9th International Workshop on Decoherence, Information, Complexity and Entropy (DICE 2018)

AA006901

Spacetime-Matter-Quantum Mechanics

Journal of Physics: Conference Series Volume 1275

Castigliocello, Italy
17 - 21 September 2018

ISBN: 978-1-5108-9950-6
ISSN: 1742-6588

Printed from e-media with permission by:

Curran Associates, Inc.
57 Morehouse Lane
Red Hook, NY 12571

CURRAN ASSOCIATES INC. proceedings.com

Some format issues inherent in the e-media version may also appear in this print version.

This work is licensed under a Creative Commons Attribution 3.0 International Licence.
Licence details: http://creativecommons.org/licenses/by/3.0/.

No changes have been made to the content of these proceedings. There may be changes to pagination and minor adjustments for aesthetics.

Printed with permission by Curran Associates, Inc. (2021)

For permission requests, please contact the Institute of Physics at the address below.

Institute of Physics
Dirac House, Temple Back
Bristol BS1 6BE UK

Phone: 44 1 17 929 7481
Fax: 44 1 17 920 0979

techtracking@iop.org

Additional copies of this publication are available from:

Curran Associates, Inc.
57 Morehouse Lane
Red Hook, NY 12571 USA
Phone: 845-758-0400
Fax: 845-758-2633
Email: curran@proceedings.com
Web: www.proceedings.com

TABLE OF CONTENTS

INVITED LECTURES

FOUNDATIONAL ISSUES RELATING SPACETIME, MATTER, AND QUANTUM MECHANICS 1
G F R Ellis

NON-LOCALITY AND QUASICLASSICAL REALITY IN KENT'S FORMULATION OF RELATIVISTIC QUANTUM THEORY 20
J Butterfield, B Marsh

A NEW RULE FOR QUANTIZATION THAT RESOLVES ALL PROBLEMS 34
John R Klauder

THE DEFORMATION PARAMETER OF THE GENERALIZED UNCERTAINTY PRINCIPLE 37
Fabio Scardigli

INFORMATION SCAN OF QUANTUM STATES BASED ON ENTROPY-POWER UNCERTAINTY RELATIONS 49
P Jizba, A Hayes, JA Dunningham

DOES GRAVITY HAVE TO BE QUANTIZED? LESSONS FROM NON-RELATIVISTIC TOY MODELS 60
Antoine Tilloy

PLANCK LENGTH CHALLENGES NON-RELATIVISTIC QUANTUM MECHANICS OF LARGE MASSES 77
Lajos Diósi

NECESSARY AND SUFFICIENT CONDITIONS FOR MACROREALISM USING TWO- AND THREE-TIME LEGGETT-GARG INEQUALITIES 82
J J Halliwell

CAUSAL FERMION SYSTEMS: DISCRETE SPACE-TIMES, CAUSATION AND FINITE PROPAGATION SPEED 94
Felix Finster

STEPS TOWARDS LORENTZIAN QUANTUM GRAVITY WITH CAUSAL SETS 108
Astrid Eichhorn

SINGULARITY AVOIDANCE IN ANISOTROPIC QUANTUM COSMOLOGY 123
Claus Kiefer, Nick Kwidzinski

ON A SOLITON-TYPE SPACETIME DEFECT 130
F R Klinkhamer

TWO ARGUMENTS FOR MORE FUNDAMENTAL BUILDING BLOCKS 146
Alfredo Iorio

NONLINEAR DETERMINISTIC-CHAOTIC COLLAPSE MODEL - PRELIMINARIES, PHILOSOPHY, LOCALITY 163
Tamás Geszti

TESTING SPONTANEOUS COLLAPSE MODELS WITH MECHANICAL EXPERIMENTS 171
Andrea Vinante

COMBINATORIAL QUANTUM GRAVITY: EMERGENCE OF GEOMETRIC SPACE FROM RANDOM GRAPHS 180
C Kelly, C A Trugenberger

ON THE HURRICANE COLLECTIVE MOLECULAR DYNAMICS 187
Gabriel Meyer, Giuseppe Vitiello

VIOLATION OF THE BELL'S TYPE INEQUALITIES AS A LOCAL EXPRESSION OF INCOMPATIBILITY 200
Andrei Khrennikov

REINFORCEMENT LEARNING BASED CONTROL OF COHERENT TRANSPORT BY ADIABATIC PASSAGE OF SPIN QUBITS 206
Riccardo Porotti, Dario Tamascelli, Marcello Restelli, Enrico Prati

QUANTUM ENHANCED CORRELATED INTERFEROMETRY FOR QUANTUM GRAVITY TESTS: THE EXPERIMENTAL SET-UP AND THE LOCKING SCHEME 215
S. T. Pradyumna, E. Losero, P. Traina, I. Ruo-Berrchera, I. Degiovanni, M. Zucco, M. Genovese

CONTRIBUTED PAPERS

CONSCIOUSNESS AS THE COLLECTIVE EXCITATION OF A BRAINWIDE WEB – UNDERSTANDING CONSCIOUSNESS FROM BELOW QUANTUM FIELDS TO ABOVE NEURONAL NETWORKS 222
Suzy Lidström, Roland E Allen

AN ALTERNATIVE TO STRING THEORY 227
Roland E Allen

FLAVOR NEUTRINO STATES FOR PEDESTRIANS 230
M Blasone, L Smaldone, G Vitiello

HEURISTIC DERIVATION OF THE CASIMIR EFFECT FROM GENERALIZED UNCERTAINTY PRINCIPLE 241
M Blasone, G Lambiase, G G Luciano, L Petruzziello, F Scardigli

COVARIANT BANDLIMITATION FROM GENERALIZED UNCERTAINTY PRINCIPLES 248
J Pye

LORENTZ BOOSTS OF BISPINOR BELL-LIKE STATES 256
Victor A. S. V. Bittencourt, Massimo Blasone

THEORY OF IMAGES AND QUANTUM MECHANICS, A COMMON PARADIGM 266
E Celeghini

VIP2 IN LNGS - TESTING THE PAULI EXCLUSION PRINCIPLE FOR ELECTRONS WITH HIGH SENSITIVITY 274
J. Marton, A. Pichler, A. Amirkhani, S. Bartalucci, M. Bazzi, S. Bertolucci, M. Bragadireanu, M. Cargnelli, A. Clozza, C. Curceanu, R. Del Grande, L. De Paolis, J.-P. Egger, C. Fiorini, C. Guaraldo, M. Iliescu, M. Laubenstein, E. Milotti, M. Milucci, D. Pietreanu, K. Piscicchia, A. Scordo, H. Shi, D. Sirghi, F. Sirghi, L. Sperandio, O. Vazquez-Doce, J. Zmeskal

EFFECT OF THE RESONANCE SPECTRA IN THE PROPAGATION OF TWO DECAYING ENTANGLED PARTICLES 280
Gastón García-Calderón, Roberto Romo, Miguel Ángel Terán

COSMIC STRINGS, DEFORMED LATTICES AND SPONTANEOUS SYMMETRY BREAKING 289
Antonino Flachi, Vincenzo Vitagliano

FEATURES OF SPINNING GRAVITY IN PARTICLE PHYSICS: SUPERSYMMETRIC CORE OF THE KERR-NEWMAN ELECTRON 294
Alexander Burinskii

PHASE-SPACE CONTINUITY EQUATIONS FOR QUANTUM DECOHERENCE, PURITY, VON NEUMANN AND RENYI ENTROPIES 308
Alex E. Bernardini, Orfeu Bertolami

SOME REMARKS ON ANALYTICAL SOLUTIONS FOR A DAMPED QUANTUM PARAMETRIC OSCILLATOR 315
Dieter Schuch

NEW REPRESENTATIONS OF POINCARE GROUP FOR CONSISTENT RELATIVISTIC PARTICLE THEORIES 323
Giuseppe Nisticò

FAST MEASUREMENTS OF ENTANGLEMENT OVER A KILOMETRIC DISTANCE TO TEST SUPERLUMINAL MODELS OF QUANTUM MECHANICS: FINAL RESULTS 331
B Cocciaro, S Faetti, L Fronzoni

SPEKKENS' TOY MODEL, FINITE FIELD QUANTUM MECHANICS, AND THE ROLE OF LINEARITY 342
Lay Nam Chang, Djordje Minic, Tatsu Takeuchi

NO TIME FOR ISOLATED CLOCKS 350
K L H Bryan, A J M Medved

QUANTUM STATISTICS IN BOHMIAN TRAJECTORY GRAVITY 356
T C Andersen

ENTANGLING QUANTUM FIELDS VIA A CLASSICAL GRAVITATIONAL INTERACTION 364
M Reginatto, M J W Hall

FROM QUANTUM MECHANICS TO RUNNING Λ COSMOLOGIES 374
A. Stachowski, M. Szydłowski, K. Urbanowski

INFLUENCE FUNCTIONALS, DECOHERENCE AND CONFORMALLY COUPLED SCALARS 387
C Burrage, C Käding, P Millington, J Minář

LINEARIZED METRIC SOLUTIONS IN GHOST-FREE NONLOCAL GRAVITY 395
Luca Buoninfante

HOW TO OBTAIN INITIAL DEGREES OF FREEDOM FOR THE EARLY UNIVERSE OF ABOUT 1000, INSTEAD OF 100, WHILE PRESERVING A UNIFORMLY INCREASING ARROW OF TIME DUE TO ENTROPY INCREASE 405
Andrew Walcott Beckwith

RECONCILING REALITY, SPACE AND TIME: A GRAVITON DRIVEN QUANTUM MECHANISM OF COSMIC EXPANSION AND CMB RADIATION 411
G. Doyen, D. Drakova

THE DISTANCE FORMULA IN ALGEBRAIC SPACETIME THEORIES 423
D Canarutto, E Minguzzi

NON-CAUCHY SURFACE FOLIATIONS AND THEIR EXPECTED CONNECTION TO QUANTUM GRAVITY THEORIES 428
Merav Hadad

GENERALLY COVARIANT QUANTUM THEORY 438
Johan Noldus

SOME ASPECTS OF THE CANONICAL ANALYSIS OF REUTER-WEYER RG IMPROVED EINSTEIN-HILBERT ACTION 448
S. J. Gionti Gabriele

BORN RULE AND SCHROEDINGER EQUATION FROM FIRST PRINCIPLES 459
Robert Oeckl

MATHEMATICS AND PHYSICS OF CONSISTENCY OF THE BORN RULE WITH NORMAL PROBABILITY DISTRIBUTION 468
Alexey A. Kryukov

HOW AN INDUCED KERR–NEWMAN BLACK HOLE RELEASES GRAVITATIONAL WAVES WITHOUT A MINI BLACK HOLE EXPLOSION 485
Andrew Walcott Beckwith

QUANTUM FIELD THEORY OF AXION-PHOTON MIXING AND VACUUM POLARIZATION 490
Antonio Capolupo

NEUTRINO NATURE, TOTAL AND GEOMETRIC PHASE 496
Antonio Capolupo, S. M. Giampaolo

PHYSICS IN ONE DIMENSION WITH PERPENDICULAR NON-LOCALITY 501
E Cerwen

LOCALIZED GRAVITONS AS ESSENCE OF DARK ENERGY: IMPLICATIONS FOR ACCELERATED COSMIC EXPANSION 509
D. Drakova, G. Doyen

QUANTUM THERMODYNAMIC INSTABILITIES IN COMPACT STARS 518
A Lavagno, D Pigato, G Gervino

SIMULATION OF QUANTUM UNIVERSE 523
Sang Pyo Kim

NO-GO THEOREMS FOR EKPYROSIS FROM TEN-DIMENSIONAL SUPERGRAVITY 531
Kunihito Uzawa

DARK MATTER IN MODIFIED COSMOLOGIES 540
G. Lambiase

MATTER-AS-CONSTRAINED-GRAVITY QUANTA ON MULTIPLE RINDLER OBSERVER GROUNDS 548
P A Mandrin

GENERALIZED UNCERTAINTY PRINCIPLE IN GRAPHENE 556
Alfredo Iorio, Pablo Pais

SOME ALGEBRAICALLY SOLVABLE TWO-DIMENSIONAL DYNAMICAL SYSTEMS WITH POLYNOMIAL INTERACTIONS 563
Francesco Calogero, Farrin Payandeh

REMARKS ON THE UNRUH EFFECT WITH MIXED NEUTRINOS 570
M Blasone, G Lambiase, G G Luciano, L Petruzziello

ON THE TIME CONTINUOUS EVOLUTION OF THE UNIVERSE IF TIME IS DISCRETE AND IRREVERSIBLE IN NATURE 577
Roland Riek

Author Index

IOP Conf. Series: Journal of Physics: Conf. Series **1275** (2019) 011001 doi:10.1088/1742-6596/1275/1/011001

DICE 2018 : Spacetime Matter Quantum Mechanics

Presented in this volume are the Invited Lectures and the Contributed Papers of the *Ninth International Workshop on Decoherence, Information, Complexity and Entropy – DICE 2018* – "From discrete structures and dynamics to top-down causation" held at **Castello Pasquini, Castiglioncello (Tuscany), September 17-21, 2018**.

These proceedings document the lively exchange of ideas during the conference for the interested public and the scientific community alike. – For further details on all aspects of the meeting, cf. the website at `http://osiris.df.unipi.it/~elze/DICE2018.html`.

Our intention has always been to unite leading researchers, advanced students, and renowned scholars from various areas, in order to stimulate new ideas and their exchange across the borders of specialization. The meetings in this series have successfully grown from DICE 2002,[1] followed by DICE 2004,[2] DICE 2006,[3] DICE 2008,[4] DICE 2010,[5] DICE 2012,[6] DICE 2014,[7] and DICE 2016.[8] About 110 participants representing almost 30 countries attended DICE 2018.

It has been a great honour and inspiration having three eminent scholars presenting Keynote Lectures: George Ellis (University of Cape Town), *"Foundational Issues Relating Spacetime, Matter, and Quantum Mechanics"*, in this volume – Steven Adler (Institute for Advanced Study, Princeton), *"Motivations for and Implications of Frame-Dependent Dark Energy"* – Pisin Chen (University of Taipei and SLAC), *"Tabletop analog black holes to investigate the information loss paradox"*.

[1] *Decoherence and Entropy in Complex Systems* ed H-T Elze *Lecture Notes in Physics* **633** (Berlin: Springer, 2004)

[2] *Proceedings of the Second International Workshop on Decoherence, Information, Complexity and Entropy – DICE 2004* ed H-T Elze *Braz. Journ. Phys.* **35**, 2A & 2B (2005) pp 205–529;
free access at: **www.sbfisica.org.br/bjp**

[3] *Proceedings of the Third International Workshop on Decoherence, Information, Complexity and Entropy – DICE 2006* eds H-T Elze, L Diósi and G Vitiello *Journal of Physics: Conference Series* **67** (2007);
free access at: **www.iop.org/EJ/toc/1742-6596/67/1**

[4] *Proceedings of the Fourth International Workshop on Decoherence, Information, Complexity and Entropy – DICE 2008* eds H-T Elze, L Diósi, L Fronzoni, J J Halliwell and G Vitiello *Journal of Physics: Conference Series* **174** (2009); free access at: **http://www.iop.org/EJ/toc/1742-6596/174/1**

[5] *Proceedings of the Fifth International Workshop on Decoherence, Information, Complexity and Entropy – DICE 2010* eds H-T Elze, L Diósi, L Fronzoni, J J Halliwell, E Prati, G Vitiello and J Yearsley *Journal of Physics: Conference Series* **306** (2011); free access at: **http://iopscience.iop.org/1742-6596/306/1**

[6] *Proceedings of the Sixth International Workshop on Decoherence, Information, Complexity and Entropy – DICE 2012* eds H-T Elze, L Diósi, L Fronzoni, J J Halliwell, E Prati, G Vitiello and J Yearsley *Journal of Physics: Conference Series* **442** (2013); free access at: **http://iopscience.iop.org/1742-6596/442/1**

[7] *Proceedings of the Seventh International Workshop on Decoherence, Information, Complexity and Entropy – DICE 2014* eds H-T Elze, L Diósi, L Fronzoni, J J Halliwell, C Kiefer, E Prati and G Vitiello *Journal of Physics: Conference Series* **626** (2015); free access at: **http://iopscience.iop.org/issue/1742-6596/626/1**

[8] *Proceedings of the Eighth International Workshop on Decoherence, Information, Complexity and Entropy – DICE 2016* eds H-T Elze, L Diósi, L Fronzoni, J J Halliwell, C Kiefer, E Prati and G Vitiello *Journal of Physics: Conference Series* **880** (2017); free access at: **http://iopscience.iop.org/issue/1742-6596/880/1**

Content from this work may be used under the terms of the Creative Commons Attribution 3.0 licence. Any further distribution of this work must maintain attribution to the author(s) and the title of the work, journal citation and DOI.
Published under licence by IOP Publishing Ltd

IOP Conf. Series: Journal of Physics: Conf. Series **1275** (2019) 011001 doi:10.1088/1742-6596/1275/1/011001

Entitled *"Spacetime - Matter - Quantum Mechanics"*, the conference took place in the very pleasant atmosphere of Castello Pasquini – in beautiful surroundings, overlooking a piece of Tuscany's coast. The 5-day program addressed aspects of the following topics:

- Emergent spacetime, matter, and symmetries
- Quantum coherence from micro- to macroscopic
- New key experiments
- Quantum mechanics interfacing with gravity
- Quantum gravity / cosmology and high-energy physics

A Roundtable Discussion dedicated to the public under the theme *"Dall'astronomia babilonese alle onde gravitazionali"* formed an integral part of the program. With the participation of F. Bonoli (Bologna), E. Prati (Milano) and A. Allocca (Pisa) [LIGO/VIRGO collaboration], the event attracted a large audience involved in the all-evening lively discussions.

Three Best Posters were voted for by the participants and awards given by MDPI were presented to A. Capolupo (Salerno - Second Prize), C. Kelly (Edinburgh - Second Prize) and L. Petruzziello (Salerno - First Prize) during a ceremony at the conference dinner.

The workshop has been organized by M Blasone (Salerno), L Diósi (Budapest), H-T Elze (Pisa, *Chair*), L Fronzoni (Pisa), J J Halliwell (London), C Kiefer (Cologne), E Prati (Milano) and G Vitiello (Salerno), with most essential help from our conference secretaries L Fratino (Paris), D Gigli (Pisa), I Pozzana (London), L Ristuccia (Pisa) and F Di Lauro (Pisa).

Several institutions and sponsors generously supported the workshop and their representatives as well as the citizens of Rosignano / Castiglioncello are warmly thanked for the kind hospitality:

- *Comune di Rosignano* – A Franchi (Sindaco di Rosignano), together with the Giunta Comunale di Rosignano, and special thanks to assessore L Montagnani.
- *Associazione Armunia* – Il Presidente e il Consiglio di Amministrazione / La direzione generale e la coordinazione artistica / Lo staff tecnico e organizzativo.
- *Solvay Italia SA, Rosignano* – A De Lorenzo (Site Communication & Total Quality Manager).

Funds made available by Università di Pisa, Domus Galilaeana (Pisa), Centro Interdisciplinare per lo Studio dei Sistemi Complessi – CISSC (Pisa), Dipartimento di Fisica (Università di Salerno), Istituto Italiano per gli Studi Filosofici – IISF (Napoli), Institute of Physics Publishing – IOP (Bristol), Springer Verlag (Heidelberg), MDPI AG (Basel), and Ericsson Telecommunicazioni (Roma) are gratefully acknowledged.

We especially thank Laura Pesce (Vitrium Galleria – http://www.laurapesce.it) for showing her artworks *"Art and Physics - a close encounter"* during DICE 2018 at Castello Pasquini.

The proceedings have been edited by H-T Elze, M Blasone, L Diósi, L Fronzoni, J J Halliwell, C Kiefer, E Prati, and G Vitiello and the accepted papers follow essentially the order of presentation during the conference, separating, however, Invited Lectures and Contributed Papers (talks and posters). Invited Lectures will also be published separately as a book. – Several lectures could not be reproduced here, partly for copyright reasons, cf. complete list and final program at `http://osiris.df.unipi.it/~elze/DICE2018.html`.

In the name of all participants, we would like to thank A Ashton with collaborators at IOP Publishing (Bristol) for friendly advice and most valuable immediate help during the editing process and, especially, for their continuing efforts to make the Journal of Physics: Conference Series available to all.

IOP Conf. Series: Journal of Physics: Conf. Series **1275** (2019) 011001 doi:10.1088/1742-6596/1275/1/011001

Pisa, Salerno, Budapest, London, Cologne, Milano — June 2019

Hans-Thomas Elze Massimo Blasone Lajos Diósi Leone Fronzoni
Jonathan Halliwell Claus Kiefer Enrico Prati Giuseppe Vitiello

IOP Conf. Series: Journal of Physics: Conf. Series **1275** (2019) 011002 doi:10.1088/1742-6596/1275/1/011002

Peer review statement

All papers published in this volume of *Journal of Physics: Conference Series* have been peer reviewed through processes administered by the proceedings Editors. Reviews were conducted by expert referees to the professional and scientific standards expected of a proceedings journal published by IOP Publishing.

Content from this work may be used under the terms of the Creative Commons Attribution 3.0 licence. Any further distribution of this work must maintain attribution to the author(s) and the title of the work, journal citation and DOI.
Published under licence by IOP Publishing Ltd

IOP Conf. Series: Journal of Physics: Conf. Series **1275** (2019) 012001 doi:10.1088/1742-6596/1275/1/012001

Foundational Issues Relating Spacetime, Matter, and Quantum Mechanics

G F R Ellis

Mathematics Department, University of Cape Town, Rondebosch, Cape Town 7700, South Africa

E-mail: george.ellis@uct.ac.za

Abstract. This article proposes the following themes: 1. Space time must be discrete at the micro level. 2. Holonomy is central to any foundational approach to relating spacetime and quantum mechanics. 3. The gravitational field equations should be trace free: gravity is essentially a conformal theory. 4. Times passes; past, present, and future are fundamentally different. 5. Causation is not only bottom-up: contextual effects occur, associated with symmetry breaking 6. Theories must adequately take account of the quantum measurement issue 7. Penrose's entropy issue is a real issue for cosmology that must be taken into account.

This paper I do not present here a coherent model of quantum gravity as a rival to any of the standard theories (for a survey see [64]). Rather I present some fundamental issues that I believe should be taken into account in any such eventual theory relating gravity and quantum physics. The sections are,

(i) space time should be discrete at the micro level (Section 1);

(ii) holonomy should be central to any foundational approach (Section 2);

(iii) the gravitational equations should be trace free: gravity is essentially a conformal theory (Section 3);

(iv) times passes; past, present, and future are fundamentally different (Section 4);

(v) contextual effects occur, with associated symmetry breaking (Section 5);

(vi) quantum gravity must adequately take account of the quantum measurement issue via contextual wave function collapse (Section 6);

(vii) cosmology must take seriously Penrose's entropy issue (Section 7).

This will lead to a growing block universe based on a discrete substrate and with loss of information as time progresses The gravitational equations will be based in holonomy in a suitable fibre bundle over a discrete foundation.The equations will be relational: they will be essentially conformal except for a matter term that will break the conformal symmetry and give scale to the whole. They will turn off gravity at very high densities so as to solve Penrose's entropy problem.

Content from this work may be used under the terms of the Creative Commons Attribution 3.0 licence. Any further distribution of this work must maintain attribution to the author(s) and the title of the work, journal citation and DOI.
Published under licence by IOP Publishing Ltd

IOP Conf. Series: Journal of Physics: Conf. Series **1275** (2019) 012001 doi:10.1088/1742-6596/1275/1/012001

1. Space time discrete at the micro level

The basic viewpoint here is that no infinities should occur in a physical theory [41], so at its foundation spacetime should emerge from a discrete structure. The point to remember is that, contrary to the way infinity is treated in many physics papers, infinity is *not* just a very large number; it is in fact an unattainable state rather than a number of any kind. It will therefore never occur in the real universe, as stated by David Hilbert [55]:

> "*The infinite is nowhere to be found in reality, no matter what experiences, observations, and knowledge are appealed to.*"

I adopt this statement, which I call *Hilbert's Golden Rule*, as a foundational philosophical principle underlying physics applications.

Physics of infinity The distinction between mathematical infinity and physical effective infinity is set out in [41]. As far as potential physical infinities are concerned, there are two essentially different kinds:

- $\infty_{(VL)}$: 'infinity' being used as a placeholder for a very large number
- $\infty_{(ESS)}$: essential infinity, where the paradoxical nature of infinity (such as 'Hilbert's Hotel' [54]) come into play.

The paradoxical nature of essential infinity is represented by the relations

$$\forall\, a, b,\ (i)\, \infty_{(ESS)} + a = \infty_{(ESS)},\ (ii)\, b \times \infty_{(ESS)} = \infty_{(ESS)}. \tag{1}$$

No finite number can satisfy both these relations, so they can be taken together as an operational characterisation of infinity. It is precisely because infinity satisfies the relations (1) that it cannot occur in physical reality; in essence, it fails to obey conservation laws. Apart from some claims about multiverses, this does not appear to play a role in physics. Similarly, for "zero", the dual of "infinity", we can distinguish two physical kinds:

- $0_{(VS)}$: 'zero' being used as a placeholder for a very small number
- $0_{(ESS)}$: essential zero, dual to $\infty_{(ESS)}$.

This is characterised by the relations

$$\forall\, a, b,\ (i)\, 0_{(ESS)} + a = a,\ (ii)\, b \times 0_{(ESS)} = 0_{(ESS)}. \tag{2}$$

The first of these characterises zero as unphysical: in essence, it fails to obey the action - reaction rule, because it affects everything by (ii), but is affected by nothing, by (i). if it were to exist in some material sense, it would have no physical effect. Thus dual to the very large number $\infty_{(ESS)}$ one can determine a very small number $0_{(VS)}$ as an effective zero in various physical contexts.

Consequence The immediate application is that there is then an unacceptable infinity in standard physics views of spacetime: They assume there is an uncountable infinity of physically distinct points points existing between my fingers when I hold them 10cm apart. This contradicts the principle just set out. This points to the need for discreteness at the foundations of spacetime (hence no such uncountable infinity of physical points characterised by local separation), Consequently spacetime must be discrete at the micro level, with an emergent spacetime continuum view emerging via coarse graining. Thus one needs something like causal set theory [23] [53] [90] but with extra structure in addition to the conformal (Section 3), or spin foam structure [72] [11]. Note that one needs discreteness at the foundations, not as an outcome, as is the case in Loop Quantum Gravity.

IOP Conf. Series: Journal of Physics: Conf. Series **1275** (2019) 012001 doi:10.1088/1742-6596/1275/1/012001

Basic statement In the *Feynman Lectures on Physics* [45], Richard Feynman said the most important statement one can make about physics is that "Matter is made of atoms" :

> "*If, in some cataclysm, all of scientific knowledge were to be destroyed, and only one sentence passed on to the next generation of creatures, what statement would contain the most information in the fewest words? I believe it is the atomic hypothesis (or the atomic fact, or whatever you wish to call it) that all things are made of atoms little particles that move around in perpetual motion, attracting each other when they are a little distance apart, but repelling upon being squeezed into one another. In that one sentence, you will see, there is an enormous amount of information about the world, if just a little imagination and thinking are applied.* " ([45]:1-2).

In parallel, one can propose that Hilbert's statement above (interpreted in this context) is:

> "*Spacetime is made of atoms*"

- an extended quantum principle! Loop Quantum Gravity [75] suggests this is indeed the case.

2. Holonomy central to any foundational approach

The proposal here is the foundation of physics should a geometric approach based in parallel transport in a suitable fibre bundle over space-time, with holonomy (variation in the effect of parallel transport round different closed loops through a point p) determining the physical outcome. That holonomy should be a foundational aspect of theory is indicated by the importance of parallel transport in Yang-Mills theories [92] and the widespread applications of Berry Phase [13] [9]. That it is a key effect in quantum physics *per se* is indicated by the Aharanov-Bohm effect [3], which has been confirmed by various experiments, e.g. [82] [83] [67] [12]. Indeed one can even claim it is the core of the Feynman path integral method [47] which with a suitable definition of parallel transport of phase, essentially involves summing of holonomy over many closed paths[1].

Yang-Mills The basis of the standard model of particle physics [73] [4] is a gauge theory [51] [81] based on a non-abelian symmetry group [92]. There is a covariant derivative, with a vector potential as the connection, and the field given by commutators of the derivatives. The Lagrangian gives equations of motion which are equivalent to Jacobi identities, and path integrals lead to Wilson loop variables [89] [49] [81] which use holonomy to determine the outcome.

Combine with discrete spacetime The previous sections suggests this holonomy should be reformulated as a theory of fundamental physics, including gravity,[2] in a discrete spacetime context. This is obviously favourable to Wilson Loop variables and so Loop Quantum Gravity; however as pointed out above, that theory is not actually based in a discrete pre-spacetime. What is needed is something like spin foam evolution of a spin network [11], involving parallel transport in a suitable bundle over discrete structures; something like like causal set theory [16] [23] [53], but the conformal symmetry has to be broken to get scale.

3. The Vacuum Energy Problem: Trace-free Gravitational equations?

A major issue for present day theory is the problem of vacuum energy. Estimates of the Quantum Field Theory ('QFT') vacuum energy density suggest a huge value for that energy, discrepant with General Relativity if that vacuum gravitates [86]. It would be an effective cosmological

[1] A forward path γ_1 from p_1 to p_2 and a reverse path γ_2 from p_2 to p_1 form a closed curve $\gamma = \gamma_2 \, o \, \gamma_1$ based in p_2

[2] But not necessarily a unification of all four forces in a single framework, as in String/M theory. Indeed from a General Relativity perspective that is a questionable goal, as at a fundamental level, gravity is not a force. It is an effective result of space-time curvature [52].

IOP Conf. Series: Journal of Physics: Conf. Series **1275** (2019) 012001 doi:10.1088/1742-6596/1275/1/012001

constant at least 80 orders of magnitude larger than indicated by the Planck date [1] [2]. This is a major problem for basic physics, because there seems to be a flagrant contradiction when we combine two of our best tested physics theories (General Relativity and QFT) in the obvious way.

One way out suggested by a number of prominent physicists is that this is solved by existence of a multiverse together with anthropic selection. I will not discuss that option here, because (i) it seems a massive violation of Ockham's Razor (the word "infinity" is thrown around with gay abandon in these discussions, in contrast to the view presented in Section 1 of this paper), (ii) the underlying proposed physics is not well established or tested, and (iii) its status as a scientific theory is disputable [42]. Rather I will look for a solution in terms of querying what the true gravitational field equations are.

3.1. Trace-free equations

The key insight is that the vacuum does not gravitate if we use the trace-free Einstein equations ("TFE") plus separate conservation equations for matter [44], giving a theory essentially equivalent to what is often labelled as "unimodular gravity" [85] [48] [84]. This then solves profound contradiction arising between QFT and EFE if we join them in the obvious way: then the vacuum does not gravitate.

Taking the trace-free part Starting with the standard Einstein Field Equations ('EFE')

$$R_{ab} - (1/2)Rg_{ab} + \Lambda g_{ab} = \kappa T_{ab} \tag{3}$$

(10 equations), these equations identically implies energy-momentum conservation:

$$T^{ab}{}_{;b} = 0 \tag{4}$$

and the vacuum energy Λ_{vac} gravitates if we regard it as an effective fluid with equation of state

$$\{\rho_{vac} = -p_{vac} = \Lambda_{vac}\} \Rightarrow \{\nabla_a \rho_{vac} = 0,\ \rho_{vac} + p_{vac} = 0\} \tag{5}$$

where the first equality on the right follows from (4). Instead, we take the trace free part of (3):

$$R_{<ab>} - (1/2)Rg_{<ab>} + \Lambda g_{<ab>} = \kappa T_{<ab>}, \tag{6}$$

where $<,>$ just means trace free part, and use these as our gravitational equations. Explicitly, because $g_{<ab>} = 0$, (6) is the Trace Free Equations ('TFE'):

$$R_{<ab>} = \kappa T_{<ab>} \Leftrightarrow R_{ab} - (1/4)Rg_{ab} = \kappa(T_{ab} - (1/4)Tg_{ab}) \tag{7}$$

(9 equations), which no longer implies energy-momentum conservation (4); we have to assume that separately. The key point is that the energy density ρ_{vac} no longer contributes to spacetime curvature via the TFE (7), because of (5). Taking the trace of (7) to get the scalar part of the gravitational equations gives the identity $0 = 0$: the vacuum energy (5) no longer gravitates, because (7) has no scalar part. Taking the divergence of (7) and using (4), on integration one recovers (3), because the divergence of the left hand side of (3) vanishes identically; but now Λ is a constant of integration that has nothing to do with vacuum energy (it is unrelated to $T := T^a{}_a$). It can take any value whatever.

The options This possibility was pointed out already by Weinberg in [86], however he did not pursue it further. Indeed Einstein tried this in 1919 [29], but used a wrong form, because at that time it was not known how to handle symmetries of tensor equations consistently. There are four possibilities:

$$G_{ab} = \kappa T_{ab}, \tag{8}$$
$$G_{<ab>} = \kappa T_{ab}, \tag{9}$$
$$G_{ab} = \kappa T_{<ab>}, \tag{10}$$
$$G_{<ab>} = \kappa T_{<ab>}. \tag{11}$$

Only the first and last are acceptable (the second and third have different index symmetries on the left and right; Einstein tried the second). The first is the standard EFE; the last is the TFE that solves the GR-QFT incompatibility. It works out acceptably in the cosmological context, and in particular as as regards inflation (see [34]), essentially because what matters in inflation is not the absolute value of the inflaton potential $V(\phi)$, but the difference $\Delta V := V(\phi_i) - V(\phi_f)$ between the values of the potential at the start and end of inflation.

This is related to unimodular gravity [48], [84],[85] where either general covariance is replaced by restricted covariance subject to $g := det\{g_{ab}\} = const$, or one considers equivalence classes of solutions as g is varied (it is no longer a dynamical variable). One can obtain variational principles for such theories, as discussed e.g. in [7].

Quantum field theory version This is classical. What does a QFT version of gravity say? Various authors have discussed this, including Feynman, Deser, Weinberg, and Zee, but they have all in one way or another assumed that the gravitational equations imply conservation (4), which then necessarily leads to the standard equations (3). However because the graviton h_{ab} is a symmetric trace free tensor:

$$h_{ab} = h_{(ab)},\ \ h^a{}_a = 0 \ \Rightarrow\ \ h_{ab} = h_{<ab>} \tag{12}$$

when one regards gravitation as a field theory, this should also give the trace free version 7 of the gravitational equations. Specifically, one should assume energy-momentum conservation (4) separate from gravity equations and then add a term to the Lagrangian of the form.

$$L = T^{ab}h_{ab} = T^{ab}h_{<ab>} = T^{<ab>}h_{<ab>} \tag{13}$$

where (12) has been used at the second equality, and the third is a simple identity. Thus the trace T of T^{ab} does not occur in these equations, so it has no ability to affect space-time curvature. One should therefore necessarily get the trace-free equations (7) from this procedure. These therefore have a good claim to be the correct equations. More sophisticated analyses (e.g. [85]) confirm this view.

3.2. Is gravity essentially a conformal theory? The Weyl tensor as variable

If the determinant g is not a dynamical variable, one has essentially a conformal theory of gravity. This is related to the possibility of using the Weyl tensor C_{abcd} (considered as the free gravitational field, in analogy to the electromagnetic field F_{ab}) as the key gravitational variable rather than the metric. Then the Bianchi Identities act as dynamical equations (analogous to Maxwell's equations) for the Weyl tensor, with the Einstein Field Equations 3 determining the matter source terms in those equations. This relates to the idea that gravity is theory of 3-d or

IOP Conf. Series: Journal of Physics: Conf. Series **1275** (2019) 012001 doi:10.1088/1742-6596/1275/1/012001

4-d conformal degree of freedom, whereas matter with mass is not: it has timelike characteristics, as opposed to the null characteristics of the gravitational field [52].

A summary of these proposals, going back particularly to Petrov, Pirani, Ehlers, and Trümper,[3], is given in [30], [59], and [40]. Relative to an observer's 4-velocity $u^a : u^a u_a = -1$, the Weyl tensor is split into 3-dimensional electric and magnetic parts:

$$E_{ac} = C_{abcd}u^b u^d = E_{<ac>},\ \ E^a{}_a = 0,\ \ E_{ab}u^b = 0, \tag{14}$$

$$H_{ac} = {}^*C_{abcd}u^b u^d = H_{<ac>},\ \ H^a{}_a = 0,\ \ H_{ab}u^b = 0, \tag{15}$$

each a spin-2 field. The Bianchi identities $R_{ab[cd;e]} = 0$ must be satisfied, and in 4 dimensions they are equivalent to

$$\nabla^d C_{abcd} = \nabla_{[a}(R_{b]c} + (1/6)Rg_{b]c}). \tag{16}$$

Using the EFE (3) to algebraically substitute for R_{bc}in terms of T_{bc}, one obtains firstly the constraints (where D^b is the spatial covariant derivative orthogonal to u^a),

$$D^b E_{ab} = -3\omega^b H_{ab} + \frac{1}{3}D_a\rho + [\sigma, H]_a \tag{17}$$

$$D^b H_{ab} = 3\omega^b E_{ab} + (\rho + p)\omega_a - [\sigma, E]_a, \tag{18}$$

in obvious analogy with the divergence Maxwell equations, and second the evolution equations:

$$\frac{dE_{<ab>}}{dt} - curl H_{ab} = -\theta E_{ab} + 3\sigma_{c<a}E^{\ c}_{b>} - \omega^c\epsilon_{cd(a}E_{b)}^{\ \ d} + 2a^c\epsilon_{cd(a}H_{b)}^{\ \ d} - \frac{1}{2}(\rho + p)\sigma_{ab} \tag{19}$$

$$\frac{dH_{<ab>}}{dt} + curl E_{ab} = -\theta H_{ab} + 3\sigma_{c<a}H^{\ c}_{b>} - \omega^c\epsilon_{cd(a}H_{b)}^{\ \ d} - 2a^c\epsilon_{cd(a}E_{b)}^{\ \ d} - \frac{1}{2}(\rho + p)\sigma_{ab} \tag{20}$$

again in analogy with the evolutionary Maxwell equations. This agrees in essence with the TFE, because in the vacuum case, $(\rho{+}p) = 0,\ D^\rho = 0)$ so the matter does not gravitate in the sense of entering as a source term in these equations for the free gravitational field C_{abcd}. This suggests that one in going to quantum gravity, one should quantize the Weyl tensor (or its potential: some connection terms) rather than a 'graviton' h_{ab} as mentioned above.

Weyl-squared theories How does this relate to conformal theories using the square of the Weyl tensor in the Lagrangian? [60] Suppose that in analogy to the way there is a term $S_{em} = -\alpha \int d^4x\sqrt{-g}\,F^{ab}F_{ab}$ in QED to represent the electromagnetic field, we replace the gravitational term $S_{grav} = -\alpha_{grav}\int d^4x\sqrt{-g}\,R$ in the Lagrangian by

$$S_{grav} = -\alpha_G \int d^4x\sqrt{-g}\,C^{abcd}C_{abcd} \tag{21}$$

Then one obtains for the gravitational sector on its own the Bach equation

$$2\nabla^a\nabla^d C_{abcd} + C^d{}_{abc}R^a{}_d = 0 \tag{22}$$

This is a field equation for C_{abcd}, which is subject to the identity (16). Matter terms have to be added in to get the full theory [60]. One can rewrite the action in terms of Ricci tensor variables because the Gauss-Bonnet invariant $\tilde{G} := \int d^4x\sqrt{-g}(R_{abcd}R^{abcd} - 4R_{ab}R^{ab} + R^2)$ vanishes for space-times topologically equivalent to Euclidean space.[4] The key point is that although these are higher derivative theories ([19]:§4.2), they reduce to the GR equations in the Ricci flat case ($R_{ab} = 0$) and the conformally flat case ($C_{abcd} = 0$) and so include as solutions both the Schwarzschild solution and the FLRW solution [60]. However, in common with other theories, the Newtonian limit is tricky (see Appendix A). Perturbation solutions in cosmology can be found [61] [8] I do not pursue this further here, but note it as a promising line to investigate.

[3] Trautmann, Bondi, Sachs, Newman, and Penrose developed the null version of these equations

[4] More generally, it is given by the Euler chracateristic $\chi(M)$ of the manifold.

IOP Conf. Series: Journal of Physics: Conf. Series **1275** (2019) 012001 doi:10.1088/1742-6596/1275/1/012001

3.3. The deep issue: The origin of scale

the deep underlying issue is that one can claim that all physics measurements are relative measurements, so that one should only use dimensionless variables in basic equations (see e.g. the literature on 'Shape Dynamics', and [79]). If so, where does scale come from? It has to be via broken symmetries [10]. At one level, the answer is that the Higgs term breaks the conformal symmetry of the Standard Model of particle physics, and so provides a scale for the matter term in the EFE through the matter source terms in (17)-(20). At a deeper level, this does not answer the problem, and still needs resolution.

4. Times passes; past, present, and future are fundamentally different.

In contrast to many claims in the literature, my view is that spacetime should such that time passes, in accord with both everyday experience and the laws of macroscopic physics and physical chemistry. Thus spacetime should be an *Evolving Block Universe* ('EBU') [33], [39], where a genuine passage of time takes place at both macro and micro levels because the future boundary of spacetime is continually advancing. This is essentially taken for granted in most cosmological analyses. Thus for example, the age of the universe was measured by the Planck Satellite team in 2015 as $t_0 = (13.813 + T_0) \times 10^9$ years where $|T_0| < 0.038$. In 2018, it is clearly 3 years older: the age of the universe is $t_0^* = t_0 + 3$. It is taken for granted by the Planck team that the Age of the Universe is a meaningful concept. This requires an EBU view of spacetime.

4.1. Flow of time: Evolving spacetime

Consider a massive object with two computer controlled rocket engines that move it right or left. Let the computer determine the outcome on the basis of measurements of decay products of excited atoms. Then because of the foundational nature of quantum uncertainty [32], as conclusively proved by the 2-slit experiment and by the decay of radioactive atoms [46], the outcome is unpredictable in principle. If the object is massive enough, it curves spacetime. The future spacetime structure is therefore not determinable or predictable from current data

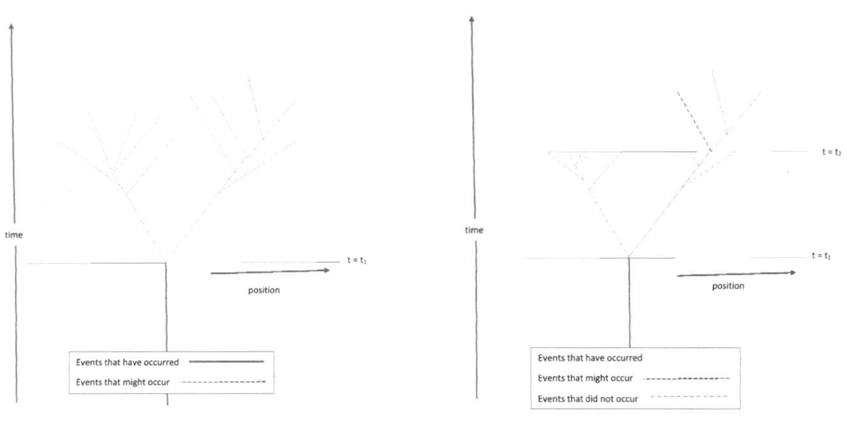

Figure 1. *The change from uncertainty to certainty: the present is where the indefinite future changes to the determined past.*

The evolving block universe grows with time: the determined past is there and fixed, the indeterminate future is not yet there and still has to be fixed (Figure 1). In this case, it is not true that the future is determined at the present. The present $t = t_0$ at any instant is where the change from possibility to definiteness takes place. It is crucially different from the past and future, and indeed separates them. At any time t_0, The future $\{\mathcal{F}\} : t > t_0$ does not exist in the

IOP Conf. Series: Journal of Physics: Conf. Series **1275** (2019) 012001 doi:10.1088/1742-6596/1275/1/012001

same sense as the past $\{\mathcal{P}\} : t < t_0$ or the present $\{\mathcal{T}_0\} : t = t_0$, because it is not yet clear what will occur there. The determinate region grows with time. Spacetime itself is growing, because its future boundary $t = t_0$ is continually extending (this is what distinguishes it from the past boundary of spacetime at the initial big-bang at time $t = 0$).

The Direction of Time This leads to the existence of the global Direction of Time, determined by the Universe as a whole. In the expanding universe, the start is a fixed initial boundary to spacetime. The present is an ever changing future boundary to spacetime: as time passes it moves to the future, in the direction indicated by the Direction of Time. At each instant spacetime exists from the start up to the present, but not to the future (because of quantum randomness, it is not yet determined what will happen in the future, so it is not a spacetime domain with definite properties), as indicated in Figure 2).

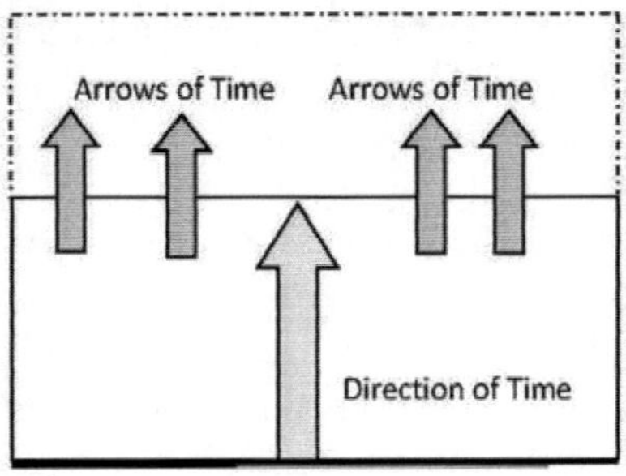

Figure 2. *The global Direction of Time is set by the expansion of the universe since its start. Local Arrows of Time are determined by local physical properties (Section 5.3).*

The time surfaces The standard objection to this proposal is that it will not work because according to Special Relativity, the surface of simultaneity $\{t = t_0\}$ depends on the motion of the observer, and so is not well defined. The response is that the big news is that Special Relativity does not describe the real universe: General Relativity does! [52]. In any realistic cosmological spacetime, there is a preferred family of observers whose motion is determined by the curved spacetime geometry as the timelike eigenvector[5] of the Ricci tensor [30] [40]. What happens physically at macro scales is determined along timelike world lines[6] rather than on spacelike surfaces (simultaneity is irrelevant to physical interactions!)

Thus the time surfaces in the EBU are secondary to uniquely defined timelike world lines (eigenlines of the stress tensor T_{ab}.). To determine the time surfaces, start at the beginning of time $t = t_0$ (here, I assume the universe had a start), and measure proper time τ along fundamental world lines from $t = 0$ to the present event $t = 0$. Then the surfaces $\{\tau = const\}$ are the surfaces of constant proper times since the start of the universe that determine the present transition time $\tau = \tau_0$ at each instant as time evolves along the fundamental world lines. these are uniquely defined, as there are no vacuum regions in a realistic model of the universe (the Cosmic background Radiation is non-zero everywhere) Thus the view is that so change happens on preferred surfaces, secondary to the timelike world lines, that need not even be spacelike.

[5] For realistic matter, this will be unique.

[6] The average motion of radiation takes place on timelike world lines: it is an almost perfect fluid with $p = /rho/3$.

IOP Conf. Series: Journal of Physics: Conf. Series **1275** (2019) 012001 doi:10.1088/1742-6596/1275/1/012001

The ADM formalism To express this in the ADM formalism [6], choose a family of spacelike surfaces with lapse function $N(x^i)$, metric $g_{ij}(x^k)$, and second fundamental form $\pi_{ij}(x^k)$, together with a family of timelike world lines with shift vector $N_i(x^j)$. The field equations for $g_{ij}(x^k)$ are as follows (where 3-dimensional quantities have the prefix (3)): four constraint equations

$$^{(3)}R + \pi^2 - \pi^{ij}\pi_{ij} = 16\pi\rho_H, \tag{23}$$

$$R^\mu := -2\pi^{\mu j}{}_{|j} = 16\pi T^\mu_0, \tag{24}$$

where $|j$ represents the covariant derivative in the 3-surfaces, and twelve evolution equations

$$\partial g_{ij}/\partial t = 2Ng^{1/2}(\pi_{ij} - \frac{1}{2}g_{ij}\pi) + N_{i|j} + N_{j|i}, \tag{25}$$

$$\partial \pi_{ij}/\partial t = -Ng^{-1/2}(^{(3)}R_{ij} - \frac{1}{2}g_{ij}{}^{(3)}R) + F_{ij}(\pi, N, N_{|i}) + 16\pi^{(3)}T_{ij}. \tag{26}$$

We choose a unique gauge by specialising the time surfaces and flow lines in that formalism to those defined above.

1. We choose the flow lines to be Ricci Eigenlines:

$$T^\mu_0 = 0 \Rightarrow R^\mu = 2\pi^{\mu j}{}_{|j} = 0 \tag{27}$$

which algebraically determines the shift vector Ni(xj), thereby solving the constraint equations (24);

2. We determine the lapse function N(xi) by the condition that the time parameter t measures proper time along the fundamental flow lines:

$$ds^2 = -d\tau^2 \Rightarrow N^2 = 1 + N^i N_i \tag{28}$$

These conditions together uniquely determine the lapse and shift.
What determines how the matter evolves? Equations of state for the matter terms $\{\rho_H, T^\mu_0, {}^{(3)}T_{ij}\}$ in (23), (24), (26) must be added, and the matter conservation equations (4) added to the set of equations to be satisfied. Then the evolution is determined:

> **ADM Theorem**: *Given the equations of state and dynamical equations for the matter, equations (25), (26) determine the time evolution of the metric in terms of proper time along the fundamental flow lines from initial data at some time t_0; if they are initially satisfied at time $\{t = t_0\}$, the constraints (23), (24) are conserved because of energy-momentum conservation (4).*

The development of spacetime with time takes place just as is the case for other physical fields, with the relevant time parameter being proper time along the fundamental flow lines. There is no problem with either the existence or the rate of flow of time. The spacetime develops accordingly via (25), (26). Times passes at the rate of 1 second per second [63], as determined by the metric tensor locally at each event. In this evolving block universe, the present is where the future changes to the past, which takes place along timelike worldlines [39]. This construction automatically provides chronology protection, because a space-time singularity will occur if these eigenlines intersect.

IOP Conf. Series: Journal of Physics: Conf. Series **1275** (2019) 012001 doi:10.1088/1742-6596/1275/1/012001

Predictability Do these equations mean the spacetime development is uniquely determined to the future and the past from initial data? That all depends on the equations of state of the matter content. For a perfect fluid with barotropic equation of state $p = p(\rho)$, the answer is yes. However one can have an explicitly time dependent equation of state:

$$p = 1/3 g^{ij}\,{}^{(3)}T_{ij}, \quad \pi_{ij} = {}^{(3)}T_{ij} - p g_{ij} = F(\tau)\Pi_{ij}(0) \tag{29}$$

where $F(\tau)$ represents local dynamics, which could possible involving random processes generated via quantum uncertainty (indeed this is the mechanism that underlies the possibility of Figure 1 above). So equations (25), (26) determine the time evolution of the spacetime, but do not guarantee predictability; this is what underlies the uncertainty of evolution depicted in Figure 1. If quantum unpredictability gets amplified to macro scales, the spacetime evolution is intrinsically undetermined till it happens (as during the generation of seed inhomogeneities in the inflationary era in the very early universe [66]).

The quantum gravity case This discussion has looked at physics occurring in an evolving classical spacetime, but with quantum fields in that spacetime. One should require the same kind of evolving spacetime structure to occur in a viable theory of quantum gravity. That seems to be the case for example for both spinfoam [72] [11], representing a growing spacetime, and causal set theory [23] [90] [91].

5. Contextual effects

Because gravity is the dominant long range force (as a consequence of gravitational mass always being positive), so larger scale effects can accumulate to influence smaller scales in crucial ways, and can even diverge.[7]. Thus top-down (contextual) effects occur ([36]: Chapter 6).

5.1. The traditional contextual effects

The traditional such top-down effects (i.e. those discussed before 1965) are [78] [17] [31]

- *Mach's Principle*: what is the origin of inertia? Newton's bucket experiment suggests it is related to the most distant matter in the universe.
- *Olber's paradox*: why is the night sky dark? A simple calculation in a static universe suggests the energy density of radiation received from all the stars in the universe should diverge (each shell with the same $\Delta R := R_2 - R_1$ adds the same amount of radiation).
- *The arrow of time*: how does a direction of time arise out of underlying time-symmetric physics? Boltzmann's derivation of the H-theorem from kinetic theory does not in fact explain the Second Law of Thermodynamics $dS/dt > 0$, because that derivation remains true if one sets $t \rightarrow t' := -t$, thus it also shows that $dS/dt' > 0$.

The issue of boundary conditions at infinity influencing local conditions led to Einstein's and Wheeler's preference for spatially closed universes, where infinity does not occur [88] (in concordance with Section 1). Olber's paradox is now resolved by our Hot Big Bang models [26] [74] which explain why the temperature of the night sky is $2.7^o K$. Mach's principle is unresolved. The arrow of time remains a key issue, discussed below (Section 5.3).

[7] Which is why Newton never succeeded in deriving a cosmological theory based in summation of gravitational forces acting on a particle in an infinite universe: this is the spatial version of Olber's paradox.

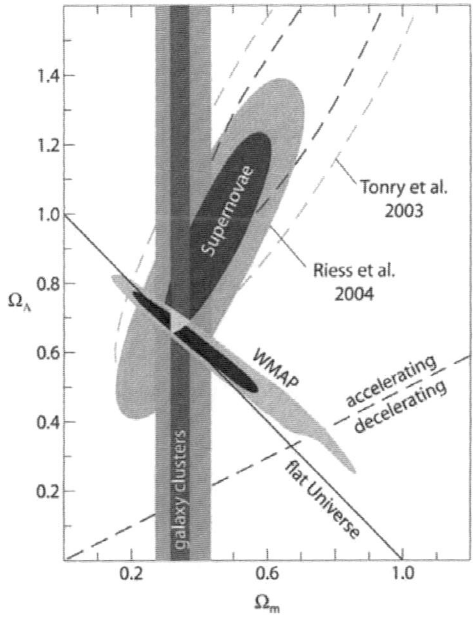

Figure 3. *Testing model by standard candles (supernovae) versus by contextually shaped outcomes of structure formation in the expanding universe (galaxy clusters), and their effect on CMB ansiotropies as measured by WMAP. The latter two top-down effects give stronger constraints on the cosmological matter density parameter Ω_m and cosmological constant parameter Ω_Λ than the direct measurements of cosmological parameters via the supernovae.* Credit: ESO

5.2. Emergence of structure: cosmology and astrophysics

The theme that the flow of causation is not only bottom up, from smaller to larger scales, but it is also top down [36], now has other important manifestations in the standard cosmology that has developed since the 1960s [22] [26] [74]. Specifically

- *Nucleosynthesis* in the early universe depends on the cosmological rate of expansion $a(t)$ at that time, which is determined by the total cosmological density $\rho(t)$ of matter and radiation then (a global parameter);
- *Structure formation* in the later universe is similar: gravitational instability develops structure (density inhomogeneities $\delta\rho$) in a way that also depends on the background expansion rate $a(t)$, which again depends on the cosmological model parameters $\rho(t), \Lambda$; for example in the pressure-free case, the comoving fractional density parameter $\mathcal{D}$ grows according to [38]

$$\ddot{\mathcal{D}} + \frac{2}{3}\theta\dot{\mathcal{D}} - \frac{1}{2}\kappa\rho\mathcal{D} = 0 \tag{30}$$

  where θ is the cosmological rate of expansion, determined by the Friedmann equation, and ρ is the cosmological density of matter.[8]
- Because these processes depend on the background cosmology, observing their outcomes is the most sensitive way of determining the background model parameters. Thus the

[8] Thus in a static universe $a(t) = a_0$, $\delta\rho$ grows exponentially, while in a universe expanding as a power law, $\delta\rho$ grows as a a power law.

perturbation equations determine the power spectrum $P(k)$ of matter density fluctuations [22], which then determine the angular power spectrum $\mathcal{C}(l)$ of CMB anisotropies [77] [26]. Measuring these matter and radiation power spectra provides the best test of cosmological parameters [1] [2], see Figure 3.

Thus the fact that the cosmology provides the context for local existence determines the nature of what can exist! This applies also to the vexed question of why the cosmos is of such a nature as to allow life to exist (the *Anthropic* issue: [35]:§6.1), which I will not pursue further here.

5.3. The direction of time

Micro physics is time symmetric, except for extremely weak time-symmetry breaking by weak interactions, which does not affect everyday life at the present time.[9] How can a difference emerge between the future and the past, on the basis of time symmetric underlying microphysics?

- How does macro physics know the direction of time?
- Why does time flow the same direction everywhere?

There is no basis for such a determination in microphysics alone: because of the symmetry $T : \{t \to -t\}$ in that underlying foundation, the H-theorem derived from it (in the classical case by Boltzmann, and in the QFT case by Weinberg [87]) applies equally in both directions of time (this has to be so, because of the T symmetry of those derivations)[10]. Microphysics cannot provide a foundation for the second law of thermodynamics with a unique arrow of time. In an EBU (section 4.1), the arrows of time arises fundamentally because the future does not yet exist: a global asymmetry in the physics context, which affects local physics in a top-down way.

Historical effects In the evolving block universe (Section 4), the past exists and is developing to the future, which does not yet exist. It is this asymmetry that leads to the arrows of time associated with historical effects. One can be influenced at the present time from many causes lying in our past, as they have already taken place and their influence can thereafter be felt today. One cannot be physically influenced by causes coming from the future, for they have not yet come into being. This is the rationale for saying the past exists but the future does not: if something can influence you, it exists. The past affects today as regards:

- Existence of elements (C,N,O, ...) on Earth, which arose in the past through primordial and stellar nucleosynthesis
- Arrival of particles (cosmic rays) due to events which took place in the past.
- Arrival of radiation from the past, both from nearby sources (e.g. the Sun) and the distnace past (e.g. the Cosmic Black Body Radiation).

We have no such effects (at the macro scale) causing effects here and now due to future events; indeed they do not yet exist, as what they will be is not yet determined, due to quantum uncertainty (Section 4.1). In summary: the future can't affect today, because it is not yet definite what it is: future possibilities exist, not specific outcomes. That is a core feature of quantum theory (Section 6).

Local Arrows of time This broken symmetry then affects microphysics. The global direction of time leads to local *Arrows of Time*, which include quantum, electrodynamic, gravitational, thermodynamic, biological, and psychological arrows of time in the local physical context. Most important are

[9] It might have played a role in the emergence of the matter-antimatter asymmetry in the early universe.
[10] Weinberg's derivation does not include the aspects of the weak force that break time symmetry.

IOP Conf. Series: Journal of Physics: Conf. Series **1275** (2019) 012001 doi:10.1088/1742-6596/1275/1/012001

- The *Second Law of Thermodynamics*, crucially affecting macrophysics, physical chemistry, and biology [28]. This arrow occurs only because of special initial conditions in the early universe (a *Past Condition* [5] [70]): the universe must start off in a low entropy (smooth) state, else entropy cannot increase. This condition is required in order that the thermodynamic arrow of time be well defined and agree with the Direction of Time.[11]
- The *electrodynamic arrow of time*: Electromagnetic radiation is received after it is sent. From a QED viewpoint, this is because the Feynman propagator can only be integrated over the past, as the future spacetime domain does not yet exist in an EBU. Advanced and retarded Feynman propagators are not on an equal footing,because of the cosmological context. The same will apply to gravitational radiation.
- A *quantum arrow of time*, discussed in the next section.

The other arrows of time (biological, psychological) will follow from these. They are all determined in a top-down way by the EBU context, together with the Past Condition.

6. Theories must adequately take account of the quantum measurement issue

It is crucial for physical theories to face the measurement problem for quantum physics, and for foundational theories to take some stance on this issue as regards quantum gravity. The key point is that we only get specific results from quantum physics, via the Born rule, if some form of collapse of the wave function to an eigenstate occurs; otherwise the probabilities represented by the wave function ψ are never instantiated, hence it has no physical meaning. In brief, an ensemble of events, to which one can apply statistics, only exists if individual events occur.

One particular area where this is relevant in cosmology is the issue of how inflationary perturbations become classical, which is ignored by almost everyone except Sudarsky and colleagues (e.g. [58]). Until this is resolved, there is a major lacuna in current cosmology.

Measurement Problem The quantum measurement process is, according to the standard view[12] ([46], [32] and references there) that the quantum to classical transition is associated with quantum *events*: collapse of the wave function to an eigenstate, which happens in a contextual way [25] (e.g. what is measured in an experiment depends on what apparatus is used and when it is turned on). If a measurement of an observable A takes place at time $t = t_*$, initially the wave function $\psi(x)$ is a linear combination of eigenfunctions $u_n(x)$ of the operator $\tilde{A}$ that represents A: for $t < t_*$, the wave function is

$$\psi_1(x) = \Sigma_n c_n u_n(x). \tag{31}$$

But immediately after the measurement has taken place, the wave function is an eigenfunction of the operator $\tilde{A}$:

$$\psi_2(x) = a_N u_N(x) \tag{32}$$

for some specific value N. The data for $t < t_*$ do not determine the index N; they just determine a probability $p_n = |c_n|^2$ for each outcome n (this is the Born rule). One can think of this process as due to the probabilistic time-irreversible collapse of the wave function; it cannot occur as a result of the unitary Schrödinger eqn. This process is irreversible: information is lost, because knowing N for $t > t_*$ does not determine the c_n for $t < t_*$. This irreversible nature of physics and loss of information at the micro scale implies an irreversible nature of physics, and loss of information at the macro level when micro uncertainty gets amplified to macro levels [71].

[11] Inflation does not solve this issue: see [70] and Section 7.
[12] There is a group of quantum cosmologists who vociferously defend the Everett ('many worlds') interpretation, but they are a very small fraction of physicists overall who work with quantum theory. The pilot wave theory also has strong proponents, but again they are a small minority of quantum theorists.

IOP Conf. Series: Journal of Physics: Conf. Series **1275** (2019) 012001 doi:10.1088/1742-6596/1275/1/012001

Contextual Wavefunction Collapse But the question is how does such a collapse process take place? In "Contextual Wavefunction Collapse: An integrated theory of quantum measurement" [25], Barbara Drossel and I propose one can solve the quantum measurement problem via the top-down effect of local physical context, with no extra *ad hoc* term introduced in the Schrödinger equation. This proposal is based in two previous papers: "On the limits of quantum theory: Contextuality and the quantum-classical cut" [32], which considers under what conditions one may expect quantum physics to describe a physical system as a whole (its component parts at the lowest levels will of course always be described by quantum theory), and "Ten reasons why a thermalized system cannot be described by a many particle wave function" [24].

These papers together make a strong case that there is no meaningful "Wave Function of the Universe" in any realistic sense (remember that quantum cosmology operates in the context of 'mini-superspace', in which all physical degrees of freedom except a few are ignored), and if there is, there is no reason to believe it will obey a unitary equation (*inter alia*, the universe contains many thermalised systems). On the contrary, physics is not causally complete at the quantum level: events take place, time passes, and information is lost at this level too.

Information loss? There is no law of information conservation in quantum physics, provided one takes measurement events into account (it was emphasized above that if we do not, the wavefunction has no meaning). On the contrary: information is lost at every "event", and in particular at every measurement. This is associated with the passage of time and the global direction of time, as discussed in [25]. The real world does not evolve in a unitary fashion! [32]. This may relate firstly to the "information loss paradox" as regards black holes (it may well be that one can regard the emission of Hawking radiation as involving wave function collapse and hence loss of information), and the issue of how quantum perturbations in inflation become classical.

7. Penrose entropy issue

The fine tuned initial state While it is often stated that inflation solves the flatness and horizon problems in the early universe, that is not in fact the case, as Penrose has pointed out in various writings that are summarised in Chapter 3 of *Fashion, Faith, and Fantasy* [70]. The essential point is that the maximum entropy of a given amount of matter is attained by collapsing it into a black hole; the most probable early universe state is a plethora of black holes. By contrast, "the Big Bang was an event of extraordinarily low entropy ... the gravitational degrees of freedom were completely suppressed" Penrose estimates the extraordinary precision that was involved in setting the initial state of the universe as $10^{10^{123}}$. This situation is not made clear by standard inflationary studies because they consider only perturbed Robertson-Walker geometries. It is not solved by inflation: rather inflation needs special initial conditions, not dominated by the large number of black holes that would correspond to maximum entropy, in order to start.

There is a counter view held by Kleban and colleagues [27] [56] [20] [21] that this is incorrect: inhomogeneous initial conditions will lead to inflation as usually claimed, and the Penrose argument is incorrect because the maximum entropy state is that given by de Sitter spacetime (because of the entropy associated with its cosmological event horizon.) I do not concur. While those models are certainly a great improvement over starting with perturbed FLRW models, they are not inhomogeneous enough to include the kind of multi-black hole situation envisaged by Penrose; and the (constant) entropy associated with that cosmological horizon, if it has any physical meaning at all, is only associated with observers in the very far future of the universe - long long after the present day, and even further to the future of the pre-inflationary situation that is Penrose' concern. It has no relevance to physical effects at those early time.

IOP Conf. Series: Journal of Physics: Conf. Series **1275** (2019) 012001 doi:10.1088/1742-6596/1275/1/012001

Gravitational entropy The issue here is the nature of gravitational entropy in general, not only in the case of black holes. The real gravitational entropy problem is the following: start off a gas in a very large container in an inhomogeneous state. Without gravity, the gas will relax to its most probable state of uniformity. Turn on Newtonian gravity, and gravitational instability will result in a non-uniform final state with matter clustered into stars. Turn on general relativity, and the final state will be a non-uniform state of matter clustered into stars and black holes ([70]:258). In the expanding universe, the latter is what indeed happens during the "dark ages" when the first stars form and black holes are not dominant, if they exist at all at that time. How this is compatible with the Second Law of Thermodynamics, which tell us the matter should tend to a uniform state? The only answer has to be that we must associate gravitational entropy in a general spacetime with the free gravitational field, that is, with the Weyl tensor (as implied by Penrose in [68], [69]:765-768).

Penrose has not given a specific formula for the gravitational entropy of a general spacetime (as opposed to a black hole, where everyone agrees on the Bekenstein-Hawking definition). However Clifton *et al* [18] propose such a formula for gravitational entropy based on the Bel-Robinson tensor: a square of the Weyl tensor that is the effective super-energy-momentum tensor of the free gravitational field. The detailed form of this entropy proposal in the Coulomb-like case shows that gravitational entropy increases as structure formation occurs in the expanding universe. This allows overall entropy (matter and gravity) to increase as structure forms, which is not true if one does not take this gravitational entropy into account. It is also in keeping with Penrose's Weyl curvature hypothesis [70] that the Weyl tensor must be zero in the limit at the initial singularity, which is a form of Past Condition to allow local thermodynamics to function the way it does (Section 5.3).

This gravitational entropy vanishes in the de Sitter universe; that spacetime therefore cannot be a state of maximum entropy, as claimed by Kleban. The same is true for any other definition of gravitational entropy based in the Weyl tensor. The most probable (highest entropy) initial gravity state is highly inhomogeneous, as claimed by Penrose.

Turning off gravity at very high densities Penrose proposes to solve this problem of how to avoid massive initial inhomogeneity by invoking a Conformal Cyclic Cosmology [70] This is a creative idea, but the mechanisms involved in realizing such a proposal are not all clear.

By contrast, consider a mechanism proposed by Greene et al [50]: suppose the strength of the gravitational force dies away at very high energies. In that case the collapse to black holes that causes the problems identified by Penrose will not occur: thermal processes will be able to overcome gravitational forces in the very early universe and prevent collapse to blackholes; the most probable state will be smooth. The gravitational degrees of freedom will not be suppressed, they will simply not dominate the degrees of freedom of ordinary matter as in the standard case considered by Penrose. Greene *et al* state [50], in response to the issue of collapse leading to highly inhomogeneous situations [68][69],

> *Our scenario does pass one crucial test, avoiding the double standard. Often natural-seeming scenarios, such as a ball rolling down a hill and coming to a stop due to friction, become completely unnatural viewed in time reverse. In our scenario, the time reverse is completely natural: gravity is strong and the universe starts in a clumpy state. Then at some point gravity turns off. The clumps start to dissolve into a smooth state. The state at all times is natural with respect to the laws operating at those times. Passing the double standard test requires that the universe spend a sufficient period of time in the weak gravity phase, otherwise there will not be sufficient time (in the time reversed sense) for a generic initial clumpy state to smooth out. In our model, we accomplish this by tuning the parameter h in the Lagrangian. The initial conditions are allowed to be generic.*

IOP Conf. Series: Journal of Physics: Conf. Series **1275** (2019) 012001 doi:10.1088/1742-6596/1275/1/012001

What is then particularly intriguing is that one can have a similar scenario with the Higgs as the inflaton [80]. This is highly desirable, as this is the only case in which the inflaton is related to testable and indeed tested particle physics [43].

Appendix A. Conformal gravity, local sources, and alternative theories of gravity
Mannheim and Kazanas [62] investigated the Newtonian limit of conformal gravity in and obtained the following exact equation of motion:[13]

$$\frac{3}{B(r)}(W^0{}_0 - W^r{}_r) = B'''' + \frac{4B'''}{r} = \frac{(rB)'''}{r} = \nabla^4 B = \frac{3}{4\alpha_g B(r)}(T^0{}_0 - T^r{}_r) = f(r) \qquad \text{(A.1)}$$

where $B(r) = -g_{00}(r)$. For a source localized to a region $r < r_0$ the solution is

$$B(r) = 1 - \frac{2\beta}{r} + \gamma r \qquad \text{(A.2)}$$

where

$$\gamma = -\frac{1}{2}\int_0^{r_0} dr' r'^2 f(r'), \qquad 2\beta = \frac{1}{6}\int_0^{r_0} dr' r'^4 f(r') \qquad \text{(A.3)}$$

They pointed out that if the source was a delta function then the coefficient β would be zero. Thus it had to have a more singular structure that would hold at the level of the microscopic sources in a star. Thus the source that gives the Newton term has to be more singular than the source that one uses in standard gravity. They also pointed out that if $f(r)$ was uniform, one would obtain $\beta \sim r_0^2\gamma$, and that could be ruled out by solar system measurements. Thus the source would have to be far from uniform.

Now in a conformal invariant theory, all mass has to be generated by dynamical symmetry breaking fields, and without them sources could not even localize into sources at rest. Thus the energy-momentum tensor of the source must contain dynamical fields and could not be approximated by a kinematic perfect fluid in which $\rho \gg p$ (a condition that would anyway violate the tracelessness of a conformal invariant energy-momentum tensor). In consequence one cannot infer the sign of the β integral by just giving it the same sign as the T_{00} term is given in GR. (In the conformal theory T_{rr} is just as big as T_{00} anyway.)

Alternative gravity Now one may or may not feel that this result rules out this particular theory. But the issue here is bigger than that. From Lovelock's Theorem [57], as nicely characterised by Clifton *et al* [19], the only alternative gravity options are that one can

(i) add other fields in addition to the metric tensor,
(ii) add higher than second order derivatives of the metric,
(iii) consider other than four spacetime dimensions,
(iv) consider non-locality, such as arises through the inverse d'Alembertian,
(v) propose emergent gravity, where the field equations do not arise from an action.

Any alternative theory of gravity (i) or (ii) that adds other fields or leads to higher derivative equations will face similar issues: they are not confined to this particular theory.

[13] I thank Prof Mannheim for the following note

References

[1] Ade, Peter AR, N. Aghanim, M. Arnaud, M. Ashdown, J. Aumont, C. Baccigalupi, A. J. Banday et al. "Planck 2015 results-xiii. cosmological parameters." Astronomy and Astrophysics 594 (2016): A13.

[2] Aghanim, N., Y. Akrami, M. Ashdown, J. Aumont, C. Baccigalupi, M. Ballardini, A. J. Banday et al. "Planck 2018 results. VI. Cosmological parameters." arXiv preprint arXiv:1807.06209 (2018).

[3] Aharonov, Y; Bohm, D (1959). "Significance of electromagnetic potentials in quantum theory". Physical Review. 115 (3): 485491

[4] Aitchison, I and A Hey (2003). Gauge Theories in Particle Physics: A Practical Introduction. Institute of Physics

[5] Albert D (2000) *Time and Chance* (Cambridge, Mass: Harvard University Press)

[6] Arnowitt R, S. Deser, C. W. Misner (1962) "The dynamics of general relativity". In: Witten, L. (ed.) *Gravitation: An Introduction to Current Research* (Wiley, New York). Reprinted as a Golden Oldie in *Gen Relativ Gravit* (2008) 40:1989

[7] Alvarez, E. (2012). The Weight of matter. Journal of Cosmology and Astroparticle Physics, 2012(07), 002. arXiv 1204.6162

[8] Amarasinghe, A., Phelps, M. G., and Mannheim, P. D. (2018). Cosmological perturbations in conformal gravity II. arXiv preprint arXiv:1805.06807.

[9] Anandan J, Jo Christian and Kazimir Wanelik (1997). "Resource Letter GPP-1: Geometric Phases in Physics". Am. J. Phys. 65 (3): 180. arXiv:quant-ph/9702011

[10] Anderson P W (1972) "More Is Different" *Science* 177:393-396.

[11] Baez, J. C. (2000). An introduction to spin foam models of BF theory and quantum gravity. In Geometry and quantum physics (pp. 25-93). Springer, Berlin, Heidelberg.

[12] Ballesteros, Miguel, and Ricardo Weder. "The AharonovBohm effect and Tonomura et al. experiments: Rigorous results." Journal of Mathematical Physics 50.12 (2009): 122108.

[13] Berry M. V. (1984). "Quantal Phase Factors Accompanying Adiabatic Changes". Proceedings of the Royal Society A. 392 (1802): 4557

[14] Bezrukov, F. and Shaposhnikov, M. "The Standard Model Higgs boson as the inflaton" *Physics Letters B* **659**:703 (2008)

[15] Bezrukov, F. "The Higgs field as an inflaton" *Classical and Quantum Gravity* **30**:214001 (2013)

[16] Bombelli L, Joohan Lee, David Meyer, and Rafael D. Sorkin (1987) Space-time as a causal set Phys. Rev. Lett. 59, 521

[17] Bondi H (1960) *Cosmology* (Cambridge: Cambridge University Press)

[18] Clifton, T., Ellis, G. F., and Tavakol, R. (2013). A gravitational entropy proposal. Classical and Quantum Gravity, 30(12), 125009. arXiv 1303.5612

[19] Clifton, T., Ferreira, P. G., Padilla, A., and Skordis, C. (2012). Modified gravity and cosmology. Physics reports, 513(1-3), 1-189.

[20] Clough, K., Lim, E. A., DiNunno, B. S., Fischler, W., Flauger, R., and Paban, S. (2017). Robustness of inflation to inhomogeneous initial conditions. Journal of Cosmology and Astroparticle Physics, 2017(09), 025. http://arxiv.org/abs/arXiv:1608.04408

[21] Clough, K., Flauger, R., and Lim, E. A. (2018). Robustness of inflation to large tensor perturbations. Journal of Cosmology and Astroparticle Physics, 2018(05), 065. http://arxiv.org/abs/arXiv:1712.07352

[22] Dodelson, S (2003) *Modern Cosmology* (Academic Press).

[23] Dowker, F. (2005). Causal sets and the deep structure of spacetime. In 100 years of relativity: Space-time structure: Einstein and beyond (pp. 445-464).

[24] Drossel, B. (2017). Ten reasons why a thermalized system cannot be described by a many-particle wave function. Studies in History and Philosophy of Science Part B: Studies in History and Philosophy of Modern Physics, 58, 12-21. https://arxiv.org/abs/1509.07275.

[25] Drossel, Barbara, and George Ellis (2018). "Contextual wavefunction collapse: An integrated theory of quantum measurement." New Journal of Physics 20.11 (2018): 113025. https://arxiv.org/abs/1807.08171

[26] Durrer R (2008) *The Cosmic Microwave Background* (Cambridge: Cambridge University Press).

[27] East, W. E., Kleban, M., Linde, A., and Senatore, L. (2016). Beginning inflation in an inhomogeneous universe. Journal of Cosmology and Astroparticle Physics, 2016(09), 010.

[28] Eddington, A. (2012). The nature of the physical world: Gifford lectures (1927). Cambridge University Press.

[29] Einstein A (1919) "Spielen Gravitationsfelder im Aufbau der materiellen Elementarteilchen eine wesentliche Rolle?" *Sitzungsber. Preuss. Akad. Wiss. Berlin* (Math. Phys.) 1919: 349-356. English translation: "Do gravitational fields play an essential part in the structure of the elementary particles of matter?" In *The Principle of Relativity* (NY: Dover Publications, 1952), 189-198.

[30] Ellis G F R (1971) "General relativity and cosmology". In *General Relativity and Cosmology*, Varenna Course No. XLVII, ed R. K. Sachs (Academic, New York). Reprinted as Golden Oldie, *General Relativity*

and Gravitation **41**: 581-660 (2009).

[31] Ellis, G. F. (2002). Cosmology and local physics. New Astronomy Reviews, 46(11), 645-657.

[32] Ellis G (2011) "On the limits of quantum theory: Contextuality and the quantum-classical cut" (Ellis:https://arxiv.org/abs/1108.5261)

[33] Ellis, G. F. R. (2013). The arrow of time and the nature of spacetime. *Studies in History and Philosophy of Science Part B: Studies in History and Philosophy of Modern Physics*, **44**:242-262. arXiv:1208.2611

[34] Ellis, George FR.(2014) "The trace-free Einstein equations and inflation." General Relativity and Gravitation 46, no. 1 : 1619.

[35] Ellis, G. F. R. (2014). On the philosophy of cosmology. Studies in History and Philosophy of Science Part B: Studies in History and Philosophy of Modern Physics, 46, 5-23.

[36] Ellis, G. (2016). How can Physics Underlie the Mind?. New York, NY, USA: Springer: Berlin/Heidelberg, Germany.

[37] Ellis G F R (2017) "Editorial note to: E. Lifshitz, On the gravitational stability of the expanding universe" *General Relativity and Gravitation* **49**:17

[38] Ellis, G. F., and Bruni, M. (1989). Covariant and gauge-invariant approach to cosmological density fluctuations. Physical Review D, 40(6), 1804.

[39] Ellis, G. F., and Goswami, R. (2014). Spacetime and the Passage of Time. In Springer Handbook of Spacetime (pp. 243-264). Springer, Berlin, Heidelberg. arXiv1407.7243

[40] Ellis, GF R, Maartens, R,and Maccallum,M A H (2012) *Relativistic Cosmology* (Cambridge: Cambridge University Press)

[41] Ellis, G. F., Meissner, K. A., and Nicolai, H. (2018). "The physics of infinity" *Nature Physics* **14**:770.

[42] Ellis, G., and Silk, J. (2014). Scientific method: Defend the integrity of physics. Nature News, 516(7531), 321.

[43] Ellis G and JP Uzan (2014) "Inflation and the Higgs particle" Astronomy and Geophysics 55:1.19-1.20

[44] Ellis, G. F., Van Elst, H., Murugan, J., and Uzan, J. P. (2011). On the trace-free Einstein equations as a viable alternative to general relativity. Classical and Quantum Gravity, 28(22), 225007.

[45] Feynman, R P, and Leighton, R. B (1964). *The Feynman Lectures on Physics Volume I* (Addison Wesley) Section 1-2, p. 1-2

[46] Feynman, R P, and Leighton, R. B. (1964). *The Feynman Lectures on Physics Volume III* (Addison Wesley).

[47] Feynman, R. P., Hibbs, A. R., and Styer, D. F. (2010). Quantum mechanics and path integrals. Courier Corporation.

[48] Finkelstein David R, Andrei A. Galiautdinov and James E. Baugh (2001) Unimodular relativity and cosmological constant J. Math. Phys. 42 340346 [arXiv:gr-qc/0009099v1].

[49] Giles, R. (1981). "Reconstruction of Gauge Potentials from Wilson loops". Physical Review D. 24 (8): 2160

[50] Greene, B., Hinterbichler, K., Judes, S., and Parikh, M. (2011). Smooth initial conditions from weak gravity. Physics Letters B, 697(3), 178-183.

[51] Gross David J. (1992) " Gauge Theory-Past, Present, and Future?" Chinese Journal of Physics **30**:955-972.

[52] Hawking S W and Ellis G F R (1973) *The Large Scale Structure of Space-Time* (Cambridge: Cambridge University Press)

[53] Henson, Joe. "The causal set approach to quantum gravity." Approaches to quantum gravity: towards a new understanding of space, time and matter 393 (2009).

[54] Hilbert D (1924) "On the infinite", in *David Hilbert's Lectures on the Foundations of Arithmetics and Logic 1917-1933*. W Ewald and W Sieg, eds. (Heidelberg: Springer-Verlag, 2013), p. 730.

[55] Hilbert D (1926) "On the infinite", Mathematische Annalen (Berlin) **95**:161-90. Reprinted in *Philosophy of mathematics: Selected readings* Ed P Benacerraf and H Putnam (Cambridge: Cambridge University Press) [available at https://math.dartmouth.edu/ matc/Readers/HowManyAngels/Philosophy/Philosophy.html]

[56] Kleban, M., and Senatore, L. (2016). Inhomogeneous anisotropic cosmology. Journal of Cosmology and Astroparticle Physics, 2016(10), 022. http://arxiv.org/abs/arXiv:1602.03520

[57] Lovelock, D. (1971). "The Einstein Tensor and Its Generalizations". Journal of Mathematical Physics. 12 (3): 498501. doi:10.1063/1.1665613

[58] Landau, Susana J., Claudia G. Scccola, and Daniel Sudarsky. "Cosmological constraints on nonstandard inflationary quantum collapse models." Physical Review D 85.12 (2012): 123001. arXiv 1112.1830v3

[59] Maartens, R., and Bassett, B. A. (1998). Gravito-electromagnetism. Classical and Quantum Gravity, 15(3), 705.

[60] Mannheim, Philip D. "Making the case for conformal gravity." Foundations of Physics 42, no. 3 (2012): 388-420. ArXiv e-print 1101.2186

[61] Mannheim, P. D. (2012). Cosmological perturbations in conformal gravity. Physical Review D, 85(12), 124008.

[62] Mannheim P D and D. Kazanas (1994), "Newtonian limit of conformal gravity and the lack of necessity of the second order Poisson equation", *General Relativity and Gravitation* **26**, 337 (1994)

IOP Conf. Series: Journal of Physics: Conf. Series **1275** (2019) 012001 doi:10.1088/1742-6596/1275/1/012001

[63] Maudlin, Tim. *Philosophy of physics: Space and time.* Princeton University Press, 2012.
[64] Murugan, J., Weltman, A., and Ellis, G. F. (Eds.) (2012). *Foundations of Space and Time: Reflections on Quantum Gravity.* Cambridge University Press.
[65] Martin J, Ringeval C, and Vennin, V (2014) "Encyclopdia inflationaris" *Physics of the Dark Universe* **5**:75-235. arXiv:1303.3787
[66] Mukhanov, V. (2005). Physical foundations of cosmology. Cambridge university press.
[67] Osakabe, N; et al. (1986). "Experimental confirmation of AharonovBohm effect using a toroidal magnetic field confined by a superconductor". Physical Review A. 34 (2): 815822.
[68] Penrose, R (1979). Singularity and Time-Asymmetry. In Hawking, S. W., and Israel, W. (Eds). *General Relativity: An Einstein Centenary Survey* (Cambridge: Cambridge UNiversity Press):581-638.
[69] Penrose, R. (2004). The road to reality: A complete guide to the physical universe. Jonathan Cape.
[70] Penrose, R (2016) Fashion, Faith, and Fantasy in the New Physics of the Universe (Princeton: Princeton UNiversity Press)
[71] Percival, I. (1991). Schrdinger's quantum cat. Nature, 351(6325), 357.
[72] Perez, A. (2013). The spin-foam approach to quantum gravity. Living Reviews in Relativity, 16(1), 3.
[73] Peskin, Michael E., and Daniel V. Schroeder. An Introduction To Quantum Field Theory (Frontiers in Physics). Westview Press Incorporated, 1995.
[74] Peter P and Uzan J-P (2013) *Primordial Cosmology* (Oxford Graduate Texts).
[75] Rovelli, C. (2008). Loop quantum gravity. Living reviews in relativity, 11(1), 5.
[76] Rubio, J. "Higgs inflation" arXiv:1807.02376 (2018)
[77] Sachs, R. K., and Wolfe (2007). Republication of: Perturbations of a cosmological model and angular variations of the microwave background. General Relativity and Gravitation, 39(11), 1929-1961.
[78] Sciama, D. W. (2012). The unity of the universe. Courier Corporation.
[79] Sloan D (2018) Dynamical Similarity Phys. Rev. D 97, 123541 [arXiv:1803.04472]
[80] Sloan D, George Ellis (2018) "Solving the Cosmological Entropy Issue with a Higgs Dilaton" arXiv preprint 1810.06522
[81] Tong D (2018) "Gauge Theory" Lecture notes, DAMTP, Cambridge
[82] Tonomura, Akira, et al "Observation of Aharonov-Bohm effect by electron holography." Physical Review Letters 48, no. 21 (1982): 1443.
[83] Tonomura, A., et al (1986). Evidence for Aharonov-Bohm effect with magnetic field completely shielded from electron wave. Physical Review Letters, 56(8), 792.
[84] Smolin L(2009) Quantization of unimodular gravity and the cosmological constant problems Phys. Rev. D 80 084003 [arXiv:0904.4841v1 [hep-th]].
[85] Unruh William G. (1989) Unimodular theory of canonical quantum gravity Phys. Rev. D 40 10481052.
[86] Weinberg S W (1989) "The Cosmological Constant Problem" *Rev. Mod. Phys.* **61**:1.
[87] Weinberg, S. (1995). The quantum theory of fields (Vol. 2). Cambridge university press.
[88] Wheeler, J A *Einsteins vision* (1968). (Berlin: Springer-Verlag).
[89] Wilson, K. (1974). "Confinement of quarks". Physical Review D. 10 (8): 2445.
[90] Wüthrich Christian (2015) Spacetime from causality: causal set theory
[91] Wüthrich, C., and Callender, C. (2016). What Becomes of a Causal Set?. The British Journal for the Philosophy of Science, 68(3), 907-925. https://arxiv.org/pdf/1502.00018
[92] Yang C. N., Mills R. L. (1954). "Conservation of Isotopic Spin and Isotopic Gauge Invariance". Phys. Rev. 96: 191195.

IOP Conf. Series: Journal of Physics: Conf. Series **1275** (2019) 012002 doi:10.1088/1742-6596/1275/1/012002

Non-locality and quasiclassical reality in Kent's formulation of relativistic quantum theory

J Butterfield[1], **B Marsh**[2]

[1] Trinity College, Cambridge CB2 1TQ, UK

[2] SITP and Department of Applied Physics, Stanford University, Stanford CA 94305, USA

E-mail: jb56@cam.ac.uk, marshbp@stanford.edu

Abstract. We report Adrian Kent's proposed framework for a realist, one-world, Lorentz-invariant formulation of quantum theory. The idea is to postulate a final boundary condition: in effect, a late-time distribution of mass-energy recording how photons scattered off macroscopic objects. Nature selects this final boundary condition with the orthodox late-time Born probability; and this defines the probability space of events, to give a realist quantum theory. We emphasize two topics. First, we consider this formulation's verdicts about traditional locality conditions, such as Outcome Independence and Parameter Independence. Second, we discuss a possible amendment to Kent's proposal that, roughly speaking, allows for the emergence of a quasiclassical history even when mass-energy is shielded or delayed from appearing in the final boundary condition.

1. Introduction

Kent proposes a solution to quantum theory's measurement problem [1, 2, 3] that is realist, one-world, and relativistic, in contrast to 'the usual suspects' of e.g. Bell's 'Six possible worlds of quantum mechanics' [4]. The idea is to postulate a final boundary condition: in effect, a late-time distribution of mass-energy recording how photons scattered off macroscopic objects. Nature selects one such final boundary condition with the late-time Born probability, and so the probability space of events is defined. Given the final boundary condition, a Lorentz-invariant rule defines a generalized field of beables in spacetime describing the quantum reality of the system. We will: (i) set the proposal in the context of quantum non-locality; (ii) explore a cousin proposal with a different Lorentz-invariant rule used to formulate beables. Our discussion will summarize and develop our recent work [5, 6, 7] in this area.

2. Peaceful coexistence?

Thirty years ago, Shimony hoped to reconcile quantum non-locality with special relativity—to secure 'peaceful coexistence'—by carefully isolating the culprit in the derivations of the Bell inequality. The main assumption of a Bell inequality is factorizability:

$$\forall\lambda; \forall x, y; \forall X, Y = \pm 1 : \ \ \mathrm{Pr}_{\lambda,x,y}(X \& Y) = \mathrm{Pr}_{\lambda,x}(X) \cdot \mathrm{Pr}_{\lambda,y}(Y) \ ; \tag{1}$$

where λ encodes a complete state, x and y are choices of possible measurements in the left and right wing of a Bell experiment, and X and Y possible outcomes. Factorizability is the

Content from this work may be used under the terms of the Creative Commons Attribution 3.0 licence. Any further distribution of this work must maintain attribution to the author(s) and the title of the work, journal citation and DOI.
Published under licence by IOP Publishing Ltd

IOP Conf. Series: Journal of Physics: Conf. Series **1275** (2019) 012002 doi:10.1088/1742-6596/1275/1/012002

conjunction of two conditions, Parameter Independence (PI) and Outcome Independence (OI):

$$\forall\lambda; x, y; X, Y = \pm 1: \ \Pr_{\lambda,x}(X) = \Pr_{\lambda,x,y}(X) := \Pr_{\lambda,x,y}(X \& Y) + \Pr_{\lambda,x,y}(X \& \neg Y)\ ; \tag{2}$$

$$\forall\lambda, x, y; X, Y = \pm 1: \ \Pr_{\lambda,x,y}(X \& Y) = \Pr_{\lambda,x,y}(X) \cdot \Pr_{\lambda,x,y}(Y)\ . \tag{3}$$

With λ as the quantum state, PI ensures that the statistics in one wing are not correlated with which quantity is chosen to be measured in the other wing. So for spacelike measurement events, PI ensures the impossibility of superluminal signaling (the information in such a signal being which quantity has been chosen to be measured). This is essentially the content of the no signalling theorem, which is derived in quantum theory from the commutation of quantities in the left wing with those on the right. So Shimony suggested that to get peaceful coexistence, it is enough to deny OI [8, 9].

On the other hand, consider the verdicts about PI and OI made by the pilot-wave theory. In its usual version, we take λ as the conjunction of the quantum state and the particles' possessed positions. OI holds thanks to the theory's determinism. But PI fails since the velocity of each particle depends instantaneously on the position of the other, thanks to $\mathbf{p}_i = \nabla_i S$, where $S \equiv S(\mathbf{x}_1, \mathbf{x}_2)$ is the phase of the wave function in configuration space. Since the pilot-wave theory (in this version) is usually considered incompatible with special relativity, these two verdicts seem to support Shimony's contention that denying OI and endorsing PI secures peaceful coexistence.

But we submit that one should not make a fundamental distinction between parameters (i.e. apparatus-settings) and outcomes (i.e. measurement results). For they are both macrophysical facts: so presumably, they are on equal terms, as regards whether a curious (i.e. unscreenable-off) correlation between examples of them at spacelike separation violates relativity theory. This seems to be the viewpoint of John Bell, especially in his last essays, 'Against measurement' [10] and 'La nouvelle cuisine' [11]. As Bell might have put it: 'surely Nature, in her causal structure, does not care whether a macrophysical fact is 'controllable' by humans, in the way a knob-setting, but not a pointer-reading, is—or at least seems to be?' More generally, the fact that the parameter/outcome distinction in PI and OI is schematic reinforces the point that having verdicts about PI and OI sufficient to avoid a Bell inequality does not amount to having a solution to the quantum measurement problem. So proposed solutions to that problem still need to be addressed.

For assessing Kent's proposal (in Section 3 et seq.), it will be important to recall that Bell's theorem predicts observable probabilities by averaging over λ. For example, letting a_1 be a left-wing setting (i.e. a value of x) and A_1 the corresponding outcome:

$$\Pr(A_1 = +1) := \int_\Lambda \Pr_{\lambda,a_1}(+1)\ d\rho\ . \tag{4}$$

Here, the fact that ρ is independent of which quantity, e.g. a_1 vs. a_2, is measured, encodes a locality assumption. Namely, there is no correlation between

(i) the causal factors influencing which quantity, e.g. a_1 vs. a_2, is measured, and

(ii) the causal factors influencing which value of λ is realized.

This *no conspiracy* assumption is very reasonable if: (i') the causal factors (i) are localized in the wings of the experiment, and (ii') the causal factors (ii) are localized in the central source. But with an eye on our later discussion of Kent's proposal, we should ask: what if λ encodes facts about a final boundary condition? In particular: what if λ encodes information in that boundary condition about traces (records) of earlier choices of measurement? The answer is surely that, if so, a probability distribution over values of λ may well depend on such choices—so that the label 'no conspiracy' is tendentious, since such dependence is in no way suspicious.

IOP Conf. Series: Journal of Physics: Conf. Series **1275** (2019) 012002 doi:10.1088/1742-6596/1275/1/012002

3. Kent's formulation of quantum theory

Kent's proposal is realist, one-world, and without state reduction. In those three features, it is like the pilot-wave theory. But it is relativistic, and its beables are very different from those of the pilot-wave theory.

Kent's main idea can be stated for Minkowski spacetime, as follows. He takes the universal quantum state to evolve unitarily, and to determine Born-rule probabilities for the outcomes of a *hypothetical* mass-energy density measurement at all points of an asymptotic late-time hypersurface S. Then he supplements the orthodox description of the quantum system given by the state, with one such mass-energy density distribution—one which "Nature selects" at random, with Born-rule probabilities prescribed by the universal quantum state at the late "time" S. Then he proposes a Lorentz-invariant rule that invokes orthodox quantum correlations with whichever distribution of mass-energy density on S "Nature has selected", so as to define beables at points of spacetime earlier than S. Thus Kent supplements quantum theory with asymptotic mass-energy distributions so as to define beables—facts about quantities' values going beyond quantum orthodoxy—at earlier points of spacetime.

The physical idea of his rule defining the beables is decoherence. In short: that a photon registers on S in one place rather than another—as encoded in the randomly selected mass-energy distribution on S—is correlated with its having earlier scattered off some macroscopic object in one way rather than another: and this scattering typically encodes the location of the macroscopic object. Thus the selected mass-energy distribution on S acts as a record of facts, like macroscopic objects' locations, at earlier times. The idea is that enough photons (or other particles) register, so as to record many definite locations of objects at earlier times: sufficiently many to imply an intuitively satisfying macroscopic world, or 'quasiclassical realm'—and so to give a realistic and one-world solution of the measurement problem. Besides, the fact that the photons travel along null (lightlike) curves makes the correlations encoded in these records suitably Lorentz-invariant.

This overview of Kent's proposal is summarized in Figure 1.

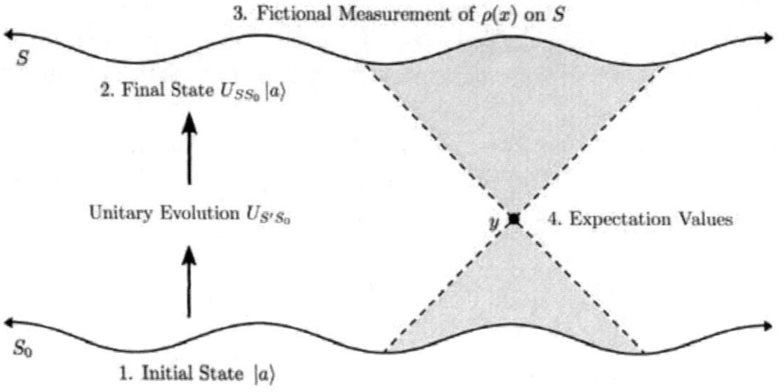

Figure 1. Overview of Kent's proposal for deriving beables for any spacetime point y between an initial spacelike hypersurface S_0 and a final spacelike hypersurface S. 1. An initial quantum state is provided on an initial spacelike hypersurface S_0. 2. The state unitarily evolves to any spacelike hypersurface S' in the future of S_0 via an appropriate unitary operator $U_{S'S_0}$ provided by the Tomonaga-Schwinger formalism [12]. 3. The final state $U_{SS_0}|a\rangle$ on S determines the Born rule probability for any given final boundary condition. 4. A single final boundary condition is used in a Lorentz-invariant rule to compute beables at spacetime point y, using only the part of the final boundary condition *outside* the future light cone of y.

We should add that this overview has simplified in two related respects. First: Kent's proposal

IOP Conf. Series: Journal of Physics: Conf. Series **1275** (2019) 012002 doi:10.1088/1742-6596/1275/1/012002

does not require Minkowski spacetime. He discusses how it can be transcribed to a large range of cosmologies, that admit late hypersurfaces on which—one can reasonably suppose—enough photons (or other particles) register, so as to record sufficiently many definite locations of objects that one intuitively wants (in a realistic, one-world solution of the measurement problem) to be definitely located. But in this paper, we can keep Minkowski spacetime in mind, as our main example. Second: the proposal does not need to postulate some single hypersurface S, the various possible mass-energy distributions on which are then the main ingredient for defining the extra beables: i.e. the facts that supplement quantum orthodoxy, analogous to the corpuscles' positions according to pilot-wave theory. After all, what matters for a realistic one-world solution of the measurement problem is a well-defined probability space of events: just *one* of which happens. Thus Kent envisages defining such a space by a sequence of ever-later hypersurfaces and their associated probability distributions: i.e. by a convergent sequence of probability distributions. But for simplicity of exposition, we will always talk about just one such late hypersurface S.

To sum up: Kent proposes to solve the measurement problem by a notional detection of photons at a late time. We are to think of a notional Lüders-rule measurement of mass-energy density at each point on S. The measured outcome distribution: (i) is correlated with where and when photons scattered off macroscopic objects at earlier times, and so: (ii) defines the probability space of real events.

In Section 3.1 et seq., we will illustrate these ideas with two of Kent's toy models of photons propagating to S after scattering off an object. But first we further clarify the proposal in general terms, with three numbered comments.

1. Supplementing decoherence

Recall the main idea of decoherence. It is not so much the destruction of coherence (i.e. of quantum probability distributions' interference terms that are characteristic of a superposition as against a mixture), but rather the *diffusion* of coherence from the system to its environment.

Thus in the paradigm case studied in e.g. [13] of photons scattering off the pointer of a macroscopic apparatus, the reduced state (density matrix) of the pointer very rapidly becomes nearly diagonal in a quantity, such as position of its centre of mass, which—to secure a definite 'quasiclassical' macroscopic world—one intuitively wants to be definite in value. But decoherence does not select a single component of the reduced state / density matrix of the system. Nor does it justify attributing a vector state to the system. For the reduced state cannot be interpreted as a matter of ignorance.

This last point is independent of the physical process of decoherence; and is of course, the source of the measurement problem, i.e. quantum orthodoxy's lack of values for quantities, e.g. macroscopic positions, that one intuitively wants to be definite. It was emphasized long ago by D'Espagnat, who suggested calling such reduced states *improper mixtures*, as against the straightforwardly ignorance-interpretable proper mixtures describing e.g. an impure beam of electrons prepared by a less-than-perfect electron-gun [14].

Indeed, it was emphasized *longer* ago by Schrödinger in his great 1935 papers, that introduced his famous cat and the word 'entanglement'. Thus in the first paper, 'The present situation in quantum mechanics' [15, 16], he says this 'is the most difficult and most interesting point of the theory'. He summarizes as follows: 'Maximal knowledge of a total system does not necessarily include total knowledge of all its parts, not even when these are fully separated from each other and at the moment are not influencing each other at all. ... Best possible knowledge of a whole does not necessarily include the same for its parts. ... The whole is in a definite state, the parts taken individually are not.' And he reinforces the point in a dialogue with an imagined interlocutor:

"How so? Surely a system must be in some sort of state."

IOP Conf. Series: Journal of Physics: Conf. Series **1275** (2019) 012002 doi:10.1088/1742-6596/1275/1/012002

"No. State is ψ-function, is maximal sum of knowledge. I didn't necessarily provide myself with this, I may have been lazy. Then the system is in no state."

"Fine, but then too the agnostic prohibition of questions is not yet in force and in our case I can tell myself: the subsystem is already in some state, I just don't know which."

"Wait. Unfortunately no. There is no 'I just don't know'. For as to the total system, maximal knowledge is at hand . . " (Section 10, pages 331-332, of [16]).

2. The light cone is respected

In Kent's proposal, the physical beables at spacetime points y in the past of S are determined only by the part of the final boundary distribution on S that is outside the future light cone of y: 'the light cone is respected'. Restriction to only the spacelike-separated region of the final boundary condition yields a Lorentz-invariant rule for the construction of the beables. Cf. Figure 2. (Here, 'determined' is to be understood in a tenseless sense: there is no backwards causation from S to y.)

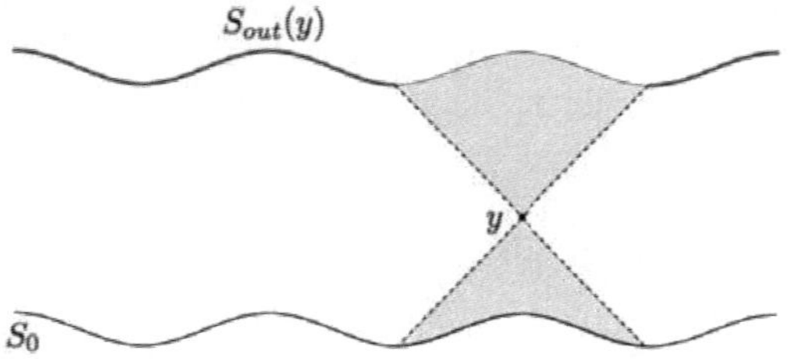

Figure 2. The spacelike hypersurface $S_{out}(y)$ in bold blue, containing the part of the final boundary condition used to determine beables at y.

3. Conditioning on a final boundary condition

We emphasize that although conditioning on a final boundary condition is of course much less common in physics than conditioning on an initial condition, it is in no way suspicious. In particular, it is a matter of logical relations between physical statements, not of causation: so it does not imply any backwards causation.

Indeed, the legitimacy of considering final as well as initial conditions is illustrated by various time-symmetric formalisms for quantum theory: in particular, by the well-known ABL rule [17], which gives the probability, conditioned on both final and initial states of a non-relativistic quantum system, of results of Lüders-rule measurements at intermediate times.

In a bit more detail: a non-relativistic system is prepared in an eigenstate $|\psi(0)\rangle = |a\rangle$ of observable A. At a later time T, observable C is measured and the eigenvalue c with eigenstate $|c\rangle$ is found. We consider an intermediate measurement of observable B with eigenvalues $\{b_i\}$, occurring at time $0 < t < T$. That is, we ask: what is the probability of obtaining outcome b_i at intermediate time t, conditioned on (i) the initial state $|\psi(0)\rangle = |a\rangle$ *and* on (ii) obtaining outcome c at the final time T?

The conditional probabilities for intermediate measurement outcomes follows as a direct result of the measurement postulate and the Born rule. For a given initial state $|a\rangle$ and final measurement outcome c, the conditional probability $\Pr(\, b_i \mid c, |a\rangle)$ of measuring b_i is prescribed by the ABL rule—see [6] for a detailed derivation. Precisely this ABL probability is used to construct the beables of Kent's non-relativistic proposals, as in Section two of his paper [1].

Agreed: Kent's relativistic proposal does not use an exact relativistic generalization of the ABL rule. For as we said in comment 2 above, only the part of S outside the light cone of a given point y is considered. The outcomes of (notional) measurements in the causal future of y are not conditioned on, as they would be in a straightforward relativistic generalization of

IOP Conf. Series: Journal of Physics: Conf. Series **1275** (2019) 012002 doi:10.1088/1742-6596/1275/1/012002

the ABL rule. However, the rule shows the legitimacy of conditioning on final as well as initial information; and is similar in spirit to Kent's proposal.

3.1. Toy model: single "photon" interaction
We now describe, with a simple semi-relativistic model in one spatial dimension, a photon interacting with a non-localized system, and later registering on a late hypersurface. We will see how Kent's invocation of a final boundary condition determines a single quasiclassical history of the relevant beable: here, the location of the system.

Consider a system initially in a superposition of two localized states:

$$\psi_0^{\text{sys}} = a\psi_1^{\text{sys}} + b\psi_2^{\text{sys}} \tag{5}$$

where ψ_i^{sys} are states localized around the points $x = x_i$, with $x_2 > x_1$. We take $|x_1 - x_2|$ to be large, so that there is negligible overlap in the wave functions, and we neglect any dynamics within and between the two systems.

For simplicity, "photons" are treated as pointlike particles following lightlike segments of spacetime. Suppose such a photon, initially unentangled with the system, propagates rightwards from the direction $x = -\infty$, so that in the absence of any interaction it would reach $x = x_1$ at $t = t_1$ and $x = x_2$ at $t = t_2 = t_1 + (x_2 - x_1)$. We take the "photon"-system interaction to instantaneously reverse the "photon"'s direction of travel, while leaving the system unaffected. Cf. Figure 3.

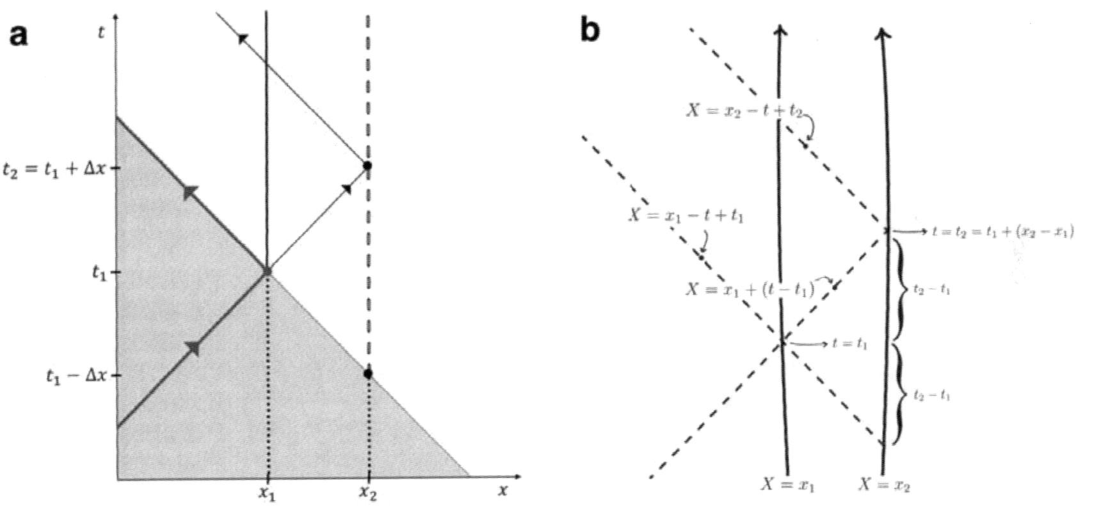

Figure 3. **a**: Single "Photon" Interaction. Diagonal lines represent possible "photon" paths, with the thick blue lines representing the actual path taken. The grey region depicts the *region of indeterminacy*, in which the final boundary condition is uncorrelated with the beables. The vertical lines at x_1 and x_2 show the temporal structure of the collapse of the wave function, and $\Delta x = x_2 - x_1$. Dotted lines represent partial mass-energy, solid vertical lines represent full mass-energy, and dashed lines represent no mass-energy. **b**: A fully labelled diagram showing the key trajectories and time intervals.

The possible outcomes of a (fictitious) mass-energy measurement at a late time $t = T \gg t_2$ are thus either: (i) finding the photon heading along the first leftward ray $X = x_1 + t_1 - t$ (cf. Figure 3b) and the system accordingly localized in the support of ψ_1^{sys}; or (ii) finding the photon

IOP Conf. Series: Journal of Physics: Conf. Series **1275** (2019) 012002 doi:10.1088/1742-6596/1275/1/012002

heading along the second leftward ray $X = x_2 + t_2 - t$ and the system accordingly localized in the support of ψ_2^{sys}.

Suppose the real world is defined by the first outcome, (i). That is: "Nature selects" a final boundary condition with the photon being detected where the late-time hypersurface intersects the first leftward ray $X = x_1 + t_1 - t$.

Then the conditional probability of finding any mass localized around the *other* massive subsystem, at x_2, becomes zero. The beables thus describe zero mass around x_2, in our frame, at all the times t for which the photon hits (while flying leftwards) the final hypersurface defined by the time slice $t = T$ outside the future light cone of (x_2, t).

It is only after t_1, i.e. when $t > t_1$, that the event of hitting the final hypersurface is outside the future light cone of (x_1, t)—thus the beables "collapse" to the full mass of a single cloud at x_1 at time t_1. Since the photon escapes the light cone from x_2 before escaping the light cone from x_1, the "collapse" to zero mass around x_2 happens, in our frame, before the "collapse" to full-mass around x_1 happens. Hence Figure 3a's triangular region of indeterminacy.

To repeat: We suppose the real world is defined by the leftmost reflection from x_1, as encoded in the final boundary condition. That is: the photon reflects from x_1 at t_1, and registers on the given time slice $t = T$ (which is much later: $T \gg t_2 > t_1$). Given this outcome, there is zero mass around x_2, in our frame, at all the times t for which the photon hits (while flying leftwards) the time slice $t = T$ outside the future light cone of (x_2, t). That is: at all times t later than $2t_1 - t_2$: a semi-infinite period. Early in this period—i.e. for $t_1 > t > 2t_1 - t_2$— the photon hitting the $t = T$ time slice is still *inside* the future light cone of (x_1, t). It is only after t_1, i.e. when $t > t_1$, that hitting the much-later slice is outside the future light cone of (x_1, t). Thus the "collapse" to zero mass around x_2 happens, in our frame, before the "collapse" to full mass around x_1 happens.

3.2. *Toy model: two "photon" interaction*

Adding a second leftward "photon" gives a similar story to that in Section 3.1's toy model. But now there is a "collapse" to full mass and zero mass across a light cone, instead of across a single light ray. In our 1+1-dimensional spacetime, this yields a triangular region of indeterminacy; cf. Figure 4.

Using x to label the position variable for the first photon, y for the system, and z for the second photon: the wave-function for times $t < t_1$ is

$$\psi(x, y, z; t) = \delta(x - (x_1 + t - t_1)) \cdot \delta(z - (x_2 - t + t_1)) \cdot (a\psi_1(y) + b\psi_2(y)) \,. \qquad (6)$$

Again, we suppose Nature selects a late-time mass-energy distribution indicating that both photons reflected from mass cloud one, i.e. at x_1. Then, as in the first toy model:

(i) the leftmost outgoing light ray for the first photon reflecting off mass cloud one encodes all the relevant information about the system;

(ii) the other light ray for the first photon reflecting off mass cloud two is inconsequential. It lags behind the leftmost light ray, thus manifesting itself at a later time in the conditional probability.

However, the rightmost outgoing ray for the second (initially-leftward) photon reflecting from mass cloud two also encodes all the relevant information about the system, in the sense that: if mass-energy is found along this light ray, the system must be found around mass cloud two—otherwise, it must be found around mass cloud 1.

Thus, we now have a "collapse" along the rightmost outgoing light ray as well as the leftmost outgoing light ray. So for any point y in spacetime which is to the future of at least one of these light rays, the relevant information about localizing the cloud is in the late-time mass-energy distribution in the surface $S_{out}(y)$ of Figure 2. So the region of spacetime which does

IOP Conf. Series: Journal of Physics: Conf. Series **1275** (2019) 012002 doi:10.1088/1742-6596/1275/1/012002

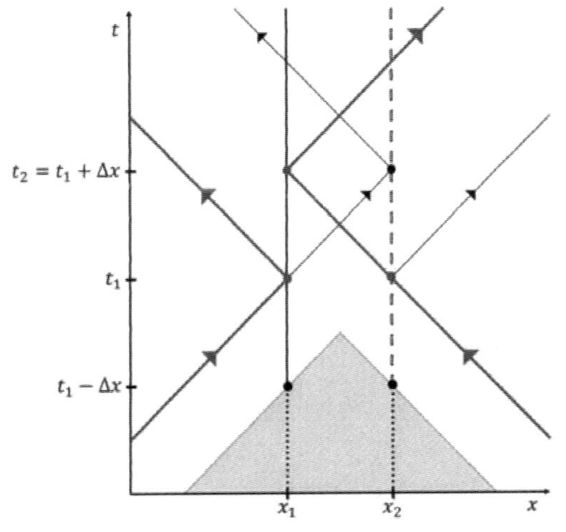

Figure 4. Two "Photon" Interaction. The same graphical representation, but now with two photons approaching from ± spatial infinity and a resulting triangular region of indeterminacy.

not "see the collapse" must lie to the past of both the leftmost and the rightmost outgoing light rays—resulting in a triangular region of indeterminacy.

4. Non-locality in Kent's proposed formulation

In this Section, we turn to the question: does Kent's proposal satisfy Shimony's conditions discussed in Section 2, viz. Outcome Independence (OI) and Parameter Independence (PI)? In short, we will find that Kent's proposal, once adapted to a Bell experiment, is much like the pilot-wave theory. (Our full analysis is in [5].)

Regarding Outcome Independence, Kent's proposal obeys OI at the 'micro-level', and outcome dependence (OD), i.e. quantum orthodoxy, at the 'observable level' (Sections 4.1 and 4.2). Here, 'micro-level' and 'observable level' have meanings analogous to their meanings in pilot-wave theory. So the idea is that the 'micro-level' includes all information: in pilot-wave theory, it includes the point-particles' positions (as well as the quantum state); and in Kent's proposal, it includes the final condition, i.e. the result of the notional measurement on the final hypersurface. On the other hand, in both theories—pilot-wave theory and Kent's proposal—'observable level' refers to orthodox probabilities prescribed by just the quantum state.

Regarding Parameter Independence, one needs to consider the recent important 'no-go' theorems of Colbeck-Renner, Landsman, and Leegwater. The gist of these theorems is that a theory supplementing orthodox quantum theory must violate: *either*

(i) Probabilities of hidden variables are independent of which quantity is measured: this is usually labelled 'no conspiracy', and so it seems mandatory: but as we said at the end of Section 2 (after equation 4) and in comment 3 of Section 3, the label is tendentious, since there need be nothing conspiratorial—or in any way suspicious—about conditioning on future information that contains traces (records) of earlier choices of measurement; *or*

(ii) Parameter Independence.

Pilot-wave theory violates (ii). We will see in Section 4.3 that at first sight, Kent's proposal is similar. But the question is open: it may violate (i) and obey PI.

4.1. Outcome Independence at the 'micro-level'

We take an outcome to be encoded by a "photon" registering on the surface S. Compare how in the pilot-wave theory, an outcome of a spin measurement by a Stern-Gerlach magnet is encoded by the point-particle's position.

IOP Conf. Series: Journal of Physics: Conf. Series **1275** (2019) 012002 doi:10.1088/1742-6596/1275/1/012002

We adapt Kent's first toy model (Section 3.1) to describe a Bell experiment by, firstly, taking the locations x_1 and x_2 ($x_2 > x_1$) to be the left-wing of the experiment. We take the two possible outcomes of a measurement on the system entering the left wing ('the L-system') to be localization of mass density around x_1 and x_2. In any run of the experiment, one outcome is definite, thanks to reflection of an initially-rightward "photon". (Alternative settings are a topic only for Parameter Independence.)

Similarly, there is a right wing and an R-system, which reflects an initially-leftward "photon". The initial joint state (with Y_L, Y_R for the two systems' spatial coordinates) is:

$$a\psi_1(Y_L)\psi_4(Y_R) + b\psi_2(Y_L)\psi_3(Y_R) \tag{7}$$

where x_3 and x_4 are located far along the x-axis, i.e. $x_3 >> x_1, x_2$ and $x_4 >> x_1, x_2$, and with $x_3 < x_4$.

Suppose that in one actual run of the experiment, the 'outer', x_1 and x_4, outcomes occur in the quasiclassical history specified by the final boundary condition, which we write as t_S. That is: (a) a photon registers on S so far to the left as to imply that it earlier reflected at x_1 rather than at x_2, while (b) another photon registers on S so far to the right as to imply that it earlier reflected at x_4 rather than at x_3.

If we condition orthodox quantum probabilities for outcomes on all this (very rich!) information in t_S, the probabilities will become trivial, i.e. 0 or 1. And so they will factorize. So we obtain probability 1 for each of the two 'outer', x_1 and x_4, outcomes. The situation is similar to the way that, once particle positions are take into account, the pilot-wave theory is past-deterministic just as much as it is future-deterministic; and so one could discuss its satisfaction of OI by invoking particles' *final* positions, instead of (as usual) their initial ones.

4.2. Outcome dependence at the observable level

Of course, we do not know the final condition t_S; and indeed, we will never know more than a tiny fraction of the actual single quasiclassical history. Kent's proposal about this is natural (and again analogous to pilot-wave theory):

(i) One should average over the possible final mass-energy distributions t_S on S, with their Born-rule probabilities as prescribed by the final quantum state $|\psi_S\rangle$.

(ii) It is these averaged probabilities that are equal, or close enough, to the orthodox probabilities.

The analogy with the pilot-wave theory's averaging over its (necessarily unknown!) particle positions, using their Born-rule probabilities, is clear. So as for the pilot-wave theory: in so far as we are confident that Kent's proposal recovers orthodox quantum probabilities at the observable level, we thus conclude that it satisfies OD at that level.

4.3. Parameter Independence?

Assessing PI, at either the micro or the observable levels, is a substantive task. One needs to model in a Kentian fashion both:

(i) different choices of apparatus-setting (parameter) on one wing, and

(ii) non-selective measurement, (i.e. defining the left-wing marginal probability by summing probabilities of the various right-wing outcomes).

A Kentian model of (i) will involve photons first scattering off the apparatus' knob that sets which quantity gets measured, and then much later, registering on the hypersurface S; so that the actual final condition encodes the setting. As we emphasized in (i) at the start of this Section: a final condition encoding a knob-setting, and probabilities dependent on such a setting, is not *ipso facto* suspicious. To model (ii), the probabilities dependent on reflection from "which knob-setting" will need to be combined with probabilities dependent on reflection from "which outcome, i.e. mass-energy lump".

We leave the development of a model addressing both (i) and (ii), to another paper. But whatever its details, one will need to consider how recent 'no go' theorems about non-locality bear on it. We will now briefly discuss this.

Colbeck and Renner [18] (made rigorous by Landsman [19]) show that, under some apparently natural extra assumptions: any theory that supplements orthodox quantum theory, in the sense of recovering orthodox Born-rule probabilities by probabilistic averaging over "hidden variables", must violate either:

(i) Probabilities of hidden variables are independent of which quantity is measured: called 'no conspiracy'—tendentiously—cf. equation (4), or

(ii) Parameter Independence.

Leegwater [20] dispenses with the extra assumptions apart from a 'no conspiracy' assumption: that the measure used to average over the hidden variables so as to recover Born-rule probabilities must be independent of which quantity, or quantities, are (chosen to be) measured.

Since Kent's probabilities are conditioned on a specific final boundary condition, they are not equal to orthodox probabilities, and thus his proposal must violate (i) or (ii). At first sight, it seems that Kent's proposal satisfies no conspiracy. For he averages over the possible final boundary conditions, with the Born-rule probabilities given by the final universal quantum state $|\psi_S\rangle$. And $|\psi_S\rangle$ is independent of which quantity, or quantities, were previously measured, even though it is the universal state.

Agreed: there is a fact about which quantity got measured (and about what the outcome was), according to Kent's overall scheme for recovering a quasiclassical world. But these facts leave no 'mark' on—have no back-reaction on—the universal quantum state, which always evolves unitarily. (This is like the pilot-wave theory's particle being guided by the ψ-wave, but having no influence on how ψ evolves.) Thus in the Schrödinger picture, $|\psi_S\rangle$ gets, at an intermediate time when an experiment is set up, components (non-zero amplitudes) for various possible choices of quantity (settings) and, soon thereafter, components for various possible outcomes. (Think of an Everettian's description of the setting-up, and performance, of a run of an experiment: the components correspond to the Everettian's branches.)

But on reflection, it is not clear that Kent's proposal satisfies no conspiracy. For the no conspiracy assumption concerns the measure used in deriving a Bell inequality. But *a priori*, this measure need not be the Born-rule probabilities given by the final universal quantum state; (cf. Section 5.1 of [5]). This means that on the question whether Kent's proposal satisfies Parameter Independence, the jury is still out.

5. Towards alternative proposals

Kent's proposal requires that photons interact with systems in superposed states, and so become entangled with the systems, and appropriately encode one or another component of the superposition—and then escape to spatial infinity in order for a quasiclassical history to emerge. So, for example, the beables for a quantum system in a superposed state which is completely contained at the centre of a large photon-absorbing box could never describe the emergence of a quasiclassical history.

Moreover, perhaps many quantum systems are not so "fortunate" as to interact with photons that then escape unimpeded to infinity. For example, photons that through entanglement record the state of macroscopic systems might only escape to infinity after several scattering events.

IOP Conf. Series: Journal of Physics: Conf. Series **1275** (2019) 012002 doi:10.1088/1742-6596/1275/1/012002

For such systems, Kent's approach will imply an *information delay*: a time-lag until "collapse", which occurs after the point of interaction.

This situation prompts us to explore modifying the Lorentz-invariant rule for constructing beables, so as to avoid any information delay. The rest of this Section builds from our ideas in [6, 7] toward this goal.

We begin by noting that in Kent's original proposal, beables at the spacetime point y are constructed via conditional expectation values for simultaneous measurements of all points x in some set $S_i \cap S$, where $\{S_i\}_i$ defines a sequence of hypersurfaces converging to the hypersurface $S_\Lambda(y)$, as defined in figure 5. The probabilities are conditional on a final boundary condition giving the actual mass-energy on the final hypersurface S.

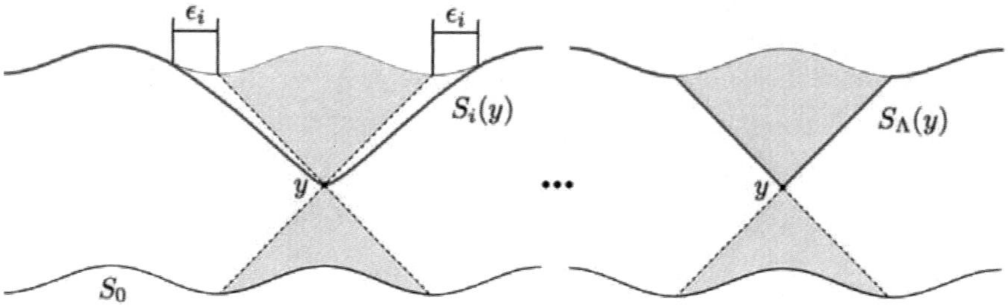

Figure 5. Spacelike hypersurfaces $S_i(y)$ (left) converging, as $\epsilon_i \to 0$, to the *effective future boundary* $S_\Lambda(y)$ (right) for a given spacetime point y between S_0 and S.

We now suggest: i) defining a sample space of mass-energy distributions on the entire spacetime, not just the final hypersurface; ii) using these distributions to formulate conditional probabilities for a point y which are conditional upon (notionally!) measuring the actual mass-energy distribution at *all* points $x \in S_i$, not just $x \in S_i \cap S$. So in the limit as $i \to \infty$, this effectively allows the conditional probabilities to be sensitive to mass-energy on the forward light cone of y, not just outside of it. Again, cf. figure 5.

Intuitively, this allows the beables to be sensitive to entangled photons crossing the boundary of their light cone, registering "pings" of energy there, in addition to entangled photons outside the light cone. This subtle addition would have significant ramifications for the beables describing physical reality, and in particular, information delay would be avoided.

In Section 3.1's single "photon" toy model, we can easily identify two possible mass-energy histories for the whole spacetime, as opposed to just the final boundary conditions. These are, similarly to Section 3.1: (i) the mass-energy distribution of the photon in the scenario where it reflects from x_1, along with the mass-energy of the massive subsystem localized at x_1; and (ii) the mass-energy distribution of the photon in the scenario where it reflects from x_2, along with the mass-energy of the massive subsystem localized at x_2.

The only spacetime points y for which the light cone of y encodes the mass-energy distribution while the hypersurface $S_{out}(y)$ does not—thus implying a difference from Kent's proposal—is the region to the left of x_1 with time coordinate greater than the incoming photon but less than the outgoing photon. At these points, the outgoing photon produces a ping on the forward light cone, thus revealing the presence or absence of the photon there in the mass-energy distribution. So the region of indeterminacy shrinks as compared to Kent's proposal: cf. figure 6. But the beables themselves do not change in this model. In this sense, the predictions of our suggested approach coincide, for this system in which no information delay is present, with Kent's predictions.

We can also compare the approaches in a scenario for which, on Kent's approach, information delay *is* present. This can be done with a simple adaptation of the single "photon" toy model. Suppose there is another massive body, fixed at $x_0 \ll x_1$, so that we can ignore any interaction

IOP Conf. Series: Journal of Physics: Conf. Series **1275** (2019) 012002 doi:10.1088/1742-6596/1275/1/012002

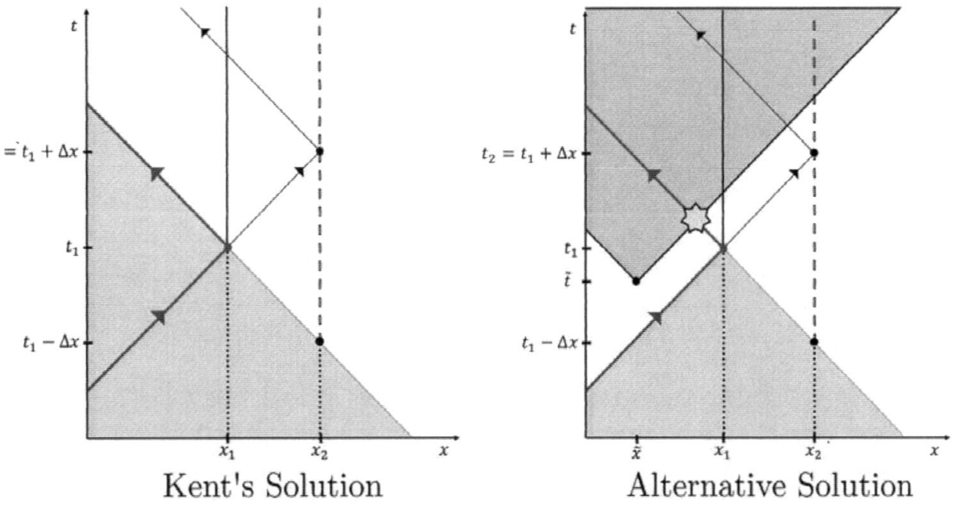

Figure 6. The single "photon" toy-model in the new approach compared to Kent's proposal. For the point $(\tilde{x}, \tilde{t})$, its forward light cone, with green interior, contains pings of mass energy from outgoing "photons". So the actual mass-energy distribution *could* contribute to determining beables at points such as $(\tilde{x}, \tilde{t})$, whereas in Kent's approach it would not. For this particular system, while the region of indeterminacy shrinks, the collapse of beables is described in exactly the same way.

between the massive body and the superposed system. We introduce the photon from a position to the left of x_1 but to the right of x_0, and suppose that the photon will reach x_1 at some time t_1. Cf. figure 7.

After reflecting from the first or second mass cloud, the photon reverses direction toward x_0 and reflects again upon reaching x_0. Technically, the photon should be trapped in the region between x_0 and x_1 or x_2 forever after reflecting from the system in superposition. However, this is clearly just an artifact of working in one spatial dimension; and we can avoid this artifact, by instead supposing the photon passes unimpeded through x_1 and x_2 upon returning for the second time. The important difference from the original toy model is that the photon undergoes one further interaction after reflecting from the system in superposition, before propagating toward spatial infinity to the right.

We again suppose that the late-time beable configuration chosen by Nature indicates that the photon reflects from the first mass cloud, at position x_1. Applying Kent's rules, there is indeed information delay: cf. figure 7 (left-hand side). The outgoing photon does not appear outside the light cone of x_1 at the time of interaction t_1, but at a later time $t_1 + 2|x_1 - x_0|/c$.

In this scenario, conditioning on the mass-energy on the future light cone provides far more information, since pings of mass-energy are registered on the light cone, as the "photon" bounces around. The region of indeterminacy also shrinks. The superposition collapses at time t_1 around x_1, and at time $t_1 - |x_2 - x_1|/c$ around x_2 as shown in figure 7 (right-hand side).

An immediate challenge for making our suggested approach rigorous is defining a probability measure of mass-energy distributions in spacetime. For we can no longer simply appeal to the Born rule on the final hypersurface. Assuming this technical challenge could be met, such a modification to Kent's framework could eliminate the information delay phenomenon.

IOP Conf. Series: Journal of Physics: Conf. Series **1275** (2019) 012002 doi:10.1088/1742-6596/1275/1/012002

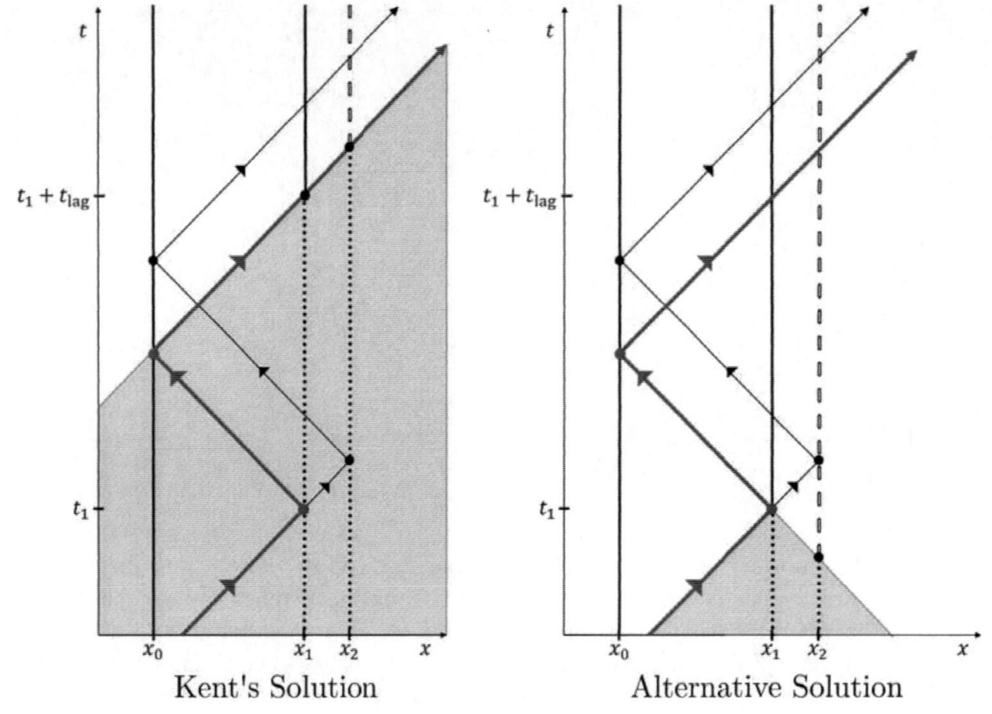

Figure 7. A comparison of Kent's approach and the alternative, for a scenario with information delay. In the alternative approach, the "superposed" state of the beables collapses, before or precisely at the time of interaction t_1.

6. Conclusion

In this paper, we have analyzed Kent's proposal for a Lorentz-invariant solution to the measurement problem from a variety of perspectives. We summarize with four points:

(i) Shimony hoped for a peaceful coexistence between quantum non-locality and special relativity by "blaming" the Bell-inequality derivation on assuming Outcome Independence, *pace* the pilot-wave theory: a viewpoint that, we urged, makes the measurement problem vivid.

(ii) Kent has proposed a solution to the problem that is: realist, one-world, relativistic, and without state reduction. The idea is: a final boundary condition makes mass-energy definite at intermediate times, via scattered photons, and this yields a single actual quasiclassical history.

(iii) Kent's proposal is like the pilot-wave theory in that: it obeys OI at the micro-level (i.e. with the final boundary condition specified)—but presumably, not at the observable level. Theorems by Colbeck, Landsman, Leegwater, and Renner imply that it also violates either: "no conspiracy", or Parameter Independence at the micro-level.

(iv) When none of the photons scattered from a macro-event (that we intuitively want to be definite) escape uninterrupted to spatial infinity from the macro-event, a quasiclassical history is, on Kent's proposal, *not* recovered. We outlined a revision. The idea is to condition on mass-energy present on the future light cone, as well as on mass-energy in the final boundary condition outside the future light cone.

Acknowledgments
We thank Adrian Kent for comments and conversations, especially about Section 5's suggested extensions of his proposal; and Gijs Leegwater for informative discussions about quantum non-locality. JB also thanks Bryan Roberts for discussions and Figure 3b. BM also acknowledges support from the National Science Foundation Graduate Research Fellowship under Grant No. DGE - 1656518.

References
[1] Kent A 2014 *Phys. Rev. A* **90**(1) 012107 arXiv:1311.0249
[2] Kent A 2015 *Philosophical Transactions of the Royal Society of London A: Mathematical, Physical and Engineering Sciences* **373** ISSN 1364-503X arXiv:1411.2957
[3] Kent A 2017 *Phys. Rev. A* **96**(6) 062121 arXiv:1608.04805
[4] Bell J S and Aspect A 2004 *Speakable and Unspeakable in Quantum Mechanics* 2nd ed (Cambridge University Press)
[5] Butterfield J 2018 *Reality and Measurement in Algebraic Quantum Theory* ed Ozawa M, Butterfield J, Halvorson H, Rédei M, Kitajima Y and Buscemi Y (Springer) arXiv:1710.07844
[6] Marsh B 2018 *PhilSci-Archive* ArXiv:1811.08950
[7] Marsh B 2019 *In preparation*
[8] Shimony A 1986 *Quantum Concepts in Space and Time* ed Penrose R and Isham C J (New York ;Oxford University Press) pp 182–203
[9] Shimony A 2017 *The Stanford Encyclopedia of Philosophy* ed Zalta E N (Metaphysics Research Lab, Stanford University) fall 2017 ed
[10] Bell J S 2004 *Speakable and Unspeakable in Quantum Mechanics* (Cambridge University Press) pp 213–231
[11] Bell J S 2004 *Speakable and Unspeakable in Quantum Mechanics* (Cambridge University Press) pp 232–248
[12] Dyson F J 1949 *Physical Review* **75** 486
[13] Schlosshauer M A 2007 *Decoherence: and the quantum-to-classical transition* (Springer Science & Business Media)
[14] d'Espagnat B 1971 *Conceptual Foundations of Quantum Mechanics* (Addison-Wesley, Advanced Book Program)
[15] Schrödinger E 1935 *Naturwissenschaften* **23** 823–828
[16] Trimmer J D 1980 *Proceedings of the American Philosophical Society* 323–338
[17] Aharonov Y, Bergmann P G and Lebowitz J L 1964 *Physical Review* **134** B1410
[18] Colbeck R and Renner R 2016 *Quantum theory: informational foundations and foils* (Springer) arXiv:1208.4123
[19] Landsman K 2015 *Journal of Mathematical Physics* **56** arXiv:1509.08498
[20] Leegwater G 2016 *Studies in History and Philosophy of Science Part B: Studies in History and Philosophy of Modern Physics* **54** 18 – 34 ISSN 1355-2198 arXiv:1705.08341

IOP Conf. Series: Journal of Physics: Conf. Series **1275** (2019) 012003 doi:10.1088/1742-6596/1275/1/012003

A new rule for quantization that resolves all problems

John R Klauder

Department of Physics and Department of Mathematics
University of Florida, Gainesville, FL 32611-8440, U.S.A.

E-mail: `john.klauder@gmail.com`

Abstract. While canonical quantization solves many problems there are some problems where it fails. A close examination of the classical/quantum connection leads to a new connection that permits quantum and classical realms to coexist, as is the case in the real world. For those problems for which conventional quantization works well, the new procedures yield identical results. However, we offer an example that fails when quantized conventionally but succeeds when quantized with the new procedures.

1. Canonical Quantization vs. Enhanced Quantization

For a single degree of freedom, canonical quantization (CQ) takes classical phase space variables, (p, q), and promotes them into Hilbert space operators, (P, Q). The classical Hamiltonian $H(p, q)$ is promoted to the quantum Hamiltonian $\mathcal{H}(P, Q)$, specifically $\mathcal{H}(P, Q) = H(P, Q) + \mathcal{O}(\hbar; P, Q)$ provided the phase space variables p and q are "Cartesian coordinates" [1] (footnote, page 114), even though phase space has no metric to determine Cartesian coordinates.

Enhanced quantization (EQ) features a different classical/quantum connection story. It starts by introducing self-adjoint operators P and Q. The quantum Hamiltonian $\mathcal{H}(P, Q)$ is chosen as a self-adjoint operator built from the basic variables. Schrödinger's equation follows from the action functional

$$A_Q = \int_0^T \langle\psi(t)|[i\hbar(\partial/\partial t) - \mathcal{H}(P, Q)]|\psi(t)\rangle \, dt \tag{1}$$

by stationary variations of the normalized Hilbert space vectors $|\psi(t)\rangle$. However, a *macroscopic* observer can only vary a subset of vectors of a *microscopic* system such as

$$|p, q\rangle \equiv e^{-iqP/\hbar} e^{ipQ/\hbar}|0\rangle \ , \tag{2}$$

where $(p, q) \in \mathbb{R}^2$, and the normalized fiducial vector $|0\rangle$ is chosen as a solution of the equation $(\omega Q + iP)|0\rangle = 0$. The reduced (R) quantum action functional is given by

$$A_{Q(R)} = \int_0^T \langle p(t), q(t)|[i\hbar(\partial/\partial t) - \mathcal{H}(P, Q)]|p(t), q(t)\rangle \, dt \ , \tag{3}$$

which leads to

$$A_{Q(R)} = \int_0^T [p(t)\dot{q}(t) - H(p(t), q(t))] \, dt \ , \tag{4}$$

Content from this work may be used under the terms of the Creative Commons Attribution 3.0 licence. Any further distribution of this work must maintain attribution to the author(s) and the title of the work, journal citation and DOI.
Published under licence by IOP Publishing Ltd

IOP Conf. Series: Journal of Physics: Conf. Series **1275** (2019) 012003 doi:10.1088/1742-6596/1275/1/012003

a natural candidate to be the classical Hamiltonian, with $\hbar > 0$, as it is in the real world! In this expression

$$\begin{aligned} H(p,q) &\equiv \langle p,q|\mathcal{H}(P,Q)|p,q\rangle \\ &= \langle 0|\mathcal{H}(P+p\mathbb{1}, Q+q\mathbb{1})|0\rangle \\ &= \mathcal{H}(p,q) + \mathcal{O}(\hbar;p,q) \ . \end{aligned} \tag{5}$$

Apart from possible $\hbar$ corrections, $\mathcal{H}(P,Q) = H(P,Q)$, which *exactly* represents the goal of seeking Cartesian coordinates. Although phase space has no metric to identify Cartesian coordinates, Hilbert space has one. It follows that a scaled form of the Fubini-Study metric [2] for ray vectors, specifically $d\sigma^2 \equiv 2\hbar[\|d|p,q\rangle\|^2 - |\langle p,q|\,d|p.q\rangle|^2]$, yields the result $d\sigma^2 = \omega^{-1}dp^2 + \omega\, dq^2$, which indeed implies Cartesian coordinates for a flat phase space. Consequently, EQ can lead to the same good results given by CQ, but EQ can do even more.

In EQ the classical and quantum Hamiltonian are related by the weak correspondence principle [3],

$$H(p,q) = \langle p,q|\mathcal{H}(P,Q)|p,q\rangle \ , \tag{6}$$

which holds for self-adjoint canonical variables. In CQ the canonical operators P and Q are required to be irreducible, and in EQ they may also be irreducible. However, from (6) it follows that the quantum variables may also be reducible. The following problem can be properly solved with reducible operators, but cannot be properly solved with irreducible operators.

2. Rotationally symmetric models

This example involves many variables, and it has been examined before [4, 5]; here we only offer an overview of the solution. The classical Hamiltonian is given by

$$H(p,q) = \tfrac{1}{2}\Sigma_{n=1}^N [p_n^2 + m_0^2 q_n^2] + \lambda_0\{\Sigma_{n=1}^N q_n^2\}^2 \ , \tag{7}$$

where $N \le \infty$, and the Poisson bracket is $\{q_k, p_l\} = \delta_{kl}$. For $\lambda_0 > 0$ and $N = \infty$ the quantum model is trivial (i.e., Gaussian) when CQ and irreducible operators are used To obtain reducible operators we introduce a second set of canonical operators $\{R_n, S_n\}$ which commute with all $\{P_n, Q_n\}$. Triviality of such models forces us to restrict the fiducial vector to be a Gaussian in a Schrödinger representation, i.e., where $Q_n|x\rangle = x_n|x\rangle$. Thus $(m_0 Q_n + iP_n)|0\rangle = 0$ for the free model, with the property that

$$\begin{aligned} (m_0 Q_n - iP_n)\cdot(m_0 Q_n + iP_n) &= P_n^2 + m_0^2 Q_n^2 + i m_0 [Q_n, P_n] \\ &=: P_n^2 + m_0^2 Q_n^2 : \ , \end{aligned} \tag{8}$$

for all n, $1 \le n \le N$, with normal ordering denoted by $:(\cdot):$. It follows, for example, that

$$\begin{aligned} &\langle p,q|\,\tfrac{1}{2}\,\Sigma_{n=1}^N (P_n^2 + m_0^2 Q_n^2) : + w \ : \{\Sigma_{n=1}^N (P_n^2 + m_0^2 Q_n^2)\}^2 : \ |p,q\rangle \\ &\quad = \tfrac{1}{2}\,\Sigma_{n=1}^N (p_n^2 + m_0^2 q_n^2) + w\{\Sigma_{n=1}^N (p_n^2 + m_0^2 g_n^2)\}^2 \ , \end{aligned} \tag{9}$$

which offers a clue of how to proceed.

For the interacting model with reducible operators this example is changed to $[m(Q_n + \zeta S_n) + iP_n]|\zeta\rangle = 0$ and $[m(S_n + \zeta Q_n) + iR_n]|\zeta\rangle = 0$, for all n, $1 \le n \le N$, where $0 < \zeta < 1$. We next introduce new reduced states as

$$|p,q;\zeta\rangle \equiv e^{-i\Sigma_{n=1}^N q_n P_n/\hbar}\, e^{i\Sigma_{n=1}^N p_n Q_n/\hbar}|\zeta\rangle \ , \tag{10}$$

which still span the relevant Hilbert space. Then it follows that

$$\begin{aligned} H(p,q) &= \langle p,q;\zeta|\,\tfrac{1}{2}\{:\Sigma_{n=1}^{N}[P_n^2+m^2(Q_n+\zeta S_n)^2]:\\ &\qquad +:\Sigma_{n=1}^{N}[R_n^2+m^2(S_n+\zeta Q_n)^2]:\}\\ &\qquad +v:\{\Sigma_{n=1}^{N}[R_n^2+m^2(S_n+\zeta Q_n)^2]\}^2:|p,q;\zeta\rangle\\ &= \tfrac{1}{2}\Sigma_{n=1}^{N}[p_n^2+m^2(1+\zeta^2)q_n^2]+v\zeta^4m^4\{\Sigma_{n=1}^{N}q_n^2\}^2\\ &\equiv \tfrac{1}{2}\Sigma_{n=1}^{N}[p_n^2+m_0^2q_n^2]+\lambda_0\{\Sigma_{n=1}^{N}q_n^2\}^2 \end{aligned} \tag{11}$$

as desired!

Further discussion of these models is available in [4, 5].

3. Additional Enhanced Quantization examples

Enhanced quantization has also been applied to other problems, among which ultralocal scalar models [6, 5] admit complete solutions and additionally suggest proposals for more complicated examples. These include covariant scalar fields [7, 8, 9, 10, 11], and quantum gravity [7, 12, 13, 14]. The use of special variables to discuss idealized cosmological models [15, 16] has led to gravitational bounces rather than an initial singularity of the universe. Further use of special variables as applied to idealized gravitational models has been given in various papers, e.g., [17, 18, 19, 20], and references therein.

References

[1] Dirac P 1958 *The Principles of Quantum Mechanics* (Oxford: Clarendon Press) p 114
[2] See, e.g., http://en.wikipedia.org/wiki/Fubini-Study metric.
[3] Klauder J 1967 Weak correspondence principle *J. Math. Phys.* **8** 2392
[4] Klauder J 1965 Rotationally-symmetric model field theories *J. Math. Phys.* **6** 1666-79
[5] Klauder J 2000 *Beyond Conventional Quantization* (Cambridge: Cambridge University Press)
[6] Klauder J 1970 Ultralocal scalar field models *Commun. Math. Phys.* **18** 307-18
[7] Klauder J 2015 *Enhanced Quantization: Particles, Fields & Gravity* (Singapore: World Scientific)
[8] Klauder J 2012 Enhanced quantization: a primer *J. Phys. A: Math. Theor.* **45** 285304-12
[9] Klauder J 2015 Enhanced quantum procedures that resolve difficult problems *Rev. Math. Phys.* **27** 1530002-45
[10] Klauder J 2015 Nontrivial quantization of ϕ_n^4, $n \geq 2$ *Theor. Math. Phys.* **182** 83-9; *do.* 2014 *Teor. Mat. Fiz.* **182** 103-11
[11] Klauder J 2011 Scalar field quantization without divergences in all spacetime dimensions *J. Phys. A: Math. Theor.* **44** 273001-31
[12] Klauder J 1999 Noncanonical quantization of gravity. I. Foundations of affine quantum gravity *J. Math. Phys.* **40** 5860
[13] Klauder J 2001 Noncanonical quantization of gravity. II. Constraints and the physical Hilbert space *J. Math. Phys.* **42** 4440
[14] Klauder J 2012 Recent results regarding affine quantum gravity *J. Math. Phys.* **53** 082501
[15] Klauder J and Aslaksen E 1970 Elementary model for quantum gravity *Phys. Rev.* D **2** 272
[16] Fanuel M and Zonetti S 2013 Affine quantization and the initial cosmological singularity *EPL* **101** 10001
[17] Bergeron H, Czuchry E, Gazeau J-P, Malkiewicz P and Piechocki W 2015 Singularity avoidance in a quantum model of the mixmaster universe *Phys. Rev.* D **92** 124018
[18] Bergeron H, Czuchry E, Gazeau J-P, Malkiewicz P and Piechocki W 2015 Smooth quantum dynamics of mixmaster universe *Phys. Rev.* D **92** 061302
[19] Bergeron H, Czuchry E, Gazeau J-P and Malkiewicz P 2016 Nonadiabatic bounce and an inflationary phase in the quantum mixmaster universe *Phys. Rev.* D **93** 124053
[20] Bergeron H, Czuchry E, Gazeau J-P and Malkiewicz P 2016 Vibronic framework for quantum mixmaster universe *Phys. Rev.* D **93** 064080

IOP Conf. Series: Journal of Physics: Conf. Series **1275** (2019) 012004 doi:10.1088/1742-6596/1275/1/012004

The deformation parameter of the generalized uncertainty principle

Fabio Scardigli[†*]

[†] Dipartimento di Matematica, Politecnico di Milano, Piazza Leonardo da Vinci 32, 20133 Milano, Italy

[*] Institute-Lorentz for Theoretical Physics, Leiden University, P.O. Box 9506, Leiden, The Netherlands

E-mail: fabio@phys.ntu.edu.tw

Abstract. After a short introduction to the generalized uncertainty principle (GUP), we review some of the physical predictions of the GUP, and we focus in particular on the bounds that present experimental tests can put on the value of the deformation parameter β. We also describe a theoretical value computed for β, and comment on the vast parameter region still unexplored, and to be probed by future experiments.

1. Introduction

Uncertainty Principle (HUP) emerged when Heisenberg [1], in a seminal paper of 1927, discussed measurement processes in Quantum Theory. The argument (for example, find the position of an electron by means of photons) is well known under the name of *Heisenberg microscope argument* [2]. In those early approaches the gravitational interaction between particles was completely neglected, although this was somehow justified by the huge weakness of gravity, when compared with other fundamental interactions. Then, in a few years, HUP became a theorem at the core of the formalism of a fully developed Quantum Mechanics [3].

However, when basic principles and elementary measurement processes are discussed in order to address fundamental questions in Nature, it seems clear that also gravity should be considered. This is what actually happened in the subsequent decades, starting from the very early attempts in generalizing HUP [4], to the more recent proposals like those of string theory, deformed special relativity, and studies of black hole physics [5, 6, 7, 8, 9, 10, 11, 12, 13].

Various revised versions of the classical Heisenberg argument have been presented, and, for example in Ref. [8], one of them is described as follows. A beam of photons of energy E can theoretically detect an object of size δx roughly given by (if we assume the dispersion relation $E = pc$)

$$\delta x \simeq \frac{\hbar c}{2E}\,, \tag{1}$$

so that increasingly large energies allow to explore decreasingly small details. As remarked above, in its original formulation, Heisenberg's gedanken experiment ignores gravity. Now, if on the contrary in this gedanken measurement process we take into account *also* the possible formation, in high energy scatterings, of micro black holes with a gravitational radius $R_S = R_S(E)$ roughly

Content from this work may be used under the terms of the Creative Commons Attribution 3.0 licence. Any further distribution of this work must maintain attribution to the author(s) and the title of the work, journal citation and DOI.
Published under licence by IOP Publishing Ltd

proportional to the (centre-of-mass) scattering energy E (see Ref. [8]), then it turns out that the usual uncertainty relation should be modified as

$$\delta x \simeq \frac{\hbar c}{2E} + \beta R_S(E) , \tag{2}$$

where β is a dimensionless parameter. Recalling that $R_S \simeq 2 G_N E/c^4 = 2 \ell_p^2 E/\hbar c$ we can write

$$\delta x \simeq \frac{\hbar c}{2E} + \beta \ell_p^2 \frac{2E}{\hbar c}, \tag{3}$$

where the Planck length is defined as $\ell_p^2 = G\hbar/c^3$, the Planck energy as $E_p \ell_p = \hbar c/2$, the Planck mass $m_p = E_p/c^2$. This kind of modification of the uncertainty principle was also proposed in Ref. [9].

Although the deforming parameter β is not in principle fixed by the theory, it is generally assumed to be of the order of unity. This happens, in particular, in some models of string theory (see again for instance Ref. [5]). An explicit analytic calculation of β in Ref. [14] has confirmed the circumstance. However, many studies have appeared in literature, with the aim to set experimental bounds on β (see, for instance, Refs. [15]).

The relation (3) can be recast in the form of an uncertainty relation, namely a deformation of the standard HUP, usually referred to as Generalized Uncertainty Principle (GUP),

$$\Delta x \, \Delta p \geq \frac{\hbar}{2}\left[1 + \beta \left(\frac{\Delta p}{m_p c}\right)^2\right] . \tag{4}$$

For mirror-symmetric states (with $\langle \hat{p} \rangle = 0$), the inequality (4) is equivalent to the commutator

$$[\hat{x}, \hat{p}] = i\hbar\left[1 + \beta \left(\frac{\hat{p}}{m_p c}\right)^2\right] , \tag{5}$$

since $\Delta x \, \Delta p \geq (1/2) \left|\langle [\hat{x}, \hat{p}] \rangle\right|$. Vice-versa, the commutator (5) implies the inequality (4) for any state. The GUP is widely studied in the context of quantum mechanics [16], quantum field theory [17], thermal effects in QFT [18, 19], and for various deformations of the quantization rules [20].

2. Tests of GUP: Deformed Quantum Mechanics

In the last decade a lively debate emerged on the *measurable* features predicted by the various kinds of Generalized Uncertainty Principles (GUPs). The discussion focused on the experimental predictions about the *size* of these modifications, and among the better elaborated experimental proposals are those, for example, of the groups of Brukner, and Marin [21, 22].

We can roughly divide in two groups the Studies on the dimensionless deforming parameter β of GUP .

In the first group, authors as Kempf, Mann, Brau, Vagenas, Nozari, etc. (Refs.[7, 15, 23, 24]), translated the GUP into a deformed commutator and developed a deformed quantum mechanics. The deformed commutator (5) is in general rewritten as

$$\left[\hat{X}, \hat{P}\right] = i\hbar\left(1 + \beta \frac{\hat{P}^2}{m_p^2 c^2}\right) , \tag{6}$$

where the fundamental variables $\hat{X}, \hat{P}$ are thought to be high energy operators valid, in particular, at or near the Planck scale. They have non linear representations, $\hat{X} = X(\hat{x})$,

$\hat{P} = P(\hat{p})$ in terms of the operators $\hat{x}$, $\hat{p}$ which are the usual position and momentum operators at low energy, obeying the standard Heisenberg commutator $[\hat{x}, \hat{p}] = i\hbar$.

This approach is used in general to extract bounds on β of non-gravitational origin by computing with the new $\hat{X}, \hat{P}$ well know physical phenomena, and comparing the results with experimental data. Explicit calculations rely on the specific transformation $\hat{X} = X(\hat{x})$, $\hat{P} = P(\hat{p})$, which is in general non linear, and also by definition non canonical, since $[\hat{x}, \hat{p}] \neq [\hat{X}, \hat{P}]$, namely the correspondent Poisson brackets are not preserved.

Authors compute corrections to quantum mechanical predictions, such as energy shifts in the spectrum of the hydrogen atom, or to the Lamb shift, the Landau levels, Scanning Tunneling Microscope, charmonium levels, etc. As we shall see, the bounds on β so obtained are quite stringent, but a debated drawback of this approach is a possible dependence of the expected shifts on the specific representation of the variables X and P in the fundamental commutator (6).

2.1. *Non gravitational bounds on β*

The following are the the most commonly discussed (atomic) phenomena, where, by applying the above discussed Deformed Quantum Mechanics, non gravitational bounds on β can be obtained.

• The Landau levels are recomputed with a GUP-corrected hamiltonian of the form

$$H = \frac{1}{2m}(\vec{p_0} - e\vec{A})^2 + \frac{\beta_0}{m}(\vec{p_0} - e\vec{A})^4 = H_0 + 4\beta_0 m H_0^2 \tag{7}$$

where $\beta_0 = \beta/(m_p c)^2$, and for an electron in a magnetic field of $10T$, $\omega_c \simeq 10^3$ GHz, authors in Refs.[23] (see also references therein) found a relative shift of the first level as

$$\frac{\Delta E_{1(GUP)}}{E_1^{(0)}} \simeq 2.30 \times 10^{-54}\beta\,. \tag{8}$$

Since we have an accuracy of one part in 10^4 in direct measurements of Landau levels using a scanning tunnel microscope (STM), the upper bound on β becomes

$$10^{-54}\beta < 10^{-4} \Rightarrow \beta < 10^{50}\,. \tag{9}$$

• For the Lamb shift, again in Refs.[23], authors get the a relative correction of

$$\frac{\Delta E_{0(GUP)}}{\Delta E_0} \simeq 0.47 \times 10^{-48}\beta \tag{10}$$

and since the current accuracy in precision measurements of Lamb shift is of about 1 part in 10^{12}, this sets an upper bound of

$$\beta < 10^{36}. \tag{11}$$

• After having computed similar GUP-corrections for the harmonic oscillator, it is then possible to use them for systems that can be well modeled by the harmonic oscillator itself, such as heavy mesons like charmonium. In so doing the correction due to GUP can be calculated at the first order in β, and in Refs.[23] authors found for the relative shift in energy

$$\frac{\Delta E_0^{(2)}}{E_0^{(0)}} \simeq 2.7 \times 10^{-39}\beta \tag{12}$$

The current accuracy in precision measurements for the case of J/ψ is at the level of 10^{-5}. Therefore they find an upper bound on β as

$$\beta < 10^{34}. \tag{13}$$

• We then have bounds obtained without relying on a representation of the fundamental commutator, but rather based directly on the deformed uncertainty relations. The first is coming from the lack of observed deviations from the standard model theory at the electroweak scale (100 GeV). In fact, Eq. (4) can be promptly cast in the form

$$\Delta x \, \Delta p \geq \frac{\hbar}{2}\left[1 + \beta\left(\frac{E_{\rm ew}}{E_p}\right)^2\right]. \tag{14}$$

The absence of deviations from the standard Heisenberg principle at the electroweak scale then implies $\beta(E_{\rm ew}/E_p)^2 \ll 1$, which means $\beta \ll 10^{34}$.

• The study of Bawaj et al. [22] opened the way of probing deformed commutators with macroscopic harmonic oscillators. The authors have checked with different devices the behavior of state-of-the-art micro and nano mechanical oscillators.

Supposing m and ω_0 are the mass and the frequency of the harmonic oscillator, then the classical standard Hamiltonian is $H = p^2/2m + m\omega_0^2 x^2/2 = \hbar\omega_0(Q^2 + P^2)/2$, where $x = \sqrt{\hbar/m\omega_0}Q$, $p = \sqrt{\hbar m\omega_0}P$, with Q and P dimensionless coordinates. The operators $\hat{x}$ and $\hat{p}$ obey Eq.(5), and therefore $[\hat{Q}, \hat{P}] = i(1 + \beta_0\hat{P}^2)$ where here $\beta_0 = \beta(\hbar m\omega_0/m_p^2c^2)$. We now introduce the transformation $Q = Q$, $P = P(\tilde{P})$, where $\tilde{P}$ is the low energy momentum, such that $[Q, \tilde{P}] = i$. Therefore at the first order in β_0 we have $P = \tilde{P}(1 + \beta_0\tilde{P}^2/3)$ and the GUP-deformed Hamiltonian for the harmonic oscillator reads

$$H = \frac{\hbar\omega_0}{2}(Q^2 + \tilde{P}^2) + \frac{\hbar\omega_0}{3}\beta_0\tilde{P}^4\,. \tag{15}$$

The solution to the first order in β_0 of the equations of motion for such oscillator reads

$$Q = Q_0\left[\sin(\tilde{\omega}t) + \frac{\beta_0}{8}Q_0^2\sin(3\tilde{\omega}t)\right] \tag{16}$$

where Q_0 is the oscillation amplitude for Q and

$$\tilde{\omega} = (1 + \frac{\beta_0}{2}Q_0^2)\omega_0\,. \tag{17}$$

Clearly, it presents two striking new effects with respect to the standard harmonic oscillator: a dependence of the oscillation frequency on the amplitude, and the appearance of a third harmonic.

Authors have examined three kinds of oscillators, with masses of respectively 10^{-4}, 10^{-7} and 10^{-11} kg. Results are summarized in Fig.1 (taken from Ref.[22]), where they are also compared with other bounds on β listed above, or discussed in the next section. The parameter β quantifies the deformation to the standard commutator between position and momentum. Its upper limits are reported on the Y-axis as a function of the mass involved in the measurement. Red dots come from the dependence of the oscillation frequency from its amplitude. They provide the best estimates of β ranging from 10^8 to 10^{19}, with an average around 10^{12}. Magenta stars come from the third harmonic distortion, not present in the standard harmonic oscillator. Light

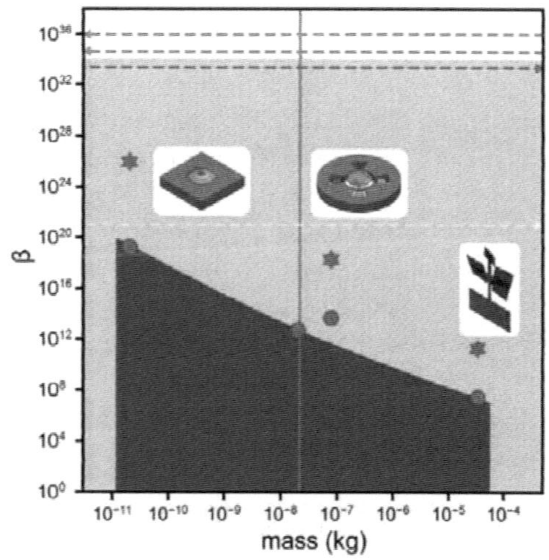

Figure 1. See comment in the text.

blue shows the area below the electroweak scale, dark blue the area that remains unexplored. Dashed lines reports some previously estimated upper limits, obtained in mass ranges outside this graph (as indicated by the arrows). Green: from high resolution spectroscopy on the hydrogen atom, considering the ground state Lamb shift (upper line) [23] and the 1S-2S level difference (lower line) [25]; on this lower green line there are also the bounds coming from the charmonium levels. Magenta: from the AURIGA detector [26]. Yellow: from the lack of violation of the equivalence principle [29]. The vertical line corresponds to the Planck mass (22 μg).

• The very recent work of P.Bushev et al. [27] is a further brilliant step in the direction of the evaluation of β with macroscopic harmonic oscillators. Actually, the sapphire mechanical resonator employed by Bushev is really "big", when compared to the standard devices used in in this field, in fact its mass amount to 0.3 kg. The Hamiltonian governing the system is essentially again given by Eq.(15),

$$H = \frac{\hat{p}^2}{2m} + \frac{1}{2}m\Omega_0^2\hat{x}^2 + \frac{\beta\hat{p}^4}{3m(m_pc)^2}, \tag{18}$$

namely a deformation in $\beta\hat{p}^4$ of the usual harmonic oscillator Hamiltonian. Experiments with the sapphire resonator have established an upper limit on β not exceeding 5.2×10^6, which is from 2 to 6 order of magnitude better than the one previously detected by Bawaj et al.. Moreover, Bushev's group is preparing a further experiment with a state of the art quartz BAW resonator, and they claim to have reasonable estimates that this would yield an even more stringent limit of $\beta < 4 \times 10^4$.

• In Ref. [28] authors deal with the broadening times of large molecular wave-packets (such as C_{60}, C_{176} and large organic molecules) and the profound effect that a modified fundamental commutator should have on the dispersion rate of free wave-packets. Authors propose experiments where the minimal length effect can be tested by measuring the timescale in which the wave-packet (associated with a particle or a system of particles) doubles its initial width. They

IOP Conf. Series: Journal of Physics: Conf. Series **1275** (2019) 012004 doi:10.1088/1742-6596/1275/1/012004

start from a fundamental commutator containing also a linear term in $\hat{p}$ (see [13])

$$[\hat{x}, \hat{p}] = i\hbar(1 - 2\alpha_0\hat{p} + 4\alpha_0^2\hat{p}^2) \tag{19}$$

where $\alpha_0 = \alpha/(m_p c)$. On comparing this with Eq.(5), we see that $\beta \sim \alpha^2$. The difference between the "doubling time" predicted by the HUP and the doubling time predicted by the GUP can be computed for a generic α and compared with experimental data. It turns out that t_{double} is proportional to the mass m of the particle considered, and that is why it is convenient to use large molecules. According to the authors, the use of organic macro molecules as the TPPF152 (consisting of 430 atoms) and of up to date atomic clocks (with an assumed precision of 10^{-15} s) would allow, in the proposed experiments, to probe values of $\alpha \sim 10^6$. Therefore this would give an upper bound for $\beta < 10^{12}$.

3. Tests of GUP: Deformed Classical Mechanics; gravitational bounds on β

In a second group of works (see, e.g., Chang [30]), classical Newtonian mechanics is deformed by modifying the standard Poisson brackets in a way that resembles the quantum commutator,

$$[\hat{x}, \hat{p}] = i\hbar\left(1 + \beta_0\,\hat{p}^2\right) \quad \Rightarrow \quad \{X, P\} = \left(1 + \beta_0\,P^2\right) , \tag{20}$$

where $\beta_0 = \beta/(m_p^2 c^2)$. In particular, in Ref. [30] authors compute the precession of the perihelion of Mercury directly from this GUP-deformed Newtonian mechanics, and interpret it as an extra contribution to the well known precession of 43"/century due to General Relativity (GR). They then compare this global result with the observational data, and the very accurate agreement between the GR prediction and observations leave them not much room for possible extra contributions to the precession. In fact, they obtain the tremendously small bound $\beta < 10^{-66}$. A problem with this approach is that this GUP-deformed Newtonian mechanics is simply superposed linearly to the usual GR theory. One may argue that a modification of GR at order β should likewise be considered, but this is however omitted in Ref. [30]. In other words, it is not clear why the two structures, GR and GUP-modified Newtonian mechanics, should coexist independently, and why the two different precession errors add into a final single precession angle. Most important, as a matter of fact, in the limit $\beta \to 0$, Ref. [30] recovers *only* the Newtonian mechanics but not GR, and GR corrections must be added as an extra structure. Clearly, the physical relevance of this approach and the bound that follows for β, remain therefore quite questionable.

Finally, there are works on the evaluation of β where authors (see for example Ghosh [29]), in a similar but slightly different way, use a covariant formalism (first defined in Minkowski space, with the metric $\eta_{\mu\nu} = \text{diag}(1, -1, -1, -1)$, and then easily generalized to curved space-times via the standard procedure $\eta_{\mu\nu} \to g_{\mu\nu}$), to introduce a deformation of classical Poisson brackets, although posited in covariant form. From deformed covariant Poisson brackets, they obtain a β-deformed geodesic equation, which leads to a violation of the Equivalence Principle. They do not deform the field equations or the metric. It is however possible to show that this violation of the Equivalence Principle is completely due to the postulate of deformed Poisson brackets, and has nothing to do with the general covariance of formalism, or with a deformation of the GR field equations or solutions. In other words, already a simple deformation of classical Newtonian (i.e. non covariant) Poisson brackets implies automatically a violation of the equivalence principle.

In fact, suppose to deform Poisson brackets in the same fashion as the quantum commutators derived from the GUP (see Refs. [29, 30]). Then, considering just a one dimensional system to keep the calculation simple, we can write the Poisson brackets of a pair of canonical variables as

$$\{q, p\} = 1 + \beta_0\,p^2 \ , \qquad \{q, q\} = \{p, p\} = 0 \ . \tag{21}$$

Then for any regular function $H(q,p)$, the following equations hold

$$\{q, H\} = (1 + \beta_0\, p^2)\frac{\partial H}{\partial p}\,, \qquad \{p, H\} = -(1 + \beta_0\, p^2)\frac{\partial H}{\partial q}\,. \tag{22}$$

A point-like particle of mass m moving in a Newtonian potential is described by the Hamiltonian (assuming $M \gg m$)

$$H = \frac{p^2}{2\,m} - \frac{G\,M\,m}{q}\,, \tag{23}$$

and evolves according to the equations of motion

$$\dot{q} = \{q, H\} = (1 + \beta_0\, p^2)\frac{p}{m}\,, \qquad \dot{p} = \{p, H\} = -(1 + \beta_0\, p^2)\frac{G\,M\,m}{q^2}\,. \tag{24}$$

To first order in β, we have therefore the equation of motion

$$\ddot{q} \simeq -(1 + 4\,\beta_0\,(m\,\dot{q})^2)\,\frac{G\,M}{q^2}\,. \tag{25}$$

Clearly, the trajectory of a test particle of mass m will depend on m, which signals a violation of the Equivalence Principle (see Refs. [33, 34]). This violation has nothing to do with covariant formalism, GR, or the geodesic equation, but strictly follows, even in simple Newtonian mechanics, from the modified Poisson brackets (21).

However, the Ghosh formalism remains covariant when $\beta \to 0$ and reproduces standard GR results in the limit $\beta \to 0$. In conclusion, the bound on β predicted by Ghosh (see Ref. [29]) from a possible violation of the universality of free fall, namely, from a violation of the (weak) equivalence principle, is $\beta < 10^{21}$.

The role of the equivalence principle in the GUP context has been discussed also in Ref. [31]. There, Tkachuk (2012) recovers Equivalence Principle also for deformed classical mechanics (i.e. deformed Poisson brackets) by considering composite bodies, and postulating that kinetic energy has the additive property, that is, it does not depend on the composition of a body but only on its total mass. The price to pay is to introduce a deformation parameter β_{0i} which depends on the mass m_i of the i-th (elementary) particle composing the macroscopic body. In other words, each (elementary) particle composing the body has its own deformation parameter β_{0i}. Then a new universal constant is introduced, $\gamma^2 = \beta_{0i} m_i^2$, which allows to recover the validity of the equivalence principle for deformed classical mechanics (by the way, to the first order in β). The construction is without doubt ingenious, however has the drawback that each specific particle would have its own specific minimal length, $\hbar\sqrt{\beta_{0i}} = \ell_i$. The effective minimal length that can be probed by a proton should be different (specifically, smaller) from that of an electron. This feature clearly seems to be at odd with the universality of gravitation, and with the fact that Planck length can be computed in a way completely independent from the particle considered (see e.g. [18]).

3.1. Gravitational bounds on β preserving the equivalence principle

By exploiting the (first) gravitational wave event ever observed, GW150914, authors of Ref. [32] have been able to set bounds on β of purely gravitational origin. They start from a representation of the deformed commutator $[X_i, P_j] = i\hbar\delta_{ij}(1 + \beta_0 P^2)$ which reads

$$X_i = x_{0i}\,, \qquad P_i = p_{0i}(1 + \beta_0 p^2) \tag{26}$$

IOP Conf. Series: Journal of Physics: Conf. Series **1275** (2019) 012004 doi:10.1088/1742-6596/1275/1/012004

where x_{0i}, p_{0i} are the low energy position and momentum operators, satisfying the canonical commutation relations $[x_{0i}, p_{0j}] = i\hbar\delta_{ij}$, and $p^2 = \eta_{ij}p^{0i}p^{0j}$. From here authors deduce a GUP-modified dispersion relation (to the first order in β)

$$E^2 = m^2c^4 + p^2c^2(1 - 2\beta_0 p^2)\,. \tag{27}$$

Therefore the group speed for massless gravitons will be

$$v_{massless} = \frac{\partial E}{\partial p} \simeq c(1 - 3\beta_0 p^2)\,. \tag{28}$$

Then, assuming for massless gravitons $E_g \simeq p_g c$, they can write $v_{massless} \simeq c(1 - 3\beta_0 E_g^2/c^2)$ where E_g is the energy of gravitons. Finally, the difference between the modified speed of gravitons and the speed of light is

$$\Delta v = c - v_{massless} = 3\beta \frac{E_g^2}{m_p^2 c^3}\,. \tag{29}$$

From the data of GW150914 we get an upper bound

$$3\beta \frac{E_g^2}{m_p^2 c^3} < 5.6 \times 10^{-12} \quad m/s\,. \tag{30}$$

Since the signal of the gravitational wave event GW150914 was peaked at 150Hz, here they can write $E_g \simeq 6.024 \times 10^{-13}$eV. Therefore, the upper bound on the GUP parameter β turns out to be

$$\beta < 2.3 \times 10^{60}\,, \tag{31}$$

which is more stringent than the gravitational bounds derived, for example, in Ref.[33]. We emphasize that in this procedure the authors, because of GUP, have modified the usual dispersion relation. Namely the propagation speed of gravitons is not the same for all the frequencies/energies, but depends on the frequencies/energies of the gravitons considered, according to Eq.(28). Note however that this procedure does not violate the Equivalence Principle (EP). In fact, we have a violation of the EP when the equation of motion contains deforming terms dependent on β and/or on the mass m of the test body, and therefore such equation does not coincide anymore with the geodesic equation. (The geodesic equation is given and fixed, once the metric is given.) In cases in which the trajectory of a test body depends on the mass of the test body, then we have a violation of EP. But of course, quite obviously, a generic trajectory/geodesic should depend on the (initial) velocity of the moving body. For example, the trajectory of a planet around the Sun ($V_{planet} \ll c$) is not the trajectory of a photon in the gravitational field of the Sun ($V_{photon} = c$). Nevertheless both trajectories are geodesics of the metric surrounding the Sun. Hence, a dispersion relation requiring different speeds for different frequencies, will imply that red light follows a path different from blue light, but however both paths will be geodesics of the metric (not trajectories different from geodesics). Therefore the EP is not violated.

A further example of bounds on β of gravitational origin, but not depending on a violation of the EP, is given in Ref.[33]. There, authors consider a deformation of the Hawking temperature of a Schwarzschild black hole when computed through the GUP. The GUP deformed Hawking temperature turns out to be (at first order in β)

$$T(\beta) = \frac{\hbar c^3}{8\pi G k_B M}\left(1 + \frac{\beta m_p^2}{4\pi^2 M^2} + \ldots\right)\,. \tag{32}$$

Such deformed temperature can be mimicked through the standard Hawking temperature of a deformed Schwarzschild black hole metric. In fact, authors show that a deformed spherically symmetric metric like

$$ds^2 = F(r)dt^2 - F(r)^{-1}dr^2 - r^2 d\Omega^2 \tag{33}$$

with

$$F(r) = 1 - \frac{2GM}{rc^2} + \varepsilon \frac{G^2M^2}{r^2c^4} \tag{34}$$

has an horizon radius $r_H = (GM/c^2)(1 + \sqrt{1-\varepsilon})$, and the correspondent standard Hawking temperature is

$$T(\varepsilon) = \frac{\hbar c}{4\pi k_B} F'(r_H) = \frac{\hbar c^3}{2\pi G k_B M} \frac{\sqrt{1-\varepsilon}}{(1+\sqrt{1-\varepsilon})^2} \,. \tag{35}$$

Imposing $T(\beta) = T(\varepsilon)$, they get a relation between β and ε that at the lowest order reads

$$\beta = -\frac{\pi^2 M^2}{4m_p^2}\varepsilon^2 \,. \tag{36}$$

Once we have a deformed metric, we can use very precise astronomical measures to set bounds on the magnitude of the deformation ε, and then transfer these bounds on β, through Eq.(36).

Incidentally, we note that the negative value of the β parameter predicted by Eq.(36), although it may appear surprising, has been already encountered in different scenarios: for an uncertainty relation formulated on a (crystal) lattice (see [20]); in the Magueijo and Smolin formulation of Doubly Special Relativity [35], where the fundamental commutator $[X, P] = i\hbar(1-E/E_p)$ exhibits a $\hbar$ that can be interpreted as depending on the energy, $\hbar(E) \to 0$ when $E \to E_{Planck}$; in the analysis of the Chandrasekhar limit [36], where Y.C.Ong shows that β should be negative in order to have a match between GUP-predicted white dwarfs masses and the astrophysical measures; in the comparison between the modification of Hawking radiation predicted by GUP and that predicted by corpuscular models of Gravity [37, 38, 39]. All these hints point towards a negative β which would implies, immediately, $[\hat{X}, \hat{P}] = i\hbar(1 + \beta\hat{P}^2/m_p^2c^2) \to 0$ for $P \to P_{Planck}$. Namely, a negative β implies the existence of a sharp, *classical world* at the Planck Scale, where $[\hat{X}, \hat{P}] = 0$. This would realize the vision formulated by G.'t Hooft in his deterministic interpretation of Quantum Mechanics, where the world is a gigantic, deterministic cellular automaton which works *CLASSICALLY* at the Planck scale [40].

To constrain ε, the best data come from the measures of the precession of Mercury. Using the deformed metric (34), we compute that the total precession after a single lap is

$$\Delta\phi = 6\pi \frac{GM}{L}\left(1 - \frac{\varepsilon}{6}\right) \tag{37}$$

to the first order in GM/L (where $L = (1-e^2)a$ is the *semilatus rectum* of Mercuty orbit). Note that for $\varepsilon \to 0$ we recover the usual, well known, General Relativity prediction. The data provided by the Messenger spacecraft [41], which orbited Mercury for two years in 2011-2013, give us a bound on ε as

$$|\varepsilon| < 1.6 \times 10^{-4} \tag{38}$$

which implies the following bound on β

$$|\beta| < 2 \times 10^{69} \,. \tag{39}$$

Although much worse than the $\beta < 10^{21}$ obtained by Ghosh, this bound has been obtained with a procedure that fully preserves the equivalence principle.

IOP Conf. Series: Journal of Physics: Conf. Series **1275** (2019) 012004 doi:10.1088/1742-6596/1275/1/012004

4. Theoretical calculation of β

By following a path similar to the one just presented in the previous section (formulae 32-36), it possible to compute an *exact* value of β (see Ref. [14]). In fact, suppose to consider a generic deformation of the spherically symmetric metric given above, as

$$F(r) = 1 - \frac{2GM}{rc^2} + \varepsilon\phi(r)\,. \tag{40}$$

Then the deformed Schwarzschild radius (to the first order in ε) reads

$$r_H \;=\; a - \frac{\varepsilon\, a\, \phi(a)}{1 \,+\, \varepsilon\,[\phi(a) \,+\, a\,\phi'(a)]}\,, \tag{41}$$

where $a = 2GM/c^2$, and the corresponding deformed standard Hawking temperature is

$$T = \frac{\hbar c}{4\pi k_B} F'(r_H) = \frac{\hbar c^3}{8\pi k_B GM}\left\{1 + \varepsilon\left[2\phi(a) + a\phi'(a)\right] + \varepsilon^2\phi(a)\left[\phi(a) - 2a\phi'(a) - a^2\phi''(a)\right] + \ldots\right\}. \tag{42}$$

Now, supposing to know the exact form of the correction term $\varepsilon\phi(r) \equiv \epsilon(r)$ in formula (40), then we can compare the two expansions (42) and (32), and from the two respective first order terms, we can extract the exact value of β as

$$\beta \;=\; \frac{4\pi^2 M^2}{m_p^2}\;\left[2\epsilon(a) + a\epsilon'(a)\right]\;. \tag{43}$$

This is what actually happens, when we consider the leading quantum corrections to the Newtonian potential computed, in an early attempt, by Duff [42], and more recently by Donoghue [43].

Donoghue reformulated General Relativity as an effective field theory. At ordinary energies gravity is a well behaved QFT. The quantum corrections at low energy, and dominant effects at large distances can be isolated, and are shown to be due to the propagation of massless particles (gravitons). The effective quantum-corrected gravitational potential generated by an heavy mass M, interacting with another heavy body m (both bodies close to rest), then reads

$$V(r) = -\frac{GM}{r}\left(1 + \frac{3GM}{rc^2}(1 + \frac{m}{M}) + \frac{41}{10\pi}\frac{\ell_p^2}{r^2}\right). \tag{44}$$

The first correction term does not contain any power of $\hbar$, so it is a classical effect, due to the non-linear nature of General Relativity. However, the second correction term, i.e., the last term of (44), is a true quantum effect, linear in $\hbar$.

Now, it is well know that for any metric of the form $ds^2 = F(r)dt^2 - F(r)^{-1}dr^2 - C(r)d\Omega^2$ an effective Newtonian potential [1] (i.e. the Newtonian limit) can be defined as

$$V(r) \;\simeq\; \frac{c^2}{2}\;(F(r) - 1) \tag{45}$$

Conversely, a metric mimicking a given Newtonian potential $V(r)$ will be

$$F(r) \;\simeq\; 1 \,+\, 2\,\frac{V(r)}{c^2}\;. \tag{46}$$

[1] The effective Newtonian potential is produced by such metric on a point particle which moves slowly, in a stationary and weak gravitational field, i.e., quasi-Minkowskian far from the source, $r \to \infty$.

IOP Conf. Series: Journal of Physics: Conf. Series **1275** (2019) 012004 doi:10.1088/1742-6596/1275/1/012004

Therefore, the metric mimicking the quantum-corrected Newtonian potential (44) will be

$$F(r) \simeq 1 + 2\frac{V(r)}{c^2} = 1 - \frac{2GM}{rc^2} - \frac{6\,G^2M^2}{r^2c^4}\left(1 + \frac{m}{M}\right) - \frac{41}{5\pi}\frac{G^3M^3}{r^3c^6}\left(\frac{\ell_p\,c^2}{GM}\right)^2 . \tag{47}$$

Identifying now

$$\epsilon(r) = -\frac{6\,G^2M^2}{r^2c^4}\left(1 + \frac{m}{M}\right) - \frac{41}{5\pi}\frac{G^3M^3}{r^3c^6}\left(\frac{\ell_p\,c^2}{GM}\right)^2 , \tag{48}$$

then, through Eq.(43), we finally arrive at

$$\beta = \frac{4\pi^2M^2}{m_p^2}\frac{41}{40\pi}\left(\frac{\ell_p c^2}{GM}\right)^2 = \frac{82\pi}{5} , \tag{49}$$

where we used $\ell_p c^2/GM = 2m_p/M$. Therefore, β results to be roughly of order 1, as expected also from other completely different approaches (among the many, see for example [5], [44]).

5. Summary and conclusions

The main results described in this paper are about the bounds and the value of the deformation parameter β of the GUP, and they can be summarized as follows.

• The best bound on β of non-gravitational origin seems to be in the range $\beta < 10^8 \div 10^{12}$ (Bawaj, Marin 2015), or probably even $\beta < 5 \times 10^6$ (Bushev 2019)

• The best bound on β of gravitational origin (if we allow for a violation of the Equivalence Principle) seems to be $\beta < 10^{21}$ (Ghosh 2014).

• The best bound on β of gravitational origin (preserving however the Equivalence Principle) seems to be $\beta < 10^{60}$ (Feng 2017).

• Different theoretical frameworks, and an explicit calculation (Ref.[14]), indicate a value of β of order 1, in particular $\beta \simeq 82\pi/5$.

So we can conclude that the gap between theoretical predictions and experimental bounds will still require a quite big leap in experimental techniques in order to probe a vast region for the parameter β. For sure, there is a lot of space for improvement!

References

[1] Heisenberg W 1927 *Z. Phys.* **43** 172
[2] Heisenberg W 1930 *Physikalische Prinzipien der Quantentheorie* (in German) Leipzig: Hirzel; 1930 *The Physical Principles of Quantum Theory* (English translation) Chicago: University of Chicago Press
[3] Robertson H P 1929 *Phys. Rev.* **34** 163; E. Schrödinger 1930 *Zum Heisenbergschen Unschrfeprinzip*, Sitzungsberichte der Preussischen Akademie der Wissenschaften, Physikalisch-mathematische Klasse **14** 296
[4] Bronstein M P 1936 *Phys. Zeitschr. der Sowjetunion* **9** 140; Snyder H S 1947 *Phys. Rev.* **71** 38; Yang C N 1947 *Phys. Rev.* **72** 874; Mead C A 1964 *Phys. Rev.* **135** B849; Karolyhazy F 1966 *Nuovo Cim. A* **42** 390
[5] Amati D, Ciafaloni M and Veneziano G 1987 *Phys. Lett. B* **197** 81; Gross D J and Mende P F 1987 *Phys. Lett. B* **197** 129
[6] Maggiore M 1993 *Phys. Lett. B* **319** 83
[7] Kempf A, Mangano G and Mann R B 1995 *Phys. Rev. D* **52** 1108
[8] Scardigli F 1999 *Phys. Lett. B* **452** 39

[9] Adler R J and Santiago D I 1999 *Mod. Phys. Lett. A* **14** 1371; Capozziello S, Lambiase G and Scarpetta G 2000 *Int. J. Theor. Phys.* **39** 15
[10] Scardigli F and Casadio R 2003 *Class. Quant. Grav.* **20** 3915
[11] Casadio R, Garattini R and Scardigli F 2009 *Phys. Lett. B* **679** 156
[12] Casadio R and Scardigli F 2014 *Eur. Phys. J. C* **74** 2685
[13] Ali A F, Das S and Vagenas E 2009 *Phys. Lett. B* **678** 497
[14] Scardigli F, Lambiase G and Vagenas E 2017 *Phys. Lett. B* **767** 242
[15] Brau F 1999 *J. Phys. A* **32** 7691; Khodadi M, Nozari K, Dey S, Bhat A and Faizal M 2016 *Sci. Rep.* **8** 1659
[16] Naimi Y 2016 *Grav. Cosmol.* **22** 382; Bernardo R C S and Esguerra J P H 2018 *Annals Phys.* **391** 293
[17] Husain V, Kothawala D and Seahra S S 2013 *Phys. Rev. D* **87** 025014
[18] Scardigli F 1995 *Nuovo Cim. B* **110** 1029
[19] Scardigli F, Blasone M, Luciano G and Casadio R 2018 *Eur. Phys. J. C* **78** 728
[20] Jizba P, Kleinert H and Scardigli F 2010 *Phys. Rev. D* **81** 084030
[21] Pikovski I, Vanner M, Brukner C et al. 2012 *Nature Physics* **8** 393
[22] Bawaj M, Biancofiore C, Marin F, et al. 2015 *Nature Comm.* **6** 7503 [arXiv:1411.6410]
[23] Das S, Vagenas E C 2008 *Phys. Rev. Lett.* **101** 221301; Ali A F, Das S, Vagenas E C 2011 *Phys. Rev. D* **84** 044013
[24] Pedram P, Nozari K and Taheri S H 2011 *JHEP* **1103** 093
[25] Quesne C, Tkachuk V M 2010 *Phys. Rev. A* **81** 012106
[26] Marin F, Cerdonio M, et al. 2013 *Nature Physics* **9** 71
[27] Bushev P, Bourhill J, Goryachev M, Kukharchyk N, Ivanov E, Galliou S, Tobar M and Danilishin S, *Testing of Quantum Gravity With Sub-Kilogram Acoustic Resonators*, [arXiv:1903.03346]
[28] Villalpando C and Modak S, *Probing Quantum Gravity with Large Molecular Wave-packets*, [arXiv:1901.09696]
[29] Ghosh S 2014 *Class. Quant. Grav.* **31** 025025
[30] Chang L N, Minic D, Okamura N and Takeuchi T 2002 *Phys. Rev. D* **66** 026003
[31] Tkachuk V M 2012 *Phys. Rev. A* **86** 062112
[32] Zhong-Wen Feng, Shu-Zheng Yang, Hui-Ling Li, Xiao-Tao Zu 2017 *Phys. Lett. B* **768** 81
[33] Scardigli F and Casadio R 2015 *Eur. Phys. J. C* **75** 425
[34] Ali A F 2011 *Class. Quantum Grav.* **28** 065013
[35] Magueijo J and Smolin L 2002 *Phys. Rev. Lett.* **88** 190403
[36] Yen Chin Ong 2018 *JCAP* **09** 015
[37] Buoninfante L, Luciano G, Petruzziello L *Generalized uncertainty principle and corpuscular gravity*, [arXiv:1903.01382]
[38] Dvali G and Gomez C 2013 *Fortsch. Phys.* **61** 742 [arXiv:1112.3359]
[39] Casadio R, Giugno A, Giusti A and Lenzi M 2017 *Phys. Rev. D* **96** 044010 [arXiv:1702.05918]; Giusti A 2019 *Int. J. Geom. Meth. Mod. Phys.* **16** 1930001
[40] 't Hooft G 2016 *The Cellular Automaton Interpretation of Quantum Mechanics* (Springer Int. Publ.)
[41] Verma A K, Fienga A, Laskar J, Manche H, and Gastineau M 2014 *Astron. Astrophys.* **561** A115
[42] Duff M J 1974 *Phys. Rev. D* **9** 1837
[43] Donoghue J F 1994 *Phys. Rev. Lett.* **72** 2996; Donoghue J F 1994 *Phys. Rev. D* **50** 3874
[44] Luciano G, Petruzziello L (2019) *Eur. Phys. J. C* **79** 283 [arXiv:1902.07059]

IOP Conf. Series: Journal of Physics: Conf. Series **1275** (2019) 012005 doi:10.1088/1742-6596/1275/1/012005

Information scan of quantum states based on entropy-power uncertainty relations

P Jizba[1], A Hayes[2] and JA Dunningham[2]

[1] FNSPE, Czech Technical University in Prague, Břehová 7, 115 19 Praha 1, Czech Republic
[2] Department of Physics and Astronomy, University of Sussex, Brighton BN1 9QH, UK

E-mail: `p.jizba@fjfi.cvut.cz`

E-mail: `Anthony.Hayes@sussex.ac.uk`

E-mail: `J.Dunningham@sussex.ac.uk`

Abstract. We use Rényi-entropy-power-based uncertainty relations to show how the information probability distribution associated with a quantum state can be reconstructed in a process that is analogous to quantum-state tomography. We illustrate our point with the so-called "cat states", which are of both fundamental interest and practical use in schemes such as quantum metrology, but are not well described by standard variance-based approaches.

1. Introduction

There has been a recent upsurge of interest in quantum-mechanical (QM) uncertainty relations (URs) catalyzed by new ideas from (quantum) information theory [1, 2, 3, 4, 5], functional analysis [6, 7] and cosmology [8, 9, 10] as well as experiments that have observed violations of Heisenberg's error-disturbance uncertainty relations [11, 12, 13, 14, 15]. Historically the most popular quantifier of quantum uncertainty has been *variance* because of its simplicity and ubiquity in probability theory. The variance determines the measure of uncertainty in terms of the fluctuation (or spread) around the mean value which, while useful for many distributions, does not provide a sensible measure of uncertainty in a number of important situations including multimodal [1, 2, 3] and heavy-tailed distributions [2, 3, 4]. To deal with this, a multitude of alternative measures of uncertainty have emerged in QM. Among these, a particularly prominent role is played by information entropies such as the Shannon entropy [16], Rényi entropy [3, 16], Tsallis entropy [17], associated differential entropies and their quantum-information generalizations [2, 3, 5].

In Ref. [2] we introduced an infinite tower of mutually distinct (generally irreducible) Rényi entropy-power-based URs (REPURs). The conventional URs based on variances (Robertson–Schrödinger URs [18, 19]) and Shannon differential entropies (Hirschman and Białynicki-Birula URs [5, 20]) naturally appeared as special cases in this hierarchy of REPURs. The concept of *entropy power* (EP) was introduced in Shannon's seminal 1948 paper [21] in order to formulate information theory for continuous random variables. Since then, the EP has proved to be essential in a number of classical and quantum information-theoretic applications ranging from interference channels to secrecy capacity [22, 23, 24, 25, 26]. Apart from its role in information theory, the EP has found wide use in pure mathematics, namely in the theory of inequalities and mathematical statistics and estimation theory [27].

Content from this work may be used under the terms of the Creative Commons Attribution 3.0 licence. Any further distribution of this work must maintain attribution to the author(s) and the title of the work, journal citation and DOI.
Published under licence by IOP Publishing Ltd

In Refs. [2, 3] we addressed the following two questions: Assuming one is able to control Rényi EPs of various orders; i) how does this set of EPs constrain the underlying state distribution and ii) how do the ensuing REPURs restrict the state distributions of conjugate variables? To answer these questions we invoked the concept of the *information distribution* associated with a given quantum state. The latter contains a complete "information scan" of the underlying state distribution. Here we concentrate on the point i) by extending the analysis of [2, 3]. In particular, we show how the information distribution associated with a given quantum state can be numerically reconstructed from EPs in a process that is akin to a quantum-state tomography.

The paper is structured as follows. We begin in Sec. 2 by introducing the concept of Rényi's EP, which is done in the context of estimation theory by generalizing the notion of Fisher information (FI) using a Rényi entropy version of De Bruijn's identity. In this regard, a particularly notable role is played by the so-called *escort distribution*, which appears naturally in the definition of higher-order *score functions* and the ensuing generalized Cramér–Rao's inequalities. After this, in Sec. 3 we introduce the concepts of information distribution and in Sec. 4 we show how cumulants of the information distribution can be obtained from knowledge of the EPs. With the cumulants at hand, one can reconstruct the underlying information distribution in a process which we call an *information scan*. Details of how this can explicitly be realized for quantum state PDFs is provided in Sec. 5. This is done with the help of a generalized Gram–Charlier A and the Edgeworth expansion. In Sec. 6 we illustrate the inner workings of the information scan using the example of a so-called *cat state*. A cat state is a superposition of the vacuum state and a coherent state of the electromagnetic field; two cases are studied comprising of different probabilistic weightings of the superposition state corresponding to *balanced* and *unbalanced* cat states. A short conclusion is presented in Sec. 7.

2. Cramér–Rao inequality and Rényi entropy powers

In Ref. [2, 3] we derived the Rényi-entropy-power-based QM uncertainty relations using the Babenko–Beckner inequality. In this section we provide another version of the proof which will hinge on the concept of FI and the (generalized) Cramér–Rao inequality. The derivation we outline fits the mathematical framework used in quantum metrology. Firstly, we recall that the Fisher information matrix $\mathbb{J}(\mathcal{X})$ of a random vector $\{\mathcal{X}_i\}$ in $\mathbb{R}^D$ with the PDF $\mathcal{F}(\boldsymbol{x})$ is

$$\mathbb{J}(\mathcal{X}) \ = \ \mathrm{cov}(\boldsymbol{V}(\mathcal{X}))\,, \tag{1}$$

where the covariance matrix is associated with the random zero-mean vector (*score* vector),

$$\boldsymbol{V}(\boldsymbol{x}) \ = \ \nabla \mathcal{F}(\boldsymbol{x})/\mathcal{F}(\boldsymbol{x})\,. \tag{2}$$

The ensuing FI $J(\mathcal{X})$ is a trace of $\mathbb{J}(\mathcal{X})$, i.e.

$$J(\mathcal{X}) \ = \ \mathrm{Tr}(\mathbb{J}(\mathcal{X})) \ = \ \mathrm{var}(\boldsymbol{V}(\mathcal{X})) \ = \ \mathbb{E}(\boldsymbol{V}^2(\mathcal{X}))\,. \tag{3}$$

Both the FI and FI matrix can be conveniently related to Shannon's differential entropy via De Bruin's identity [28].

De Bruin's identity: Let $\{\mathcal{X}_i\}$ be a random vector in $\mathbb{R}^D$ with the PDF $\mathcal{F}(\boldsymbol{x})$ and let $\{\mathcal{Z}_i^G\}$ be a Gaussian noise vector with zero mean and unit-covariance matrix, independent of $\{\mathcal{X}_i\}$. Then

$$\frac{d}{d\epsilon}\mathcal{H}(\mathcal{X} + \sqrt{\epsilon}\,\mathcal{Z}^G)|_{\epsilon=0} \ = \ \frac{1}{2}J(\mathcal{X})\,, \tag{4}$$

where $\mathcal{H} = -\int_{\mathbb{R}^D}\mathcal{F}(\boldsymbol{x})\log\mathcal{F}(\boldsymbol{x})\,d\boldsymbol{x}$ is Shannon's differential entropy (measured in *nats*). In the case when the independent additive noise $\{\mathcal{Z}_i\}$ is non-Gaussian with zero mean and covariance matrix $\boldsymbol{\Sigma} = \mathrm{cov}(\mathcal{Z})$ then the following generalization holds

$$\frac{d}{d\epsilon}\mathcal{H}(\mathcal{X} + \sqrt{\epsilon}\,\mathcal{Z})|_{\epsilon=0} \ = \ \frac{1}{2}\mathrm{Tr}\,(\mathbb{J}(\mathcal{X})\boldsymbol{\Sigma})\,. \tag{5}$$

The crux of De Bruin's identity is that it provides a very useful intuitive interpretation of the Fisher information, namely that it quantifies the sensitivity of transmitted (Shannon) information to an arbitrary independent additive noise. An important aspect to note is that this quantifier of sensitivity depends only on the covariance of the noise vector and so it is independent of the shape of the noise distribution. This is because De Bruin's identity remains unchanged for both Gaussian and non-Gaussian additive noise with the same covariance matrix. We now prove that the shape of the noise distribution can be further quantified by the Rényi-entropy-based FI matrix. Indeed, the following statement holds:

Generalized De Bruin's identity: Let $\{\mathcal{X}_i\}$ be a random vector in $\mathbb{R}^D$ with the PDF $\mathcal{F}(\boldsymbol{x})$ and let $\{\mathcal{Z}_i\}$ be an independent (generally non-Gaussian) noise vector with the zero mean and covariance matrix $\boldsymbol{\Sigma} = \text{cov}(\mathcal{Z})$, then

$$\frac{d}{d\epsilon}\mathcal{I}_q(\mathcal{X} + \sqrt{\epsilon}\,\mathcal{Z})|_{\epsilon=0} \;=\; \frac{1}{2q}\text{Tr}\left(\mathbb{J}_q(\mathcal{X})\boldsymbol{\Sigma}\right), \tag{6}$$

where $\mathcal{I}_q = \frac{1}{1-q}\log\int_{\mathbb{R}^D}\mathcal{F}^q(\boldsymbol{x})d\boldsymbol{x}$ is Rényi's differential entropy (measured in *nats*) with $\mathcal{I}_1 = \mathcal{H}$. The generalized FI matrix of order q has the explicit form

$$\mathbb{J}_q(\mathcal{X}) \;=\; \text{cov}_q(\boldsymbol{V}_q(\mathcal{X}))\,, \tag{7}$$

with the score vector

$$\boldsymbol{V}_q(\boldsymbol{x}) \;=\; \nabla\rho_q(\boldsymbol{x})/\rho_q(\boldsymbol{x}) \;=\; q\nabla\mathcal{F}(\boldsymbol{x})/\mathcal{F}(\boldsymbol{x}) \;=\; q\boldsymbol{V}(\boldsymbol{x})\,. \tag{8}$$

Here $\rho_q = \mathcal{F}^q/\int_{\mathbb{R}^D}\mathcal{F}^q d\boldsymbol{x}$ is the so-called escort distribution [29]. The "cov_q" denotes the covariance matrix computed with respect to ρ_q. A proof of this generalized De Bruin's identity is provided in [30]. Finally, as in the Shannon case we define the FI of order q — denoted as $J_q(\mathcal{X})$, via De Bruin's identity with a unit noise covariance matrix. This gives

$$\text{Tr}\left(\mathbb{J}_q(\mathcal{X})\right) \;\equiv\; J_q(\mathcal{X})\,. \tag{9}$$

As for Shannon's information theory, there also exists a close connection between the FI matrix $\mathbb{J}_q(\mathcal{X})$ and the corresponding Rényi entropy power $N_p(\mathcal{X})$. Rényi's entropy power is defined as the solution of the equation

$$\mathcal{I}_p\left(\mathcal{X}\right) \;=\; \mathcal{I}_p\left(\sqrt{N_p(\mathcal{X})}\cdot\mathcal{Z}^G\right), \tag{10}$$

where $\{\mathcal{Z}_i^G\}$ represents a Gaussian random vector with zero mean and unit covariance matrix. So, $N_p(\mathcal{X})$ denotes the variance of a would be Gaussian distribution that has the same Rényi information content as the random vector $\{\mathcal{X}_i\}$ described by the PDF $\mathcal{F}(\boldsymbol{x})$. Expression (10) was studied in [2, 3, 31] where it was shown that the only class of solutions of (10) is

$$N_p(\mathcal{X}) \;=\; \frac{1}{2\pi}p^{-p'/p}\exp\left(\frac{2}{D}\,\mathcal{I}_p(\mathcal{X})\right), \tag{11}$$

with $1/p + 1/p' = 1$ and $p \in \mathbb{R}^+$. In addition, when $p \to 1_+$ one has $N_p(\mathcal{X}) \to N(\mathcal{X})$, where $N(\mathcal{X})$ is the conventional Shannon entropy power [32].

Generalized isoperimetric inequality: Let $\{\mathcal{X}_i\}$ be a random vector in $\mathbb{R}^D$ with the PDF $\mathcal{F}(\boldsymbol{x})$. Then

$$\frac{1}{D}N_q(\mathcal{X})J_q(\mathcal{X}) \;\geq\; N_q(\mathcal{X})[\det(\mathbb{J}_q(\mathcal{X}))]^{1/D} \;\geq\; 1\,, \tag{12}$$

IOP Conf. Series: Journal of Physics: Conf. Series **1275** (2019) 012005 doi:10.1088/1742-6596/1275/1/012005

where the Rényi parameter $q \geq 1$. A simple proof of the generalized isoperimetric inequality is provided in [30].

It is also worth noting that the relation (12) implies another important inequality. By using the fact that the Shannon entropy is maximized (among all PDF's with identical covariance matrix $\mathbf{\Sigma}$) by the Gaussian distribution we have $N_1(\mathcal{X}) \leq \det(\mathbf{\Sigma})^{1/D}$ (see, e.g. [33]). If we further employ that $\mathcal{I}_q$ is a monotonously decreasing function of q, see, e.g. [36, 34], we can write (recall that $q \geq 1$)

$$\frac{q^{1/(q-1)}}{e} N_q \leq N_1 = \frac{\exp(\frac{2}{D}\mathcal{I}_1)}{2\pi e} \leq \det(\mathbf{\Sigma})^{1/D}. \tag{13}$$

The isoperimetric inequality (12) then implies

$$\det(\mathbf{\Sigma}(\mathcal{X})) \geq \frac{\left(q^{1/(q-1)}\right)^D}{e^D \det(\mathbb{J}_q(\mathcal{X}))} \geq \frac{1}{e^D \det(\mathbb{J}_q(\mathcal{X}))}. \tag{14}$$

We can further use the inequality

$$\frac{1}{D}\mathrm{Tr}(\mathbb{A}) \geq [\det(\mathbb{A})]^{1/D}, \tag{15}$$

(valid for any positive semi-definite $D \times D$ matrix $\mathbb{A}$) to write

$$\sigma^2(\mathcal{X}) = \frac{1}{D}\mathrm{Tr}(\mathbf{\Sigma}(\mathcal{X})) \geq \frac{Dq^{1/(q-1)}}{eJ_q(\mathcal{X})} \geq \frac{D}{eJ_q(\mathcal{X})}, \tag{16}$$

where σ^2 is an average variance per component.

Relations (14)-(16) represent the q-generalizations of the celebrated Cramér–Rao information inequality. In the limit of $q \to 1$ we recover the standard Cramér–Rao inequality which is widely used in statistical inference theory [37]. A final logical step needed to complete the proof of REPURs is represented by the so called generalized Stam inequality. To this end we first define the concept of *conjugate random variables*. We say that random vectors $\{\mathcal{X}_i\}$ and $\{\mathcal{Y}_i\}$ in $\mathbb{R}^D$ are conjugate if their respective PDF's $\mathcal{F}(\boldsymbol{x})$ and $\mathcal{G}(\boldsymbol{y})$ can be written as

$$\mathcal{F}(\boldsymbol{x}) = |\psi_{\mathcal{F}}(\boldsymbol{x})|^2/\|\psi_{\mathcal{F}}\|_2^2, \quad \mathcal{G}(\boldsymbol{y}) = |\psi_{\mathcal{G}}(\boldsymbol{y})|^2/\|\psi_{\mathcal{G}}\|_2^2, \tag{17}$$

where the (generally complex) probability amplitudes $\psi_{\mathcal{F}}(\boldsymbol{x}) \in L_2(\mathbb{R}^D)$ and $\psi_{\mathcal{G}}(\boldsymbol{y}) \in L_2(\mathbb{R}^D)$ are mutual Fourier images, i.e., $\psi_{\mathcal{F}}(\boldsymbol{x}) = \hat{\psi}_{\mathcal{G}}(\boldsymbol{x})$ and $\psi_{\mathcal{G}}(\boldsymbol{y}) = \hat{\psi}_{\mathcal{F}}(\boldsymbol{y})$. With this we can state the generalized Stam inequality.

Generalized Stam inequality: Let $\{\mathcal{X}_i\}$ and $\{\mathcal{Y}_i\}$ be conjugate random vectors in $\mathbb{R}^D$. Then

$$16\pi^2 N_q(\mathcal{Y}) \geq [\det(\mathbb{J}_r(\mathcal{X}))]^{1/D}, \tag{18}$$

valid for any $r \in [1, \infty)$ and $q \in [1/2, 1]$ that are connected via relation $1/r + 1/q = 2$. A proof of the generalized Stam inequality is provided in [30].

Combining the isoperimetric inequality (12) together with the generalized Stam inequality (18) we obtain a one-parameter class of REP-based inequalities

$$N_{p/2}(\mathcal{X})N_{q/2}(\mathcal{Y}) = N_{p/2}(\mathcal{X})\frac{[\det(\mathbb{J}_{p/2}(\mathcal{X}))]^{1/D}}{[\det(\mathbb{J}_{p/2}(\mathcal{X}))]^{1/D}}N_{q/2}(\mathcal{Y}) \geq \frac{N_{q/2}(\mathcal{Y})}{[\det(\mathbb{J}_{p/2}(\mathcal{X}))]^{1/D}} \geq \frac{1}{16\pi^2}, \tag{19}$$

where p and q form now a Hölder double. By symmetry the role of q and p can be reversed. In Ref. [2] we presented an alternative (though more abstract) derivation of above REPURs which was based on the Beckner–Babenko theorem. There it was also proved that the inequality saturates if only if the distributions involved are Gaussian. Only exception to this rule is for the asymptotic values $p = 1$ and $q = \infty$ (or vice versa) where the saturation happens whenever the peak of $\mathcal{F}(\boldsymbol{x})$ and tail of $\mathcal{G}(\boldsymbol{y})$ (or vice versa) are Gaussian.

Importantly, since the REs are, in principle, measurable [38, 39, 40], the associated REPs are experimentally accessible. In addition, REPs of various orders are often used as convenient measures of entanglement — e.g., N_2 represents *tangle* τ (with $\sqrt{\tau}$ being *concurence*) [41], $N_{1/2}$ is related to both *fidelity* F and *robustness* R of a pure state [42], N_∞ quantifies the Bures distance to the closest separable pure state [16], etc. For some recent applications of REs and REPs in quantum theory see, e.g., [43, 44, 45, 39].

3. Information distribution

Let $\mathcal{F}(\boldsymbol{x})$ be the PDF for the random variable $\mathcal{X}$. We define the *information random variable* $i_{\mathcal{X}}(\mathcal{X})$ so that $i_{\mathcal{X}}(\boldsymbol{x}) = \log_2 1/\mathcal{F}(\boldsymbol{x})$. In other words $i_{\mathcal{X}}(\boldsymbol{x})$ represents the information in $\boldsymbol{x}$ with respect to $\mathcal{F}(\boldsymbol{x})$. In this connection it is expedient to introduce the cumulative distribution function for $i_{\mathcal{X}}(\mathcal{X})$ as

$$\wp(y) \;=\; \int_{-\infty}^{y} d\wp(i_{\mathcal{X}}) \;=\; \int_{\mathbb{R}^D} \mathcal{F}(\boldsymbol{x})\theta(\log_2 \mathcal{F}(\boldsymbol{x}) + y)d\boldsymbol{x}\,. \tag{20}$$

The function $\wp(y)$ thus represents the probability that the random variable $i_{\mathcal{X}}(\mathcal{X})$ is less or equal than y. We have denoted the corresponding probability measure as $d\wp(i_{\mathcal{X}})$. Taking the Laplace transform of both sides of (20), we get

$$\mathcal{L}\{\wp\}(s) \;=\; \int_{\mathbb{R}^D} \mathcal{F}(\boldsymbol{x})\,\frac{e^{s\log_2 \mathcal{F}(\boldsymbol{x})}}{s}\,d\boldsymbol{x} \;=\; \frac{\mathbb{E}\left[e^{s\log_2 \mathcal{F}}\right]}{s}\,, \tag{21}$$

where $\mathbb{E}\left[\cdots\right]$ denotes the mean value with respect to $\mathcal{F}$. By assuming that $\wp(x)$ is smooth then the PDF associated with $i_{\mathcal{X}}(\mathcal{X})$ — the so-called *information PDF* – is

$$g(y) \;=\; \frac{d\wp(y)}{dy} \;=\; \mathcal{L}^{-1}\left\{\mathbb{E}\left[e^{s\log_2 \mathcal{F}}\right]\right\}(y)\,. \tag{22}$$

Setting $s = (p-1)\log 2$ we have

$$\mathcal{L}\{g\}(s = (p-1)\log 2) \;=\; \mathbb{E}\left[2^{(1-p)i_{\mathcal{X}}}\right]. \tag{23}$$

The mean here is taken with respect to the PDF g. Eq. (23) can also be written explicitly as

$$\int_{\mathbb{R}^D} d\boldsymbol{x}\,\mathcal{F}^p(\boldsymbol{x}) \;=\; \int_{\mathbb{R}} g(y)2^{(1-p)y}dy\,. \tag{24}$$

Note that when $\mathcal{F}^p$ is integrable for $p \in [1, 2]$ then (24) ensures that the moment-generating function for $g(x)$ PDF exists. So in particular, the moment-generating function exists when $\mathcal{F}(\boldsymbol{x})$ represents Lévy α-stable distributions, including the heavy-tailed stable distributions (i.e, PDFs with the Lévy stability parameter $\alpha \in (0, 2]$). The same holds for $\hat{\mathcal{F}}$ and $p' \in [2, \infty)$ due to the Beckner–Babenko theorem [2, 3, 46, 47].

4. Reconstruction theorem

Since $\mathcal{L}\{g\}(s)$ is the *moment-generating function* of the random variable $i_{\mathcal{X}}(\mathcal{X})$ one can generate all moments of the PDF $g(x)$ (if they exist) by taking the derivatives of $\mathcal{L}\{g\}$ with respect to s. From a conceptual standpoint, it is often more useful to work with cumulants rather than moments. Using the fact that the *cumulant generating function* is the logarithm of the moment-generating function, we see from (24) that the differential RE is a reparametrized version of the cumulant generating function of the random variable $i_{\mathcal{X}}(\mathcal{X})$. In fact, from (23) we have

$$\mathcal{I}_p(\mathcal{X}) \;=\; \frac{1}{(1-p)}\log_2 \mathbb{E}\left[2^{(1-p)i_{\mathcal{X}}}\right]. \tag{25}$$

To understand the meaning of REPURs we begin with the cumulant expansion (25), i.e.

$$p\mathcal{I}_{1-p}(\mathcal{X}) \;=\; \log_2 e \sum_{n=1}^{\infty}\frac{\kappa_n(\mathcal{X})}{n!}\left(\frac{p}{\log_2 e}\right)^n, \tag{26}$$

where $\kappa_n(\mathcal{X}) \equiv \kappa_n(i_{\mathcal{X}})$ denotes the n-th cumulant of the information random variable $i_{\mathcal{X}}(\mathcal{X})$ (in units of *bits*^{n}). We note that

$$\kappa_1(\mathcal{X}) \;=\; \mathbb{E}\left[i_{\mathcal{X}}(\mathcal{X})\right] \;=\; \mathcal{H}(\mathcal{X}), \quad \kappa_2(\mathcal{X}) \;=\; \mathbb{E}\left[i_{\mathcal{X}}(\mathcal{X})^2\right] - \left(\mathbb{E}\left[i_{\mathcal{X}}(\mathcal{X})\right]\right)^2, \tag{27}$$

i.e., they represent the Shannon entropy and *varentropy*, respectively. By employing the identity

$$\mathcal{I}_{1-p}(\mathcal{X}) \;=\; \frac{D}{2}\log_2\left[2\pi(1-p)^{-1/p}N_{1-p}(\mathcal{X})\right], \tag{28}$$

we can rewrite (26) in the form

$$\log_2\left[N_{1-p}(\mathcal{X})\right] \;=\; \log_2\left[\frac{(1-p)^{1/p}}{2\pi}\right] + \frac{2}{D}\sum_{n=1}^{\infty}\frac{\kappa_n(\mathcal{X})}{n!}\left(\frac{p}{\log_2 e}\right)^{n-1}. \tag{29}$$

From (29) one can see that

$$\kappa_n(\mathcal{X}) = \frac{nD}{2}(\log_2 e)^{n-1}\left.\frac{d^{n-1}\log_2\left[N_{1-p}(\mathcal{X})\right]}{dp^{n-1}}\right|_{p=0} + \frac{D}{2}(\log_2 e)^n\left[(n-1)! + \delta_{1n}\log 2\pi\right], \tag{30}$$

which, in terms of the Grünwald–Letnikov derivative formula (GLDF) [48], allows us to write

$$\begin{aligned}\kappa_n(\mathcal{X}) \;&=\; \lim_{\Delta\to 0}\frac{nD}{2}\frac{(\log_2 e)^n}{\Delta^{n-1}}\sum_{k=0}^{n-1}(-1)^k\binom{n-1}{k}\log\left[N_{1+k\Delta}(\mathcal{X})\right] \\ &\quad + \frac{D}{2}(\log_2 e)^n\left[(n-1)! \;+\; \delta_{1n}\log 2\pi\right].\end{aligned} \tag{31}$$

So, in order to determine the first m cumulants of $i_{\mathcal{X}}(\mathcal{X})$ we need to know all $N_1, N_{1+\Delta}, \ldots, N_{1+(m-1)\Delta}$ entropy powers. In practice Δ corresponds to a characteristic resolution scale for the entropy index which will be chosen appropriately for the task at hand, but is typically of the order 10^{-2}. Note that the last term in (30) and (31) can be also written

$$\frac{D}{2}(\log_2 e)^n\left[(n-1)! + \delta_{1n}\log 2\pi\right] \;=\; \kappa_n(\mathcal{Z}_G^{\mathbb{1}}) \;\equiv\; \kappa_n(i_{\mathcal{Y}}), \tag{32}$$

with $\mathcal{Y}$ being the random variable distributed with respect to the *Gaussian* distribution $\mathcal{Z}_G^{\mathbb{I}}$ with the *unit* covariance matrix.

When all the cumulants exist then the problem of recovering the underlying PDF for $i_{\mathcal{X}}(\mathcal{X})$ is equivalent to the *Stieltjes* moment problem [49]. Using this connection, there are a number of ways to proceed; the PDF in question can be reconstructed e.g., in terms of sums involving orthogonal polynomials (e.g., the Gram–Charlier A series or the Edgeworth series [50]), the inverse Mellin transform [51] or via various maximum entropy techniques [52]. Pertaining to this, the theorem of Marcinkiewicz [53] implies that there are no PDFs for which $\kappa_m = \kappa_{m+1} = \ldots = 0$ for $m \geq 3$. In other words, the cumulant generating function cannot be a finite-order polynomial of degree greater than 2. The only exceptions to Marcinkiewicz's theorem, are the *Gaussian* PDFs which can have the first two cumulants nontrivial and $\kappa_3 = \kappa_4 = \ldots = 0$. Thus, apart from the special case of Gaussian PDFs where only N_1 and $N_{1+\Delta}$ are required, one needs to work with as many entropy powers $N_{1+k\Delta}, k \in \mathbb{N}$ as possible to receive as much information as possible about the structure of the underlying PDF. In theory, the whole infinite tower of REPURs would be required to uniquely specify a system's information PDF. Note, that for *Gaussian* information PDFs one needs only N_1 and $N_{1+\Delta}$ to reconstruct the PDF uniquely. From (29) and (31) we see that knowledge of N_1 corresponds to $\kappa_1(\mathcal{X}) = \mathcal{H}(\mathcal{X})$ while $N_{1+\Delta}$ further determines κ_2, i.e. the varentropy. Since N_1 is involved [via (31)] in the determination of all cumulants, it is the most important entropy power in the tower.

We should stress, that the focus of the reconstruction theorem we present is on cumulants κ_n which can be directly used for a shape estimation of $g(x)$ but not $\mathcal{F}(\boldsymbol{x})$. However, by knowing $g(y)$ we have a complete "information scan" of $\mathcal{F}(\boldsymbol{x})$. Such an information scan is, however, not unique, indeed two PDFs that are rearrangements of each other – i.e., *equimeasurable* PDFs, have identical $\wp(y)$ and $g(y)$. Even though equimeasurable PDFs cannot be distinguished via their entropy powers, they can be, as a rule, distinguished via their respective momentum-space PDFs and associated entropy powers. So the information scan has a tomographic flavor to it. From the multi-peak structure of $g(y)$ one can determine the *number* and *height* of the stationary points. These are invariant characteristics of a given family of equimeasurable PDFs.

5. Information scan of the state PDF

With knowledge of the entropy powers, the question now is how we can reconstruct the information distribution $g(x)$. The inner workings of this will be now explicitly illustrated with the (generalized) Gram-Charlier A expansion. However, other – often more efficient methods – are also available [50]. Let κ_n be cumulants obtained from entropy powers and let $G(x)$ be some reference PDF whose cumulants are γ_k. The information PDF $g(x)$ can be then written as [50]

$$g(x) \;=\; \exp\left[\sum_{k=1}^{\infty}(\kappa_k - \gamma_k)(-1)^k \frac{(d^k/dx^k)}{k!}\right] G(x)\,. \tag{33}$$

With the hindsight we choose the reference PDF $G(x)$ to be a shifted gamma PDF, i.e.

$$G(x) \;\equiv\; \mathcal{G}(x|a,\alpha,\beta) \;=\; \frac{e^{-(x-a)/\beta}(x-a)^{\alpha-1}}{\beta^{\alpha}\Gamma[\alpha]}\,, \tag{34}$$

with $a < x < \infty$, $\beta > 0$, $\alpha > 0$. In doing so, we have implicitly assumed that the $\mathcal{F}(y)$ PDF is in the first approximation equimeasurable with the Gaussian PDF. To reach a corresponding matching we should choose $a = \log_2(2\pi\sigma^2)/2$, $\alpha = 1/2$ and $\beta = \log_2 e$. Using the fact that [54]

$$\beta^{k+1/2}\,\frac{d^k\,\mathcal{G}(x|a,1/2,\beta)}{k!\,dx^k} = \left(\frac{x-a}{\beta}\right)^{-k} L_k^{(-1/2-k)}\left(\frac{x-a}{\beta}\right)\mathcal{G}(x|a,1/2,\beta)\,, \tag{35}$$

IOP Conf. Series: Journal of Physics: Conf. Series **1275** (2019) 012005 doi:10.1088/1742-6596/1275/1/012005

(L_k^δ is an associated Laguerre polynomial of order k with parameter δ). Given that $\kappa_1 = \gamma_1 = \alpha\beta + a = \log_2(2\pi\sigma^2 e)/2$, and $\gamma_k = \Gamma(k)\alpha\beta^k = (\log_2 e)^k/2$ for $k > 1$ we can write (33) as

$$\begin{aligned} g(x) &= \mathcal{G}(x|a, 1/2, \beta)\left[1 + \frac{(\kappa_2 - \gamma_2)}{\beta^{1/2}(x-a)^2} L_2^{(-5/2)}\left(\frac{x-a}{\beta}\right)\right. \\ &\quad \left. - \frac{(\kappa_3 - \gamma_3)}{\beta^{1/2}(x-a)^3} L_3^{(-7/2)}\left(\frac{x-a}{\beta}\right) + \cdots\right]. \end{aligned} \tag{36}$$

For the Gram–Charlier A expansion various formal convergence criteria exist (see, e.g., [50]). In particular, the expansion for nearly Gaussian equimeasurable PDFs $\mathcal{F}(y)$ converges quite rapidly and the series can be truncated fairly quickly. Since in this case one needs fewer κ_k's in order to determine $g(x)$, only EPs in the small neighborhood of the index 1 are needed. On the other hand, the further the $\mathcal{F}(y)$ is from Gaussian the higher orders of κ_k are required to determine $g(x)$, and hence a wider neighborhood of the index 1 will be needed for EPs.

6. (Un)Balanced Cat State and reconstruction theorem

We now demonstrate an example of the reconstruction in the context of QM. Specifically, we consider cat states that are often considered in the foundations of quantum physics as well as in various applications, including solid state physics [55] and quantum metrology [56]. The form of the state we consider is $|\psi\rangle = \mathcal{N}(|0\rangle + \nu|\alpha/\nu\rangle)$ where, $\mathcal{N} = [1 + 2\nu\exp(-\alpha^2/2\nu^2) + \nu^2]^{-1/2}$ is the normalization factor, $|0\rangle$ is the vacuum state, $\nu \in \mathbb{R}$ a weighting factor and $|\alpha\rangle$ is the coherent state given by ($\alpha \in \mathbb{R}$)

$$|\alpha\rangle = e^{-\alpha^2/2}\sum_{n=0}^{\infty}\frac{\alpha^n}{\sqrt{n!}}|n\rangle\,, \tag{37}$$

For $\nu = 1$ we refer to the state as a *balanced cat state* (BCS) and for $\nu \neq 1$, as an *unbalanced cat state* (UCS). Changing the basis of $|\psi\rangle$ to the eigenstates of the general quadrature operator

$$\hat{Y}_\theta = \frac{1}{\sqrt{2}}\left(\hat{a}e^{-i\theta} + \hat{a}^\dagger e^{i\theta}\right), \tag{38}$$

where, $\hat{a}$ and $\hat{a}^\dagger$ are the creation and annihilation operators of the electromagnetic field, we find the PDF for the general quadrature variable y_θ to be

$$\mathcal{F}(y_\theta) = \mathcal{N}^2\pi^{-\frac{1}{2}}e^{-y_\theta^2}\left|1 + \nu\exp\left[-\frac{\alpha^2}{\nu^2 2}\left(1 + e^{2i\theta} - 2\sqrt{2}e^{i\theta}\frac{\nu}{\alpha}y_\theta\right)\right]\right|^2, \tag{39}$$

Setting $\theta = 0$ and $\nu = 1$ returns the PDF of the BCS for the position-like variable y_0. With this, the Rényi EPs $N_{1-p}(\chi)$ are calculated and found to be constant across varying p. This is because $\mathcal{F}(y_0)$ for the BCS is in fact a piecewise rearrangement of a Gaussian PDF (yet has an overall non-Gaussian structure), thus $N_{1-p}(\chi) = \sigma^2$ for all p, where σ^2 is the variance of the 'would be Gaussian'. Taking the reference PDF to be $G(x) = \mathcal{G}(x|a, \alpha, \beta)$, with $a = \log_2(2\pi\sigma^2)/2$, $\alpha = 1/2$ and $\beta = \log_2(e)$, it is evident that $(\kappa_k - \gamma_k) = 0$ for all $k \geq 1$, and from the Gram–Charlier A series (33), a perfect matching in the reconstruction is achieved. Furthermore, it can be shown that the variance of (39) increases with α, i.e. the variance increases as the peaks of the PDF diverge, which is in stark contrast to the Rényi EPs which remain constant for increasing α. This reveals the shortcomings of variance as a measure of uncertainty for non-Gaussian PDFs.

The peaks, located at $\mathcal{F}(y_\theta) = 2^{-a_j^+}$, where j is an index labelling the distinct peaks, gives rise to sharp singularities in the target $g(x)$. With regard to the BCS position PDF, distributions of the conjugate parameter $\mathcal{F}(y_{\pi/2})$ distinguish $\mathcal{F}(y_0)$ from it's equimeasurable Gaussian PDF and hence the Rényi EPs also distinguish the different cases. The number of available cumulants k is computationally limited but as this grows, information about the singularities will be recovered in the reconstruction. In the following, we show how the tail convergence and location of a singularity for $g(x)$ can be reconstructed using $k = 5$. We consider the case of a UCS with $\nu = 0.97$, $\alpha = 10$ and we take $\theta = 0$ in equation (39) to find the PDF in the y_0 quadrature which is non-Gaussian for all piecewise rearrangements. As such, all REPs N_{1-p} vary with p and consequently all cumulants κ_k carry information on $g(x)$. Here we choose to reconstruct the UCS information distribution by means of the Edgeworth series [50] so that

$$g(x) = \exp\left[n\sum_{j=2}^{\infty}(\kappa_j - \gamma_j)\frac{(-1)^j}{j!}\frac{d^j}{dx^j}n^{-j/2}\right]G(x), \tag{40}$$

where the reference PDF $G(x)$ is again the shifted gamma distribution. Using the Edgeworth series, the information PDF is approximated by expanding in orders of n, which has the advantage over the Gram–Charlier A expansion discussed above of bounding the errors of the approximation. For the particular UCS of interest, expanding to order $n^{-3/2}$ reveals convergence toward the analytic form of the information PDF shown as the target line in Fig. 1. This shows that, for a given characteristic resolution, control over the first five Rényi EPs can be enough for a useful information scan of a quantum state with an underlying non-Gaussian PDF. In the example shown in Fig. 1 we see that the information scan accurately predicts the tail behavior as well as the location of the singularity, which corresponds to the second (lower) peak of $\mathcal{F}(y_0)$.

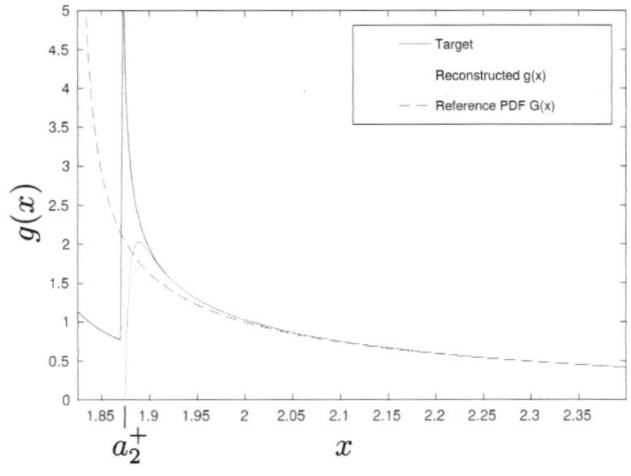

Figure 1. Reconstructed information distribution of an unbalanced cat state with $\nu = 0.97$ and $\alpha = 10$. The Edgeworth expansion has been used here to order $n^{-3/2}$ requiring control of the first five REPs. Good convergence of the tail behaviour is evident as well as the location of the singularity corresponding to the second peak; a_2^+ corresponds to the value of x at the point of intersection with the second (lower) peak of $\mathcal{F}(y_0)$.

7. Conclusions

In the first part of this paper we have presented a new proof of a one-parameter class of Rényi-entropy-power-based URs for pairs of observables in an infinite-dimensional Hilbert space. This was done with the help of the generalized isoperimetric and Stam inequality. This proof substantiates our earlier version in [2, 3] which was based on the Babenko–Beckner inequality.

IOP Conf. Series: Journal of Physics: Conf. Series **1275** (2019) 012005 doi:10.1088/1742-6596/1275/1/012005

The mathematical language employed in the new proof (i.e., Fisher information matrix, Cramér–Rao inequality, etc.) is much closer to the language familiar in quantum information theory and, in particular, quantum metrology, and so makes stronger connections to these fields.

In the second part we present the method for reconstructing the underlying information PDF associated with a given quantum state from known entropy powers. The details of this reconstruction procedure have been explained in terms of what we call an information scan and mathematically formalized in a reconstruction theorem. A numerical implementation of the reconstruction theorem was discussed and we demonstrated its utility in a proof-of-principle analysis by reconstructing the information PDF of a quantum state described by a non-Gaussian probability distribution. In this case it was found that control of the first five REPs gave enough information for a meaningful reconstruction of the information PDF. In particular we demonstrated the superiority of REPs over variance as a measure of uncertainty for the important class of non-Gaussian states — (un)balanced cat states.

REPURs, together with the reconstruction theorem that has been established and discussed here, can be used for various problems in the theory of quantum information and metrology. Specific example include separability conditions and the characterization of multipartite entanglement or improved witnesses of quantum entanglement and (weak) measurements in the presence of quantum memory. These avenues are currently being actively pursued.

Acknowledgments

P.J. was supported by the Czech Science Foundation Grant No. 19-16066S. J.D. acknowledges support from the UK EPSRC through the NQIT Quantum Technology Hub (EP/M013243/1).

References

[1] Bialynicki-Birula I 2007 *AIP Conference Proceedings* **889** 52
[2] Jizba P, Ma Y, Hayes A and Dunningham J A 2016 *Phys. Rev.* E**93** 060104
[3] Jizba P, Dunningham J A and Joo J 2015 *Annals of Physics* **355** 87
[4] Maassen H and Uffink J B M 1988 *Phys. Rev. Lett.* **60** 1103
[5] Bialynicki-Birula I and Mycielski J 1975 *Commun. Math. Phys.* **44** 129
[6] Dang P, Deng G T and Qian T 2013 *J. Func. Analysis* **265** 2239.
[7] Ozawa T and Yuasa K 2017 *J. Math. Anal. Appl.* **445** 998
[8] Jizba P, Kleinert H and Scardigli F 2010 *Phys. Rev.* D**81** 084030
[9] Marin F et al. 2013 *Nature Physics* **9** 71
[10] Pikovski I. et al. 2012 *Nature Physics* **8** 393
[11] Erhart J et al. 2012 *Nature Physics* **8** 185
[12] Sulyok G et al. 2013 *Phys. Rev.* A**88** 022110
[13] Baek S Y et al. 2013 *Scientific Reports* **3** 2221
[14] Dressel J and Nori F 2014 *Phys. Rev.* A**89** 022106
[15] Busch P, Lahti P and Werner R F 2013 *Phys. Rev. Lett.* **111** 160405
[16] Bengtsson I and Życzkowski K 2006 *Geometry of Quantum States. An Introduction to Quantum Entanglement* (Cambridge: Cambridge Univ. Press)
[17] Wilk G and Włodarczyk Z 2009 *Phys. Rev.* A**79** 062108
[18] Schrödinger E 1930 *Sitzungsber. Preuss. Acad. Wiss.* **24** 296
[19] Robertson H P 1929 *Phys. Rev.* **34** 163
[20] Hirschman Jr. I I 1957 *Am. J. Math.* **79** 152
[21] Shannon C E 1948 *Bell Syst. Tech. J.* **27** 623
[22] Liu R, Liu T, Poor H V and Shamai S 2010 *IEEE Trans. Inf. Theory* **56** 1865
[23] Costa M H 1985*IEEE Trans. Inf. Theory* **31** 607
[24] Polyanskiy Y and Wu Y 2016 *IEEE Trans. Inf. Theory* **62** 3992
[25] Bagherikaram G, Motahari A S and Khandani A K 2013 *IEEE Trans. Inf. Theory* **59** 2673
[26] De Palma G, Mari A, Lloyd S and Giovannetti V 2015 *Phys. Rev.* A**91** 032320
[27] Barron A 1986 *Ann. Probab.* **14** 336
[28] Rioul O 2011 *IEEE Trans. Inf. Theory* **57** 33
[29] Beck C and Schlögl F 1993 *Thermodynamics of Chaotic Systems* (Cambridge: Cambridge University Press)
[30] Jizba P, Dunningham J A and Joo J 2019 *Phys. Rev.* E to apper

IOP Conf. Series: Journal of Physics: Conf. Series **1275** (2019) 012005 doi:10.1088/1742-6596/1275/1/012005

[31] Gardner R J 2002 *Bull. Amer. Math. Soc.* **39** 355
[32] Shannon C E 1948 *Bell Syst. Tech. J.* **27** 379
[33] Stam A 1959 *Inform. Contr.* **2** 101
[34] Rényi A 1970 *Probability Theory* (Amsterdam: North-Holland)
[35] Rényi A 1976 *Selected Papers of Alfred Rényi* vol 2 (Budapest: Akadémia Kiado)
[36] Jizba P and Arimitsu T 2004 *Ann. Phys.* **312** 17
[37] Cramér H 1946 *Mathematical Methods of Statistics* (Princeton: Princeton University Press)
[38] Campbell L L 1965 *Inf. Control* **8** 423
[39] Jizba P and Arimitsu T 2004 *Phys. Rev.* **E69** 026128
[40] Elben A, Vermersch B, Dalmonte M, Cirac J I, and Zoller P 2018 *Phys. Rev. Lett.* **120** 050406
[41] Minter F, Kuś M and Buchleitner A 2004 *Phys. Rev. Lett.* **92** 167902
[42] Vidal G and Tarrach R 1999 *Phys. Rev.* A**59** 141
[43] Bacco D, Canale M, Laurenti N, Vallone G and Villoresi P 2013 *Nat. Commun.* **4** 2363
[44] Müller-Lennert M, Dupuis F, Szehr O, Fehr S and Tomamichel M 2013 *J. Math. Phys.* **54** 122203
[45] Coles P J, Colbeck R, Yu L and Zwolak M 2012 *Phys. Rev. Lett.* **108** 210405
[46] Beckner W 1975 *Ann. Math.* **102** 159
[47] Babenko K I 1962 *Amer. Math. Soc. Transl.* **44** 115
[48] Samko S G, Kilbas A A and Marichev O I 1993 *Fractional Integrals and Derivatives: Theory and Applications* (New York: Gordon and Breach)
[49] Reed M and Simon B 1975 *Methods of Modern Mathematical Physics* vol XI (New York: Academic Press)
[50] Wallace D L 1958 *Ann. Math. Stat.* **29** 635
[51] Zolotarev V M 1957 *Teor. Veroyatnost. i Primenen.* **2** 444
[52] Tagliani A 1998 *J. Comp. Appl. Math.* **90** 157
[53] Lukacs E 1970 *Characteristic Functions* (London: Charles Griffin)
[54] Pal N, Jin C and Lim W K 2005 *Handbook of Exponential and Related Distributions for Engineers and Scientists* (New York: Taylor & Francis Group)
[55] Kira M, Koch S W, Smith R P, Hunter A E and Cundiff S T 2011 *Nature Physics* **7** 799
[56] Knott P A, Cooling J P, Hayes A, Proctor T J and Dunningham J A 2016 *Phys. Rev.* A**93** 033859

IOP Conf. Series: Journal of Physics: Conf. Series **1275** (2019) 012006 doi:10.1088/1742-6596/1275/1/012006

Does gravity have to be quantized? Lessons from non-relativistic toy models

Antoine Tilloy
Max-Planck-Institut für Quantenoptik, Hans-Kopfermann-Straße 1, 85748 Garching, Germany
E-mail: antoine.tilloy@mpq.mpg.de

Abstract. It is often argued that gravity has to be a quantum theory simply because a fundamentally semiclassical approach would necessarily be inconsistent. Here I review recent Newtonian toy models of (stochastic) semiclassical gravity. They provide one option to implement a force semiclassically without getting into the known problems associated with mean-field. These models are not complete theories and should not be considered too seriously, but their consistency shows that semiclassical gravity is hard to dismiss on purely theoretical grounds.

1. Introduction

Do we have to quantize gravity? Is it merely possible that gravity is not quantum? Are there theoretical arguments or experimental results cornering us to the quantization route? Is it already known that gravity must be a quantum theory and that only its specific form is yet to be understood?

My point, which is not particularly original, is simply that we are not sure. The situation is not as clear cut as one might think and the arguments in favor of the quantization of gravity are so far weak. Naturally, I do not aim to demonstrate that gravity is not quantum with certainty, but merely that this is a possibility one ought to consider. I am obviously not the first one to defend this line, and the possibility that gravity is ultimately not quantum has abundantly been discussed in the literature [1–3]. My objective here is simply to review recent constructions of consistent Newtonian toy models [4–11] that exemplify how standard arguments against semiclassical gravity can be bypassed.

The present contribution to these proceedings does not contain anything fundamentally new on the mathematical side, but puts together all the arguments and derivations, in an hopefully selfcontained way. It follows to a large extent, although in a more detailed manner, the rather informal discussion in [9].

1.1. Semiclassical and fully quantum gravity

So what is the alternative to a fully quantum theory of gravity? Because the words *classical*, *semiclassical*, and *quantum* have been used to mean different things in this context, a clarification is perhaps not superfluous. In a classical theory of gravity, one has a spacetime, classical (well localized) and its properties are encoded in a metric $g(t, \mathbf{x})$. Matter is classical, one has particles with well defined trajectories or *e.g.* a scalar field $\phi(t, \mathbf{x})$. The two evolve jointly: the dynamics of matter is influenced by the metric (essentially through the substitution $\partial_\mu \to D_\mu$), and matter

Content from this work may be used under the terms of the Creative Commons Attribution 3.0 licence. Any further distribution of this work must maintain attribution to the author(s) and the title of the work, journal citation and DOI.
Published under licence by IOP Publishing Ltd

IOP Conf. Series: Journal of Physics: Conf. Series **1275** (2019) 012006 doi:10.1088/1742-6596/1275/1/012006

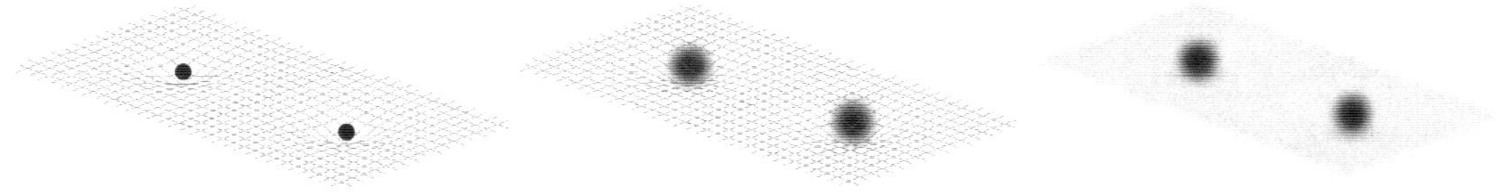

Figure 1. In a classical theory of gravity (left), matter and spacetime are classical. In a semiclassical theory of gravity (center), matter is quantum but there is a single classical spacetime; all particles feel the same. Finally, in a fully quantum theory of gravity (right), matter and spacetime are quantum. While the need for having a quantum description of matter is experimentally demonstrated, it is so far unclear if this is needed for spacetime.

in turn curves space time via Einstein's equation $G_{\mu\nu} \propto T_{\mu\nu}$. In a fully quantum theory of gravity, say canonical quantum gravity to fix the ideas, both sides become quantum and a wave function $\psi_t(\phi, g)$ is needed to describe the couple matter + spacetime.

In a semiclassical theory of gravity, spacetime is still classical, with a well defined metric $g(\mathbf{x}, t)$. Matter is quantum, and is described at an intuitive and naive level by some wave-function $\psi_t(\phi)$. Its dynamics is given by Quantum Field Theory in curved spacetime (QFTCST), which is naively given by the transformation $\partial_\mu \to D_\mu$ in Lorentzian QFT, and is a refined version of ordinary quantum mechanics in an external classical field. The difficult part is then to find a law for the backreaction of quantum matter on a classical spacetime: when there are superpositions of matter, how is spacetime curved? While QFTCST for a fixed metric $g(\mathbf{x}, t)$ is uncontroversial and even experimentally well tested in the non-relativistic limit (e.g. by letting quantum stuff fall), the backreaction problem is really where wild speculations are allowed. A particular proposal, sometimes called semiclassical Einstein equations, Møller-Rosenfeld semiclassical gravity [12,13], or sometimes simply semiclassical gravity, is to have spacetime be curved by the quantum expectation values, i.e. $G_{\mu\nu} \propto \langle \hat{T}_{\mu\nu} \rangle$. Heuristically, at the non-relativistic level, it means having the modulus squared of the wave-function source the gravitational field: if the wavefunction for a single particle has two blobs, they attract each other. Note again that this is just *one* option to curve a classical spacetime with quantum matter.

Semiclassical gravity is sometimes used in a rather restrictive sense, to mean QFTCST alone –that is without any backreaction of matter on spacetime–, to mean the specific Møller-Rosenfeld backreaction proposal, or finally to designate certain classes of *approximations* to a more fundamental quantum theory of gravity. Rather, I want to consider a potentially fundamental theory in which quantum matter curves a classical spacetime (in whatever way) and in which matter moves in this spacetime following (at least approximately) QFTCST[1]. Is this possible? Can it be made consistent? If so, is it already falsified or soon falsifiable?

1.2. In the Newtonian limit

The difference between a fully quantum theory of gravity and semiclassical gravity is perhaps best illustrated in the Newtonian limit. Consider the Schrödinger equation for a matter state $|\psi_t\rangle$:

$$i\hbar \frac{\mathrm{d}}{\mathrm{d}t} |\psi_t\rangle = \left(\hat{H}_0 + \hat{V}\right) |\psi_t\rangle \tag{1}$$

where $\hat{H}_0$ is the Hamiltonian in the absence of gravity and $\hat{V}$ is the potential of gravitational origin. If gravity is fully quantum, then $\hat{V}$ is simply a pair potential, in the same way the

[1] This latter requirement is used to prevent labelling "semiclassical" a Bohmian quantum theory of gravity. In such a theory, there would be a well defined trajectory for the metric, but the dynamics for the matter part would not be given by QFTCST with the Bohmian metric.

IOP Conf. Series: Journal of Physics: Conf. Series **1275** (2019) 012006 doi:10.1088/1742-6596/1275/1/012006

Coulomb force is obtained in the non-relativistic limit of Quantum Electrodynamics (QED). On the other hand, for a semiclassical theory, all particles see the same classical (here scalar) field $\Phi(\mathbf{x})$. Namely:

$$\text{Quantum gravity} \quad \hat{V} = -\frac{G}{2}\int_{\mathbb{R}^3} \mathrm{dx} \int_{\mathbb{R}^3} \mathrm{dy}\, \frac{\hat{M}(\mathbf{x})\hat{M}(\mathbf{y})}{|\mathbf{x}-\mathbf{y}|} \tag{2}$$

$$\text{Semiclassical gravity} \quad \hat{V} = \int_{\mathbb{R}^3} \mathrm{dx}\, \Phi(\mathbf{x})\, \hat{M}(\mathbf{x}) \tag{3}$$

where $\hat{M}(\mathbf{x})$ is the mass density operator, which reads *e.g.*

$$\hat{M}(\mathbf{x}) = \sum_{\ell=1}^{n} m_\ell\, a_\ell^\dagger(\mathbf{x})\, a_\ell(\mathbf{x}), \tag{4}$$

for n different species of particles of mass m_ℓ, with $a_\ell^\dagger(\mathbf{x})$ and $a_\ell(\mathbf{x})$ the local creation and annihilation operators of particles of type ℓ. For a semiclassical theory, equation (3), which is uncontroversial, only tells how matter moves once the gravitational field is given. One needs to complement it with a specification for backreaction, which means identifying what object $M(\mathbf{x})$ sources the field Φ in the Poisson equation:

$$\nabla^2\Phi(\mathbf{x}) = 4\pi G M(\mathbf{x}). \tag{5}$$

In this context, defining a theory of semiclassical gravity requires providing a prescription for $M(\mathbf{x})$ and typically its dependence on the quantum state (or on extra variables): this is the non-trivial part. However, whatever choice one makes for the $M(\mathbf{x})$ and thus Φ, it is not possible to recover the quantum evolution (2) from the semiclassical prescription (3), as it would require an operator valued field $\Phi \to \hat{\Phi}/2$.[2]

2. The standard approach and its problems

We now discuss the standard approach to semiclassical gravity and its legitimate shortcomings.

2.1. Møller-Rosenfeld semiclassical gravity

To source the gravitational field out of quantum matter, one needs a prescription to extract some classical quantity out of something quantum. A natural option, historically due to Møller and Rosenfeld [12, 13], is to use quantum expectation values to turn operators into scalars and use the result as source in Einstein's equation, *i.e.* to posit that:

$$R_{\mu\nu} - \frac{1}{2}R g_{\mu\nu} = 8\pi G\, \langle\psi|\hat{T}_{\mu\nu}|\psi\rangle. \tag{6}$$

Let us note already that it is non trivial to make sense of this equation in the general case. One technical difficulty, among others, is that the r.h.s. is generically divergent, even for the vacuum, and needs to be renormalized [14, 15]. The technical difficulties of the fully relativistic setting may have overshadowed the conceptual issues that can be seen already in the Newtonian limit. In my opinion, these are the most serious difficulties and the ones that most urgently need to be addressed if one wants to explore the option that gravity is not fully quantum.

[2] This factor of $1/2$ is often missed in the literature where the semiclassical approach is compared with the purely quantum one. One does not get the correct quantum prescription simply with the naive procedure of putting hats in the Poisson equation and using (3). One needs to take into account the field Hamiltonian, which is not zero in the non-relativistic limit, and which precisely contributes to $-1/2$ times what the matter-field coupling yields, giving the overall $1/2$ in (2).

IOP Conf. Series: Journal of Physics: Conf. Series **1275** (2019) 012006 doi:10.1088/1742-6596/1275/1/012006

2.2. Newtonian limit

If (6) is taken as fundamental[3], then its non-relativistic limit is simply given by the Poisson equation [18], where field degrees of freedom are slaved to matter:

$$\nabla^2\Phi(\mathbf{x},t) = 4\pi G\,\langle\psi_t|\hat{M}(\mathbf{x})|\psi_t\rangle. \tag{7}$$

Once this classical field is given, which is the non-trivial part, it acts on quantum matter in the standard way, as an external field:

$$\frac{\mathrm{d}}{\mathrm{d}t}|\psi\rangle = -\frac{i}{\hbar}\left(H_0 + \int \mathrm{d}\mathbf{x}\,\Phi(\mathbf{x},t)\hat{M}(\mathbf{x})\right)|\psi_t\rangle. \tag{8}$$

Inverting the Poisson equation (7), to replace the field by its matter source, then yields the celebrated Schrödinger-Newton equation [16, 19]:

$$\frac{\mathrm{d}}{\mathrm{d}t}|\psi_t\rangle = -\frac{i}{\hbar}H_0|\psi_t\rangle + i\frac{G}{\hbar}\int \mathrm{d}\mathbf{x}\,\mathrm{d}\mathbf{y}\,\frac{\langle\psi_t|\hat{M}(\mathbf{x})|\psi_t\rangle\,\hat{M}(\mathbf{y})}{|\mathbf{x}-\mathbf{y}|}|\psi_t\rangle. \tag{9}$$

For a single (otherwise free) particle of wave function ψ, it takes the perhaps more familiar form:

$$\frac{\mathrm{d}}{\mathrm{d}t}\psi(t,\mathbf{x}) = \frac{i}{2\hbar m}\nabla^2\psi(t,\mathbf{x}) + i\frac{Gm^2}{\hbar}\int \mathrm{d}\mathbf{y}\,\frac{|\psi(\mathbf{y},t)|^2}{|\mathbf{x}-\mathbf{y}|}\,\psi(t,\mathbf{x}), \tag{10}$$

which is cubic in the wavefunction. With such an evolution, a single particle weakly interacts with itself.

2.3. The problem of non-linear back-reaction

The deterministic non-linear character of the Schrödinger-Newton brings well known difficulties [20–23] which are hard to bypass. It is worth discussing these conceptual problems because arguments against semiclassical gravity often leverage them in one way or another.

Discussing the consequences (even only operational) of a non-linear modification of the Schrödinger equation is not as easy as one may think. Indeed, non-linearities break the empirical equivalence "for all practical purposes" between decoherence and collapse. Consequently different interpretations of quantum theory make different predictions and should be discussed separately.

The first possibility is to consider an interpretation of quantum mechanics in which there is no fundamental collapse process, such as the Many-World interpretation or de Broglie-Bohm (dBB) theory. In such approaches, the problem of the Schrödinger-Newton equation is one of *empirical inadequacy*. Indeed, the non-linear gravitational term couples even fully orthogonal branches of the universe wave-function. Decoherence no longer achieves the decoupling role it has in standard linear quantum mechanics: Schrödinger's dead and live cat(s), even fully decohered with the environment, attract each other[4]. Hence for interpretations of quantum mechanics without fundamental collapse, the theory is well defined but makes predictions that seem completely at odds with our experience.

[3] The Poisson equation (7) is only a good non-relativistic approximation of (6) if the latter is taken to be fundamental [16]. Naturally, if (6) is aimed to provide a mean-field approximation of a more fundamental theory of gravity, then the Poisson equation (7) for a system of a few particles will be a very poor approximation of the full quantum description [17], but this is not the scenario we discuss here.

[4] The initial hope was that the manifest clash with our experience this implies would be avoided because the non-linear addition would also dynamically collapse macroscopic superpositions precisely in the way one would want, according to the Born rule. Unfortunately, this is not the case [16].

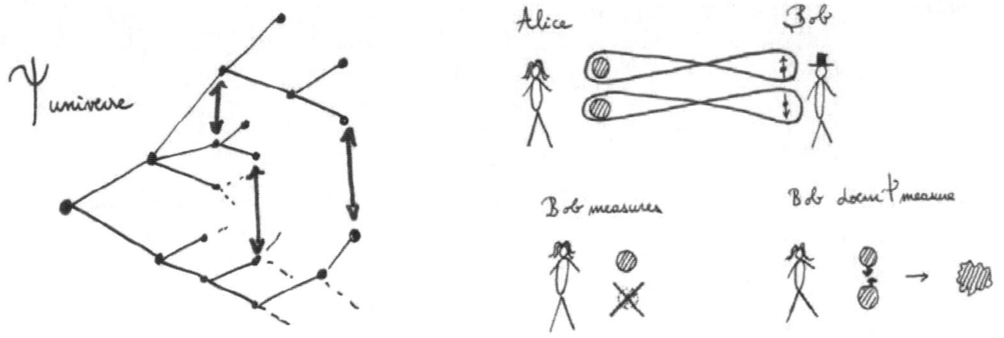

Figure 2. *Left*: without collapse, orthogonal branches of the universe wavefunction attract each other via the non-linear gravitational interaction. *Right*: Alice and Bob share an entangled pair with a massive superposition on Alice's side. In linear quantum mechanics, the statistics on Alice's side do not depend on what Bob does on his side. In a non-linear theory, Alice can know, via measurements on *her* side, if Bob carried a measurement on his side. Note that this information is not transmitted by gravity from Bob to Alice. Rather, it is the local non-linearity of gravity that allows to witness the non-local collapse of the wave function.

The second option is to have an interpretation in which collapse is fundamental and happens (not necessarily exclusively) in measurement situations. In that case, the difficulty is a bit more subtle and lies in the breakdown of the statistical interpretation of the quantum state. Consider a situation in which Alice and Bob share an Einstein-Podolski-Rosen (EPR) pair, with a massive superposition on Alice's side (see Fig. 2). If the wave-function collapses upon measurement on Bob's side, then Alice's mass is localized in a given position and nothing happens to it. If Bob carries no measurement and the wave-function does not collapse, Alice's mass stays superposed and evolves with the non-linear self gravitation term. Alice can easily distinguish the two situations, e.g. by making a measurement in the region where most the mass would lie in the second situation. Not only would this generically allow faster than light signalling (at least in one frame), the Born rule would break down when used even on well isolated subsystems. Alice's density matrix is insufficient to determine statistics on her side.

In practice, the kind of massive superpositions required to make this example clean may very well be suppressed by an intrinsic collapse process, but the conclusion remains that it would in general be impossible to use the statistical interpretation of the state vector and that there would be residual faster than light signalling effects.

2.4. Conclusion on standard semiclassical gravity

The standard (Møller-Rosenfeld) approach to semiclassical gravity is not completely killed by the previous arguments. Especially in the second case, for interpretations of quantum mechanics with an objective collapse process, one may hope that violations of the statistical interpretation of the quantum state do not amplify from micro to macro, remaining generically small and under control. Further, faster-than-light effects could be so small that no practical consequences would ensue, and restricted to a single frame such that no theoretical inconsistency could be derived. Hence, it is technically possible to still defend this approach provided one is ready to pay the price of an auxiliary collapse process and of serious practical complications (the Born rule becoming only approximately valid, without any good control on the approximation).

Nonetheless, I think it is fair to say that if this were the only representative of semiclassical theories, the case for fundamental semiclassical gravity would be rather slim. However, one should remember that, while simple, the *choice* of making expectation values the classical stuff that sources the gravitational field is not unique nor even natural. At a heuristic level, the Møller-Rosenfeld prescription assumes that $|\psi|^2$ is what is "real", that it is the tangible thing that bends spacetime. But we are free to explore other options.

3. Discrete models restoring linearity

3.1. Basic idea

A natural thing to do if one wants to get something classical from quantum matter is to go to an interpretation of quantum mechanics where there is such a thing as classical stuff by construction. For example, one could use Bohmian particles as classical sources of the gravitational field. In fact, this has been tried by Struyve [24, 25]. However, while such a theory could behave in a better way than the previous approach, and importantly not require an additional collapse process to make sense, it still suffers from the breakdown of the statistical interpretation and the theoretical possibility to send information faster-than-light. In that case, these unappealing features emerge from the breakdown of equivariance which usually insures that the distribution of Bohmian particles is provided by the Born rule.

Hence it seems that rather generically, when one extracts a classical quantity from quantum matter, through the wave-function directly (with quantum expectation values) or through additional variables (Bohmian positions), one breaks the simplicity of the formalism, i.e. the ability to extract predictions from a simple statistical prescription.

However, this difficulty can be avoided and it seems at present that there is essentially only one simple way to do it. The intuition comes from (orthodox) quantum measurement and feedback. In a lab, an experimentalist measures a quantum systems, obtains a certain result, processes it, and applies a control pulse depending on this result. This complete procedure of measurement and feedback does not break the formalism as the latter was precisely constructed to describe such situations. Further, the measurement result is a tangible thing, which after some (classical) processing is used to act on the quantum state. Measurement and feedback thus allow the consistent coupling of classical and quantum systems. This intuition can be used to construct (potentially) fundamental semiclassical theories that are without conceptual problems [26–28]. In the context of gravity, this intuition was first explored in a toy model by Kafri *et al.* [4] where it was called "classical channel gravity".

Of course, this measurement and feedback intuition is only used to construct a model with good properties. It should not be understood "ontologically" as meaning there are actual detectors carrying the measurement task.

3.2. Intuition in the discrete

Let us consider a rather general weak measurement with n outcomes, parameterized by n matrices $\hat{N}_k$ which verify as only constraint $\sum_k \hat{N}_k^\dagger \hat{N}_k = \mathbb{1}$. Such a measurement acts on a quantum state in the following way:

$$|\psi\rangle \xrightarrow{\text{measurement}} |\psi_k\rangle = \frac{\hat{N}_k|\psi\rangle}{\sqrt{\langle\psi|\hat{N}_k^\dagger \hat{N}_k|\psi\rangle}}, \tag{11}$$

with probability $p_k = \langle\psi|\hat{N}_k^\dagger N_k|\psi\rangle$ where k labels the measurement outcome. This is the first step of any measurement + feedback scheme.

Now, "feeding back" the result k just amounts to apply a k-dependent unitary operator $\hat{U}_k$ on the state. That is, for a measurement immediately followed by a feedback (or neglecting any intrinsic system dynamics happening in the meantime), one has:

$$|\psi\rangle \xrightarrow[\text{+ feedback}]{\text{measurement}} |\psi_k^{\text{f.b.}}\rangle = \hat{U}_k|\psi_k\rangle = \hat{U}_k \frac{\hat{N}_k|\psi\rangle}{\sqrt{\langle\psi|\hat{N}_k^\dagger \hat{N}_k|\psi\rangle}}, \tag{12}$$

with the same probability p_k. Notice that the full evolution is still a legitimate generalized measurement with the substitution $\hat{N}_k \to \hat{B}_k = \hat{U}_k \hat{N}_k$.

By construction, this measurement and feedback scheme does not break the standard formalism despite the fact that the measurement step is non-linear. Indeed this non-linearity vanishes upon averaging over measurement results:

$$\rho^{\text{f.b.}} \equiv \mathbb{E}\left[|\psi^{\text{f.b.}}\rangle\langle\psi^{\text{f.b.}}|\right] = \sum_{k=1}^{n} p_k \frac{\hat{B}_k|\psi\rangle\langle\psi|\hat{B}_k^\dagger}{\langle\psi|\hat{N}_k^\dagger\hat{N}_k|\psi\rangle} \equiv \sum_{k=1}^{n} \hat{B}_k \underbrace{|\psi\rangle\langle\psi|}_{\rho} \hat{B}_k^\dagger, \tag{13}$$

where $\mathbb{E}[\,\cdot\,]$ denotes the statistical average. Any modification of the measurement rule (12), which specifies how the classical variable k is extracted from the quantum state, would generically preclude this cancellation of non-linearities. Unless there is a particular reason why linearity should hold, it typically does not.

3.3. Simplest discrete model for gravity

With this simple understanding of feedback, one can already construct a toy model of semiclassical gravity with discrete Poisson distributed measurement and feedback events [8].

Let us consider for convenience N distinguishable particles with coordinates $\{\mathbf{x}_1, \cdots, \mathbf{x}_N\}$, masses $\{m_1, \cdots, m_N\}$, and without spin. The N-particle system is described by the wave-function $\psi_t(\mathbf{x}_1, \cdots, \mathbf{x}_N)$ with $\mathbf{x}_\ell \in \mathbb{R}^3$. Our objective is to implement a consistent semiclassical gravitational force between them. We consider the following weak position measurement of particle ℓ:

$$\psi_t \longrightarrow \frac{\hat{L}_\ell(\mathbf{x}_f)\psi_t}{\|\hat{L}_\ell(\mathbf{x}_f)\psi_t\|}, \tag{14}$$

with the operator

$$\hat{L}_\ell(\mathbf{x}_f) = \frac{1}{(\pi r_c^2)^{3/4}} \mathrm{e}^{-(\hat{\mathbf{x}}_\ell - \mathbf{x}_f)^2/(2r_c^2)}, \tag{15}$$

where $\hat{\mathbf{x}}_\ell$ is the position operator associated to particle ℓ. The probability for the outcome $\mathbf{x}_f$ is $P(\mathbf{x}_f) = \|\hat{L}_k(\mathbf{x}_f)\psi_t\|^2$. This is as just before in eq. (12) with discrete indices labeling results promoted to continuous position variables $k \to \mathbf{x}_f$ and $\hat{N}_k \to \hat{L}_\ell(\mathbf{x}_f)$. The parameter r_c encodes the sharpness of the position measurement.

We now assume that each particle can be hit by this spontaneous weak measurement process independently of the others and with a distribution uniform in time with intensity λ. The astute reader will have noticed that this random dynamics for the quantum state is nothing but the Ghirardi-Rimini-Weber (GRW) model [29–31], the simplest dynamical reduction model. However, its measurement interpretation is fundamental in that it guides us to the proper choice of variables to use as sources for the gravitational field.

Now assume that a collapse (or weak measurement) with center (or result) $\mathbf{x}_f$ has hit particle ℓ at time t_f. We now take this space-time event (or "flash") $(t_f, \mathbf{x}_f)$ as source of the gravitational field[5]:

$$\nabla^2\Phi(\mathbf{x}, t) = 4\pi G m_\ell \lambda^{-1} \delta(t - t_f)\delta(\mathbf{x} - \mathbf{x}_f). \tag{16}$$

More generally, one could replace these Dirac distributions by a generic causal spacetime smearing function f such that:

$$\int_{\mathbb{R}^4} \mathrm{d}t\,\mathrm{d}\mathbf{x}\, f(t, \mathbf{x}) = 1 \qquad f \geq 0 \qquad \forall t < 0,\;\; f(t, \mathbf{x}) = 0. \tag{17}$$

[5] This possibility was mentioned, although not explored further, by Derakhshani [32].

Since the measurement step has resolution r_c, it would *e.g.* also make sense to have f spatially smear the point source with a Gaussian of width r_c. The objective here is just to provide the simplest model and thus we will just consider punctual sources. The gravitational potential created by this event is:

$$\hat{V} = \int \mathrm{d}\mathbf{x}\Phi(\mathbf{x})\hat{M}(\mathbf{x}), \tag{18}$$

where the mass density operator is simply $\hat{M}(\mathbf{x}) = \sum_{\ell=1}^{N} m_\ell\, \delta(\mathbf{x} - \hat{\mathbf{x}}_\ell)$ in this first quantized representation. Inverting the Poisson equation (16) yields:

$$\hat{V}(t) = -\delta(t - t_f)\, G\lambda^{-1} \sum_{\ell'=1}^{N} \frac{m_\ell m_{\ell'}}{|\mathbf{x}_f - \hat{\mathbf{x}}_{\ell'}|}. \tag{19}$$

This time-singular potential may be worrying at first sight, but once put in the Schrödinger equation it simply implements a unitary transformation:

$$\hat{U}_\ell(\mathbf{x}_f) = \exp\left(i\frac{G}{\lambda\hbar} \sum_{\ell'=1}^{N} \frac{m_\ell m_{\ell'}}{|\mathbf{x}_f - \hat{\mathbf{x}}_{\ell'}|}\right), \tag{20}$$

which happens immediately *after* the collapse. As before, it just changes the collapse operators $\hat{L}_\ell(\mathbf{x}_f) \to \hat{C}_\ell(\mathbf{x}_f) = \hat{U}_\ell(\mathbf{x}_f)\hat{L}_\ell(\mathbf{x}_f)$. This fully specifies the model. In a nutshell, the position of each particle is weakly measured and the measurement result is used to create a unitary gravitational kick. In the language of the GRW model, we just made one possible choice of local beables [33] (or primitive ontology [34–36]), the *flashes*, the source of the gravitational field [8].

The master equation for this model is very simple and reads:

$$\frac{\mathrm{d}}{\mathrm{d}t}\rho_t = -\frac{i}{\hbar}[\hat{H}_0, \rho_t] + \lambda \sum_{\ell=1}^{N} \left\{ \int_{\mathbb{R}^3} \hat{C}_\ell(\mathbf{x}_f)\rho_t \hat{C}_\ell^\dagger(\mathbf{x}_f)\, \mathrm{d}\mathbf{x}_f - \rho_t \right\}, \tag{21}$$

where $\hat{H}_0$ is just the standard Schrödinger Hamiltonian containing all interactions but gravity. Importantly, it is linear, and thus the statistical intepretation of the density matrix survives. This model has a slightly more complicated phenomenology than continuous alternatives, and the fact that gravity acts through punctual kicks creates a new length scale. Indeed, assuming all the masses are the same $m_k = m$:

$$\hat{U}_\ell(\mathbf{x}_f) = \exp\left(i \sum_{\ell'=1}^{N} \frac{r_G}{|\mathbf{x}_f - \hat{\mathbf{x}}_{\ell'}|}\right), \tag{22}$$

with $r_G = Gm^2/(\hbar\lambda)$. For $\lambda = 10^{-16}\mathrm{s}^{-1}$ (historical value of λ for the GRW model that is not yet experimentally falsified) we get $r_G \simeq 1.8 \times 10^{-14}$m for protons and $r_G \simeq 5.3 \times 10^{-21}$m for electrons. In both cases, $r_G \ll r_c$, the smearing length of the measurement. If we consider interparticle distances r much larger than r_G we can expand the unitary operator to first order in r_G/r and we obtain:

$$\frac{\mathrm{d}}{\mathrm{d}t}\rho_t \simeq -\frac{i}{\hbar}[\hat{H}_0 +, \rho_t] - \frac{i}{\hbar}\hat{V}_{\text{eff}} \cdot \rho_t + \lambda \sum_{\ell=1}^{N} \left\{ \int_{\mathbb{R}^3} \hat{L}_\ell(\mathbf{x}_f)\rho_t \hat{L}_\ell^\dagger(\mathbf{x}_f)\, \mathrm{d}\mathbf{x}_f - \rho_t \right\}, \tag{23}$$

with

$$\hat{V}_{\text{eff}} \cdot \rho = \sum_{\ell=1}^{N} \sum_{\ell'=1}^{N} \int_{\mathbb{R}^3} \mathrm{d}\mathbf{x}_f \left[-\frac{Gm_\ell m_{\ell'}}{|\mathbf{x}_f - \hat{\mathbf{x}}_{\ell'}|}, \hat{L}_\ell(\mathbf{x}_f)\rho \hat{L}_\ell^\dagger(\mathbf{x}_f) \right], \tag{24}$$

which looks a lot like the purely quantum pair potential up to some smearing around ρ. The way the expected pair potential is approximately recovered will be clearer for continuous models, but it can already be seen here for a simple situation. Consider N particles separated from each other by an average distance $r \gg r_c$. Writing $\rho^{(1)} = \mathrm{tr}_{N-1}[\rho]$ the reduced density matrix of the first particle, one has [8]:

$$\mathrm{tr}_{N-1}\left[\hat{V}_{\mathrm{eff}} \cdot \rho\right] \simeq \mathrm{tr}_{N-1}\left(\left[-\sum_{\ell=2}^{N} \frac{Gm_1 m_\ell}{|\hat{\mathbf{x}}_1 - \hat{\mathbf{x}}_\ell|}, \rho\right]\right). \tag{25}$$

Further assuming that the $N-1$ particles traced over are localized over a distance much smaller than their distance to particle 1 yields:

$$\mathrm{tr}_{N-1}\left[\hat{V}_{\mathrm{eff}} \cdot \rho\right] \simeq \left[-\sum_{\ell=2}^{N} \frac{Gm_1 m_\ell}{|\hat{\mathbf{x}}_1 - \mathbf{x}_\ell|}, \rho^{(1)}\right], \tag{26}$$

which is precisely the potential one would have put by hand in a scenario where a quantum particle interacts with $N-1$ fixed classical particles.

In this model, gravity behaves in a rather peculiar way, through discrete kicks. Yet, it is not obviously falsified by experiments probing gravity in the Newtonian limit [8].

4. Continuous models restoring linearity

The previous model is easy to define, and the reason why it preserves linearity at the master equation level is particularly understandable and straightforward to verify. However, its phenomenology is not completely trivial, and the fact that gravity acts through punctual kicks makes everything more complicted. This motivates the discussion of continuous models, which were actually introduced first.

4.1. Continuous measurement and feedback

When infinitely weak measurements are repeated infinitely frequently, one can obtain a continuous measurement described by a diffusive stochastic differential equation [37, 38]. The real-time result (or signal) from the continuous measurement of a self-adjoint operator $\mathcal{O}$ reads:

$$\mathscr{S}_{\mathcal{O}}(t) = \langle \mathcal{O} \rangle_t + \eta_t, \tag{27}$$

where $\langle \mathcal{O} \rangle_t = \langle \psi_t | \mathcal{O} | \psi_t \rangle$ and η is white noise:

$$\mathbb{E}\left[\eta_t \eta_s\right] = \frac{1}{\gamma}\delta(t-s) \qquad \mathbb{E}[\eta_t] = 0, \tag{28}$$

where γ encodes the measurement strength. The signal given by equation (27) is the equivalent of the result k and its probability distribution $P(k) = \langle \psi | N_k^\dagger N_k | \psi \rangle$ in the discrete.

The corresponding non-linear backaction on the quantum state reads:

$$\frac{\mathrm{d}}{\mathrm{d}t}|\psi\rangle_t = \left[\frac{\gamma}{2}\left(\mathcal{O} - \langle \mathcal{O} \rangle_t\right)\eta_t - \frac{\gamma}{4}\left(\mathcal{O} - \langle \mathcal{O} \rangle_t\right)^2\right]|\psi_t\rangle. \tag{29}$$

This is a stochastic differential equation with multiplicative white noise, and it should be understood in the Itô convention. In this context, one needs to be particularly careful with stochastic calculus conventions as corrections to the standard differentiation rules coming from Itô's formula are necessary to get the physically correct results. This has sometimes led to confusions in the literature. Hence, to mark the Itô convention, and at the risk of seeming overly pedantic, I now switch to the Itô notation and rewrite equations (27) and (29):

$$\mathrm{d}R_t = \langle \mathcal{O} \rangle_t \, \mathrm{d}t + \mathrm{d}W_t \quad \text{“ = ”} \quad \mathscr{S}_{\mathcal{O}}(t)\, \mathrm{d}t \tag{30}$$

$$\mathrm{d}|\psi\rangle_t = \left[\frac{\gamma}{2}\left(\mathcal{O} - \langle \mathcal{O}\rangle_t\right) \mathrm{d}W_t - \frac{\gamma}{4}\left(\mathcal{O} - \langle \mathcal{O}\rangle_t\right)^2 \mathrm{d}t\right] |\psi_t\rangle, \tag{31}$$

with the heuristic Itô rule $\mathrm{d}W_t \mathrm{d}W_t = \gamma^{-1}\,\mathrm{d}t$. For all practical purposes, one obtains the Itô formula for differentiation simply by expanding all differentials to second order in $\mathrm{d}W$ and using this heuristic rule. For example, one can derive the stochastic master equation for the (pure) density matrix $\rho_t = |\psi_t\rangle\langle\psi_t|$ writing:

$$\mathrm{d}\rho_t = \mathrm{d}|\psi_t\rangle\, \langle\psi_t| + |\psi_t\rangle\, \mathrm{d}\langle\psi_t| + \underbrace{\mathrm{d}|\psi_t\rangle \mathrm{d}\langle\psi_t|}_{\text{Itô correction}} \tag{32}$$

$$= -\frac{\gamma}{8}\left[\mathcal{O}, \left[\mathcal{O}, \rho_t\right]\right] \mathrm{d}t + \frac{\gamma}{2}\left(\mathcal{O}\rho_t + \rho_t \mathcal{O} - 2\langle \mathcal{O}\rangle_t \rho_t\right) \mathrm{d}W_t, \tag{33}$$

the latter containing exactly the same information as (31). Because ρ_t is what shows up when computing expectation values, one often works with it directly instead of the state $|\psi_t\rangle$ even if ρ_t is of rank 1. An important property of stochastic differential equations in the Itô convention is that in integrals of the form $\int f_t \mathrm{d}W_t$, f_t and $\mathrm{d}W_t$ are uncorrelated. As a result, taking the average of a stochastic differential equation simply removes the noise term. Writing $\bar{\rho} = \mathbb{E}\left[\rho\right]$ we thus get the master equation

$$\frac{\mathrm{d}}{\mathrm{d}t}\bar{\rho} = -\frac{\gamma}{8}\left[\mathcal{O}, \left[\mathcal{O}, \bar{\rho}\right]\right]. \tag{34}$$

As expected, this master equation is linear: continuous measurements, just like discrete ones, bring no non-linearities at the master equation level.

Let us now go back to the stochastic master equation without averaging and introduce a feedback term depending on the real-time signal. Namely, let us heuristically consider the addition of a potential proportional to the signal

$$\hat{V}^{\text{f.b.}}(t) \quad \text{“ = ”} \quad \mathscr{S}_{\mathcal{O}}(t)\, \hat{K} \tag{35}$$

in the Schrödinger equation, where $\hat{K}$ is self-adjoint. Because this potential contains white noise, one needs to be careful to implement it. Importantly, this potential does not act at the exact same instant as the measurement from which the result is extracted, but infinitesimally *after*. This is analogous to the discrete situation where the feedback unitary $\hat{U}_k$ is applied just after the measurement operator $\hat{N}_k$. Hence we have to translate the informal procedure (35) into the rigorous prescription

$$\rho_t + \mathrm{d}\rho_t^{\text{f.b.}} = \exp(-i\hat{K}\mathrm{d}R_t/\hbar)(\rho_t + \mathrm{d}\rho_t)\exp(i\hat{K}\,\mathrm{d}R_t/\hbar), \tag{36}$$

where $\mathrm{d}\rho_t$ is the infinitesimal evolution of ρ_t given by pure continuous measurement (31). Expanding all terms up to second order in $\mathrm{d}W$ and using the heuristic Itô rule yields

$$\mathrm{d}\rho_t^{\text{f.b.}} = \mathrm{d}\rho_t - \frac{i}{\hbar}[\hat{K}, \rho_t]\,\mathrm{d}W_t - \frac{i}{2}\left[\hat{K}, \{\mathcal{O}, \rho_t\}\right] \mathrm{d}t - \frac{1}{2\gamma\hbar^2}\left[\hat{K}\left[\hat{K}, \rho_t\right]\right] \mathrm{d}t. \tag{37}$$

Note that one gets the same stochastic master equation with the more physical procedure of regulating the signal with a low pass filter first, and subsequently sending the frequency cutoff

IOP Conf. Series: Journal of Physics: Conf. Series **1275** (2019) 012006 doi:10.1088/1742-6596/1275/1/012006

of the filter to infinity (or equivalently the associated smearing timescale to zero) [38]. One may again average over the measurement randomness to get the linear master equation:

$$\frac{\mathrm{d}}{\mathrm{d}t}\bar{\rho} = -\frac{i}{2\hbar}\left[\hat{K},\{\mathcal{O},\bar{\rho}\}\right] - \frac{\gamma}{8}\left[\mathcal{O},[\mathcal{O},\bar{\rho}]\right] - \frac{1}{2\gamma^{-}\hbar}\left[\hat{K}\left[\hat{K},\bar{\rho}\right]\right]. \tag{38}$$

At the master equation level, this feedback has thus added a new decoherence term $\propto [\hat{K},[\hat{K},\rho]]$ and a so far hard to interpret term $\propto -i\left[\hat{K},\{\mathcal{O},\bar{\rho}\}\right]$ which effectively acts like a potential up to an additional non-trivial dissipation. We will need a generalization of this procedure to the continuous simultaneous measurement of n self-adjoint operators $\mathcal{O}^j$. In that case, a fairly general consistent continuous measurement model can be written:

$$\mathrm{d}R_t^j = \langle\mathcal{O}^j\rangle_t + \mathrm{d}W_t^j \tag{39}$$

$$\mathrm{d}\rho_t = -\frac{1}{8}\Gamma_{jk}\left[\mathcal{O}^j,\left[\mathcal{O}^k,\rho_t\right]\right]\,\mathrm{d}t + \frac{1}{2}\Gamma_{jk}\left(\mathcal{O}^j\rho_t + \rho_t\mathcal{O}^j - 2\langle\mathcal{O}^j\rangle_t\rho_t\right)\mathrm{d}W_t^k, \tag{40}$$

with the Itô rule $\mathrm{d}W_t^j\mathrm{d}W_t^k = (\Gamma^{-1})^{kj}\,\mathrm{d}t$ where Γ is a *real* symmetric positive definite matrix and we have used implied summation on repeated indices. Intuitively, in a measurement situation, this Γ matrix encodes the correlation of the detectors and its eigenvalues encode the measurement strength. In this more general setup, a general real time feedback prescription reads:

$$\mathrm{d}\hat{V}^{\text{f.b.}} = \hat{K}_j\,\mathrm{d}R_t^j, \tag{41}$$

which yields a stochastic master equation for ρ_t that can be derived in the same way as in equations (36) and (37). Upon averaging over the measurement randomness, one finally gets a master equation generalizing (38):

$$\frac{\mathrm{d}}{\mathrm{d}t}\bar{\rho} = -\frac{i}{2\hbar}\left[\hat{K}_j,\left\{\mathcal{O}^j,\bar{\rho}\right\}\right] - \frac{\Gamma_{jk}}{8}\left[\mathcal{O}^j,\left[\mathcal{O}^k,\bar{\rho}\right]\right] - \frac{(\Gamma^{-1})^{jk}}{2\hbar^2}\left[\hat{K}_j\left[\hat{K}_k,\bar{\rho}\right]\right]. \tag{42}$$

The first term in this master equation, which contains a mix of unitary and dissipative effects, may again seem mysterious. It is simpler to interpret when $\hat{K}_j\otimes\mathcal{O}^j = \mathcal{O}^j\otimes\hat{K}_j$, in which case there is no dissipation and we are left with a pure potential term:

$$-\frac{i}{2\hbar}\left[\hat{K}_j,\left\{\mathcal{O}^j,\bar{\rho}\right\}\right] = -\frac{i}{4\hbar}\left[\left\{\mathcal{O}^j,\hat{K}_j\right\},\bar{\rho}\right]. \tag{43}$$

In that case, which is the one we will be considering for gravity, measurement + feedback allows to implement a purely quantum potential at the price of additional decoherence coming from the measurement itself and from the feedback noise. These two sources of decoherence, by design, make sure that the interaction is in fact purely classical despite what the presence of the quantum potential (43) in the master equation (42) may suggest.

4.2. General continuous mass density measurement and feedback

We are now equipped to implement gravity by a continuous measurement of the mass density and feedback of the corresponding gravitational potential. To this end, we start from a continuous measurement model of (regularized) local mass density in every point of space. The corresponding signal reads:

$$\mathrm{d}R(\mathbf{x},t) = \langle\hat{M}_{r_c}(\mathbf{x})\rangle_t + \mathrm{d}W_t(\mathbf{x}) \quad \text{“}=\text{''} \quad \mathscr{S}_{M_{r_c}(\mathbf{x})}(t)\,\mathrm{d}t \tag{44}$$

$$\text{with}\quad \mathrm{d}W_t(\mathbf{x})\mathrm{d}W_t(\mathbf{y}) = \Gamma^{-1}(\mathbf{x},\mathbf{y})\,\mathrm{d}t. \tag{45}$$

IOP Conf. Series: Journal of Physics: Conf. Series **1275** (2019) 012006 doi:10.1088/1742-6596/1275/1/012006

The corresponding backaction on the state is

$$\begin{aligned} \mathrm{d}\rho_t = &-\frac{1}{8}\int \mathrm{d}\mathbf{x}\mathrm{d}\mathbf{y}\ \Gamma(\mathbf{x},\mathbf{y}) \left[\hat{M}_{r_c}(\mathbf{x})\left[\hat{M}_{r_c}(\mathbf{y}),\rho_t\right]\right]\mathrm{d}t \\ &+\frac{1}{2}\int \mathrm{d}\mathbf{x}\mathrm{d}\mathbf{y}\ \Gamma(\mathbf{x},\mathbf{y})\left(\hat{M}_{r_c}(\mathbf{x})\rho_t + \rho_t\hat{M}_{r_c}(\mathbf{x}) - 2\langle\hat{M}_{r_c}(\mathbf{x})\rangle_t\rho_t\right)\mathrm{d}W_t(\mathbf{y}), \end{aligned} \tag{46}$$

with $\hat{M}_{r_c}(\mathbf{x})$ a regularized mass density operator:

$$\hat{M}_{r_c}(\mathbf{x}) = \left[g_{r_c} * \hat{M}\right](\mathbf{x}) = g_{r_c} * \sum_{\ell=1}^{n} m_\ell\, a_\ell^\dagger(\mathbf{x})\, a_\ell(\mathbf{x}), \tag{47}$$

where "$*$" denotes the convolution product and g_{r_c} is a Gaussian smearing of width r_c

$$g_{r_c}(\mathbf{x}) = e^{-\mathbf{x}^2/(2r_c^2)}/(\sqrt{2\pi r_c^2})^3. \tag{48}$$

This continuous measurement model of the regularized mass density everywhere in space, if taken as fundamental, is simply a continuous collapse model. More precisely, one obtains:

(i) The Continuous Spontaneous Localization (CSL) model [39, 40] for:

$$\Gamma^{\mathrm{CSL}}(\mathbf{x},\mathbf{y}) = \frac{4\gamma}{m_N^2} \times \delta(\mathbf{x}-\mathbf{y}), \tag{49}$$

where m_N is the mass of a nucleon and γ is the collapse "strength". It is a rate $\times$ distance^3, the corresponding rate is $\lambda_{\mathrm{CSL}} \equiv \gamma/(4\pi r_c^2)^{3/2}$ historically fixed at $\lambda_{\mathrm{CSL}} \simeq 10^{-16}\mathrm{s}^{-1}$ (the so called "GRW" value).

(ii) The Diósi-Penrose (DP) model [41, 42] for:

$$\Gamma^{\mathrm{DP}}(\mathbf{x}-\mathbf{y}) = \frac{2G}{\hbar} \times \frac{1}{|\mathbf{x}-\mathbf{y}|}. \tag{50}$$

The interest of knowing that, at a mathematical level, this model has a continuous measurement interpretation, is that it allows us to introduce the "signal" $\mathscr{S}_{\hat{M}_{rc}(\mathbf{x})}$ (44). This is not a natural object to consider in the collapse model context, but it is fundamental to implement a consistent semiclassical interaction.

As in the discrete, we can implement gravity by sourcing the gravitational field from the mass density signal:

$$\nabla^2\Phi_t(\mathbf{x}) = 4\pi G\mathscr{S}_{M_{r_c}(\mathbf{x})}(t). \tag{51}$$

As before, this expression is a bit formal because the signal has white noise fluctuations. We nonetheless keep on formally, and invert the Poisson equation to get:

$$\Phi_t(\mathbf{x}) = -G\int \mathrm{d}\mathbf{y}\ \frac{\mathscr{S}_{M_{r_c}(\mathbf{x}))}(t)}{|\mathbf{x}-\mathbf{y}|}. \tag{52}$$

The corresponding signal dependent feedback potential to introduce in the Schrödinger equation thus reads (still formally):

$$\hat{V}^{\mathrm{f.b.}} = \int \mathrm{d}\mathbf{x}\,\Phi_t(\mathbf{x})\hat{M}_{r_c}(\mathbf{x}). \tag{53}$$

This is equivalent with feeding back directly the operator $\hat{\Phi}_{r_c}(\mathbf{x})$ multiplied by the mass density signal with

$$\hat{\Phi}_{r_c}(\mathbf{x}) := -G\int \mathrm{d}\mathbf{y}\ \frac{\hat{M}_{r_c}(\mathbf{y})}{|\mathbf{x}-\mathbf{y}|}. \tag{54}$$

IOP Conf. Series: Journal of Physics: Conf. Series **1275** (2019) 012006 doi:10.1088/1742-6596/1275/1/012006

Now more rigorously, we can write the elementary feedback step as in (36):

$$\rho_t + \mathrm{d}\rho_t^{\text{f.b.}} = \exp(-i\mathrm{d}\hat{V}^{\text{f.b.}}/\hbar)(\rho_t + \mathrm{d}\rho_t)\exp(i\mathrm{d}\hat{V}^{\text{f.b.}}/\hbar) \tag{55}$$

with

$$\mathrm{d}\hat{V}^{\text{f.b.}} = -G\int \mathrm{d}\mathbf{x}\mathrm{d}\mathbf{y}\,\frac{\hat{M}_{r_c}(\mathbf{x})\,\mathrm{d}R_t(\mathbf{y})}{|\mathbf{x}-\mathbf{y}|} = \int \mathrm{d}\mathbf{x}\,\hat{\Phi}_{r_c}(\mathbf{x})\,\mathrm{d}R_t(\mathbf{x}). \tag{56}$$

Following the same steps as before, it yields the master equation for matter

$$\frac{\mathrm{d}}{\mathrm{d}t}\bar{\rho} = -\frac{i}{\hbar}\left[\hat{V},\bar{\rho}\right] - \underbrace{\int \mathrm{d}\mathbf{x}\mathrm{d}\mathbf{y}\frac{\Gamma(\mathbf{x},\mathbf{y})}{8}\left[\hat{M}_{r_c}(\mathbf{x})\left[\hat{M}_{r_c}(\mathbf{y}),\bar{\rho}\right]\right]}_{\text{measurement decoherence}} - \underbrace{\int \mathrm{d}\mathbf{x}\mathrm{d}\mathbf{y}\frac{\Gamma^{-1}(\mathbf{x},\mathbf{y})}{2\hbar^2}\left[\hat{\Phi}_{r_c}(\mathbf{x})\left[\hat{\Phi}_{r_c}(\mathbf{y}),\bar{\rho}\right]\right]}_{\text{feedback induced decoherence}} \tag{57}$$

with the pair potential

$$\hat{V} = -\frac{G}{2}\int \mathrm{d}\mathbf{x}\mathrm{d}\mathbf{y}\frac{\hat{M}_{r_c}(\mathbf{x})\hat{M}_{r_c}(\mathbf{y})}{|\mathbf{x}-\mathbf{y}|}. \tag{58}$$

This pair potential is exactly the quantum pair potential one would have obtained at the non-relativistic limit of a fully quantum theory of gravity, up to the smearing of the interaction for distances shorter than r_c. The additional decoherence terms in (57) precisely insure that the resulting interaction remains classical. They would have been non-trivial to guess starting directly from a master equation. This decomposition of semiclassical gravity into 3 decoupled terms –a purely quantum potential, "intrinsic" measurement decoherence, and "gravitational" feedback decoherence – is a major advantage of the continuum. The master equation (21) we obtained for the discrete model where flashes source the gravitational field, while simpler to derive, had no such simple decomposition.

4.3. Minimum decoherence prescription

We constructed a class of non-relativistic theories of semiclassical gravity parameterized by a kernel $\Gamma(\mathbf{x},\mathbf{y})$. Should we take the CSL kernel which looks simpler, as it is local, or the DP kernel, which is non-local but requires one parameter less? Should we pick yet anoter kernel? Following [7], we may want to find a principle allowing us to single out one kernel. A natural principle is to ask that the total decoherence induced by the model be minimal. Decoherence coming from the measurement and feedback parts can be grouped in a single kernel $\Gamma^{\text{f.b.}}$

$$\frac{\mathrm{d}}{\mathrm{d}t}\bar{\rho} = -\frac{i}{\hbar}\left[\hat{V},\bar{\rho}\right] - \frac{1}{8}\int \mathrm{d}\mathbf{x}\mathrm{d}\mathbf{y}\;\Gamma^{\text{f.b.}}(\mathbf{x},\mathbf{y})\left[\hat{M}_{r_c}(\mathbf{x})\left[\hat{M}_{r_c}(\mathbf{y}),\bar{\rho}\right]\right] \tag{59}$$

where

$$\Gamma^{\text{f.b.}} = \Gamma + 4\,(4\pi G/\hbar)^2\,\frac{1}{\nabla^2}\,\Gamma^{-1}\,\frac{1}{\nabla^2}. \tag{60}$$

There are many ways to define the minimum for such a kernel. Assuming a translation invariant kernel $\Gamma(\mathbf{x},\mathbf{y}) = \Gamma(\mathbf{x}-\mathbf{y})$ we can go to Fourier space where all terms in (60) are diagonal. A natural prescription is then to minimize decoherence Fourier mode by Fourier mode, or eingenvalue per eigenvalue. This will insure that the kernel we find is indeed the minimum for most commonly used norms. Going back to real space, this yields a minimal total decoherence kernel $\Gamma^{\text{f.b.}}_{\text{min}}$:

$$\Gamma^{\text{f.b.}}_{\text{min}}(\mathbf{x},\mathbf{y}) = 2\,\Gamma_{\text{min}}(\mathbf{x},\mathbf{y}) = \frac{4G}{\hbar}\frac{1}{|\mathbf{x}-\mathbf{y}|}. \tag{61}$$

IOP Conf. Series: Journal of Physics: Conf. Series **1275** (2019) 012006 doi:10.1088/1742-6596/1275/1/012006

This is precisely the decoherence kernel of the Diósi-Penrose model (up to a factor 2)[6]. At the minimum of decoherence, exactly as much decoherence is provided by the continuous measurement and feedback parts.

5. Consequences and limitations of "measurement and feedback" inspired semiclassical models

5.1. Non-linear stochastic description and the measurement problem

An important byproduct of the semiclassical approach we have considered is that it is intimately tied to a solution of the measurement problem. Starting from the will to implement gravity in a semiclassical manner, one is lead to the need to have a quantity formally behaving like a measurement result. Extracting such a result cannot be done without a backreaction on the quantum state, acting in kicks or continuously, that kills off macroscopic superpositions. Requiring that a force is implemented semiclassically, formally as a measurement + feedback scheme naturally leads to objective collapse models. Conversly, taking a collapse model for granted, one gets a way to implement consistent semiclassical interactions "for free".

We have focused on feedback master equations averaged over measurement outcomes (or collapse events), which are indeed sufficient to get all possible predictions in all possible experimental situations. But the stochastic description with pure states is important: it is necessary to understand why this approach solves the measurement problem by collapsing macroscopic superpositions. Further, without the initial stochastic state description, there is no way to implement the feedback in the first place, and the empirical content *does depend* on the measurement stochastic Schrödinger equation one starts with. Different unravelings of the measurement master equation yield different feedback master equations.

Is it generic that a fundamentally semiclassical theory of gravity is tied to a solution of the measurement problem? I think so. In any reasonable semiclassical theory of gravity, there is a classical configuration associated to spacetime. This classical configuration can be directly given the status of local beable or used, via Einstein's equation, to define a configuration for matter beables. In any case, one has something tangible which can, for example, yield definite measurement outcomes upon coarse graining.

5.2. Regularization and complete falsifiability

An undoubtedly unappealing feature of the present approach is the need for an arbitrary spacetime regularization of the mass-density measurement. Sending the smearing length-scale r_c to zero yields a divergent model even for the kernel $\Gamma^{\text{f.b.}}_{\text{min}}$ yielding minimal decoherence (Diósi-Penrose kernel). For standard collapse models, this regularization freedom in principle allows to escape falsifiability indefinitely. In practice, if regularization length-scales get too large, collapse models no longer achieve their purpose of collapsing macroscopic superpositions. This gives a rather soft metaphysical bound that one can easily move by orders of magnitude (see e.g. [43]).

In contrast here, the gravitational potential $\hat{V}$ is inevitably smeared at the same lengthscale as decoherence. Making the regularization lengthscale too large destroys the Newtonian pair potential at short distances. This means that r_c can be experimentally lower bounded by probing intrinsic decoherence and upper bounded by probing gravity at short distances. The semiclassical models we presented are thus falsifiable for all values of r_c, although they are so far *not* falsified.

[6] The present approach gives more than the initial Diósi-Penrose model. First the derivation of the kernel is not heuristic but motivated by a precise principle: the fact that we get a kernel $\propto 1/|\mathbf{x}|$ is not by analogy with gravity, it comes naturally. Second, our model does contain an effective gravitational interaction $\hat{V}$, which is not in the historical DP model.

5.3. Difficult relativistic extension

The main limitation of the approach I have discussed here is that the construction is made directly in the Newtonian limit. There is so far no complete relativistic theory from which one obtains the Newtonian limit after a robust derivation. What goes in the way of such a construction?

First, constructing a fully relativistic theory of continuous quantum measurement seems difficult. Intuitively, it seems necessary to continuously measure local operators to preserve Lorentz invariance, as was *e.g.* proposed in [44]. The difficulty is that such theories are divergent in a way that does not seem easily curable except with an explicit smearing like the one we used for non-relativistic models. In the Lorentzian context, a space smearing necessarily goes hand in hand with a time smearing, i.e. some form of non-Markovianity. While the theory of non-Markovian stochastic Schrödinger equations is well developed [45–49], the latter have no real-time continuous measurement interpretation [50–53] and cannot be used to implement a consistent semiclassical interaction (at least not in any obvious way). There have nonetheless been recent attempts at making general relativistic collapse models with discrete jumps [54] that may be usable to extend the first toy model we presented.

Once a proper continuous measurement model or collapse model is available in the general relativistic setting, there is still a difficulty. Generically, the signal obtained will not verify the conservation laws the corresponding field would obey in the classical theory. For example, the "signal" stress-energy tensor will not be conserved, which makes it unusable as source in Einstein's equations. This lack of conservation can be dealt with by going to unimodular gravity [55], but the precise consequences in a complete model are yet to be explored.

A promising avenue of research, in the foreseeable future, is cosmology, where having a complete relativistic model is not necessary. The cosmological consequences of a semiclassical coupling in the spirit of the one we presented here, but for the scale factor, have been explored already by Altamirano *et al.* in [56, 57].

6. Conclusion

The core of the argument in favor of merely considering the option that gravity could be fundamentally semiclassical is that one can construct semiclassical models that are not obviously falsified and that bypass the main no-go argument against semiclassical theories. Admittedly, these models are not the most appealing on the market, one might even say that they are ugly. Further, they have been made precise only in Newtonian settings: while this is the setting in which criticisms of semiclassical gravity have been formulated, and thus where it is fair to refute them, this is a strong limitation. As a result, the semiclassical models which have been explored in the last 5 years and which I presented here should not be taken too seriously as an alternative to more developed programs like String Theory or Asymptotically safe gravity (to name only a few), but rather as counter examples, as toys, simply reminding us to remain open minded. It is unlikely that semiclassical gravity can be eliminated exploiting purely theoretical arguments, and a hint about the nature of gravity will most likely have to be found in tabletop experiments [58, 59].

Acknowledgments

I thank the participants of the DICE 2018 meeting for interesting and enjoyable discussions. I am indebted to Lajos Diósi with whom I have worked on many of the ideas presented here. This work was supported by a postdoctoral fellowship of the Alexander von Humboldt foundation.

References

[1] Carlip S 2001 *Rep. Prog. Phys.* **64** 885

[2] Mattingly J 2005 *The Universe of General Relativity* ed Kox A J and Eisenstaedt J (Boston, MA: Birkhäuser Boston) pp 327–338 ISBN 978-0-8176-4454-3
[3] Wüthrich C 2005 *Philosophy of Science* **72** 777–788
[4] Kafri D, Taylor J M and Milburn G J 2014 *New J. Phys.* **16** 065020
[5] Kafri D, Milburn G J and Taylor J M 2015 *New J. Phys.* **17** 015006
[6] Tilloy A and Diósi L 2016 *Phys. Rev. D* **93**(2) 024026
[7] Tilloy A and Diósi L 2017 *Phys. Rev. D* **96**(10) 104045
[8] Tilloy A 2018 *Phys. Rev. D* **97**(2) 021502
[9] Tilloy A 2018 *Found. Phys.* **48** 1753–1769 ISSN 1572-9516
[10] Khosla K E and Altamirano N 2017 *Phys. Rev. A* **95**(5) 052116
[11] Altamirano N, Corona-Ugalde P, Mann R B and Zych M 2018 *Class. Quantum Grav.* **35** 145005
[12] Møller C 1962 *Colloques Internationaux CNRS* **91** 1–96
[13] Rosenfeld L 1963 *Nuclear Physics* **40** 353–356
[14] Wald R M 1978 *Phys. Rev. D* **17**(6) 1477–1484
[15] Wald R M 1994 *Quantum field theory in curved spacetime and black hole thermodynamics* (University of Chicago Press)
[16] Bahrami M, Großardt A, Donadi S and Bassi A 2014 *New J. Phys.* **16** 115007
[17] Anastopoulos C and Hu B L 2014 *New J. Phys.* **16** 085007
[18] Giulini D and Großardt A 2012 *Class. Quantum Grav.* **29** 215010
[19] Diósi L 1984 *Phys. Lett. A* **105** 199 – 202 ISSN 0375-9601
[20] Gisin N 1989 *Helv. Phys. Acta* **62** 363–371
[21] Gisin N 1990 *Phys. Lett. A* **143** 1–2
[22] Polchinski J 1991 *Phys. Rev. Lett.* **66**(4) 397–400
[23] Bassi A and Hejazi K 2015 *Eur. J. Phys.* **36** 055027
[24] Struyve W 2015 *Preprint* arXiv:1507.04771
[25] Struyve W 2017 *The Philosophy of Cosmology* (Cambridge University Press) chap 18, p 356
[26] Diósi L and Halliwell J J 1998 *Phys. Rev. Lett.* **81**(14) 2846–2849
[27] Halliwell J J 1999 *Int. J. Theor. Phys.* **38** 2969–2986 ISSN 1572-9575
[28] Diósi L and Tilloy A 2017 *Open Systems & Information Dynamics* **24** 1740020
[29] Ghirardi G C, Rimini A and Weber T 1986 *Phys. Rev. D* **34**(2) 470–491
[30] Bassi A and Ghirardi G 2003 *Phys. Rep.* **379** 257 – 426 ISSN 0370-1573
[31] Bassi A, Lochan K, Satin S, Singh T P and Ulbricht H 2013 *Rev. Mod. Phys.* **85**(2) 471–527
[32] Derakhshani M 2014 *Phys. Lett. A* **378** 990 – 998 ISSN 0375-9601
[33] Bell J S 1976 *Epistemological Letters* **9** URL `http://cds.cern.ch/record/980036/files/197508125.pdf`
[34] Allori V, Goldstein S, Tumulka R and Zanghì N 2014 *Brit. J. Philos. Sci.* **65** 323–352
[35] Allori V 2015 *International Journal of Quantum Foundations* **1** 107–122 URL `http://www.ijqf.org/archives/2394`
[36] Tumulka R 2011 *Preprint* arXiv:1102.5767
[37] Jacobs K and Steck D A 2006 *Contemp. Phys.* **47** 279–303
[38] Wiseman H M and Milburn G J 2009 *Quantum measurement and control* (Cambridge university press, Cambridge UK)
[39] Pearle P 1989 *Phys. Rev. A* **39**(5) 2277–2289
[40] Ghirardi G C, Pearle P and Rimini A 1990 *Phys. Rev. A* **42**(1) 78–89
[41] Diósi L 1987 *Phys. Lett. A* **120** 377 – 381 ISSN 0375-9601
[42] Penrose R 1996 *Gen. Relativ. Gravit.* **28** 581–600 ISSN 1572-9532
[43] Feldmann W and Tumulka R 2012 *J. Phys. A: Math. Theor.* **45** 065304
[44] Diósi L 1990 *Phys. Rev. A* **42**(9) 5086–5092
[45] Strunz W T 1996 *Phys. Lett. A* **224** 25 – 30 ISSN 0375-9601
[46] Diósi L and Strunz W T 1997 *Phys. Lett. A* **235** 569–573
[47] Gambetta J and Wiseman H M 2002 *Phys. Rev. A* **66**(1) 012108
[48] Diósi L and Ferialdi L 2014 *Phys. Rev. Lett.* **113**(20) 200403
[49] Tilloy A 2017 *Quantum* **1** 29 ISSN 2521-327X
[50] Gambetta J and Wiseman H M 2003 *Phys. Rev. A* **68**(6) 062104
[51] Diósi L 2008 *Phys. Rev. Lett.* **100**(8) 080401
[52] Diósi L 2008 *Phys. Rev. Lett.* **101**(14) 149902
[53] Wiseman H M and Gambetta J M 2008 *Phys. Rev. Lett.* **101**(14) 140401
[54] Juárez-Aubry B A, Kay B S and Sudarsky D 2018 *Phys. Rev. D* **97**(2) 025010
[55] Josset T, Perez A and Sudarsky D 2017 *Phys. Rev. Lett.* **118**(2) 021102
[56] Altamirano N, Corona-Ugalde P, Khosla K E, Milburn G J and Mann R B 2017 *Class. Quantum Grav.* **34**

115007
[57] Altamirano N, Pascalie R and Mann R B 2018 *Gen. Relativ. Gravit.* **50** 120 ISSN 1572-9532
[58] Bose S, Mazumdar A, Morley G W, Ulbricht H, Toroš M, Paternostro M, Geraci A A, Barker P F, Kim M S and Milburn G 2017 *Phys. Rev. Lett.* **119**(24) 240401
[59] Carney D, Stamp P C E and Taylor J M 2019 *Class. Quantum Grav.* **36** 034001

IOP Conf. Series: Journal of Physics: Conf. Series **1275** (2019) 012007 doi:10.1088/1742-6596/1275/1/012007

Planck length challenges non-relativistic quantum mechanics of large masses

Lajos Diósi

Wigner Research Centre for Physics, H-1525 Budapest 114, P.O.Box 49, Hungary

E-mail: diosi.lajos@wigner.mta.hu

Abstract. With the simplest proof ever, we justify the significance of quantum-gravity in non-relativistic quantum mechanics together with the related theories and experiments. Since the de Broglie wave length is inverse proportional to the mass, it would descend towards and below the Planck scale 10^{-33} cm for large masses even at slow non-relativistic motion. The tricky relationship between gravity and quantum mechanics —well-known in the relativistic case— shows up in non-relativistic motion of massive objects. Hence the gravity-related modification of their Schrödinger equation is *mandatory*. We also recall the option of an autonomous Newtonian quantum-gravity, a theory parametrized by $\hbar$ and G. On cancellation of c from the Newtonian limit of Planck scale metric fluctuations is given a new hint.

1. Introduction

Quantum-gravitational limitations of our very notion of space-time were conjectured by Bronstein in 1936 [1], the same issue was famously characterized by Wheeler's foamy structure of space-time [2] In the back-ground of quantized matter, the classical notion of space-time continuum can not be maintained at short distances. In the vicinity of the Planck length

$$\ell_{\rm Pl} = \sqrt{\frac{\hbar G}{c^3}} = 1.6 \times 10^{-33} cm, \tag{1}$$

the quantization of space-time would become unavoidable according to mainstream views. Efforts to construct a satisfactory theory of quantum-gravity have reached partial successes only. There is no consensual model of distances about $\ell_{\rm Pl}$ [3, 4].

Direct observation of distances like 10^{-33} cm would require incredible high precision — beyond imaginations. Their indirect test would be possible around the extreme high Planck energy $E_{\rm Pl} = 1.2 \times 10^{19}$ GeV per particle, which existed right after the Big Bang only. At low energies, Planck scale effects are too small to be testable. That's the mainstream wisdom at least.

"... it certainly is not a viable option to simply wait until the Planck scale becomes accessible experimentally in order to have an indication on how to proceed theoretically, as this will require a long time" — says the recent review Ref. [5] arguing for low-energy gravity-related decoherence theories and experiments. There are many discussions of quantum-gravity related effects that might occur at non-relativistically low energies, cf., e.g. [4, 5, 6, 7, 8].

But, what is the reason that the extreme-relativistic Planck scale effects, suppressed by a factor 10^{-28} at low energies (a number used in [6]), may amplify (or accumulate) to the level of laboratory testability? Here we are going to show the trivial answer lies in non-relativistic

Content from this work may be used under the terms of the Creative Commons Attribution 3.0 licence. Any further distribution of this work must maintain attribution to the author(s) and the title of the work, journal citation and DOI.
Published under licence by IOP Publishing Ltd

IOP Conf. Series: Journal of Physics: Conf. Series **1275** (2019) 012007 doi:10.1088/1742-6596/1275/1/012007

quantized motion of massive objects, and the argument is no doubt simpler than any earlier ones.

2. Non-relativistic quantized motion reaches Planck length

Let us consider and object of mass m and compare its de Broglie wave length $\lambda_{\rm dB}$ with the Planck length $\ell_{\rm Pl}$:

$$\underbrace{\lambda_{\rm deBroglie}}_{\frac{2\pi\hbar}{mv}\sqrt{1-v^2/c^2}} \sim \underbrace{\ell_{\rm Planck}}_{\sqrt{\frac{\hbar G}{c^3}}}$$

where v is the object's velocity. Now, $\lambda_{\rm dB}$ can sink to $\ell_{\rm Pl}$ in two ways. First: velocity v approaches c. Obviously, the required energy is $E_{\rm Pl}$, too high to occur apart from the Big Bang. Second: mass m grows macroscopic keeping $v \ll c$. That's the way we go.

2.1. Periods of the non-relativistic plane wave

Construct the non-relativistic time-dependent plane wave $\Psi(x,t)$, at velocity $v \ll c$, along the coordinate x:

$$\begin{aligned}\Psi(x,t) &= \exp\left(-iEt/\hbar + ipx/\hbar\right)\\ &= \exp\left(-i\tfrac{mv^2}{2\hbar}t + i\tfrac{mv}{\hbar}x\right)\\ &= \exp\left(-2\pi i\tfrac{t}{\tau} + 2\pi i\tfrac{x}{\lambda}\right),\end{aligned} \tag{2}$$

where E, p are the energy and momentum, respectively. The plane wave is periodic both in time and space, with the corresponding periods τ and $\lambda = \lambda_{\rm dB}$:

$$\tau = (4\pi\hbar/mv^2), \tag{3a}$$
$$\lambda_{\rm dB} = (2\pi\hbar/mv). \tag{3b}$$

These periods should be compared to the Planck time $t_{\rm Pl}$ and length $\ell_{\rm Pl}$:

$$t_{\rm Pl} = \sqrt{\hbar G/c^5} \sim 10^{-43}s, \tag{4a}$$
$$\ell_{\rm Pl} = \sqrt{\hbar G/c^3} \sim 10^{-33}cm. \tag{4b}$$

$\Psi(x,t)$ is legitimate non-relativistic wave function as long as its periods are much longer than the Planck time and length, respectively: $\tau \gg t_{\rm Pl}$ and $\lambda_{\rm dB} \gg \ell_{\rm Pl}$, discussed similarly in [9]. That's the case for atomic masses m, where $\tau/t_{\rm Pl} \sim 10^{18}$ and $\lambda_{\rm dB}/\ell_{\rm Pl} \sim 10^{18}$. But much larger masses will push the structure of $\Psi(x,t)$ towards the Planck scales, such role of the mass is what our work aims to emphasize.

2.2. How large should the mass be?

Masses M, requested for a Planck scale structure of $\Psi(x,t)$ non-relativistically, are much greater than the Planck mass $m_{\rm Pl} = 2.18 \times 10^{-5}$ g. The time-period τ would sink to the Planck time $t_{\rm Pl}$ if the mass is

$$M|_{\tau=t_{\rm Pl}} = \frac{c^2}{v^2}m_{\rm Pl} \sim \frac{c^2}{v^2}10^{-5}g, \tag{5}$$

whereas the spatial period $\lambda_{\rm dB}$ drops down to $\ell_{\rm Pl}$ provided

$$M|_{\lambda_{\rm dB}=\ell_{\rm Pl}} = \frac{c}{v}m_{\rm Pl} \sim \frac{c}{v}10^{-5}g. \tag{6}$$

Remember that we keep $v << c$. Interestingly, the condition $\tau = t_{\rm Pl}$ is equivalent with $Mv^2/2 = E_{\rm Pl}/2$, i.e., the center-of-mass wave function temporal structure approaches the Planck scale $\ell_{\rm Pl}$ provided the non-relativistic kinetic energy approaches, due to the large M, the Planck energy $E_{\rm Pl}$ (see related observation in [9]). Btw, the incoherent preparation of macroscopic objects, unlike of elementary particles, of Planckian kinetic energy $E_{\rm Pl}$ would not be an issue.

But, for the spatial structure, the bell rings earlier at much smaller mass M and kinetic energy $Mv^2/2 = (v/c)E_{\rm Pl}$. Through examples, we illustrate that the system can be perfectly non-relativistic, though apparently not available by state-of-the-art coherent preparation of massive objects. Our first example is mass $M = 10$ g at velocity $v = 10$ km/s, yielding $\lambda_{\rm dB} = 4.2\times10^{-33}$ cm which is of the order of $\ell_{\rm Pl}$! Of course, the single plane wave is not necessarily illegitimate in itself since its Planck scale structure may be the artifact of the chosen frame of reference. So, we consider the superposition of two opposite plane waves of the mass $M = 10$ g:

$$|+10km/s\rangle + |-10km/s\rangle, \tag{7}$$

which is a Schrödinger cat in momentum. Its wave function has a Planck scale spatial structure in any reference frame hence it is not legitimate.

Our second example is a solid mass $M = 10$ kg, with an elastic vibrational mode of frequency $\omega = 100$ kHz. If the amplitude is $a = 0.01$ cm, then $\lambda_{\rm dB}$ oscillates between infinity and $\lambda_{\rm dB} = (2\pi\hbar/ma\omega) = 4.2\times10^{-33}$ cm. Again, standard Schrödinger equation becomes illegitimate for this massive (macroscopic) non-relativistic oscillator.

2.3. How can tiny Planckian effects accumulate for massive objects?

As we mentioned, in non-relativistic atomic quantum systems the Planck scale effects are suppressed by an incredible small factor guessed deliberately by 10^{-28} which is the ratio of typical atomic energies (eV's) to $E_{\rm Pl} \sim 10^{19}$ GeV [6]. We can illustrate how such small effects can accumulate in macroscopic massive systems. Our example is a minimum toy model of the uncertainty u of the classical spatial continuum at the Planckian distances. We suppose the replacement

$$x \Rightarrow x + u \tag{8}$$

where u is a global uncertainty of the coordinate x. It can be quantum or just classically random, we choose the latter option for its simplicity. Let u be a Gaussian random number of squared spread $\mathbb{E}u^2 = \ell_{\rm Pl}^2$. Consider the many-body wave function of a macroscopic object. Due to the random uncertainty u, it should be replaced randomly as

$$\Psi(x_1, x_2, \ldots, x_{10^{23}}) \Rightarrow \Psi(x_1 + u, x_2 + u, \ldots, x_{10^{23}} + u). \tag{9}$$

The shifts $u \sim \ell_{\rm Pl}$ are irrelevant for the non-relativistic individual constituents, but their effects accumulate for the center-of-mass. The mechanism of accumulation is best seen in momentum representation of Eq. (9):

$$\widetilde{\Psi}(p_1,p_2,\ldots,p_{10^{23}}) \Rightarrow \exp\left(\frac{i}{\hbar}u(\underbrace{p_1+p_2+\ldots+p_{10^{23}}}_{P})\right)\widetilde{\Psi}(p_1,p_2,\ldots,p_{10^{23}}). \tag{10}$$

Observe the appearance of the phase $u/\hbar$ times the total momentum P of the center-of-mass motion of the macroscopic object. P is the "large" number to compensate the the smallness of u. We can derive closed expression for the influence of the Planckian coordinate uncertainty u on the center-of-mass density matrix if we take the average over u:

$$\rho(P,P') = \mathbb{E}\exp\left(\frac{i}{\hbar}u(P-P'\right)\rho(P,P') = \exp\left(-\frac{\ell_{\rm Pl}^2}{2\hbar^2}(P-P')^2\right)\rho(P,P'). \tag{11}$$

IOP Conf. Series: Journal of Physics: Conf. Series **1275** (2019) 012007 doi:10.1088/1742-6596/1275/1/012007

In our toy model, the Planck scale coordinate uncertainty decoheres the center-of-mass momentum superpositions of massive objects. As to our Schrödinger cat (7), the interference term $\rho(mv, -mv)$ is suppressed by the factor

$$\exp\left(-\frac{\ell_{\rm Pl}^2}{2\hbar^2}(2mv)^2\right) \approx e^{-1.2}. \tag{12}$$

The cat (7) becomes partially decohered. The atom-wise ignorable Planck scale uncertainty grew testable indeed non-relativistically in the massive degree of freedom. Yet a consistent theory (advancing our minimum toy model) of Planck scale uncertainties should either eliminate the illegitimate cat states or make them legitimate.

To improve the phenomenology of our minimum toy model, the unnatural globality and constancy of u should be replaced by some correlated structure of local space-time randomnesses. They result in much stronger decoherence effects whose significance extends for masses M much below $m_{\rm Pl}$ towards mesoscopic masses. Cat states (7) become eliminated at fast decoherence rates. Numerous different choices of space-time fluctuations have been summarized recently in Ref. [10], we discuss the particular DP-model below.

3. Newtonian quantum gravity: $\hbar$ + G, no c

So far we argued how quantum-gravity, relativistic by its definition, should influence non-relativistic quantum mechanics. The extreme-relativistic quantum-gravity mechanism must have its footprints on non-relativistic quantum mechanics of massive objects. Now, this opens a remarkable perspective: what if velocity of light c cancels from the said footprints? Then there would be an autonomous theory of Newtonian quantum-gravity, coined so by [11]. And, apparently, this shall be the case. A first indication was the non-relativistic limit (cf. footprint) of quantum-gravity in the semi-classical approximation which led to the Newton-Schrödinger equation, containing G and $\hbar$ only, and predicting a regime of significance for quantum dynamics of meso- and macroscopic masses [12, 13]. The other instance is the DP-model [13, 14] of gravity-related decoherence and wave function collapse in massive degrees of freedom, parametrized again by G and $\hbar$ only. It is based on a conjectured ultimate uncertainty of space-time but neither the relationship to the Planck length $\ell_{\rm Pl}$ (sketched in [15]) nor the cancellation of c are explained in concrete forms.

The question itself was explicitly asked in Ref. [18]: whether the space-time Newtonian unpredictabilities/fluctuations *´a la*DP [14] are the non-relativistic limit of the Planckian's? A possible, affirmative, answer may come from Unruh's intuitive quantization of space-time [16], as mentioned in [14, 17, 18]. Below we show a direct affirmative derivation, still starting from a phenomenology of Planckian fluctuations. Let us represent the uncertainty around the Minkowski metric by introducing a perturbative conformal factor $1 + h$ (at $|h| \ll 1$) where the random fluctuations h are proportional to $\ell_{\rm Pl}$. We choose the formal expression of relativistic covariance:

$$\mathbb{E}h(x)h(y) = \text{const.} \times \frac{\ell_{\rm Pl}^2}{(2\pi)^4}\int e^{-ik(x-y)}\frac{\theta(-k^2)}{-k^2}dk, \tag{13}$$

ignoring the issues of regularizing $\theta(-k^2)/k^2$. Now we write h into the form $h = 2\Phi/c^2$ anticipating that Φ plays the role of a (random) Newton potential for non-relativistic matter. In the limit $c \to \infty$, the good thing about the formal relativistic correlator (13) is that it leads to a convergent correlator for Φ:

$$\mathbb{E}\Phi(t,\mathbf{x})\Phi(s,\mathbf{y}) = \text{const.} \times \frac{\hbar G}{|\mathbf{x}-\mathbf{y}|}\delta(t-s). \tag{14}$$

The conform uncertainties, proportional to $\ell_{\rm Pl}$, have become equivalent non-relativistically with the (Newtonian) space-time uncertainties proposed for the DP-model [14]. The way c has gone is being demonstrated directly, tentatively though, for the first time.

4. Summary

We wished relaxing the belief, based mainly on quantum-field-theory considerations, that the significance of quantum-gravity is restricted for the regime of extreme high energies. We simply argue that the Planck length $\ell_{\rm Pl}$ invokes a *mandatory* —gravity-related, yet to be specified— modification of non-relativistic quantum mechanics of massive (macroscopic) degrees of freedom because their de Broglie wave lengths can descend into the vicinity of $\ell_{\rm Pl}$ non-relativistically. The modification depends on the yet also unknown behavior of space-time at the Planckian scale.

No matter whether space-time is classical or quantized, some ultimate statistical or quantum uncertainties (fluctuations) are attributed to the background space-time, to result in some particular decoherence of massive degrees of freedom. We explained the accumulation of the tiny extreme-relativistic effects in massive degrees of freedom. Such non-relativistic footprints of the Planck scale fluctuations do, in general, depend on G, $\hbar$, and c. However, the footprints according to the DP-model are fully non-relativistic, the light velocity c cancels from them. A mechanism of this cancellation has been outlined.

Acknowledgments

The author thanks the National Research Development and Innovation Office of Hungary Projects Nos. 2017-1.2.1-NKP-2017-00001 and K12435, and the EU COST Action CA15220 for support.

References

[1] Bronstein M 1936 *Phys. Z. Sowjetunion* **9** 140
[2] Wheeler J A 1962 *Geometrodynamics* (New York: Academic Press)
[3] Garay L J 1995 *Int. J. Mod. Phys. A* **10** 145
[4] Hossenfelder S 2013 *Living Rev. Rel.* **16** 2
[5] Bassi A, Grossardt A and Ulbricht H 2017 *Class. Quantum Grav.* **34** 193002
[6] Amelino-Camelia G, Lämmerzahl C, Macias A and Müller H 2005 *AIP Conf. Proc.* **758** 30
[7] Pikovski I, Vanner M R, Aspelmeyer M, Kim M S and Brukner Č 2012 *Nat. Phys. 8*
[8] Howl R, Hackermüler L, Bruschi D E and Fuentes I 2018 *Advances in Physics: X* **3**, 1383184
[9] Gao S 2017 *Meaning of the Wave Function* (Cambridge, UK: Cambridge University Press)
[10] Kwon O and Hogan C J 2016 *Class. Quantum Grav.* **33** 105004
[11] Diósi L and Lukács B 1987 *Annln. Phys.* **499** 488
[12] Diósi L 1984 *Phys. Lett. A* **105** 199
[13] Penrose R 1996 *Gen. Relativ. Gravit.* **28** 581
[14] Diósi L 1987 *Phys. Lett. A* **120** 377
[15] Penrose R 2016 *Fashion, Faith, and Fantasy in the New Physics of the Universe* (Princeton, NJ: Princeton University Press) p 365
[16] Unruh W G 1984 *Quantum Theory of Gravity*, ed S M Christensen (Bristol, UK: Adam Hilger Ltd, Bristol) p 234
[17] Tilloy A and Diósi L 2017 *Phys. Rev. D* **96** 104045
[18] Diósi L 2018 *Entropy* **20** 496

IOP Conf. Series: Journal of Physics: Conf. Series **1275** (2019) 012008 doi:10.1088/1742-6596/1275/1/012008

Necessary and sufficient conditions for macrorealism using two- and three-time Leggett-Garg inequalities

J J Halliwell
Blackett Laboratory, Imperial College, London SW7 2BZ, UK
E-mail: j.halliwell@imperial.ac.uk

Abstract. The Leggett-Garg (LG) inequalities were introduced, as a temporal parallel of the Bell inequalities, to test macroscopic realism – the view that a macroscopic system evolving in time possesses definite properties which can be determined without disturbing the future or past state. The original LG inequalities are only a necessary condition for macrorealism, and are therefore not a decisive test. We argue, for the case of measurements of a single dichotomic variable Q, that when the original four three-time LG inequalities are augmented with a set of twelve two-time inequalities also of the LG form, Fine's theorem applies and these augmented conditions are then both necessary and sufficient. A comparison is carried out with the alternative necessary and sufficient conditions for macrorealism based on no-signaling in time conditions which ensure that all probabilities for Q at one and two times are independent of whether earlier or intermediate measurements are made. We argue that the two tests differ in their implementation of the key requirement of non-invasive measurability so are testing different notions of macrorealism, and these notions are elucidated.

1. Introduction

The purpose of this paper is to develop, discuss and compare various conditions that have been proposed to test macrorealism, the idea that a time-evolving macroscopic system can possess definite properties at a number of times uninfluenced by measurements of it. The aim in particular is to present conditions for macrorealism which are necessary and sufficient, and also to elucidate the different, inequivalent, notions of non-invasiveness involved in such tests. This paper is in part a more streamlined statement of a formulation given in Refs.[1, 2] and removes some inessential assumptions contained therein.

1.1. Macrorealism and the Leggett-Garg inequalities

Macrorealism (MR) was proposed by way of analogy to the notion of local realism for spatially entangled systems and leads to a set of inequalities obeyed by the temporal correlation functions of a single system, the Leggett-Garg inequalities [3, 4, 5]. Most investigations to date focus on a single dichomotic variable Q which is measured in various ways at three (or more) times leading to the determination of the temporal correlation functions of the form,

$$C_{12} = \langle Q(t_1)Q(t_2)\rangle. \tag{1}$$

These are argued, for a macrorealistic theory, to obey the Leggett-Garg (LG) inequalities,

$$1 + C_{12} + C_{23} + C_{13} \geq 0, \tag{2}$$

Content from this work may be used under the terms of the Creative Commons Attribution 3.0 licence. Any further distribution of this work must maintain attribution to the author(s) and the title of the work, journal citation and DOI.
Published under licence by IOP Publishing Ltd

IOP Conf. Series: Journal of Physics: Conf. Series **1275** (2019) 012008 doi:10.1088/1742-6596/1275/1/012008

$$1 - C_{12} - C_{23} + C_{13} \geq 0, \tag{3}$$
$$1 + C_{12} - C_{23} - C_{13} \geq 0, \tag{4}$$
$$1 - C_{12} + C_{23} - C_{13} \geq 0, \tag{5}$$

which are identical in mathematical form to the Bell inequalities. Measurements at four times lead to a set of eight LG inequalities identical in mathematical form to the CHSH inequalities.

To derive these inequalities, the notion of macrorealism is broken down into three separate assumptions. These are:
1. Macrorealism *per se* (MRps): the system is in one of the states available to it at each moment of time.
2. Non-invasive measurability (NIM): it is possible in principle to determine the state of the system without disturbing the subsequent dynamics.
3. Induction (Ind): future measurements cannot affect the present state.

These assumptions ensure the existence of an underlying joint probability distribution from which the LG inequalities readily follow. Any experimental test thus tests the combination of these assumptions. Induction is always taken for granted so what is being tested is the combination of MRps and NIM. To ensure NIM, Leggett and Garg proposed that the measurement of the correlation functions be carried out using ideal negative measurements, in which the detector is coupled to, say, only the $Q = +1$ state, at the first time, and a null result then permits us to deduce that the system is in the $Q = -1$ state but without any interaction taking place, from the macrorealistic perspective. This procedure rules out alternative classical explanations of the correlation functions [6] analogous to the way in which signaling is ruled out in Bell experiments and has been successfully implemented in a number of recent experiments [7, 8, 9, 10]. Many other experimental tests of the LG inequalities have also been carried out, on a variety of different physical systems (see the extensive list of references in Refs.[2, 5]).

Maroney and Timpson [12] have argued that MRps actually comes in three different varieties, only one of which can be ruled out by LG test, namely "operational eigenstate mixture macrorealism", which includes spontaneous collapse models, such as those of the GRW type [13]. Almost all LG tests are focused on this variety of MRps and we do so here. Of the other two varieties of MRps, the most significant one is realist theories of the Bohmian type, which would need some sort of locality arguments in order to be ruled out [12, 14]. The remaining type have the form of restricted Bohmian theories, and include the Kochen-Specker model [15] for two-dimensional Hilbert spaces, but for Hilbert spaces of dimension greater than 3 these can be ruled out [16, 17].

1.2. Different interpretations of NIM

In the various papers on tests of macrorealism, NIM appears to be interpreted in a number of different ways depending on exactly how the measurements are carried out. Most experimental tests of the LG inequalities measure the three correlation functions in three different experiments, analogous to the Bell case, with each involving measurements at just two times and non-invasiveness is required only in each separate experiment. Furthermore, even within each experiment involving two times, there are different choices, as we shall see in more detail. However, some approaches assume a stronger reading of NIM – that it should not make any difference if one, two or three sequential measurements are made in the same experiment. Both of these versions of NIM permit access to the information required to determine whether MRps holds, but since MR is defined to be the conjunction of both MRps and NIM, it means that there are *different versions* of MR depending on how the NIM requirement is implemented. These differences will be important for what follows.

For convenience we will denote the stronger version of NIM, entailing non-invasiveness for sequential measurements at three times, by NIM_{seq}, and the weaker one, in which NIM is only

IOP Conf. Series: Journal of Physics: Conf. Series **1275** (2019) 012008 doi:10.1088/1742-6596/1275/1/012008

satisfied in a piecewise (pw) way in each one of a number of experiments, by $\mathrm{NIM}_{\mathrm{pw}}$. These different characterizations will be further refined as required.

1.3. Parallels with Bell experiments

The LG framework for testing macrorealism was designed by analogy with Bell experiments and it does indeed have some genuinely close parallels (reviewed in Ref.[2]). However, the analogy fails at a number of points. As Maroney and Timpson have argued [12], Bell and LG tests are not methodologically on a par since the notion of non-invasiveness typically carries some model-dependent assumptions so is difficult to motivate as a general feature, unlike local causality in Bell experiments.

This general point relates to our main question of interest, which is that of sufficient conditions for macrorealism. The LG inequalities are necessary conditions but they are not sufficient, as has been noted by a number of authors [18, 19]. By contrast in Bell experiments, Fine's theorem [20, 21] guarantees that the Bell [22] or CHSH [23] inequalities are both necessary and sufficient conditions for the existence of an underlying probability matching the given correlation functions, and so are necessary and sufficient conditions for local realism. This means that the Bell or CHSH inequalities are a decisive test.

The point at which Fine's theorem fails to apply to the LG framework relates to the description of the system at two moments of time. Suppose we carry out sequential measurements of Q at t_1 and t_2, yielding probability $p_{12}(s_1, s_2)$ and in a separate experiment, measure Q at t_2 only, yielding probability $p_2(s_2)$. For Fine's theorem to hold, it would be necessary for conditions of the form

$$\sum_{s_1} p_{12}(s_1, s_2) = p_2(s_2), \tag{6}$$

to hold, since such conditions ensure that the pairwise probabilities such as $p_{12}(s_1, s_2)$, $p_{23}(s_2, s_3)$ etc are compatible with each other. Conditions of the form Eq.(6) are referred to as "no signaling in time" (NSIT) conditions [18, 19]. They were proposed as a clear analogy to no-signaling in Bell experiments but are not satisfied in general (except for special parameter choices).

However, this difficulty clearly relates to the particular way in which NIM is implemented. Eq.(6) is naturally linked to $\mathrm{NIM}_{\mathrm{seq}}$, but one could also contemplate a less direct measurement scheme for the two-time probabilities, consistent with $\mathrm{NIM}_{\mathrm{pw}}$, which yields a different two-time probability compatible with Eq.(6). From a quantum-mechanical perspective this corresponds to the fact that probabilities for non-commuting quantities can often be defined but they are not unique, with different measurements procedures giving different candidate probabilities, and furthermore, their interpretation as probabilities comes with restrictions.

1.4. Necessary and sufficient conditions for macrorealism

We now outline two different ways in which necessary and sufficient conditions for MR can be defined, making use of the two different versions of NIM outlined above. The first way is to work with the weaker form of non-invasiveness, $\mathrm{NIM}_{\mathrm{pw}}$, and stay as close as possible to the original LG framework, in which the three correlation functions are determined in a number of different runs. To sidestep the failure of Eq.(6) we seek another way of determining a candidate probability for the system at two times which does not involve sequential measurements. Our approach is to use a number of different non-invasive experiments to determine the moments of the system at two times and then use a set of two-time LG inequalities to determine indirectly whether or not a probability distribution at two times exsists. This approach yields a set of two-time and three-time LG inequalities which, when the two-time LG inequalities hold, have a mathematical form identical to that of the Bell system and are therefore necessary and sufficient conditions for MR. Hence, the desired parallel with the Bell system and a decisive test for MR is achieved using an augmented set of LG inequalities measured in a judiciously chosen set of

IOP Conf. Series: Journal of Physics: Conf. Series **1275** (2019) 012008 doi:10.1088/1742-6596/1275/1/012008

runs. The measurements are non-invasive, by design, so this protocol is, under ideal conditions, most accurately thought of as a direct test of MRps.

The second way is to follow the much stronger reading of NIM outlined above, NIM_{seq}, and restrict to initial states and other parameter ranges so that relationships of the form Eq.(6) hold for sequential measurements. In particular, Clemente and Kofler [19] proposed a scheme in which the underlying three-time probability $p_{123}(s_1, s_2, s_3)$ is determined in a single experiment by sequential measurements at all three times, subject to a set of two- and three-time NSIT conditions, similar to Eq.(6). We review this work. When these conditions hold, the probability $p_{123}(s_1, s_2, s_3)$ is a properly defined probability for a set of three independent variables, and hence the set of NSIT conditions are a necessary and sufficient condition for MR. The LG inequalities are not involved in this sort of test, but are clearly implied by the set of NSIT conditions. These conditions test a combination of NIM_{seq} and MRps (and induction). From a quantum-mechanical point of view, it is very different to a Bell test since it involves sequential measurement of incompatible quantities, but this test is of the same type as some of the "coherence witness" tests proposed recently [8, 24, 25].

These two possibilities clearly delineate two extremes in terms of how strongly or weakly NIM is implemented. We will also find intermediate possibilities that involve combinations of both. (We also note here a possible connnection with the so-called Wigner Leggett-Garg inequalities, which lie midway between the LG inequalities and no-signaling conditions [26]).

Note also that in talking about measurements which we refer to as "non-invasive", we have in mind a theoretical ideal situation. In practice experimental clumsiness is difficult to eliminate and this leaves loopholes for alternative explanations of the results [27]. See Refs.[7, 8] for further discussions of how this may be handled in specific experiments.

Tests of macrorealism involving the augmented LG inequalities are described in Section 2, and tests involving NSIT conditions are described in Section 3. Some quantum-mechanical aspects of the LG and NSIT approaches are briefly discussed in Section 4 along with a simple property of coherence witnesses. We summarize and conclude in Section 5.

2. Necessary and sufficient conditions for macroealism using Leggett-Garg inequalities

We now exhibit a set of necessary and sufficient conditions for macrorealism using an augmented set of Leggett-Garg inequalities and using NIM_{pw}. We suppose that we can carry out a set of experiments to determine the averages and second order correlation functions of $Q(t)$ at three times. NIM_{pw} is the requirement that all quantities are measured non-invasively, but the measurements need be non-invasive only within each experiment – there is no requirement that non-invasiveness should persist if different experiments are combined. This is a natural parallel with the Bell case [2]. The clearest way to accomplish this is to do a total of six different experiments, in which the three correlation functions C_{12}, C_{23}, C_{13} are determined in three different experiments, as in traditional LG tests, using a suitable non-invasive measurement protocol; and $\langle Q_1 \rangle$, $\langle Q_2 \rangle$, $\langle Q_3 \rangle$ are determined in three further experiments, in which there is clearly no issue of invasiveness since only a single measurement is made in each case. (We use the convenient notation $Q_i = Q(t_i)$ for $i = 1, 2, 3$.)

Non-invasive protocols exist in which the correlation function *and nothing else* is measured [28, 29], which is why it is reasonable to suppose that $\langle Q_1 \rangle$ and C_{12} are determined in genuinely seperate experiments. Having said that, when we measure correlation functions such as C_{12} with ideal negative measurements (as suggested in the original LG framework), this usually entails a single experiment in which the full two-time probability $p(s_1, s_2)$ is measured, from which one could also read off $\langle Q_1 \rangle$. Hence it may be possible to reduce the six different experiments to three, as long as there is a clear argument for non-invasiveness. However, for logical clarity, we make no assumptions about which pairs of moments can be measured in the same experiment

IOP Conf. Series: Journal of Physics: Conf. Series **1275** (2019) 012008 doi:10.1088/1742-6596/1275/1/012008

(in contrast to Ref.[2] which invokes quantum mechanics to argue that certain pairs of moments are compatible).

Macrorealism is the question as to whether or not there exists an underlying probability $p(s_1, s_2, s_3)$ matching the six moments measured in the way described above. We build this up in three steps.

First, the single time probabilties are given by

$$p(s_i) = \frac{1}{2}\left(1 + s_i \langle Q_i \rangle\right), \tag{7}$$

for $i = 1, 2, 3$, and these are non-negative by construction. Second, there are three two-time probabilties, given by

$$p(s_i, s_j) = \frac{1}{4}\left(1 + s_i \langle Q_i \rangle + s_j \langle Q_j \rangle + s_i s_j C_{ij}\right), \tag{8}$$

where $ij = 12, 13, 23$. (See Refs.[30, 31] for more information on these useful moment expansions). Because the averages $\langle Q_i \rangle$ are measured in such a way that there can be no disturbance from an earlier measurement, these two-time probabilities automatically obey all compatibility conditions of the form,

$$\sum_{s_i} p(s_i, s_j) = p(s_j) = \sum_{s_k} p(s_j, s_k). \tag{9}$$

These relations are of course mathematically identical to the NSIT conditions, Eq.(6), but there is no sense in which they indicate the absence of "signaling", since the two-time probabilities are assembled indirectly from different experiments, not measured sequentially in a single experiment. Instead these relations are simply the compatibility relations between the two-time probabilities that are required for Fine's theorem to apply.

In a macrorealist theory in which the averages and correlation function are non-invasively measured, the two-time probabilities Eq.(8) are guaranteed to be non-negative. This follows very easily from a simple argument similar to the derivation of the LG inequalities: we have

$$(1 + s_i Q_i)(1 + s_j Q_j) \geq 0, \tag{10}$$

and averaging this, we obtain

$$1 + s_i \langle Q_i \rangle + s_j \langle Q_j \rangle + s_i s_j C_{ij} \geq 0. \tag{11}$$

These twelve conditions, which we will call *two-time Leggett-Garg inequalities* are necessary conditions for macrorealism at the two-time level. They are also sufficient because if satisfied, the left-hand side of Eq.(11), multiplied by $\frac{1}{4}$, are precisely the probabilities Eq.(8) matching the given averages and correlation functions.

Finally, the most general possible form of the desired three-time probability is

$$\begin{aligned} p(s_1, s_2, s_3) &= \frac{1}{8}\left(1 + s_1 \langle Q_1 \rangle + s_2 \langle Q_2 \rangle + s_3 \langle Q_3 \rangle\right. \\ &+ \left. s_1 s_2 C_{12} + s_2 s_3 C_{23} + s_1 s_3 C_{13} + s_1 s_2 s_3 D\right). \end{aligned} \tag{12}$$

It involves a coefficient D, essentially the triple correlator, which is *not* measured in the experiment. The question is whether there is any possible value of D for which

$$p(s_1, s_2, s_3) \geq 0. \tag{13}$$

IOP Conf. Series: Journal of Physics: Conf. Series **1275** (2019) 012008 doi:10.1088/1742-6596/1275/1/012008

Fine's theorem guarantees that this is indeed possible under the following conditions: the twelve two-time LG inequalities Eq.(11) hold; the compatibility conditions Eq.(9) hold; and the four three-time LG inequalities, Eqs.(2)-(5) hold. The proof of this result is spelled out in detail in Ref.[21].

The new feature in this protocol, compared to standard LG tests, are the twelve two-time LG inequalities, Eq.(11). It is these that fill the shortfall in the usual three-time LG inequalities and lead to conditions for MR which are both necessary and sufficient.

Concisely summarized, the protocol just described tests a specific definition of MR, consisting three sets of two-time LG inequalities, one set of three-time inequalities, together with induction and piecewise non-invasive measurability. This is definition of MR is arguably the weakest one possible, and we write,

$$\mathrm{MR_{weak}} = \mathrm{NIM_{pw}} \wedge \mathrm{LG_{12}} \wedge \mathrm{LG_{23}} \wedge \ \mathrm{LG_{13}} \wedge \mathrm{LG_{123}} \wedge \mathrm{Ind}. \tag{14}$$

Like the Bell and CHSH inequalities, it may be satisfied in the face of non-zero interferences, as long as they are not too large [2].

The twelve two-time and four three-time LG inequalities can be readily simplified by a particular choice of initial state. Suppose that we choose the initial state of the system to be at time t_1 to be $Q_1 = +1$. Then $C_{12} = \langle Q_2 \rangle$ and $C_{13} = \langle Q_3 \rangle$. The four three-time LG inequalities Eq.(2)-(5), then read,

$$1 + \langle Q_2 \rangle + \langle Q_3 \rangle + C_{23} \geq 0, \tag{15}$$
$$1 - \langle Q_2 \rangle - \langle Q_3 \rangle + C_{23} \geq 0, \tag{16}$$
$$1 + \langle Q_2 \rangle - \langle Q_3 \rangle - C_{23} \geq 0, \tag{17}$$
$$1 - \langle Q_2 \rangle + \langle Q_3 \rangle - C_{23} \geq 0, \tag{18}$$

which therefore coincide with four of the twelve two-time LG inequalities. The remaining eight two-time LG inequalities consist of trivially satisfied conditions of the form $|\langle Q_i \rangle| \leq 1$. Hence in this simplied situation the four inequalities Eq.(15)-(18) are necessary and sufficient conditions for $\mathrm{MR_{weak}}$. Inequalities of this general form have been tested experimentally [10, 11, 25].

The above protocol is readily extended to the four-time situation, for which we find,

$$\mathrm{MR_{weak}} = \mathrm{NIM_{pw}} \wedge \mathrm{LG_{12}} \wedge \mathrm{LG_{23}} \wedge \ \mathrm{LG_{34}} \wedge \mathrm{LG_{14}} \wedge \mathrm{LG_{1234}} \wedge \mathrm{Ind}. \tag{19}$$

That is, there are four two-time LG inequalities together with the eight four-time LG inequalities, which have the form

$$-2 \leq C_{12} + C_{23} + C_{34} - C_{14} \leq 2, \tag{20}$$

plus the three more pairs of inequalities obtained by moving the minus sign to the other three possible positions.

3. Necessary and sufficient conditions for macrorealism using no-signaling in time

We now review very different conditions for macrorealism which make use of no-signaling in time conditions and do not involve the LG inequalities at all. The most comprehensive version of this approach is that of Clemente and Kofler [19] which is followed here. (Coherence witness conditions [8, 24, 25] are simpler examples of this approach and conditions similar to the ones that follow have been given by Maroney and Timpson [12]). They suppose that the system is measured using sequential measurements at three times, with all measurements done in the same experiment and then conditions are imposed to ensure that these measurements are non-invasive, hence we are working with $\mathrm{NIM_{seq}}$. This procedure accesses the underlying three-time probability $p(s_1, s_2, s_3)$ directly but the nature of the measurements means that the procedure works only under conditions considerably stricter than those required in the augmented LG tests.

In the face of potentially invasive sequential measurements, the most general possible form for the three-time probability is

$$\begin{aligned} p_{123}(s_1, s_2, s_3) &= \frac{1}{8}\Big(1 + s_1\langle Q_1\rangle + s_2\langle Q_2^{(1)}\rangle + s_3\langle Q_3^{(12)}\rangle \\ &+ s_1 s_2 C_{12} + s_2 s_3 C_{23}^{(1)} + s_1 s_3 C_{13}^{(2)} + s_1 s_2 s_3 D\Big). \end{aligned} \tag{21}$$

Here, the superscripts again acknowledge that the values of averages and correlation functions can depend on whether earlier or intermediate measurements are made. So for example, $\langle Q_3^{(12)}\rangle$ can depend on whether measurements were made at both t_1 and t_2, and $C_{13}^{(2)}$ can depend on whether a measurement is made at the intermediate time t_2. In contrast to the three-time probability discussed in the LG framework, Eq.(12), here, the triple correlator D is determined by the measurement process. We assume induction throughout so there is no possibility of dependence on later measurements.

Eq.(21) is non-negative by definition since it is a measurement probability. However, because of the possible dependencies of its components on the context of the measurement, it is not the probability of an independent set of variables, so is not yet the sought after description of macrorealism we seek. Clemente and Kofler therefore imposed a set of NSIT conditions at two and three times to ensure this.

We consider the related measurement probabilities in which measurements are made at only two times, or just one time:

$$p_{13}(s_1, s_3) = \frac{1}{4}\Big(1 + s_1\langle Q_1\rangle + s_3\langle Q_3^{(1)}\rangle + s_1 s_3 C_{13}\Big), \tag{22}$$

$$p_{23}(s_2, s_3) = \frac{1}{4}\Big(1 + s_2\langle Q_2\rangle + s_3\langle Q_3^{(2)}\rangle + s_2 s_3 C_{23}\Big), \tag{23}$$

$$p_{12}(s_1, s_2) = \frac{1}{4}\Big(1 + s_1\langle Q_1\rangle + s_2\langle Q_2^{(1)}\rangle + s_1 s_2 C_{12}\Big), \tag{24}$$

$$p_3(s_2) = \frac{1}{2}\left(1 + \langle Q_3\rangle\right). \tag{25}$$

The NSIT condition

$$\sum_{s_2} p_{23}(s_2, s_3) = p_3(s_3), \tag{26}$$

conveniently denoted $\text{NSIT}_{(2)3}$, implies that $\langle Q_3^{(2)}\rangle = \langle Q_3\rangle$. The NSIT condition

$$\sum_{s_1} p_{123}(s_1, s_2, s_3) = p_{23}(s_2, s_3), \tag{27}$$

which we denote $\text{NSIT}_{(1)23}$, implies that $C_{23}^{(1)} = C_{23}$, $\langle Q_2^{(1)}\rangle = \langle Q_2\rangle$, and $\langle Q_3^{(12)}\rangle = \langle Q_3^{(2)}\rangle$ (which therefore equals $\langle Q_3\rangle$). Finally, the NSIT condition

$$\sum_{s_2} p_{123}(s_1, s_2, s_3) = p_{13}(s_1, s_3) \tag{28}$$

which we denote $\text{NSIT}_{1(2)3}$, implies $C_{13}^{(2)} = C_{13}$ and $\langle Q_3^{(12)}\rangle = \langle Q_3^{(2)}\rangle$ (and so they are both equal to $\langle Q_3\rangle$). These three NSIT conditions therefore establish that all averages and correlation functions take values independent of whether earlier or intermediate measurements were performed, and the three time probability may then be written:

$$\begin{aligned} p_{123}(s_1, s_2, s_3) &= \frac{1}{8}(1 + s_1\langle Q_1\rangle + s_2\langle Q_2\rangle + s_3\langle Q_3\rangle \\ &+ s_1 s_2 C_{12} + s_2 s_3 C_{23} + s_1 s_3 C_{13} + s_1 s_2 s_3 D)\,. \end{aligned} \tag{29}$$

Hence this combination of NSIT conditions are necessary and sufficient conditions for a variety of macrorealism that is clearly stronger than that described in the previous section, and we write,

$$\mathrm{MR}_{\mathrm{strong}} = \mathrm{NSIT}_{(2)3} \wedge \mathrm{NSIT}_{(1)23} \wedge \mathrm{NSIT}_{1(2)3} \wedge \mathrm{Ind}. \tag{30}$$

From the quantum-mechanical point of view, the NSIT conditions can only hold if the interferences are zero [2].

In contrast to the LG case, where the measurements are non-invasive by design, the sequential measurements used in these NSIT conditions are invasive in general. In any experimental test it is therefore necessary to adjust the initial state and measurement times (and perhaps other parameters too) to ensure that the NSIT conditions are satisfied. This is why this definition of MR appears to involve far more restrictive conditions than in the augmented LG case, i.e. equalities, rather than inequalities [19]. The NSIT conditions are primarily measures of NIM for sequential measurements, whereas NIM is already taken to be satisfied, by design, in the augmented LG case. Of course, the values of the averages and correlation functions in Eq.(29) must be the same as those determined in the augmented LG protocol, in Eq.(12), but the conditions under which they can be determined are different in each case: in Eq.(29) they can be determined only if the equalities consisting of the NSIT conditions hold, whereas in Eq.(12), no such restrictions are required.

The two different types of protocols described in this and the last section are not the only possibilities and clearly delineate the two extremes. A third, intermediate option naturally arises, which is to do three experiments with sequential measurements made at only two times in each case, and then require that the three measured two-time probabilities all satisfy two-time NSIT conditions, of the form, Eq.(26); in addition, we then require that the three-time LG inequalities are satisfied. This therefore tests the following version of MR:

$$\mathrm{MR}_{\mathrm{int}} = \mathrm{NSIT}_{(1)2} \wedge \mathrm{NSIT}_{(1)3} \wedge \mathrm{NSIT}_{(2)3} \wedge \mathrm{LG}_{123} \wedge \mathrm{Ind} \tag{31}$$

Like the augmented LG protocol, it stays close to the spirit of the original LG framework and clearly supplies necessary and sufficient conditions for macrorealism. It requires zero coherence at the two-time level but allows non-zero coherences at the three-time level, as long as they are suitably bounded. This protocol readily extends to the four times, analagous to the augmented LG case for four times.

All the two-time NSIT conditions can be satisfied quite easily, in a quantum-mechanical description, by choosing an initial state such at $\langle\hat{Q}(t)\rangle = 0$ at all three times. Also, $\mathrm{NSIT}_{(1)2}$ and $\mathrm{NSIT}_{(1)3}$ can be satisfied by choosing an initial state at t_1 diagonal in $\hat{Q}_1$. Furthermore, in practice, the NSIT condtitions in $\mathrm{MR}_{\mathrm{int}}$ will only be satisfied approximately and it is then necessary to develop extended forms of the LG inequalities appropriate to the case in which there is some signaling. This extension has been carried out by Dzhafarov and Kujala [32] (and is also briefly reviewed in Ref.[28]).

Clemente and Kofler gave a simple general argument that NIM implies MRps [19] and the detailed conditions reviewed in this section exemplify this. (Although note that this argument needs to be adjoined with the assumption that the measurements are repeatable [17]). However, in view of the different types of NIM characterized in this paper, it is clear that this argument can only involve NIM_{seq} and not NIM_{pw}. Their argument refers to measurements made in a single experiment, and not to combinations of results obtained from different experiments. Hence in the LG tests involved in MR_{weak}, MRps can be violated when NIM_{pw} holds.

4. NSIT vs two-time LG inequalities in a quantum-mechanical description

In a quantum-mechanical description, a direct comparison may be made between the NSIT conditions and LG inequalities for two-times using explicit measurement formulae. The probability for two sequential projective measurements at times t_1, t_2 is,

$$p(s_1, s_2) = \text{Tr}\left(P_{s_2}(t_2) P_{s_1}(t_1) \rho P_{s_1}(t_1)\right), \tag{32}$$

where the projection operators $P_s(t)$ are defined by $P_s(t) = e^{iHt} P_s e^{-iHt}$ and

$$P_s = \frac{1}{2}\left(1 + s\hat{Q}\right). \tag{33}$$

By contrast, the two-time probabilities Eq.(8) correspond in quantum mechanics to the quantities,

$$q(s_1, s_2) = \frac{1}{2}\text{Tr}\left(\left(P_{s_2}(t_2) P_{s_1}(t_1) + P_{s_1}(t_1) P_{s_2}(t_2)\right)\rho\right), \tag{34}$$

and the two-time LG inequalities Eq.(11) are simply then $q(s_1, s_2) \geq 0$. Eq.(34) may be measured either by measuring the averages and correlation function in various runs, as described, or, as argued in Ref.[1], more directly, using sequential measurements in which the first one is a weak measurement. Eqs.(32) and (34) have the same correlation function and same $\langle \hat{Q}_1 \rangle$, but differ in the average of $\hat{Q}(t)$ at the second time. The sequential measurement probability Eq.(32) does not satisfy the NSIT conditions Eq(6) in general. By contrast, Eq.(34) formally satisfies NSIT, but can be negative.

The relation between these two measurement formulae is given by

$$p(s_1, s_2) = q(s_1, s_2) + \frac{1}{8}\langle[\hat{Q}(t_1), \hat{Q}(t_2)]\hat{Q}(t_1)\rangle s_2. \tag{35}$$

The extra term on the right-hand side, which vanishes for commuting measurements, represents interferences (as shown more explicitly in Ref.[1]). If we impose NSIT on $p(s_1, s_2)$ this clearly implies that the interference term is zero and hence that $p(s_1, s_2) = q(s_1, s_2)$. This also means that $q(s_1, s_2) \geq 0$, which is equivalent to the two-time LG inequalities, Eq.(11).

However, the converse is not true: $q(s_1, s_2) \geq 0$ clearly does not imply NSIT for $p(s_1, s_2)$. Furthermore, since $p(s_1, s_2)$ is always non-negative, the two-time LG inequalities $q(s_1, s_2) \geq 0$ will be satisfied if the interference term is bounded:

$$\frac{1}{8}\left|\langle[\hat{Q}(t_1), \hat{Q}(t_2)]\hat{Q}(t_1)\rangle\right| \leq p(s_1, s_2) \tag{36}$$

This confirms in this case the general story mentioned earlier (and described at greater length in Ref.[2]): NSIT conditions require zero interference but the LG inequalities, like the CHSH case, require only bounded interference.

NSIT for $p(s_1, s_2)$ and $q(s_1, s_2) \geq 0$ are both conditions for MR at two times but they are different types of conditions. NIM is assumed to hold already in the measurement of $q(s_1, s_2)$ and $q(s_1, s_2) \geq 0$ may therefore be thought as a direct measure of MRps. By contrast, NSIT for $p(s_1, s_2)$ measures a combination of NIM and MRps, without being able to distinguish between them.

As an aside, we note the following. Although the role of the quasi-probability $q(s_1, s_2)$ is primarily as a measure of quantumness, when positive, it has the mathematical form of a true probability, and it is therefore natural to ask if it has any interpretation as a probability in the frequentist sense. The sequentially measured probability $p(s_1, s_2)$ clearly has a frequentist interpretation, but $q(s_1, s_2)$ is determined indirectly. However, it can be argued that when NSIT fails but $q(s_1, s_2) \geq 0$, then this can be interpreted as a situation in which $p(s_1, s_2)$ is the probability obtained when classically disturbing measurements are used to measure an underlying probability $q(s_1, s_2)$ [33]. Differently put, it is the situation in which the interference effects in sequential measurements which cause NSIT to fail can be modelled as a classical disturbance.

Finally, with the above explicit formulae in hand, we note a connection to coherence witness conditions [8, 24, 25]. One can define a witness $W(s_2)$ measuring the degree to which NSIT is violated:

$$W(s_2) = \left| \sum_{s_1} p_{12}(s_1, s_2) - p_2(s_2) \right|. \tag{37}$$

This is easily seen to be proportional to the interference term,

$$W(s_2) = \frac{1}{4} \left| \langle [\hat{Q}(t_1), \hat{Q}(t_2)] \hat{Q}(t_1) \rangle \right|. \tag{38}$$

There is therefore a simple relation between the degree of violation of NSIT and the two-time LG inequalities. If the witness is bounded according to

$$\frac{1}{2} W(s_2) \leq p(s_1, s_2), \tag{39}$$

then $q(s_1, s_2) \geq 0$. Hence witness conditions, which are usually used to check NSIT, can also be used to check the two-time LG inequalities. This result is also in keeping with the observation in Ref.[26] that violations of NSIT have to reach a threshold value before the LG inequalities are violated.

5. Summary and conclusions

The purpose of this paper was to elucidate and compare two very different sets of necessary and sufficient conditions for macrorealism which differ in the way in which they implement the notion of non-invasive measurability. In the first, the weaker form, measurements are made using a number of different experiments, analogous to Bell tests, and then the underlying probability, when it exists, is assembled indirectly. The measurements are non-invasive in a piecewise way, denoted $\mathrm{NIM_{pw}}$. The probability then exists provided that a set of two and three time LG inequalities hold (and also induction). This leads to a weak notion of macrorealism, which we write,

$$\mathrm{MR_{weak}} = \mathrm{NIM_{pw}} \wedge \mathrm{LG}_{12} \wedge \mathrm{LG}_{23} \wedge \mathrm{LG}_{13} \wedge \mathrm{LG}_{123} \wedge \mathrm{Ind}. \tag{40}$$

In the second, stronger form, proposed by Clemente and Kofler [19], macrorealism is defined by a series of NSIT conditions for sequential measurements in which all three measurements are made in the same experiment, together with induction:

$$\mathrm{MR_{strong}} = \mathrm{NSIT}_{(2)3} \wedge \mathrm{NSIT}_{(1)23} \wedge \mathrm{NSIT}_{1(2)3} \wedge \mathrm{Ind}. \tag{41}$$

An intermediate notion of MR also naturally arises, in which there are three pairwise experiments with NSIT satisfied for each pair, with all correlation functions required to satisfy the three-time LG inequalities (and induction):

$$\mathrm{MR_{int}} = \mathrm{NSIT}_{(1)2} \wedge \mathrm{NSIT}_{(1)3} \wedge \mathrm{NSIT}_{(2)3} \wedge \mathrm{LG}_{123} \wedge \mathrm{Ind} \tag{42}$$

These three conditions have a clear logical connection,

$$\mathrm{MR}_{\mathrm{strong}} \Rightarrow \mathrm{MR}_{\mathrm{int}} \Rightarrow \mathrm{MR}_{\mathrm{weak}}, \tag{43}$$

but the converse implications clearly do not hold. The relation between the NSIT conditions at two times and the two-time LG inequalities was spelled out explicitly in the quantum-mechanical analysis in Section 4. We also noted a relation between the two-time LG inequalities and the degree of violation of coherence witness conditions.

$\mathrm{MR}_{\mathrm{strong}}$ is primarily a measure of non-invasiveness and in the quantum case is satisfied only when the interferences are zero, so is essentially the same type of condition as a number of coherence witness conditions. By contrast, $\mathrm{MR}_{\mathrm{weak}}$ allows non-zero interferences. The measurements are non-invasive by design and hence $\mathrm{MR}_{\mathrm{weak}}$ is in effect a direct test of MRps (subject of course to loopholes due to experimental clumsiness [27]). Both of these types of macrorealism have been discussed and tested, at least in part, in a number of previous works. The purpose of the present work has been to make clear that these are different notions of macrorealism, due to the different ways in which NIM is implemented, although each clearly of interest to explore and test.

From the perpsective of the consistent histories approach to quantum mechanics these different notions of macrorealism correspond to the fact that there exists a hierarchy of classicality conditions. This is utilized and explored in Ref.[34].

For all of the protocols described in this paper, it would clearly be of interest to check experimentally a full set of necessary and sufficient conditions for macrorealism. This should not be difficult to accomplish with a modest extension of recent experiments: two and three-time LG inequalities have been tested in many different experiments, and likewise two-time NSIT conditions. What is required is an experiment which tests the appropriate combination of such conditions.

Acknowledgments
I am very grateful Clive Emary, George Knee, Johannes Kofler, Owen Maroney and James Yearsley for many useful discussions and email exchanges about the Leggett-Garg inequalities. I am particularly grateful as always to Thomas Elze for putting together such an excellent conference.

References
[1] Halliwell J J 2016 *Phys. Rev.* A **93** 022123
[2] Halliwell J J 2017 *Phys. Rev.* A **96** 012121
[3] Leggett A J and Garg A 1985 *Phys. Rev. Lett.* **54** 857
[4] Leggett A J 1988 *Found. Phys.* **18** 939; Leggett A J 2008 *Rep. Prog. Phys.* **71** 022001
[5] Emary C *et al* 2014 *Rep. Prog. Phys* **77** 016001
[6] Montina A 2012 *Phys. Rev. Lett.* **108** 160501
[7] Knee G C *et al* 2012 *Nat. Commun.* **3** 606
[8] Robens C *et al* 2015 *Phys. Rev.* X **5** 011003
[9] Katiyar H *et al* 2016 *Preprint* arxiv:1606.07151
[10] Wang K *et al* 2017 *Preprint* arxiv:1701.02454
[11] Goggin M E *et al* 2011 *Proc. Natl. Acad. Sci.* **108** 1256; Robens C *et al* 2015 *Phys. Rev.* X **5** 011003; Lambert N *et al* 2016 *Phys. Rev.* A **94** 012105
[12] Maroney O J E and Timpson C G 2014 *Preprint* arxiv:1412.6139
[13] Ghirardi G C *et al* 1986 *Phys. Rev.* D **34** 470
[14] Bacciagalupi G 2014 *Preprint* arxiv:1409.4104
[15] Kochen S and Specker E P 1967 *Journal of Mathematics and Mechanics* **17** 59
[16] Allen J-M A *et al* 2016 *Preprint* arxiv:1610.00022
[17] Hermens R and Maroney O J E 2017 *Preprint* arxiv:1706.01641
[18] Kofler J and Brukner C 2013 *Phys. Rev.* A **87** 052115

IOP Conf. Series: Journal of Physics: Conf. Series **1275** (2019) 012008 doi:10.1088/1742-6596/1275/1/012008

[19] Clemente L and Kofler J 2015 *Phys. Rev.* A **91** 062103; Clemente L and Kofler J 2016 *Phys. Rev. Lett.* **116** 150401
[20] Fine A 1982 *J. Math. Phys* **23** 1306; Fine A 1982 *Phys. Rev. Lett.* **48** 291
[21] Halliwell J J 2014 *Phys. Lett* A **378** 2945
[22] Bell J S 1987 *Speakable and Unspeakable in Quantum Mechanics* (Cambridge: Cambridge University Press)
[23] Clauser J F *et al* 1982 *Phys.Re v. Lett.* **23** 1306
[24] Li C-M *et al* 2012 *Sci. Rep.* **2** 885; Schild G and Emary C 2015 *Phys. Rev.* A **92** 032101; Wang K *et al* 2017 *Phys. Rev.* A **95** 032122
[25] Emary C 2017 *Phys. Rev.* A **96** 042102
[26] Saha D *et al* 2015 *Phys. Rev.* A **91** 032117; Kumari S and Pan A K 2017 *Phys. Rev.* A **96** 042107
[27] Wilde M M and Mizel A 2012 *Found. Phys.* **42** 256
[28] Halliwell J J 2016 *Phys. Rev.* A **94** 052131
[29] Halliwell J J 2016 *Phys. Rev.* A **94** 052114
[30] Halliwell J J and Yearsley J M 2013 *Phys. Rev.* A **87** 022114
[31] Klyshko D N 1996 *Phys. Lett.* A **218** 119
[32] Dzhafarov E N and Kujala J V 2014 *Preprint* arxiv:1407.2886
[33] Halliwell J J *et al* 2018 *Preprint* arxiv:1810.10976
[34] Halliwell J J 2017 *Phys. Rev.* A **96** 012123

Causal Fermion Systems: Discrete Space-Times, Causation and Finite Propagation Speed

Felix Finster
Fakultät für Mathematik, Universität Regensburg, D-93040, Germany
E-mail: finster@ur.de

Abstract. The theory of causal fermion systems is a recent approach to fundamental physics. Giving quantum mechanics, general relativity and quantum field theory as limiting cases, it is a candidate for a unified physical theory. The dynamics is described by a novel variational principle, the so-called causal action principle. The causal action principle does not rely on a presupposed space-time structure. Instead, it is a variational principle for space-time itself as well as for all structures in space-time (like particles, fields, etc.).

After a general motivation and introduction, we report on mathematical results for two-particle causal fermion systems which state that every minimizer describes a discrete space-time. We explain and make precise that on scales which are much larger than the scale of the microscopic space-time structures, the dynamics of a causal fermion system respects causality with a finite speed of propagation.

1. From a lattice system to causal fermion systems

The focus of this conference is on discrete space-time structures. Already in previous talks the motivation for thinking about discrete space-times was mentioned: The ultraviolet divergences of quantum field theory indicate that the structure of space-time should be modified on a microscopic scale, and the Planck length gives a natural length scale for such modifications and for new physics. To my opinion, instead of taking sides in favor of discrete or continuous microstructures, it is more promising to consider the following more general question:

What is the structure of space-time on the Planck scale?

I shall present a concise proposal for what the mathematical structure of microscopic space-time could be. The mathematical structures are quite general and allow for the description of both discrete and continuum space-times. Nevertheless, as we shall see, there are results in favor of a discrete microstructure.

1.1. Example: A two-dimensional lattice system

For the motivation, I would like to begin with the familiar example of a space-time lattice. In order to keep the setting as simple as possible, we consider a two-dimensional lattice (one space and one time dimension), but higher-dimensional lattices can be described similarly. Thus let $\mathscr{M} \subset \mathbb{R}^{1,1}$ be a lattice in two-dimensional Minkowski space. We denote the spacing in time direction by Δt and in spatial direction by Δx (see Figure 1). The usual procedure for setting up equations on a lattice is to replace derivatives by difference quotients, giving rise to an evolution equation which can be solved time step by time step according to deterministic rules.

Content from this work may be used under the terms of the Creative Commons Attribution 3.0 licence. Any further distribution of this work must maintain attribution to the author(s) and the title of the work, journal citation and DOI.
Published under licence by IOP Publishing Ltd

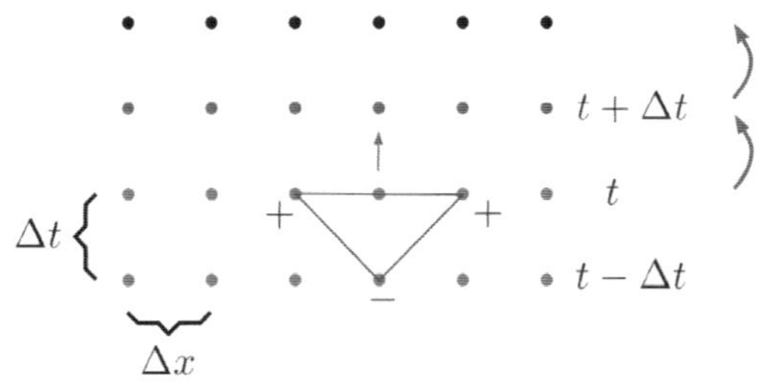

Figure 1. Time evolution of a lattice system $\mathscr{M} \subset \mathbb{R}^{1,1}$.

As a concrete example, let us consider a discretization of the two-dimensional wave equation on the lattice,

$$\begin{aligned} 0 = \Box\phi(t,x) := \frac{1}{(\Delta t)^2}\Big(\phi(t+\Delta t,x) - 2\phi(t,x) + \phi(t-\Delta t,x)\Big) \\ - \frac{1}{(\Delta x)^2}\Big(\phi(t,x+\Delta x) - 2\phi(t,x) + \phi(t,x-\Delta x)\Big) . \end{aligned} \tag{1}$$

Solving this equation for $\phi(t+\Delta t,x)$ gives a rule for computing $\phi(t+\Delta t,x)$ from the values of ϕ at earlier times t and $t-\Delta t$ (see again Figure 1).

While this method is very simple and gives well-defined evolution equations, it also has several drawbacks:

- The above method of discretizing the continuum equations is very *ad hoc*. Why do we choose a regular lattice, why do we work with difference quotients? There are many other ways of discretizing the wave equation.
- The method is *not background-free*. In order to speak of the "lattice spacing," the lattice must be thought of as being embedded in a two-dimensional ambient space-time.
- The concept of a space-time lattice is not invariant under general coordinate transformations. In other words, the assumption of a space-time lattice is *not compatible with the equivalence principle.*

In view of these shortcomings, the following basic question arises:

> Can one formulate equations without referring to the nearest neighbor relation and the lattice spacing?

The answer to this question is yes, and we will now see how this can be done in the example of our two-dimensional lattice system. Although our example is somewhat oversimplified, this consideration will lead us quite naturally to the setting of causal fermion systems.

We consider complex-valued wave functions $\psi_1, \ldots, \psi_f$ on the lattice $\mathscr{M}$ (for simplicity a finite number, i.e. $f < \infty$). These wave functions do not need to satisfy the wave equation. On the complex vector space spanned by these wave functions we introduce a scalar product $\langle .|. \rangle_{\mathcal{H}}$ and assume that the wave functions are orthonormal, i.e.

$$\langle \psi_k | \psi_l \rangle_{\mathcal{H}} = \delta_{kl} . \tag{2}$$

We thus obtain an f-dimensional Hilbert space $(\mathcal{H}, \langle .|. \rangle_{\mathcal{H}})$. Note that the scalar product is given abstractly (meaning that it has no representation in terms of the wave functions as a sum over lattice points). Next, for any lattice point $(t,x) \in \mathscr{M}$ we introduce the so-called *local correlation operator* $F(t,x) : \mathcal{H} \to \mathcal{H}$ as a matrix with elements given by

$$(F(t,x))^j_k = \overline{\psi_j(t,x)}\psi_k(t,x) . \tag{3}$$

IOP Conf. Series: Journal of Physics: Conf. Series **1275** (2019) 012009 doi:10.1088/1742-6596/1275/1/012009

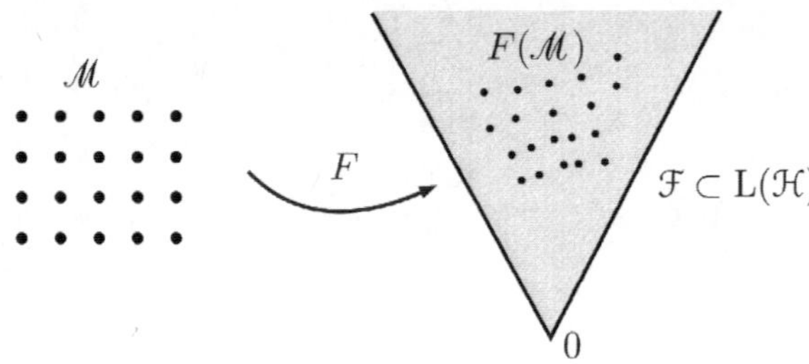

Figure 2. Embedding in $\mathcal{F}$.

The diagonal elements of this matrix are the absolute squares $|\psi_k(t,x)|^2$ of the corresponding wave functions. The off-diagonal elements, on the other hand, tell us about the correlation of the jth and k^{th} wave functions at the lattice point (t,x). This is the reason for the name "local correlation operator." This operator can also be characterized in a basis-invariant way by the relations

$$\langle \psi, F(t,x)\,\phi \rangle_{\mathcal{H}} = \overline{\psi(t,x)}\phi(t,x)\,,$$

to be satisfied for all $\psi, \phi \in \mathcal{H}$. Taking the complex conjugate, one sees immediately that the matrix defined by (3) is Hermitian. Stated equivalently independent of bases, the local correlation operator is a *selfadjoint* linear operator on $\mathcal{H}$. Moreover, a local correlation operator has *rank at most one* and is *positive semi-definite*. This can be seen by writing it as

$$F(t,x) = e^* e \qquad \text{with} \qquad e : \mathcal{H} \to \mathbb{C}\,, \quad \psi \mapsto \psi(x)\,. \tag{4}$$

It is useful to denote the set of all operators with the above properties by $\mathcal{F}$,

$$\mathcal{F} := \big\{ F \in \mathrm{L}(\mathcal{H}) \mid F \text{ is selfadjoint, positive semi-definite and has rank at most one} \big\}\,.$$

Varying the lattice point, we obtain a mapping (see Figure 2)

$$F : \mathcal{M} \to \mathcal{F}\,, \qquad (t,x) \mapsto F(t,x)\,.$$

For clarity, we note that the set $\mathcal{F}$ is *not* a vector space, because the linear combination of operators in $\mathcal{F}$ will in general have rank bigger than one. But it is a *conical* set in the sense that a positive multiple of any operator in $\mathcal{F}$ is again in $\mathcal{F}$ (this is why in Figure 2 the set $\mathcal{F}$ is depicted as a cone).

We point out that the local correlation operators do not involve the lattice spacing or the nearest neighbor relation; instead they contain information only on the local correlations of the wave functions at each lattice point. With this in mind, our strategy for formulating equations which do not involve the structures of the lattice is to work exclusively with the local correlation operators, i.e. with the subset $F(\mathcal{M}) \subset \mathcal{F}$. In other words, in Figure 2 we want to disregard the lattice on the left and work only with the objects on the right.

How can one set up equations purely in terms of the local correlation operators? In order to explain the general procedure, we consider a finite number of operators $F_1, \ldots, F_L$ of $\mathcal{F}$. Each of these operators can be thought of as giving information on the local correlations of the wave functions at a space-time point. However, this "space-time point" is no longer a lattice point, but at the moment it is merely a point without additional structure. In order to obtain a "space-time" in the usual sense, one needs additional structures and relations between the space-time points. Such relations can be obtained by multiplying the operators. Indeed, the operator product $F_i F_j$ tells us about correlations of the wave functions at different space-time points. Taking the trace of this operator product gives a real number. Our method for formulating physical equations is to set up a variational principle. This variational formulation has the advantage that symmetries give rise to conservation laws by Noether's theorem (as will be explained in Section 6). Therefore, we want to minimize an action $\mathcal{S}$. A simple example is to

$$\text{minimize} \qquad \mathcal{S}(F_1, \ldots, F_n) := \sum_{i,j=1}^{L} \text{Tr}(F_i\, F_j)^2 \tag{5}$$

under variations of the points $F_1, \ldots, F_L \in \mathcal{F}$. In order to obtain a mathematically sensible variational principle, one needs to impose certain constraints. Here we do not enter the details, because the present example is a bit too simple. Instead, we merely use it as a motivation for the general setting of causal fermion systems, which we now introduce.

1.2. The setting of causal fermion systems

In order to get from our example to the general setting of causal fermion systems, we must extend the above constructions in several steps:

(a) The previous example works similarly in higher dimensions, in particular for a lattice $\mathscr{M} \subset \mathbb{R}^{1,3}$ in four-dimensional Minkowski space. This has no effect on the resulting structure of a finite number of distinguished operators $F_1, \ldots, F_L \in \mathcal{F}$.

(b) Suppose that on the lattice we consider multi-component wave functions $\psi(t,x) \in \mathbb{C}^N$. Then the pointwise product on the right side of (3) must be replaced by a complex inner product, which we denote by $\prec .|. \succ$ (in mathematical terms, this inner product is a non-degenerate sesquilinear form; we always use the convention that the wave function on the left is complex conjugated). Thus the definition of the local correlation operator (3) is replaced by
$$(F(t,x))^j_k = -\prec \psi_j(t,x) | \psi_k(t,x) \succ$$
(the minus sign compared to (3) merely is a useful convention). The resulting local correlation operator is no longer an operator of rank at most one, but it has rank at most N (as can be seen for example by writing it similar to (4) in the form $F(t,x) = -e^* e$ with $e : \mathcal{H} \to \mathbb{C}^N$). If the inner product $\prec .|. \succ$ on $\mathbb{C}^N$ is positive definite, then the operator $F(t,x)$ is negative semi-definite. However, in the physical applications, this inner product will *not* be positive definite. Indeed, a typical example in mind is that of four-component Dirac spinors. The Lorentz invariant inner product $\overline{\psi}\phi$ on Dirac spinors in Minkowski space (with the usual adjoint spinor $\overline{\psi} := \psi^\dagger \gamma^0$) is indefinite of signature $(2,2)$. In order to describe systems involving leptons and quarks, one must take direct sums of Dirac spinors, giving the signature (n,n) with $n \in 2\mathbb{N}$. With this in mind, we assume more generally that
$$\prec .|. \succ \quad \text{has signature } (n,n) \text{ with } n \in \mathbb{N}\,.$$
Then the resulting local correlation operators are selfadjoint operators of rank at most $2n$, which (counting multiplicities) have at most n positive and at most n negative eigenvalues.

(c) Finally, it is useful to generalize the setting such as to allow for continuous space-times and for space-times which have both continuous and discrete components. In preparation, we note that the sums over the operators $F_1, \ldots, F_L$ in (5) can be written as integrals,
$$\mathcal{S}(\rho) = \int_{\mathcal{F}} d\rho(x) \int_{\mathcal{F}} d\rho(y)\; \text{Tr}(F_i F_j)^2\,, \tag{6}$$
if ρ is a measure on $\mathcal{F}$ chosen as the sum of Dirac measures supported at these operators,
$$\rho = \sum_{i=1}^{L} \delta_{F_i}\,. \tag{7}$$

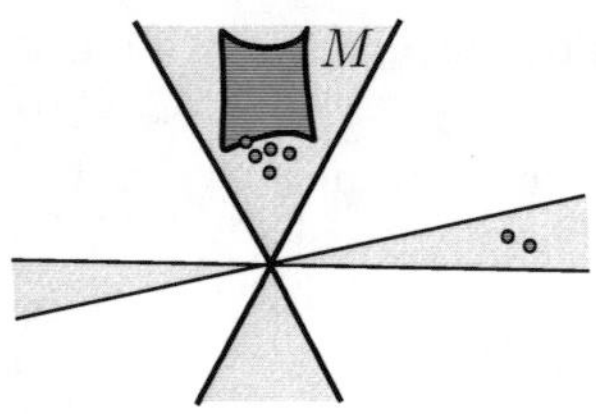

Figure 3. A causal fermion system.

In this formulation, the measure plays a double role: First, it distinguishes the points $F_1, \ldots, F_L$ as those points where the measure is non-zero, as is made mathematically precise by the notion of the *support* of the measure, i.e.

$$\operatorname{supp} \rho = \{F_1, \ldots, F_L\} . \tag{8}$$

Second, a measure makes it possible to integrate over its support, an operation which for the measure (7) reduces to the sum over $F_1, \ldots, F_L$.
Now one can extend the setting simply by considering (6) for more general measures on $\mathcal{F}$ (like for example regular Borel measures). The main advantage of working with measures is that we get into a mathematical setting in which variational principles like (5) can be studied with powerful analytic methods. In particular, as we shall see in Section 3, in this setting the question whether minimizers give rise to discrete or continuous space-time structures can be analyzed and answered.

These generalizations lead us to the following definition:

Definition 1.1 *(causal fermion system)* Given a separable complex Hilbert space $\mathcal{H}$ with scalar product $\langle .|. \rangle_{\mathcal{H}}$ and a parameter $n \in \mathbb{N}$ (the *"spin dimension"*), we let $\mathcal{F} \subset \mathrm{L}(\mathcal{H})$ be the set of all selfadjoint operators on $\mathcal{H}$ of finite rank, which (counting multiplicities) have at most n positive and at most n negative eigenvalues. On $\mathcal{F}$ we are given a positive measure ρ (defined on a σ-algebra of subsets of $\mathcal{F}$), the so-called *universal measure*. We refer to $(\mathcal{H}, \mathcal{F}, \rho)$ as a *causal fermion system*.

This definition is illustrated in Figure 3. Now the set $\mathcal{F}$ is invariant in addition under the transformation where an operator is multiplied by a negative number, as is indicated in the figure by the double cones. The support of the measure, denoted by

$$M := \operatorname{supp} \rho , \tag{9}$$

is referred to as *space-time*. In generalization of the example of the lattice system, where space-time consisted of discrete points (8), now the measure ρ can also have continuous components.

The Hilbert space $\mathcal{H}$ can be understood as being spanned by all wave functions of the system, similar as explained for our lattice system after (2). Indeed, starting from the abstract definition, to every vector $u \in \mathcal{H}$ one can associate a corresponding wave function in space-time, the so-called *physical wave function* (for details see [1, §1.1.4]). In the applications, the physical wave functions describe fermions. This is the reason for the name "causal *fermion* system." The dimension of $\mathcal{H}$ can be regarded as the number of fermionic particles in the system (where in the physical applications we also count the particles of the so-called Dirac sea; see [2] for further explanations of this point).

The above notions evolved over several years. Based on preparations in [3], the present formulation was first given in [4]. For the general background we refer to the non-technical introduction for physicists in [5] or to the textbooks [1, 6].

2. The causal action principle

Having explained the general definition of a causal fermion system (see Definition 1.1), we can now introduce the variational principle used to describe the dynamics of a causal fermion system, the so-called *causal action principle.* The mathematical structure of the causal action is similar to the action (5) given in our example of the lattice system. Its detailed form, however, is far from obvious and is the result of many computations and long considerations. The causal action was first proposed in [3, Section 3.5].

For any $x, y \in \mathcal{F}$, the product xy is an operator of rank at most $2n$. However, in general it is no longer a selfadjoint operator because $(xy)^* = yx$, and this is different from xy unless x and y commute. As a consequence, the eigenvalues of the operator xy are in general complex. We denote these eigenvalues counting algebraic multiplicities by $\lambda_1^{xy}, \ldots, \lambda_{2n}^{xy} \in \mathbb{C}$ (more specifically, denoting the rank of xy by $k \leq 2n$, we choose $\lambda_1^{xy}, \ldots, \lambda_k^{xy}$ as all the non-zero eigenvalues and set $\lambda_{k+1}^{xy}, \ldots, \lambda_{2n}^{xy} = 0$). We introduce the Lagrangian and the causal action by

$$\textit{Lagrangian:} \qquad \mathcal{L}(x,y) = \frac{1}{4n} \sum_{i,j=1}^{2n} \Big(\big|\lambda_i^{xy}\big| - \big|\lambda_j^{xy}\big| \Big)^2 \tag{10}$$

$$\textit{causal action:} \qquad \mathcal{S}(\rho) = \iint_{\mathcal{F}\times\mathcal{F}} \mathcal{L}(x,y)\, d\rho(x)\, d\rho(y)\,. \tag{11}$$

The *causal action principle* is to minimize $\mathcal{S}$ by varying the measure ρ under the following constraints:

$$\textit{volume constraint:} \qquad \rho(\mathcal{F}) = \text{const} \tag{12}$$

$$\textit{trace constraint:} \qquad \int_{\mathcal{F}} \text{tr}(x)\, d\rho(x) = \text{const} \tag{13}$$

$$\textit{boundedness constraint:} \qquad \iint_{\mathcal{F}\times\mathcal{F}} \bigg(\sum_{i=1}^{2n} \big|\lambda_j^{xy}\big|^2 \bigg)\, d\rho(x)\, d\rho(y) \leq C\,, \tag{14}$$

where C is a given parameter (and tr denotes the trace of a linear operator on $\mathcal{H}$). This variational principle is mathematically well-posed if $\mathcal{H}$ is finite-dimensional and if one varies the measure in the class of regular Borel measures (with respect to the topology on $\text{L}(\mathcal{H})$ induced by the operator norm). For the existence theory and the analysis of general properties of minimizing measures we refer to [7, 8, 9].

Given a minimizing measure ρ, we define *space-time* again as the support of the measure (9). Thus the space-time points are selfadjoint linear operators on $\mathcal{H}$.

The fact that the eigenvalues $\lambda_1^{xy}, \ldots, \lambda_{2n}^{xy}$ are complex makes it possible to introduce the following notion of causality:

Definition 2.1 *(causal structure)* The points x and y are called *spacelike* separated if all the eigenvalues λ_j^{xy} have the same absolute value. They are said to be *timelike* separated if the λ_j^{xy} are all real and do not all have the same absolute value. In all other cases (i.e. if the λ_j^{xy} are not all real and do not all have the same absolute value), the points x and y are said to be *lightlike* separated.

Restricting the causal structure of $\mathcal{F}$ to M, we get causal relations in space-time.

The Lagrangian (10) is compatible with the above notion of causality in the following sense. Suppose that two points $x, y \in \mathcal{F}$ are spacelike separated. Then the eigenvalues λ_i^{xy} all have the same absolute value. Therefore, the Lagrangian (10) vanishes. Thus pairs of points with spacelike separation do not enter the action. This can be seen in analogy to the usual notion

IOP Conf. Series: Journal of Physics: Conf. Series **1275** (2019) 012009 doi:10.1088/1742-6596/1275/1/012009

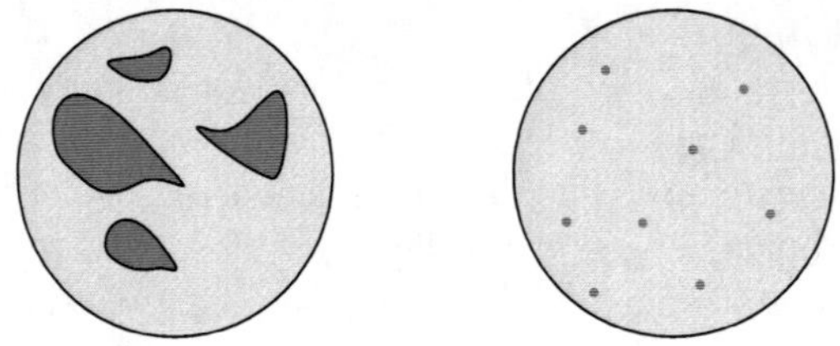

Figure 4. The causal variational principle on the sphere.

of causality where points with spacelike separation cannot influence each other. This analogy is the reason for the notion *causal* in "causal fermion system" and "causal action principle."

We point out that the fact that pairs of points with spacelike separation do not enter the action is much weaker than the usual concept of causation which states that the present can only influence the future. Indeed, the structure of the causal Lagrangian leaves the possibility that the future could influence the present or past. In order to analyze whether this really happens, one must analyze the initial value problem. If for given initial data there is a unique solution to the future, we can conclude that causation holds. We will come back to this important point in Section 8.

3. Discreteness results

Clearly, the general setting of causal fermion systems is quite abstract. In order to explain the causal action principle more concretely, we now consider the simplest interesting example. To this end, we choose the spin dimension $n = 1$ and the Hilbert space $\mathcal{H} = \mathbb{C}^2$. Then $\mathcal{F}$ consists of all Hermitian 2×2-matrices F which have at most one positive and at most one negative eigenvalue. Such a matrix can be represented with the help of Pauli matrices as

$$F = \alpha\,1 + \vec{v}\vec{\sigma} \qquad \text{with} \qquad \alpha \in \mathbb{R},\ \vec{v} \in \mathbb{R}^3 \text{ and } |\vec{v}| \geq |\alpha|\,.$$

In order to simplify the constraints (13) and (14), we replace them by the condition that the matrix F should have eigenvalues $1 \pm \tau$, where $\tau \geq 1$ is a parameter of the model (which plays a similar role as the parameter C in the boundedness constraint (14)). The set of all matrices with these properties can be written as

$$\mathcal{F} = \big\{ F = \tau\,\vec{x}\vec{\sigma} + \mathbb{1} \quad \text{with} \quad \vec{x} \in S^2 \subset \mathbb{R}^3 \big\}\,.$$

Thus the set $\mathcal{F}$ can be identified with the unit sphere S^2, which also simplifies illustrations (see Figure 4). The volume constraint (12) can be implemented most easily by restricting attention to *normalized* regular Borel measures on $\mathcal{F}$ (i.e. measures with $\rho(\mathcal{F}) = 1$). Such a measure can be both continuous, discrete or a mixture. Examples of *continuous measures* are obtained by multiplying the Lebesgue measure on the sphere by a non-negative smooth function. By a *discrete measure*, on the other hand, we here mean a *weighted counting measure*, i.e. a measure obtained by inserting weight factors into (7),

$$\rho = \sum_{i=1}^{L} c_i\,\delta_{x_i} \qquad \text{with} \qquad x_i \in \mathcal{F}\,, \quad c_i \geq 0 \quad \text{and} \quad \sum_{i=1}^{L} c_i = 1\,. \tag{15}$$

In this setting, a straightforward computation yields for the Lagrangian (10)

$$\begin{aligned} \mathcal{L}(x,y) &= \max\big(0, \mathcal{D}(x,y)\big) \qquad \text{with} \\ \mathcal{D}(x,y) &= 2\tau^2\,\big(1 + \langle x,y\rangle\big)\Big(2 - \tau^2\,\big(1 - \langle x,y\rangle\big)\Big)\,, \end{aligned}$$

IOP Conf. Series: Journal of Physics: Conf. Series **1275** (2019) 012009 doi:10.1088/1742-6596/1275/1/012009

and $\langle x, y \rangle$ denotes the Euclidean scalar product of two unit vectors $x, y \in S^2 \subset \mathbb{R}^3$ (thus $\langle x, y \rangle = \cos \vartheta$, where ϑ is the angle between x and y). The resulting so-called *causal variational principle on the sphere* was introduced in [8, Chapter 1] and analyzed in [10, Sections 2 and 5] and more recently in [11]. We now explain a few results from these papers.

First of all, the causal variational principle on the sphere is well-posed, meaning that the minimum is attained in the class of all normalized regular Borel measures. The minimizing measure is not unique; indeed, there are typically many minimizers. The study in [10, Section 2] gives the following

> *numerical result*: If $\tau > \sqrt{2}$, every minimizing measure is a weighted counting measure (15).

Thus, although we minimize over all regular Borel measures (i.e. measures which can have discrete and continuous components), a minimizing measure always describes a discrete space-time consisting of a finite number of space-time points (as shown on the right of Figure 4). To some extent, this numerical result could be underpinned by analytic results. First, it was proven in [10, Section 5.1] that the support has an empty interior:

Theorem 3.1 *If $\tau > \sqrt{2}$, the support of any minimizing measure does not contain an open subset of S^2.*

Intuitively speaking, this result shows that the space-time points are a subset of the sphere of dimension strictly smaller than two. More recently, it was shown in [11] that the dimension of the support is even strictly smaller than one:

Theorem 3.2 *In the case $\tau > \sqrt{6}$, the support of any minimizing measure is totally disconnected and has Hausdorff dimension at most $6/7$.*

For brevity, we cannot enter the proof of these theorems. But we point out that the method of proof gives a good understanding of the underlying mechanism which has the effect that minimizing measures tend to be supported on low-dimensional subsets of $\mathcal{F}$. This mechanism applies not only to the causal variational principle on the sphere, but similarly also to the general setting in higher dimensions (see [10, Section 3.3]). The gap between the analytic and numerical results comes about mainly due to shortcomings of the mathematical methods.

To summarize, the above mathematical results indicate that minimizing measures of the causal action should be discrete, and the methods of proof reveal the reason why this should be the case.

4. Linearized fields

We again point out that the above results give an *indication* that minimizers of the causal action principle should correspond to discrete space-times. However, the theorems apply in too special situations for giving a definitive answer for general causal fermion systems. Therefore, as mentioned at the beginning of this talk, we do not want to take sides in favor of discrete or continuous space-times. Moreover, this distinction does not seem essential to me. My message is that the fundamental structure is a measure on $\mathcal{F}$ which minimizes the causal action. It is an interesting question whether this measure is discrete or continuous, but this question seems of secondary importance.

Nevertheless, in the remainder of my talk, I will restrict attention to discrete measures for two reasons: First, because this facilitates the comparison to other discrete approaches at this conference (like causal sets or cellular automata). Second, it has the advantage that integrals reduce to sums, and that the presentation simplifies because space-time M can be depicted as a discrete set of points $x_1, x_2, \ldots \in \mathcal{F}$ (see Figure 5). We now proceed by explaining a few recent structural results on causal fermion systems, specialized to discrete measures.

IOP Conf. Series: Journal of Physics: Conf. Series **1275** (2019) 012009 doi:10.1088/1742-6596/1275/1/012009

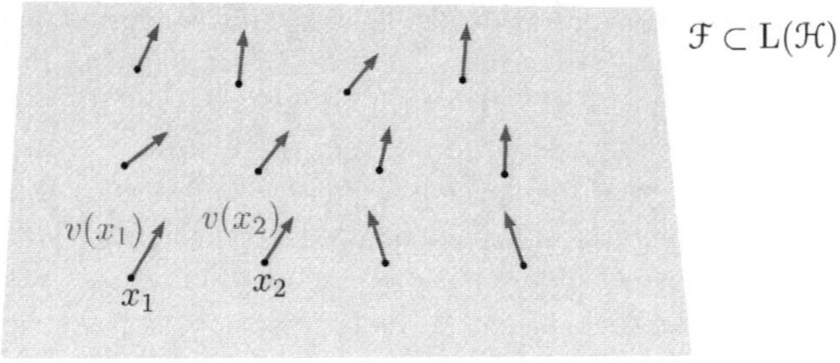

Figure 5. Jets in a discrete space-time.

Let us assume that ρ is a minimizing measure which is discrete in the sense that its support is a discrete point set $M = \{x_1, x_2, \ldots\} \subset \mathcal{F}$ (the set M could be finite or countable, which for our presentation makes no difference). The fact that ρ is a minimizer of the causal action poses many conditions on this measure. In more physical terms, these conditions tell us about the interaction and the dynamics of the system. It is an important and difficult task to analyze what these conditions actually mean, for example by rewriting them in a form familiar from classical field theory or quantum theory. In more general terms, the question is:

> What is the dynamics of a causal fermion system as described by the causal action principle?

In order to get detailed and explicit insight into this question, the following strategy turns out to be fruitful: A class of minimizing measures which are relatively easy to construct are minimizers which have certain symmetries and can therefore be regarded as describing the vacuum. Now one can consider small perturbations of the measure ρ and analyze their dynamics. Considering linear perturbations [12, 13], this procedure is similar to analyzing linearized fields in Minkowski space. Linear perturbations are also a suitable starting point for analyzing nonlinear fields, for example perturbatively [14]. The procedure of considering small perturbations of a discrete system bears some similarity to the procedure in solid states physics, where perturbations of the positions of the atoms of a solid give rise to lattice oscillations described by phonons. But in our case, of course, the "ambient space" in which the discrete points "live" is not the three-dimensional Euclidean space or space-time, but instead it is the set $\mathcal{F}$ of linear operators on the Hilbert space $\mathcal{H}$.

Perturbations of the measure ρ are described by so-called *jets*, which we now introduce. There are two ways to perturb a weighted counting measure (15): one can perturb the weight factors c_i or the positions x_i of the space-time points. The first class of perturbations is described by a real-valued function b on the space-time points. The perturbation of the positions x_i, on the other hand, is described by a vector v which is tangential to $\mathcal{F}$ at the point x_i (see Figure 5). The jet $\mathfrak{v}$ is defined as the pair of the real-valued function and the vector field,

$$\mathfrak{v} = (b, v) .$$

The vector field v appearing in the jet can also be described more concretely in terms of wave functions and fields in space-time. In order to get the connection, we first recall that in the example of the lattice system, we constructed the points in $\mathcal{F}$ as local correlation operators of wave functions in space-time (3). A similar construction works for general causal fermion systems, with the result that every point $x \in M$ can be recovered as the local correlation operator of so-called *physical wave functions* ψ_k, i.e.

$$x = F(x) \qquad \text{with} \qquad F(x)^j_k = -\prec \psi_j(x) | \psi_k(x) \succ_x ,$$

where $\prec.|.\succ_x$ is the so-called spin scalar product (for details see [1, §1.1.4]). A vector field v at x is a first variation of $F(x)$, i.e.

$$v(x) = \delta F(x) = -\prec \delta\psi_j(x) \,|\, \psi_k(x) \succ_x - \prec \psi_j(x) \,|\, \delta\psi_k(x) \succ_x .$$

In this way, the vector field v can be described by linear variations of the physical wave functions. Now one can consider different types of such variations:

(a) One can vary individual physical wave functions. This gives rise to the so-called *fermionic jets*.

(b) Alternatively, one can vary all physical wave functions collectively. In physical applications when the physical wave functions satisfy the Dirac equation, such variations can be described by bosonic fields, like for example

$$\delta\psi_j(x) = -(s_m \, A_j \gamma^j \, \psi_j)(x) \,,$$

where s_m is a Dirac Green's operator and A is the electromagnetic potential. The corresponding jets are referred to as *bosonic jets*.

More detailed explanations on these constructions can be found in [14, Sections 6 and 7].

We remark that the connection to bosonic and fermionic fields is worked out in detail in the so-called *continuum limit*, which can be thought of as the limiting case of a lattice system in four-dimensional Minkowski space when the lattice spacing tends to zero. For brevity, we cannot enter the methods and results of the continuum limit analysis, but we refer instead to the textbook [1] or the short survey in [5, Section 6].

5. The jet dynamics

The dynamics of linearized fields is described by equations in space-time, referred to as the *linearized field equations.* We now outline how these equations are derived and say a few words on their general structure. Following the causal action principle, a measure ρ describing physical space-time should be a minimizer of the causal action. This implies that ρ must satisfy corresponding *Euler-Lagrange equations.* These equations take the form (more precisely these are the so-called weak EL equations; for details [12, Sections 3.1 or 4.1])

$$\nabla_{\mathfrak{u}} \bigg(\int_{\mathcal{F}} \mathcal{L}(x,y) \, d\rho(y) - \frac{\nu}{2} \bigg) = 0 \,, \qquad (16)$$

which must hold for all jets $\mathfrak{u}$ (the so-called *test jets*) and all $x \in M$, where ∇ denotes the combination of multiplication and differentiation

$$\nabla_{\mathfrak{v}} \eta(x) = b(x) \, \eta(x) + D_v \eta(x) \,.$$

In simple terms, the function in the brackets in (16) as well as its derivatives must vanish at all space-time points.

We next assume that ρ describes the vacuum and consider linear perturbations of the measure ρ described by a jet $\mathfrak{v}$. These perturbations should be physical in the sense that the perturbed measures should also satisfy the Euler-Lagrange equations. In other words, the jet $\mathfrak{v}$ should generate a one-parameter family of solutions of (16). The linearized field equations are obtained by differentiating the Euler-Lagrange equations with respect to this parameter. They take the form (for details see [12, Section 4.2])

$$\langle \mathfrak{u}, \Delta \mathfrak{v} \rangle(x) = 0 \qquad \text{for all } x \in M \,,$$

IOP Conf. Series: Journal of Physics: Conf. Series **1275** (2019) 012009 doi:10.1088/1742-6596/1275/1/012009

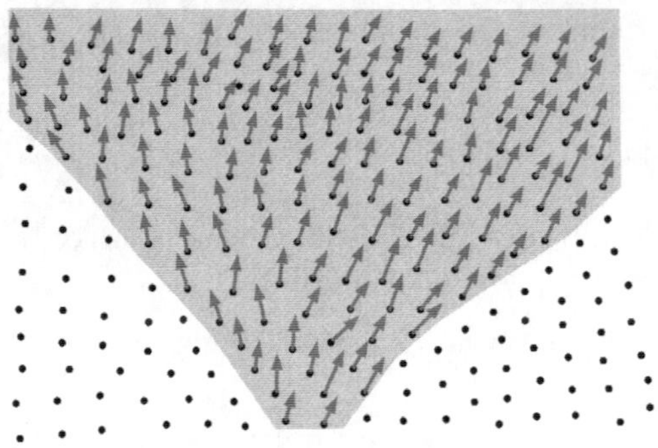

Figure 6. Finite propagation speed.

where the "Laplacian" is defined by

$$\langle \mathfrak{u}, \Delta \mathfrak{v} \rangle(x) := \nabla_{\mathfrak{u}} \bigg(\int_M \big(\nabla_{1,\mathfrak{v}} + \nabla_{2,\mathfrak{v}} \big) \mathcal{L}(x,y) \, d\rho(y) - \nabla_{\mathfrak{v}} \frac{\nu}{2} \bigg) . \tag{17}$$

For a discrete measure, the last integral reduces to a sum over the space-time points $x_1, x_2, \ldots \in \mathcal{F}$. Therefore, similar to the wave equation on the lattice (1), the linearized field equations involve a sum over lattice points. But, in contrast to (1), we do not sum only over the nearest neighbors, but the equation (17) is *nonlocal* in the sense that it typically involves a sum over many lattice points.

Despite this nonlocality, the linearized field equations describe a dynamics with *finite propagation speed* (as shown schematically in Figure 6). Moreover, *causation* holds in the sense that the initial value problem is well-posed. It is important to note, however, that these statements hold only "macroscopically" on a scale which is much larger than the length scale of the discretization. This is made mathematically precise in the recent paper [13]. Before explaining in Section 8 how these statements come about and what precisely we mean by "macroscopic," we need to discuss a few other structures which arise for minimizers of causal variational principles.

6. Surface layer integrals

In analogy to the situation for the classical Noether theorem, symmetries of minimizers of the causal action give rise to corresponding conservation laws. In the resulting *Noether-like theorems* [15], the conserved quantities are expressed in terms of so-called *surface layer integrals* which have the following general structure. We choose a subset Ω of space-time which can be thought of as the past of an equal time hypersurface of an observer (see Figure 7). We now form a double integral where one variable x is integrated over Ω and the other variable y over the complement of Ω,

$$\int_\Omega d\rho(x) \int_{M \setminus \Omega} d\rho(y) \, (\cdots) \mathcal{L}(x,y) \, .$$

Here $(\cdots)$ stands for a an operator acting on the Lagrangian which typically involves jet derivatives. For specific choices of this operator, one recovers conservation laws for charge, energy and angular momentum. The structure of a surface integral can be understood most easily if one keeps in mind that in the applications, the Lagrangian $\mathcal{L}(x,y)$ as well as its jet derivatives are of short range. More precisely, these functions decay on the Compton scale m^{-1}. As a consequence, the relevant contribution to the surface layer integral is obtained when both x and y are close to the boundary of Ω. As a result, a surface layer integral can be thought of as an integral over a "time strip" of size m^{-1} (see Figure 8).

IOP Conf. Series: Journal of Physics: Conf. Series **1275** (2019) 012009 doi:10.1088/1742-6596/1275/1/012009

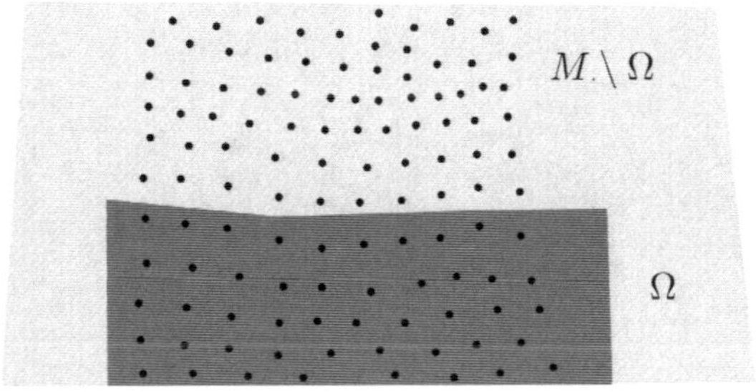

Figure 7. A surface layer integral.

There are also conservation laws which are not related to symmetries of space-time. Instead, they are a direct consequence of the linearized field equations. Again, the corresponding conserved quantities are expressed in terms of surface layer integrals. From the general class of conservation laws found in [16], the most useful surface layer integrals for the applications are the *symplectic form* σ_Ω (see [12, Sections 3.3 and 4.3])

$$\sigma_\Omega(\mathfrak{u}, \mathfrak{v}) = \int_\Omega d\rho(x) \int_{M\setminus\Omega} d\rho(y)\, \big(\nabla_{1,\mathfrak{u}}\nabla_{2,\mathfrak{v}} - \nabla_{1,\mathfrak{v}}\nabla_{2,\mathfrak{u}}\big)\, \mathcal{L}(x,y) \tag{18}$$

and the *surface layer inner product* $(.,.)_\Omega$ (see [16, Theorem 1.1])

$$(\mathfrak{u}, \mathfrak{v})_\Omega = \int_\Omega d\rho(x) \int_{M\setminus\Omega} d\rho(y)\, \big(\nabla_{1,\mathfrak{u}}\nabla_{1,\mathfrak{v}} - \nabla_{2,\mathfrak{u}}\nabla_{2,\mathfrak{v}}\big)\, \mathcal{L}(x,y)\,. \tag{19}$$

Note that the surface layer inner product is symmetric in its two arguments. In [17] it is shown by explicit computation that it is positive definite for Dirac sea configurations in Minkowski space. Therefore, it defines a scalar product on the real vector space of linearized solutions.

7. Unitary time evolution on Fock spaces

Based on the above surface layer integrals, the nonlinear dynamics as described by the causal action principle can be rewritten in terms of a unitary time evolution on Fock spaces, as we now outline (for details see [18]): Assuming that the surface layer inner product (19) is positive definite and combining it with the symplectic form (18), one can endow the space of linearized solutions with the structure of a complex Hilbert space, denoted by $(\mathfrak{h}, \langle .|.\rangle)$. The scalar product on $\mathfrak{h}$ has the property that its real part coincides with the scalar product $(.,.)_\Omega$. In the application of Dirac sea configurations in Minkowski space [17], one sees that this scalar product agrees with the usual scalar product used in quantum theory. In particular, it is compatible with the probabilistic interpretation. The conservation laws for the surface layer integrals imply that the linearized fields preserve the complex structure and the scalar product. In other words, the linearized dynamics is described by a unitary operator U on $\mathfrak{h}$ (see Figure 9).

In order to describe the interacting dynamics, one makes essential use of the perturbation theory developed in [14], which makes it possible to express the dynamics of the physical system as described by the causal action principle in terms of a nonlinear time evolution on jet spaces, which can be expanded in a linear, a quadratic, a cubic contribution, and so on. Rewriting the resulting multilinear operators as linear operators on the symmetric tensor product, one obtains a *linear dynamics on bosonic Fock spaces* (for details see [18]). Again making use of conservation laws for surface layer integrals, this dynamics can be written with a norm-preserving

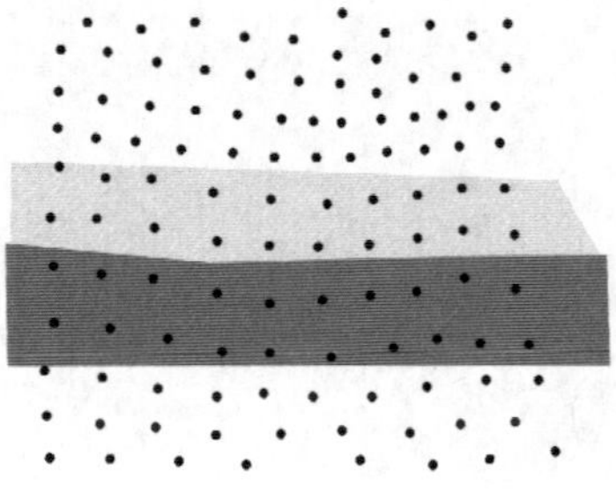

Figure 8. Localization of a surface layer integral.

complex-linear time evolution operator acting on the bosonic Fock space tensored with its dual. A *unitary time evolution* on the bosonic Fock space is obtained in the so-called holomorphic approximation. In [18] the error of this approximation is quantified, and it is argued that in the physical applications in mind, this error is very small.

8. Hyperbolicity, causation and finite propagation speed

Having surface layer integrals to our disposal, we can finally explain in which sense causation holds and what we mean by finite propagation speed (see the end of Section 4 and Figure 6). The fact that certain surface layer integrals are conserved for solutions of the linearized field equations can also be used for proving existence of solutions and for analyzing their properties. Generally speaking, one works with surface layer integrals which are positive. These positivity properties are subsumed in so-called *hyperbolicity conditions*. Then one derives *energy estimates*, which basically show that the surface layer integrals cannot grow too fast in time. Once these energy estimates have been established, one can mimic the procedure for hyperbolic partial differential equations to prove existence and uniqueness of solutions of the initial value problem. These solutions respect *causation* in the sense that the present has an influence only on the future. Moreover, one gets *finite propagation speed*. But since in the surface layer integrals the jets are integrated over regions of space-time of the size of the Compton scale, the energy methods give us information on the jet dynamics only when taking "averages" over macroscopic space-time regions. This is what at the end of Section 5 we meant by the statement that causation and finite propagation speed hold "only macroscopically." For precise mathematical statements we refer to [13].

Summarizing from a conceptual point of view, in a causal fermion system there is no strict causation. Nevertheless, the usual notions of causality, future and past are well-defined on scales larger than the Compton length. Moreover, on this macroscopic scale, the dynamics is compatible with these notions, and linearized fields propagate with finite speed.

Acknowledgments

I would like to thank the organizers of the International Workshop "DICE2018: Spacetime - Matter - Quantum Mechanics" for the kind invitation to this inspiring conference. I am grateful to Maximilian Jokel and Niky Kamran for helpful comments on the manuscript.

References

[1] Finster F 2016 *The Continuum Limit of Causal Fermion Systems* (*arXiv:1605.04742 [math-ph]*, *Fundamental Theories of Physics* vol 186) (Springer)

[2] Finster F 2011 *arXiv:0911.2102 [hep-th]*, *Lett. Math. Phys.* **97** 165–183

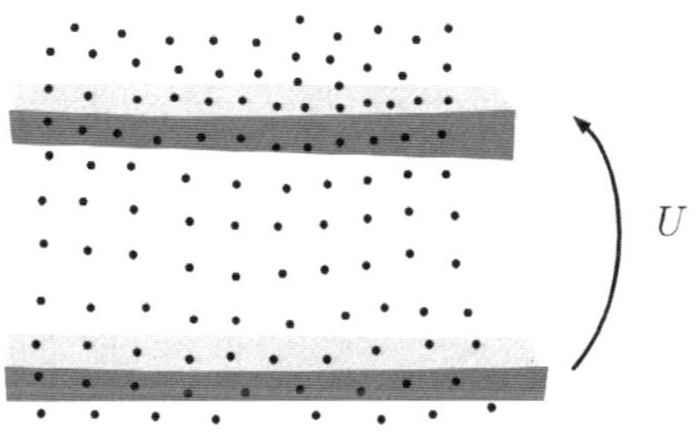

Figure 9. Unitary time evolution.

[3] Finster F 2006 *The Principle of the Fermionic Projector* (*hep-th/0001048, hep-th/0202059, hep-th/0210121, AMS/IP Studies in Advanced Mathematics* vol 35) (Providence, RI: American Mathematical Society)

[4] Finster F, Grotz A and Schiefeneder D 2012 Causal fermion systems: A quantum space-time emerging from an action principle *arXiv:1102.2585 [math-ph], Quantum Field Theory and Gravity* ed Finster F, Müller O, Nardmann M, Tolksdorf J and Zeidler E (Basel: Birkhäuser Verlag) pp 157–182

[5] Finster F and Kleiner J 2015 *arXiv:1502.03587 [math-ph], J. Phys.: Conf. Ser.* **626** 012020

[6] Finster F, Kleiner J and Treude J H *in preparation*

[7] Finster F 2007 *arXiv:math-ph/0503069, Calc. Var. Partial Differential Equations* **29** 431–453

[8] Finster F 2010 *arXiv:0811.2666 [math-ph], J. Reine Angew. Math.* **646** 141–194

[9] Bernard Y and Finster F 2014 *arXiv:1205.0403 [math-ph], Adv. Calc. Var.* **7** 27–57

[10] Finster F and Schiefeneder D 2013 *arXiv:1012.1589 [math-ph], Arch. Ration. Mech. Anal.* **210** 321–364

[11] Bäuml L, Finster F, von der Mosel H and Schiefeneder D 2018 *arXiv:1808.09754 [math.CA]*

[12] Finster F and Kleiner J 2017 *arXiv:1612.07192 [math-ph], Calc. Var. Partial Differential Equations* **56:73** 33

[13] Dappiaggi C and Finster F 2018 *arXiv:1811.10587 [math-ph]*

[14] Finster F 2017 *arXiv:1703.05059 [math-ph]*

[15] Finster F and Kleiner J 2016 *arXiv:1506.09076 [math-ph], Calc. Var. Partial Differential Equations* **55:35** 41

[16] Finster F and Kleiner J 2019 *arXiv:1801.08715 [math-ph], to appear in Calc. Var. Partial Differential Equations*

[17] Finster F 2017 *arXiv:1711.07058 [math-ph]*

[18] Finster F and Kamran N 2018 *arXiv:1808.03177 [math-ph]*

IOP Conf. Series: Journal of Physics: Conf. Series **1275** (2019) 012010 doi:10.1088/1742-6596/1275/1/012010

Steps towards Lorentzian quantum gravity with causal sets

Astrid Eichhorn

CP3-Origins, University of Southern Denmark, Campusvej 55, DK-5230 Odense M, Denmark and

Institute for Theoretical Physics, University of Heidelberg, Philosophenweg 16, 69120 Heidelberg, Germany

E-mail: eichhorn@sdu.dk

Abstract. Causal set quantum gravity is a Lorentzian approach to quantum gravity, based on the causal structure of spacetime. It models each spacetime configuration as a discrete causal network of spacetime points. As such, key questions of the approach include whether and how a reconstruction of a sufficiently coarse-grained spacetime geometry is possible from a causal set. As an example for the recovery of spatial geometry from discrete causal structure, the construction of a spatial distance function for causal sets is reviewed. Secondly, it is an open question whether the path sum over all causal sets gives rise to an expectation value for the causal set that corresponds to a cosmologically viable spacetime. To provide a tool to tackle the path sum over causal sets, the derivation of a flow equation for the effective action for causal sets in matrix-model language is reviewed. This could provide a way to coarse-grain discrete networks in a background-independent way. Finally, a short roadmap to test the asymptotic-safety conjecture in Lorentzian quantum gravity using causal sets is sketched. [1]

1. Lorentzian quantum gravity

Causal set quantum gravity is an approach to the path integral over spacetimes. Unlike many other approaches to quantum gravity, it emphasizes the spatiotemporal nature of gravity, i.e., it is intrinsically Lorentzian, and cannot be defined in a Riemannian setting. Therefore its starting point is a manifold with a Lorentzian metric, $(\mathcal{M}, g)$. Yet, due to diffeomorphism invariance not all components of the metric are actually physical. Instead, the physical content of the metric consists of the causal structure (locally corresponding to the lightcone at a point), providing the partial order relation $\prec$ ("precedes") and a local scale (encoded in the conformal factor), i.e., the local volume element ϵ. The order relation $\prec$ on physical spacetimes should satisfy

(a) transitivity $x \prec y,\, y \prec w \;\Rightarrow x \prec w$

(b) acyclicity (no closed timelike curves) $x \prec y,\, y \prec x \;\Rightarrow x = y$,

for spacetime points x, y, w. As causal order does not relate points at spacelike proper distance, it has no Riemannian equivalent and generates a partial instead of a total order in the Lorentzian case.

A manifold endowed with a causal structure and a local volume element, $(\mathcal{M}, \prec, \epsilon)$ is physically

[1] Based on a talk given at the Ninth International Workshop DICE2018 Castello Pasquini/Castiglioncello (Tuscany), September 17-21, 2018 Spacetime - Matter - Quantum Mechanics.

Content from this work may be used under the terms of the Creative Commons Attribution 3.0 licence. Any further distribution of this work must maintain attribution to the author(s) and the title of the work, journal citation and DOI.
Published under licence by IOP Publishing Ltd

IOP Conf. Series: Journal of Physics: Conf. Series **1275** (2019) 012010 doi:10.1088/1742-6596/1275/1/012010

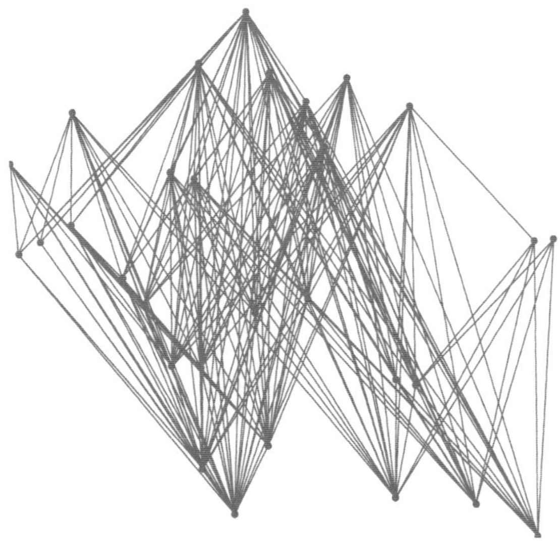

Figure 1. A sprinkling into the interval $[0,1] \times [0,1]$ in $\mathbb{M}^2$ with the red dots indicating the coordinates of the causal set elements in the embedding spacetime, and the blue lines indicating the existence of a causal relation. The embedding is not part of the fundamental definition of a causal set.

equivalent to $(\mathcal{M}, g)$, but only keeps the physical information encoded in the metric g, [1, 2], making $(\mathcal{M}, \prec, \epsilon)$ an arguably more suitable starting point for quantum gravity than $(\mathcal{M}, g)$. It underlies the causal-set approach to quantum gravity [3], see [4–9] for reviews. Causal set quantum gravity is based on the additional assumption of spatiotemporal discreteness. Accordingly, a causal set is a set of spacetime points, together with the order relation $\prec$, which in addition to the conditions (a) and (b) satisfies local finiteness

(c) $\forall x, y \in \mathcal{C}$: $\text{card}\{z \in \mathbb{C}, s.t.\ x \prec z \prec y\} < \infty$.

The latter is the discretization condition. One can impose it in two conceptually very distinct ways: One can view it as a central part of the definition of the physical theory, as is actually done in causal set quantum gravity. In this view, geometric structures at sub-discretization scales on the continuum manifold have no physical meaning, and are therefore not encoded in the corresponding causal set. Alternatively, one could regard condition (c) as a *regularization* that enables the study of the regularized gravitational path integral, e.g., using numerical simulations or matrix-model techniques. In the second setting, the continuum limit is actually the physically relevant limit, cf. Sec. 3.2.

The last missing ingredient to approximate $(\mathcal{M}, \prec, \epsilon)$ by a causal set is the discrete counterpart of the scale ϵ: A causal set approximated by a manifold encodes the information on the local scale in the number of elements: By associating a spacetime volume of Planckian size on average to each causal set element, the number of elements provides the spacetime volume in Planckian units. This gives rise to the causal-set-"slogan" that "order +number = geometry", coined by Rafael Sorkin, and underlies an argument that provided a prediction of the value of the cosmological constant [10]. Specifically, a causal set is approximated by a manifold, if the causal set has a high probability to arise through a sprinkling process. A sprinkling is a random selection of spacetime points according to a Poisson distribution. The random nature of the sprinkling is crucial to avoid a violation of Lorentz invariance despite the discreteness. An illustration of a

IOP Conf. Series: Journal of Physics: Conf. Series **1275** (2019) 012010 doi:10.1088/1742-6596/1275/1/012010

causal set approximated by a region of 1+1-dimensional Minkowski spacetime is shown in Fig. 1.

2. Key challenges of causal sets: Recovering geometry and making sense out of the path sum over all causal sets

2.1. Reconstructing geometry from causal structure

The minimalistic structure of causal sets motivates the question whether a causal set can indeed encode all sufficiently coarse-grained geometric information of a continuum spacetime. The Hawking-King-McCarthy/Malament-theorem indeed suggests that this could be the case. It applies to the continuum setting, and states that under certain causality conditions, the conformal geometry is completely determined by the causal space $(M, \prec)$. Therefore one could indeed expect that the causal relations together with the number-to-volume correspondence could be sufficient to recover a sufficiently coarse-grained geometry from a causal set. Explicit examples that information on continuum geometry can be reconstructed explicitly include the spacetime dimension [11–14], timelike and spacelike distances [15], the spatial topology [16–18], the scalar d'Alembertian [19–21] and the scalar curvature in arbitrary dimensions [22, 23].

Due to the intrinsic spatiotemporal nature of causal sets one might expect the spatial geometry in a foliated spacetime to be most challenging to reconstruct. In [24], a proposal to extract the spatial distance between points in a causal set was put forward and tested numerically. In fact, in accordance with the Hawking-King-McCarthy/Malament-theorem, the causal structure allows to recover the spatial geometry. The key idea of the construction for a continuum setting is the following:

Given two points p, q on a flat spacelike hypersurface Σ in $\mathbb{M}^d$, they have a common causal future, i.e., a set of spacetime points that is preceded by both p and q. For every point r in that set, its causal past $J^-(r) = \{s|s \prec r\}$ has a non-empty intersection with the future of Σ, $J^+(\Sigma) = \cup_{s\in\Sigma} J^+(s)$. This intersection is defined as $J(\Sigma, r) = J^+(\Sigma) \cap J^-(r)$ and has spacetime volume $V(r) = \mathrm{vol}(J(\Sigma, r))$. The intersection $J(\Sigma, r)$ essentially constitutes an inverted cone, with the tip at the point r and the base on the hypersurface Σ. This suspended spacetime volume provides a distance function on Σ. Specifically, the distance can be extracted from $V(\mathbf{r}_m)$, where $\mathbf{r}_m$ is the point that minimizes $V(r)$. For the case of d-dimensional Minkowski spacetime $\mathbb{M}^d$ with a flat hypersurface Σ, the spatial distance between two points is accordingly given by

$$\widetilde{d}(p, q) = 2\left(\frac{V(r)}{\zeta_d}\right)^{\frac{1}{d}}, \tag{1}$$

where

$$\zeta_d = \frac{\pi^{\frac{d-1}{2}}}{d\,\Gamma\left(\frac{d+1}{2}\right)}. \tag{2}$$

This construction of the spatial distance makes no reference to the induced spatial metric on the hypersurface. Instead, it is extracted from the flat four-metric, by making use of the causal structure to the future of the hypersurface [2].

In the presence of intrinsic and/or extrinsic curvature, such a direct relation between the suspended volume and the distance no longer exists, as curvature corrections have to be taken into account. Yet, at distances sufficiently smaller than the smallest curvature scale l_K, where the metric becomes locally well-approximated by the Minkowski metric, the construction works. Therefore, a distance function between widely separated p, q can still be defined using the relation Eq. (1) by summing the distances in a piecewise fashion along a path between p and q, and then minimizing over all possible paths, while limiting the maximum distance of each contributing

[2] An equivalent construction is possible by making use of the past of the spatial hypersurface Σ. In this case $\mathbf{r}_m$ constitutes the "minimal" point in the common past of two points on Σ.

IOP Conf. Series: Journal of Physics: Conf. Series **1275** (2019) 012010 doi:10.1088/1742-6596/1275/1/012010

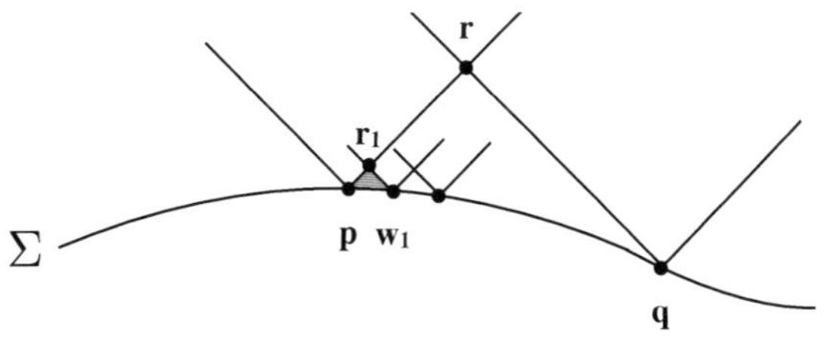

Figure 2. Illustration of an extrinsically curved hypersurface Σ with the two points $\mathbf{p}, \mathbf{q}$. There distance is not well-approximated in terms of the suspended volume $V(r)$. Instead, a piecewise construction along a path w_1... is done. For instance, the suspended volume $V(\mathbf{r}_1)$ (marked by the blue triangle) provides the correct distance between $\mathbf{p}, \mathbf{w}_1$.

piece. Specifically, we first define a predistance function $\widetilde{d}$ by using Eq. (1). In the presence of curvature, unlike in the case of a flat hypersurface embedded in $\mathbb{M}^d$, this is not yet the distance. For instance, in the presence of negative (positive) extrinsic curvature, $\widetilde{d}$ significantly underestimates (overestimates) the distance. In particular, this can lead to violations of the triangle inequality.
Accordingly we construct the spatial distance in a piecewise fashion. We define discretized paths in Σ, i.e., $W_k = (w_0, w_1, ..., w_k) \subset \Sigma$, such that $w_0 = p$, $w_k = q$. The path distance along such a path is defined as

$$d_{W_k}(p, q) = \sum_{i=0}^{k-1} \widetilde{d}(w_i, w_{i+1}). \tag{3}$$

Herein, it is crucial to impose a mesoscale cutoff ℓ, such that only such paths are allowed for which

$$\widetilde{d}(w_i, w_{i+1}) < \ell. \tag{4}$$

The mesoscale cutoff has to be chosen sufficiently smaller than all curvature scales (extrinsic and intrinsic), i.e., $\ell \ll l_K$. The distance function is the minimum of the path distance over all possible paths for which all steps satisfy Eq. (4),

$$d(p, q) = \min_{W_k} d_{W_k}(p, q) \tag{5}$$

The construction is illustrated in Fig. 2.

In summary, this provides a spatial distance purely from causal information, constituting one explicit example of the general insight of the Hawking-King-McCarthy/Malament-theorem.

This construction is easily adapted to the discrete causal-set case. The discrete analogue of a spatial hypersurface is an inextendible antichain $\mathcal{A}$, i.e., a set of unrelated elements, such that no further element can be added that is not related to an element already included in $\mathcal{A}$, [25]. The suspended volume between a point in the future of the antichain and the antichain itself can be obtained by counting the number of elements in the corresponding causal interval. This allows to directly transfer the above definitions to the causal-set case: The predistance between two elements p, q of the antichain is given by the minimum of the suspended volume, taken over all points in their common future $\mathcal{F}(p, q)$,

$$\widetilde{\mathbf{d}}(p, q) = \min_{\mathbf{e} \in \mathcal{F}(\mathbf{p}, \mathbf{q})} 2 \left(\frac{\mathbf{V}(\mathbf{e})}{\zeta_d} \right)^{\frac{1}{d}}. \tag{6}$$

IOP Conf. Series: Journal of Physics: Conf. Series **1275** (2019) 012010 doi:10.1088/1742-6596/1275/1/012010

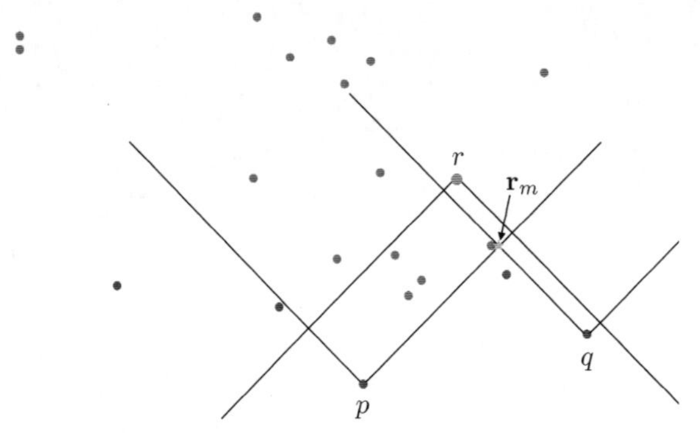

Figure 3. The antichain at the bottom of the causal set shown in Fig. 1 is marked in blue. In the continuum, the spacetime point $\mathbf{r}_m$, marked by a yellow star, would provide the distance between p and q. There is no causal-set element at this point. Instead, the causal-set element marked in green at the spacetime point r is the first in the common future of p, q. $V(r)$ is significantly larger than $V(\mathbf{r}_m)$, leading to an overestimation of the distance between p and q.

Herein, $\mathbf{V}(\mathbf{e})$ is obtained by counting causal-set elements, and the dimension d of the spacetime that the causal set embeds into must be known.

In the antichain, there is a finite number of possible paths between two elements p, q. Specifically, one can form the set $\mathcal{W}(p,q)$ of all ordered subsets χ_k of $\mathcal{A}$ that contain p and q, i.e., all paths between p and q. The predistance can be calculated between all pairs of elements in any χ_k. Summing these provides the predistance between p and q along a path χ_k, directly mimicking the continuum construction:

$$d_{\chi_k}(p,q) = \sum_{i=0}^{k-1} \widetilde{\mathbf{d}}(e_i, e_{i+1}), \tag{7}$$

where $\chi_k = \{p, e_1, ...e_{k-1}, q\}$. Subsequently, all paths are discarded where the predistance in at least one step is larger than the mesoscale cutoff as in Eq. (4), such that $\widetilde{\mathbf{d}}(e_i, e_{i+1}) < \ell$. The distance is defined by minimizing the path-distance over all remaining paths in the antichain $\mathcal{A}$, i.e.,

$$\mathbf{d}(p,q) = \min_{\chi \in \mathcal{W}(p,q)} d_{\chi_k}(p,q). \tag{8}$$

Here, it is crucial to impose the mesoscale cutoff on all pieces of the paths χ_k, otherwise the minimization can result in violations of the triangle inequality.

In the limit of infinitely densely sprinkled causal-set elements, i.e., the discreteness scale going to zero, one obviously recovers the continuum construction and therefore a calculation of the spatial distance from the information on the causal structure. At finite discreteness scale, the interplay of the mesoscale cutoff and the discreteness scale leads to an additional effect that is not present in the continuum.

This important difference between the discrete causal-set case and the continuum case is the presence of discrete asymptotic silence [27, 28]. Given a causal set that is well-approximated by $\mathbb{M}^d$, with an antichain that is well-approximated by a flat hypersurface and contains the points p and q, the causal set does almost certainly not contain an element corresponding to the point $\mathbf{r}_m$, cf. Fig. 3. Instead, the first point r in the common future of p and q lies "further in the future", i.e., its suspended volume is larger than $V(\mathbf{r}_m)$. This effect can be understood

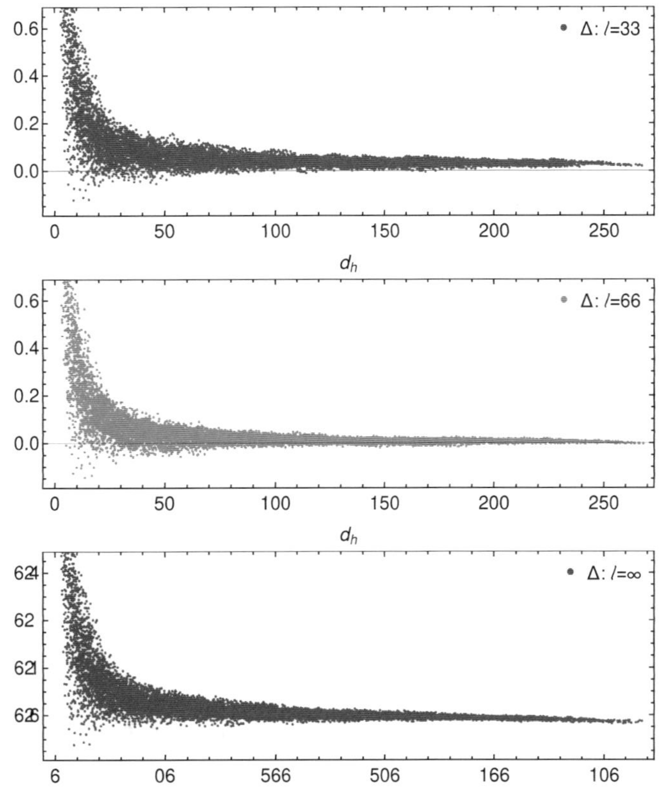

Figure 4. To check how well the construction of the discrete distance $\mathbf{d}(p,q)$ compares to the continuum quantity $d_h(p,q)$ calculated in terms of the induced metric h on Σ, we evaluate $\Delta = (\mathbf{d}(p,q) - d_h(p,q))/d_h(p,q)$ for sprinklings into $\mathbb{M}^2$. In the above example, the hypersurface Σ has constant extrinsic curvature characterized by the curvature scale $l_k = 1000$, measured in units of the discreteness scale. For small distances $d_h \leq l_{DAS}$, the distance is overestimated, $\Delta >> 0$. If the mesoscale cutoff is chosen too small, e.g., $\ell = 33$ (upper panel), all distances are overestimated as a consequence of discrete asymptotic silence, $\Delta > 0$ at all d_h. If the mesoscale cutoff is chosen too large, e.g., $\ell = \infty$ (lower panel), the distance is underestimated at larger distances due to the presence of the curvature scale. For this setting, the hierarchy of scales $l_{DAS} << \ell << l_K$, can be implemented, such that $\Delta \to 0$ for $d_h > l_{DAS}$ (middle panel). The data is taken from [24].

as a form of decoupling of spatially nearby points: the causal interval relevant for the causal-set construction, $J(\Sigma, r)$ is larger than the causal interval that provides the distance in the continuum case, $J(\Sigma, \mathbf{r}_m)$. As this is similar to the effect of asymptotic silence [26] in classical gravity in the vicinity of spacetime singularities, the effect has been called discrete asymptotic silence. It is a direct consequence of the kinematically discrete nature of a causal set and vanishes in the limit of infinite sprinkling density.
The effects of discrete asymptotic silence set in at the length scale l_{DAS} which lies about one or two orders of magnitude above the discreteness scale [28]. Accordingly, the mesoscale cutoff ℓ should not be smaller than l_{DAS}, since distances $d \lesssim l_{DAS}$ are overestimated due to discrete asymptotic silence. If the piecewise distances along a path between p and q are overestimated due to this effect, then also $\mathbf{d}(p,q)$ will be overestimated.

IOP Conf. Series: Journal of Physics: Conf. Series **1275** (2019) 012010 doi:10.1088/1742-6596/1275/1/012010

The construction of the spatial distance for causal sets accordingly requires a separation of scales according to

$$l_{DAS} << \ell << l_K. \tag{9}$$

To determine how much of a separation is needed, numerical simulations are required. They indicate that $\ell \gtrsim 60\,\ell_{\rm disc}$, where $\ell_{\rm disc}$ is the discreteness scale, is sufficient. Numerical simulations in $\mathbb{M}^d$ with $d = 2, 3$ with flat and extrinsically curved hypersurfaces in [24] confirm that the causal information in a causal set is sufficient to reconstruct the spatial distance of points in the hypersurface, cf. Fig. 4.

This provides another example that the intuition, based on the Hawking-King-McCarthy/Malament-theorem, that a causal set indeed encodes the sufficiently coarse-grained geometry of a continuum spacetime, could be justified.

2.2. The path integral for causal sets as a Renormalization Group equation

Recovering geometric information from appropriate causal sets and reconstructing a manifold is not sufficient to guarantee that the sum over all causal sets is well-defined and phenomenologically viable. This in particular requires that the expectation value taken over all causal sets provides a causal set approximated by a deSitter like spacetime, suitable for a description of our universe. A second challenge of the causal set program therefore consists in the evaluation of the path sum over all causal sets. In particular, the set of all causal sets contains many causal sets which are not approximated by a Lorentzian manifold. Examples include the so-called Kleitman-Rothschild orders [29], which consist of only three "layers", with $N/4$ elements in the upper-most and lowest layer and $N/2$ elements in the middle. Each element in the middle layer has probability $1/2$ to have share a link with each individual element in the upper and the lower layer. Clearly this structure cannot be embedded in an extended Lorentzian manifold. Entropically, i.e., based on a counting argument, these causal sets dominate the set of all causal sets starting already at relatively low numbers of elements [30]. The path-sum over all causal sets is therefore not at all guaranteed to give rise to an effective Lorentzian spacetime manifold at a sufficiently coarse-grained scale. Moreover, the existence of non-manifoldlike causal sets also makes it more challenging to evaluate the path sum, e.g., using numerical simulations, as it very significantly enlarges the configuration space that must be sampled.

In [31] it was proposed to tackle the path integral over causal sets by writing it as a specific matrix model using the link matrix. Background-independent coarse-graining techniques for matrix models were introduced in [32], see [33] for a review, and can be applied to the causal-set case. More broadly speaking, this tool constitutes a way of coarse graining causal networks, irrespective of whether these are given a spacetime interpretation. As a first step, one rewrites causal sets as a matrix model: Each causal set comes with a matrix associated to it, the link matrix [34]. Links are the irreducible relations in a causal set. For two elements e_i, e_j, the corresponding entry in the link matrix is 1 if e_i and e_j share a link, i.e., $e_i \prec e_j$ and $\{e_k \in C \,|\, e_i \prec e_k \prec e_j\} = \emptyset$. Otherwise the corresponding entry in the link matrix is zero. The link matrix can be written down once the elements in the causal set are labelled. As the label cannot carry any physical meaning, different link matrices that arise by a relabelling of the same causal set are physically equivalent. The label independence is the discrete remainder of diffeomorphism invariance.

In the path sum, one can thus sum over all link matrices (modulo relabellings) instead of all causal sets. To implement the path sum as a matrix model, also the action has to be written fully in terms of the elements of the link matrix. That this must be possible is clear, since the link matrix encodes the full structure of a causal set. To arrive at a background-independent dynamics, no other structure than the causal set itself can be used. Accordingly, nothing but the entries of the link matrix can occur in the dynamics. In particular, the number of m-paths, i.e., the number of paths with a total length of $m+1$ elements, between two elements e_i

and e_j is encoded in $(L^m)_{ij}$. To respect background-independence and discrete diffeomorphism invariance, the full dynamics can only depend on $(L^m)_{ij}$. Specifically, one can also convince oneself explicitly that, e.g., the discrete analogue of the Einstein-Hilbert action can indeed be written purely as a function of the link matrix, e.g., the two-dimensional Benincasa-Dowker action [22] evaluated on a causal set $\mathcal{C}$ with N' elements takes the form

$$S^{(2)}_{\mathrm{BD}}[\mathcal{C}] = N' - \sum_{i,j} \left(2L_{ij} - 4L_{ij}^2 + 2\left(L_{ij}^3 + \delta_{L_{ij}^2,2}\right)\right). \tag{10}$$

The generating functional to be evaluated therefore takes the form

$$Z = \int \mathcal{D}L\, e^{iS[L]}, \tag{11}$$

where the sum goes over all link matrices of size $N' \times N'$, and the measure is understood to account for the discrete relabelling symmetry. Introducing a source term J_{ij} as well as a parameter α that allows for an analytical continuation, we define

$$Z[J] = \int \mathcal{D}L\, e^{-\alpha S[L] + \sum_{i,j} J_{ij} L_{ij}}. \tag{12}$$

The key idea of Renormalization Group techniques is to not perform this integral all at once but in steps. The decomposition into stepwise integrations should be done in such a way that the effective configurations that arise during the integration procedure capture physically relevant degrees of freedom. For instance, in local quantum field theories, one decomposes the integration into such steps that field configurations are averaged over in local patches, and long-wavelength fluctuations are integrated over in the very last steps. This provides coarse-grained effective configurations, which do indeed capture the relevant physics at large scales. In the absence of a background, which is the setting for several quantum gravity models, including causal sets, such a local coarse-graining cannot be defined. Instead, the number of degrees of freedom, i.e., the causal-set size, provides a scale according to which the integration can be done in a stepwise manner. This implements a coarse-graining procedure for causal sets that is rather intuitive: At the microscopic level, one starts with causal sets of size N', and a corresponding microscopic dynamics. After a coarse-graining step, the same physics is encoded in causal sets of size $N'' < N'$ (i.e., *coarser* causal sets), and an effective dynamics that includes the effect of the degrees of freedom that have been integrated out. This procedure can be implemented in the form of a formally exact equation that can be derived in analogy to a similar construction in the continuum [35–37]: One generalizes the generating functional in Eq. (12) to a family of N-dependent generating functionals with $N < N'$ as follows:

$$Z_N[J] = \int \mathcal{D}L\, e^{-\alpha S[L] + \sum_{ij} J_{ij} L_{ij} + \frac{1}{2}\sum_{ijkl} L_{ij} R_N(i,j,k,l) L_{kl}}, \tag{13}$$

where

$$R_N(i,j,k,l) \begin{cases} > 0 \ \text{for} \ (i+j+k+l)/N < 1 \\ = 0 \ \text{for} \ (i+j+k+l)/N > 1. \end{cases} \tag{14}$$

This ensures that only entries of the link matrix with large index value are summed over, as the path sum over these entries is unchanged. In contrast, the finite value of R_N for the low-index entries strongly suppresses their contribution in the path sum. Note that according to this definition, N is an infrared cutoff scale in the generating functional, whereas N' is a UV cutoff.

In a next step, it is convenient to express the dynamics in terms of the scale-dependent effective action, which is defined via a (modified) Legendre transform,

$$\Gamma_N[\bar{L}] = \sum_{ij} J_{ij} L_{ij} - \ln Z_N[J] - \frac{1}{2} \sum_{ijkl} \bar{L}_{ij} R_N(i,j,k,l) \bar{L}_{kl}, \tag{15}$$

where $\bar{L}_{ij} = \langle L_{ij} \rangle$ is the expectation value of the link matrix, i.e., the "classical" link matrix. The advantage of the transition to the scale-dependent effective action is that a differential equation can be derived that encodes the change of Γ_N purely in terms of the second derivative of Γ_N, and no longer makes any reference to the microscopic action S. The derivation makes use of

$$\sum_{k,l} \left(\langle L_{ij} L_{kl} \rangle_N - \bar{L}_{ij} \bar{L}_{kl} \right) \left(\frac{d^2 \Gamma_N}{dL_{kl} dL_{mn}} + R_N(k,l,m,n) \right) = \delta_{im} \delta_{jn}, \tag{16}$$

which says that the (regularized) propagator of the theory is related to the inverse of the second derivative of the scale-dependent effective action. Taking a derivative of the effective action in Eq. (15) and using Eq. (16) results in

$$\partial_N \Gamma_N = \frac{1}{2} \sum_{ijkl} \partial_N R_N(i,j,k,l) \left(\frac{d^2 \Gamma_N}{dL_{kl} dL_{mn}} + R_N(k,l,m,n) \right)^{-1}. \tag{17}$$

This equation encodes how the effective dynamics changes as one integrates out entries in the link matrix, such that the physics that is encoded in the family of effective actions Γ_N stays the same, even though the degrees of freedom become coarser in each step. This is just the usual idea underlying the Renormalization Group, adapted to a setting which lacks a local notion of scale, and instead comes with a background-independent notion of scale, namely N, the number of degrees of freedom.

Note that Eq. (17) is a formal equation which has not been tested in practical calculations yet. In the following we briefly discuss to what end studies based on Eq. (17) could be useful.

Plugging in a form of the microscopic dynamics provides the initial condition that is necessary to integrate Eq. (17) and arrive at $\Gamma_{N \to 0}$. This is the full effective action that encodes the information on all correlation functions, and most importantly provides the quantum equation of motion

$$\frac{d\Gamma_{N \to 0}}{d\bar{L}_{ij}} = 0. \tag{18}$$

The solution provides the physical expectation value of L_{ij}. For causal set quantum gravity to be a candidate for a description of quantum spacetime in our universe, the expectation value should be a sprinkling into a Friedmann-Robertson-Walker spacetime with a tiny cosmological constant in units of the Planck mass. Eq. (17) could be useful to tackle this question, because it could allow to calculate (an approximation to) $\Gamma_{N \to 0}$ in practise. While the full calculation of the effective action amounts to solving the full path sum over all causal sets, Eq. (17) provides a way of tackling this by studying a differential equation (instead of a sum/integral). Specifically, starting with a form of the dynamics at some initial scale, in addition to a scale-dependence of the couplings in this dynamics, new interaction terms are expected to be generated. Approximations to $\Gamma_{N \to 0}$ are obtained by taking these terms into account in successive enlargements of the truncation, starting by keeping only those interaction terms that are already present in the microscopic dynamics. Whether such truncations converge relatively quickly in the case of causal sets is presumably linked to the question whether the coarse-graining procedure reviewed above is based on physically relevant degrees of freedom.

IOP Conf. Series: Journal of Physics: Conf. Series **1275** (2019) 012010 doi:10.1088/1742-6596/1275/1/012010

3. Discreteness, the continuum limit and asymptotic safety in causal sets

3.1. Discreteness in quantum gravity

As already highlighted, the discretization condition (c) can be understood either as part of the physical definition of the theory or as a regularization condition. The latter appears to contradict the commonly made assumption that quantum spacetime should be discrete. In particular, there are arguments, e.g., based on the finiteness[3] of black-hole entropy [43] as well as on the expected production of black holes in the measurement of Planckian distances, that quantum gravity should come with a discreteness scale which should have the simultaneous interpretation of a minimum length/mininum volume, see [44] for a review. Sometimes the presence of Landau poles in the Standard Model of particle physics is advanced as an additional argument for the existence of a fundamental minimal length scale, although this is of course only one possible manifestation of the new physics that is required to render the Standard Model ultraviolet complete. Based on this collection of arguments one might be led to expect that configurations of quantum spacetime must necessarily be discrete, and continuum approaches to quantum spacetime are automatically ruled out. This would imply that the continuum limit in causal sets cannot have a physical meaning, and is at best of purely mathematical interest.

Yet, discreteness in quantum gravity could arise in different ways. In particular, models defined in the continuum (limit) could nevertheless exhibit physical discreteness. In other words, there is no automatic connection between kinematical and physical discreteness. This is also highlighted by models which are discrete at the kinematical level but nevertheless exhibit continuous dynamics.

An example for the first point is Loop Quantum Gravity, where the physical limit is the continuum limit. Nevertheless, the area and spatial volume operators exhibit discrete spectra, at least in the kinematical Hilbert space [45–47].

Moreover, one might at a first glance not expect that a model that is defined in the continuum can exhibit a minimal length scale. Asymptotically safe quantum gravity could constitute a counterexample: It exhibits a transition scale $l_{\rm tr}$, and is scale invariant at all distance scales $l/l_{\rm tr} < 1$. This implies that all dimensionful quantities, such as the Planck length, scale canonically in that regime. Accordingly, the Planck length acts as a particular form of minimum length scale [48], as the ratio of the "resolution" scale l to the Planck length remains constant as l is lowered in the regime $l < l_{\rm tr}$.

Examples for continuous dynamics in discrete settings are given by perfect actions. These mimick continuum dynamics on a lattice even at finite lattice spacing. While these have been pioneered for non-Abelian gauge theories, [49], their application in a quantum-gravity context has been discussed, e.g., in [50]. In both cases, even though the model is kinematically discrete (i.e., each field configuration is discrete), the physics appears continuous.

Lastly, discrete configurations, i.e., kinematically discrete models could give rise to a continuum dynamics by being summed over in the path integral. For instance, in spin foams, this idea goes under the slogan that "summing = refining" [51].

These examples highlight that requiring a theory to show geometric discreteness (or conversely a geometric continuum) at the level of its physics content does *not* imply that there must be discreteness (a continuum) at the kinematical level. Accordingly, even if we aim at developing a theory of quantum gravity that shows geometric discreteness, we can still choose a route via the continuum limit and view kinematical discreteness as a regularization.

In particular, simple discrete models, where discreteness is interpreted as a cutoff to be imposed on configurations *and* dynamics could suffer from a lack of predictivity near the cutoff scale. This is similar to a generic effective field theory, the dynamics of which are only constrained by symmetry, and accordingly contain infinitely many terms. In the infrared, all but finitely

[3] Note that the entanglement entropy in QFT across the horizon is actually finite once the scale-dependence of the Newton coupling is accounted for, see, e.g., [38–42]

IOP Conf. Series: Journal of Physics: Conf. Series **1275** (2019) 012010 doi:10.1088/1742-6596/1275/1/012010

many of these terms are suppressed and have a vanishing impact on measurable quantities. Yet, at high energy scales, all infinitely many terms contribute. Therefore, infinitely many parameters, in the form of the couplings of the infinitely many interaction monomials, enter the dynamics. Then, one needs to impose a principle that provides relations between these couplings, such that only a finite number of free parameters remain. Demanding the existence of a universal[4] continuum limit is one possible way of achieving this. The reason lies in the fact that a universal continuum limit can be taken at a fixed point of the Renormalization Group flow: At such a fixed point, scale invariance is realized, allowing to "zoom in" to arbitrarily small distances. Following RG trajectories away from the fixed point (towards finite cutoff scales) is only possible along finitely many directions in coupling space. Accordingly a model which allows one to take the continuum limit is (typically) specified by only a finite number of free parameters. Conversely, if discreteness is from the outset imposed on configurations and dynamics (as opposed to being emergent in a mathematically continuous setting), then any choice in this infinite-dimensional space of models is a priori viable for the dynamics at the cutoff scale. Just as in the effective field theory setting, predictivity is (expected to be) restored at scales far below the cutoff, where most couplings are suppressed in a canonical power counting [5]. Yet, close to the cutoff scale – which in quantum gravity one would usually associate with the Planck scale – predictivity is lost, unless a new principle is found that can be imposed to provide infinitely many relations between the various couplings.
For causal sets, discreteness is imposed on the configurations. Yet, the question whether the physics arising from causal sets will exhibit a discreteness scale is not yet clarified. For instance, just as in spin foams, the summation over all discrete configurations might potentially even give rise to continuum physics. Conversely, it is not excluded that it is actually necessary to demand the existence of a continuum limit to render the path integral for causal sets predictive and fix the freedom in the choice of microscopic dynamics. This necessitates the search for a Renormalization Group fixed point in the setting where all causal sets, also all non-manifoldlike ones, are taken into account. As argued above, this does not a priori preclude the emergence of physical discreteness, e.g., in the form of discreteness at the level of geometric observables.

3.2. Sketch of a roadmap towards Lorentzian asymptotic safety from causal sets

In what follows, we view causal sets not as "fundamental" configurations of spacetime, but rather as regularizations of the underlying continuum geometry. This differs from the discussion in the previous subsection, as here we focus only on the path sum over sprinklings and exclude, e.g., KR-orders, as these are not discretizations of Lorentzian manifolds and accordingly not part of the relevant configuration space.

This setup provides a novel angle from which to explore the asymptotic-safety conjecture for quantum gravity [52, 53]. The asymptotic-safety program proposes that the path integral for local continuum quantum gravity could make sense at and beyond the Planck scale, where the usual effective-field theory treatment breaks down. The extension beyond the Planck scale becomes possible because of the interacting Reuter fixed point, which underlies a scale-invariant regime that is realized asymptotically. Intuitively speaking, this allows us to "zoom in" into the dynamics up to arbitrarily small scales without running into any inconsistencies. Compelling indications for the Reuter fixed point exist in a Riemannian setting, [54–63], see [64–68] for reviews and [69] for a recent account of quantum scale invariance. As no straightforward Wick rotation is available in quantum gravity, this does not imply the existence of an analogous fixed point in the Lorentzian case. Moreover, important questions such as whether quantum gravity breaks global symmetries due to the presence of virtual black-hole configurations [70], as well

[4] Universality in this context means that the continuum physics does not depend on the details of the regularization/discretization.

[5] Whether this expectation from local QFTs also holds in, e.g., the non-local causal-set setting is an open question.

as whether there is a UV-IR-interplay through black-hole formation, cannot be conclusively tackled in the Riemannian case, as only the spacetime exterior to the black-hole horizon can be continued analytically, see, e.g., [71]. Accordingly, exploring asymptotic safety in a Lorentzian setting is a key outstanding challenge. In continuum functional RG studies, this is technically challenging; for a first study see [72]. In [31], it was instead proposed that causal sets could serve as a framework to search for Lorentzian asymptotic safety. Specifically, sprinklings provide discretizations of all configurations of spacetime that enter the path integral. Accordingly, a regularized form of the Lorentzian path integral over all metrics is provided by the path sum over all sprinklings $\mathcal{C}_{sp}$, i.e.,

$$\int \mathcal{D}\mathcal{C}_{sp}\, e^{i\, S^{(d)}[\mathcal{C}_{sp}]} \to \int \mathcal{D}g_{\mu\nu} e^{i\, S[g_{\mu\nu}]}, \tag{19}$$

where $S^{(d)}[\mathcal{C}_{sp}]$ is a d-dimensional microscopic dynamics selected by the asymptotic-safety conjecture. If the Lorentzian path integral over all metrics is well-defined, then the continuum limit can be taken if $S^{(d)}[\mathcal{C}_{sp}]$ is used to determine the phase of configurations. This is expected to be equivalent to asymptotic safety, since the perturbative non-renormalizability of gravity prohibits a well-defined continuum limit at the free fixed point. Therefore, searching for the existence of a continuum limit in the path sum over all sprinklings provides insight into whether asymptotic safety is realized in Lorentzian gravity.

To search for such a universal continuum limit, one has (at least) two options. The first is to use the Renormalization Group equation discussed above. In analogy to matrix and tensor models, one might expect that a universal continuum limit is visible as a fixed point in the large-N limit. A fixed point at an arbitrary point in the space of couplings can be searched for with the flow equation Eq. (17), as it does not make any reference to a microscopic action any more. Instead, the flow equation encodes a local vector field in the space of couplings, as becomes clear by expanding Γ_N in terms of the couplings and the corresponding interaction monomials. A zero of this vector field corresponds to a fixed point. This fixed point ultimately determines $S^{(d)}[\mathcal{C}_{sp}]$ and can a priori lie anywhere in the space of couplings. Accordingly, for a search for Lorentzian asymptotic safety with causal sets it is most likely not sufficient to map out the phase diagram in the space spanned by the Newton coupling and the cosmological constant, i.e., the two couplings of the classical Einstein-Hilbert action. Instead, one is presumably required to include further couplings, corresponding to higher-order terms in the gravitational dynamics. For instance, the construction in [22] can presumably be generalized to provide the causal-set analogue of $\int d^dx\, \sqrt{-g}\, (R^2 + \Box R)$ as well as higher orders in the curvature. As a first step towards asymptotic safety, one could simply work in a theory space spanned by all possible functions of the elements of the link matrix. Their geometric interpretation need not be known in order to determine whether Lorentzian gravity is asymptotically safe.

An important advantage of causal sets in contrast to a discretization in the form of a regular lattice is that a random sprinkling does not break Lorentz-invariance as it does not select a preferred frame. This presumably has important implications for the continuum limit, as typically it requires the tuning of additional parameters to recover symmetries broken by a lattice discretization. In a Renormalization-Group language, the breaking of symmetries leads to an enlarged theory space in which fixed points acquire additional relevant directions.

As an alternative tool to search for a universal scaling regime, one can evaluate the path sum over all sprinklings by Monte Carlo simulations. Just as the use of the flow equation, this requires to introduce a parameter in the generating functional that allows for an analytical continuation. The search for a universal continuum limit proceeds by mapping out the phase diagram in terms of the microscopic couplings. The existence of a universal continuum limit implies the existence of a second-order phase transition in the space of couplings. In the vicinity of a second-order phase transition, a universal scaling regime sets in, linking the physics in the

vicinity of the phase transition to fixed-point behavior.
Although the corresponding studies are based on a different configuration space, it can be viewed as an encouraging result that [73] finds a second-order phase transition. This would enable a continuum limit in the restricted class of two-orders, which are a subset of all causal sets. Moreover, the addition of a simple toy model of matter leads to additional second-order phase transitions [74]. Extensions of these results to larger classes of causal sets might be feasible making use of the algorithms developed in [75]. First steps along a similar direction, exploring asymptotic safety numerically using discrete random graphs have been made in [76].

This example highlights that, as in many other difficult questions in quantum gravity, there could be significant untapped potential hiding in the unexplored connections between different approaches to quantum gravity.

Acknowledgments
I thank R. Sorkin, S. Surya and F. Versteegen for helpful discussions, S. Surya and F. Versteegen for collaboration on [24], and the organizers of the Ninth International Workshop DICE2018 Castello Pasquini/Castiglioncello (Tuscany), Spacetime - Matter - Quantum Mechanics for the invitation. I acknowledge support by the DFG through an Emmy-Noether fellowship under grant no. Ei/1037-1.

References

[1] S. W. Hawking, A. R. King and P. J. Mccarthy, J. Math. Phys. **17**, 174 (1976). doi:10.1063/1.522874
[2] D. B. Malament, J. Math. Phys. **18**, 1399-1404 (1977)
[3] L. Bombelli, J. Lee, D. Meyer and R. Sorkin, Phys. Rev. Lett. **59**, 521 (1987).
[4] R. D. Sorkin, gr-qc/0309009.
[5] F. Dowker, doi:10.1142/9789812700988-0016 gr-qc/0508109.
[6] J. Henson, In *Oriti, D. (ed.): Approaches to quantum gravity* 393-413 [gr-qc/0601121].
[7] J. Henson, arXiv:1003.5890 [gr-qc].
[8] S. Surya, arXiv:1103.6272 [gr-qc], in Recent Research in Quantum Gravity, edited by A. Dasgupta (Nova Science Publishers NY).
[9] F. Dowker, Gen. Rel. Grav. **45**, no. 9, 1651 (2013). doi:10.1007/s10714-013-1569-y
[10] Rafael D. Sorkin. Spacetime and Causal Sets - World Scientific. in J.C. D'Olivo, E. Nahmad-Achar, M. Rosenbaum, M.P. Ryan, L.F. Urrutia and F. Zertuche (eds.) Relativity and Gravitation: Classical and Quantum (Proceedings of the SILARG VII Conference, held Cocoyoc, Mexico, December, 1990), pages 150-173 Singapore, 1991). http://128.230.72.124/ sorkin/some.papers/66.cocoyoc.pdf.
[11] J. Myrheim, CERN-TH-2538.
[12] D. A. Meyer, PhD thesis, M.I.T., (1988).
[13] D. D. Reid, Phys. Rev. D **67** (2003) 024034 [gr-qc/0207103].
[14] L. Glaser and S. Surya, Phys. Rev. D **88**, no. 12, 124026 (2013) doi:10.1103/PhysRevD.88.124026 [arXiv:1309.3403 [gr-qc]].
[15] D. Rideout and P. Wallden, Class. Quant. Grav. **26**, 155013 (2009) doi:10.1088/0264-9381/26/15/155013 [arXiv:0810.1768 [gr-qc]].
[16] S. Major, D. Rideout and S. Surya, J. Math. Phys. **48**, 032501 (2007) doi:10.1063/1.2435599 [gr-qc/0604124].
[17] S. Major, D. Rideout and S. Surya, Class. Quant. Grav. **26**, 175008 (2009) doi:10.1088/0264-9381/26/17/175008 [arXiv:0902.0434 [gr-qc]].
[18] R. D. Sorkin, Y. K. Yazdi and N. Zwane, arXiv:1811.09765 [gr-qc].
[19] D. M. T. Benincasa, F. Dowker and B. Schmitzer, Class. Quant. Grav. **28**, 105018 (2011) doi:10.1088/0264-9381/28/10/105018 [arXiv:1011.5191 [gr-qc]].
[20] S. Aslanbeigi, M. Saravani and R. D. Sorkin, JHEP **1406**, 024 (2014) doi:10.1007/JHEP06(2014)024 [arXiv:1403.1622 [hep-th]].
[21] A. Belenchia, D. M. T. Benincasa and F. Dowker, Class. Quant. Grav. **33**, no. 24, 245018 (2016) doi:10.1088/0264-9381/33/24/245018 [arXiv:1510.04656 [gr-qc]].
[22] D. M. T. Benincasa and F. Dowker, Phys. Rev. Lett. **104**, 181301 (2010) doi:10.1103/PhysRevLett.104.181301 [arXiv:1001.2725 [gr-qc]].
[23] F. Dowker and L. Glaser, Class. Quant. Grav. **30**, 195016 (2013) doi:10.1088/0264-9381/30/19/195016 [arXiv:1305.2588 [gr-qc]].

[24] A. Eichhorn, S. Surya and F. Versteegen, arXiv:1809.06192 [gr-qc].
[25] S. Major, D. Rideout and S. Surya, Class. Quant. Grav. **23**, 4743 (2006) doi:10.1088/0264-9381/23/14/011 [gr-qc/0506133].
[26] V. A. Belinsky, I. M. Khalatnikov and E. M. Lifshitz, Adv. Phys. **19**, 525 (1970). doi:10.1080/00018737000101171
[27] S. Carlip, Class. Quant. Grav. **32**, no. 23, 232001 (2015) doi:10.1088/0264-9381/32/23/232001 [arXiv:1506.08775 [gr-qc]].
[28] A. Eichhorn, S. Mizera and S. Surya, Class. Quant. Grav. **34**, no. 16, 16LT01 (2017) doi:10.1088/1361-6382/aa7d1b [arXiv:1703.08454 [gr-qc]].
[29] D. Kleitman and B. Rothschild, Trans. Amer. Math. Society **205**, 205 (1975).
[30] J. Henson, D. P. Rideout, R. D. Sorkin and S. Surya, arXiv:1504.05902 [math.CO].
[31] A. Eichhorn, Class. Quant. Grav. **35**, no. 4, 044001 (2018) doi:10.1088/1361-6382/aaa0a3 [arXiv:1709.10419 [gr-qc]].
[32] A. Eichhorn and T. Koslowski, Phys. Rev. D **88**, 084016 (2013) doi:10.1103/PhysRevD.88.084016 [arXiv:1309.1690 [gr-qc]].
[33] A. Eichhorn, T. Koslowski and A. D. Pereira, arXiv:1811.12909 [gr-qc].
[34] S. P. Johnston, arXiv:1010.5514 [hep-th].
[35] C. Wetterich, Phys. Lett. B **301**, 90 (1993) doi:10.1016/0370-2693(93)90726-X [arXiv:1710.05815 [hep-th]].
[36] U. Ellwanger, Z. Phys. C **62**, 503 (1994) doi:10.1007/BF01555911 [hep-ph/9308260].
[37] T. R. Morris, Int. J. Mod. Phys. A **9**, 2411 (1994) doi:10.1142/S0217751X94000972 [hep-ph/9308265].
[38] L. Susskind and J. Uglum, Phys. Rev. D **50**, 2700 (1994) doi:10.1103/PhysRevD.50.2700 [hep-th/9401070].
[39] T. Jacobson, gr-qc/9404039.
[40] J. G. Demers, R. Lafrance and R. C. Myers, Phys. Rev. D **52**, 2245 (1995) doi:10.1103/PhysRevD.52.2245 [gr-qc/9503003].
[41] F. Larsen and F. Wilczek, Nucl. Phys. B **458**, 249 (1996) doi:10.1016/0550-3213(95)00548-X [hep-th/9506066].
[42] C. Pagani and M. Reuter, JHEP **1807**, 039 (2018) doi:10.1007/JHEP07(2018)039 [arXiv:1804.02162 [gr-qc]].
[43] R. D. Sorkin, Stud. Hist. Phil. Sci. B **36**, 291 (2005) doi:10.1016/j.shpsb.2005.02.002 [hep-th/0504037].
[44] S. Hossenfelder, Living Rev. Rel. **16**, 2 (2013) doi:10.12942/lrr-2013-2 [arXiv:1203.6191 [gr-qc]].
[45] C. Rovelli and L. Smolin, Nucl. Phys. B **442**, 593 (1995) Erratum: [Nucl. Phys. B **456**, 753 (1995)] doi:10.1016/0550-3213(95)00150-Q, 10.1016/0550-3213(95)00550-5 [gr-qc/9411005].
[46] A. Ashtekar and J. Lewandowski, Class. Quant. Grav. **14**, A55 (1997) doi:10.1088/0264-9381/14/1A/006 [gr-qc/9602046].
[47] A. Ashtekar and J. Lewandowski, Adv. Theor. Math. Phys. **1**, 388 (1998) doi:10.4310/ATMP.1997.v1.n2.a8 [gr-qc/9711031].
[48] R. Percacci and G. P. Vacca, Class. Quant. Grav. **27**, 245026 (2010) doi:10.1088/0264-9381/27/24/245026 [arXiv:1008.3621 [hep-th]].
[49] P. Hasenfratz and F. Niedermayer, Nucl. Phys. B **414**, 785 (1994) doi:10.1016/0550-3213(94)90261-5 [hep-lat/9308004].
[50] B. Bahr and B. Dittrich, Phys. Rev. D **80**, 124030 (2009) doi:10.1103/PhysRevD.80.124030 [arXiv:0907.4323 [gr-qc]].
[51] C. Rovelli and M. Smerlak, Class. Quant. Grav. **29**, 055004 (2012) doi:10.1088/0264-9381/29/5/055004 [arXiv:1010.5437 [gr-qc]].
[52] S. Weinberg, *In *Hawking, S.W., Israel, W.: General Relativity*, 790-831* (Cambridge University Press, Cambridge, 1980).
[53] M. Reuter, Phys. Rev. D **57**, 971 (1998).
[54] M. Reuter and F. Saueressig, Phys. Rev. D **65**, 065016 (2002) doi:10.1103/PhysRevD.65.065016 [hep-th/0110054].
[55] D. F. Litim, Phys. Rev. Lett. **92**, 201301 (2004) doi:10.1103/PhysRevLett.92.201301;
[56] A. Codello, R. Percacci and C. Rahmede, Annals Phys. **324**, 414 (2009) doi:10.1016/j.aop.2008.08.008 [arXiv:0805.2909 [hep-th]].
[57] D. Benedetti, P. F. Machado and F. Saueressig, Mod. Phys. Lett. A **24**, 2233 (2009);
[58] K. Falls, D. F. Litim, K. Nikolakopoulos and C. Rahmede, arXiv:1301.4191 [hep-th];
[59] P. Donà, A. Eichhorn and R. Percacci, Phys. Rev. D **89**, no. 8, 084035 (2014) doi:10.1103/PhysRevD.89.084035 [arXiv:1311.2898 [hep-th]].
[60] D. Becker and M. Reuter, Annals Phys. **350**, 225 (2014) doi:10.1016/j.aop.2014.07.023;
[61] H. Gies, B. Knorr, S. Lippoldt and F. Saueressig, Phys. Rev. Lett. **116**, no. 21, 211302 (2016);
[62] T. Denz, J. M. Pawlowski and M. Reichert, Eur. Phys. J. C **78**, no. 4, 336 (2018) doi:10.1140/epjc/s10052-018-5806-0 [arXiv:1612.07315 [hep-th]].

[63] N. Christiansen, K. Falls, J. M. Pawlowski and M. Reichert, Phys. Rev. D **97**, no. 4, 046007 (2018) doi:10.1103/PhysRevD.97.046007 [arXiv:1711.09259 [hep-th]].
[64] M. Niedermaier and M. Reuter, Living Rev. Rel. **9**, 5 (2006).
[65] M. Reuter and F. Saueressig, New J. Phys. **14**, 055022 (2012).
[66] R. Percacci, "An Introduction to Covariant Quantum Gravity and Asymptotic Safety," World Scientific Publishing (2017).
[67] A. Eichhorn, Found. Phys. **48**, no. 10, 1407 (2018) doi:10.1007/s10701-018-0196-6 [arXiv:1709.03696 [gr-qc]].
[68] A. Eichhorn, arXiv:1810.07615 [hep-th].
[69] C. Wetterich, arXiv:1901.04741 [hep-th].
[70] R. Kallosh, A. D. Linde, D. A. Linde and L. Susskind, Phys. Rev. D **52**, 912 (1995) doi:10.1103/PhysRevD.52.912 [hep-th/9502069].
[71] A. Baldazzi, R. Percacci and V. Skrinjar, arXiv:1811.03369 [gr-qc].
[72] E. Manrique, S. Rechenberger and F. Saueressig, Phys. Rev. Lett. **106**, 251302 (2011) doi:10.1103/PhysRevLett.106.251302 [arXiv:1102.5012 [hep-th]].
[73] S. Surya, Class. Quant. Grav. **29**, 132001 (2012) doi:10.1088/0264-9381/29/13/132001 [arXiv:1110.6244 [gr-qc]].
[74] L. Glaser, Class. Quant. Grav. **35**, no. 8, 084001 (2018) doi:10.1088/1361-6382/aab139 [arXiv:1802.02519 [gr-qc]].
[75] W. J. Cunningham and D. Krioukov, Comput. Phys. Commun. **233**, 123 (2018) doi:10.1016/j.cpc.2018.06.008 [arXiv:1709.03013 [gr-qc]].
[76] C. Kelly and C. A. Trugenberger, arXiv:1811.12905 [hep-th].

IOP Conf. Series: Journal of Physics: Conf. Series **1275** (2019) 012011 doi:10.1088/1742-6596/1275/1/012011

Singularity avoidance in anisotropic quantum cosmology

Claus Kiefer and Nick Kwidzinski

Institut für Theoretische Physik, Universität zu Köln, Zülpicher Straße 77, 50937 Köln, Germany

Abstract. We discuss the fate of the classical singularities in quantum cosmological models. We state our criteria of singularity avoidance and apply them to Friedmann-Lemaître models models as well as, in more detail, to the anisotropic case of a Bianchi I universe. We find that the classical singularities are generally avoided in these cases.

1. Introduction

A major issue in any quantum theory is the fate of the classical singularities. Since so far no final theory of quantum gravity exists, this problem can only be addressed within particular approaches [1]. In our contribution, we shall address singularity avoidance in the framework of canonical (metrical) quantum general relativity (quantum geometrodynamics).

In Section 2, we state and discuss our general criteria for singularity avoidance. We employ a generalization of the DeWitt criterion that demands the quantum gravitational state to approach zero in the region of the classical singularity. In Section 3, we briefly review the situation for Friedmann-Lemaître models. In Section 4, we extend the disucssion to an anisotropic model – the Bianchi I universe. Sections 2 and 4 are based on our recent paper [2]. Section 5 contains our conclusions and an outlook.

2. Criteria for singularity avoidance

It is well known that singularities are ubiquitous in general relativity. In their general form, the singularity theorems do not display the physical nature of singularity. They usually state incompleteness for timelike or null geodesics. Since cases of physical importance mostly deal with curvature singularities, it seems appropriate to include them explicitly in the definition. In the words of Ellis *et al.* ([3], p. 145):

> We shall define a singularity as a boundary of spacetime where either the curvatures diverges ... or geodesic incompleteness occurs.

We shall adopt this definition here. (Scalar) curvature singularities are either Ricci tensor divergences or Weyl tensor divergences or a mixture of the two. The first are related to unbounded matter (or field) densities, while the second are related to gravitational field divergences in the vacuum. Both cases will be of relevance here.

In general relativity, singularity theorems can be formulated and proved in a mathematically rigorous manner. This is possible because the theory itself has a sound mathematical foundation. This is not the case in quantum gravity. First, there exist various approaches to such a theory, without any consensus so far [1]. Second, even within particular approaches, the mathematical foundation is limited. Nevertheless, it is of importance to study the fate of classical singularities in quantum gravity approaches, even if this is only possible *heuristically*. The reason for this is to gain hints on the viability of approaches and to get an idea about the conceptual nature of singularity avoidance.

Content from this work may be used under the terms of the Creative Commons Attribution 3.0 licence. Any further distribution of this work must maintain attribution to the author(s) and the title of the work, journal citation and DOI.
Published under licence by IOP Publishing Ltd

IOP Conf. Series: Journal of Physics: Conf. Series **1275** (2019) 012011 doi:10.1088/1742-6596/1275/1/012011

Our approach here is quantum geometrodynamics. This is, in a sense, the most conservative approach, because its central equations (Wheeler-DeWitt equation and quantum momentum constraints) are obtained by formulating quantum wave equations that lead to the Einstein equations[1] in the semiclassical (WKB) limit. The kinematic quantity here is the wave functional of all three-geometries (and non-gravitational fields). This configuration space is called superspace.

In his pioneering paper on canonical quantum gravity, DeWitt came up with a heuristic proposal for singularity avoidance [4]. Assuming that the wave functional Ψ in quantum gravity is at least loosely related to the probability interpretation in ordinary quantum theory, he suggested to demand that $\Psi = 0$ in regions where classical singularities are situated. He writes ([4], p. 1129):

> *Provided it does not turn out to be ultimately inconsistent*, this condition,[2] ... yields two important results. Firstly, it makes the probability amplitude for catastrophic 3-geometries vanish, and hence gets the physicist out of his classical predicament. Secondly, it may permit the Cauchy problem for the "wave equation" ... to be handled in a manner very similar to that of the ordinary Schrödinger equation.

In his second point, DeWitt expresses the hope that $\Psi = 0$ together with the specification of Ψ on *one* three-geometry (instead of two, as is natural for a second-order equation) suffices to determine Ψ for all three-geometries. This is, however, mathematically far from clear. It is even imaginable that only the trivial solutions remains from these two boundary conditions. The specification of Ψ on one boundary only is related, in spirit, to the no-boundary proposal by Hartle and Hawking (see e.g. [1], p. 279).

The DeWitt condition has frequently been used in the context of Friedmann-Lemaître cosmology, where the configuration space is finite-dimensional and called minisuperspace [1, 6]. It also arises as a *consequence* of unitary evolution for quantum collapsing dust shells; see, for example, [5] in which it is shown how a wave packet representating a collapsing shell will develop into a superposition of collapsing and expanding shell, thereby avoiding the black and while hole singularities.

In the rest of this section, we shall formulate the DeWitt criterion for a general anisotropic model in quantum cosmology. In this, we make use of the fact that the quantum cosmological wave function is, in fact, defined over a conformal manifold, which leads us to a generalization of the original DeWitt criterion [2]. This generalization becomes relevant for models with dimension of configuration space greater than two; see below the example for a Bianchi I universe.

To be specific, let us consider the action for general (diagonal) Bianchi class A models with unspecified matter part (see for the following [2] and the references therein):

$$S_{\text{EH}} + S_{\text{m}} = \int \text{d}t\; N\text{e}^{3\alpha} \left[\frac{-\dot{\alpha}^2 + \dot{\beta}_+^2 + \dot{\beta}_-^2}{2N^2} + \frac{{}^{(3)}R}{12} \right] + S_{\text{m}}. \tag{1}$$

The minisuperspace, $\mathcal{M}$, of these models is parametrized here by the coordinates $\mathbf{q} = \{\alpha, \beta_+, \beta_-, \phi\}$, where $\alpha \equiv \ln a$, β_+, and β_- are the 'Misner variables', and ϕ denotes matter field degrees of freedom; ${}^{(3)}R$ denotes the three-dimensional Ricci scalar. Units are chosen such that $3c^6 V_0/4\pi G = 1$, where V_0 is the volume of three-dimensional space (assumed to be compact here).

The central equation of the canonical theory is the Hamiltonian constraint, which follows from (1) by variation with respect to the lapse function N,

$$\mathcal{H} = \frac{1}{2}\mathcal{G}^{IJ} p_I p_J + \mathcal{V} = 0. \tag{2}$$

Here, $\mathcal{G}_{IJ}$ are the components of the DeWitt metric, $\mathcal{G}^{IJ}$ the components of its inverse, and $\mathcal{V}$ is the minisuperspace potential which contains contributions from the three-curvature and from the matter part. The p_I are the momenta canonically conjugate to the $\mathbf{q}$ above.

[1] More precisely, the four Einsteins equations which are constraints.
[2] He refers to $\Psi = 0$.

IOP Conf. Series: Journal of Physics: Conf. Series **1275** (2019) 012011 doi:10.1088/1742-6596/1275/1/012011

Under a re-scaling of the lapse, $N \to \widetilde{N} = \Omega^2 N$, the transformation of the Hamiltonian constraint follows from the invariance of the total Hamiltonian H,

$$H = N\mathcal{H} = \widetilde{N}\widetilde{\mathcal{H}} = \widetilde{N}\left(\frac{1}{2}\Omega^{-2}\mathcal{G}^{IJ}p_I p_J + \Omega^{-2}\mathcal{V}\right) \equiv \widetilde{N}\left(\frac{1}{2}\widetilde{\mathcal{G}}^{IJ}p_I p_J + \widetilde{\mathcal{V}}\right). \tag{3}$$

For this reason, we can interpret the minisuperspace as a *conformal manifold* $(\mathcal{M}, [\mathcal{G}_{IJ}\mathrm{d}q^I \otimes \mathrm{d}q^J])$. This conformal structure motivates the choice of a factor ordering that renders the Wheeler–DeWitt equation conformally covariant. Explicitly,

$$\left[-\frac{\hbar^2}{2}(\Box - \xi\mathcal{R}) + \mathcal{V}\right]\Psi = 0, \tag{4}$$

where $\mathcal{R}$ denotes the Ricci scalar constructed from $\mathcal{G}_{IJ}$ and $\xi = \frac{d-2}{4(d-1)}$, with $d = \dim(\mathcal{M})$. For the equation to be conformally covariant the wave function has to transform as $\Psi \to \widetilde{\Psi} = \Omega^{-\frac{d-2}{2}}\Psi$. This is why the original DeWitt criterion ($\Psi \to 0$) is only a good criterion when dealing with two-dimensional minisuperspaces.

We are now interested in the formulation of a conformally invariant DeWitt criterion. For this we note that the following scalar density is invariant under a conformal transformation in minisuperspace:

$$\star|\Psi|^{\frac{2d}{d-2}} = |\Psi|^{\frac{2d}{d-2}}\mathrm{dvol}. \tag{5}$$

Here, dvol contains the square root of the (absolute value of the) determinant of the DeWitt metric, and $\star$ denotes the Hodge star. The generalized DeWitt criterion then reads:

A singularity is said to be avoided if $\star|\Psi|^{\frac{2d}{d-2}} \to 0$ *in the vicinity of the singularity.*

We emphasize that this criterion, as the original one, is a sufficient criterion only, not a necessary one. As one knows from quantum mechanics (e.g. from the solutions for the energy eigenstates of the hydrogen atom from the Dirac equation), one can have singularity avoidance even for diverging wave functions.

One can think of other criteria. In the first paper along this direction, [7], singularity avoidance was infered from the breakdown of the semiclassical approximation when approaching the region of the singularity. Then, the notion of a classical trajectory loses its meaning and the classical theorems cannot be applied. In a sense, the DeWitt criterion also obeys this criterion, because the vanishing of the wave function signals the breakdown of the semiclassical approximation.

In [2], we have also applied the criterion that implies singularity avoidance if the Klein-Gordon flux vanishes in the vicinity of the classical singularity[3]. In the explicit examples discussed there, this leads in general to results different from the DeWitt criterion. Because some aspects of the Klein-Gordon flux are problematic in this context (e.g. it is not positive definite), we shall not consider this criterion here. It must, however, be remarked that the validity of the DeWitt criterion usually demands the superposition of different semiclassical components for the Klein-Gordon current (see Eq. (6.29) in [4]).

3. Singularity avoidance in Friedmann-Lemaître cosmology

Friedmann-Lemaître models are homogeneous and isotropic. From the gravitational side, the scale factor $a(t)$ is the only dynamical quantity. In order to represent matter and render the model non-trivial, usually a scalar field $\phi(t)$ with an appropriate potential $V(\phi)$ is added. We thus have a two-dimensional minisuperspace. The metric is given by

$$\mathrm{d}s^2 = -N^2(t)\mathrm{d}t^2 + a^2(t)\mathrm{d}\Omega_3^2, \tag{6}$$

[3] This was motivated by the fact that the Klein-Gordon current is a conformally invariant $(d-1)$-form.

where $\mathrm{d}\Omega_3^2$ is the three-dimensional spatial line element (here chosen to be the three-sphere). After quantization, the Wheeler-DeWitt equation reads

$$\frac{1}{2}\left(\frac{\hbar^2}{a^2}\frac{\partial}{\partial a}\left(a\frac{\partial}{\partial a}\right)-\frac{\hbar^2}{a^3}\frac{\partial^2}{\partial\phi^2}-a+\frac{\Lambda a^3}{3}+2a^3V(\phi)\right)\Psi(a,\phi)=0. \tag{7}$$

Here, the Laplace-Beltrami factor ordering is chosen, which for a two-dimensional configuration space is equivalent to the above conformal factor ordering.

The standard singularities in general relativity are big bang and (for a recollapsing universe) big crunch. Motivated by the observation of a currently accelerating universe, one can introduce a scalar field with a suitable potential as the origin of such an acceleration. The presence of such a 'dark energy' can lead to further singularities, such as big rip, big brake, big freeze, and others; see, for example, [8] and the references therein. In a series of papers (see e.g. [9] and the references therein) it was possible to investigate the applicability of the DeWitt criterion in detail. The main result is that this is in generally possible: the Wheeler-DeWitt equation possesses solutions with $\Psi = 0$ in the region of the classical singularity. Let us quote one specific example. In [10], the avoidance of a classical big-brake singularity[4] was demonstrated. The result is illustrated in Figure 1.

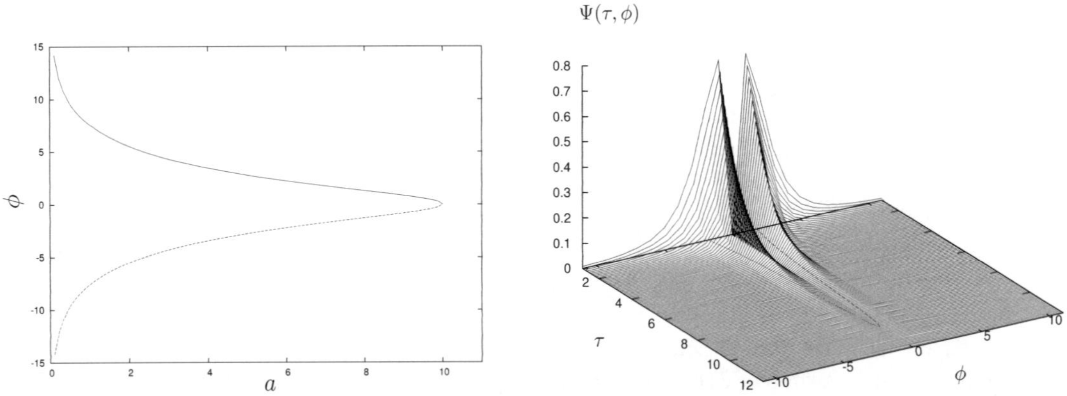

(a) Classical trajectory. From [10]. (b) Wave packet; here, $\tau = a^6$. From [10].

Figure 1: The avoidance of a classical singularity in quantum cosmology.

The left part (a) shows the classical trajectory $\phi(a)$ in configuration space. The big brake occurs at $\phi = 0$. The right part b) shows a plot of an exact solution of the Wheeler-DeWitt equation close to $\phi = 0$; this solution goes to zero there and thus satisfies the DeWitt criterion.

4. Singularity avoidance in Bianchi I quantum cosmology

The simplest anisotropic model is the vacuum Bianchi I model, also called Kasner universe. The metric can be written as

$$\mathrm{d}s^2 = -\mathrm{d}t^2 + t^{2p_x}\mathrm{d}x^2 + t^{2p_y}\mathrm{d}y^2 + t^{2p_z}\mathrm{d}z^2 \quad \text{with} \tag{8}$$

$$p_x^2 + p_y^2 + p_z^2 = 1 \quad \text{and} \quad p_x + p_y + p_z = 1. \tag{9}$$

[4] In a big-brake singularity the universe comes to an abrupt halt in the future, with the density ρ finite, but the pressure p diverging.

Details on this model (and other anisotropic models) can be found, for example, in [3]. In the Kasner universe one has a singularity which can be described by the divergence of the Weyl squared scalar. After quantization, one obtains the Wheeler-DeWitt equation, which in the above conformal factor ordering reads (with $\hbar = 1$),

$$\left[-\frac{\partial^2}{\partial\alpha^2} + \frac{\partial^2}{\partial\beta_+^2} + \frac{\partial^2}{\partial\beta_-^2}\right]\Psi(\alpha, \beta_+, \beta_-) = 0. \tag{10}$$

This is identical, in its form, to the classical wave equation in $d = 1 + 2$ dimensions. Using decay rate estimates for such an equation, one finds that the DeWitt criterion is fulfilled for a small universe ($\alpha \to -\infty$) as well as for a big universe ($\alpha \to +\infty$):

$$\star|\Psi|^6 \to 0 \quad \text{as } \alpha \to \pm\infty. \tag{11}$$

The initial singularity is thus avoided by this criterion. The decay of the amplitude $|\Psi|$ is induced by a spreading of the wave packet as shown in Figure 2. It is at first glance surprising that this criterion applies

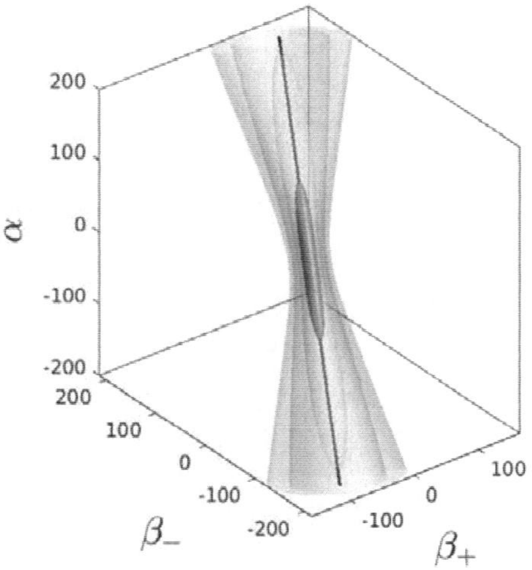

Figure 2: Plot of equipotential surfaces of $|\Psi|$ for a wave packet Ψ solving the Wheeler-DeWitt equation (10). The thin black line is the classical trajectory.

to the future evolution, too: whereas classically the universe expands forever, quantum cosmologically the wave packets cease to evolve, and the semiclassical approximation breaks down. This demonstrates that quantum effects can, in principle, occur at any scale, as a mere consequence of the superposition principle. Other examples are the above mentioned quantum effects near the big brake as well as quantum effects near the turning point of a classically recollapsing universe [11].

Let us now add matter to this model. The simplest way is to treat matter by an effective potential only, without considering a kinetic term for it [2]. Using the ansatz for a barotropic fluid, one finds that the situation for a small universe is similar to the quantum Kasner model, and that the DeWitt criterion holds as before: the initial singularity is avoided. For a big universe, one can have classically an ever expanding universe or a big-rip singularity, depending on the exact form of the equation of state. In both cases, the DeWitt criterion holds, so the wave packets do not reflect either of these two cases. A plot of a corresponding wave packet is shown in Figure 3.

At a fundamental level, matter should be described dynamically, for example by using a scalar field. Let us here only address the case of a phantom field, that is, a field with negative kinetic term that can mimic dark energy [2]. Classically, this leads to a big-rip singularity in the future (universe expanding

IOP Conf. Series: Journal of Physics: Conf. Series **1275** (2019) 012011 doi:10.1088/1742-6596/1275/1/012011

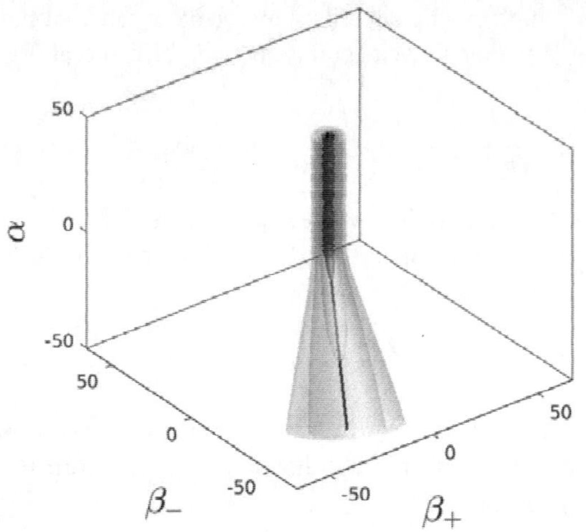

Figure 3: Plot of equipotential surfaces of $|D^{-1/2}\Psi|$ for a wave packet Ψ solving the Wheeler-DeWitt equation with an effective dust potential. The factor $D^{-1/2}$ provides a suitable rescaling for the purpose of visualization. The thin black line is the classical trajectory, which describes an isotropizing Bianchi I universe. The wave packet spreads in the region close to the big-bang singularity similarly to the Kasner model.

to infinity in a finite time). Unlike the situation shown in Figure 1, the field ϕ diverges in this limit. The conformally covariant Wheeler–DeWitt equation in the limit approaching the big rip is found to read

$$\left[\frac{\partial^2}{\partial\alpha^2} - \frac{\partial^2}{\partial\beta_+^2} - \frac{\partial^2}{\partial\beta_-^2} + \frac{\partial^2}{\partial\phi^2} + V_0 e^{6(\alpha+|\phi|)}\right]\Psi(\alpha,\beta_+,\beta_-,\phi) = 0. \tag{12}$$

A detailed analysis of this equation and its solutions reveals that the DeWitt criterion is fulfilled and the big rip is thus avoided [2].

Figure 4 shows a plot of a wave packet using numerical integration. It gives the results for $\alpha = 10$ and increasing values of $|\phi|$. For increasing $|\phi|$, the wave packet assumes an annular shape and propagates outwards with decreasing amplitude; it is peaked in the direction of negative $\beta_\pm$.

5. Conclusion and Outlook

In this contribution, we have discussed the fate of classical singularities in Friedmann-Lemaître as well as Bianchi I models of quantum cosmology. Employing a generalized form of the DeWitt criterion, we have seen that these singularities can generally be avoided.

Singularity avoidance is also discussed in other approaches such as loop quantum cosmology. There, the resolution is usually treated as a non-singular bounce in quantum-corrected cosmological models; see, for example, the recent discussion of Bianchi models in [12]. It would, of course, be of great interest to perform a detailed comparison of this approach with the present one.

Future work should address the issue of singularity avoidance for models of increasing complexity. The Bianchi I models discussed here belong to the class of asymptotically velocity dominated (AVTD) models, for which the kinetic terms become dominant in the vicinity of the initial singularity. Our results for these models should thus be representative for more general models which admit such an AVTD behaviour.

The general approach to a classical spacelike singularity is been assumed to be of the Belinsky-Khalatnikov-Lifshits (BKL) type; see, for example, [13] for a recent discussion. This corresponds to

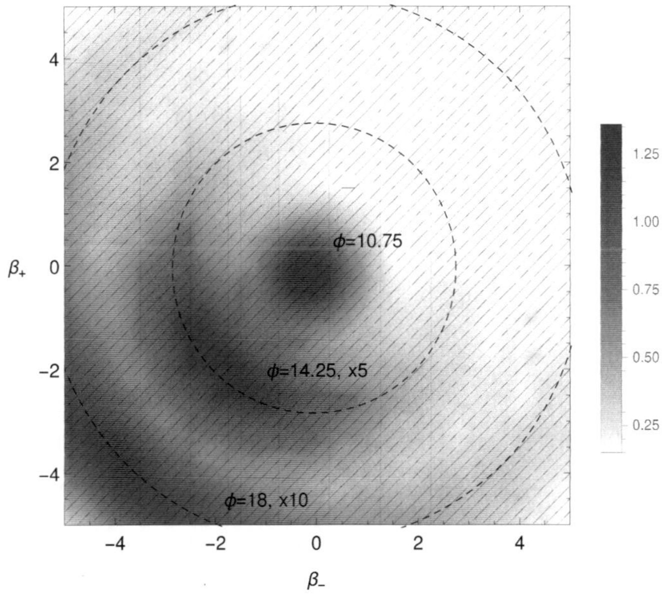

Figure 4: Density plot for $|\Psi|e^{\frac{3}{2}(\alpha+|\phi|)}$, showing the position of the wave packet for different values of $|\phi|$, each with a different scaling to visualize the decaying wave in a single graphic. From [2].

an evolution that behaves as if a separate mixmaster model were attached to each spatial point. The discussion of the Wheeler-DeWitt equation for such a situation should give a quantum picture of the BKL behaviour and should answer the question whether the singularities are avoided in the generic case. If this could be achieved, one would have gained a complete picture of the situation in quantum geometrodynamics, given our present level of understanding.

Acknowledgments
We are grateful to Dennis Piontek for collaboration on the quantum Bianchi I model.

References

[1] Kiefer C 2012 *Quantum Gravity* third edition (Oxford: Oxford University Press)
[2] Kiefer C, Kwidzinski N and Piontek D 2019 Singularity avoidance in Bianchi I quantum cosmology *arXiv:1903.04391 [gr-qc]*
[3] Ellis G F R, Maartens R and MacCallum M A H 2012 *Relativistic Cosmology* (Cambridge: Cambridge University Press)
[4] DeWitt B S 1967 Quantum Theory of Gravity. 1. The Canonical Theory *Phys. Rev.* **160** 1113
[5] Hájíček P and Kiefer C 2001 Singularity avoidance by collapsing shells in quantum gravity *Int. J. Mod. Phys. D* **10** 775; Kiefer C 2015 Quantum black hole without singularity *arXiv:1512.08346 [gr-qc]*
[6] Halliwell J J 1991 Introductory lectures on quantum cosmology, in *Quantum cosmology and baby universes* ed S Coleman, J B Hartle, T Piran and S Weinberg (Singapore: World Scientific) pp 159–243
[7] Dąbrowski M, Kiefer C and Sandhöfer B 2006 Quantum phantom cosmology *Phys. Rev. D* **74** 044022
[8] Kiefer C 2010 On the avoidance of classical singularities in quantum cosmology *J. Phys.: Conf. Ser.* **222** 012049
[9] Albarran I, Bouhmadi-López M, Kiefer C, Marto J and Vargas Moniz P 2016 Classical and quantum cosmology of the little rip abrupt event *Phys. Rev. D* **94** 063536
[10] Kamenshchik A, Kiefer C and Sandhöfer B 2007 Quantum cosmology with big-brake singularity *Phys. Rev. D* **76** 064032
[11] Kiefer C and Zeh H D 1995 Arrow of time in a recollapsing quantum universe *Phys. Rev. D* **51** 4145
[12] Wilson-Ewing E 2018 The loop quantum cosmology bounce as a Kasner transition *Class. Quantum Grav.* **35** 065005
[13] Kiefer C, Kwidzinski N, and Piechocki W 2018 On the dynamics of the general Bianchi IX spacetime near the singularity *Eur. Phys. J. C* **78** 691

IOP Conf. Series: Journal of Physics: Conf. Series **1275** (2019) 012012 doi:10.1088/1742-6596/1275/1/012012

On a soliton-type spacetime defect

F R Klinkhamer

Institute for Theoretical Physics, Karlsruhe Institute of Technology (KIT), 76128 Karlsruhe, Germany

E-mail: `frans.klinkhamer@kit.edu`

Abstract. We review the construction of a particular soliton-type solution of the classical Einstein and matter-field equations. This localized finite-energy static classical solution can be interpreted as a single spacetime defect embedded in Minkowski spacetime and may give rise to several new effects. For a Skyrme-type theory with small enough matter-field energy scale compared to the Planck energy scale and for a sufficiently small defect length scale, the existence of a globally regular solution requires a negative active gravitational mass, so that the defect repels a distant test particle ("antigravity"). There also exist "stealth defects" which have a vanishing asymptotic gravitational mass. These stealth defects are, however, not entirely invisible as they bring about a new type of gravitational lensing.

1. Introduction

One hypothesis is that the Universe started out in some form of "quantum phase" which gave rise to classical spacetime and gravity, as described by Einstein's General Theory of Relativity [1]. It is then possible that this process is analogous to the cooling of a liquid, which produces an atomic crystal. But, if the cooling of the latter process is rapid, the resulting crystal will be imperfect, containing crystallographic defects. For the above-mentioned quantum phase and the resulting classical spacetime, the analogy suggests the possibility of having "spacetime defects" (i.e., imperfections in the fabric of spacetime). Remark that, historically, some of the earliest ideas on a foam-like structure of spacetime go back to Wheeler in the 1950s (cf. Sec. 43.4 and Box 44.3 in Ref. [1] and Chap. 6 in Ref. [2]).

Little is known for sure about the quantum phase of spacetime. Loop quantum gravity, for example, does have something like "atoms of space," but the emergence of a classical spacetime has not been established (cf. Refs. [3, 4, 5] and references therein).

Here, we will stay on the classical side of the problem and use the framework of Einstein's General Relativity (GR). Specifically, we will obtain a soliton-type classical solution to describe a single spacetime defect and we will investigate certain novel effects which this soliton-type solution produces.

The aim of the present contribution is to provide a more or less self-contained discussion of one particular soliton-type classical solution, namely a Skyrmion spacetime defect solution [6] and to explain the origin of certain new phenomena [7, 8, 9]. In Sec. 2, we first review some background material on the manifold considered (see also Refs. [10, 11, 12, 13]), then define the fields and their action, and, finally, present the details of the Skyrmion spacetime defect solution. In Sec. 3, we explain why certain Skyrmion spacetime defects produce "antigravity," that is, repulsion of a distant test particle. In Sec. 4, we consider other Skyrmion spacetime

Content from this work may be used under the terms of the Creative Commons Attribution 3.0 licence. Any further distribution of this work must maintain attribution to the author(s) and the title of the work, journal citation and DOI.
Published under licence by IOP Publishing Ltd

IOP Conf. Series: Journal of Physics: Conf. Series **1275** (2019) 012012 doi:10.1088/1742-6596/1275/1/012012

defects which have a positive energy density of the matter fields but a vanishing asymptotic gravitational mass, so that these defects may be called "stealth defects." In Sec. 5, we show that such stealth defects are not entirely invisible, as they give lensing of light (this lensing is different from standard gravitational lensing). In Sec. 6, we, first, compare our soliton-type spacetime defect with another kind of spacetime defect [14, 15] and, then, review the general properties of the Skyrmion spacetime defect solution.

Before we start with the technical discussion of the next section, we emphasize that our Skyrmion spacetime defect is a genuine solution of standard GR, as long as we allow for degenerate metrics. The hope is that the equations of GR know about the "edge of the theory" and that they may provide some insight into the nature of possible defects of spacetime.

2. Skyrmion spacetime defect

2.1. *Basic idea*

Let us start by presenting a rough description of our spacetime defect solution [6], with mathematical details to follow later. The basic idea is to obtain a *nonsingular* finite-energy static defect solution of the Einstein field equation, with a length parameter $b > 0$ and topology as suggested by the sketch in Fig. 1.

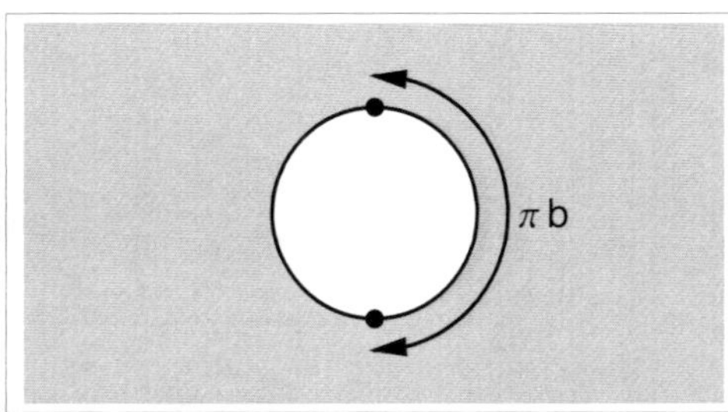

Figure 1. Three-space $\widetilde{M}_3$ obtained by surgery from the Euclidean 3-space E_3 : interior of a ball with radius b removed and antipodal points on the boundary of the ball identified (as indicated by the dots). The "long distance" between identified points on the defect surface equals $\pi\, b$.

An important ingredient of such a regular solution is the choice of *appropriate coordinates.* The standard Cartesian coordinates of Euclidean 3-space are unsatisfactory, as a single point may have different coordinates. Taking the origin of the Cartesian coordinate system to coincide with the center of the 3-ball in Fig. 1, the coordinates $(x^1, x^2, x^3) = (0, b, 0)$ and $(x^1, x^2, x^3) = (0, -b, 0)$, for example, correspond to the same point (the identified dots in Fig. 1). Instead of a single chart of Cartesian coordinates, it is possible to use three overlapping charts of coordinates, each one centered on one of the three Cartesian coordinate axes [12, 13]. The promised mathematical details now follow.

2.2. *Manifold*

The four-dimensional spacetime manifold considered has topology

$$\widetilde{M}_4 = \mathbb{R} \times \widetilde{M}_3 , \tag{1}$$

where $\widetilde{M}_3$ is a noncompact, orientable, nonsimply-connected manifold without boundary. Up to a point, $\widetilde{M}_3$ is homeomorphic to the 3-dimensional real-projective plane,

$$\widetilde{M}_3 \simeq \mathbb{R}P^3 - p_\infty , \tag{2}$$

IOP Conf. Series: Journal of Physics: Conf. Series **1275** (2019) 012012 doi:10.1088/1742-6596/1275/1/012012

where p_∞ corresponds to the "point at spatial infinity."

For the explicit construction of $\widetilde{M}_3$, we perform *local surgery* (Fig. 1) on the 3-dimensional Euclidean space $E_3 = \left(\mathbb{R}^3,\, \delta_{mn}\right)$. Recall the standard Cartesian and spherical coordinates on $\mathbb{R}^3$,

$$\vec{x} \;\equiv\; |\vec{x}|\,\widehat{x} = (x^1,\, x^2,\, x^3) = (r\sin\theta\cos\phi,\, r\sin\theta\sin\phi,\, r\cos\theta)\,, \tag{3a}$$

with ranges

$$x^m \;\in\; (-\infty,\, +\infty)\,, \quad r \geq 0\,, \quad \theta \in [0,\, \pi]\,, \quad \phi \in [0,\, 2\pi)\,. \tag{3b}$$

Now, $\widetilde{M}_3$ is obtained from $\mathbb{R}^3$ by removing the interior of the ball B_b with radius b and identifying antipodal points on its boundary $S_b \equiv \partial B_b$. With point reflection denoted $P(\vec{x}) = -\vec{x}$, the 3-space $\widetilde{M}_3$ is given by

$$\widetilde{M}_3 = \Big\{\vec{x} \in \mathbb{R}^3\, :\, \Big(|\vec{x}| \geq b > 0\Big) \wedge \Big(P(\vec{x}) \,\widehat{=}\, \vec{x} \;\;\text{for}\;\; |\vec{x}| = b\Big)\Big\}\,, \tag{4}$$

where "$\widehat{=}$" stands for point-wise identification.

As mentioned before, a relatively simple covering of $\widetilde{M}_3$ uses three charts of coordinates, labeled by $n = 1, 2, 3$. Each chart surrounds one of the three Cartesian coordinate axes [12, 13], as illustrated by Fig. 2.

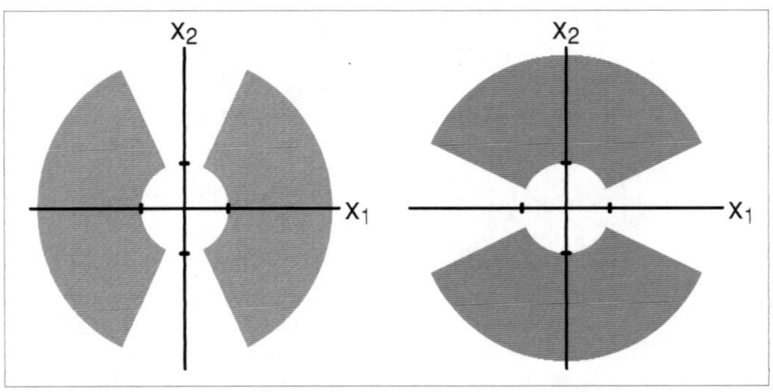

Figure 2. Slice $x_3 = 0$ of the manifold $\widetilde{\mathcal{M}}_3$ with the domains of the chart-1 coordinates (left) and the chart-2 coordinates (right). The tick marks on the x_1 and x_2 axes correspond to the values $\pm b$. The 3-dimensional domains are obtained by revolution around the x_1-axis (left) or the x_2-axis (right). The domain of the chart-3 coordinates is defined similarly.

These coordinate charts are denoted

$$(X_n,\, Y_n,\, Z_n)\,, \quad \text{for} \quad n = 1,\, 2,\, 3\,. \tag{5}$$

Note that the triples (5) are, despite appearances, non-Cartesian coordinates. Specifically, the set of coordinates surrounding the x^2-axis segments with $|x^2| \geq b$ is given by (cf. Sec. II D in Ref. [11])

$$X_2 \;=\; \begin{cases} \phi\,, & \text{for} \;\; 0 < \phi < \pi\,, \\ \phi - \pi\,, & \text{for} \;\; \pi < \phi < 2\pi\,, \end{cases} \tag{6a}$$

$$Y_2 \;=\; \begin{cases} +\sqrt{r^2 - b^2}\,, & \text{for} \;\; 0 < \phi < \pi\,, \\ -\sqrt{r^2 - b^2}\,, & \text{for} \;\; \pi < \phi < 2\pi\,, \end{cases} \tag{6b}$$

$$Z_2 \;=\; \begin{cases} \theta\,, & \text{for} \;\; 0 < \phi < \pi\,, \\ \pi - \theta\,, & \text{for} \;\; \pi < \phi < 2\pi\,, \end{cases} \tag{6c}$$

with ranges

$$X_2 \in (0,\, \pi)\,, \quad Y_2 \in (-\infty,\, \infty)\,, \quad Z_2 \in (0,\, \pi)\,. \tag{6d}$$

The two other coordinate sets, $(X_1,\, Y_1,\, Z_1)$ and $(X_3,\, Y_3,\, Z_3)$, are defined similarly.

In the following, we explicitly discuss only one coordinate chart, which we take to be (6). Furthermore, the notation is simplified as follows:

$$(T,\, X,\, Y,\, Z) \equiv (T,\, X_2,\, Y_2,\, Z_2)\,, \tag{7}$$

where the time coordinate T has been added in order to describe the spacetime manifold $\widetilde{M}_4$.

2.3. Fields and action

Consider a Skyrme-type scalar field $\Omega(X) \in SO(3)$, which propagates over the spacetime manifold (1) and has the following action ($c = \hbar = 1$):

$$S = \int_{\widetilde{M}_4} d^4X\, \sqrt{-g}\,\Big(\mathcal{L}_{\text{grav}} + \mathcal{L}_{\text{mat}}\Big)\,, \tag{8a}$$

$$\mathcal{L}_{\text{grav}} = \frac{1}{16\pi\, G_N}\, R\,, \tag{8b}$$

$$\mathcal{L}_{\text{mat}} = \frac{f^2}{4}\,\mathrm{tr}\Big(\omega_\mu\, \omega^\mu\Big) + \frac{1}{16\, e^2}\,\mathrm{tr}\Big(\,[\omega_\mu,\, \omega_\nu]\,[\omega^\mu,\, \omega^\nu]\Big) + \frac{1}{2}\, m^2\, f^2\,\mathrm{tr}\Big(\Omega - \mathbb{1}_3\Big)\,, \tag{8c}$$

$$\omega_\mu \equiv \Omega^{-1}\,\partial_\mu\,\Omega\,, \tag{8d}$$

with $g \equiv \det g_{\mu\nu}$ and the Ricci curvature scalar $R \equiv R_{\kappa\lambda\mu\nu}\, g^{\kappa\mu}\, g^{\lambda\nu}$. Defining "pions" π^a by

$$\Omega(X) = \exp\Big[S^a\, \pi^a(X)/f\Big]\,, \tag{9}$$

for three 3×3 matrices S^a given by

$$S^1 \equiv \begin{pmatrix} 0 & 0 & 0 \\ 0 & 0 & 1 \\ 0 & -1 & 0 \end{pmatrix}, \quad S^2 \equiv \begin{pmatrix} 0 & 0 & -1 \\ 0 & 0 & 0 \\ 1 & 0 & 0 \end{pmatrix}, \quad S^3 \equiv \begin{pmatrix} 0 & 1 & 0 \\ -1 & 0 & 0 \\ 0 & 0 & 0 \end{pmatrix}, \tag{10}$$

we have

$$\mathcal{L}_{\text{mat}} = -\frac{1}{2}\,\partial_\mu\pi^a\partial^\mu\pi^a - \frac{1}{2}\, m^2\, \pi^a\pi^a + \cdots\,. \tag{11}$$

The theory (8) has three dimensional parameters and a single dimensionless parameter e,

$$G_N \geq 0\,, \tag{12a}$$

$$f > 0\,, \tag{12b}$$

$$m \geq 0\,, \tag{12c}$$

$$e > 0\,. \tag{12d}$$

From these parameters, we obtain the following two dimensionless parameters:

$$\widetilde{\eta} \equiv 8\pi\, G_N\, f^2 \geq 0\,, \tag{13a}$$

$$\widetilde{m} \equiv \frac{m}{e\, f} \geq 0\,. \tag{13b}$$

IOP Conf. Series: Journal of Physics: Conf. Series **1275** (2019) 012012 doi:10.1088/1742-6596/1275/1/012012

2.4. *Ansätze*

The self-consistent *Ansätze* for the metric and the $SO(3)$ matter field are [6]

$$ds^2\Big|_{\widetilde{M}_4\,,\,\text{chart}-2} = -\big[\mu(W)\big]^2 dT^2 + \big(1 - b^2/W\big)\big[\sigma(W)\big]^2 (dY)^2 + W\Big[(dZ)^2 + \sin^2 Z\,(dX)^2\Big]\,, \tag{14a}$$

$$\Omega(X) = \cos\big[F(W)\big]\,\mathbb{1}_3 - \sin\big[F(W)\big]\,\widehat{x}\cdot\vec{S} + \big(1-\cos\big[F(W)\big]\big)\,\widehat{x}\otimes\widehat{x}\,, \tag{14b}$$

$$W \equiv b^2 + Y^2\,, \tag{14c}$$

with a unit 3-vector $\widehat{x} \equiv \vec{x}/|\vec{x}|$ from the Cartesian coordinates $\vec{x}$ defined in terms of the chart-2 coordinates X, Y, and Z [see (3), (6), and (7) above]. Note that the sine-F term in (14b) displays the well-known hedgehog behavior, linking the "isospin" dependance of the matter field $\Omega(X)$ to its spatial dependance.

The boundary conditions on the three *Ansatz* functions are

$$F(b^2) = \pi\,, \quad F(\infty) = 0\,, \tag{15a}$$

$$\sigma(b^2) \in (0,\,\infty)\,, \tag{15b}$$

$$\mu(b^2) \in (0,\,\infty)\,. \tag{15c}$$

The $W = b^2$ boundary condition (15a) for (14b) is consistent with the topology of $\widetilde{M}_3$ [identified antipodal points in Fig. 1] and the boundary conditions (15b)–(15c) will be discussed later.

Two remarks on these *Ansätze* are in order. First, with finite *Ansatz* functions $\mu(W)$ and $\sigma(W)$, the metric from (14a) is **degenerate** at the defect surface $Y = 0$ (or $W = b^2$),

$$\det\big[g_{\mu\nu}(X)\big]\Big|_{W=b^2} = 0\,, \tag{16}$$

and the standard elementary-flatness property does not apply (cf. App. D in Ref. [10]). Further discussion on the degeneracy property is relegated to Sec. 6.2.

Second, the matter-field *Ansatz* (14b) corresponds to a topologically nontrivial scalar field configuration, a Skyrmion-like configuration [16, 17] with **unit winding number**,

$$N \equiv \deg[\Omega] = -\frac{2}{\pi}\int_{\pi}^{0} dF\,\sin^2(F/2) = 1\,, \tag{17}$$

where the endpoints of the integral correspond to the boundary conditions (15a). Again, further discussion is relegated to Sec. 6.2.

2.5. *Reduced field equations*

In our numerical analysis, we will use the following dimensionless variables:

$$y \equiv e\,f\,Y \in \big(-\infty,\,\infty\big)\,, \tag{18a}$$

$$y_0 \equiv e\,f\,b \in \big(0,\,\infty\big)\,, \tag{18b}$$

$$w \equiv (e\,f)^2\,W \equiv y_0^2 + y^2 \in \big[y_0^2,\,\infty\big)\,, \tag{18c}$$

and the *Ansatz* functions are simply written as $\sigma(w)$, $\mu(w)$, and $F(w)$. The reduced field equations are three ordinary differential equations (ODEs). With the auxiliary functions

$$A(w) \equiv 2\sin^2\frac{F(w)}{2}\left(\sin^2\frac{F(w)}{2} + w\right)\,, \quad C(w) \equiv 4\sin^2\frac{F(w)}{2} + w\,, \tag{19}$$

these ODEs are [8, 13]

$$
\begin{aligned}
4\,w\,\sigma'(w) &= +\sigma(w)\,\left[\left[1-\sigma^2(w)\right]+\widetilde{\eta}\,\frac{2}{w}\left(A(w)\,\sigma^2(w)+C(w)\,\left[w\,F'(w)\right]^2\right)\right] \\
&\quad +2\,w\,\widetilde{m}^2\,\widetilde{\eta}\,\sigma^3(w)\,\sin^2\frac{F(w)}{2}\,, \qquad (20a)\\
4\,w\,\mu'(w) &= -\mu(w)\,\left[\left[1-\sigma^2(w)\right]+\widetilde{\eta}\,\frac{2}{w}\left(A(w)\,\sigma^2(w)-C(w)\,\left[w\,F'(w)\right]^2\right)\right] \\
&\quad -2\,w\,\widetilde{m}^2\,\widetilde{\eta}\,\sigma^2(w)\,\sin^2\frac{F(w)}{2}\,, \qquad (20b)\\
C(w)\,w^2\,F''(w) &= +\sigma^2(w)\,\sin F(w)\left(\sin^2\frac{F(w)}{2}+\frac{w}{2}\right)-\frac{1}{2}\,C(w)\,\sigma^2(w)\,w\,F'(w) \\
&\quad \times\left[1-4\,\widetilde{\eta}\,\frac{1}{w}\,\sin^2\frac{F(w)}{2}\left(\sin^2\frac{F(w)}{2}+w\right)\right] \\
&\quad -w\,F'(w)\left[w\,F'(w)\,\sin F(w)+w\right]+\frac{\widetilde{m}^2}{2}\,w^2\,\sigma^2(w)\,\sin\frac{F(w)}{2} \\
&\quad \times\left[\cos\frac{F(w)}{2}+2\,\widetilde{\eta}\,C(w)\,\sin\frac{F(w)}{2}\,F'(w)\right]\,, \qquad (20c)
\end{aligned}
$$

where the prime stands for differentiation with respect to w.

2.6. *Matter energy density and gravitational mass*

For later use, we also give the reduced expression for the 00-component of the energy-momentum tensor $T^{\mu}{}_{\nu}(w)$ which corresponds to the negative of the energy density $\rho(w)$,

$$
\begin{aligned}
T^{0}{}_{0}(w) &= -f^2\,(e\,f)^2\,\frac{2}{w^2\,\sigma^2(w)} \\
&\quad \times\left(A(w)\,\sigma^2(w)+C(w)\,\left[w\,F'(w)\right]^2+\widetilde{m}^2\,w^2\,\sigma^2(w)\,\sin^2\frac{F(w)}{2}\right). \qquad (21)
\end{aligned}
$$

The expression in the large brackets on the right-hand side of (21) has already appeared on the right-hand side of (20a), which resulted from the 00-component of the Einstein equation.

A further useful quantity is the dimensionless length scale $l(w)$, defined by

$$
l(w)\equiv\sqrt{w}\,\left[1-\frac{1}{\sigma^2(w)}\right]\,, \qquad (22)
$$

which corresponds to a Schwarzschild-type behavior of the square of the metric function,

$$
\sigma^2(w)=\frac{1}{1-l(w)/\sqrt{w}}\,. \qquad (23)
$$

The σ–ODE (20a) gives immediately

$$
l'(w) = \widetilde{\eta}\,w^{-3/2}\left(A(w)+C(w)\,\frac{\left[w\,F'(w)\right]^2}{\sigma^2(w)}\right)\,, \qquad (24)
$$

where the prime stands for the derivative d/dw and the auxiliary functions $A(w)$ and $C(w)$ are given by (19). As the right-hand-side of (24) is manifestly nonnegative, the interpretation is

IOP Conf. Series: Journal of Physics: Conf. Series **1275** (2019) 012012 doi:10.1088/1742-6596/1275/1/012012

that the Schwarzschild-type length scale $l(w)$, corresponding to a w-dependent effective mass, can only increase by the addition of positive energy density from the matter fields [7]. The asymptotic value of $l(w)$ defines the Arnowitt–Deser–Misner (ADM) mass in our context,

$$M_{\mathsf{ADM}} = l_\infty / \big(2\, G_N\, e\, f\big)\,, \quad l_\infty \equiv \lim_{w\to\infty} l(w)\,, \tag{25}$$

which corresponds to the active gravitational mass of the soliton-type solution [12, 13].

2.7. Numerical solutions

The ODEs (20) can be solved numerically with boundary conditions from (15). Specifically, we have $F(y_0^2) = \pi$ and $F(\infty) = 0$ for the matter-field *Ansatz* function $F(w)$ and we, first, take $\sigma(y_0^2) = 1$ for the metric *Ansatz* function $\sigma(w)$ [the value of $\mu(y_0^2)$ can be rescaled arbitrarily].

Figure 3 shows the *Ansatz* functions $F(w)$, $\sigma(w)$, and $\mu(w)$ of one particular numerical solution. Also displayed are the dimensionless Ricci curvature scalar $R(w)$, the dimensionless Kretschmann curvature scalar $K(w) \equiv R_{\kappa\lambda\mu\nu}(w)\, R^{\kappa\lambda\mu\nu}(w)$, and the negative of the 00 component of the dimensionless Einstein tensor $E^{\mu}{}_{\nu}(w) \equiv R^{\mu}{}_{\nu}(w) - (1/2)\, R(w)\, \delta^{\mu}{}_{\nu}$ [note that $-E^{0}{}_{0}(w)$ is proportional to the energy density $\rho(w)$ by the Einstein equation]. This Fig. 3 shows, moreover, the behavior of the dimensionless Schwarzschild-type length scale $l(w)$ defined by (22), which, as mentioned in Sec. 2.6, stays constant or increases with increasing w. All physical quantities are well-behaved at the defect surface $w = y_0^2$, as shown by Fig. 9 in Ref. [7].

The boundary condition $\sigma(y_0^2) = 1$ may be called the "standard" boundary condition, because the limit $b \to 0$ then connects to the standard Minkowski spacetime manifold. But with $b \neq 0$ and the nontrivial topology $\mathbb{R}P^3$ from (2), the boundary condition on the metric *Ansatz* function can be generalized,

$$\sigma\big(y_0^2\big) \in (0,\, \infty)\,, \tag{26}$$

where the value zero has been excluded, in order that the field equations be well-defined at the $w = y_0^2$ defect surface [13].

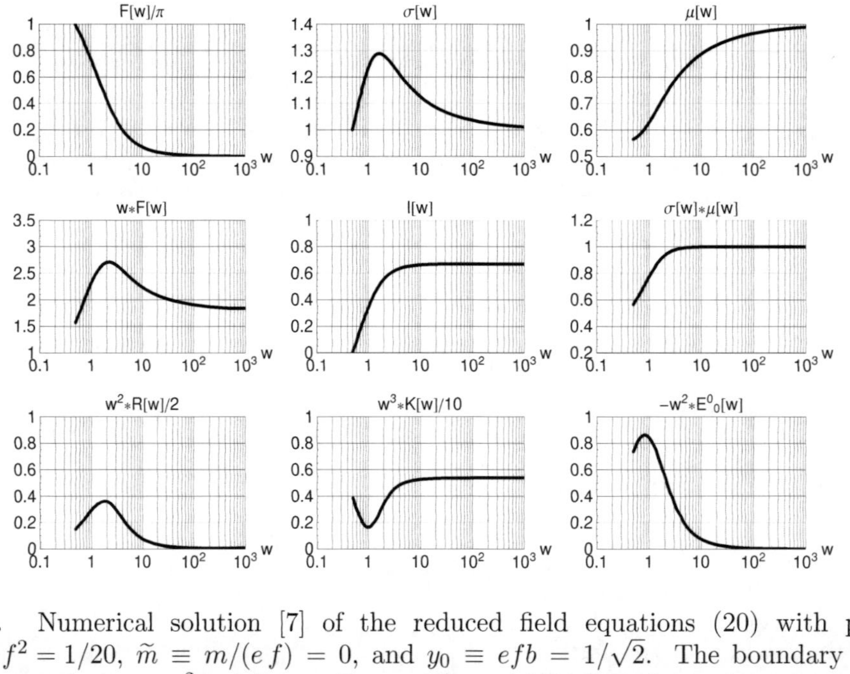

Figure 3. Numerical solution [7] of the reduced field equations (20) with parameters $\widetilde{\eta} \equiv 8\pi\, G_N\, f^2 = 1/20$, $\widetilde{m} \equiv m/(e\, f) = 0$, and $y_0 \equiv efb = 1/\sqrt{2}$. The boundary conditions at the defect surface $w = y_0^2 = 1/2$ are: $F = \pi$, $F' = -1.9718377138$, $\sigma = 1$, and $\mu = 0.564337$.

IOP Conf. Series: Journal of Physics: Conf. Series **1275** (2019) 012012 doi:10.1088/1742-6596/1275/1/012012

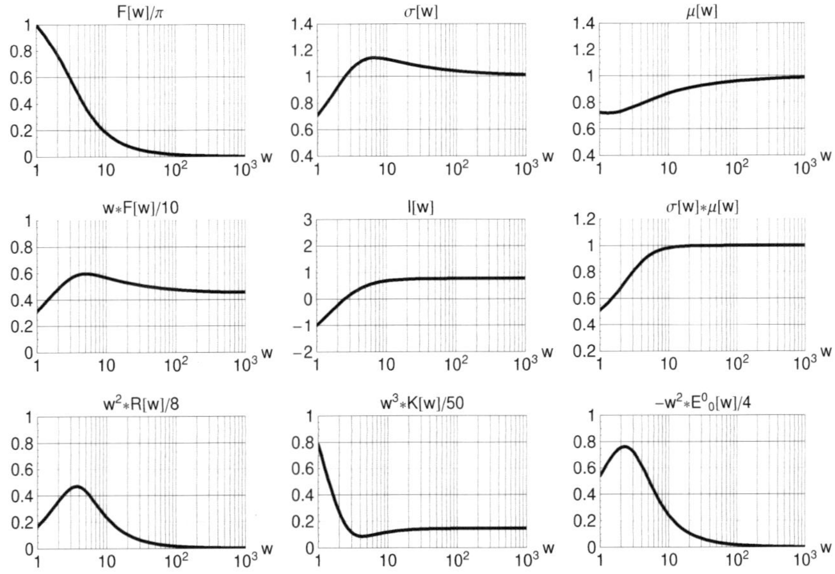

Figure 4. Numerical solution [7] of the ODEs (20) with parameters $\widetilde{\eta} = 1/10$, $\widetilde{m} = 0$, and $y_0 = 1$. The boundary conditions at $w = y_0^2 = 1$ are: $F = \pi$, $F' = -0.82561881304$, $\sigma = 1/\sqrt{2}$, and $\mu = 0.725818$. This numerical solution with $l_\infty \approx 0.8$ has a positive ADM mass (25).

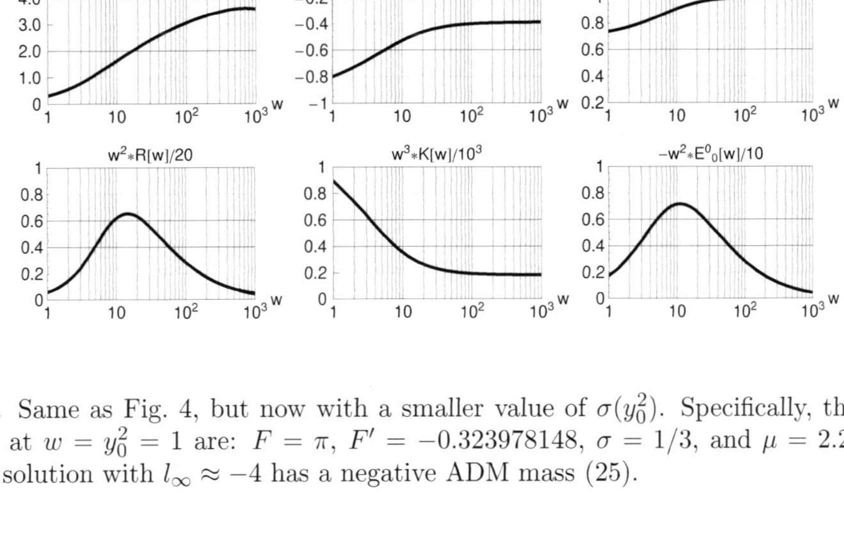

Figure 5. Same as Fig. 4, but now with a smaller value of $\sigma(y_0^2)$. Specifically, the boundary conditions at $w = y_0^2 = 1$ are: $F = \pi$, $F' = -0.323978148$, $\sigma = 1/3$, and $\mu = 2.21176$. This numerical solution with $l_\infty \approx -4$ has a negative ADM mass (25).

IOP Conf. Series: Journal of Physics: Conf. Series **1275** (2019) 012012 doi:10.1088/1742-6596/1275/1/012012

Figures 4 and 5 give numerical solutions for two different values of $\sigma(y_0^2) < 1$. The results of Figs. 3–5 are to be understood as having *different* values of the boundary condition $\sigma(y_0^2)$ for *equal* boundary conditions $F(y_0^2) = \pi$ and $F(\infty) = 0$ [the numerical values for $F'(y_0^2)$ in the figure captions are mentioned purely for technical reasons, as the numerical solutions are easier to obtain with all boundary conditions at one side, $w = y_0^2$]. The qualitative behavior of the $\sigma(w)$ curves in these figures changes: the $\sigma(w)$ curves of Figs. 3 and 4 approach unity from above as $w \to \infty$ [resulting in $M_{\text{ADM}} > 0$, according to (22) and (25)], while the $\sigma(w)$ curve of Fig. 5 approaches unity from below as $w \to \infty$ [resulting in $M_{\text{ADM}} < 0$].

3. Antigravity

3.1. Origin of a new phenomenon

Any localized object made of ponderable matter (e.g., quarks and leptons of the Standard Model) **attracts** a distant test particle. This phenomenon is called **gravity** and was first studied by Newton in his *Principia* [1687].

With the solution of Fig. 5, we have a localized object which **repels** a distant test particle. The phenomenon may be called "**antigravity**." The crucial ingredients of this particular object are, in the framework of Einstein's General Relativity [1915], the nontrivial topology of space [here, $\mathbb{R}P^3$] and the nontrivial gravitational fields at the defect surface [here, $\sigma(b^2) < 1$ for the *Ansatz* (14a)].

Figures 4 and 5 suggest that the Skyrmion spacetime defect can have either positive or negative gravitational mass, but we have a further result: *a sufficiently small defect solution exists only if it has negative gravitational mass* [a more precise formulation will be given later].

Consider, first, the nature of the solutions with "standard" boundary condition $\sigma(b^2) = 1$. It is, then, found that the solution collapses if it becomes too small (loosely speaking, if b is of the order of or less than the effective Schwarzschild radius); see Fig. 6 for numerical results at one particular value of the coupling constant $\widetilde{\eta}$. In fact, there is a *critical curve* in the $(b, \widetilde{\eta})$ plane, above which (or, to the left of which) there are no globally regular solutions with $\sigma(b^2) = 1$; see Fig. 7 for a numerical approximation of this critical curve.

Let us now discuss the heuristics for obtaining an anti-gravitating spacetime defect. In order to get a globally regular solution in the region *above* the critical curve of Fig. 7, we need to arrange for a sufficiently negative effective mass at the defect surface ($y = 0$). Now, the dimensionless effective mass is given by (22) with $w \equiv y_0^2 + y^2$. Then, an effective mass $l(y_0^2) < 0$ results from a nonstandard boundary condition $\sigma(y_0^2) < 1$.

For a fixed positive value of $\widetilde{\eta} \equiv 8\pi\, G_N\, f^2$, a sufficiently small globally regular defect solution thus requires a sufficiently negative effective mass at the defect surface $w = y_0^2$ from a nonstandard boundary condition on the *Ansatz* function for the yy component of the metric, namely, a positive value of $\sigma(y_0^2)$ sufficiently far below unity. For a small enough value of the coupling constant $\widetilde{\eta}$, this boundary condition at the defect surface directly gives a negative ADM mass at spatial infinity [see (24) with a near-zero right-hand-side from $\widetilde{\eta} \sim 0$ and the definition (25) of M_{ADM}].

3.2. Antigravity from a Planck-scale defect

Let us give an explicit example [7] of a negative-gravitational-mass Skyrmion spacetime defect. With ζ a number of order 1 or larger, the model parameters are taken as follows:

$$f^2 \ll \left(E_{\text{planck}}\right)^2 \equiv 1/(8\pi G_N) \approx \left(2.44 \times 10^{18}\ \text{GeV}\right)^2, \tag{27a}$$

$$e \leq 1/\zeta, \tag{27b}$$

where the first inequality corresponds to $\widetilde{\eta} = \left(f/E_{\text{planck}}\right)^2 \ll 1$. The defect is considered to be

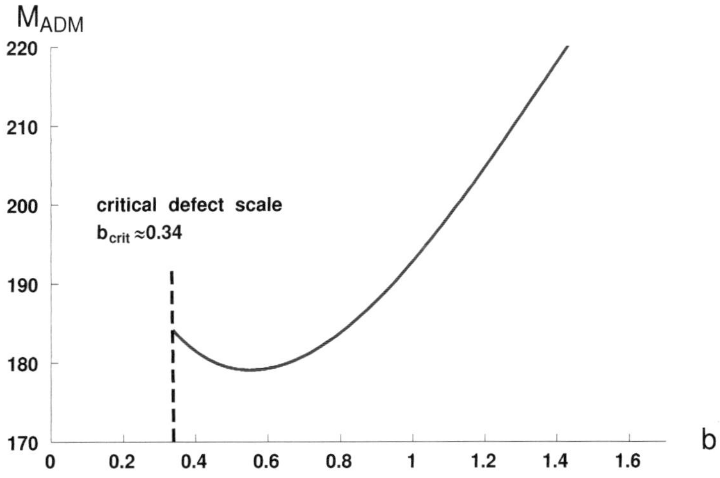

Figure 6. Gravitational mass M_{ADM} [in units f/e] vs. defect length scale b [in units $1/(ef)$] from the numerical solutions [7] of the ODEs (20) with boundary condition $\sigma(b^2) = 1$, for model parameters $\widetilde{\eta} = 0.033$ and $\widetilde{m} = 0$.

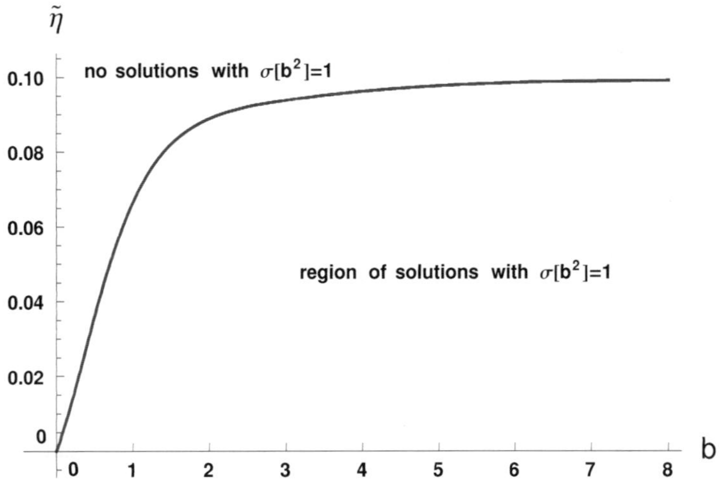

Figure 7. Curve of the critical defect length scale b_{crit} [in units of $1/(ef)$] and the corresponding critical coupling constant $\widetilde{\eta}_{\text{crit}}$, obtained from the numerical solutions [7] of the ODEs (20) with boundary condition $\sigma(b^2) = 1$ and model parameter $\widetilde{m} = 0$.

a remnant of a quantum-spacetime phase and the defect length scale is given by

$$b_{\text{remnant}} = \zeta \, l_{\text{planck}} \,, \tag{28a}$$

$$l_{\text{planck}} \equiv \sqrt{8\pi G_N \, \hbar / c^3} \approx 8.10 \times 10^{-35} \ \text{m} \,, \tag{28b}$$

with $\zeta \gtrsim 1$. The quantum nature of b_{remnant} has been emphasized by displaying $\hbar$ in (28b).

The bounds (27) and the defect length scale (28) imply that the solution corresponds to a point above the critical curve of Fig. 7, so that a negative effective mass at the defect surface is required in order to prevent collapse. The resulting gravitational mass of the defect [obtained from (25) with $l_\infty \sim -1$] is

$$M_{\mathsf{ADM}} \sim -\frac{4\pi}{\widetilde{\eta}}\,\frac{f}{e} = -\left(\frac{4\pi}{e}\right)\left(\frac{E_{\mathsf{planck}}}{f}\right) E_{\mathsf{planck}}\,. \tag{29}$$

According to the inequalities (27a) and (27b), the absolute value $|M_{\mathsf{ADM}}|$ from (29) typically is much larger than $E_{\mathsf{planck}} \approx 4.34 \times 10^{-6}\,\mathrm{g}$.

4. Stealth defect

We have seen that certain soliton-type defect solutions can have positive gravitational mass but also negative gravitational mass. As the gravitational mass of such a spacetime-defect solution is a continuous variable, there must be special spacetime defects with vanishing gravitational mass. These defects with positive energy density of the matter fields and zero asymptotic gravitational mass will be called "stealth defects" [8]. An explicit solution with vanishing gravitational mass and an exponentially-vanishing energy density of the matter fields is given in Fig. 8.

Now, assume that all matter fields have some form of non-gravitational interaction with each other. If so, there will be some interaction between the "pions" of the theory considered in (8) and the elementary particles of the Standard Model. Then, consider what happens with the head-on collision of a stealth defect from Fig. 8 and a human observer made of Standard Model particles (mostly up and down quarks, gluons, and electrons). In close approximation, the observer will have no idea of what is going to happen, until he/she is within a distance of order $\hbar/(mc)$ from the defect, where m is the "pion" mass scale from the matter action (8c). What happens during the collision itself and afterwards depends on the details of the setup, for example, the size of the observer compared to the defect length scale b.

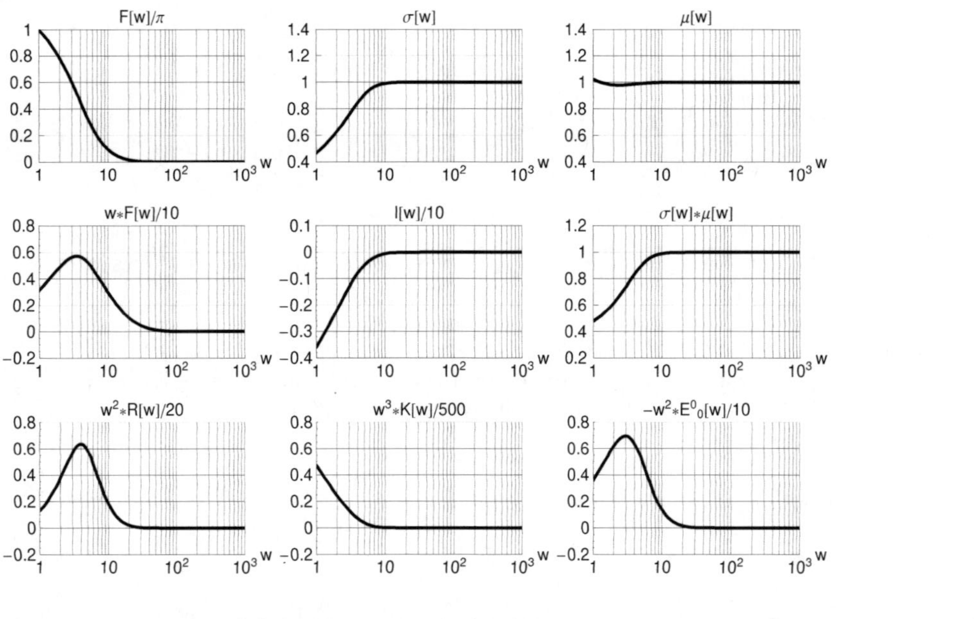

Figure 8. Numerical solution [8] of the ODEs (20) with parameters $\widetilde{\eta} = 1/10$, $\widetilde{m} = 1$, and $y_0 = 1$. The boundary conditions at $w = 1$ are: $F/\pi = 1.00000$, $F' = -0.752388$, $\sigma = 0.466343$, and $\mu = 1.02282$. The value of $|l(10^3)|$ is less than 10^{-11}, which gives an essentially vanishing ADM mass (25). The matter energy density $\rho(w)$ is proportional to $-{E^0}_0(w)$ by the Einstein equation. The energy density $\rho(w)$ of this numerical solution vanishes exponentially as $w \to \infty$.

IOP Conf. Series: Journal of Physics: Conf. Series **1275** (2019) 012012 doi:10.1088/1742-6596/1275/1/012012

5. Lensing by a stealth defect

Consider the behavior of light rays propagating over a spacetime-defect manifold. It is, in fact, possible to give a simplified discussion [9] by use of an exact vacuum solution [10],

$$ds^2\Big|^{\text{(vac. sol.)}}_{\widetilde{M}_4\,,\,\text{chart-2}} = -\left(1-\widehat{l}/\sqrt{w}\right)(dt)^2 + \frac{1-y_0^2/w}{1-\widehat{l}/\sqrt{w}}\,(dy)^2 + w\left[(dz)^2+\sin^2 z\,(dx)^2\right], \tag{30a}$$

$$w \equiv y_0^2+y^2\,,\quad y_0 \equiv e\,f\,b>0\,,\quad \widehat{l}\in(-\infty,\,y_0)\,. \tag{30b}$$

In the notation of (14a) with dimensionless variables, we have $[\mu(w)]^2 = 1/[\sigma(w)]^2 = 1-\widehat{l}/\sqrt{w}$. In our setup, the vacuum metric (30) can arise in two ways: first, from the solution of Sec. 2 with nontrivial matter-field boundary conditions (15a) in the limit $G_N \to 0$ for fixed energy scale $f>0$ or, second, from the metric of Sec. 2 with a trivial matter field $F(W)=0$, so that the matter energy density (21) vanishes identically.

Next, consider the special case of a stealth-defect vacuum solution (30) with

$$\widehat{l} = 0\,, \tag{31}$$

which has a flat spacetime with vanishing curvature scalars, $R(w) = K(w) = 0$. We consider this “extreme” case, in order to emphasize the difference with standard gravitational lensing [18] which is due to the curvature of spacetime resulting from a nonvanishing matter distribution.

Remark that *exact* multi-defect solutions of the vacuum Einstein equation can be obtained by superposition of these static $\widehat{l}=0$ defects, provided the individual defect surfaces do not intersect. The resulting “gas” of static defects violates Lorentz invariance (see Sec. 6 in Ref. [9] for further details). In principle, we can also obtain an exact multi-defect solution of the vacuum Einstein equation which is approximately Lorentz invariant, if we superpose quasi-randomly positioned and quasi-randomly moving $\widehat{l}=0$ defects (arranged to be nonintersecting initially).

Returning to the single $\widehat{l}=0$ defect with metric (30), the geodesics are readily calculated:

- straight lines in the ambient Euclidean 3-space, if there are no intersections with the defect surface;
- matching straight-line segments in the ambient Euclidean 3-space, if there are intersections with the defect surface.

Figure 9 gives an example of a geodesic staying away from the defect surface, while Figs. 10 and 11 show geodesics crossing the defect surface, with or without parallel shift of the straight-line segments in the ambient Euclidean 3-space (the solid line in Fig. 11, for example, has a parallel shift but the dot-dashed line not).

Due to the parallel shifts at the defect surface, there is a lensing effect, as shown in Fig. 12. This lensing of the flat-spacetime defect results in image formation, as illustrated by Fig. 13. A few remarks are in order:

(i) The image in Fig. 13 is located at the reflection point on the other side of the defect.

(ii) The image is inverted and the image size is equal to the object size. Note that this is also the case if an object in Minkowski spacetime is located at a $2f$ distance from a thin double-convex lens, where f is the focal length of the lens.

(iii) The irradiance of the image (defined as the power per unit receiving area) depends on the defect length scale b and the location of the object: the irradiance of the image will be larger if b is increased for unchanged object position or if the object is brought closer to the defect for unchanged b.

(iv) If a permanent pointlike light source is placed at point P of Fig. 12, then an observer at point P' in the same figure will see a *luminous disk* (different from the *Einstein ring* [18] which the observer would see if the defect were replaced by a patch of Minkowski spacetime with a static spherical star at the center).

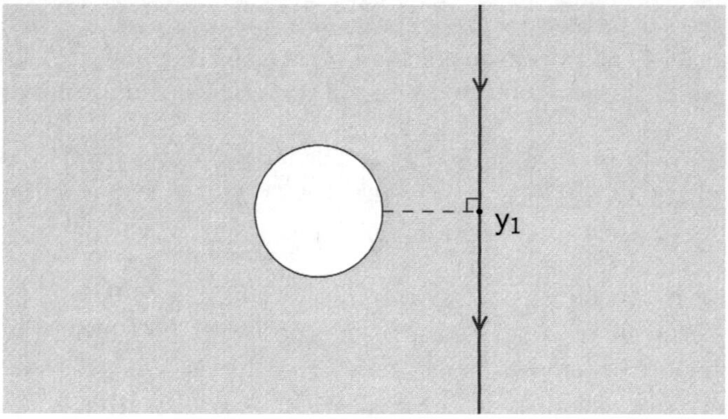

Figure 9. Geodesic which does not cross the defect surface, with part of the 3-space manifold (30)–(31) indicated by the shaded area. The dimensionless quasi-radial coordinate y_1 corresponds to an "impact parameter."

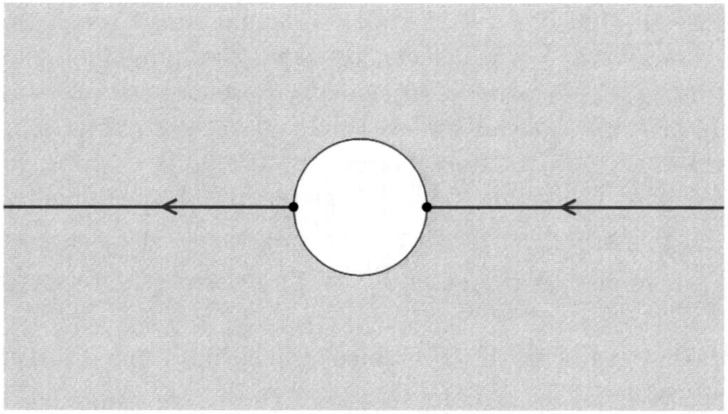

Figure 10. Radial geodesic which crosses the defect surface, where antipodal points (dots) on the defect surface are identified (cf. Fig. 1).

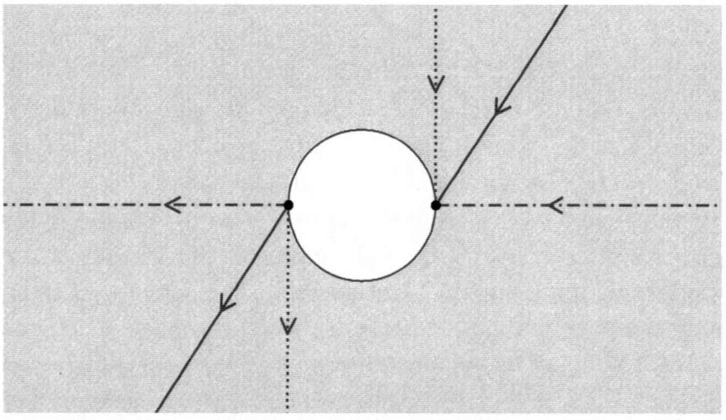

Figure 11. A family of geodesics crossing the defect surface.

IOP Conf. Series: Journal of Physics: Conf. Series **1275** (2019) 012012 doi:10.1088/1742-6596/1275/1/012012

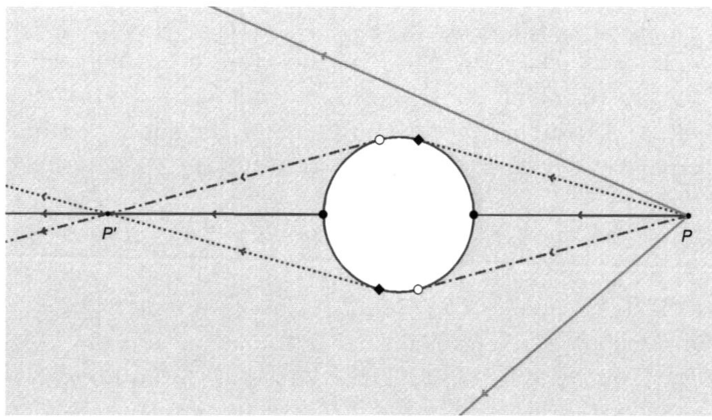

Figure 12. Geodesics with intersection points P and P'.

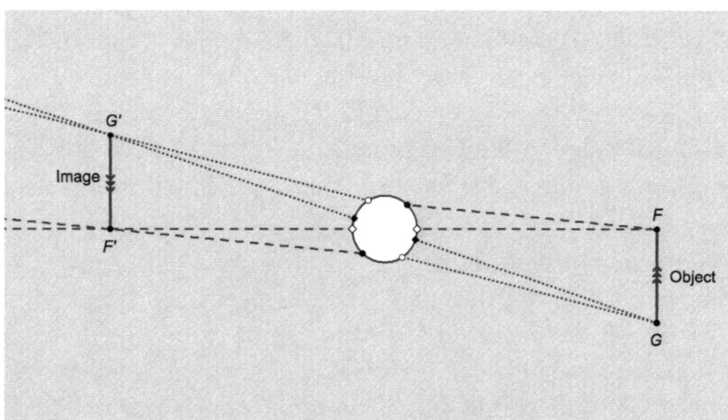

Figure 13. Image formation by the stealth defect.

6. Discussion

Before we review the main characteristics of our soliton-type spacetime defect, we wish to compare our approach to another approach which is more or less orthogonal.

6.1. Comparison to another kind of spacetime defect

It is, in principle, possible to define a local spacetime defect not by what it *is* but rather by what it *does*. Precisely this approach has been adopted in Refs. [14, 15], where the effect considered is the violation of energy-momentum conservation. More precisely, a single particle moves in the ambient spacetime (i.e., the smooth spacetime away from the localized defect) and is arranged to "collide" with a pointlike spacetime defect, which randomly changes the 4-momentum of the particle. The problem with this approach is how to implement it into a consistent theory, without knowing the details of the defect.

In the flat-spacetime model of Ref. [14], the suggestion is to modify the gauge-covariant derivative appearing in the Lagrange density. But, then, gauge invariance is violated, which will introduce an unacceptable violation of unitarity for the particles in the ambient spacetime.

In the curved-spacetime model of Ref. [15], the standard Levi-Civita connection is modified by the addition of a rank-3 tensor $Q^{\lambda}_{\mu\nu} = Q^{\lambda}_{\nu\mu}$, which is thought to be defined on a discrete set of spacetime points (corresponding to the pointlike spacetime defects). The problem, now, is to obtain the modified Einstein equation, which somehow involves derivatives of $Q^{\lambda}_{\mu\nu}$, even though these $Q^{\lambda}_{\mu\nu}$ are only defined on a discrete set of points. It is perhaps possible to consider some smeared-out version of $Q^{\lambda}_{\mu\nu}$, but this is not really satisfactory for a fundamental theory.

Another possibility is simply to view the model of Ref. [15] as an effective theory over a smooth manifold, with pointlike defects replaced by finite-size soliton-type defects. Instead of static soliton-type defects, we then need genuine time-dependent solutions of the classical field equations. Alternatively, we may consider our Skyrmion spacetime defect as a first approximation, where the defect size is considered to be time-dependent: $b = b(t)$ with an approximately constant value $b(t) \sim b_c > 0$ for $|t| < \Delta t$ and an approximately vanishing value $b(t) \sim 0^+$ for $|t| > \Delta t$. This last suggestion for $b(t)$ would correspond to "topology change without topology change" (cf. Sec. 6.6.4 in Ref. [2]). The issue of topology change will also be discussed at the end of Sec. 6.2.

Expanding on the last paragraph, we remark that it is actually possible to calculate the scattering of a "pion" [as defined in (9)] by the $SO(3)$ Skyrmion spacetime defect and to obtain the recoil of the defect with the corresponding energy-momentum change of the "pion." The main observation is that our Skyrmion spacetime defect has a finite-width shell of nonvanishing pion-density, as follows from setting $F(w) \sim \pi$ in (21) for w values between y_0^2 and $y_0^2 + (\Delta y)^2$. This finite matter-density shell then scatters a single incoming "pion." In fact, pion-nucleon scattering [19, 20] has been discussed for the standard $SU(2)$ Skyrmion in Minkowski spacetime and the recoil of this Skyrmion (a finite matter-density ball) has been obtained in Ref. [21, 22]. A similar calculation of the recoil would appear to be feasible for our $SO(3)$ Skyrmion spacetime defect (a finite matter-density shell). This calculated recoil would provide an explicit realization for the defect-induced energy-momentum change [15] of the incoming particle (here, a "pion").

6.2. Remarks on the Skyrmion spacetime defect

The following general remarks aim to put our Skyrmion spacetime defect solution in perspective. First, the new type of Skyrmion solution is rather interesting by itself, as it combines the nontrivial topology of spacetime with the nontrivial topology of field-configuration space. Indeed, the nontrivial topology of the underlying space manifold allows the internal $SO(3)$ space to be covered only once, $N = 1$ in (17).

Second, it remains to be proved that the solution obtained is stable. The scalar fields by themselves would be stable because of the topological charge $N = 1$, but there could still be more branches of solutions with even lower values of the ADM mass (the two known branches are shown in Fig. 2 of Ref. [7]).

Third, the Skyrmion-spacetime-defect metric from (14) is degenerate: $\det g_{\mu\nu} = 0$ at the defect surface $Y = 0$, which corresponds to a submanifold $\mathbb{R}P^2 \sim S^2/\mathbb{Z}_2$ (cf. Fig. 1). A heuristic argument for the necessity of a degenerate metric is as follows: the submanifold $\mathbb{R}P^2$ cannot be differentially embedded in $\mathbb{R}^3$ and this fact implies that a nonsingular solution requires a vanishing metric component g_{YY} at $Y = 0$, making for a vanishing determinant at $Y = 0$ (more details can be found in the third remark of Sec. VI in Ref. [7]).

Fourth, this degenerate metric makes that the Gannon singularity theorem [23] and the Schoen–Yau positive-mass theorem [24] are not directly applicable; we refer to Sec. 3.1.5 in Ref. [13] for further discussion. The special feature of the Skyrmion spacetime defect solution is that certain geodesics at the $\mathbb{R}P^2$ defect surface cannot be continued uniquely: compare the full curve of Fig. 9 in the limit of $y_1 \to 0^+$ with the dotted curve of Fig. 11.

Fifth, the negative ADM mass found for small enough defect length scale b (at a given value of $\widetilde{\eta}$) is not due to ponderable matter but to nontrivial gravitational fields at the $\mathbb{R}P^2$ defect surface with area $2\pi b^2$. Specifically, the boundary value $\sigma(b^2) \in (0, 1)$ determines the approach to zero of the metric component g_{YY}, according to the metric *Ansatz* (14a).

Sixth, having a negative-gravitational-mass spacetime defect [notably a vacuum solution (30a) with $\widehat{l} < 0$] prompts us to reconsider the stability of Minkowski spacetime, especially as classical topology change appears to be possible if degenerate metrics are allowed [25].

Seventh, the crucial open question is the origin and role of nontrivial spacetime topology. For our Skyrmion spacetime defect solution, two specific questions are

(i) what sets the constant defect length scale b?

(ii) can the defect length scale b become a dynamic variable?

There are, of course, many further questions, such as what happens if two defects collide?

The two specific questions listed in the last remark have also been raised in the penultimate paragraph of Sec. 6.1. The issue of (genuine or effective) topology change is essential for a proper understanding of the small-scale structure of spacetime, which brings us back to the "quantum phase" of the first sentence in Sec. 1.

Acknowledgments
The author thanks the organizers of this interesting workshop, in particular Hans-Thomas Elze. He also thanks J.M. Queiruga and Z.L. Wang for useful comments on the manuscript.

References
[1] Misner C W, Thorne K S and Wheeler J A 2017 *Gravitation* (Princeton NJ: Princeton University Press)
[2] Visser M 1995 *Lorentzian Wormholes: From Einstein to Hawking* (New York NY: Springer)
[3] Ashtekar A and Lewandowski J 1997 Quantum theory of geometry. 1. Area operators *Class. Quant. Grav.* **14** A55 (Preprint arXiv:gr-qc/9602046)
[4] Ashtekar A and Lewandowski J 1998 Quantum theory of geometry. 2. Volume operators *Adv. Theor. Math. Phys.* **1** 388 (Preprint arXiv:gr-qc/9711031)
[5] Rovelli C 2008 Loop quantum gravity *Living Rev. Rel.* **11** 5
[6] Klinkhamer F R 2014 Skyrmion spacetime defect *Phys. Rev.* D **90** 024007 (Preprint arXiv:1402.7048)
[7] Klinkhamer F R and Queiruga J M 2018 Antigravity from a spacetime defect *Phys. Rev.* D **97** 124047 (Preprint arXiv:1803.09736)
[8] Klinkhamer F R and Queiruga J M 2018 A stealth defect of spacetime *Mod. Phys. Lett.* A **33** 1850127 (Preprint arXiv:1805.04091)
[9] Klinkhamer F R and Wang Z L 2019 Lensing and imaging by a stealth defect of spacetime *Mod. Phys. Lett.* A **34** 1950026 (Preprint arXiv:1808.02465)
[10] Klinkhamer F R 2014 A new type of nonsingular black-hole solution in general relativity *Mod. Phys. Lett.* A **29** 1430018 (Preprint arXiv:1309.7011)
[11] Klinkhamer F R and Sorba F 2014 Comparison of spacetime defects which are homeomorphic but not diffeomorphic *J. Math. Phys.* **55** 112503 (Preprint arXiv:1404.2901)
[12] Schwarz M 2010 Nontrivial spacetime topology, modified dispersion relations, and an $SO(3)$-Skyrme model *PhD Thesis, Karlsruhe Institute of Technology* (Munich, Germany: Verlag Dr. Hut) [`ISBN 9783868536232`]
[13] Guenther M 2017 Skyrmion spacetime defect, degenerate metric, and negative gravitational mass *Master Thesis, Karlsruhe Institute of Technology* [`https://www.itp.kit.edu/en/publications/diploma`]
[14] Hossenfelder S 2013 Phenomenology of space-time imperfection. II. Local defects *Phys. Rev.* D **88** 124031 (Preprint arXiv:1309.0314)
[15] Hossenfelder S and Gallego Torromé R 2018 General relativity with space-time defects *Class. Quant. Grav.* **35** 175014 (Preprint arXiv:1709.02657)
[16] Skyrme T H R 1961 A nonlinear field theory *Proc. Roy. Soc.* A **260** 127
[17] Manton N S and Sutcliffe P 2004 *Topological Solitons* (Cambridge, UK: Cambridge University Press)
[18] Schneider P, Ehlers J and Falco E E 1992 *Gravitational Lenses* (Berlin: Springer-Verlag)
[19] Hayashi A, Eckart G, Holzwarth G and Walliser H 1984 Pion nucleon scattering phase shifts in the Skyrme model *Phys. Lett.* B **147** 5
[20] Mattis M P and Peskin M E 1985 Systematics of πN scattering in chiral soliton models *Phys. Rev.* D **32** 58
[21] Uehara M, Hayashi A and Saito S 1991 Meson-soliton scattering with full recoil in standard collective coordinate quantization *Nucl. Phys.* A **534** 680
[22] Hughes J and Mathews G J 1992 Skyrmion recoil in pion nucleon scattering *Phys. Rev.* D **46** 970
[23] Gannon D 1975 Singularities in nonsimply connected spacetimes *J. Math. Phys.* **16** 2364
[24] Schoen R and Yau S T 1979 On the proof of the positive mass conjecture in general relativity *Commun. Math. Phys.* **65** 45
[25] Horowitz G T 1991 Topology change in classical and quantum gravity *Class. Quant. Grav.* **8** 587

Two arguments for more fundamental building blocks

Alfredo Iorio

Faculty of Mathematics and Physics, Charles University, V Holešovičkách 2, 18000 Prague 8, Czech Republic

E-mail: alfredo.iorio@mff.cuni.cz

Abstract. We present two lines of reasoning, leading to elementary constituents more fundamental than the ones we know. One such arguments is new, and based on the holographic maximal bound for the number of degrees of freedom of any system. In this case, both matter and space are emergent. The other argument is old, and was given by Richard Feynman as a possible explanation of why analog systems do describe the same physics. The former argument naturally points to a solution of the information paradox. The latter argument elevates analogs from mere curiosities, to reliable tests of fundamental theories. Amusingly, the names given to this fundamental level, both by Feynman and by some of the modern quantum gravity researchers, e.g., Jacob Bekenstein, resemble each others: "Xons" (Feynman) vs "level X" (Bekenstein).

1. Introduction

Here we elaborate on the work of [1], where it is proposed that, at our energies, the fields and spacetimes they live in are entangled. This entanglement is there because both, matter and space, emerge from the dynamics of the same more fundamental objects, whose existence can be inferred from the celebrated upper bound on the entropy of any system, conjectured by Bekenstein [2, 3]. Quoting Feynman [4], and paraphrasing Bekenstein [5], here we call those objects "Xons".

The most noticeable result of [1] is that the evaporation of a black hole (BH) can only lead to an information loss, in the sense that, in general, there is a nonzero entanglement entropy associated to the final products of the evaporation. In what follows, we recollect that result, next to suggestions on how to realize the model on certain analog systems. This gives us the chance to discuss briefly analogs in general, and to offer a reflection on Feynman's famous (but not well-known...) lecture "Electrostatic Analogs", that is chapter twelve of the second volume of [4].

The structure of the paper is as follows. In Section 2 we explain why the Bekenstein bound could mean that Xons exist, we present the quasiparticle picture stemming from this, and we show that it could be the key to the solution of the information-loss puzzle. In Section 3 we discuss how the above scenario could be realized in analogs, then we face the issue of analogs in general, and we close that Section by reflecting on Feynman's conclusions on why analogs do work, leading him to evoke Xons, as the ultimate elementary constituents of matter[1]. That

[1] As we shall show later, Feynman went very close to postulate Xons as elementary constituents of *space*, as well as matter. Indeed, it was the analysis of intimate properties of space itself, that lead him to conjecture the existence of these objects. Nonetheless, there is no clear statement he makes in this respect.

Content from this work may be used under the terms of the Creative Commons Attribution 3.0 licence. Any further distribution of this work must maintain attribution to the author(s) and the title of the work, journal citation and DOI.
Published under licence by IOP Publishing Ltd

striking argument is simply based on the observation of what we daily have around. It was given decades before the much more sophisticated arguments of quantum gravity. The final Section is devoted to our conclusions.

2. The new story: Bekenstein's Xons and the unavoidable information loss

Bekenstein proposed [2] that the entropy of any system, in a given volume V, can never be bigger than the entropy of the BH whose event horizon coincides with the boundary of V

$$S \leq S_{BH} \,. \tag{1}$$

In the proper units (and forgetting log corrections), one has [6] $S_{BH} = A/4$, where $A = \partial V$. The implications of the *holographic* aspects of (1) go far and wide, see, e.g., [7, 8, 9, 10]. We focus here, instead, on the fact that (1) is a universal *upper bound.*

That formula means that the Hilbert space associated to anything contained in V has finite dimension $\sim e^{S_{BH}}$. The world at our energy scales is well described by fields and by the space they live in. Fields, as we know them, have infinite-dimensional Hilbert spaces, to which one should add the degrees of freedom surely carried by (the quanta of) space itself. How can then be that the *ultimate* Hilbert space, that must include *all* degrees of freedom, is not only separable, like for a single harmonic oscillator, but it is actually finite dimensional?

This simple logic points to the existence of something new, more fundamental. Something making both, fields and space. Objects in terms of which particles like the electron, the muon, etc. are, in fact, *quasi-particles*, emerging from the interaction with some sort of *lattice*, whose emergent picture is space itself. Inspired by Feynman (see [4] and later here) we call these objects: *Xons.* To access the Xons one would need resolutions of the order of the Planck length, which might be not only technically unfeasible, but actually impossible (see, e.g., [11]).

2.1. The universal quasiparticle picture

Emergent, nonequivalent descriptions of the same underlying dynamics are ubiquitous in quantum field theory (QFT) [12, 13, 14], as, in general, the vacuum has a nontrivial structure [15] with nonequivalent "phases". That is, for a given basic dynamics one should expect several different Hilbert spaces, representing different "phases" of the system with distinct physical properties. Distinct excitations play the role of the elementary excitations for the given "phase" [16], but their general character is that of the quasiparticles of condensed matter [17, 18]. Noticeably, in theories with certain *dualities* [19, 20], elementary and collective excitations are interchangeable pictures. On this we also comment in Section 3.

What we add here to that QFT picture are, essentially, two important ingredients:

- the finiteness of the degrees of freedom, forcing to interpret fields as emergent;
- the necessity to include the spacetime into such emergent picture[2].

Taking this view, the continuum of fields and space is then only the result of a limiting process. In general, there must be (many!) microscopic configurations of the Xons giving raise to the *same* emergent geometry, but to *different/non-equivalent* field configurations.

Let us now summarize a possible way to formalize the above, for more details see [1]. If H indicates the fundamental Hilbert space of the Xons, what discussed above means that there

[2] That gravity could be an emergent phenomenon is an old idea of Sakharov [21, 22], with many present ramifications, such as, e.g., the "world crystal" model of Kleinert [23], the hydrodynamics approach of Jacobson [24], and many more, see, e.g., [25, 26, 27]. Proposals of particular interest for our discussion are that of [28], where spacetime emerges from the quantum entanglement between the actual fundamental degrees of freedom, see also [29], while in "quantum graphity" [30], fundamental degrees of freedom and their interactions are represented by a complete graph with dynamical structure.

is some mapping, $\mathsf{P}_{\mathrm{G}} : \mathsf{H} \mapsto \mathsf{H}_{\mathsf{G}}$, which assigns to a microscopic state in H the corresponding "classical geometry". On the other hand, there must be some mapping, $\mathsf{P}_{\mathrm{F}} : \mathsf{H} \mapsto \mathsf{H}_{\mathsf{F}}$, which extracts the "field content" of a state in H. The states of the H can be interpreted as states with some classical geometry, $g^{(a)}$, via the mapping P_{G}, and states of the quantum field, ϕ, via the mapping P_{F}:

$$\begin{array}{ccccc} & & |\psi\rangle \in \mathsf{H} & & \\ & \overset{\mathsf{P}_{\mathrm{G}}}{\swarrow} & & \overset{\mathsf{P}_{\mathrm{F}}}{\searrow} & \\ |g^{(a)}\rangle \in \mathsf{H}_{\mathsf{G}} & & & & |\phi\rangle \in \mathsf{H}_{\mathsf{F}} \end{array}$$

In other words, the Hilbert space H_G represents the states which appear as classical spacetimes, on the emergent level, while H_F is the Hilbert space representing states that appear as states of quantum fields, on the emergent level. Henceforth, the generic state in H can be labelled by the values of the coarse-grained variables, $|\psi\rangle = |g^{(a)}, \phi\rangle$.

We can also split H into a direct sum of the subspaces $T_{(i)}$,

$$\mathsf{H} = \bigoplus_{i=1}^{N_T} T_{(i)}, \qquad \dim \mathsf{H} = N_T\, N, \tag{2}$$

where each $T_{(i)}$ has a fixed dimension N and a specific "topology", i.e., a specific distribution of fundamental degrees of freedom among the $|g^{(a)}\rangle$ and the $|\phi\rangle$, and a specific arrangement. N_T is then the number of different topologies. By assumption, each $T_{(i)}$ has a structure

$$T_{(i)} = \mathsf{H}_{\mathrm{G}}^{p_i} \otimes \mathsf{H}_{\mathrm{F}}^{q_i}, \qquad p_i\, q_i = N, \tag{3}$$

where $\mathsf{H}_{\mathrm{G}}^{p}$ ($\mathsf{H}_{\mathrm{F}}^{q}$) is a Hilbert space of dimension p (q) representing possible microscopic states of the geometry (fields). With this in mind, the generic state $|\psi\rangle \in \mathsf{H}$ can be written as

$$|\psi\rangle = \bigoplus_{i=1}^{N_T} \sum_{I=1}^{p_i} \sum_{n=0}^{q_i-1} c_{In}^{(i)} |I_i\rangle \otimes |n_i\rangle, \tag{4}$$

where the vectors $|I_i\rangle$ and $|n_i\rangle$ form a basis of $\mathsf{H}_{\mathrm{G}}^{p_i}$ and $\mathsf{H}_{\mathrm{F}}^{q_i}$, respectively, and $c_{In}^{(i)}$ are numerical coefficients.

By denoting with $\mathsf{P}_{(i)} : \mathsf{H} \mapsto T_{(i)}$ a projector onto $T_{(i)}$, the squared norm of $\mathsf{P}_{(i)}|\psi\rangle$ is the probability $p_{(i)}$ of finding the system in the state with the topology $T_{(i)}$, $p_{(i)} = ||\mathsf{P}_{(i)}|\psi\rangle||^2$.

Importantly, the generic state of $T_{(i)}$ has entanglement between the geometry and the field, in the sense that its decomposition reads

$$\mathsf{P}_{(i)}|\psi\rangle = \sum_{I,n} c_{In}^{(i)} |I_i\rangle \otimes |n_i\rangle \,. \tag{5}$$

The associated density matrix, representing the state of the field, is

$$\rho_{(i)} = \mathrm{Tr}_{\mathsf{H}_{\mathrm{G}}^{p_i}} |\psi\rangle_i \langle\psi|_i, \tag{6}$$

where we first define the normalized state $|\psi\rangle_i = p_{(i)}^{-1/2} \mathsf{P}_{(i)}|\psi\rangle$, and then trace over the degrees of freedom of the gravitational field. Correspondingly, the entropy of geometry-fields entanglement, for a given topology of the lattice is then the usual expression

$$S_{(i)} = -\mathrm{Tr}_{\mathsf{H}_{\mathrm{F}}^{q_i}} \rho_{(i)} \ln \rho_{(i)} \,. \tag{7}$$

As it is impossible to distinguish the space corresponding to different topologies of the lattice, the expected value of the entanglement between the fields and the geometrical degrees of freedom is

$$\langle S\rangle = \sum_i p_{(i)} S_{(i)} \,, \tag{8}$$

This picture needs to be compared to the standard QFT picture, recalled earlier, of the non-equivalent field configurations, or "phases". It is very tempting to consider, in particular, the Thermo-Field Dynamics (TFD) [31] framework, with its tilde degrees of freedom playing the role of (a coarse-grained and effective description of) the degrees of freedom of the geometry to trace away. Indeed, the vacuum of TFD can be written as [31]

$$|0(\theta)\rangle = \sum_n \sqrt{w_n} |n, \tilde{n}\rangle, \tag{9}$$

where θ is a physical order parameter, such as temperature, magnetization, etc., (but also acceleration, surface gravity, etc., in gravitational contests [32, 33]) labeling the different "'phases", w_n are probabilities such that $\sum_n w_n = 1$, and the states $|n, \tilde{n}\rangle$ (infinite in number) are the components of the condensate, each made of pairs of n quanta and their n mirror counterparts ($\tilde{n}$). Therefore, such vacuum is clearly an entangled state. Notice that [14]

$$\langle 0(\theta)|0(\theta')\rangle \to 0, \tag{10}$$

in the field limit, that formalizes the inequivalence we have discussed. Notice also that, if one fixes θ, there is no unitary evolution to disentangle the vacuum, as the interaction with the environment and non-unitarity are the basis for the generation and the stability of such entanglement [33].

The expected value of *field's* observables, O, are obtained by tracing away the *mirror* modes, $\tilde{n}$. In the TFD formalism this corresponds to taking the vacuum expectation value over the vacuum (9)

$$\langle O\rangle \equiv \langle 0(\theta)|O|0(\theta)\rangle = Z^{-1}(\theta) \sum_n e^{\theta E_n} \langle n|O|n\rangle \,, \tag{11}$$

where Z is a partition function.

In particular, there is always an *entanglement entropy* associated to any field, given by, e.g.,

$$\langle S\rangle = \sum_k \left[n_k \ln n_k + (1 - n_k) \ln(1 - n_k)\right] \,. \tag{12}$$

where $n_k = \langle N_k\rangle$ is the expected value of the number operator for the given (fermionic, in this example) mode k. The analogy of (12) with (8) and (7) is stronger, if we think that in TFD the process of taking statistical averages through tracing is replaced, by construction [31], by taking vevs over the vacuum (9).

In this comparison, the mirror (tilde) image of the field mimics the effects of the entanglement with space where the field lives, even when the space is flat. This happens on a level that is both emergent and effective. This would have far reaching consequences, surely worth a serious exploration. For instance, the entanglement entropy associated to any field, would never be zero. Furthermore, this would explain why quantizing gravity as we quantize the matter fields, cannot make much sense.

IOP Conf. Series: Journal of Physics: Conf. Series **1275** (2019) 012013 doi:10.1088/1742-6596/1275/1/012013

Table 1: Three possible choices of the of the parameters, for two topologies, $N_T = 2$, thirty geometries, $N_G = 30$, and no "field degeneracy", $R_F = 1$.

case	i	R^i_G	$p_i = 30R^i_G$	q_i	$N = p_i q_i$	$dim\mathsf{H} = N_T N$
I	1	1	30	200	6000	12000
	2	5	150	40		
II	1	2	60	200	12000	24000
	2	10	300	40		
III	1	4	120	200	24000	48000
	2	20	600	40		

2.2. Effects on black hole evaporation

When applied to BH evaporation, the immediate consequence of the above, is that it is impossible that after the evaporation we can retrieve the very same "phase" we had before the BH was formed. Hence, the information associated to the quantum fields before the formation of the BH is, in general, lost after the BH has evaporated, due to the fields-geometry entanglement.

Even when the emergent spaces, before the formation and after the evaporation, are the same, say Minkowski, the emergent fields belong, in general, to non-equivalent Hilbert spaces. Therefore, even assuming unitary evolution at the X level, the initial and final Hilbert spaces of fields cannot be the same. There is always a *relic* field-geometry entanglement entropy.

We shall now try to go all the way to actually computing these relic entropies, for an idealized scenario of BH evaporation, that we now describe.

We might think that initially matter, described by a quantum field, is in an almost flat space. Such matter then collapses[3], and eventually forms a BH of mass M_0. In our picture, the BH evaporates in a discrete manner, and we take each emitted quantum with the same energy ε, so that $M_0 = N_G \varepsilon$ for some integer N_G. At the end of the evaporation, the space becomes almost flat again, but the field is in a different state with N_G quanta.

Putting together (2) and (3), the Hilbert space can be written as

$$\mathsf{H} = \bigoplus_{i=1}^{N_T} \mathsf{H}_G^{p_i} \otimes \mathsf{H}_F^{q_i} \tag{13}$$

where N^T is, as said, the number of topologies, and we now introduce measures, R_Fs,R_Gs, of the "degenaracies", i.e.,

$$p_i \quad = \quad N_G \, R^i_G. \tag{14}$$

with N_G classical geometries available (they represent the BH with mass $M^{(a)} = a\,\varepsilon$, where $a = 0, 1, \ldots N_G - 1$), and each classical geometry can be realized by R^i_G microstates, while

$$q_i \quad = \quad N_F \, R^i_F. \tag{15}$$

that is, each emergent field state can be realized by R^i_F indistinguishable microstates.

When the BH starts to evaporate, we assume continuous unitary evolution of the state in H but take the "snapshots" of the system, when the expected values are

$$\langle M \rangle = (N_G - 1 - k)\,\varepsilon, \quad \langle n \rangle = k, \tag{16}$$

[3] The mechanism for BH formation could be less immediate when the spacetime is of lower dimension. For instance, in three dimensions, although a solid and well known BH solution of Einstein gravity exists [34], its degrees of freedom are of topological origin, i.e., due to specific boundary conditions (see, e.g. [35]), and collapse is not as immediate as in four dimensions.

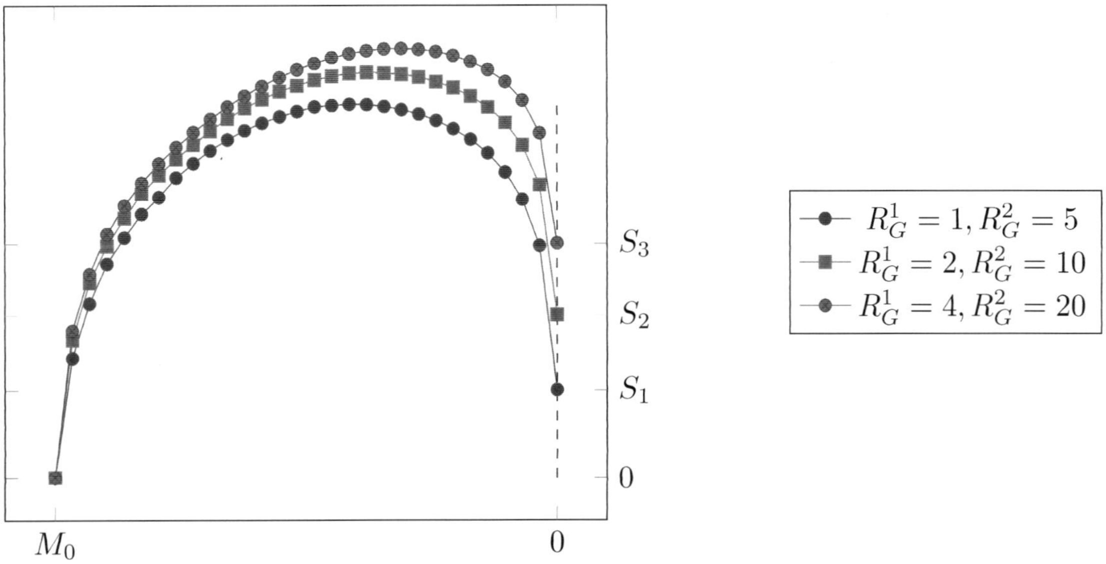

Figure 1: Field-geometry entanglement entropy, as a function of the decreasing mass of the evaporating BH. The initial and final points of this curve are in exact correspondence with the initial and final points of the Page curve. The plot here is for two topologies and three cases. The more microscopic realizations of macroscopic classical geometries are allowed, the higher the residual entropies. Here: $S_1 = 0.77$, $S_2 = 1.43$, $S_3 = 2.06$. Taken from [1].

where k acquires discrete values $k = 0, 1, \dots N_G - 1$. Here $k = 0$ corresponds to BH of maximal mass $M_0 = M^{(N_G-1)} = (N_G - 1)\varepsilon$ and vacuum outside the BH. That does not necessarily mean that the Hilbert space for the field outside the BH has dimension 1, because in our model such dimension depends on the topology. However, since there is only one vacuum state in all topologies, the field is disentangled from the geometry. Hence, our starting point coincides with the starting point of the standard analysis of Page [36]: expected entanglement vanishes at the beginning of the evaporation.

At the k-th step, the BH already emitted k quanta of the field, so its mass decreased by the value $k\,\varepsilon$, while the field is in the state with k quanta outside. When the BH is fully evaporated for $k = N_G - 1$, the field is in the state with $N_G - 1$ quanta.

The actual analytic computations of the entanglement entropy are a heavy tall, so in [1] we proceed with easier (but not trivial...) numerical computations. The case we want to report here is for the following choice of $N_G = 30$, $N_T = 2$, and $R_F^i = 1$, for each topology. We present in Table 1 three cases, I, II, and III, and plot in Fig. 1 the corresponding entanglement entropies, as functions of the discrete parameter k.

As can be seen from the figure, the residual entropies are never zero, and are given by

$$S_1 = 0.77, \quad S_2 = 1.43, \quad S_3 = 2.06\,, \tag{17}$$

for the cases I, II, and III, respectively. As it must be, the more microscopic realizations of the same macroscopic geometry (i.e. the bigger the degeneracy R_G), the higher the relic entanglement entropy.

The fact that, at the end of the evaporation, the entanglement entropy remains finite signals a dramatic departure from the *information conservation* scenario of the famous Page curve [36, 37], presented here in Fig. 2. To fully appreciate this, we need to give the tools to compare the figures obtained within the quasiparticle non-unitaty picture of [1], with that obtained in

IOP Conf. Series: Journal of Physics: Conf. Series **1275** (2019) 012013 doi:10.1088/1742-6596/1275/1/012013

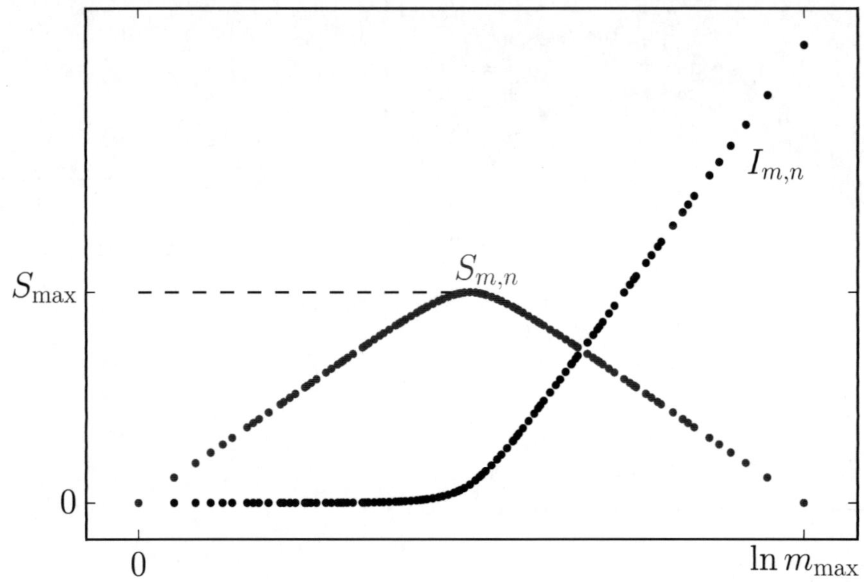

Figure 2: Field entropy of entanglement between modes inside the BH and modes of the radiation vs the dimension of the Hilbert space of the radiation, obtained in [37] by Page via a unitary evolution. The point $m = 0$ corresponds to the initial mass of the BH, $M = M_0$, while m_{max} corresponds a fully evaporated BH, $M = 0$.

the fully unitary picture of [36, 37].

In Page's scenario only fields are considered, and there is no explicit room for the degrees of freedom associated to the geometry. The Hilbert space represents the states of the field "inside" and "outside" the BH horizon. Thus, this is as for any standard bipartite system, $\mathsf{H} = \mathsf{H}_A^m \otimes \mathsf{H}_B^n$, where superscripts m and n indicate the dimension of the corresponding Hilbert space, so that $\dim \mathsf{H} = m\,n$. Then, picking an arbitrary fixed state $|\psi_0\rangle \in \mathsf{H}$, and a random unitary matrix U, $U|\psi_0\rangle$ is a random state in H. By tracing away the subsystem B, to such state one associates the density matrix $\rho_A(U)$, and the corresponding entanglement entropy $S_{m,n}(U)$. Averaging through U we get the average entanglement entropy of the subsystem A, $S_{m,n} = \langle S_{m,n}(U)\rangle$ and the average *information* contained in A, defined as $I_{m,n} = \ln m - S_{m,n}$. It is found [36], see also [38], that

$$I_{m,n} = \ln m + \frac{m-1}{2n} - \sum_{k=n+1}^{mn} \frac{1}{k}, \tag{18}$$

for $m < n$.

These results, applied to BH evaporation, give the curves in Fig.2. There we see that, when the BH is formed, there is no Hawking radiation outside, hence, $n = 1$ and $m = \dim \mathsf{H}$. $S_{m,n}$ is trivially zero. As the BH evaporates, n increases while m decreases, keeping $m\,n$ constant. Since the emitted photons are entangled with the particles under the horizon, $S_{m,n}$ increases, but only up to, approximately, half of the evaporation process. There, the information stored below the horizon starts to leak from the BH, so that $S_{m,n}$ decreases untill full evaporation, hence $m = 1$ and $n = \dim \mathsf{H}$ and $S_{m,n}$ returns to zero.

To conclude this Section, we may say that, even if one takes a conservative view, for which the Xons evolve unitarily, nonunitarity is unavoidable. Indeed,

- the unitary evolution may as well be only formally possible, but physically

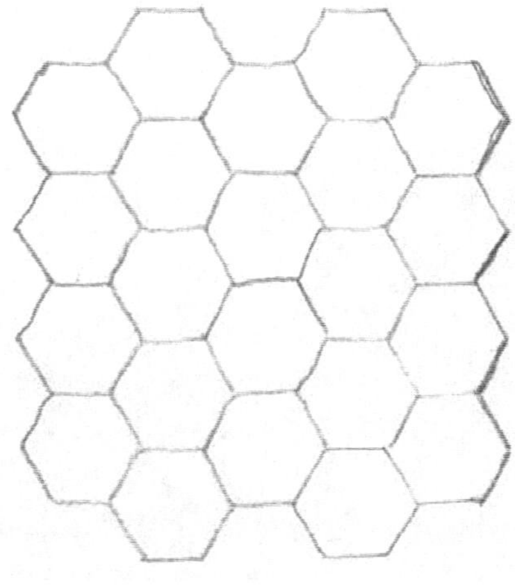

Figure 3: An hexagonal lattice, common to many Dirac materials.

impossible to spot, for a generalized uncertainty forbidding the necessary Planck scale localization/resolution (see, e.g., [11]);

- the emergent description of the evolution is that of the combined system gravity+matter, hence there is inevitably information loss, due to the relic entanglement of the matter field with the geometry;
- this description should apply also to standard nonunitary features of QFT, and we evoke here the possibility that the tilde degrees of freedom of TFD could be interpreted as "how the emergent fields see the degrees of freedom of geometry with which they are entangled".

Notice that our description does allow for an arbitrary number of different fields, hence naturally includes the possibility of yet unknown kinds of matter.

3. The old story: Feynman's Xons and the elevation of analogs' status

It is extremely important to bring the long lasting open questions on the information loss, and on many other fundamental issues, under the judgement of experiments. In fact, we believe that there is probably no way to solve these issues solely on the basis of logical consistency of no matter which theory. This is also witnessed by the fact that, over the years, many have changed their views on the information loss and related issues, including Stephen Hawking himself[4].

For instance, it would be extremely interesting to perform an analog experiment on BH evaporation, to see which picture is correct, Page's or ours. We do not suggest here a specific set-up for the experiment, but suggest, instead, how certain scenarios, evoked in the quasiparticle picture of last Section, could find analog realizations in condensed matter. In fact, such analog scenarios inspired the theoretical model of [1] in the first place.

Examples of quasiparticles are many, some more, some less cogent as possible tests of our picture. Quasiparticles are the Cooper pairs of type II superconductors [39, 40] that are bosonic free excitations emerging from the basic dynamics of the electrons interacting with the lattice. Quasiparticles are the more recently discovered low energy excitations of graphene (see, e.g., the review [41]), that are massless Dirac pseudo-relativistic fermions (the matter ψ), emerging

[4] I cannot refrain to report here a funny anecdote. At the keynote lecture of a conference in Vienna, Roger Penrose, in a large and packed lecture-hall, wanted to defend his persistent view that information loss indeed takes place during BH evaporation. As he wanted Stephen Hawking on his side he said "I agree with the old Hawking, and by *old* I mean the young one".

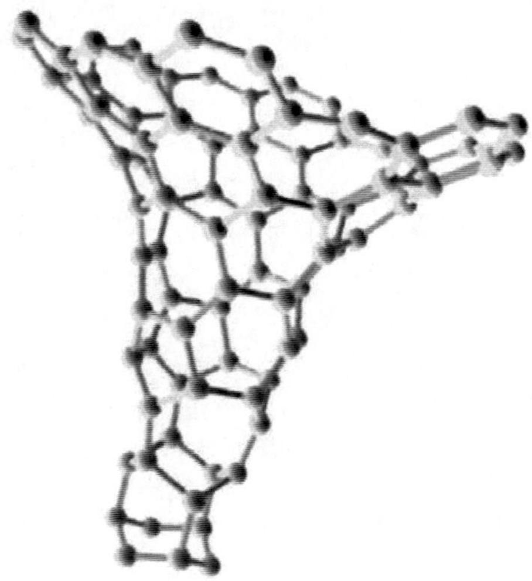

Figure 4: A configuration with the necessary minimal number (six) of heptagonal defects, in order to have the intrinsic Gaussian curvature of a Beltrami pseudospehere [48, 49]. In the appropriate continuum limit, that is for a large number of atoms, and for large enough wavelengths of the π-electrons, this configuration realizes a spacetime conformal to a BH.

from the dynamics of electrons propagating in a carbon two-dimensional honeycomb lattice, whose emergent (continuum limit) description is that of a pseudo-Riemannian (2+1)-dimensional spacetime $g_{\mu\nu}$. Here the role of the Xons is played by carbon's electrons, ions and photons.

Indeed, we know from previous research [42, 43, 44, 45] (see also the review [46]), that the emergence of intrinsic as well as extrinsic curvature in graphene can be used to probe the fundamental physics of the quantum Dirac field theory in the presence of a variety of spacetimes, or even quantum gravity scenarios [47].

3.1. Black hole evaporation in graphene and graphene-like materials

Let us start with an hexagonal lattice of ions, hold together by the σ bonds. On this lattice π-electrons are the quasiparticles, whose properties are determined by a combination of the effects of the hexagonal geometry, and of the quantum interaction with the ions. One can associate a "flat world", (η, ψ), to the configuration of Fig. 3, that is free of inelastic defects. The world (η, ψ) is made of two components: the flat spacetime η, that is the emergent description of the lattice, and the Dirac field ψ, that is the emergent description of the π-electrons. In general, one should also add a $U(1)$ field that is the strain-induced gauge field with components,

IOP Conf. Series: Journal of Physics: Conf. Series **1275** (2019) 012013 doi:10.1088/1742-6596/1275/1/012013

$A_x \sim u_{xx} + u_{yy}$ and $A_y \sim u_{xy}$, where u_{ij} is the strain tensor, but for simplicity, we do not consider that here.

Then defects set in, according to specific arrangements [48, 49], and make shapes like those of Fig.4 , that, in certain specific limits, recreate conditions analog, through a conformal mapping, to those of a three-dimensional BH [42, 43]. In the example of Fig.4 there are exactly 6 heptagonal defects, each carrying a Gaussian curvature $K_7 = -\pi/6$ [48].

In the limit of a large number of atoms (a necessary condition for the emergent description, along with wavelength of π-electron much larger than the lattice spacing), more and more defects, in the form of chains pentagons-heptagons, set in. Each such chain carries either a surplus of one heptagon, or even a net zero curvature. The role of these chains, usually referred to as *scars*, is to distribute the curvature as evenly as possible on the surface. The emergent description of this is a "BH world", that we indicate with $(g^{(a)}_{\rm BH}, \psi_{\rm BH})$.

In a process that carries away the excess defects, that would mimic the BH evaporation through the running of the parameter a, one should arrive again at a "flat world", that is a membrane with overall zero curvature and zero torsion. Such flat world would still have η, as emergent description for spacetime, and some sort of flat space Dirac field as emergent description for the field.

What is interesting here is that, the final product of this process, in general, could have defects still trapped into its texture but, if arranged in specific manners, the emergent spacetimes are indistinguishable from η.

Three such examples are shown in Figs. 5, 6 and 7. These defects configurations are routinely found in real materials [50], and include the famous Stone-Thrower-Wales (STW) defect [51, 52], shown in Fig. 6.

They all have some nonzero local Gaussian curvature, at the location of the defect, but due to the arrangement of such defects, the curvature cannot propagate, giving an overall zero total curtavure. For Fig.5 one has $K_{tot} = 2K_5 + K_8 = 2(\pi/6) - 2\pi/6 = 0$, for the STW defect of Fig.6 one has $K_{tot} = 2K_5 + 2K_7 = 2(\pi/6) + 2(-\pi/6) = 0$, and, finally, for Fig.7 one has $K_{tot} = 3K_5 + 3K_7 = 3(\pi/6) + 3(-\pi/6) = 0$. Furthermore, as is clear from the figures, these configurations do not introduce any nonzero Burger vector beyond the localized deformation, hence also torsion is absent from the emergent picture [45]. Summarizing, the emergent spacetimes, in all three cases, are indeed η spacetimes.

On the field side, though, the situation is each time different, becasue the spectrum of the π-electrons is actually modified by the presence of such defects [50]. Hence, a correct emergent description of the three cases should be (η, ψ'), (η, ψ'') and (η, ψ'''). The geometry part is the same, η, while the matter part, $\psi, \psi', \psi'', \psi'''$, being sensitive to number, location and orientation of the defects, is each time different[5].

[5] Clearly, these defects carry with them some strain, that makes the membrane a little deformed as compared to the pristine case, and introduce the emergent A_μ we already discussed. As said, this is not included in the present picture, nonetheless it is easy to do so by enlarging the "matter content" of the theory by including $A_\mu, A'_\mu, A''_\mu, A'''_\mu$.

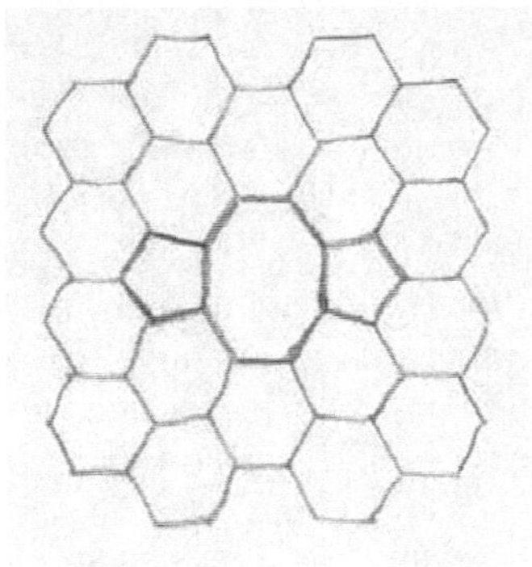

Figure 5: (η, ψ')

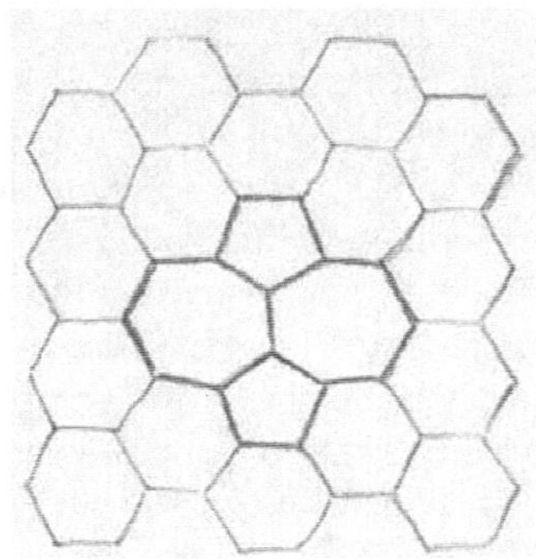

Figure 6: (η, ψ'') STW

Figure 7: (η, ψ''')

$$\begin{array}{ccc} & Xons & (\eta, \psi') \text{ or} \\ (\eta, \psi) \rightarrow & (g^{(a)}_{\rm BH}, \psi_{\rm BH}) \rightarrow & (\eta, \psi'') \text{ or} \\ & & (\eta, \psi''') \end{array} \tag{19}$$

This scenario should then reproduce, with Dirac materials, the reshuffling of the fundamental degrees of freedom of the Xons (carbon's electrons, ions and photons, in the example of graphene), supposed to be at work during the evaporation as described in the previous Section. A schematic view is reported in (19).

3.2. The importance of analogs

The field of analogs is, by now, an old and well established one [53, 54]. Furthermore, due to our deeper understanding and experimental control of condensed matter systems, it is becoming increasingly popular to reproduce aspects of fundamental physics in analog systems. Examples are the cited Hawking phenomenon on graphene [42, 43] and in Bose-Einstein condensates [55], gravitational and axial anomalies in Weyl semimetals [56], and more (for a recent result see, e.g., [57]). Actually, that does not apply only to condensed matter, as can be seen in the interpretation of hadronization in high energy collisions as a Unruh phenomenon [58] (see also [59] and [60]).

Despite that, the importance of analogs for fundamental theoretical physics is still not appreciated. This is witnessed by the ongoing philosophical debates (see, e.g., [61]), and by the fact that theorists look down at experiments on analogs as mere *divertissements*, and never use any of those results to fix any fundamental theory. The rest of this paper is devoted to discuss why this is wrong, and to advocate the necessity to raise the status of analogs to the indirect experimental tests of fundamental theories, otherwise not-testable.

First, let us say that *analogy* (from the Greek $\alpha\nu\alpha - \lambda o\gamma o\varsigma$, , equal – relation), fully qualifies to be a member of an important family of *relations* among different descriptions, or different theories, or else, whose scope is that of a unified view, and points to an underlying invariant structure of nature.

The members of the family we could identify are four

- *Symmetry.* Here we have a theory, with action $\mathcal{A}(\Phi)$, mapped into itself

$$\mathcal{A}(\Phi) \rightarrow \mathcal{A}(\Phi') = \mathcal{A}(\Phi) \tag{20}$$

when $\Phi \rightarrow \Phi'$ is a mapping with some algebraic (Lie group, usually, but not escluseively) structure. The advantages are two. One is that there are conserved charges, pointing, in the case of spatiotemporal symmetries, to certain properties of spacetime (e.g., homogeneity and isotropy). The second advantage is that, as long as Φ, Φ', ..., Φ'' are all given by the same mapping, we can use any of the different descriptions, $\mathcal{A}(\Phi)$, $\mathcal{A}(\Phi')$, ..., $\mathcal{A}(\Phi'')$, at our convenience, to describe the *same* physics. The descriptions, albeit equivalent, can be dramatically different.

- *Duality.* This time, in general a theory is mapped to a different theory, or to a different regime of the same theory
$$\mathcal{A}(\Phi) \to \mathcal{A}_D(\Phi_D) \neq \mathcal{A}(\Phi) \tag{21}$$
when $\Phi \to \Phi_D$. There might be some algebraic structure behind the mapping, think of the $SL(2,R)$ structure behind the electromagnetic duality of certain supersymmetric theories [62, 63]. There might be no structure at all. Surely, here one cannot identify conserved charges. This is an important departure from the case of symmetry, because it is not easy to define what is the fixed entity we are describing in different and related, but not equivalent fashions. For symmetry, one can say that the fixed (invariant) entities are the conserved (invariant) charges, and the various descriptions only differ because different variables are used. Here the entities even change their nature, namely, collective solitonic excitations are mapped into elementary excitations, and *viceversa*. What this relation shares with symmetry is the possibility to use different descriptions, $\mathcal{A}(\Phi)$ and $\mathcal{A}_D(\Phi_D)$, at one's convenience. The descriptions are not equivalent, but we have a mapping that allows to translate results in one setting, into results in the other setting. This last occurrence, and the fact that elementary and collective excitations are interchangeable concepts, make one think that there could be an underlying dynamics, common to both (or to more than two) dual formulations, of fundamental entities, in terms of which everything at the higher level of complexity, is just collective/quasiparticle.
- *Correspondence.* Although someone might classify this relation as some form of the previous one(e.g., sometimes Juan Maldacena calls his AdS/CFT correspondence [64] a "holographic duality" [65]), we better dedicate to it a separate entry. In fact, this relation is more loose than the previous two, as it might include both, in certain limits, but surely is not a map from a theory into itself (symmetry) (or into a dual description) that works all the time. All that is necessary to have a correspondence is that certain key quantities, such as quantum correlators, are related. Therefore, this relation is one step ahead in the generalization of the first two relations of this family. It shares with them that there is a *criterion* behind the relation. The criterion can be inspired by physics (e.g., the conservation of energy, the electromagnetic duality, holography), and again it points to an underlying structure common to all descriptions.
- *Analogy.* Here we move a bigger step further, but, according to Feynman (see later), we are still in the same family. We are in the presence of an analogy between different physical phenomena when the equations governing such phenomena are identical. Therefore, the formal developments of the theory of one of them are valid for the others, provided we pay due attention to the fact that: a) the symbols used in the equations have, in general, different meaning, and b) the boundary conditions for one system may make no sense for an analog system, or could be very hard to realize in practice. Here there is no theory turning into itself (symmetry). Perhaps one can look at that as a theory mapped into another theory, so some sort of duality or correspondence, but in any case there is no *criterion* for this relation. That is, what is missing is a reason, based on some physical argument (like conservation of energy, of charge, etc.), or inspired by some speculation (like electromagnetic duality, or holography, etc), for establishing this relation. Missing a criterion, we do not have a specific *mapping*. We have, instead, a *vocabulary* of the meaning of symbols, on both sides of the relation: fields, couplings, sources, etc.. It is then very difficult to

identify any common underlying structure.

The apparent lack of a criterion to establish an analogy, and the absence of arguments in favor of a common underlying structure, is what probably made the misfortune of analogs among pure theorists. Theorists use with no limitations all sort of combinations, spontaneous breaking, deformations, etc., of the first three members of this family, but look down to the last member as a curious coincidence, but nothing more. They, usually, do not know what to do with this, and, do not include analogs as possible experimental ways to tests their models. On this we shall comment in the following, with the help of Feynman, who had a different attitude towards analogs.

3.3. Feynman's lecture and lesson on analogs

In this last part of the paper, we offer some reflections on a lecture given by Feynman titled "Electrostatic Analogs", available in [4]. This lecture is very famous, but we are not sure that it is equally well known...

By this we mean that, probably, everyone has heard of the title of the first Section there, saying "*The same equations have the same solutions*", but Feynman, in that Section and in the whole lecture, offers much more than that sentence, and we doubt that this made any serious impact.

This is a pity, because Feynman's extraordinary intuition and freedom of thoughts has proved pivotal in disparate areas of physics. Quantum computing is a prominent example of his visionary power. Thus, let us summarize here what is in that lecture, and let us comment on it.

The first Section, whose title is so famous, deals with the possibility for a physicist to be able to retain a broad knowledge as opposed to the apparent necessity for specialization. The reasons in favor of the broad knowledge, mentioned by Feynman, all refer to some underlying unity: general principles that always apply (like conservation of energy); complicated phenomena based on an underlying quantum electrodynamics (taken here as a prototype of a fundamental theory); and finally, the "[...] *most remarkable coincidence: The equations for many different physical situations have exactly the same appearance*" says Feynman.

He goes on with "*Of course, the symbols may be different - one letter is substituted for another - but the mathematical form of the equations is the same. This means that having studied one subject, we immediately have a great deal of direct and precise knowledge about the solutions of the equations of another*".

We disagree only with the use of the word "direct" there, as "indirect" is probably more correct. That said, though, Feynman put it so clearly that it is really surprising that analogs are still looked down by (too) many. But let us keep going.

After having said what quoted above, in that Section he recalls the basic equations of electrostatics, summarized by

$$\vec{\nabla} \cdot (\kappa \vec{\nabla} \phi) = -\frac{\rho_{free}}{\epsilon_0}, \tag{22}$$

with standard notation. The point he wants to make here is that "[...] *there are many physics problems whose mathematical equations have the same form. There is a potential (ϕ) whose gradient multiplied by a scalar function (κ) has a divergence equal to another scalar function (ρ_{free}/ϵ_0)). Whatever we know about electrostatics can immediately be carried over into that other subject, and vice versa*".

There are then five beautiful Sections dedicated to five such different physical problems, governed by (22): The flow of heat (a point source near an infinite plane boundary) – The stretched membrane – The diffusion of neutrons (a uniform spherical source in a homogeneous medium) – Irrotational fluid flow (the flow past a sphere) – Illumination (the uniform lighting of a plane).

For instance, the Section on the stretched membrane considers a thin rubber sheet stretched over a large horizontal frame. Feynman shows that the displacement $u(x, y)$, obtained by pushing the membrane up in one place and down in another, obeys

$$\nabla^2 u(x,y) = -\frac{f(x,y)}{\tau}\,, \tag{23}$$

where f is the force with which we push and pull, and τ is the surface tension. This is just the equation (22), with u corresponding to ϕ, and f/τ corresponding to ρ_{free}/ϵ_0 (here τ is constant, that corresponds to constant κ there). Therefore, having some fixed values of u in some places is "[...] *analog of having a definite potential at the corresponding places in an electrical situation. So, for instance, we may make a positive 'potential' by pushing up on the membrane with an object having the cross-sectional shape of the corresponding cylindrical conductor*".

Here Feynman says that not only one can use electrostatics (a "fundamental" theory) to solve the elasticity (an "emergent" theory) problem, but also the reverse is true! He says: "*Various rods and bars are pushed against the sheet to heights that correspond to the potentials of a set of electrodes. Measurements of the height then give the electrical potential for the electrical situation. The analogy has been carried even further. If little balls are placed on the membrane, their motion corresponds approximately to the motion of electrons in the corresponding electric field. One can actually watch the 'electrons' move on their trajectories. This method was used to design the complicated geometry of many photomultiplier tubes (such as the ones used for scintillation counters, and the one used for controlling the headlight beams on Cadillacs)*".

What strikes here is that one can design specific shapes of the membrane, corresponding to electric potentials very difficult to obtain, and then let the physics of the membrane follow its paths, governed by equation (23). The results are valid for the other system, governed by the same equation, just with different meaning of the symbols, (22). Any further warning makes no sense. Actually, as noticed, Feynman says that one gets "direct" knowledge of system A, by studying the analog system B. We would say "indirect", instead, but that is the only place where we differ (and it is only a matter of wording). The fact is that there is no reason whatsoever for doubting that the results obtained with B would be valid for A, and *viceversa*.

Our question here is: so, why we are not fully exploiting the analogs to acquire with accessible systems knowledge of unaccessible systems? Although it might seem that analogs are very popular these days, this popularity is still not what we advocate here. Analogs are popular as a sort of amusement, but they are not considered as experimental tests of the fundamental theories.

For instance, with reference to Section 2 of this paper, it would be truly important to settle the information loss vs conservation puzzle through experiments. Although some analogs can be proposed (and we offered one in Section 2), the general attitude is that, these not being true BHs, they cannot be used as test-beds of the theory. We think the opposite, as said already in other occasions [66, 67].

On the other hand, Feynman question here is: "*Why are the equations from different phenomena so similar*"? This means to look for the criterion behind the analogy, we absence of which we have pointed to in the discussion about the family of four relations. By evoking the "*underlying unity of nature*", Feynman starts elaborating on the possibility that "*everything is made out of the same stuff, and therefore obeys the same equations*". Then concludes that "*[...] the thing which is common to all the phenomena is the space, the framework into which the physics is put. As long as things are reasonably smooth in space, then the important things that will be involved will be the rates of change of quantities with position in space. That is why we always get an equation with a gradient. The derivatives must appear in the form of a gradient or a divergence; because the laws of physics are independent of direction, they must be expressible in vector form. The equations of electrostatics are the simplest vector equations that*

IOP Conf. Series: Journal of Physics: Conf. Series **1275** (2019) 012013 doi:10.1088/1742-6596/1275/1/012013

one can get which involve only the spatial derivatives of quantities. Any other simple problem - or simplification of a complicated problem - must look like electrostatics. What is common to all our problems is that they involve space and that we have imitated what is actually a complicated phenomenon by a simple differential equation."

Eventually, simply on the basis of what just quoted, he arrives at other questions, leading straight to the Xons that make (nearly) everything: "*Are they* [the electrostatic equations, ed] *also correct only as a smoothed-out imitation of a really much more complicated microscopic world? Could it be that the real world consists of little Xons which can be seen only at very tiny distances? And that in our measurements we are always observing on such a large scale that we cant see these little Xons, and that is why we get the differential equations?*"

Amazingly, the fact that systems A, B, C, etc are all analog, in a smoothed, approximated (emergent, in today's jargon) fashion of a system D (electrostatics), that surely was deemed elementary at that time, lead Feynman look the way a physicist looks, i.e. in a democratic fashion: all systems, including D, are emergent descriptions of one more fundamental thing.

Although Feynman took brave steps into including space into the picture, as a fundamental criterion for the analogy, he really meant the properties of space (isotropy, in this case). So he only made half of the journey of Section 2 of this paper. That is, Feynman space is not really made of Xons: they need be there, but they do not make space itself.

Thus the last part of his comments, where he throws negative statements on the finding of a consistent theory of Xons, should be now confronted with explanations of the fundamental world where space also dramatically changes its nature at the ultimate X level. For instance, Conne's noncommutative geometry [68], appears to reproduces the *standard* Standard Model of particle physics.

4. Conclusions

We elaborated on the hypothesis that both matter, and the geometry where it lives, emerge from the dynamics of more fundamental objects, as can be inferred from the Bekenstein bound on the entropy of any system.

A natural consequence is that, at our energies, fields and spaces are generally entangled entities. This leads in [1] to the noticeable result that BH evaporation is unavoidably a nonunitary process, hence with information loss. The final stage of the famous Page curve is then analyzed in this framework, and the conclusion is that there is always a nonzero entanglement entropy associated to the final products of the evaporation.

As this description should apply also to standard inequivalent representations of QFT, we evoke here a fascinating possibility for the tilde degrees of freedom of TFD. The question is whether such mirror images of the physical degrees of freedom could be interpreted as the way emergent fields see the degrees of freedom of geometry with which they are entangled.

We then suggest how certain scenarios of the quasiparticle picture could find analog realizations in condensed matter. In particular we propose graphene and graphene-like materials as possible analogs, with the necessary reshuffling of the fundamental degrees of freedom, between geometry and fields, at work during BH evaporation. In the example of graphene, the fundamental constituents, that we call Xons, are the carbon's electrons, ions and photons. A schematic view of how the evaporation mechanism should be reproduced with these systems is discussed.

As analogs popped up, we took the chance to discuss them in general. First by trying to include analogy into a family of relations (symmetry, duality, correspondence, analogy), and then by scrutinizing certain intriguing arguments of Feynman that elevate analogs from mere curiosities, to reliable tests of fundamental theories. Feynman's arguments point to the existence of fundamental constituents, that we see as predecessors of the Xons stemming from the Bekenstein bound, although the latter make also space not only matter.

IOP Conf. Series: Journal of Physics: Conf. Series **1275** (2019) 012013 doi:10.1088/1742-6596/1275/1/012013

The discussion of this paper adds even more reasons to construct an experimental facility, where to systematically test fundamental theories with analogs, as already advocated in various occasions [66, 67].

Acknowledgments

The author is indebted to Giovanni Acquaviva and Martin Scholtz for their collaboration on these matters (and geometries), and for managing to bring the Xons (aka "dots") under the judgment of computations. He also thanks Georgios Lukas-Gerakopulos and Pablo Pais for many interesting and fruitful discussions.

References

[1] Acquaviva G, Iorio A and Scholtz M 2017 *Ann. Phys.* **387** 317
[2] Bekenstein J D 1981 *Phys. Rev. D* **23** 287
[3] Bekenstein J D 2014 *Phys. Rev. E* **89** 1
[4] Feynman R, Leighton R, Sands M and Gottlieb M 2006 *The Feynman Lectures on Physics* (Pearson/Addison-Wesley)
[5] Bekenstein J D 2003 *Sci. Am.* **289** 58
[6] Bekenstein J D 1973 *Phys. Rev. D* **7** 2333
[7] 't Hooft G 1993 *Conf. Proc.* **C930308** 284 (*Preprint* gr-qc/9310026)
[8] Susskind L 1995 *J. Math. Phys.* **36** 6377
[9] Bousso R 1999 *J. High Energy Phys.* **06** 24
[10] Bousso R 2002 *Rev. Mod. Phys.* **74**(3) 825
[11] Doplicher S, Fredenhagen K and Roberts J E 1995 *Comm. Math. Phys.* **172** 187
[12] Haag R 1992 *Local quantum physics: fields, particles, algebras* (Springer-Verlag)
[13] Dirac P A M 1966 *Lectures on Quantum Field Theory* (Yeshiva University)
[14] Umezawa H, Matsumoto H and Tachiki M 1982 *Thermo field dynamics and condensed states* (North-Holland Pub. Co.)
[15] Milonni P W 2013 *The Quantum Vacuum: An Introduction to Quantum Electrodynamics* (Elsevier Science)
[16] Umezawa H 1993 *Advanced field theory: micro, macro, and thermal physics* (American Institute of Physics)
[17] Landau L D, Lifshitz, E M and Pitaevski, L P 1984 *Electrodynamics of continuous media* (Pergamon)
[18] Landau L D, Lifshitz, E M and Pitaevski, L P 1984 *Statistical Physics* 2 (Pergamon)
[19] Montonen C and Olive D 1977 *Phys. Lett. B* **72** 117
[20] Castellani E 2017 *Stud. Hist. Philos. Sci. B* **59**
[21] Sakharov A D 1967 *Dokl. Akad. Nauk, SSSR, reprinted in Gen. Rel. Grav. (2000)* **32** 70
[22] Visser M 2002 *Mod. Phys. Lett. A* **17** 977
[23] Kleinert H 1987 *Ann. Phys.* **499** 117
[24] Jacobson T 1995 *Phys. Rev. Lett.* **75** 1260
[25] Baez J C 1999 *Lect. Notes Phys.* **543** 55 (*Preprint* gr-qc/9905087)
[26] Lombard J 2017 *Phys. Rev. D* **95** 024001
[27] Oriti D 2014 *Stud. Hist. Philos. Sci. B* **46** 186 (*Preprint* 1302.2849)
[28] van Raamsdonk M 2010 *Gen. Rel. Grav.* **42** 2323
[29] Cao C, Carroll S M and Michalakis S 2017 *Phys. Rev. D* **95** 024031
[30] Konopka T, Markopoulou F and Severini S 2008 *Phys. Rev. D* **77** 104029
[31] Takahashi Y and Umezawa H 1996 *Int. J. Mod. Phys. B* **10** 1755
[32] Iorio A, Lambiase G and Vitiello G 2001 *Ann. Phys.* **294** 234
[33] Iorio A, Lambiase G and Vitiello G 2004 *Ann. Phys.* **309** 151
[34] Banados M, Teitelboim C and Zanelli J 1992 *Phys. Rev. Lett.* **69** 1849
[35] Carlip S 2003 *Quantum gravity in 2+1 dimensions* (Cambridge University Press)
[36] Page D N 1993 *Phys. Rev. Lett.* **71** 1291
[37] Page D N 1993 *Phys. Rev. Lett.* **71** 3743
[38] Harlow D 2016 *Rev. Mod. Phys.* **88** 015002
[39] Bardeen J, Cooper L N and Schrieffer J R 1957 *Phys. Rev.* **108** 1175
[40] Altland A and Simons B 2010 *Condensed Matter Field Theory* (Cambridge University Press)
[41] Castro Neto A H, Guinea F, Peres N M R, Novoselov K S and Geim A K 2009 *Rev. Mod. Phys.* **81** 109
[42] Iorio A and Lambiase G 2012 *Phys. Lett.* **B716** 334
[43] Iorio A and Lambiase G 2014 *Phys. Rev.* **D90** 025006
[44] Iorio A and Pais P 2015 *Phys. Rev.* **D92** 125005

[45] Iorio A and Pais P 2018 *Ann. Phys.* **398** 265
[46] Iorio A 2015 *Int. J. Mod. Phys.* **D24** 1530013 (*Preprint* 1412.4554)
[47] Iorio A, Pais P, Elmashad I A, Ali A F, Faizal M and Abou-Salem L I 2018 *Int. J. Mod. Phys.* **D27** 1850080
[48] Iorio A 2013 *J. Phys.: Conf. Ser.* **442** 012056
[49] Taioli S, Gabbrielli R, Simonucci S, Pugno N M and Iorio A 2016 *J. Phys.: Cond. Matt.* **28** 13LT01
[50] Araujo P T, Terrones M and Dresselhaus M S 2012 *Materials Today* **15** 98
[51] Thrower P A 1969 *Chem. Phys. of Carbon* **5** 217
[52] Stone A and Wales D 1986 *Chem. Phys. Lett.* **128** 501
[53] Volovik G E 2006 *Int. Ser. Monogr. Phys.* **117** 1–526
[54] Barceló C, Liberati S and Visser M 2011 *Liv. Rev. Rel.* **14** 3
[55] Steinhauer J 2016 *Nature Phys.* **12** 959
[56] Gooth J *et al.* 2017 *Nature* **547** 324
[57] Drori J, Rosenberg Y, Bermudez D, Silberberg Y and Leonhardt U 2019 *Phys. Rev. Lett.* **122** 010404
[58] Castorina P, Kharzeev D and Satz H 2007 *Eur. Phys. J.* **C52** 187
[59] Castorina P, Grumiller D and Iorio A 2008 *Phys. Rev.* **D77** 124034
[60] Castorina P, Iorio A and Satz H 2015 *Int. J. Mod. Phys.* **E24** 1550056
[61] Dardashti R 2016 *Challenging Scientic Methodology: Theory Assessment and Development in Modern Fundamental Physics* (LMU PhD thesis, München)
[62] Iorio A 2000 *Phys. Lett. B* **487** 171
[63] Iorio A, O'Raifeartaigh L and Wolf S 2001 *Ann. Phys.* **290** 156
[64] Maldacena J 1999 *Int. J. Theor. Phys.* **38** 1113
[65] Maldacena J 2005 *Sci. Am.* **293** 56
[66] Iorio A 2015 *Front. Mat.* **1** 36
[67] Iorio A 2015 *J. Phys.: Conf. Ser.* **626** 012035
[68] Connes A 1994 *Noncommutative geometry* (Academic Press)

IOP Conf. Series: Journal of Physics: Conf. Series **1275** (2019) 012014 doi:10.1088/1742-6596/1275/1/012014

Nonlinear deterministic-chaotic collapse model - preliminaries, philosophy, locality

Tamás Geszti

Department of Physics of Complex Systems, Eötvös Loránd University, Budapest, Hungary

E-mail: geszti@elte.hu

Abstract. This is an extended discussion of Ref. [1], presenting a nonlinear dynamical model of quantum collapse, with randomness emerging from self-generated noise. Here we focus on a few issues: 1) the way chaos theory explains "deterministic but unpredictable" as a generic feature of nonlinear dynamics; 2) a new argument about why Bell-CHSH and GHZ experiments confirm quantum mechanics; 3) a discussion of why the heating effect predicted by CSL-type theories is not expected to happen according to our approach; 4) in looking for direct-product solutions of our nonlinear Von Neumann equation, we use a parametrization different from that used in Ref. [1], allowing better insight to the locality issue.

1. Introduction: unpredictable determinism

Quantum measurement is a nonlinear phenomenon: an entering wave is getting separated into a superposition of partial waves, each facing a detector; finally, one of the detectors, apparently randomly chosen, gives a signal, the others not. That requires *pumping* weights of a superposition from one partial wave into another; a nonlinear process, since in linear evolution those weights remain constant. On the contrary, quantum phenomena are routinely described by the linear Schrödinger equation, offering an extremely accurate description of whatever happens before the measurement.

That contradiction is obviously violating our longstanding belief that physics is based on measurements, observing objective properties of the world around us, furnishing data independent of the act of measurement. A successful escape is the Copenhagen interpretation: a special status is assigned to the measurement process, meant to offer random data at probabilities determined by the solutions of the Schrödinger equation through Born's rule.

As recognized by Einstein [2, 3], whether measurement results reflect properties of the observed system independent of the measuring apparatus ("reality"), is open to experimental tests by observing detector-detector correlations. Further analysis focused experimental tests on two-particle correlations [4, 5] with a flexible control on detector setting; later extended to three-particle correlations [6]. An enormous amount of experimental evidence accumulated since that time [7] demonstrates that the expectations are *not* satisfied: measurement outcomes are not determined by any known or unknown properties of the incoming particles ("local reality"); rather, individual outcomes emerge in the process of measurement; only their statistics can be predicted through Born's rule. This is in clear contradiction with our classicality-based prejudices about apparatus-independent reality, and suggests we should accept that the actual

Content from this work may be used under the terms of the Creative Commons Attribution 3.0 licence. Any further distribution of this work must maintain attribution to the author(s) and the title of the work, journal citation and DOI.
Published under licence by IOP Publishing Ltd

IOP Conf. Series: Journal of Physics: Conf. Series **1275** (2019) 012014 doi:10.1088/1742-6596/1275/1/012014

state of the apparatus, including all its microscopic details, is part of the reality displayed in measurement outcomes.

On the contrary, this is not in contradiction with full determinism, as we learned in the 1970's from the rise of nonlinear chaotic dynamics [8]; the trivial example of which is throwing dice. The key idea is *sensitivity to initial conditions,* making the evolution unpredictable, even if deterministic, since one is unable to control initial conditions accurately enough to decide between markedly different outcomes. As foreseen by Lyapunov's groundbreaking treatment of nonlinear differential equations, and subsequently identified on low-dimension models of chaos [8], this is a decisive property of macroscopic bodies, like a particle detector, deliberately prepared in a metastable initial state [9, 10].

All that points into the direction of looking for a nonlinear dynamical equation, reproducing the collapse phenomenology, including Born's rule. The treatment should focus on finding a description for the events of pumping; the route to Born's rule is then open through Pearle's "gambler's ruin" mechanism [11].

Generically, nonlinear dynamics carries the danger of superluminal signaling [12, 13]; however, a shortcut is offered by the observation that fully reproducing quantum predictions implies no signaling [14].

Taking that route, in Ref. [1] we presented a deterministic evolution equation for the density matrix, containing a minimal nonlinear extension of the Von Neumann equation. As briefly summarized below, straightforward analysis of that equation is capable to fully reproduce the collapse phenomenology of quantum measurement, with random outcomes emerging from nonlinear dynamics, at probabilities determined by Born's rule. That would resolve the signaling issue, with one contradiction remaining: our treatment excludes signaling through statistics, not in individual events, as explicitely seen in the equations. Although quantum predictions refer to statistics only, individual events are part of the reality around us - this is a logical catch that our treatment is unable to avoid.

Our treatment is close to the widely studied model class called CSL [15], with important differences: those models postulate an omnipresent external noise, fundamentally connected to the actual quantum state through Born's rule, whereas we consider self-generated noise emerging in detector dynamics, reproducing Born's rule only in measurement situations. Analysing non-measurement scenarios on the quantum-classical border, amply discussed in terms of CSL-type collapse models, may require a different approach; as mentioned in Ref. [1], a possible explanation of $1/f$ noise is one of the obvious scopes of further research.

2. The nonlinear equation

As described in detail in Ref. [1], we are looking for strictly deterministic evolution of an individual system, composed of a microsystem and a delocalized set of detectors. Evolution should be nonlinear, which means that the Hamiltonian $\hat{H}$ should depend on the actual quantum state of the composed system. Our first task is to specify the form of that dependence.

The requirement of gauge invariance excludes linear dependence on the state vector $|\mathbf{\Psi}\rangle$; the obvious alternative is a Hamiltonian $\hat{H}(\hat{\varrho})$ depending on the density matrix $\hat{\varrho}$. Since nothing physically disappears during collapse, only weights of a superposition - constant in linear dynamics - are pumped by nonlinearity from one quantum state into another; their sum remaining unchanged. Accordingly, the evolution should be unitary, and $\hat{H}$ Hermitian. To account for the accuracy of linear dynamics on the microscopic level, we are looking for a minimal nonlinearity, with $\hat{H}(\hat{\varrho})$ depending linearly on $\hat{\varrho}$. The obvious choice is then to add nonlinearity in the form of a single Kraus operator $\hat{M}\hat{\varrho}\hat{M}^{\dagger}$ to the linear Hamiltonian $\hat{H}_0$. Accordingly, we

IOP Conf. Series: Journal of Physics: Conf. Series **1275** (2019) 012014 doi:10.1088/1742-6596/1275/1/012014

are going to study evolution governed by a nonlinear Von Neumann equation

$$\begin{aligned}\partial_t \hat{\varrho} &= -\frac{i}{\hbar}\left[\hat{H}(\hat{\varrho}), \hat{\varrho}\right] \\ &= -\frac{i}{\hbar}\left[\left(\hat{H}_0 + \zeta \hat{M}\hat{\varrho}\hat{M}^\dagger\right), \hat{\varrho}\right].\end{aligned} \tag{1}$$

Following Ref. [1], we focus on typical measurement situations, and describe collapse as pumping weights among terms of a superposition. Quantum measurement is done by a set of remote detectors. It starts by a Stern-Gerlach-type separation process of incoming particles, open to control by changing beamsplitter orientations and detector positions. Separation results in a superposition of the form $\sum_k c_k|k\rangle$, expanded on an orthogonal basis of non-overlapping partial waves. Each basis state is defined by a different combination of those detectors which are hit by particles, and those left quiet - a distinctive feature of measurement situations, shared by Stern-Gerlach, Bell-CHSH and GHZ[1]. Our derivation below demonstrates that this is sufficient to arrive at Born's rule; we conjecture it is necessary as well, and in non-measurement situations non-Born statistical features may emerge.

Interaction with the detectors at time $t = 0$, through fast linear quantum dynamics, creates an entangled state of particles and detectors in the form

$$|\Psi\rangle = \sum_k c_k |k\rangle |\Phi_k\rangle, \tag{2}$$

which becomes the initial state for subsequent temporal changes concluding in collapse. $|\Phi_k(t)\rangle$ denotes a multi-detector pure state vector, with local environments of each detector included.

The density matrix corresponding to state vector (2) takes the form

$$\hat{\varrho} = |\Psi\rangle\langle\Psi| = \sum_{k,l} |k\rangle\langle l| \, c_k c_l^* \, \hat{R}_{kl}(t), \tag{3}$$

with

$$\hat{R}_{kl} = |\Phi_k\rangle\langle\Phi_l|. \tag{4}$$

In coordinate representation, the operator $\hat{R}_{kl}$ appears as a Hermitian matrix $R_{kl}(x, x'; t) = R_{lk}(x', x; t)^*$, acting on the configuration space of all detectors, including their local environments.

The weight of state $|k\rangle|\Phi_k\rangle$ in the superposition is given by

$$w_k(t) = |c_k|^2 \ \mathrm{Tr}_x \, \hat{R}_{kk}(t) = |c_k|^2 \ \langle\Phi_k(t)|\Phi_k(t)\rangle. \tag{5}$$

Unlike in Ref. [1], now we postulate $|c_k|^2 = const$, and follow the evolution of $w_k(t)$ - pumping from one state to the other, ending in collapse - through the nonlinear dynamics of $\hat{R}_{kl}(t)$. Apart from the initial normalization

$$w_k(t=0) = |c_k|^2 \quad \Longrightarrow \quad \mathrm{Tr}_x \, \hat{R}_{kk}(t=0) = 1 \quad \forall k, \tag{6}$$

the possible initial states of the macroscopic measurement apparatus are of enormous variety, serving as the multi-local hidden parameters mentioned in the Introduction.

[1] As an example, in a Bell measurement, in basis state $|k\rangle = |+-\rangle$ of two entangled particles from a common source, left-going particle is directed towards detector left-up, right-going particle towards detector right-down. It should be noted that these basis states are not what is called "Bell states".

IOP Conf. Series: Journal of Physics: Conf. Series **1275** (2019) 012014 doi:10.1088/1742-6596/1275/1/012014

Since we expect nonlinearity to dominate dynamics of the macroscopic apparatus, we postulate the Kraus operator $\hat{M}$ to act on the configuration space coordinates, not on indices k. Then from Eqs. (5) and (1) we immediately obtain

$$\dot{w}_k = -i\frac{\zeta}{\hbar}\sum_m |c_k|^2|c_m|^2\,\mathrm{Tr}_x\left(\hat{M}\hat{R}_{km}\hat{M}^\dagger\hat{R}_{mk} - \hat{R}_{km}\hat{M}\hat{R}_{mk}\hat{M}^\dagger\right). \tag{7}$$

By cyclic invariance of the trace, this cancels to 0 if $\hat{M}$ is self-adjoint, so to describe pumping, we have to postulate a non-Hermitian $\hat{M} = \hat{M}_1 + i\hat{M}_2$, with $\hat{M}_1$ and $\hat{M}_2$ non-zero Hermitian operators. Then dropping the non-pumping combinations $\hat{M}_1\hat{\varrho}\hat{M}_1$ and $\hat{M}_2\hat{\varrho}\hat{M}_2$, what remains is a term $i\zeta(\hat{M}_1\hat{\varrho}\hat{M}_2 - \hat{M}_2\hat{\varrho}\hat{M}_1)$ in the Hamiltonian.

Trying to give some physical identity to $\hat{M}_1$ and $\hat{M}_2$, while keeping the full nonlinear Hamiltonian scalar, by intuition we choose $\hat{x}$ and $\hat{p}$ as configuration-space vectors, and the whole combination a scalar product over the configuration space. Accordingly, in the following we use Equation (1) in the final form

$$\partial_t\hat{\varrho} = -\frac{i}{\hbar}\left[\hat{H}_0, \hat{\varrho}\right] + \frac{\zeta}{\hbar}\left[\left(\hat{x}\hat{\varrho}\hat{p} - \hat{p}\hat{\varrho}\hat{x}\right), \hat{\varrho}\right] \tag{8}$$

where $\hat{H}_0$ describes linear evolution of the incoming particles and all detectors interacting with them, including environments of the whole measurement setup.[2]

The constant ζ characterizes the strength of nonlinearity. Since the combination $\hat{x}\hat{\varrho}\hat{p} - \hat{p}\hat{\varrho}\hat{x}$ has the dimension of action, measured in quantum units by the factor $\hbar^{-1}$, ζ is of dimensionality t^{-1}. The immediate insight gained therefrom is to connect the quantum-classical border to time scales: for microscopic systems, moving $\mathcal{O}(1)$ amount of action, quantum phenomena go to the end without noticeable signature of nonlinearity; macroscopic amount of action makes nonlinearity immediately dominant. In Ref. [1], $\zeta \approx 10^{-10}\ s^{-1}$, $\zeta^{-1} \approx 1/2\ yr$ is proposed as a rough estimate, based on known parameters of an avalanche photon detector.

3. The route to Born's rule

In what follows, we use the interaction picture and look for a solution of Eq. (8) in the form of Eq. (3). As a preparation, we evaluate the mean value of a configuration-space operator $\hat{A}_x$ acting on the configuration space coordinates, not on indices k; with the result

$$\langle\hat{A}_x\rangle = \sum_k |c_k|^2\,\langle\Phi_k|\hat{A}_x|\Phi_k\rangle = \sum_k w_k\,\langle\hat{A}_x\rangle_k, \tag{9}$$

where

$$\langle\hat{A}_x\rangle_k = \frac{\langle\Phi_k|A_x|\Phi_k\rangle}{\langle\Phi_k|\Phi_k\rangle} = \frac{\mathrm{Tr}_x\left(\hat{R}_{kk}\hat{A}_x\right)}{\mathrm{Tr}_x\ \hat{R}_{kk}} \tag{10}$$

is the mean value of $\hat{A}_x$ in basis state k.

Considering that the vector operators $\hat{x}$ and $\hat{p}$ are configuration-space operators as defined above, and using Eq. (10), in straightforward steps one arrives at the system of equations

$$\frac{\partial\hat{R}_{kl}}{\partial t} = \frac{\zeta}{\hbar}\sum_m w_m\left(\langle\hat{p}\rangle_m\cdot\{\hat{x}, \hat{R}_{kl}\} - \langle\hat{x}\rangle_m\cdot\{\hat{p}, \hat{R}_{kl}\}\right) \tag{11}$$

[2] Rejecting the non-pumping terms is not fully trivial, since the model should be tested in non-measurement situations too. In that context, we notice that rejecting $\hat{x}\hat{\varrho}\hat{x}$ is trivial, since that term would violate translation invariance. On the contrary, a term $\hat{p}\hat{\varrho}\hat{p}$ would add a kind of kinetic energy with a rest mass depending on the quantum state to the Hamiltonian, which is not absurd; anyway, that would introduce a length scale into the framework.

IOP Conf. Series: Journal of Physics: Conf. Series **1275** (2019) 012014 doi:10.1088/1742-6596/1275/1/012014

where $\{...,...\}$ denotes the anticommutator.

To calculate the pumping rates, we focus on $l = k$. Starting from Eq. (5), using cyclic invariance of the trace, one finally obtains the pumping rates in the form

$$\dot{w}_k = \zeta \sum_m w_k w_m A_{km}, \tag{12}$$

where

$$A_{km}(t) = \frac{2}{\hbar} \Big(\langle \hat{x} \rangle_k \cdot \langle \hat{p} \rangle_m - \langle \hat{x} \rangle_m \cdot \langle \hat{p} \rangle_k \Big). \tag{13}$$

Eqs. (12) with (13) constitute a system of balance equations, with the obvious antisymmetry property

$$A_{km} = -A_{mk}, \tag{14}$$

granting $\sum_k \dot{w}_k = 0$, i.e., conservation of the total weight, traced back to self-adjointness of the Hamiltonian.

It is the dimensionless quantities A_{km} defined by Eq. (13) which - although obeying deterministic dynamics under control of Hamiltonian $\hat{H}(\hat{\varrho})$ - behave like a random noise, since the dynamics of metastable detectors is chaotic, and thereby sensitive to details of the initial states of the detectors, impossible to control. That quasi-randomness of $A_{km}(t)$ makes the process described by Eq. (12) similar to a stochastic game of the "gambler's ruin" type [11].

Eq. (12) is immediately seen to describe some kind of dynamics in which weights w_k are vanishing one after the other; once vanished, that weight never reappears; finally, one "winner" remains alive at $w_m = 1$, all the rest disappearing, in accordance with the "gambler's ruin" model. Randomness in quantum measurements emerges from chaotic nonlinear dynamics of the macroscopic detectors biased to a metastable state, giving $A_{km}(t)$ the character of a noise, self-generated by deterministic dynamics.

As discussed in detail in Ref. [1], the no-drift property $\langle A_{km}(t) \rangle = 0$, equivalent to the "fair play" or "martingale" property in the language of games, is sufficient to assure that the probability of a given outcome m is equal to the initial weight $|c_m|^2$, which is Born's rule. By modeling $A_{km}(t)$ as white noise, all that can be followed in detail.

4. Bell, CHSH, GHZ

Testing two-particle correlations in the Bell-CHSH setup [4, 5], as well as three-particle correlations in the GHZ setup [6], were a milestone in rendering quantum foundations a reliable object of research. The common scheme is to test entangled particles prepared in a reproducibly identical way, using a set of detectors in different settings. The assumption that identical preparation of particles implies identical values of hidden parameters determining measurement outcomes ("local realism") leads to predictions which are in disagreement with experiments.

In our scheme that is the natural thing to happen.[3] If hidden parameters are the initial states of detectors, then different detector settings imply different hidden parameters. To strengthen that statement, macroscopically different detector settings imply orthogonality of the corresponding detector quantum states - a trivial case of Anderson's "orthogonality catastrophe" [16]; therefore the postulate of identical values of hidden parameters is not satisfied. Instead, statistical predictions of quantum mechanics are reproduced through the "gambler's ruin" dynamics described by Eq. (12), in agreement with experiments.

[3] This discussion has been inspired by correspondance with Lev Vaidman.

5. No heating

An interesting prediction of noise-based collapse models, heating due to explicit breaking of time translation symmetry by external white noise [17, 18, 19], is absent in the present scheme: noise generation being attached to strongly chaotic dynamics of detectors in the pre-firing period, nothing would break time translation symmetry, leaving no reason for heating in non-measurement situations. In particular, this should be true for the possibility of nucleon decay through bound state excitation [20, 21].[4]

A formal argument to demonstrate that heating is absent in our approach is this: in the interaction picture, using cyclic invariance of the trace, the mean value of the Hamiltonian is

$$\langle \hat{H}(\hat{\varrho}(t))\rangle \;=\; i\zeta\,\mathrm{Tr}\left[\left(\hat{x}(t)\hat{\varrho}(t)\hat{p}(t) - \hat{p}(t)\hat{\varrho}(t)\hat{x}(t)\right)\hat{\varrho}(t)\right] \;=\; 0 \quad \forall t, \tag{15}$$

which excludes that energy, conserved in linear quantum dynamics, might be changed by nonlinearity.

6. Locality and direct product structure

Although the emblematic case of Bell-CHSH and GHZ correlations is handled by the above argument, more insight to what is local and what is not in the dynamics of a multi-detector setup can be obtained by analyzing the structure of solutions of Eq. (8).

The separability of the scalar product offers a convenient starting point to study the issue of locality, since the Hamiltonian governing the evolution according to Eq. (8) is decomposed into a sum over detectors:

$$\hat{H}(\hat{\varrho}) = \sum_d \left(\hat{H}_0^d + i\zeta\left(\hat{x}_d\hat{\varrho}\hat{p}_d - \hat{p}_d\hat{\varrho}\hat{x}_d\right)\right). \tag{16}$$

Here $\hat{H}_0^d$ describes coupled linear evolution of detector d, its interacting environment, as well as the respective partial wave of incoming particles initially interacting with that detector. As an immediate consequence, Eq. (11) takes the form

$$\frac{\partial \hat{R}_{kl}}{\partial t} = \frac{\zeta}{\hbar}\sum_d\sum_m w_m\left(\langle\hat{p}_d\rangle_m\cdot\{\hat{x}_d,\hat{R}_{kl}\} - \langle\hat{x}_d\rangle_m\cdot\{\hat{p}_d,\hat{R}_{kl}\}\right). \tag{17}$$

The pumping rate equation (12) remains unchanged, with Eq. (13) separating into the sum of one-detector terms:

$$A_{km}(t) = \frac{2}{\hbar}\sum_d\left(\langle\hat{x}_d\rangle_k\cdot\langle\hat{p}_d\rangle_m \;-\; \langle\hat{x}_d\rangle_m\cdot\langle\hat{p}_d\rangle_k\right). \tag{18}$$

An important property of the relevant solutions of Eq. (17) emerges if we assume that before measurement starts, remote detectors are uncorrelated. That imposes that following interaction with the separated partial waves of incoming particles from a common source, the multi-detector density matrices $\hat{R}_{kl}$ in Eq. (3) take the direct product structure

$$\hat{R}_{kl}(t) \;=\; \bigotimes_d \hat{r}_{kl}^d(t) \tag{19}$$

with $\hat{r}_{kl}^d = |\varphi_k^d\rangle\langle\varphi_l^d|$.[5] In accordance with our expectations, that structure is preserved during subsequent evolution. Substituting Eqs. (19) and (16) into Eq. (8), then following steps

[4] I thank S. Adler for calling my attention to that work.

[5] In Ref. [1] the one-detector density matrix elements were assumed to be built of normalized one-detector states, varying local weights being incorporated into a single global one; that gave rise to ill-defined equations, as noted in the Corrigendum. In the present treatment the norms of each $|\varphi_k^d\rangle$ are changing by dynamics.

preceding Eq. (11), including use of the interaction picture, one obtains a sum over d of the analogous one-detector equations

$$\frac{\partial \hat{r}^d_{kl}}{\partial t} = \frac{\zeta}{\hbar} \sum_m w_m \left(\langle \hat{p}_d \rangle_m \cdot \{\hat{x}_d, \hat{r}^d_{kl}\} - \langle \hat{x}_d \rangle_m \cdot \{\hat{p}_d, \hat{r}^d_{kl}\} \right), \tag{20}$$

for each d multiplied by a factor $\bigotimes_{d' \neq d} \hat{r}^{d'}_{kl}$.

We introduce the one-detector weights $w^d_k(t)$ as

$$w^d_k(t) = \mathrm{Tr}_d \, \hat{r}^d_{kk}(t), \tag{21}$$

connected to the global weights w_k through Eqs. (19) and (5) as

$$w_k(t) = |c_k|^2 \prod_d w^d_k(t); \quad \frac{\dot{w}_k}{w_k} = \sum_d \frac{\dot{w}^d_k}{w^d_k}. \tag{22}$$

Using Eqs. (20) and (21), each term on the r.h.s. obeys a one-detector dynamical equation

$$\frac{\dot{w}^d_k}{w^d_k} = \frac{2\zeta}{\hbar} \sum_m w_m \Big(\langle \hat{x}_d \rangle_k \cdot \langle \hat{p}_d \rangle_m - \langle \hat{x}_d \rangle_m \cdot \langle \hat{p}_d \rangle_k \Big). \tag{23}$$

Finally, we see that the pumping equations (12) to (14) remain valid, however, the product structure expressed in Eq. (22) implies a strange combination of single-detector dynamics: $w_k \to 0$, as soon as $w^d_k \to 0$ at *anyone* of the detectors. This assigns an unexpectedly strong meaning to the short-living one-detector weights w^d_k, to be explored by further research. Further insight is obtained by noticing that for any operator $\hat{A}_d$ acting on the configuration subspace of detector d, like $\hat{x}_d$ and $\hat{p}_d$, in Eq. (10) the $d' \neq d$ multipliers cancel, therefore all one-detector averages appearing in Eq. (23) are fully determined locally, and can be calculated by $\hat{r}^d_{kk}$:

$$\langle \hat{A}_d \rangle_k = \frac{\langle \varphi_k | \hat{A}_d | \varphi_k \rangle}{\langle \varphi^d_k | \varphi^d_k \rangle} = \frac{\mathrm{Tr}_d \left(\hat{r}^d_{kk} \hat{A}_d \right)}{\mathrm{Tr}_d \; \hat{r}^d_{kk}}. \tag{24}$$

As already emphasized in Ref. [1], it is important to note that whereas Eqs. (20) are quasi-local, since each of them refers to a single detector d, they include the global weights w_m, representing remote entanglement in the scenario. As mentioned in the Introduction, this does not allow signaling through statistics, as granted by the results summarized in Ref. [14]: since our treatment reproduces Born's-rule-based quantum statistical predictions, it remains a subset of non-signaling theories. Thereby, the performance of our model is equivalent to that of CSL-type models [15], with the advantage that here noise appears as a consequence of multilocal dynamics of the measurement apparatus. However, "spooky action-at-a-distance" in individual events, not accessible through measurement statistics but still part of reality, remains with us.

7. Final remarks

As discussed in more detail in Ref. [1], our treatment demonstrates that a deterministic, nonlinear dynamical equation can be capable to reproduce Born's rule quantum statistics, without signaling. Unlike CSL-type theories [15], no external noise is needed for that, randomness emerging from chaotic dynamics. That is demonstrated for genuine measurement situations; in non-measurement contexts different statistics, eventually, 1/f noise [22] can emerge.

Nonlinearity challanges our feeling of comfort offered by linear quantum mechanics. In particular, the equivalence of averaging a density matrix over a Gibbs ensemble of different

IOP Conf. Series: Journal of Physics: Conf. Series **1275** (2019) 012014 doi:10.1088/1742-6596/1275/1/012014

states, and obtaining the same density matrix by tracing out over a big environment, is lost by nonlinearity. In the present case, it is easy to check that the derivation of Born's rule is insensitive to that difference; however, fine details of collapse dynamics, as well as some details of non-measurement dynamics, can show up observable consequences of the loss of that linearity-based equivalence.

Acknowledgment. It is my pleasure to acknowledge endless discussions over many years with Lajos Diósi; special thanks for a critical reading of an earlier version of the present manuscript. I thank helpful recent correspondance with Philip Pearle, Lev Vaidman and Stephen Adler, and stimulating discussions with Michael Vanner at DICE2018.

References

[1] T. Geszti, J. Phys. A: Math. Theor. **51,** 175308 (2018); Corrigendum: *ibid.* **51,** 469501 (2018).
[2] A. Einstein, *Fifth Solvay Conference, 1927,* discussed in: M. Jammer, *The Philosophy of Quantum Mechanics,* (Wiley, New York, 1974), pp. 115-117.
[3] A. Einstein, B. Podolsky, and N. Rosen. Physical Review **47,** 777 (1935).
[4] J. S. Bell, Physics **1,** 195 (1964).
[5] J. F. Clauser, M. A. Horne, A. Shimony, R. A. Holt, Phys. Rev. Lett. **23,** 15 (1969).
[6] D. M. Greenberger, M. Horne, A. Zeilinger, in: *Bell's theorem, Quantum Theory, and Conceptions of the Universe,* ed. M. Kafatos, Kluwer, Dordrecht 69 (1989).
[7] B. Hensen *et al.*, Nature **526**, 682 (2015); L. K. Shalm et al., Phys. Rev. Lett. **115,** 250402 (2015); M. Giustina et al., Phys. Rev. Lett. **115,** 250401 (2015).
[8] See e.g. H. G. Schuster: *Deterministic Chaos: An Introduction* (4th ed., Wiley-VCH 2006).
[9] N. G. van Kampen, Physica A **153,** 97 (1988).
[10] An early attempt to explore detector metastability in a Newton-Schrödinger context appears in T. Geszti, Phys. Rev. A **69**, 032110 (2004).
[11] P. Pearle, Found Phys. **12,** 249 (1982). For background, see W. Feller, *An Introduction to Probability Theory and its Applications*, 3rd ed. (Wiley, New York 1968), Vol. 1, Chap.14. An earlier treatment in several aspects anticipating ours is P. Pearle, Phys. Rev. D **13,** 857 (1976).
[12] N. Gisin, Helv. Phys. Acta **62**, 363 (1989).
[13] J. Polchinski, Phys. Rev. Lett. **66,** 397 (1991); C. Simon, V. Bužek and N. Gisin, Phys. Rev Lett. **87**, 170405 (2001); Ll. Masanes, A. Acin, and N. Gisin, Phys. Rev. A **73**, 012112 (2006); A. Bassi and K. Hejazi, European Journal of Physics, **36**, No. 5 (2015); N. Gisin, Phys. Lett. A **143,** 1 (1990); L. Diósi, J. Phys. Conf. Ser. **701,** 012019 (2016); an early appearance is L. Jánossy, Acta Phys. Hung. **1**, 423 (1952).
[14] For a recent review, see N. Brunner *et al.*, Rev. Mod. Phys. **86**, 419 (2014).
[15] G. Ghirardi, A. Rimini and T. Weber, Phys. Rev. D **34,** 470 (1986) is the primary version; the CSL elaboration appears in P. Pearle, Phys. Rev. A **39,** 2277 (1989); see also Ref. [12] as well as L. Diósi, Phys. Rev. A **40**, 1165 (1989); a recent review is A. Bassi et al., Rev. Mod. Phys. **85**, 471 (2013).
[16] P. W. Anderson, Phys. Rev. Lett. **18,** 1049 (1967); F. Tonielli et al., Phys. Rev. Lett. **122,** 040604 (2019).
[17] L. Diósi, Phys. Rev. Lett. **114**, 050403 (2015).
[18] A. Vinante *et al.*, Phys. Rev. Lett. **116**, 090402 (2016).
[19] S. Adler *et al.*, arXiv:1901.10963 (2019).
[20] P. Pearle and E. Squires, Phys. Rev. Lett. **73,** 1 (1994).
[21] S. N. Mayburov, arXiv:1812.01455 (2018).
[22] See e.g. M. Niemann, H. Kantz and E. Barkai, Phys. Rev. Lett. **110**, 140603 (2013).

IOP Conf. Series: Journal of Physics: Conf. Series **1275** (2019) 012015 doi:10.1088/1742-6596/1275/1/012015

Testing spontaneous collapse models with mechanical experiments

Andrea Vinante

Department of Physics and Astronomy, University of Southampton, SO17 1BJ, United Kingdom

E-mail: `A.Vinante@soton.ac.uk`

Abstract. Spontaneous collapse models (CM) have been proposed as a possible solution of the quantum measurement problem. In CM the unitary evolution of the wave function is modified by stochastic corrections which lead to a dynamical reduction at macroscopic level. Unlike interpretations of quantum mechanics, CM predict departures from standard theory that can be experimentally tested. Here, we focus on one of these predictions, a universal force noise acting on any mechanical system. In particular, we discuss recent bounds on collapse models set by ultracold cantilevers experiments and gravitational wave detectors. First notable results are a partial exclusion of the CSL parameters proposed by Adler and a full exclusion of an early model based on quantum gravity ideas.

1. Introduction

Spontaneous wave function collapse (or dynamical reduction) models (CM) [1, 2, 3, 4, 5] have been proposed to reconcile the linear and deterministic evolution of quantum mechanics with the nonlinearity and stochasticity of the measurement process. According to CM, random collapses in space (i.e. localizations) of the wave function of any system occur spontaneously, independently of measurement processes, leading to a progressive spatial localization. The collapse rate scales with the size of the system, in such a way as to produce rapid localization of any macroscopic system, while giving no measurable effect at the microscopic level, where standard quantum mechanics holds. Importantly, collapse models lead to a natural solution of the measurement problem, by predicting the emergence of well-defined outcomes in any measurement process in agreement with the Born rule, provided that the measurement system is macroscopic.

The standard collapse model described in literature is the Continuous Spontaneous Localization (CSL) [2, 3]. It can be regarded as a continuous version of the earliest collapse model introduced in 1986 by Ghirardi, Rimini and Weber (GRW) [1], and solves some technical issues of the latter. CSL is characterized by two phenomenological constants, a collapse rate λ and a characteristic length r_C, which characterize respectively the intensity and the spatial resolution of the spontaneous collapse process. λ and r_C are free parameters which should be derived, or bounded, by experiments. Other relevant collapse models are the Diosi-Penrose gravitational model [6, 7, 8, 9] and quantum gravity inspired models [10, 11].

The direct effect of collapse models such as CSL is to destroy quantum superpositions, resulting in a loss of coherence in interferometric matter-wave experiments [12, 13, 14]. In

Content from this work may be used under the terms of the Creative Commons Attribution 3.0 licence. Any further distribution of this work must maintain attribution to the author(s) and the title of the work, journal citation and DOI.
Published under licence by IOP Publishing Ltd

addition, the noise field associated with the collapse leads, unavoidably, to the violation of the energy conservation. So called non-interferometric tests have been proposed to look for these effects. Such tests include spontaneous emission of x-rays [15, 16], force noise in mechanical systems [17, 18, 19, 20, 21, 22, 23, 24] and spontaneous heating of bulk matter [25] or ultracold atoms [26, 27].

In this paper, I will review recent experimental progress in testing collapse models by means of non-interferometric mechanical experiments and discuss possible future developments.

2. Models

2.1. The CSL model

The CSL master equation can be written as [4]:

$$\frac{\mathrm{d}}{\mathrm{d}t}\hat{\rho}(t) = -\frac{i}{\hbar}\left[\hat{H}, \hat{\rho}(t)\right] + \mathcal{L}_{\text{CSL}}[\hat{\rho}(t)], \tag{1}$$

where $\hat{H}$ describes the free evolution of the system and $\mathcal{L}[\hat{\rho}(t)]$ is a Lindblad term. In the case of a rigid body undergoing small displacements in a given direction z, with the motion described in terms of the center of mass coordinate, the CSL term can be written as [21]:

$$\mathcal{L}_{\text{CSL}}[\hat{\rho}_{\text{CM}}(t)] = -\eta\,[\hat{q}, [\hat{q}, \hat{\rho}_{\text{CM}}(t)]], \tag{2}$$

$\hat{q} = \hat{z}$ being the position operator of the center-of-mass, and:

$$\eta = \frac{2\lambda}{m_0^2}\iint \mathrm{d}^3\mathbf{r}\,\mathrm{d}^3\mathbf{r}'\,\exp\left(-\frac{|\mathbf{r}-\mathbf{r}'|^2}{4r_C^2}\right)\frac{\partial\varrho(\mathbf{r})}{\partial z}\frac{\partial\varrho(\mathbf{r}')}{\partial z'} \tag{3}$$

$$= \frac{(4\pi)^{\frac{3}{2}}\,\lambda\, r_C^3}{m_0^2}\int\frac{\mathrm{d}^3\mathbf{k}}{(2\pi)^3}\,k_z^2\,e^{-\mathbf{k}^2 r_C^2}\,|\tilde{\varrho}(\mathbf{k})|^2 \tag{4}$$

with $\mathbf{k} = (k_x, k_y, k_z)$, $\tilde{\varrho}(\mathbf{k}) = \int \mathrm{d}^3\mathbf{x}\, e^{i\mathbf{k}\cdot\mathbf{r}}\,\varrho(\mathbf{r})$ and $\varrho(\mathbf{r})$ the mass density distribution of the system. In the latter expressions m_0 is the nucleon mass, λ and r_C are free parameters. The CSL term causes a collapse of the wave function in the coordinate space, i.e. a localization of the center of mass, and therefore a loss of quantum coherence as an effect of the collapse process.

The CSL dynamics is characterized by two phenomenological parameters: the collapse rate λ, which represents the rate of localization of a single nucleon, and the characteristic length r_C, which is related to a smearing of the mass-density function. These are free parameters of the model which should be derived, or bounded, by experiments. The standard conservative values suggested by GRW are $\lambda \simeq 10^{-16}$ s^{-1} and $r_C = 10^{-7}$ m [1, 2] and are sufficient to guarantee almost instantaneous localization of macroscopic objects. A strongly enhanced value for the collapse rate has been suggested by Adler [28], motivated by the requirement of making the wave function collapse effective at the level of latent image formation in photographic process. The values suggested by Adler are $\sim 10^{9\pm2}$ times larger than standard values at $r_C = 10^{-7}$ m, and $\sim 10^{11\pm2}$ times larger at $r_C = 10^{-6}$ m.

Besides causing localization, the CSL term expressed by Eq. (2) describes a diffusion of the center of mass, and η can be interpreted as diffusion constant [17, 18]. In other words, the center of mass of a rigid body will undergo a brownian motion due to the action of the continuous collapse-localization process. This consideration is at heart of so called non-interferometric methods to look for CSL. More generally, the existence of a fundamental CSL collapse mechanism necessarily implies the appearance of a universal force noise, acting on any massive object, with (one-sided) power spectral density $S_{ff} = 2\hbar^2\eta$. A simple consequence is that for a mechanical resonator in equilibrium with a thermal bath at temperature T, the mean energy is given by [21, 22]:

$$\langle E \rangle = k_B T + \frac{\hbar^2 \eta Q}{2m\omega_0}, \tag{5}$$

i.e. CSL predicts a net excess of energy with respect to the classical equipartition value.

Therefore, one can look for CSL effects by continuously monitoring a mechanical resonator and measuring its mean energy as a function of temperature. The equipartition term $k_B T$ can be characterized and subtracted, and the hypothetical CSL term can be estimated or bounded. A more general procedure, valid also for free masses, is to monitor the free dynamics of the system and infer the power spectral density of the force noise acting on it. This is given, in the thermal limit $k_B T \gg \hbar\omega$ and in absence of technical noise, by:

$$S_{ff} = \frac{4k_B T m\omega}{Q} + 2\hbar^2 \eta. \tag{6}$$

Accurate measurements as a function of the temperature T and the damping factor $\gamma = m\omega/Q$ allow to isolate the thermal term and to estimate or bound the collapse-noise term.

In both situations, in order to increase the sensitivity to CSL noise, one has to maximize the ratio between the CSL-induced noise and the thermal noise. In practice this means lowest possible temperature T, highest possible time constant $\tau = Q/\omega$, and highest possible η/m ratio. While the first two conditions merely express the requirement of lowest possible rate of energy exchange with the thermal bath, the third condition leads to consequences which are peculiar to the model details. For CSL we can distinguish two relevant limits. When the characteristic size L of the system is small, $L \ll r_C$, then $\eta \propto m^2$ and thus $\eta/m \propto m$. When the characteristic length of the system (along the direction of motion) L is large, $L \gg r_C$, then $\eta \propto \rho^2 A$ and thus $\eta/m \propto \rho/L$, where ρ is the mass density and A is the area transverse to the motion [21, 22]. The expressions in the two limits imply that, for a well defined characteristic length r_C, the optimal system is a plate or disk with thickness $L \sim r_C$ and the largest possible density ρ.

2.2. The Diosi-Penrose model

Besides CSL, other models have been proposed, attempting to provide a connection of the wave function collapse to specific mechanisms, in particular to gravity. On the one hand, gravity is the only known force that we have not been able to merge with quantum theory, despite enormous theoretical efforts. Indeed, there is no compelling evidence that gravity must be quantized. On the other hand, collapse models suggest that any breakdown of the quantum superposition principle has to be related to a mass-proportional localizing mechanism, pointing to gravity as a natural candidate.

Among the supporters of this point of view, Penrose argues that the quantum superposition principle is in fundamental contradiction with the general covariance principle of general relativity [6, 7]. Therefore, he suggests that superpositions of a massive object in different spatial positions can survive only for a finite time $\tau \sim \hbar/\Delta E$ where ΔE is a properly defined difference between the self-gravitational energies associated to the two states. In other words, the quantum superposition principle must be broken at fundamental level by gravity.

Another attempt in his direction was done by Diosi [8]. His model is similar to CSL, with a different kernel for the the localization integral. The intensity of the collapse rate would be determined by gravity, while a phenomenological length scale r_{DP}, analogous to r_C, is necessary to avoid exceedingly large violations of energy conservation [9]. Experimentally, values of r_{DP} as low as the nuclear size 10^{-15} m are allowed, although values $r_{DP} \sim r_C = 10^{-7}$ m have been suggested as more reasonable [9]. Notably, the collapse time scale predicted by Diosi coincides with the one heuristically suggested by Penrose [6, 7]. This is the reason motivating the denomination of Diosi-Penrose (DP) model in literature.

The DP model leads to localization and diffusion similar to CSL. The diffusion constant η_{DP} is given by [22]:

$$\eta_{DP} = \frac{G\rho m}{6\sqrt{\pi}\hbar}\left(\frac{a}{r_{DP}}\right)^3, \tag{7}$$

where a is the lattice constant and G is the gravitational constant.

Experimentally, Eqs. (5) and (6) hold exactly in the DP model as in CSL, by replacing η with η_{DP}. The requirements of lowest possible T and longest possible time constant $\tau = Q/\omega$ remain unchanged. A notable difference from CSL is that the diffusion constant η_{DP} does not show any shape dependence. In particular, for the relevant case of large size $L \gg r_{DP}$, the ratio η_{DP}/m, which determines the experimental sensitivity, depends only on the density. In other words there is no optimal size or shape.

Another peculiarity of the DP model is that it is characterized by a single free parameter. While a null result experiment will exclude a given region of the CSL two-parameter space $\lambda - r_C$, in the DP case it will set a lower bound on the r_{DP} parameter.

2.3. The Ellis model

Another model was proposed by Ellis and coworkers [10, 11], based on quantum gravity considerations. The wave function of a massive object would be effectively localized by the interaction with a bath of quantum wormholes which characterize the spacetime structure at the Planck length scale. As such, the Ellis model is not strictly speaking a collapse model, rather it is a decoherence model, where decoherence is induced by practically unaccessible degrees of freedom.

As long as the wavelength of wormholes is much longer than the characteristic magnitude of the motion of the system, decoherence can be effectively described by Eq. (2), with diffusion constant [29]:

$$\eta_E = \frac{(m_0 c)^4 m^2}{(\hbar m_{Pl})^3}, \tag{8}$$

where m_{Pl} is the Planck mass and c is the speed of light. Compared to CSL and DP models, the Ellis model is parameter-free, which allows for unambiguous experimental falsifiability.

3. Experiments

3.1. Overview

We consider here two classes of experiments which have been shown to be suitable in testing collapse models: (a) micro- and nano-mechanical systems developed in the context of force microscopy technologies and (b) macroscopic experiments to detect weak forces, such as gravitational wave detectors.

Force microscopy technologies are nowadays a standard tool in applied physics and surface science. The invention of atomic force microscopy (AFM) [30] has been awarded by the Nobel Prize. AFM is based on very sensitive cantilevers, featuring atomically sharp tip, which can be used to map the surface of a sample with atomic resolution by detecting the tip-surface force. One of the most extreme versions of force microscopy is the so called magnetic resonance force microscopy (MRFM), which combines magnetic resonance imaging with ultrasensitive microcantilevers to yield three-dimensional maps of a sample with nanometer resolution. This technique is very demanding, and has led to the development of very sensitive nanocantilevers with relatively low frequency in the kHz range, high quality factor, and to their operation at very low temperature down to the millikelvin regime [31]. According to Eqs. (5) and (6) these systems are natural candidates to detect collapse-induced noise. Furthemore, their size compares favourably with the theoretically suggested CSL length scale [2, 28].

IOP Conf. Series: Journal of Physics: Conf. Series **1275** (2019) 012015 doi:10.1088/1742-6596/1275/1/012015

Macroscopic experiments such as torsion pendulums and gravitational wave detectors [32, 33] are by design force sensors with exquisitely low force noise. These experiments consist of macroscopic mechanical systems with high quality factor and typically very low frequency, in the Hz or sub-Hz range. According to Eqs. (5) and (6) they are obvious candidates to detect or bound collapse-induced force noise. The main strong feature with respect to other systems is the operation at low frequency, which requires extreme care in mechanical isolation from environmental noise. The only limitations come from being macroscopic and thus far from the theoretically suggested CSL length scale [2, 28], and from the fact that these experiments are normally operated at room temperature. However, resonant bar gravitational wave detectors [33] and prototype of future interferometric detectors have been operated at cryogenic temperature as well.

3.2. Ultracold cantilevers

The first serious attempt to exploit mechanical resonators to bound the parameters of collapse models, in particular the CSL model, was done by Vinante et al. in 2016 [34]. The system described in the latter work is a MRFM nanocantilever cooled to temperatures down to 10 mK. A magnetic microsphere was attached to the cantilever free end to allow low noise magnetic detection using a SQUID. The original motivation of this solution was to provide a way to operate a MRFM nanocantilever at millikelvin temperature avoiding the overheating typical of laser-based detection. This goal was successfully achieved, as the cantilever thermal noise was found in good agreement with the equipartition theorem down to 25 mK [31]. From this observation an upper bound on any non-thermal contribution, including hypothetical CSL noise, was estimated. An exclusion curve on the CSL parameter space was then derived (Fig. 1, orange line) [34].

The main notable result was the first partial exclusion of the CSL parameters suggested by Adler, in particular the full exclusion of the parameters suggested at $r_C = 10^{-6}$. However the experiment was far from being optimized. The spherical geometry is not the one that maximizes the effect of CSL-induced noise, and the density of the material is three times lower than the densest known materials, such as osmium or iridium. Moreover, the quality factor in this experiment, $Q \sim 10^5$ was not really high.

A second similar experiment was reported later [35] with cantilever and magnetic sphere roughly 10 times larger than the previous one. The cantilever was a standard AFM cantilever, featuring a very large quality factor, up to 10^7 at 20 mK. Very accurate measurements of the force noise were performed and, according to Eq. (6), the thermal contribution was characterized and subtracted. In contrast with the first experiment, a residual force noise was revealed. The ultimate origin of this noise is still unexplained. Further measurements have shown that it is not related to back-action from the SQUID and likely not due to vibrational noise.

For the purpose of testing CSL, the experiment is still able to enlarge the exclusion region. The exclusion line is shown in red in Fig. 1. If the source of excess noise were explained, the bound would be improved by one order of magnitude (dashed red line in Fig. 1). Of course, the excess noise could be in principle explained as CSL noise as well, and it would be compatible with the CSL effect predicted by Adler. However, given the latest data from LISA Pathfinder (see next section), this explanation seems unlikely.

3.3. Macroscopic experiments: LISA Pathfinder

The bounds on collapse models that can be inferred by macroscopic experiments have been recently discussed by Carlesso et al. [29]. Different experiments have been considered: cryogenic resonant bar detectors of gravitational waves, interferometric detectors of gravitational waves (LIGO), torsion pendulums. It was finally shown that the strongest bounds on collapse models

IOP Conf. Series: Journal of Physics: Conf. Series **1275** (2019) 012015 doi:10.1088/1742-6596/1275/1/012015

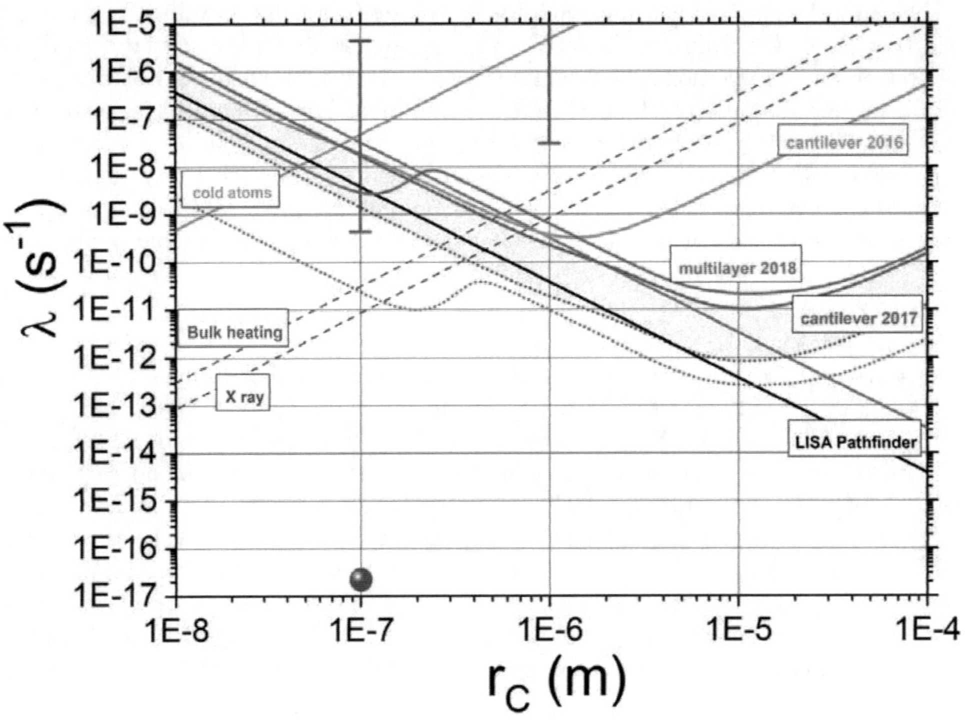

Figure 1. Upper limits on the CSL model collapse rate from different non-interferometric experiments. For each curve the region above the curve is excluded. Solid lines refer to actual experimental bounds, dashed lines to experimental bounds probing the CSL field at very high frequency, dotted lines to hypothetical bounds. Specifically, x-ray spontaneous emission (dashed blue) [16], bulk heating in solid matter (dashed dark green) [25], cold atoms (light green) [27], cantilever 2016 experiment (orange) [34], cantilever 2017 experiment (red) [35], LISA Pathfinder early data (gray) [29, 36] and final data (black) [37]. We also plot the potential improvement from the cantilever 2017 experiment if the excess noise were removed (dotted red line), and a very preliminary bound (magenta) and projected upper limit (magenta dotted) from ongoing cantilever experiments with optimized multilayer test mass. The theoretical parameters suggested by Adler [28] and GRW [1] are also shown.

are set by the recently launched space mission LISA Pathfinder (LPF) [36], the technological demonstrator of the space-born gravitational wave detector LISA.

In LPF, two identical AuPt cubic test masses are nominally in free-fall in space, protected by a spacecraft which follows the masses in such way to minimize the residual coupling. The goal of the mission was to assess the achievable accuracy of free-fall, by optically measuring the residual relative acceleration noise (or equivalently force noise) between the two test masses. Thanks to the very quiet space environment, it was possible to achieve an impressive low acceleration noise of 5.2 $\mathrm{fm/(m/s^2)/\sqrt{Hz}}$ [36], eventually improved by a factor of three following a progressive reduction of residual gas pressure with time [37].

LPF and the future LISA mission are optimized for extremely low frequencies in the range from 10^{-4} Hz to 10^{-1} Hz. In this frequency range, LPF outperforms ground-based torsion pendulums by several orders of magnitude in force power spectral density. Ground based gravitational wave detectors such as Advanced LIGO, despite the impressive mechanical attenuation, are outperformed too by LPF in terms of force noise, mainly because of the higher characteristic frequency. In fact the minimum of the force noise in LIGO is around 30 Hz [29].

The bounds on the CSL model that can be set using LPF data are easily calculated using Eq. (6). As in LPF there is no absolute calibration of the thermal noise, one assumes the worst case condition and attributes the all measured noise to CSL. The gray and black curves in Fig. 1 refer respectively to the first data released officially [36] and the final data [37].

3.4. Discussion

The bounds on CSL obtained from ultracold cantilevers and LPF are reported in Fig. 1, together with other relevant bounds from other experimental techniques, such as spontaneous emission of x-ray [16], spontaneous heating of bulk matter [25] and spontaneous heating of ultracold atoms [27]. In addition, the CSL parameters suggested by Adler (dark green bars) [28] and GRW (violet dot) [1] are shown.

We remark that the bound from x-ray relies on the strong assumption that the CSL noise field extends up to 10^{18} Hz. The bound may be easily evaded by a CSL noise field with a frequency cutoff. Similarly, bounds from bulk matter heating require noise extending up to 10^{11} Hz. In contrast the bounds from mechanical systems and cold atoms probe the CSL noise field at low frequencies, and are therefore considered more significant. On the other hand, bounds from matter-wave interferometry, which represent direct test of the quantum superposition principle, are so far are at the level of $\lambda \sim 10^{-6}$ s^{-1} [38], thus far weaker than those inferred from mechanical resonators.

As shown in Fig. 1, current mechanical bounds are seriously challenging the predictions of the collapse rate by Adler, based on the assumption that the wave function collapse is effective already at the mesoscopic scale [28]. In contrast, the more conservative prediction by GRW is still many orders of magnitude off, and its testing will require impressive improvements on the experimental side.

It is an interesting fact that, at the end of the day, experiments with so huge difference in the relevant parameters, such as ultracold cantilevers and LPF, provide a comparable bound on CSL. We can explain this remarkable coincidence with the help of Eqs. (6). As discussed in the theory section, in the relevant limit $L \gg r_C$, the CSL over thermal noise ratio scales as $\rho Q/(T\omega L)$. Referring to the cantilever experiment, the LPF experiment provides roughly a 10^1 times larger effective Q and 10^6 times lower ω, giving 7 orders of magnitude improvement. However, the temperature T and size L are respectively 10^4 and 10^3 times larger, resulting in a nearly exact compensation by 7 orders of magnitude. This gives also a clear indication of how the ultimate experiment to test CSL ought to be: very low frequency as in LPF, ultralow temperature and micrometer size as in the cantilever case. This calls for an experiment to be performed at ultralow temperature in a space environment.

Let us consider the DP model. In this case, there is no optimal length scale, and consistently it is found that the strongest bound on the characterstic length r_{DP} is set by LPF, roughly one order of magnitude better than cantilevers [39]. Specifically, using the final data of LPF [37], the lower bound $r_{DP} > 83$ fm is found, implying that r_{DP} must be larger than the size of any stable nuclei. This result does not appear so significant, for instance it is very far from the normalization length suggested by Ghirardi et al [9] of 10^{-7} m. However, it rules out some earlier speculation that the nuclear size could possibly set the DP model length scale [22].

Finally, let us consider the Ellis model. It is possible to show that current LPF data, but data from other macroscopic experiments as well, are inconsistent with the Ellis model, as they would have already been able to point out the associated noise expressed by Eq. (8) [29]. Remarkably, the same conclusion has been recently obtained by analyzing atom interferometry experiments [40]. This is perhaps the first example of a collapse model which is completely ruled out by experiments.

4. Recent updates and outlook

We conclude with an update on ongoing measurements and an outlook of possible short and mid term progress.

Concerning macroscopic experiments, a significant improvement within the next 15 years seems unlikely. The LISA Pathfinder mission is over, and the LISA mission is scheduled to be launched in the early 2030s. Moreover, as the LPF mission has already achieved the free-fall requirement for LISA it is not clear if LISA will be able to improve over LPF force noise significantly. On the other hand, ground based experiment, such as advanced LIGO or similar detectors, are quite far from being significant with respect to bounding collapse models. Specifically, the upper limit on the CSL model is worse then the one from LPF by almost 4 orders of magnitude with the initial advanced LIGO data and 2 orders of magnitude with the final projected design sensitivity. However, the third generation of gravitational wave interferometers is expected to significantly improve the low frequency force noise, by means of cryogenic suspension and underground operation. This might perhaps allow to approach performance close to LPF.

In contrast, experiments with ultracold cantilevers or different micromechanical resonators have a better potential for improvement in the short-mid term, as they are far from being optimized. Improvements by 2-3 orders of magnitude appear possible in the near future by a combination of two main strategies. The first is moving to lower frequency, by means of softer resonators or larger mass. This will require a much better mechanical isolation, but entirely within reach. The second is an optimization of the test mass. In the first cantilever experiments the test mass was a magnetic sphere made of NdFeB. Far better results can be obtained by using disk geometries and high density materials such as Au or Pt as in LPF. Recent calculations show that the sensitivity to an hypothetical CSL noise field can be enhanced by using a multilayer test mass composed of alternate materials with high density contrast [41].

We have recently started a preliminary experiment with a multilayer test mass made of 47 layers of W and WO_3 of thickness 300 μm, attached on the same cantilever used in Ref. [35]. The first very preliminary measurements show a larger excess noise as in the simple cantilever experiment. Nevertheless, the bound on CSL at $r_C = 10^{-7}$ m, shown as magenta curve in Fig. 1, is already better than any other mechanical experiment, including LPF, although the improvement is marginal. An additional hypothetical curve is shown (magenta dotted curve) which assumes a further optimized multilayered mass with Au or W as heavy material and excess noise comparable with the one of Ref. [35]. This experiment would completely rule out the CSL collapse rate suggested by Adler, and would outperform LPF over the whole relevant range of r_C up to 10^{-5} m. In addition, it would be able to completely rule out the excess noise observed in [35] as a possible CSL effect. Needless to say, any further reduction of the force noise, for instance by removing the excess noise or reducing the effective T/Q factor, would further improve the bounds.

A more radical approach towards a huge improvement is to replace the cantilever with a optical, electrical or magnetic trap, i.e. by levitating a nano or micromechanical test mass. Operation under extreme vacuum and ultralow temperature conditions might in principle allow effective quality factors exceeding 10^{12}. However, we are far from a demonstration of this figure. Optomechanical levitation of nanoparticles is a hot topic of the latest years. For operation at very low temperature, electrical levitation by Paul traps or by magnetic levitation with superconductors appears more suitable. This approach is being explored by the TEQ Project, recently funded by EU. In order to exploit the extreme performance that may be achievable using levitated particles, space environment may be eventually necessary, as in the LPF case. An ambitious program in this direction is the proposed space mission MAQRO [42].

Acknowledgments
The author acknowledges support from the EU H2020 FET project TEQ (Grant No. 766900).

References
[1] Ghirardi G, Rimini A and Weber T 1986 *Phys. Rev.* D **34** 470
[2] Ghirardi G C, Pearle P and Rimini A 1990 *Phys. Rev.* A **42** 78
[3] Ghirardi G C, Grassi R and Benatti F 1995 *Found. Phys.* **25** 5
[4] Bassi A and Ghirardi G C 2003 *Phys. Rep.* **379** 257
[5] Bassi A, Lochan K, Satin S, Singh T P and Ulbricht H 2013 *Rev. Mod. Phys.* **85** 471
[6] Penrose R 1996 *Gen. Relativ. Gravit.* **28** 581
[7] Penrose R 1998 *Phil. Trans. R. Soc.* A **356** 1927
[8] Diosi L 1989 *Phis. Rev.* A **40** 1165
[9] Ghirardi G C, Grassi R and Rimini A 1990 *Phys. Rev.* A **42** 1057
[10] Ellis J, Nanopoulos D V and Mohanty S 1984 *Nucl. Phys.* B **241** 381
[11] Ellis J, Mohanty S and Nanopoulos D V 1989 *Phys. Lett.* B **221** 113
[12] Hornberger K, Gerlich S, Haslinger P, Nimmrichter S and Arndt M 2012 *Rev. Mod. Phys.* **84** 157
[13] Juffmann T, Ulbricht H and Arndt M 2013 *Rep. Prog. Phys.* **76** 086402
[14] Arndt M and Hornberger K 2014 *Nat. Phys.* **10** 271
[15] Adler S L and Ramanazoglu F 2007 *J. Phys.* A **40** 13395
[16] Curceanu C, Hiesmayr B C and Piscicchia K 2015 *J. Adv. Phys.* **4** 263
[17] Collett B and Pearle P 2003 *Found. Phys.* **33** 1495
[18] Adler S L 2005 *J. Phys.* A **38** 2729
[19] Bassi A, Ippoliti E and Adler S L 2005 *Phys. Rev. Lett.* **94** 030401
[20] Bahrami M, Paternostro M, Bassi A and Ulbricht H 2014 *Phys. Rev. Lett.* **112** 210404
[21] Nimmrichter S, Hornberger K and Hammerer K 2014 *Phys. Rev. Lett.* **113** 020405
[22] Diosi L 2016 *Phys. Rev. Lett.* **114** 050102
[23] Goldwater D, Paternostro M and Barker P F 2016 *Phys. Rev.* A **94** 010104
[24] Li J, Zippilli S, Zhang J and Vitali D 2016 *Phys. Rev.* A **93** 050102
[25] Adler S L and Vinante A 2018 *Physical Review* A **97** 052219
[26] Kovachy T, Hogan J M, Sugarbaker A, Dickerson S M, Connelly C A, Overstreet C and Kasevich M A 2015 *Phys. Rev. Lett* **114** 143004
[27] Bilardello M, Donadi S, Vinante A and Bassi A 2016 *Physica* A **462** 764
[28] Adler S L 2007 *J. Phys.* A **40** 2935
[29] Carlesso M, Bassi A, Falferi P and Vinante A 2016 *Phys. Rev.* D **94** 124036
[30] Binnig G, Quate C F and Gerber C 1986 *Phys. Rev. Lett.* **56** 930
[31] Usenko O, Vinante A, Wijts G and Oosterkamp T H 2011 *Appl. Phys. Lett.* **98** 133105
[32] Abbott B P *et al.* (LIGO Collaboration) 2016 *Phys. Rev. Lett.* **116** 061102
[33] Vinante A *et al.* (AURIGA Collaboration) 2006 *Class. Quantum Grav.* **23** 103
[34] Vinante A, Bahrami M, Bassi A, Usenko O, Wijts G and Oosterkamp T H 2016 *Phys. Rev. Lett.* **116** 090402
[35] Vinante A, Mezzena R, Falferi P, Carlesso M and Bassi A 2017 *Phys. Rev. Lett.* **119** 110401
[36] Armano M *et al.* 2017 *Phys. Rev. Lett.* **116** 231101
[37] Armano M *et al.* 2018 *Phys. Rev. Lett.* **120** 061101
[38] Toros M and Bassi A 2018 *J. Phys. A: Math Theor.* **51** 115302
[39] Helou B, Slagmolen B, McClelland D E and Chen Y 2017 *Phys. Rev.* D **95** 084054
[40] Minar J, Sekatski P and Sangouard N 2016 *Phys. Rev.* A **94** 062111
[41] Carlesso M, Vinante A and Bassi A 2018 *Phys. Rev.* A **98** 022122
[42] Kaltenbaek R *et al.* 2012 *Exp. Astron.* **34** 123

Combinatorial Quantum Gravity: Emergence of Geometric Space from Random Graphs

C Kelly[1] and C A Trugenberger[2]

[1] School of Engineering and Physical Sciences, Heriot-Watt University, Edinburgh EH14 4AS, UK

E-mail: ckk1@hw.ac.uk

[2] SwissScientific Technologies SA, rue due Rhone 59, CH-1204 Genève, Switzerland

E-mail: c.trugenberger@swissscientific.com

Abstract. We review and extend the recently proposed model of combinatorial quantum gravity. Contrary to previous discrete approaches, this model is defined on (regular) random graphs and is driven by a purely combinatorial version of Ricci curvature, the Ollivier curvature, defined on generic metric spaces equipped with a Markov chain. It dispenses thus of notions such as simplicial complexes and Regge calculus and is ideally suited to extend quantum gravity to combinatorial structures which have a priori nothing to do with geometry. Indeed, our results show that geometry and general relativity emerge from random structures in a second-order phase transition due to the condensation of cycles on random graphs, a critical point that defines quantum gravity non-perturbatively according to asymptotic safety. In combinatorial quantum gravity the entropy area law emerges naturally as a consequence of infinite-dimensional critical behaviour on networks rather than on lattices. We propose thus that the entropy area law is a signature of the random graph nature of space-(time) on the smallest scales.

1. Introduction

Ultraviolet (UV) fixed point of statistical mechanics models define renormalizable quantum field theories via the Wilson renormalization group [1]. Here we review the evidence, first presented in [2], that quantum gravity is defined by an UV fixed point for a graph model [3]. The asymptotic safety scenario [4] would thus be realized on networks, rather than traditional statistical mechanics models.

In the traditional discrete approach to quantum gravity [5] a smooth background is assumed, which is then approximated by piecewise flat geometries on which curvature is computed by Regge calculus [6]. In [2], instead, one of us first posited that the fundamental structure on Planckian scales are not smooth but, rather, graphs, on which even notions like Regge curvature are lost. Random graphs are generic metric spaces. When equipped with a Markov chain, like a probability measure, a purely combinatorial notion of Ricci curvature, first introduced by Ollivier [7, 8, 9], can be defined on such structures. This was used in [2] to define a model of purely combinatorial quantum gravity. This approach was subsequently pursued in [10], where a modified version of the Ollivier curvature was introduced.

Albeit in a simplified model, we will provide strong evidence that geometric space emerges from random graphs at a second-order phase transition driven by a combinatorial version of

Content from this work may be used under the terms of the Creative Commons Attribution 3.0 licence. Any further distribution of this work must maintain attribution to the author(s) and the title of the work, journal citation and DOI.
Published under licence by IOP Publishing Ltd

the Einstein-Hilbert action and corresponding to the condensation of elementary loops on the graphs. In the geometric phase the combinatorial Einstein-Hilbert action becomes its standard continuum version. One notable result is that, in this model, the entropy of quantum space automatically follows an area law. The posited critical point on graphs could thus be the origin of the famed area law for the entropy in quantum gravity. Note that a relation between geometry and the density of triangles (the so-called clustering coefficient [3]) has been also noted in the network literature [11]. In the simplified model considered here we will be dealing with squares but the general case can also be treated [12].

2. The Ollivier curvature and the combinatorial Einstein-Hilbert action

The continuum Ricci curvature is associated with two (infinitely) close points on a manifold, defining a tangent vector. It can be thought of as a measure of how much (infinitesimal) spheres around these points are, on average, closer (positive Ricci curvature) or more distant (negative Ricci curvature) than the two points at their centres. Its combinatorial version, the Ollivier curvature [7, 8, 9], is a discrete version of the same measure. Consider two vertices i and $j = i + e_{ij}$ separated by edge e_{ij} on a graph. The Ollivier curvature compares the Wasserstein (or earth-mover) distance $W(\mu_i, \mu_j)$ between two uniform probability measures $\mu_{i,j}$ on the unit spheres around i and j to the distance $d(i, j)$ on the graph and is defined as

$$\kappa(i,j) = 1 - \frac{W(\mu_i, \mu_j)}{d(i,j)} \ . \tag{1}$$

The Wasserstein distance between two probability measures μ_1 and μ_2 on the graph is defined as

$$W(\mu_1, \mu_2) = \inf \sum_{i,j} \xi(i,j) d(i,j) \ , \tag{2}$$

where the infimum has to be taken over all couplings (or transference plans) $\xi(i, j)$ i.e. over all plans on how to transport a unit mass distributed according to μ_1 around i to the same mass distributed according to μ_2 around j without losses,

$$\sum_j \xi(i,j) = \mu_1(i) \ , \qquad \sum_i \xi(i,j) = \mu_2(j) \ . \tag{3}$$

The Ollivier curvature is very intuitive but, in general not easy to compute and work with. Fortunately, it becomes much simpler for bipartite graphs [9], which have no odd cycles. Since the Ollivier curvature of an edge depends only on the triangles, squares and pentagons supported on that edge (a discrete form of locality) [8] and there are no triangles and pentagons on bipartite graphs, one can use for all practical purposes the simpler version of the Ollivier curvature for bipartite regular graphs [9]:

$$\begin{aligned} \kappa(i,j) \quad &= -\frac{1}{d}\Big[(2d-2) - |N_1(j)| \\ &+ \sum_a \left(|L_a(j)| - |U_a(i)|\right) \times \mathbf{1}_{\{|U_a(i)|<|L_a(j)|\}}\Big]_+ \ , \end{aligned} \tag{4}$$

where $N_1(i)$ denotes the set of neighbours of i which are on a 4-cycle supported on (ij), $\mathbf{1}$ denotes the indicator function (1 if the corresponding condition is satisfied, 0 otherwise) and the undescript "+" denotes $z_+ = Max(z, 0)$ so that the Ollivier Ricci curvature for bipartite graphs is always zero or negative. The definition of U and L is as follows: suppose that $R(i, j)$ is the subgraph induced by $N_1(i) \cup N_1(j)$ and $R_1(i, j)...R_q(i, j)$are the connected components of $R(i, j)$. Then $U_a(i) = R_a(i, j) \cap N_1(i)$ and $L_a(j) = R_a(i, j) \cap N_1(j)$ for $a = 1 \ldots q$.

IOP Conf. Series: Journal of Physics: Conf. Series **1275** (2019) 012016 doi:10.1088/1742-6596/1275/1/012016

Equipped with a combinatorial version of Ricci curvature we can now formulate a purely combinatorial version of the Einstein Hilbert action. First we define the combinatorial Ricci scalar as

$$\kappa(i) = \sum_{j \sim i} \kappa(i,j) \ , \tag{5}$$

where $\sim$ denotes the neighbours of i on the graph. Then we obtain the combinatorial Einstein-Hilbert action simply as

$$S_{\rm EH} = -\frac{1}{g} \sum_i \kappa(i) \ , \tag{6}$$

where the sum runs over all the vertices of the graph and g is a coupling constant with dimension 1/action. This expression for the Ollivier curvature still looks forbidding. However, as we will show in a moment, it will become extremely simple on the physical configuration space.

3. Configuration space and combinatorial quantum gravity

To fully specify a combinatorial quantum gravity model we need, in addition to the action, a configuration space over which to sum in the partition function. The configuration space in our simplified model consists of all random bipartite graphs. There is, however a further restriction that must be taken into account. As mentioned above we will model the emergence of geometry by the condensation of elementary loops on the graphs. We have thus to remind ourselves that even the Bose condensation of point particles is not well defined in absence of interactions, because of the infinite compressibility of the condensate. In exactly the same way, the condensation of "non-interacting" loops is unstable, since it leads to crumpling and disconnected graphs (baby universes). We will thus follow the same route as for point particles and introduce, as the simplest stability mechanism, a hard-core condition for elementary loops. However, while for point particles the meaning of a hard core condition is unequivocal, for loops we must define what exactly we have in mind. The definition we will use is that elementary squares on the graphs will be allowed to share one edge but not more. Note that two squares can share two edges without being identical: it is exactly these configurations that we exclude.

When the hard-core condition is implemented, the Ollivier combinatorial curvature becomes really simple. Indeed, it is easy to convince oneself that the second term in (4), involving the sum of connected components of a subgraph, only contributes for squares that share 2 edges. Indeed, for an isolated square $|N_1| = 1$ for all vertices on the square. If an edge supports N_s squares which do not share another edge, then $|N_1(i)| = |N_1(j)| = N_s$ and $|U_a(i)| = |L_a(j)|$ since all the vertices within $N_1(i)$ and $N_1(j)$ are disconnected because of the absence of triangles in a bipartite graph and all the vertices of $N_1(i)$ are disconnected from those in $N_1(j)$ since, by assumption, the edge does not support two different squares. The Ollivier combinatorial curvature reduces thus simply to

$$\kappa(i,j) = -\frac{1}{d}[(2d-2) - N_s(ij)]_+ \ , \tag{7}$$

where $N_s(ij)$ is the total number of squares supported on edge (ij). The full model of combinatorial quantum gravity can thus be specified as

$$\mathcal{Z} = \sum_{CF} \exp\left(-\frac{1}{g\hbar} \sum_i \kappa(i)\right) \ , \tag{8}$$

where CF denotes the configuration space of random regular bipartite graphs with squares satisfying the hard-core condition and $\kappa(i)$ given by (5) and (7).

IOP Conf. Series: Journal of Physics: Conf. Series **1275** (2019) 012016 doi:10.1088/1742-6596/1275/1/012016

4. The classical limit and the mean field action

The classical limit $\hbar \to 0$ corresponds to the weak coupling limit of small g. In the quantum regime $\hbar g \gg 1$ the Boltzmann probability becomes uniform over all configuration space of random regular bipartite graphs. Random regular bipartite graphs are locally tree-like, with very sparse short cycles governed by a Poisson distribution with mean $(2d-1)^l/$ for cycles of length l on $2d$-regular graphs [13]. The quantity $(2d-2)$, instead, is the number of squares supported on an edge in a $\mathbb{Z}^d$ lattice. The combinatorial Einstein-Hilbert action (6) thus favours the formation of squares on the graph until the amount corresponding to a $\mathbb{Z}^d$ lattice is reached, after which it vanishes. The number of squares based on an edge, however, can be larger than $(2d-2)$ even for graphs with hard core squares, indeed it can reach up to $(2d-1)$. This is the maximum that the "mean field" version of the action (7) without subscript + [2] would favour in the classical limit. It can be shown, however, that configurations attaining this maximum split into large quantities of disconnected baby universes while the classical limit of the exact action avoids this fate and approaches a regular $\mathbb{Z}^d$ lattice [12]. The hard core condition is thus sufficient to stabilize space.

On a $\mathbb{Z}^d$ lattice the optimal coupling for the Wasserstein distance between the two vertices at the extremities of an edge is the translation along the lattice links connecting the unit balls around the two vertices. As derived in [7], the Ollivier curvature then becomes the average of the sectional curvatures of the planes defined by the original edge and each of the links defining the unit ball around one of the vertices. If we assign a length ℓ to each link of the lattice and scale this as $\ell = \ell_0 N^{-1/d}$, with ℓ_0 a renormalization constant related to the Planck length, we obtain the formal continuum limit

$$\frac{1}{\hbar g} \sum_i \kappa(i) \to \frac{1}{2(d+2)\ell_0^{d-2}} \frac{N^{1-2/d}}{\hbar g} \int d\text{Vol}\ R\ . \tag{9}$$

Both sides do of course vanish since the $\mathbb{Z}^d$ lattice is Ollivier flat and the Euclidean space $\mathbb{R}^d$ it approximates is Ricci flat. However, this formal continuum limit shows, first, that the combinatorial Einstein-Hilbert action goes over into the continuum Einstein-Hilbert action and, secondly, that this requires a scaling of $g\hbar \sim N^{1-2/d}$. We will return to this all-important scaling below. For the moment let us retain that the the combinatorial quantum gravity model has the correct formal continuum limit.

Having established that the hard-core condition is sufficient to stabilize space and obtain the correct (formal) continuum limit we can adopt the simpler mean field action expressed in terms of the total number of squares provided we explicitly exclude configurations with more than $(2d-2)$ squares per edge. To this end we compute

$$\begin{aligned} \sum_i \kappa(i) &= -(4d-4)N + \frac{1}{d} \sum_i \sum_{e_i} N_s(e_i) \\ &= \frac{-8}{d} \left[\frac{d(d-1)}{2} N - N_s \right] , \end{aligned} \tag{10}$$

which gives the final result

$$S_{EH}^{mf} = \frac{4d-4}{g} N[1-\zeta] , \tag{11}$$

where $\zeta = 2N_s/(d(d-1)N)$ $(0 \le \zeta \le 1)$ is the density of squares.

5. A continuous network phase transition and the entropy area law

Let us consider the free energy of the model (divided by the "temperature" $\hbar g$)

$$F = \frac{4d-4}{\hbar g} N[1-\zeta] - S(N) , \tag{12}$$

where $S(N)$ is the entropy of the graphs. In traditional statistical mechanics models the degrees of freedom, typically (but not necessarily) living on the vertices of a lattice, interact with a fixed number of their neighbours. As a consequence, both the energy and the entropy are extensive quantities, scaling like the volume N (number of vertices of the lattice). Phase transitions, thus, show up when the external intensive parameter temperature T (or coupling constant in case of quantum phase transitions) crosses a critical value T_c where energy and entropy exactly compensate. This is not so in the statistical mechanics of networks [14]. On networks, interactions are represented by edges. Each vertex can thus interact with a number of other vertices that diverges in the limit $N \to \infty$: contrary to traditional statistical mechanics models on lattices, network are infinite-dimensional. Moreover, there is no a priori notion of locality on networks. On random graphs, e.g., there can be an edge between any two vertices. As explained in detail in [3], the infinite-dimensionality of networks has the consequence that the phase structure of networks is determined by *critical functions* of N rather than critical values, e.g.

$$\begin{aligned} \text{phase 1 if} \quad & \lim_{N\to\infty} \left(\frac{T(N)}{T_c(N)} \right) = 0 \ , \\ \text{phase 2 if} \quad & \lim_{N\to\infty} \left(\frac{T(N)}{T_c(N)} \right) = \infty \ . \end{aligned} \tag{13}$$

When the temperature (or the coupling in the quantum case) is chosen to scale exactly as the critical function,

$$\lim_{N\to\infty} \left(\frac{T(N)}{T_c(N)} \right) = t \ , \tag{14}$$

the phase transition appears typically as a traditional one in terms of the rescaled coupling t, one phase appearing for $t < t_c$ and the other for $t > t_c$ with the difference that energy, entropy and free energy need not be extensive quantities anymore but can have a different scaling in terms of the volume N (see e.g. the mean field solution of the two-star model in [14]).

This is exactly what seems to be realized in the present model of combinatorial quantum gravity. The formal continuum limit (9) is well-defined only if the coupling $\hbar g$ scales as $\hbar g \sim N^{1-2/d}$. And remarkably, exactly when this scaling is chosen, the order parameter $\zeta = 2N_s/(d(d-1)N)$ representing the density of squares collapses onto an N-independent function suggestive of a traditional second-order phase transition, as shown in Fig. 1 for the case $d = 4$.

Second order phase transitions are characterized by the divergence of the correlation length at the critical point. We can define a correlation length also on graphs. To this end we define the local order parameter $\zeta(i) = 2N_s(i)/(d(d-1))$ characterizing the density of squares at each vertex i and we compute the correlation function

$$C\left(d(ij)\right) = \frac{< (\zeta(i) - \bar{\zeta})(\zeta(j) - \bar{\zeta}) >}{\sigma(i)\sigma(j)} \ , \tag{15}$$

as a function of the graph distance $d(ij)$ between vertices. Here $\bar{\zeta}$ denotes the expectation value whereas $\sigma(i)$ is the standard deviation of $\zeta(i)$. Expressing, as usual, this correlation function as

$$C\left(d(ij)\right) = \exp\left(-\frac{d(ij)}{\xi} \right) \ , \tag{16}$$

defines the correlation length ξ. The correlation length in units of the graph diameter, averaged over different graph sizes, is plotted as a function of the rescaled coupling $G = \hbar g$ in Fig. 2 for the case $d = 3$.

IOP Conf. Series: Journal of Physics: Conf. Series **1275** (2019) 012016 doi:10.1088/1742-6596/1275/1/012016

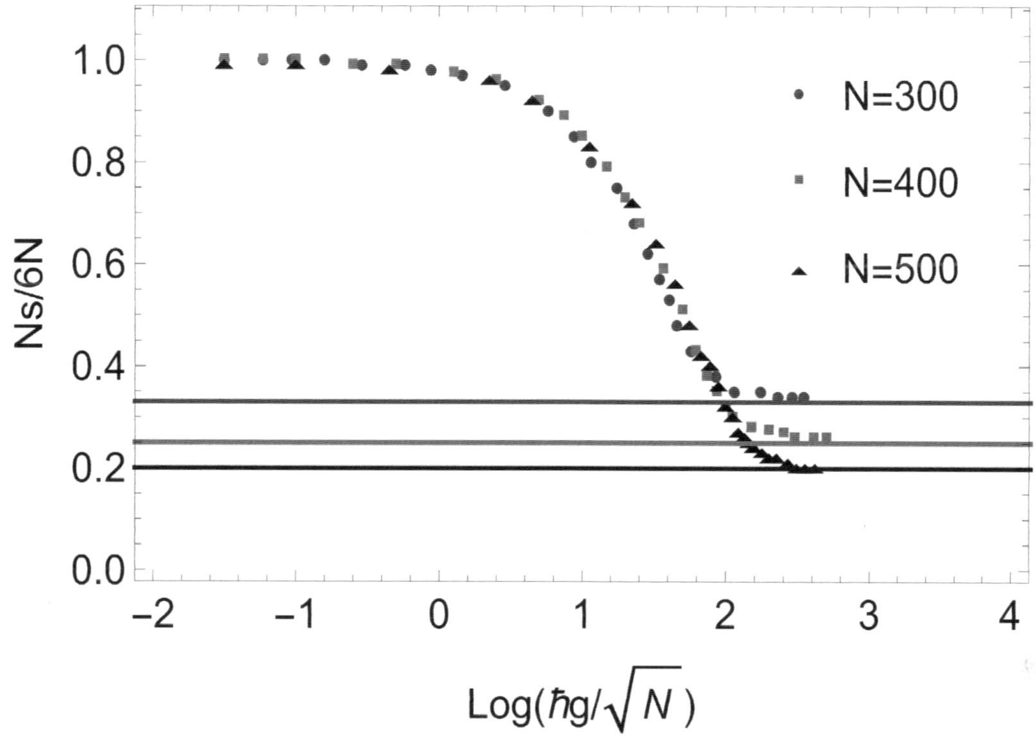

Figure 1. Monte Carlo simulation of the average number of squares for $d = 4$ and $N = 300$, $N = 400$ and $N = 500$ as a function of the rescaled coupling $\hbar g/\sqrt{N}$. Random regular graphs with sparse squares $N_s \sim$ Poisson (600.25) and logarithmic distance scaling at large values of the coupling constant turn into $\mathbb{Z}^4$ lattices with the maximum number of squares $N_s = 6N$ and power-law distances when gravitation becomes weak.The horizontal lines correspond to the expected number of squares for random regular graphs of the corresponding volume N.

Remarkably, this plot indeed shows the typical behaviour of the correlation length at a second order phase transition, the divergence being of course cut-off by finite size effects. This second order phase transition, if confirmed by further studies, marks the emergence of geometric space, in the form of flat tori locally isomorphic to $\mathbb{Z}^d$ from random regular graphs as a consequence of the condensation of the shortest cycles, squares. This critical point would provide a non-perturbative definition of quantum gravity according to the asymptotic safety scenario [4], in which the critical value $G_c \simeq 400$ ($d = 4$) (see Fig.1) of the rescaled coupling $\hbar g/N^{1-2/d}$ defines Newton's gravitational constant via the relation $G_{\text{Newton}} = (3/4\pi)G_c\ell_0^2/\hbar$.

A further important consequence of this second-order phase transition is derived from the expression (12) of the free energy. As explained above, the scaling $\hbar g \sim N^{1-2/d}$ is exactly the critical function of the graph model. The existence of a second-order phase transition with this scaling would imply the existence of a balance critical value between energy and entropy, which would further imply that, with this scaling, energy and entropy have themselves the same scaling behaviour with the volume N. This immediately leads to the scaling behaviour $S(N) \sim N^{2/d}$ for the entropy of the graphs, in both phases of course. Since N represents the volume, this scaling law is a combinatorial version of the entropy area law in quantum gravity [15].

IOP Conf. Series: Journal of Physics: Conf. Series **1275** (2019) 012016 doi:10.1088/1742-6596/1275/1/012016

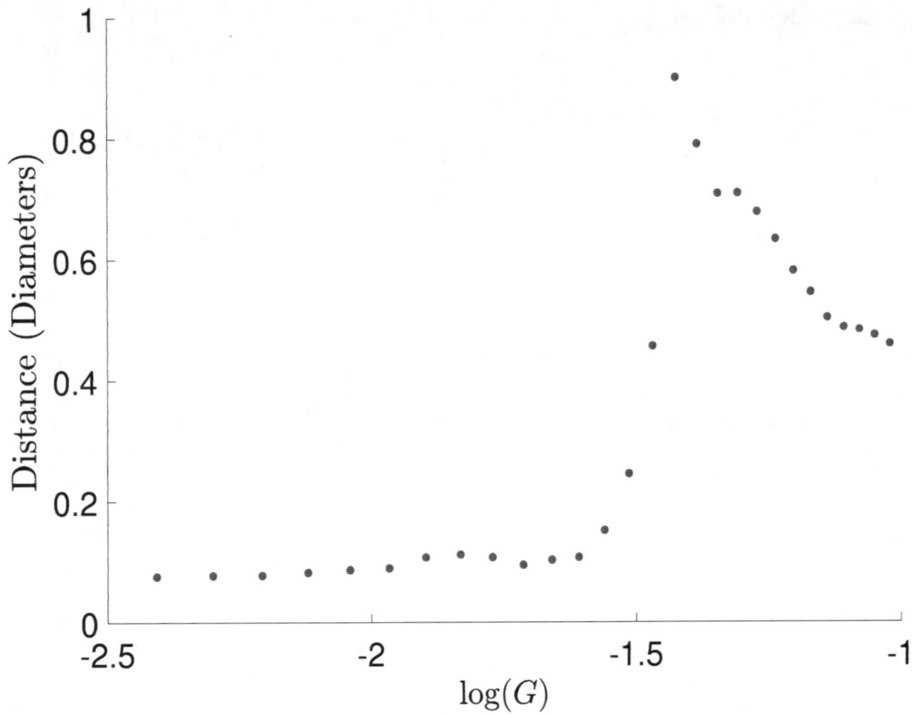

Figure 2. The correlation length in units of the graph diameter, averaged over different graph sizes as a function of the rescaled coupling $G = \hbar g$ in Fig. 2 for the case $d = 3$.

References

[1] Zinn-Justin J 1989 *Quantum Field Theory and Critical Phenomena* (Oxford: Clarendon Press)
[2] Trugenberger C A 2017 *JHEP* **1709** 45
[3] Albert R and Barabasi L 2002 *Rev. Mod. Phys.* **74** 47
[4] Eichhorn A 2017 *Preprint* arXiv:1709.03696
[5] Ambjorn J, Görlich A, Jurkiewicz J and Loll R 2012 *Phy. Rep.* **58** 127
[6] Regge T 1961 *Nuovo Cim.* **19** 558; Williams R M and Tuckey P A 1992 *Class. Quant. Gravity* **9** 1409
[7] Ollivier, Y 2009 *J. Funct. Anal.* **256** 810; Ollivier, Y 2010 *Adv. Stud. Pure Math.* **57** 343
[8] Linn, Y, Lu L and Yau, S T 2011 *Tohoku Math. J.* **63** 605; Loisel B and Romon P 2014 *Axioms* **3** 119; Jost J and Liu S 2014 *Discrete Comput. Geom.* **51** 300
[9] Bhattacharya B B and Mukherjee S 2015 *Discrete Mathematics* **338** 23
[10] Klitgaard N and Loll R 2018 *Phys. Rev* D **97** 046008; *do.* 2018 *Preprint* arXiv:1802.10524
[11] Dall J and Christensen M 2002 *Phys. Rev.* E **66** 016121; Krioukov D 2016 *Phys. Rev. Lett.* **116** 208302
[12] Kelly C, Trugenberger C A and Biancalana F 2018, in preparation
[13] O'Neil P E 1969 *Bull. Am. Math. Soc.* **75** 1276; Wormald N C 1999 *Surveys in Combinatorics*, LMS Lectures Note Series **267**, Lamb J D and Pierce D A eds. 239
[14] Park J and Newman M J 2004 *Phys. Rev.* **E70** 066117-1
[15] Bekenstein J D 1973 *Phys. Rev.* **D7** 2333; Jacobson T 1995 *Phys. Rev. Lett.* **75** 1260, Jacobson T 2016 *Phys. Rev. Lett.* **116** 201101

IOP Conf. Series: Journal of Physics: Conf. Series **1275** (2019) 012017 doi:10.1088/1742-6596/1275/1/012017

On the hurricane collective molecular dynamics

Gabriel Meyer[1] **and Giuseppe Vitiello**[2]
[1] Ecole Normale Suprieure de Lyon, 46 Allée d'Italie 69007 Lyon
[2] Dipartimento di Fisica "E.R. Caianiello" and Istituto Nazionale di Fisica Nucleare
Universitá di Salerno, I-84100 Fisciano (Salerno), Italy
E-mail: [1] `gabriel.meyer@neel.cnrs.fr`
E-mail: [2] `vitiello@sa.infn.it`

Abstract. We study the molecular dynamics underlying the formation and the evolution of hurricanes. The giant global structure of hurricanes cannot be solely the results of random molecular kinematics. We show that the non-stationary, topologically non-trivial hurricane behavior is the manifestation of the collective dynamics at the molecular level. This is described in terms of spontaneous breakdown of the rotational symmetry of the water molecule electrical dipoles. The long range molecular correlations generated as a consequence of the spontaneous breakdown of symmetry are responsible of the large scale collective motion resulting in the hurricane structure. The critical non-stationary dynamics is analyzed in terms of the non-stationary time dependent Ginzburg-Landau equation leading to the vortex equation. Phase transition, criticality, temperature dependence, energy exchanges and entropy are studied. Energy storage and dissipation processes in the coherent state dynamics are analyzed. Finally the self-similar fractal structure of the hurricane is shown to be the manifestation of the coherence of the collective molecular dynamics.

1. Introduction

In this paper we report some results obtained in the study of the molecular dynamics in the hurricane formation and interaction with its environment [1]. In tropical regions, large-scale storms, called hurricanes in the Atlantic ocean, form over warm ocean waters. A general consensus on the physical processes involved in cyclone formation and intensification or weakening is still lacking, although it is clear that the temperature of the ocean surface, the troposphere humidity and vorticities play crucial roles in their genesis [2, 3, 4, 5, 6]. Of great interest is for obvious reasons the understanding of the mechanisms by which they may be getting stronger or weaker on their way in crossing the ocean. Heat and angular momentum are anyway main sources of energy for tropical cyclones [3, 7]. Indeed the hurricane may get more heat in the regions where more water will evaporate from the ocean, due to the warmer ocean's surface in those regions. Flowing upward, the vapor cools down and latent heat is released from its condensation. It is also thought that an increase of the hurricane angular momentum may occur when it captures wind currents blowing in the proper direction (in this connection see the so called balanced vortex models and other numerical models [2, 8]).

Evidently, the hurricane formation and evolution result from highly non-trivial non-equilibrium thermodynamic and hydrodynamic processes and the complexity of the problem requires heavy use of numerical simulations. It is, however, also evident that the hurricane cannot be solely the result of random molecular kinematics; a giant collective molecular dynamics must

Content from this work may be used under the terms of the Creative Commons Attribution 3.0 licence. Any further distribution of this work must maintain attribution to the author(s) and the title of the work, journal citation and DOI.
Published under licence by IOP Publishing Ltd

IOP Conf. Series: Journal of Physics: Conf. Series **1275** (2019) 012017 doi:10.1088/1742-6596/1275/1/012017

be at the basis of the hurricane formation and evolution. Aim of our work is in fact the study of the molecular dynamics responsible of large scale collective molecular behaviors turning into the hurricane topologically non-trivial structure, its energy storage and dissipation, its propagation speed as a macroscopic system, the criticality of its non-equilibrium dynamics, temperature dependence, etc. In Section 2 we introduce few elements of the fluid dynamics underlying the hurricane Burgers model description. In Section 3 we show how the central features of the Burgers model can be obtained in the canonical formalism for molecular dynamics. In Section 4 we discuss the collective molecular dynamics in the quantum field theory (QFT) framework. Criticality, vorticity and non-stationary properties are analyzed in Section 5 by using the time dependent Ginzburg-Landau equation. In Section 6 the fractal self-similarity of the logarithmic spiral structure of hurricane is shown to be the macroscopic manifestation of the coherence of the collective molecular dynamics. Conclusions and final comments on energy storage in coherent systems are presented in Section 7.

2. The hurricane fluid dynamics

In fluid dynamics studies of the hurricane phenomenon, one considers the Navier-Stokes equation for a fluid in a given space region with fixed density ρ and viscosity ν:

$$\frac{\partial \mathbf{u}}{\partial t} + (\mathbf{u} \cdot \nabla)\mathbf{u} = \frac{1}{\rho}\nabla p + \nu \Delta \mathbf{u}. \tag{1}$$

$\mathbf{u}$ is the velocity field and p the pressure. The rotor of (1) gives:

$$\frac{\partial \omega}{\partial t} - \nabla \times (\mathbf{u} \times \omega) = \nu \Delta \omega. \tag{2}$$

where $\omega = \nabla \times \mathbf{u}$ is the vorticity. Since in the considered region the fluid is assumed to be incompressible, the continuity equation reduces to $\nabla \cdot \mathbf{u} = 0$ and the vorticity equation is:

$$\frac{\partial \omega}{\partial t} + (\mathbf{u} \cdot \nabla)\omega = (\omega \cdot \nabla)\mathbf{u} + \nu \Delta \omega. \tag{3}$$

Within such a frame, we consider the Burgers model [9] describing how a vortex may stretch so as to concentrate its vorticity in a smaller region (see the Appendix A of [1] for details). We use cylindrical coordinates with the z-axis along the axis of the vortex and we look for a flow such that the vertical component of the velocity is $u_z = 2Cz$ (with C a constant), and the radial component u_r and u_θ are functions of r only. Then, one obtains that the radial velocity must be equal to $u_r = -Cr + A/r$. We will choose $A = 0$ for convenience and thus we have

$$u_z = 2Cz; \qquad u_r = -Cr. \tag{4}$$

In general C depends on several space-time dependent variables, such as temperature T, pressure, density and viscosity of the medium, its molecular composition, the density and charge of present particles, atoms, molecules, ions, etc.. Thus, C actually denotes its effective value $C(\mathbf{r}, t)$ resulting from its dependence on the said variables. However, for definiteness in the model it is considered the approximation of constant C over a relatively wide space region and relatively long time intervals. Putting $z = 0$ at the ground and limiting ourselves to the region $z > 0$, we see that $C > 0$ corresponds to a radially inward and vertically upward flow, whereas $C < 0$ corresponds to a radially outward and vertically downward flow (flow reversal). As mentioned, in the core of a hurricane, there is a vertical current of warm and wet air going upward. Unless differently stated, in the following the case $C > 0$ is considered.

IOP Conf. Series: Journal of Physics: Conf. Series **1275** (2019) 012017 doi:10.1088/1742-6596/1275/1/012017

Note that $C = 0$ in the absence of incoming or outgoing flux. On the other hand, the fact that for positive non-vanishing C the velocity increases without bond with the distance from the origin indicates that this model is only locally valid .

The conditions imposed on the flow imply that the vorticity is parallel to the z-axis: $\omega(\mathbf{r}, t) = \omega(r, t)u_z$. The vorticity equation becomes:

$$\frac{\partial \omega}{\partial t} = 2C\omega + Cr\frac{\partial \omega}{\partial r} + \frac{\nu}{r}\frac{\partial}{\partial r}(r\frac{\partial \omega}{\partial r}). \tag{5}$$

Observe that for $C = 0$ Eq. (5) reduces to the diffusion equation:

$$\frac{\partial \omega}{\partial t} = \nu \Delta \omega, \tag{6}$$

which has the solution:

$$\omega(r, t) = \frac{\Gamma}{4\pi\nu t}\exp(\frac{-r^2}{4\nu t}). \tag{7}$$

Here Γ denotes the circulation along a horizontal circle with infinite radius or, equivalently, to the flux of vorticity across the entire xy-plane :

$$\Gamma = \lim_{r\to\infty}\oint_{C_r} u_\theta \, r \, d\theta = \int_{xy\text{-plane}} \omega dS \tag{8}$$

In order to solve the case $C \neq 0$, we look for a Gaussian shaped solution of the form

$$\omega(r, t) = \frac{1}{r}\frac{\partial}{\partial r}[ru_\theta(r, t)] = \frac{\Gamma}{2\pi\delta^2(t)}\exp(\frac{-r^2}{2\delta^2(t)}), \tag{9}$$

Eq. (9) used in (5) yields:

$$(\delta(t)\dot{\delta}(t) + C\delta^2(t) - \nu)(2 - \frac{r^2}{\delta^2(t)}) = 0. \tag{10}$$

We look for a solution $\delta(t)$ independent of r. For $2 - r^2/\delta^2(t) \neq 0$ we obtain the first order linear differential equation for $\delta^2(t)$ and find the central result:

$$\delta^2(t) = \frac{\nu}{C} + (\delta_0^2 - \frac{\nu}{C})e^{-2Ct}, \tag{11}$$

where δ_0 is the initial value of δ, and $\delta^2(t = \infty) = \nu/C$. The sign of $\delta^2(t) - \nu/C$ is determined by the sign of $\delta_0^2 - \nu/C$ (in principle δ_0^2 can be larger or smaller than ν/C). Using (9) we finally obtain [1]

$$u_\theta = \frac{\Gamma}{2\pi r}[1 - \exp(\frac{-r^2}{2\delta^2(t)})]. \tag{12}$$

Summarizing, we have obtained

$$\begin{cases} u_z = 2Cz \\ u_r = -Cr \\ u_\theta = \frac{\Gamma}{2\pi r}[1 - \exp(\frac{-r^2}{2\delta^2(t)})] \end{cases} \tag{13}$$

The flow is therefore a combination of two motions. An irrotational part controlled by the parameter C, with radial and vertical velocity components u_r and u_z and a rotational part

IOP Conf. Series: Journal of Physics: Conf. Series **1275** (2019) 012017 doi:10.1088/1742-6596/1275/1/012017

characterized by the angular velocity u_θ. It describes an axis-symmetric, vertically uniform vortex flow with strength and size controlled by the parameters Γ and δ respectively.

The energy associated to the vortex flow (per vertical unit) is :

$$E_v = \int u_\theta^2 \pi r \, dr = \frac{\Gamma^2}{4\pi} \int_0^\delta \frac{1}{r}[1 - \exp(\frac{-r^2}{2\delta^2})]^2 \, dr \tag{14}$$

This integral is well defined in the limit $r \to 0$ but it diverges as $\ln(r)$ for $r \to \infty$. Therefore, the natural cutoff δ is introduced in the integration. The necessity of the cutoff confirms that this model is only locally valid. We also recognize the relevance of the role played by δ.

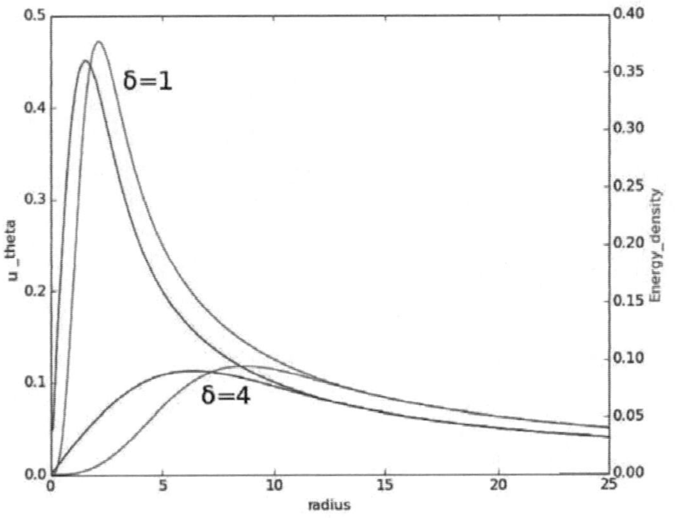

Figure 1. u_θ and $E_{density}$ is plotted for two different values of δ. The energy maximum occurs for a slightly larger r-value than the velocity maximum.

We verified numerically how the energy (14) depends on δ. In figure 2, for two values of δ, the velocity and energy distribution are shown as a function of the distance from the axis of the vortex. The maximum of energy and the radius at which this maximum occurs are shown as a function of the vortex size in figure 2. The integrated energy is clearly independent of δ. The energy density instead depends on δ as follows:

$$E_{\text{density}} = \frac{\Gamma^2}{4\pi r}[1 - \exp(\frac{-r^2}{2\delta^2})]^2 \, . \tag{15}$$

The storage or dissipation of energy by hurricanes during their time evolution is of course a question of great practical interest.

3. The hurricane molecular dynamics

We want now to study the underlying molecular dynamics in the hurricane-environment interactions. As already observed, δ plays a crucial role in the dynamics described by the Navier-Stokes equation. It evolves in time according to Eq. (11), which shows that there is a competition between the viscosity, which tends to make the vortex diffuse, and the stretching flow, which tends to make the vortex more concentrated around its axis.

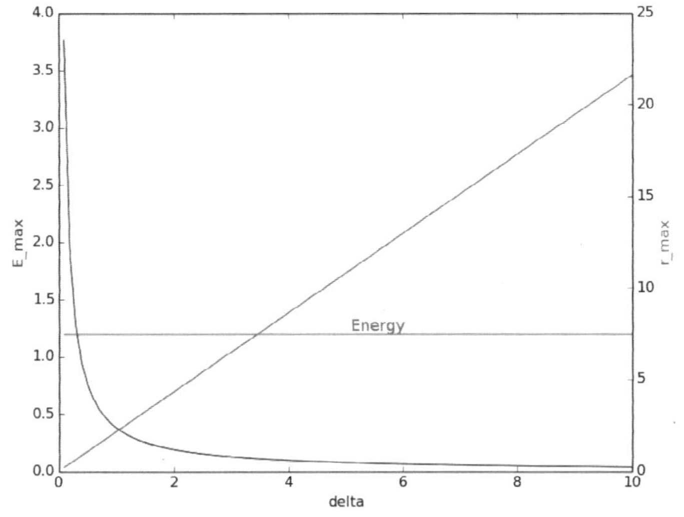

Figure 2. Plot of the energy maximum (blue) and radius where the maximum occurs (green) for different values of δ. The maximum of energy increases without bond for small δ and the radius at the maximum increases linearly with δ (like $\approx 2\delta$). The energy integrated over five δ (red) is shown to be clearly independent of δ.

Define $\varrho^2(t) = |\delta^2(t) - \nu/C|$, $\varrho_0^2 = |\delta_0^2 - \nu/C|$. Eq. (11) is then written as

$$\varrho(\theta) = \varrho_0 \, e^{-\theta} \tag{16}$$

where we put $\theta \equiv \theta(t) = C\,t$. The associated parametric equations in polar coordinates are

$$\xi = \varrho(\theta) \cos\theta = \varrho_0 \, e^{-\theta} \cos\theta \,, \tag{17}$$
$$\eta = \varrho(\theta) \sin\theta = -\varrho_0 \, e^{-\theta} \sin\theta \,. \tag{18}$$

The completeness of the (hyperbolic) basis $\{e^{-\theta}, e^{+\theta}\}$ requires that both elements $q = e^{\pm\theta}$ must be taken into account in the complex z-plane $z = \xi + i\,\eta$.

We thus consider both the points $z_1 = \varrho_0 \, e^{-\theta} \, e^{-i\theta}$ and $z_2 = \varrho_0 \, e^{+\theta} \, e^{+i\theta}$, and for convenience both signs also for the imaginary exponent $\pm i\,\theta$.

We note that $z_1(t) = \varrho_0 \; e^{-i\Omega t} \, e^{-Ct}$ and $z_2(t) = \varrho_0 \; e^{+i\Omega t} \, e^{+Ct}$ solve the equations

$$m\,\ddot{z}_1 + \gamma\,\dot{z}_1 + \kappa\,z_1 = 0 \,, \tag{19}$$
$$m\,\ddot{z}_2 - \gamma\,\dot{z}_2 + \kappa\,z_2 = 0 \,, \tag{20}$$

respectively, with positive real m, γ and κ, $C \equiv \gamma/2\,m$; $\Omega^2 \equiv C^2 = (1/m)(\kappa - \gamma^2/4m)$, with $\kappa > \gamma^2/4m$; "dot" denotes derivative with respect to t. Note that no new parameters have been introduced. In fact the ratios γ/m and κ/m are proportional to C and C^2, respectively.

By putting $x(t) \equiv [z_1(t) + z_2^*(-t)]/2$ and $y(t) \equiv [z_1^*(-t) + z_2(t)]/2$, Eqs. (19) reduce to the couple of damped and amplified harmonic oscillators

$$m\ddot{x} + \gamma\dot{x} + kx = 0 \,, \tag{21}$$
$$m\ddot{y} - \gamma\dot{y} + ky = 0 \,, \tag{22}$$

respectively. Note that $C = \gamma/2m$ controls the dissipation (amplification) terms. Eqs. (21) and (22) are one the time-reversed image of the other one ($\gamma \to -\gamma$) (the $\pm$ sign of C is taken into account and shifted to the $\pm$ of time t). The physical meaning of this is that total energy is conserved. The global system (x, y) is indeed a closed system. The canonical formalism can only describe closed systems. The y system plays indeed the role of the environment for the x system (or vice-versa). This shows thus how important it is to consider the role of the interaction with the environment, as indeed stressed in hurricane studies [3].

We remark that the domain of surface $\varrho^2(t) = |\delta^2(t) - \nu/C|$ has blurred borders, i.e. not sharply defined linear dimension $\varrho(t)$, due to fluctuating molecular movements arising from the hurricane interaction with the environment molecules. The hurricane exchanges indeed energy by interacting at a molecular level with the atmosphere molecules (mostly nitrogen (N_2) and oxygen (O_2), but also carbon dioxide (CO_2) and Argon (Ar)) and water molecules and droplets largely present in the environment due to moist convection, cumulus and clouds. These brownian-like molecular motions may be represented by two coordinates $x_+(t)$ (going forward in time) and $x_-(t)$ (going backward in time)

$$x_\pm \equiv x \pm \frac{y}{2} . \tag{23}$$

Eqs. (21) and (22) are then rewritten as

$$m\ddot{x}_+ + \gamma\dot{x}_- + kx_+ = 0 \; , \tag{24}$$
$$m\ddot{x}_- + \gamma\dot{x}_+ + kx_- = 0 \; . \tag{25}$$

The Lagrangian from which they can be derived and the Hamiltonian are [10, 11]

$$L = \frac{m}{2}\dot{x}_+^2 - \frac{k}{2}x_+^2 - \frac{m}{2}\dot{x}_-^2 + \frac{k}{2}x_-^2 + \frac{\gamma}{2}\left(\dot{x}_+x_- - \dot{x}_-x_+\right), \tag{26}$$

$$H = H_+ - H_- = \frac{1}{2m}(p_+ - \frac{\gamma}{2}x_-)^2 + \frac{k}{2}x_+^2 - \frac{1}{2m}(p_- + \frac{\gamma}{2}x_+)^2 - \frac{k}{2}x_-^2. \tag{27}$$

respectively, with conjugate momenta $p_+ = m\dot{x}_+ + (\gamma/2)x_-$, $p_- = -m\dot{x}_- - (\gamma/2)x_+$. It is interesting that the damping manifests itself as a correction in the kinetic energy and that the dissipative constant acts as a coupling between x_+ and x_- (cf. Eq. (26)).

From Eq. (27), the forward and backward in time velocities $v_\pm$ are

$$v_\pm = \frac{\partial H}{\partial p_\pm} = \pm\frac{1}{m}(p_\pm \mp \frac{\gamma}{2}\, x_\mp) \; . \tag{28}$$

By using as usual $p_\pm = -i\hbar d/dx_\pm$, we have

$$[v_+, v_-] = i\hbar\,\frac{\gamma}{m^2} \; . \tag{29}$$

Provided that we set $\gamma \equiv q\,B_3/c$, Eq. (29) is similar to the usual commutators for the velocities $\mathbf{v} = (\mathbf{p} - q\mathbf{A}/c)/m$ of a charged particle moving in a magnetic field $\mathbf{B}$; i.e., $[v_1, v_2] = (i\hbar\, q\, B_3/m^2c)$ (also analogous to the Bohm-Aharonov phase interference for charged particles). The commutator (29) implies a topologically non-trivial (noncommutative) geometry in the (v_+, v_-) plane, reflecting of course the non-trivial topology of the hurricane vortex.

These remarks and the non-vanishing density of ionized particles suggest to us to write H as

$$H = H_+ - H_- = \frac{1}{2m}(p_+ - \frac{q_+}{c}A_+)^2 + q_+\Phi_+ - \frac{1}{2m}(p_- + \frac{q_-}{c}A_-)^2 + q_-\Phi_-, \tag{30}$$

IOP Conf. Series: Journal of Physics: Conf. Series **1275** (2019) 012017 doi:10.1088/1742-6596/1275/1/012017

with

$$A_+ \equiv \frac{B}{2}x_- \ , \qquad A_- \equiv \frac{B}{2}x_+ \ , \qquad B \equiv \frac{c}{q}\gamma, \tag{31}$$

namely, $(\gamma/2)x_\pm$ in Eq. (27) are represented as the components $A_+ \equiv A_1$ and $A_- \equiv A_2$ of the em vector potential $\mathbf{A}$ [11, 12, 13] with the magnetic field $\mathbf{B} = \nabla \times \mathbf{A} = -B\hat{\mathbf{3}}$.

The Hamiltonian H then describes two particles with opposite charge $q_+ = -q_- = q$ in the (oscillator) potential $\Phi \equiv (k/2q)(x_+{}^2 - x_-{}^2) \equiv \Phi_+ - \Phi_-$, $\Phi_\pm \equiv (k/2q)x_\pm{}^2$, and in the constant magnetic field $\mathbf{B}$. One particle thus moves in the em field whose source is the other particle. Eqs.(24) and (25) are recognized to describe nothing but the Lorentz forces on particles with charge $q_+ = -q_- = q$, in the magnetic field $\mathbf{B} = -B\hat{\mathbf{3}}$ and in the electric field $\mathbf{E} = -\nabla\Phi$:

$$\mathcal{F}_i = m\ddot{x}_i = q_i[E_i + \frac{1}{c}(\mathbf{v} \times \mathbf{B})_i] \ , \tag{32}$$

$i = 1, 2 \equiv +, -$, no sum on i, respectively, and $\mathbf{v} = (\dot{x}_+, \dot{x}_-, 0)$.

In conclusion, we have expressed the dynamical content of Eq. (11) of the Burgers model by the above Lagrangian and Hamiltonian which account for the dynamics of molecular motion and, with the identification (31), also of charged particles motion in the hurricane-environment interaction.

4. Molecular collective modes

The role played by cloud water droplets, cumulus and moist convection in hurricane genesis and evolution has been much discussed in the literature by the help of numerical and analytical studies [2, 3]. Our standpoint is that in the region where the hurricane is generated, a weak perturbation, e.g. the em field generated by ionized particles, or perturbations of other kind, mechanical, chemical, etc., may break the rotational symmetry of the electric dipole of the water molecules present in that region. The molecular dynamics will be then conditioned by the spontaneous breakdown of the dipole rotational symmetry (SBS). As a consequence, the system components undergo an *in phase*, or coherent collective motion resulting in ordered states. The ordering is generated by long range correlations among them (the Goldstone theorem in QFT) [14, 15]. Ion-dipole interactions may be indeed observed to induce ordering of water molecular electric dipole even at very low ion concentrations over coherent domains of nanoscopic length scales [16, 17, 18] with observable effects in surface tension experiments at the macroscopic scale. The degree of ordering is described by the "order parameter", characterizing the macroscopic behavior of the system. The change of scale, from the dynamics of the elementary components to the system's macroscopic behavior, is thus dynamically obtained and is expressed by classical field equations and observables. In the following we will always refer to such a classical level of description, which emerges out of the molecular QFT dynamics.

We thus consider the non-homogeneous polarization density $\mathcal{P}(\mathbf{r}, t)$ that arises as a dynamical effect of the SBS. The consequent collective correlations among the in phase oscillating molecular dipoles are described by dipole fields (the Nambu-Goldstone (NG) fields) whose coherent condensation in the ground state manifests itself in the classical, macroscopic properties and behavior of the system.

Let $\rho(\mathbf{r}, t)$ denote the charge density and ℓ the (average) dipole length. The symmetry under rotations by a constant phase λ around the dipole axis (the global U(1) symmetry) is broken since it is impossible to change by a constant amount the phase of the dipole field simultaneously at every space-point. Then, one can show [19] that under quite general conditions the order parameter $v(\mathbf{r}, t)$ is proportional to $\mathcal{P}(\mathbf{r}, t)$: $|v(\mathbf{r}, t)|^2 \propto 2\mathcal{P}(\mathbf{r}, t) = \rho(\mathbf{r}, t)\,\ell$.

By adopting a standard recipe [14, 15, 20], the order parameter is described in terms of the charge density wave function $\sigma(\mathbf{r}, t)$:

$$\sigma(\mathbf{r}, t) = \sqrt{\rho(\mathbf{r}, t)}\, e^{i\varphi(\mathbf{r}, t)} \ , \tag{33}$$

where the real phase $\varphi(\mathbf{r},t)$ field, plays the role of the NG quantum field. The phase transformation for $\sigma(\mathbf{r},t)$, i.e. $\varphi(\mathbf{r},t) \to \varphi(\mathbf{r},t) + (q/\hbar c)\,\lambda(\mathbf{r},t)$, describes the coherent NG boson condensation process. In full generality,

$$\varphi(\mathbf{r},t) \to \varphi(\mathbf{r},t) + f(\mathbf{r},t), \tag{34}$$

with $f(\mathbf{r},t)$ playing the role of a form factor in the φ non-homogeneous (i.e. space-time dependent) condensation. A topologically non-trivial structure is obtained when $f(\mathbf{r},t)$ carries topological singularities [21, 22]. For example, vortices appear by using

$$f(x) = \arctan\left(\frac{x_2}{x_1}\right)\,, \tag{35}$$

which indeed carries a singularity on the line $r = 0$, with $r^2 = x_1^2 + x_2^2$.

In order to ensure the invariance under local phase transformation, as usual in the canonical formalism one considers the em field gauge transformation

$$\mathbf{A}(\mathbf{r},t) \to \mathbf{A}' = \mathbf{A}(\mathbf{r},t) + \boldsymbol{\nabla}\lambda(\mathbf{r},t). \tag{36}$$

When using φ as gauge function in $\mathbf{A} \to \mathbf{A}' = \mathbf{A} + (\hbar c/q)\boldsymbol{\nabla}\varphi$, the transformation $\mathbf{A}' \to \mathbf{A}' + \boldsymbol{\nabla}\lambda$ is induced by $\varphi(\mathbf{r},t) \to \varphi(\mathbf{r},t) + (q/\hbar c)\,\lambda(\mathbf{r},t)$.

We adopt the gauge condition $\boldsymbol{\nabla}\cdot\mathbf{A} = 0$ ($\boldsymbol{\nabla}\cdot\mathbf{A}' = 0$), which requires $\boldsymbol{\nabla}^2\varphi(\mathbf{r},t) = 0$ and $\boldsymbol{\nabla}^2\lambda(\mathbf{r},t) = 0$ (and $\boldsymbol{\nabla}^2 f(\mathbf{r},t) = 0$). When the polarization current $\mathbf{J}_p = \rho(\mathbf{r},t)\mathbf{v}$, with $\mathbf{v} = \partial\boldsymbol{\ell}/\partial t$, is assumed to be the only relevant current, the Maxwell equation $\Box\mathbf{A}(\mathbf{r},t) = \mathbf{J}_p(\mathbf{r},t)$ is

$$(\Box + M^2)\mathbf{A}(x) = \frac{2\,\mathcal{P}}{m\,\ell}\,\hbar\,\boldsymbol{\nabla}\varphi(\mathbf{r},t). \tag{37}$$

where $M^2 \equiv 2q\,\mathcal{P}/(m\,\ell\,c)$. In obtaining (37) we used

$$\mathbf{J}_p(\mathbf{r},t) = \frac{1}{m}\rho(\mathbf{r},t)(\hbar\,\boldsymbol{\nabla}\varphi(\mathbf{r},t) - \frac{q}{c}\,\mathbf{A}(\mathbf{r},t)), \tag{38}$$

derived in a standard fashion by using (33) in the definition of $\mathbf{J}_p$ and the usual minimal coupling derivative $(-i\hbar\boldsymbol{\nabla} - q\mathbf{A})$. From (38) we see that $m\,\mathbf{v} = \hbar\,\boldsymbol{\nabla}\varphi - (q/c)\,\mathbf{A}$, with $\boldsymbol{\nabla}\cdot\mathbf{v} = 0$ and $\boldsymbol{\nabla}\cdot\mathbf{J}_p = 0$. In condensed matter physics, $(q/mc)\,\rho(\mathbf{r},t)\mathbf{A}(\mathbf{r},\mathbf{t}))$ is the Meissner current term and $(1/m)\rho(\mathbf{r},t)\hbar\,\boldsymbol{\nabla}\varphi(\mathbf{r},t)$ is the boson current.

The SBS thus dynamically produces the non-vanishing mass M for the field $\mathbf{A}$ [21], implying its damping by a factor $\propto \exp(-c\,M\,s/\hbar)$ and its self-focusing propagation with a transversal size of the order of $s \propto 1/M$ [20].

Self-focusing propagation of the em field is a well studied phenomenon in classical nonlinear optics [23, 24, 25] and it can be shown [20] that the present analysis accounts for phenomenological aspects of em wave propagation in nonlinear non-homogeneous media, as it is indeed the medium of the hurricane.

The dipole of the water molecule is due to the displacement of a pair of electrons; thus $q = 2e$ and $m = 2m_e$ in the above expression for M^2. Then $\mathcal{P} = 2\,e\,\ell/c \times n$ with n the number of oriented dipoles. In the extreme case where all dipoles are oriented, $n = N/18$, where N is the Avogadro number. We then obtain [20] $M_H\,c^2 = \hbar\,c\,(16\,\pi\,r_0\,n)^{1/2} \approx 13.60$ eV, with $r_0 = (1/4\,\pi\,\epsilon_0)(e^2/m_e\,c^2) = 2,8 \times 10^{-15}$ m the "classical radius" of the electron, $\hbar = 0,6582 \times 10^{-15}$ eV sec; c the light velocity and 1 m^3 the unit volume. We thus see that the mass acquired by the self-focusing em field in such a case corresponds approximately to the hydrogen ionization energy.

A single photon with energy $\hbar\nu < M_H c^2$ is thus not able to propagate through the correlated medium. A highly energetic photon ($\hbar\nu \gg M_H c^2$) would instead produce the breakdown of the correlation modes, thus restoring the masslessness of em field (symmetry restoration). In general the polarization density values are smaller than the maximum value above considered. This allows, however, an interesting mechanism of energy storage in coherent states.

Energies from different processes (mechanical, chemical, electromagnetic), less than the hydrogen ionization energy, could be stored indeed as polarization modes in the coherent state, provided that they can produce excitations in the correlated medium. When the accumulated energy is enough to reach the ionization value, the whole polarization mode would manifest as an em field propagating as in (37) with the corresponding energy value $M c^2$.

Indeed, in the quasi-stationary approximation, the minimization of the free energy, $d\mathcal{F} = dE - \frac{1}{\beta}dS = 0$, with $\beta = 1/k_B T$, implies that the rate of change of the number N_κ of condensate polarization modes is related to the changes of the 'internal energy' dE and entropy dS [10, 14, 15]

$$dE = \sum_\kappa E_\kappa \dot{N}_\kappa(t)dt = \frac{1}{\beta}dS \;, \tag{39}$$

with $\dot{N}_\kappa \equiv dN_\kappa/dt$. The exchanged heat is $dQ = (1/\beta)dS$, which enters in the balance of the hurricane heating/cooling.

Most interesting is, however, the non-equilibrium dynamics characterized by criticality and phase transitions (assumed anyway to occur between thermodynamic equilibrium states to which the fluctuation theorem applies [26, 27, 28]).

5. Criticality, the Ginzburg-Landau non-stationary regime and the vortex equation

In order to study the variations of the charge density wave function $\sigma(\mathbf{r},t)$ occurring in the non-equilibrium phase transition processes, we start by considering the free energy density Ginzburg-Landau (GL) functional $F(\sigma,\sigma^*,\mathbf{A})$ [29]. Extremizing $F(\sigma,\sigma^*,\mathbf{A})$ with respect to $\sigma^*(\mathbf{r},t)$ gives the stationary Ginzburg-Landau (GL) equation: $\partial F/\partial\sigma^* = 0$. In full generality, we may write:

$$\frac{\partial F}{\partial \sigma^*} \equiv \left[\frac{1}{2m}(-i\hbar\boldsymbol{\nabla} - \frac{q}{c}\mathbf{A})^2 + \mu^2 + \lambda|\sigma(\mathbf{r},t)|^2\right]\sigma(\mathbf{r},t), \tag{40}$$

where the explicit expressions of $\mu^2(T)$ and $\lambda(T)$ depend on the system under study. They act as mass term and positive coupling constant term, respectively. The quantity $(\mu^2 + \lambda|\sigma|^2)\sigma$ may be considered as derived from the potential $V(\sigma,\sigma^*) = -(\lambda/2)(|\sigma|^2 + \mu^2/\lambda)^2$.

The value of σ minimizing $V(\sigma,\sigma^*)$ is zero (disordered or symmetric ground state) for $\mu^2 \geq 0$, and non-zero for $\mu^2 < 0$, with $|\sigma|^2 = -\mu^2/\lambda \neq 0$, playing the role of the order parameter (the ordered or asymmetric ground state). The dependence on a critical temperature T_C may be taken into account through the dependence of μ^2 on T_C in a form, e.g., proportional to $(T - T_C)/T$, assuming T varies linearly with time near the transition region.

By tuning μ^2, the pump parameter in the Haken interpretation [30], the system may be carried in the lasering process between the disordered and the ordered (laser) state (the phase transition), $|\sigma|^2$ changing in time from zero to a non-vanishing value proportional to μ^2.

In the non-stationary regime, $\partial F/\partial\sigma^*$ is non-vanishing and expresses the rate at which $\sigma(\mathbf{r},t)$ approaches its stationary value at the minimum of the free energy. The generalized time dependent GL (TDGL) equation is:

$$i\hbar\frac{\partial\sigma}{\partial t} = \hat{H}\sigma - \frac{i}{\gamma}\frac{\partial F}{\partial\sigma^*} \;. \tag{41}$$

The second term in the r.h.s. of Eq. (41) describes the dissipative contribution coming from incoherent relaxation processes with γ the relaxation parameter [29, 31].

In the limit of the stationary regime $(1/\gamma)\partial F/\partial\sigma^* \to 0$, the equilibrium wave function is denoted by $\sigma^\circ(t) = |\sigma^\circ|\exp(-i\epsilon t/\hbar)$, with $\hat{H}\sigma^\circ = \epsilon\sigma^\circ$.

By using $\sigma = \sqrt{n}\exp(i\varphi)$, one can show that the TDGL equation becomes

$$\frac{1}{D_{\rm GL}}\frac{d\chi}{dt} = \nabla^2\chi + \frac{1}{\xi_{\rm GL}^2}\left[(1-\chi^2) - \xi_{\rm GL}^2\left(\frac{mv}{\hbar}\right)^2\right]\chi\,, \tag{42}$$

where $\chi = |\sigma|/|\sigma^\circ| \equiv (n/n^\circ)^{1/2}$ is the normalized wave function, $D_{GL} \equiv \xi_{GL}^2/\tau_{GL}$, is the diffusion coefficient, with ξ_{GL} and τ_{GL} denoting the GL correlation length and the GL relaxation life-time, respectively (see [1] for details).

Eq. (42), for $(1/D_{\rm GL})d\chi/dt = \hbar\gamma\, d\chi/dt \approx 0$ is indeed the vortex equation [32]. It is associated with the singularity at $r = 0$.

For numerical analysis and simulations of Eq. (42) see, e.g. [33] and refs. therein quoted.

One may derive from the TDGL equation that the vortex core size is in the first approximation of the order $\xi_{\rm GL}^2 \propto |\mu^2|^{-1}$, and thus it increases as T increases, approaching to T_C from below; the vortex envelope disappears at T_C. Symmetry restoration is obtained for temperature rising above T_C. Vice-versa, going back to T_C from above the unstructured (fully symmetric) background activity exhibits an undefined phase at T_C, i.e. the singularity at the center line $r = 0$. The critical regime starts as T goes below T_C, vortices appear, and their core shrinks as temperature decreases. The vortex formation starts form the singularity, at the vortex core, in the process of phase transition due to SBS ("where the vortex comes from") [32].

The cooling itself is a manifestation of the process of symmetry breakdown. The non-homogeneous condensation and ordered localized patterns are lower in energy and separated by an energy gap from the symmetric state. The phase transition is non-instantaneous and begins with cooling [34].

In the TDGL formalism it is assumed that a time-dependent temperature might still be defined in non-instantaneous phase transition processes. In the general picture the transition is assumed to start at the critical temperature T_C. Time evolution leads then to the stable configuration at the so-called "Ginzburg temperature" T_G, with $T_G < T_C$.

The critical regime is defined to be the one evolving in time between T_C and T_G, and is supposed to be always a transition between different thermodynamic equilibrium states, regardless of the non-equilibrium processes occurring in the course of the transition process.

Thus the general thermodynamic description derived from the fluctuation theorem [26] applies to the 'critical dynamics' of our system. As already mentioned, in the process of reaching new equilibrium configurations below the critical temperature T_C, local exchanges of heat with environments turn into entropy changes of the vortex, with consequent rearrangements of its configurations and (internal) energy density contents.

We remark that the departure from the stationary regime (at T_C), namely the start of the critical regime, is driven by fluctuations which can trigger a phase transition, if the necessary transition energy is provided by some external input [35]. These ground state fluctuations turn into temperature fluctuations since in our dissipative model the ground state is in fact a thermal state [32, 36]. At the end of the critical regime (at T_G) the system arrives at a "new" ground state configuration and the phase transition is thus completed. The system thus undergoes a continuous sequence of phase transitions going through a path, or trajectory, through the (infinitely) many coherent ground states (in each of them free energy is minimized). Remarkably, these trajectories in the manifold of the coherent states turn out to be classical chaotic trajectories [37], which shows the sensibility of the molecular dynamics and of the hurricane evolution in general to slight changes in the environmental conditions.

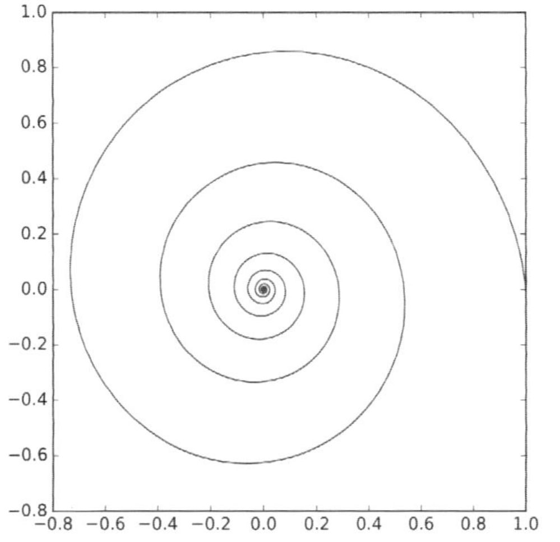

Figure 3. Eq. (16): the clockwise logarithmic spiral.

6. The hurricane fractal structure

Our discussion in the previous Sections has shown that coherent state structures characterize the molecular dynamics out of which hurricanes originate. Since an isomorphism exists between (deformed) coherent states and fractals [38, 39], this suggests to us to search for fractal self-similarity properties of the hurricane. We observe indeed that Eq. (16) may be regarded as the logarithmic spiral equation in polar coordinates [40] (see Fig. 3).

In order to see this, suppose that Ω may be scaled by tuning a real parameter, say d, so that $\Omega = C/d$. By putting $\alpha = \theta/d$, we write then $\xi = \varrho_0\, e^{-\theta} \cos\alpha$, $\eta = -\varrho_0\, e^{-\theta} \sin\alpha$ (cf. Eqs. (17)) and Eq. (16) is written as $\varrho(\alpha) = \varrho_0\, e^{-d\alpha}$, which is represented in a log-log plot with abscissa $\alpha = \ln e^{\alpha}$ by the straight line of slope d:

$$\ln \frac{\varrho}{\varrho_0} = -d\,\alpha\ . \tag{43}$$

The constancy of the angular coefficient $\tan^{-1} d$ represents the self-similarity property of ϱ.

The parameter d is called the fractal or self-similarity dimension. Rescaling $t \to n\,t$ changes $\varrho/\varrho_0 = e^{-d\alpha}$ by the power $(\varrho/\varrho_0)^n$. At time $\tau = 2\,\pi\,d/C$ it is $(C/d)\,\tau = 2\,\pi$ and at $t = n\,\tau$, $z_1 = \varrho_0\,(e^{-2\,\pi d})^n$, $z_2 = \varrho_0\,(e^{2\,\pi d})^n$, with integer $n = 1, 2, 3...$

Such a macroscopic self-similarity feature of the logarithmic-spiral-like hurricane finds its correspondence in the d-deformed coherent states structure at the molecular level previously discussed. For brevity we do not show here how such an isomorphism is proved. See [38] for details.

7. Conclusions

We have shown that central results of the Burgers model for hurricanes may be derived in the QFT of the collective molecular dynamics in the hurricane-environment interaction. Macroscopic hurricane behaviors and properties, such as vorticity, transitions between different dynamical regimes (phase transitions), criticality and chaotic behavior, energy storage and dissipation, free energy and entropy changes, etc. are thus the result not of random kinetic molecular motions, but of collective coherent molecular dynamics of water molecules in the hurricane interaction with moist convection flows, cumulus and cloud water droplets.

The QFT spontaneous breakdown of dipole rotational symmetry of water molecules is induced by weak perturbations due, e.g., to ionized particles, or to other sources, mechanical, chemical, etc.. The consequent generation of long range molecular correlations is responsible of collective coherent molecular motions characterized by non-trivial topology and fractal self-similar structures which manifest at the hurricane macroscopic level as vortices and logarithmic spiral configurations. The time dependent Ginzburg-Landau equation has been used to describe non-stationary processes.

Energy storage and dissipation by hurricanes has been related with the properties of the coherent state molecular structure. Energy can be accumulated on the coherent net of long range correlations and when it reaches a threshold (energy gap) depending on the system under study, it may be then released, as discussed in Section 4. For example, the total hamiltonian in normal modes of a net of harmonically coupled point masses, when subject to the action of an external potential linearly dependent on space, with a force of Fourier components F_k, has the form $H_T = H_0 + H_F$, with $H_0 = \sum_k \hbar\omega_k(c_k^\dagger c_k + 1/2)$ and H_F the contribution (the energy gap ΔE) due to the external force [15, 41]. Consider the coherent state $|\alpha\rangle = \Pi_k|\alpha_k\rangle$, where $|\alpha_k\rangle$ is the k-mode coherent state, $|\alpha_k\rangle = \exp(-|\alpha_k|^2/2)\sum_{n_k}(1/n_k!)(\alpha_k)^{n_k}(c_k^\dagger)^{n_k}|0\rangle$, and α_k is the k-mode coherence strength of $|\alpha_k\rangle$. $|0\rangle$ denotes the vacuum, or ground state, for the c_k-oscillators, $c_k|0\rangle = 0$ for any k. The number of coherently condensed c_k-modes in $|\alpha\rangle$ is $\langle\alpha|c_k^\dagger c_k|\alpha\rangle = |\alpha_k|^2$. The expectation value of H_T in $|\alpha\rangle$ is $\langle\alpha|H_T|\alpha\rangle = \sum_k[\hbar\omega_k|\alpha_k|^2 - F_k\alpha - F_k^*\alpha^*]$, without considering the zero point energy. It is minimized for $(1/\hbar\omega_k)F_k = \alpha_k^*$, namely when the external action matches the coherence phase of the net. The energy "reservoir" in the coherent state is thus $-2\sum_k \hbar\omega_k|\alpha_k|^2$, which is clearly not due to random kinematical motion, but to the coherent dynamics. Energy supplied to the coherent net (below of course a critical threshold) does not thermalize the net components increasing their kinetic energy, it is instead stored "in the net".

Note that divergences arise in the case $\omega_k = 0$. This corresponds to the absence of restoring forces opposing to F_k with increasingly unbounded acceleration for the system as a whole [41].

As already observed in the previous Sections, coherence provides also the bridge to go from microscopic (quantum) to macroscopic (classical) dynamics, namely from the quantum operator fields c_k to the c-number (classical) α_k functions. For example, the probability of finding n_k quantum modes in the coherent state $|\alpha\rangle$ is given by the classical distribution $|\langle n_k|\alpha\rangle|^2 = \exp(-|\alpha_k|^2)(1/n_k!)|\alpha_k|^{2n_k}$, which is recognized to be the Poisson distribution.

The above remarks shed some light on the fact that the hurricane can deliver its giant amounts of energy as a whole, not just as the uncorrelated molecular energy bits delivered by the non-coherent molecular masses of a "normal" or "simple", although energetic and speedy, wind. In this last case, damages are produced by the intensity of the kinematical (uncorrelated) molecular wind flows (currents).

In conclusion, we have shown that QFT may reproduce features of the Burgers model; however, classical fluid dynamics alone cannot account for the coherent molecular quantum dynamics. Hurricanes appear to be macroscopic quantum systems, in the sense, indeed, that some of their properties cannot be accounted without recourse to quantum dynamics.

References

[1] Meyer G and Vitiello G 2018 *Phys. Lett.* A **382** 1441
[2] Ooyama K V 1982 *J. Meteor. Soc. Japan* **60** 369
[3] Holland G J and Merrill R T 1984 *Quart. J . R. Met. Soc.* **110** 723
[4] Montgomery M T and Farrell B F 1992 *Amer. Meteor. Soc.* **50** 285
[5] Montgomery M T and Smith R K 2014 *Austr. Meteor. Ocean. J.* **64** 37

IOP Conf. Series: Journal of Physics: Conf. Series **1275** (2019) 012017 doi:10.1088/1742-6596/1275/1/012017

[6] Tory K J and Franck M W 2010 *Global Perspectives on Tropical Cyclones* Chapter 2 (Series on Asia-Pacific Weather and Climate) vol 4 ed J. C. L. Chan and J. D. Kepert (Singapore: World Scientific)
[7] Gray W M 1975 *Amospheric Science Paper No. 234* (Fort Collins: Colorado State University, Colorado)
[8] Eliassen A 1951 *Astrophys. Norv.* **5** 19
[9] Burgers J M 1948 *Advances in Applied Mechanics* **1** 171
[10] Celeghini E, Rasetti M and Vitiello G 1992 *Ann. Phys.* **215** 156
[11] Sivasubramanian S, Srivastava Y N, Vitiello G and Widom A 2003 *Phys. Lett.* A **311** 97
[12] Blasone M, Graziano E, Pashaev O K and Vitiello G 1996 *Ann. Phys.* **252**, 115
[13] Tsue Y, Kuriyama A and Yamamura M 1994 *Progr. Theor. Phys.* **91** 469
[14] Umezawa H 1993 *Advanced field theory: Micro, macro, and thermal physics* (New York: AIP)
[15] Blasone M, Jizba Pand Vitiello G 2011 *Quantum Field Theory and its macroscopic manifestations* (London: Imperial College Press)
[16] Chen Y, Okur H I, Gomopoulos N, Macias-Romero C, Cremer P S, Petersen P B, Tocci G, Wilkins D M, Liang C, Ceriotti M and Roke S 2016 *Sci. Adv.* **2** e1501891
[17] De Ninno A, De Francesco M 2018 *Chemical Physics Letters* **705** 7
[18] De Ninno A, De Francesco M 2018 *Chemical Physics* **513** 266
[19] Del Giudice E, Doglia S, Milani M and Vitiello G 1985 *Nucl. Phys.* B **251** (FS 13) 375
[20] Del Giudice E, Doglia S, Milani M and Vitiello G 1986 *Nucl. Phys.* B **275** (FS 17) 185
[21] Matsumoto H, Papastamatiou N J, Umezawa H and Vitiello G 1975 *Nucl. Phys.* B **97** 61
[22] Matsumoto H, Papastamatiou N J, Umezawa H 1975 *Nucl. Phys.* B **97**, 90
[23] Marburgher J H 1975 *Prog. Quant. Electr.* **4** 35
[24] Chiao R Y, Gustafson T K and Kelley P L 2009 *Self-Focusing of Optical Beams* (Self Focusing: Past and Present) ed R W Boyd, S G Lukishova and Y R Shen (New York: Springer) p. 129
[25] Shen Y R 1975 *Prog. Quant. Electr.* **4**, 1
[26] Crooks G E 1998 *J. Stat. Phys.* **90** 1481
[27] Jarzynski C 1997 *Phys. Rev. Lett.* **78** 2690
[28] Jarzynski C 1997 *Phys. Rev.* E **56** 5018
[29] Freeman W J, Livi R, Obinata M and Vitiello G 2012 *Int. J. of Mod. Phys.* B **26**(6) 1250035
[30] Haken H 1984 *Laser theory* (Berlin: Springer-Verlag)
[31] Barybin A A 2011 *Advances in Condensed Matter Physics* 425328
[32] Mańka R and Vitiello G 1990 *Ann. Phys.* **199** 61
[33] Antunes N A, Bettencourt L M A and Zurek W H 2000 *Phys. Rev. Lett.* **82** 2824
[34] Alfinito E and Vitiello G 2002 *Phys. Rev.* B **65** 054105
[35] Zurek W H 1997 *Phys. Rep.* **276** 177 and Refs. therein quoted.
[36] Mańka M, Kuczyński J and Vitiello G 1986 *Nucl. Phys.* B **276** 533
[37] Vitiello G 2004 *Int. J. Mod. Phys.* B **18** 785
[38] Vitiello G 2012 *Phys. Lett.* A **376** 2527 (2012).
[39] Vitiello G 2014 *Systems* **2014 2** 203
[40] Peitgen H O, Jürgens H and Saupe D 1986 *Chaos and fractals. New frontiers of Science* (Berlin: Springer-Verlag)
[41] Greiner W and Reinhardt J 1996 *Field quantization* (Berlin: Springer-Verlag)

9th International Workshop DICE2018 : Spacetime - Matter - Quantum Mechanics IOP Publishing
IOP Conf. Series: Journal of Physics: Conf. Series **1275** (2019) 012018 doi:10.1088/1742-6596/1275/1/012018

Violation of the Bell's type inequalities as a local expression of incompatibility

Andrei Khrennikov
Linnaeus University, International Center for Mathematical Modeling in Physics and Cognitive Sciences Växjö, SE-351 95

E-mail: Andrei.Khrennikov@lnu.se

Abstract. By filtering out the philosophic component we can be said that the EPR-paper was directed against the straightforward interpretation of the Heisenberg's uncertainty principle or more generally the Bohr's complementarity principle. The latter expresses contextuality of quantum measurements: dependence of measurement's output on the complete experimental arrangement. However, Bell restructured the EPR-argument against complementarity to justify nonlocal theories with hidden variables of the Bohmian mechanics' type. Then this Bell's kind of nonlocality - *subquantum nonlocality* - was lifted to the level of quantum theory - up to the terminology *"quantum nonlocality"*. The aim of this short note is to explain that Bell's test is simply a special *test of local incompatibility of quantum observables*, similar to interference experiments, e.g., the two-slit experiment.

1. Introduction

In this note I repeat the message presented in papers [1]-[3]. This note was motivated by the recent INTERNET discussion on the Bell inequality. This discussion involved both sides: those who support the conventional interpretation of a violation of the Bell type inequalities and those who present a variety of non-conventional positions. As always happens in such discussions on the Bell inequality, the diversity of for and against arguments was really amazing, chaotic, and mainly misleading.[1] My attempt to clarify the situation by replying R. Gill generated a new burst of chaos. In parallel, I had the email exchange with W. De Muynck and his sharing my views on the Bell inequality interpretation stimulated me to write this note (with hope to stop further misinterpreting of the Bell inequality).

The aim of this short note is to explain that Bell's test is simply a special *test of local incompatibility of quantum observables*, similar to interference experiments, e.g., the two slit experiment. In particular, Bell's test has nothing to do with nonlocality, neither subquantum nor quantum. I guess that the fathers of the Copenhagen interpretation, e.g., Bohr and Heisenberg, would not be interested in such a test since the principle of complementarity was well supported by interference experiments.

[1] I (as well as R. Gill and T. Nieuwenhuizen) participate in similar discussions since year 2000 and the situation did not change so much. It is interesting that the recent (2015) loophole free experiments did not have any impact on "non-conventional" part of discussants. But A. Aspect told me around year 2000 that closing loopholes would not close the foundational discussion.

Content from this work may be used under the terms of the Creative Commons Attribution 3.0 licence. Any further distribution of this work must maintain attribution to the author(s) and the title of the work, journal citation and DOI.
Published under licence by IOP Publishing Ltd

IOP Conf. Series: Journal of Physics: Conf. Series **1275** (2019) 012018 doi:10.1088/1742-6596/1275/1/012018

The main message of this paper is presented in a few words and formulas in the next section. Only the readers who are deeply involved into quantum foundations can be interested in the discussions presented in appendixes.[2]

2. CHSH inequality is violated only under the condition of local incompatibility

The mathematical counterpart of the story which I plan to tell is well known and rather trivial. The main issue is the interpretation. Here we proceed by using mathematics from article [4]. This article is based on the quantum formalism (cf., e.g., [1] - [3], [5]-[8]). [3]

Consider the Bohm-Bell type experiments. There are considered four observables A_1, A_2, B_1, B_2 taking values ± 1. It is assumed that the pairs of observables $(A_i, B_j), i, j = 1, 2$, can be measured jointly, i.e., A-observables are compatible with B-observables. However, the observables in pairs A_1, A_2 and B_1, B_2 are incompatible, i.e., they cannot be jointly measured. Thus probability distributions $p_{A_i B_j}$ are well defined theoretically by quantum mechanics and they can be verified experimentally; probability distributions $p_{A_1 A_2}$ and $p_{B_1 B_2}$ are not defined by quantum mechanics and, hence, the question of their experimental verification does not arise.

For spatially separated systems S_1 and S_2, incompatibility of the A-observables on S_1 and the B-observables on S_2 is natural to call *local incompatibility*

Now we proceed generally, i.e., without the local incompatibility assumption. In the quantum formalism observables are presented by Hermitian operators which will be denoted by the same symbols as the observables. The quantum representation of the CHSH-correlations can be obtained with the aid of the "CHSH-operator":

$$C = \frac{1}{2}[A_1(B_1 + B_2) + A_2(B_1 - B_2)]. \tag{1}$$

As was shown in article [4], for given observables, the CHSH inequality holds for all quantum states iff

$$C^2 \leq I, \tag{2}$$

where I is the unit operator.

It is easy to show that

$$C^2 = 1 + (1/4)[A_1, A_2][B_1, B_2]. \tag{3}$$

Thus if *at least one of the commutators* equals to zero, i.e.,

$$[A_1, A_2] = 0, \tag{4}$$

or

$$[B_1, B_2] = 0, \tag{5}$$

then the CHSH inequality holds for all quantum states. It is impossible to violate this inequality, if at least at one side (say at the A-side) observables are compatible. We emphasize that *both these conditions are local.* Condition (4) is about observables on system S_1 and condition (4) is about observables on system S_2. (How can one find any trace of nonlocality here?)

[2] Appendix 1 (coupling the EPR-argument and the Bohr's complementarity principle, Bohr's reply to Einstein) and in Appendix 2 (the contextual nature of the Bohr's principle of complementarity, so to say the principle of *contextuality-complementarity*), Appendix 3 (the role of the existence of the irreducible quantum of action, the Planck constant).

[3] The direct appeal to the quantum formalism makes the presentation clearer for so to say "real physicists". Suppose I met a physicist who is not so much interested in quantum foundations, who believes in the validity of quantum theory and its completeness, following, for example, Bohr, Heisenberg, Fock, Landau, and who is not interested in anything "beyond quantum mechanics". Suppose he asked me: "What does a violation of the Bell's inequality mean?" He is not interested in an answer related to hidden variables, or foundations of probability theory.

IOP Conf. Series: Journal of Physics: Conf. Series **1275** (2019) 012018 doi:10.1088/1742-6596/1275/1/012018

3. Conclusion

In the light of the previous consideration, it is clear that the common claim that the Bell story about nonlocality is really misleading. (Any rationally thinking expert in quantum foundations should dismiss the use of the term "nonlocality".) The Bell-tests are just special tests of incompatibility. The natural question arises:

Do the Bell type experiments add some foundational value to the interference type experiments?

Thus if one hope to go beyond quantum mechanics, then he should *search for possible violations of the complementarity principle.* However, such experiment have nothing to do with spatial separation of systems.

Appendix 1: The role of complementarity principle in the EPR-reasoning

I start with the remark that one has to separate the Bell inequality as the mathematical result and its interpretation, including conditions of its derivation. So, there is no problem with the "Bell theorem", the problem is the conventional interpretation of this mathematical statement. Thus people who claim that they found mistakes in Bell's mathematics do very bad job for quantum foundations: they discredit the rational critique of the conventional interpretation of the Bell inequality.

The second important remark is that using the notion *"local realism"* in interpreting violations of the Bell type inequalities is really misleading. It seems that Bell have never used this notion. At least in his book [9], the collection of his main works, "local realism" appears only in the title of the reference to the of one philosopher.

Bell started his project as being on the realist position [9] (by following directly the EPR-paper [10]). Then he claimed that realism can be saved only through the rejection of locality. Finally, he understood that

"the result of an observation may reasonably depend not only upon the state of the system (including the hidden variables) but also on the complete disposition of the apparatus" [11].

Shimony [12] stressed that this is the first statement about contextuality (although Bell did not use this terminology):

"John Stewart Bell (1928-90) gave a new lease on life to the program of hidden variables by proposing contextuality. In the physical example just considered the complete state λ in a contextual hidden variables model would indeed ascribe an antecedent element of physical reality to each squared spin component s_n^2 but in a complex manner: the outcome of the measurement of s_n^2 is a function $s_n^2(\lambda, C)$ of the hidden variable λ and the context C, which is the set of quantities measured along with s_n^2. ... a minimum constraint on the context C is that it consist of quantities that are quantum mechanically compatible, that is represented by self-adjoint operators which commute with each other. ..."

However, this is nothing else as Bohr's statement that measurement's output depends on the complete experimental arrangement (see Appendix 2). In principle, there is just one step from understanding of the role of such type of contextuality t o Bohr's complementarity principle and to understanding that violations of the Bell-type inequalities is a consequence of this principle (Appendix 2). Unfortunately, Bell was addicted to the nonlocality issue and this final step has never been done.

One may say that Bohr did not consider the possibility to go beyond the quantum theory, e.g., by introducing hidden variables. As I learned from Arkady Plotnitsky who is the author of a series of monographs about Bohr, the complementarity principle, and the interpretations in the *Spirit of Copenhagen* (see, e.g., [13, 14] and our joint paper [15]), in principle Bohr did not reject the possibility of construction of mathematical subquantum models. He was not excited by

IOP Conf. Series: Journal of Physics: Conf. Series **1275** (2019) 012018 doi:10.1088/1742-6596/1275/1/012018

mathematical no-go theorems. For him the possibility to go beyond quantum theory was blocked by the principle of complementarity. Some people think that the latter is a purely philosophical principle. However, for Bohr and other Copenhagenists, this principle was a physical principle such as Newton laws (see Appendix 2).

Bell's behavior has its root in the EPR-paper [10] in that the anti-complementarity issue was covered by the "philosophic souse" with "elements of reality" as its basic component. Let us clean the EPR-argument from the "philosophic souse". Such a rafined argument has the following structure:

- The Copenhagen interpretation is heavily based on the complementarity principle.[4]
- This principle was (straightforwardly) interpreted as the rejection of the joint *existence* of the position and momentum of a quantum system.
- The existence of the EPR-states exhibiting perfect correlations contradicts to the complementarity principle.

The natural conclusion from the EPR-argument could be formulated as follows:

Bohr's principle of complementarity should be rejected!

However, Einstein, Podolsky, and Rosen combined the above reasoning on physical foundations of quantum mechanics with philosophic argumentation based on the elements of reality. The above conclusion was not clearly stated in [10].

This is the good place to mention the Bohr reply [16] to the EPR-paper. During conferences on quantum foundations, speakers permanently say that it is impossible to understand Bohr's paper. I do not think so. The message of this paper is very clear:

The principle of complementarity is not about the joint existence. It is about the possibility of joint measurement!

In any event, Bell did not understand the Bohr reply and tried to justify the EPR-argument by using subquantum mathematical models of the hidden variables type. As was already stressed, ignoring the crucial role played by the complementarity principle in the EPR-paper led him to the emphasis of nonlocality. We recall that in the EPR-paper [10] nonlocality was mentioned as an absurd alternative to incompleteness of quantum mechanics. (And we interpret the latter as the possibility to violate the complementarity principle.)

Appendix 2: Bohr's complementarity principle as a principle of contextuality

Bohr's complementarity principle is not just a kind of the impossibility principle. In the modern terminology the essence of the complementarity principle is contextuality. But Bohr by himself did not operate with the notion "experimental context"; he wrote about experimental conditions [17]:

> "Strictly speaking, the mathematical formalism of quantum mechanics and electrodynamics merely offers rules of calculation for the deduction of expectations pertaining to observations obtained under well-defined experimental conditions specified by classical physical concepts."

We now present the basic futures of Born's principle of contextuality-complementarity [3]:

- (B1): There exists the fundamental quantum of action given by the Planck constant h.
- (B2): The presence of h prevents approaching internal features of quantum systems.
- (B3): Therefore it is meaningless (from the viewpoint of physics) to build scientific theories about such features.

[4] It seems that in the EPR-paper it was identified with Heisenberg's uncertainty principle. But at that time such identification was common.

- (B4): An output of any observable is composed of contributions from a system under measurement and the measurement device.
- (B5): Therefore the complete experimental arrangement (context) has to be taken into account.
- (B6): There is no reason to expect that all experimental contexts can be combined. Therefore there is no reason to expect that all observables can be measured jointly. This is the essence of *the principle of complementarity.* Hence, there exist incompatible observables.

We emphasize that the principle (B6) is just a consequence of (B4) and (B5). And Bohr understood the complementarity principle as the combination of (B1-B6). It is the good place to remark that statement (B6) is very natural in any experimental situation. The existence of incompatible experimental contexts is not surprising. Non-existence would be really surprising.

Unfortunately, the name "complementarity principle" led to the common understanding of this principle as just an impossibility prtinciple. I think that in the light of the above consideration, the name of Bohr's "complementarity principle" should be changed to

The Bohr's contextuality principle or the contextuality-complementarity principle.

Of course, such renaming would be a delicate operation since nowadays the terminology "contextuality" is widely used in foundational discussion on the Bell type inequalities. In such discussions, the meaning of the notion "contextuality" does not coincide with so to say "Bohr's contextuality", as taking into account the experimental context to explain the mechanism of generating the values of quantum observables. From this viewpoint, any single quantum observable is contextual. Roughly speaking contextuality is present everywhere. There is no need in consideration joint measurements of compatible observables, A, B_1 and A, B_2, to find its exhibitions.

Appendix 3: A role of the Planck quantum of action

As was already noted, the Bohr's reasoning leading to the complementarity principle (B6) was fundamentally based on the principle (B1): the existence in nature of the fundamental quantum of action given by the Planck constant h. It seems that nowadays this fundamental "Bohr-Planck postulate" is totally forgotten. I have not seen any its trace in nowadays foundational discussions on the complementarity principle. Moreover, the Planck constant is not present at all in modern discussions: neither the Bell's type no-go theorems nor theorems on "quantum contextuality" do not containn h (nor the Kochen-Specker theorem). I think that the essence of the complementarity principle (and hence the role of the EPR-counterargument) can be understood only through the careful analysis of the Bohr-Planck postulate. Such an analysis with attraction of new ideas can be done only by real physicist. The mathematical games around no-go theorem would not help so much. In principle, we cannot exclude that Bohr was wrong and the existence of the fundamental quantum of action does not imply contextuality-complementarity. However, if he were right, then, as we have seen in section 2, all manipulations with classical probabilistic constraints have not so much relation to physics.

References

[1] Khrennikov A 1999 *Interpretations of Probability* (Utrecht/Tokyo: VSP Int. Sc. Publ.; Berlin: De Gruyter, 2009, second edition)
[2] Khrennikov A 2009 *Contextual Approach to Quantum Formalism* (Berlin-Heidelberg-New York: Springer)
[3] Khrennikov, A 2017 *Open Phys.* **15**, 734
[4] Wolf M M, Perez-Garcia D, Fernandez, C 2009 *Phys. Rev. Lett.* **103**, 230402
[5] De Muynck D 2002 *Foundations of Quantum Mechanics, an Empiricist Approach* (Dordrecht: Springer Netherlands)
[6] Nieuwenhuizen T M 2011 *Found. Phys.* **41**, 580

IOP Conf. Series: Journal of Physics: Conf. Series **1275** (2019) 012018 doi:10.1088/1742-6596/1275/1/012018

[7] Kupczynski M 2017 *Phil. Trans. Royal Soc.* A **375,** 2016039
[8] Kupczynski M 2018 *Entropy* **20**, 877
[9] Bell J S 2004 *Speakable and Unspeakable in Quantum Mechanics* (Cambridge: Cambridge Univ. Press)
[10] Einstein A, Podolsky B, Rosen N 1935 *Phys. Rev.* **47**, 777
[11] Bell J S 1966 *Rev. Mod. Phys.* **38**, 450
[12] Shimony A 2009 Hidden-variables models of quantum mechanics (Noncontextual and contextual). Ed D. Greenberger, K. Hentschel, F. Weinert *Compendium of Quantum Physics* (Berlin-Heidelberg: Springer) pp 287–291
[13] Plotnitsky A 2009 *Epistemology and Probability: Bohr, Heisenberg, Schrödinger and the Nature of Quantum-Theoretical Thinking* (Berlin-New York: Springer)
[14] Plotnitsky A 2012 *Niels Bohr and Complementarity: An Introduction* (Berlin-New York: Springer)
[15] Plotnitsky A and Khrennikov A 2015 *Found. Phys.* **45**, 1269.
[16] Bohr N 1935 *Phys. Rev.* **48**, 696
[17] Bohr N 1987 *The philosophical writings of Niels Bohr* (Woodbridge, Conn.: Ox Bow Press).

IOP Conf. Series: Journal of Physics: Conf. Series **1275** (2019) 012019 doi:10.1088/1742-6596/1275/1/012019

Reinforcement Learning Based Control of Coherent Transport by Adiabatic Passage of Spin Qubits

Riccardo Porotti[1,2], **Dario Tamascelli**[4], **Marcello Restelli**[5] and **Enrico Prati**[1]

[1] Istituto di Fotonica e Nanotecnologie, Consiglio Nazionale delle Ricerche, Piazza Leonardo da Vinci 32, I-20133 Milano, Italy
[2] Dipartimento di Fisica "Aldo Pontremoli", Università degli Studi di Milano, via Celoria 16, I-20133 Milano, Italy
[4] Quantum Technology Lab, Dipartimento di Fisica "Aldo Pontremoli", Università degli Studi di Milano, via Celoria 16, I-20133 Milano, Italy
[5] Politecnico di Milano, Piazza Leonardo da Vinci 32, I-20133 Milano, Italy

E-mail: enrico.prati@cnr.it

Abstract. Several tasks involving the determination of the time evolution of a system of solid state qubits require stochastic methods in order to identify the best sequence of gates and the time of interaction among the qubits. The major success of deep learning in several scientific disciplines has suggested its application to quantum information as well. Thanks to its capability to identify best strategy in those problems involving a competition between the short term and the long term rewards, reinforcement learning (RL) method has been successfully applied, for instance, to discover sequences of quantum gate operations minimizing the information loss. In order to extend the application of RL to the transfer of quantum information, we focus on Coherent Transport by Adiabatic Passage (CTAP) on a chain of three semiconductor quantum dots (QD). This task is usually performed by the so called counter-intuitive sequence of gate pulses. Such sequence is capable of coherently transfer an electronic population from the first to the last site of an odd chain of QDs, by leaving the central QD unpopulated. We apply a technique to find nearly optimal gate pulse sequence without explicitly give any prior knowledge of the underlying physical system to the RL agent. Using the advantage actor-critic algorithm, with a small neural net as function approximator, we trained a RL agent to choose the best action at every time step of the physical evolution to achieve the same results previously found only by *ansatz* solutions.

1. Introduction

We generate coherent adiabatic passage suitable for transferring quantum states by using a reinforcement learning approach, thus recovering counter-intuitive pulse sequences with no prior knowledge of the system. Such result demonstrates that quantum information processing can be optimized and controlled by reinforcement learning technique, to discover non-trivial strategies.

Coherent transport by adiabatic passage [1] has been introduced in semiconductor quantum dots [2] and later discussed for ab-initio simulations and extended to multiple spin qubits as well [3, 4, 5] . Contrary to supervised learning [6, 7], which has been applied to quantum information related systems earlier [8] , reinforcement learning has been introduced only very recently

Content from this work may be used under the terms of the Creative Commons Attribution 3.0 licence. Any further distribution of this work must maintain attribution to the author(s) and the title of the work, journal citation and DOI.
Published under licence by IOP Publishing Ltd

IOP Conf. Series: Journal of Physics: Conf. Series **1275** (2019) 012019 doi:10.1088/1742-6596/1275/1/012019

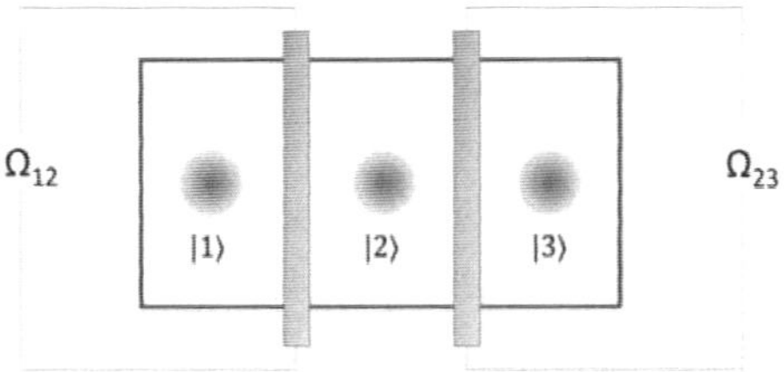

Figure 1. The triple quantum dot system. Top control gates of the three quantum dots not shown. The system is occupied by an individual spin at the time, so that it can move along the three quantum dots. The coupling gates can be controlled separately by changing Ω_{12} and Ω_{23} respectively.

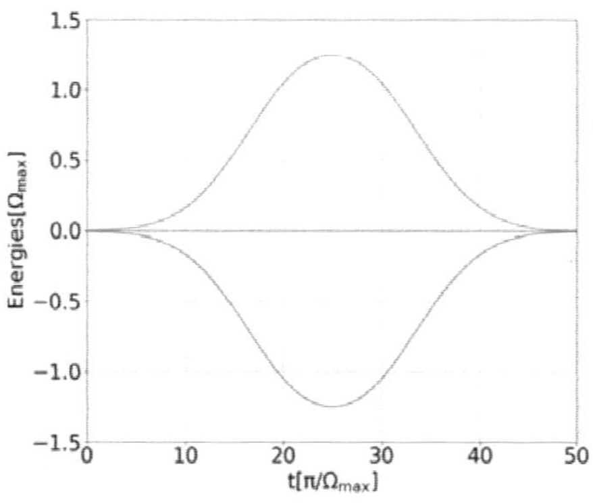

Figure 2. The eigenvalues of the triple quantum dot system. The orange value corresponds to the $|D_0\rangle$ eigenvector.

[9, 10]. In the past, some of us have developed genetic algorithms to generate universal logic port silicon spin qubits [11] , CMOS silicon logical qubit design for both Steane and surface codes respectively [12] and CTAP for the hybrid qubits. CTAP provides an excellent example of non-trivial strategy to find a gate pulse sequence to achieve optimum by moving in a quantum framework. The coherent transfer is indeed achieved by a counter-intuitive combination of gate pulsing sequence, i. e. the near and the far gates from the initially occupied quantum dot are opened second and first respectively. By adopting reinforcement learning, we show that such a non-trivial pulse sequence can be discovered with no prior knowledge of the system by applying reinforcement learning method by a neural network.

First, we show that our software based on QuTIP library [13] is able to recover all the classical results of prior literature. Next, the reinforcement learning algorithm is sized to the problem in order to have a reasonable neural network consisting of three layers (input, hidden and output layers respectively), in terms of number of neurons and its implementation. Like for the case of the artificial intelligence learning in the classical Atari game environment [14], the reinforcement learning routine here interacts with the QuTIP simulation of the CTAP implementing the Hilbert space and the time evolution of the system, and exploits the information retrieved by the feedback given by such time evolution generated by the time-varying Hamiltonian. By appropriately defining a simple reward function which constrains the system to avoid occupation of the central quantum dot and to reward occupation of the third quantum dot, the neural network progressively learns how to shape the gate pulses in order to prevent the occupation of the central dot to increase and next to maximize the occupation of the third, thanks of the structure of the eigenvectors of the Hamiltonian. We therefore conclude that reinforcement learning can be successfully applied to quantum information processing to discover highly non-trivial sequences of actions.

2. Standard CTAP

Figure 1 shows the typical configuration of three semiconductor quantum dots used to investigate CTAP. Each quantum dot is controlled by a control gate (not shown) from the top to set the ground state energy relatively to the Fermi energy of some external reservoir of electrons. Two coupling gates control the coupling between adjacent dots. Ground state are set by $E_1 = E_2 = E_3 = 0$. If we define the coupling between i^{th} QD and j^{th} QD as Ω_{ij}, the Hamiltonian

IOP Conf. Series: Journal of Physics: Conf. Series **1275** (2019) 012019 doi:10.1088/1742-6596/1275/1/012019

reads:

$$H = \begin{pmatrix} E_1 & -\Omega_{12} & 0 \\ -\Omega_{12} & E_2 & -\Omega_{23} \\ 0 & -\Omega_{23} & E_3 \end{pmatrix}. \tag{1}$$

The eigenstates of H, written in the QD base (with the respective energies), are [1] :

$$\begin{aligned} |D_+\rangle &= \sin\Theta_1 \sin\Theta_2 \, |1\rangle + \cos\Theta_2 \, |2\rangle + \cos\Theta_1 \sin\Theta_2 \, |3\rangle \\ |D_-\rangle &= \sin\Theta_1 \cos\Theta_2 \, |1\rangle - \sin\Theta_2 \, |2\rangle + \cos\Theta_1 \cos\Theta_2 \, |3\rangle \\ |D_0\rangle &= \cos\Theta_1 \qquad |1\rangle + 0 \qquad |2\rangle - \sin\Theta_1 \qquad |3\rangle \end{aligned}$$

with

$$\Theta_1 = \arctan\left(\frac{\Omega_{12}}{\Omega_{23}}\right). \tag{2}$$

The Hamiltonian dynamics is generated by a routine based on QuTIP. This section is dedicated to show that our implementation recovers all the previous findings known from literature, in particular to Ref.[1]. The energies of the eigenstates are plotted in Figure 2. By changing the eigenstate parameters in time, $|D_0\rangle$ can transform from $|1\rangle$ at $t = 0$ to $|3\rangle$ at $t = t_{max}$. If the Hamiltonian is prepared in $|D_0\rangle$ at $t = 0$, it will remain in the same eigenstate if the adiabaticity criterion is met, that is:

$$|\epsilon_0 - \epsilon_\pm| \gg |\left\langle \dot{D}_0 \middle| D_\pm \right\rangle|. \tag{3}$$

Consequently Ω_{12} and Ω_{23} pulses of Gaussian form can achieve coherent transport with high fidelity, if $t_{max} \geq \frac{10\pi}{\Omega_{max}}$. The remarking fact is that the two pulses must be applied in the so-called *counter-intuitive* sequence, as shown in Fig. 2. In Fig. 3a, CTAP for a 3-QD system is shown with a $t_{max} \geq \frac{50\pi}{\Omega_{max}}$.
Time evolution is governed by a master equation involving the density matrix ρ:

$$\dot{\rho} = \frac{1}{i\hbar}[H, \rho]. \tag{4}$$

Therefore the occupation of the first dot is provided by $\rho_{11}(t)$, that of the central by $\rho_{22}(t)$ and the final by $\rho_{33}(t)$.
Figure 3 shows the transfer of population from the state $|1\rangle$ to the state $|3\rangle$ with no passage in the state $|2\rangle$. This is achieved by a sequence of pulse called counter-intuitive, as the second gate is pulsed before the first one, closest to the $|1\rangle$ quantum dot. Figure 4 reproduces the results of the detuning of the ground state by $\Delta_{12} = E_2 - E_1$ and $\Delta_{13} = E_3 - E_1$ shown in Ref. [1]. In Figure 5 we further exploit our software to show the pulse region for which $\rho_{33}(t_f) = 1$ at the end of the pulse sequence, complemented with the integral of $\rho_{22}(t)$ during the process. The CTAP is successful only when both $\rho_{33} = 1$ and $\rho_{22} = 0.$ for the whole time evolution. The pulses are centered at α_1 and α_2 respectively, being their time evolution:

$$\begin{aligned} \Omega_{12} &= \Omega^{max} e^{[-(t - \frac{t_{max}+\sigma}{2})^2/(2\sigma^2)]} \\ \Omega_{23} &= \Omega^{max} e^{[-(t - \frac{t_{max}-\sigma}{2})^2/(2\sigma^2)]} \\ \sigma &= t_{max}/8. \end{aligned} \tag{5}$$

There is a wide region of counter-intuitive sequence of pulses allowing to achieve $\rho_{33}(t_f) = 1$. This value is achieved also for some combinations for the "intuitive sequence", i.e. when the closest gate is pulsed first. The integral of ρ_{22} during the process rules out those pulse sequences where the left gate is controlled first, as $\rho_{22}(t)$ does not equal zero for the whole time evolution.

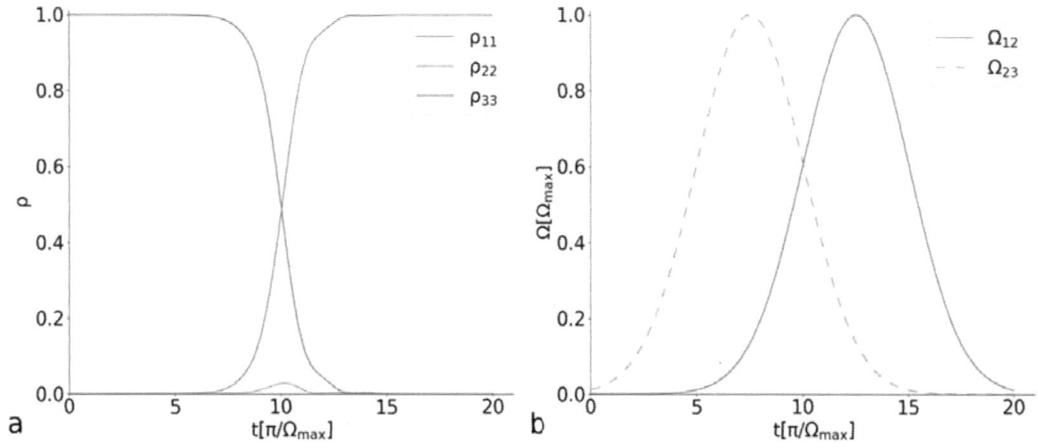

Figure 3. a) The CTAP from QD 1 to QD 3 with no occupation of the central QD 2 generated by QuTIP. b) the counter-intuitive pulse sequence of the coupling gates. Blue pulse of the gate voltage between QD 1 and QD 2 opens second.

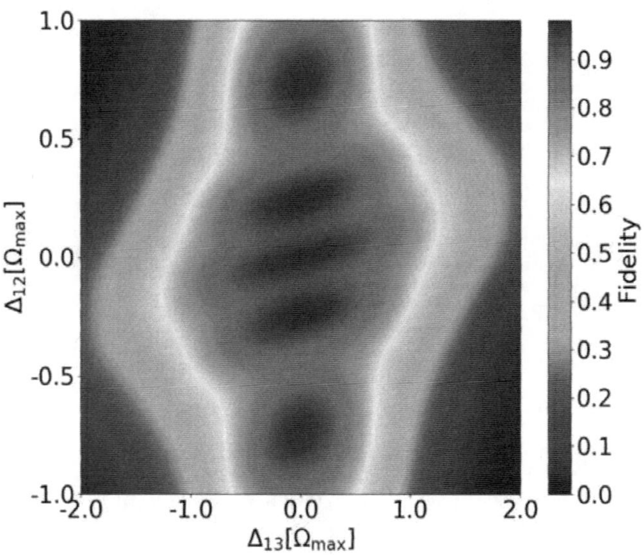

Figure 4. CTAP as a function of the detuning of the ground state energy of QD 1 with respect to QD 2 and QD 3 like in [1].

3. Description of the Reinforcement Learning algorithm

To rediscover the best pulses for CTAP by exploiting an artificial intelligence agent, we used an Advantage Actor-Critic (A2C) [10, 11] model. The algorithm finds the shape of the pulses of the two control gates without providing any prior knowledge of the system. From now on, we impose the system to behave consistently with episodic Markov Decision Processes (MDPs), so there exists a final state s_f that terminates the episode. A2C mixes two well-known paradigms of RL, namely value-based and policy-based algorithms. In a value-based framework, the expected

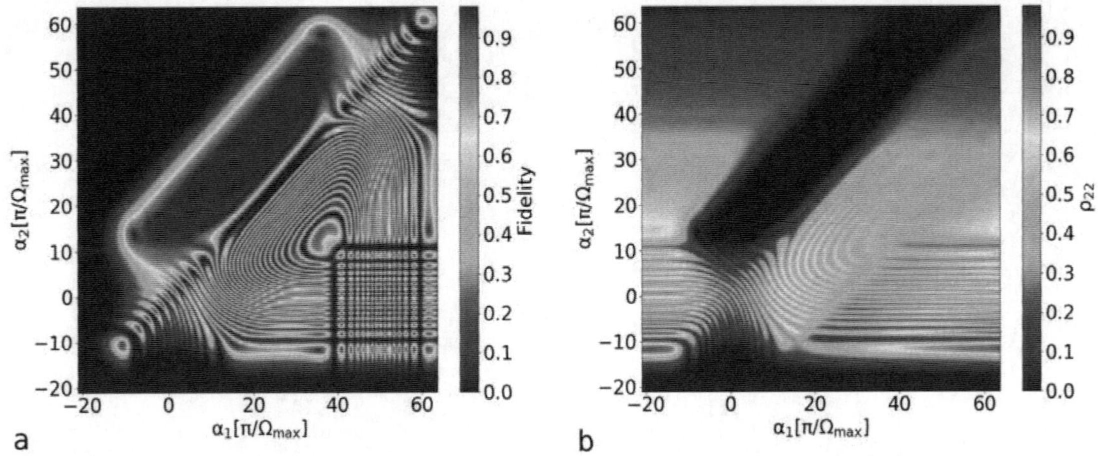

Figure 5. a) Value of $\rho_{33}(t_f)$ at the final time t_f after two Gaussian pulses centered at α_1 and α_2 respectively. b) The integral of ρ_{22} during the process.

return value of a state given an action is determined following a given policy π, called $Q^{\pi}(s,a)$. Then, if $Q^{\pi^*}(s,a)$, where π^* is the optimal policy:

$$\pi^*(s) = \underset{a}{argmax}\left[Q^{\pi}(s,a)\right]. \tag{6}$$

In a policy-based framework, instead, one is interested in directly finding the optimal policy, by minimizing:

$$\boldsymbol{\nabla}_{\theta} J(\theta) = \mathbb{E}_{\pi}\left[G_t \boldsymbol{\nabla}_{\theta} \log \pi_{\theta}\right] \tag{7}$$

where J is a proper loss function and G_t is:

$$G_t = \sum_{k=0}^{\infty} \gamma^k R_{t+k+1} \tag{8}$$

where γ is the discount factor and R_t is the reward at time t.

In A2C, the so-called advantage function $Q^{\pi^*}(s,a)$ replaces R_t. The latter method reduces the variance in the results and it could be also used to upgrade value-based techniques to non-episodic problems. The algorithm operates as in the following. The time evolution of H is divided in N timesteps, with $t \in [0, t_{max}]$. At each time step, the A2C agent can choose 9 discrete actions corresponding to 3 possible actions of both the control gates, namely increase, hold and decrease by an arbitrary quantity $\Delta\Omega$:

$$\Omega_{ij}(t+1) = \begin{cases} \Omega_{ij}(t) - \Delta\Omega \\ \Omega_{ij}(t) \\ \Omega_{ij}(t) + \Delta\Omega \end{cases}. \tag{9}$$

As the next step, the master equation solver implemented in QuTiP is used to evolve the quantum state at time t with the Hamiltonian calculated at step $t+1$, by providing as input the updated pulse values as decided by the agent. During each time step, the Hamiltonian is constant. As both actor and critic, we chose a neural network as a function approximator. The actor network is trained with policy gradient (PG) and it involves a Softmax output layer

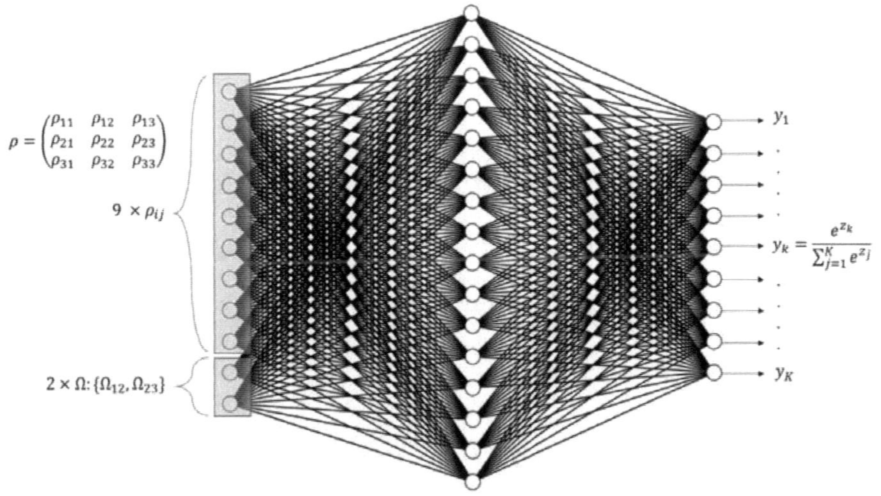

Figure 6. The network used to implement the advantage actor-critic algorithm. One hidden layer (*hl*) of 16 neurons is used. The 9 output neurons generate the combinations of possible actions (upward, hold, downward) of the two neurons.

consisting of 9 neurons. The input states of both networks are two 11-dimension vectors (Figure 6), in which the first 9 elements are populated by the density matrix elements, and the last 2 are the values of the two gate pulses during the last time step. By choosing an appropriate reward and by looking at the 3x3 density matrix of the system, the agent can learn which is the next best action. The pseudocode is shown in Table 1.

Indent	Functions
1	for each episode e:
2	initialize s0=[1.,0.,0.] total reward=0 for each time step t:
3	choose action a with prob. p evolve the quantum state total reward+=reward if done:
4	train act. & critic nets
5	break

Table 1. Pseudocode

4. Results

We developed an OpenAI-like environment [12] to perform the simulations. For the RL part of the problem, we used Keras with Tensorflow backend [13]. The parameters of the neural networks and the optimizers are listed in Table 2.

The simulations were run on an Intel Core i7-8700K @ 3.7 GHz and a Nvidia Titan Xp 12

IOP Conf. Series: Journal of Physics: Conf. Series **1275** (2019) 012019 doi:10.1088/1742-6596/1275/1/012019

Architecture and Settings	**Neural Network**	
	Actor	***Critic***
Layers	(11, 16, 9), Dense	(11,16,1), Dense
Activation function	ReLU for the *hl*, Softmax for the output	ReLU for the *hl*, Linear for the output
Optimizer	Adam (LR[a] =0.001, default [15] elsewhere)	Adam (LR[a] =0.001, default [15] elsewhere)
Loss Function	Categorical Crossentropy	Mean Squared Error (MSE)
Initialization	Xavier uniform	He uniform

[a] LR stands for Learning Rate

Table 2. Actor Critic Neural Networks

GB. The reward function used for the results shown in Figure 7 consists of 5 parts:

$$\begin{aligned} R_{t+1} = & (\rho_{33,t} + \rho_{11,t})\Theta(\rho_{33,t} - \rho_{33,t-2}) \\ & -\Theta(\rho_{22,t} - 0.1) \\ & -\Theta(\Omega_{12,t} - \Omega_{12,t-4})\Theta(t - \tfrac{3}{4}t_f) \\ & -\Theta(\Omega_{23,t} - \Omega_{23,t-4})\Theta(t - \tfrac{3}{4}t_f) \\ & +100\,\Theta(\rho_{33,t} - 0.9)\delta_{t,t_f} \end{aligned} \tag{10}$$

where Θ is the Heaviside function and δ is Kronecker delta.

The agent gets a positive reward if ρ_{33} rises and if it reaches a fidelity higher than 0.9 in the final step of the evolution. Instead, it gets a punishment when the two pulses keep getting higher near the end of the simulation. We chose $N = 300$ steps for the time evolution, where the last step corresponds to $t = t_f$. When an episode ends, the network's weights are updated. In Figure 7 the artificial intelligence-based analogue of the Figure 3 is shown, consisting of the CTAP of the population and the pulse shape obtained by the neural network. There, we chose a $t_{max} = \frac{20\pi}{\Omega_{max}}$ and $\Delta\Omega = 5 \cdot 10^{-4}\Omega_{max}$. After training of 60000 simulations ρ_{22} remains below 0.035 and ρ_{33} reaches 0.99 with approximately 10 hours of computation by a single thread. Figure 8 shows the results during the training after 30, 300, 20000 and 57000 epochs respectively. The artificial intelligence-based method discovers two pulses obeying the counter-intuitive sequence, like those found by human intuition in the seminal paper of Vitanov et al. [14]. In other words, if no one guessed such method for coherent adiabatic transfer of quantum states, the RL would be able to discover it from scratch by exploring the Hilbert space.

5. Conclusion

We have shown how to apply advantage actor-critic (A2C) reinforcement learning methods to achieve control of the coherent transport by adiabatic passage of qubits. We have built an environment consisting of a software simulating the CTAP, which returns known results if the pulses are Gaussian and follows the counter-intuitive sequence, as from literature. Next, we use such an environment to train a network by reinforcement learning, giving no prior knowledge to the network. The A2C method exploits a neural network with a hidden layer of 16 neurons

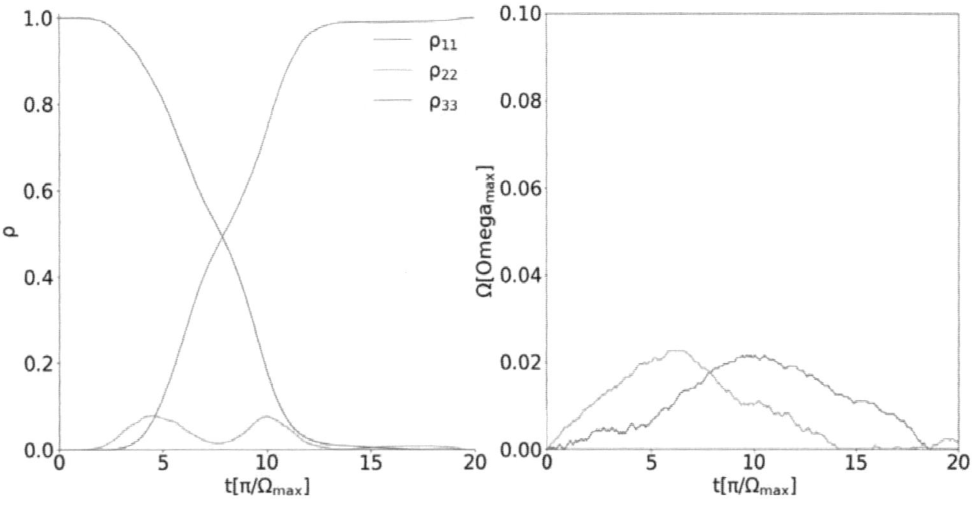

Figure 7. a) The CTAP found by reinforcement learning after 60000 epochs. b) The gate pulses found by the actor-critic RL algorithm to achieve the CTAP. The artificial intelligence discovers that the time evolution of the gates must follow a Gaussian-like shape and that the sequence is counter-intuitive.

to maximize the reward consisting of achieving maximum population transfer by keeping null occupation of the central dot in the meanwhile. We therefore demonstrated that RL is capable to discover non-trivial sequences of control gates in quantum information processing, opening the way of calibrating complex and non-ideal systems of qubits.

Acknowledgments

EP gratefully acknowledges the support of Nvidia Corporation with the donation of the Titan Xp GPU used for this research.

References

[1] Greentree A D, Cole J H, Hamilton A R, Hollenberg L C L 2004 *Phys. Rev. B* **70** 235317
[2] Maurand R, Jehl X, Kotekar-Patil D, Corna A, Bohuslavskyi H, Laviéville R, Hutin L, Barraud S, Vinet M, Sanquer M, De Franceschi S 2016 *Nat. Comm.* **7** 13575
[3] Ferraro E, De Michielis M, Fanciulli M, Prati E 2015 *Phys. Rev. B* **91** 075435
[4] Prati E, Rotta D, Sebastiano F, Charbon E 2017 *In 2017 IEEE International Conference on Rebooting Computing (ICRC)* 1-4IEEE
[5] Rotta D, De Michielis M, Ferraro E, Fanciulli M, Prati E 2016 *Quant. Inf. Proc.* **15**(6) 2253-2274.
[6] Prati E 2016 *Int. J. of Nanotech.* **13** 7 509
[7] Prati E 2017 *J. of Phys.: Conference Series* **880** 1 012018
[8] Sentís G, Calsamiglia, Munoz-Tapia R, Bagan E 2012 *Sci. Rep.* *2* **708**
[9] Porotti R, Tamascelli D, Restelli M, Prati E 2019 arXiv:1901.06603
[10] Fösel T, Tighineau P, Weiss T, Marquardt F 2018 *Phys. Rev. X* **8** 031084
[11] De Michielis M, Ferraro E, Fanciulli M, Prati E 2015 *Journal of Physics A: Math. Theor.* **48** 065304
[12] Rotta D, Sebastiano F, Charbon E, Prati E 2017 *npj Quantum Information* **3** 42
[13] Johansson J R, Nation P D, Nori F 2012 *Comp. Phys. Comm.* **183** 1760
[14] Mnih V *et al* 2015 *Nature* **518** 7540
[15] Sutton R, Barto A 2017 *Reinforcement Learning: An Introduction* (MIT Press)
[16] Szepesvari C 2010 *Algorithms for Reinforcement Learning* (Morgan and Claypool Publishers)
[17] Brockman G, Cheung V, Pettersson L, Schneider J, Schulman J, Tang J, Zaremba W 2016 OpenAI Gym (*Preprint* cs.LG/1606.01540)
[18] Chollet F *et al* 2015 *Keras* https://keras.io

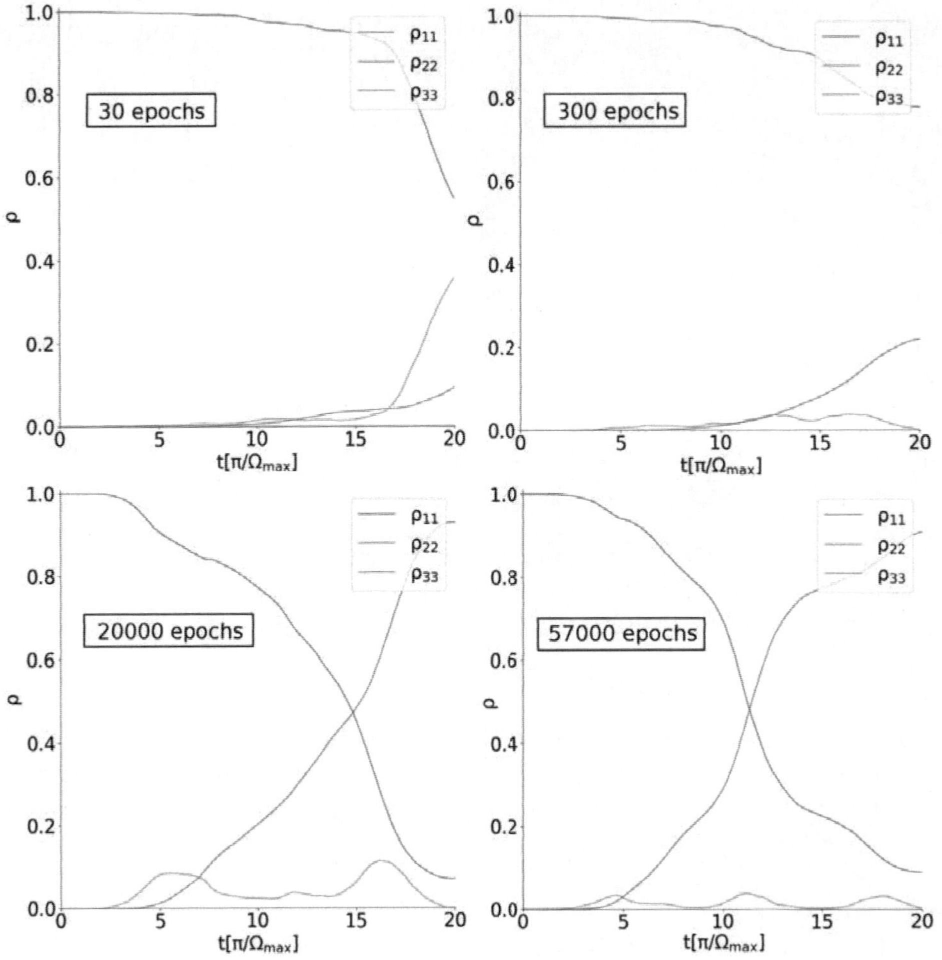

Figure 8. Time evolution of the trace elements of the density matrix during the training, after 30, 300, 20000 and 57000 epochs respectively. The training ended at 60000 epochs.

[19] Vitanov N V, Halfmann T, Shore B W, Bergmann K 2001 *Ann. Rev. of Phys. Chem.* **52**
[20] Kingma D P, Ba J 2015 Adam: A Method for Stochastic Optimization (*Preprint* cs.LG/1412.6980)

IOP Conf. Series: Journal of Physics: Conf. Series **1275** (2019) 012020 doi:10.1088/1742-6596/1275/1/012020

Quantum enhanced correlated interferometry for quantum gravity tests: the experimental set-up and the locking scheme

S.T. Pradyumna [1,2], E. Losero[1,2], P. Traina[1], I. Ruo-Berrchera [1], I. Degiovanni[1], M. Zucco[1], M. Genovese[1,3]

[1]INRIM, Strada delle Cacce 91, 10135, Torino, Italy
[2]DET, Politecnico di Torino, Corso Duca degli Abruzzi 24, 10129, Torino, Italy
[3]INFN, sezione di Torino, via P. Giuria 1, 10125 Torino, Italy

E-mail: m.genovese@inrim.it

Abstract. The search for Planck scale scale effect is one of holy grails of physics. In Fermilab a double Michelson interferometer was built for this purpose: the holometer. In this proceedings we review our effort to build a quantum enhanced holometer.

1. Introduction

The search for Planck scale scale effect is one of holy grails of physics. In Fermilab a double Michelson interferometer was built for this purpose: the holometer [1]. In this proceedings we review our effort to build a quantum enhanced holometer. In particular, we study in detail a system of two power-recycling Michelson interferometers aimed at detecting extremely faint phase fluctuations. This system can represent a breakthrough for detecting a faint correlated signal that would remain otherwise undetectable even using the most sensitive individual interferometric devices. A faint correlated phase noise acting in both the interferometers can emerge, for example, by performing the cross-correlation of their outputs whilst shot noise being uncorrelated in the two interferometers is rejected. Injecting quantum light in the system through the antisymmetric port of each interferometer can further reduce the uncertainty in the measurement of phase correlation. The improvement introduced by quantum light is theoretically discussed in [2, 3], in particular two cases are considered: two single-mode squeezing and twin-beam state. An experiment demonstrating the advantage of using quantum light in a holometer-like set-up has been developed and is at its final stage. This paper describes in details the experimental set-up.

The general idea is represented in Fig. 1

Content from this work may be used under the terms of the Creative Commons Attribution 3.0 licence. Any further distribution of this work must maintain attribution to the author(s) and the title of the work, journal citation and DOI.
Published under licence by IOP Publishing Ltd

IOP Conf. Series: Journal of Physics: Conf. Series **1275** (2019) 012020 doi:10.1088/1742-6596/1275/1/012020

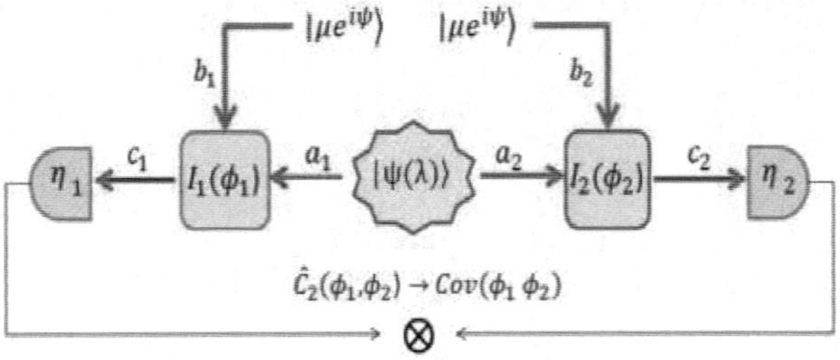

Figure 1: Scheme representing the general idea of a quantum enhanced correlation measurement. $I_1(\phi_1)$ and $I_2(\phi_2)$ are two power-recycling Michelson intereferometer fed by the coherent state $|\mu e^{i\psi}\rangle$. A quantum bi-partite state $|\Psi(\lambda)\rangle$ can be sent in the interferometers from the classically unused port. From measuring the observable $\hat{C}(\phi_1, \phi_2)$ the correlation of the two interferometers phases ϕ_1 and ϕ_2 can be estimated. The use of quantum light can enhance the sensitivity of this measurement. Picture from [2]

Let us consider two Michelson interferometers placed closed to each other fed with two coherent states $|\mu e^{i\phi}\rangle$ in b_k inputs. The quantum state $|\Psi\rangle$ (it can be the product of two independent squeezed states or a two-modes twin-beam state) is injected from the antisymmetric ports a_1 and a_2 of the interferometers. The observable $\hat{C}(\phi_1, \phi_2)$ between the outputs of the interferometer is measured. From $\hat{C}(\phi_1, \phi_2)$ it is possible to retrieve the correlation in the phases fluctuation, i.e. $\text{Cov}(\phi_1, \phi_2)$ (see [2, 3] for details). The observable $\hat{C}$ is different for the two input quantum states considered, in order to minimize the uncertainty on $\text{Cov}(\phi_1, \phi_2)$.

1.1. Single mode squeezed states

When independent squeezed vacuum states are injected into the interferometer, $\hat{C}(\phi_1, \phi_2)$ can be chosen as the product of the photon number operators $\hat{N}_1\hat{N}_2$. However to remove the dc component, we evaluate

$$\hat{C}(\phi_1, \phi_2) = (\hat{N}_1 - \langle \hat{N}_1 \rangle)(\hat{N}_2 - \langle \hat{N}_2 \rangle). \tag{1}$$

It can be demonstrated that in ideal situations when losses are zero the uncertainty on the covariance is $\mathscr{U}^{(0)}(\delta\phi_1\delta\phi_2) \propto e^{-r_1}e^{-r_2}$, being r_1 and r_2 the two squeezing factors [2, 3].

1.2. Twin-beam state

The second quantum state considered is the twin-beam state: in this case each mode of the twin-beam state is injected into each interferometer. The general operator $\hat{C}(\phi 1, \phi 2)$, differently from the case of independent squeezed vacuum, is defined as $(\hat{N}_1 - \hat{N}_2)^2$. The results of the theoretical uncertainty reduction in correlated phase detection is thoroughly discussed in [2, 3] It comes out that the uncertainty reduction is divided into two regimes: quadrature correlations regime and photon number entanglement regime. For extremely high detection efficiencies, together with the parameter condition $T\lambda \gg (1-T)\mu$ the photon number correlation dominates and leads to a dramatic reduction in the uncertainty. $T = cos^2(\phi)$ is the interferometer equivalent transmission of the input port (mode) a_j to the output port (mode) c_j (j=1,2), λ and μ are the mean number of photons in the coherent and squeezed mode respectively. While for the opposite condition $T\lambda \ll (1-T)\mu$ and slightly away from dark fringe, twin-beam quadrature correlations can be exploited and the uncertainty behaves similarly to the independent squeezed

IOP Conf. Series: Journal of Physics: Conf. Series **1275** (2019) 012020 doi:10.1088/1742-6596/1275/1/012020

states case. Twin-beam state can be generated by superimposing two single mode squeezed states at a BS, however a twin-beam-like state which has the same non-classical behavior, in terms of quadratures correlation (although only in one quadrature angle), can be generated by splitting a single mode squeezer on a 50-50 beam splitter. This simpler option is implemented in the experiment.

2. Classical part: power-recycling MI

Our system consists of two power-recycling Michelson interferometers placed close to each other with small separation between the two beam-splitters.

These are the types of interferometer currently operated at larger scale gravitational-wave detectors. They are based on the combination of two fundamental interferometers : the Fabry-Perot interferometer and the Michelson interferometer.

A Fabry-Perot cavity is a linear cavity formed by two partially reflecting mirrors. When light is incident on this linear cavity, some part is transmitted and some part is reflected. The beam is transmitted through the cavity only when the length of the cavity L matches the multiple of wavelength λ of the light incident. In correspondence, the optical power inside the cavity is enhanced (optical gain) because of constructive interference of repeated reflections. The transmission in the cavity is maximum when 2kL is multiple of 2π which implies (whenever the detuning frequency ν is a multiple of c/2L) the transmission of the cavity is maximum. This spacing of the detuning between two successive maxima or two successive minima of the transmission (and/or optical gain occurrence) of the cavity $\Delta\nu = c/2L$ is called Free Spectral Range (FSR). Another property, which defines the losses in the cavity, is the finesse. It is defined as the ratio of Free Spectral Range of the cavity to the Full-Width-at-Half-Maximum (FWHM) of the transmitted peak. It defines the quality of the cavity. In general, the optical losses of the cavity widen the peak, decreasing the value of the finesse.

A Michelson interferometer consists of a 50-50 beam splitter and two piezo transducer driven end-mirrors. The laser field entering from the input port propagates in each arm of the interferometer, before reflecting back towards the beamsplitter. The two beams interfere at the beam splitter: the intensity of the transmission and reflection depends on the differential arm length (DARM) $(L - L')$ and can be sensed by measuring the power at the anti-symmetric port, using a photodetector. The DARM point corresponding to constructive interference at the anti-symmetric port is called bright fringe while the one corresponding to destructive interference is called dark fringe.

The figure of merit to quantify the contrast between the dark and bright fringe is the interferometric visibility and is given by:

$$V = \frac{(P_{max} - P_{min})}{(P_{max} + P_{min})} \tag{2}$$

When a Michelson interferometer is operated near the dark fringe, most of the power is reflected back towards the laser. This reflected power could be recycled by placing a partially transmitting mirror called power-recycling mirror (PRM) and forming a resonant cavity with the interferometer. In this sense the power-recycling interferometer is combination of a Fabry-Perot cavity and Michelson interferometer. With this technique, as the optical power P in the interferometer is increased by the power-recycling, the shot noise reduces as the uncertainty is proportional to $(P)^{-1/2}$. The Michelson interferometer operated at the dark fringe can be considered as a high-reflective compound mirror. The recycling mirror forms a recycling cavity with this compound mirror of equivalent length $CARM = \frac{L+L'}{2}$; the power in the cavity is enhanced when the cavity is resonant with the input laser beam.

Fig. 2 depicts the simplified schematic of our experimental setup, concerning the classical part.

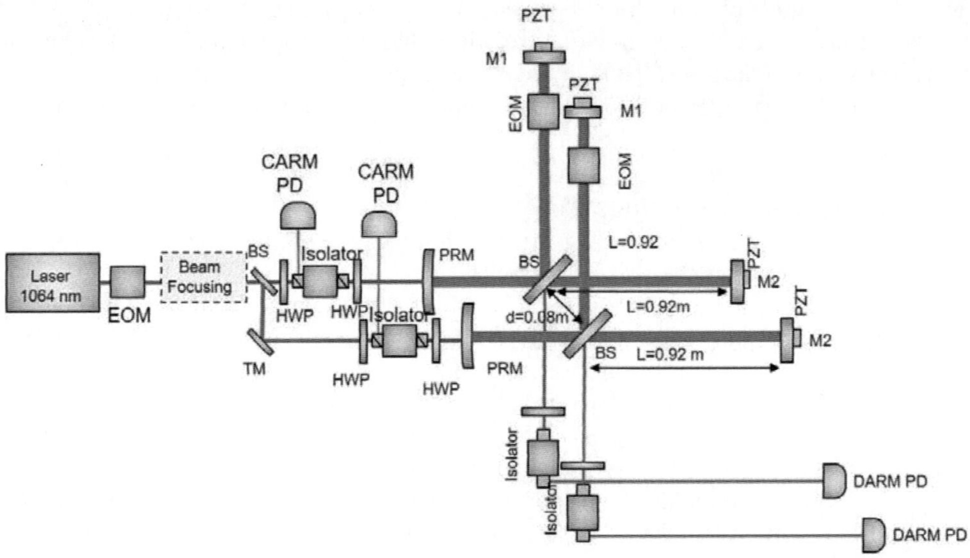

Figure 2: Simplified schematic of the double-interferometer setup. Two Michelson Interferometers with arm length $L = 0.92m$ were co-located, with a distance between the two balanced beam splitters (BSs) of around 8cm. M1,M2: piezo-actuated high-reflectivity (99.9%) end-mirrors. PRM: partially reflecting (90%) power-recycling mirror, radius of curvature $r_c = 1.5m.$; HWP:Half waveplate; PD: Photodetector; BS: Beam splitter

A Nd-YAG laser with an output power of 300 mW at wavelength 1064nm is used. Part of the light is sent to the generation of the squeezed light setup. For having an efficient TEM00 input mode for the interferometers, the light is fiber coupled. A power of 1.5 mW enters each interferometer.

Each interferometer is formed by a power-recycling mirror having the radius of curvature of 1.5m, a 50-50 beam splitter, and two plain end-mirrors M_1 and M_2 (reflectivity around 99.9 %). M_1 and M_2 are stabilized by a PZT actuators. The power-recycling mirror's reflectivity is chosen to be of 90 % to form an over-coupled cavity with end-mirrors when locked near the dark fringe.

The length of each arm of the interferometer is $\sim$ 92 cm. The distance between the two beam splitters is $\sim$ 8 cm. A set of lenses is placed before the interferometers for mode matching the power-recycling interferometer such that the waist of the beam is at the end-mirror. A resonant Electro Optic Modulator (EOM) is placed before the interfereometers for generating the proper error signal for locking the CARM (see next section) while another phase modulator, which consists of RTP crystal(4mm x 5mm area) is placed inside the interferometer for locking the DARM. A photodetector with quantum Efficiency 99% is placed at the output port: it is used both for locking the DARM and acquiring the data. Another photodetector, used for locking the CARM, collect the light reflected from the power-recycling cavity.

2.1. Locking schemes

Fig. 3 depicts the detailed locking scheme of each interferometer. It is necessary to lock the system to the proper working point, in order to get rid of environmental noise and reach the desired sensitivity. In particular we work close to the dark fringe and at the resonance. The first condition implies for the DARM: $L - L' = n\lambda + \delta$ while the second condition: CARM $L_{eff} = \frac{L+L'}{2} = m\lambda$.

To lock the cavity to the resonance, the Pound-Drever-Hall locking technique is utilized [4]. For this technique it is necessary to induce a periodic phase modulation of the laser beam, applying a periodic electronic signal with a EOM ($\omega_C = 20MHz$). In this way two fields, called sidebands, separated in frequency from the original field by the modulation frequency are generated. The signal reflected from the cavity is collected by the CARM photodiode, demodulated at ω_C and opportunely processed in the lockbox. We have used a set of lock boxes made of Proportional-Integral controller with inbuilt low pass filters) to generate an error signal. This error signal is then sent equally (phase concordance) to both the end-mirrors passing through a high voltage amplifier (HVA).

For locking the DARM at a desired working point different techniques are possible. In our case, being interested in locking close to the dark fringe the internal modulation technique is chosen. As in the PDH technique the generation of sidebands is necessary. In this case the EOM is placed inside the interferometer to generate sidebands at a frequency $\omega_D = 7.6MHz$. The DARM photodiode signal is demodulated by a mixer at the frequency ω_D and then sent to the lockbox. The dark fringe corresponds to the zero of the error signal and corresponds to its maximum slope. This error signal is then sent differentially (i.e with a relative phase of π) to both the mirrors with a help of a High Voltage Amplifier. The phase between the electronic local oscillator (LO) and the AC output signal from the photodetector has to be adjusted properly to maximize the slope of the error signal. Sending the error signal equally for the CARM locking and differentially for the DARM locking allows to decouple the control of the two degree of freedom, i.e. the DARM error signal does not affect the CARM and viceversa.

The locking for the squeezed beam is described in the next section.

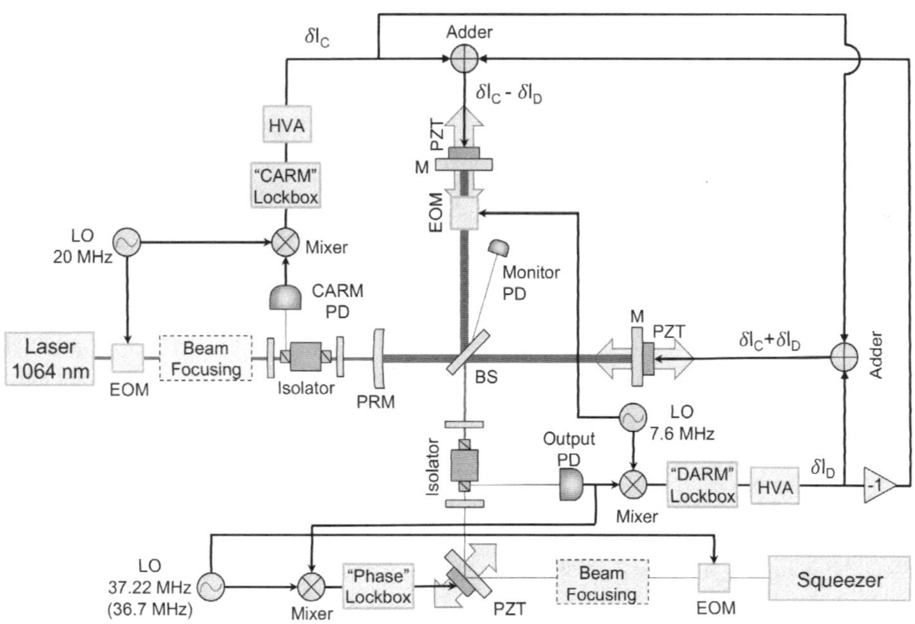

Figure 3: Detailed locking schematic of each interferometer. PZT: piezo-electric actuator. EOM: electro-optical modulator. HVA: high-voltage amplifier. LO: local oscillator. PD: photo-diode, InGaAs with high quantum-efficiency photodiodes (99%) and low noise (Noise Equivalent Power $1.2 \cdot 10^{-11} W/\sqrt{Hz}$) are used at the read-out port. Beam-Focusing: set of lenses in order to have Gaussian Optical mode TEM_{00}. $\delta l_C, \delta l_D$: correction signals from the CARM and DARM lockbox respectively. The DARM was locked using the internal modulation technique while the CARM was locked using the Pound-Drever-Hall technique

3. Squeezing generation set-up

Fig. 4 depicts, on the left-hand side, the experimental scheme for preparation of squeezed vacuum state, on the right-hand side, the two quantum states considered and their implementation. In both cases the squeezed light is injected into the interferomenters through the anti-symmetric port.

Each squeezer comprises of a ppKTP (periodically poled Potassium Titanyl Phosphate) crystal placed inside a semi-monolithic cavity formed by the edges of the crystal and transparent mirror. A separate 1064nm beam dubbed control beam was sent through a phase modulation of 37.22 MHz and 36.5 MHz respectively for each squeezer cavity and is used to lock each of the squeezer cavities using Pound-Drever-Hall technique. An internal module for second harmonic generation provided the 532 nm radiation as a pump for the generation of squeezed light. The pump beam is then sent to a mode-cleaner to have clean TEM00 mode. The output of the mode cleaner is then incident on the crystal for having parametric down conversion inside the cavity. A Peltier element and a thermistor attached to the crystal was used along with a temperature controller for having an efficient phasematching between the control beam and incoming pump beam. The phase relation between pump and control beam determines the orientation of the squeezing ellipse. A PZT attached to mirror is used to lock this phase. The control beam perfectly overlaps spatially to the generated squeezed light at 1064nm. Thus it is used to align the squeezed light into the interferometer and to adjust the spatial mode matching between the beam in the interferometer cavity and the squeezed beam. A dichroic beam splitter (DBS) is used to reflect the 1064 nm and would not allow any 532 nm to be injected into the interferometer. Since squeezer is very sensitive to loss, a very high reflective turning mirrors (R$\sim$ 99%) were used to inject into interferometers.

The squeezed light after it passes through DBS is injected into the interferometer, previously locked near to the dark fringe. The control beam, with output power maintained at 5μW, enters the interferometer together with the squeezed light. When the squeezed light enters the interferometer, the DARM photodetector is demodulated at 37.22 MHz/36.5 MHz with an inbuilt mixer. The error signal is generated such that the point of squeezing quadrature corresponds to the maximum slope part of the error signal. The interferometer is then locked to the squeezing quadrature by help of a phase shifter with PZT attached.

The TEM 00 interferometer output mode has to be modematched to the squeezed light. This is done by reversing the injection of interferometer mode into the squeezer cavity. Appropriate lenses are placed to improve the mode-matching between the interferometer output mode and squeezed light mode. The squeezer cavity which has a PZT attached to it is scanned and modes of the cavity are observed. The TEM 00 mode was enhanced by turning mirrors placed for aligning the squeezed light to the interferometer. At the end of the procedure the modematching was around 99%.

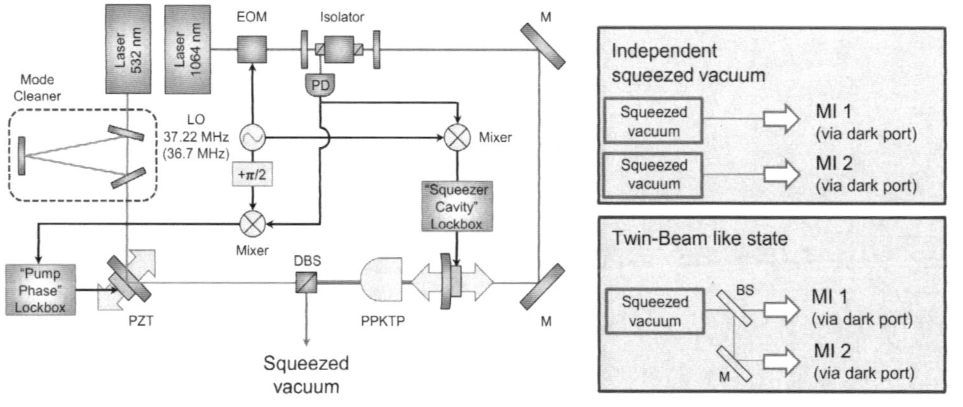

Figure 4: On the left-hand side a scheme of the squeezer generation set-up is presented, see text for the details. On the right-hand side the quantum states tested are summarized: independent single mode squeezed vacuum and twin-beam like state.

The twin-beam state is entangled in number of photons and presents also non classical quadrature correlations between the two optical modes. Although photon number entanglement would provide in principle a dramatic enhancement of the sensitivity, its practical realization is extremely challenging. On the other side exploiting non-classical quadrature correlations is definitely feasible with some slight modification of the previous configuration.

As anticipated in the introduction we use an approximate twin-beam-like state in our experiment. Its implementation is straight forward having implemented the two single mode squeezing configuration. The quantum noise locking is similar to that of the double squeezer setup previously described, in this case the sidebands are both at 36.5 MHz in the two modes. The main difference respect the two single-mode squeezed beams is in the data elaboration: the variance of difference of the output signal of two interferometers is considered instead that the correlation.

References

[1] Chou AS, Gustafson R, Hogan C, Kamai B, Kwon O, Lanza R, McCuller L, Meyer SS, Richardson J, Stoughton C and Tomlin R 2016 First measurements of high frequency cross-spectra from a pair of large Michelson interferometers. Physical Review Letters **117** 111102

[2] Ruo-Berchera I, Degiovanni IP, Olivares S, Samantaray N, Traina P and Genovese M 2015 One-and two-mode squeezed light in correlated interferometry. Physical Review A **92** 5 053821

[3] Ruo-Berchera I, Degiovanni IP, Olivares S and Genovese M 2013 Quantum light in coupled interferometers for quantum gravity tests. Physical Review Letters **110** 21 213601

[4] Black ED 2001 An introduction to Pound–Drever–Hall laser frequency stabilization. American Journal of Physics **69** 1 79-87

IOP Conf. Series: Journal of Physics: Conf. Series **1275** (2019) 012021 doi:10.1088/1742-6596/1275/1/012021

Consciousness as the collective excitation of a brainwide web – understanding consciousness from below quantum fields to above neuronal networks

Suzy Lidström and Roland E. Allen

Department of Physics and Astronomy, Texas A&M University, College Station, Texas 77843-4242

E-mail: suzy.lidstrom@gmail.com, allen@tamu.edu

Abstract. In the spirit of previous ideas in the neuroscience community, but with a more physics-oriented perspective, we propose that consciousness can be described as the collective excitation of a *brainwide web* of neurones. This picture is inspired by the fact that, in all major areas of physics, a collective excitation has just as much physical reality as a particle or other localized object. The brainwide web extends into those regions (neuronal networks) where processed information is received from the senses, memories, etc. (emerging out of unconscious processes in prior networks). It unifies those regions (plus motor control regions) via the vast complexity of the neuronal interactions that it spans. A crude analogy is the worldwide web, which extends into servers which have been prepared by external agents. Other crude analogies are the many collective excitations in physics. True understanding of consciousness must rely mostly on experiment, but since probes of living brains have limited precision, there is an important role for large-scale simulations in revealing details and complementing experiment. Petascale computational facilities should make it possible to perform simulations with realistic complexity.

In discussing consciousness, we need a vocabulary that prevents confusion of different concepts. Continuing with a notation introduced elsewhere [1], we use the subscripts O and P to refer respectively to the phenomena of nature accessible to an external *observer* and the inner reality of nature experienced by a *participant*, with direct conscious experiences being only one example. The "hard problem" of David Chalmers [2] is the problem of explaining consciousness$_{\mathbf{P}}$: Why do we have the specific inner experiences of seeing red, feeling pain, etc.? As argued elsewhere [1], it is not clear that this problem has meaning for observers or participants embedded within nature. Like the question "Why is there something rather than nothing?" [3], it appears either to be meaningless as an answerable problem or else to require a deeper level of understanding than we can now imagine.

Here we focus on understanding consciousness$_{\mathbf{O}}$, which can potentially be addressed at various levels:

1. Quantum fields, which are the ultimate reality of present-day physics. Presumably biological consciousness primarily involves the electron and photon fields, since the quark, gluon, ... fields are perturbed mostly at inappropriate distance and energy scales. However, nuclear motion is of central importance in processes like the release of neurotransmitters, and

Content from this work may be used under the terms of the Creative Commons Attribution 3.0 licence. Any further distribution of this work must maintain attribution to the author(s) and the title of the work, journal citation and DOI.
Published under licence by IOP Publishing Ltd

IOP Conf. Series: Journal of Physics: Conf. Series **1275** (2019) 012021 doi:10.1088/1742-6596/1275/1/012021

nuclear spins are of central importance in experimental techniques like functional magnetic resonance imaging (fMRI).

2. Particles, which are excitations of quantum fields (analogous to the quantized excitations of a harmonic oscillator). In a quantum description, one can form collective excitations that involve only one species of particle (e.g. plasma oscillations) or several species, as in polaritons (electron-photon hybrid modes) or polarons (electron-phonon hybrid modes, with phonons being the quanta of nuclear motion). A central point is that a collective excitation has just as much physical reality as a particle or other localized object. If one reverts to level 1 above, excitations of quantum fields can be defined in different ways according to the physical situation. For example, an observer far from a black hole and one who is nearby will define (level 2) particles differently, using different vacua and particle detectors, even though the (level 1) quantum fields in curved spacetime are well-defined. This is the origin of the Hawking effect [4]. Similarly, one can choose different descriptions of quantum excitations in material systems – for example, by performing a canonical transformation from atomic to normal mode coordinates for the vibrations of a molecule or material. In this case, a phonon (quantum of vibration) has just as much physical reality as an atom.

3. Atoms, and then molecules (with a macromolecule often consisting of an enormous number of atoms). At this level one still needs quantum mechanics, but the most nuanced features of quantum theory, such as entanglement, are very unlikely to be relevant, because a normal body temperature implies extremely short decoherence times [5]. Thermal agitation also implies that the states discussed below are to be interpreted as thermal averages over the microscopic states accessible at thermal energies.

4. The biochemistry and biophysics of synapses, ion channels, etc. At this point we are already into an area of biological science which is rich in potential for important discoveries. (Of course, this is also true for some aspects of the preceding level, such as protein structure and protein folding.)

5. The architecture of neuronal networks. In the brain there are many kinds of neurones, with axons of various lengths, thousands of dendrites per neurone, and differences in other attributes, comprising networks of astounding intricacy. In the formal model proposed here, there is one well-defined global network – the *brainwide web* discussed below – which can be identified as the physical system in which consciousness is located.

6. The time-dependent states of these networks, which are arrayed in a hierarchy, again of astounding complexity. These states must be viewed as collective modes of neurones. At the top of this hierarchy is the brainwide web proposed here, which includes those networks where final representations are formed of what is perceived, remembered, ... The networks which feed information into these components in the visual cortex etc. are not themselves part of the brainwide web (i.e. the conscious brain), so all their enormously sophisticated processing is unconscious.

Since physics has come so far during the past 120 years, it is quite plausible that new physics may lie ahead, with yet another level:

0. A deeper understanding which explains quantum fields, and which may even render the "hard problem" meaningful.

However, level 0 physics will not invalidate quantum field theory, which has been demonstrated to describe the world with remarkable success and accuracy. Even gravity must ultimately be consistent with quantum theory, as can be demonstrated by a thought experiment

IOP Conf. Series: Journal of Physics: Conf. Series **1275** (2019) 012021 doi:10.1088/1742-6596/1275/1/012021

involving a "Schrödinger planet": Imagine that the trajectory of a planet can be perturbed by a radioactive decay or other quantum event or measurement. The gravitational field must follow the planet, so one has two outcomes for the trajectory of the planet's gravitational field in the same way that one has two outcomes for the state of Schrödinger's cat. In the path integral formulation of quantum mechanics, the gravitational field, as well as the matter fields and other force fields, must follow all allowed trajectories.

Since consciousness is one aspect of physical reality, and physical reality is described with great success by quantum fields, we conclude that consciousness must be expressible in terms of quantum fields. In other words, we conclude that consciousness can be regarded as a state $|\Psi(t)\rangle$ of these fields. (Again, this state must properly be regarded as a nearly classical thermal average over specific quantum states, and interference phenomena for specific states are precluded by rapid decoherence.) This is the most fundamental physics perspective currently available.

Here we are using the Dirac bra-ket notation [6], with a general state represented by a ket like $|\Psi\rangle$, and the amplitude for this state being in a specific state like $|\psi\rangle$ given by its projection onto this state (the bra-ket), $\langle\psi\,|\Psi\rangle$. An example is the wavefunction $\phi(\vec{r},t) = \langle\vec{r}\,|\phi(t)\rangle$ for a particle in a state $|\phi(t)\rangle$, where $|\vec{r}\rangle$ is the state of a particle located exactly at position $\vec{r}$.

There are two aspects of $|\Psi(t)\rangle$ that are immediately revealed just by introspection: It has boundaries, which means it must be localized in a *coherent domain of fields* within nature. And it spans a very large number of degrees of freedom x, with time-dependent amplitudes

$$\Psi\left(x,t\right) = \langle x\,|\Psi(t)\rangle \ . \tag{1}$$

The content of consciousness is constantly shifting from one centre of attention to another – from one object to another in the visual field, from one memory to another, from one sound to another, ..., or equally often from a visual object to a sound or a thought, ... This is the time dependence of the state $|\Psi(t)\rangle$. Let $|X\rangle$ represent some set of degrees of freedom – for example a perceived face or a remembered melody. The amplitude for $|X\rangle$ being in the consciousness at time t is $\langle X|\Psi(t)\rangle$. I.e., this is the degree to which one is conscious of $|X\rangle$.

From a biology perspective [7–15] – based on a vast number of brilliant experimental studies which it is impossible to cite here – it is clear that biological consciousness must involve *neuronal networks* in specific ways. The coherence of consciousness implies that it must be localized in a *coherent domain of neurones*, which we call the brainwide web. A crude analogy is the worldwide web, which extends into servers which have been prepared by external agents, in the same way that the brainwide web extends into regions that have been prepared by unconscious processes in other networks lower in the hierarchy. With the brainwide web regarded as a well-defined (but open) system, its time-dependent state $|\Psi(t)\rangle$ can be represented by whatever neuronal states $|\psi_n(t)\rangle$ are most appropriate for experimental study or simulation:

$$|\Psi(t)\rangle = \sum_n a_n(t)\,|\psi_n(t)\rangle \ . \tag{2}$$

If the $|\psi_n(t)\rangle$ are chosen to be orthonormal – i.e., $\langle\psi_n(t)|\psi_{n'}(t)\rangle = \delta_{n\,n'}$, where $\delta_{n\,n'} = 0$ if $n' \neq n$ and 1 if $n' = n$ – then the amplitudes satisfy

$$a_n(t) = \langle\psi_n(t)|\Psi(t)\rangle \ . \tag{3}$$

The $a_n(t)$ might be amplitudes for experiences localized in particular regions – in the visual cortex, in memory, etc. They are clearly driven by many influences which are both outside and within consciousness, from the barking of a dog to a worrisome memory. We represent this by the purely formal equation

$$a_n(t) = F_n[\psi, a, \chi] \tag{4}$$

where ψ, a, and χ respectively represent the neuronal states $|\psi_n(t)\rangle$ accessible to consciousness, the amplitudes $a_n(t)$ themselves, and direct influences χ_i outside consciousness (as opposed to indirect influences mediated by the $|\psi_n(t)\rangle$). A trivial example of a χ_i would be a stimulant, depressant, or hallucinogen entering the brain. The more obvious influences are the networks

IOP Conf. Series: Journal of Physics: Conf. Series **1275** (2019) 012021 doi:10.1088/1742-6596/1275/1/012021

within the brainwide web which acquire information processed by unconscious networks. It is they that interact to focus attention on inputs that may be enticing, threatening, ...

If $|N\rangle$ is the state of a specific neural network N, the amplitude of consciousness at a given moment t in that network is $\langle N\,|\Psi(t)\rangle$, and this is a quantity that is in principle accessible to experiment.

The functionals $F_n[\psi, a, \chi]$ are currently even more unknown and mysterious than the one in density-functional theory [16], which is treated with semiempirical models. A corresponding model for consciousness and other aspects of neuroscience will have to be vastly more sophisticated and complex.

As an aside, we note that the word consciousness as used here applies to any sufficiently complex *coherent domain* which has the qualities implicit in the functional $F_n[\psi, a, \chi]$ written above – including richness of experience and a capacity for self-awareness. A bat [17], bee [18], or creature of the remote future with artificial intelligence will certainly have a different consciousness$_{\mathbf{P}}$ than human beings – with e.g. different inner experiences red$_{\mathbf{P}}$ for the same colour. And they will have varying levels of consciousness. But the word "consciousness" is not misapplied to any biological or nonbiological creature that satisfies the conditions implied here. On the other hand, it is certainly a misuse of language to say that an electron or a stone or a nebula is conscious. This misusage reflects confusion between consciousness and the mysterious inner reality of nature. There is a reality$_{\mathbf{O}}$ accessible to science and a reality$_{\mathbf{P}}$ which we cannot directly access (with any science that is currently imaginable), but very little of reality$_{\mathbf{P}}$ is conscious.

We also note that our picture is similar to those proposed by others – for example, in Refs. [7] and [13] and by Tononi [19] – but here we place more emphasis on the physics setting and the physical reality of conscious experience.

Understanding of consciousness will come primarily from increasingly sophisticated experiments, employing methods that largely are developed by the physics community. In Fig. 1, we show an example of a promising new technique [20].

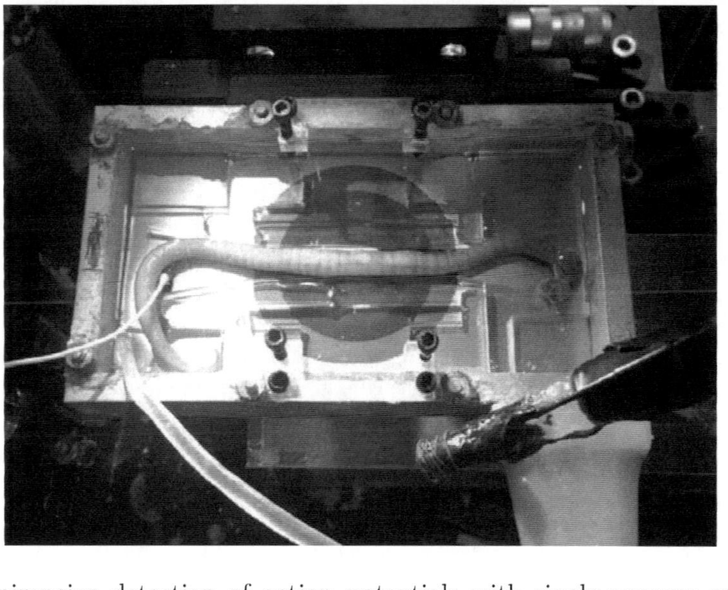

Figure 1. Noninvasive detection of action potentials with single-neurone sensitivity, using a sensor composed of quantum defects (nitrogen-vacancy colour centres in a diamond chip), with no observed adverse effect on the animal (marine fanworm Myxicola infundibulum). After Ref. [20]. Credit: John Barry, Matthew Turner, Jennifer Schloss, and Ronald Walsworth.

IOP Conf. Series: Journal of Physics: Conf. Series **1275** (2019) 012021 doi:10.1088/1742-6596/1275/1/012021

However, there are two points which demonstrate that very large-scale *simulations* can have an important role in complementing experiment:

- Unless a dramatically more effective technique is invented for observing deep inside a living (and normal) human brain with atomic-scale precision, neuroscience experiments are limited to rather gross observations from the outside. This means that realistic (and therefore extremely large-scale) simulations are needed to study what neurones do at the microscopic scale, in organisms whose neural complexities transcend that of Fig. 1 by many orders of magnitude. An analogy is a supernova, which also can only be observed from the outside, with no possibility of getting at the detailed processes where the explosion actually occurs. In both cases, the exterior observations serve as boundary conditions on simulations of the interior processes. If the underlying theory is good enough (e.g. nuclear physics and fluid dynamics for the supernova, biochemistry and biophysics for the brain) well-designed simulations can provide information that complements the experiments or observations.
- We have entered the era of petascale computing. The connections in the human brain are also petascale. This means that we can finally perform realistic simulations of the neuronal processes in the brain, within models that simplify the dynamics to include only the neural connections. But well-chosen models of these processes, guided by experimental discoveries of how neurones work, should be sufficient to capture the essential processes behind consciousness. And, of course, revealing how the brain works will certainly have a strong impact on medical and other biological applications.

Understanding consciousness is just as exciting a goal for our time as understanding the origin of the universe and its laws.

References

[1] Alexander G et al 2019 The sounds of science – a symphony for many instruments and voices, submitted; see the section by Allen R E and Lidström S: What is consciousness, and is there really a "hard problem"?
[2] Chalmers D 1995 Facing up to the problem of consciousness *Journal of Consciousness Studies* **2** 200
[3] Wippel J F 2007 Thomas Aquinas on the ultimate why question: why is there anything at all rather than nothing whatsoever? *The Review of Metaphysics* **60** 731
[4] Hawking S W 1976 Black holes and thermodynamics *Phys. Rev. D* **13** 191
[5] Tegmark M 2000 Importance of quantum decoherence in brain processes *Phys. Rev. E* **61** 4194
[6] Dirac P A M 1939 A new notation for quantum mechanics *Mathematical Proceedings of the Cambridge Philosophical Society* **35** 416
[7] Baars B J 1989 *A cognitive theory of consciousness* (Cambridge: Cambridge University Press)
[8] Crick F 1994 *The astonishing hypothesis – The scientific search for the soul* (USA: Touchstone)
[9] Kandel E R, Schwartz J H, and Jessell T M 2000 *Principles of neural science* (USA: McGraw-Hill Companies Inc)
[10] Koch C 2004 *The Quest for Consciousness – A neurobiological approach* (USA: Roberts and Company)
[11] Edelman G M 2006 *Second nature – brain science and human knowledge* (USA: Yale University Press)
[12] Northoff G, Heinzel A, de Greck M, Bermpohl F, Dobrowolny H, and Panksepp J 2006 Self-referential processing in our brain – A meta-analysis of imaging studies on the self *NeuroImage* **31** 440 and references therein
[13] Dehaene S 2014 *Consciousness and the brain: Deciphering how the brain codes our thoughts* (USA: Penguin)
[14] Koch C 2017 *Consciousness – Confessions of a romantic reductionist* (USA: MIT Press)
[15] Demirtas M et al. 2019 Hierarchical Heterogeneity across Human Cortex Shapes Large-Scale Neural Dynamics *Neuron* **101** 1-14 and references therein
[16] Kohn W 1964 Inhomogeneous Electron Gas *Phys. Rev.* **136** B864
[17] Nagel T 1974 What is it like to be a bat? *The Philosophical Review* **83** 435
[18] Feinberg T E and Mallatt J M 2018 *Consciousness Demystified* (USA: MIT Press)
[19] Tononi G 2004 An information integration theory of consciousness *BMC Neurosci.* **5** 42
[20] Barry J F, Turner M J, Schloss J M, Glenn D R, Song Y, Lukin M D, Park H, and Walsworth R L 2016 Optical magnetic detection of single-neuron action potential using quantum defects in diamond *P.N.A.S.* **113** 14133

IOP Conf. Series: Journal of Physics: Conf. Series **1275** (2019) 012022 doi:10.1088/1742-6596/1275/1/012022

An alternative to string theory

Roland E Allen

Department of Physics and Astronomy, Texas A&M University, College Station, Texas 77843-4242

E-mail: allen@tamu.edu

Abstract. String theory has many compelling features, including higher dimensions, supersymmetry, and topological defects of central importance. Here we describe a theory which also contains these features, but which is much more ambitious than string theory and also much closer to experiment. Since the mathematical details are given in a much longer paper, we summarize only the most important results, which include $SO(N)$ gauge unification, vanishing of the usual cosmological constant, and a credible dark matter candidate.

1. Introduction

Although string theory is now half a century old [1], it has not yet been successful in confronting experiment or naturally resolving problems like an enormous cosmological constant. Here we argue that a likely reason is its conservatism: It is basically an extension of field theory from world lines to world sheets, with (higher-dimensional) spacetime, Lorentz invariance etc. assumed from the beginning.

In a longer paper [2], a theory is offered with a radically different starting point – in fact, the simplest imaginable starting point. Spacetime, Lorentz invariance, etc. then emerge at a later stage, rather than being put in by hand at the beginning.

Some successes of the theory will be enumerated below and emphasized with bullets. It is the author's experience that essentially everyone finds Ref. [2] difficult to read, because of the density of new ideas and level of abstraction (and perhaps also the many mathematical details). The present paper therefore represents the opposite extreme, with only a summary of the most central conclusions and almost no mathematics.

John Wheeler conjectured that physics is fundamentally based on bits and the act of observation: "It from bit" [3]. The theory of Ref. [2] differs by starting with dits, which can assume any of d states, and which have physical reality apart from any observer. Giving rise to all of Nature, there is a fundamental system which consists of an arbitrary number of dits, with each dit in any of the many states available to it. D of the states correspond to coordinates x^M and the rest to fields.

- The fact that fields and coordinates have essentially the same fundamental status then finds a natural explanation. (String theory, for example, shows this when coordinates are interpreted as fields on a 2-dimensional manifold.)
- A point that may seem trivial but is not: Since the theory is initially based on counting arguments involving only integers (with the continuum defined merely as an approximation), it is manifestly mathematically consistent.

Content from this work may be used under the terms of the Creative Commons Attribution 3.0 licence. Any further distribution of this work must maintain attribution to the author(s) and the title of the work, journal citation and DOI.
Published under licence by IOP Publishing Ltd

IOP Conf. Series: Journal of Physics: Conf. Series **1275** (2019) 012022 doi:10.1088/1742-6596/1275/1/012022

As described in detail in Ref. [2], one can then proceed through a set of arguments to define an entropy S which ultimately becomes the familiar Lorentzian action for bosonic and fermionic fields in a path integral description.

- There is then a natural explanation of the similarity of statistical partition functions and quantum path integrals.

The following aspects of nature emerge automatically:

- bosonic fields
- fermionic fields
- path-integral quantization (which can then be converted to canonical quantization)
- 4-dimensional spacetime
- a metric tensor with one time coordinate
- Lorentz invariance
- the vierbein of general relativity
- correct coupling of the bosonic and fermionic fields to the vierbein
- supersymmetry (without which the theory cannot even be formulated).

The most natural cosmological model involves a vortex-like topological singularity in 4-dimensional spacetime, corresponding to the Big Bang.

There is also an internal space with $D-4$ dimensions, again containing a vortex-like topological singularity, and from this the following features emerge:

- gauge fields
- correct coupling of the original bosonic and fermionic fields to the gauge fields
- an unavoidable $SO(N)$ symmetry for the gauge fields, with $SO(10)$ being an excellent candidate for grand unification of gauge fields (and $N > 10$ possibly yielding family structure).

Except for one term which is relevant only at high energy, the theory is both Lorentz invariant and supersymmetric, with a Lagrangian density for fermions and scalar bosons of the form

$$\mathcal{L} = \psi_f^\dagger(x)\, ie_\alpha^\mu \sigma^\alpha D_\mu \psi_f(x) - g^{\mu\nu}\left(D_\mu\phi(x)\right)^\dagger D_\nu\phi(x) + F^\dagger(x)\,F(x) \tag{1}$$

where

$$g^{\mu\nu} = \eta^{\alpha\beta} e_\alpha^\mu e_\beta^\nu \tag{2}$$

and F contains auxiliary fields which emerge automatically. The notation is standard and explained in Ref. [2]. The spin 1/2 fermion fields in ψ_f, the scalar boson fields in ϕ, and the auxiliary fields in F span the various physical representations of the fundamental gauge group, which must be $SO(D-4)$ in the present theory (or more precisely $Spin(D-4)$).

The fact that there is not a factor of $e = \left|\det e_\mu^\alpha\right|$ in (1) is discussed at length in Ref. [2]. This is one of two features which, in conjunction, imply vanishing of the usual enormous cosmological constant. (The other feature is normal ordering of field operators in the quantized action for gauge fields.)

We conclude with three predictions discussed in recent papers:

- a dark matter particle with many favorable features, including well-defined couplings and a mass of ≤ 125 GeV/c^2 [4–6]
- many other neutral and charged new particles at higher energy [4–6]

IOP Conf. Series: Journal of Physics: Conf. Series **1275** (2019) 012022 doi:10.1088/1742-6596/1275/1/012022

- a natural origin of the Bekenstein-Hawking entropy for all black holes, which is based on a remarkable result of Gibbons and Hawking [8].

The above claims may appear to be beyond outrageous, so we refer the reader to the arguments in Refs. [2, 4–7], and we also reiterate that many aspects – including the actions of the gauge and gravitational fields, and the topological defects accounting for curvature – are far from complete at this point and need further work. We also await experimental tests.

References

[1] Schwarz J H 1985 *Superstrings: The First 15 Years of Superstring Theory, Volumes 1 & 2* (Singapore: World Scientific)
[2] Allen R E 2017 Predictions of a fundamental statistical picture (arXiv:1101.0586v9 [hep-th])
[3] Wheeler J A 1989 Information, physics, quantum: the search for links *Proc. 3rd Int. Symp. Foundations of Quantum Mechanics* (Tokyo: Physical Society of Japan) pp. 354-368
[4] Allen R E and Saha A 2017 Dark matter candidate with well-defined mass and couplings *Mod. Phys. Lett. A* **32** 1730022 (arXiv:1706.00882 [hep-ph])
[5] Allen R E 2019 Saving supersymmetry and dark matter WIMPs – a new kind of dark matter candidate with well-defined mass and couplings *Phys. Scr.* **94** 014010 (arXiv:1811.00670 [hep-ph])
[6] Throm M, Thornberry R, Killough J, Sun B, Abdulla G, and Allen R E 2019 Two natural scenarios for dark matter particles coexisting with supersymmetry *Mod. Phys. Lett. A* **34** 1930001 (arXiv:1901.02781 [hep-ph])
[7] Allen R E 2019 Black hole entropy, the black hole information paradox, and time travel paradoxes from a new perspective *J. Mod. Opt.* in press (arXiv:1901.09096 [physics.gen-ph])
[8] Gibbons G W and Hawking S W 1977 Action integrals and partition functions in quantum gravity *Phys. Rev.* **15** 2752

IOP Conf. Series: Journal of Physics: Conf. Series **1275** (2019) 012023 doi:10.1088/1742-6596/1275/1/012023

Flavor neutrino states for pedestrians

M Blasone[1], L Smaldone[1] and G Vitiello[1]

[1] Dipartimento di Fisica, Università di Salerno, Via Giovanni Paolo II, 132 84084 Fisciano, Italy & INFN Sezione di Napoli, Gruppo collegato di Salerno, Italy

E-mail: blasone@sa.infn.it

E-mail: lsmaldone@sa.infn.it

E-mail: vitiello@sa.infn.it

Abstract. In this paper we discuss the ontology of flavor states of oscillating neutrinos. While an heuristic approach to this subject, experimentally successful in the high energy regime, is generally adopted, a logically consistent definition of flavor states describing neutrinos produced and detected in weak processes is still desirable and essential from a theoretical perspective. Here we briefly review basic facts and present some arguments which suggest that the definition of flavor states as eigenstates of flavor charges is the most reasonable one.

1. Introduction

Neutrinos are strange beasts which changed identity several times during their history: they were firstly theorized by Pauli to explain the continuous spectrum of electrons in the β decay[1]. They were observed for the first time, after long efforts, by F. Reines and C. Cowan [2]. Then it was proved that neutrinos exist in three different *flavors* [3], in correspondence with the three flavors of charged leptons.

For almost three decades people thought that neutrinos were exactly massless, and then they were described as left-handed Weyl fermions [4]. This assumption is also contained in the original Standard Model (SM) by Weinberg and Salam [5]. However, the results of the Homestake experiment [6], and many others along the years [7], inescapably proved that they must possess small masses. Actually, the nature itself of neutrino masses, Dirac [8] or Majorana [9] is not clear at the moment and different extensions of SM [10, 11, 12] were proposed.

Neutrinos also possess flavor mixing, a feature which is at the basis of the phenomenon of neutrino oscillations. This was theorized by Pontecorvo [13], Maki, Nakagawa and Sakata [14], and it is nowadays universally accepted [15, 16].

Although quantum mechanical (QM) Pontecorvo theory works extremely well in the actual experiments where neutrinos, due to their small masses, are ultrarelativistic, it is not satisfactory from a theoretical point of view, and deviations in low energy experiments are expected in different scenarios based on quantum field theory (QFT) [17, 18, 19, 20, 21]. In particular, the correct definition of flavor states is somewhat mysterious. The reason is that, because of the flavor oscillation phenomenon, these cannot be constructed as the usual *in* or *out* states [22]. A good review of the different approaches can be found in Ref. [23].

[1] The name *neutrino* was introduced by E. Fermi, to distinguish it from the neutron [1].

Content from this work may be used under the terms of the Creative Commons Attribution 3.0 licence. Any further distribution of this work must maintain attribution to the author(s) and the title of the work, journal citation and DOI.
Published under licence by IOP Publishing Ltd

IOP Conf. Series: Journal of Physics: Conf. Series **1275** (2019) 012023 doi:10.1088/1742-6596/1275/1/012023

In the present paper we review few basic facts on neutrino flavor states. In doing so we will limit to the case of Dirac neutrinos. We also show some paradoxes and mistakes that may occur in an oversimplified treatment of neutrino mixing and oscillations and propose some arguments which suggests that a correct description of neutrino flavor states would be the one developed in Refs. [19, 21].

2. Pontecorvo flavor states

The mixing transformation is

$$\nu_\sigma(x) \;=\; \sum_j U_{\sigma\, j} \nu_j(x)\,, \tag{1}$$

where ν_σ are the *flavor fields*, involved in the weak interaction, ν_j are the mass fields, which describe neutrinos with definite masses and U is the unitary mixing matrix.

The problem of constructing a Fock space for flavor neutrinos was studied many times in literature [17, 18, 19, 21, 24, 25]. There is general agreement [18, 19, 21, 24, 25, 26] on the existence of such a Fock space in the ultra-relativistic limit $m_j/|\mathbf{k}| \to 0$, where m_j are the neutrino masses. In this case annihilation operators are defined as

$$\tilde{\alpha}^r_{\mathbf{k},\sigma} \;=\; \sum_j U^*_{\sigma\, j}\, \alpha^r_{\mathbf{k},j} \tag{2}$$

where $\alpha^r_{\mathbf{k},j}$ are the annihilation operator of fields with definite mass:

$$\nu_j(x) = \sum_r \int \frac{\mathrm{d}^3 k}{(2\pi)^{\frac{3}{2}}} \left[u^r_{\mathbf{k},j}(t)\, \alpha^r_{\mathbf{k},j} + \, v^r_{-\mathbf{k},j}(t)\, \beta^{r\dagger}_{-\mathbf{k},j}\right] e^{i\mathbf{k}\cdot\mathbf{x}}\,. \tag{3}$$

Similar relations hold for $\beta^r_{\mathbf{k},j}$. Flavor states can be thus constructed as:

$$|\nu^r_{\mathbf{k},\sigma}\rangle_P \;\equiv\; \tilde{\alpha}^{r\dagger}_{\mathbf{k},\sigma}|0\rangle\,, \tag{4}$$

where $|0\rangle$ is the vacuum state, which is annihilated by $\alpha^r_{\mathbf{k},j}$ and $\beta^r_{\mathbf{k},j}$ (*mass vacuum*). These states can be recognized as the flavor states originally introduced by Pontecorvo [13]:

$$|\nu^r_{\mathbf{k},\sigma}\rangle_P \;=\; \sum_j U^*_{\sigma\, j}\, |\nu^r_{\mathbf{k},j}\rangle\,. \tag{5}$$

In the relativistic limit these are eigenstates of flavor charges at fixed time:

$$\lim_{m_i/|\mathbf{k}|\to 0} \; Q_{\nu_\sigma}(0)|\nu^r_{\mathbf{k},\sigma}\rangle_P \;=\; |\nu^r_{\mathbf{k},\sigma}\rangle_P\,, \tag{6}$$

where [27, 29] (see Section 3)

$$Q_{\nu_\sigma}(t) \;\equiv\; \int \mathrm{d}^3 x\, \nu^\dagger_\sigma(x)\, \nu_\sigma(x)\,. \tag{7}$$

However, this is not true at all energy scales. To see this, let us explicitly consider the 2-flavor case ($\sigma = e, \mu$ and $j = 1, 2$). The mixing matrix is given by:

$$U \;=\; \begin{pmatrix} \cos\theta & \sin\theta \\ -\sin\theta & \cos\theta \end{pmatrix}\,. \tag{8}$$

IOP Conf. Series: Journal of Physics: Conf. Series **1275** (2019) 012023 doi:10.1088/1742-6596/1275/1/012023

We can evaluate the oscillation formula as the expectation value of the flavor charge on a reference neutrino state [27, 28, 29]:

$$\widetilde{\mathcal{P}}_{e\to\mu}(t) \equiv {}_P\langle \nu^r_{\mathbf{k},e}|Q_{\nu_\mu}(t)|\nu^r_{\mathbf{k},e}\rangle_P = \frac{\sin^2(2\theta)}{2}\Big\{1-|U_{\mathbf{k}}|\cos[(\omega_{\mathbf{k},1}-\omega_{\mathbf{k},2})t]\Big\}, \tag{9}$$

where $|U_{\mathbf{k}}| \equiv u^{r\dagger}_{\mathbf{k},1}u^r_{\mathbf{k},2}$. In the ultra-relativistic case $|U_{\mathbf{k}}|\to 1$, and we get the standard oscillation formula:

$$\mathcal{P}_{e\to\mu}(t) = \sin^2(2\theta)\,\sin^2\left(\frac{\omega_{\mathbf{k},1}-\omega_{\mathbf{k},2}}{2}t\right). \tag{10}$$

Note in particular that

$$\widetilde{\mathcal{P}}_{e\to\mu}(0) = \frac{\sin^2(2\theta)}{2}\left(1-|U_{\mathbf{k}}|\right), \tag{11}$$

which is unacceptable because it tells us that flavor is undefined even at $t=0$ [29, 30]. A similar paradox was also pointed out in Ref. [28], in connection with the study of Feynman propagator for flavor fields.

3. Lepton number conservation

3.1. Basic considerations

In Ref. [18] it was pointed out that the amplitude of the neutrino detection process $\nu_\sigma + X_i \to e^- + X_f$, where X_i and X_f are the initial and the final particles, respectively, and e^- is the electron, is generally different from zero if $\sigma \neq e$, if we use the Pontecorvo states. In fact, for low-energy weak processes (where we can use the four-fermion Fermi interaction):

$$\langle e^s_{\mathbf{q},-}|\bar{e}(x)\,\gamma^\mu\,(1-\gamma^5)\,\nu_e(x)|\nu^r_{\mathbf{k},\sigma}\rangle_P\,h_\mu(x) = \sum_j U_{ej}\,U^*_{\sigma j}\langle e^s_{\mathbf{q},-}|\bar{e}(x)\,\gamma^\mu\,(1-\gamma^5)\,\nu_j(x)|\nu^r_{\mathbf{k},j}\rangle\,h_\mu(x)\,, \tag{12}$$

where h_μ are the matrix elements of the X part. This is generally different from $\delta_{\sigma e}$. This seems to be inconsistent because the flavor of the neutrino is *defined* by the flavor of the associated charged lepton in the (lepton-neutrino) doublet (see also Ref. [31]). From this observation, in the attempt of developing a consistent QFT approach, *weak process states* were introduced[2] [18]:

$$|\nu^r_\sigma\rangle_{WP} = \frac{1}{\sqrt{\sum_j |\mathcal{A}_{\sigma j}|^2}}\sum_j \mathcal{A}_{\sigma j}\,|\nu^r_{\mathbf{k}_j,j}\rangle\,, \tag{13}$$

where

$$\mathcal{A}_{\sigma j} = \langle \nu^r_{\mathbf{k}_j,j} X^{r'}_{\mathbf{k}',i}|S^\dagger|l^{s,-}_{\mathbf{q},\sigma}\,X^{s'}_{\mathbf{q}',f}\rangle\,. \tag{14}$$

Here l^-_α indicates the charged lepton and S is the S-matrix. If we consider, as before, a low-energy weak process, we can write:

$$\mathcal{A}_{\sigma j} = U^*_{\sigma j}\,\mathcal{M}_{\sigma j}\,, \tag{15}$$

with

$$\mathcal{M}_{\sigma j} \equiv -i\frac{G_F}{\sqrt{2}}\int d^4x\,\langle \nu^r_{\mathbf{k}_j,j}X^{r'}_{\mathbf{k}',i}|\bar{\nu}_j(x)\,\gamma^\mu\,(1-\gamma^5)\,l_\sigma(x)J_\mu(x)|l^{s,-}_{\mathbf{q},\sigma}X^{s'}_{f,\mathbf{q}'}\rangle\,, \tag{16}$$

[2] Here we are presenting the so called *detection states*. Analogously *production* states, can be defined [18].

where $J_\mu(x)$ is the X current. It is important to stress that the definition (13) explicitly depends on the details of the detection process. However, the inconsistency of Eq.(12) was not solved by this new definition. In fact:

$$\langle e^s_{\mathbf{q},-}|\bar{e}(x)\,\gamma^\mu\,(1-\gamma^5)\,\nu_e(x)|\nu^r_\sigma\rangle_{WP}\,h_\mu(x) = \sum_j \frac{U_{ej}\,\mathcal{A}_{\sigma j}}{\sqrt{\sum_j|\mathcal{A}_{\sigma j}|^2}}\langle e^s_{\mathbf{q},-}|\bar{e}(x)\,\gamma^\mu\,(1-\gamma^5)\,\nu_j(x)|\nu^r_{\mathbf{k}_j,j}\rangle\,h_\mu(x)\,, \tag{17}$$

which is also generally different from $\delta_{\sigma e}$.

We now have to do an important remark. The S matrix is defined so it connects *in* and *out* states [22]:

$$S_{AB} \;\equiv\; \langle A; out|B; in\rangle\,. \tag{18}$$

The asymptotic states are defined far before and after the interaction, so that neutrino oscillation occurs leading to the violation of family lepton number conservation. This is natural and expected. The previous example are thus not pathological in this sense. However, lepton number has to be conserved in the production and detection vertexes (at tree level), where flavor oscillation can be neglected[3]. Because this point created some discussion in literature [32, 33, 34], we think it is important to expose it in more detail, starting from basic elementary considerations in order to avoid to incur in the contradictions of oversimplified treatments.

3.2. Flavor charge conservation in the vertex

Let us consider a charged weak decay $W^+ \to e^+ + \nu_e$. The relevant part of SM Lagrangian is $\mathcal{L} = \mathcal{L}_0 + \mathcal{L}_{int}$ with

$$\mathcal{L}_0(x) = \overline{\nu}(x)\,(i\gamma_\mu\partial^\mu - M_\nu)\,\nu(x) \,+\, \overline{l}(x)\,(i\gamma_\mu\partial^\mu - M_l)\,l(x)\,, \tag{19}$$

$$\mathcal{L}_{int}(x) = \frac{g}{2\sqrt{2}}\left[W^+_\mu\,\overline{\nu}(x)\,\gamma^\mu\,(1-\gamma^5)\,l(x) + h.c.\right]\,, \tag{20}$$

where $\nu = (\nu_e, \nu_\mu)^T$, $l = (e, \mu)^T$, and

$$M_\nu \;=\; \begin{pmatrix} m_e & m_{e\mu} \\ m_{e\mu} & m_\mu \end{pmatrix}, \qquad M_l \;=\; \begin{pmatrix} \tilde{m}_e & 0 \\ 0 & \tilde{m}_\mu \end{pmatrix}. \tag{21}$$

The Lagrangian $\mathcal{L}$ is invariant under the global $U(1)$ transformations $\nu \to e^{i\alpha}\nu$ and $l \to e^{i\alpha}l$ leading to the conservation of the total flavor charge Q^{tot}_l corresponding to the total lepton-number conservation [15, 35]. This can be written in terms of the flavor charges for neutrinos and charged leptons [27]

$$Q^{tot}_l \;=\; \sum_{\sigma=e,\mu} Q^{tot}_\sigma(t)\,, \quad Q^{tot}_\sigma(t) = Q_{\nu_\sigma}(t) + Q_\sigma\,, \tag{22}$$

with

$$\begin{aligned} Q_e &= \int d^3x\, e^\dagger(x)e(x)\,, & Q_{\nu_e}(t) &= \int d^3x\, \nu_e^\dagger(x)\nu_e(x)\,, \\ Q_\mu &= \int d^3x\, \mu^\dagger(x)\mu(x)\,, & Q_{\nu_\mu}(t) &= \int d^3x\, \nu_\mu^\dagger(x)\nu_\mu(x)\,. \end{aligned} \tag{23}$$

[3] Obviously loop diagrams can produce violation of lepton number in the production and detection vertexes, but these contributions are negligible in the following discussion.

IOP Conf. Series: Journal of Physics: Conf. Series **1275** (2019) 012023 doi:10.1088/1742-6596/1275/1/012023

The above charges can be derived via Noether's theorem [36] from the Lagrangian (20). Note the time dependence of the neutrino charges, due to the non-diagonal mass matrix M_ν.

By observing that $[\mathcal{L}_{int}(\mathbf{x},t), Q^{tot}_\sigma(t)] = 0$, we see that a neutrino flavor state is well defined in the production vertex as an eigenstate of the corresponding flavor charge [30]. Let us now consider the (Pontecorvo states) amplitude of the above mentioned process

$$\mathcal{A}^P_{W^+\to e^+\,\nu_e} \;=\; {}_P\langle \nu^r_{\mathbf{k},e}| \otimes \langle e^s_{\mathbf{q}}| \left[-i \int_{x^0_{in}}^{x^0_{out}} d^4x\, \mathcal{H}^e_{int}(x) \right] |W^+_{\mathbf{p},\lambda}\rangle\,. \tag{24}$$

The interaction Hamiltonian density is

$$\mathcal{H}^e_{int}(x) \;=\; -\frac{g}{2\sqrt{2}}\, W^+_\mu(x)\, J^\mu_e(x) + h.c.\,, \tag{25}$$

and

$$J^\mu_e(x) \;=\; \bar{\nu}_e(x)\, \gamma^\mu\, (1-\gamma^5) e(x)\,, \tag{26}$$

as it can be deduced from Eq.(20). The usual amplitude is obtained by taking the asymptotic limit $x^0_{out} \to +\infty$, $x^0_{in} \to -\infty$. However, as mentioned, the flavor states are not asymptotic stable states, and we want to investigate the short-time behavior of the amplitude (around the interaction time $x_0 = 0$). Explicit calculations (cfr. Ref.[33]) give:

$$\mathcal{A}^P_{W^+\to e^+\,\nu_e} \;=\; \frac{ig}{2\sqrt{4\pi}} \frac{\varepsilon_{\mathbf{p},\mu,\lambda}}{\sqrt{2E^W_p}}\, \delta^3(\mathbf{p}-\mathbf{q}-\mathbf{k})$$
$$\times\; \sum_{j=1}^{2} U^2_{ej} \int_{x^0_{in}}^{x^0_{out}} dx^0\, e^{-i\omega_{\mathbf{k},j}\, x^0_{out}}\, \bar{u}^r_{\mathbf{k},j}\, \gamma^\mu (1-\gamma^5)\, v^s_{\mathbf{q},e}\, e^{-i(E^W_p - E^e_q - \omega_{\mathbf{k},j})x^0}\,, \tag{27}$$

where $\varepsilon_{\mathbf{p},\mu,\lambda}$ is the polarization vector of W^+ and $v^s_{\mathbf{q},e}$ is the positron wave function. We take $x^0_{in} = -\Delta t/2$ and $x^0_{out} = \Delta t/2$, when $\tau_W \ll \Delta t \ll t_{osc}$, where τ_W is the W^+ lifetime, while t_{osc} is the oscillation time. Under this condition, we can expand the amplitude at the leading order in Δt, obtaining:

$$\mathcal{A}^P_{W^+\to e^+\,\nu_e} \;\approx\; \frac{ig}{2\sqrt{4\pi}} \frac{\varepsilon_{\mathbf{p},\mu,\lambda}}{\sqrt{2E^W_p}}\, \delta^3(\mathbf{p}-\mathbf{q}-\mathbf{k})\Delta t \sum_{j=1}^{2} U^2_{ej}\; \bar{u}^r_{\mathbf{k},j}\, \gamma^\mu (1-\gamma^5)\, v^s_{\mathbf{q},e}\,. \tag{28}$$

In the same way, one can evaluate the "wrong" amplitude

$$\mathcal{A}^P_{W^+\to e^+\,\nu_\mu} \;=\; {}_P\langle \nu^r_{\mathbf{k},\mu}| \otimes \langle e^s_{\mathbf{q}}| \left[-i \int_{x^0_{in}}^{x^0_{out}} d^4x\, \mathcal{H}^e_{int}(x) \right] |W^+_{\mathbf{p},\lambda}\rangle\,. \tag{29}$$

Proceeding as before we get:

$$\mathcal{A}^P_{W^+\to e^+\,\nu_\mu} \;=\; \frac{ig}{2\sqrt{4\pi}} \frac{\varepsilon_{\mathbf{p},\mu,\lambda}}{\sqrt{2E^W_p}}\, \delta^3(\mathbf{p}-\mathbf{q}-\mathbf{k})$$
$$\times\; \sum_{j=1}^{2} U_{\mu j}\, U_{ej} \int_{x^0_{in}}^{x^0_{out}} dx^0\, e^{-i\omega_{\mathbf{k},j}\, x^0_{out}}\, \bar{u}^r_{\mathbf{k},j}\, \gamma^\mu (1-\gamma^5)\, v^s_{\mathbf{q},e}\, e^{-i(E^W_p - E^e_q - \omega_{\mathbf{k},j})x^0}\,, \tag{30}$$

which in the short time limit becomes

$$\mathcal{A}^{P}_{W^+\to e^+\nu_\mu} \approx \frac{ig}{2\sqrt{4\pi}} \frac{\varepsilon_{\mathbf{p},\mu,\lambda}}{\sqrt{2E^W_p}} \delta^3(\mathbf{p}-\mathbf{q}-\mathbf{k})\Delta t \sum_{j=1}^{2} U_{\mu j}\, U_{ej}\;\bar{u}^r_{\mathbf{k},j}\,\gamma^\mu(1-\gamma^5)\,v^s_{\mathbf{q},e}\,. \tag{31}$$

This is clearly different from zero, which is inconsistent.

Let us notice that the same procedure for the weak process states above defined would lead to

$$\mathcal{A}^{WP}_{W^+\to e^+\nu_\mu} \approx \frac{ig}{2\sqrt{4\pi}} \frac{\varepsilon_{\mathbf{p},\mu,\lambda}}{\sqrt{2E^W_p}} \Delta t \sum_{j=1}^{2} \frac{\delta^3(\mathbf{p}-\mathbf{q}-\mathbf{k}_j)\,\mathcal{A}^*_{\mu j}}{\sqrt{\sum_{j=1}^{2}|\mathcal{A}_{\mu j}|^2}} U_{ej}\;\bar{u}^r_{\mathbf{k}_j,j}\,\gamma^\mu(1-\gamma^5)\,v^s_{\mathbf{q},e}\,. \tag{32}$$

The main reason behind this ambiguity is that Eq.(6) does not hold in the non-relativistic case. The same is true for weak process states: these are *not* eigenstates of the flavor charges which commute with the interaction vertex. Therefore it is obvious that the use of these states leads to violation of the family lepton number at tree level, in the production (and detection) vertex.

4. First quantized oscillation formula and Dirac equation

In this section, following Ref. [37] we review the correct derivation of the oscillation formula in relativistic quantum mechanics. Flavor wavefunction satisfies the Dirac equation:

$$(i\gamma^\mu\,\partial_\mu \otimes 1\!\!1_2 - 1\!\!1_4 \otimes M_\nu)\;\Psi(x) \;=\; 0\,, \tag{33}$$

where $1\!\!1_n$ indicates the $n\times n$ identity matrix and M_ν is the mass matrix introduced in Eq.(21).

For simplicity we limit to study the problem in one spatial dimension (along the z-axis). By introducing wavefunctions of neutrino with definite masses:

$$\left(i\gamma^0\partial_0 + i\gamma^3\partial_3 - m_j\right)\,\psi_j(z,t) \;=\; 0\,, \qquad j=1,2\,, \tag{34}$$

one can express a neutrino wavepacket Ψ as:

$$\begin{aligned}\Psi(z,t) \;&=\; \cos\theta\,\psi_1(z,t)\,\otimes\,\nu_1 \,+\, \sin\theta\,\psi_2(z,t)\,\otimes\,\nu_2\\ &=\; \left[\psi_1(z,t)\,\cos^2\theta \,+\, \psi_2(z,t)\,\sin^2\theta\right]\,\otimes\,\nu_\sigma + \sin\theta\,\cos\theta\,\left[\psi_1(z,t) - \psi_2(z,t)\right]\,\otimes\,\nu_\rho\\ &\equiv\; \psi_\sigma(z,t)\,\otimes\,\nu_\sigma \,+\, \psi_\rho(z,t)\,\otimes\,\nu_\rho\,,\end{aligned} \tag{35}$$

where $\psi_j(z,t)$ are the wavepackets describing neutrinos with definite masses, ν_1,ν_2 are the eigenstates of M_ν and ν_σ,ν_ρ are flavor eigenstates. A neutrino is produced as a flavor eigenstate if $\psi_1(z,0)=\psi_2(z,0)=\psi_\sigma(z,0)$, with $\sigma=e,\mu$. The oscillation probability will be given by

$$P_{\nu_\sigma\to\nu_\rho} \;=\; \int_{-\infty}^{+\infty} \mathrm{d}z\,\psi^\dagger_\rho(z,t)\,\psi_\rho(z,t)\,. \tag{36}$$

By using Eq.(35) we can derive:

$$P_{\nu_\sigma\to\nu_\rho} \;=\; \frac{\sin^2 2\theta}{2}\,\left[1 - I_{12}(t)\right]\,, \tag{37}$$

where the interference term is given by

$$I_{12}(t) \;=\; \Re e\left[\int_{-\infty}^{+\infty}\mathrm{d}z\,\psi^\dagger_1(z,t)\,\psi_2(z,t)\right]\,. \tag{38}$$

IOP Conf. Series: Journal of Physics: Conf. Series **1275** (2019) 012023 doi:10.1088/1742-6596/1275/1/012023

Note that respect to the usual treatment, where only positive frequency modes are included, this analysis explicitly shows that also negative frequency contribution have to be involved in the computation of the interference term (38). In fact, let us consider the Fourier expansion of $\psi_j(z,t)$:

$$\psi_j(x) = \sum_r \int_{-\infty}^{+\infty} \frac{\mathrm{d}p_z}{2\pi} \left[u^r_{p_z,j}\, \alpha^r_{p_z,j}\, e^{-i\,\omega_{p_z,j}\,t} + v^r_{-p_z,j} \beta^{r*}_{-p_z,j}\, e^{i\,\omega_{p_z,j}\,t}\right] e^{i\,p_z\,z}\,, \quad j = 1,2\,. \tag{39}$$

The requirement that neutrino is produced with definite flavor, assumes the form:

$$u^r_{p_z,j}\, \alpha^r_{p_z,j} + v^r_{-p_z,j} \beta^{r*}_{-p_z,j} \;=\; \varphi_\sigma(p_z - p_0)\, w\,, \tag{40}$$

where $\varphi_\sigma(p_z - p_0)$ is the flavor neutrino distribution in the momentum space, at $t = 0$, p_0 is the mean momentum of mass wavepackets and w is a constant spinor, satisfying $w^\dagger w = 1$. By using orthogonality conditions of Dirac spinors we derive the relations

$$\alpha^r_{p_z,j} \;=\; \varphi_\sigma(p_z - p_0)\, u^{r\dagger}_{p_z,j}\, w\,, \tag{41}$$

$$\beta^{r*}_{-p_z,j} \;=\; \varphi_\sigma(p_z - p_0)\, v^{r\dagger}_{-p_z,j}\, w\,. \tag{42}$$

Substituting in Eq.(39) and then in Eq.(38) we finally arrive at [29, 37]:

$$I_{12}(t) \;=\; \int_{-\infty}^{+\infty} \frac{\mathrm{d}p_z}{2\pi}\, \varphi^2_\sigma(p_z - p_0) \left(|U_{p_z}|^2 \cos(\Omega^-_{p_z} t) + |V_{p_z}|^2 \cos(\Omega^+_{p_z} t)\right)\,, \tag{43}$$

where

$$\Omega^\pm_{p_z} \;=\; \omega_{p_z,1} \pm \omega_{p_z,2}\,, \tag{44}$$

$$|V_{p_z}|^2 \;=\; 1 - |U_{p_z}|^2 \;=\; \frac{\omega_{p_z,1}\,\omega_{p_z,2} - p_z^2 - m_1 m_2}{2\omega_{p_z,1}\,\omega_{p_z,2}}\,. \tag{45}$$

The notation here is slightly different respect to Refs.[29, 37], in order to get in touch with the next section. For plane waves, $\varphi_\sigma(p_z - p_0) \;=\; \delta(p_z - p_0)$. The oscillation probability thus reads

$$P_{\nu_\sigma \to \nu_\rho} \;=\; \sin^2 2\theta \left[|U_{p_0}|^2 \sin^2\left(\frac{\Omega^-_{p_z}}{2} t\right) + |V_{p_0}|^2 \sin^2\left(\frac{\Omega^+_{p_0}}{2} t\right)\right]. \tag{46}$$

The main difference with respect to the standard oscillation formula Eq.(10) is the presence of a fast oscillating term, with frequency $\Omega^+_{p_0}/2$. This is analogous to the *Zitterbewegugng* encountered in atomic physics, which leads to the Darwin contribution to fine structure of hydrogen atom. In our case this effect is very small when $p_0 \gg \sqrt{m_1 m_2}$, i.e. for ultrarelativistic neutrinos. In that regime $|U_{p_0}|^2 \to 1$ and $|V_{p_0}|^2 \to 0$, and the oscillation probability recovers its standard form (10).

Let us remark that the usual treatment [13, 15, 16], not including negative frequency terms, is very similar to the *rotating wave approximation* usually encountered in quantum optics and atomic physics [38, 39], where fast oscillating terms in the Hamiltonian are neglected in order to find exact solutions of the eigenvalue problem. However, in the case of neutrino oscillations, there are no reasons to neglect the contribution with $\Omega^+_{p_0}$, apart from ultrarelativistic case.

5. Flavor eigenstates

Let us now come back to the problem of constructing flavor states in QFT. Resuming to our previous considerations, we aim to construct states which are eigenstates of the flavor charges. To this end, we limit ourselves to the two flavor case. The mixing relations for neutrino fields are

$$\begin{pmatrix} \nu_e(x) \\ \nu_\mu(x) \end{pmatrix} = \begin{pmatrix} \cos\theta & \sin\theta \\ -\sin\theta & \cos\theta \end{pmatrix} \begin{pmatrix} \nu_1(x) \\ \nu_2(x) \end{pmatrix}, \tag{47}$$

with $\tan 2\theta = 2m_{e\mu}/(m_\mu - m_e)$. Eq. (47) can be equivalently rewritten as [21]

$$\nu_\sigma(x) = G_\theta^{-1}(t)\, \nu_j(x)\, G_\theta(t)\,, \tag{48}$$

with $(\sigma, j) = (e, 1), (\mu, 2)$ and $G_\theta(t)$ given by

$$G_\theta(t) = \exp\left[\theta \int d^3x \left(\nu_1^\dagger(x)\nu_2(x) - \nu_2^\dagger(x)\nu_1(x)\right)\right]. \tag{49}$$

From (3) and (48) it follows that flavor fields are:

$$\nu_\sigma(x) \;=\; \sum_r \int \frac{d^3k}{(2\pi)^{\frac{3}{2}}} \left[u^r_{\mathbf{k},j}(t)\, \alpha^r_{\mathbf{k},\sigma}(t) \,+\, v^r_{-\mathbf{k},j}(t)\, \beta^{r\dagger}_{-\mathbf{k},\sigma}(t)\right] e^{i\mathbf{k}\cdot\mathbf{x}}, \tag{50}$$

with $(\sigma, j) \;=\; (e, 1), (\mu, 2)$, and flavor ladder operators are given by

$$\begin{pmatrix} \alpha^r_{\mathbf{k},\sigma}(t) \\ \beta^r_{-\mathbf{k},\sigma}(t) \end{pmatrix} \;=\; G_\theta^{-1}(t) \begin{pmatrix} \alpha^r_{\mathbf{k},j}(t) \\ \beta^r_{-\mathbf{k},j}(t) \end{pmatrix} G_\theta(t)\,. \tag{51}$$

The *flavor vacuum* is defined as [19, 21]:

$$|0\rangle_{e,\mu} = G_\theta^{-1}(0)\; |0\rangle_{1,2}\,, \tag{52}$$

where $|0\rangle_{1,2}$ denotes the mass vacuum (cf. Eq.(4)) for the two-flavor case. One can easily verify that it is annihilated by the flavor operators defined in Eq. (51). Moreover, one can prove that [19, 21]

$$\lim_{V\to\infty} {}_{1,2}\langle 0|0\rangle_{e,\mu} = \lim_{V\to\infty} e^{\frac{V}{(2\pi)^3}\int d^3k \ln(1-\sin^2\theta|V_{\mathbf{k}}|^2)^2} = 0\,, \tag{53}$$

where

$$|V_{\mathbf{k}}| = \frac{|\mathbf{k}|}{\sqrt{4\omega_{\mathbf{k},1}\omega_{\mathbf{k},1}}}\left(\sqrt{\frac{\omega_{\mathbf{k},2}+m_2}{\omega_{\mathbf{k},1}+m_1}} - \sqrt{\frac{\omega_{\mathbf{k},1}+m_1}{\omega_{\mathbf{k},2}+m_2}}\right), \tag{54}$$

i.e. flavor and massive fields belong to unitarily inequivalent representations of the anticommutation relations [40, 41].

True flavor eigenstates can be explicitly constructed as

$$|\nu^r_{\mathbf{k},\sigma}\rangle \;=\; \alpha^{r\dagger}_{\mathbf{k},\sigma}|0\rangle_{e,\mu}\,. \tag{55}$$

where flavor operators are taken at reference time $t = 0$. One can prove that:

$$Q_{\nu_\sigma}(0)|\nu^r_{\mathbf{k},\sigma}\rangle \;=\; |\nu^r_{\mathbf{k},\sigma}\rangle\,. \tag{56}$$

IOP Conf. Series: Journal of Physics: Conf. Series **1275** (2019) 012023 doi:10.1088/1742-6596/1275/1/012023

The corresponding oscillation formula can be found by taking the expectation value of the flavor charges [28]

$$\mathcal{Q}_{\sigma\to\rho}(t) \;=\; \langle Q_{\nu_\rho}(t)\rangle_\sigma\,, \tag{57}$$

where $\langle\cdots\rangle_\sigma = \langle\nu^r_{\mathbf{k},\sigma}|\cdots|\nu^r_{\mathbf{k},\sigma}\rangle$, which gives

$$\begin{aligned}\mathcal{Q}_{\sigma\to\rho}(t) &= \sin^2(2\theta)\Big[|U_{\mathbf{k}}|^2\sin^2\Big(\frac{\Omega^-_{\mathbf{k}}}{2}t\Big)+|V_{\mathbf{k}}|^2\sin^2\Big(\frac{\Omega^+_{\mathbf{k}}}{2}t\Big)\Big],\\ \mathcal{Q}_{\sigma\to\sigma}(t) &= 1\;-\;\mathcal{Q}_{\sigma\to\rho}(t)\,,\quad \sigma\neq\rho\,,\end{aligned} \tag{58}$$

where now $\Omega^\pm_{\mathbf{k}}\equiv\omega_{\mathbf{k},2}\pm\omega_{\mathbf{k},1}$ and $|U_{\mathbf{k}}|^2=1-|V_{\mathbf{k}}|^2$. This is nothing but the formula (46), obtained in the previous section from Dirac equation in the quantum mechanical wave packet formalism.

By using the flavor states (55) we can also evaluate the amplitudes of the decay $W^+\to e^+\,\nu_e$ [33]:

$$\begin{aligned}\mathcal{A}_{W^+\to e^+\,\nu_e} \;=\;& \frac{ig}{2\sqrt{2}(2\pi)^{\frac{3}{2}}}\,\delta^3(\mathbf{p}-\mathbf{q}-\mathbf{k})\int_{x^0_{in}}^{x^0_{out}}\mathrm{d}x^0\,\frac{\varepsilon_{\mathbf{p},\mu,\lambda}}{\sqrt{2E^W_p}}\,\delta^3(\mathbf{p}-\mathbf{q}-\mathbf{k})\\ \times\;& \Big\{\cos^2\theta\, e^{-i\omega_{\mathbf{k},1}x^0_{in}}\,\bar{u}^r_{\mathbf{k},1}\gamma^\mu(1-\gamma^5)\,v^s_{\mathbf{q},e}\,e^{-i(E^W_p-E^e_q-\omega_{\mathbf{k},1})x^0}\\ +\;& \sin^2\theta\,\Big[|U_{\mathbf{k}}|\,e^{-i\omega_{\mathbf{k},2}x^0_{in}}\,\bar{u}^r_{\mathbf{k},2}\gamma^\mu(1-\gamma^5)\,v^s_{\mathbf{q},e}\,e^{-i(E^W_p-E^e_q-\omega_{\mathbf{k},2})x^0}\\ +\;& \varepsilon^r|V_{\mathbf{k}}|\,e^{i\omega_{\mathbf{k},2}x^0_{in}}\,\bar{v}^r_{-\mathbf{k},2}\gamma^\mu(1-\gamma^5)\,v^s_{\mathbf{q},e}\,e^{-i(E^W_p-E^e_q+\omega_{\mathbf{k},2})x^0}\Big]\Big\}\,,\end{aligned} \tag{59}$$

with $\varepsilon^r\equiv(-1)^r$, and the "wrong" one, $W^+\to e^+\,\nu_\mu$:

$$\begin{aligned}\mathcal{A}_{W^+\to e^+\,\nu_\mu} \;=\;& \sin\theta\,\cos\theta\,\frac{ig}{2\sqrt{2}(2\pi)^{\frac{3}{2}}}\,\delta^3(\mathbf{p}-\mathbf{q}-\mathbf{k})\int_{x^0_{in}}^{x^0_{out}}\mathrm{d}x^0\,\frac{\varepsilon_{\mathbf{p},\mu,\lambda}}{\sqrt{2E^W_p}}\,\delta^3(\mathbf{p}-\mathbf{q}-\mathbf{k})\\ \times\;& \Big\{e^{-i\omega_{\mathbf{k},2}x^0_{in}}\,\bar{u}^r_{\mathbf{k},2}\gamma^\mu(1-\gamma^5)\,v^s_{\mathbf{q},e}\,e^{-i(E^W_p-E^e_q-\omega_{\mathbf{k},2})x^0}\\ -\;& \Big[|U_{\mathbf{k}}|\,e^{-i\omega_{\mathbf{k},1}x^0_{in}}\,\bar{u}^r_{\mathbf{k},1}\gamma^\mu(1-\gamma^5)\,v^s_{\mathbf{q},e}\,e^{-i(E^W_p-E^e_q-\omega_{\mathbf{k},1})x^0}\\ +\;& \varepsilon^r|V_{\mathbf{k}}|\,e^{i\omega_{\mathbf{k},1}x^0_{in}}\,\bar{v}^r_{-\mathbf{k},1}\gamma^\mu(1-\gamma^5)\,v^s_{\mathbf{q},e}\,e^{-i(E^W_p-E^e_q+\omega_{\mathbf{k},1})x^0}\Big]\Big\}\,.\end{aligned} \tag{60}$$

Looking at the case $\tau_W\ll\Delta t\ll t_{osc}$, we find [33]:

$$\begin{aligned}\mathcal{A}_{W^+\to e^+\,\nu_e} \;\approx\;& \frac{ig}{2\sqrt{2}(2\pi)^{\frac{3}{2}}}\,\delta^3(\mathbf{p}-\mathbf{q}-\mathbf{k})\frac{\varepsilon_{\mathbf{p},\mu,\lambda}}{\sqrt{2E^W_p}}\,\delta^3(\mathbf{p}-\mathbf{q}-\mathbf{k})\,\Delta t\\ \times\;& \Big\{\cos^2\theta\,\bar{u}^r_{\mathbf{k},2}+\sin^2\theta\,\big[|U_{\mathbf{k}}|\bar{u}^r_{\mathbf{k},2}+\varepsilon^r|V_{\mathbf{k}}|\,\bar{v}^r_{-\mathbf{k},2}\big]\Big\}\,\gamma^\mu(1-\gamma^5)\,v^s_{\mathbf{q},e}\,,\end{aligned} \tag{61}$$

and

$$\mathcal{A}_{W^+\to e^+\,\nu_\mu} \;\approx\; 0\,. \tag{62}$$

which is the expected result.

6. Conclusions

We have reviewed some basic facts on neutrino flavor states, and put in evidence some inconsistencies, coming out from the use of weak states which are not eigenstates of the flavor charges. The definition of neutrino flavor states as exact eigenstates of flavor charges [21] instead provides the solution to overcome such inconsistencies and recovers the correct results, as the exact oscillation formula Eqs. (46),(58) and family lepton number conservation (at tree level), in the production/detection vertex Eqs.(61),(62).

Another independent argument in favor of flavor states as "ontological" entities, comes from the study of the proton decay in accelerated frames in presence of neutrino mixing [43, 44]. From general covariance, the proton decay rates calculated in the inertial and comoving frame should agree: this leads to the necessity of Unruh radiation but also to the use of flavor neutrino states in the decay vertex.

We finally point out that recently [35], the use of flavor states led to the interpretation of flavor neutrinos as unstable particles, which periodically *decay* into different neutrino species. This view is also compatible with the formal statement, based on the study of Schwinger–Dyson equations, according to which flavor neutrinos can be formally thought as *single-particle bound states*, i.e. bound states in the flavor space.

Acknowledgments

L.S. would like to thank G. Fiore, M. Kurkov and F. Lizzi for interesting discussions on the subject.

References

[1] Segrè E 1970 *Enrico Fermi: Physicist* (Chicago: Univ. Chicago Press)
[2] Reines F and Cowan C L, *Phys. Rev.* 1953 **92** 830; Cowan C L, Reines F, Harrison F B, Kruse H W and McGuire A D 1956 *Science* **124** 103
[3] Danby G et al. 1962 *Phys. Rev. Lett.* **9** 36; M. L. Perl et al. 1975 *Phys. Rev. Lett.* **35** 1489
[4] Weyl H 1929 *Z. Phys.* **56** 5
[5] Weinberg S W 1967 *Phys. Rev. Lett.* **19** 1264; Salam A 1968 *Conf. Proc.* C **680519** 367
[6] Davis R 1994 *Prog. Part. Nucl. Phys.* **32** 13
[7] Smy M B (Super-Kamiokande) 2003 *Nucl. Phys. Proc. Suppl.* **118** 25; Hallin A L et al. 2003 *Nucl. Phys. Proc. Suppl.* **118** 3; Eguchi K et al. (KamLAND) 2003 *Phys. Rev. Lett.* **90** 021802; Hampel W et al. (GALLEX) 1999 *Phys. Lett.* B **447** 127; Altmann M et al. (GNO) 2000 *Phys. Lett.* B **490** 16
[8] Dirac P A M and Fowler R H 1928 *Roy. Soc. Proc.* A **117** 610
[9] Majorana E 1937 *Nuovo Cim.* **5** 171.
[10] Minkowski P 1977 *Phys. Lett. B* **67** 421; Yanagida T 1980 *Prog. Theor. Phys.* **64** 1103
[11] Cheng T P and Li L F 1980 *Phys. Rev. D* **22** 2860
[12] King S F 2008 *J. Phys. Conf. Ser.* **136** 022038
[13] Pontecorvo B 1957 *Sov. Phys. JETP* **6** 429; *Sov. Phys. JETP* **7** 172-02; Gribov V and Pontecorvo B 1969 *Physics Letters* B **28** 493; Bilenky S and Pontecorvo B 1978 *Phys. Rep.* **41** 225
[14] Z. Maki, M. Nakagawa and S. Sakata, Prog. Theor. Phys. **28**, 870 (1962).
[15] Bilenky S M and Petcov S T 1987 *Rev. Mod. Phys.* **59** 671
[16] Giunti C and Kim C W 2007 *Fundamentals of Neutrino Physics and Astrophysics* (Oxford: Oxford Univ. Press); Kleinert H 2016 *Particles and Quantum fields* (Singapore, World Scientific); Strumia A and Vissani F *Preprint* arXiv:hep-ph/0606054v3.
[17] Mannheim P T 1988 *Phys. Rev.* D **37** 1935
[18] Giunti C, Kim C W and Lee U W 1992 *Phys. Rev.* D **45** 2414
[19] Blasone M and Vitiello G 1995 *Ann. Phys.* **244** 283
[20] Giunti C, Kim C W, Lee J A and Lee U W 1993 *Phys. Rev.* D **48** 4310; Grimus W and Stockinger P 1996 *Phys. Rev.* D **54** 3414; Cohen A G, Glashow S L and Ligeti Z 2009 *Phys. Lett.* B **678** 191 Akhmedov E K and Smirnov A Y 2009 *Phys. Atom. Nucl.* **72** 1363
[21] Fujii K, Habe C and Yabuki T 1999 *Phys. Rev.* D **59** 113003; 2001 *Phys. Rev.* D **64** 013011; Hannabuss K C and Latimer D C 2000 *J. Phys.* A **33** 1369; 2003 *J. Phys.* A **36** L69; Ji C R and Mishchenko Y 2002 *Phys. Rev.* D **65** 096015; 2005 *Ann. Phys.* **315** 488; Blasone M, Gargiulo M V and Vitiello G 2016 *Phys. Lett.* B **716** 104

IOP Conf. Series: Journal of Physics: Conf. Series **1275** (2019) 012023 doi:10.1088/1742-6596/1275/1/012023

[22] Lehmann H, Symanzik K and Zimmerman W 1955 Nuovo Cimento **1** 205
[23] Beuthe M 2003 *Phys. Rep.* **375** 105; Kruppke D 2007 *On Theories of Neutrino Oscillations: A Summary and Characterisation of the Problematic Aspects*, Bielefeld Univ.
[24] Ho C M 2012 *JHEP* **1212** 022
[25] Lobanov A E 2019 *Ann. Phys.* **403** 82
[26] Fantini G, Gallo Rosso A, Vissani F and Zema V 2018 *Adv. Ser. Direct. High Energy Phys.* **28** 37
[27] Blasone M, Jizba P and Vitiello G 2001 *Phys. Lett.* B **517** 471
[28] Blasone M, Henning P A and Vitiello G 1999 *Phys. Lett.* B **451** 140
[29] Nishi C C 2006 *Phys. Rev.* D **73** 053013; 2008 *Phys. Rev.* D **78** 113007
[30] Blasone M, Capolupo A, Ji C R and Vitiello G 2009 *Nucl. Phys. Proc. Suppl.* **188** 37
[31] Bilenky S M and Giunti C 2001 *Int. J. Mod. Phys.* A **16** 3931
[32] Li Y F and Liu Q Y 2006 *JHEP* **0610** 048
[33] Blasone M, Capolupo A, Ji C R and Vitiello G 2010 *Int. J. Mod. Phys.* A **25** 4179
[34] Lee C Y *Preprint* arXiv:1709.06306v3 [hep-ph]
[35] Blasone M, Jizba P and Smaldone L 2019 *Phys. Rev.* D **99** 016014
[36] Noether E 1918 *Nachrichten von der Gesellschaft der Wissenschaften zu Gttingen, Mathematisch-Physikalische Klasse* **1918** 235
[37] Bernardini A E and De Leo 2004 *Eur. Phys. J.* C **37** 471; 2005 *Phys. Rev.* D **71** 076008
[38] Kurcz A, Capolupo A, Beige A, Del Giudice E and Vitiello G 2010 *Phys. Lett.* A **374** 3726
[39] Fuji K 2017 *J. Mod. Phys.* **8** 2042
[40] Umezawa H, Matsumoto H and Tachiki M 1982 *Thermo Field Dynamics And Condensed States* (Amsterdam: North-holland); Umezawa H 1993 *Advanced Field Theory: Micro,Macro and Thermal Physics* (New York: AIP)
[41] Blasone M, Jizba P and Vitiello G 2011 *Quantum Field Theory and its Macroscopic Manifestations* (London: World Scientific & ICP)
[42] Blasone M, Jizba P and Smaldone L 2018 *J. Phys. Conf. Ser.* **1071** 012003
[43] Blasone M, Lambiase G, Luciano G G and Petruzziello L 2018 *Phys. Rev.* D **97** 105008
[44] Blasone M, Lambiase G, Luciano G G and Petruzziello L 2018 *PoS CORFU* **2017** 198

IOP Conf. Series: Journal of Physics: Conf. Series **1275** (2019) 012024 doi:10.1088/1742-6596/1275/1/012024

Heuristic derivation of the Casimir effect from Generalized Uncertainty Principle

M Blasone[†1], **G Lambiase**[†2], **G G Luciano**[†3], **L Petruzziello**[†4] **and F Scardigli**[‡*5]

[†] INFN and Università di Salerno, Via Giovanni Paolo II, 132 – 84084 Fisciano (SA), Italy
[‡] Dipartimento di Matematica, Politecnico di Milano, Piazza Leonardo da Vinci 32, 20133 Milano, Italy
[*] Institute-Lorentz for Theoretical Physics, Leiden University, P.O. Box 9506, Leiden, The Netherlands

Abstract. After a short introduction to the generalized uncertainty principle (GUP), we discuss heuristic derivations of the Casimir effect, first from the usual Heisenberg uncertainty principle (HUP), and then from GUP. Results are compared with those obtained from more standard calculations in Quantum Field Theory (QFT).

1. Introduction

In his seminal paper of 1927, Heisenberg [1] introduced the Uncertainty Principle (HUP) in Quantum Theory by discussing measurement processes in which the gravitational interaction between particles was completely neglected. The argument (for example, find the position of an electron by means of photons) is well known under the name of *Heisenberg microscope argument* [2]. Of course, this procedure was fully justified by the huge weakness of gravity, when compared with the other fundamental interactions. Then things evolved and, in a few years, the HUP became a theorem in the framework of a fully developed Quantum Mechanics [3].

However, if we want to address fundamental questions in Nature, it seems clear that also gravity should be considered when we discuss basic principles and elementary measurement processes. This, in fact, is what historically happened, from the very early attempts in generalizing HUP [4], to the more recent proposals like those of string theory, loop quantum gravity, deformed special relativity, non-commutative geometry, and studies of black hole physics [5, 6, 7, 8, 9, 10, 11, 12].

A revised version of the classical Heisenberg argument has been presented in [8] and can be described as follows. The size δx of the smallest detail of an object, theoretically detectable with a beam of photons of energy E, is roughly given by (if we assume the dispersion relation

[1] blasone@sa.infn.it
[2] lambiase@sa.infn.it
[3] gluciano@sa.infn.it
[4] lpetruzziello@na.infn.it
[5] fabio@phys.ntu.edu.tw

Content from this work may be used under the terms of the Creative Commons Attribution 3.0 licence. Any further distribution of this work must maintain attribution to the author(s) and the title of the work, journal citation and DOI.
Published under licence by IOP Publishing Ltd

IOP Conf. Series: Journal of Physics: Conf. Series **1275** (2019) 012024 doi:10.1088/1742-6596/1275/1/012024

$E = pc$)

$$\delta x \simeq \frac{\hbar c}{2\,E}\,, \tag{1}$$

so that increasingly large energies are required to explore decreasingly small details. As remarked above, in its original formulation, Heisenberg's gedanken experiment ignores gravity. Nevertheless, gedanken experiments taking into account the possible formation, in high energy scatterings, of micro black holes with a gravitational radius $R_S = R_S(E)$ roughly proportional to the (centre-of-mass) scattering energy E (see Ref. [8]), suggest that the usual uncertainty relation should be modified as

$$\delta x \;\simeq\; \frac{\hbar c}{2\,E} \;+\; \beta\, R_S(E)\;, \tag{2}$$

where β is a dimensionless parameter. Recalling that $R_S \simeq 2\,G_N\,E/c^4 = 2\,\ell_p^2\,E/\hbar c$, we can write

$$\delta x \;\simeq\; \frac{\hbar c}{2\,E} \;+\; 2\beta\,\ell_p^2\,\frac{E}{\hbar c}, \tag{3}$$

where the Planck length is defined as $\ell_p^2 = G\hbar/c^3$, the Planck energy as $E_p \ell_p = \hbar c/2$, the Planck mass $m_p = E_p/c^2$. This kind of modification of the uncertainty principle was also proposed in Ref. [9].

The dimensionless deforming parameter β is not (in principle) fixed by the theory, although it is generally assumed to be of the order of unity. This happens, in particular, in some models of string theory (see again for instance Ref. [5]), and has been confirmed by an explicit calculation in Ref. [13]. However, many studies have appeared in literature, with the aim to set bounds on β (see, for instance, Refs. [14]).

The relation (3) can be recast in the form of an uncertainty relation, namely a deformation of the standard HUP, usually referred to as Generalized Uncertainty Principle (GUP),

$$\Delta x\, \Delta p \geq \frac{\hbar}{2}\left[1 + \beta\left(\frac{\Delta p}{m_p c}\right)^2\right]\,. \tag{4}$$

For mirror-symmetric states (with $\langle \hat{p} \rangle = 0$), the inequality (4) is equivalent to the commutator

$$[\hat{x}, \hat{p}] = i\hbar\left[1 + \beta\left(\frac{\hat{p}}{m_p c}\right)^2\right]\,, \tag{5}$$

since $\Delta x\, \Delta p \geq (1/2)\,|\langle[\hat{x},\hat{p}]\rangle|$. Vice-versa, the commutator (5) implies the inequality (4) for any state. The GUP is widely studied in the context of quantum mechanics [15], quantum field theory [16], thermal effects in QFT [17, 18], and for various deformations of the quantization rules [19].

Casimir effect as well has a long history. It is a direct manifestation of the non-trivial nature of quantum vacuum [20], which occurs whenever a quantum field is bounded in a finite region of space; such a confinement gives rise to a net attractive force between the confining plates, whose intensity has been successfully measured [21]. Since the foundational paper [22], the Casimir effect has been largely investigated in both flat [23] and curved [24] background, and in particular in quadratic theories of gravity [25], which have been studied also in other frameworks [26]. Further interesting applications have been addressed in the context of Lorentz symmetry breaking [27] and flavor mixing of fields [28].

Among the many applications of GUP, some papers have been recently devoted to the study of Casimir effect in the framework of modified uncertainty relations and/or minimal length.

IOP Conf. Series: Journal of Physics: Conf. Series **1275** (2019) 012024 doi:10.1088/1742-6596/1275/1/012024

In particular, in Refs. [29, 30] the corrections to the standard Casimir effect are computed for different GUP scenarios. Such calculations are performed within the framework of QFT, using the formalism of maximally localized states in Hilbert space representation proposed in particular in Refs. [7, 31], which implements various types of deformed commutators.

In the case of GUP involving βp^2 term only, as in Eq. (4), the authors of Refs. [29, 30] find corrections to the energy density of the form:

$$\Delta\mathcal{E} = -\frac{\pi^{2-}hc}{720d^3}\left(1+\widetilde{\beta}\frac{2\pi^{2-}\hbar^{2}}{3d^2}\right), \tag{6}$$

where $\widetilde{\beta} = \beta c^2/E_p^2$.

On the other hand, recently an interesting derivation of Casimir effect from HUP has been given by Giné [32] by means of heuristic arguments only. Motivated by the utility of heuristic procedures, which help to develop physical intuition, in this paper we consider similar derivation of the Casimir effect from GUP. We do this for the case of βp^2 model and in one (space) dimension.

The work is organized as follows: in Section 2 we briefly review the treatment of Ref. [32]. The heuristic derivation of Casimir effect from GUP is given in Section 3. Conclusions are given in Section 4.

2. Casimir effect and Uncertainty Principle

In Ref. [32], an heuristic argument is proposed for deriving the Casimir effect from the HUP. Such a derivation motivated us to look for similar arguments in the case of GUP.

Here, we briefly review the discussion of Ref. [32]. The Casimir effect is there derived from the idea that the contribution to the energy density at a given point of a plate is modified by the presence of the other plate.

The author considers virtual photons produced by vacuum fluctuations arriving at a point P_0 of the plate. In principle, the uncertainty Δx about the distance between the photon production point and P_0 can have any value within a sphere of (large) radius R enclosing the plates (eventually, such a radius will be sent to infinity).

The energy density will originate from the contributions of photons from a volume $A\Delta x$, where A is the area of the plate of side L.

Now, if there is only one plate, the relevant volume will be the one of entire sphere $V_T = 4/3\pi R^3$ (we take here $R \gg L$). The situation is different when another identical parallel plate is present at distance d from the first one: indeed, in such a case the photons originating from behind the second plate cannot reach the point P_0 and, as a result, a reduced volume $V_T - V_c$ will be associated to the energy density, where V_c is the volume shielded by the second plate (see Ref. [32] for details).

In the case of infinite plates, or even better when $L/(2d) \to \infty$, one obtains

$$A\Delta x = V_T - V_c \simeq \frac{2}{3}\pi R^3 . \tag{7}$$

In the above setting, no length scale is present, so Δx goes to infinity as R increases. At this point, the author of Ref. [32] introduces a radius r_e, representing the effective distance beyond which photons have a very small probability to reach the plate.

Therefore, the above equation becomes

$$A\Delta x = V_T - V_c \simeq \frac{2}{3}\pi r_e^3 . \tag{8}$$

IOP Conf. Series: Journal of Physics: Conf. Series **1275** (2019) 012024 doi:10.1088/1742-6596/1275/1/012024

The uncertainty principle in the form $\Delta x \Delta E \simeq \hbar c/2$, with $E = pc$, is then used to obtain

$$\Delta E = \frac{3\hbar c A}{4\pi r_e^3}. \tag{9}$$

Finally, the above formula is identified with the exact expression for the Casimir energy obtained in QFT:

$$|\Delta E| = \frac{\pi^2 \hbar c A}{720 d^3}, \tag{10}$$

which allows to fix $r_e^3 = 540 d^3/\pi^3$, i.e. $r_e \simeq 2,6\, d$.

3. Modified Casimir effect from GUP

The procedure described in the previous Section leading to the Casimir effect from the HUP, appears to be interesting although rather obscure in some points. In particular, in order to clarify the origin of the effective radius r_e above introduced, and to derive the Casimir effect in the GUP framework, we consider here the one-dimensional case.

Starting from the expression of the GUP

$$\Delta E \Delta x \simeq \frac{\hbar c}{2}\left[1 + \beta\left(\frac{\Delta E}{E_p}\right)^2\right], \tag{11}$$

we can certainly write $\Delta x = \alpha d$ (being d the only physical scale of the system), where α is a parameter which will be fixed below.

Eq. (11) can then be rewritten as

$$\Delta E = \frac{E_p^2 \alpha d}{\beta \hbar c}\left(1 - \sqrt{1 - \beta\left(\frac{\hbar c}{E_p \alpha d}\right)^2}\right). \tag{12}$$

This can be expanded in $\beta\left(\frac{\hbar c}{E_p \alpha d}\right)^2 \ll 1$, thus obtaining, up to the first order,

$$\Delta E = \frac{\hbar c}{2\alpha d}\left[1 + \beta\left(\frac{\hbar c}{2E_p \alpha d}\right)^2\right]. \tag{13}$$

Now we fix the parameter α by taking $\beta \to 0$ in the above equation and comparing the result with the one-dimensional standard QFT expression for the Casimir energy [32]:

$$\Delta E = -\frac{\pi \hbar c}{12 d}, \tag{14}$$

from which we get $\alpha = -6/\pi$. Inserting this back into Eq. (13), we finally obtain

$$\Delta E = -\frac{\pi \hbar c}{12 d}\left[1 + \beta\left(\frac{\pi \hbar c}{12 E_p d}\right)^2\right]. \tag{15}$$

This is to be compared with the expression (6) of the Casimir energy computed with GUP in QFT. In spite of our minimal setting, we can see that the behavior of GUP correction is approximately recovered[6].

[6] Strictly speaking, in order to compare our one-dimensional treatment with the three-dimensional one of Ref. [29, 30], we should divide Eq. (15) by d^2. This is the most direct procedure that can be adopted; in the case of one-dimensional confinement of the photon field, indeed, it is impossible to refer to an energy density per unit area since the binding plates reduce to mere points. In spite of these technicalities, however, we stress that the GUP correction to the Casimir energy derived via a heuristic approach exhibits the same dependence on d as the more rigorous field theoretical expression in Eq. (6).

IOP Conf. Series: Journal of Physics: Conf. Series **1275** (2019) 012024 doi:10.1088/1742-6596/1275/1/012024

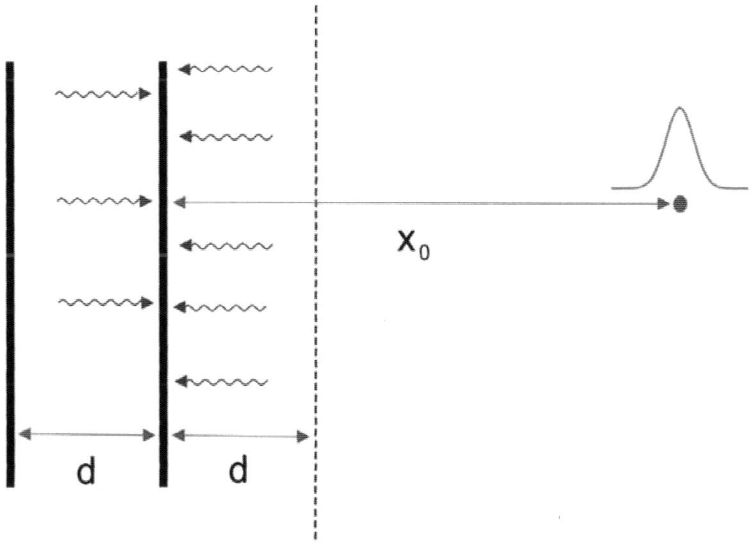

Figure 1. Setup for the calculation of the Casimir effect: two infinite parallel plates (bold lines), at distance d. Photons act on the right plate, coming from right and left regions. See text for detailed explanation.

4. Conclusions and perspectives

Let us now try to better understand the meaning of the effective radius r_e introduced in Section 2. It is clear from the Heisenberg uncertainty relation that large fluctuations in energy live for very short times and thus very energetic photons of energy ΔE can travel only very short distances, of order $\hbar c/(\Delta E)$. Therefore, energetic photons originating far away from a plate will contribute very little to the energy around the plate, Eqs. (9), (14) and (15).

With reference to Fig. 1, let us focus on the right plate. It is evident that photons originating between the plates, i.e. within a distance d from one plate, will not contribute because of the symmetric action due to the photons originating in the strip of width d on the other side of the plate. On the other hand, photons coming from the right region, from a distance greater than d, will not be compensated by photons coming from the left region, whose action is screened by the (infinite) left plate. Therefore only photons coming from a distance greater than d, from the right region, can be responsible for a resultant non-zero action on the right plate. Of course, this argument applies symmetrically to the left plate, and here we have a qualitative explanation of the origin of Casimir force.

Now, consider a point at a distance $x_0 > d$ from the plate. Photons are created by quantum fluctuations in a (small) region around that point. The region cannot be smaller than a Compton length of the electron, $\lambda_C = \hbar/m_e c$, otherwise the energy amplitude of the fluctuation would be larger than the threshold ($m_e c^2$) for pairs creation of electrons. Photons can reach the plate (and therefore contribute to the Casimir force) only if their energy is in the range $0 < E < E_0$, where $E_0 = \hbar c/x_0$. Photons of energies E greater than E_0 can only travel paths shorter than x_0, before recombining ($x = \hbar c/E < \hbar c/E_0 = x_0$), thus they do not arrive to touch the plate. Photons at x_0 originate from fluctuations of energy E with a probability which can be assumed to behave like a Boltzmann factor $e^{-E/m_e c^2}$.[7] Therefore the total linear energy density (energy

[7] Note that such an assumption is not fundamental for the validity of the result (15). See the discussion below for more details.

IOP Conf. Series: Journal of Physics: Conf. Series **1275** (2019) 012024 doi:10.1088/1742-6596/1275/1/012024

per unit length) arriving on the plate from a small region around the point x_0 will be

$$\Delta\varepsilon\,(E_0) = \int_0^{E_0} \left(\frac{E}{m_e c^2}\right) e^{-E/m_e c^2} \frac{dE}{\lambda_C} \;=\; \int_0^{E_0} \left(\frac{E}{\hbar c}\right) e^{-E/m_e c^2} dE\,, \tag{16}$$

where, since we deal with the electromagnetic field, we introduced the natural threshold of the electron mass/energy $m_e c^2$. In terms of x_0 the above integral reads

$$\Delta\varepsilon(x_0) = \int_{x_0}^{+\infty} \frac{\hbar c}{x^3} e^{-\hbar/m_e c x} dx\,. \tag{17}$$

Finally, summing over all the points x_0 such that $d < x_0 < +\infty$, we get the whole contribution to the energy from all the photons able to arrive to the plate as

$$\Delta E(d) = \int_d^{+\infty} \Delta\varepsilon(x_0)\, dx_0\,. \tag{18}$$

Now, since for x large enough, $e^{-\hbar/m_e c x} \simeq 1$, then Eq. (17) yields

$$\Delta\varepsilon(x_0) \simeq \frac{\hbar c}{2x_0^2}\,, \tag{19}$$

and hence the total energy reads

$$\Delta E(d) \simeq \frac{\hbar c}{2d}\,. \tag{20}$$

We find therefore again for the energy ΔE the behavior expressed by Eq. (14).

Distributions like those given in Eqs. (18), (19) allow us to easily define an "effective" radius, or distance from the plate, below which the vast majority of photons are responsible for the large part of the Casimir force. In fact, we can define $r_e > d$ as the distance up to which a fraction γ $(0 < \gamma < 1)$ of the total energy is responsible for the effect. This means

$$\int_d^{r_e} \frac{\hbar c}{2x^2} dx = \gamma \Delta E(d) \tag{21}$$

which yields

$$r_e = \frac{d}{1-\gamma}\,. \tag{22}$$

A final important note is in order at this point. The above equations maintain essentially their validity also for different choices of the distribution function

$$f(E) = e^{-E/m_e c^2}\,, \tag{23}$$

provided $0 < f(E) < 1$. In particular, the approximation is as better as the function $f(E)$ is closer to 1. This is important, since a perusal of current literature has revealed that the search for distributions of the energy fluctuations of the quantum vacuum is not at all a closed chapter, but on the contrary, a hot and magmatic research topic (see for example the important papers of Larry Ford [33] and references therein).

References

[1] Heisenberg W 1927 *Z. Phys.* **43** 172

[2] Heisenberg W 1930 *Physikalische Prinzipien der Quantentheorie* (in German) Leipzig: Hirzel; 1930 *The Physical Principles of Quantum Theory* (English translation) Chicago: University of Chicago Press

[3] Robertson H P 1929 *Phys. Rev.* **34** 163; E. Schrödinger 1930 *Zum Heisenbergschen Unschrfeprinzip*, Sitzungsberichte der Preussischen Akademie der Wissenschaften, Physikalisch-mathematische Klasse **14** 296

[4] Bronstein M P 1936 *Phys. Zeitschr. der Sowjetunion* **9** 140; Snyder H S 1947 *Phys. Rev.* **71** 38; Yang C N 1947 *Phys. Rev.* **72** 874; Mead C A 1964 *Phys. Rev.* **135** B849; Karolyhazy F 1966 *Nuovo Cim. A* **42** 390

[5] Amati D, Ciafaloni M and Veneziano G 1987 *Phys. Lett. B* **197** 81; Gross D J and Mende P F 1987 *Phys. Lett. B* **197** 129

[6] Maggiore M 1993 *Phys. Lett. B* **319** 83

[7] Kempf A, Mangano G and Mann R B 1995 *Phys. Rev. D* **52** 1108

[8] Scardigli F 1999 *Phys. Lett. B* **452** 39

[9] Adler R J and Santiago D I 1999 *Mod. Phys. Lett. A* **14** 1371; Capozziello S, Lambiase G and Scarpetta G 2000 *Int. J. Theor. Phys.* **39** 15; Kanazawa T, Lambiase G, Vilasi G and Yoshioka A 2019 *Eur. Phys. J. C* **79** 95

[10] Scardigli F and Casadio R 2003 *Class. Quant. Grav.* **20** 3915

[11] Casadio R, Garattini R and Scardigli F 2009 *Phys. Lett. B* **679** 156

[12] Casadio R and Scardigli F 2014 *Eur. Phys. J. C* **74** 2685

[13] Scardigli F, Lambiase G and Vagenas E 2017 *Phys. Lett. B* **767** 242

[14] Brau F 1999 *J. Phys. A* **32** 7691; Pedram P, Nozari K and Taheri S H 2011 *JHEP* **1103** 093; Scardigli F and Casadio R 2015 *Eur. Phys. J. C* **75** 425; Khodadi M, Nozari K, Dey S, Bhat A and Faizal M 2016 *Sci. Rep.* **8** 1659

[15] Naimi Y 2016 *Grav. Cosmol.* **22** 382; Bernardo R C S and Esguerra J P H 2018 *Annals Phys.* **391** 293

[16] Husain V, Kothawala D and Seahra S S 2013 *Phys. Rev. D* **87** 025014

[17] Scardigli F 1995 *Nuovo Cim. B* **110** 1029

[18] Scardigli F, Blasone M, Luciano G and Casadio R 2018 *Eur. Phys. J. C* **78** 728

[19] Jizba P, Kleinert H and Scardigli F 2010 *Phys. Rev. D* **81** 084030

[20] Blasone M and Vitiello G 1995 *Annals Phys.* **244** 283; Blasone M, Henning P A and Vitiello G 1999 *Phys. Lett.* B **451** 140; Blasone M, Lambiase G and Luciano G G 2017 *Phys. Rev. D* **96** 025023; Blasone M, Jizba P and Smaldone L 2017 *Annals Phys.* **383** 207; Blasone M, Lambiase G and Luciano G G 2018 *J. Phys. Conf. Ser.* **956** 01202; Blasone M, Lambiase G, Luciano G G and Petruzziello L 2018 *Phys. Rev. D* **97** 105008; Blasone M, Jizba P and Luciano G G 2018 *Annals Phys.* **397** 213; Blasone M, Lambiase G, Luciano G G and Petruzziello 2018 *EPL* **124** 51001

[21] Mohideen U and Roy A 1998 *Phys. Rev. Lett.* **81** 4549; Bressi G, Carugno G, Onofrio R and Ruoso G 2002 *Phys. Rev. Lett.* **88** 041804

[22] Casimir H B G 1948 *Indag. Math.* **10** 261; 1948 *Kon. Ned. Akad. Wetensch. Proc.* **51** 793; 1987 *Front. Phys.* **65** 342; 1997 *Kon. Ned. Akad. Wetensch. Proc.* **100N3-4** 61

[23] Lambiase G, Nesterenko V V and Bordag M 1999 *J. Math. Phys.* **40**; Nesterenko V V, Lambiase G and Scarpetta G 2001 *J. Math. Phys.* **42** 1974; Nesterenko V V, Lambiase G and Scarpetta G 2002 *Annals Phys.* **298** 403

[24] Sorge F 2005 *Class. Quant. Grav.* **22** 5109; Fulling S A, Milton K A, Parashar P, Romeo A, Shajesh K V and Wagner J 2007 *Phys. Rev. D* **76** 025004; Lambiase G, Stabile A and Stabile An 2017 *Phys. Rev. D* **95** 084019

[25] Buoninfante L, Lambiase G, Petruzziello L and Stabile A 2019 *Eur. Phys. J. C* **79** 41

[26] Buoninfante L, Harmsen G, Maheshwari S and Mazumdar A 2018 *Phys. Rev. D* **98** 084009; Buoninfante L, Cornell A S, Harmsen G, Koshelev A S, Lambiase G, Marto J and Mazumdar A 2018 *Phys. Rev. D* **98** 084041; Buoninfante L, Koshelev A S, Lambiase G and Mazumdar A 2018 *JCAP* **1809** 034

[27] Blasone M, Lambiase G, Petruzziello L and Stabile An 2018 *Eur. Phys. J. C* **78** 976

[28] Blasone M, Luciano G G, Petruzziello L and Smaldone L 2018 *Phys. Lett. B* **786** 278

[29] Nouicer K 2005 *J. Phys. A* **38** 10027

[30] Frassino A M and Panella O 2012 *Phys. Rev. D* **85** 045030

[31] Detournay S, Gabriel C and Spindel P 2002 *Phys. Rev. D* **66** 125004

[32] Giné J 2018 *Mod. Phys. Lett. A* **33** 1850140

[33] Fewster C J, Ford L H and Roman T A 2012 *Phys. Rev. D* **85** 125038; Fewster C J, Ford L H and Roman T A 2014 *Springer Proc. Phys.* **157** 489

Covariant bandlimitation from Generalized Uncertainty Principles

J Pye

Department of Applied Mathematics, University of Waterloo, Waterloo, Ontario, N2L 3G1, Canada
Institute for Quantum Computing, University of Waterloo, Waterloo, Ontario, N2L 3G1, Canada
Perimeter Institute for Theoretical Physics, 31 Caroline St. N., Waterloo, Ontario, N2L 2Y5, Canada

E-mail: j2pye@uwaterloo.ca

Abstract. It is widely believed that combining the uncertainty principle with gravity will lead to an effective minimum length scale. A particular challenge is to specify this scale in a coordinate-independent manner so that covariance is not broken. Here we examine a class of Lorentz-covariant generalizations of the uncertainty principle which aim to provide an effective low-energy model for a Lorentz-invariant minimum length. We show how this modification leads to a covariant bandlimitation of quantum field theory. However, we argue that this does not yield an adequate regulator for many quantities of interest, e.g., the entanglement entropy between spatial regions. The possibility remains open that it could aid in regulating interactions.

1. Introduction

In 1936, Bronstein [1] suggested a fundamental limit to the precision of gravitational field measurements. Ensuing thought experiments have indicated that in the presence of gravity, there should generally be a limitation to the precision of distance measurements in the form of a lower bound on the uncertainty, given by the Planck length: $\Delta X \geq \ell_P := \sqrt{\hbar G/c^3} \sim 10^{-35}m$ [2,3]. It is widely believed this indicates that the classical conception of spacetime breaks down at the Planck scale. As Bronstein [1,4] writes: "Without a deep revision of classical notions it seems hardly possible to extend the quantum theory of gravity also to this domain."

Intuitively, the presence of a minimum length suggests that spacetime may be fundamentally discrete, rather than an infinitely-divisible continuum. Naively, if one were to imagine this in analogy with a crystal lattice, one would seem to be forced to abandon the continuous spacetime symmetries fundamental to, for example, the structure of the Standard Model. The spirit of the work presented in this paper is to examine whether one must abandon continuous symmetries by adopting a minimum length. As it happens, this forfeiture is not generally required.

This endeavor is inspired by a model for a minimum length scale called the Generalized Uncertainty Principle (GUP) [5,6], which directly introduces Planck-scale corrections to the ordinary quantum-mechanical uncertainty principle. A distinguishing feature of this model is that it introduces a minimum length while preserving Euclidean symmetries (translation and rotation invariance). Given this fact, it is natural to ask whether an extension of the GUP could be made to also include Lorentz symmetry.

Content from this work may be used under the terms of the Creative Commons Attribution 3.0 licence. Any further distribution of this work must maintain attribution to the author(s) and the title of the work, journal citation and DOI.
Published under licence by IOP Publishing Ltd

IOP Conf. Series: Journal of Physics: Conf. Series **1275** (2019) 012025 doi:10.1088/1742-6596/1275/1/012025

Without a full working theory of quantum gravity, it is difficult to obtain a clear picture of the nature of this minimum length. The goal here is to begin in the well-established territory of quantum field theory (QFT) and deform it toward the Planck scale while retaining Lorentz-invariance. Our particular focus will be on developing Lorentz-covariant versions of the GUP and apply it to QFT. The purpose is to clarify the role of the minimum length as a covariant regulator for QFT [7, 8]. Of particular interest here is the entanglement entropy of the field, with the ultimate aim of studying black hole entropy.

In this paper, we will begin in Section 2 by reviewing the original version of the GUP. In Section 3, we will show how the GUP relates to bandlimitation and sampling theory, which clearly demonstrates the possibility of discrete representations while retaining translation invariance. In Section 4, we introduce Lorentz-covariant extensions of the GUP. We then proceed to show in Section 5 that one can arrive at a covariant analogue of bandlimitation, but it does not yield discrete spacetime representations as in the Euclidean case. Lorentz-covariant bandlimitation has been considered before [9–11], in direct analogy with the Euclidean case. The contribution we provide here is to show that one is lead to the same conclusion beginning with a covariant adaptation of the GUP. In particular, we show that covariant bandlimitation is the only non-trivial possibility arising from the Lorentz-covariant GUP we study here. We then provide further discussion of these results in Section 6.

Throughout we will work in Planck units, for which $\hbar = c = G = 1$.

2. Generalized Uncertainty Principles

The idea behind the Generalized Uncertainty Principle (GUP) is to restrict the set of allowable states by directly modifying the ordinary uncertainty principle to impose a minimum position uncertainty, $\Delta X \geq \ell_P$. (For purposes of simplicity, here we will discuss the one-dimensional version of the GUP, although it can be extended to higher dimensions [5,6].) The allowable states in ordinary quantum mechanics satisfy $\Delta X \Delta P \geq \frac{1}{2}$. A Planck-scale modification should become significant when the momentum uncertainty is large, since this will cause large uncertainties in the geometry through the gravitational field equations, which will in turn generate an increase in position uncertainty. The GUP achieves this by modifying the uncertainty principle to contain gravitational corrections, for example, to lowest order of the form [5, 6]:

$$\Delta X \Delta P \geq \tfrac{1}{2}(1 + \ell_P^2 \Delta P^2). \tag{1}$$

The physical states must satisfy this modified uncertainty principle. It is easy to show that the above inequality implies $\Delta X \geq \ell_P$. There are many other modifications which yield a finite minimum uncertainty, but generically each exhibits the same qualitative behavior [12].

One can arrive at a GUP of the above form through a modification of the commutation relation between the operators X and P [5, 6]:

$$[X, P] = i(1 + \ell_P^2 P^2). \tag{2}$$

This implies an uncertainty principle of the above form (up to a term proportional to $\langle P\rangle^2$, although this does not significantly change the above conclusion). Hence, if we consider states in a representation of this modified Heisenberg algebra, they will automatically exhibit the feature of a minimum uncertainty in position. For example, one such representation is:

$$(X\tilde{\psi})(p) = i(1 + \ell_P^2 p^2)\frac{d}{dp}\tilde{\psi}(p), \qquad (P\tilde{\psi})(p) = p\tilde{\psi}(p), \tag{3}$$

with inner product, $\langle\tilde{\phi}|\tilde{\psi}\rangle = \int \frac{dp}{2\pi}(1+\ell_P^2 p^2)^{-1}\tilde{\phi}^*(p)\tilde{\psi}(p)$. The physical states are those in a dense subset of this Hilbert space (typically the common domain of $[X, P]$ and various powers of X, P, and their products).

There have been many papers studying various implications of GUPs (e.g., see [12–18]). Our interest in this paper is to study Lorentz-covariant modifications of QFT as one approaches the Planck scale. In order to apply the GUP to QFT, one must first identify how the ordinary uncertainty principle of non-relativistic quantum mechanics manifests itself in QFT. After making this identification and performing the appropriate generalization to the GUP, we will show how such a modification can be interpreted in terms of bandlimitation and sampling theory.

3. Bandlimitation and sampling theory

The classical space of field configurations (on flat spacetime) is taken to be a representation space of the Poincaré group. From the Poincaré generators $L_{\mu\nu}$ and P^λ, one can build momentum space representations of the fields, $\tilde{\phi}(p) \equiv (p|\tilde{\phi})$, with $P^\lambda|p) = p^\lambda|p)$. (Note that we use round brackets to emphasize the distinction from quantum mechanical wavefunctions.) The position space or spacetime representation of the fields is then naturally obtained via a Fourier transform. The assumption that position and momentum space are Fourier-related presumes an underlying Heisenberg algebra structure, $[X_\mu, P^\nu] = i\delta_\mu{}^\nu$, with position representation $\phi(x) \equiv (x|\phi)$, where $X^\mu|x) = x^\mu|x)$. This is formally similar to the typical account in quantum mechanics. Naturally, this is the step at which one can introduce a modification of the Heisenberg algebra. That is, one can continue to build momentum space representations for the Poincaré group, but a modified Heisenberg algebra may alter the relationship between position and momentum space.

Notice that in the position representation, x is simply a label for an element in the joint spectrum of the X^μ operators. Hence, field configurations in the spacetime representation reside on the joint spectra of these operators. A modification of the Heisenberg algebra may change the structure of these spectra. For example, if the spectra become discrete after a modification, then these field configurations could be thought of as living on a lattice.

Now we will consider more explicitly such a modification using the GUP discussed in Section 2. For simplicity, we will continue with the one-dimensional case at a fixed time (however, the Lorentz-covariant GUP in Section 4 will apply to arbitrary spacetime dimensions). In the previous section, we presented the momentum space representation of the GUP corresponding to $[X, P] = i(1 + \ell_P^2 P^2)$. In order to find the position space representation, we must find the eigenvalues and eigenvectors of X. However, the finite minimum uncertainty, $\Delta X \geq \ell_P$, implies that X has no eigenvectors in the domain where the GUP is represented, since the uncertainty for an eigenvector would vanish (or in the case of a continuous spectrum could be made arbitrarily close to zero).

This situation can be made more clear by performing a momentum space diffeomorphism. Using the functional calculus of P, we define an operator, $K := \frac{1}{\ell_P}\arctan(\ell_P P)$, which can be easily shown satisfies the canonical commutation relation with X: $[X, K] = i$. It may appear that this simply reverts the GUP to the ordinary uncertainty principle, with x-space and k-space being Fourier-related. However, because arctan has a finite range, the fields in k-space have finite support (such fields are called *bandlimited*). The k-space representation consists of fields on the interval $[-\frac{\pi}{2\ell_P}, \frac{\pi}{2\ell_P}]$ obeying Dirichlet boundary conditions (and with the usual flat measure $dk/2\pi$). Therefore, by focusing solely on the x- and k-space representations, we can view the deformation as equivalent to simply restricting to a subset of k-space. In other words, we are restricting the spectrum of the K operator (ordinarily represented in position space as the derivative operator $-id/dx$). Note that one can also perform this diffeomorphism when considering the GUP for wavefunctions in quantum mechanics, and think of the resulting system as a particle in a box in momentum space.

Do we still have finite minimum uncertainty in position? Does the position operator continue to lack eigenvectors? Of course this should be the case, since we have not changed the abstract algebraic properties enforcing these features. Although the operators X and K are canonically related, the bandlimitation implies a maximum uncertainty in K. Therefore, the uncertainty

IOP Conf. Series: Journal of Physics: Conf. Series **1275** (2019) 012025 doi:10.1088/1742-6596/1275/1/012025

in X continues to exhibit a finite minimum since $\Delta X \geq 1/2\Delta K \geq 1/2\Delta K_{\text{max}}$. The situation involving the eigenvectors is more subtle, for it turns out that X is symmetric on the physical domain, but not essentially self-adjoint [5,6]. Although the formal eigenvectors of X in k-space are e^{-ikx}, these do not obey the appropriate Dirichlet boundary conditions. In order to obtain a position space representation, one needs to extend the domain of physical states to a domain on which X is self-adjoint. However, in this case the choice of extension is not unique; indeed, there is a one-parameter family of self-adjoint extensions. These correspond to the periodic boundary conditions up to a phase $e^{-2\pi i\alpha}$, where $\alpha \in [0,1)$ parametrizes the family of extensions. For each extension α, one obtains a discrete spectrum $\{x_n^{(\alpha)} = 2\ell_P(n+\alpha)\}_{n\in\mathbb{Z}}$ with corresponding eigenvectors. Because the physical domain is contained within the domain of each of these extensions, any one of the set of eigenvectors can be used as a basis in which one can represent the physical fields: $|\phi) = \sum_{n\in\mathbb{Z}} |x_n^{(\alpha)})(x_n^{(\alpha)}|\phi)$. Therefore, these fields admit a family of discrete representations in position space. Furthermore, although we have discrete representations, we also retain full translation invariance in space because none of the lattices are preferred.

This situation can be illustrated in a much simpler scenario. If we introduce an ultraviolet cutoff for the space of L_2 functions on a circle, then we obtain a finite-dimensional function space (finitely many Fourier modes). Suppose we then pick a set of sample points on the circle, equal in number to the number of Fourier modes. Generically, we can use the function values at these points to completely specify a function in the space. We can then view this as a representation of the functions on this lattice of sample points, although we have not changed the fact that the underlying space remains continuous.

The discreteness of the position representations of the bandlimited fields is analogous to this example, except extended to the real line and an infinite-dimensional function space. Notice that the sample points $x_n^{(\alpha)} = 2\ell_P(n+\alpha)$ cover the real line exactly once as one runs through both $n \in \mathbb{Z}$ and $\alpha \in [0,1)$. Hence, we can consider a field value at an arbitrary $x \in \mathbb{R}$ as $\phi(x) \equiv (x|\phi)$, since $|x) = |x_n^{(\alpha)})$ for some n and α. Therefore, we can consider position space to be the entire real line, with the fields admitting discrete representations on these sampling lattices. The field value at any other point on the real line can be obtained via the following reconstruction formula:

$$\phi(x) = (x|\phi) = \sum_{n\in\mathbb{Z}} (x|x_n^{(\alpha)})(x_n^{(\alpha)}|\phi) = \sum_{n\in\mathbb{Z}} \frac{\sin[\pi(x - x_n^{(\alpha)})/2\ell_P]}{\pi(x - x_n^{(\alpha)})/2\ell_P}\phi(x_n^{(\alpha)}), \tag{4}$$

which is the basic result of Shannon sampling theory [19–21]. The quantization of these bandlimited fields is relatively straightforward [22–25]. Also, many of these results have been extended to higher-dimensional spaces [26, 27] as well as curved Riemannian manifolds [9, 28].

The main message of sampling theory is that one can have discrete representations while retaining continuous Euclidean spatial symmetries. Armed with this encouragement, we will now examine whether this can be extended to Lorentzian spacetime symmetries.

4. Lorentz-covariant Generalized Uncertainty Principles

The strategy of the GUP was to introduce a minimum length scale by modifying the commutation relation between X and P to give a finite minimum uncertainty in position. Now we will attempt to do this Lorentz-covariantly. By *Lorentz-covariant*, we mean that the coordinates, and not necessarily the momenta, transform appropriately under an infinitesimal Lorentz transformation. This is because we are primarily concerned with symmetries of spacetime, rather than the space described by the momenta. Although there are known problems with deformations of momentum space (e.g., the so-called "soccer-ball problem" [3, 29]), we will not address these issues here.

We will continue to focus primarily on the kinematical structure of the fields (although dynamics will be briefly discussed in Section 6). Indeed, if we are interested in a regulator

IOP Conf. Series: Journal of Physics: Conf. Series **1275** (2019) 012025 doi:10.1088/1742-6596/1275/1/012025

for the entanglement entropy, it should arise from the kinematics, as entanglement is typically thought of as between regions of space at a fixed time.

We will consider a somewhat general set of deformations of the Heisenberg algebra:

$$[X_I, P^\mu] = i\theta_I{}^\mu(P), \qquad [X_I, X_J] = 0, \qquad [P^\mu, P^\nu] = 0. \tag{5}$$

where we assume $\theta_I{}^\mu$ is a smooth function of the P^μ operators, and as a matrix is pointwise-invertible (i.e., $\det\theta \neq 0$). (Note that $\theta_I{}^\mu$ should be a function of the dimensionless quantity $\ell_P P^\mu$, however, to keep the notation compact, we will omit an explicit reference to ℓ_P.) We take $\theta_I{}^\mu$ to be a function of the P^μ operators and not the X_I operators for both simplicity and in analogy with the one-dimensional GUP (where X-dependence yields an infrared, rather than an ultraviolet, cutoff). Both sets of indices traverse all spacetime dimensions, $I, \mu = 0, 1, 2, \ldots, n-1$. We label the X_I operators and P^μ operators differently since we only require the upper-case Latin indices to transform appropriately under Lorentz transformations.

The requirements of commuting coordinates and commuting momenta are made solely as simplifying assumptions in order to make conclusive statements. If this prevents the model from producing a suitable cutoff, then one could take this as an indication that these assumptions (such as the commutative geometry assumption) should be relaxed in order to obtain a cutoff in a model of this kind. Indeed, we are attempting to determine whether one can obtain a minimum length without resorting to non-commutative geometry.

Of course, we will also assume that the Jacobi identities are satisfied. This introduces the following restriction for θ: $[X_I, \theta_J{}^\mu(P)] = [X_J, \theta_I{}^\mu(P)]$. With this structure, it is simple to build a p-space representation for this algebra as:

$$(X_I\phi)(p) = i\theta_I{}^\mu(p)\frac{\partial}{\partial p^\mu}\phi(p), \qquad (P^\mu\phi)(p) = p^\mu\phi(p), \tag{6}$$

with inner product: $(\phi|\psi) = \int \frac{d^n p}{(2\pi)^n}(\det\theta(p))^{-1}\phi^*(p)\psi(p)$. Note that throughout we assume the summation convention. In this representation, the condition for the Jacobi identity becomes: $\theta_I{}^\nu\partial_\nu\theta_J{}^\mu = \theta_J{}^\nu\partial_\nu\theta_I{}^\mu$. This ensures that the coordinates commute and are symmetric operators. One can write the Lorentz generators in this representation as:

$$L_{IJ} := X_I \int^p dp'^\mu(\theta^{-1})_{\mu J}(p') - X_J \int^p dp'^\mu(\theta^{-1})_{\mu I}(p'), \tag{7}$$

where the upper-case Latin indices are lowered using the Minkowski metric, $\eta_{IJ} := \mathrm{diag}(+1, -1, -1, \ldots, -1)$. It is straightforward to show that these indeed satisfy the Lorentz algebra, and that the X_I operators transform appropriately as: $[X_K, L_{IJ}] = i\eta_{JK}X_I - i\eta_{IK}X_J$.

In the one-dimensional case, we made a simplification by enacting a coordinate change in momentum space to recover canonical commutation relations, but with a restriction of the spectrum of the new momentum operator. We can also establish such a diffeomorphism here for an arbitrary deformation $\theta_I{}^\mu(P)$ (with the already-mentioned assumptions). Notice that the operators $-iX_I$ provide a basis of vector fields on p-space. Because these vector fields commute with each other, they can be written as a coordinate basis after a suitable diffeomorphism (see, e.g., [30]). That is, there is some diffeomorphism $k^I(p)$ for which we can write $X_I\phi(k) = i(\partial/\partial k^I)\phi(k)$. The functions parametrizing the deformation of the Heisenberg algebra can then simply be associated with the (inverse) Jacobian of this diffeomorphism, $\theta_I{}^\mu(p) = (\partial k^I(p)/\partial p^\mu)^{-1}$. We can define multiplication operators in k-space as $K^I\phi(k) = k^I\phi(k)$ (or in p-space as $K^I\phi(p) = k^I(p)\phi(p)$), which satisfy: $[X_I, K^J] = i\delta_I{}^J$. Notice that with the K^I operators, the Lorentz generators take their familiar form: $L_{IJ} = X_I K_J - X_J K_I$.

5. Covariant bandlimitation

We have established that a deformation of the commutator between X_I and P^μ in the Heisenberg algebra can be reverted to the canonical commutation relations (provided the coordinate and momentum operators remain commuting among themselves). Therefore, in regards to the kinematical structure of the classical field configurations, we are only left with the possibility of modifying the global properties of momentum space. Can we restrict to a subset of k-space in order to achieve a kind of bandlimitation, as in the one-dimensional case?

Of course, we cannot make an arbitrary restriction. A faithful representation of the Lorentz group in k-space requires the hypersurfaces $k^2 = \text{const}$ to remain untouched. This leaves only the possibility of restricting k-space to a subset of these hypersurfaces, i.e., we can restrict the set of admissible mass shells. For example, we can constrain the quantity $|k^2| = |k_0^2 - \boldsymbol{k}^2| < 1/\ell_P^2$. This can be achieved, for example, by the following deformation:

$$\theta_I{}^J(p) = f(p^2)\delta_I{}^J + p_I p^J \frac{2f(p^2)f'(p^2)}{f(p^2) - 2p^2 f'(p^2)}, \tag{8}$$

where $p^2 := p_0^2 - \boldsymbol{p}^2$, with diffeomorphism $k^I = p^I/f(p^2)$ and inverse denoted $p^I = b(k^2)k^I$. To achieve the desired restriction of k^2, we can choose $b(k^2) = \sum_{n=0}^{\infty}(\ell_P^2 k^2)^{2n}$. Due to the finite radius of convergence of this series, fields in the domain of P^I are confined to the region $|k^2| < 1/\ell_P^2$. One can check that this indeed defines a diffeomorphism in this region, and by writing $\theta_I{}^J(p(k)) = b(k^2)\delta_I{}^J + 2k_I k^J b'(k^2)$, that θ satisfies the required properties.

In the literature, such a restriction of momentum space has been called "covariant bandlimitation" [9–11]. In these previous works, this was considered as a restriction of the spectrum of the covariant d'Alembertian operator, in analogy with the restriction of the spectrum of $-id/dx$ in the one-dimensional case. Here we have shown that one is also led to covariant bandlimitation by a class of deformations of the Heisenberg algebra.

Despite the analogy with the Euclidean case, covariant bandlimitation does not exhibit the discrete spacetime representations characteristic of Shannon sampling theory. The fact that this cannot occur is a consequence of a theorem of Landau [26], which states that the required density of sampling points is bounded below by the volume of the support of the space of functions in momentum space. If we require a faithful representation of the Lorentz group in this space, this volume must be non-compact. This immediately implies an infinite density of degrees of freedom. (It is possible to derive a type of reconstruction formula for covariantly bandlimited fields, although the features are qualitatively different than the Euclidean case [9, 10].)

6. Discussion and conclusion

Despite the infinite density of degrees of freedom, what are the implications of a covariant bandlimit? Does it provide a regulator for field theory? It is clear that any on-shell quantities, such as the Wightman function, commutator, and correlation functions, are unaffected by the covariant bandlimit (for masses smaller than the cutoff) since these mass shells remain unchanged. Hence, for a free field theory, it has limited impact as many physical quantities can be calculated on-shell. For example, the entanglement entropy in the ground state between spatial regions can be calculated from the commutator and correlation functions (see, e.g., [31–33]).

Of course, the notion of "on-shell" presumes a form for the dynamics, such as that associated with the Klein-Gordon action, $S[\phi] = \frac{1}{2}(\phi|K_I K^I \phi) - \frac{1}{2}m^2(\phi|\phi)$. However, one could also specify an action in terms of the additional set of momentum operators, P^μ. For example, a natural generalization could be $S[\phi] = (\phi|\sigma(P, p_0)\phi) - (\phi|\sigma(p_m, p_0)\phi)$, where σ is the Synge world function in p-space, and we define p_0 and p_m as points in p-space which satisfy $k^I(p_0) = 0$ and $k_I(p_m)k^I(p_m) = m^2$. Due to the coordinate-invariance of σ, if we choose the p-space metric to be the pullback of the Minkowski metric on k-space, that is, $g_{\mu\nu}(p) = \eta_{IJ}(\theta^{-1})_\mu{}^I(\theta^{-1})_\nu{}^J$, then this simply reduces to the Klein-Gordon action. One could alternatively introduce another metric on p-space for this purpose. For example, with the deformation used in the previous section (with $p^I = b(k^2)k^I$), one could use η_{IJ} in p-space to obtain the action: $S[\phi] = \frac{1}{2}(\phi|P_I P^I \phi) - \frac{1}{2}b(m^2)^2 m^2(\phi|\phi)$. This can be viewed as a replacement of the d'Alembertian operator $\Box \to \Box/(1 - \ell_P^4 \Box^2)^2$ along with a re-scaling of the mass. However, regardless of the choice, to retain covariance in x-space one is restricted to using functions of the d'Alembertian operator in the action. A locally analytic function whose power series has a finite radius of convergence will yield covariant bandlimitation [9]. Viewed in this way, the fact that covariant bandlimitation does not regulate the entanglement entropy is consistent with previous investigations in non-local field theories and modified dispersion relations [34, 35]. (Also, certain deformations considered here could be viewed as equivalent to non-local d'Alembertian operators arising from averaging over causal sets [36, 37].)

If we focus solely on the effect of covariant bandlimitation (and retain the usual Klein-Gordon action), then on-shell quantities are not modified, but the covariant bandlimit can be interpreted as cutting off allowable virtual masses. Therefore, this will modify the inhomogeneous Green's functions of the theory, such as the Feynman propagator [10, 11]. Covariant bandlimitation may then be of interest in interacting theories. For example, a simple interacting theory one could consider is that between the field and an accelerating detector, in which one can study the Unruh effect [38]. However, because the Unruh effect corresponds to a first-order response, there are no internal field lines which could be affected by the covariant bandlimit. Therefore, there is no modification to the usual leading-order thermal response of an accelerated detector. One could also consider interacting fields, although bandlimited functions do not form an algebra with pointwise multiplication in spacetime. Thus, to retain a consistent mathematical description of interacting bandlimited fields, one should alter the multiplication by modifying the vertex.

The conclusions regarding covariant bandlimitation also hold on other spacetimes. On a general pseudo-Riemannian manifold, covariant bandlimitation can be analogously defined in terms of cutting off the spectrum of a covariant differential operator, such as the Laplace-Beltrami operator, $\Box_g = |g|^{-1/2}\partial_\mu(|g|^{1/2}g^{\mu\nu}\partial_\nu\cdot)$ [9–11]. However, even in this case, the eigenvalues correspond to allowable masses in the Klein-Gordon equation, $\Box_g\phi = -m^2\phi$. Therefore, restricting the allowable masses has no effect on the homogeneous solutions to the Klein-Gordon equation, and thus cannot be used to regulate the entanglement entropy (for example, between the interior and exterior regions of a black hole). This last observation is also consistent with the investigations of [34, 35].

We noted above that the obstruction to obtaining discrete spacetime representations was due to the infinite volume of momentum space. If we consider restricting to a finite volume of momentum space (through whatever means), then in spacetime one obtains a space of analytic functions (see, e.g., [39]). Analytic functions exhibit a severe non-locality since all of the degrees of freedom are contained in a small volume around any point in spacetime (as the Taylor series can be evaluated in a neighborhood). (For a related discussion, see also [25].) Hence it seems that anything akin to ordinary sampling theory would allow for superluminal signalling, thus would be inappropriate for the configuration space of a relativistic field theory. In practical applications of sampling theory (e.g., in communications engineering [19–21]) this non-locality does not appear as realistic signals are only approximately bandlimited. In any case, practically one could not fully evaluate the Taylor series coefficients to arbitrary precision. Therefore, instead of attempting to model the fundamental configuration space of the field theory, one could adopt these analytic functions as an approximation, and perhaps prevent superluminal effects in the model by introducing other practical limitations (such as finite precision of measurements).

To conclude, although the original GUP yields a sampling theory for bandlimited fields, we have shown that in a class of Lorentz-covariant GUPs one does not obtain a finite density of degrees of freedom in spacetime. Consequently, quantities such as entanglement between

IOP Conf. Series: Journal of Physics: Conf. Series **1275** (2019) 012025 doi:10.1088/1742-6596/1275/1/012025

spatial subsystems is not regulated, although covariant bandlimitation may still be pertinent in interacting theories. In order to obtain a stronger regulator, perhaps certain assumptions should be relaxed, for example, the commutativity of the coordinates (see, e.g., the reviews [40, 41]).

Acknowledgments
The author would like to thank Achim Kempf for his supervision, continuing support, and many invaluable discussions. The author also acknowledges support from the Natural Sciences and Engineering Research Council (NSERC) of Canada through the Doctoral Canada Graduate Scholarship (CGS) program, as well as from the Ontario Graduate Scholarship (OGS) program. This research was supported in part by Perimeter Institute for Theoretical Physics. Research at Perimeter Institute is supported by the Government of Canada through the Department of Innovation, Science and Economic Development and by the Province of Ontario through the Ministry of Research and Innovation.

References

[1] Bronstein M P 1936 *Phys. Z. Sowjetunion* **9** 140–157 (Republished in [4])
[2] Mead C A 1964 *Phys. Rev.* **135** B849
[3] Hossenfelder S 2013 *Liv. Rev. Rel.* **16** 2
[4] Bronstein M 2012 *Gen. Rel. and Grav.* **44** 267–283
[5] Kempf A 1994 *J. Math. Phys.* **35** 4483–4496
[6] Kempf A, Mangano G and Mann R B 1995 *Phys. Rev.* D **52** 1108
[7] Deser S 1957 *Rev. Mod. Phys.* **29** 417
[8] DeWitt B S 1964 *Phys. Rev. Lett.* **13** 114
[9] Kempf A 2004 *Phys. Rev. Lett.* **92** 221301
[10] Kempf A, Chatwin-Davies A and Martin R T W 2013 *J. Math. Phys.* **54** 022301
[11] Chatwin-Davies A, Kempf A and Martin R T W 2017 *Phys. Rev. Lett.* **119** 031301
[12] Kempf A 1997 *J. Phys.* A **30** 2093
[13] Brau F 1999 *J. Phys.* A **32** 7691
[14] Hossenfelder S, Bleicher M, Hofmann S, Ruppert J, Scherer S and Stöcker H 2003 *Phys. Lett.* B **575** 85–99
[15] Das S and Vagenas E C 2008 *Phys. Rev. Lett.* **101** 221301
[16] Pikovski I, Vanner M R, Aspelmeyer M, Kim M and Brukner Č 2012 *Nature Phys.* **8** 393
[17] Scardigli F and Casadio R 2015 *Eur. Phys. J.* C **75** 425
[18] Scardigli F, Blasone M, Luciano G and Casadio R 2018 *Eur. Phys. J.* C **78** 728
[19] Shannon C E 1948 *Bell Sys. Tech. J.* **27** 379–423
[20] Shannon C E 1949 *Proc. IRE* **37** 10–21
[21] Nyquist H 1928 *Trans. IEEE* **47** 617–644
[22] Kempf A 1994 *Czech. J. Phys.* **44** 1041–1048
[23] Kempf A 1997 *J. Math. Phys.* **38** 1347–1372
[24] Kempf A 2000 *Phys. Rev. Lett.* **85** 2873
[25] Pye J, Donnelly W and Kempf A 2015 *Phys. Rev.* D **92** 105022
[26] Landau H 1967 *Acta Math.* **117** 37–52
[27] Jerri A J 1977 *Proc. IEEE* **65** 1565–1596
[28] Kempf A and Martin R 2008 *Phys. Rev. Lett.* **100** 021304
[29] Hossenfelder S 2014 *SIGMA* **10** 8
[30] Lee J M 2002 *Introduction to Smooth Manifolds* (Springer)
[31] Sorkin R D 1983 *Tenth Int. Conf. Gen. Rel. Grav.* vol 2 ed Bertotti B, de Felice F and Pascolini A pp 734–736
[32] Bombelli L, Koul R K, Lee J and Sorkin R D 1986 *Phys. Rev.* D **34** 373
[33] Serafini A 2017 *Quantum Continuous Variables: A Primer of Theoretical Methods* (CRC Press)
[34] Nesterov D and Solodukhin S N 2011 *Nucl. Phys.* B **842** 141–171
[35] Solodukhin S N 2011 *Liv. Rev. Rel.* **14** 8
[36] Aslanbeigi S, Saravani M and Sorkin R D 2014 *J. High Energy Phys.* **2014** 24
[37] Belenchia A, Benincasa D M and Liberati S 2015 *J. High Energy Phys.* **2015** 36
[38] Unruh W G 1976 *Phys. Rev.* D **14** 870
[39] Rudin W 1987 *Real and Complex Analysis* 3rd ed (McGraw-Hill)
[40] Douglas M R and Nekrasov N A 2001 *Rev. Mod. Phys.* **73** 977–1029
[41] Szabo R J 2003 *Phys. Rep.* **378** 207–299

IOP Conf. Series: Journal of Physics: Conf. Series **1275** (2019) 012026 doi:10.1088/1742-6596/1275/1/012026

Lorentz boosts of bispinor Bell-like states

Victor A. S. V. Bittencourt[1] **and Massimo Blasone** [2]

[1]Max Planck Institute for the Science of Light, Staudtstraße 2, PLZ 91058, Erlangen, Germany
[2]Dipartimento di Fisica, Università degli Studi di Salerno, Via Giovanni Paolo II, 132 84084 Fisciano, Italy & INFN Sezione di Napoli, Gruppo collegato di Salerno, Italy

E-mail: `victor.bittencourt@mpl.mpg.de; blasone@sa.infn.it`

Abstract. We describe in this paper the effects of Lorentz boost on the quantum entanglement encoded in two-particle Dirac bispinor Bell-like states. Each particle composing the system described in this formalism has three degrees of freedom: spin, chirality, and momentum, and the joint state can be interpreted as a 6 qubit state. Given the transformation law of bispinor under boosts, we compute the change of the Meyer-Wallach global measure of quantum entanglement due to the frame transformation and study its equivalence to the results obtained for the relativistic spin 1/2 Bell-like states, constructed in the framework of the irreducible representations of the Lorentz group. We verify that the monotonic increase of the global entanglement under boosts for ultra-relativistic states is solely due to an increasing of the entanglement associated with the spins subsystems. For such ultra-relativistic states, the entanglement related to the chirality degrees of freedom is invariant, and the variation of the global entanglement of bispinor states is the same as the one calculated for relativistic spin 1/2 states. We also show that the particle-particle entanglement is invariant under boosts for any Bell-like state.

1. Introduction

Relativistic quantum mechanics and quantum information theory have been the framework to devise information protocols in relativistic setups, such as clock synchronization [1–3] and position verification [4]. In this context, transformation properties of quantum entanglement under frame transformations, in particular under Lorentz boosts, have been addressed and discussed in the last two decades [5–15]. The focus of many studies were on the transformation properties of spin $\frac{1}{2}$ entangled states [5–9], since spin states can be used to encode qubits, the basic quantum information unit.

In the relativistic quantum information framework, spin $\frac{1}{2}$ states are described by the irreducible representations (irreps) of the Poincarè group and a Lorentz boost acts on the spin degree of freedom as a momentum dependent rotation, the Wigner rotation [5–9]. Such transformation property is the basis of many interesting results, such as the generation of spin-momentum entanglement by frame transformations and non-invariance of the entanglement encoded in a pair of spin $\frac{1}{2}$ particles [5–9]. Despite the richness of this description, in the covariant Hamiltonian formulation of relativistic quantum mechanics of the Dirac equation, the inclusion of mass requires the consideration of the parity symmetry [16,17]. Spatial parity couples positive and negative parity states with positive and negative helicities and to have invariance under this symmetry one needs to consider the irreps of the *extended Poincarè* group [17]. In this case, spin $\frac{1}{2}$ is carried by Dirac four-component spinors (Dirac bispinors) satisfying the Dirac equation.

Content from this work may be used under the terms of the Creative Commons Attribution 3.0 licence. Any further distribution of this work must maintain attribution to the author(s) and the title of the work, journal citation and DOI.
Published under licence by IOP Publishing Ltd

IOP Conf. Series: Journal of Physics: Conf. Series **1275** (2019) 012026 doi:10.1088/1742-6596/1275/1/012026

Such bispinorial structure can be used not only to describe charged massive fermions (quarks, electrons, muons, etc.) but also for investigating properties of Dirac-like systems, such as bilayer graphene [18–21]

Dirac bispinors are described in terms of a representation supported by a subgroup of the $SL(2,C) \otimes SL(2,C)$, the $SU(2) \otimes SU(2)$, accordingly carrying two discrete degrees of freedom (DoFs): spin and intrinsic parity [17], that are in general entangled [22–24]. In the scope of relativistic quantum information theory, the effects of Lorentz boosts on quantum entanglement encoded in a pair of bispinorial particles have been described in connection with Wigner rotations [25] and in the context of Fouldy-Wouthuysen (FW) spin operator [10]. Moreover, the role of the intrinsic parity-spin entanglement in the change of the total amount of entanglement shared by two bispinorial particles was addressed for general boosts scenarios and for anti-symmetric states [13, 14].

In this paper we consider the effects of Lorentz boosts on entangled two-particle states constructed as Bell-like superpositions of helicity eigenstates of the Dirac equation, including momentum superposition. We consider a scenario where two particles, described as helicity eigenstates of the free Dirac Hamiltonian, are in an entangled Bell-like state with respect to an inertial frame $\mathcal{S}$ and we study the effects of a Lorentz boost on the correlational content of the state. The momentum degree of freedom is included as a discrete DoF and treated as an additional qubit carried by each particle. We verify that the particle-particle entanglement, encoded in the partition {all DoFs of A; all DoFs of B}, is invariant under boosts. Otherwise, for ultra-relativistic states the variation of the global entanglement, quantified via the Meyer-Wallach measure [26] is the same as obtained with the formalism using the irreps of the Lorentz group as presented in [8]. In this case, the amount of entanglement encoded in the partition (chirality vs all other DoFs) is invariant therefore not contributing for the change of the global entanglement.

2. Composite structure of Dirac equation and bispinor Bell-like states

The free Dirac equation that describes the dynamics of free fermionic particles reads (in momentum space) [27]

$$i\frac{\partial \psi}{\partial t} = \hat{H}_D \psi = (\boldsymbol{p} \cdot \hat{\boldsymbol{\alpha}} + m\hat{\beta})\psi, \tag{1}$$

where $\boldsymbol{p}$ is the particle momentum and m the particle mass. The 4×4 Dirac matrices $\hat{\boldsymbol{\alpha}} = \{\hat{\alpha}_x, \hat{\alpha}_y, \hat{\alpha}_z\}$ and $\hat{\beta}$ satisfy the anti-commutation relations $\hat{\alpha}_i\hat{\alpha}_j + \hat{\alpha}_j\hat{\alpha}_i = 2\delta_{i,j}\hat{I}_4$, $\hat{\alpha}_i\hat{\beta} + \hat{\beta}\hat{\alpha}_i = 0$ (for $i = x, y, z$) and $\hat{\beta}^2 = \hat{I}_4$. We have adopted natural units $\hbar = c = 1$, and from hereafter bold letters represents vector quantities with modulus $v = \sqrt{\boldsymbol{v} \cdot \boldsymbol{v}}$ and circumflexes denotes operators. $\hat{I}_N$ denotes the N dimensional identity operator. Among the different representations of the Dirac matrices satisfying the required anti commutation relations, we adopt in this paper the Chiral (or Weyl) representation, where the matrix $\hat{\gamma}_5 = i\hat{\beta}\hat{\alpha}_x\hat{\alpha}_y\hat{\alpha}_z$ is diagonal and given by $\hat{\gamma}_5 = \text{diag}\{\hat{I}_2, -\hat{I}_2\}$ [1]. In this representation the Dirac matrices are given by

$$\hat{\beta} = \begin{bmatrix} 0 & \hat{I}_2 \\ \hat{I}_2 & 0 \end{bmatrix}, \quad \hat{\boldsymbol{\alpha}} = \begin{bmatrix} \hat{\boldsymbol{\sigma}} & 0 \\ 0 & -\hat{\boldsymbol{\sigma}} \end{bmatrix}. \tag{2}$$

The Dirac Hamiltonian exhibit symmetries that are supported by a group structure given in terms of the direct product of two algebras composing a subset of the $SL(2,C) \otimes SL(2,C)$, the

[1] The chiral and the so called usual (or Dirac) representation of the Dirac matrices, where $\hat{\beta} = \text{diag}\{\hat{I}_2, -\hat{I}_2\}$, are related by the unitary transformation

$$\hat{T} = \frac{1}{\sqrt{2}} \begin{bmatrix} \hat{I}_2 & \hat{I}_2 \\ \hat{I}_2 & -\hat{I}_2 \end{bmatrix}.$$

IOP Conf. Series: Journal of Physics: Conf. Series **1275** (2019) 012026 doi:10.1088/1742-6596/1275/1/012026

$SU(2) \otimes SU(2)$. The origin of such structure is related to the mass term of Dirac equation that requires the consideration of spatial parity symmetry, which couples two different irreducible representations of the Lorentz group, each one given in terms of the $SU(2)$ group structure [16, 17](for more details see [14]). Correspondingly, the Hamiltonian (1) supported by such structure describes the dynamics of two discrete degrees of freedom: the chirality, associated to the Hilbert space $\mathcal{H}_C$, and the spin, associated with the Hilbert space $\mathcal{H}_S$ [2]. The joint dynamics is therefore described in terms of a composite Hilbert space $\mathcal{H} = \mathcal{H}_C \otimes \mathcal{H}_S$, with $\dim[\mathcal{H}_C] = \dim[\mathcal{H}_S] = 2$. This structure becomes evident through the decomposition of the Dirac matrices in terms of tensor products of operators acting in each Hilbert space that for the chiral representation reads

$$\hat{\alpha} = \hat{\sigma}_z^{(C)} \otimes \hat{\boldsymbol{\sigma}}^{(S)}, \quad \hat{\beta} = \hat{\sigma}_x^{(C)} \otimes \hat{I}_2^{(S)}, \tag{3}$$

where the subscripts $C\,(S)$ indicates the chirality (spin) Hilbert space. In this framework, the Dirac Hamiltonian can be interpreted as a two-qubit operator and the particles described by such dynamics carry two discrete degrees of freedom with dynamics governed by a continuous variable, position or momentum.

Within the above prescription, the Dirac Hamiltonian eigenstates can be written as two-qubit states describing a bipartite system. For instance, the two degenerate positive energy eigenstate of the Dirac Hamiltonian (1) $|u_s(p)\rangle \in \mathcal{H}_C \otimes \mathcal{H}_S$ (with $s = 0, 1$) are given by

$$|u_s(\boldsymbol{p})\rangle = \frac{1}{2}\sqrt{\frac{E_p + m}{E_p}} \left[\left(1 + \frac{\boldsymbol{p} \cdot \hat{\boldsymbol{\sigma}}^{(S)}}{E_p + m}\right) |0\rangle_C \otimes |\eta_s(\boldsymbol{p})\rangle_S + \left(1 - \frac{\boldsymbol{p} \cdot \hat{\boldsymbol{\sigma}}^{(S)}}{E_p + m}\right) |1\rangle_C \otimes |\eta_s(\boldsymbol{p})\rangle_S \right], \tag{4}$$

where $E_p = \sqrt{p^2 + m^2}$ and $|\eta_s(\boldsymbol{p})\rangle_S$ is a two-component spinor that depends on the spin polarization of the particle, and we have indicated the positive (negative) eigenstates of $\hat{\sigma}_z^{(C)}$ by $|0\rangle$ ($|1\rangle$). The solutions of the Dirac equation (4) cannot be decomposed as $|\phi\rangle_C \otimes |\phi\rangle_S$, thus they are entangled states. The entanglement content of the Dirac equation solutions was characterized including external potentials [24] and in connection with low-energy systems [20, 21]. A convenient description of Dirac equations solutions that was adopted in previous studies [13, 14] are helicity eigenstates:

$$\hat{I}^{(C)} \otimes \hat{\boldsymbol{\sigma}}^{(S)} \cdot \frac{\boldsymbol{p}}{p} |u_s(\boldsymbol{p})\rangle = (-1)^s |u_s(\boldsymbol{p})\rangle, \tag{5}$$

thus from now on the label $s = 0\,(1)$ denotes positive (negative) helicity. The helicity eigenstates are given explicitly as

$$|u_s(\boldsymbol{p})\rangle = \frac{1}{2}\sqrt{\frac{E_p + m}{E_p}} \left[\left(1 + (-1)^s \frac{p}{E_p + m}\right) |0\rangle_C + \left(1 - (-1)^s \frac{p}{E_p + m}\right) |1\rangle_C \right] \otimes |\eta_s(\boldsymbol{p})\rangle_S, \tag{6}$$

which, different from the general solutions (4), are separable states.

With such particular solutions of the free Dirac equation, two-particle states are constructed as arbitrary superpositions of bispinors. Designating the two particles as A and B, a general two-particle state reads

$$|\psi_{A,B}\rangle = \frac{1}{N} \sum_i^n c_i |u_{s_i}(\boldsymbol{p}_i)\rangle_A \otimes |u_{r_i}(\boldsymbol{q}_i)\rangle_B. \tag{7}$$

[2] This is equivalent to the intrinsic parity-spin description of the Dirac equation reported previously [13, 14, 20, 21, 24], but here we choose to label one of the states with the chirality DoF, since we are working with the chiral representation of the Dirac matrices.

IOP Conf. Series: Journal of Physics: Conf. Series **1275** (2019) 012026 doi:10.1088/1742-6596/1275/1/012026

Such type of states can be created for example in quantum electrodynamical processes [28]. Since each of the particles are associated with a two-qubit state, the two-particle state is a 4 qubit state with the following DoFs: spins of particle A and B and chiralities of particle A and B. By further generalizing the bispinor superpositions (7) we can construct "Bell-like" states. A two-spins $\frac{1}{2}$ Bell-like state has the form

$$|\phi_{\text{Bell}}\rangle = \big(\cos\alpha|\boldsymbol{p}\rangle_A\otimes|\boldsymbol{q}\rangle_B+\sin\alpha|\boldsymbol{q}\rangle_A\otimes|\boldsymbol{p}\rangle_B\big)\otimes\big(\cos\beta|+\rangle_A\otimes|-\rangle_B+\sin\beta|-\rangle_A\otimes|+\rangle_B\big), \quad (8)$$

whose entanglement properties under relativistic transformations where studied in terms of Wigner rotations [8]. A bispinor Bell-like state equivalent to (8) is constructed by including in the superposition (7) the momentum DoF. Adopting the shorthand notation $|u_s(\boldsymbol{p})\rangle\otimes|\boldsymbol{p}\rangle\equiv|u_s(\boldsymbol{p}),\boldsymbol{p}\rangle$, a bispinor Bell-like state is written as

$$\begin{aligned}|\psi_{\text{Bell}}\rangle &= \cos\alpha\,\cos\beta|u_0(\boldsymbol{p}),\boldsymbol{p}\rangle_{\text{A}}\otimes|u_0(\boldsymbol{q}),\boldsymbol{q}\rangle_{\text{B}}+\cos\alpha\,\sin\beta|u_1(\boldsymbol{p}),\boldsymbol{p}\rangle_{\text{A}}\otimes|u_1(\boldsymbol{q}),\boldsymbol{q}\rangle_{\text{B}}\\ &+ \sin\alpha\,\cos\beta|u_1(\boldsymbol{q}),\boldsymbol{q}\rangle_{\text{A}}\otimes|u_1(\boldsymbol{p}),\boldsymbol{p}\rangle_{\text{B}}+\sin\alpha\,\sin\beta|u_0(\boldsymbol{q}),\boldsymbol{q}\rangle_{\text{A}}\otimes|u_0(\boldsymbol{p}),\boldsymbol{p}\rangle_{\text{B}}. \quad (9)\end{aligned}$$

The angles α and β parametrize, as in (8), the degree of superposition between the momenta and between the helicities, respectively. The momentum eigenstates $|p\rangle$ are normalized as

$$\int\frac{d^3p}{2E_p}\langle\boldsymbol{p}|\boldsymbol{q}\rangle = 1. \quad (10)$$

Since each bispinor carries two DoFs (chirality and spin), the system described by (9) has 6 DoFs: the chiralities of A and of B (indicated by $C_{A(B)}$), the spins of A and B ($S_{A(B)}$) and the momenta of A B ($P_{A(B)}$). Moreover, from now on the letter ψ is used to refer to a state at the bispinor level, while the letter ϕ will always refer to states in the spin $\frac{1}{2}$ level of the irreps of the Lorentz group such as (8). We indicate the density matrix associated to a state $|\xi\rangle$ by $\rho_\xi=|\xi\rangle\langle\xi|$

From now on we consider that the momenta distributions are sharply concentrated around their mean value, such that the overlap between $|\boldsymbol{p}\rangle$ and $|\boldsymbol{q}\rangle$ can be ignored. This supports an effective description of the momentum as a dichotomic variable, associated to an additional qubit of the systems, that is $|\boldsymbol{p}\rangle\equiv|0\rangle$ and $|\boldsymbol{q}\rangle\equiv|1\rangle$. This simplifying hypothesis was used in several investigations of transformation properties of quantum entanglement [8] and will be used through this paper. Moreover we will assume that the particles propagate in the $\boldsymbol{e}_z$ direction such that p and q are opposite, that is, $\boldsymbol{p}=-\boldsymbol{q}=p\boldsymbol{e}_z$. For this momenta configuration the helicity eigenstates are given by

$$|\eta_0(\boldsymbol{p})\rangle=|\eta_1(\boldsymbol{q})\rangle=|+\rangle,\quad |\eta_1(\boldsymbol{p})\rangle=|\eta_0(\boldsymbol{q})\rangle=|-\rangle, \quad (11)$$

where $|\pm\rangle$ are the positive (negative) eigenstates of $\hat{\sigma}_z^{(S)}$.

Following the above considerations we verify that the state (9) is closely related to its counterpart (8) in the non-relativistic limit $p\ll m$. In this case the bispinor helicity eigenstates (6) read

$$|u_s(p)\rangle\rightarrow\frac{|0\rangle_C+|1\rangle_C}{\sqrt{2}}\otimes|\eta_s\rangle_S, \quad (12)$$

and since for the considered momentum configuration $|\eta_s\rangle_S$ is given by (11), one has

$$|\psi_{\text{Bell}}\rangle\rightarrow|\psi_{\text{NR}}\rangle=\frac{(|0\rangle+|1\rangle)_{C_A}\otimes(|0\rangle+|1\rangle)_{C_B}}{2}\otimes|\phi_{\text{Bell}}\rangle. \quad (13)$$

Therefore, the subsystem associated with the DoFs $\{S_A,P_A;S_B,P_B\}$ is described by the reduced density matrix $\text{Tr}_{\text{C}_\text{A},\text{C}_\text{B}}\Big[|\psi_{\text{NR}}\rangle\langle\psi_{\text{NR}}|\Big]=|\phi_{\text{Bell}}\rangle\langle\phi_{\text{Bell}}|$. One also notices that for $m\gg p$ the

IOP Conf. Series: Journal of Physics: Conf. Series **1275** (2019) 012026 doi:10.1088/1742-6596/1275/1/012026

bispinors $|u_s(\boldsymbol{p})\rangle$ are eigenstates of the operator $\hat{\sigma}_x^{(C)} \otimes \hat{I}^{(S)}$, which is the intrinsic parity operator in the chiral representation.

Under the approximation of the momentum as a dichotomic variable, the bispinor Bell-like state (9) can be interpreted as a 6 qubit state. Different from the entanglement encoded in states containing only two DoFs (bipartite states) whose characterization is supported by the Schmidt decomposition theorem [29], the description of quantum entanglement for multipartite states, such as (9) and (8), is an involved task [29]. The subsystems of a multipartite state can be correlated in nonequivalent ways such that multipartite entanglement can be studied following different approaches. Since the joint state (9) is a pure state, the multipartite entanglement shared among the different DoFs of the system can be computed via the Meyer-Wallach global measure of entanglement E_G [26]. For a system described by a pure state ρ and composed of N subsystem labeled as $\{\alpha_1, \alpha_2, ..., \alpha_N\}$, E_G is given by

$$E_G[\rho] = \bar{E}[\rho^{\alpha_k}] = \frac{1}{N}\sum_{i=1}^{N} E_L[\rho^{\alpha_i}], \tag{14}$$

with the linear entropies E_L given by

$$E_L[\rho] = \frac{d}{d-1}(1 - \mathrm{Tr}[\rho^2]), \tag{15}$$

where d is the dimension of the Hilbert space in which ρ acts, and ρ^{α_i} is the reduced density matrix of the subsystem α_i, obtained by tracing out all other subsystems $\rho^{\alpha_i} = \mathrm{Tr}_{\{\alpha_k\}\neq\alpha_i}[\rho]$. The linear entropy $E_L[\rho^{\alpha_i}]$ quantifies the amount of entanglement between the subsystem α_i and the subsystem composed by all α_k with $k \neq i$, therefore E_G is the mean value of the entanglement shared in the partitions of the form $\{\alpha_i; \{\alpha_{k\neq i}\}\}$. The global entanglement of Bell-like state (9) is given by

$$E_G[\rho_{\psi_{\mathrm{Bell}}}] = \frac{1}{6}\left[2 - \cos 4\alpha - \cos 4\beta + \frac{p^2}{2E_p^2}\left(3 - \cos 4\alpha - \cos 4\beta - \cos 4\alpha \cos 4\beta\right)\right]. \tag{16}$$

The last term in the above expression is the contribution due to the chirality degree of freedom, which vanishes for $p \ll m$. In this limit, as highlighted by Eq. (12), the chirality factorizes out of the bispinorial structure, and therefore it is not entangled with all other subsystems. Moreover, for such non-relativistic limit $E_G[\rho_{\psi_{\mathrm{Bell}}}] \to \frac{2}{3}E_G[\rho_{\phi_{\mathrm{Bell}}}]$, and therefore the global entanglement properties of (9) are the same of (8).

3. Global entanglement of bispinor Bell-like states under Lorentz boosts

Once the general framework is set, we now proceed to describe the effects of Lorentz boosts on the entanglement content of bispinor Bell-like states (9) and its connections to the results obtained in the scope of Wigner rotations and spin $\frac{1}{2}$ states in the form (8).

Under a given Lorentz transformations Λ connecting two inertial frames $\mathcal{S}$ and $\mathcal{S}'$, a state $|\psi(x)\rangle$ transforms as [16, 17]

$$|\psi(x)\rangle \to \hat{S}[\Lambda]|\psi(\Lambda^{-1}x)\rangle, \tag{17}$$

where $\hat{S}[\Lambda]$ is an operator describing how Λ acts on the state, which depends on the representation of the Lorentz group to which the state belongs and on the specific transformation. A Lorentz boost describes the transformation from a inertial frame $\mathcal{S}$ to a frame $\mathcal{S}'$ moving with constant speed $\boldsymbol{v}$ with respect to $\mathcal{S}$. The operator representing the action of Lorentz Boosts on bispinor states $\hat{S}_{\mathrm{Boost}}$ reads [16, 17, 27]

$$\hat{S}_{\mathrm{Boost}}[\omega] = \cosh\left(\frac{\omega}{2}\right)\hat{I}_4 + \sinh\left(\frac{\omega}{2}\right)\boldsymbol{n}\cdot\hat{\boldsymbol{\alpha}}, \tag{18}$$

where $\boldsymbol{n} = \boldsymbol{v}/v$ and $\omega = \text{arctanh}\left[\frac{v}{\sqrt{1-v^2}}\right]$ is the boost rapidity. The state constructed with the positive energy solutions of the free Dirac equation in momentum space $|u_s(\boldsymbol{p}), \boldsymbol{p}\rangle$ transforms under boosts as

$$|u_s(\boldsymbol{p})\rangle \otimes |\boldsymbol{p}\rangle \to \hat{\mathcal{A}}[\omega]\Big[|u_s(\boldsymbol{p})\rangle \otimes |\boldsymbol{p}\rangle\Big] = \left(\frac{1}{N_{\boldsymbol{p}}}\hat{S}_{\text{Boost}}[\omega]|u_s(\Lambda^{-1}\boldsymbol{p}')\rangle\right) \otimes |\boldsymbol{p}'\rangle, \tag{19}$$

where $N_{\boldsymbol{p}}$ is a normalization factor and $\boldsymbol{p}'$ is the momentum with respect to $\mathcal{S}'$. To be in correspondence with the scenario described in the previous section, we consider now a boost in the direction $\boldsymbol{n} = (\sin\theta, 0, \cos\theta)$ in the $x - z$ plane as depicted in Fig. 1. In this case, for the helicity eigenstates (Eq. 6) the normalization factors are given by

$$\begin{aligned} N_{\boldsymbol{p}} &= \sqrt{\cosh\omega + \frac{p}{E_p}\sinh\omega\cos\theta} \qquad \text{for } \boldsymbol{p} = p\boldsymbol{e}_z, \\ N_{\boldsymbol{q}} &= \sqrt{\cosh\omega - \frac{p}{E_p}\sinh\omega\cos\theta} \qquad \text{for } \boldsymbol{q} = -p\boldsymbol{e}_z. \end{aligned}$$

We assume that the orthonormalization of the momentum qubits is preserved under boosts: $\langle\boldsymbol{p}|\boldsymbol{q}\rangle \to \langle\boldsymbol{p}'|\boldsymbol{q}'\rangle = \delta_{\boldsymbol{p}',\boldsymbol{q}'}$, such that for the joint bispinor-momentum helicity eigenstates

$$\langle u_s(\boldsymbol{p}), \boldsymbol{p}|u_r(\boldsymbol{q}), \boldsymbol{q}\rangle \to \frac{1}{N_{\boldsymbol{p}}N_{\boldsymbol{q}}}\langle u_s(\Lambda^{-1}\boldsymbol{p}'), \boldsymbol{p}'|\hat{S}^2_{\text{Boost}}[\omega]|u_r(\Lambda^{-1}\boldsymbol{q}'), \boldsymbol{q}'\rangle = \delta_{sr}\delta_{\boldsymbol{p}',\boldsymbol{q}'}. \tag{20}$$

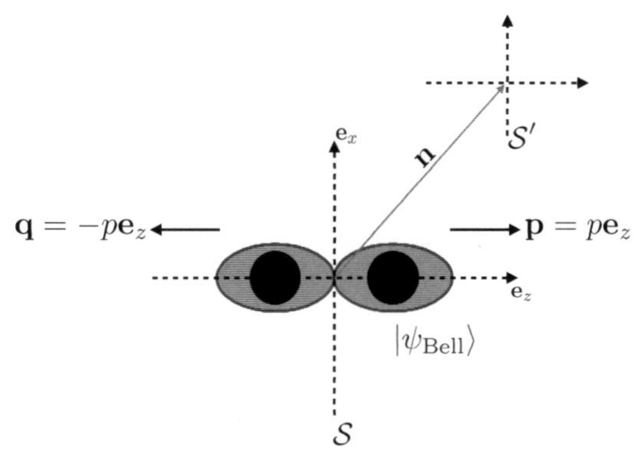

Figure 1. Schematic representation of the boost scenario. The Lorentz boost is performed to describe a frame transformation from $\mathcal{S}$ to $\mathcal{S}'$, which moves uniformly in the $\boldsymbol{n} = \sin\theta\boldsymbol{e}_x + \cos\theta\boldsymbol{e}_z$ direction with respect to $\mathcal{S}$. The transformation of the two-particle state (9) is described by the operator $\hat{\mathcal{A}}^{(A)} \otimes \hat{\mathcal{A}}^{(B)}$ whose action in the bispinors are defined in (18), which induces non trivial changes in the quantum entanglement shared among the different DoFs of the system.

The action of boosts on two-particle states is given by $\hat{\mathcal{A}}^{(A)}[\omega] \otimes \hat{\mathcal{A}}^{(B)}[\omega]$ and the bispinor Bell-like state transforms as

$$|\psi_{\text{Bell}}\rangle \to |\psi'_{\text{Bell}}\rangle = \hat{\mathcal{A}}^{(A)}[\omega] \otimes \hat{\mathcal{A}}^{(B)}[\omega]|\psi_{\text{Bell}}\rangle, \tag{21}$$

modifying in a non trivial way the superposition and consequently inducing changes in the entanglement encoded in the different partitions of the system. In particular, the entanglement

IOP Conf. Series: Journal of Physics: Conf. Series **1275** (2019) 012026 doi:10.1088/1742-6596/1275/1/012026

between all DoFs of particle A and all DoFs of particle B, which we call *particle-particle entanglement*, is invariant. This can be verified by considering the reduced density matrix ρ^{A} of subsystem A obtained by tracing our all DoFs of B:

$$\begin{aligned}\rho^{\mathrm{A}}_{\psi_{\mathrm{Bell}}} = \mathrm{Tr}_{\mathrm{B}}[\rho_{\psi_{\mathrm{Bell}}}] &= \cos^2\alpha\left(\cos^2\beta|u_0(\boldsymbol{p}),\boldsymbol{p}\rangle\langle u_0(\boldsymbol{p}),\boldsymbol{p}| + \sin^2\beta|u_1(\boldsymbol{p}),\boldsymbol{p}\rangle\langle u_1(\boldsymbol{p}),\boldsymbol{p}|\right) \\ &+ \sin^2\alpha\left(\cos^2\beta|u_1(\boldsymbol{q}),\boldsymbol{q}\rangle\langle u_1(\boldsymbol{q}),\boldsymbol{q}| + \sin^2\beta|u_0(\boldsymbol{q}),\boldsymbol{q}\rangle\langle u_0(\boldsymbol{q}),\boldsymbol{q}|\right) \quad (22)\end{aligned}$$

Under a Lorentz boosts the orthonormalization of the bispinors is preserved (see Eq. 20), and thus

$$\mathrm{Tr}_{\mathrm{A}}[(\rho^{\mathrm{A}}_{\psi_{\mathrm{Bell}}})^2] = \mathrm{Tr}_{\mathrm{A}}[(\rho^{\mathrm{A}}_{\psi'_{\mathrm{Bell}}})^2]. \quad (23)$$

Therefore the linear entropy $E_L[\rho^{\mathrm{A}}_{\psi_{\mathrm{Bell}}}] = \frac{6}{5}(1 - \mathrm{Tr}_{\mathrm{A}}[(\rho^{\mathrm{A}}_{\psi_{\mathrm{Bell}}})^2])$ quantifying the entanglement in the partition {all DoFs of A; all DoFs of B} is invariant under boosts, in agreement with the invariance of particle-particle entanglement of spin $\frac{1}{2}$ Bell-like states (8) [8]. The linear entropies of the momenta subsystems $E_L[\rho^{P_{A(B)}}_{\psi_{\mathrm{Bell}}}]$ are also invariant under boosts, which is again in correspondence with the behavior reported for spin $\frac{1}{2}$ states $|\phi_{\mathrm{Bell}}\rangle$. On the other hand, as verified in previous studies, bispinor superpositions like (7) exhibit invariance of particle-particle entanglement only if A and B have definite momenta [13].

Although the bispinor Bell-like state exhibits some invariant quantum correlations, the global measure of entanglement (14) is not invariant under boosts, with variation depending not only on the superposition and boost parameters but also on the energy regime of the state, encoded in the parameter p/m. Figure 2 shows the plot of the global entanglement of the transformed state $E_G[\rho_{\psi'_{\mathrm{Bell}}}]$ for a boost in the $\boldsymbol{e}_x$ in function of the boost rapidity for several superposition angles for a non-relativistic state (left plot) and for a state with $p/m = 10$ (right plot). Non-relativistic states show an monotonic increase in the global entanglement. For states with $p \gg m$ the increase on E_G is exclusively due to the variation in the linear entropy associated to the spins subsystems, since in this limit

$$E_L[\rho^{C_{A(B)}}_{\psi'_{\mathrm{Bell}}}] = \frac{1}{4}(3 - \cos 4\alpha - \cos 4\beta - \cos 4\alpha \cos 4\beta) = E_L[\rho^{C_{A(B)}}_{\psi_{\mathrm{Bell}}}]. \quad (24)$$

For such ultra-relativistic states, $|u_s(\boldsymbol{p})\rangle \to |s\rangle_C \otimes |\eta_s\rangle_S$ and because chirality is invariant under boosts, the contribution of the chirality subsystems entropy to the global entanglement, $E_L[\rho^{C_{A(B)}}_{\psi'_{\mathrm{Bell}}}]$, is invariant. Additionally, for specific superposition angles the state has invariant E_G and we also notice that the overall behavior of E_G under boosts is similar to the one reported for anti-symmetric superpositions of the form (9), which do not include momentum entanglement.

The discussion of the previous paragraph sets the framework to relate the changes of the global entanglement of (9) under boosts with those of spin $\frac{1}{2}$ states (8). Although the states have the same amount of global entanglement for $p \ll m$, a Lorentz boost will increase the chirality subsystems entropy of bispinor Bell-like states, and the changes on the quantum entanglement are affected in a different way from the ones expected for spin $\frac{1}{2}$ states. Otherwise, for $p \gg m$ (24) holds and, although $|\psi_{\mathrm{Bell}}\rangle$ and $|\phi_{\mathrm{Bell}}\rangle$ have different amounts of global entanglement, the variation of E_G induced by boosts is the same in both states. In fact, for $p \gg m$ one has

$$\rho^{S_{A(B)}}_{\psi'_{\mathrm{Bell}}} = \rho^{S_{A(B)}}_{\phi'_{\mathrm{Bell}}}, \quad (25)$$

as can be verified by considering the transformation properties of the spin $\frac{1}{2}$ states given in terms of Wigner rotations, with general formula given elsewhere (see for instance [9]). Therefore $E_L[\rho^{S_{A(B)}}_{\psi'_{\mathrm{Bell}}}] = E_L[\rho^{S_{A(B)}}_{\phi'_{\mathrm{Bell}}}]$ and thus, since for the bispinor state in the limit $p \gg m$ (24) holds, the variations of E_G due to Lorentz boosts are the same for $|\psi_{\mathrm{Bell}}\rangle$ and for $|\phi_{\mathrm{Bell}}\rangle$.

IOP Conf. Series: Journal of Physics: Conf. Series **1275** (2019) 012026 doi:10.1088/1742-6596/1275/1/012026

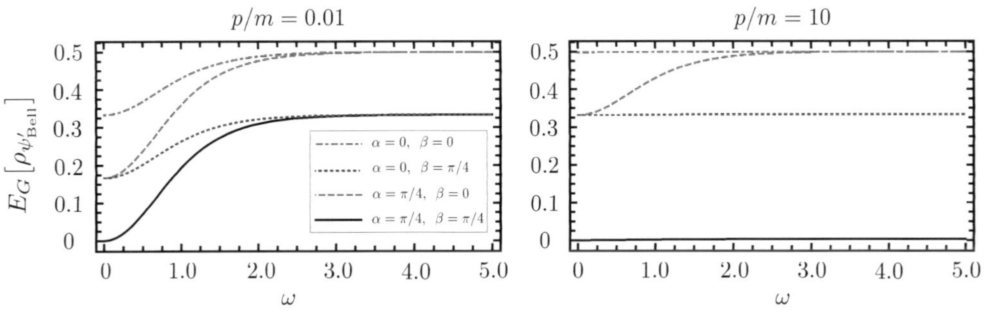

Figure 2. Global entanglement of the bispinor Bell-like state (9) under a Lorentz boost in the $\boldsymbol{e}_x$ direction in function of the boost rapidity ω, for non-relativistic states (left plot) and ultra-relativistic states (right plot) in the unboosted frame and for several superposition angles as indicated in the legend. The change on E_G of ultra-relativistic state is due to an increase of the reduced entropy of only the spin subsystems, since for such states the entropies associated with the chiralities and with the momenta are invariant.

Taking as an example the Bell-like state with $\alpha = \pi/4$ and $\beta = 0$ (the red curve in Fig. 2), Figure 3 shows the difference between the variations of global entanglement

$$\delta E_G = 6\Delta E_G[\rho_{\psi_{\rm Bell}}] - 4\Delta E_G[\rho_{\phi_{\rm Bell}}] \tag{26}$$

where $\Delta E_G[\rho_\xi] = E_G[\rho_{\xi'}] - E_G[\rho_\xi]$ is the variation of the global entanglement induced by a Boost in a given pure state ρ_ξ. The first row shows $|\delta E_G|$ as a function of both p/m and the boost rapidity ω while the second row depicts it as a function of p/m for several values of ω and for boost in two directions. For $p \gg m$, $\delta E_G \to 0$, and the variation of E_G of the bispinor Bell state is the same of spin $\frac{1}{2}$ Bell state (8). We also notice that for small ω and small p/m, $\delta E_G \to 0$ since, as discussed in Sec. II, in this limit both states are equivalent and the chiralities are separable from the other subsystems.

4. Conclusions

In summary, we have described the effects of Lorentz boosts on the quantum entanglement encoded in two-particle states constructed as Dirac bispinors Bell-like superpositions. In this scenario, given the composite structure of the Dirac bispinors, each particle has 3 DoFs: spin, chirality and momentum, and we have considered the latter as a dichotomic variable, such that the joint two-particle state is interpreted as a 6 qubit state. Considering the non-relativistic regime, the constructed superposition has the same correlation properties of the relativistic spin $\frac{1}{2}$ state described in [8].

We verified that Lorentz boosts affect the Meyer-Wallach global measure of entanglement [26] shared among the different DoFs of the system. In general the frame transformation increases the amount of the global entanglement encoded in the state in a similar fashion to what was reported for bispinor superpositions not including the momentum DoF [14]. In correspondence to the boost effects on spin $\frac{1}{2}$ states as shown in [8], the entanglement in the partition { all DoFs of A; all DoFs of B} is invariant under boosts. Moreover, for ultra-relativistic states the variation of the global entanglement is the same as the one calculated for spin $\frac{1}{2}$ states in the framework of Wigner rotations. In this limit, the entanglement of the chiralities with all other DoFs of the two-particle state is invariant and all changes in the global entanglement are due to an increase in the spins entanglement.

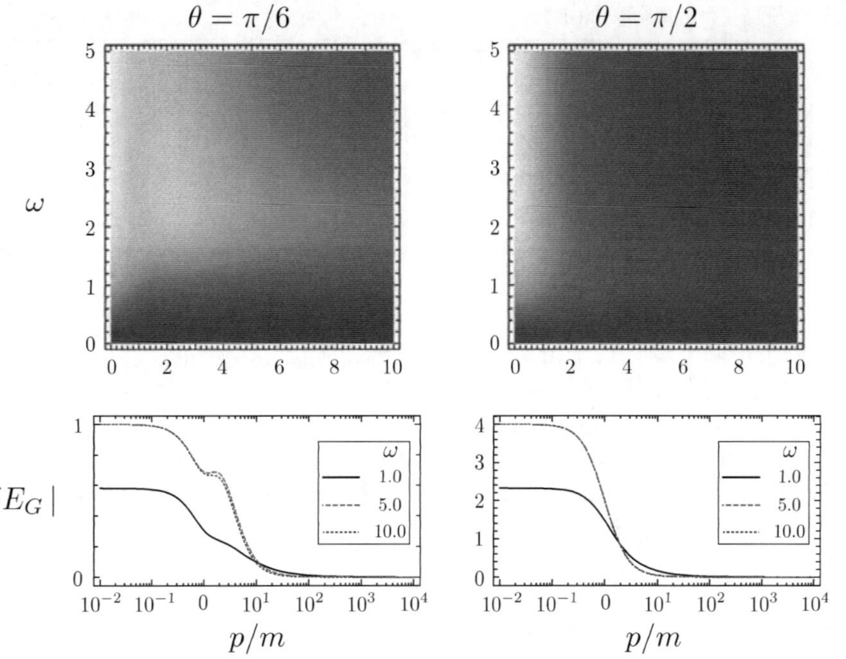

Figure 3. Absolute difference $|\delta E_G|$ between the variation of the global entanglement in bispinor Bell-like states (9) and in spin $\frac{1}{2}$ Bell-like states (8) under a Lorentz boost in the direction $\boldsymbol{n} = \sin\theta \boldsymbol{e}_x + \cos\theta \boldsymbol{e}_x$, as schematically depicted in Fig. 1. The plots are in function of p/m and the boost rapidity ω (top row), and in function of p/m for different values of ω (bottom row). The plots are for two different boost directions: $\pi/6$ (left column) and $\pi/2$ (right column). For ultra-relativistic states $p \gg m$ and the variation in both states are the same as a consequence of the invariance of the chirality subsystem entropy.

Our results generalize the preliminar discussion of [13] connecting the description of the transformation properties of quantum entanglement encoded in spin $\frac{1}{2}$ states, constructed in the framework of the irreps of the (proper) Lorentz group [5–9], with those of bispinor states, described via irreps of the *complete* Lorentz group [13, 14], which includes the parity symmetry related to the mass term of the Dirac equation. The role played by the mass with respect to such connection between the frameworks is two-folded: although the states considered here have equal entanglement properties in the non-relativistic regime, the variation of the global entanglement under boosts is the same only in the ultra-relativistic regime, for which the bispinor states have definite chirality.

Acknowledgments

VASVB acknowledge financial support from the Max Planck Gesellschaft through an Independent Max Planck Research Group.

References

[1] Jozsa R, Abrams D S, Dowling J P and Williams C P 2000 Phys. Rev. Lett. **85** 2010
[2] Giovanetti V, Lloyd S and Maccone L 2001 Nature **412** 417
[3] Yurtsever U and Dowling J P 2002 Phys. Rev. A **65** 052317
[4] Kent A, Munro W J and Spiller T P 2011 Phys. Rev. A **84** 012326
[5] Gingrich R M and Adami C 2002 Phys. Rev. Lett. **89** 270402

IOP Conf. Series: Journal of Physics: Conf. Series **1275** (2019) 012026 doi:10.1088/1742-6596/1275/1/012026

[6] Ahn D, Lee H J, Moon Y H and Hwang S W 2003 Phys. Rev. A **67** 012103
[7] Jordan T F, Shaji A and Sudarshan E C G 2007 Phys. Rev. A **75** 022101
[8] Friis N, Bertlmann R A, Huber M and Hiesmayr B C 2010 Phys. Rev. A **81** 042114
[9] Cafaro C, Capozziello S and Mancini S 2012 Int. J. Theor. Phys. **51** 2313
[10] Caban P and Rembielinski 2006 Phys. Rev. A **74** 042103
[11] Choi T, Hur J and Kim J 2011 Phys. Rev. A **84** 012334
[12] Choi T 2013 J. Korean Phys. Soc. **62** 1085
[13] Bittencourt V A S V, Bernardini A E and Blasone M 2018 J. Phys.: Conf. Ser. **1071** 012001
[14] Bittencourt V A S V, Bernardini A E and Blasone M 2018 Phys. Rev. A **97** 032106
[15] Rembieliński J, Caban P, Smoliński K A and Khrapko S 2018 Quantum Inf. Proc. **17** 262
[16] Fonda L and Ghirardi G C 1970 *Symmetry Principles in Quantum Physics* (New York: Marcel Dekker INC)
[17] Tung W K 2003 *Group Theory* (London: World Scientific Publishing)
[18] Castro Neto A H, Guinea F, Peres R, Novoselov K S and Geim A K 2009 Rev. Mod. Phys. **81** 109
[19] Rozhkov A V, Sboychakov O, Rakhmanov A L and Nori F 2016 Phys. Rep. **648**
[20] Bittencourt V A S V and Bernardini A E 2017 Phys. Rev. B **95** 195145
[21] Bittencourt V A S V, Blasone M and Bernardini A E 2018 Phys. Rev. B **97** 125435
[22] Mizrahi S S 2009 Physica Scripta **T135** 014007
[23] Bernardini A E and Mizrahi S S 2014 Physica Scripta **89** 075105
[24] Bittencourt V A S V and Bernardini A E 2016 Ann. Phys. **364** 182
[25] Alsing P M and Milburn G J 2002 Quantum Inf. Comput. **2** 487
[26] Meyer D A and Wallach N R 2002 J. Math. Phys. **43** 4273
[27] Greiner W 2000 *Relativistic Quantum Mechanics: Wave equations* (Berlin: Springer)
[28] Pachos J and Solano E 2002 Quantum Inf. Comput. **3** 115
[29] Horodecki R, Horodecki P, Horodecki M and Horodecki K 2009 Rev. Mod. Phys. **81** 865

IOP Conf. Series: Journal of Physics: Conf. Series **1275** (2019) 012027 doi:10.1088/1742-6596/1275/1/012027

Theory of Images and Quantum Mechanics, a common paradigm

E Celeghini

Dipartimento di Fisica, Università di Firenze, I50019 Sesto Fiorentino, Firenze, Italy
Departamento de Fisica Téorica, Universidad de Valladolid, E-47005, Valladolid, Spain

E-mail: `celeghini@fi.infn.it`

Abstract.

The Paradigms introduced in philosophy of science one century ago are shown to be quite more satisfactory of that introduced by Galileo. This is particularly evident in the physics based on Hilbert Spaces and related mathematical structures that we apply in this paper to Quantum Mechanics and to Theory of Images. An exhaustive discussion, that include the algebraic analysis of the operators acting on them, exhibits that the Hilbert Spaces -that have fixed dimension- must be generalized to the Rigged Hilbert Spaces that contain spaces with continuous and discrete dimensions. This is the property of Rigged Hilbert Spaces that allows a consistent formal description of the physics we are considering. Theory of Quantum Mechanics and of Images are similar and the fundamental difference between them comes from the definition of measure that is outside the theory of the spaces: while in Quantum Mechanics the measure is a probabilistic action, in Images it is a classical functional.

1. Introduction

The paradigm in philosophy of science is the *Gestalt*, the general frame, that define the fundamental attitudes of scientists active in similar fields.

The first general paradigm in science has been stated by Galileo: it could be summarized in the statement that experience is necessary and sufficient to find the "Truth".

The paradigms of the philosophy of science of the XX century are less arrogant and more problematic. The author point of view can be summed up in the Wittgenstein statement: *"The procedure of induction consists in accepting as true the simplest law that can be reconciled with our experience"* [1]. This means that a law is considered a recipe that worked -up to now- every time has been taken into account and, between the working recipes, is the *"simplest"* one.

The role of the general paradigm is relevant in physics, where we have at least three conflicting sectorial paradigms that, at the moment, we have not been able to reduce to a single one. They are General Relativity, Classical Physics and theories related to Hilbert Spaces with their operators. It is normally assumed that in the appropriate limit all of them are equivalent but, for the time being, this is more an act of faith that a scientific truth.

Moreover in physics the word *"Paradigm"* is almost always substituted by the word *"Theory"* that is less general but suggests a strong connection with *"Truth"* and the philosophical attitudes of XX century are not too popular since the analysis follows meanly the mental framework of

Content from this work may be used under the terms of the Creative Commons Attribution 3.0 licence. Any further distribution of this work must maintain attribution to the author(s) and the title of the work, journal citation and DOI.
Published under licence by IOP Publishing Ltd

Galileo: the physicists tend to refuse to discuss about the meaning of the theories believing that physical concepts have in themselves a content of reality.

In this paper we consider only the Hilbert Space paradigm and its applications to Quantum Mechanics (QM) and Theory of Images (TI). We discuss as the Hilbert spaces (HS) -where the dimension is fixed- are insufficient to describe consistently these applications where operators with spectrum of dimension $\aleph_0$ (discrete) and $\aleph_1$ (continuous) appear together. We thus introduce a generalization of HS, the Rigged Hilbert Spaces (RHS), that contains simultaneously spaces of dimensions $\aleph_0$ and $\aleph_1$[2, 3, 4, 5, 6].

RHS -and not HS- are indeed, in the opinion of the author, the more appropriate specific scheme where QM and TI can be formulated. To read this paper the copious mathematical literature on RHS (see, for instance, ref.[2]) is not necessary: at the level of applicative aspects RHS are quite similar to HS, except for the fact that inside RHS operators and bases of dimension $\aleph_0$ and $\aleph_1$ can be considered together[7].

Summarizing, this paper is focused on the applications of RHS to QM and TI. The conclusion is that both applications need similar RHS. A fundamental difference between QM and TI exists not at the level of the spaces but at the level of something added to the scheme from outside RHS: it is the definition of the measure that is probabilistic in QM and described by deterministic functionals in TI.

2. Quantum Mechanics Hilbert Space

Hilbert Space is a generalization of inner product vector space. It has fixed number of dimensions and this is the reason we consider in this paper the extension to RHS. The dimensions normally considered in Quantum Mechanics are $\aleph_0$ (the dimension of natural numbers) and $\aleph_1$ (the dimension of the continuum).

Essentials of Quantum Mechanics are:

1) Any physical state can be exhaustively characterized by a vector.

2) All observables correspond to Hermitian operators.

3) The measure process is absolutely different from all others processes.

While the points 1) and 2) are the same in QM and TI, the point 3) is absolutely dissimilar, because QM is a quantum theory and TI a classical one. Let us thus consider this point 3) of QM in slight more detail.

3a) The value of one observable on one of the eigenstates of its corresponding operator is equal to the corresponding eigenvalue.

3b) The value of one observable on a generic state is not fixed but probabilistic and restricted to the eigenvalues of the corresponding operator. The probability of each eigenvalue is proportional to the modulus square of the related component.

As stated by Bohr the measure is something added to the theory of HS. The definition of measure in QM is probabilistic and independent from the formal theory of HS because it describes a peculiar action that implies an interaction of the physical microscopic state with a macroscopic object. All the rest of QM has nothing to do with probability.

Theory of Quantum Mechanics, as it is, is not actually considered too satisfactory. Let us recall the Feynman's authoritative citations:

"I think I can safely say that nobody understands Quantum Mechanics"[8].

The point is that physicists pose a lot of questions like:

- Is the wave function a real object?
- Do measurement generates many-worlds?
- What is the Copenhagen statistical collapse?
- What means the Born rule for repeated measurement?

IOP Conf. Series: Journal of Physics: Conf. Series **1275** (2019) 012027 doi:10.1088/1742-6596/1275/1/012027

All these questions are perfectly reasonable inside the philosophy of the XVI-XVIII centuries but a complete *nonsense* in the spirit of the contemporary theory of science (see, for instance, [1, 9, 10]) where a law, a physical law, is related to the known experimental results and isn't an absolute. Therefore in this spirit Quantum Mechanics is a magnificent recipe because it describes our electronic apparatuses and it has been -up to now- in agreement with all the experimental results, but it is only a recipe: there is not any ontological content in its postulates.

To the author opinion, the formal limit of standard QM based on HS is that -to consider a particular case- it is unable to describe together, in a mathematically acceptable way, energy and position operators of the harmonic oscillator. In general we believe that the essential problem of HS in QM is that they have fixed dimensions, while we have to consider together operators with spectrum of different dimensions. This point will be discussed in next Sections and the proposed solution is to move from HS to the *"simpler"*[1] RHS.

3. Rigged Hilbert Space on the line Quantum Mechanics

Rigged Hilbert Space on the line $\mathbb{R}$ is defined by[5, 6]:

a) Two continuous bases $\{|\,x\,\rangle\}$ and $\{|\,p\,\rangle\}$ in $\mathbb{R}$ with properties:

$$\langle\, x\,|\,x'\,\rangle = \sqrt{2\pi}\;\delta(x-x')\,, \qquad \frac{1}{\sqrt{2\pi}}\int_{-\infty}^{+\infty} |x\rangle\, dx\, \langle x| = \mathbb{I}\,,$$

$$|p\rangle := \frac{1}{\sqrt{2\pi}}\int_{-\infty}^{+\infty} dx\; e^{+\mathbf{i}px}\,|x\rangle \qquad \Rightarrow \qquad \langle\, p\,|\,x\,\rangle = e^{-\mathbf{i}px}\,,$$

$$\langle\, p\,|\,p'\,\rangle = \sqrt{2\pi}\,\delta(p-p')\,, \qquad \frac{1}{\sqrt{2\pi}}\int_{-\infty}^{+\infty} |p\rangle\, dp\, \langle p| = \mathbb{I}\,.$$

b) Two bases in $L^2(\mathbb{R})$ -the space of square integrable functions defined in $\mathbb{R}$- the Hermite functions $\{\psi_n(x)\}$ and $\{\psi_n(p)\}$:

$$\psi_n(x) := \frac{e^{-x^2/2}}{\sqrt{2^n n! \sqrt{\pi}}}\, H_n(x)\,,$$

$$\psi_n(p) := (-\mathbf{i})^n [FT]\, \psi_n(x) = \frac{(-\mathbf{i})^n}{\sqrt{2\pi}}\int_{-\infty}^{\infty} dx\; e^{\,\mathbf{i}px}\, \psi_n(x)$$

(where H_n are the Hermite polynomials). The $\{\psi_n\}$ have indeed the following properties:

$$\int_{-\infty}^{\infty} \psi_n(x)\, \psi_{n'}(x)\, dx = \delta_{n,n'}\,, \qquad \sum_{n=0}^{\infty} \psi_n(x)\; \psi_n(x') = \delta(x-x')\,,$$

$$\int_{-\infty}^{\infty} \psi_n(p)\, \psi_{n'}(p)\, dp = \delta_{n,n'}\,, \qquad \sum_{n=0}^{\infty} \psi_n(p)\; \psi_n(p') = \delta(p-p')\,.$$

Note that the Kronecker delta and the Dirac delta appear together showing that both the dimensions $\aleph_0$ and $\aleph_1$ are present in $L^2(\mathbb{R})$ that thus cannot be a HS.

IOP Conf. Series: Journal of Physics: Conf. Series **1275** (2019) 012027 doi:10.1088/1742-6596/1275/1/012027

c) One discrete basis in the line $\{|n\rangle\}$:

Let us define the vector $|n\rangle \in \mathbb{R}$

$$|n\rangle \;:=\; (2\pi)^{-1/4} \int_{-\infty}^{\infty} dx\; \psi_n(x)\, |x\rangle \;.$$

By inspection, the discrete set $\{|n\rangle\}$ is orthonormal and complete

$$\langle\, n \,|\, n' \,\rangle = \delta_{n\, n'} \;, \qquad\qquad \sum_{n=0}^{\infty} |n\rangle\, \langle n| \;=\; \mathbb{I}' \,.$$

$\{|n\rangle\}$ is thus another basis on the real line $\mathbb{R}$, but with cardinality $\aleph_0$. As

$$\langle n|x\rangle \;=\; \langle x|n\rangle \;=\; (2\pi)^{1/4}\, \psi_n(x)\,,$$

$$\langle n|p\rangle \;=\; \mathbf{i}^n\, (2\pi)^{1/4}\, \psi_n(p)\,, \qquad\qquad \langle p|n\rangle \;=\; (-\mathbf{i})^n\, (2\pi)^{1/4}\, \psi_n(p)\,,$$

(where $\mathbf{i}$ is the imaginary unit) the Hermite functions are nothing else that the transformation matrices between the continuous bases and the discrete one in $\mathbb{R}$.

d) The Weyl-Heisenberg algebra (WHA):

Let us introduce in the space $\{\psi(x)\}$ the operators X, P, N (that read position, momentum and order of the Hermite functions) and the identity **I**. Taking into account the recurrence relation of the Hermite polynomials, we can write

$$N\; \psi_n(x) \;=\; n\; \psi_n(x)\,, \qquad\quad \mathbf{I}\; \psi_n(x) \;=\; \psi_n(x)\,,$$

$$a\; \psi_n(x) \;=\; \sqrt{n}\; \psi_{n-1}(x)\,, \qquad\quad a^\dagger\; \psi_n(x) = \sqrt{n+1}\; \psi_{n+1}(x)\;;$$

where

$$a = \frac{1}{\sqrt{2}}(X + \mathbf{i}P)\;, \qquad a^\dagger = \frac{1}{\sqrt{2}}(X - \mathbf{i}P).$$

By inspection $N, a, a^\dagger$ and $\mathbf{I}$ close the Weyl-Heisenberg algebra

$$[N, a] = -a\;, \qquad [N, a^\dagger] = \; a^\dagger\;, \qquad [a, a^\dagger] = \mathbf{I}\;, \qquad [\mathbf{I}, \bullet] = 0\;;$$

with Casimir invariant $\mathbb{C} = (N + 1/2)\mathbf{I} - \frac{1}{2}\{a, a^\dagger\}$. $\{\psi_n(x)\}$ is isomorphic to $\{|n\rangle\}$ and define the representation of WHA with Casimir $\mathbb{C} = 0$ that is irreducible: the isomorphism between the Universal Enveloping Algebra of the WHA, $UEA[h(1)]$, and the space of the linear operators acting on $\{|n\rangle\}$ can be extended to $\{\psi_n\}$ i.e. every linear operators on $\mathbb{R}$ and $L^2(\mathbb{R})$ can be written (with α, β, γ integers and $c_{\alpha,\beta,\gamma}$ complex numbers) as

$$\mathcal{O} \;=\; \sum_{\alpha,\beta,\gamma} c_{\alpha,\beta,\gamma}\; (a^\dagger)^\alpha\, N^\beta\, a^\gamma \;\;.$$

The representation is also unitary: every transformation of $\{\psi_n\}$ under the group is a basis.

IOP Conf. Series: Journal of Physics: Conf. Series **1275** (2019) 012027 doi:10.1088/1742-6596/1275/1/012027

4. Rigged Hilbert Space on the disk and Theory of Images

Theory of Images is quite similar to the theory of Quantum Mechanics exposed above. Again we have a RHS, constructed now on the disk $\mathbb{D}$. As told before, the fundamental difference is not at the level of Rigged Hilbert Space, but at the level of measure that is defined in a completely different way: as Theory of Images is a classical theory, the measure doesn't operate on images but is realized by means of functionals.

Rigged Hilbert Space of Images is defined on the disk $\mathbb{D}$ by[7, 11]:

a) A continuous basis in the disk $\{|r,\phi\rangle\}$

A point in the disk is characterized by two coordinates. By means of the two continuous variables $\{r,\phi\}$ we define a continuous basis $\{|r,\phi\rangle\}$ (that corresponds to the $\{|x\rangle\}$ of $\mathbb{R}$):

$$\langle r_1,\phi_1|r_2,\phi_2\rangle = 2\pi\,\delta({r_1}^2-{r_2}^2)\,\delta(\phi_1-\phi_2)\,, \qquad \frac{1}{\pi}\int_{-\pi}^{+\pi} d\phi \int_0^1 dr\ r\ |r\,\phi\rangle\langle r,\phi| = \mathbb{I}\,,$$

where the ranges are $(0 \le r \le 1,\ -\pi \le \phi < \pi)$.

b) A Zernike basis in the space of square integrable functions $L^2(\mathbb{D})$ defined in $\mathbb{D}$.

Like in QM we find that the Hermite functions are a basis in $L^2(\mathbb{R})$, starting from the Zernike polynomials $\{R_n^m(r)\}$ we find that the Zernike functions $\{R_n^m(r)\,e^{\mathbf{i}m\phi}\}$ [12] are a basis in the space of images $L^2(\mathbb{D})$.

Following Dunkl [13] we change the Born and Wolf parametrization $\{n,m\}$ to $\{u,v\}$:

$$u := \frac{n+m}{2} \qquad v := \frac{n-m}{2}$$

that have ranges simpler then $\{n,m\}$, as they are independent natural numbers i.e. $\{u = 0,1,2,\ldots,\ \ v = 0,1,2,\ldots\}$. We introduce also a normalization factor to have the explicit orthonormalization we are used in QM and we redefine the Zernike functions as:

$$Z_{u,v}(r,\phi) := \sqrt{u+v+1}\ R_{u+v}^{u-v}(r)\ e^{\mathbf{i}(u-v)\phi}\,.$$

$\{Z_{u,v}(r,\phi)\}$ have the following properties:

1. $Z_{u,v}(r,\phi)$ are square integrable in $L^2(\mathbb{D},\, r\,dr\,d\phi) \equiv L^2(\mathbb{D})$.
2. They have the symmetries:

$$Z_{v,u}(r,\phi) = Z^*_{u,v}(r,\phi) = Z_{u,v}(r,-\phi)\,.$$

3. They are orthonormal:

$$\langle Z_{u',v'}|Z_{u,v}\rangle = \frac{1}{\pi}\int_{-\pi}^{+\pi} d\phi \int_0^1 dr\ r\ Z^*_{u,v}(r,\phi)\ Z_{u',v'}(r,\phi) = \delta_{u,u'}\,\delta_{v,v'}\,.$$

4. They satisfy the completeness relation:

$$\sum_{u,v=0}^{\infty} Z^*_{u,v}(r_1,\phi_1)\ Z_{u,v}(r_2,\phi_2) = \pi\ \delta(r_1^2-r_2^2)\ \delta(\phi_1-\phi_2)\,.$$

Thus the Zernike functions $\{Z_{u,v}(r,\phi)\}$ are an orthonormal basis for the square integrable functions of the disk $L^2(\mathbb{D})$ exactly as the Hermite functions $\{\psi_n\}$ are an orthonormal basis

IOP Conf. Series: Journal of Physics: Conf. Series **1275** (2019) 012027 doi:10.1088/1742-6596/1275/1/012027

for the square integrable functions defined in $\mathbb{R}$. Any function $f(r,\phi) \in L^2(\mathbb{D})$ can be written, in the sense of convergence on the Hilbert space $L^2(\mathbb{D})$, as superposition of $Z_{u,v}(r,\phi)$:

$$f(r,\phi) = \sum_{u,v=0}^{\infty} f_{u,v}\, Z_{u,v}(r,\phi)$$

where $f_{u,v}$ are complex numbers given by

$$f_{u,v} = \frac{1}{\pi}\int_{-\pi}^{+\pi} d\phi \int_0^1 dr\; r\; Z^*_{u,v}(r,\phi)\; f(r,\phi).$$

Moreover

$$\langle f|f\rangle = \frac{1}{\pi}\int_{-\pi}^{+\pi} d\phi \int_0^1 dr\; r\; |f(r,\phi)|^2 = \sum_{u,v=0}^{\infty} |f_{u,v}|^2 < \infty\,.$$

Note that we have assumed $\langle f|f\rangle < \infty$ but not $\langle f|f\rangle = 1$. If we restrict $f(r,\phi)$ to be real (as it is usual in optics) we have $f_{v,u} = f^*_{u,v}$.

c) A discrete basis in the disk $\{|u,v\rangle\}$.

We follow the same procedure of the line and introduce in the disk $\mathbb{D}$ the discrete set $\{|u,v\rangle\}$:

$$|u,v\rangle := \frac{1}{\pi}\int_{-\pi}^{+\pi} d\phi \int_0^1 dr\; r\; |r,\phi\rangle\; Z_{u,v}(r,\phi) \qquad (u,v = 0,1,2,\ldots).$$

Like in QM, we have the orthonormal discrete relations:

$$\langle u,v|u',v'\rangle = \delta_{uu'}\,\delta_{vv'}\,, \qquad \sum_{u,v}^{\infty} |u,v\rangle\langle u,v| = \mathbb{I}'\,,$$

where, as $\{|u,v\rangle\}$ is a discrete basis and $\{|r,\phi\rangle\}$ a continuous basis, the two identities $\mathbb{I}$ and $\mathbb{I}'$ are identities in different spaces as it happens on the line. Likewise $\{\psi_n\}$ in QM, the Zernike functions are the transition matrices from continuous and discrete basis:

$$Z_{u,v}(r,\phi) = \langle r,\phi|u,v\rangle\,.$$

d) the disk $\mathbb{D}$ is an algebra representation of $su(1,1) \oplus su(1,1)$.

Analogously to the line, we introduce the discrete operators U and V such that

$$U\,Z_{u,v}(r,\phi) = u\,Z_{u,v}(r,\phi)\,, \qquad V\,Z_{u,v}(r,\phi) = v\,Z_{u,v}(r,\phi)$$

and the continuous ones R, D_r and Φ such that

$$R\,Z_{u,v}(r,\phi) = r\,Z_{u,v}(r,\phi), \quad D_r\,Z_{u,v}(r,\phi) = \frac{\partial Z_{u,v}(r,\phi)}{\partial r}, \quad \Phi\,Z_{u,v}(r,\phi) = \phi\,Z_{u,v}(r,\phi)\,.$$

Four additional operators on the $\{Z_{u,v}(r,\phi)\}$ define the recurrence relations:

$$A_+\,Z_{u,v}(r,\phi) = (u+1)\;Z_{u+1,v}(r,\phi)\,, \qquad A_-\,Z_{u,v}(r,\phi) = u\;Z_{u-1,v}(r,\phi)\,,$$

$$B_+\,Z_{u,v}(r,\phi) = (v+1)\;Z_{u,v+1}(r,\phi)\,, \qquad B_-\,Z_{u,v}(r,\phi) = v\;Z_{u,v-1}(r,\phi)\,;$$

where

IOP Conf. Series: Journal of Physics: Conf. Series **1275** (2019) 012027 doi:10.1088/1742-6596/1275/1/012027

$$A_+ := \frac{e^{+i\Phi}}{2}\sqrt{\frac{U+V+1}{U+V}}\left[-(1-R^2)D_R + R(U+V+2) + \frac{1}{R}(U-V)\right],$$

$$A_- := \frac{e^{-i\Phi}}{2}\sqrt{\frac{U+V+1}{U+V}}\left[+(1-R^2)D_R + R(U+V) + \frac{1}{R}(U-V)\right],$$

$$B_+ := \frac{e^{-i\Phi}}{2}\sqrt{\frac{U+V+1}{U+V}}\left[-(1-R^2)D_R + R(U+V+2) - \frac{1}{R}(U-V)\right],$$

$$B_- := \frac{e^{+i\Phi}}{2}\sqrt{\frac{U+V+1}{U+V}}\left[+(1-R^2)D_R + R(U+V) - \frac{1}{R}(U-V)\right].$$

The relevant commutation relations are :

$$[U,V]=0, \quad [U,A_\pm]=\pm A_\pm, \quad [U,B_\pm]=0,$$

$$[V,A_\pm]=0, \quad [V,B_\pm]=\pm B_\pm, \quad [A_\pm,B_\pm]=0, \quad [A_\pm,B_\mp]=0\,;$$

and, on the $\{Z_{u,v}(r,\phi)\}$:

$$[A_+,A_-]\,Z_{u,v}(r,\phi) \;=\; -2(u+1/2)\,Z_{u,v}(r,\phi)\,;$$

defining therefore $A_3 := U+1/2$ we find that $\{A_+,\,A_3,\,A_-\}$ are the generators of one algebra $su(1,1)$, we call $su_A(1,1)$:

$$[A_+,\,A_-] \;=\; -2A_3 \qquad [A_3,\,A_\pm] \;=\; \pm A_\pm\,.$$

Analogously

$$[B_+,B_-]\,Z_{u,v}(r,\phi) = -2(v+1/2)\,Z_{u,v}(r,\phi)$$

exhibits that $\{B_+,\,B_3 := V+1/2\,,\,B_-\}$ are the generators of one other algebra $su(1,1)$, we call $su_B(1,1)$:

$$[B_+,\,B_-] \;=\; -2B_3 \qquad [B_3,\,B_\pm] \;=\; \pm B_\pm\,.$$

Finally, as

$$[A_i,B_j]=0\,,$$

we obtain on the disk $\mathbb{D}$ a differential representation of the 6 dimensional Lie algebra $su_A(1,1)\oplus su_B(1,1)$.

We can now calculate the values of the Casimir invariants:

$$\mathbb{C}_A\,Z_{u,v}(r,\phi) \;=\; \left[A_3^2 - \frac{1}{2}\{A_+,A_-\}\right]\,Z_{u,v}(r,\phi) \;=\; -\frac{1}{4}\,Z_{u,v}(r,\phi)\,,$$

$$\mathbb{C}_B\,Z_{u,v}(r,\phi) \;=\; \left[B_3^2 - \frac{1}{2}\{B_+,B_-\}\right]\,Z_{u,v}(r,\phi) \;=\; -\frac{1}{4}\,Z_{u,v}(r,\phi)\,.$$

As the Casimir of the discrete principal series of $su(1,1)$ is $j(j+1)$ with $j=-1/2,-3/2,\ldots$, the RHS $\{Z_{u,v}(r,\phi)\}$ is isomorphic to the fundamental unitary irreducible representation $D^+_{-1/2}\otimes D^+_{-1/2}$ of $SU(1,1)\otimes SU(1,1)$ where the eigenvalues of U and V are the natural numbers and the eigenvalues of A_3 and B_3 are the half-integers numbers $1/2,\,3/2,\,5/2,\ldots$ [14].

IOP Conf. Series: Journal of Physics: Conf. Series **1275** (2019) 012027 doi:10.1088/1742-6596/1275/1/012027

5. Operators, UEA and Lie group in the space of Images

The Universal Enveloping Algebra of $su(1,1) \oplus su(1,1)$ UEA$[su(1,1) \oplus su(1,1)]$ is the algebra constructed on the ordered monomials $A_+^\alpha A_3^\beta A_-^\gamma B_+^\delta B_3^\epsilon B_-^\zeta$ where $\alpha, \beta, \gamma, \delta, \epsilon$ and ζ are natural numbers. Analogously to $\{\psi_n\}$:

- $\{Z_{u,v}(r,\phi)\}$ is an irreducible representation so that every linear operator $\mathcal{O}$ acting on $\mathbb{D}$ and $L^2(\mathbb{D})$ belongs to UEA$[su(1,1) \oplus su(1,1)]$.

- $\{Z_{u,v}(r,\phi)\}$ is a unitary representation so that every base in the space $L^2(\mathbb{D})$ has the form $\{g\, W_{u,v}(r,\phi)\}$ where g is an arbitrary element of the group $SU(1,1) \otimes SU(1,1)$.

6. Measure in the space of Images

Measure does not belong to the theory of the spaces. It is something added in applications to the mathematical formalism. HS and RHS are completely independent from the measure and different measures describe different physics. We have discussed the probabilistic QM measure. As Theory of Images is a classical theory its measure is deterministic: mathematically is a functional i.e. a mapping $\mathcal{M}$ from the images into the real numbers:

$$\mathcal{M}\,[|f\rangle] = m\,[|f\rangle] \in \mathbf{R}\,.$$

For example, the luminosity $\mathcal{L}$ of the image $|f\rangle$ could be defined as the number

$$\mathcal{L}\,[|f\rangle] \;:=\; \frac{1}{\pi} \int_{-\pi}^{+\pi} d\phi \int_0^1 dr\; r\; |f(r,\phi)|^2\,.$$

7. Conclusions

- RHS describe both quantum and classical theories.
- In RHS we have continuous and discrete variables that, through the related lowering and rising operators, define a Lie algebra.
- The space of the wave functions is an unitary irreducible representation of this algebra.
- Elements of the group define the bases of this space.
- UEA is a basis in the space of operators acting on the system.
- Classical measures are described by functionals.

References

[1] Wittgenstein L 1922 *Tractatus Logico-Philosophicus* (London: Kegan)
[2] Reed M and Simon B 2008 *Methods of Modern Mathematical Physics 1, 2* (London: Acad Press)
[3] Böhm A 1978 *The Rigged Hilbert Space and Quantum Mechanics* (Berlin: Springer-Verlag)
[4] Gadella M and Gomez F 2003 *Int J Theor Phys* **42** 2225
[5] Celeghini E 2015 *J Phys: Conf Ser* **626** 012047
[6] Celeghini E, Gadella M and del Olmo M A 2016 *J Math Phys* **57** 072105
[7] Celeghini E 2017 *J Phys: Conf Ser* **880** 012055
[8] Feynman R 1967 *The Character of Physical Law* (Cambridge, Mass: The MIT Press)
[9] Popper K 1992 *The Logic of Scientific Discovery* (London: Routledge)
[10] Bhaskar R 1975 *A Realistic Theory of Science* (London: Verso)
[11] Celeghini E, Gadella M and del Olmo M A 2019 *to be published*
[12] Born M and Wolf E 2005 *Principles of Optics* 7^{th}ed (Cambridge: Univ Press)
[13] Dunkl C F 2001 *Enciclopaedia of Mathematics* **Suppl 3** (Dordrecht: Kluwert), pag 454
[14] Bargmann V 1954 *Ann of Math* **59** 1

IOP Conf. Series: Journal of Physics: Conf. Series **1275** (2019) 012028 doi:10.1088/1742-6596/1275/1/012028

VIP2 in LNGS - Testing the Pauli Exclusion Principle for electrons with high sensitivity

J. Marton[1], A. Pichler[1], A. Amirkhani[4], S. Bartalucci[2], M. Bazzi[2], S. Bertolucci[6], M. Bragadireanu[2,3], M. Cargnelli[1], A. Clozza[2], C. Curceanu[2,3,9], R. Del Grande[2], L. De Paolis[2], J.-P. Egger[7], C. Fiorini[4], C. Guaraldo[2], M. Iliescu[2], M. Laubenstein[8], E. Milotti[5], M. Milucci[2], D. Pietreanu[2,3], K. Piscicchia[9,2], A. Scordo[2], H. Shi[10], D. Sirghi[2,3], F. Sirghi[2,3], L. Sperandio[2], O. Vazquez-Doce[2,11] and J. Zmeskal[1]

[1] Stefan Meyer Institute for subatomic physics, Boltzmanngasse 3, 1090 Vienna, Austria
[2] INFN, Laboratori Nazionali di Frascati, CP 13, Via E. Fermi 40, I-00044, Frascati (Roma), Italy
[3] Horia Hulubei National Institute of Physics and Nuclear Engineering, Str. Atomistilor no. 407, P.O. Box MG-6, Bucharest - Magurele, Romania
[4] Politecnico di Milano, Dipartimento di Elettronica, Informazione e Bioingegneria and INFN Sezione di Milano, Milano, Italy
[5] Dipartimento di Fisica, Università di Trieste and INFN Sezione di Trieste, Via Valerio, 2, I-34127 Trieste, Italy
[6] University and INFN Bologna, Via Irnerio 46, I-40126, Bologna, Italy
[7] Institut de Physique, Université de Neuchâtel 1 rue A.-L. Breguet, CH-2000 Neuchâtel, Switzerland
[8] INFN, Laboratori Nazionali del Gran Sasso, S.S. 17/bis, I-67010 Assergi (AQ), Italy
[9] Centro Fermi - Museo Storico della Fisica e Centro Studi e Ricerche "Enrico Fermi", Roma, Italy
[10] Institut für Hochenergiephysik, Nikolsdorfergasse 18, 1050 Vienna, Austria
[11] Excellence Cluster Universe, Technische Universität München, Boltzmannstraße 2, D-85748 Garching, Germany

E-mail: `johann.marton@oeaw.ac.at`

Abstract. The VIP2 (VIolation of the Pauli Exclusion Principle) experiment at the Gran Sasso underground laboratory (LNGS) is searching for possible violations of standard quantum mechanics predictions in atoms at very high sensitivity. We investigate atomic transitions with precision X-ray spectroscopy in order to test the Pauli Exclusion Principle (PEP) and therefore the related spin-statistics theorem. We will present our experimental method for the search for "anomalous" (i.e. Pauli-forbidden) X-ray transitions in copper atoms, produced by "new" electrons, which could have tiny probability to undergo Pauli-forbidden transition to the ground state already occupied by two electrons. We will describe the VIP2 experimental setup, which is taking data at LNGS presently. The goal of VIP2 is to test the PEP for electrons with unprecedented accuracy, down to a limit in the probability that PEP is violated at the level of 10^{-31}. We will present current experimental results and discuss implications of a possible violation.

Content from this work may be used under the terms of the Creative Commons Attribution 3.0 licence. Any further distribution of this work must maintain attribution to the author(s) and the title of the work, journal citation and DOI.
Published under licence by IOP Publishing Ltd

IOP Conf. Series: Journal of Physics: Conf. Series **1275** (2019) 012028 doi:10.1088/1742-6596/1275/1/012028

1. Introduction

The Pauli Exclusion Principle (PEP) was proposed by Wolfgang Pauli [1] who received the Nobel prize for this finding. It is regarded as a consequence of the spin-statistics theorem. In his Nobel prize lecture Pauli noticed [2] *we want to stress here a law of Nature which is generally valid, namely, the connection between spin and symmetry class. A half-integer value of the spin quantum number is always connected with antisymmetrical states (exclusion principle), an integer spin with symmetrical states.* First the PEP was found for electrons and later extended to all fermions (half-integer spin systems). Thus the nature is divided into bosonic (symmetric) and fermionic (antisymmetric) systems. *Is anything between, with mixed symmetry?* is an obvious question. Up-to-now no contradiction with the validity of PEP was found. Therefore, the PEP is extremely successful in explaining many fundamental features of nature. Here some examples: periodic table of the elements, the existence of neutron stars and the stability of matter [3]. Therefore the spin plays a decisive role, dividing nature in half-integer and integer particles and systems, i.e. in fermions and bosons respectively. Nevertheless, to probe the PEP validity with utmost sensitivity is extremely important because of its role in foundations of quantum physics. Even a tiny violation would point to new physics. The combination of high sensitivity X-ray spectroscopy and the large background suppression in an underground laboratory like LNGS is employed to test PEP in the VIP2 experiment.

2. Experimental tests of PEP

In the past searches for both PEP-violating transitions and PEP-forbidden system were carried out. Such investigations were performed with atoms and nuclei but also astrophysical bounds on the PEP were searched. The accuracy of tests of PEP for electrons has gradually incresed over time (see table 1). The quantity $\beta^2/2$ - widely used in the literature - gives the upper limit of PEP violation and can be traced back to a model introduced by Ignatiev and Kuzmin [5]. We are using this quantity to enable a comparison of the results employing different approaches. The Messiah and Greenberg superselection rule [6] constrains the possible transitions as it forbids symmetry changes in stable systems. Experiments investigating stable systems reached extremely low upper limits for PEP violation [7, 8]. Regarding atomic transitions the first experimental test of PEP for electrons can be traced back to a pioneering experiment by Goldhaber and Scharff-Goldhaber more than half a century ago. In that experiment new electrons from beta decay were used as probes of the validity of PEP. About 40 years later the validity of PEP for electrons was examined in an experiment by Ramberg an Snow using a current to bring new electrons into the system resulted in an upper limit for PEP violation in the order of 10^{-26}. The VIP2 experiment follows the Ramberg-Snow approach with largely improved instrumentation and in a low background LNGS environment.

3. The VIP2 experiment at LNGS

3.1. Experimental concept of VIP2

The VIP2 experiment at the Laboratori Nazionali del Gran Sasso (LNGS) searches for Pauli-forbidden X-ray transitions in copper at the utmost sensivity (see fig.1). If an atomic transition from the 2p state to the already filled 1s ground state (K_α transition) takes place - thus violating PEP - then the transition energy is lower by about 300 eV (i.e. the transition energy for the PEP violating transition is 7747 eV instead of 8048 eV) which can be determined by modern X-ray spectroscopy employing solid state X-ray detectors. The *new* electrons as probes of PEP are introduced by an electric current circulating through a copper foil. Therefore, periods with current (possible PEP-violating transitions) and without current (no PEP violating transitions) are alternating in VIP2.

IOP Conf. Series: Journal of Physics: Conf. Series **1275** (2019) 012028 doi:10.1088/1742-6596/1275/1/012028

Table 1. Upper bounds on $\beta^2/2$ from experimental tests of the PEP validity searching for PEP-forbidden atomic transitions. Only experiments in accordance with the Messiah-Greenberg superselection rule are given. The VIP experiment repeatedly succeeded in improving the bound by optimization of the experimental setup [9].

Year	experiment	Supply of new electrons	$\beta^2/2$	reference
1948	Goldhaber/Scharff-Goldhaber	beta-rays	$3\text{x}10^{-3}$	[10]
1990	Ramberg/Snow	electric current	$1.7\text{x}10^{-26}$	[11]
2006	VIP	electric current	$4.5\text{x}10^{-28}$	[12]
2011	VIP2	electric current	$4.7\text{x}10^{-29}$	[13]
2018	VIP2	electric current	$3.4\text{x}10^{-29}$	[14, 15]

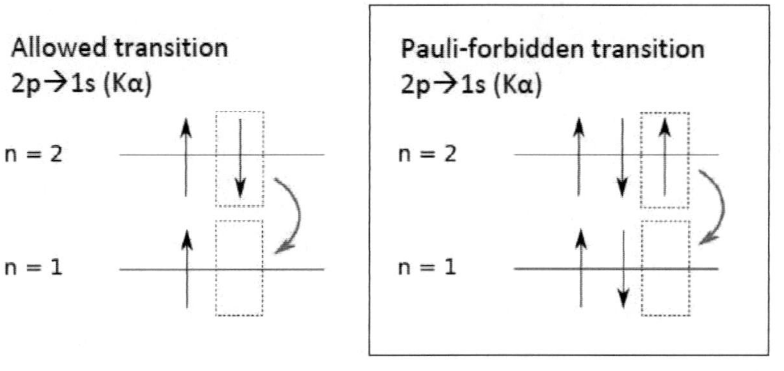

Figure 1. The VIP2 experiment is searching for Pauli-forbidden the atomic transition to the already filled groundstate n=1 (K_α) displayd in the right picture.

3.2. VIP2 setup

The VIP experiment used charge-coupled devices (CCDs) for the X-ray detection. CCDs provide excellent energy resolution but the background reduction is limited because of the lack of timing information. Subsequently the VIP2 experiment employs Silicon Drift Detectors (SDDs) (see fig. 2). These semiconductor detectors are ideally suited for soft X-ray spectroscopy combining large active area, small capacity, arrangements in arrays and provide excellent energy resolution and timing capability. The X-rays generate free electrons, which drift to the anode with an applied electric field. The energy spectrum of the X-rays can be deduced from the electrons arriving at the anode. This current is amplified and read out by an ADC. The timing performance is sufficient (below 1 μs) for anticoincidence with scintillation detectors in order to suppress actively background. Their larger depletion depth leads to a higher detection efficiency of X-rays in the region of interest. The timing capability of SDDs enables the use of an active shielding with scintillators against external radiation. For this purpose, 32 scintillator bars read out by silicon photomultipliers are installed around the SDDs. In addition to the passive shielding around the VIP2 apparatus their veto signal helps to reduce further the background.

Figure 2. SDD detector (left) consisting of an array of 2x4 SDDs and the arrangement in the vacuum box of VIP2 (right). The white circle indicates the position of the X-ray detectors positioned on both sides of the water-cooled copper target conducting a current of about 100A.

3.3. VIP2 Shielding

The shielding of the VIP2 apparatus employs active and passive shielding. The active shielding is using plastic bars readout by silicon photomultipliers. The efficiency of the active shielding depends on the lower threshold of the scintillation detectors. Presently tests and optimization of the active shielding system is in progress. The passive shielding consists of copper and lead blocks surrounding the outer part of the VIP2 apparatus. This method of passive shielding was successfully employed in the VIP-CCD experiment.

4. Measurements and Results

Recently the VIP2 experimental setup was upgraded. In 2018 the new Silicon Drift Detectors [16] (see fig 2) were mounted in the VIP2 setup in the underground laboratory Gran Sasso (LNGS). These detectors consist of 4 arrays with 8 single cell SDDs per array. Each square cell has 8 x 8 mm^2 surface area. In total the active detector arrea is around 20 cm^2. Compared to the formerly employed SDDs the active area is increased by a factor larger than 3. Therefore, the detection efficiency for possible photons from Pauli Exclusion Principle (PEP) violating transitions is increased by a factor greater than 3. An energy resolution of 150 eV -160 eV (FWHM) at 6 keV could be verified for all channels. The experiment is taking data with and without current under stable conditions with the new configuration (see fig.3). Furthermore Monte Carlo simulations conducted with GEANT 4, which have the environmental gamma radiation spectrum as input, verified the background rate detected with this new configuration.

The progress of the VIP2 experiment has been reported in [9, 15, 17, 18, 19]. In 2016 we collected data in a time period of $\sim$70 without current and $\sim$40 days with 100 A current. In fig.3 a typical X-ray spectrum measured with the SDD detector system is displayed.

5. Summary and Outlook

The experimental program for testing a possible PEP violation for electrons made great progress so far. The use of a new type of SDDs as X-ray detectors will further enhance the sensitivity by providing larger sensitive area. Concerning the reduction of the X-ray background we installed a part of the passive shielding with copper and lead which will be completed in 2019. Given a running time of 3 years and alternating measurement with and without current we might either find a tiny violation or we can lower the upper limit of PEP violation by about two orders of magnitude (see fig.4). On the theoretical side studies are made to investigate the concept of new electrons provided by a current in the copper bulk material [20].

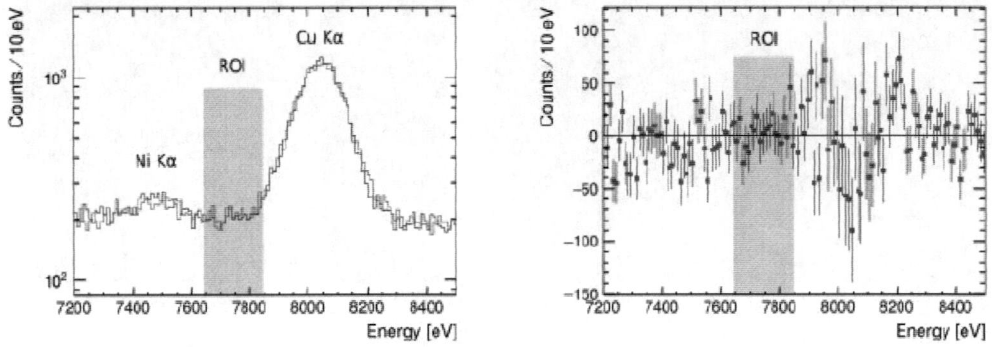

Figure 3. The X-ray spectrum measured with current and without current in VIP2 (left). The region-of interest (ROI) is shaded. After subtraction of the background events the residual X-ray spectrum is shown in the right panel.

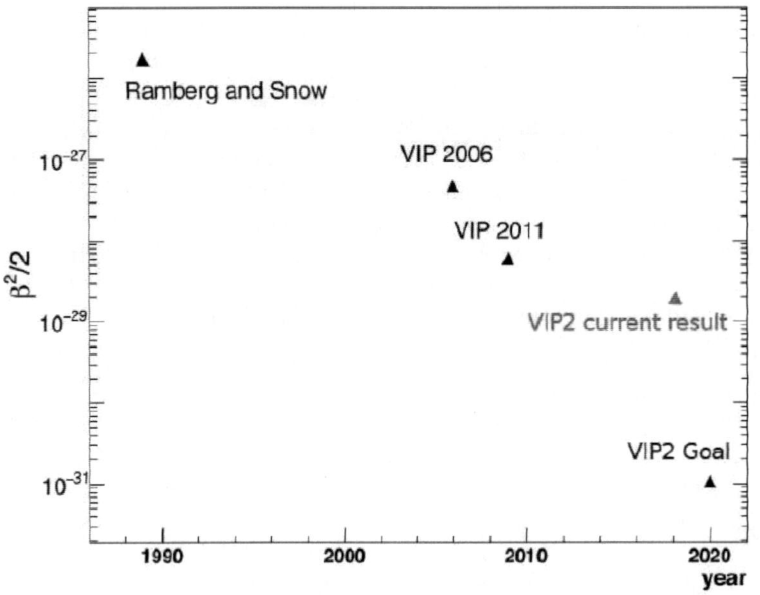

Figure 4. Progress in lowering the upper limit of PEP violation in atomic transitions (respecting the Messiah-Greenberg superselection rule) in the last 20 years. The goal of the new VIP2 experiment with a running time of about 3 year is indicated.

Acknowledgments
We thank H. Schneider, L. Stohwasser, and D. Pristauz-Telsnigg from Stefan-Meyer-Institut for their fundamental contribution in designing and building the VIP2 setup. We acknowledge the very important assistance of the INFN-LNGS laboratory. The support from the EU COST Action CA 15220 and of the EU FET project TEQ (grant agreement 766900) is gratefully acknowledged. We thank the Austrian Science Foundation (FWF) which supports the VIP2 project with the grants P25529-N20, project P 30635-N36 and W1252-N27 (doctoral college particles and interactions) and Centro Fermi for the grant Problemi Aperti nella Meccanica Quantistica. Furthermore, these studies were made possible through the support of a grant from the Foundational Questions Institute, FOXi (and a grant from the John Templeton Foundation (ID 58158). The opinions expressed in this publication are those of the authors and do not necessarily respect the views of the John Templeton Foundation.

References
[1] Pauli W 1925 *Z. Phys.* **31** 765
[2] Pauli W 1945 Wolfgang Pauli Nobel Lecture
[3] Lieb E. H. 1976 The stability of matter *Rev. Mod. Phys.* **48** 553
[4] Curceanu C, Marton J, Budker D, Milotti E and Hall E Putting the Pauli exclusion principle on trial *Cern Courier* 2018
[5] Ignatiev A Yu and Kuzmin V A 1987 *Yad. Fiz.* **46** 786
[6] Messiah A M I and Greenberg O W 1964 *Phys. Rev. B* **136** 248
[7] Bernabei R *et al* 2009 New search for processes violating the Pauli exclusion principle in sodium and in iodine *Eur. Phys. J. C* **62** 327
[8] Back O. *et al* (2004 New experimental limits on violations of the Pauli exclusion principle obtained with the Borexino Counting Test Facility *European Physical Journal C* **37** 421, 10.1140/epjc/s2004-01991-1.
[9] Pichler A 2018 Test of the Pauli Exclusion Principle for electrons in the Gran Sasso underground laboratory *PhD Thesis, TU Vienna* to be published
[10] Goldhaber M and Scharff-Goldhaber G 1948 Identification of Beta-Rays with Atomic Electrons. *Physical Review* **73(12)** 472
[11] Ramberg E and Snow G A 1990 Experimental limit on a small violation of the Pauli principle *Physics Letters B* **238(2)** 438
[12] Bartalucci S *et al* 2006 *Phys. Lett. B* **641** 18
[13] Curceanu C *et al* 2011 Experimental tests of quantum mechanics: Pauli Exclusion Principle Violation (the VIP experiment) and future perspectives *Journal of Physics: Conference Series* **306(1)** 012036
[14] Shi H *et al* 2018 Experimental search for the violation of Pauli Exclusion Principle *Eur. Phys. J C* **78** 319. https://doi.org/10.1140/epjc/s10052-018-5802-4
[15] Shi H *et al* 2018 Search for the violation of Pauli Exclusion Principle at LNGS *EPJ Web of Conferences* **volume 182** 02118
[16] Quaglia, R *et al* 2015 Silicon Drift Detectors and CUBE preamplifiers for high-resolution X-ray spectroscopy *IEEE Transactions on Nuclear Science* **62 (1)** 221
[17] Marton J *et al* 2018 Underground Test of Quantum Mechanics: The VIP2 Experiment *Quantum Foundations , Probability and Information* 155
[18] Curceanu C *et al* 2017 *Entropy* **19 (7)** 300
[19] C. Curceanu *et al* 2017 Quantum mechanics under X-rays in the Gran Sasso underground laboratory *Int. J. Quant. Inf.* **15** 1740004 doi:10.1142/S0219749917400044
[20] Milotti E *et al* 2018 On the Importance of Electron Diffusion in a Bulk-Matter Test of the Pauli Exclusion Principle *Entropy* **20(7)** 515

IOP Conf. Series: Journal of Physics: Conf. Series **1275** (2019) 012029 doi:10.1088/1742-6596/1275/1/012029

Effect of the resonance spectra in the propagation of two decaying entangled particles

Gastón García-Calderón[1], Roberto Romo[2] and Miguel Ángel Terán[1]
[1]Instituto de Física, Universidad Nacional Autónoma de México, 04510 Ciudad de México, Mexico
[2]Facultad de Ciencias, Universidad Autónoma de Baja California, 22800 Ensenada, Baja California, Mexico

E-mail: gaston@fisica.unam.mx

Abstract. An exact analytical solution of the decaying wave function of two identical noninteracting particles, which are entangled by spatial symmetry, is used to analyze the effect of the resonance spectra in the propagation of the decaying probability density outside the interaction potential region. We find, using exactly solvable problems, that a usual approximation that considers the two resonance levels associated with the initial states, is affected substantially in the case of sharp high energy resonances by disrupting the pure exponential decaying regime exhibited by the two resonance level approximation, whereas for broad high energy resonances, we find that the probability density profile is well described by the two resonance approximation.

1. Introduction

In recent work it is shown rigorously, using the analytical properties of the outgoing Green's function for single particle potentials that vanish exactly beyond a distance, that the non-Hermitian formulation using resonance states and the Hermitian formulation in terms of continuum wave solutions to the time dependent Schrödinger equation for tunneling decay, yield identical numerical results [1]. See also [2, 3]. The above refers to the dynamics of a class of open quantum systems that deal with the full Hamiltonian to the system in which a particle initially confined within an interaction region decays by tunneling to the outside. This should not be confused with approaches where the Hamiltonian is separated into a part corresponding to a closed system and a part responsible for the decay which usually is treated to some order of perturbation theory, as in the old work by Weisskopf and Wigner to describe the decay of an excited atom interacting with a quantized radiation field [4], that has become a standard procedure in some decay problems where perturbation theory may be justified.

Artificial quantum systems at the nanometric scale provide interesting examples of decay of open quantum systems, as electronic decay in double-barrier resonant tunneling semiconductor structures at low temperatures [5] and the decay of ultracold atoms initially confined by light [6]. The formalism of resonance states has been used to investigate the conditions for nonexponential decay in these systems [7, 8].

In particular, studies involving ultracold atoms have opened the way to study the decay of an arbitrary number of atoms. In particular, for two atoms, one may control the interaction strength between the atoms to reach the limit of no interaction between them. These may allow

Content from this work may be used under the terms of the Creative Commons Attribution 3.0 licence. Any further distribution of this work must maintain attribution to the author(s) and the title of the work, journal citation and DOI.
Published under licence by IOP Publishing Ltd

IOP Conf. Series: Journal of Physics: Conf. Series **1275** (2019) 012029 doi:10.1088/1742-6596/1275/1/012029

to study the decay of identical non interacting atoms entangled by spatial symmetry. In recent work, García-Calderón and Mendoza-Luna showed, using the formalism of resonant states, that the survival and nonescape probabilities for non interacting identical entangled symmetric and entangled antisymmetric initial states evolve in a distinct form along the exponentially decaying and nonexponential regimes [9]. These probabilities, however, are defined within the potential interaction region both for single and two particle decay, as reviewed in [15], and hence omit the relevant case of propagation of decaying particles beyond the range of the potential.

In this work we investigate the probability density for decay of two identical entangled particles by spatial symmetry as they propagate along the external region of an interaction potential. In particular we address the issue of the effect of the resonance spectra in the time evolution of the decaying probability density.

The paper is organized as follows. Section 2 reviews briefly the time evolution of decay of a single particle. Section 3 deals with the decay of symmetric and antisymmetric identical particles and shows that it may be expressed in terms of combinations of single particle decaying density probabilities. Section 4 discusses some model calculations, and finally, section 5 presents some concluding remarks.

2. Decay of a single particle

In this section we briefly recall the relevant points of the derivation of the decaying wave solution for a single particle confined initially, at $t = 0$, along the internal region of a real spherically symmetrical potential that vanishes beyond a distance, *i.e.*, $V(r) = 0$ for $r > a$. We choose the natural units $\hbar = 2m = 1$ and for simplicity of the description we refer to s waves. Hence, the solution to the time-dependent Schrödinger equation, as an initial value problem, may be written at time $t > 0$ in terms of the retarded Green's function $g(r, r'; t)$ of the problem as [10],

$$\Psi(r,t) = \int_0^a g(r,r',t)\Psi(r',0)\,\mathrm{d}r', \tag{1}$$

where $\Psi(r, 0)$ stands for the arbitrary state initially confined within the internal interaction region. Since the decay refers to tunneling into the continuum, for the sake of simplicity it is assumed that the potential does not possess bound nor antibound states. It is convenient to express the retarded time-dependent Green's function in terms of the outgoing Green's function $G^+(r, r'; k)$ of the problem. Both quantities are related by a Laplace transformation. The Bromwich contour in the k complex plane corresponds to a hyperbolic contour along the first quadrant that may be deformed to a contour that goes from $-\infty$ to ∞ along the real k axis,

$$g(r,r';t) = \frac{i}{2\pi}\int_{-\infty}^{\infty} G^+(r,r';k)\mathrm{e}^{-ik^2t}\,2kdk. \tag{2}$$

This allows to make use of the resonant expansion of the outgoing Green's function [10]

$$G^+(r,r';k) = \frac{1}{2k}\sum_{n=-\infty}^{\infty}\frac{u_n(r)u_n(r')}{k-\kappa_n}, \quad (r,r')^\dagger \le a \tag{3}$$

where the notation $(r, r')^\dagger$ means that the point $r = r' = a$ is excluded in the above expansion (otherwise it diverges) and the set of functions $\{u_n(r)\}$ refer to the resonance states to the problem, which are obtained from the residues at the complex poles $\{\kappa_n\}$ of the outgoing Green's function that also provide its normalization condition [11, 10],

$$\int_0^a u_n^2(r)dr + i\frac{u_n^2(a)}{2\kappa_n} = 1. \tag{4}$$

IOP Conf. Series: Journal of Physics: Conf. Series **1275** (2019) 012029 doi:10.1088/1742-6596/1275/1/012029

Resonance states satisfy the Schrödinger equation of the problem $[\kappa_n^2 - H]u_n(r) = 0$ with outgoing (radiative) boundary conditions $u_n(0) = 0$, $[du_n(r)/dr]_{r=a} = i\kappa_n(a)$. The complex energy eigenvalues are $\kappa_n^2 = E_n = \mathcal{E}_n - i\Gamma_n/2$, where $\mathcal{E}_n$ stands for the resonance energy of the decaying fragment and Γ_n stands for the resonance width, which is related to the lifetime as, recalling that $\hbar = 1$, $\tau_n = 1/\Gamma_n$. The lifetime of the system is defined by the longest lifetime. The complex poles $\kappa_n = v_n - i\gamma_n$ are distributed along the third and fourth quadrants of the complex k plane in a well known fashion [12].

One should note that for $r' < a$ and $r \geq a$ one may write $G^+(r,r';k)$ as [1, 10],

$$G^+(r,r';k) = G^+(r',a;k)\mathrm{e}^{ik(r-a)}, \quad r' < a, \quad r \geq a. \tag{5}$$

The above expression permits to get a resonance expansion for $G^+(r,r';k)$ which is valid along the external region of the interaction. On the other hand, it is easily shown that the representation of $G^+(r,r';k)$ given by (3) satisfies the closure relation [1, 10],

$$\mathrm{Re}\left\{\sum_{n=1}^{\infty} u_n(r)u_n(r')\right\} = \delta(r-r'), \quad (r,r')^\dagger \leq a. \tag{6}$$

The above results allows us to write the decaying time-dependent wave function as [1, 10],

$$\Psi(r,t) = \sum_{n=-\infty}^{\infty} \begin{cases} C_n u_n(r) M(y_n^\circ), & r \leq a \\ C_n u_n(a) M(y_n), & r \geq a, \end{cases} \tag{7}$$

where the coefficients C_n are defined as,

$$C_n = \int_0^a \Psi(r,0)u_n(r)dr. \tag{8}$$

and the functions $M(y_n)$, the so named Moshinsky functions, are defined as [10]

$$M(y_n) = \frac{i}{2\pi}\int_{-\infty}^{\infty} \frac{\mathrm{e}^{ikr}\mathrm{e}^{-ik^2t}}{k-\kappa_n}dk = \frac{1}{2}\mathrm{e}^{(ir^2/4t)}w(iy_n), \tag{9}$$

where

$$y_n = \mathrm{e}^{-i\pi/4}(1/4t)^{1/2}[(r-a) - 2\kappa_n t], \tag{10}$$

and the function $w(z) = \exp(-z^2)\mathrm{erfc}(-\mathrm{iz})$ in (9) stands for the Faddeyeva or complex error function [13] for which there exist efficient computational tools to calculate it [14]. The argument y_n° of the functions $M(y_n^0)$ in (7) is that of y_n given by (10) with $r = a$, namely,

$$y_n^\circ = -\mathrm{e}^{-i\pi/4}\kappa_n t^{1/2}. \tag{11}$$

Assuming that the initial state $\Psi(r,0)$ is normalized to unity, it follows from the closure relation (6) that,

$$\mathrm{Re}\sum_{n=1}^{\infty}\{C_n\bar{C}_n\} = 1, \tag{12}$$

where

$$\bar{C}_n = \int_0^a \Psi^*(r,0)u_n(r)dr. \tag{13}$$

Equation (12) indicates that $\mathrm{Re}\,\{C_n\bar{C}_n\}$ cannot be interpreted as a probability, since in general it is not a positive definite quantity. Nevertheless, one may see that it represents the 'strength' or 'weight' of the initial state in the corresponding resonant state. One might see the coefficients $\mathrm{Re}\,\{C_n\bar{C}_n\}$ as some sort of quasi-probabilities.

The solution $\Psi(r,t)$ for $r \leq a$, given by the first equation in (7), is the relevant ingredient to calculate the survival and nonescape probabilities, as discussed in [1, 2, 10]. For $r \geq a$, the solution $\Psi(r,t)$, given by the second equation in (7), describes the propagation of a single decaying particle along the external region. This has been discussed in [1, 3, 10].

3. Decay of two identical noninteracting particles

For identical non interacting particles, it is well known that the Hamiltonian H must be symmetric under the permutation of the indices of the particles so the exchange operator and H necessarily commute. It is enough to impose the appropriate symmetry or antisymmetry conditions on the initial state $\Psi(y_1,y_2,0)$ since symmetry is conserved as time evolves. As a consequence, the time evolution for decay of two identical particles may be written as,

$$\Psi(r_1,r_2,t) = \int_0^a \int_0^a g(r_1,y_1,t)g(r_2,y_2,t)\Psi(y_1,y_2,0)\, dy_1\, dy_2. \tag{14}$$

A simple choice corresponding to a symmetric state, is given by the product of single particle states $\psi_\alpha(y_1,0)$ and $\psi_\alpha(y_2,0)$, with α denoting the corresponding initial state,

$$\Psi_{\alpha,\alpha}(y_1,y_2,0) = \psi_\alpha(y_1,0)\psi_\alpha(y_2,0). \tag{15}$$

Substitution of (15) into (14) allows us to write the decaying *factorized symmetric* state as a product of two evolving single particle decaying states,

$$\Psi_{\alpha,\alpha}(r_1,r_2,t) = \psi_\alpha(r_1,t)\psi_\alpha(r_2,t). \tag{16}$$

Using (7) one may write the resonance expansion for $\Psi(r_1,r_2,t)$ as,

$$\Psi_{\alpha,\alpha}(r_1,r_2,t) = \sum_{p,q=-\infty}^{\infty} \begin{cases} \left[C_{p,\alpha}u_p(r_1)M(y_p^\circ)\right]\left[C_{q,\alpha}u_q(r_2)M(y_q^\circ)\right], & (r_1,r_2) \leq a \\ \left[C_{p,\alpha}u_p(a)M(y_p)\right]\left[C_{q,\alpha}u_q(a)M(y_q)\right], & (r_1,r_2) \geq a, \end{cases} \tag{17}$$

where $C_{n,\alpha}$, with $n = p$ or q, is given by,

$$C_{n,\alpha} = \int_0^a u_n(y)\psi_\alpha(y)\, \mathrm{d}y. \tag{18}$$

For different initial states for each particle, it follows from the Pauli Principle that initial two-particle state must consist of a linear combination of single-particle states corresponding to the symmetric and antisymmetric combinations. Denoting by $\psi_s(y_1,0)$ and $\psi_s(y_2,0)$ the initial states, with $s = \alpha$ or β, one may write,

$$\Psi_{\alpha,\beta}(y_1,y_2,0) = \frac{1}{\sqrt{2}}\left[\psi_\alpha(y_1,0)\psi_\beta(y_2,0) \pm \psi_\beta(y_1,0)\psi_\alpha(y_2,0)\right], \tag{19}$$

where respectively, the plus sign refers to *entangled symmetric* and the minus sign to *entangled antisymmetric* states. It follows immediately that the corresponding time evolved state reads,

$$\Psi_{\alpha,\beta}(r_1,r_2,t) = \frac{1}{\sqrt{2}}\left[\psi_\alpha(r_1,t)\psi_\beta(r_2,t) \pm \psi_\beta(r_1,t)\psi_\alpha(r_2,t)\right]. \tag{20}$$

Then, using (7) one may write the two-particle decaying wave function as

$$\Psi_{\alpha,\beta}(r_1,r_2,t) = \sum_{p,q=-\infty}^{\infty} \begin{cases} [C_{p,\alpha}C_{q,\beta} \pm C_{p,\beta}C_{q,\alpha}]\, u_p(r_1)u_q(r_2)M(y_p^\circ)M(y_q^\circ), & (r_1,r_2) \leq a \\ [C_{p,\alpha}C_{q,\beta} \pm C_{p,\beta}C_{q,\alpha}]\, u_p(r_1)u_q(a)M(y_p^\circ)M(y_q), & r_1 \leq a,\, r_2 \geq a \\ [C_{p,\alpha}C_{q,\beta} \pm C_{p,\beta}C_{q,\alpha}]\, u_p(a)u_q(a)M(y_p)M(y_q), & (r_1,r_2) \geq a, \end{cases} \tag{21}$$

where the coefficients $C_{n,s}$ with $n = (p,q)$ or $s = (\alpha,\beta)$ may be figured out from (18). It is worth recalling that the coefficients $\{C_{n,s}\}$, which involve only single-particle states, fulfill the relationship [10]

$$\mathrm{Re}\left(\sum_{n=1}^{\infty} C_{n,s}\bar{C}_{n,s}\right) = 1. \tag{22}$$

In the above expression $\bar{C}_{n,s}$ is defined as (18) with $\psi_s(y)$ substituted by $\psi_s^*(y)$. Hence for real initial states, $\bar{C}_{n,s} = C_{n,s}$.

The two-particle solution given by the first equation in (21) permits to calculate the corresponding two-particle expressions for the survival and nonescape probabilities [9, 16]. The situation described by the second equation in (21), which may be used to obtain the probability density where one particle remains within the interaction region whereas the other propagates along the external region of the interaction is, since both particles decay, presumably a very small quantity. In any case, here we shall be concerned with the situation provided by the last equation in (21) to obtain the probability density where both decaying particles propagate along the external region of the interaction.

4. Models

In this section we consider the s-wave delta-shell potential and the double barrier resonant potential to illustrate the profile of the probability density for two identical non interacting decaying particles as a function of time and fixed values of the corresponding positions. In particular, we are interested in how the resonance spectra affects the profile of the probability density.

4.1. Delta-shell potential

The delta-shell potential is defined as,

$$V(r) = \lambda\delta(r-a), \tag{23}$$

where λ stands for the intensity of the potential and a for the radius. We model the initially confined state by a given infinite box state,

$$\Psi(r,0) = \sqrt{\frac{2}{a}}\,\sin\left(\frac{q\pi}{a}r\right), \quad q = 1,2,3... \tag{24}$$

The resonance solutions to the problem with complex energy eigenvalues $\kappa_n^2 = \mathcal{E}_n - i\Gamma_n/2$ read,

$$u_n(r) = \begin{cases} A_n \sin(\kappa_n r) & r \leq a \\ B_n\, e^{i\kappa_n r}, & r \geq a, \end{cases} \tag{25}$$

where $\kappa_n = \upsilon_n - i\delta_n$. From the continuity of the above solutions and the discontinuity of its derivatives with respect to r (due to the δ-function interaction) at the boundary value $r = a$, it is obtained that the set of κ_n's satisfy the equation,

$$2i\kappa_n + \lambda(e^{2i\kappa_n a} - 1) = 0. \tag{26}$$

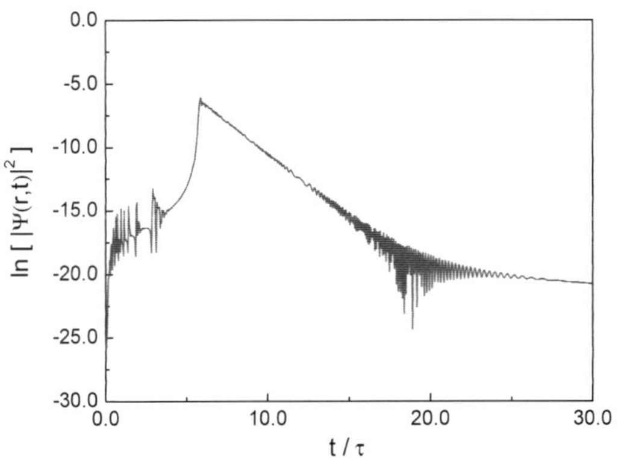

Figure 1. $\ln[|\Psi(r,t)|^2]$ vs time in lifetime units for single-particle decay. See text.

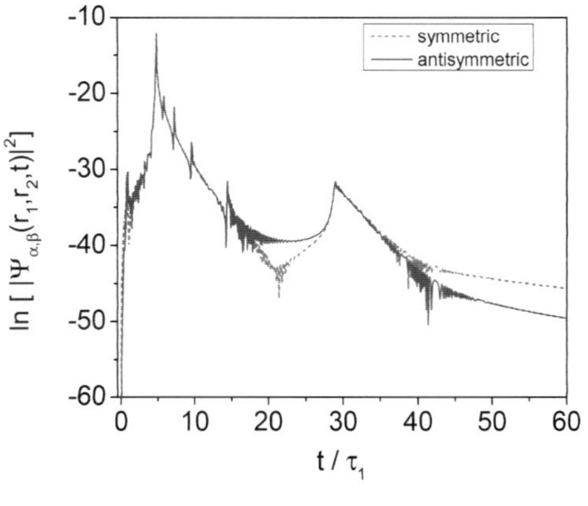

Figure 2. $\ln[|\Psi_{\alpha,\beta}(r,t)|^2]$ vs time in lifetime units for two-particle decay of symmetric and antisymmetric states. See text.

For $\lambda > 1$ one may write the approximate analytical solutions to Eq. (26) as [10]

$$\kappa_n \approx \frac{n\pi}{a}\left(1-\frac{1}{\lambda a}\right) - i\frac{1}{a}\left(\frac{n\pi}{\lambda a}\right)^2. \tag{27}$$

Using iterative procedures as the Newton-Rapshon method, one may obtain the poles κ_n with the desired degree of approximation. It is worth noticing that as the intensity of the potential $\lambda \to \infty$, the complex poles $\kappa_n = v_n - i\delta_n$ tend to the real infinite box eigenfunctions, and similarly, the resonance eigenfunctions $u_n(r)$ tend to the infinite box model eigenfunctions. This implies that for a large finite value of the intensity λ, an initial infinite box state $\Psi(r,0)$ with $q = m$, has a larger overlap with the resonant state $u_m(r)$ than with any other resonant state. For a given value of the intensity λ and a radius a of the δ-potential, one may then evaluate the corresponding set of complex poles $\{\kappa_n\}$, the set of normalized resonance states $\{u_n(r)\}$, and the expansion coefficients $\{C_n\}$ to the problem.

Figure 1 exhibits a plot of $\ln[|\Psi(r,t)|^2]$, given by the second equation of equation of (7), as a function of time in lifetime units, calculated at the distance $r = 3000\,a$. Here we choose $q = 1$, and hence the real part of the overlap of the initial state with the resonant state with $n = 1$, in view of (12), is of the order of unity. We choose the potential parameters $\lambda = 100$ and $a = 1$. As time evolves we observe a buildup of the probability density characterized by resonance peaks that are due to high resonance contributions until it reaches the highest peak which corresponds to the dominating contribution that arises from the resonance term with $n = 1$ in the second equation of (7). This is corroborated by the fact that the highest peak is given by $t_1 \approx r/2v_1$, which follows by inspection of the argument (10) of the Moshinsky function. As time evolves further, the probability density mimics the behavior of the survival or nonescape probabilities. There is a regime of exponential decay which is followed by the transition to nonexponential decay where the probability density evolves as the inverse power of time t^{-3} [3].

Figure 2 displays the plot of $\ln[|\Psi_{\alpha,\beta}(r,t)|^2]$ as a function of time in lifetime units for the two-particle probability density given by the third equation of (21) with $r_1 = 2400\,a$ and $r_2 = 15000\,a$, for the delta-shell potential with parameters $\lambda = 100$ and $a = 1$ with $\alpha = 1$, and $\beta = 6$, which correspond, respectively to $q = 1$ and $q = 6$ of equation (24). One sees that the probability density profile is very similar for symmetric and antisymmetric states. This seems to be a

IOP Conf. Series: Journal of Physics: Conf. Series **1275** (2019) 012029 doi:10.1088/1742-6596/1275/1/012029

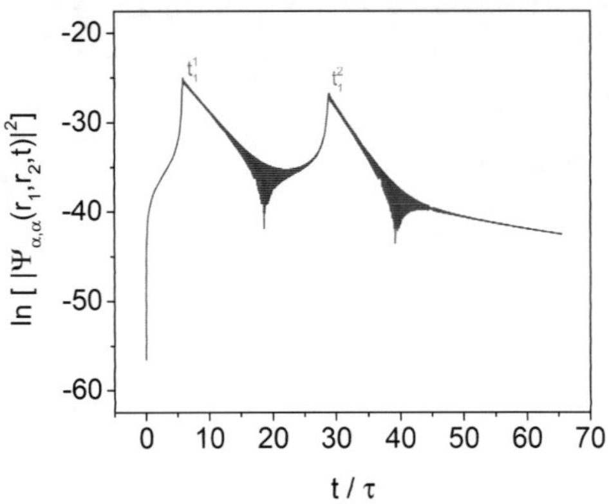

Figure 3. $\ln[|\Psi_{\alpha,\alpha}(r,t)|^2]$ vs time in lifetime units for single-particle decay. See text.

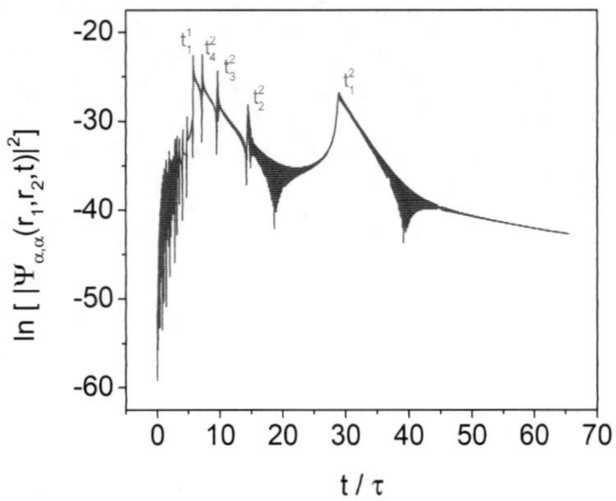

Figure 4. $\ln[|\Psi_{\alpha,\alpha}(r,t)|^2]$ vs time in lifetime units for two-particle decay. See text.

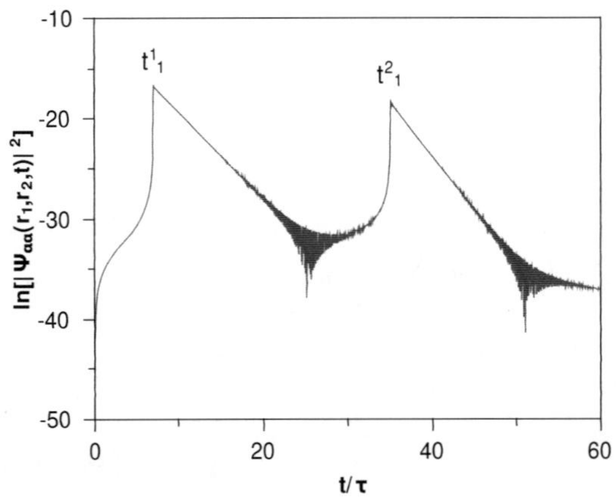

Figure 5. $\ln[|\Psi_{\alpha,\alpha}(r,t)|^2]$ vs time in lifetime units for single-particle decay. See text.

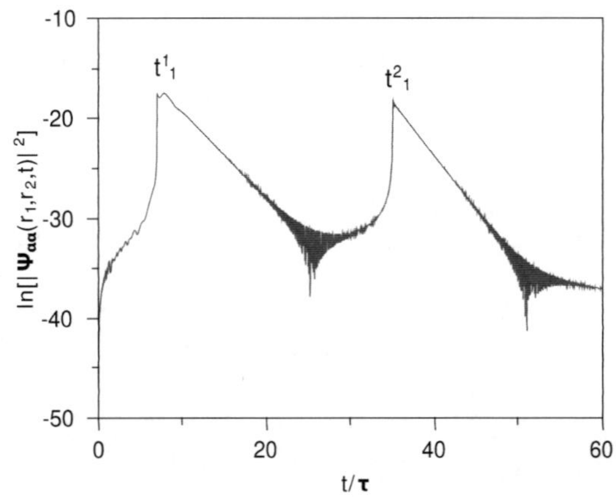

Figure 6. $\ln[|\Psi_{\alpha,\alpha}(r,t)|^2]$ vs time in lifetime units for two-particle decay. See text.

common feature of the propagating solutions. This is not so along the internal interaction region [9]. Essentially the symmetric and antisymmetric quantities differ in the valley region between the two broad peaks and in the corresponding asymptotic behavior at long times. It turns out that the symmetric state goes as t^{-6} and the antisymmetric one as t^{-10}, as obtained along the internal region [9]. Comparison of figures 1 and 2 suggests that the two broad resonance peaks are related to the decay of the particles in the initial states $\alpha = 1$ and $\beta = 6$. This is corroborated by calculating the corresponding peak valu in time domain for the state $\alpha = 1$, $t_1^1 = r_1/2v_1$, and for the state $\beta = 6$, given by $t_6^2 = r_2/2v_6$. Further inspection of figure 2 suggests that the resonance peaks in time domain that seem to disrupt the decaying part of the first broad peak correspond to high energy resonances of the system.

This is further corroborated by figures 3 and 4, which display the plot of the $\ln[|\Psi_{\alpha,\alpha}(r,t)|^2]$ as a function of time in lifetime units for a symmetric state with $\alpha = 1$ with $r_1 = 3000\,a$ and

$r_2 = 5r_1$. In figure 3 both states correspond to α_1, namely, (24) with $q = 1$. One sees two broad peaks, one peaked at $t_1^1 = r_1/2v_1$ and the other at $t_1^2 = r_2/2v_1$. In fact in lifetime units $t_1^1 = 5.73$ with and $t_1^2 = 28.68$ with $v_1 = 3.11052$ The lifetime is $\tau = 82.058$. The probability density is larger around these peaked values. On the other hand, figure 4 refers to the same system with the contribution of 1000 resonance poles. One sees that a number of peaks that arise from higher resonance terms at t_2^2, t_3^2 and t_4^2, disrupt the exponential decaying regime corresponding to the first broad peak. In lifetime units these times are given by: $t_2^2 = 14.34$, $t_3^2 = 9.56$ and $t_4^2 = 7.17$. The corresponding v's read:$v_2 = 6.2213$, $v_3 = 9.3325$,$v_4 = 12.4444$.

4.2. Double barrier system

The delta potential in general corresponds to a situation that involves many sharp isolated resonances poles, namely, $v_n \gg \delta_n$. It is good model in systems with these characteristics. However, there are systems with just a few sharp resonances, may be two, which refer to particles trapped by barrier of finite height, where the high energy resonances are not trapped and hence are broad and presumably overlapping. A model with these characteristics is a double barrier with a well in between. The calculation of the complex poles have been discussed elsewhere [7] and there is no space to present it here. It follows a procedure analogous to the previous model. It should be noted that since the potential is one dimensional, the normalization has an additional term.

As in the previous two figures, our aim is to make a comparison with a situation where the double barrier system holds two sharp resonances and the remaining, the higher resonance energies, are situated above the barrier height. The system parameters are: barrier heights $V = 40$, well width $w = 1$, and barrier widths $b = 1$. Figures 5 and 6 exhibit each a plot of $\ln[|\Psi_{\alpha,\alpha}(r,t)|^2]$ as a function of time in lifetime units with $r_1 = 600000$ and $r_2 = 5r_1$. The Since there are only two resonance states, it is not surprising to obtain a profile for the decaying probability density similar to that in figure 3. However, in contrast to figure 4, the many resonance calculation, up to 50 states presented in figure 6 remains almost identical to that involving just the two sharp resonances situated below the barrier heights. In this case, in lifetime units $t_1^1 = 7.0$ and $t_2^1 = 35.0$. The corresponding pole is $v_1 = 2.3725$ and the lifetime $\tau = 18067.23$. The above result seems to imply that broad overlapping resonances do not play a relevant role in the profile of the decaying probability density for two particles.

5. Concluding remarks

We find of interest the result that a multilevel sharp resonance spectra may disrupt the exponential decaying regime which occurs in the case of a multilevel broad resonance spectra. Recent experiments involving ultracold gases [6] may be used to test this novel nonexponential behavior in the time evolution of decay of identical noninteracting particles. Another result worth noticing is that symmetric and antisymmetric states evolve almost identically except in the small valley region between the two broad peaks related to the initial resonance states and in the long time asymptotic regime.

Acknowledgments

G.G-C acknowledges the partial financial support of UNAM-DGAPA-PAPIIT IN105618, Mexico. R.R. acknowledges financial support from PRODEP under the program *Apoyo para Estancias Cortas de Investigación*, Convocatoria 2018, Mexico. R.R. also thanks Instituto de Física of UNAM for hospitality.

References

[1] García-Calderón G, Máttar A and Villavicencio J 2012 *Phys. Scr.* **T151** 1
[2] García-Calderón G, Maldonado I and Villavicencio J 2007 *Phys. Rev. A* **76** 012103

IOP Conf. Series: Journal of Physics: Conf. Series **1275** (2019) 012029 doi:10.1088/1742-6596/1275/1/012029

[3] García-Calderón G, Maldonado I and Villavicencio J 2013 *Phys Rev A* **88** 052114
[4] Weisskopf V F and Wigner E P 1930 *Z. Phys.* **65** 54
[5] Tsuchiya M, T. Matsusue T and Sakaki H 1987 *Phys. Rev. Lett.* **59** 2356
[6] Serwane F, Zürn G , Lompe T, Ottenstein T B, Wenz A N and Jochim S 2011 *Science* **332** 336
[7] García-Calderón G and Villavicencio J 2006 *Phys. Rev A* **73** 062115
[8] García-Calderón G and Romo R 2016 *Phys Rev. A* **93** 022118
[9] García-Calderón G and Mendoza-Luna L G 2011 *Phys. Rev. A* **84** 032106
[10] García-Calderón G 2010 *Adv. in Quant. Chem.* **60** 407
[11] García-Calderón G and Peierls R 1976 *Nucl.Phys. A* **265** 443
[12] Newton R G 2002 *Scattering Theory of Waves and Particles* (New York: Dover Publications) chapter 12
[13] Abramowitz M and Stegun A I 1964 *Handbook of Mathematical Functions* (New York: Dover Publications)
[14] Poppe G M P and Wijers C M J 1990 *ACM Trans. Math. Softw.* **16** 38
[15] García-Calderón G 2015 *J. Phys. Conf. Series* **626** 012064 (arXiv:quant-ph/1812.09319)
[16] García-Calderón G and Romo R 2017 *Phys. Rev. A* **96** 062124

IOP Conf. Series: Journal of Physics: Conf. Series **1275** (2019) 012030 doi:10.1088/1742-6596/1275/1/012030

Cosmic strings, deformed lattices and spontaneous symmetry breaking

Antonino Flachi and Vincenzo Vitagliano

Department of Physics & Research and Education Center for Natural Sciences, Keio University, 4-1-1 Hiyoshi, Kanagawa 223-8521, Japan

E-mail: flachi@phys-h.keio.ac.jp, vincenzo.vitagliano@keio.jp

Abstract. External conditions have a dramatic impact on the way dynamical symmetry breaking occurs. In a curved background, the natural expectation is that curvature works toward the restoration of the internal symmetry. We show instead that, for topological defects, the competing action of the locally induced curvature and boundary conditions generated by the non-trivial topology allows configurations where symmetries can be spontaneously broken close to the core. Inspired by the effect of geometrical deformations on 2D lattices, we then propose a novel mechanism to induce a superconducting phase by triggering condensation along cosmic strings.

1. Introduction

Most of the revolutionary results of modern theoretical physics have been achieved using tools and techniques provided by quantum-mechanical perturbation theory: if the strength of the interaction among particles is weak (namely the coupling constant is small), then it is possible to extract information about how the coupling-induced small corrections affect a system.

Nevertheless, *contemporary* theoretical physics is looking forward. Typically, as energy scales become higher, non-perturbative effects get into the game, and the system is said to have entered the strongly coupled regime. Remarkable examples of strongly coupled systems include QCD and the theory of quarks, high-temperature superconductors and the very primordial plasma filling the universe a few instants after the big bang, but also the theory of the propagation of conducting electrons in graphene, the contemporary high-tech superstar material.

The study of the dynamics of such systems is a formidable task, and clear statements on the physics at strong coupling scales are only possible at the cost of demanding numerical simulations. On the other hand, general phenomenological guidelines for the study of strongly coupled systems can be precisely drafted exploiting mathematical considerations on the underlying symmetries. The physics of symmetry breaking, in fact, notably relies on exact mathematical statements which are intimately non-perturbative, as in the case of the Goldstone theorem.

A seminal paper by Coleman and Weinberg [1] showed how symmetries can be spontaneously broken due to quantum fluctuations modifying the structure of the vacuum. This mechanism of *dynamical* symmetry breaking, initially developed in the context of scalar field self-interactions, can be naturally extended to fermions.

Content from this work may be used under the terms of the Creative Commons Attribution 3.0 licence. Any further distribution of this work must maintain attribution to the author(s) and the title of the work, journal citation and DOI.
Published under licence by IOP Publishing Ltd

IOP Conf. Series: Journal of Physics: Conf. Series **1275** (2019) 012030 doi:10.1088/1742-6596/1275/1/012030

To better understand the picture, let's introduce some concepts for the unfamiliar reader. In a field theory with massless fermions, the interactions between particles preserve helicity: the polarised left- and right-handed sectors evolve separately. However, in D-dimensional four fermion effective theories[1],

$$S_{\rm 4fET} = \int d^D x \sqrt{g} \left\{ \bar{\psi} i \gamma^\mu \nabla_\mu \psi + \frac{\lambda}{2N} (\bar{\psi}\psi)^2 + ... \right\} , \tag{1}$$

(like the celebrated Nambu–Jona-Lasinio model [2, 3], and in its progeny of models for strong interactions) such symmetry is spontaneously broken by a non-vanishing order parameter: when the composite operator $\phi \sim \bar{\psi}\psi$ acquires a non-zero vacuum expectation value, then a dynamical effective mass for the fermions, $M_{\rm eff} \sim \langle \bar{\psi}\psi \rangle$, is generated. This mechanism, analogous to the spontaneous magnetisation, is responsible for most of the observed mass of hadrons.

Transitions between broken symmetry and restored symmetry phases are generally triggered by changes in the external conditions. In particular, a nonzero temperature leads to temperature-dependent mass generation and restoration of broken symmetries once a certain critical value of the temperature is reached. Other factors modifying the phase diagram of self-interacting theories span from finite density effects to the action of a non-vanishing chemical potential or of an external gauge field. Here we instead concentrate on geometrical effects, and how these challenge the vacuum stability of a theory with four fermions interactions [4].

Many are the configurations in which geometry affects the symmetry breaking of strongly interacting systems: in flat spacetime with $R^3 \times S^1$ topology and periodic boundary conditions, for example, the consequences of the non-trivial topology are very similar to those of nonzero temperature [5]; on the other hand, in curved spacetime the effects of external gravitational fields resemble those of an effective extra mass [6, 7, 8, 9]. The combination of these external factors (in particular topology, nonzero temperature, and curvature) acting on self-interacting theories is likely to have been of considerable importance in the early stages of the evolution of the universe. During those eras a spontaneous breaking of an internal symmetry group results in the production of topological defects – the well-known Kibble–Zurek mechanism [10, 11].

[1] Here, λ is the coupling constant and N the total number of fermion degrees of freedom.

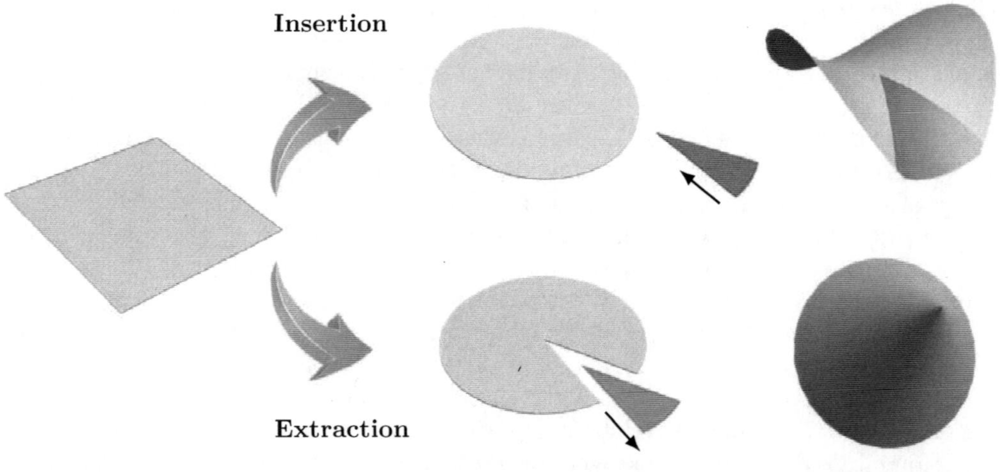

Figure 1. The flat space is modified through the insertion or the extraction of a piece of lattice. Adding a section with a given angle, one finds a saddle; subtracting the same section, and then sewing on the cut, a cone.

IOP Conf. Series: Journal of Physics: Conf. Series **1275** (2019) 012030 doi:10.1088/1742-6596/1275/1/012030

2. Cosmic defects and crystal lattices

Suppose the dynamical symmetry breaking for some model to be at the origin of the formation of a static straight cosmic string [12, 13, 14, 15] lying along the z-axis, namely an infinitely long thin tube of false vacuum generated in the sudden temperature-driven transition from a phase to another (here, the transverse size of the cosmic string is neglected while 'sudden transition' means a transition with a rate that is fast if compared with the size of the system). Away from the defect, the spacetime associated with the gravitating string is accurately described by the vacuum Einstein equations. It turns out that at large distance from the string, the geometry is locally flat, $ds^2_{\rm con} = dt^2 - dz^2 - dr^2 - r^2 d\theta^2$, but with an important *caveat*: it is not globally Euclidean, since the angular coordinate does not run on the entire 2π circle; instead, $0 \leq \theta < 2\pi - \Delta$, with $\Delta > 0$ ($\Delta < 0$) being the deficit (excess) angle: surfaces at constant t and z are cones, not planes.

This is a remembrance of defects insertion in crystal lattices: starting from a (locally) flat lattice (see Fig. 1), the subtraction (or addition) of atoms is equivalent to the extraction (or insertion) of sections of the lattice with a given angle. The procedure results in an ice-cream-shaped lattice (or a saddle) which is locally flat everywhere apart for the apex. Note yet that the spacetime surgery does not come for free: the price to pay is the implementation of some non-trivial new boundary conditions along the cut, which will be a reminder of the deficit (excess) angle at the apex. What is the role of the background geometry in modifying the vacuum structure? Surprisingly enough, we can now show that curvature, localised close to a topological defect, enhances, rather than inhibits, the condensation along the defect itself. In order to do so, we will consider the lesson coming from a $(2+1)$–dimensional system borrowed from condensed matter [16], a honeycomb lattice characterised by a bipartite symmetry, that is, whose hexagonal structure is obtained by the superposition of two triangular sublattices (see Fig. 2): the breakdown of this discrete symmetry is behind the phenomenon of condensation we describe here.

The dynamics of the delocalized electrons on such a lattice is often described in terms of a generalisation of the Hubbard model, that in the continuum limit is mapped onto a field theory with nine different couplings [17]; however, considering the limit for a large number of fermion flavours N and after bosonization, the Hubbard model for the honeycomb lattice acquires the form of a bosonized (2+1) Gross–Neveu model,

$$\mathcal{L}_{\rm GN} = \bar{\psi}_\sigma \imath \partial\!\!\!/ \psi_\sigma + \sigma \bar{\psi}_\sigma \phi \psi_\sigma + \frac{\phi^2}{2\lambda}, \qquad (2)$$

where $\sigma = \pm$ is a spin index on which one sums. The final goal is to study the behaviour of the order parameter ϕ when moving toward the apex of the cone.

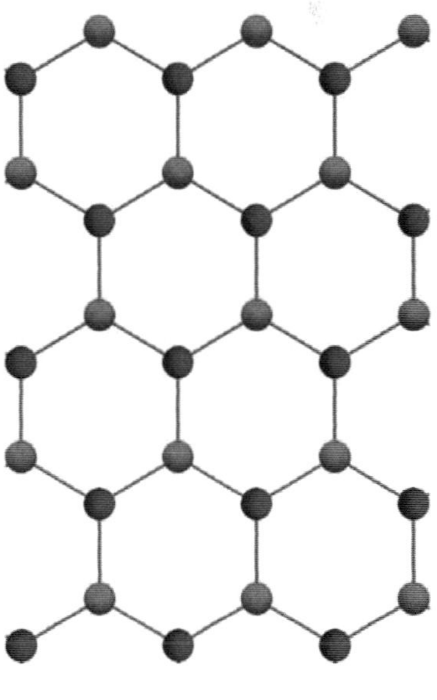

Figure 2. The hexagonal honeycomb lattice with the two triangular sublattices (blue and red spheres)

Nature abhors (or discretely hides) singularities: the sympathetic reader will then not complain about a regularization of the space generated by the topological 'stringy' defect with a family of smoothed versions of the conical solution (in Euclidean time), $ds^2 = d\tau^2 + f_\epsilon(r)dr^2 + r^2 d\theta^2$, where $f_\epsilon(r)$ is a regularising function: in the limit $\epsilon \to 0$ one might recover the singular cone. Using standard technique (Schrödinger–Lichnerowicz–Weitzenböck formula) to square the Dirac operator, the effective action reads

$$\Gamma[\phi] = -\int d^3x\sqrt{g}\frac{\phi^2}{2\lambda} + \frac{1}{2}\sum_{p=\pm}\log\det\left(\Box + \frac{R}{4} + \phi_p^2\right), \qquad \phi_\pm^2 = \phi^2 \pm \sqrt{g^{rr}}\phi' \tag{3}$$

where the metric here employed is the smoothed one, ds^2, rather than ds^2_{con}. As previously mentioned, the change in the topology has a prominent role in altering the boundary conditions on the glued side of the lattice; the response to this change is captured by the employment of a modified covariant derivative, $D_\mu = \nabla_\mu + \imath\mathcal{A}_\mu$, acting on spinors and encompassing an effective non-dynamical gauge field, $\mathcal{A}_\mu$, that depends on the deficit angle[18]. Expressing the functional determinant of the (squared) Dirac operator in curved space in terms of its heat kernel expansion [19, 20], and using zeta function regularization, it is possible to calculate the effective action (3) [21] and find out the effective equations of motion of the order parameter ϕ, whose solutions can be reconstructed numerically. Although the presence of curvature (which acts, as for the chiral gap effect [22, 23], as an effective mass term) is supposed to enhance a phase of symmetry restoration, the presence of the extra effective gauge field, another reminder of the geometry of the system, catalyse the formation of a bubble of condensed particles close the vertex of the cone, (see Fig. 3, the saddle case is similar). A well known duality between defective crystal configurations and differential geometry [24, 25, 26] raises the question about how generalised non-Riemannian structures can contribute to the mechanism described so far. Similar issues would enter the game in the case of presence of inhomogeneities [27, 28].

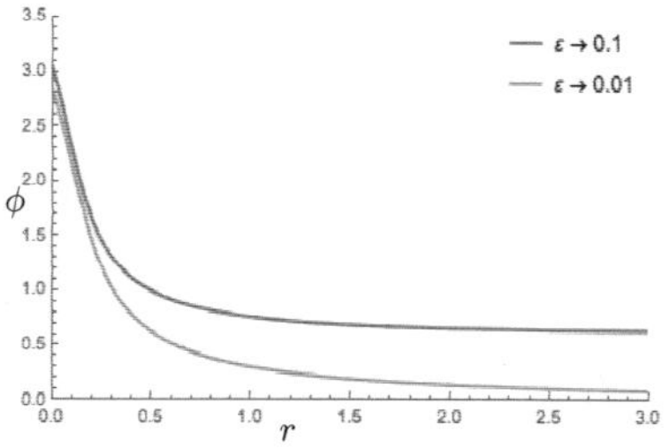

Figure 3. Profile of the condensate ϕ as a function of the distance from the apex, r. Different colours refer to different values of the ϵ parameter regularising the cone:

3. Conclusions

For relativistic cosmic strings, the possibility of condensation in the region surrounding the string core due only to configurational elements is a novel mechanism to provide a superconducting phase around the defect, whose phenomenological potentiality is completely to explore. However, it is worth mentioning that in a more realistic setup, phase transitions in the early universe did not seed the formation of a single straight string, but of a network of cosmic strings, which renders eventually even more striking and intriguing the connection with a crystal lattice, where a distribution of defects is more natural to occur. Another interesting aspect is the following: in order to simplify the discussion we have here considered the spontaneous symmetry breaking for the Gross–Neveu model catalysed by an external string defect, namely originated by the breakdown of some symmetry of a different field. Different would be the situation in which the responsible for the defect formation is the very same field, in which case one might take into account possible backreaction effects.

Acknowledgments

We are thankful to E. V. Castro and P. Ribeiro for the collaboration that led to the deformed lattice calculation. AF acknowledges the support of JSPS through the KAKENHI Grant No. 18K03626 and the support of MEXT Program for the Strategic Research Foundation at Private Universities 'Topological Science' (Grant No. S1511006). VV is supported by JSPS (Grant No. P17763).

IOP Conf. Series: Journal of Physics: Conf. Series **1275** (2019) 012030 doi:10.1088/1742-6596/1275/1/012030

References

[1] Coleman SR and Weinberg EJ 1973 *Phys. Rev.* D **7** 1888
[2] Nambu Y and Jona-Lasinio G 1961 *Phys. Rev.* **122** 345-358
[3] Nambu Y and Jona-Lasinio G 1961 *Phys. Rev.* **124** 246-254
[4] Inagaki T, Muta T and Odintsov SD 1997 *Prog. Theor. Phys. Suppl.* **127** 93
[5] Dolan L and Jackiw R 1974 *Phys. Rev.* D **9** 3320
[6] Flachi A and Tanaka T 2011 *J. High Energy Phys.* JHEP02(2011)026
[7] Flachi A and Tanaka T 2011 *Phys. Rev.* D **84** 061503
[8] Flachi A 2012 *J. High Energy Phys.* JHEP01(2012)023
[9] Flachi A 2012 *Phys. Rev.* D **86** 104047
[10] Kibble TWB 1976 *J. Phys.* A **9** 1387
[11] Zurek WH 1985 *Nature* **317** 505
[12] Vilenkin A and Shellard EPS 2000 *Cosmic strings and other topological defects* (Cambridge, UK: Cambridge University Press)
[13] Brandenberger RH, Huang T, Yang K and Zhang X 1997 *Phys. Rev.* D **55** 4511
[14] Brandenberger RH 1998 *Pramana* **51** 191
[15] Russell IH and Toms DJ 1989 *Class. Quant. Grav.* **6** 1343
[16] Castro EV, Flachi A, Ribeiro P and Vitagliano V 2018 *Phys. Rev. Lett.* **121** 221601
[17] Herbut IF 2006 *Phys. Rev. Lett.* **97** 146401
[18] Sitenko YA and Vlasii ND 2007 *Nucl. Phys.* B **787** 241
[19] Parker LE and Toms DJ 2009 *Quantum Field Theory in Curved Spacetime: Quantized Fields and Gravity* (Cambridge, UK: Cambridge University Press)
[20] Avramidi IG 2000 *Heat Kernel and Quantum Gravity* (Berlin: Springer)
[21] Buchbinder IL, Odintsov SD and Shapiro IL 1992 *Effective action in quantum gravity* (Bristol, UK: IOP)
[22] Flachi A and Fukushima K 2014 *Phys. Rev. Lett.* **113** 091102
[23] Flachi A, Fukushima K and Vitagliano V 2015 *Phys. Rev. Lett.* **114** 181601
[24] Vitagliano V 2014 *Class. Quant. Grav.* **31** 045006
[25] Luz P and Vitagliano V 2017 *Phys. Rev.* D **96** 024021
[26] de Juan F, Cortijo A and Vozmediano M 2010 *Nucl. Phys.* B **828** 625
[27] Faraoni V, Vitagliano V, Sotiriou TP and Liberati S 2012 *Phys. Rev.* D **86** 064040
[28] Prain A, Vitagliano V, Faraoni V and Lapierre-Leonard M 2016 *Class. Quant. Grav.* **33** 145008

IOP Conf. Series: Journal of Physics: Conf. Series **1275** (2019) 012031 doi:10.1088/1742-6596/1275/1/012031

Features of spinning gravity in particle physics: supersymmetric core of the Kerr-Newman electron

Alexander Burinskii

Theor. Phys. Laboratory, Nuclear Safety Institute, Russian Academy of Sciences, B.Tulskaya 52, 115191, Moscow,
Russia

E-mail: `burinskii@mail.ru`

Abstract. The model of electron based on regularized Kerr-Newman solution shows that spinning gravity is consistently united with particle physics. Extremely high spin of the electron creates topological deformation of space on the Compton scale, and we show that supersymmetric Higgs model regularizes this space, forming a nonperturbative solution as the bag-like core of the regularized Kerr-Newman electron model. Conflict between gravity and quantum interior of the bag is eliminated by the domain wall boundary of the bag which interpolates between the external exact Kerr-Newman solution and the free from gravity superconducting interior of the bag, forming vacuum state necessary for action of the perturbative quantum theory. Similar to typical bag models, the Kerr-Newman bag is deformable and creates the controlled by gravity stringy structure.

We show that contrary to the widely-discussed confrontation, spinning gravity cooperates with quantum theory, playing fundamental role in the structure of the dressed electron.

1. Introduction

Unification of Gravity with Quantum theory represents principal problem of the modern theoretical physics. The both theories are confirmed experimentally with great precision. Nevertheless, they contradict and cannot be combined in a unified theory.

In this paper we show how this problem is solved for the Kerr-Newman (KN) solution, which is the known solution for a charged and rotating black hole. Fifty years ago, in 1968, Carter noticed that KN solution has gyromagnetic ratio $g = 2$ just the same as that of the Dirac electron [1], which gave rise to study of the electron model based on the KN solution [1, 2, 3, 4, 5, 6] and so on.

The KN model of electron is consistent with gravity by nature, and there appears the question, how the known insuperable contradictions between gravity and quantum physics is solved in this model.

In this paper, and in the series of previous works [7, 8], we show that the reason of this discrepancy lies in the delusion about weakness of gravity, and in underestimation of the role of spin in gravitational interactions.

Treatment of the electron model becomes principal for two reasons i) the electron is the most studied of the particles from theoretical and experimental point of view, ii) the KN electron model displays natural consistency with gravity, shedding the light on the features of spinning gravity and on the related problems of interpretation of the quantum theory.

Content from this work may be used under the terms of the Creative Commons Attribution 3.0 licence. Any further distribution of this work must maintain attribution to the author(s) and the title of the work, journal citation and DOI.
Published under licence by IOP Publishing Ltd

IOP Conf. Series: Journal of Physics: Conf. Series **1275** (2019) 012031 doi:10.1088/1742-6596/1275/1/012031

1.1. Compton scale of the spinning Kerr gravity

Usually, it is recognized that the cause of the contradiction lies in gravity, which is very weak, but not subordinated to quantum theory and does not allow to be quantized.

However, the classical gravity, which has its own high self-esteem as the space-time arena for action of quantum theory, does not accept the quantum conception of the particles, in which the electron is considered as pointlike and even structureless. Gravity requires an extended description of the particle in configuration space in terms of the real fields, in particular, as a nonperturbative solution of the field equations. It needs a stress-energy tensor for the right hide side of the Einstein equations, $G_{\mu\nu} = 8\pi T_{\mu\nu}$.

The modern theories of electron do not satisfy this requirement. In the Dirac theory, QED and in the Standard model the electron model is pointlike. Although, the point-like electron is confirmed experimentally at the high energy scattering, it does not exclude the existence of some thin structure (similar to strings) which cannot be seen at the high energy scattering, and we show further that this is indeed the case, and the KN model is extended as a "dressed" electron on the Compton region in the form of a circular string model. The development of the KN electron model along the line of publications [1, 2, 3, 4, 5, 6] leads us to a Supersymmetric Bag model [9, 10] which contains the important elements of the stringy structure.

The transfer from pointlike particles to strings was revolutionary step in the problem of unification, and so far as the string theory raised a claim on unification with gravity, this step was correlated with weakness of gravity. However, this stringy structure and the related extra dimensions are "invisible" experimentally, and following to the old argumentations of the five-dimensional Kaluza-Klein theory (which is predecessor of the string theory), the extra dimensions are "invisible" because they are extremely small, and we read in [11] "*...The Planck scale is the natural first guess for a rough estimate of the fundamental string length scale as well as the characteristic size of compact extra dimensions. Experiments at energies far below the Planck energy cannot resolve distances as short as the Planck length. Thus, at such energies, strings can be accurately approximated by point particles.*" It means that particles are not point-like indeed, but their extended structure is "invisible" above the Planck scale $\approx 10^{-33} cm$.

However, as it was mentioned by John Schwarz, [12], "*...Since 1974 superstring theory stopped to be considered as particle physics...* " and "*... a realistic model of elementary particles still appears to be a distant dream ...* "

The extended particles, representing localized concentration of energy are described as solitons. These are just what is necessary for association with gravitation. However, gravity is ignored usually in the solitonic models, as it is conventionally assumed as very weak and not essential on the scale of electroweak interactions.

Indeed, mass of the elementary particles is very small concerning the scale of masses in cosmology. However almost all of the elementary particles have spin, which distort space as spinning Kerr's gravity. Spin has only discrete values of order $\hbar$, and typically, its value with respect to masses of elementary particles is extremely high, about $10^{20} - 10^{22}$ (in the dimensionless plank's units $G = \hbar = c = 1$).

Another very plausible arguments about weakness of gravity were given John C Baez, who is based on the Schwarzschild solution and doesn't take into account Kerr's distortion of the space caused by spin of elementary particles [13]: "*... general relativity says that associated to any mass m there is a length called the Schwarzschild radius, l_s, such that compressing an object of mass m to a size smaller than this results in the formation of a black hole. The Schwarzschild radius is roughly the distance scale at which general relativity becomes crucial for understanding the behavior of an object of a given mass. Now, ignoring some numerical factors, we have*

$$l_c = \frac{^-h}{mc} \tag{1}$$

IOP Conf. Series: Journal of Physics: Conf. Series **1275** (2019) 012031 doi:10.1088/1742-6596/1275/1/012031

and

$$l_s = \frac{Gm}{c^2}. \tag{2}$$

These two lengths become equal when m is the Planck mass. And when this happens, they both equal the Planck length..."

In this sentence l_c is the Compton length of the dressed quantum particle, and comparison of the lengths l_c and l_s is indeed comparison of the influence of quantum theory and Schwarzschild gravity. Author shows that these lengths become equal on the Planck scale. This comparison is well known and very popular, but similar comparison based on the Kerr solution alters this conclusion resolutely.

Gravitational field of the Kerr solution becomes strong at the Kerr singular ring, which has the radius $a = \frac{J}{mc}$, where J is angular momentum of the Kerr solution. For parameters of the spinning particles one has to use $J \approx \bar{}h$,which gives for characteristic length of the Kerr solution

$$l_{Kerr} = a \approx \frac{\bar{}h}{mc} = l_c \tag{3}$$

which shows that l_{Kerr} and l_c are approximately equivalent for any values of the mass m, and i) the given by John C Baez arguments on the exclusive role of the Planck mass and the Planck length do not work for the spinning Kerr gravity, 2)equality (3) shows explicitly that l_{Kerr} corresponds to the Compton length scale, which for electron is about $10^{-11}cm$, what is much more than the Planck length $l_p \approx 10^{-33}cm$.

1.2. Two-sheeted Kerr space as Einstein-Rosen bridge

The Kerr singular ring is the branch line of the Kerr space into two sheets, and the Compton scale is the characteristic length at which Kerr's gravity creates two-sheeted topology.

Note that the Kerr solution with parameters of elementary particles is not black hole because in the dimensionless units we have $a \gg m$, which is condition for disappearance of the horizons. The Kerr singular ring and topologically two-sheeted space are naked – not covered by horizon, and there appeared the attempts to consider the two-sheeted Kerr solution as a "wormhole" or the famous Einstein-Rosen bridge [14, 15] – the old proposals on the related with gravity models of elementary particles.

By the condition (3) the Kerr gravity and Quantum theory are at the same scale and the conflict related with their incompatibility must be increased.

For electron, the two-sheeted Kerr's background forms in space a hole of the Compton radius $a = \hbar/2m$. Size of this hole corresponds to the region of the quantum virtual processes – the Compton zone of dressed electron. However, this space is singular and topologically different from the required Minkowski space, so that neither the Dirac theory nor perturbative QED can be applied, and to resolve this conflict, some special mechanism of regularization is necessary. Over the last fifty years many works were devoted to the problem of source of Kerr solution, and we note here only several of them, which are relevant to our direction of the research [3, 4, 5, 6, 14, 18, 19, 20, 21, 22]. Moving along this line, we arrive at the bag model of the KN source [9, 16, 17], which unifies the bubble and solitonic models of the KN source with string-like models, and also allows us to resolve the conflict between gravity and quantum theory *without modification of the Einstein-Maxwell equations,* by using a *supersymmetric scheme of phase transition* which is equivalent to *supersymmetric Higgs model of symmetry breaking.* Note, that supersymmetry becomes really necessary here in the form of the supersymmetric Landau-Ginzburg or Wess-Zumino field model (the concept of superparticles is not used here). In this model singular region of KN solution is replaced by the flat internal space of the Compton size. This region is free from gravity and forms interior of the bag, such so the external exact Kerr-Newman gravitational solution leaves untouched.

IOP Conf. Series: Journal of Physics: Conf. Series **1275** (2019) 012031 doi:10.1088/1742-6596/1275/1/012031

In our previous treatments [9, 16, 17] we found the corresponding non-perturbative BPS-saturated solution in frame of the generalized Landau-Ginzburg (LG) field, which is very close to true supersymmetric LG model, but nevertheless has some features (we use here term "generalized Landau-Ginzburg theory" for the case when the superpotential is not holomorphic but depends also on the complex conjugate anti-chiral fields). Boundary of the bag is formed by the domain wall (DW) interpolating between the external KN gravity and superconducting vacuum state inside the bag. Similar to the known MIT and SLAC bag models [23, 24], the KN bag becomes deformable, displaying consistency with external gravitational and electromagnetic (EM) KN field. In fact, the corresponding solutions of the Einstein-Maxwell fields equations determine the size, shape and dynamics of the bag. As a result, the mass, spin and charge of the electron fixe boundary of the bag as a highly oblate ellipsoidal surface. Bag forms a superconducting disk of the Compton radius, and the KN vector potential forms a closed Wilson loop, inducing the closed string of superficial currents placed at the sharp border of the oblate superconducting disk.

In this work we generalize the previous treatments by using the true supersymmetric Landau-Ginzburg model, which is equivalent to supersymmetric Higgs model and specifies the associated with bag stringy structure. This supersymmetric field model becomes equivalent to the Wess-Zumino SuperQED model [25], revealing the bridge between the non-perturbative bag-like solutions and the conventional perturbative technics used in QED.

Contrarily to general point of view that gravitation conflicts with quantum theory, we obtain that Kerr's spinning gravity actively interacts with quantum theory, playing a fundamental role in the structure of elementary particles.

2. Basic features of the Kerr-Newman solution

In the Kerr-Schild coordinates, metric of the KN solutions is [2]

$$g_{\mu\nu} = \eta_{\mu\nu} + 2H k_\mu k_\nu, \tag{4}$$

where $\eta_{\mu\nu}$ is metric of an auxiliary Minkowski space M^4, (signature $(- + ++)$), and H is the scalar function which for the KN solution takes the form

$$H_{KN} = \frac{mr - e^2/2}{r^2 + a^2 \cos^2\theta}, \tag{5}$$

and the KN vector potential is

$$A_\mu = \frac{-er}{(r^2 + a^2\cos^2\theta)} k_\mu . \tag{6}$$

The non-linear part of the KN metric along with vector potential are collinear with null vector field k_μ, $(k_\mu k^\mu = 0)$, forming Principal Null Congruence of the algebraically special solutions [2].

The used in [2] oblate spheroidal coordinates r, θ and ϕ_K are related to Cartesian coordinates as follows

$$\begin{aligned} x + iy &= (r + ia)\exp\{i\phi_K\}\sin\theta, \\ z &= r\cos\theta, \quad \rho = r - t, \end{aligned} \tag{7}$$

where $\rho = r - t$ is the light cone coordinate adapted to null direction

$$k_\mu dx^\mu = dr - dt - a\sin^2\theta d\phi_K, \tag{8}$$

Kerr's congruence k^μ forms a vortex of the null field which propagates analytically from negative sheet of Kerr metric, $r < 0$, to positive one, $r > 0$, where the ingoing field k_μ^- turns

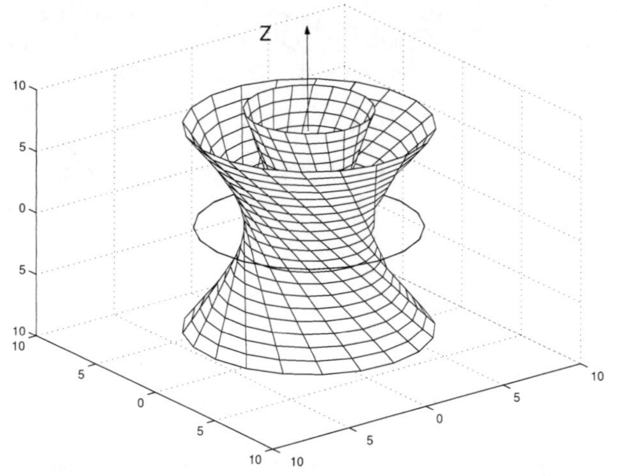

Figure 1. Einstein-Rosen bridge: null lines of the Kerr congruence k^μ forms a vortex field which propagates analytically from negative sheet of Kerr metric, $r < 0$, to positive one, $r > 0$.

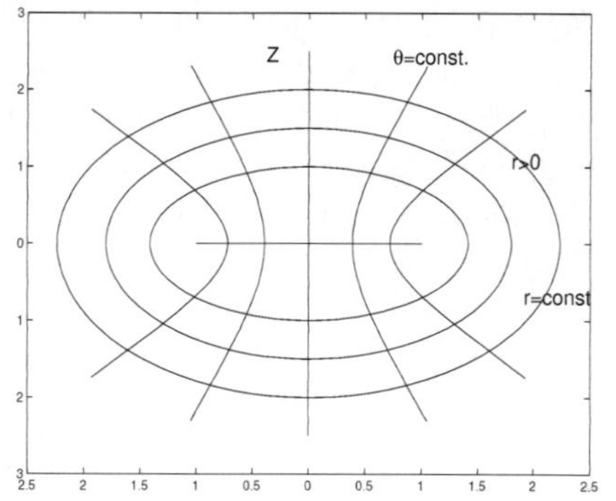

Figure 2. Kerr's oblate spheroidal coordinate system with two-sheeted "radial" coordinate – ellipsoidal surfaces $r = const.$, ($r \geq 0$ and $r \leq 0$) and the orthogonal system of the confocal hyperboloids $\theta = const.$

into outgoing k^+_μ. In equatorial plane, $\cos\theta = 0$, lines of the Kerr congruence are focus on the Kerr singular ring $r = 0$.

The metrics on the ingoing and outgoing ($\pm$)-sheets become different$g^\pm_{\mu\nu} = \eta_{\mu\nu} + 2Hk^\pm_\mu k^\pm_\nu$, and there is a freedom in choice of the k^+_μ or k^-_μ for our "basic" space [26].

We take as the "basic" the outgoing field k^+_μ, that corresponds to the classical physical picture with the retarded electromagnetic field, and we chose as positive rotation the anticlockwise direction of the Kerr congruence near the Kerr singular ring.

Since electron and other spinning particles have the great spin/mass ratio $J/m = \hbar/2m$ (about $10^{20} - 10^{22}$ in the dimensionless units), the black hole horizons of the Kerr solution disappear, and the Kerr singular ring becomes naked.

Acting as a branch line of the Kerr space, the Kerr singular ring spoils space topologically, creating in space-time hole of the Compton radius $a = J/m$ (for seldom exclusion we use in this paper the dimensionless units $G = c = \hbar = 1$). This hole opens the pass to another sheet of the Kerr space, which was interpreted recently as a model of the worm-hole or the Einstein-Rosen bridge [14, 15].

It seems that emergence of this hole in the background intensifies the conflict between gravity and quantum theory and makes their coexistence absolutely incompatible. However, as it was shown in [9, 16] this confrontation can be removed by supersymmetry, more precisely, by the nonperturbative bag model based on the supersymmetric scheme of phase transition. Still more, this confrontation turns into a cooperation, and this gravitational hole emerges as an important part of the KN electron model.

3. Bag-like source of Kerr-Newman solution as a semi-classical electron model

Supersymmetry resolves conflict between the broken by Kerr's gravity background and quantum theory *without modifications of the Einstein-Maxwell equations.* In the supersymmetric bag model, singular region of KN solution is covered by flat interior of the bag, which is formed by a supersymmetric vacuum state of the energy

$$< 0|\mathcal{H}|0 >= 0. \qquad (9)$$

IOP Conf. Series: Journal of Physics: Conf. Series **1275** (2019) 012031 doi:10.1088/1742-6596/1275/1/012031

The corresponding nonperturbative bag-like solution is built as a domain wall phase transition interpolating between the supersymmetric vacuum state inside the bag and the external gravitational KN solution.

3.1. Emergence of the Compton wave length l_c, classical radius r_e and fine structure constant α

Typical features of the bag models are the softness and deformability of the bags under influence of the field distributions, and in particular, under rotation. The KN super-bag is also soft and elastic, displaying a compliance with the external gravitational and electromagnetic KN field.

The Compton wave length appears as radius of the KN spinning particle automatically from basic relation of the Kerr geometry [26]

$$J = ma. \tag{10}$$

Setting for an electron $J = \frac{\hbar}{2}$, we obtain that radius of the Kerr singular ring corresponds to the reduced Compton wave length

$$a = \hbar/2mc \tag{11}$$

which could be wrapped twice around the Kerr singular ring.

The shape and dynamics of the bag are fully defined by matching of the bag boundary with a special surface (López, 1986 [4]), defined by condition

$$H_{KN}(r) = 0. \tag{12}$$

Near this surface metric becomes flat, $g_{\mu\nu} = \eta_{\mu\nu}$, and external gravitational field can be matched continuously with flat space inside the bag.

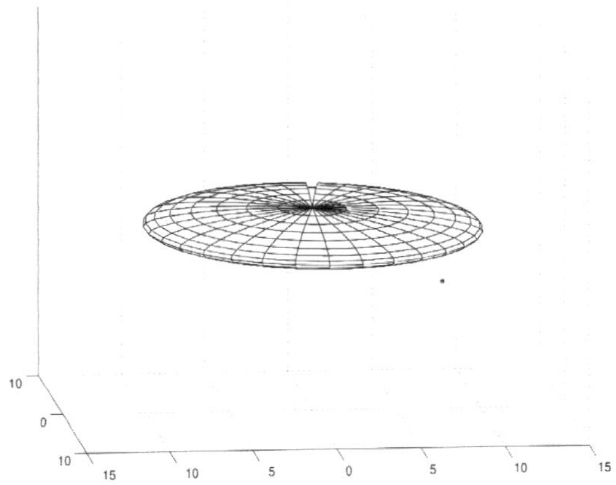

Figure 3. Ellipsoidal shape of the Kerr-Newman bag defined by the boundary surface $r = r_e = e^2/2m$.

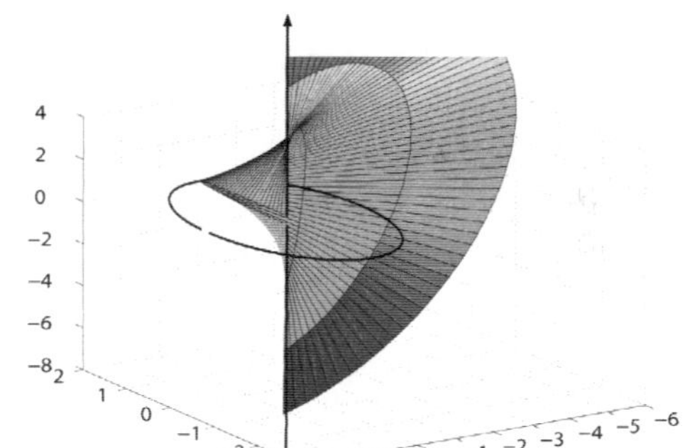

Figure 4. Kerr's coordinate $\phi_K = const$. Singular ring drags the KN vector potential, forming a closed Wilson loop along edge border of the bag.

It gives the additional parameter – classical electron radius. According (5), it defines position of the bag border at

$$r = r_e = e^2/2m, \tag{13}$$

and the relations (7) show that this bag takes the form of a thin ellipsoidal disk of the thickness $2r_e = e^2/m$, and radius

$$r_c = \sqrt{r_e^2 + a^2}, \tag{14}$$

IOP Conf. Series: Journal of Physics: Conf. Series **1275** (2019) 012031 doi:10.1088/1742-6596/1275/1/012031

what is slightly more than the reduced Compton wave length $a = \hbar/2mc$. So that the ratio

$$r_e/a = \alpha \approx 1/137, \tag{15}$$

becomes the fine structure constant, and the ratio $r_e/r_c = \frac{r_e}{a}(1 + r_e^2/a^2)^{-1/2} = \alpha(1 + \alpha^2)^{-1/2}$, which characterizes the degree of oblateness of the bag, is very close to $\alpha \approx 1/137$. Bag covers singular ring, and its flat interior regularizes KN space, where the Cartesian distance δ works as cutoff for the KN electromagnetic and gravitational field, see Fig.6.

3.2. Quantization of the angular momentum

Vector-potential of the external KN solution (6)

$$A_\mu dx^\mu = -\Re e \frac{er}{r^2 + a^2 \cos^2\theta}(dr - dt - a\sin^2\theta d\phi) \tag{16}$$

grows near the core and takes maximal value at the Kerr radial distance $r = r_e = e^2/2m$, $\cos\theta = 0$,

$$A_\mu^{max} dx^\mu = -\frac{2m}{e}(dr - dt - ad\phi). \tag{17}$$

Note, that the component A_r is a perfect differential (see for example [2]) and can be ignored. At the boundary$r = r_e$, the dragged by the Kerr congruence component A_ϕ^{max} is

$$A_\phi^{max} d\phi = \frac{2m}{e} ad\phi. \tag{18}$$

Taking the Wilson loop integral along the bag border for the period $\phi \in [0, 2\pi]$, we obtain

$$e \oint A_\phi^{max} d\phi = 4\pi ma = 4\pi J, \tag{19}$$

where the last equality corresponds to the basic relation $J = ma$ (10). Therefore, for the solutions with integer or half-integer angular momentum, $J = \frac{1}{2}, 1, \frac{3}{2}...$, vector potential (17) forms closed Wilson loop along border of the bag.

3.3. Stringy structure created by spinning bag model

It was noticed long ago [5, 6], that the Kerr singular ring can be considered as a string, which is similar to string-like filament appearing in superconductivity [27] and in dual string models [28]. Kerr's string forms a waveguide for traveling electromagnetic and gravitational waves, and it was shown in [29], that the structure of the electromagnetic and gravitational fields near the Kerr singular ring is close to structure of the Sen heterotic string solutions to low energy string theory [30].

Bag models are relatives of the string models. Bags are soft and flexible, and emergence of the string-like structures in the bag models is usually related with their deformations under influence of strong rotation [31, 32]. Typically, the rotating bag takes the form of a flux-tube with lines of the electric force stretched between the charges (quarks) placed at the ends. In toroidal models of the bags [33], the flux-tubes form closed loops of the electric lines.

These strings have much in common with the known Nambu-Goto string model, and also with solitonic string models [34] and pp-wave solutions [35]. These strings differ from the famous higher-dimensional critical strings of the superstring theory because they are four-dimensional and also because their longitudinal excitations play important role. The string of the KN bag model is formed by the Higgs mechanism of symmetry breaking. It is similar to the Nielsen-Olesen string model, formed as a vortex line in superconducting media [27], and to the Vilenkin-Witten superconducting cosmic string model [36], which was developed by Morris in [37] to a supersymmetric string model.

Similar to other bag models, the bag-like source of the KN solution is also deformed under influence of rotation and takes the shape of a very thin disk of the ellipsoidal form. Circular string emerges in equatorial plane ($\cos\theta = 0$) on the edge of the superconducting disk, instead of the Kerr singular ring and very close to its former position.

As consequence of the strong concentration of the electromagnetic field, the superficial circular current appear at the border of the superconducting bag.

The Kerr string is regular and is build of the longitudinal mode of the vector potential, which forms the closed Wilson loop wrapped around the border. Structure of the Kerr-Newman electromagnetic field shows that there is also the transverse electromagnetic field in the form of traveling waves.

According to Carter [1], the original KN solution has magnetic momentum $\mu = \mu_{Dir} = ea$, which is in full correspondence with the Dirac electron model. The Kerr-Newman solution has singular ring, and therefore, it can be considered as a non-regularized electron model. So far as $\mu = \mu_{Dir}$, the anomalous magnetic momentum in the exact KN solution is zero. So far as the regularized bag model of the KN solution deforms initial model, and we can expect the emergence of the anomalous magnetic moment, which true value is $\alpha/2\pi$. We show that the effect of order α emerges in the considered regular bag model.

The related with regularization string is created very close to the lightlike Kerr singular ring, which grads the KN vector potential in the lightlike circular direction. In our previous works [29, 22, 9, 17] we considered approximate model, in which this string was perfectly lightlike and described in the auxiliary Minkowski background as the surface $x^\mu(\phi, t)$ parametrized by the cylindric coordinates ϕ, t.

The Cartesian distance δ between border of the bag and singular ring is very small with respect to radius of the ring a, and this string can be considered as tangential to the Kerr singular ring with the lightlike direction

$$k^+_\mu dx^\mu = (dt - a d\phi), \tag{20}$$

In this case δ works as a cut-off parameter. However, there is one more essential effect – the tangent to string direction $d\phi$ deviates from the lightlike direction of the Kerr congruence k^+_μ, and the real velocity of the bag border becomes smaller than the speed of the light.

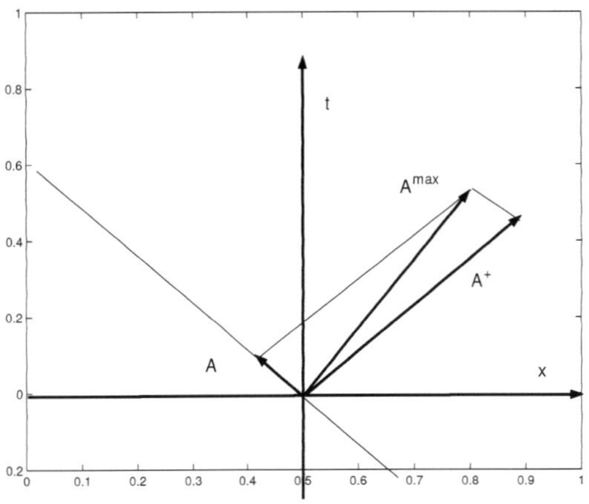

Figure 5. Decomposition A^{max} in the left and right modes A^+ and A^-. Contribution of the small lightlike right mode A^- reduces speed and increases length of the relativistic string.

This deviation can be described as a small lightlike component A^-_μ in direction $k^-_\mu dx^\mu = (dt + ad\phi)$, which is opposite (antipodal) to k^+_μ.

When the initiate basic potential

$$A^{max}_\mu dx^\mu = A^-_\mu|_{r_e} dx^\mu = -\frac{2m}{e}(dt - ad\phi), \tag{21}$$

is completed by antipodal components A^-, the resulting potential

$$A^{max}_\mu dx^\mu = A^+_\mu dx^\mu + A^-_\mu dx^\mu \tag{22}$$

is composed proportionally to contributions of the amplitudes A^+ and A^-, where according Fig.5

$$A^+_\mu dx^\mu = -\frac{2m}{e}\cos\theta(dt + ad\phi), \quad A^-_\mu dx^\mu = -\frac{2m}{e}\sin\theta(dt - ad\phi). \tag{23}$$

On the other hand, kinematic relations presented in Fig.6 show that the angle is

$$\theta_r = \arctan\frac{r_e}{a} = \arctan\alpha, \tag{24}$$

which is approximately equal to the fine structure constant α.

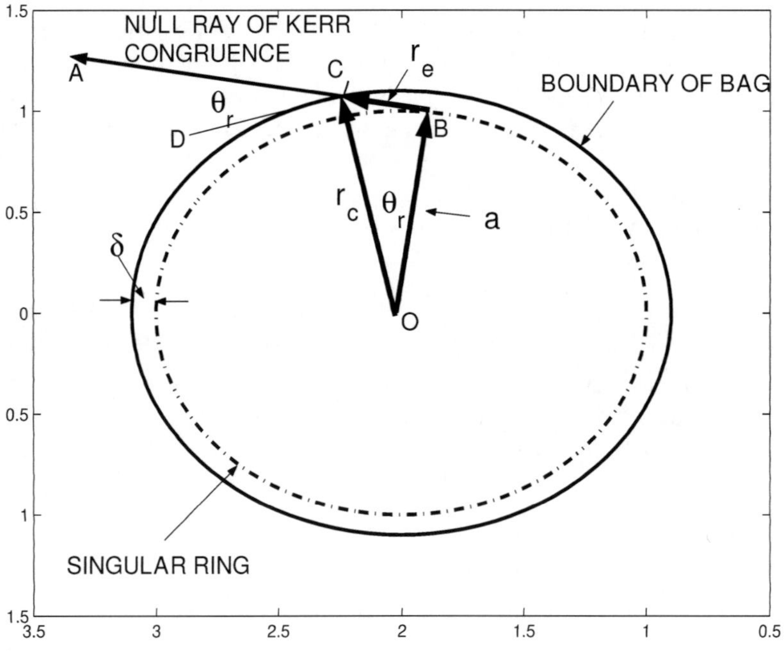

Figure 6. Kinematic relation in the equatorial plane $\cos\theta = 0$. The tangent to Kerr singular ring light-like beam of the Kerr congruence crosses the edge of the bag at the angle $\theta = \arctan r_e/a \approx \alpha$.

We see that influence of the oppositely directed mode A^-_μ is proportional to

$$\tan\theta_r = r_e/a = \alpha. \tag{25}$$

Physically, it leads to deceleration and to the corresponding lengthening of the string and expansion of the relevant current loop. The fields $A^+_\mu(t+\sigma)$ and $A^-_\mu(t-\sigma))$ can be considered now as the left and right modes which complete the full stringy system.

Deviation from the basic lightlike direction k^+_μ, caused by additional contribution from the direction k^-_μ is small, because this mode is directed against the principal null congruence of the KN solution, which is prohibited by the KN gravitational field. It can impact only in the zone near the bag border where the regularized space is flat, $g_{\mu\nu} \approx \eta_{\mu\nu}$. Velocity of the originally lightlike string is reduced, and its length $l = 2\pi a$ increases proportionally to $\theta_r = \arctan r_e/a \approx \alpha$.

Identification of the angles θ and θ_r could give solution of the old problem – origin of the anomalous magnetic moment of the electron [38]. In the report to 12th Solvay Congress Feynman noted [39]: "We have no physical picture by which we can easily see that the correction is roughly $\alpha/2\pi$, in fact, we do not even know why the sign is positive (other than by computing it). " Unfortunately, we have not found so far solution of this problem. We see definitely that θ is proportional to θ_r, but we didn't explained so far the emergence of the factor $\frac{1}{2\pi}$.

We see that modes $A^+_\mu(t+\sigma)$ and $A^-_\mu(t-\sigma))$ have opposite direction of rotation, which corresponds to the opposite sign of a and is required by supersymmetry (see later (38)). Geometric origin of this difference is nontrivial.

We note, that although the supersymmetric phase transition regularized the Kerr-Newman space-time background, and the negative sheet of the Kerr metric is removed, the electromagnetic field remained two-valued, as far as it is described by the two-valued Kerr coordinate system (7). One can expect that emergence of the the factor $\frac{1}{2\pi}$ can be related with the pole in the potential, which must be bypassed in the process of integration over the domain wall layers $r \in [-r_e, +r_e]$.

From (6) and (8) we see that potential has the form

$$A_\mu dx^\mu = -e\frac{er}{r^2 + a^2\cos^2\theta}(dr - dt - a\sin^2\theta d\phi_K), \tag{26}$$

and the (t, ϕ) components are extended smoothly to negative values of r, where the potential changes the sign and direction. However, at the "negative" border of the bag, $r = -r_e$, potential forms oppositely oriented loop. Orientation of the string is retained when this transformation is completed by the replacement $e \to -e$. The resulting string retains orientation, but changes direction of the rotation a. The potential is continued analytically to negative r, where the boundary of the bag-like source is completed by an anti-domain wall boundary. Therefore, the string mode formed by the potential A^- should be placed on the "negative" border of the anti-domain wall boundary. Such two-valued structure of the Kerr source, forming the "Domain Wall – Anti-Domain Wall" combination similar to the known kink-antikink solutions, sometimes was called as "breather". In the source of the KN solution it was at first noticed in [17].

4. Supersymmetric phase transition

It was noticed in several works that source of the KN solution must be superconducting [4, 19, 21], similar to many soliton models, the famous Nielsen-Olesen dual string model [27] and the bag models [23, 24]. Superconductivity is close related with Higgs mechanism of symmetry breaking [27] and in the simplest case it is described by the "minimal" Landau-Ginzburg model

$$\mathcal{L}_{NO} = -\frac{1}{4}F_{\mu\nu}F^{\mu\nu} - \frac{1}{2}(\mathcal{D}_\mu\Phi)(\mathcal{D}^\mu\Phi)^* - V(|\Phi|), \tag{27}$$

where $\mathcal{D}_\mu = \nabla_\mu + ieA_\mu$, and $F_{\mu\nu} = A_{\mu,\nu} - A_{\nu,\mu}$.

The typical form of the scalar potential $V = \lambda(\Phi^\dagger\Phi - \eta^2)^2$, cannot be matched with gravity, because it breaks gauge symmetry in external space. In particular, in the most of superconducting string models and in the bag models, strings are created as a vortex in superconductor [27], as well as the bags appear as "cavities in superconductor" [23, 24]. The consistency with gravity requires opposite construction – the bag must be superconducting itself, leaving unbroken the external gravitational and electromagnetic fields. As it was first noticed by Witten [36], this cannot be realized with one Higgs field, and a special form of the potential is

IOP Conf. Series: Journal of Physics: Conf. Series **1275** (2019) 012031 doi:10.1088/1742-6596/1275/1/012031

required which has a few Higgs-like fields. Morris [37] suggested to use "pseudo-supersymmetric" model where potential V is formed through superpotential

$$W(\Phi_i) = Z(\Sigma\bar{\Sigma} - \eta^2) + (Z + \mu)H\bar{H}, \tag{28}$$

depending on several chiral fields Φ_i, (μ and η are real constants). The notations $(H, Z, \Sigma) \equiv (\Phi_1, \Phi_2, \Phi_3)$ identify the complex Higgs field $H = |H|e^{i\chi}$, as the first field Φ_1.

The scalar potential is formed as follows [25] (we use the sign "star" for complex conjugation along with the sign "bar")

$$V(r) = \sum_i F_i F_i^*, \tag{29}$$

where

$$F_i = \partial W/\partial \Phi_i \equiv \partial_i W. \tag{30}$$

$\mathcal{D}_{1\mu} = \nabla_{1\mu} + ieA_\mu$, where A_μ is the external KN vector field. The fields Φ_2 and Φ_3 are assumed uncharged, and $\mathcal{D}_{i\mu} = \nabla_{i\mu}$ for $i = 2, 3$. The corresponding Lagrangian differs from (27) only by summation over the fields Φ_i. This model was used in our previous works [9, 16, 17]. It is called as "pseudo-supersymmetric", because the true superpotential must indeed be holomorphic function of the all superfields [25, 37],

$$W(\Phi_i) = Z(\Sigma^+\Sigma^- - \eta^2) + (Z + \mu)H^+H^-, \tag{31}$$

and therefore, the number of independent chiral fields Φ_i must be increased to five:

$$(H^+, H^-, Z, \Sigma^+, \Sigma^-) \equiv (\Phi_1, \Phi_2, \Phi_3, \Phi_4, \Phi_5). \tag{32}$$

The auxiliary fields $F_i^* = \partial W/\partial \Phi_i = (F_+^*, F_-^*, F_{\Sigma+}^*, F_{\Sigma-}^*, F_Z^*)$ take the form

$$F_\pm^* = (Z + \mu)H^\mp, \tag{33}$$
$$F_{\Sigma\pm}^* = Z\Sigma^\mp, \tag{34}$$
$$F_Z^* = \Sigma^+\Sigma^- + H^+H^- - \eta^2, \tag{35}$$

Vacuum expectation values of fields Φ_i for which $F_i^* = 0$ give minima of the potential $V = 0$ corresponding to supersymmetric vacuum states. Just as in case (28), we obtain Domain Wall with two isolated vacua

Internal: $H^-H^+ = \eta^2, \quad Z = -\mu, \quad \Sigma^+ = \Sigma^- = 0$, and

External: $H^- = H^+ = 0,\ Z = 0,\ \Sigma^+\Sigma^- = \eta^2$,

which are separated by Domain Wall – zone **(R)**, formed by of the positive scalar potential

$$V = |\Sigma^+\Sigma^- + H^+H^- - \eta^2|^2 + |(Z + \mu)H^+|^2 + |(Z + \mu)H^-|^2 + |Z|^2(|\Sigma^+|^2 + |\Sigma^-|^2). \tag{36}$$

As in the considered previously cases [9, 16, 17] the domain wall forms ellipsoidal boundary of the bag. The both Higgs fields H^- and H^+ are concentrated inside the bag.

The most important difference between the supersymmetric model and the considered earlier pseudo-supersymmetric Landau-Ginzburg model is the relation

$$H^-H^+ = \eta^2, \tag{37}$$

which shows that the chiral Higgs fields $H^- = |H^-|e^{i\chi^-}$ and $H^+ = |H^+|e^{i\chi^+}$ can have different amplitudes $|H^-| \neq |H^+|$, but their phases must be correlated

$$\chi^+ = -\chi^- = \chi, \tag{38}$$

which is necessary to form Domain Wall corresponding to profile of the curved bag boundary $r = r_e$ (see Fig.4.) and provide BPS stability, similar to results in [9, 16, 17].

5. Superficial currents

According supersymmetric Landau-Ginzburg model, inside the bag the chiral fields $\Sigma^{\pm}$ vanish, (4), and potential on the border of the bag (17) interacts only with the Higgs fields H^+ and H^-, through covariant derivatives $\mathcal{D}^+_\mu = \nabla_\mu + ieA^+_\mu$ and $\mathcal{D}^-_\mu = \nabla_\mu - ieA^-_\mu$, which appropriate to opposite $U(1)$ charges and correspond to the Wess-Zumino field model of supersymmetric QED [25]. One can expect that the summarized potential (22), $A^{max}_\mu dx^\mu = A^+_\mu dx^\mu + A^-_\mu dx^\mu$ creates on the border of the bag two surface currents

$$\Box A^{max}_\mu = J^+_\mu + J^-_\mu , \tag{39}$$

where

$$J^{\pm}_\mu = e^{\pm}|H^{\pm}|^2(\chi^{\pm},_\mu - e^{\pm}A^{\pm}_\mu). \tag{40}$$

The right mode A^-_μ is placed at the "negative" border $r = -r_e$, where is to be used opposite sign of the charge $e^- = -e^+ = -e$, and sign of the phase $\chi^- = -\chi^+ = -\chi$ of the Higgs fields $H^{\pm}$, according to (38).

In the core of superconducting disk we have $J^+_\mu = 0$, and the equation $\chi^+,_\mu - eA^+_\mu = 0$ gives

$$\chi^+,_t - eA^+_t = 0, \quad \chi^+,_\phi - eA^+_\phi = 0. \tag{41}$$

From (23) we obtain $\chi^+,_t = eA^+_t = 2m\cos\theta$, $\chi^+,_\phi = eA^+_\phi = 2ma\cos\theta$, and integration gives

$$\chi^+ = 2m\cos\theta(t - t_0 + a\phi), \tag{42}$$

where t_0 is the constant of integration which is omitted later.

Similar calculations for $\chi^-,_\mu + eA^-_\mu = 0$ (with change $e \to -e$ and (23)) give in components

$$\chi^-,_t = -eA^-_t = -2m\sin\theta, \qquad \chi^-,_\phi = -eA^-_\phi = -2ma\sin\theta, \tag{43}$$

and integration gives

$$\chi^- = -2m\sin\theta(t - a\phi). \tag{44}$$

Since $\chi^+ = -\chi^-$, we have to set for the left and right modes equivalent frequencies and opposite directions of rotation, which demands to use the different parameters $a = a^+ = -a^-$, and the different masses $m^+ = m/\cos\theta$, $m^- = m/\sin\theta$. Using (42) and (44), we obtain

$$\chi^+ = 2m^+\cos\theta(t + a^+\phi) = -\chi^- = 2m^-\sin\theta(t - a^-\phi), \tag{45}$$

6. SuperBag as nonperturbative solution of the SuperQED model

Two oppositely charged Higgs fields H^+ and H^- give rise to kinetic part of the Wess-Zumino SuperQED model [25],

$$-\frac{1}{4}W^aW_a + \Phi^+_+ e^{eV}\Phi_+|_{\theta\theta\bar\theta\bar\theta} + \Phi^+_- e^{-eV}\Phi_-|_{\theta\theta\bar\theta\bar\theta}, \tag{46}$$

where V is vector superfield, and $W^a = -\frac{1}{4}\bar D\bar D D_\alpha V$. In the same time, the potential part (29), (30) and (31) corresponds to the most general renormalizable supersymmetric Lagrangian and gives rise to nonperturbative generalization of the SuperQED model.

The chiral superfields $\Phi_\pm$, can be expressed in the component form

$$\Phi_\pm(y) = H^{\pm}(y^\mu) + \sqrt{2}\theta\psi_\pm(y^\mu) + \theta\theta F_\pm(y^\mu), \tag{47}$$

as functions of the chiral coordinates $y^\mu = x^\mu + i\theta\sigma^\mu\bar\theta$ and θ.

Interaction of the oppositely charged Higgs fields $H^{\pm}$ with vector potential in zone **(I)** is defined by (39), (40) and yields

$$H^{\pm} = |H^{\pm}|e^{\pm i\chi}, \ \bar{H}^{\pm} = |H_{\pm}|e^{\mp i\chi}, \ \chi = 2mt + 2am\phi, \tag{48}$$

where the fields $\bar{H}^{\pm}$ are scalar components of the antichiral fields

$$\Phi^{+}_{\pm}(y^{+}) = \bar{H}^{\pm}(y^{+\mu}) + \sqrt{2}\bar{\theta}\bar{\psi}_{\pm}(y^{+\mu}) + \bar{\theta}\bar{\theta}\bar{F}_{\pm}(y^{+\mu}), \tag{49}$$

as functions of the antichiral coordinates $y^{+\mu} = x^{\mu} - i\theta\sigma^{\mu}\bar{\theta}$ and $\bar{\theta}$. The corresponding nonperturbative solution with doubled Higgs fields (48) can be obtained similar to [16].

In the Wess-Zumino SuperQED model, the two Weyl spinors $\psi^{\pm}$ in (47) combine into one massive Dirac spinor of the electron – superpartner of the Higgs doublet $H^{\pm}$, [25].

Super-bag solution generates in the core of spinning particle the free from gravity Compton zone, which is flat and supersymmetric, representing necessary conditions for work of the perturbative SuperQED model, while the "miracleous cancellations" of the component super-graphs [25] forms the link to perturbative QED.

7. Conclusion

We have considered principal features of the spinning Kerr-Schild geometry which specify the bag model as a consistent with gravity way to particle physics and electroweak sector of the Standard Model. Two of these features are principally new, relative to the widespread belief:

– the spinning KN gravity is not weak, and becomes very strong at the Compton scale of the particle physics,

– compatibility between Quantum and Gravity can be achieved by means of supersymmetric generalization of the matter sector, without modification of the Einstein-Maxwell equations.

Due to extreme high spin/mass ratio, impact of the gravitational KN field on the structure of space-time becomes very strong, and the necessary for regularization nonperturbative bag-like solution becomes very sensible to the external Einstein-Maxwell field. As a result, the Kerr gravitational field determines many important parameters of the electron, such as the Compton wave length, the classical radius of the electron and the fine structure constant, creating the bag as a thin superconducting disk of the Compton radius with the thickness equal to classical radius of the electron. Simultaneously, the consistent with gravity bag forms the necessary stringy structure together with the quantum Wilson loop, which defines quantization of the angular momentum.

All that, along with the wonderful consistency of the Kerr geometry with solutions of the Dirac equation in the Weyl form [16, 40] and the known Carter result [1] that the exact Kerr-Newman solution has the gyromagnetic ratio of the Dirac theory, indicate that Kerr's gravity participates actively in formation of the electron structure, contrary to wide-spread opinion that it is only a weak field theory which for some reason resists to be quantized. The spinning Kerr gravity deforms background topologically, as it is necessary to create quantum particle.

In conclusion we must answer the typical question: Why this bag-like and string structures are not seen experimentally, and electron looks like a point? Answer follows from the features of the Lorentz contraction for the relativistically rotating systems [41]. *The large left mode of the relativistic Kerr string shrinks to point by Lorentz contraction* [42].

Acknowledgments

Author wishes to acknowledge the useful discussions on supersymmetry with K.V Stepanyants and also with the colleagues in the Theor. Phys. Lab. of the NSI RAS R. V. Arutunian, P.S. Kondratenko, L.V. Matweev and Yu.N. Obukhov.

References

[1] Carter B 1968 *Phys. Rev.* **174** 1559
[2] Debney G C, Kerr R P and Schild A 1969 *J. Math. Phys.* **10** 1842
[3] Israel W 1970 Source of the Kerr metric *Phys. Rev.* D **2** 641
[4] López C A 1984 An Extended Model Of The Electron In General Relativity *Phys. Rev.* D **30** 313
[5] Burinskii A Ya 1974 Microgeons with spin *Sov. Phys. JETP* **39** 193
[6] Ivanenko D D and Burinskii A Ya 1975 Gravitational strings in the models of elementary particles *Izv. Vuz. Fiz.* **5** 135
[7] Burinskii A 2017 Weakness of gravity as illusion which hides true path to unification of gravity with particle physics *Int. J. Mod. Phys.* D **26** 1743022
[8] Burinskii A 2017 Illusion about weakness of gravity hides new way to unify gravity with particle physics *J. Phys.: Conf. Series* **942** 012006
[9] Burinskii A 2015 Gravitating lepton bag model *JETP (Zh. Eksp. Teor. Fiz.)* **148**(8) 228 [arXiv:1505.03439]
[10] Burinskii A 2018 Supersymmetric bag model for unification of gravity with spinning particles *Phys. of Part. and Nucley* **49**(5) 958
[11] Becker K, Becker M and Schwarz J 2007 *String Theory and M-Theory: A Modern Introduction* Cambridge University Press
[12] Schwarz J 2012 The Early History of String Theory and Supersymmetry Preprint CALT-68-2858 (arXiv:1201.0981)
[13] Baez J C 2001 Higher Dimensional Algebra and Planck Scale Physics, In *Physics Meets Philosophy at the Planck Length*, eds. Craig Callender and Nick Huggett, Cambridge U. Press pp 177-195, [arXiv:gr-qc/9902017]
[14] Burinskii A 2014 Kerr-Newman electron as spinning soliton, *Int J. of Mod.Phys.* A **29** 1450133 (Preprint arXiv:1410.2888) .
[15] Gibbons G V and Volkov M S 2017 Zero mass limit of Kerr spacetime is a wormhole *Phys. Rev.* D **96** 024053
[16] Burinskii A 2015 Stability of the lepton bag model based on the Kerr-Newman solution *JETP (Zh. Eksp. Teor. Fiz.)* **148** (11) 937 [arXiv:1706.02979]
[17] Burinskii A 2016 Source of the Kerr-Newman solution as a supersymmetric domain-wall bubble: 50 years of the problem *Phys. Lett* B **754** 99
[18] Gürses M and Gürsey F 1975 *J. Math. Phys.* **16** 2385
[19] Tiomno J 1973 Electromagnetic field of rotating charged bodies *Phys. Rev.*D **7** 992
[20] Hamity V 1976 An "interior" of the Kerr metric *Phys. Lett.* A **56** 77
[21] Dymnikova I 2006 Spinning superconducting electrovacuum soliton *Phys. Lett.* B **639** 368
[22] Burinskii A 2010 Regularized Kerr-Newman Solution as a Gravitating Soliton *J. Phys. A: Math. Theor.* **43** 392001
[23] Chodos A et al. 1974 New extended model of hadrons *Phys. Rev.* D **9** 3471
[24] Bardeen W A at al. 1974 Heavy quarks and strong binding: A field theory of hadron structure *Phys. Rev.* D **11** 1094
[25] Wess J and Bagger J *Supersymmetry and Supergravity*, Princeton Univ. Press, New Jersey 1983.
[26] Misner Ch W, Thorne K S and Wheeler J A 1973 *Gravitation* v.3, San Francisco: W. Freeman
[27] Nielsen H B and Olesen P 1973 Vortex-line models for dual strings *Nucl. Phys.* B **61** 45
[28] Burinskii A 2003 Orientifold D-String in the Source of the Kerr Spinning Particle, *Phys. Rev.* D **68** 105004
[29] Burinskii A 1995 Some properties of the Kerr solution to low-energy string theory. *Phys. Rev.* D **52** 5826
[30] Sen A 1992 Rotating charged black hole soltion in heterotic string theory, *Phys. Rev. Lett.* **69** 1006
[31] Giles R C 1976 Semiclassical dynamics of the "SLAC bag", *Phys. Rev.* **D 70** 1670
[32] Johnson K and Thorn C B 1976 Stringlike solutions of the bag model *Phys. Rev.* **D 13** 1934
[33] Tye S-H H 1976 Quark-binding string *Phys. Rev.* **D 13**, 3416
[34] Dabholkar A et al. 1996 Strings as solitons and black holes as strings *Nucl.Phys.* B **474**, hep-th/9511053.
[35] Horowitz G and Steif A 1990 Spacetime Singularities in String Theory. *Phys. Rev. Lett.* **64**, 260
[36] Witten E 1985 Superconducting strings *Nucl.Phys.* B **249** 557
[37] Morris J R 1996 *Phys. Rev.* **D 53** 2078, [arXiv:hep-ph/9511293]
[38] Drell S R and Pagels H R 1965 *Phys. Rev.* **140** B 397
[39] Feynman R P 1961 in *The Quantum Theory of Fields* (Interscience Publishers, Inc., New York)
[40] Burinskii A. 2015 Emergence of the Dirac Equation in the Solitonic Source of the Kerr Spinning Particle. *Grav. Cosmol.* **21 (1)** 28 , arXiv:1404.5947
[41] Landau L D and Lifshitz E M *Course of Theoretical Physics (Volume 4): Quantum Electrodynamics*, 2nd ed. ISBN 10: 0750633719
[42] Punsly B 1985 *J. Math. Phys.* **26** 1728

9th International Workshop DICE2018 : Spacetime - Matter - Quantum Mechanics IOP Publishing
IOP Conf. Series: Journal of Physics: Conf. Series **1275** (2019) 012032 doi:10.1088/1742-6596/1275/1/012032

Phase-space continuity equations for quantum decoherence, purity, von Neumann and Rényi entropies

Alex E. Bernardini

Departamento de Física, Universidade Federal de São Carlos, PO Box 676, 13565-905, São Carlos, SP, Brasil.

E-mail: `alexeb@ufscar.br`

Orfeu Bertolami

Departamento de Física e Astronomia, Faculdade de Ciências da Universidade do Porto, Rua do Campo Alegre, 687, 4169-007, Porto, Portugal

E-mail: `orfeu.bertolami@fc.up.pt`

Abstract. Phase-space features of the Wigner flow are examined so to provide a set of continuity equations that describe the flux of quantum information in the phase-space. The reported results suggest that the non-classicality profile of anharmonic (periodic) quantum systems can be consistently obtained in terms of the fluxes of probability, purity and von Neumann entropy. Extensions of the such phase-space quantifiers are also investigated in the context of the so-called Rényi entropy.

1. Introduction

The formulation of the phase-space dynamics of quantum systems according to the Weyl-Wigner quantum mechanics [1, 2] provides elementary tools to connect the predictive features of quantum mechanics with the investigation of quantum information. As it has been shown, the Weyl-Wigner phase-space formulation [3, 4, 5] can provide a wider understanding of the limits between classical and quantum regimes of physical systems.

The Wigner function, $W(q,p)$, encompasses the information from coordinate and momentum Schrödinger wave function representation of quantum mechanics in a convolutional representation of the wave function described by the Weyl transform of a quantum operator, $\hat{Q}$,

$$Q^W(q,p) = \int_{-\infty}^{+\infty} ds\, \exp\left[2\,i\,p\,s/\hbar\right] \langle q-s|\hat{Q}|q+s\rangle = \int_{-\infty}^{+\infty} dr\, \exp\left[-2\,i\,q\,r/\hbar\right] \langle p-r|\hat{Q}|p+r\rangle, \quad (1)$$

such that

$$h^{-1}\hat{\rho} \to W(q,p) = (\pi\hbar)^{-1} \int_{-\infty}^{+\infty} ds\, \exp\left[2\,i\,p\,s/\hbar\right] \psi(q-s)\,\psi^*(q+s), \quad (2)$$

the Weyl transform of a density matrix, $\hat{\rho} = |\psi\rangle\langle\psi|$, is interpreted as the Fourier transform of the off-diagonal elements of $\hat{\rho}$.

Content from this work may be used under the terms of the Creative Commons Attribution 3.0 licence. Any further distribution of this work must maintain attribution to the author(s) and the title of the work, journal citation and DOI.
Published under licence by IOP Publishing Ltd

In spite of being consistent with the Moyal's picture of quantum mechanics [6], where the noncommutative nature of canonical operators supports the Moyal *star*-product definition, the above definition does not yield a real-valued probability distribution, given that it admits local negative values, which is inconsistent with the positive-valued probability interpretation. This provides a *quasi*-probability interpretation which admits an enhanced phase-space framework and eventually fixes this issue [7].

Keeping in mind that the Weyl transform and the Wigner function connect quantum observables, $\hat{Q}$, with their expectation values, one notices that the trace of the product of the operators, $\hat{\rho}$ and $\hat{Q}$, can be evaluated from the $dq\,dp$ integral of the product of their Weyl transforms [1, 2], [1]

$$Tr_{\{q,p\}}\left[\hat{\rho}\hat{Q}\right] \to \langle Q\rangle = \int_{-\infty}^{+\infty}\int_{-\infty}^{+\infty} dq\,dp\,W(q,p)\,Q^W(q,p), \tag{4}$$

which allows for the probability distribution interpretation, with the normalization condition established by $\hat{\rho}$, $Tr_{\{q,p\}}[\hat{\rho}] = 1$. The Weyl-Wigner definitions also admit extensions from pure states to statistical mixtures. In particular, the purity,

$$Tr_{\{q,p\}}[\hat{\rho}^2] = 2\pi \int_{-\infty}^{+\infty}\int_{-\infty}^{+\infty} dq\,dp\,W(q,p)^2, \tag{5}$$

is constrained by the conditions of $Tr_{\{q,p\}}[\hat{\rho}^2] = Tr_{\{q,p\}}[\hat{\rho}] = 1$ for pure states. In the investigation of classical and quantum regimes of one-dimensional physical systems, the statistical aspects related to the nature of the density matrix quantum operators play a relevant role in the computation of entropy and information quantifiers.

Considering these theoretical foundations, a fluid-analog of the phase-space quantum entropy (information) flux in the phase-space can be recovered from the Weyl-Wigner formalism of quantum mechanics [8]. In this contribution, phase-space continuity equations for quantum decoherence, purity and von Neumann entropy are recast in a dimensionless framework in order to be useful for discussing a broad set of Hamiltonian quantum systems. In particular, extensions of such phase-space quantifiers are also investigated in the context of the Rényi entropy definition. The conclusion is that, the set of phase-space continuity equations discussed below emphasizes the relevancy of the Wigner formalism in discussing the flux of quantum information in the phase-space.

2. Phase-space flow analysis and continuity equations

The dynamics of the Wigner function, $W(q,\,p;\,t)$ can be described in terms of a vector flux, $\mathbf{J}(q,\,p;\,t)$, which represents the flow of $W(q,\,p;\,t)$ in the phase-space [4, 9, 10, 11]. In this case, the flow field, $\mathbf{J}(q,\,p;\,t)$, is decomposed into coordinate and momentum components, $\mathbf{J} = J_q\,\hat{q} + J_q\,\hat{q}$, with $\hat{p} = \hat{p}_q$, through which one can identify the quantum version of the Liouville equation, that is a continuity equation given by [2, 7, 4, 8],

$$\frac{\partial W}{\partial t} + \frac{\partial J_q}{\partial q} + \frac{\partial J_p}{\partial p} \equiv \frac{\partial W}{\partial t} + \boldsymbol{\nabla}\cdot\mathbf{J} = 0, \tag{6}$$

where

$$J_q(q,\,p;\,t) = \frac{p}{m}\,W(q,\,p;\,t), \tag{7}$$

[1] In general one has

$$Tr_{\{q,p\}}\left[\hat{Q}_1\hat{Q}_2\right] = \int_{-\infty}^{+\infty}\int_{-\infty}^{+\infty} dq\,dp\,Q_1^W(q,p)\,Q_2^W(q,p), \tag{3}$$

for any product of two operators, $\hat{Q}_1$ and $\hat{Q}_2$.

IOP Conf. Series: Journal of Physics: Conf. Series **1275** (2019) 012032 doi:10.1088/1742-6596/1275/1/012032

$$J_p(q,\,p;\,t) = -\sum_{\nu=0}^{\infty}\left(\frac{i\,\hbar}{2}\right)^{2\nu}\frac{1}{(2\nu+1)!}\left[\left(\frac{\partial}{\partial q}\right)^{2\nu+1}V(q)\right]\left(\frac{\partial}{\partial p}\right)^{2\nu}W(q,\,p;\,t), \tag{8}$$

and $V(q)$ is the potential for a non-relativistic quantum system described by the Hamiltonian

$$H(q,\,p) = \frac{p^2}{2m} + V(q), \tag{9}$$

Notice from Eq. (8) that the contributions from $j \geq 1$ in the series expansion quantify the distortion due to the quantum corrections with respect to the classical Liouvillian phase-space trajectories.

In particular, for the generalized discussion of non-relativistic quantum systems where m is simply a mass scale, generalized results can be obtained when $H(q,\,p)$ is put into a dimensionless form, $\mathcal{H}(x,\,k) = k^2/2 + \mathcal{U}(x)$. The dimensionless variables, $x = \left(m\,\omega\,\hbar^{-1}\right)^{1/2} q$ and $k = (m\,\omega\,\hbar)^{-1/2} p$, provide the identification of $\mathcal{H} = (\hbar\omega)^{-1} H$ and $\mathcal{U}(x) = (\hbar\omega)^{-1} V\left(\left(m\,\omega\,\hbar^{-1}\right)^{-1/2} x\right)$, where ω^{-1} is the time scale.

The corresponding dimensionless form of the Wigner function and of the corresponding Wigner current are thus obtained from

$$\begin{aligned}
\mathcal{W}(x,\,k;\,\tau) &\equiv (m\omega\,\hbar)^{1/2}\,W(q,\,p;\,t), &(10)\\
\mathcal{J}_x(x,\,k;\,\tau) &\equiv m\,J_q(q,\,p;\,t), &(11)\\
\mathcal{J}_k(x,\,k;\,\tau) &\equiv \omega^{-1}\,J_p(q,\,p;\,t), &(12)
\end{aligned}$$

which can be recast in the form [14]

$$\begin{aligned}
\mathcal{W}(x,\,k;\,\tau) &= \pi^{-1}\int_{-\infty}^{+\infty} dy\,\exp\left[2\,i\,k\,y\right]\varphi(x-y;\,\tau)\,\varphi^*(x+y;\,\tau), \quad \text{with } y = \left(m\,\omega\,\hbar^{-1}\right)^{1/2} s, &(13)\\
\mathcal{J}_x(x,\,k;\,\tau) &= k\,\mathcal{W}(x,\,k;\,\tau), &(14)\\
\mathcal{J}_k(x,\,k;\,\tau) &= -\sum_{\nu=0}^{\infty}\left(\frac{i}{2}\right)^{2\nu}\frac{1}{(2\nu+1)!}\left[\left(\frac{\partial}{\partial x}\right)^{2\nu+1}\mathcal{U}(x)\right]\left(\frac{\partial}{\partial k}\right)^{2\nu}\mathcal{W}(x,\,k;\,\tau), &(15)
\end{aligned}$$

where $\tau = \omega t$ is a dimensionless quantity[2]. From this framework, the phase-space information flux associated to extensions of the above continuity equation can be obtained [12, 8].

To implement a self-contained procedure for obtaining the continuity equations derived from the fluid-analog properties of the Wigner functions, one notices that locally and globally conservative properties associated to a point in the phase-space, $\boldsymbol{\xi}$, and to a phase-space *volume integral bounded by a comoving closed surface*, $V = \int_V dx\,dk$, are connected by a convective (or substantial) derivative [4, 8] operator defined by

$$\frac{D}{D\tau}\int_V dV\,\mathcal{W} \equiv \int_V dV\left[\frac{D\mathcal{W}}{D\tau} + \mathcal{W}\boldsymbol{\nabla}_\xi\cdot\mathbf{v}_\xi\right], \tag{18}$$

[2] Notice that $\varphi(x,\,\tau)$ is consistent with the normalization condition given by

$$\int_{-\infty}^{+\infty} dx\,|\varphi(x;\,\tau)|^2 = \int_{-\infty}^{+\infty} dq\,|\psi(q;\,t)|^2 = 1, \tag{16}$$

and the Eq. (6) can be multiplied by $(m\,\hbar/\omega)^{1/2}$ so to yield the dimensionless continuity equation,

$$\frac{\partial\mathcal{W}}{\partial\tau} + \frac{\partial\mathcal{J}_x}{\partial x} + \frac{\partial\mathcal{J}_k}{\partial k} = \frac{\partial\mathcal{W}}{\partial\tau} + \boldsymbol{\nabla}_\xi\cdot\boldsymbol{\mathcal{J}} = 0, \tag{17}$$

where the phase-space coordinate vector, $\boldsymbol{\xi} = (x,\,k)$, is defined.

with $dV \equiv dx\,dk$, and where $\mathbf{v}_\xi$ corresponds to the classical phase-space vector velocity, $\mathbf{v}_{\xi(\mathcal{C})} = (k, -\partial\mathcal{U}/\partial x)$. Assuming that a two-dimensional classical path, $\mathcal{C}$, can be identified, from the definition of $\mathbf{v}_{\xi(\mathcal{C})}$, one sets

$$\frac{D\mathcal{W}}{D\tau} = -\mathcal{W}\,\boldsymbol{\nabla}_\xi \cdot \mathbf{v}_{\xi(\mathcal{C})}, \tag{19}$$

which implies into a conservation law, $D\mathcal{W}/D\tau = 0$. In this case, the divergenceless behavior of the classical fluid-analog associated to the Wigner function is expressed by $\boldsymbol{\nabla}_\xi \cdot \mathbf{v}_{\xi(\mathcal{C})} = 0$, which defines the Liouvillian behavior.

To implement the above properties onto a quantum scenario, one has to notice that $\boldsymbol{\mathcal{J}}$ can be cast in the form of product $\mathbf{w}\,\mathcal{W}$, where the now introduced Wigner phase-velocity, $\mathbf{w}$, satisfies the constraint equation given by

$$\boldsymbol{\nabla}_\xi \cdot \mathbf{w} = \frac{\mathcal{W}\,\boldsymbol{\nabla}_\xi \cdot \boldsymbol{\mathcal{J}} - \boldsymbol{\mathcal{J}} \cdot \boldsymbol{\nabla}_\xi \mathcal{W}}{\mathcal{W}^2} \neq 0, \tag{20}$$

that can be used to quantify the departure from the Liouvillian behavior [4, 8].

For Hamiltonians describing periodic motions (in particular, for those driven by symmetric quantum wells), the parameterization to the two-dimensional volume enclosed by the classical path, $\mathcal{C}$, results into

$$\sigma_{(\mathcal{C})} = \int_{V_\mathcal{C}} dV\,\mathcal{W}, \tag{21}$$

where $\sigma_{(\mathcal{C})}$ corresponds to the phase-space volume integrated probability flux.

Using the properties of $\mathbf{w}$, after some manipulation involving Eqs. (18)-(19), one obtains

$$\frac{D}{D\tau}\sigma_{(\mathcal{C})} = \frac{D}{D\tau}\int_{V_\mathcal{C}} dV\,\mathcal{W} = \int_{V_\mathcal{C}} dV\,\left[\boldsymbol{\nabla}_\xi \cdot (\mathbf{v}_{\xi(\mathcal{C})}\mathcal{W}) - \boldsymbol{\nabla}_\xi \cdot \boldsymbol{\mathcal{J}}\right], \tag{22}$$

which allows for identifying the role of the quantum corrections given in terms of $\Delta\boldsymbol{\mathcal{J}} = \boldsymbol{\mathcal{J}} - \mathbf{v}_{\xi(\mathcal{C})}\mathcal{W}$, which effectively drives the volume variation of $\sigma_{(\mathcal{C})}$ in terms of a path integral given by

$$\frac{D}{D\tau}\sigma_{(\mathcal{C})} = -\int_{V_\mathcal{C}} dV\,\boldsymbol{\nabla}_\xi \cdot \Delta\boldsymbol{\mathcal{J}} = -\oint_\mathcal{C} d\ell\,\Delta\boldsymbol{\mathcal{J}} \cdot \mathbf{n} \equiv -\oint_\mathcal{C} d\ell\,\boldsymbol{\mathcal{J}} \cdot \mathbf{n}, \tag{23}$$

where the unitary vector $\mathbf{n}$ is defined by $\mathbf{n} = (-dk_\mathcal{C}/d\tau, dx_\mathcal{C}/d\tau)|\mathbf{v}_{\xi(\mathcal{C})}|^{-1}$ and $\mathbf{n} \cdot \mathbf{v}_{\xi(\mathcal{C})} = 0$ has been used in the last step. One thus has a parametric integral given by

$$\left.\frac{D}{D\tau}\sigma_{(\mathcal{C})}\right|_{\tau=T} = -\oint_\mathcal{C} d\ell\,\Delta\boldsymbol{\mathcal{J}} \cdot \mathbf{n} = -\int_0^T d\tau\,\Delta\mathcal{J}_p(x_\mathcal{C}(\tau), k_\mathcal{C}(\tau); \tau)\,\frac{d}{d\tau}x_\mathcal{C}(\tau), \tag{24}$$

where, in the first step, the line element, ℓ, is set as $d\ell \equiv |\mathbf{v}_{\xi(\mathcal{C})}|d\tau$, and, in the second step, one has $x_\mathcal{C}(\tau)$ and $k_\mathcal{C}(\tau)$ as typical solutions of the classical Hamiltonian problem. In addition, $T = 2\pi$ is the dimensionless period of the classical motion, and $\Delta\mathcal{J}_p(x, p; \tau)$ is identified by the series expansion from Eq. (8) for $\nu \geq 1$.

The above result straightforwardly quantifies the loss of information related to quantum decoherence. The same analysis can be applied to the informational content related to the Wigner associated von Neumann entropy and purity expressed respectively by [8]

$$S_{vN} = -\int_V dV\,\mathcal{W}\,\ln|\mathcal{W}|, \tag{25}$$

and

$$\mathcal{P} = 2\pi \int_V dV\, \mathcal{W}^2. \tag{26}$$

In these cases, some suitable mathematical manipulations involving Eq. (20)[3] result into

$$\begin{aligned}
\frac{DS_{vN}}{D\tau} &= -\frac{D}{D\tau}\left(\int_V dV\, \mathcal{W}\ln(\mathcal{W})\right)\\
&= -\int_V dV\left[\frac{D}{D\tau}(\mathcal{W}\ln(\mathcal{W})) + \mathcal{W}\ln(\mathcal{W})\boldsymbol{\nabla}_\xi\cdot\mathbf{v}_{\xi(\mathcal{C})}\right]\\
&= -\int_V dV\left[\frac{\partial}{\partial\tau}(\mathcal{W}\ln(\mathcal{W})) + \boldsymbol{\nabla}_\xi\cdot(\mathbf{v}_{\xi(\mathcal{C})}\mathcal{W}\ln(\mathcal{W}))\right]\\
&= \int_V dV\left[\mathcal{W}\,\boldsymbol{\nabla}_\xi\cdot\mathbf{w} + \boldsymbol{\nabla}_\xi\cdot\left(\boldsymbol{\mathcal{J}}\ln(\mathcal{W}) - \mathbf{v}_{\xi(\mathcal{C})}\mathcal{W}\ln(\mathcal{W})\right)\right]\\
&= \int_V dV\,\mathcal{W}\,\boldsymbol{\nabla}_\xi\cdot\mathbf{w} + \oint d\ell\,\ln(\mathcal{W})\,(\boldsymbol{\Delta\mathcal{J}}\cdot\mathbf{n})\\
&\overset{V\to\infty}{=} \langle\boldsymbol{\nabla}_\xi\cdot\mathbf{w}\rangle,
\end{aligned} \tag{27}$$

and

$$\begin{aligned}
\frac{1}{2\pi}\frac{D\mathcal{P}}{D\tau} &= \frac{D}{D\tau}\left(\int_V dV\,\mathcal{W}^2\right)\\
&= \int_V dV\left[\frac{D}{D\tau}\mathcal{W}^2 + \mathcal{W}^2\boldsymbol{\nabla}_\xi\cdot\mathbf{v}_{\xi(\mathcal{C})}\right]\\
&= \int_V dV\left[\frac{\partial}{\partial\tau}\mathcal{W}^2 + \boldsymbol{\nabla}_\xi\cdot(\mathbf{v}_{\xi(\mathcal{C})}\mathcal{W}^2)\right]\\
&= -\int_V dV\left[\mathcal{W}^2\,\boldsymbol{\nabla}_\xi\cdot\mathbf{w} + \boldsymbol{\nabla}_\xi\cdot(\boldsymbol{\mathcal{J}}\mathcal{W} - \mathbf{v}_{\xi(\mathcal{C})}\mathcal{W}^2)\right]\\
&= \int_V dV\,\mathcal{W}^2\,\boldsymbol{\nabla}_\xi\cdot\mathbf{w} + \oint d\ell\,\mathcal{W}\,(\boldsymbol{\Delta\mathcal{J}}\cdot\mathbf{n})\\
&\overset{V\to\infty}{=} \langle\mathcal{W}\boldsymbol{\nabla}_\xi\cdot\mathbf{w}\rangle,
\end{aligned} \tag{28}$$

for von Neumann entropy and purity, respectively, where $\langle\ldots\rangle = Tr_{\{x,k\}}\left[\hat{\rho}(\ldots)\right]$ and, in both cases, it has been used that $\partial\mathcal{W}/\partial\tau = -\boldsymbol{\nabla}_\xi\cdot\boldsymbol{\mathcal{J}} = -\boldsymbol{\nabla}_\xi\cdot(\mathbf{w}\,\mathcal{W})$.

Notice that the surface terms have been consistently suppressed in the limit where $V\to\infty$, which leads to the results from Refs. [8, 14]. However, when a classical surface, $\mathcal{C}$, encloses a finite phase-space volume, $V_\mathcal{C}$, the surface term must be taken into account. In fact, given the definitions from Eqs. (13)-(15), it is possible to verify that $\boldsymbol{\nabla}_\xi\cdot\mathbf{w}$ is proportional to

$$\sum_{\nu=1}^{\infty}\left(\frac{i}{2}\right)^{2\nu}\frac{1}{(2\nu+1)!}\left[\left(\frac{\partial}{\partial x}\right)^{2\nu+1}\mathcal{U}\right]\mathcal{W}^2\frac{\partial}{\partial k}\left((1/\mathcal{W})\frac{\partial}{\partial k}\right)^{2\nu}\mathcal{W},$$

and for parity symmetric potentials, $\mathcal{U}(x) = \mathcal{U}(-x)$ which lead to periodic anharmonic motions, the above obtained averaged terms vanish. Therefore, the obtained continuity equations can be recast in the form of [8, 14]

$$\begin{aligned}
\left.\frac{D}{D\tau}S_{vN(\mathcal{C})}\right|_{\tau=T} &= \oint_\mathcal{C} d\ell\,\ln(\mathcal{W})\,(\Delta\boldsymbol{\mathcal{J}}\cdot\mathbf{n})\\
&= \int_0^T d\tau\,\ln(\mathcal{W}(x_\mathcal{C}(\tau), k_\mathcal{C}(\tau);\tau))\,\Delta\mathcal{J}_p(x_\mathcal{C}(\tau), k_\mathcal{C}(\tau);\tau)\,\frac{d}{d\tau}x_\mathcal{C}(\tau),
\end{aligned} \tag{29}$$

[3] In particular, if one notices that $\boldsymbol{\nabla}_\xi\cdot\boldsymbol{\mathcal{J}} = \mathcal{W}\,\boldsymbol{\nabla}_\xi\cdot\mathbf{w} + \mathbf{w}\cdot\boldsymbol{\nabla}_\xi\mathcal{W}$.

and

$$\begin{aligned}\left.\frac{D}{D\tau}\mathcal{P}_{(C)}\right|_{\tau=T} &= -\oint_C d\ell\,\mathcal{W}\,\Delta\boldsymbol{\mathcal{J}}\cdot\mathbf{n}\\ &= -\int_0^T d\tau\,\mathcal{W}(x_C(\tau),\,k_C(\tau);\tau)\,\Delta\mathcal{J}_p(x_C(\tau),\,k_C(\tau);\tau)\,\frac{d}{d\tau}x_C(\tau),\end{aligned} \tag{30}$$

for von Neumann entropy and purity, respectively, through which one can quantify the role of quantum fluctuations driven by the contributions from $\Delta\boldsymbol{\mathcal{J}}\cdot\mathbf{n} \equiv \boldsymbol{\mathcal{J}}\cdot\mathbf{n}$[4].

Of course, as emphasized, the integrals from Eqs. (23), (29) and (33) vanish in the classical limit, i.e. for $\boldsymbol{\mathcal{J}} \sim \mathbf{v}_{\xi(C)}\mathcal{W}$. Therefore the obtained results work as an optimized quantifier of non-classicality for a pletora of Wigner functions.

3. Rényi entropy continuity equation

The Rényi entropy, R_β, generalizes the Hartley entropy, the Shannon entropy, the collision entropy and the minimal entropy [13]. Once contextualized in the phase-space formulation, it can be written as

$$R_\beta = \frac{1}{1-\beta}\log\left[\int_{-\infty}^{+\infty}\int_{-\infty}^{+\infty} dx\,dk\,\mathcal{W}^\beta_{(x,k)}\right], \tag{31}$$

where the index β is introduced to denote an estimative of the fractal dimension. By following the same derivative structure reported in the previous section, one could notice that

$$\begin{aligned}\left.e^{[(1-\beta)R_\beta]}\frac{D}{D\tau}R_\beta\right|_{\tau=T} &= -\frac{D}{D\tau}\left(\int_V dV\,\mathcal{W}^\beta\right)\\ &= -\int_V dV\left[\frac{D}{D\tau}(\mathcal{W}^\beta) + \mathcal{W}^\beta\boldsymbol{\nabla}_\xi\cdot\mathbf{v}_{\xi(C)}\right]\\ &= -\int_V dV\left[\frac{\partial}{\partial\tau}(\mathcal{W}^\beta) + \boldsymbol{\nabla}_\xi\cdot(\mathbf{v}_{\xi(C)}\mathcal{W}^\beta)\right]\\ &= -\int_V dV\left[-\beta\mathcal{W}^{\beta-1}\boldsymbol{\nabla}_\xi\cdot\boldsymbol{\mathcal{J}} + \boldsymbol{\nabla}_\xi\cdot(\mathbf{v}_{\xi(C)}\mathcal{W}^\beta)\right]\\ &= \int_V dV\left[(\beta-1)\mathcal{W}^\beta\,\boldsymbol{\nabla}_\xi\cdot\mathbf{w} + \boldsymbol{\nabla}_\xi\cdot(\boldsymbol{\mathcal{J}}\mathcal{W}^{\beta-1} - \mathbf{v}_{\xi(C)}\mathcal{W}^\beta)\right]\\ &= \int_V dV\,(\beta-1)\mathcal{W}^\beta\,\boldsymbol{\nabla}_\xi\cdot\mathbf{w} + \oint d\ell\,\mathcal{W}^{\beta-1}\,(\boldsymbol{\Delta\mathcal{J}}\cdot\mathbf{n})\\ &\overset{V\to\infty}{=} (\beta-1)\langle\mathcal{W}^{\beta-1}\boldsymbol{\nabla}_\xi\cdot\mathbf{w}\rangle,\end{aligned} \tag{32}$$

and that

$$\begin{aligned}\left.e^{[(1-\beta)R_\beta]}\frac{D}{D\tau}R_\beta\right|_{\tau=T} &= -\oint_C d\ell\,\mathcal{W}^{\beta-1}\,\Delta\boldsymbol{\mathcal{J}}\cdot\mathbf{n}\\ &= -\int_0^T d\tau\,\mathcal{W}^{\beta-1}(x_C(\tau),\,k_C(\tau);\tau)\,\Delta\mathcal{J}_p(x_C(\tau),\,k_C(\tau);\tau)\,\frac{d}{d\tau}x_C(\tau),\end{aligned} \tag{33}$$

which is consistent with the previously obtained continuity equations.

[4] For quantum systems which account for the all order corrections from Eq. (15), quantumness and classicality can be quantified through the above obtained continuity equation framework through the results from Eqs. (24), (29), and (33) [8, 12, 14].

IOP Conf. Series: Journal of Physics: Conf. Series **1275** (2019) 012032 doi:10.1088/1742-6596/1275/1/012032

4. Conclusions

In this work, it has been shown that the Weyl transform quantum states lead to phase-space continuity equations which allow for evaluating departures from Liouvillian properties of Wigner currents and fluxes. The results provide the tools for discussing quantumness and classicality in terms of phase-space quantum decoherence, purity and von Neumann entropy fluxes, in the context of (an)harmonic periodic quantum systems. The informational content analysis was extended in order to provide an equivalent formulation for the phase-space Rényi entropy, in terms of an additional continuity equation. Our results suggest that the Wigner flow framework can be universally applied to distinguish how quantum and classical regimes are quantitatively distinguished from each other [8, 12, 14].

Acknowledgments

The work of A. E. B. was supported by the Brazilian agencies FAPESP (grant 2018/03960-9) and CNPq (grants 451081/2018-8 and 300831/2016-1). The work of O. B. was partially supported by the COST action MP1405 Quantum Structure of Spacetime (QSPACE).

References

[1] Wigner E 1932 *Phys. Rev.* **40** 749
[2] Case W B 2008 *Am. J. Phys.* **76** 937
[3] Bernardini A E and Chinaglia M 2015 *Mod. Phys. Lett.* **A30** 1550118
[4] Steuernagel O, Kakofengitis D and Ritter G 2013 *Phys. Rev. Lett.* **110** 030401
[5] Meekhof D M, Monroe C, King B E, Itano W M and Wineland D J 1996 *Phys. Rev. Lett.* **76** 1796
[6] Moyal J 1949 *Proc. Camb. Phil. Soc.* **45** 99
[7] Ballentine L E 1998 *Quantum Mechanics: a Modern Development* pp. 633 (World Scientific).
[8] Bernardini A E and Bertolami O 2017 *EPL* **120** 20002
[9] Ferraro A and Paris M G A 2012 *Phys. Rev. Lett.* **108** 260403
[10] Donoso A and Martens C C 2001 *Phys. Rev. Lett.* **87** 223202
[11] Domcke W, Hänggi P and Tannor D 1997 *Chem. Phys.* **217** 117
[12] Bernardini A E, Leal P and Bertolami O 2018 *JCAP* **02** 025
[13] Jizba P and Arimitsu T 2004 *Annals of Physics* **312** 17
[14] Bernardini A E 2018 *Phys. Rev.* **A98** 052128

Some remarks on analytical solutions for a damped quantum parametric oscillator

Dieter Schuch

Institut für Theoretische Physik, J.W. Goethe-Universität Frankfurt am Main, Max-von-Laue-Str. 1, D-60438 Frankfurt am Main, Germany

E-mail: `Schuch@em.uni-frankfurt.de`

Abstract. The time-dependent Schrödinger equation for quadratic Hamiltonians has Gaussian wave packets as exact solutions. For the parametric oscillator with frequency $\omega(t)$, the width of these wave packets must be time-dependent. This time-dependence can be determined by solving a complex nonlinear Riccati equation or an equivalent real nonlinear Ermakov equation. All quantum dynamical properties of the system can easily be constructed from these solutions, e.g., uncertainties of position and momentum, their correlations, ground state energies, etc. In addition, the link to the corresponding classical dynamics is supplied by linearizing the Riccati equation to a complex Newtonian equation, actually representing two equations of the same kind: one for the real and one for the imaginary part. If the solution of one part is known, the missing (linear independent) solution of the other can be obtained via a conservation law for the motion in the complex plane. Knowing these two solutions, the solution of the Ermakov equation can be determined immediately plus the explicit expressions for all the quantum dynamical properties mentioned above. The effect of a dissipative, linear velocity dependent friction force on these systems is discussed.

1. Introduction

A parametric oscillator, i.e., an oscillator with time-dependent (TD) frequency $\omega(t)$ is a non-trivial problem even in classical mechanics. Whether or not there are analytical solutions strongly depends on the time-dependence of the frequency. However, in most cases, at least a numerical solution should be possible.

Extending the problem to quantum dynamics adds an additional aspect, the dynamics of the typical quantum fluctuations, i.e., the uncertainties of position, momentum and their correlation.

For the harmonic oscillator (HO) (or any other at-most-quadratic Hamiltonian operator), exact analytic solutions in the form of Gaussian wave packets (WPs) exist that are uniquely determined by the maximum and the width of the Gaussian function. The maximum follows the classical trajectory and so its time-evolution is determined by the solution of the corresponding Newtonian equation of motion. This also applies to the parametric oscillator.

The dynamics of the WP-width is determined by a complex nonlinear Riccati equation and is proportional to the time-dependence of the position uncertainty. Once this is known, also the momentum uncertainty and the correlation between these uncertainties can be determined explicitly. In principle, this is possible not only for the HO, but also for the parametric oscillator.

For the HO it can be shown that the solution for the WP-width and, consequently, the solution of the complex Riccati equation, can also be obtained by solving the equation for the

Content from this work may be used under the terms of the Creative Commons Attribution 3.0 licence. Any further distribution of this work must maintain attribution to the author(s) and the title of the work, journal citation and DOI.
Published under licence by IOP Publishing Ltd

IOP Conf. Series: Journal of Physics: Conf. Series **1275** (2019) 012033 doi:10.1088/1742-6596/1275/1/012033

WP-maximum.

In Section 2, the relationship between the classical equations of motion and the dynamics of the WP-width will be given in an overview.

In Section 3, analytical solutions for the classical parametric oscillator with frequency inversely proportional to time are examined, including an extension that incorporates the quantum mechanical uncertainty aspect.

The extension to effective descriptions of dissipative systems and the consequences for the afore-mentioned relations are discussed in Section 4.

A generalization from the damped HO to the damped parametric oscillator is considered and specific differences and corresponding difficulties are pointed out in Section 5.

Finally, a summary and perspectives of future developments are given.

2. Interrelations between the classical and quantum aspects of the Gaussian wave packet dynamics

For Hamiltonians that are at most quadratic in position and momentum variables, the Schrödinger equation (SE) possesses exact analytical solutions in the form of Gaussian WPs. In one dimension, to which the following discussion will be restricted, such a solution can be written in the form [1, 2, 3, 4, 5]

$$\Psi_{\mathrm{WP}}(x,t) = N(t) \exp\left\{ \mathrm{i} \left[y(t)\, \tilde{x}^2 + \frac{\langle p \rangle}{\hbar}\, \tilde{x} + K(t) \right] \right\} \tag{1}$$

with the purely TD normalization factor $N(t)$ (and the TD function $K(t)$ that will not be relevant to the following discussion). The variable $\tilde{x} = x - \langle x \rangle = x - \eta(t)$ is so defined that the maximum of the Gaussian function is at $\langle x \rangle$, the mean value of position, that (according to Ehrenfest) coincides with the classical trajectory $\eta(t)$ and is calculated via $\langle x \rangle = \int_{-\infty}^{+\infty} dx \Psi^* x \Psi = \eta(t)$; similarly, $\langle p \rangle = m\dot{\eta}$ is the mean value of momentum. Important for the quantum mechanical aspect of the system's dynamics is the generally complex coefficient $y(t)$ of $\tilde{x}^2$ in the exponent, $y(t) = y_{\mathrm{R}}(t) + \mathrm{i} y_{\mathrm{I}}(t)$. The imaginary part is related to the position uncertainty via $y_{\mathrm{I}} = \frac{1}{4\langle \tilde{x}^2 \rangle}$ with the mean square deviation of position $\langle \tilde{x}^2 \rangle = \langle x^2 \rangle - \langle x \rangle^2$. The square root of $\langle \tilde{x}^2 \rangle$ essentially determines the width of the Gaussian function. This function is completely determined by two parameters, the width and the maximum of the bell-shaped curve. In this quantum mechanical context, both parameters can be TD. The equations of motion that determine this time-dependence can be obtained by inserting the WP(1) into the TDSE,

$$\mathrm{i}\hbar \frac{\partial}{\partial t} \Psi_{\mathrm{WP}}(x,t) = \left\{ -\frac{\hbar}{2m} \frac{\partial^2}{\partial x^2} + V(x,t) \right\} \Psi_{\mathrm{WP}}(x,t) \ , \tag{2}$$

in the following for $V(x,t) = \frac{m}{2}\omega^2 x^2$ with the general case $\omega = \omega(t)$ and the particular cases $\omega = \omega_0 = const.$ (HO) and $\omega = 0$ (free motion). For the motion of the maximum one obtains, as expected, the Newtonian equation of motion (divided by mass m)

$$\ddot{\eta} + \omega^2(t)\eta = 0 \ . \tag{3}$$

Defining the complex variable $\mathcal{C}(t) = \frac{2\hbar}{m} y(t)$, the terms proportional to $\tilde{x}^2$ yield the complex nonlinear Riccati equation

$$\dot{\mathcal{C}} + \mathcal{C}^2 + \omega^2(t) = 0 \tag{4}$$

that determines the time-evolution of the WP-width, as $\mathcal{C}_{\mathrm{I}} = \frac{2\hbar}{m} y_{\mathrm{I}} = \frac{\hbar}{2m\langle \tilde{x}^2 \rangle}$. Therefore, in order to obtain the information about the typical quantum mechanical aspects of the system's dynamics, this Riccati equation must be solved. Possible ways of doing this have been discussed, e.g., in [1, 2, 3, 4, 5].

For our discussion, two possible treatments of the Riccati equation (4) will be of relevance.

First, one can introduce a new variable $\alpha(t)$ via the definition of $\mathcal{C}_{\rm I}(t)$ and consequently, from the imaginary part of Eq. (4), directly follows the real part $\mathcal{C}_{\rm R}(t)$ as

$$\mathcal{C}_{\rm I}(t) = \frac{1}{\alpha^2} \quad , \quad \mathcal{C}_{\rm R} = \frac{\dot{\alpha}}{\alpha} \, . \tag{5}$$

According to $\alpha = \sqrt{\frac{2m}{\hbar}\langle\tilde{x}^2\rangle}$, $\alpha(t)$ is directly proportional to the WP-width; the real part $\mathcal{C}_{\rm R}$ vanishes only for constant width, i.e., $\dot{\alpha} = 0$. In all other cases, particularly for the parametric oscillator, the Riccati variable $\mathcal{C}(t)$ is complex. From the real part of Eq. (4), one finally obtains the equation for $\alpha(t)$,

$$\ddot{\alpha} + \omega^2(t)\alpha = \frac{1}{\alpha^3} \, , \tag{6}$$

known in the literature as Ermakov equation [6].

In a second treatment, the Riccati equation (4) can be linearized using the logarithmic derivative of a new complex variable $\lambda(t)$ via $\mathcal{C} = \frac{\dot{\lambda}}{\lambda}$ to yield the complex Newtonian equation

$$\ddot{\lambda} + \omega^2(t)\lambda = 0 \tag{7}$$

with $\lambda(t) = u + {\rm i}\, z = \alpha\, {\rm e}^{{\rm i}\varphi}$.

Real and imaginary parts, or phase and amplitude, of the complex quantity λ are not independent of each other but coupled via the conservation law

$$\dot{z}u - \dot{u}z = \alpha^2\, \dot{\varphi} = 1 \tag{8}$$

which is equivalent to the Wronskian determinant of two linear independent solutions u and z of the complex Newtonian equation (7).

Knowing one solution (e.g., $z(t)$), it is possible to calculate the other via

$$u = -z \int^{t} \frac{1}{z^2}(t)\, dt' \, . \tag{9}$$

Knowing both solutions of the classical Newtonian equation immediately provides the solution for the dynamics of the WP width via $\alpha^2 = u^2 + z^2$. If α is known, all typical quantum mechanical properties like uncertainties and ground state energies can be determined as they can be expressed in terms of this quantity in the form

$$\begin{aligned} \langle\tilde{x}^2\rangle &= \frac{\hbar}{2m}\alpha^2 \, , \; \langle\tilde{p}^2\rangle = \frac{m\hbar}{2}\left[\dot{\alpha}^2 + \frac{1}{\alpha^2}\right] \, , \; \langle[\tilde{x},\tilde{p}]_+\rangle = \hbar\, \alpha\, \dot{\alpha} \, , \\ U = \langle\tilde{x}^2\rangle\langle\tilde{p}^2\rangle &= \frac{\hbar^2}{4}\left\{1 + (\alpha\,\dot{\alpha})^2\right\} , \; \tilde{E} = \frac{\langle\tilde{p}^2\rangle}{2m} + \frac{m}{2}\omega^2\langle\tilde{p}^2\rangle = \frac{\hbar}{4}\left\{\dot{\alpha}^2 + \frac{1}{\alpha^2} + \omega^2\alpha^2\right\} . \end{aligned} \tag{10}$$

Further, with the help of the TD Green function (also called Feynman kernel), it can be shown that the imaginary part of $\lambda(t)$ is (up to a constant factor) identical to the classical trajectory, i.e.,

$$z(t) = \frac{m}{\alpha_0\, p_0}\, \eta(t) \tag{11}$$

IOP Conf. Series: Journal of Physics: Conf. Series **1275** (2019) 012033 doi:10.1088/1742-6596/1275/1/012033

with the initial WP-width α_0 and the initial classical momentum $p_0 = m\dot{\eta}(t_0)$ (for details, see e.g. [1, 2, 3, 4, 5]).

Therefore, with the solutions of the classical Newtonian equation (analytical or numerical) it is possible to construct the solution of the Ermakov equation that describes the dynamics of the uncertainties and thus the quantum mechanical aspect. In other words, if the dynamics of the WP-maximum is known, the dynamics of the WP-width (for given initial width α_0) can be determined in the way shown above.

3. Analytical solution for a parametric oscillator with frequency inversely proportional to time

Comparing the WP with the Gaussian solution of the diffusion equation it has been shown in [7] that an analytical solution of the Newtonian equation

$$\ddot{z} + \omega^2(t)\; z = 0 \tag{12}$$

can be found for

$$\omega(t) \;=\; \frac{1}{2\;(t+b)} \tag{13}$$

with $b = constant \neq 0$, having the dimension "*time*".

In this case, the two linear-independent solutions of the (complex) Newtonian equation (7) are

$$z = (t+b)^{\frac{1}{2}} \quad \rightarrow \quad u = -(t+b)^{\frac{1}{2}}\; \ln(t+b)\;, \tag{14}$$

and, consequently, the solution of the Ermakov equation is obtained from

$$\alpha^2(t) = (t+b)\;\left[1 + \ln^2(t+b)\right]\;. \tag{15}$$

Using relations (10), the analytic form of the quantum mechanical properties can be given explicitly as

$$\begin{aligned}
\langle \tilde{x}^2 \rangle \;&=\; \frac{\hbar}{2m}(t+b)\left[1+\ln^2(t+b)\right],\; \langle \tilde{p}^2 \rangle \;=\frac{m\hbar}{2}\,\frac{1}{(t+b)}\left\{\frac{1}{4}\,[1+\ln(t+b)]^2 + \frac{1}{2}\ln(t+b)+1\right\},\\
\langle [\tilde{x},\tilde{p}]_+ \rangle \;&=\; \frac{\hbar}{2}\,[1+\ln(t+b)]^2\,,\; U(t) = \frac{\hbar}{4}\left\{1+\frac{1}{4}\,[1+\ln(t+b)]^4\right\},\\
\tilde{E} \;&=\; \frac{\hbar}{4}\left\{\frac{1}{(t+b)}\left[\frac{1}{2}\,(1+\ln(t+b))^2+1\right]\right\}\;.
\end{aligned} \tag{16}$$

Obviously, the position uncertainty and thus the WP width are diverging for $t \to \infty$, as in the case of the free motion, which is reasonable as in this limit $\omega(t)$, and thus the potential, vanishes. The momentum uncertainty decreases and approaches zero, whereas the uncertainty product is going to infinity for $t \to \infty$. The energy contribution $\tilde{E}(t)$ is decreasing and approaches zero, similar to a (classical) dissipative system (without fluctuating force).

According to (11), $z(t)$ is proportional to the classical trajectory $\eta(t)$. Therefore, using Eq. (12), the solution of the classical Newtonian equation (3) for the frequency (13) and consequently the corresponding velocity can immediately be given in the form

$$\eta(t) = v_0\;\tau^{\frac{1}{2}}(t+b)^{\frac{1}{2}}\;\;,\;\; \dot{\eta}(t) = \frac{1}{2}\;v_0\;\tau^{\frac{1}{2}}(t+b)^{-\frac{1}{2}} = v_0\;\frac{\tau^{\frac{1}{2}}}{2\;(t+b)^{\frac{1}{2}}} \tag{17}$$

with $\eta_0 = v_0\tau^{\frac{1}{2}}b^{\frac{1}{2}}$ and $\dot{\eta}_0 = v_0\;\frac{\tau^{\frac{1}{2}}}{2\;b^{\frac{1}{2}}}$ where the constant τ with dimension time has been introduced to get the correct dimension for $\eta(t)$. From $\dot{\eta}_0 = v_0$ follows $b = \frac{\tau}{4}$.

The classical energy of the system then acquires the form

$$E = T + V = \frac{m}{2}\,\dot{\eta}^2 + \frac{m}{2}\left(\frac{1}{2(t+b)}\right)^2 \eta^2 = \frac{m}{2}\,v_0^2\,\frac{\tau}{2(t+b)} \propto \frac{1}{t} \tag{18}$$

and decreases inversely proportional to time t. The initial energy E_0 is given by $E_0 = \frac{m}{2}\,v_0^2\,\frac{\tau}{2b} = m\,v_0^2$, the final energy for $t \to \infty$ approaches zero, $E_\infty \to 0$, like in a dissipative system.

To compare this parametric oscillator with decreasing energy with the damped HO with a friction force linear proportional to velocity, obeying the equation of motion

$$\ddot{\eta} + \gamma\,\dot{\eta} + \omega_0^2\eta = 0 \tag{19}$$

with friction coefficient γ and constant frequency ω_0, the decrease in energy of both systems is compared.

For the parametric oscillator this decrease is given by

$$\dot{E} = -\frac{m}{2}\,v_0^2\,\frac{\tau}{2}\left(\frac{1}{t+b}\right)^2 = -2\omega(t)E = -4\,\omega(t)\;T < 0\;, \tag{20}$$

for the damped HO it is

$$\dot{E} = -2\gamma\,T = -4\frac{\gamma}{2}\,T\;. \tag{21}$$

So, the TD frequency would correspond to half of the friction parameter γ, $\omega(t) \mathrel{\hat{=}} \frac{\gamma}{2}$.

It can also be shown (details will be provided in a forthcoming work) that analytical solutions are possible for a more general parametric oscillator, namely one with the frequency

$$\omega(t) = \frac{1}{a(t+b)}\quad. \tag{22}$$

However, qualitatively different behaviour of dynamics and energetics is obtained for different values of a. As for the damped HO, three different regimes can be distinguished. The case of $a = 2$ (discussed above) corresponds to the *aperiodic limit*, i.e., $\omega_0 = \frac{\gamma}{2}$; the case $a > 2$ to *overcritical damping*, i.e., $\omega_0 > \frac{\gamma}{2}$; and the case $a < 2$ to *undercritical damping*, $\omega_0 < \frac{\gamma}{2}$. Only in the last case the damped HO, as well as the corresponding parametric oscillator provide oscillatory solutions.

Looking into the combined effect of TD frequency and damping force on the WP dynamics should be interesting. For this purpose, different approaches for the description of dissipative quantum systems will be presented subsequently that lead to the same equations of motion for the quantum mechanical properties of interest.

4. Effective descriptions of dissipative quantum systems

There are several approaches in the literature to describe open quantum systems in an effective way, i.e., without taking into account the environmental degrees of freedom explicitly, but only include the effect of the environment via modification of the TDSE. Essentially two different types of approaches can be distinguished that shall be called *canonical* and *physical*.

In the *canonical* methods (based on modifications of classical Hamiltonian mechanics) the corresponding Hamiltonian operator (after canonical quantization) might be explicitly TD, but the resulting TDSE still remains linear. However, the connection between the canonical variables of these open Hamiltonian systems and the physical position and momentum is given classically by a *non-canonical* transformation, corresponding to a *non-unitary* transformation [8] for the respective WP-solution of the SE (but still allowing normalization of the WPs).

The *physical* methods keep the definitions of position- and momentum-operators unchanged, as in conventional quantum mechanics, but modify the Hamiltonian operator in a way that, according to Ehrenfest's theorem, the equation of motion for the mean value of position contains a linear velocity-dependent friction force as given in Eq. (19). The price to be paid is the occurrence of nonlinear terms in the corresponding Hamiltonian operators. Another similar approach [9], changing the reversible continuity equation for the probability density $\varrho = \Psi^*\Psi$ to an irreversible Fokker–Planck-type equation with diffusion term, also leads, via separation, to a NLSE.

Four of these methods are now presented that can be transformed into each other in a unique way and all possess Gaussian WP solutions of the corresponding TDSE where the maximum and width of these WPs obey the same set of equations of motion.

On the canonical level (where canonical quantities are characterized by a hat) the approach of Caldirola [10] and Kanai [11] leads to a TDSE with explicitly TD Hamiltonian (where $\hat{x} = x$ is still the physical position variable)

$$i\hbar\frac{\partial}{\partial t}\hat{\Psi}(x,t) = \left\{-\frac{\hbar}{2m}\,\mathrm{e}^{-\gamma t}\,\frac{\partial^2}{\partial x^2} + \frac{m}{2}\,\omega^2 x^2 \mathrm{e}^{\gamma t}\right\}\hat{\Psi}(x,t) \ . \qquad (23)$$

This time-dependence of the Hamiltonian can be eliminated by changing to an exponentially expanding position variable via $\hat{Q} = x\mathrm{e}^{\frac{\gamma}{2}t}$, leading to the TDSE

$$i\hbar\frac{\partial}{\partial t}\hat{\Psi}(\hat{Q},t) = \left\{-\frac{\hbar}{2m}\,\frac{\partial^2}{\partial \hat{Q}^2} + \frac{m}{2}\Omega^2\,\hat{Q}^2\right\}\hat{\Psi}(\hat{Q},t) \qquad (24)$$

with $\Omega^2 = \omega^2 - \frac{\gamma^2}{4}$, looking like an undamped HO with frequency Ω. However, to avoid unphysical results, the important point is the difference between the canonical wave function $\hat{\Psi}$ and the physical one, Ψ. The connection between both is given by the *non-unitary* transformation [8]

$$\ln\hat{\Psi} = \mathrm{e}^{\gamma t}\ln\Psi \ . \qquad (25)$$

On the physical level, based on Ehrenfest's theorem, a modification of Hasse's NLSE [12] yields

$$i\hbar\frac{\partial}{\partial t}\Psi(x,t) = \left\{H_L + \frac{\gamma}{4}\left([\tilde{x}, p + \langle p\rangle]_+ - \langle[\tilde{x}, p + \langle p\rangle]_+\rangle\right)\right\}\Psi(x,t) \qquad (26)$$

where H_L is the conventional Hamiltonian of the the linear theory, while the TDSE (26) is nonlinear due to the occurrence of the mean values $\langle p\rangle$ and $\langle x\rangle$ (in $\tilde{x} = x - \langle x\rangle$).

Finally, via separation of a Fokker–Planck-type equation for the probability density [5, 9, 13], one obtains the logarithmic NLSE

$$i\hbar\frac{\partial}{\partial t}\Psi(x,t) = \left\{H_L + \gamma\,\frac{\hbar}{i}(\ln\Psi - \langle\ln\Psi\rangle)\right\}\Psi(x,t) \ . \qquad (27)$$

In all cases, the two parameters of the Gaussian WP solution obey the equations of motion

$$\text{Maximum}: \quad \ddot{\eta} + \gamma\dot{\eta} + \omega^2(t)\eta = 0\ , \quad \text{Width}: \quad \ddot{\alpha} + \left(\omega^2(t) - \frac{\gamma^2}{4}\right)\alpha = \frac{1}{\alpha^3}\ , \qquad (28)$$

and, expressed in terms of the expanding coordinates, the maximum obeys

$$\ddot{\hat{Q}} + \left(\omega^2(t) - \frac{\gamma^2}{4}\right)\hat{Q} = 0 \qquad (29)$$

with $\eta(t) = \hat{Q}\ \mathrm{e}^{-\frac{\gamma}{2}t}$, i.e., the same as for the conservative HO, only $\omega(t)$ is replaced by $\Omega(t) = \left(\omega^2(t) - \frac{\gamma^2}{4}\right)^{\frac{1}{2}}$.

The complex Riccati equation corresponding to the Ermakov equation in (28) is $\dot{\mathcal{C}} + \gamma\ \mathcal{C} + \mathcal{C}^2 + \omega^2(t) = 0$ with $\mathcal{C}_{\mathrm{I}} = \frac{1}{\alpha^2}$, $\mathcal{C}_{\mathrm{R}} = \frac{\dot{\alpha}}{\alpha} - \frac{\gamma}{2}$. Via $\mathcal{C} = \frac{\dot{\lambda}}{\lambda}$, this equation can be linearized to $\ddot{\tilde{\lambda}} + \gamma\dot{\tilde{\lambda}} + \omega^2(t)\tilde{\lambda} = 0$ with $\tilde{\lambda} = \mathrm{e}^{-\frac{\gamma}{2}t}\lambda$, in particular $\tilde{z} = \mathrm{e}^{-\frac{\gamma}{2}t}z \propto \eta(t)$.

5. Damped parametric oscillator

Considering the damped HO obeying Eqs. (28) and (29), it appears that the transition from the undamped system to the damped one can simply be achieved by:

1) replacing $\omega(t)$ in η by $\Omega(t) = \sqrt{\omega^2(t) - \frac{\gamma^2}{4}} \ \rightarrow\ \hat{Q}(t)$;
2) multiplying $\hat{Q}(t)$ by $\mathrm{e}^{-\frac{\gamma}{2}t}\ \rightarrow$ dissipative trajectory η_{dis};
3) dissipative velocity: $\dot{\eta}_{dis}\ \rightarrow$ classical dissipation of energy;
4) $\hat{Q} \propto z\ \rightarrow u$ in analogy with (9), leading to $\alpha^2 = u^2 + z^2 \rightarrow$ quantum mechanical properties.

However, this procedure is too naive, because η changes qualitatively as time proceeds. This is caused by the time-dependence of

$$\Omega^2(t) = \left(\frac{1}{4(t+b)^2} - \frac{\gamma^2}{4}\right) \tag{30}$$

that is positive for $t < \frac{1}{\gamma} - b$ (like for the undercritically damped HO), vanishes for $t = \frac{1}{\gamma} - b$ (like for the aperiodic limit) and becomes negative for $t > \frac{1}{\gamma} - b$ (like for overcritical damping). That means, the attractive potential in Eq. (29) turns into a repulsive one for $t > \frac{1}{\gamma} - b$, thus changing the solution of Eq. (29) qualitatively and likewise the behaviour of $\eta(t)$ at and after this time.

Therefore, the proper procedure is to solve directly the equation of motion for the expanding coordinate $\hat{Q}$,

$$\frac{d^2}{dt^2}\hat{Q} + \left(-\frac{\gamma^2}{4} + \frac{\frac{1}{4}}{(t+b)^2}\right)\hat{Q}\ =\ 0 \tag{31}$$

which is formally equivalent to Whittaker's equation

$$\frac{d}{dt^2}W + \left(-\frac{1}{4} + \frac{k}{t} + \frac{\frac{1}{4} - m^2}{t^2}\right)W\ =\ 0 \tag{32}$$

for particular choices of parameters k, m (e.g., $k = 0$ leads to Bessel functions).

Knowing the solution of (31) enables one to obtain the classical trajectory of the system via $\eta_{\mathrm{dis}}(t) = \hat{Q}\mathrm{e}^{-\frac{\gamma}{2}t}$. However, the particular form of the solution strongly depends on the parameters. If one solution $\hat{Q}_1$ is known, the second linear independent one, $\hat{Q}_2$, can be obtained using $\dot{\hat{Q}}_1\hat{Q}_2 - \dot{\hat{Q}}_2\hat{Q}_1 = 1$ in analogy with (9). With both solutions, $\alpha^2 = \hat{Q}_1^2 + \hat{Q}_2^2$ is known and thus all quatum mechanical properties of the system can be determined via

$$\langle\tilde{x}^2\rangle(t) = \frac{\hbar}{2m}\alpha^2\ ,\ \langle\tilde{p}^2\rangle(t) = \frac{m\hbar}{2}\left[\left(\dot{\alpha} - \frac{\gamma}{2}\alpha\right)^2 + \frac{1}{\alpha^2}\right]\ ,\ \langle[\tilde{x},\tilde{p}]_+\rangle(t) = \hbar\ \left[\alpha\ \dot{\alpha} - \frac{\gamma}{2}\alpha^2\right]\ ,$$

$$U = \frac{\hbar}{4}\left\{1 + \left(\alpha\ \dot{\alpha} - \frac{\gamma}{2}\alpha^2\right)^2\right\}\ ,\ \tilde{E} = \frac{\hbar}{4}\left\{\left(\dot{\alpha} - \frac{\gamma}{2}\alpha\right)^2 + \frac{1}{\alpha^2} + \omega^2(t)\alpha^2\right\}\ . \tag{33}$$

Alternatively, $\alpha(t)$, and thus all quantities in (33), can also be obtained by solving the Ermakov equation (6) for frequency (30).

IOP Conf. Series: Journal of Physics: Conf. Series **1275** (2019) 012033 doi:10.1088/1742-6596/1275/1/012033

6. Conclusions

The TDSE for potentials that are at most quadratic in position variable possesses exact solutions that have the form of Gaussian WPs. These functions are uniquely determined by two parameters, the maximum and the width. In this case, both parameters are TD. The time-dependence of the maximum obeys the classical Newtonian equation of motion for the given potential; the time-evolution of the width is determined by a complex nonlinear Riccati equation or, equivalently, by a real nonlinear Ermakov equation. For TD frequency $\omega = \omega(t)$, it is the explicit time-dependence of $\omega(t)$ that determines whether analytic solutions exist or not.

Independent of the particular $\omega(t)$, the solutions of the Riccati or Ermakov equations provide the entire information about the quantum dynamics of the system.

Furthermore, the complex Riccati equation can also be linearized to a complex Newtonian equation where real and imaginary parts of the complex variable $\lambda(t)$ fulfilling this equation are two linear-independent solutions of the second-order Newtonian differential equation. However, these two solutions are coupled via a kind of conservation law. Essentially, the solution of the classical Newtonian equation makes it possible to obtain a second linear-independent solution with both providing the solution of the Ermakov equation and thus all information about the quantum aspect.

The TD frequency $\omega(t) = \frac{1}{2(t+b)}$ enables analytical solutions of the WP maximum and width. Comparison of the (classical) energy decrease of this oscillator with that of the damped HO shows similarity with the aperiodic limit in the damped case. A more general frequency, $\omega(t) = \frac{1}{a(t+b)}$, also allowing analytical solutions, resembles for $a < 2$ the undercritical and for $a > 2$ the overcritical damped HO. A first attempt at solving the classical equation (29) by simply comparing it with the equations describing the damped HO fails because the damped frequency $\Omega(t)$ changes sign as time passes, thus changing the potential from an attractive to a repulsive one. Finally, a Whittaker-type equation must be solved that depends on the values of the occurring parameters. Explicit cases shall be investigated in future work.

References

[1] Schuch D 1989 Connections between Newton- and Schrödinger-type equations in the description of reversible and irreversible dynamics. *Int. J. Quant. Chem., Quant. Chem. Symp.* **36 S23** 59-72

[2] Schuch D and Moshinsky M 2006 Connection between quantum-mechanical and classical time evolution via a dynamical invariant. *Phys. Rev.* A **73** 063111

[3] Schuch D 2008 Riccati and Ermakov Equations in Time-Dependent and Time-Independent Quantum Systems. *SIGMA* **4** 043

[4] Cruz H, Schuch D, Castaños O and Rosas-Ortiz O 2015 Time-evolution of quantum systems via a complex nonlinear Riccati equation I. Conservative systems with time-independent Hamiltonians. *Ann. Physics* **360** 44-60

[5] Schuch D 2018 *Quantum Theory from a Nonlinear Perspective – Riccati Equations in Fundamental Physics* (Heidelberg: Springer)

[6] Ermakov V P 1880 Second-Order Differential Equations: Conditions of Complete Integrability. *Univ. Jzv. Kiev* **20 No 9** 1-25; Harin A O (English translation) 2008 *Appl. Anal. Discrete Math.* **2** 123-145

[7] Schuch D 2018 Analytical solutions for the quantum parametric oscillator from corresponding classical dynamics via a complex Riccati equation *J. Phys.: Conf. Ser.* **1071** 012020

[8] Schuch D 1997 Nonunitary connection between explicitly time-dependent and nonlinear approaches for the description of dissipative quantum systems. *Phys. Rev.* A **55** 935

[9] Schuch D, Chung K-M and Hartmann H 1983 Nonlinear Schrödinger-type field equation for the description of dissipative systems. I. Derivation of the nonlinear field equation and one-dimensional example. *J. Math. Phys.* **24** 1652

[10] Caldirola P 1941 Forze non conservative nella meccanica quantistica. *Nuovo Cimento* **18** 393-400

[11] Kanai E 1948 On the Quantization of the Dissipative Systems. *Progr. Theor. Phys.* **3** 440-442

[12] Hasse R W 1975 On the quantum mechanical treatment of dissipative systems. *J. Math. Phys* **16** 2005

[13] Cruz H, Schuch D, Castaños O and Rosas-Ortiz O 2016 Time-evolution of quantum systems via a complex nonlinear Riccati equation. II. Dissipative systems. *Ann. Physics* **373** 609-630

IOP Conf. Series: Journal of Physics: Conf. Series **1275** (2019) 012034 doi:10.1088/1742-6596/1275/1/012034

New representations of Poincaré group for consistent Relativistic Particle Theories

Giuseppe Nisticò

Dipartimento di Matematica e Informatica, Università della Calabria, Italy and
INFN - gruppo collegato di Cosenza, Italy

E-mail: giuseppe.nistico@unical.it

Abstract. The class of representations of the Poincaré group for the formulation of relativistic quantum theories of single particle is consitently extended. In so doing new species of such representations are singled out. By making use of the enlarged class quantum theories of Klein-Gordon free particle are formulated without the problems suffered by the early theory. Furthermore, new species of consistent theories can be developed.

1. Introduction

The symmetry character of Poincaré transformations for a free elementary particle implies that the Hilbert space $\mathcal{H}$ of its quantum theory must admit [1] an irreducible representation U of the Poincaré group $\mathcal{P}$ such that the quantum transformation $A \to S_g[A]$ of each observable A, operated by a Poincaré transformation g, is obtained according to the relation $S_g[A] = U_g A U_g^{-1}$. Therefore, the identification of the irreducible representations of $\mathcal{P}$ lays the groundwork for the formulation of relativistic quantum theories of one free particle. Elsewhere [2] we have shown that the collection of the irreducible representations of $\mathcal{P}$ considered in the literature does not exhaust the irreducible representations of $\mathcal{P}$ a quantum theory of an elementary particle can be coherently based on. In particular, the irreducible representations with anti-unitary space inversion operator are currently discarded [3], [4] but in fact quantum theories of a particle characterized by anti-unitary space inversion operator can be consistently developed [2]; in particular, these further representations turn out to be successful for formulating quantum theories of a Klein Gordon particle without the inconsistencies that plagued the early theory [4].

For this reason it is important to redetermine the possible irreducible representations of $\mathcal{P}$ without *a priori* preclusions. In section 2 we outline such a redetermination, limited to *positive mass* and *spin 0* irreducible representations just for reason of space. In so doing, for every value of the of mass four irreducible representations are identified besides those considered by the current literature. Furthermore we show a further possibility of forming *irreducible* representations of $\mathcal{P}$ from *reducible* representations of the subgroup $\mathcal{P}_+^\uparrow$, completely neglected so far.

In section 3 we show that making use of this redetermination, it is possible to formulate particle theories that do not suffer the problems of early theories, and also *new species* of theories.

Content from this work may be used under the terms of the Creative Commons Attribution 3.0 licence. Any further distribution of this work must maintain attribution to the author(s) and the title of the work, journal citation and DOI.
Published under licence by IOP Publishing Ltd

IOP Conf. Series: Journal of Physics: Conf. Series **1275** (2019) 012034 doi:10.1088/1742-6596/1275/1/012034

2. Positive mass and spin 0 irreducible representations of $\mathcal{P}$

2.1. Notation and mathematical prerequisites

2.1.1. Poincaré group. Given any vector $\underline{x} = (x_0, \mathbf{x}) \in \mathbb{R}^4$, we call x_0 the *time component* of $\underline{x}$ and $\mathbf{x} = (x_1, x_2, x_3)$ the *spatial component* of $\underline{x}$. The proper orthochronous Poincaré group $\mathcal{P}_+^\uparrow$ is the separable locally compact group of all transformations of $\mathbb{R}^4$ generated by the ten one-parameter sub-groups $\mathcal{T}_0$, $\mathcal{T}_j, \mathcal{R}_j$, $\mathcal{B}_j$, $j = 1, 2, 3$, of, respectively, time translations, spatial translation, proper spatial rotations and Lorentz boosts relative to axis x_j. The Euclidean group $\mathcal{E}$ is the sub-group generated by all $\mathcal{T}_j$ and $\mathcal{R}_j$. The sub-group generated by $\mathcal{R}_j$, $\mathcal{B}_j$ is the proper orthochronous Lorentz group $\mathcal{L}_+^\uparrow$ [5]. It does not include time reversal $\mathfrak{t}$ and space inversion $\mathfrak{s}$. Time reversal $\mathfrak{t}$ transforms $\underline{x} = (x_0, \mathbf{x})$ into $(-x_0, \mathbf{x})$; space inversion $\mathfrak{s}$ transforms $\underline{x} = (x_0, \mathbf{x})$ into $(x_0, -\mathbf{x})$. The group generated by $\{\mathcal{P}_+^\uparrow, \mathfrak{t}, \mathfrak{s}\}$ is the separable and locally compact Poincaré group $\mathcal{P}$. By $\mathcal{L}_+$ we denote the subgroup generated by $\mathcal{L}_+^\uparrow$ and $\mathfrak{t}$, while $\mathcal{L}^\uparrow$ denotes the subgroup generated by $\mathcal{L}_+^\uparrow$ and $\mathfrak{s}$; analogously, $\mathcal{P}_+$ denotes the subgroup generated by $\mathcal{P}_+^\uparrow$ and $\mathfrak{t}$, while $\mathcal{P}^\uparrow$ is the subgroup generated by $\mathcal{P}_+^\uparrow$ and $\mathfrak{s}$.

2.1.2. Mathematical structures. The following mathematical structures, based on a complex and separable Hilbert space $\mathcal{H}$, are of general interest in quantum theory.

- The set $\Omega(\mathcal{H})$ of all self-adjoint operators of $\mathcal{H}$; in a quantum theory these operators represent *quantum observables.*
- The lattice $\Pi(\mathcal{H})$ of all projections operators of $\mathcal{H}$; in a quantum theory they represent observables with spectrum $\{0, 1\}$.
- The set $\Pi_1(\mathcal{H})$ of all rank one orthogonal projections of $\mathcal{H}$.
- The set $\mathcal{S}(\mathcal{H})$ of all density operators of $\mathcal{H}$; in a quantum theory these operators represent *quantum states.*
- The set $\mathcal{V}(\mathcal{H})$ of all unitary or anti-unitary operators of the Hilbert space $\mathcal{H}$.
- The set $\mathcal{U}(\mathcal{H})$ of all unitary operators of $\mathcal{H}$; trivially, $\mathcal{U}(\mathcal{H}) \subseteq \mathcal{V}(\mathcal{H})$ holds.

2.1.3. Generalized representations of groups. The following definition introduces generalized notions of group representation.

Definition 2.1. *Let G be a separable, locally compact group with identity element e. A correspondence $U : G \to \mathcal{V}(\mathcal{H})$, $g \to U_g$, with $U_e = \mathbb{I}$, is a generalized projective representation (shortly ġprep') of G if the following conditions are satisfied.*

i) *A complex function $\sigma : G \times G \to \mathbb{C}$, called multiplier, exists such that $U_{g_1 g_2} = \sigma(g_1, g_2) U_{g_1} U_{g_2}$; the modulus $|\sigma(g_1, g_2)|$ is always 1, of course;*

ii) *for all $\phi, \psi \in \mathcal{H}$, the mapping $g \to \langle U_g \phi \mid \psi \rangle$ is a Borel function in g.*

Whenever U_g is unitary for all $g \in G$, U is called projective representation (shortly ṗrep'), or σ-representation.

A generalized projective representation is said to be continuous if for any fixed $\psi \in \mathcal{H}$ the mapping $g \to U_g \psi$ from G to $\mathcal{H}$ is continuous with respect to g.

If $g \to U_g$ is a gprep and $\theta(g) \in \mathbb{R}$, then $g \to \tilde{U}_g = e^{i\theta(g)} U_g$ is a gprep *equivalent* [6] to $g \to U_g$.

In [7] we have proved that the following statement holds.

Proposition 2.1. *If G is a connected group, then every continuous generalized projective representation of G is a projective representation, i.e. $U_g \in \mathcal{U}(\mathcal{H})$, for all $g \in G$.*

2.2. Generalized representations of the Poincaré group $\mathcal{P}$
All sub-groups $\mathcal{T}_0$, $\mathcal{T}_j, \mathcal{R}_j$, $\mathcal{B}_j$ of $\mathcal{P}_+^\uparrow$ of Poincaré group $\mathcal{P}$ are additive; in fact, $\mathcal{B}_j$ is not additive with respect to the parameter *relative velocity* u, but it is additive with respect to the parameter $\varphi(u) = \frac{1}{2}\ln\frac{1+u}{1-u}$. Then, according to Stone's theorem [9], for every continuous prep of $\mathcal{P}_+^\uparrow$, an equivalent prep U exists for which there are ten self-adjoint generators P_0, P_j, J_j, K_j, $j = 1,2,3$, of the ten one-parameter unitary subgroups $\{e^{iP_0t}\}$, $\{e^{-iP_ja_j}, a \in \mathbb{R}\}$, $\{e^{-iJ_j\theta_j}, \theta_j \in \mathbb{R}\}$, $\{e^{-iK_j\varphi(u_j)}, u_j \in \mathbb{R}\}$ of $\mathcal{U}(\mathcal{H})$ that represent the one-parameter sub-groups $\mathcal{T}_0$, $\mathcal{T}_j, \mathcal{R}_j$, $\mathcal{B}_j$ according to the projective representation $g \to U_g$ of the Poincaré group $\mathcal{P}_+^\uparrow$.

2.2.1. Commutation relations. The structural properties of $\mathcal{P}_+^\uparrow$ as a Lie group imply that every continuous prep of $\mathcal{P}_+^\uparrow$ admits an equivalent prep U such that the following commutation relations [10] hold for its generators.

$$\begin{array}{lll}
(i)\ [P_j,P_k]=\mathbb{O}, & (ii)\ [J_j,P_k]=i\hat{\epsilon}_{jkl}P_l, & (iii)\ [J_j,J_k]=i\hat{\epsilon}_{jkl}J_l,\\
(iv)\ [J_j,K_k]=i\hat{\epsilon}_{jkl}K_l, & (v)\ [K_j,K_k]=-i\delta_{j,k}J_l, & (vi)\ [K_j,P_k]=i\delta_{jk}P_0,\\
(vii)\ [P_j,P_0]=\mathbb{O}, & (viii)\ [J_j,P_0]=\mathbb{O}, & (ix)\ [K_j,P_0]=iP_0,
\end{array} \quad (1)$$

where $\hat{\epsilon}_{jkl}$ is the Levi-Civita symbol ϵ_{jkl} restricted by the condition $j \neq l \neq k$.

Let $U : \mathcal{P} \to \mathcal{V}(\mathcal{H})$ be a gprep of $\mathcal{P}$ whose restriction to $\mathcal{P}_+^\uparrow$ is continuous. Since time reversal $\mathfrak{t}$ and space inversion $\mathfrak{s}$ are *not connected* with the identity transformation $e \in \mathcal{P}$, the operators $\mathsf{T} = U_{\mathfrak{t}}$ and $\mathsf{S} = U_{\mathfrak{s}}$ can be unitary or anti-unitary; The phase factor $e^{i\theta(g)}$ can be chosen in such a way that the following statements hold in the equivalent gprep [10].

If S is unitary, then $[\mathsf{S},P_0]=\mathbb{O}$, $\mathsf{S}P_j=-P_j\mathsf{S}$, $[\mathsf{S},J_k]=\mathbb{O}$, $\mathsf{S}K_j=-K_j\mathsf{S}$; $\mathsf{S}^2=\mathbb{I}$; (2)

If S is anti-unitary, then $\mathsf{S}P_0=-P_0\mathsf{S}$, $[\mathsf{S},P_j]=\mathbb{O}$, $\mathsf{S}J_k=-J_k\mathsf{S}$, $\mathsf{S}K_j=K_j\mathsf{S}$; (3)
$\mathsf{S}^2=c\mathbb{I}$, *so that* $\mathsf{S}^{-1}=c\mathsf{S}$, *where* $c=1$ *or* $c=-1$.

If T is unitary, then $\mathsf{T}P_0=-P_0\mathsf{T}$, $[\mathsf{T},P_j]=\mathbb{O}$, $[\mathsf{T},J_k]=\mathbb{O}$, $\mathsf{T}K_j=-K_j\mathsf{T}$; $\mathsf{T}^2=\mathbb{I}$ (4)

If T is anti-unitary, then $\mathsf{T}P_0=P_0\mathsf{T}$, $\mathsf{T}P_j=-P_j\mathsf{T}$, $\mathsf{T}J_k=-J_k\mathsf{T}$, $\mathsf{T}K_j=K_j\mathsf{T}$. (5)
$\mathsf{T}^2=c\mathbb{I}$, *so that* $\mathsf{T}^{-1}=c\mathsf{T}$, *either* $c=1$ *or* $c=-1$ *must hold.*

The commutation condition for the pair S,T, *is* $\mathsf{S}\mathsf{T}=\omega\mathsf{T}\mathsf{S}$, *with* $\omega\in\mathbb{C}$ *and* $|\omega|=1$. (6)

2.3. Classifying and identifying irreducible representations of $\mathcal{P}$
From now on we specialize to *irreducible* generalized projective representations of $\mathcal{P}$, with $U\mid_{\mathcal{P}_+^\uparrow}$ continuous, where (1)-(6) hold. The following proposition holds [8].

Proposition 2.2. *Properties (1)-(6) imply that if a gprep U of $\mathcal{P}$ is irreducible, then two real numbers η, ϖ exist such that the following equalities hold.*

$$(i)\ P_0^2 - \mathbf{P}^2 = \eta\mathbb{I} \quad \text{and} \quad (ii)\ W^2 \equiv W_0^2 - (W_1^2 + W_2^2 + W_3^2) = \varpi\mathbb{I} \quad (7)$$

where $W_0 = \mathbf{P}\cdot\mathbf{J}$ *and* $W_j = P_0J_j - (\mathbf{P}\times\mathbf{K})_j$.

Therefore every irreducible gprep of $\mathcal{P}$ is characterized by the real constant η, ϖ. We restrict our investigation to those irreducible gprep of $\mathcal{P}$ for which $\eta > 0$, so that $\eta = \mu^2$, with $\mu > 0$; with this restriction it can be proved that $s \in \frac{1}{2}\mathbb{N}$ exists such that $\varpi = \mu s(s+1)$.

The following proposition [2] establishes that the spectrum $\sigma(P_0)$ of P_0 must be one of three definite subsets I_μ^+, I_μ^-, $I_\mu^+ \cup I_\mu^-$ of $\mathbb{R}$, where $I_\mu^+ = [\mu,\infty)$ and $I_\mu^- = (-\infty,-\mu]$. The different possibilities are ruled over by the following proposition.

Proposition 2.3. *If $U : \mathcal{P} \to \mathcal{V}(\mathcal{H})$ is an irreducible generalized projective representation, then there are only the following mutually exclusive possibilities for the spectrum $\sigma(P_0)$ of P_0.*

(**u**) $\sigma(P_0) = I_\mu^+$ and $\sigma(P_0) = [\mu, \infty)$, *up spectrum;*

(**d**) $\sigma(P_0) = I_\mu^-$ and $\sigma(P_0) = (-\infty, -\mu]$, *down spectrum;*

(**s**) $\sigma(P_0) = I_\mu^+ \cup I_\mu^-$ *and* $\sigma(P_0) = [\mu, \infty) \cup (-\infty, -\mu]$, *symmetrical spectrum.*

If $\triangleleft T$ *is anti-unitary and* $\mathcal{S}$ *is unitary, then either* $\sigma(P_0) = I_\mu^+$ *or* $\sigma(P_0) = I_\mu^-$, *and hence* $\sigma(P_0) = I_\mu^+ \cup I_\mu^-$ *cannot occur.*
If $\triangleleft T$ *is unitary then* $\sigma(P_0) = I_\mu^+ \cup I_\mu^-$, *independently of* $\mathcal{S}$.
If $\mathcal{S}$ *is anti unitary then* $\sigma(P_0) = I_\mu^+ \cup I_\mu^-$, *independently of* $\triangleleft T$.

Given an irreducible gprep U of $\mathcal{P}$, we define the projection operators $E^\pm = \int_{I_\mu^\pm} p_0 dE_{p_0}^{(0)}$ where $E_{p_0}^{(0)}$ is the resolution of the identity of P_0. In [2] we prove the following proposition.

Proposition 2.4. *In an irreducible gprep U of $\mathcal{P}$ the relation $[E^\pm, U_g] = \mathbb{O}$ holds for all $g \in \mathcal{P}_+^\uparrow$.*

According to Prop. 2.4, in the case of symmetrical spectrum $\sigma(P_0) = I_\mu^+ \cup I_\mu^-$, the restriction $U\mid_{\mathcal{P}_+^\uparrow}$ is always reduced by E^+ into $U^+\mid_{\mathcal{P}_+^\uparrow} = E^+ U\mid_{\mathcal{P}_+^\uparrow} E^+$ and by $U^-\mid_{\mathcal{P}_+^\uparrow} = E^- U\mid_{\mathcal{P}_+^\uparrow} E^-$.
If $\sigma(P_0) = I_\mu^+$ (resp., $\sigma(P_0) = I_\mu^-$), then $U\mid_{\mathcal{P}_+^\uparrow} = U^+\mid_{\mathcal{P}_+^\uparrow}$ (resp., $U\mid_{\mathcal{P}_+^\uparrow} = U^-\mid_{\mathcal{P}_+^\uparrow}$).

In any case, given an irreducible gprep $U : \mathcal{P} \to \mathcal{U}(\mathcal{H})$, once established which of the three conditions (**u**), (**d**), (**s**) is satisfied, $U^+\mid_{\mathcal{P}_+^\uparrow}$ or $U^-\mid_{\mathcal{P}_+^\uparrow}$ can be reducible or not, though the "mother" representation U is irreducible. In sections 2.3.1 and 2.3.2 we completely identify the possible irreducible gpreps U of $\mathcal{P}$ for which $U^+\mid_{\mathcal{P}_+^\uparrow}$ and $U^-\mid_{\mathcal{P}_+^\uparrow}$ are *irreducible*. For reasons of space, we restrict ourselves to the case $s = \varpi = 0$; nevertheless, the identification is extendable to all values of s [2].

In section 2.4 we show that also irreducible gpreps of $\mathcal{P}$ exist with $U^+\mid_{\mathcal{P}_+^\uparrow}$ (or $U^-\mid_{\mathcal{P}_+^\uparrow}$) reducible exist, though they are never considered by the literature.

2.3.1. The case $\sigma(P_0) = I_\mu^\pm$ with $U^\pm\mid_{\mathcal{P}_+^\uparrow}$ irreducible. For each value $\mu = \sqrt{\eta} > 0$, *modulo unitary isomorphisms* there is only one irreducible projective representation of $\mathcal{P}$ with $\sigma(P_0) = I_\mu^+$ and only one with $\sigma(P_0) = I_\mu^-$, that we briefly present. Once fixed μ, the Hilbert space of the projective representation is the space $L_2(\mathbb{R}^3, d\nu)$ of all complex functions $\psi : \mathbb{R}^3 \to \mathbb{C}$, $\mathbf{p} \to \psi(\mathbf{p})$, square integrable with respect to the measure $d\nu(\mathbf{p}) = \frac{dp_1 dp_2 dp_3}{\sqrt{\mu^2 + \mathbf{p}^2}}$.

For the well known [10] irreducible preps with $\sigma(P_0) = I_\mu^+$ the following statements hold.

- The generators P_j are the multiplication operators defined by $(P_j\psi)(\mathbf{p}) = p_j\psi(\mathbf{p})$;
- $(P_0\psi)(\mathbf{p}) = p_0\psi(\mathbf{p})$ where $p_0 = +\sqrt{\mu^2 + \mathbf{p}^2}$; this P_0 has a positive spectrum;
- $J_k = i\left(p_l \frac{\partial}{\partial p_j} - p_j \frac{\partial}{\partial p_l}\right)$, (k, l, j) being a cyclic permutation of $(1, 2, 3)$;
- the generators K_j are given by $K_j = ip_0 \frac{\partial}{\partial p_j}$;
- the space inversion operator is $\mathcal{S} = \Upsilon$ and the time reversal operator is $\triangleleft T = \mathcal{K}\Upsilon$, where
  - Υ is the unitary operator defined by $(\Upsilon\psi)(\mathbf{p}) = \psi(-\mathbf{p})$,
  - $\mathcal{K}$ is the anti-unitary complex conjugation operator defined by $\mathcal{K}\psi(\mathbf{p}) = \overline{\psi(\mathbf{p})}$.

The irreducible projective representation with $\sigma(P_0) = I_\mu^-$ differs only for P_0 and $\mathbf{K}$:

$(P_0\psi)(\mathbf{p}) = -p_0\psi(\mathbf{p})$, and $K_j = -ip_0 \frac{\partial}{\partial p_j}$.

2.3.2. The case $\sigma(P_0) = I_\mu^+ \cup I_\mu^-$ with $U^+ |_{\mathcal{P}_+^\uparrow}$ irreducible. For every value $\mu > 0$ there are six inequivalent irreducible generalized projective representations $U^{(l)}$, $l = 1, 2, \ldots, 6$ of $\mathcal{P}$; modulo unitary isomorphisms, the Hilbert space of each $U^{(l)}$ is $L_2(\mathbb{R}^3, d\nu) \oplus L_2(\mathbb{R}^3, d\nu)$. If each vector $\psi \in \mathcal{H}$ is represented as a column vector $\psi = \begin{bmatrix} \psi^+ \\ \psi^- \end{bmatrix}$, with $\psi^\pm \in L_2(\mathbb{R}^3, d\nu)$, then $E^+ = \begin{bmatrix} 1 & 0 \\ 0 & 0 \end{bmatrix}$, $E^- = \begin{bmatrix} 0 & 0 \\ 0 & 1 \end{bmatrix}$, and the generators of $U^{(l)} |_{\mathcal{P}_+^\uparrow}$ are represented by the following matrices, known as the *canonical form*.

$$P_j = \begin{bmatrix} p_j & 0 \\ 0 & -p_j \end{bmatrix}, \quad P_0 = \begin{bmatrix} p_0 & 0 \\ 0 & -p_0 \end{bmatrix}, \quad J_k = \begin{bmatrix} \mathsf{j}_k & 0 \\ 0 & \mathsf{j}_k \end{bmatrix}, \quad K_j = \begin{bmatrix} \mathsf{k}_j & 0 \\ 0 & -\mathsf{k}_j \end{bmatrix}, \tag{8}$$

where $$\mathsf{j}_k = i\left(p_l \frac{\partial}{\partial p_j} - p_j \frac{\partial}{\partial p_l}\right) \text{ and } \mathsf{k}_j = i p_0 \frac{\partial}{\partial p_j}. \tag{9}$$

The six representations $U^{(l)}$, $l = 1, 2, \ldots, 6$ differ just for the different combinations of time reversal and space inversion operators, according to the following list.

$U^{(1)}$ has unitary $\mathrm{T} = \begin{bmatrix} 0 & 1 \\ 1 & 0 \end{bmatrix}$ and unitary $\mathcal{S} = \Upsilon \begin{bmatrix} 1 & 0 \\ 0 & 1 \end{bmatrix}$;

$U^{(2)}$ has unitary $\mathrm{T} = \begin{bmatrix} 0 & 1 \\ 1 & 0 \end{bmatrix}$ and unitary $\mathcal{S} = \Upsilon \begin{bmatrix} 1 & 0 \\ 0 & -1 \end{bmatrix}$;

$U^{(3)}$ has unitary $\mathrm{T} = \begin{bmatrix} 0 & 1 \\ 1 & 0 \end{bmatrix}$ and anti-unitary $\mathcal{S} = \begin{bmatrix} 0 & 1 \\ 1 & 0 \end{bmatrix} \mathcal{K}$;

$U^{(4)}$ has unitary $\mathrm{T} = \begin{bmatrix} 0 & 1 \\ 1 & 0 \end{bmatrix}$ and anti-unitary $\mathcal{S} = \begin{bmatrix} 0 & 1 \\ -1 & 0 \end{bmatrix} \mathcal{K}$;

$U^{(5)}$ has anti-unitary $\mathrm{T} = 1\mathcal{K}\Upsilon \begin{bmatrix} 0 & 1 \\ 1 & 0 \end{bmatrix}$ and anti-unitary $\mathcal{S} = \begin{bmatrix} 0 & 1 \\ 1 & 0 \end{bmatrix} \mathcal{K}$;

$U^{(6)}$ has anti-unitary $\mathrm{T} = 1\mathcal{K}\Upsilon \begin{bmatrix} 0 & 1 \\ 1 & 0 \end{bmatrix}$ and anti-unitary $\mathcal{S} = \begin{bmatrix} 0 & 1 \\ -1 & 0 \end{bmatrix} \mathcal{K}$.

Thus we have identified, besides the well known [10] irreducible gpreps $U^{(1)}$, $U^{(2)}$, the gpreps $U^{(3)}$, $U^{(4)}$ with *T anti-unitary and $\mathcal{S}$ unitary*, $U^{(5)}$, $U^{(6)}$or with *T anti-unitary and $\mathcal{S}$ anti-unitary*, which are neglected by the literature about elementary particles theory, at the best of our knowledge.

2.4. New species of gpreps: irreducible $U : \mathcal{P} \to \mathcal{U}(\mathcal{H})$ with $U^\pm |_{\mathcal{P}_+^\uparrow}$ reducible

Hence, for each $\mu > 0$ there are eight inequivalent irreducible gpreps of $\mathcal{P}$. The class of all such octets, however, does not exhaust the class $\mathcal{I}_\mathcal{P}$ of all irreducible gpreps of $\mathcal{P}$ with $s = 0$, because the class $\mathcal{I}_\mathcal{P}(U^\pm\text{red.})$ of the irreducible representations with $U^+ |_{\mathcal{P}_+^\uparrow}$ or $U^- |_{\mathcal{P}_+^\uparrow}$ *reducible* is not empty.

2.4.1. The case $\sigma(P_0) = I_\mu^\pm$. Fixed any $\mu > 0$, let us consider the Hilbert space $\mathcal{H} = \mathcal{H}^{(1)} \oplus \mathcal{H}^{(2)}$, where $\mathcal{H}^{(1)} \equiv \mathcal{H}^{(2)} = L_2(\mathbb{R}^3, d\nu)$. We represent every vector $\psi = \psi_1 + \psi_2$ in $\mathcal{H}$, with $\psi_n \in \mathcal{H}^{(n)}$, as the column vector $\psi \equiv \begin{bmatrix} \psi_1 \\ \psi_2 \end{bmatrix}$, so that every linear (resp., anti-linear) operator A of $\mathcal{H}$ can be represented by a matrix $\begin{bmatrix} A_{11} & A_{12} \\ A_{21} & A_{22} \end{bmatrix}$, where A_{mn} is a linear (resp., anti-linear) operator from $\mathcal{H}^{(m)}$ to $\mathcal{H}^{(n)}$, and $A\psi = \begin{bmatrix} A_{11} & A_{12} \\ A_{21} & A_{22} \end{bmatrix} \begin{bmatrix} \psi_1 \\ \psi_2 \end{bmatrix} = \begin{bmatrix} A_{11}\psi_1 + A_{12}\psi_2 \\ A_{21}\psi_1 + A_{22}\psi_2 \end{bmatrix}$.

Let us define $P_0 = \begin{bmatrix} p_0 & 0 \\ 0 & p_0 \end{bmatrix}$, $P_j = \begin{bmatrix} p_j & 0 \\ 0 & p_j \end{bmatrix}$, $J_k = \begin{bmatrix} \mathsf{j}_k & 0 \\ 0 & \mathsf{j}_k \end{bmatrix}$, $J_j = \begin{bmatrix} \mathsf{k}_j & 0 \\ 0 & \mathsf{k}_j \end{bmatrix}$, they satisfy (1).

In such a way, a *reducible* prep $U : \mathcal{P}_+^\uparrow \to L_2(\mathbb{R}^3, d\nu) \oplus L_2(\mathbb{R}^3, d\nu)$ is completely determined, where $E^+ = \mathbb{1}$, so that $U^+ = U$. The possible extensions to the whole $\mathcal{P}$ are obtained by adding a time reversal operator T and a space inversion operator S that satisfy (2)-(6). Now we show that some of these extensions are *irreducible* gpreps of $\mathcal{P}$. Indeed, the operators

$$\mathsf{S} = \Upsilon \begin{bmatrix} 0 & 1 \\ 1 & 0 \end{bmatrix} \quad \text{and} \quad \mathsf{T} = \Upsilon \mathcal{K} \begin{bmatrix} 0 & 1 \\ -1 & 0 \end{bmatrix} \tag{10}$$

satisfy all conditions (2)-(6). Therefore, by defining $U_{\mathsf{S}} = \Upsilon \begin{bmatrix} 0 & 1 \\ 1 & 0 \end{bmatrix}$ and $U_{\mathsf{t}} = \Upsilon \mathcal{K} \begin{bmatrix} 0 & 1 \\ -1 & 0 \end{bmatrix}$ the projective representation U of $\mathcal{P}_+^\uparrow$ is extended to a generalized projective representation of $\mathcal{P}$. Moreover, if $A = \begin{bmatrix} A_{11} & A_{12} \\ A_{21} & A_{22} \end{bmatrix}$ is any self-adjoint operator of $\mathcal{H} = L_2(\mathbb{R}^3, d\nu) \oplus L_2(\mathbb{R}^3, d\nu)$, then the conditions $[A, P_0] = [A, P_j] = [A, J_k] = [A, K_j] = [A, \mathsf{S}] = [A, \mathsf{T}] = \mathbb{O}$ imply $A = \begin{bmatrix} a & 0 \\ 0 & a \end{bmatrix} \equiv a\mathbb{1}$. Thus, this generalized projective representation U of $\mathcal{P}$ is irreducible, while $U^+ |_{\mathcal{P}_+^\uparrow}$ is reducible.

2.4.2. The case $\sigma(P_0) = I_\mu^+ \cup I_\mu^-$. Let us introduce the Hilbert space $\mathcal{H} = L_2(\mathbb{R}^3, d\nu) \oplus L_2(\mathbb{R}^3, d\nu) \oplus L_2(\mathbb{R}^3, d\nu) \oplus L_2(\mathbb{R}^3, d\nu)$. Every vector $\psi \in \mathcal{H}$ is represented as a column vector $\psi = \begin{bmatrix} \psi_+^{(1)} \\ \psi_-^{(1)} \\ \psi_+^{(2)} \\ \psi_-^{(2)} \end{bmatrix}$, with $\psi_\pm^{(m)} \in L_2(\mathbb{R}^3, d\nu)$. To identify a gprep U of $\mathcal{P}$ with $U^+ |_{\mathcal{P}_+^\uparrow}$ reducible, we first define the self-adjoint generators of a prep $U : \mathcal{P}_+^\uparrow \to \mathcal{U}(\mathcal{H})$ as

$$P_0 = \begin{bmatrix} p_0 & 0 & 0 & 0 \\ 0 & -p_0 & 0 & 0 \\ 0 & 0 & p_0 & 0 \\ 0 & 0 & 0 & -p_0 \end{bmatrix}, \quad P_j = \begin{bmatrix} p_j & 0 & 0 & 0 \\ 0 & p_j & 0 & 0 \\ 0 & 0 & p_j & 0 \\ 0 & 0 & 0 & p_j \end{bmatrix},$$

$$J_k = \begin{bmatrix} \mathsf{j}_k & 0 & 0 & 0 \\ 0 & \mathsf{j}_k & 0 & 0 \\ 0 & 0 & \mathsf{j}_k & 0 \\ 0 & 0 & 0 & \mathsf{j}_k \end{bmatrix}, \quad K_j = \begin{bmatrix} \mathsf{k}_j & 0 & 0 & 0 \\ 0 & -\mathsf{k}_j & 0 & 0 \\ 0 & 0 & \mathsf{k}_j & 0 \\ 0 & 0 & 0 & -\mathsf{k}_j \end{bmatrix}.$$

So, a *reducible* projective representation of $\mathcal{P}_+^\uparrow$ is defined. To extend it to an *irreducible* gprep of $\mathcal{P}$ we introduce $\mathsf{T} = \begin{bmatrix} 0 & 1 & 0 & 0 \\ 1 & 0 & 0 & 0 \\ 0 & 0 & 0 & 1 \\ 0 & 0 & 1 & 0 \end{bmatrix}$ (unitary) and $\mathsf{S} = \mathcal{K} \begin{bmatrix} 0 & 0 & 0 & 1 \\ 0 & 0 & 1 & 0 \\ 0 & -1 & 0 & 0 \\ -1 & 0 & 0 & 0 \end{bmatrix}$ (anti-unitary), that satisfy (2)-(6).

Now, if $A = \begin{bmatrix} A_{11} & A_{12} & A_{13} & A_{14} \\ A_{21} & A_{22} & A_{23} & A_{24} \\ A_{31} & A_{32} & A_{33} & A_{34} \\ A_{41} & A_{42} & A_{43} & A_{44} \end{bmatrix}$ is any self-adjoint operator of $\mathcal{H}$, then the conditions $[A, P_0] = [A, P_j] = [A, J_k] = [A, K_j] = [A, \mathsf{T}] = [A, \mathsf{S}] = \mathbb{O}$ are satisfied if and only if $A = \begin{bmatrix} a & 0 & 0 & 0 \\ 0 & a & 0 & 0 \\ 0 & 0 & a & 0 \\ 0 & 0 & 0 & a \end{bmatrix} \equiv a\mathbb{1}$ with $a \in \mathbb{R}$. Thus, a new species of irreducible gprep of $\mathcal{P}$ is identified.

IOP Conf. Series: Journal of Physics: Conf. Series **1275** (2019) 012034 doi:10.1088/1742-6596/1275/1/012034

3. Consistent relativistic quantum theories of elementary spin-0 particle

By *localizable free particle*, shortly *free particle*, we mean an isolated system whose quantum theory is endowed with a unique triple $(Q_1, Q_2, Q_3) \equiv \mathbf{Q}$ of quantum observables, called *position* operator, such that

(Q.1) $[Q_j, Q_k] = \mathbb{O}$, for all $j, k \in \{1, 2, 3\}$. This condition requires that a measurement of position yields all three values of the coordinates of the particle position.

(Q.2) The triple $(Q_1, Q_2, Q_3) \equiv \mathbf{Q}$ is characterized by the specific properties of transformation of position with respect to the group $\mathcal{P}$; hence, in particular,

(Q.2 .a) $S_{\mathsf{t}}[\mathbf{Q}] = \mathbf{Q}$ and $S_{\mathsf{s}}[\mathbf{Q}] = -\mathbf{Q}$, equivalent to $\mathsf{T}\mathbf{Q} = \mathbf{Q}\mathsf{T}$ and $\mathsf{S}\mathbf{Q} = -\mathbf{Q}\mathsf{S}$.

(Q.2 .b) If $g \in \mathcal{E}$ then $S_g[\mathbf{Q}] = U_g \mathbf{Q} U_g^{-1} = \mathbf{g}(\mathbf{Q})$, where $\mathbf{x} \to \mathbf{g}(\mathbf{x})$ is the function that realizes g.

A free particle is said *elementary* if the gprep U for which $S_g[A] = U_g A U_g^{-1}$ is irreducible. Accordingly, by selecting the irreducible gpreps U of $\mathcal{P}$, satisfying (1)-(6), that admit such a triple $\mathbf{Q}$ satisfying (Q.1) and (Q.2) we identify the possible theories of an elementary free particle.

3.1. Elementary particle theories with $s = 0$, and $U^{\pm}|_{\mathcal{P}_+^{\uparrow}}$ irreducible

For the irreducible gpreps with $\sigma(P_0) = I_\mu^{\pm}$ of section 2.3.1, conditions (Q.1) and (Q2.a,b) are sufficient [2] to univocally determine $\mathbf{Q}$ as $Q_j = F_j$, where $F_j = i\frac{\partial}{\partial p_j} - \frac{i}{2p_0^2}p_j$ are the Newton and Wigner operators [11].

If $\sigma(P_0) = I_\mu^+ \cup I_\mu^-$, ($Q$.1) and ($Q$2.a,b) are sufficient [2] to univocally determine $\mathbf{Q}$ only for $U^{(5)}$ and $U^{(6)}$ in section 2.3.1, where $\mathbf{Q} = \hat{\mathbf{F}} = \begin{bmatrix} F_j & 0 \\ 0 & F_j \end{bmatrix}$. So we have four complete theories. Thus, our approach proves that anti-unitary space inversion operators, discarded in the literature, are consistent in a particle theory with $U+|_{\mathcal{P}_+^{\uparrow}}$ irreducible and $\sigma(P_0)$ symmetrical.

The early theory for such a kind of particle is Klein-Gordon theory [12]-[14], that suffered serious problems.

A first problem is that the wave equation of Klein-Gordon theory is second order in time, while according to the general laws of quantum theory it should be first order.

Furthermore, in Klein-Gordon thoery $\hat{\rho}(t, \mathbf{x}) = \frac{i}{2m}\left(\overline{\psi_t}\frac{\partial}{\partial t}\psi_t - \psi_t\frac{\partial}{\partial t}\overline{\psi_t}\right)$ was interpreted as the *probability of position* density, and $\hat{\mathbf{j}}(t, \mathbf{x}) = \frac{i}{2m}\left(\psi_t\nabla\overline{\psi_t} - \overline{\psi_t}\nabla\psi_t\right)$ as its *current* density. This interpretation is at the basis of the Dirac concern that, due to the time derivative of ψ, such position probability density can be negative. A way to overcome the difficulty was proposed by Feshbach and Villars [15]. They derive an equivalent form of Klein-Gordon equation as a first order equation $i\frac{\partial}{\partial t}\Psi_t = H\Psi_t$ for the state vector $\Psi_t = \begin{bmatrix} \phi_t \\ \chi_t \end{bmatrix}$, where $\phi_t = \frac{1}{\sqrt{2}}(\psi_t + \frac{1}{m}\frac{\partial}{\partial t}\psi_t)$, $\chi_t = \frac{1}{\sqrt{2}}(\psi_t - \frac{1}{m}\frac{\partial}{\partial t}\psi_t)$, and $H = (\sigma_3 + \sigma_2)\frac{1}{2m}(\nabla + m\sigma_3)$, ψ_t being the Klein-Gordon wave function and the σ_j's are the Pauli matrices; in this representation $\hat{\rho} = |\phi_t|^2 - |\chi_t|^2$, without time derivatives. The minus sign in $\hat{\rho}$ forbids to interpret it as probability density of position; Feshbach and Villars proposed to interpret it as *density probability of charge*. Nevertheless, according to Barut and Malin [16], covariance with respect to boosts should imply that $\hat{\rho}$ must be the time component of a four-vector. Barut and Malin proved that is not the case!

In order to check our theories with respect to these problems, the theories based on $U^{(5)}$ and $U^{(6)}$ are reformulated in equivalent forms, obtained by means of unitary transformations operated by the unitary operator $Z = Z_1 Z_2$, where $Z_2 = \frac{1}{\sqrt{p_0}}\mathbb{I}$ and Z_1 is the inverse of the *Fourier-Plancherel* operator, that transforms $\psi(\mathbf{p})$ into $(Z\psi)(\mathbf{x}) \equiv (\hat{\psi})(\mathbf{x})$. In the so reformulated theories

the Hilbert space for both turs out to be $\mathcal{H} = Z\left(L_2(\mathbb{R}^3, d\nu) \oplus L_2(\mathbb{R}^3, d\nu)\right) \equiv L_2(\mathbb{R}^3) \oplus L_2(\mathbb{R}^3)$; the new self-adjoint generators are $\hat{P}_j = \begin{bmatrix} -i\frac{\partial}{\partial x_j} & 0 \\ 0 & -i\frac{\partial}{\partial x_j} \end{bmatrix}$, $\hat{P}_0 = \sqrt{\mu^2 - \nabla^2}\begin{bmatrix} 1 & 0 \\ 0 & -1 \end{bmatrix}$, $\hat{J}_k = -i\left(x_l\frac{\partial}{\partial x_j} - x_j\frac{\partial}{\partial x_l}\right)\begin{bmatrix} 1 & 0 \\ 0 & 1 \end{bmatrix}$; $\hat{K}_j = \frac{1}{2}\left(x_j\sqrt{\mu^2 - \nabla^2} + \sqrt{\mu^2 - \nabla^2}x_j\right)\begin{bmatrix} 1 & 0 \\ 0 & -1 \end{bmatrix}$.
The wave equation trivially is $i\frac{\partial}{\partial t}\psi_t = P_0\psi_t$, that is first order.

The position operator is $\hat{Q}_j = \begin{bmatrix} x_j & 0 \\ 0 & x_j \end{bmatrix}$, so that also the other problems disappear. Indeed, the position is represented by the *multiplication* operator; therefore, the probability density of position must necessarily be given by the non negative function $\rho(t, \mathbf{x}) = |\psi_t^+(\mathbf{x})|^2 + |\psi_t^-(\mathbf{x})|^2$. On the other hand, the covariance properties with respect to boosts, according to $(Q.2)$, are explicitly expressed by $S_g[\mathbf{Q}] = e^{iK_j\varphi(u)}\mathbf{Q}e^{-iK_j\varphi(u)}$, being K_j and $\mathbf{Q}$ explicitly known, and there is no need of a four-density concept.

3.2. New species of particle theories

Conditions $(Q.1)$, $(Q.2.a,b)$ univocally determine [2] the position operator also for the irreducible gprep of section 2.3.3 with $\sigma(P_0) = I_\mu^\pm$ [2] where $\mathcal{H} = L_2(\mathbb{R}^3, d\nu) \oplus L_2(\mathbb{R}^3, d\nu)$, and again $\hat{Q}_j = \begin{bmatrix} F_j & 0 \\ 0 & F_j \end{bmatrix}$.

Thus, complete theories of an elementary free particle turn out to be identified, which corresponds to none of the early theories. Only Nature establishes whether the zoo of all existing particles hosts also particles described by these theories.

[1] Wigner EP. 1959 *Group Theory and its Applications to the Quantum Theory of Atomic Spectra* (Boston: Academic Press)
[2] Nisticò G. 2018 Arxiv:1811.01546
[3] Costa G, Fogli G. 2012 *Lecture Notes in Physics* **823** (New York: Springer)
[4] Weinberg S. 1995 *The quantum theory of fields. Vol.I* (Cambridge: Cambridge University press)
[5] Barut AO, Racza R. 1986 *Theory of group repesentations and applications* (Singapore: World Scientific)
[6] Mackey GW. 1968 *Induced Representations of Group and Quantum Mechanics* (New York: Benjamin Inc.)
[7] Nisticò G. 2017 Group theoritical derivation of the minimum coupling principle. *Proc. R. Soc.* **A473** 20160629
[8] Bargmann V, Wigner EP. 1948 *Nat.Ac.Sc.* **34** 211
[9] Simon B. 1976 in *Studies in Mathematical Physics: Essays in Honor of Valentine Bargmann* ed Lieb E H, Simon B and Wightman A S (Princeton: Princeton Un. Press) p. 327
[10] Foldy LL. 1956 *Phys. Rev.* **102** 568
[11] Newton TD, Wigner EP. 1949 *Rev. Mod. Phys.* **21** 400
[12] Klein O. 1926, *Z. Physik* **37** 895
[13] Fock V. 1926 *Z. Physik* **37** 242
[14] Gordon W. 1926 *Z. Physik* **40** 117
[15] Felshbach H, Villars F. 1958 *Rev. Mod. Phys.* **30** 24
[16] Barut AO, Malin S. 1968 *Rev. Mod. Phys.* **40** 632

IOP Conf. Series: Journal of Physics: Conf. Series **1275** (2019) 012035 doi:10.1088/1742-6596/1275/1/012035

Fast measurements of entanglement over a kilometric distance to test superluminal models of Quantum Mechanics: final results.

B Cocciaro[1], **S Faetti**[2] **and L Fronzoni**[2]

[1] High School XXV Aprile, Via Milano 2, I-56025 Pontedera (Pisa), Italy
[2] Department of Physics Enrico Fermi, Largo Pontecorvo 3, I-56127 Pisa, Italy

E-mail: `b.cocciaro@comeg.it`, `sandro.faetti@unipi.it`, `leone.fronzoni@unipi.it`

Abstract. Quantum mechanics is a non local theory. To recover the locality principle some physicists suggested that the correlations between entangled particles could be established by communications propagating with a velocity $v_t > c$ in a preferred frame. In the previous DICE conference we reported the preliminary results of a high sensitivity experiment to test the superluminal models. Here we report the final results. No breaking of the quantum predictions has been found. Then, we infer that the velocity of the quantum communications has to be higher than a lower bound of the order of some millions of times the light velocity.

1. Introduction

1.1. The non-locality of quantum mechanics and the superluminal models.

In 1935 Einstein, Podolsky and Rosen (EPR-paradox) [1] showed that orthodox quantum mechanics is a non-local theory. Consider photons a and b in Figure 1 that are in the maximally entangled state

$$|\psi\rangle = \frac{1}{\sqrt{2}}\left(|H,H\rangle + |V,V\rangle\right), \tag{1}$$

where H and V denote horizontal and vertical polarization, respectively. Photons a and b impinge on two absorption polarizers P_A and P_B with the polarization axes that make the angles α and ξ with respect to the horizontal axis. Quantum mechanics predicts that the probability that photon a (or b) passes through a polarizer is $p = 1/2$ for any polarizer orientation. Furthermore, if one photon passes first through polarizer P_A having the polarization axis at angle α and is detected by the photon counter, the entangled state in eq.(1) collapses *instantaneously* to the state $|\alpha_a, \alpha_b\rangle = |\alpha_a\rangle\,|\alpha_b\rangle$ where $|\alpha_a\rangle$ and $|\alpha_b\rangle$ denote single photon states linearly polarized at the same angle α. This means that the polarization measurement at point A affects instantaneously the result of the polarization measurement at point B (*spooky action at a distance*). The probability that both photons pass through the polarizers is $P(\alpha,\xi) = \frac{1}{2}\cos^2(\alpha - \xi)$. Einstein, Podolsky and Rosen[1] concluded that quantum mechanics is an incomplete theory and suggested that a complete local theory should contain some additive local variables ("hidden variables" in the literature). Before 1964, the choice between orthodox quantum mechanics and local variables models was just a matter of personal taste but, in 1964, J. Bell [2] demonstrated that any theory

Content from this work may be used under the terms of the Creative Commons Attribution 3.0 licence. Any further distribution of this work must maintain attribution to the author(s) and the title of the work, journal citation and DOI.
Published under licence by IOP Publishing Ltd

based on local variables must satisfy an inequality that is violated by Quantum Mechanics. Analogous inequalities have been proposed by Clauser et al. [3, 4](the Bell-Clauser-Horne-Shimony-Holt inequality). Then, a direct experimental comparison between quantum mechanics and local variables models became possible.

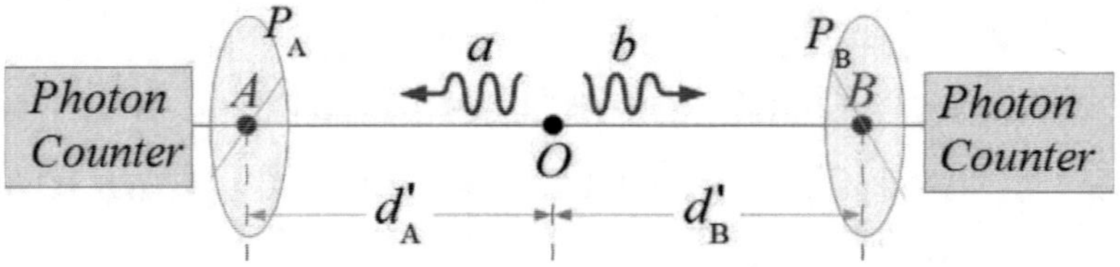

Figure 1. Two entangled photons a and b are generated at O, get the absorption polarizing films P_A (Alice polarize) and P_B(Bob polarizer) and are collected by photon counting modules.

In 1982 Aspect et al. [5] found that the Bell inequality is violated and also showed that the quantum correlations at a distance cannot be induced by luminal communications. Other successive experiments confirmed the Aspect et al. results and also closed some residual loopholes [6, 7, 8, 9]. Despite the remarkable successes of orthodox quantum mechanics, some physicists are very reluctant to abandon the principle of locality and are still looking for alternative local theories. Newton himself in a letter to R. Bentley [10] expressed a strong criticism on action at distance and wrote

"*That gravity should be innate, inherent, and essential to matter, so that one body may act upon another at a distance through a vacuum, without the mediation of anything else, by and through which their action and force may be conveyed from one to another, is to me so great an absurdity that I believe no man who has in philosophical matters a competent faculty of thinking can ever fall into it* ".

The Newton reluctance to accept the idea of a nonlocal behavior of the Nature found a strong support in the successive history of Physics. In fact, at the beginning of the 20th century Einstein showed that the gravitational interactions are not instantaneous but are established by exchange of gravitational waves that have been recently detected[11]. A similar behavior characterized the history of the electric interactions. For this reason, some physicists think that suitable communications could explain the apparent nonlocality of the quantum correlations. After that the Aspect et al. experiment [5] showed that these eventual communications cannot travel with luminal or subluminal velocities, J. Bell, in a famous interview, said: " *in these EPR experiments there is the suggestion that behind the scenes something is going faster than light*" (see page 49 in [12]). To avoid causal paradoxes, the superluminal communications must propagate in a preferred frame at a velocity $v_t = \beta_t\, c$ higher than the light speed where $\beta_t = v_t/c$ is the adimensional velocity. The existence of a preferred frame is needed to avoid causal paradoxes but Refs. [13, 14] strongly stressed that it is not in contrast with the relativity theory. Note that a preferred frame has been already observed: it is the Cosmic Microwave Background frame (CMB frame) that moves with the adimensional velocity $\beta \approx 10^{-3}$ with respect to the Earth geocentric frame. Of course, the CMB frame could be a good candidate as a preferred frame for the superluminal communications. Eberhard[15] and Bohm and Hiley[16] developed specific superluminal models based on this idea (v-causal models in the current literature [17]). Superluminal communications between only two entangled particles could not be used to exchange information in the macroscopic world (*signalling*), but it was recently shown[18, 19] that superluminal communications between three or four entangled particles allow signalling. Although one of us believes that signalling is not in a strict conflict with the relativity theory [13, 14], most physicists think that there is complete incompatibility between signalling and relativity. According to them, the experimental evidence of signalling would need a revision of the relativity theory. Anyway, we stress here the Physics is an experimental matter and only the experiment can give a definitive answer.

1.2 The method to test the superluminal models.
It is important to emphasize that in our experiment we use absorption polarizers and not polarizing beamsplitters. When a photon impinges on an absorption polarizer, it is either absorbed or transmitted by the polarizer and, thus, we think that the outgoing state is not a pure quantum state but is a quantum mixture of a vacuum state (absorbed photon) and a linearly polarized state (transmitted photon). If a successive detection of the photon occurs, one can infer that the photon has not previously absorbed and that it was already linearly polarized just at the exit of the polarizer. Then, we make here the assumption that the collapse of the polarization state of the photon occurs at the absorption polarizer and not at the detector. Here we consider all the possible superluminal models, as well as the Eberhard one, where a first event (the passage of a photon through a polarizer) affects a second event (the passage of the other photon through the second polarizer) via a superluminal communication. If photon a goes first (in the preferred frame) through polarizer P_A, it does not induce the instantaneous collapse of the entangled state to the state $|\alpha_a\rangle\,|\alpha_b\rangle$ but only the local collapse of the state of photon a to $|\alpha_a\rangle$. The collapse of the second photon to the linear polarized state $|\alpha_b\rangle$ will occur only when a collapsing superluminal wave reaches it. The collapsing wave propagates isotropically in the preferred frame at the adimensional superluminal velocity β_t (β_t >1) and is created where (and when) the first event occurs (e.g. a photon hits a polarizer). The first event is the one which occurs temporally before in the preferred frame. For $\beta_t \to \infty$, the collapse occurs instantaneously everywhere and the superluminal model leads to the same predictions of quantum mechanics. If polarizers P_A and P_B are at the same distances d'_A and d'_B from the source O of the entangled photons in the preferred frame, the two photons get to the polarizers simultaneously (the wavepackets width is disregarded here) and no communication is possible if β_t has a finite value. Then, probability $P(\alpha,\xi)$ must satisfy the local Bell-Clauser-Horne-Shimony-Holt inequality and the correlation parameter

$$S_{max} = P_0 - \sum_{i=1}^{3} P_i, \tag{2}$$

with $P_0 = P\,(45°, 67.5°)$, $P_1 = P\,(0°, 67.5°)$, $P_2 = P\,(45°, 112.5°)$ and $P_3 = P\,(90°, 22.5°)$ must satisfy the inequality $S_{max} \leq 0$ [20, 21] while quantum mechanics predicts $S_{max} = {(\sqrt{2}-1)}/{2} \approx 0.2071$. Probabilities $P(\alpha,\xi)$ can be experimentally obtained using the relation

$$P(\alpha,\xi) = \frac{N\,(\alpha,\xi)}{N_{tot}}, \tag{3}$$

where $N(\alpha,\xi)$ are the coincidences between entangled photons passing through the polarizers during the acquisition time $\delta_a t$ and N_{tot} is the total number of entangled photons couples that can be also written in the form:

$$N_{tot} = \sum_{i=0}^{3} N_i, \tag{4}$$

where $N_0 = N\,(0°, 0°)$, $N_1 = N\,(0°, 90°)$, $N_2 = N\,(90°, 0°)$ and $N_3 = N\,(90°, 90°)$. Due to the experimental uncertainty, the arrival times of the photons at the polarizers could differ by a quantity $\Delta t' = {\Delta d'}/{c}$ where $\Delta d'$ is the uncertainty of the *effective optical paths difference* that takes into account the uncertainty of the optical path difference but also any other source of spatial uncertainty [for a detailed analysis of the error sources see Eq.(7) in [21]]. In such a case, a superluminal communication would be impossible only if $\Delta t'$ is smaller than the time $\Delta t'' = d'_{AB}/(\beta_t c)$ that is needed to establish a communication between points A and B, where d'_{AB} is the optical path from A to B in the preferred frame (see Figure 1). The condition $\Delta t' < \Delta t''$ is satisfied if β_t is lower than a maximum detectable value $\beta_{t,max} = {d'_{AB}}/{\Delta d'}$. Therefore, a breakdown of the quantum predictions ($S_{max} < 0$) could be observed in the preferred frame only if $\beta_t < \beta_{t,max}$.

On the Earth laboratory frame the analysis is more complex. The time difference $\Delta t'$ in the preferred frame is related to that in the laboratory frame by the Lorentz relation:

$$\Delta t' = \gamma(\Delta t - \vec{\beta} \cdot \overrightarrow{AB}), \tag{5}$$

where $\vec{\beta}$ is the unknown adimensional relative velocity vector of the preferred frame with respect to the laboratory frame and $\overrightarrow{AB}$ is the vector that connects points A and B in the laboratory frame. The detection of the superluminal effects requires that $\Delta t'$ is as small as possible to satisfy the condition $\Delta t' < \Delta t''$. However a small value of the time uncertainty in the laboratory frame ($\Delta t \approx 0$) does not imply $\Delta t' \approx 0$ except if $\vec{\beta} \cdot \overrightarrow{AB} \approx 0$ in Eq.(5). If the AB segment is east-west aligned, due to the earth's rotation around its axis, there are always two times t_1 and t_2 for each sidereal day where $\overrightarrow{AB}$ becomes orthogonal to $\vec{\beta}$ [22, 23]. Our experiment is performed inside the so called east-west gallery of the European Gravitational Observatory (EGO)[24] that is not aligned along the east-west axis but makes the angle $\gamma = 18° = \pi/10$ with it. In this case, the condition $\vec{\beta} \cdot \overrightarrow{AB} = 0$ is still verified at two times t_1 and t_2 each sidereal day except if the velocity vector of the preferred frame makes a polar angle $\chi < |\gamma| = \pi/10$ or $\chi > \pi - |\gamma| = 9\pi/10$ with respect to the Earth rotation axis (see Figure (2)). This means that our experiment is virtually insensitive to the fraction $\Omega/(4\pi) = \int_0^{\gamma} \sin\theta d\theta < 5\%$ of all the possible alignments of the preferred frame velocity vector. Note that the reference frame of the Cosmic Microwave Background ($\chi \approx 97°$) is accessible to our experiment.

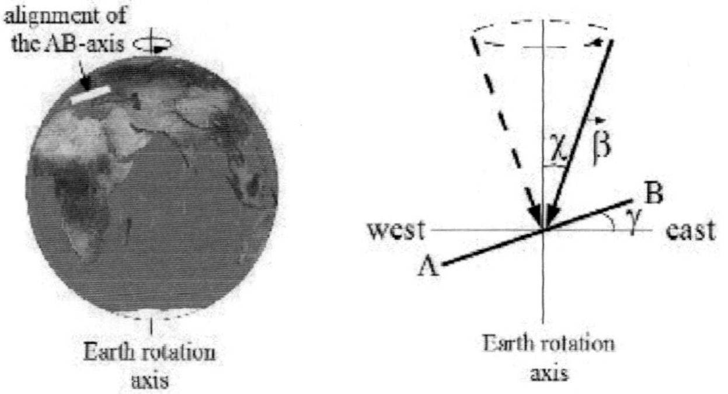

Figure 2. The velocity vector $\vec{\beta}$ with respect to the laboratory frame rotates on a cone of angle χ around the Earth axis (the small contributions due to the rotation of the Earth around the Sun and to the precession and nutation of the Earth axis are disregarded here). The angle between vector $\vec{\beta}$ and segment AB varies between $\pi/2-\gamma-\chi$ and $\pi/2-\gamma+\chi$. Then, $\vec{\beta}$ become orthogonal to AB only if $\pi - |\gamma| \geq \chi \geq |\gamma|$.

For $\pi - |\gamma| \geq \chi \geq |\gamma|$, the maximum measurable superluminal velocity is[22] (see also Eqs.(6) and (7) in Ref.[25]):

$$\beta_{t,max} = \sqrt{1 + \frac{(1-\beta^2)\left[1-\rho^2\right]}{\left[\rho + A\,\frac{2\pi\beta\delta_a t}{T}\right]^2}}\,, \tag{6}$$

where

$$A = \sqrt{\sin^2\chi\cos^2\gamma - \cos^2\chi\sin^2\gamma}\,. \tag{7}$$

T is the earth rotation day and $\rho = {}^{d_{AB}}/_{\Delta d}$, where d_{AB} is the optical path between points A and B in the laboratory frame. Parameter $\beta_{t,max}$ greatly decreases out of the interval $\pi - |\gamma| \geq \chi \geq |\gamma|$[22]. If $\beta_t < \beta_{t,max}$, each day close to times t_1 and t_2 the parameter S_{max} should jump from the quantum value $S_{max} = 0.2071$ to the Bell-Clauser-Horne-Shimony-Holt value $S_{max} \leq 0$.

Some experimental tests of the superluminal models have been performed but no violation of the quantum mechanics predictions has been found and only lower bounds $\beta_{t,max}$ for the velocity β_t of the superluminal signals have been established [22, 23, 26, 21]. In the previous Conference we reported some preliminary results of an improved experiment and here we report the final results[25] where the maximum detectable velocity of the superluminal communications is increased by about two orders of magnitude.

2. Experiment.

2.1. The experimental apparatus and procedures.

To increase the value of the maximum detectable adimensional velocity $\beta_{t,max}$, one has to make the experimental parameters ρ and $\delta_a t$ in Eq.(6) as small as possible. To get a small value of $\rho = {}^{d_{AB}}/_{\Delta d}$ we exploit the large distances available in the EGO gallery ($d_{AB} \approx 1200$ m) and we use an interferometric method to equalize the optical paths d_A and d_B within 5 μm($d_A \approx d_B \approx 600$ m). The final uncertainty $\Delta d \approx 0.22$ mm of the equality of the effective optical paths is due to many error sources including (in order of relevance) the finite thickness of the polarizing layers, the air dispersion, the uncertainty of the interferometric measurement, and the wavepackets width (the analysis of the error sources is given in[21]). To get a very small acquisition time $\delta_a t$ we develop a high intensity source of entangled photons and we minimize photons losses using a suitable optical apparatus and feedback procedures to stabilize the photons paths.

The experimental apparatus has been already described in detail in Ref. [21] and, thus, we show only a very simplified scheme in Figure3 where the four reference optical beams used to stabilize the propagation of the entangled photons beams and to equalize the optical paths are not shown. Figure3 a) shows the system used to generate a high intensity beam of entangled photons with a sufficient purity (Fidelity > 0.95). A pump laser beam is polarized at 45° with respect to the horizontal axis by a polarizer, passes through a $\lambda/2$ plate, a Babinet-Soleil compensator, a compensating plate C and is focused (spot diameter equal to 0.6 mm) at the center of two adjacent BBO nonlinear optical crystal plates (29.05° tilt angle, 0.56 mm-thickness) cut for type I phase matching [27]. The BBO plates have the optical axes lying in the horizontal and vertical plane, respectively. Down-conversion generates two outgoing beams of entangled photons at the average wavelength $\lambda = 2\,\lambda_p = 812.6\,\mathrm{nm}$ that mainly propagate at two symmetric angles (±2.42°) with respect to the normal to the crossed BBO plates. The entangled beams are deviated in opposite directions along the EGO gallery by two right-angle prisms, pass through the diaphragmed compensating plates C_A and C_B and exit from the central table. Plates C, C_A and C_B are used to increase the purity of the entangled state[28, 29, 30]. In such a way we get a high-intensity source of entangled photons ($N_{tot} \approx 23000$ coincidences/s) in an entangled state of sufficient purity. The beams outgoing from the central table are collected by two large achromatic lenses L_A and L_B having a 6 m focal length and a 15 cm diameter as shown in Figure3 b). The

IOP Conf. Series: Journal of Physics: Conf. Series **1275** (2019) 012035 doi:10.1088/1742-6596/1275/1/012035

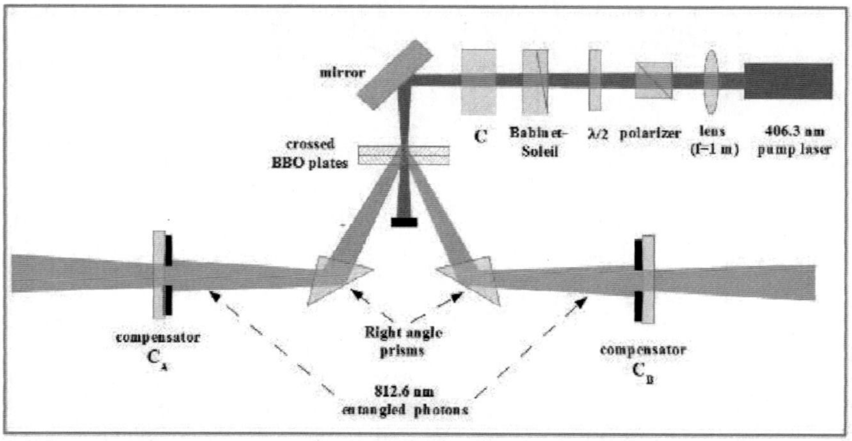

a)

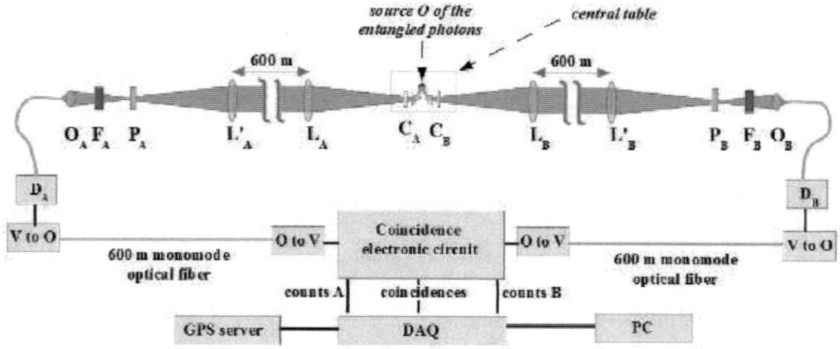

b)

Figure 3. **a)** The optical apparatus to generate high intensity beams of entangled photons. C, C_A and C_B are anisotropic compensator plates used to get a sufficient fidelity of the entangled state. The pump beam is focused at the centre of two crossed adjacent BBO plates (29.05° tilt angle) where entangled photons are generated and emitted at the average angles ±2.42° with respect to the pump laser beam. The whole system is thermostated within 0.1°C. **b)** The optical scheme used to produce a 1:1 real image of the source O of the entangled photons on polarizers P_A and P_B at about 600 m from O. The figure is not to scale. L_A, L_B, L'_A and L'_B are specially designed 15 cm-diameter achromatic lenses aligned along the EGO gallery and having a 6.00 m. P_A and P_B are absorption polarizing filters. F_A and F_B are sets of adjacent optical filters, O_A, O_B are optical lenses, D_A and D_B are photon counting detectors. V to O are electronic systems that transform the output voltage pulses into optical pulses, whilst O to V transform the optical pulses into voltage pulses. DAQ is a National Instruments CompactDAC that provides a real time acquisition of coincidences.

transmitted beams propagate along the gallery and impinge on two identical achromatic lenses L'_A and L'_B at a distance approximately 600 m from the source O in the figure. Two real 1:1 images (0.6 mm wide) of the source of the entangled photons are produced at the centers of the linear absorption polarizers P_A and P_B (Thorlab LPNIR). Four reference beams at wavelength λ_R =681 nm (not shown in the Figure but extensively described in Ref.[21]) are utilized to align the optical system, to equalize the optical paths and to keep under control the deviations of the entangled beams due to the air turbulence and to the vertical gradients of the air refractive index. In this way the spots of the entangled beams on the polarizers remained fixed within ± 0.4 mm during the whole 8 days measurement time. Each of the entangled photons beams outgoing from the two polarizers P_A or P_B passes through a filtering set (F_A or F_B in Figure3 b)) made by two long-pass optical filters ($\lambda_c = 765$ nm) that stop the reference 681-nm beams and a bandpass filter ($\lambda = 810\,\text{nm} \pm 20\,\text{nm}$). Then, each beam is focused by a system of optical lenses (O_A or O_B) on a multimode optical fiber having a large numerical aperture (0.39) and a core diameter of 200 μm connected to a Perkin Elmer photons-counter module. The output pulses of the photons counters are transformed into optical pulses (using the LCM155EW4932-64 modules of Nortel Networks) that propagate in two monomode optical fibers toward the central optical table where the entangled photons are generated. Finally, the optical pulses are transformed again into electric pulses and sent to an electronic coincidence circuit. An electronic counter connected to a National Instruments CompactDAQ counts the Alice pulses N_A, the Bob pulses N_B, and the coincidences pulses N.

2.2. The fast acquisition procedure

In the previous Conference DICE 2016[21] we reported some preliminary results where the measurements of the probabilities $P_i = P(\alpha_i, \xi_i)$ (i = 0-3) appearing in eq.(2) were made sequentially. The procedure needed some consecutive rotations of the polarizers before a single value of S_{max} was obtained leading to a long acquisition time $\delta_a t \approx$100 s for each measurement of S_{max}. In the final experiment the acquisition time $\delta_a t$ has been greatly reduced measuring each of the four probabilities P_i in successive daily experimental runs. The polarization angles α_i and ξ_i are set before the beginning of each i-th run made of 2^{19} successive acquisitions of the coincidences $N(\alpha_i, \xi_i)$ and the corresponding probabilities $P_i = N(\alpha_i, \xi_i)/N_{tot}$ are obtained. The procedure is fully automatized using a National Instruments CompactDAQ where a real time LABVIEW program runs. The real time procedure ensures a full continuity of the acquisitions and a constant acquisition time. The values of parameters ρ and $\delta_a t$ are

$$\rho = 1.83 \times 10^{-7}, \; \delta_a t = 0.247\, s. \qquad (8)$$

A Global Positioning System (GPS) Network Time Server (TM2000A) provides the actual Coordinated Universal Time (UTC) [31, 32] with an absolute accuracy better than 1 ms. Since the investigated phenomenon is related to the Earth's rotation, we synchronize the acquisitions with the Earth's rotation time $t = \theta \times 240\,\text{s}$ where θ is the Earth Rotation Angle(ERA) [33] expressed in degrees. The synchronization procedure is fully automatized and it ensures that each acquisition run starts at the UTC time that corresponds to the zero Greenwich Earth rotation time[25]. Four acquisition runs each with a duration of 36 Earth rotation hours (about 35 h, 54 min, and 7 s in the standard UTC time) are performed to obtain the four probabilities P_i with i = 0,1,2,3 that appear in Eq.(2). The spurious statistical coincidences are subtracted.

3. Results and conclusions.

Figures 4(a)-4(d) show the probabilities $P_i = P(\alpha_i, \xi_i) = {}^{N(\alpha_i,\xi_i)}/_{N_{tot}}$ obtained in the successive runs (i = 0,1,2,3). Due to the large number of points (2^{19}) acquired in each run, the experimental measured values are distributed within a black region having a thickness of several standard

IOP Conf. Series: Journal of Physics: Conf. Series **1275** (2019) 012035 doi:10.1088/1742-6596/1275/1/012035

deviations. Note the great stability of our equipment that ensured the full continuity of the measurements without any interruption during the whole measuring time (about 8 days). The solid green line represents the average value, and the green dotted line represents the value predicted by quantum mechanics for the pure maximally entangled state in Eq.(1) (fidelity $F = 1$). The discrepancy between the solid and dotted lines indicates that either our entangled state is not completely pure ($F \approx 0.95$) or some systematic noise is present. The experimental results in Figures 4(a)-4(d) do not exhibit any evident peak outgoing from the noise region and, thus, no apparent breakdown of the quantum mechanics predictions. Assuming that the investigated phenomenon has the same periodicity as the Earth rotation, one can calculate the S_{max} value at each Earth rotation time by substituting into the theoretical expression of S_{max} in Eq.(2) the P_i contributions of Figure 4 measured at the same Earth rotation time t in the successive runs.

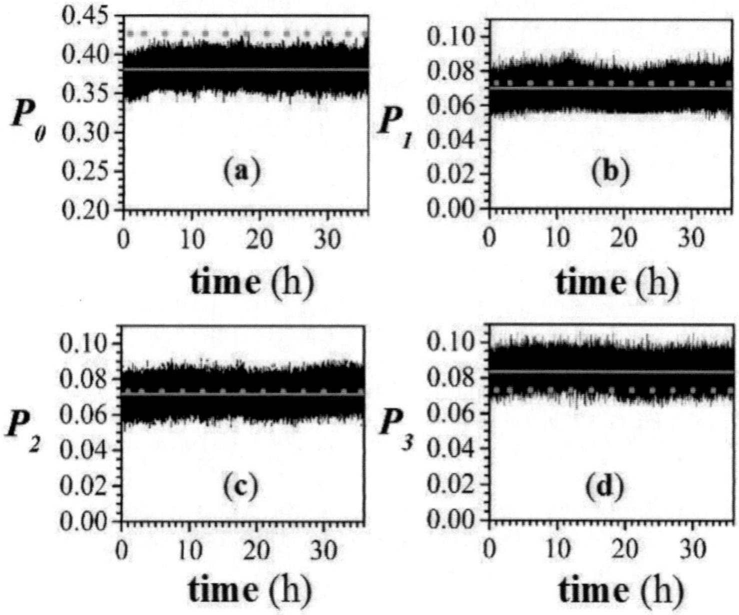

Figure 4. Probabilities ***a)*** P_0, ***b)*** P_1, ***c)*** P_2 and ***d)*** P_3 measured in successive runs versus the Greenwich earth rotation time. The 2^{19} values measured in each run are connected by straight lines leading to the resulting black regions in the figure. The acquisition time is $\delta_a t \approx 0.247$ s. The green solid lines represent the average values of the measured probabilities: $< P_0 >= 0.38087$, $< P_1 >= 0.06999$, $< P_2 >= 0.07187$, $< P_3 >= 0.08378$. The green dotted lines correspond to the values predicted by quantum mechanics for a pure maximally entangled state: $P_0 = 0.4267$, and $P_1 = P_2 = P_3 = 0.0732$.

With this procedure we get the results shown in Figure 5(a) (black region) and the corresponding frequency distribution ρ_0 shown in Figure 5(b) where black points represent the experimental results and the solid green line is the best fit with the Gaussian function $A\exp\left[-(S_{max}-<S_{max}>)^2/(2\sigma^2)\right]$ with standard deviation $\sigma = 0.01272$ and $< S_{max} >= 0.15523$. The green solid line in Figure 5(a) shows the average value $< S_{max} >$ and the green dotted line is the quantum mechanics value $S_{max} = 0.2071$ for the maximally entangled state in Eq.(1)($F = 1$). Unfortunately, the phenomenon that is investigated here has not exactly the same periodicity

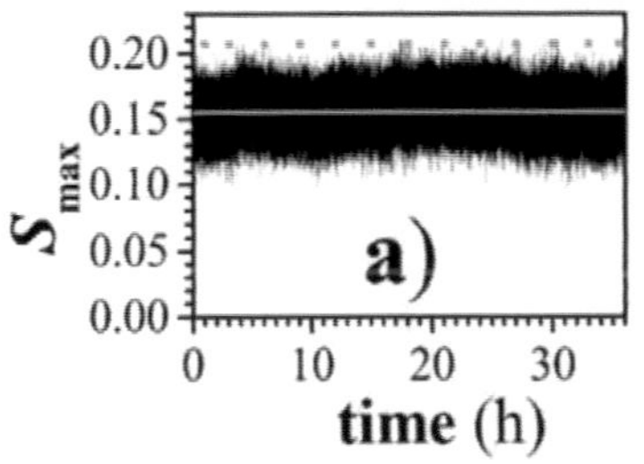

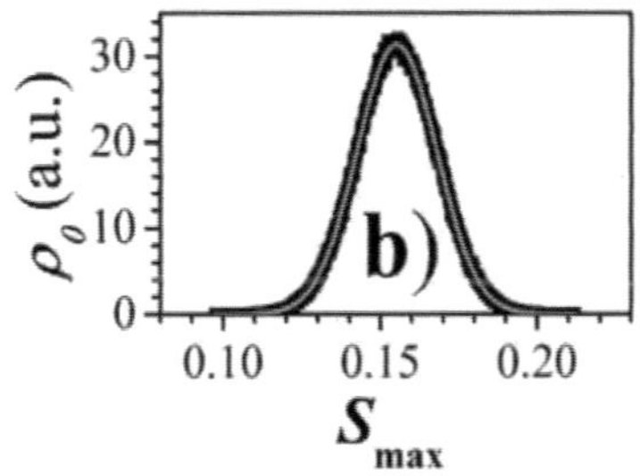

Figure 5. **a)** Parameter S_{max} versus the earth rotation time obtained using the relation $S_{max}(t) = P_0(t) - P_1(t) - P_2(t) - P_3(t)$. The green solid line is the average value $< S_{max} >= 0.15523$, while the green dotted line represents the quantum mechanics average value $< S_{max} >= 0.207$ predicted for the pure maximally entangled state in Eq.(1). **b)** Frequency distribution ρ_0 of the 2^{19} measured values of S_{max} in arbitrary units. The solid green curve is the Gaussian fit with standard deviation $\sigma = 0.01272$ and $< S_{max} >= 0.15523$.

as the Earth rotation because the breaking of quantum mechanics should occur when vector $\vec{\beta}$ becomes orthogonal to vector $\overrightarrow{AB}$. Due to the revolution motion of the earth around the sun and other minor motions (precession and nutation of the Earth's axis), vector $\vec{\beta}$ does not return exactly to the same orientation with respect to the laboratory frame after one Earth rotation day. Denote by t_{i1} and t_{i2} the successive two unknown times during the i-th measurement run $(i = 0-3)$ where the orthogonality condition $\vec{\beta} \cdot \overrightarrow{AB} = 0$ is satisfied and by $P_i(t_{ij})$, with $i = 0-3$ and $j = 1, 2$, the probabilities P_0, P_1, P_2 and P_3 measured at these times in the successive runs $(i = 0-3)$. If $\beta_t < \beta_{t,max}$, all or some of these probabilities should be different from the quantum mechanics values and the correlation parameters

$$S_{max}(j) = P_0(t_{0j}) - \sum_{i=1}^{3} P_i(t_{ij}) \qquad (9)$$

should satisfy the Bell-Clauser-Horne-Shimony-Holt inequality.

We do not know what are the times t_{ij} and we cannot calculate $S_{max}(j)$ but it is obvious that $S_{max}(j) \geq S = MIN(P_0) - MAX(P_1) - MAX(P_2) - MAX(P_3)$ where $MIN(P_i)$ and $MAX(P_i)$ are the absolute minimum and maximum measured values of P_i during each acquisition run, respectively. From the experimental data in Figure 4 we get $S = 0.04237$ and thus $S_{max}(j) \geq 0.04237 \approx 3.3\,\sigma$. The probability that a value of $S_{max}(j)$ lower than or equal to zero could be compatible with our measured values is $p \leq \frac{1}{2}\mathrm{erfc}\left[0.04237/(\sqrt{2}\sigma)\right] = 4.3 \cdot 10^{-4}$, where $\mathrm{erfc}(x)$ is the complementary error function. At least two breakdowns of S_{max} should occur during the time of 36 h and the probability P that both these breakdowns happen here is $P = p^2 \sim 2 \times 10^{-7}$. Then, we can conclude that there is no evidence of a breakdown of the quantum mechanics predictions and that the velocity of the superluminal signals has to be higher than the maximum measurable value $\beta_{t,max}$.

Substituting $\rho = 1.83 \times 10^{-7}$ and $\delta_a t = 0.247\,\mathrm{s}$ in Eq.(6) we obtain the lower bound $\beta_{t,max}$ of the superluminal velocities as a function of the unknown adimensional velocity β $(\beta < 1)$ of the preferred frame and of the angle χ. According to Eq.(6), $\beta_{t,max}$ reaches the maximum value at the borders $\chi = |\gamma|$ and $\chi = \pi - |\gamma|$ of the region accessible to our experiment and the minimum value at $\chi = \pi/2$. Curve-a in Figure 6 shows our $\beta_{t,max}$ versus the unknown adimensional velocity β of the preferred frame in the unfavorable case $\chi = \pi/2$. In the case of the Cosmic

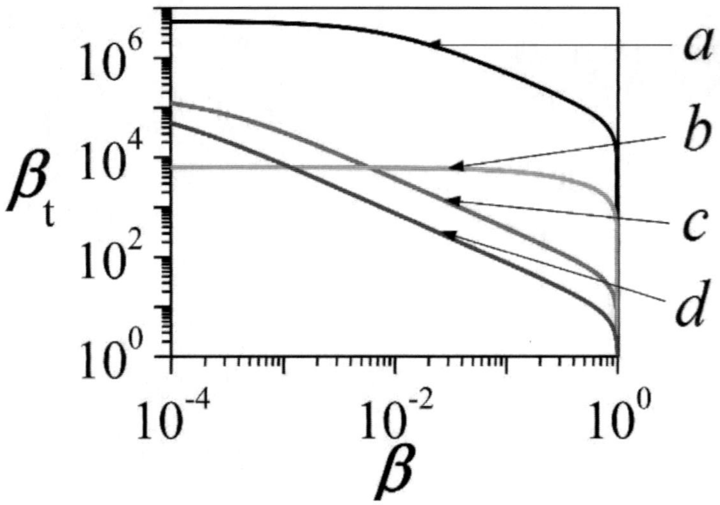

Figure 6. Curve a shows the $\beta_{t,max}$ values obtained in our experiment using Eq.(6) ($\rho = 1.83 \times 10^{-7}$, $\delta_a t = 0.247\ s$, and $\gamma = 18°$) versus the unknown adimensional velocity β of the preferred frame for $\chi = \pi/2$; curve b is the result obtained in Ref. [23] ($\rho = 1.6 \times 10^{-4}$, $\delta_a t = 4\ s$ and $\gamma = 0°$); curve c is the result obtained in Ref. [22] ($\rho = 5.4 \times 10^{-6}$, $\delta_a t = 360\ s$ and $\gamma = 5.9°$) and curve d is the result obtained in Ref. [26] ($\rho = 7.3 \times 10^{-6}$, $\delta_a t = 1800\ s$ and $\gamma = 0°$). In the case of curve d also the locality and the freedom-of-choice loopholes were addressed.

Microwave Background frame ($\beta \approx 10^{-3}, \chi \approx 97°$) we get $\beta_{t,max} \approx 5 \times 10^6$. The other curves represent the experimental values of $\beta_{t,max}$ obtained in the previous experiments [22, 23, 26].

In conclusion, no evidence for the occurrence of superluminal communications has been found in our experiment and a higher lower bound for the speed of the superluminal has been established. We remind here that, due to the inclination of our apparatus with respect to the east-west direction, we are not sensitive to about a 5% of all the possible orientations of the velocity vector of the preferred frame and that our experimental results have been obtained under the assumption that the collapse of the polarization state occurs at the absorption polarizers.

Acknowledgements

We acknowledge the Fondazione Pisa for financial support. We acknowledge the EGO and VIRGO staff that made possible the experiment and, in particular, F. Ferrini, F. Carbognani, A. Paoli and C.Fabozzi. We are especially grateful to M. Bianucci (Pisa Physics Department) and to S. Cortese (VIRGO) for their invaluable and continuous contribution to the solution of many electronic and informatic problems. Finally, we acknowledge N. Gisin for helpful comments and T. Faetti for helpful suggestions on real-time procedures.

References

[1] Einstein A, Podolsky B and Rosen N 1935 *Phys. Rev.* **47**(10) 777–780
[2] Bell J S 1964 *Physics* **1** 195–200
[3] Clauser J F, Horne M A, AShimony and Holt R 1969 *Phys. Rev. Lett.* **23**(2) 880
[4] Clauser J F and Horne M A 1974 *Phys. Rev. D* **10**(2) 526

IOP Conf. Series: Journal of Physics: Conf. Series **1275** (2019) 012035 doi:10.1088/1742-6596/1275/1/012035

[5] Aspect A, Dalibard J and Roger G 1982 *Phys. Rev. Lett.* **49**(25) 1804–1807

[6] Hensen B *et al.* 2015 *Nature* **526** 682

[7] Giustina M *et al.* 2015 *Phys. Rev. Lett.* **115**(25) 250401

[8] Shalm L K *et al.* 2015 *Phys. Rev. Lett.* **115**(25) 250402

[9] Rosenfeld W, Burchardt D, Garthoff R, Redeker K, Ortegel N, Rau M and Weinfurter H 2017 *Phys. Rev. Lett.* **119**(1) 010402

[10] Newton, letter to R Bentley, 1692/3,189R447, f 6, Trinity College Library, Cambridge, UK

[11] Abbot B P *et al.* 2016 *Phys. Rev. Lett.* **116**(25) 061102

[12] Davies P C W and Brown J R 1986 *The ghost in the atom: a discussion of the mysteries of quantum physics* canto ed. ed (Cambridge; New York: Cambridge University Press)

[13] Cocciaro B 2013 *Physics Essays* **26** 531–547

[14] Cocciaro B 2015 *Journal of Physics: Conference Series* **626** 012054

[15] Eberhard P H 1989 A realistic model for quantum theory with a locality property *Quantum theory and pictures of reality: foundations, interpretations, and new aspects* ed Schommers W (Berlin; New York: Springer-Verlag) pp 169–2016

[16] Bohm D and Hiley B J 1993 *The undivided universe: an ontological interpretation of quantum mechanics* (Routledge,London,)

[17] Gisin N 2014 *Quantum Correlations in Newtonian Space and Time: Faster than Light Communication or Nonlocality* (Milano: Springer Milan) pp 185–203

[18] Bancal J, Pironio S, Acín A, Liang Y, Scarani V and Gisin N 2012 *Nat. Phys.* **8** 867–870

[19] Cocciaro B, Faetti S and Fronzoni L 2018 *Phys. Rev. A* **97**(5) 052124

[20] Aspect A 2002 Bell's theorem: The naive view of an experimentalist *Quantum (Un)speakables: From Bell to Quantum Information* ed Bertlmann R A and Zeilinger A (Springer)

[21] Cocciaro B, Faetti S and Fronzoni L 2017 *Journal of Physics: Conference Series* **880** 012036

[22] Salart D, Baas A, Branciard C, Gisin N and Zbinden H 2008 *Nature* **454** 861–864

[23] Cocciaro B, Faetti S and Fronzoni L 2011 *Phys. Lett.* A **375** 379–384

[24] European Gravitational Observatory,https://wwwego-gwit/ URL `https://www.ego-gw.it`

[25] Barnea T J, Bancal J D, Liang Y C and Gisin N 2013 *Phys. Rev. A* **88**(2) 022123

[26] Yin J *et al.* 2013 *Phys. Rev. Lett.* **110**(26) 260407

[27] Kwiat P G, Waks E, White A G, Appelbaum I and Eberhard P H 1999 *Phys. Rev. A* **60**(2) R773–R776

[28] Altepeter J, Jeffrey E and Kwiat P 2005 *Opt. Express* **13** 8951–8959

[29] Akselrod G M, Altepeter J B, Jeffrey E R and Kwiat P G 2007 *Opt. Express* **15** 5260–5261

[30] Rangarajan R, Goggin M and Kwiat P 2009 *Opt. Express* **17** 18920–18933

[31] PK Seidelmann B Guinot L D 1992 *Explanatory Supplement to the Astronomical Almanac,cp.2* (US Naval Observatory,University Science books,Mill Walley,CA)

[32] National Radio Astronomy Observatory, Rick Fisher homepage, `https://www.cv.nrao.edu/~rfisher/Ephemerides/times.htm`

[33] I R 2012 *A Dictionary of Astronomy,2nd.ed* (Oxford: Oxford University Pressl)

IOP Conf. Series: Journal of Physics: Conf. Series **1275** (2019) 012036 doi:10.1088/1742-6596/1275/1/012036

Spekkens' Toy Model, Finite Field Quantum Mechanics, and the Role of Linearity

Lay Nam Chang, Djordje Minic, and Tatsu Takeuchi[1]

Department of Physics, Virginia Tech, 850 West Campus Drive, Blacksburg 24061 USA

E-mail: laynam@vt.edu, dminic@vt.edu, takeuchi@vt.edu

Abstract. We map Spekkens' toy model to a quantum mechanics defined over the finite field $\mathbb{F}_5$. This allows us to define arbitrary linear combinations of the epistemic states in the model. For Spekkens' elementary system with only $2^2 = 4$ ontic states, the mapping is exact and the two models agree completely. However, for a pair of elementary systems there exist interesting differences between the entangled states of the two models.

1. Introduction

The interpretation of Quantum Mechanics (QM) remains a subject of intense investigation and heated debate, as well as the source of puzzlement and intellectual discomfort for many. To deepen our understanding of QM, various ontological models which posit some type of underlying physical "reality" to QM have been proposed. An illuminating discussion on ontological models of QM can be found in Harrigan and Spekkens [1], in which the models are classified into two broad categories: ψ-ontic, and ψ-epistemic. ψ-ontic refers to models in which different wave-functions ψ correspond to different physical states of the system itself, while ψ-epistemic refers to models in which different ψ's correspond to different states of our (limited) knowledge about the system. Both types of models have their pros and cons, and are often unable to completely reproduce the predictions of canonical QM. Nevertheless, scrutinizing where they succeed and where they fail will help us identify how or whether the models can be improved, clarify the limitations of our "classical" thinking, and guide us to think more "quantum-ly" about QM.

In this note, we will look at a simple ψ-epsitemic ontological model proposed by Spekkens in Ref. [2]. The most elementary version of the model succeeds in reproducing the predictions of canonical QM of spin restricted to the six states of $|0\rangle$ (spin up), $|1\rangle$ (spin down), $|\pm\rangle = \frac{1}{\sqrt{2}}(|0\rangle \pm |1\rangle)$, and $|\pm i\rangle = \frac{1}{\sqrt{2}}(|0\rangle \pm i|1\rangle)$. Furthermore, the two-"spin" system in Spekkens' model possesses the analog of entanglement. However, one aspect of QM which is missing from Spekkens' model is *linearity*. Linearity, or the superposition principle, is what leads to the various mysteries of canonical QM, so it is a property one would like to impart on any model of QM. We argue that this can be done to Spekkens' model by mapping it to a QM defined over the finite field $\mathbb{F}_5$ [3, 4, 5, 6, 7].

2. Spekkens' Toy Model

We begin with a quick review of Spekkens' toy model [2].

[1] Presenting Author

Content from this work may be used under the terms of the Creative Commons Attribution 3.0 licence. Any further distribution of this work must maintain attribution to the author(s) and the title of the work, journal citation and DOI.
Published under licence by IOP Publishing Ltd

2.1. Ontic & Epsitemic States

Spekkens' *knowledge balance principle* states that for a system which requires $2N$ bits of information to completely specify, only N bits of information may be known about the system at any time. The state of the system is called the *ontic state*, while the state of our (limited) knowledge about the system is called the *epistemic state*.

The most elementary system onto which Spekkens' knowledge balance principle could be applied would be such that require 2 bits of information to completely specify. Such an elementary system would have $2^2 = 4$ ontic states which can be represented graphically as

$$\boxed{1}\boxed{2}\boxed{3}\boxed{4} \tag{1}$$

The elementary system inhabits one and only one of these four ontic states at any time.

The epistemic states each represent 1 bit of knowledge about this system, which would only allow us to narrow down the possible ontic states to two: $a \vee b$ where $a, b \in \{1, 2, 3, 4\}$ and $a \neq b$. $a \vee b$ and $b \vee a$ represent the same knowledge, so we will use the convention that $a < b$ for all epistemic states $a \vee b$.[2] Therefore, an elementary system possesses six epistemic states and these can be represented by shading the boxes where the actual ontic state may be:

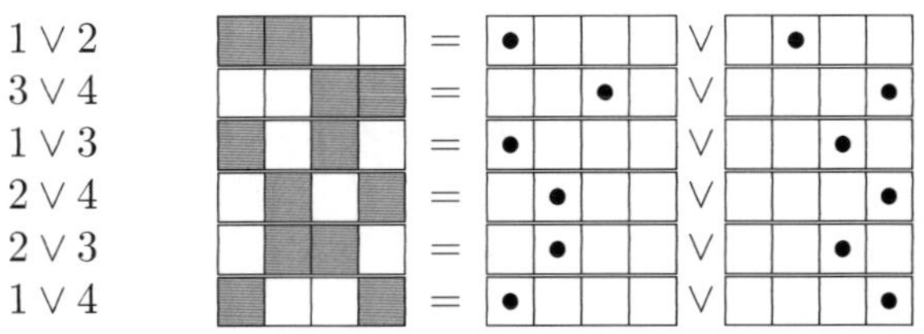

(2)

For each epistemic state $a \vee b$, we assume that the probability that the actual ontic state is a or b is $\frac{1}{2}$ each.

The subset of ontic states $\{a, b\}$ that appear in the epistemic state $a \vee b$ is called the *ontic support* of the epistemic state. When the ontic supports of two epistemic states are disjoint, then the two epistemic states are said to be *disjoint*. $1 \vee 2$ and $3 \vee 4$, $1 \vee 3$ and $2 \vee 4$, $2 \vee 3$ and $1 \vee 4$ are disjoint pairs of epistemic states. We will also call one member of these disjoint pairs the *disjoint* of the other.

2.2. Measurements, Observables, & Eigenstates

The only allowed *measurements* in Spekkens' model are yes-no inquiries on whether one of the $a \vee b$ statements is true or not. For instance, if the system is in ontic state 1, the result of a measurement of $1 \vee 2$ would result in "yes," while a measurement of $3 \vee 4$ would result in "no." It is clear that an "yes" answer for "$1 \vee 2$?" is equivalent to a "no" answer for "$3 \vee 4$?", yielding the same information on whether the system is in state $1 \vee 2$ or $3 \vee 4$. Each measurement provides information on whether the system is in one epistemic state or its disjoint. Therefore, *observables* can be defined as pairs of disjoint epistemic states, which are $X = \{1 \vee 3, 2 \vee 4\}$, $Y = \{2 \vee 3, 1 \vee 4\}$, and $Z = \{1 \vee 2, 3 \vee 4\}$.

The three observables can be represented graphically as

$$\begin{array}{ccc} X & : & \boxed{+}\boxed{-}\boxed{+}\boxed{-} \\ Y & : & \boxed{-}\boxed{+}\boxed{+}\boxed{-} \\ Z & : & \boxed{+}\boxed{+}\boxed{-}\boxed{-} \end{array} \tag{3}$$

where we denote the two possible outcomes of a measurement of each observable for a given underlying ontic state with $+1$ and -1. The measurement of Z, for instance, on the epistemic

[2] The symbol $\vee$ represents an OR.

IOP Conf. Series: Journal of Physics: Conf. Series **1275** (2019) 012036 doi:10.1088/1742-6596/1275/1/012036

state $1 \vee 2$ always yields $+1$, while on the epistemic state $3 \vee 4$ always yields -1. On the other hand, the measurement of X on the epistemic state $1 \vee 2$, which is in ontic state 1 with probability $\frac{1}{2}$ and in ontic state 2 with probability $\frac{1}{2}$, will yield ± 1 with probability $\frac{1}{2}$ each.

Let us call the epistemic states comprising each observable the *eigenstates* of the observable. For instance, $1 \vee 3$ and $2 \vee 4$ are the eigenstates of X. A measurement of X on $1 \vee 3$ will always yield $+1$ while a measurement of X on $2 \vee 4$ will always yield -1.

2.3. Ontic State Update/Epistemic State Collapse

The knowledge balance principle demands that repeated measurements cannot be allowed to narrow down the possible underlying ontic state. The ontic state must therefore be perturbed by each measurement and update accordingly. For instance, a $+1$ result on the measurement of X tells us that the system is in ontic state 1 or 3 with $\frac{1}{2}$ probability each. If the actual ontic state were 1, and did not update as a result of the measurement of X, a followup measurement of Z would yield $+1$ with probability 1 and allow us to deduce what the ontic state was. Therefore, the measurement of X must knock the ontic state 1 into ontic state 3 with $\frac{1}{2}$ probability so that while another measurement of X will yield $+1$ as before, a measurement of Z (or Y) will yield ± 1 with probability $\frac{1}{2}$ each.

A measurement will also update the epistemic state. A measurement of X yielding $+1$ will update the epistemic state to $1 \vee 3$ while a -1 outcome will update it to $2 \vee 4$. In other words, a measurement of an observable will *collapse* the epistemic state to one of the observable's eigenstates.

2.4. Correspondence to Qubit/Spin States

Spekkens' six epistemic states introduced above correspond nicely to qubit/spin states as

$$\begin{array}{lcl} \{1 \vee 2, 3 \vee 4\} & \Leftrightarrow & \{|\,0\,\rangle, |\,1\,\rangle\}\,, \\ \{1 \vee 3, 2 \vee 4\} & \Leftrightarrow & \{|+\rangle, |-\rangle\}\,, \\ \{2 \vee 3, 1 \vee 4\} & \Leftrightarrow & \{|+i\,\rangle, |-i\,\rangle\}\,, \end{array} \tag{4}$$

where $|\,0\,\rangle$ (spin up) and $|\,1\,\rangle$ (spin down) are the computational basis states of the qubit (eigenstates of $\hat{Z} = \sigma_z$) while

$$|\pm\rangle \;=\; \frac{1}{\sqrt{2}}\Big(|\,0\,\rangle \pm |\,1\,\rangle\Big)\,, \qquad |\pm i\,\rangle \;=\; \frac{1}{\sqrt{2}}\Big(|\,0\,\rangle \pm i\,|\,1\,\rangle\Big)\,. \tag{5}$$

The states $|\pm\rangle$ and $|\pm i\,\rangle$ are respectively the eigenstates of $\hat{X} = \sigma_x$ and $\hat{Y} = \sigma_y$ with eigenvalues ± 1. Given this correspondence, it is straightforward to show that the measurement outcomes of X, Y, Z on Spekkens' six epistemic states correspond exactly to the measurement outcomes of $\hat{X}$, $\hat{Y}$, $\hat{Z}$ on the six qubit/spin states. In other words, Spekkens' toy model in its most elementary form succeeds in reproducing the predictions of a spin-$\frac{1}{2}$ quantum mechanical system albeit restricted to the six states listed here.

2.5. Entanglement of two "Qubits/Spins"

A pair of elementary systems, each with $2^2 = 4$ ontic states, will have $2^4 = 16$ ontic states. A complete specification of the ontic state will require 4 bits of information, whereas the knowledge balance principle demands that only 2 bits of information may be known about the system at any time. Furthermore, the knowledge balance principle continues to apply to each of the 2-bit subsystems. This restricts the epistemic states of the elementary-system pair to the types depicted graphically in Figure 1: there are $6^2 = 36$ product states, and $4! = 24$ entangled states. The entangled states must have one shaded box in each row and each column. Each measurement of either system 1 or system 2 will update the ontic and epistemic states as in the one-system case. Any measurement on an entangled state will collapse it to a product state.

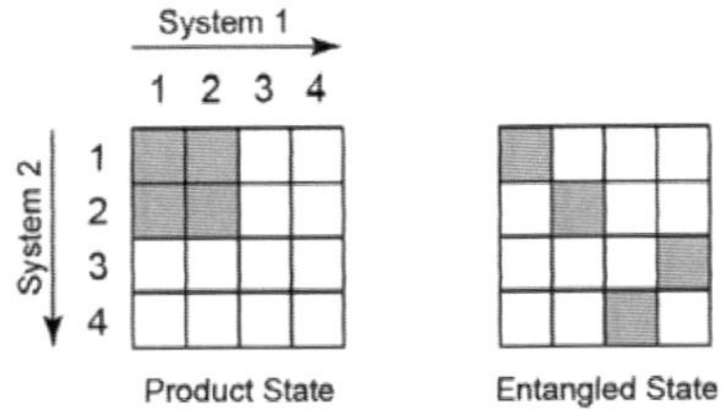

Figure 1. Examples of product and entangled states in Spekkens' model.

2.6. Linearity

In Ref. [2], Spekkens defines four types of "coherent superpositions" as

$$\begin{aligned}(a\vee b)+_1(c\vee d) &= a\vee c\,,\\ (a\vee b)+_2(c\vee d) &= b\vee d\,,\\ (a\vee b)+_3(c\vee d) &= b\vee c\,,\\ (a\vee b)+_4(c\vee d) &= a\vee d\,.\end{aligned}\tag{6}$$

If applied to $(1\vee 2)$ and $(3\vee 4)$, we have

$$\begin{aligned}(1\vee 2)+_1(3\vee 4) &= 1\vee 3\,,\\ (1\vee 2)+_2(3\vee 4) &= 2\vee 4\,,\\ (1\vee 2)+_3(3\vee 4) &= 2\vee 3\,,\\ (1\vee 2)+_4(3\vee 4) &= 1\vee 4\,,\end{aligned}\tag{7}$$

in exact parallel to Eq. (5). Thus, these four different sums are analogous to addition but with different phases multiplying the second state of the sum.

However, note that these sums are only defined for pairs of disjoint epistemic states. The sum of $1\vee 2$ and $1\vee 3$, for instance, is not defined. Moreover, the sums are dependent on the ordering of the ontic-state labels. One could try to identify $a\vee b$ and $b\vee a$ with the same qubit/spin state but with different phases, but it can be shown that such an identification would not work.

Since one cannot define arbitrary sums of the six qubit/spin states $|0\rangle$, $|1\rangle$, $|\pm\rangle$, and $|\pm i\rangle$ in canonical QM either without going outside of this set, one could argue that the inability to define sums of arbitrary pairs of epistemic states in Spekkens' model is not really a problem. Nevertheless, the lack of linearity suggests that an important aspect of QM has not been completely captured within the model.

3. Finite Field Quantum Mechanics

The missing linearity of Spekkens' toy model can be supplied by mapping it to a QM defined over the finite field $\mathbb{F}_5$. There are multiple ways to define QM on finite fields as discussed in Refs. [3, 4, 5, 6, 7]. The formalism presented here is closest to Refs. [3, 4].

The finite field $\mathbb{F}_5=\{\underline{0},\pm\underline{1},\pm\underline{2}\}$ consists of five elements with addition and multiplication defined modulo 5. We underline elements of $\mathbb{F}_5$ to distinguish them from elements of $\mathbb{Z}$. Note that $\underline{2}^2=(-\underline{2})^2=-\underline{1}$, that is:

$$\pm\underline{2} = \pm\sqrt{-\underline{1}}\,.\tag{8}$$

In analogy to the qubit/spin states in canonical QM, which are vectors in the complex projective space $CP^1=P^1(\mathbb{C})$ (the Bloch sphere), the "qubit/spin" states in $\mathbb{F}_5$ QM are vectors in the $\mathbb{F}_5$ projective space $P^1(\mathbb{F}_5)$. This is basically the 2D vector space $\mathbb{F}_5^2$ with vectors differing by

IOP Conf. Series: Journal of Physics: Conf. Series **1275** (2019) 012036 doi:10.1088/1742-6596/1275/1/012036

overall non-zero multiplicative constants identified as representing the same state. Consequently, in $P^1(\mathbb{F}_5)$ there are $(5^2-1)/4 = 6$ inequivalent states and these can be taken to be

$$|a\rangle = \begin{bmatrix} \underline{1} \\ \underline{0} \end{bmatrix}, \ |b\rangle = \begin{bmatrix} \underline{0} \\ \underline{1} \end{bmatrix}, \ |c\rangle = \begin{bmatrix} \underline{1} \\ \underline{1} \end{bmatrix}, \ |d\rangle = \begin{bmatrix} \underline{1} \\ -\underline{1} \end{bmatrix}, \ |e\rangle = \begin{bmatrix} \underline{1} \\ \underline{2} \end{bmatrix}, \ |f\rangle = \begin{bmatrix} \underline{1} \\ -\underline{2} \end{bmatrix}. \tag{9}$$

The inequivalent dual-vectors can be taken to be

$$\begin{aligned} \langle a| &= \begin{bmatrix} \underline{1} & \underline{0} \end{bmatrix}, \quad \langle b| = \begin{bmatrix} \underline{0} & \underline{1} \end{bmatrix}, \quad \langle c| = \begin{bmatrix} -\underline{2} & -\underline{2} \end{bmatrix}, \\ \langle d| &= \begin{bmatrix} -\underline{2} & \underline{2} \end{bmatrix}, \quad \langle e| = \begin{bmatrix} -\underline{2} & -\underline{1} \end{bmatrix}, \quad \langle f| = \begin{bmatrix} -\underline{2} & \underline{1} \end{bmatrix}. \end{aligned} \tag{10}$$

The actions of these dual-vectors on the vectors are:

	$\|a\rangle$	$\|b\rangle$	$\|c\rangle$	$\|d\rangle$	$\|e\rangle$	$\|f\rangle$
$\langle a\|$	$\underline{1}$	$\underline{0}$	$\underline{1}$	$\underline{1}$	$\underline{1}$	$\underline{1}$
$\langle b\|$	$\underline{0}$	$\underline{1}$	$\underline{1}$	$-\underline{1}$	$\underline{2}$	$-\underline{2}$
$\langle c\|$	$-\underline{2}$	$-\underline{2}$	$\underline{1}$	$\underline{0}$	$-\underline{1}$	$\underline{2}$
$\langle d\|$	$-\underline{2}$	$\underline{2}$	$\underline{0}$	$\underline{1}$	$\underline{2}$	$-\underline{1}$
$\langle e\|$	$-\underline{2}$	$-\underline{1}$	$\underline{2}$	$-\underline{1}$	$\underline{1}$	$\underline{0}$
$\langle f\|$	$-\underline{2}$	$\underline{1}$	$-\underline{1}$	$\underline{2}$	$\underline{0}$	$\underline{1}$

Observables are defined by pairs of dual-vectors:

$$\underline{X} = \{\langle c|, \langle d|\}\,, \qquad \underline{Y} = \{\langle e|, \langle f|\}\,, \qquad \underline{Z} = \{\langle a|, \langle b|\}\,, \tag{11}$$

where the first (second) dual-vector of the pair corresponds to the outcome $+1$ (-1). The probability of obtaining outcome $+1$ (-1) when observable $O = \{\langle x|, \langle y|\}$ is measured on state $|\psi\rangle$ is defined as

$$P(+1, O, \psi) = \frac{|\langle x|\psi\rangle|^2}{|\langle x|\psi\rangle|^2 + |\langle y|\psi\rangle|^2}\,, \qquad P(-1, O, \psi) = \frac{|\langle y|\psi\rangle|^2}{|\langle x|\psi\rangle|^2 + |\langle y|\psi\rangle|^2}\,, \tag{12}$$

where the absolute value of $\underline{k} \in \mathbb{F}_5$ is defined to be $|\underline{k}| = 0$ if $\underline{k} = \underline{0}$, and $|\underline{k}| = 1$ otherwise. It is straightforward to demonstrate that the predictions of $\mathbb{F}_5$ QM agrees with those of canonical QM and Spekkens' toy model via the correspondence

$$\begin{array}{lllll} \{1 \vee 2, 3 \vee 4\} & \Leftrightarrow & \{|\,0\,\rangle, |\,1\,\rangle\} & \Leftrightarrow & \{|a\rangle, |b\rangle\}\,, \\ \{1 \vee 3, 2 \vee 4\} & \Leftrightarrow & \{|+\rangle, |-\rangle\} & \Leftrightarrow & \{|c\rangle, |d\rangle\}\,, \\ \{2 \vee 3, 1 \vee 4\} & \Leftrightarrow & \{|+i\,\rangle, |-i\,\rangle\} & \Leftrightarrow & \{|e\rangle, |f\rangle\}\,. \end{array} \tag{13}$$

However, in $\mathbb{F}_5$ QM, the sums of arbitrary pairs of states are well defined. In addition to

$$\begin{aligned} |c\rangle &= |a\rangle + |b\rangle\,, \\ |d\rangle &= |a\rangle - |b\rangle\,, \\ |e\rangle &= |a\rangle + \underline{2}\,|b\rangle = |a\rangle + \sqrt{-\underline{1}}\,|b\rangle\,, \\ |f\rangle &= |a\rangle - \underline{2}\,|b\rangle = |a\rangle - \sqrt{-\underline{1}}\,|b\rangle\,, \end{aligned} \tag{14}$$

we also have, for instance:

$$\begin{aligned} |a\rangle + |c\rangle &= \underline{2}|f\rangle\,, \\ |a\rangle - |c\rangle &= -|b\rangle\,, \\ |a\rangle + \underline{2}\,|c\rangle &= -\underline{2}|d\rangle\,, \\ |a\rangle - \underline{2}\,|c\rangle &= -|e\rangle\,, \end{aligned} \tag{15}$$

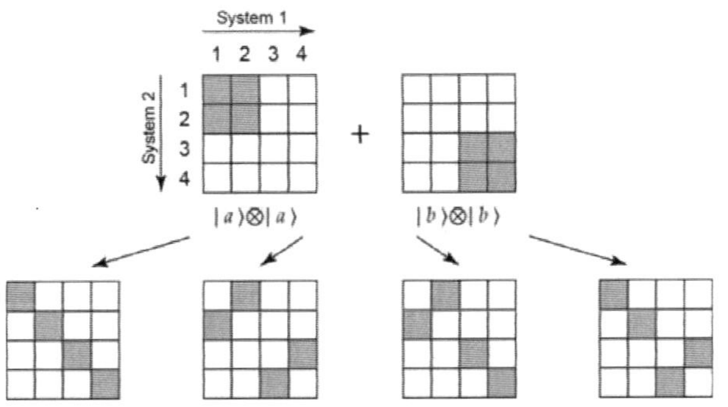

Figure 2. Possible "coherent superpositions" of disjoint epistemic states.

which corresponds to

$$\begin{aligned}(1 \vee 2) +_1 (1 \vee 3) &= (1 \vee 4)\,, \\ (1 \vee 2) +_2 (1 \vee 3) &= (3 \vee 4)\,, \\ (1 \vee 2) +_3 (1 \vee 3) &= (2 \vee 4)\,, \\ (1 \vee 2) +_4 (1 \vee 3) &= (2 \vee 3)\,. \end{aligned} \tag{16}$$

In this fashion, one can introduce sums of non-disjoint epistemic states into Spekkens' toy model via the correspondence Eq. (13).

4. Entanglement in $\mathbb{F}_5$ QM and in Spekkens' Toy Model

Since $\mathbb{F}_5$ QM is equipped with a vector space, the two-qubit/spin system state space is obtained by tensoring the one-qubit/spin system state space. This gives us $P^3(\mathbb{F}_5)$, namely $\mathbb{F}_5^4$ with vectors differing by overall non-zero multiplicative constants identified. The number of inequivalent states in this space is $(5^4 - 1)/4 = 156$, of which $6^2 = 36$ are product states and $156 - 36 = 120$ are entangled. All 120 entangled states can be expressed as superpositions of two product states without a common factor.

There is clearly a mismatch in the number of entangled states between $\mathbb{F}_5$ QM and Spekkens' model. Thus, imposing linearity on Spekkens' model seems to require many more entangled epistemic states to be included in the model than were originally allowed. So which ones are they? And can we prevent the inclusion of such states from violating the knowledge balance principle, perhaps by modifying the ontic-state update rule upon measurement?

Before pursuing these questions further, let us first identify which of the 120 $\mathbb{F}_5$ entangled states correspond to the 24 Spekkens' entangled states. Do all Spekkens' entangled states have an $\mathbb{F}_5$ analog? Considering how "coherent superpositions" were originally defined for the elementary system, one guess would be that the Spekkens' entangled states would result from sums of disjoint product states as depicted in Figure 2. Indeed, for the disjoint pair $\{|a\rangle \otimes |a\rangle, |b\rangle \otimes |b\rangle\}$ one finds

$$\begin{aligned} |a\rangle\otimes|a\rangle + |b\rangle\otimes|b\rangle &= -2(|c\rangle\otimes|c\rangle + |d\rangle\otimes|d\rangle) &&= -2(|e\rangle\otimes|f\rangle + |f\rangle\otimes|e\rangle)\,, \\ |a\rangle\otimes|a\rangle - |b\rangle\otimes|b\rangle &= -2(|c\rangle\otimes|d\rangle + |d\rangle\otimes|c\rangle) &&= -2(|e\rangle\otimes|e\rangle + |f\rangle\otimes|f\rangle)\,, \\ |a\rangle\otimes|a\rangle + \underline{2}|b\rangle\otimes|b\rangle &= -2(|c\rangle\otimes|e\rangle + |d\rangle\otimes|f\rangle) &&= -2(|e\rangle\otimes|c\rangle + |f\rangle\otimes|d\rangle)\,, \\ |a\rangle\otimes|a\rangle - \underline{2}|b\rangle\otimes|b\rangle &= -2(|c\rangle\otimes|f\rangle + |d\rangle\otimes|e\rangle) &&= -2(|f\rangle\otimes|c\rangle + |e\rangle\otimes|d\rangle)\,. \end{aligned} \tag{17}$$

Comparing the result of various measurements on these states with those of Spekkens' states, one finds that $|a\rangle \otimes |a\rangle \pm \underline{2}\,|b\rangle \otimes |b\rangle$ have Spekkens' analogs, cf. Figure 3, but $|a\rangle \otimes |a\rangle \pm |b\rangle \otimes |b\rangle$

IOP Conf. Series: Journal of Physics: Conf. Series **1275** (2019) 012036 doi:10.1088/1742-6596/1275/1/012036

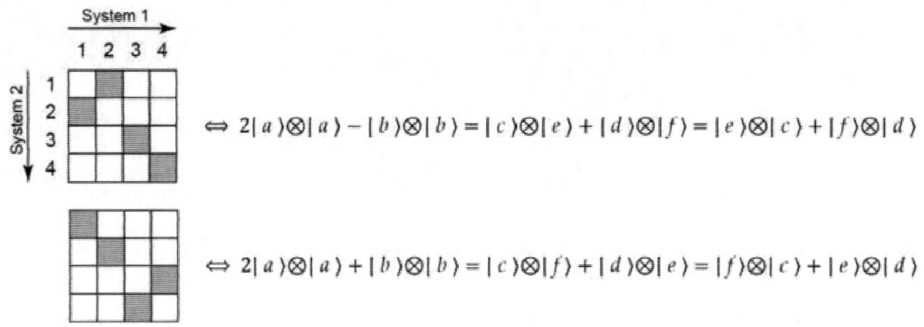

Figure 3. Spekkens' entangled states and the corresponding $\mathbb{F}_5$ entangled states.

do not. For instance, the Spekkens' state

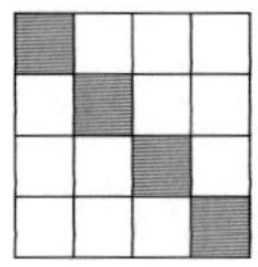
(18)

will collapse to the product state $|a\rangle \otimes |a\rangle$ or $|b\rangle \otimes |b\rangle$ if Z is measured on system 1 or 2, $|c\rangle \otimes |c\rangle$ or $|d\rangle \otimes |d\rangle$ if X is measured on system 1 or 2, and $|e\rangle \otimes |e\rangle$ or $|f\rangle \otimes |f\rangle$ if Y is measured on system 1 or 2. However, as we have demonstrated in Eq. (17), no superposition of $|a\rangle \otimes |a\rangle$ and $|b\rangle \otimes |b\rangle$ is also a superposition of $|c\rangle \otimes |c\rangle$ and $|d\rangle \otimes |d\rangle$, and a superposition of $|e\rangle \otimes |e\rangle$ and $|f\rangle \otimes |f\rangle$ simultaneously. Indeed, of the $4! = 24$ entangled states of Spekkens' model, 12 do not have $\mathbb{F}_5$ analogs that give the same predictions, cf. Figure 4.

The reason for this mismatch can be traced to the fact that the correspondence of Eq. (13) does not respect the symmetries of the respective models. Spekken's toy model is invariant under any relabelling of the four ontic states. In other words, it has an S_4 symmetry. On the other hand, the qubit/spin states of both canonical QM and $\mathbb{F}_5$ QM transform under 2D

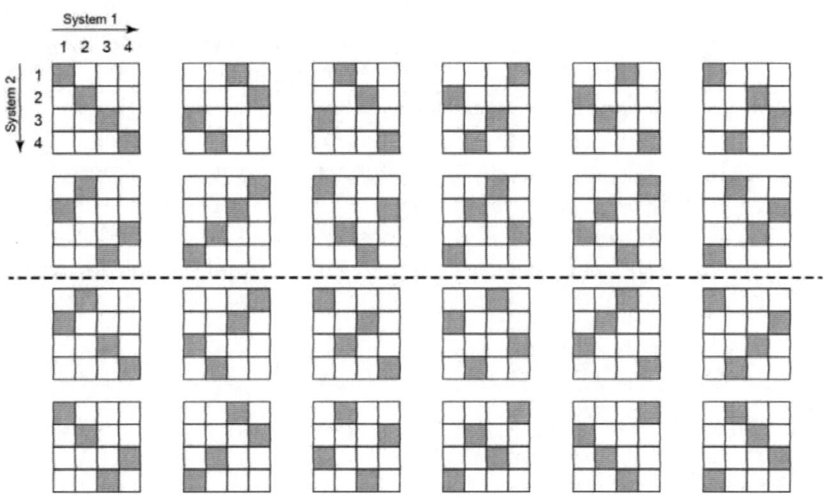

Figure 4. The $4! = 24$ entangled states of the Spekkens' model. The 12 states above the dashed line do not have $\mathbb{F}_5$ analogs.

projective representations of $SO(3)$. Though S_4 can be embedded into $SO(3)$ [8], Eq. (13) demands that all the even permutations of S_4 (the A_4 subgroup) be mapped to rotations, but all the odd permutations be mapped to reflections, which would involve a complex conjugation. In fact, this mismatch was already evident in Ref. [2] where it was noted that the coherent superpositions of Eq. (6) do not produce the expected results for sums of $1 \vee 3$ and $2 \vee 4$.

Thus, we have identified a rather serious problem if one wishes to utilize Spekkens' model as a model of qubits/spin. While the linearity requirement demanding more entangled epistemic states to be added to the mix is interesting in itself, the symmetry mismatch suggests we could be comparing apples and oranges. One possible way out may be to embed the S_4 permutations into 4D projective representations of $SO(3)$ in canonical and $\mathbb{F}_5$ QM. This will allow the odd permutations to be also represented as rotations, but it will also double the number of states available leading to a different set of complications. These, and other questions will be addressed in more detail in a forthcoming paper.

Acknowledgments

Takeuchi would like to thank the Institute of Advanced Studies at Nanyang Technological University, Singapore, for their hospitality during the QM90 Conference (23-26 January 2017), and Sixia Yu for bringing Spekkens' model to his attention. We would also like to thank Chen Sun and Matthew Henriques for their respective contributions to this project.

References

[1] Harrigan N and Spekkens R W 2010 *Found. Phys.* **40** 125-157
[2] Spekkens R W 2007 *Phys. Rev.* A **75** 032110
[3] Chang L N, Lewis Z, Minic D and Takeuchi T 2013 *Mod. Phys. Lett.* B **27** 1350064
[4] Chang L N, Lewis Z, Minic D and Takeuchi T 2013 *J. Phys.* A **46** 065304
[5] Chang L N, Lewis Z, Minic D and Takeuchi T 2013 *J. Phys.* A **46** 485306
[6] Chang L N, Lewis Z, Minic D and Takeuchi T 2014 *Int. J. Mod. Phys.* A **29** 1430006
[7] Chang L N, Lewis Z, Minic D and Takeuchi T 2014 *J. Phys.* A **47** 405304
[8] Klein F 1956 *Lectures on the Icosahedron* (New York, Dover Publications)

IOP Conf. Series: Journal of Physics: Conf. Series **1275** (2019) 012037 doi:10.1088/1742-6596/1275/1/012037

No time for isolated clocks

K L H Bryan$^{(1)}$, A J M Medved$^{(1,2)}$

(1) Department of Physics and Electronics, Rhodes University, Grahamstown, South Africa
(2) National Institute for Theoretical Physics (NITheP), Western Cape 7602, South Africa

E-mail: g08b1231@gmail.com, j.medved@ru.ac.za

Abstract. The Wheeler-DeWitt equation provides a model of the Universe as a timeless, isolated system. Page and Wootters developed the conditional probability interpretation (CPI) to account for the experience of time within this timeless model by identifying part of the Universe as the clock. In resolving criticisms of the CPI, some investigators have since concluded that the clock must be completely isolated from the remainder of the Universe. This isolation has also been used to reinforce the conclusion that time must be an illusion. However, some interactions must exist between all subsystems of the Universe as gravitational effects, however weak, cannot be shielded. Here, we present the results of an investigation which used a toy model to consider the implications of including interactions. Although counter intuitive, their inclusion is seen to directly correlate with the accuracy of measurements of the clock state. The case is then made that, if interactions are indeed a necessary feature of the CPI, the viability of treating time as an illusion needs to be reconsidered. We conclude that the experience of time can be interpreted as a real phenomenon, which naturally incorporates an arrow of time, once interactions are taken seriously in the CPI paradigm.

1. Introduction

Time is often taken for granted but, yet, can be a sensitive subject in some fields of physics. Although the concept of time is used within the mathematical framework of physical theories, its description in different theories does not always agree. The most prominent case in point is the disagreement between quantum mechanics and general relativity. Both of these versions of time could be expected to emerge from a single description of time which is contained in a more fundamental theory of quantum gravity. If so, the more fundamental theory requires a single description that is apparently consistent with (at least) two contradictory versions of emergent time. This issue encapsulates one aspect of the infamous 'problem of time'. For a detailed overview of both this aspect of the problem and various others, see, for example, [1].

A mathematical framework that nicely incorporates both quantum mechanics and general relativity, albeit superficially, is provided by the Wheeler–DeWitt equation [2]. The equation, which is meant to model the quantum nature of the Universe, is expressible as

$$\hat{H}\,|\Psi\rangle \;=\; 0\,, \tag{1}$$

where $|\Psi\rangle$ can be regarded as the wavefunction of the Universe and the Hamiltonian operator $\hat{H}$ is a suitably quantized version of the Hamiltonian constraint from general relativity. The right-hand side of equation (1) implies that no energy is being exchanged with systems that are (somehow) external to the Universe; in other words, the Universe is, at least effectively, an isolated system.

Content from this work may be used under the terms of the Creative Commons Attribution 3.0 licence. Any further distribution of this work must maintain attribution to the author(s) and the title of the work, journal citation and DOI.
Published under licence by IOP Publishing Ltd

IOP Conf. Series: Journal of Physics: Conf. Series **1275** (2019) 012037 doi:10.1088/1742-6596/1275/1/012037

Irrespective of the legitimacy of the Wheeler–DeWitt equation, an isolated model for the Universe has important ramifications for the interpretation of time. Equation (1) does, however, help to clarify the point by showing that any physical state is annihilated by the Hamiltonian, implying that the Universe as a whole cannot be functionally dependent on time. The conclusion is one of 'timelessness': The Universe cannot experience any change in its quantum state, even in the Schrödinger picture.

This timelessness is problematic: If the Universe as a whole includes no notion of time evolution, then why should any smaller subsystem? A solution, which is often referred to as the conditional probability interpretation (CPI) of time, was offered quite a while ago by Page and Wootters [3]. Given that the Universe can be divided into a pair of maximally entangled, weakly interacting subsystems, then one of these – which is typically identified as the clock C — will necessarily provide a time parameter for the remaining subsystem, which is normally referred to as R. This emergent notion of time circumvents the need for the 'global time' parameter t that is commonly adopted in standard quantum mechanics, and so a means for describing the evolution of subsystems can be recovered.

The CPI has been confronted with a variety of criticisms. The most prominent of these was the concern that, while one pair of states for C and R could be described at some given time, this description could not be extended in a meaningful way so as to connect a pair of states at one time to a second pair at a later time [4]. This matter has since been resolved by some different approaches; see, for instance, [5] and [6].

Another concern is the so-called clock-ambiguity problem [7]. To explain, the CPI dynamics will in general be arbitrary, as one can expect a multitude of different ways to partition the Universe into a suitable pair of subsystems and that each such partition would provide its own dynamical description. A proposed resolution to alleviate this concern is to restrict the choice of partitions to subsystems which are perfectly isolated from one another [8]. This is a stronger constraint than that prescribed by the CPI, which stipulates that C and R be weakly interacting systems but need not be cut off entirely. The proposal works because, given that C and R are indeed isolated, any other partition is related to the first by a local unitary transformation if and only if the resultant subsystems are similarly isolated. There is, however, a cost: An isolated clock leads to the interpretation that any experience of time within the Universe is an illusion. We elaborate on this reasoning in Section 3.

It is our contention that interactions between C and R have to be present and then duly accounted for. Doing so, one finds the (perhaps) counter-intuitive outcome that the inclusion of interactions actually improves the accuracy of the clock. Moreover, and as just alluded to, this inclusion is necessary if one is to recover a non-illusionary description of the experience of time. But, irrespective of these perks, the choice of non-interacting subsystems is not even feasible as a strict matter of principle. This is because there is no way to shield them against the effects of gravity.

In Section 2, we present the results of an investigation into the effectiveness of interacting clock systems. This is followed in Section 3 by an assessment of the interpretive issues. We conclude in Section 4 with a brief summary and comment on the outlook.

2. A realistic and interacting clock

A refined version of the CPI [5] (also see [9]) can be applied to physically realistic clocks and their respective environments [10]. [1] In [11], we employed this refined CPI to study just such a clock system, while adopting the viewpoint that the clock and its environment are toy models of a partitioned form of the isolated Universe. This analysis revealed that the idealized case of

[1] With regard to the application, an essential refinement is the replacement of the abstract CPI time parameter — the eigenvalue of the conjugate to the clock's reduced Hamiltonian — with a more tangible parameter such as the position of the clock's center of mass.

an isolated clock does not make for a particularly accurate timepiece. The rest of this section recalls this investigation and our conclusions thereof.

For purposes of simplicity, we adopted the assumption that the clock C was much smaller than the environment (or remaining system) R. As such, the effects of the interactions — which are regarded as weak as mandated by the CPI — would have no bearing on R and could be directly incorporated into the dynamics of C. [2]

As a clock, we selected a coherent-state description of a (weakly) damped harmonic oscillator [13, 14], with the position of the oscillator playing the role of the time. This would provide a time parameter for R in an analogous fashion to how the position of the hands on a clock indicates the time.

We were able to find a Gaussian expression for the probability of finding the clock in a particular position x at some value of abstract clock time n (see below). The associated uncertainty, which could be identified as the width of the Gaussian and regarded as a measure of the inaccuracy of the clock, was found to be of the form

$$\delta = e^{-rn/2}\sqrt{\frac{\hbar}{2m\omega}}\,, \tag{2}$$

where r represents the strength of the damping (*i.e.*, the oscillations are damped by a factor of e^{-rn}), m is the mass and ω is the undamped frequency.

As for the abstract clock time n, it initially appears as the external time parameter (*i.e.*, the t in standard quantum mechanics) but ultimately plays the role of an integration variable in the refined CPI formalism. As far as R is concerned, the time parameter is the clock position x, as n never appears in any of the final (conditional) probability expressions and so cannot be viewed as a measurable time. Rather, n acts as a 'synchronization' tool, whose purpose is to keep track of how C and R are evolving in relation to one another.

In terms of this damped harmonic oscillator, n is also relevant to the condition of weak damping because of the oscillator's damping factor of e^{-rn}. Hence, n must be small but, on the other hand, one would like n to be as large as possible so as to ensure a lengthy 'run time' for the clock. (This does not refer to a physical run time but rather to a breakdown of the formalism as the clock progresses through more and more states.) These two conflicting considerations leads one to an operational range of $0 < rn < 1$, where the right-hand side really means a number on the order of unity.

From an inspection of equation (2), it becomes clear that the uncertainty is minimized when the damping parameter r approaches infinity. However, the previous constraint restricts the size of r with respect to n. Meaning that, in order for the clock to function effectively for as long as possible, one arrives at the choice of $r = \frac{1}{n}$. This is a essentially a sweet spot for which the operation of the clock is optimized.

What is evident is that the optimal setting for r is not zero, as this would have maximized the uncertainty in equation (2). In short, a completely isolated clock may be ideal (to some) but it is hardly optimal. It appears that the condition of a weakly interacting clock must be taken more literally: the interactions must be *weak* but never removed completely. This conclusion is further supported by an investigation in [15], which produced a very similar result for the 'more quantum' case of an atomic clock. This leaves us with the task of reassessing the role of interactions in the original (Universal) version of the CPI.

[2] Investigations into the inclusion of interactions can be found elsewhere; *e.g.*, [12], where the interactions have been formally inserted into $\hat{H}$.

IOP Conf. Series: Journal of Physics: Conf. Series **1275** (2019) 012037 doi:10.1088/1742-6596/1275/1/012037

3. Implications of interactions

The discussion to follow is based on parts of [11, 15], where more thorough treatments can be found. We once again regard C and R as the maximally entangled subsystems of a partitioned form of the isolated Universe.

Let us suppose that C and R are completely isolated as is apparently needed to resolve the clock-ambiguity problem as in [8]. The basic problem now is that this isolated clock must be as timeless as the isolated Universe. This is because of the elimination of any energy transfer to or from the clock, ensuring that its state cannot change, even in the Schrödinger picture, and so the clock can have no inherent sense of time. Nor can it gain a sense of time from its surroundings inasmuch as there are no interactions between C and R.

Let us clarify a somewhat subtle point: In standard quantum mechanics, a time-independent system, even if completely isolated, is presumed to exist within a larger system that does indeed experience time. In this manner, the state of the smaller system can 'inherit' time, allowing for features such as a time-dependent phase. The CPI prohibits this feature out of hand, as the larger system, the isolated Universe, cannot experience time, leaving no sense of time for the clock to inherit.

There is then no inherent 'flow of time' for the isolated clock, nor for its similarly isolated complement R. Rather than each subsystem tracking the other as they run through a succession of synchronized pairs of states, the interpretation now becomes that every state of C exists both 'simultaneously' and 'eternally' as does every state of R. [3] This is much the same as a series of pages making up a book because every such page must still exist alongside all of the others. A momentary state of C forms a part of the evolution of C but, as each other state exists alongside of it, the notion of change must be an illusion: No system ever actually 'moves' from one state to another. One is thus left to conclude that the flow of time, as it emerges in the isolated Universe with isolated subsystems, must be an illusionary effect.

Given that real change is indeed a mirage, why is it that the experience of time proceeds in such an orderly fashion, whereby the states of any given system follow a particular order along a distinct arrow towards the future? An argument in favor of the timeless interpretation that addresses this question can be found in [8], where an *ad hoc* memory system is introduced in order to explain the ordered series of moments that one experiences. With reference to the book analogy, the memory system essentially provides a way to gather up loose pages and arrange them into a story. However, since every possible page is always represented, what one really has is an infinite pile of loose pages rather than a single bound book. A given path through the pile would represent the 'experience' of a single book.

Various concerns with this perspective have been raised by the current authors in [17]. Our focus at present will be on the difficulties that non-trivial or irreversible interactions pose when time is regarded as illusional. Fortuitously, the inclusion of such interactions can resolve the clock-ambiguity problem by singling out thermodynamic time and, in this way, altogether nullify what is (in our opinion) the main motivation for an illusionary interpretation of time. This alternative resolution of the clock-ambiguity problem is somewhat involved; see [17] for a detailed account.

Let us suppose that irreversible interactions are now included and consider the state of some system just before and after an interaction takes place. Can this inclusion be incorporated into the above interpretation, written onto the respective pages of the book as it were? The answer is not necessarily. When interpreted as an intervention of one system on another, an interaction requires an associated process of change. This change can be thought of as one system 'learning' about another by transferring information about their respective states through an appropriate conduit. The time-is-an-illusion viewpoint, on the other hand, implies that each state (or page)

[3] This picture is essentially describing a quantum version of the Block Universe. For further discussion, see, *e.g.*, [16].

IOP Conf. Series: Journal of Physics: Conf. Series **1275** (2019) 012037 doi:10.1088/1742-6596/1275/1/012037

of each system (or book) is locked into an 'eternal moment' and, as such, each state is itself essentially an independent system. If one state were somehow removed or changed, there is nothing that is built into this picture to suggest that the remaining states would be affected. In other words, two systems cannot interact in a meaningful way because they are already locked into particular states at whatever 'moment' is meant to represent that just after the interaction.

Irreversible interactions also imply an arrow of time. As would clearly be the case for the damped clock in the previous section, there is a continual increase in entropy as a system interacts with its environment. This is another feature which appears to contradict the notion that time is an illusion, as the lack of real change is also suggestive of complete reversibility. One is supposed to get around this contradiction through yet another *ad hoc* assumption; namely that each state of the system contains the information, or 'memory', which would enable it to create the illusion of an arrow of time. [4] This picture necessarily requires each state to 'know' enough about the other states so as to produce the illusion of a consistent arrow of time. It remains an open question as to how such a convoluted picture could come to fruition or, indeed, why is it that Nature would so diligently construct such an illusion.

Rather than trying to square these circles, we are inclined to take irreversible interactions seriously and arrive at a different interpretation of the CPI: The experience of time as an emergent feature of the isolated Universe is a real phenomenon. Given that one system has the opportunity to learn about others through a process of change, then each state of the system is directly responsible for the state which follows it. There is then a contiguity in time such that one can identify the final state of a system as having evolved from the initial one.

It is the arrow of time, which is now aimed in the direction of increasing entropy, enables one to differentiate the past from the future. If the experience of time is real, this distinction between past and future can be expected to have physical meaning. Systems would be 'transforming' from their present states into future states, while their memories (*cf*, footnote 4) would be referring to states from the past. What is implicit in this interpretation is a postulate of causality. Given this, the intervention of one system on another would be the cause and the resulting transformation would be the effect. In spite of its fundamental nature, causality has fallen out favor in some areas of physics but is currently experiencing something of a revival. See, for example, [18].

There are still many issues to resolve under the interpretation that time is real. To name just a few, a definitive description of *how* change occurs, why there should be a low entropy point in the past to evolve from and to clarify the role of causality. Nonetheless, these are not insurmountable challenges; certainly no more so than those facing the interpretation that time is an illusion.

4. Discussion

The presence of interactions is required in principle when dealing with gravitational systems. Our quantitative results in Section 2 reinforce the notion that interacting clocks are a crucial aspect of the CPI and should not be dismissed from the outset. Qualitative arguments, such as those in Section 3, lend further credence to this sentiment.

Even if the experience of time is regarded as real, as would be necessary to accommodate the inclusion of interactions, there are still many interpretive issues to be resolved. For example, causality plays a prominent role in the experience of time but its origin is not really understood. We discuss this sensitive topic in more detail in [17]. In essence, whatever it is that places the future events before us and the past ones behind us must either be put in by hand or else deferred to a more fundamental treatment. But, regardless of how causality does fit into the grand scheme of things, our conclusion remains that interactions are a key ingredient if emergent time is to be regarded as a real phenomenon in an isolated model of the Universe.

[4] Memory does not have to imply consciousness in this context. For instance, a sea-side rock can carry the memory of the tide via its pattern of erosion.

Acknowledgments

The research of AJMM received support from an NRF Incentive Funding Grant 85353 and NRF Competitive Programme Grant 93595. KLHB is supported by an NRF bursary through Competitive Programme Grant 93595 and a Henderson Scholarship from Rhodes University. This work is based on the research also supported in part by the National Research Foundation of South Africa (Grant Numbers: 111616).

References

[1] Isham C 1993 *Integrable systems, quantum groups, and quantum field theories* (Springer) pp 157–287
[2] DeWitt B 1967 *Phys. Rev. D* **160** 1113
[3] Page D and William W 1983 *Phys. Rev.* D **27** 2885
[4] Kuchar K 1992 *4th Canadian conference on general relatibity and relativistic astrophysics* p 211
[5] Dolby C 2004 The conditional probability interpretation of the hamiltonian constraint (*Preprint* gr-qc/0406034)
[6] Giovannetti V, Lloyd S and Maccone L 2015 *Phys. Rev. D* **92** 045033
[7] Albrecth A and Iglesias A 2008 *Phys. Rev. D* **77** 063506
[8] Marletto C and Vedral V 2017 *Phys. Rev. D* **95** 043510
[9] Gambini R, Porto R and Pullin J 2004 *New Journ. of Phys.* **6** 45
[10] Corbin V and Cornish N 2009 *Found. of Phys.* **39** 474–485
[11] Bryan K and Medved A 2018 *Found. of Phys.* **48** 48–59
[12] Smith A and Ahmadi M 2017 Quantizing time: Interacting clocks and systems (*Preprint* quant-ph/1712.00081)
[13] Yeon K, Um C and George T 1987 *Phys. Rev. A* **36** 5287
[14] Um C, Yeon K and George T 2002 *Phys. Rep.* **362** 63–192
[15] Bryan K and Medved A 2018 Requiem for an ideal clock (*Preprint* quant-ph/1803.02045)
[16] Dainton B 2016 *Time and space* (Routledge)
[17] Bryan K and Medved A 2018 The problem with 'the problem of time' (*Preprint* quant-ph/1811.09660)
[18] DAriano G 2018 *Phil. Trans. R. Soc. A* **376**

IOP Conf. Series: Journal of Physics: Conf. Series **1275** (2019) 012038 doi:10.1088/1742-6596/1275/1/012038

Quantum statistics in Bohmian trajectory gravity

T C Andersen

NSCIR - 046516 Meaford, Ontario N4L 1W7, Canada

E-mail: tandersen@nscir.ca

Abstract. The recent experimental proposals by Bose et al. and Marletto et al. (BMV) outline a way to test for the quantum nature of gravity by measuring gravitationally induced differential phase accumulation over the superposed paths of two $\sim 10^{-14}kg$ masses. These authors outline the expected outcome of these experiments for semi-classical, quantum gravity and collapse models. It is found that both semi-classical and collapse models predict a lack of entanglement in the experimental results. This work predicts the outcome of the BMV experiment in Bohmian trajectory gravity - where classical gravity is assumed to couple to the particle configuration in each Bohmian path, as opposed to semi-classical gravity where gravity couples to the expectation value of the wave function, or of quantized gravity, where the gravitational field is itself in a quantum superposition. In the case of the BMV experiment, Bohmian trajectory gravity predicts that there will be quantum entanglement. This is surprising as the gravitational field is treated classically. A discussion of how Bohmian trajectory gravity can induce quantum entanglement for a non superposed gravitational field is put forward.

1. Introduction

In papers published in 2017, Bose et al. [1], and Marletto and Vedral [2] (BMV) describe an experimental proposal to test for quantum gravity . Christodoulou and Rovelli [3] summarize these papers and clarify the quantum-gravity interaction in these proposals. Christodoulou and Rovelli develop a simple common language to describe the experiment, using an assumption that while perhaps not possible in an actual experiment, simplifies discussion and understanding of the effect. The notation and terminology of the Christodoulou and Rovelli paper will be used here.

We will first review the BMV experimental setup, defining base states and terminology. Then the quantized gravity, semi-classical and collapse predictions will be recovered.

A section follows on how Bohmian trajectory gravity would be interpreted for the BMV experiment. The experimental predictions for Bohmian quantum gravity are then explored.

2. The BMV experiment

The in the BMV experimental proposal, Bose [1] and Marletto and Vedral [2] propose, in different experimental setups to measure the same effect, namely the expected behavior of quantum systems when particles interact only via mutual gravitational attraction. Christodoulou and Rovelli [3] point out that the effect can be understood as the time dilation effects of one mass on the other while each particle is in a spatial superposition. Although the time dilation effect is tiny, the experimental effect is amplified as the experimenters can take advantage of the extremely high Compton frequency of the massive (in the quantum particle sense) particle.

Content from this work may be used under the terms of the Creative Commons Attribution 3.0 licence. Any further distribution of this work must maintain attribution to the author(s) and the title of the work, journal citation and DOI.
Published under licence by IOP Publishing Ltd

The experimenters only need to measure the *difference* in phase accumulation along different paths, and once that phase difference is $\sim \pi$ the experimenters will detect that the state has become quantum entangled. For details on how a heavy quantum particle with spin ready for a Stern-Gerlach device can be made and kept in a superposition, see Bose et al. [1]

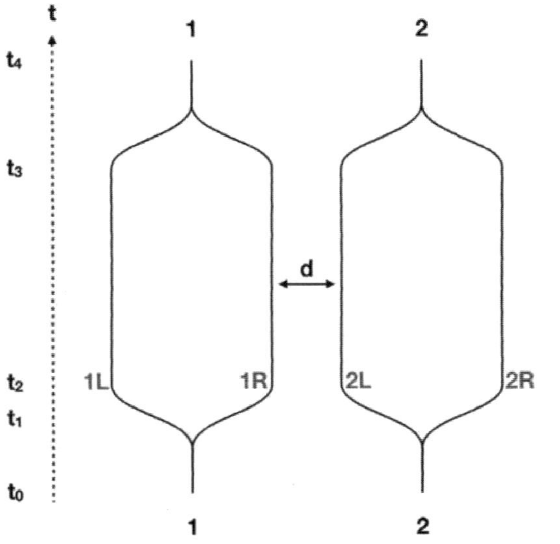

Figure 1. Particles 1 and 2 start off in the spin up $|u\rangle$ (defined as +z) direction, and are spatially split into superposed $|L\rangle+|R\rangle$ states by a Stern-Gerlach mechanism. The particles fly through the experiment at the same time, and then through a second Stern-Gerlach where they are brought back into up/down state. If there is negligible gravitational interaction, (for example with particles such as silver atoms) the final state will be $|u\rangle\otimes|u\rangle$ (times an overall phase factor) and the particles are not entangled.

3. Measuring quantum entanglement

We wish to determine the quantum entanglement of the final state in the quantum gravity models considered in this paper. In some cases such as semi-classical and quantized gravity case the final state is pure, while in others a mixed state is the result. For example a mixed state will result from environmental collapse. Quantum entanglement cannot be measured using partial traces and von Neumann entropy on mixed states [4]. Rather, a well chosen entanglement witness is a good measure. We will in this paper use a witness similar to that of Bose et al. [1]. Witness:

$$\mathcal{W} = \langle \sigma_x^{(1)} \otimes \sigma_z^{(2)} \rangle + \langle \sigma_y^{(1)} \otimes \sigma_y^{(2)} \rangle \tag{1}$$

which shows entanglement when

$$\mathcal{W} \leq 0. \tag{2}$$

We use the convention

$$u = \begin{bmatrix} 1 \\ 0 \end{bmatrix}, \quad d = \begin{bmatrix} 0 \\ 1 \end{bmatrix}, \tag{3}$$

$$\sigma_x = \begin{bmatrix} 0 & 1 \\ 1 & 0 \end{bmatrix}, \quad \sigma_y = \begin{bmatrix} 0 & -i \\ i & 0 \end{bmatrix}, \quad \sigma_z = \begin{bmatrix} 1 & 0 \\ 0 & -1 \end{bmatrix}. \tag{4}$$

IOP Conf. Series: Journal of Physics: Conf. Series **1275** (2019) 012038 doi:10.1088/1742-6596/1275/1/012038

So that the operator W of the witness $\mathcal{W}$ is

$$W = \sigma_x^{(1)} \otimes \sigma_z^{(2)} + \sigma_y^{(1)} \otimes \sigma_y^{(2)} = \begin{bmatrix} 0 & 0 & 1 & -1 \\ 0 & 0 & 1 & -1 \\ 1 & 1 & 0 & 0 \\ -1 & -1 & 0 & 0 \end{bmatrix}. \tag{5}$$

See for instance chapter 2 of *Entanglement and decoherence: foundations and modern trends* [5]. The witness $\mathcal{W}$ (1) is thus in the range [-2, 2].

The authors also found the online entanglement calculator [6] useful. The mathematics package Maple was employed to handle density matrix and witness calculations.

In this paper we will always express the final state with a density matrix, even if the final state is pure. This simplifies the analysis. The density matrix will be given in the $|u\rangle, |d\rangle$ basis. With a density matrix ρ, the expression to evaluate the entanglement witness is simply:

$$\mathcal{W}(\rho) = Tr(\rho W). \tag{6}$$

4. Quantized gravity prediction

In quantized gravity approaches to quantum gravity, the gravitational field can/will be in a superposition. Theories like canonical, Wheeler-DeWitt, and loop quantum gravity all predict that the gravitational field will be in a superposition, and that at low energy, (i.e. perturbatively) will behave the same way, as a quantized gravity. Thus the BMV experiment will not be able to distinguish between these quantized gravity models. Rather the expectation is that the results of the BMV experiment will show that gravity is quantized.

The initial state is, with u representing a $+z$ spin direction,

$$|\Psi_{t_0}\rangle = |u_1\rangle \otimes |u_2\rangle \otimes |g\rangle. \tag{7}$$

Following Christodoulou and Rovelli [3]:

> We assume that $|g\rangle$ belongs to a Hilbert space that contains semi-classical states that approximate classical geometries g, *but also linear superposition of these.* This is the key property needed to derive the BMV effect.

Each particle is passed through a typical Stern-Gerlach device, where L and R represent spins in the $\pm x$ direction.

$$|u_1\rangle = \frac{|L_1\rangle + |R_1\rangle}{\sqrt{2}} \quad , \quad |u_2\rangle = \frac{|L_2\rangle + |R_2\rangle}{\sqrt{2}}. \tag{8}$$

So that at t_1 we have

$$|\Psi_{t_1}\rangle = \frac{1}{2}\Big(|L_1\rangle + |R_1\rangle) \otimes (|L_2\rangle + |R_2\rangle\Big) \otimes |g\rangle \tag{9}$$

$$= \frac{1}{2}\Big(|LL\rangle + |RR\rangle + |LR\rangle + |RL\rangle\Big) \otimes |g\rangle. \tag{10}$$

At this point the gravitational field is felt differently by all 4 branches, and the state quickly becomes

$$|\Psi_{t_2}\rangle = \frac{1}{2}\Big(|LL\rangle \otimes |g_{d_{LL}}\rangle + |RR\rangle \otimes |g_{d_{RR}}\rangle + |LR\rangle \otimes |g_{d_{LR}}\rangle + |RL\rangle \otimes |g_{d_{RL}}\rangle\Big). \tag{11}$$

Christodoulou and Rovelli [3] assume the experiment is set up such that only $|g_{d_{RL}}\rangle$ is of physical consequence, the other gravitational states being too weak to affect their respective branches.

The state evolves over time to t_3 where the additional phase $e^{i\frac{Gm^2t}{\hbar d}}$ appears on the closest approach branch, while the huge overall phase evolution on all 4 branches is $e^{-i\frac{mc^2t}{\hbar}}$, where $t = t_3 - t_2$ is the time that the particles spend gravitationally interacting:

$$|\Psi_{t_3}\rangle = \frac{e^{-i\frac{mc^2t}{\hbar}}}{2}\Big(|LL\, g_{d_{LL}}\rangle + |RR\, g_{d_{RR}}\rangle + |LR\, g_{d_{LR}}\rangle + e^{i\frac{Gm^2t}{\hbar d}}|RL\, g_{d_{RL}}\rangle\Big). \tag{12}$$

Dropping the overall phase factor as it does not change the physical state, and at t_4 recombining the branches with another Stern-Gerlach device, the gravitational field is again in one state, with $t = t_3 - t_2$ chosen such that $e^{i\frac{Gm^2t}{\hbar d}} = -1$, the state becomes:

$$|\Psi_{t_4}\rangle_{qg} = \frac{1}{2}\Big(|LL\rangle + |RR\rangle + |LR\rangle - |RL\rangle\Big) \otimes |g\rangle. \tag{13}$$

Christodoulou and Rovelli[3] trace out the gravity and final state of one particle, arriving at the conclusion that the state (13) is a maximally entangled pure state.

Here we arrive at the same conclusion by first calculating the density matrix for the final state (13):

$$\rho_{qg} = \frac{1}{4}\begin{bmatrix} 1 & 1 & -1 & 1 \\ 1 & 1 & -1 & 1 \\ -1 & -1 & 1 & -1 \\ 1 & 1 & -1 & 1 \end{bmatrix}. \tag{14}$$

Calculating the witness (1) on the density matrix ρ_{qg} shows:

$$\mathcal{W}(\rho_{qg}) = Tr(\rho_{qg}W) = -2 \quad (\textit{quantum gravity}). \tag{15}$$

5. Semi-classical gravity prediction

While semi-classical gravity is mostly thought of as an aid to calculate the effects of gravity on quantum systems, there are some researchers [7] who look at semi-classical gravity as a real possibility. In semi-classical gravity the gravitational field of a particle in a quantum state is taken to be the expectation value of the position of the particle.

It's worthwhile to recall the physical model of semi-classical gravity:

$$R_{\mu\nu} - \frac{1}{2}g_{\mu\nu}R = \frac{8\pi G}{c^4}\langle\Psi|T_{\mu\nu}|\Psi\rangle \tag{16}$$

We see that Einstein's gravity connects to the expectation value of the matter-energy configuration T. Thus in this scenario, gravity is somehow able to violate the measurement postulates of quantum mechanics, as the gravitational field always knows the zeroth moment of the position operator on an arbitrary wave function.

In the case of the BMV experiment, in the configuration considered here, with the distance d in Figure 2 small in relation to the $L - R$ split distance on each particle, the experimental outcome is not changed from the 'light' particle prediction, and the final state is the same as the initial state:

$$|\Psi_{t_4}\rangle_{sc} \;=\; \frac{1}{2}\Big(|LL\rangle + |RR\rangle + |LR\rangle + |RL\rangle\Big), \tag{17}$$

which is of course separable, and can be written as

$$|\Psi_{t_4}\rangle_{sc} \;= |u\rangle \otimes |u\rangle, \tag{18}$$

$$\rho_{sc} = \begin{bmatrix} 1 & 0 & 0 & 0 \\ 0 & 0 & 0 & 0 \\ 0 & 0 & 0 & 0 \\ 0 & 0 & 0 & 0 \end{bmatrix}. \tag{19}$$

As a check, running the witness:

$$\mathcal{W}(\rho_{sc}) = Tr(\rho_{sc}W) = 0 \quad (semi-classical\ gravity). \tag{20}$$

6. Collapsed state prediction

The BMV experiment is not simple to build or execute, and it is entirely possible that the superposed states will collapse into L or R for each of the particles irrespective of the other particle. There are also collapse models where particles as massive as that used in the proposed BMV experiment will collapse the wave function [8]. In these cases the final state will be a mixed state where we know the source of mixing and result will be one of in equal mixture:

$$|LL\rangle, |RR\rangle, |LR\rangle, |RL\rangle, \tag{21}$$

represented with a density matrix,

$$\rho_{col} = \frac{1}{4}\begin{bmatrix} 1 & 0 & 0 & 0 \\ 0 & 1 & 0 & 0 \\ 0 & 0 & 1 & 0 \\ 0 & 0 & 0 & 1 \end{bmatrix}, \tag{22}$$

with the witness

$$\mathcal{W}(\rho_{col}) = Tr(\rho_{col}W) = 0 \quad (collapse). \tag{23}$$

7. Bohmian trajectory prediction

The de Broglie Bohm mechanics, as begun by Louis de Broglie and formalized by David Bohm [9] uses non local hidden variables to create an interpretation of quantum mechanics where particles have trajectories, with an ensemble of trajectories traversing the entire wave function.

If one were to use a Bohmian description of quantized gravity, then the results would be identical to those of the quantized gravity predictions and equation (13) would describe the final state. Instead, what is assumed for Bohmian trajectory gravity is that as Struyve writes: [10]

IOP Conf. Series: Journal of Physics: Conf. Series **1275** (2019) 012038 doi:10.1088/1742-6596/1275/1/012038

> Bohmian mechanics solves the measurement problem by introducing an actual configuration (particle positions in the non-relativistic domain, particle positions or fields in the relativistic domain) that evolves under the influence of the wave function. According to this approach, instead of coupling classical gravity to the wave function, it is natural to couple it to the actual matter configuration.

In equation form, comparing to the semi-classical form (16) above, we have:

$$R_{\mu\nu} - \frac{1}{2} g_{\mu\nu} R = 8\pi G T_{\mu\nu}(\varphi_B, g) \,. \tag{24}$$

Where φ_B is the trajectory of one member of the ensemble of Bohmian trajectories, and $T_{\mu\nu}(\varphi_B, g)$ is the stress energy tensor as determined by the specific trajectory in combination with perhaps other (e.g. earth's gravity) sources. A simple interpretation of a quantum - gravity coupling such as (24) has theoretical problems such as energy violation, which it shares with other similar approaches [11].

In order to interpret what happens in the BMV experiment assuming Bohmian trajectory gravity governs the quantum-gravity coupling consider

Superposition: Since gravity couples directly to each particle position, gravity is not in a superposition, instead the configuration of the gravitational field changes for each run of the experiment, with 4 possible configurations of particle trajectories.

No interaction: With the BMV experiment set up as outlined in Figure 2, in three of the four configurations there is no substantial gravitational interaction.

Interaction: Only when particle 1 takes the right track and particle 2 takes the left track will there be a gravitational interaction.

Measurement: There is no measurement taking place, so there is no collapse. The effect on the quantum state when the particle trajectories take the R and L paths is for the wave function to be altered in the same way as in the quantized gravity solution.

The end result is that at t_4 there are two possible states, we have a mixed state. 3/4 of the time, there will be no gravitational interaction and the final state will be the same as the starting state at t_0:

$$|\Psi_{t_4}\rangle_{bt} \quad = \quad \frac{1}{2}\Big(|LL\rangle + |RR\rangle + |LR\rangle + |RL\rangle\Big). \tag{25}$$

1/4 of the time there is gravitational time dilation and the final state will be the same as that for quantized gravity. The final state is not simply $-|RL\rangle$ as even though 'we know' the particles interacted, the wave function is not collapsed, and interference between the branches will still occur, guiding the particles with the Bohmian guidance equation:

$$|\Psi_{t_4}\rangle_{bt} \quad = \quad \frac{1}{2}\Big(|LL\rangle + |RR\rangle + |LR\rangle - |RL\rangle\Big). \tag{26}$$

In this scenario we have somehow turned a pure state into a mixed state with no measurement (accidental or otherwise). This is to be expected, as gravity is not quantized and the Bohmian trajectory interpretation interacts outside the rules of quantum mechanics.

Another way of considering this effect: Gravity is linked to a hidden variable, and thus can peer into the quantum wave function. Of course the hidden Bohmian trajectory variable is not hidden from gravity in this case and thus the state can be evolved based on this 'no longer

IOP Conf. Series: Journal of Physics: Conf. Series **1275** (2019) 012038 doi:10.1088/1742-6596/1275/1/012038

hidden' variable. The density matrix for this mixed state is

$$\rho_{bt} = \frac{1}{16}\begin{bmatrix} 13 & 1 & -1 & 1 \\ 1 & 1 & -1 & 1 \\ -1 & -1 & 1 & -1 \\ 1 & 1 & -1 & 1 \end{bmatrix}. \tag{27}$$

Evaluating the witness:

$$\mathcal{W}(\rho_{bt}) = Tr(\rho_{bt} W) = -\frac{1}{2} \quad (Bohmian\ trajectory). \tag{28}$$

The state is entangled. The witness shows a result 1/4 of that of the maximally entangled quantized gravity case, which makes sense - the density matrix in this case is 3/4 plus 1/4-times a non-entangled state.

Thus we have created entanglement with a classical single valued field. This seems in violation of the LOCC theorems. But LOCC is built on quantum mechanics - and cannot have any bearing over a scheme such as Bohmian trajectory gravity which violates quantum mechanics. Chitambar et al. in *Everything You Always Wanted to Know About LOCC (But Were Afraid to Ask)* [12] describes LOCC and summarizes:

> Quantum operations implemented in such a manner are known as LOCC (local operations with classical communication), and we can think of LOCC as a special subset of all physically realizable operations on the global system.

Other authors [13] discussing the role of gravity with respect to LOCC also share this view.

8. Discussion

The BMV (and related experiments) provide a good probe of the nature of the interaction of gravity and quantum mechanics. The proposed experiment while very challenging to perform, is simple to study theoretically.

The result prediction for Bohmian trajectory gravity shows entanglement with a witness of $-\frac{1}{2}$, while the quantized gravity approach has an entanglement witness value of -2. The semi-classical and collapse models, while having different final states, do not show any entanglement.

Hall and Reginatto [14] note that the LOCC theorems assume Koopman-type dynamics, and that there are other classical-quantum interaction schemes. They prove that the result obtained above for semi-classical interactions is generic in that semi-classical mean field models will never create quantum entanglement. They find that there exist other interaction schemes which can create entanglement.

It is interesting to compare the two schemes considered here that connect a classical field of gravity to a quantum particle. The difference between the two interaction schemes is in some sense small. In both the semi-classical and Bohmian trajectory quantum gravity schemes the classical gravitational field must still interact with a quantum entity. In both semi-classical and Bohmian trajectory gravity, the connection lies outside quantum mechanics (non Koopman-style [14] interactions). This is in contrast to schemes like the KTM [8] and others [15] where a way for a classical field to interact with a massive quantum particle is outlined that follows the rules of quantum mechanics. In [8] and [15] weak measurements both slowly measure the centre of mass of the particle and also automatically collapse the wave function.

In some sense semi-classical gravity seems more complicated than Bohmian trajectory gravity, as in semi-classical gravity the gravitational field has to somehow integrate the entire position space of the wave function (a non local entity) in real time (via the Schrödinger - Newton equation), in order to continuously use the expectation value as a source for the gravitational field. In Bohmian mechanics, the gravitational field connects directly to an existing 'hidden' particle position, which is conceptually simpler.

While this paper concerns the BMV experiment in particular, it is not hard to envision applying Bohmian trajectory gravity to other experimental predictions. Applying the method to a BMV experiment where all the branch distances take part is one future extension.

Of course the canonical viewpoint [3] is that the BMV experiment if done without environmental state collapse will side with quantized gravity and show an entangled final state with the entanglement witness approaching -2 as the experiment is performed with increasing accuracy.

Acknowledgements
The author wishes to thank Hans-Thomas Elze and the other organizers and attendees of DICE2018 Spacetime - Matter - Quantum Mechanics Ninth International Workshop in Castiglioncello Italy, for the chance to speak on the matter presented in this paper.

References
[1] Bose S, Mazumdar A, Morley G W, Ulbricht H, Toroš M, Paternostro M, Geraci A A, Barker P F, Kim M S and Milburn G 2017 *Phys. Rev. Lett.* **119** 240401
[2] Marletto C and Vedral V 2017 *Phys. Rev. Lett.* **119** 240402
[3] Christodoulou M and Rovelli C *Preprint arXiv:* **1808.05842**
[4] Vedral V, Plenio M B, Rippin M A and Knight P L 1997 *Phys. Rev. Lett.* **78** 2275–2279
[5] Buchleitner A, Viviescas C and Tiersch M 2008 *Entanglement and decoherence: foundations and modern trends* vol 768 (Springer Science & Business Media)
[6] Technion State Separator (V2.0) URL `https://physics.technion.ac.il/stateseparator/`
[7] Großardt A 2017 *Proceedings of Science* **Corfu Summ**
[8] Kafri D, Taylor J M and Milburn G J 2014 *New J. Phys.* **16** 1–11
[9] Bohm D 1952 *Phys. Rev.* **85** 166–179
[10] Struyve W 2017 Towards a Novel Approach to Semi-Classical Gravity *The Philosophy of Cosmology* (Cambridge University Press) pp 356–374 ISBN 9781316535783
[11] Pinto-Neto N and Struyve W 2018 *Preprint arXiv:* **1801.03353**
[12] Chitambar E, Leung D, Mancinska L, Ozols M and Winter A 2014 *Commun. Math. Phys.* **238** 303–326 (*Preprint* `1210.4583`)
[13] Altamirano N, Corona-Ugalde P, Mann R B and Zych M 2018 *Class. Quant. Grav.* **35** 145005 (*Preprint* `arXiv:1612.07735v2`)
[14] Hall M J and Reginatto M 2018 *J. Phys. A: Math. Theor.* **51** 1–18 (*Preprint* `1707.07974`)
[15] Khosla K E and Nimmrichter S 2018 *Preprint arXiv:* **1812.03118**

IOP Conf. Series: Journal of Physics: Conf. Series **1275** (2019) 012039 doi:10.1088/1742-6596/1275/1/012039

Entangling quantum fields via a classical gravitational interaction

M Reginatto[1] **and M J W Hall**[2]

[1]Physikalisch-Technische Bundesanstalt, Bundesallee 100, 38116 Braunschweig, Germany
[2]Department of Theoretical Physics, Research School of Physics and Engineering, Australian National University, Canberra ACT 0200, Australia

E-mail: marcel.reginatto@ptb.de

Abstract. We consider the coupling of quantum fields to classical gravity in the formalism of ensembles on configuration space, a model that allows a consistent formulation of interacting classical and quantum systems. Explicit calculations show that there are solutions for which two initially independent quantum fields evolve into an entangled state, even though their interaction occurs solely via a common classical gravitational field. Thus in contrast to recent suggestions, an observed generation of entanglement would not provide a definitive test of the nonclassical nature of gravity.

1. Introduction

The publication of proposals for witnessing nonclassical features of gravity, by Bose *et al.* [1] and by Marletto and Vedral [2], has encouraged a discussion about the possibility of generating entanglement between quantum systems which only interact via a classical gravitational field, as well as new proposals for looking for evidence of quantum gravity in laboratory experiments [3, 4, 5, 6, 7, 8, 9]. Whether entanglement is possible under these circumstances depends on *which* hybrid model is used to describe the interaction of classical and quantum sectors [7]. While some hybrid models of classical-quantum interactions seem to exclude entanglement, other models, in particular the formalism of ensembles on configuration space [10], allow for it. Thus an observed generation of entanglement cannot provide a definitive test of the nonclassicality of gravity without additional assumptions concerning the nature of classical-quantum interactions.

The purpose of this paper is to provide fully relativistic calculations showing that (a) there are solutions for which two quantum fields are in an entangled state even though their interaction occurs solely via a common classical gravitational field, and (b) such entangled solutions can evolve from initially unentangled ones. These calculations are carried out using the formalism of ensembles on configuration space.

The paper is organized as follows. In section 2, we give a brief introduction to the basic aspects of the configuration ensemble approach. Section 3 is more general in nature, consisting of a brief overview on consistency requirements for models of quantum-classical interactions and a summary of previous results on entanglement production for some specific models. In section 4, we return to the configuration ensemble approach and discuss its application to quantum matter fields coupled to a classical gravitational field. We focus on the midisuperspace formulation of spherical gravity, in particular on the case where the quantum sector consists of two quantized

Content from this work may be used under the terms of the Creative Commons Attribution 3.0 licence. Any further distribution of this work must maintain attribution to the author(s) and the title of the work, journal citation and DOI.
Published under licence by IOP Publishing Ltd

IOP Conf. Series: Journal of Physics: Conf. Series **1275** (2019) 012039 doi:10.1088/1742-6596/1275/1/012039

scalar fields. In section 5 we consider a perturbative approach and show that there are solutions for which the two quantum fields are in an entangled state, and in section 6 we discuss how entangled solutions can evolve from initially unentangled ones. Finally, in section 7, we provide some concluding remarks.

2. Ensembles on configuration space describing classical, quantum and mixed classical-quantum systems

The formalism of ensembles on configuration is a general framework that can describe classical, quantum and hybrid systems [10, 11, 12]. It forms a natural starting platform for several axiomatic approaches to reconstructing quantum theory [10, 13, 14, 15, 16, 17] and it can describe the coupling of ensembles of quantum fields to classical spacetimes [10, 11, 18, 19].

2.1. Classical, quantum, and mixed classical-quantum systems

Start from the assumption that the configuration of a physical system is an inherently statistical concept. The system will then be described by an ensemble of configurations, with probability density P, with $P \geq 0$ and $\int dx\, P(x,t) = 1$. To describe dynamics, introduce an *ensemble Hamiltonian* $H[P,S]$, where S is an auxiliary field that is canonically conjugate to P. The equations of motion take the form

$$\frac{\partial P}{\partial t} = \frac{\delta H}{\delta S}, \qquad \frac{\partial S}{\partial t} = -\frac{\delta H}{\delta P}, \tag{1}$$

where $\delta/\delta P$ ($\delta/\delta S$) denotes the functional derivative with respect to P (S).

The following ensemble Hamiltonians lead to equations that describe the evolution of quantum and classical non-relativistic particles of mass m:

$$H_C[P,S] = \int dx\, P\left[\frac{|\nabla S|^2}{2m} + V(x)\right], \qquad H_Q[P,S] = H_C[P,S] + \frac{\hbar^2}{4}\int dx\, P\frac{|\nabla \log P|^2}{2m}. \tag{2}$$

For example, the equations of motion derived from $H_Q[P,S]$ are given by

$$\frac{\partial P}{\partial t} + \nabla.\left(P\frac{\nabla S}{m}\right) = 0, \qquad \frac{\partial S}{\partial t} + \frac{|\nabla S|^2}{2m} + V - \frac{\hbar^2}{2m}\frac{\nabla^2 P^{1/2}}{P^{1/2}} = 0 \tag{3}$$

while the equations of motion derived from $H_C[P,S]$ are the same as Eq. (3) but with $\hbar = 0$. The first equation in Eq. (3) is a continuity equation, the second equation is the classical Hamilton-Jacobi equation when $\hbar = 0$ and a modified Hamilton-Jacobi equation when $\hbar \neq 0$. Defining $\psi = \sqrt{P}\, e^{iS/\hbar}$, Eq. (3) takes the form

$$i\hbar\frac{\partial \psi}{\partial t} = \frac{-\hbar^2}{2m}\nabla^2\psi + V\psi, \tag{4}$$

which is the usual form of the Schrödinger equation. It is straightforward to extend the formalism in a natural way to allow for mixed quantum-classical systems. A mixed quantum-classical ensemble Hamiltonian is given by [10, 11]

$$H_{QC}[P,S] = \int dq\, dx\, P\left[\frac{|\nabla_x S|^2}{2M} + \frac{|\nabla_q S|^2}{2m} + \frac{\hbar^2}{4}\frac{|\nabla_q \log P|^2}{2m} + V(q,x,t)\right]. \tag{5}$$

Here q denotes the configuration space coordinate of a quantum particle of mass m and x that of a classical particle of mass M, and $V(q,x,t)$ is a potential energy function describing the

IOP Conf. Series: Journal of Physics: Conf. Series **1275** (2019) 012039 doi:10.1088/1742-6596/1275/1/012039

quantum-classical interaction. The equations of motion for the joint probability density $P(q,x)$ and its conjugate $S(q,x)$ follow from H_{QC} as

$$\frac{\partial P}{\partial t} = -\nabla_q \cdot \left(P\frac{\nabla_q S}{m}\right) - \nabla_x \cdot \left(P\frac{\nabla_x S}{M}\right), \qquad \frac{\partial S}{\partial t} = -\frac{|\nabla_q S|^2}{2m} - \frac{|\nabla_x S|^2}{2M} - V + \frac{\hbar^2}{2m}\frac{\nabla_q^2 P^{1/2}}{P^{1/2}}. \quad (6)$$

These can be rewritten as a nonlinear Schrödinger equation for the 'hybrid' wave function $\psi = \sqrt{P}e^{iS/\hbar}$, with a similar nonlinear equation for the case of two classical particles. Such nonlinearity does not automatically lead to difficulties in either case, essentially because the form of classical observables is fundamentally different to that of quantum ones, see below.

2.2. *Observables*

The *state* of a system is determined by specifying P and S. *Observables* are defined as suitable functionals of P and S (arbitrary functionals $A[P,S]$ are not necessarily observables because these have to satisfy certain mild requirements, see [10] for a detailed discussion). Given observables $A[P,S]$ and $B[P,S]$, define their Poisson bracket

$$\{A,B\}_{PB} = \int dx \left(\frac{\delta A}{\delta P}\frac{\delta B}{\delta S} - \frac{\delta A}{\delta S}\frac{\delta B}{\delta P}\right). \quad (7)$$

This gives us an *algebra of obervables.* A critical physical distinction between classical and quantum systems (or classical and quantum components of a composite hybrid system) is that they have quite different sets of observables, and distinct algebras for these observables [10].

For a purely classical configuration space labelled by position x, the classical observable C_f corresponding to the phase space function $f(x,k)$ (where k is the momentum) is defined by the functional

$$C_f[P,S] := \int dx\, P\, f(x, \nabla_x S). \quad (8)$$

For a purely quantum configuration space labelled by the possible outcomes q of some complete basis set $\{|q\rangle\}$ of a Hilbert space $\mathcal{H}$ (i.e., $\int dq\, |q\rangle\langle q| = \hat{1}$, with integration replaced by summation for discrete ranges of q), the quantum observable $Q_{\hat{M}}$ corresponding to the Hermitian operator $\hat{M}$ is defined by the functional

$$Q_{\hat{M}}[P,S] := \langle\psi|\hat{M}|\psi\rangle, \quad (9)$$

where $|\psi\rangle \in \mathcal{H}$ is the wave function defined via $\langle q|\psi\rangle = \sqrt{P(q)}e^{iS(q)/\hbar}$.

Evaluating the Poisson bracket of any pair of classical observables C_f, C_g or quantum observables $Q_{\hat{M}}, Q_{\hat{N}}$ via Eq. (7) yields

$$\{C_f, C_g\} = C_{\{f,g\}}, \qquad \{Q_{\hat{M}}, Q_{\hat{N}}\} = Q_{[\hat{M},\hat{N}]/(i\hbar)} \quad (10)$$

where $\{f,g\} = \sum_i \left(\frac{\partial f}{\partial x_i}\frac{\partial g}{\partial k_i} - \frac{\partial f}{\partial k_i}\frac{\partial g}{\partial x_i}\right)$ denotes the usual phase space bracket and $[\hat{M},\hat{N}]$ the usual commutator. Thus *Eq. (8) is an isomorphism between the algebra of observables C_f on configuration space and the algebra of observables f on classical phase space, and Eq. (9) is an isomorphism between the algebra of observables $Q_{\hat{M}}$ on configuration space and the algebra of quantum observables $\hat{M}$*. As the Poisson bracket properties (10) remain unchanged for mixed classical-quantum systems of the type described by the ensemble Hamiltonian of Eq. (5), this statement remains true even under interactions between the classical and quantum components. In particular, *the classical/quantum distinction is always maintained.*

IOP Conf. Series: Journal of Physics: Conf. Series **1275** (2019) 012039 doi:10.1088/1742-6596/1275/1/012039

2.3. Entanglement

Two ensembles with respective configuration spaces X and Y are defined to be *independent* if $P(x,y)$ and $S(x,y)$ satisfy $P(x,y) = P_X(x)P_Y(y)$ and $S(x,y) = S_X(x) + S_Y(y)$ (with the latter only required to hold up to some additive constant) [10, 11]. For quantum ensembles, note that independence is equivalent to a factorisable wave function $\psi = \sqrt{P}e^{iS/\hbar}$ and hence *any two quantum ensembles are either independent or entangled.*

The concept of entanglement remains meaningful in the general case [10]. However, it is important to note that the notion of 'entanglement' referred to here is not in the strong sense of Bell inequality violation, but in Schrödinger's original weaker sense that the properties of a joint ensemble cannot be decomposed into properties of the individual ensembles [20].

This can be understood by looking at a simple classical example. Consider a classical joint ensemble, corresponding to two classical particles described by respective configuration spaces X and Y, with probability density $P(x,y)$ and conjugate quantity $S(x,y)$. The product of two classical phase space functions $f(x,p_x)$ and $g(y,p_y)$ is itself a classical phase space function and the expectation value of this product corresponds to the classical observable [10]

$$C_{fg} = \langle fg \rangle = \int dxdy\, P(x,y)\, f(x,\partial_x S)\, g(y,\partial_y S). \tag{11}$$

Now, there is clearly a trivial hidden variable for any such observable. In particular, defining $\lambda := [x,y,S(x,y)]$, $P(\lambda) := P(x,y)$, $F(\lambda) := f(x,\partial_x S)$, and $G(\lambda) := g(y,\partial_y S)$, one has

$$\langle fg \rangle = \int d\lambda\, P(\lambda)\, F(\lambda)\, G(\lambda). \tag{12}$$

Hence, no Bell inequality can be violated via such observables [21].

Nevertheless, if the independence condition $S(x,y) = S_X(x)+S_Y(y)$ is *not* satisfied, then the 'hidden value' of the observable $f(x,p_x)$ for the *first* particle, i.e., $F(\lambda)$, will in general depend on the position of the *second* particle, via $p_x = \partial_x S(x,y)$. That is, while knowledge of the position and momentum of the first particle at a given time is sufficient to determine all observables for the particle at that time, it will not be sufficient to determine them at any later time: one needs to know the evolution of the joint quantity $\partial_x S(x,y)$. Moreover, if one locally perturbs the position of the second particle, from y to y', the corresponding perturbation of $S(x,y)$ to $S(x,y')$ will typically perturb the value of p_x in this model. Hence, *a kind of nonlocality, or inseparability, can be associated even with classical configuration space ensembles.* We will, by analogy with Schrödinger's original discussion [20], refer to this property as 'entanglement'. This leads to the following general definition which applies to *all* configuration space ensembles [10]:

> *A joint ensemble is entangled with respect to the joint configuration space $X \times Y$ if and only if $S(x,y) \not\equiv S_X(x) + S_Y(y)$ (up to some additive constant).*

For further details on entanglement for quantum, classical and hybrid systems, see Ref. [10].

3. On models of quantum-classical interactions, consistency requirements, and the generation of entanglement in some particular models

The task of finding a physically consistent approach for modelling interactions between classical and quantum systems is highly non-trivial. Thus there are many possible models of classical-quantum interactions in the literature, and these typically have mathematical or physical difficulties associated with them [7, 10, 11].

IOP Conf. Series: Journal of Physics: Conf. Series **1275** (2019) 012039 doi:10.1088/1742-6596/1275/1/012039

3.1. Consistency requirements

A number of consistency requirements have been proposed to evaluate whether a particular approach to classical-quantum interactions provides a viable model or not (see Ref. [10] for a discussion on various attempts at solving this problem). Almost all of the approaches which have been proposed run into difficulties with these requirements, but two types of models have been shown to satisfy them. Consistency of the model based on ensembles on configuration space is shown in Refs. [10, 12]; for models based on a mean-field approach [22, 23], it follows from the consistency of the model due to Elze, who showed that his model satisfies the requirements [24] and therefore that all equivalent mean-field models are consistent. We will not go over the consistency conditions here but instead refer the reader to these publications and to the references therein.

The particular approach considered by Bose *et al.* [1] and Marletto and Vedral [2] is a third type of model, based on formally embedding a classical system into a diagonal basis of some quantum system; i.e., equivalent to describing classical-quantum interactions via Koopman-type dynamics [25, 26, 27]. While this approach has some interesting properties, it has been shown that it already fails for simple examples, thus some of the basic consistency requirements proposed in the literature are not satisfied. For example, classical observables remain classical for a limited class of interactions only, which do not include some of the textbook examples of experimental setups like the standard Stern-Gerlach measurement interaction [28]. Peres and Terno have further shown that this approach does not reproduce the correct classical limit for classical-quantum oscillators, and indeed may result in a runaway increase of the classical oscillator amplitude [27, 29]. We see then that the approach already runs into serious difficulties with the simple problem of classical-quantum oscillators, and one would expect even greater difficulties if one were to fully work out the details of the interaction of quantum matter with a classical gravitational field using this model (which, to the best of our knowledge, has not been done yet). Further, predictions made by such models in the limit of perturbative Newtonian gravity appear to conflict with observational data [30].

3.2. Previous analyses of entanglement in mixed classical-quantum systems

We restrict our discussion of previous analyses of entanglement to the first two models discussed above as these are known to satisfy consistency requirements. The creation of entanglement for two quantum subsystems via a classical interaction has already been discussed in both cases.

For the case of mixed classical-quantum systems described by ensembles on configuration space, it has been shown that one can construct explicit examples in which entanglement between quantum subsystems is created via an interaction with a classical subsystem [7]. The results presented in sections 5 and 6 of this paper complement these earlier results with further examples involving two quantum fields which are entangled via a common classical gravitational field.

For the approach of Elze, detailed calculations are available for the entanglement dynamics of a system of two qubits and one classical oscillator [31]. In particular, it is shown that the concurrence remains constant if the two qubits have some initial entanglement and therefore that no additional entanglement is generated in this model. This result is consistent with the observation in Ref. [7] concerning the absence of entanglement creation in mean-field approaches, of which the Elze model is a particular formulation.

These results regarding entanglement creation reflect fundamental differences between the two models [10]. In particular, in a mean-field approach the classical particle follows a deterministic trajectory in phase space, rather than being described by an ensemble on configuration space. Thus, unlike the configuration ensemble approach, the mean-field approach cannot couple quantum fluctuations into correlated classical observables.

This reinforces a point made earlier, that the possibility of entanglement generation via a classical mediator depends on the particular approach used to model the interaction between

IOP Conf. Series: Journal of Physics: Conf. Series **1275** (2019) 012039 doi:10.1088/1742-6596/1275/1/012039

classical and quantum systems. While some hybrid models of classical-quantum interactions seem to exclude entanglement, other models, in particular the formalism of ensembles on configuration space, allow for it.

4. The coupling of scalar quantum fields to classical gravity in the spherically symmetric case

The analysis of section 2 for interacting particles can be extended to interacting fields, in particular to the coupling of scalar quantum fields to classical gravity. A detailed description of the formalism is given in [10]. Here we consider only the case of spherically symmetric spacetimes and the corresponding midisuperspace formulation of general relativity known as spherical gravity. For the case of spherical symmetry, the line element may be written in the form

$$g_{\mu\nu}dx^{\mu}dx^{\nu} = -N^2dt^2 + \Lambda^2\left(dr + N_r dt\right)^2 + R^2 d\Omega^2, \tag{13}$$

where the lapse function N and the shift function N_r are functions of the radial coordinate r and the time coordinate t. Thus the configuration space for the gravitational field consists of two fields, R and Λ. Spherically symmetric gravity is discussed in detail in a number of papers, mostly in reference to the canonical quantization of black hole spacetimes [32, 33, 34, 35, 36].

4.1. The case of vacuum gravity

The most direct way of introducing a classical configuration space ensemble for gravity is to start from the Einstein-Hamilton-Jacobi (EHJ) equation [37, 38]. For the case of spherical gravity without matter fields, it takes the form $H_{\Lambda R} = 0$ with

$$H_{\Lambda R} = -\frac{1}{R}\frac{\delta S}{\delta R}\frac{\delta S}{\delta \Lambda} + \frac{\Lambda}{2R^2}\left(\frac{\delta S}{\delta \Lambda}\right)^2 + V, \qquad V = \frac{RR''}{\Lambda} - \frac{RR'\Lambda'}{\Lambda^2} + \frac{R'^2}{2\Lambda} - \frac{\Lambda}{2} \tag{14}$$

where we have set $c = G = \hbar = 1$. In spherical gravity, the momentum constraints of the full theory are replaced by a single (radial) diffeomorphism constraint, $(\delta S/\delta R)R' - \Lambda\,(\delta S/\delta\Lambda)' = 0$, where primes indicate derivatives with respect to r [38]. We will require that S be invariant under diffeomorphisms so that it automatically solves the momentum constraint.

An appropriate ensemble Hamiltonian for spherically symmetric gravity is given by

$$\mathcal{H} = \int dr\, N \int \mathrm{D}R\,\mathrm{D}\Lambda\, P H_{\Lambda R}. \tag{15}$$

where P is a probability density function (which, like S, is assumed to satisfy the diffeomorphism constraint) and $\mathrm{D}R\,\mathrm{D}\Lambda$ is an appropriate measure [10]. Assuming the constraints $\frac{\partial S}{\partial t} = \frac{\partial P}{\partial t} = 0$, the ensemble Hamiltonian of Eq. (15) leads to two equations for S and P, the EHJ equation of spherical gravity,

$$\int dr\, N \left[-\frac{1}{R}\frac{\delta S}{\delta R}\frac{\delta S}{\delta \Lambda} + \frac{\Lambda}{2R^2}\left(\frac{\delta S}{\delta \Lambda}\right)^2 + V\right] = 0, \tag{16}$$

and the continuity equation

$$\int dr\, N \left[\frac{\delta}{\delta R}\left(P\frac{1}{R}\frac{\delta S}{\delta \Lambda}\right) + \frac{\delta}{\delta \Lambda}\left(P\frac{1}{R}\frac{\delta S}{\delta R} - P\frac{\Lambda}{R^2}\frac{\delta S}{\delta \Lambda}\right)\right] = 0. \tag{17}$$

4.2. The addition of quantum scalar fields

The ensemble Hamiltonian of a hybrid system where matter is in the form of n minimally coupled quantized radially symmetric scalar fields ϕ_a of mass m is given by

$$\mathcal{H}_{\phi\Lambda R} = \int dr \int \mathrm{D}\phi\mathrm{D}\Lambda\mathrm{D}R\; P\,N \left[H^C_{\phi\Lambda R} + \frac{1}{8\Lambda R^2}\sum_{a=1}^{n}\left(\frac{\delta \log P}{\delta \phi_a}\right)^2\right], \tag{18}$$

where

$$H^C_{\phi\Lambda R} = H^C_{\Lambda R} + \sum_{a=1}^{n} \left[\frac{1}{2\Lambda R^2} \left(\frac{\delta S}{\delta \phi_a} \right)^2 + \frac{R^2}{2\Lambda} \phi_a'^2 + \frac{\Lambda R^2 m_a^2}{2} \phi_a^2 \right], \tag{19}$$

is a purely classical term which now includes the coupling to scalar fields ϕ_a and the last term in Eq. (18) is an additional, non-classical term that must be included in the ensemble Hamiltonian when the scalar field is quantized (recall we have set $\hbar = 1$). Assuming again constraints $\frac{\partial S}{\partial t} = \frac{\partial P}{\partial t} = 0$, the corresponding equations are

$$\int dr\, N \left[H^C_{\phi\Lambda R} - \frac{1}{2\Lambda R^2} \sum_{a=1}^{n} \left(\frac{1}{\sqrt{P}} \frac{\delta^2 \sqrt{P}}{\delta \phi_a^2} \right) \right] = 0, \tag{20}$$

and the continuity equation

$$\int dr\, N \left[\frac{\delta}{\delta R} \left(P \frac{1}{R} \frac{\delta S}{\delta \Lambda} \right) + \frac{\delta}{\delta \Lambda} \left(P \frac{1}{R} \frac{\delta S}{\delta R} - P \frac{\Lambda}{R^2} \frac{\delta S}{\delta \Lambda} \right) - \sum_{a=1}^{n} \frac{\delta}{\delta \phi_a} \left(P \frac{1}{\Lambda R^2} \frac{\delta S}{\delta \phi_a} \right) \right] = 0. \tag{21}$$

5. Black hole with two scalar quantum fields in spherical gravity: entangled solutions

We now apply the formalism presented in the previous section to the case of two quantized scalar fields, ϕ_1 and ϕ_2, in the space-time of a classical black hole. We assume that the quantum fields act as a perturbation to the space-time; i.e., that the contribution to the gravitational field from the mass of the black hole is much larger than that of the quantum matter fields. Under these circumstances, it is appropriate to search for an approximate perturbative solution of Eqs. (20) and (21) based on an expansion in powers of ϕ_a [39]. The advantage of using such an approach is that it is possible to solve the equations iteratively, term by term, as is clear from the equations below. We will use the notation of [39] where $S^{(n)}$ stands for a functional of order $(\phi_a)^n$. While the term $S^{(0)}$ can be chosen freely, the higher order terms depend on the previous ones.

To carry out the calculation, it will be convenient to write the expression for $P[R, \Lambda, \phi_1, \phi_2]$ in the form

$$P = e^{-\left(F^{(0)}[R,\Lambda]\right)}\, e^{-\left(\sum_{n>0} F^{(n)}[R,\Lambda,\phi_1,\phi_2]\right)} =: P_A[R, \Lambda]\, P_B[R, \Lambda, \phi_1, \phi_2]. \tag{22}$$

P_A depends on the gravitational degrees of freedom only while P_B depends on both gravitational and scalar field degrees of freedom. Furthermore, we will require that, *up to order* $(\phi_a)^2$,

$$\frac{\delta P_B}{\delta \phi_a} = -\frac{\delta F^{(2)}}{\delta \phi_a} P_B, \qquad \frac{\delta P_B}{\delta h_{ij}} = -\frac{\delta F^{(2)}}{\delta h_{ij}} P_B, \qquad \frac{1}{\sqrt{P_B}} \frac{\delta^2 \sqrt{P_B}}{\delta \phi_a^2} = -\frac{1}{2} \frac{\delta^2 F^{(2)}}{\delta \phi_a^2} + \frac{1}{4} \left(\frac{\delta F^{(2)}}{\delta \phi_a} \right)^2, \tag{23}$$

(note the terms on the right of the last equation are of order $(\phi_a)^0$ and $(\phi_a)^2$) respectively).

Our ansatz then is that the odd terms vanish, i.e., $F^{(1)} = F^{(3)} = 0$, so that P_B is to a first approximation a *Gaussian functional* with respect to the ϕ_a. While not an essential assumption, this choice seems physically reasonable as it implies a solution of the quantum sector that is in some respect close to the simplest solution for quantum field theory in curved space time (i.e., the ground state functional). The expression $\delta^2 F^{(2)}/\delta\phi_a^2$ needs to be regularized (such a term appears also in solutions of the Schrödinger functional equation). We will not consider the regularization problem here, we will simply assume that this term has been regularized and that it is finite. We will assume $\delta^2 F^{(2)}/\delta\phi_1^2 = \delta^2 F^{(2)}/\delta\phi_2^2$ and introduce the notation $C_F[R, \Lambda] := \frac{1}{2}\delta^2 F^{(2)}/\delta\phi_a^2$ for this term.

IOP Conf. Series: Journal of Physics: Conf. Series **1275** (2019) 012039 doi:10.1088/1742-6596/1275/1/012039

We now give explicit solutions for $S^{(n)}$ for the first two terms in the expansion. This is already sufficient to demonstrate the existence of entanglement. The first equation in the expansion is

$$\left[-\frac{1}{R}\frac{\delta S^{(0)}}{\delta R}\frac{\delta S^{(0)}}{\delta \Lambda}+\frac{\Lambda}{2R^2}\left(\frac{\delta S^{(0)}}{\delta \Lambda}\right)^2+V\right]+\sum_{a=1}^{2}\left\{\frac{1}{2\Lambda R^2}\left(\frac{\delta S^{(1)}}{\delta \phi_a}\right)^2-\frac{C_F}{8\Lambda R^2}\right\}=0. \tag{24}$$

We choose the $S^{(0)}$ that solves the classical EHJ equation for a black hole; i.e, that makes the terms in square brackets equal to zero. The solution is well known [32, 36]. Thus, to zeroth order, we are dealing with a black hole space-time. With this choice of $S^{(0)}$, it is straightforward to find a solution for $S^{(1)}$,

$$S^{(1)}=\int dr\,\frac{\phi_1+\phi_2}{2}\sqrt{\frac{C_F}{2}}\;+\;C^{(1)}[R,\Lambda], \tag{25}$$

where $C^{(1)}$ is an arbitrary functional of the gravitational degrees of freedom. The next equation in the expansion is

$$-\frac{1}{R}\left(\frac{\delta S^{(0)}}{\delta R}\frac{\delta S^{(1)}}{\delta \Lambda}+\frac{\delta S^{(0)}}{\delta \Lambda}\frac{\delta S^{(1)}}{\delta R}\right)+\frac{\Lambda}{R^2}\frac{\delta S^{(0)}}{\delta \Lambda}\frac{\delta S^{(1)}}{\delta \Lambda}+\sum_{a=1}^{2}\left\{\frac{1}{2\Lambda R^2}\frac{\delta S^{(1)}}{\delta \phi_a}\frac{\delta S^{(2)}}{\delta \phi_a}\right\}=0. \tag{26}$$

The equation is linear in $S^{(2)}$. Except for $\delta S^{(2)}/\delta\phi_a$, all terms are known and they depend on $\phi_1+\phi_2$ only, so it is straightforward to solve for $S^{(2)}$. It is given by

$$\begin{aligned} S^{(2)} &= \frac{1}{2}\int dr\,(\phi_1+\phi_2)\left[-\frac{1}{R}\left(\frac{\delta S^{(0)}}{\delta R}\frac{\delta S^{(1)}}{\delta \Lambda}+\frac{\delta S^{(0)}}{\delta \Lambda}\frac{\delta S^{(1)}}{\delta R}\right)+\frac{\Lambda}{R^2}\frac{\delta S^{(0)}}{\delta \Lambda}\frac{\delta S^{(1)}}{\delta \Lambda}\right]\frac{2\Lambda R^2}{\sqrt{C_F/2}} \\ &\quad +C^{(2)}[R,\Lambda,\phi_1-\phi_2] \end{aligned} \tag{27}$$

where $C^{(2)}=[R,\Lambda,\phi_1-\phi_2]$ is a quadratic but otherwise arbitrary functional of $\phi_1-\phi_2$. *Notice that in general* $S^{(2)}\neq S_1^{(2)}[R,\Lambda,\phi_1]+S_2^{(2)}[R,\Lambda,\phi_2]$, *which implies the entanglement of* ϕ_1 *and* ϕ_2, as per the discussion in section 2.3. This is the main result of this section.

The next term in the expansion is discussed in Ref. [40]. A perturbative solution requires also solving the continuity equation, Eq. (20), to the same order (the relevant equations are given in Ref. [40]). We do not carry out this step here, as the main purpose of the exercise, which is to show the existence of entangled states, is already accomplished with the solution of the EHJ equation.

6. The emergence of time and entanglement

Time did not play a role in the formalism used in the previous section. However, one would like to see whether entanglement can arise when such a hybrid system evolves. This *is* possible: since the gravitational field is treated classically, one may introduce a well defined *gravitational time* and derive a time-dependent equation for the quantum fields. Although a solution $S[\Lambda,R]$ of Eq. (16) is a functional of Λ and R, the two-dimensional space-time of spherical gravity can always be reconstructed by means of the rate equations [10, 33]

$$\dot{R}=-\frac{N}{R}\frac{\delta S}{\delta \Lambda}+N_rR',\qquad \dot{\Lambda}=-\frac{N}{R}\frac{\delta S}{\delta R}+\frac{\Lambda}{R^2}\frac{\delta S}{\delta \Lambda}+(\Lambda N_r)', \tag{28}$$

where N is the lapse function and N_r is the shift function in Eq. (13). The introduction of a gravitational time in this way has been discussed in the context of the semi-classical approximation to quantum geometrodynamics [38, 41, 42]).

IOP Conf. Series: Journal of Physics: Conf. Series **1275** (2019) 012039 doi:10.1088/1742-6596/1275/1/012039

We once more set $S = S_A[R,\Lambda] + S_B[R,\Lambda,\phi_a]$, $P = P_A[R,\Lambda]\, P_B[R,\Lambda,\phi_a]$, but instead of a perturbative expansion, we introduce a suitable physical approximation. We choose S_A and P_A to be black hole solutions of the EHJ [32, 36] and continuity equations. Then, defining $\Psi[R,\Lambda,\phi_a;t] := \sqrt{P_B}\exp(iS_B)$ and assuming a weak fields limit (i.e., $\delta S_A/\delta R >> \delta S_B/\delta R$ and $\delta S_A/\delta\Lambda >> \delta S_B/\delta\Lambda$, see [40]), we derive the non-linear Schrödinger functional equation

$$i\hbar\dot{\Psi} = \hat{H}_f\Psi = \int dr\left[\sum_{a=1}^{2}\left\{\frac{1}{2\Lambda R^2}\frac{\delta^2}{\delta\phi_a^2} + \frac{R^2}{2\Lambda}(\phi_a')^2 + \frac{\Lambda R^2 m^2}{2}\phi_a^2\right\} + \Delta\right]\Psi, \tag{29}$$

where the non-linear correction term Δ is given by

$$\Delta = -\frac{1}{R}\frac{\delta S_B}{\delta R}\frac{\delta S_B}{\delta\Lambda} + \frac{\Lambda}{2R^2}\left(\frac{\delta S_B}{\delta\Lambda}\right)^2, \tag{30}$$

which may also be written in terms of Ψ and $\bar{\Psi}$. The term Δ is a new "correction" term that distinguishes the time evolution as evaluated by quantum field theory in curved space-time (where this term is absent) from the time evolution in the theory of ensembles on configuration space. We can now consider the following question: suppose that initially the two quantum fields ϕ_1 and ϕ_2 are *not* entangled. Can this non-linear time-dependent functional Schrödinger equation lead to their entanglement? The crucial point here is that there is no reason to believe that the term Δ will preserve non-entanglement of states, as it is quadratic in the functional derivatives of $S_B[R,\Lambda,\phi_1,\phi_2]$ with respect to R and Λ. One can argue as follows. Consider calculating the time evolution of the wavefunctional Ψ after an infinitesimally small time interval δt. If the initial state is not entangled so that

$$S_B[R,\Lambda,\phi_1,\phi_2;t=0) = S_B^1[R,\Lambda,\phi_1;t=0) + S_B^2[R,\Lambda,\phi_2;t=0), \tag{31}$$

the initial Δ will have in general mixed terms in ϕ_1 and ϕ_2 which will generically lead to entanglement, so that one would expect at time δt that

$$S_B[R,\Lambda,\phi_1,\phi_2;t=\delta t) \neq S_B^1[R,\Lambda,\phi_1;t=\delta t) + S_B^2[R,\Lambda,\phi_2;t=\delta t). \tag{32}$$

This suggests then that the interaction of ϕ_1 and ϕ_2 via a common gravitational field will in general create entanglement between them.

7. Discussion

Our main result is that entanglement between quantum fields may be generated via a classical gravitational interaction (section 6). This result is based on the configuration ensemble formalism (which is able to describe the coupling of quantum and classical systems more generally), with explicit calculations made for the case of black-hole spacetimes in spherical gravity under a weak-field approximation. The effective evolution equation for the quantum fields, Eq.(29), is defined with respect to a gravitational time.

The above result strongly supports the arguments made in [7], that observation of entanglement *per se*, in the experiments proposed by Bose *et al.* and by Marletto and Vedral [1, 2], does not necessarily imply that gravity is nonclassical in nature. Such an observation can only rule out some classical models of gravity, such as Koopmanian and mean-field models [7], but not all. In particular, entanglement appears to be ubiquitous in the configuration-ensemble model, as exemplified by the spherical gravity solutions in section 5 and the approximate evolution equation in section 6.

IOP Conf. Series: Journal of Physics: Conf. Series **1275** (2019) 012039 doi:10.1088/1742-6596/1275/1/012039

References

[1] Bose S, Mazumdar A, Morley G W, Ulbricht H, Toro M, Paternostro M, Geraci A A, Barker P F, Kim M S and Milburn G 2017 *Phys. Rev. Lett.* **119** 240401
[2] Marletto C and Vedral V 2017 *Phys. Rev. Lett.* **119** 240402
[3] Anastopoulos C and Hu B-L 2018 *arXiv:* 1804.11315v1
[4] Belenchia A, Wald R, Giacomini F, Castro-Ruiz E, Brukner C and Aspelmeyer M 2018 *Phys. Rev. D* **98** 126009
[5] Christodoulou M and Rovelli C 2018 *arXiv:* 1808.05842v1
[6] Christodoulou M and Rovelli C 2018 *arXiv:* 1812.01542v2
[7] Hall M J W. and Reginatto M 2018 *J. Phys. A: Math. Theor.* **51** 085303
[8] Marletto C and Vedral V 2018 *Phys. Rev. D* **98** 046001
[9] Marletto C, Vedral V and Deutsch D (2018) *New J. Phys* **20** 083011
[10] Hall M J W and Reginatto M 2016 *Ensembles on configuration space: classical, quantum and beyond* (Switzerland: Springer)
[11] Hall M J W and Reginatto M 2005 *Phys. Rev. A* **72** 062109
[12] Hall M J W 2008 *Phys. Rev. A* **78** 042104
[13] Hall M J W and Reginatto M 2002 *J. Phys A* **35**, 3289-3303
[14] Hall M J W, Kumar K and Reginatto M 2003 *J. Phys A* **36** 9779-9794
[15] Hall M J W and Reginatto M 2012 *AIP Conf. Proc.* **1443** 96-103
[16] Hall M J W and Reginatto M 2013 *AIP Conf. Proc.* **1553** 246-253
[17] Reginatto M 2014 *J. Phys.: Conf. Ser.* **538** 012018
[18] Albers M, Kiefer C and Reginatto M 2008 *Phys. Rev. D* **78** 064051
[19] Reginatto M 2013 *J. Phys.: Conf. Ser.* **442** 012009
[20] Schrödinger E 1935 *Proc. Camb. Phil. Soc.* **31** 555-563
[21] Horodecki R, Horodecki P, Horodecki M and Horodecki K 2009 *Rev. Mod. Phys.* **81** 865-942
[22] Boucher W and Traschen J 1988 *Phys. Rev. D* **37** 3522-3532
[23] Elze H-T 2012 *Phys. Rev. A* **85** 052109
[24] Elze H-T 2013 *J. Phys.: Conf. Ser.* **442** 012007
[25] Koopman B O 1931 *Proc. Natl. Acad. Sci. U.S.A.* **17** 315-318
[26] Sudarshan E C G 1976 *Pramana* **6** 117-126
[27] Peres A and Terno D R 2001 *Phys. Rev. A* **63** 022101
[28] Sherry T N and Sudarshan E C G 1979 *Phys. Rev. D* **20** 857-868
[29] Terno D R 2006 *Found. Phys.* **36** 102-111
[30] Altamirano N, Corona-Ugalde P, Mann R B and Zych M 2016 *Class. Quantum Grav.* **35** 145005
[31] Fratino L, Lampo A and Elze H-T 2014 *Phys. Scr.* **T163** 014005
[32] Fischler W, Morgan D and Polchinski J 1990 *Phys. Rev. D* **42** 4042-4055
[33] Kuchař K V 1994 *Phys. Rev. D* **50** 3961-3981
[34] Romano J D 1995 *arXiv:* gr-qc/9501015v1
[35] Lau S R 1996 *Class. Quantum Grav.* **13** 1541-1570
[36] Brotz T and Kiefer C 1997 *Phys. Rev. D* **55** 2186-2191
[37] Misner C W, Thorne K S and Wheeler J A 1973 *Gravitation* (San Francisco: Freeman)
[38] Kiefer C 2012 *Quantum Gravity* (Oxford: Oxford University Press)
[39] Salopek D S *Phys. Rev. D* **56** 2057-2064
[40] Reginatto M and Hall M J W 2018 *arXiv:* gr-qc/1809.04989v1
[41] Kiefer C 1994 *Canonical gravity: from classical to quantum* ed Ehlers, J and Friedrich H (Berlin: Springer)
[42] Kiefer C and Singh T P 1991 *Phys. Rev D* **44** 1067-1076

From Quantum Mechanics to running Λ cosmologies

A. Stachowski[1], **M. Szydłowski**[2]
Astronomical Observatory, Jagiellonian University, Orla 171, 30-244 Krakow, Poland
E-mail: [1]`aleksander.stachowski@doctoral.uj.edu.pl`, [2]`marek.szydlowski@uj.edu.pl`

K. Urbanowski
Institute of Physics, University of Zielona Góra, ul. Prof. Z. Szafrana 4a, 65-516 Zielona Góra, Poland
E-mail: `K.Urbanowski@if.uz.zgora.pl`

Abstract. The cosmological model with running dark energy is considered. We assume that the vacuum of the Universe is in a meta–stable state and decays into a true (bare) vacuum state with increasing time. We use quantum theory of decay processes to find energy of the system in the unstable state and apply obtained results to a description of the running dark energy identified with vacuum energy. We also estimate model parameters using astronomical data. From the astronomical point of view our model is in good agreement with data. Moreover, the framework of this model one can explain naturally, smallness of the cosmological constant parameter Λ_{bare}.

1. Introduction

The two most important problems of the contemporary cosmology are: 1) The cosmological constant problem, that is the difference between the measured value of the vacuum energy and the value calculated using quantum field theory methods [1], and 2) The problem why the current densities of matter and dark energy in the Universe have the same order of magnitude [2]. (This question is called the coincidence problem). These mentioned cosmological problems can be considered in the framework of cosmological models where the cosmological constant is running and, in result, its value is changing during the cosmic evolution.

The current observations indicate that the Universe is in an accelerated phase [3] which can be explained by the existence of the dark energy. The analysis of astronomical observations shows that there is a tension between local and primordial measurements of cosmological parameters [3]. A possible explanation of this tension is the dark energy changing in time [4].

In this talk we will assume that the dark energy is metastable and depends on time t, $\rho_{\text{de}} = \rho_{\text{de}}(t)$ and decays with the increasing time t to ρ_{bare} ($\rho_{\text{de}}(t) \to \rho_{\text{bare}} \neq 0$ when $t \to \infty$). Decaying vacuum energy was considered in many papers (see e.g. [5, 6] and also [7, 8]). Shafieloo et al. [9] assumed that $\rho_{\text{de}}(t)$ decays according to the radioactive exponential decay law. Such an assumption seems to be insufficient when one wants to explain the evolution in time of the Universe with decaying dark energy. It is because the creation of the Universe is a quantum process. Therefore the metastable dark energy should be considered as the value of a scalar field

Content from this work may be used under the terms of the Creative Commons Attribution 3.0 licence. Any further distribution of this work must maintain attribution to the author(s) and the title of the work, journal citation and DOI.
Published under licence by IOP Publishing Ltd

in the false vacuum state and the process of a decay of the dark energy should be considered as a quantum decay process.

The quantum decay processes consist of the following phases [10, 11]: (i) The early time initial phase, (ii) The canonical or exponential phase, and (iii) The late time non-exponential phase. Consequently, the first phase and the third one are missed in the case of the radioactive decay law only. The theoretical analysis of quantum decaying processes shows that for the late time, the survival probability of the system to be in its initial metastable state (i.e. the decay law), should tend to zero as $t \to \infty$ much more slowly than any exponential function of time, and that, as a function of time, it has the inverse power–like form at this regime of time [10, 12]. So, all consequences of the decay process of the dark energy as a quantum decay process can be found only if one uses a quantum decay law to describe decaying metastable dark energy. (This idea was considered in [13]). It also appears that energy of the system is different in each phase. In particular at the late time phase it is much smaller than at the canonical phase and this property will be discussed in this talk.

2. Decay of a dark energy as a quantum decay process

Studying quantum unstable systems, one usually uses the survival probability (the decay law)

$$\mathcal{P}(t) = |\mathcal{A}(t)|^2, \tag{1}$$

to describe changes in time of the unstable system. This means these properties of the survival amplitudes

$$\mathcal{A}(t) = \langle \phi | \phi(t) \rangle \tag{2}$$

are analyzed in this case. Here, a vector $|\phi\rangle$ represents the unstable state of the considered system and $|\phi(t)\rangle$ is the solution gotten from the Schrödinger equation

$$i\hbar \frac{\partial}{\partial t} |\phi(t)\rangle = \mathfrak{H} |\phi(t)\rangle. \tag{3}$$

For the case, the initial condition for Eq. (3) is usually assumed as

$$|\phi(t = t_0 \equiv 0)\rangle \stackrel{\text{def}}{=} |\phi\rangle, \text{ or equivalently, } \mathcal{A}(0) = 1. \tag{4}$$

The $\mathfrak{H}$ symbol denotes the complete (full) self-adjoint Hamiltonian of the system in Eq. (3).

Using a basis in $\mathcal{H}$, which is build from normalized eigenvectors $|E\rangle$ of $\mathfrak{H}$ (where $E \in \sigma_c(\mathfrak{H})$), we can expand $|\phi\rangle$ in this basis, and then we can express the amplitude $\mathcal{A}(t)$ as the Fourier integral

$$\mathcal{A}(t) \equiv \mathcal{A}(t - t_0) = \int_{E_{\min}}^{\infty} \omega(E)\, e^{-\frac{i}{\hbar} E\,(t - t_0)}\, dE, \tag{5}$$

where $\omega(E) = \omega(E)^*$ and $\omega(E) > 0$ is the probability to find the energy of the system for the state $|\phi\rangle$ between E and $E + dE$. The above relation (5) denotes that the survival amplitude $\mathcal{A}(t)$ is given by a Fourier transform of an absolute integrable function $\omega(E)$. When we apply the Riemann-Lebesgue Lemma to the integral (5) then we can conclude that there must be $\mathcal{A}(t) \to 0$ as $t \to \infty$. This property and relation (5) are an essence of the Fock–Krylov theory of unstable states [14, 15].

In this approach, the amplitude $\mathcal{A}(t)$ and, consequently, the decay law $\mathcal{P}(t)$ of the unstable state $|\phi\rangle$, are totally determined by the density of the energy distribution $\omega(E)$ for the system for this state [14, 15] (see also [10, 16], and so on). (This approach is also used in Quantum Field Theory models [17]).

Note that the amplitude $\mathcal{A}(t)$ contains information about the decay law $\mathcal{P}(t)$ for the state $|\phi\rangle$: About the decay rate γ_ϕ for this state and the energy E_ϕ of the system for this state. This information can be extracted from $\mathcal{A}(t)$. It can be done using the rigorous equation governing the time evolution in the subspace of unstable states, $\mathcal{H}_{\parallel} \ni |\phi\rangle_{\parallel} \equiv |\phi\rangle$. Such an equation results from the Schrödinger equation (3) for the total state space $\mathcal{H}$.

When we use the Schrödinger equation (3), then we can find that within the problem considered

$$i\hbar \frac{\partial}{\partial t} \langle\phi|\phi(t)\rangle = \langle\phi|\mathfrak{H}|\phi(t)\rangle. \tag{6}$$

From the above relation it follows that the amplitude $\mathcal{A}(t)$ satisfies the following equation

$$i\hbar \frac{\partial \mathcal{A}(t)}{\partial t} = h(t)\, \mathcal{A}(t), \tag{7}$$

where

$$h(t) = \frac{\langle\phi|\mathfrak{H}|\phi(t)\rangle}{\mathcal{A}(t)} \equiv \frac{i\hbar}{\mathcal{A}(t)} \frac{\partial \mathcal{A}(t)}{\partial t}, \tag{8}$$

$h(t)$ is the effective Hamiltonian which governs the time evolution in the one–dimensional subspace of unstable states $\mathcal{H}_{\parallel} = \mathbb{P}\mathcal{H}$, where $\mathbb{P} = |\phi\rangle\langle\phi|$ (see [18] and also [19, 20] and references therein). The subspace $\mathcal{H} \ominus \mathcal{H}_{\parallel} = \mathcal{H}_{\perp} \equiv \mathbb{Q}\mathcal{H}$ is a subspace of decay products. Here, we have $\mathbb{Q} = \mathbb{I} - \mathbb{P}$. One meets the effective Hamiltonian $h(t)$ if one starts with the Schrödinger equation for the total state space $\mathcal{H}$ and looks for the rigorous evolution equation for a distinguished subspace of states $\mathcal{H}_{\parallel} \subset \mathcal{H}$ [18, 21, 22]. Generally, $h(t)$ is a complex function of time but in the case of $\mathcal{H}_{\parallel}$ of dimension two or more, the effective Hamiltonian which governs the time evolution in $\mathcal{H}_{\parallel}$ is a non-hermitian matrix $H_{\parallel}$ or a non-hermitian operator. There is

$$h(t) = E_{\phi}(t) - \frac{i}{2}\gamma_{\phi}(t), \tag{9}$$

where $E_{\phi}(t) = \Re\,[h(t)]$, $\gamma_{\phi}(t) = -2\,\Im\,[h(t)]$, are respectively instantaneous energy (mass) $E_{\phi}(t)$ and the instantaneous decay rate, $\gamma_{\phi}(t)$. (Here $\Re\,(z)$ and $\Im\,(z)$ mean the real and imaginary parts of z). The quantity $\gamma_{\phi}(t) = -2\,\Im\,[h(t)]$ can be interpreted as the decay rate since it satisfies the definition of the decay rate which is used in quantum theory: When we use (8) it is easy to check that

$$\frac{\gamma_{\phi}(t)}{\hbar} \stackrel{\text{def}}{=} -\frac{1}{\mathcal{P}(t)} \frac{\partial \mathcal{P}(t)}{\partial t} = -\frac{1}{|\mathcal{A}(t)|^2} \frac{\partial |\mathcal{A}(t)|^2}{\partial t} \equiv -\frac{2}{\hbar}\,\Im\,[h(t)]. \tag{10}$$

We have the relation $|\phi(t)\rangle = \exp\,[-\frac{i}{\hbar}t\mathfrak{H}]|\phi\rangle$. It is easy to see that this property and hermiticity of $\mathfrak{H}$ lead to the conclusion that [10]

$$(\mathcal{A}(t))^{*} = \mathcal{A}(-t). \tag{11}$$

The conclusion that results from the above property and from the relation (8) is

$$h(-t) = (h(t))^{*}\,. \tag{12}$$

That's why there should be:

$$E_{\phi}(-t) = E_{\phi}(t), \quad \text{and} \quad \gamma_{\phi}(-t) = -\gamma_{\phi}(t), \tag{13}$$

So, the instantaneous energy $E_{\phi}(t) = \Re\,[h(t)]$ is an even function of time t and the instantaneous decay rate $\gamma_{\phi}(t) = -2\,\Im\,[h(t)]$ is an odd function of t.

In the large literature, in many cases quantum unstable systems are studied using the Fock–Krylov theory for the Breit–Wigner energy density distribution function $\omega_{BW}(E)$. The use $\omega_{BW}(E)$ is convenient since it describes relatively well a large class of unstable systems and permits to find an analytical form of the survival amplitude $\mathcal{A}(t)$ (see, eg. [23] and other papers). It is because the decay curves which are obtained in this simplest case are very similar in the form to the curves which are calculated for the more general $\omega(E)$, (see [16] and analysis in [10]). Therefore looking for the most typical properties of the decay process it is enough to make the relevant calculations for $\omega(E)$ which is modeled by the Breit–Wigner distribution of the energy density $\omega_{BW}(E)$. For this energy density distribution, we can find relatively easy an analytical form of $\mathcal{A}(t)$ for very late times as well as an analytical asymptotic form of $h(t)$, $E(t)$ and $\gamma(t)$ for such times. So, we assume that

$$\omega(E) \equiv \omega_{\rm BW}(E) \stackrel{\rm def}{=} \frac{N}{2\pi}\,\Theta(E - E_{\rm min})\;\frac{\Gamma_0}{(E-E_0)^2 + (\frac{\Gamma_0}{2})^2}, \tag{14}$$

where N is a normalization constant. The parameters E_0 and Γ_0 correspond to energy of the system for the unstable state $|\phi\rangle$ and its decay rate at the exponential (or canonical) regime of the decay process. The $E_{\rm min}$ parameter is the minimal (the lowest) energy of the system. When we insert $\omega_{\rm BW}(E)$ into formula (5) for the amplitude $\mathcal{A}(t)$ and assume for simplicity that $t_0 = 0$, after some algebra we can find that

$$\mathcal{A}(t) = \frac{N}{2\pi}\,e^{-\frac{i}{\hbar}E_0 t}\,I_\beta\left(\frac{\Gamma_0 t}{\hbar}\right), \tag{15}$$

where

$$I_\beta(\tau) \stackrel{\rm def}{=} \int_{-\beta}^{\infty} \frac{1}{\eta^2 + \frac{1}{4}}\,e^{-i\eta\tau}\,d\eta. \tag{16}$$

Here, $\tau = \frac{\Gamma_0\,t}{\hbar} \equiv \frac{t}{\tau_0}$, τ_0 is the lifetime, $\tau_0 = \frac{\hbar}{\Gamma_0}$, and $\beta = \frac{E_0 - E_{min}}{\Gamma_0} > 0$. The integral $I_\beta(\tau)$ can be decomposed as follows

$$I_\beta(\tau) = I_\beta^{\rm pole}(\tau) + I_\beta^{L}(\tau), \tag{17}$$

where

$$I_\beta^{\rm pole}(\tau) = \int_{-\infty}^{\infty} \frac{1}{\eta^2 + \frac{1}{4}}\;e^{-i\eta\tau}\;d\eta \equiv 2\pi\,e^{-\frac{\tau}{2}}, \tag{18}$$

and

$$I_\beta^{L}(\tau) = -\int_{+\beta}^{\infty} \frac{1}{\eta^2 + \frac{1}{4}}\;e^{+i\eta\tau}\;d\eta. \tag{19}$$

(The integral $I_\beta^L(\tau)$ can be rewritten in terms of the integral–exponential function [23, 19, 20, 24] (for a definition, see [25, 26])). The above relation (17) implies a natural decomposition of the survival amplitude $\mathcal{A}(t)$ into two parts

$$\mathcal{A}(t) = \mathcal{A}_c(t) + \mathcal{A}_L(t), \tag{20}$$

where

$$\mathcal{A}_c(t) = \frac{N}{2\pi}\,e^{-\frac{i}{\hbar}E_0 t}\,I_\beta^{\rm pole}\left(\frac{\Gamma_0 t}{\hbar}\right) \equiv N\,e^{-\frac{i}{\hbar}E_0 t}\,e^{-\frac{\Gamma_0\,t}{2}}, \tag{21}$$

and

$$\mathcal{A}_L(t) = \frac{N}{2\pi}\,e^{-\frac{i}{\hbar}E_0 t}\,I_\beta^{L}\left(\frac{\Gamma_0 t}{\hbar}\right), \tag{22}$$

$\mathcal{A}_c(t)$ is the canonical part of the amplitude $\mathcal{A}(t)$ which describes the pole contribution into $\mathcal{A}(t)$ and $\mathcal{A}_L(t)$ represents the remaining part of $\mathcal{A}(t)$.

From the decomposition (20), we have that in the general case within the considered model, the survival probability (1) contains the following parts

$$\begin{aligned}\mathcal{P}(t) = |\mathcal{A}(t)|^2 \;&\equiv\; |\mathcal{A}_c(t) + \mathcal{A}_L(t)|^2 \\ &=\; |\mathcal{A}_c(t)|^2 \,+\, 2\,\Re\left[\mathcal{A}_c(t)\,(\mathcal{A}_L(t))^*\right] \,+\, |\mathcal{A}_L(t)|^2.\end{aligned} \tag{23}$$

IOP Conf. Series: Journal of Physics: Conf. Series **1275** (2019) 012040 doi:10.1088/1742-6596/1275/1/012040

The above relation is especially useful when one looks for a contribution of late time properties of the quantum unstable system into the survival amplitude.

The late time form of the integral $I_{\beta}^{L}(\tau)$ and hence the late time form of the amplitude $\mathcal{A}_L(t)$ is relatively easy to find performing the integration by parts in (19). It appears that for $t \to \infty$ (or $\tau \to \infty$) the leading term of the late time asymptotic expansion of the integral $I_{\beta}^{L}(\tau)$ takes the following form [24],

$$\begin{aligned} I_{\beta}^{L}(\tau) \simeq \frac{i}{\tau}\,\frac{e^{i\beta\tau}}{\beta^2+\frac{1}{4}}\Big\{ -1 &+ \frac{2\beta}{\beta^2+\frac{1}{4}}\,\frac{i}{\tau} + \frac{2}{\beta^2+\frac{1}{4}}\Big[1-\frac{4\beta^2}{\beta^2+\frac{1}{4}}\Big]\Big(\frac{i}{\tau}\Big)^2 \\ &+ \frac{24\beta}{(\beta^2+\frac{1}{4})^2}\Big[\frac{2\beta^2}{\beta^2+\frac{1}{4}}-1\Big]\Big(\frac{i}{\tau}\Big)^3 \\ &+ \frac{24}{(\beta^2+\frac{1}{4})^2}\Big[-\frac{16\beta^4}{(\beta^2+\frac{1}{4})^2}+\frac{12\beta^2}{\beta^2+\frac{1}{4}}-1\Big]\Big(\frac{i}{\tau}\Big)^4 + \ldots\Big\}. \end{aligned} \tag{24}$$

So, when we insert (24) into (22) then we can find a late time form of $\mathcal{A}_L(t)$. The typical decay curve obtained numerically for $\omega(E) = \omega_{BW}(E)$ is presented in Fig. (1).

Now we can analyze properties of the instantaneous energy $E(t)$ and instantaneous decay rate $\gamma_{\phi}(t)$ for the considered model. These quantities are defined by using the effective Hamiltonian $h(t)$. Looking for $h(t)$ we need the quantity $i\hbar\,\frac{\partial \mathcal{A}(t)}{\partial t}$ (see (8)). From (15) we can find that

$$i\hbar\frac{\partial \mathcal{A}(t)}{\partial t} = E_0\,\mathcal{A}(t) + \Gamma_0\,\frac{N}{2\pi}\,e^{-\frac{i}{\hbar}E_0 t}\,J_{\beta}(\tau(t)), \tag{25}$$

where

$$J_{\beta}(\tau) = \int_{-\beta}^{\infty}\frac{x}{x^2+\frac{1}{4}}\,e^{-ix\tau}\,dx \equiv i\frac{\partial I_{\beta}(\tau)}{\partial \tau}. \tag{26}$$

If to use (15), (25) and (8) the conclusion follows,

$$h(t) = i\hbar\frac{1}{\mathcal{A}(t)}\,\frac{\partial \mathcal{A}(t)}{\partial t} = E_0 + \Gamma_0\,\frac{J_{\beta}(\tau(t))}{I_{\beta}(\tau(t))}, \tag{27}$$

which means that

$$E(t) = \Re\,[h(t)] = E_0 + \Gamma_0\,\Re\left[\frac{J_{\beta}(\tau(t))}{I_{\beta}(\tau(t))}\right]. \tag{28}$$

For a visualization of properties of $E(t)$ it is convenient to use the following function:

$$\kappa(t) \stackrel{\text{def}}{=} \frac{E(t)-E_{min}}{E_0-E_{min}}. \tag{29}$$

The use of the relation (28) gives

$$E(t) - E_{min} = E_0 - E_{min} + \Gamma_0\,\Re\Big[\frac{J_{\beta}(\tau)}{I_{\beta}(\tau)}\Big], \tag{30}$$

When to divide two sides of the above equation by $E_0 - E_{min}$ then we obtain the function $\kappa(t)$ (see (29)) which we are looking for:

$$\kappa(\tau(t)) = 1 + \frac{1}{\beta}\,\Re\Big[\frac{J_{\beta}(\tau(t))}{I_{\beta}(\tau(t))}\Big]. \tag{31}$$

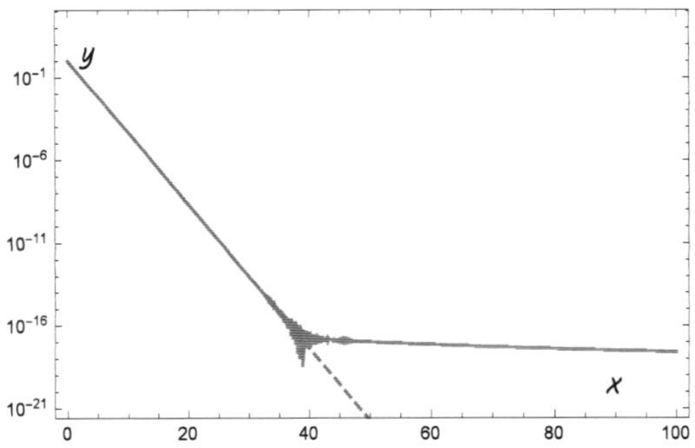

Figure 1. Decay curves obtained for $\omega_{BW}(E)$ given by Eq. (14). Axes: $x = t/\tau_0$ — time t is measured in lifetimes τ_0, y — survival probabilities on a logarithmic scale. The solid line: the decay curve $\mathcal{P}(t) = |\mathcal{A}(t)|^2$; The dotted line: the canonical decay curve $\mathcal{P}_c(t) = |\mathcal{A}_c(t)|^2$. The case $\beta = 1000$.

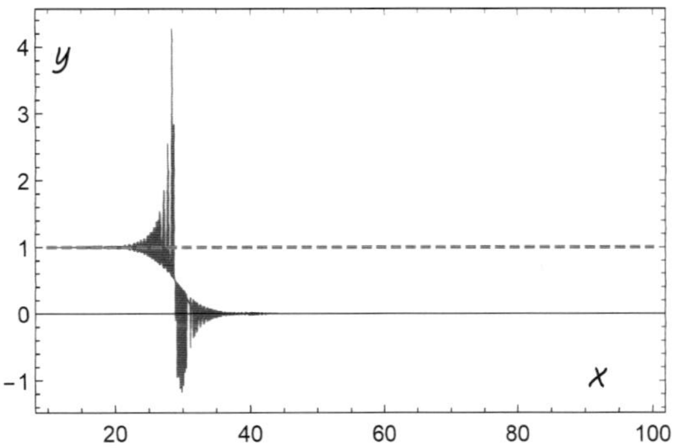

Figure 2. The instantaneous energy $E(t)$ as a function of time t. Axes: $y = \kappa(t) = \frac{E(t)-E_{min}}{E_0-E_{min}}$, $x = t/\tau_0$. The horizontal red dashed line denotes $y = \kappa(t) = 1$, that is $E(t) = E_0$. The case $\beta = 50$.

The typical behavior of the instantaneous energy $E(t)$ changing in time t is shown in Fig. (2).

To find the asymptotic form of $h(t)$ we need to know the late time form for $J_\beta(\tau)$. Using similar methods as in the case of $I_\beta^L(\tau)$ we find for $\tau \to \infty$ [24]

$$\begin{aligned} J_\beta(\tau) \simeq \frac{i}{\tau}\frac{e^{i\beta\tau}}{\beta^2+\frac{1}{4}}\Big\{\beta + \Big[1 - \frac{2\beta^2}{\beta^2+\frac{1}{4}}\Big]\frac{i}{\tau} + \frac{2\beta}{\beta^2+\frac{1}{4}}\Big[\frac{4\beta^2}{\beta^2+\frac{1}{4}} - 3\Big]\Big(\frac{i}{\tau}\Big)^2 \\ + \frac{6}{\beta^2+\frac{1}{4}}\Big[-\frac{8\beta^4}{(\beta^2+\frac{1}{4})^2} + \frac{8\beta^2}{\beta^2+\frac{1}{4}} - 1\Big]\Big(\frac{i}{\tau}\Big)^3 \\ + \frac{24\beta}{(\beta^2+\frac{1}{4})^2}\Big[\frac{16\beta^4}{(\beta^2+\frac{1}{4})^2} - \frac{20\beta^2}{\beta^2+\frac{1}{4}} + 5\Big]\Big(\frac{i}{\tau}\Big)^4 \ldots\Big\}. \end{aligned} \quad (32)$$

Starting from the asymptotic expression (32) and using formula (27) we can find the late time asymptotic form of $h(t)$ and hence of $E(t)$ and $\gamma_\phi(t)$ for the considered model [24],

$$\begin{aligned} E(t)|_{t\to\infty} &= \Re\,[h(t)]|_{t\to\infty} \quad (33) \\ &\simeq E_{\min} - 2\frac{E_0 - E_{min}}{\Gamma_0^2\,(\beta^2+\frac{1}{4})}\Big(\frac{\hbar}{t}\Big)^2 \\ &+ \frac{1}{4}\frac{6 - 21\beta + 48\beta^2 - 64\beta^3 - 288\beta^4 + 464\beta^5}{\Gamma_0^3\,(\beta^2+\frac{1}{4})^4}\Big(\frac{\hbar}{t}\Big)^4 + \ldots. \end{aligned}$$

This last relation is valid for $t > T_C$, where T_C is the cross–over time, i.e. the time when the canonical exponential and late time inverse power law contributions to the survival amplitude start to be comparable.

In the general case we get for $t \to \infty$,

$$E(t)|_{t\to\infty} \simeq E_{\min} + \sum_{k\geq 1} f_{2k}\,(\frac{\hbar}{t})^{2k}, \quad (34)$$

where $f_{2k} = (f_{2k})^*$. This expansion is consistent with the property (13), which says that $E(t)|_{t\to\infty}$ must be an even function of time t.

All the above described properties of the quantum unstable system are general one and are independent on the specific form of interactions which force the decay process or a mechanism responsible for such a process. General properties of the decay law are the same in all quantum decay processes. The decay law always has three phases: an initial non–exponential phase, the exponential (canonical) phase and the late time non-exponential phase. This concerns also such a quantum process like the decaying dark energy.

When we analyze the temporal behavior of the decaying dark energy as a function of time t then the following strategies can be used: we can analyze the decay process as the quantum decay process which is governed by the quantum mechanical decay law $\mathcal{P}(t)$, or we can consider the dark energy as the energy density in the meta–stable false vacuum state and analyze temporal properties of this energy density similarly to the properties of the instantaneous energy of the system for the unstable state $|\phi\rangle$.

In the case of the first strategy using

$$\tilde{\rho}_{\rm de}(t) \stackrel{\rm def}{=} \rho_{\rm de}(t) - \rho_{\rm bare}, \tag{35}$$

we have $\lim_{t\to\infty}(\rho_{\rm de}(t) - \rho_{\rm bare}) = 0$ and we can describe such a process assuming that (see [13]), $\tilde{\rho}_{\rm de}(t) = \tilde{\rho}_{\rm de}(t_0)\,\mathcal{P}(t)$ where $\mathcal{P}(t)$ is given by the relation (23). Such an approach can be deliberated as a generalization of the idea studied by Shafieloo *et al.* in [9]. (This was done in the paper [13]).

Considering the second strategy: As it was mentioned earlier the general properties of the quantum decay process shortly described above are independent on the mechanism responsible for a decay. The inherent mechanism of quantum decay processes presented earlier in Fig. (2) is a reduction of the energy of the system for the meta–stable state $|\phi\rangle$ from the large values measured at exponential decays time region to much smaller values for the asymptotically late times region. Our hypothesis is that this pure quantum effect occurs also in the case of the decaying dark energy.

Now we will consider the cosmological scenario for which the false vacuum decays at the inflationary stage of the Universe. This corresponds with the hypothesis which was analyzed by Krauss and Dent in [5, 6]. (The hypothesis in which some false vacuum regions do survive well up to the time T_C or later was formulated in these papers). Let $|\phi\rangle = |0\rangle^{false}$, be a false state, $|0\rangle^{true}$ – a true state of vacuum and $E_0 = E_0^{false}$ be energy of a state corresponding to the false vacuum measured at the canonical decay time, which leads to the vacuum energy density calculated by means of quantum field theory methods. Similarly, let E_0^{true} be the energy of the true vacuum (i.e. the true ground state of the system).

From the results presented earlier we get that the energy of those false vacuum regions which survived up to T_C and much later differs from E_0^{false}. Now, when we assume that $E_0^{true} \equiv E_{min}$ and $E_0^{false} = E_0$ and take into account results described above then we can conclude that energy of the system in the false vacuum state has the following form at asymptotically late times $t \gg T_C$.

$$E^{false}(t) \simeq E_0^{true} + f_2\frac{\hbar^2}{t^2} + f_4\frac{\hbar^4}{t^4}\ldots \neq E_0^{false}. \tag{36}$$

Next we will identify $\rho_{\rm de}(t_0)$ with energy E_0 of the unstable system divided by the volume V_0 (where V_0 is the volume of the system at $t = t_0$): $\rho_{\rm de}(t_0) \equiv \rho_{\rm de}^{\rm qft} \stackrel{\rm def}{=} \rho_{\rm de}^0 = \frac{E_0}{V_0}$ and $\rho_{\rm bare} = \frac{E_{\rm min}}{V_0}$, ($\rho_{\rm de}^{\rm qft}$ is the vacuum energy density calculated by the use of quantum field theory methods). Hence at times $t \gg T_C$,

$$\rho_{\rm de}(t) = \rho_0^{false}(t) \simeq \rho_{bare} + \frac{d_2}{t^2} + \frac{d_4}{t^4} + \ldots, \quad (t \gg T_C), \tag{37}$$

where $d_{2k} = d_{2k}^*$. Analogous relations are for $\Lambda(t) = \frac{8\pi G}{c^2}\,\rho(t)$, or $\Lambda(t) = 8\pi G\,\rho(t)$ where we assume $\hbar = c = 1$ units:

$$\Lambda(t) \simeq \Lambda_{bare} + \frac{\alpha_2}{t^2} + \frac{\alpha_4}{t^4} + \ldots, \quad (t \gg T_C). \tag{38}$$

The coefficients d_{2k} and α_{2k} are functions of β and Γ_0 (see (33)) and in this case we get that

$$\beta = \frac{E_0 - E_{\rm min}}{\Gamma_0} \equiv \frac{\rho_{\rm de}^0 - \rho_{\rm bare}}{\gamma_0} > 0, \tag{39}$$

(where $\gamma_0 = \Gamma_0/V_0$), or equivalently, $\Gamma_0/V_0 \equiv \frac{\rho_{\rm de}^0 - \rho_{\rm bare}}{\beta}$.

We get the good approximation of Eq. (38) replacing the cosmological time t with the Hubble cosmological scale time $t_{\rm H} = \frac{1}{H}$. As the result, instead of (38) we get

$$\Lambda(t) = \Lambda(H(t)) \simeq \Lambda_{bare} + \alpha_2\,(H(t))^2 + \alpha_4\,(H(t))^4 + \ldots\,, \tag{40}$$

(where $H(t)$ is the Hubble function), which is exactly the parametrization considered in [27, 28] and in many papers of these and other authors.

Note that the form of $\kappa(t)$ is unchanged when one passes from energies $E(t)$, E_0, E_{min} to the above defined energy density ρ and Λ:

$$\begin{aligned} \kappa(t) &= \frac{E(t) - E_{\rm min}}{E_0 - E_{\rm min}} \equiv \frac{\frac{E(t)}{V_0} - \frac{E_{\rm min}}{V_0}}{\frac{E_0}{V_0} - \frac{E_{\rm min}}{V_0}} \\ &= \frac{\rho_{\rm de}(t) - \rho_{\rm bare}}{\rho_{\rm de}^0 - \rho_{\rm bare}} = \frac{\Lambda(t) - \Lambda_{\rm bare}}{\Lambda_{\rm qft} - \Lambda_{\rm bare}}, \end{aligned} \tag{41}$$

This means that time evolution of $\rho_{\rm de}(t)$ and $\Lambda(t)$ should the same form as it is presented earlier in Fig. (2).

3. Cosmological implications of decaying vacuum

Now we will consider cosmological equations with $\rho_{\rm de} = \Lambda$. It describes the density of dark energy. Here, the form of Λ has been derived in the previous section and is given by

$$\Lambda \equiv \Lambda_{\rm eff}(t) = \Lambda_{\rm bare} + \delta\Lambda(t) = \rho_{\rm de}, \tag{42}$$

where $\delta\Lambda(t)$ describes quantum corrections given by a series with respect to $\frac{1}{t}$, i.e.

$$\delta\Lambda(t) = \sum_{n=1}^{\infty} \alpha_{2n} \left(\frac{1}{t}\right)^{2n}, \tag{43}$$

and t is the cosmological scale time. The function $\delta\Lambda(t)$ has a reflection symmetry with respect to the cosmological time ($\delta\Lambda(-t) = \delta\Lambda(t)$).

The cosmological equations has the following form:

$$3H(t)^2 = \rho_{\rm m}(t) + \rho_{\rm de}(t), \tag{44}$$

$$\dot{\rho}_{\rm m} = -3H(t)\rho_{\rm m}(t) - \dot{\rho}_{\rm de}(t), \tag{45}$$

where $H(t) \equiv \frac{\dot{a}}{a}$ is the Hubble function, the dot denotes the derivative with respect to time: $\dot{} \equiv \frac{d}{dt}$, and a is the scale factor. The equation (44) is called the Friedmann equation and the equation (45) is called the continuous equation.

IOP Conf. Series: Journal of Physics: Conf. Series **1275** (2019) 012040 doi:10.1088/1742-6596/1275/1/012040

The equations of state for matter and dark energy are assumed to have the form $p_{\rm m} = 0$ and $p_{\rm de} = -\rho_{\rm de}$ respectively, where $p_{\rm m}$ is a pressure of matter and $p_{\rm de}$ is a pressure of the dark energy. The Einstein field equation for the FRW metric is given by

$$\frac{dH(t)}{dt} = -\frac{1}{2}(\rho_{\rm eff}(t) + p_{\rm eff}(t)) = -\frac{1}{2}\rho_{\rm m}(t) = -\frac{1}{2}\left(3H(t)^2 - \Lambda_{\rm bare} - \delta\Lambda(t)\right), \tag{46}$$

where $\rho_{\rm m}$ is the density of the matter, $\rho_{\rm eff} = \rho_{\rm m} + \rho_{\rm de}$ and $p_{\rm eff} = p_{\rm m} + p_{\rm de}$.

Now we cut off the series (43) on the second term. As a result, we get

$$\delta\Lambda(t) = \frac{\alpha_2}{t^2} + \frac{\alpha_4}{t^4}. \tag{47}$$

In this case, the Friedmann equation has the following form

$$3H^2 = \rho_{\rm m} + \Lambda_{\rm bare} + \frac{\alpha_2}{t^2} + \frac{\alpha_4}{t^4}. \tag{48}$$

For the case $\delta\Lambda(t) = \frac{\alpha_2}{t^2}$ ($\alpha_4 = 0$), the solution of Eq. (46) looks as follows

$$H(t) = \frac{1-2n}{3t} + \sqrt{\frac{\Lambda_{\rm bare}}{3}}\,\frac{I_{n-1}\left(\frac{\sqrt{3\Lambda_{\rm bare}}}{2}t\right)}{I_n\left(\frac{\sqrt{3\Lambda_{\rm bare}}}{2}t\right)}, \tag{49}$$

where I_n denotes the Bessel function of the first kind. The parameter n is equal to $\frac{1}{2}\sqrt{1+3\alpha_2}$.

Now we can obtain from the Friedmann equation the formula for $\rho_{\rm m}(t)$ for the case $\alpha_4 = 0$. The result is:

$$\rho_{\rm m}(t) = \left(\frac{1-2n}{\sqrt{3}t} + \sqrt{\Lambda_{\rm bare}}\,\frac{I_{n-1}\left(\frac{\sqrt{3\Lambda_{\rm bare}}}{2}t\right)}{I_n\left(\frac{\sqrt{3\Lambda_{\rm bare}}}{2}t\right)}\right)^2 - \Lambda_{\rm bare} - \frac{\alpha_2}{t^2}. \tag{50}$$

If $\alpha_4 \neq 0$ then $\delta\Lambda(t)$ has the form given by (47) and the Friedmann equation rewritten in dimensionless terms looks as follows:

$$\frac{H(t)^2}{H_0^2} = \Omega_{{\rm m},0} f(t) + \Omega_{\Lambda_{\rm bare},0} + \Omega_{\alpha_2,0}\frac{T_0^2}{t^2} + \Omega_{\alpha_4,0}\frac{T_0^4}{t^4}, \tag{51}$$

where $\Omega_{{\rm m},0} = \frac{\rho_{{\rm m},0}}{3H_0^2}$, $\Omega_{\Lambda_{\rm bare},0} = \frac{\Lambda_{\rm bare}}{3H_0^2}$, $\Omega_{\alpha_2,0} = \frac{\alpha_2}{3H_0^2T_0^2}$, $\Omega_{\alpha_4,0} = \frac{\alpha_4}{3H_0^2T_0^4}$, H_0 is the today's value of the Hubble constant, $f(t) = \frac{\rho_{\rm m}(t)}{\rho_{{\rm m},0}}$, and T_0 is the age of the Universe. Here, $\rho_{{\rm m},0}$ parameter denotes the present day value of the energy density of matter.

In our model we observe a small deviation from the canonical scaling of the energy density of matter ($\rho_{\rm m} \propto a^{-3}$) [27, 28]. This deviation is defined as

$$\lambda(t) = \frac{\ln\frac{\rho_{\rm m}(t)}{\rho_{{\rm m},0}}}{\ln a(t)} + 3. \tag{52}$$

As result we get

$$\rho_{\rm m} = \rho_{{\rm m},0}a^{-3+\lambda}. \tag{53}$$

Our model has five parameters: H_0, $\Omega_{{\rm m},0}$, $\Omega_{\Lambda_{\rm bare},0}$, $\Omega_{\alpha_2,0}$, and $\Omega_{\alpha_4,0}$. But, for the present age of the Universe, from the above equation, we obtain the following condition

$$1 = \Omega_{{\rm m},0} + \Omega_{\Lambda_{\rm bare},0} + \Omega_{\alpha_2,0} + \Omega_{\alpha_4,0}, \tag{54}$$

IOP Conf. Series: Journal of Physics: Conf. Series **1275** (2019) 012040 doi:10.1088/1742-6596/1275/1/012040

which reduces the number of independent parameters from five to four. Now we need to determine all these four parameters using observational data. Having all these parameters we can use the above presented equations to test our model. In order to do this we used parameter obtained from the astronomical data: supernovae of type Ia, Baryon Acoustic Oscillations (Sloan Digital Sky Survey Release 7, 6dF Galaxy Redshift Survey, WiggleZ), measurements of the Hubble parameter $H(z)$ of galaxies, the Alcock-Paczynski test and measurements of Cosmic Microwave Background (CMB) and lensing by Planck satellite and low-ℓ polarization from the WMAP (WP) (the acoustic scale, the shift parameter).

We took the data of supernovae of type Ia from the Union 2.1 dataset [29]. In this case, the likelihood function is given by the following equation

$$\ln L_{\text{SNIa}} = -\frac{1}{2}[A - B^2/C + \log(C/(2\pi))], \tag{55}$$

where $A = (\mu^{\text{obs}} - \mu^{\text{th}})\mathbb{C}^{-1}(\mu^{\text{obs}} - \mu^{\text{th}})$, $B = \mathbb{C}^{-1}(\mu^{\text{obs}} - \mu^{\text{th}})$, $C = \text{Tr}\mathbb{C}^{-1}$ and $\mathbb{C}$ is a covariance matrix for SNIa. The observer distance modulus μ^{obs} is given by equation $\mu^{\text{obs}} = m - M$ (where m is the apparent magnitude and M is the absolute magnitude of SNIa). The theoretical distance modulus is defined by equation $\mu^{\text{th}} = 5\log_{10} D_L + 25$ (where the luminosity distance has the form $D_L = c(1+z)\int_0^z \frac{dz'}{H(z)}$).

For our estimation, we got the BAO data: Sloan Digital Sky Survey Release 7 (SDSS DR7) dataset at $z = 0.275$ [30], 6dF Galaxy Redshift Survey measurements at redshift $z = 0.1$ [31], and WiggleZ measurements at redshift $z = 0.44, 0.60, 0.73$ [32]. In this case the likelihood function is given by the equation

$$\ln L_{\text{BAO}} = -\frac{1}{2}\left(\mathbf{d}^{\text{obs}} - \frac{r_s(z_d)}{D_V(\mathbf{z})}\right)\mathbb{C}^{-1}\left(\mathbf{d}^{\text{obs}} - \frac{r_s(z_d)}{D_V(\mathbf{z})}\right), \tag{56}$$

where $r_s(z_d)$ is the sound horizon for the drag epoch [33, 34].

We took measurements of the Hubble parameter $H(z)$ of galaxies from [35, 36, 37]. In this case, the likelihood function is defined by the following formula

$$\ln L_{H(z)} = -\frac{1}{2}\sum_{i=1}^{N}\left(\frac{H(z_i)^{\text{obs}} - H(z_i)^{\text{th}}}{\sigma_i}\right)^2. \tag{57}$$

The likelihood function for the Alcock-Paczynski test [38, 39] looks as follows

$$\ln L_{AP} = -\frac{1}{2}\sum_i \frac{\left(AP^{th}(z_i) - AP^{obs}(z_i)\right)^2}{\sigma^2}. \tag{58}$$

where $AP(z)^{\text{th}} \equiv \frac{H(z)}{z}\int_0^z \frac{dz'}{H(z')}$ and $AP(z_i)^{\text{obs}}$ are observational data [40, 41, 42, 43, 44, 45, 46, 47, 48].

We use the likelihood function for the measurements of CMB [3] and lensing by Planck, and low-ℓ polarization from the WMAP (WP) which can be written in the following form

$$\ln L_{\text{CMB+lensing}} = -\frac{1}{2}(\mathbf{x}^{\text{th}} - \mathbf{x}^{\text{obs}})\mathbb{C}^{-1}(\mathbf{x}^{\text{th}} - \mathbf{x}^{\text{obs}}), \tag{59}$$

where $\mathbb{C}$ is the covariance matrix describing the errors and $\mathbf{x}$ is a vector having the following components: the acoustic scale l_A, the shift parameter R and $\Omega_b h^2$,

$$l_A = \frac{\pi}{r_s(z^*)}c\int_0^{z^*}\frac{dz'}{H(z')} \tag{60}$$

$$R = \sqrt{\Omega_{\text{m},0}H_0^2}\int_0^{z^*}\frac{dz'}{H(z')}, \tag{61}$$

IOP Conf. Series: Journal of Physics: Conf. Series **1275** (2019) 012040 doi:10.1088/1742-6596/1275/1/012040

where z^* is the redshift at the epoch of the recombination [33].

The final formula for likelihood function has the following form

$$L_{\rm tot} = L_{\rm SNIa} L_{\rm BAO} L_{\rm AP} L_{H(z)} L_{\rm CMB+lensing}. \tag{62}$$

Performing the statistical analysis we used our own code CosmoDarkBox. where Metropolis-Hastings algorithm [49, 50] is used.

From (33), we conclude that α_2 parameter is negative so we can we assume for the statistical analysis that $\Omega_{\alpha_2,0} < 0$. Results of our estimations are presented below in Tab. 1. Figs 3, 4, 5, and 6 show the predictions of our model based on the values of parameters resulting from the fitting process.

Table 1. The best fit and errors for the estimated model with H_0 assumed as $67.74 \frac{\rm km}{\rm s\ Mpc}$, $\Omega_{\rm m,0}$, $\Omega_{\alpha_2,0}$, $\Omega_{\alpha_4,0}$ and $\Omega_{\rm b,0}$ assumed as 0.048468. The value of χ^2 per degree of freedom is equal to 0.094 for our model. For ΛCDM model, χ^2 per degree of freedom is equal to 0.099.

parameter	best fit	68% CL	95% CL
$\Omega_{\rm m,0}$	0.3045	$+0.0045$ -0.0043	$+0.0089$ -0.0087
$\Omega_{\alpha_2,0}$	-0.0056	$+0.0056$ -0.0257	$+0.0056$ -0.0391
$\Omega_{\alpha_4,0}$	4.1×10^{-6}	$+0.000032$ -6.2×10^{-6}	$+0.000059$ -6.8×10^{-6}

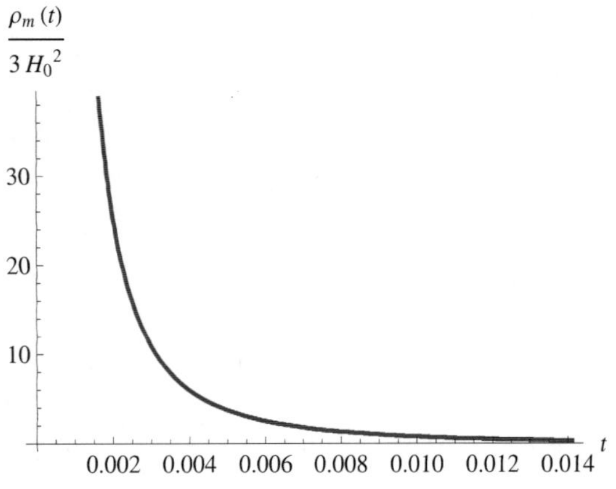

Figure 3. The diagram presents the evolution of $\frac{\rho_{\rm m}(t)}{3H_0^2}$, for the best fit values of model parameters. The cosmological time t is expressed in $\frac{\rm s\ Mpc}{\rm km}$. Here, the present age of the Universe is equal to $0.014 \frac{\rm s\ Mpc}{\rm km}$.

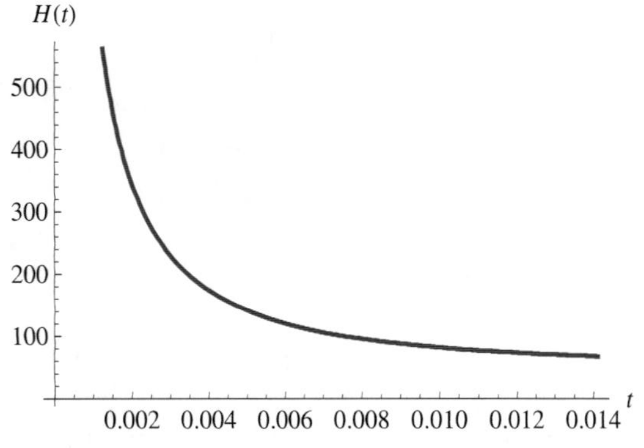

Figure 4. The diagram presents the evolution of $H(t)$, for the best fit values of model parameters. The cosmological time t is expressed in $\frac{\rm s\ Mpc}{\rm km}$. Here, the present age of the Universe is equal to $0.014 \frac{\rm s\ Mpc}{\rm km}$.

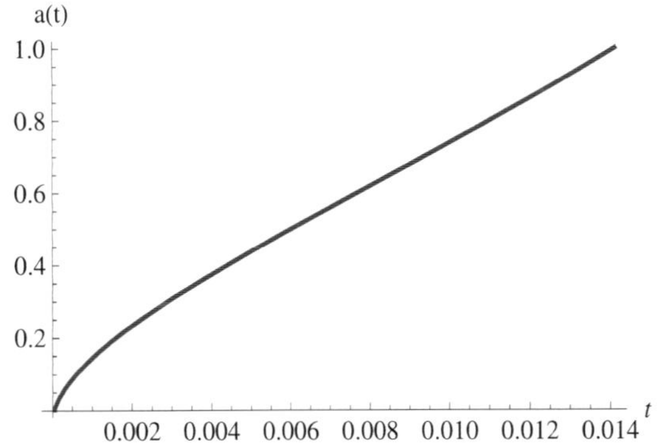

Figure 5. The diagram presents the evolution of $a(t)$, for the best fit values of model parameters. The cosmological time t is expressed in $\frac{\text{s Mpc}}{\text{km}}$. Here, the present age of the Universe is equal to $0.014\frac{\text{s Mpc}}{\text{km}}$.

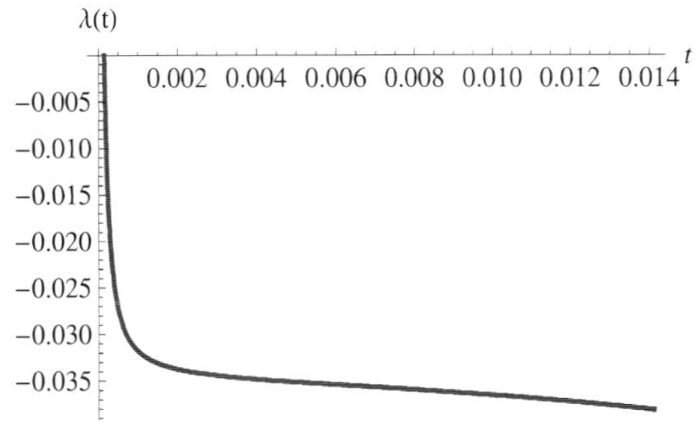

Figure 6. The diagram presents the evolution of $\lambda(t)$, for the best fit values of model parameters. The cosmological time t is expressed in $\frac{\text{s Mpc}}{\text{km}}$. Here, the present age of the Universe is equal to $0.014\frac{\text{s Mpc}}{\text{km}}$.

4. Summary

From statistical analysis, we get that the best fit of values of dimensionless density parameters $\Omega_{\alpha_2,0}$ and $\Omega_{\alpha_4,0}$

$$\Omega_{\alpha_2,0} = -0.0056, \quad \Omega_{\alpha_4,0} = 4\times 10^{-6}, \quad \Omega_{\text{m},0} = 0.3045,$$
$$\Omega_{\text{bare}} = 1 - \Omega_{\alpha_2,0} - \Omega_{\alpha_4,0} - \Omega_{\text{m},0} = 0.701096.$$

These parameters characterize the phenomenological properties of decaying quantum vacuum. While there exists theoretical prior for negative value Ω_{α_2} there is no prior for Ω_{α_4}. Having values of these parameters one can calculate another characteristic of the process of the decaying dark energy.

The considered model is in good agreement with the astronomical data. This shows that $\frac{\alpha_2}{t^2}$ term dominates in formula for $\delta\Lambda(t)$ and $\frac{\alpha_4}{t^4}$ term is negligible comparing to the term $\frac{\alpha_2}{t^2}$. This means that the function $\delta\Lambda(t)$ is negative in the present epoch. As the result, we get that the value of the energy density of dark matter increases for the our epoch and for the late times. In consequence, we get the quantum process of the annihilation of particles of dark matter for the late Universe.

We found a deviation from the standard scaling of the energy density of matter. We can explain the source of this deviation showing that the energy density of matter decreases more rapidly or slowly due to the energy transfer from dark matter to dark energy sector or in the opposite direction. The direction of the energy transfer crucially depends on the sign of $\delta\Lambda(t)$ in our model. For the best fit value, we get that the deviation speeds up decreasing of the energy density of matter.

We found that for the best fit value of the model parameters, the evolution of the Universe in our model and in the standard cosmological model ΛCDM are similar in the present epoch and for late epochs. In this model Λ decays into Λ_{bare}. As a result, we can treated the Λ term as a constant for our epochs. The process of decaying of dark energy has not a big influence for the evolution of the current Universe. Therefore, this model can be considered as a simple extension of the ΛCDM model. The deviation of this model from the ΛCDM model is less than 1σ today. Note that our model explains naturally the smallness of the cosmological constant parameter Λ_{bare}: In the model the running $\Lambda = \Lambda(t)$ term, which is very, very large near the initial singularity (i.e., for very small times t), tends to much, much smaller Λ_{bare} when $t \to \infty$.

References

[1] Weinberg S 1989 *Rev. Mod. Phys.* **61**, 1–23

[2] Weinberg S 2000 in *Sources and detection of dark matter and dark energy in the universe. Proceedings, 4th International Symposium, DM 2000, Marina del Rey, USA, February 23-25, 2000*, pp. 18–26

[3] Ade P A R *et al.* 2016 *Astron. Astrophys.* **594**, A14

[4] Valentino E, Linder E V and Melchiorri A 2018 *Phys. Rev.* **D97**, 043528

[5] Krauss L M and Dent J 2008 *Phys. Rev. Lett.* **100**, 171301

[6] L. M. Krauss L M, Dent J and Starkman G D 2009 *Int. J. Mod. Phys.* **D17**, 2501

[7] Urbanowski K 2011 *Phys. Rev. Letters*, **107**, 209001

[8] Urbanowski K 2017 *Theor. Math. Phys.* **190**, 458–469

[9] Shafieloo A, Hazara B K, Sahni V and Starobinsky A A 2018 *Mon. Not. Roy. Astron. Soc.* **473**, 2760

[10] Fonda L, Ghirardi G C and Rimini A 1978 *Rep. Prog. Phys.* **41**, 587

[11] Peshkin M, Volya A and Zelevinsky V 2014 *Europhys. Lett.* **107**, 40001

[12] Khalfin L A 1957 *Zh. Exp. Teor. Fiz.* **33**, 1371; Soviet Physics JETP, 6, 1053 (1958)

[13] Szydlowski M, Stachowski A and Urbanowski K 2017 *Eur. Phys. J.* **C77**, 902

[14] Krylov N S and Fock V A 1947 *Zh. Eksp. Teor. Fiz.* **17**, 93

[15] Fock V A 1978 *Fundamentals of Quantum Mechanics* (Moscow: Mir Publishers)

[16] Kelkar N G and Nowakowski M 2010 *J. Phys.* **A43**, 385308

[17] Giacosa F 2018, *Advances in High Energy Physics*, Volume 2018, Article ID 4672051, 7

[18] Urbanowski K 1994 *Phys. Rev.* **A50**, 2847

[19] Urbanowski K 2008 *Eur. Phys. J.* **C58**, 151

[20] Urbanowski K 2009 *Cent. Eur. J. Phys.* **7**, 696

[21] Giraldi F 2015 *Eur. Phys. J.* **D69**, 5

[22] Giraldi F 2016 Eur. Phys. J. **D70**, 229

[23] Sluis K M and Gislason E A 1991 *Phys. Rev.* **A43**, 4581

[24] Raczyńska K and Urbanowski K 2018 *Acta Physica Polonica* **B49**, 1683

[25] Olver W J, Lozier D W, Boisvert R F and Clark C W eds. 2010 *NIST Handbook of Mathematical Functions* (Cambridge: Cambridge University Press)

[26] Abramowitz M and Stegun I A, eds. 1964 *Handbook of Mathematical Functions*, National Bureau of Standards, Appl. Math. Ser. No. 55 (Washington, DC: U.S. G.P.O.)

[27] Shapiro I L, Sola J, Espana–Bonet C and Ruiz–Lapuente P 2003 *Phys. Lett.* **B574**, 149

[28] Espana-Bonet C, Ruiz-Lapuente P, Shapiro I L and Sola J 2004 *JCAP* **0402**, 006

[29] Suzuki N, et al. 2012 *Astrophys. J.* **746**, 85

[30] Percival W J, et al. 2010 *Mon. Not. Roy. Astron. Soc.* **401**, 2148

[31] Beutler F, Blake C, Colless M, Jones D H, Staveley-Smith L, Campbell L, Parker Q, Saunders W, Watson F 2011 *Mon. Not. Roy. Astron. Soc.* **416**, 3017

[32] Blake C, et al. 2012 *Mon. Not. Roy. Astron. Soc.* **425**, 405

[33] Hu W, Sugiyama N 1996 *Astrophys. J.* **471**, 542

[34] Eisenstein D J, Hu W 1998 *Astrophys. J.* **496**, 605

[35] Simon J, Verde L, Jimenez R 2005 *Phys. Rev.* **D71**, 123001

[36] Stern D, Jimenez R, Verde L, Kamionkowski M, Stanford S A 2010 *JCAP* **1002**, 008

[37] Moresco M, et al. 2012 *JCAP* **1208**, 006

[38] Alcock C, Paczynski B 1979 *Nature* **281**, 358

[39] Lopez-Corredoira M 2014 *Astrophys. J.* **781**(2), 96

[40] Sutter P M, Lavaux G, Wandelt B D, Weinberg D H 2012 *Astrophys. J.* **761**, 187

[41] Blake C, et al. 2011 *Mon. Not. Roy. Astron. Soc.* **418**, 1725

[42] Ross N P, et al. 2007 *Mon. Not. Roy. Astron. Soc.* **381**, 573

[43] Marinoni C, Buzzi A 2010 *Nature* **468**(7323), 539

[44] da Angela J, Outram P J, Shanks T 2005 *Mon. Not. Roy. Astron. Soc.* **361**, 879

[45] Outram P J, Shanks T, Boyle B J, Croom S M, Hoyle F, Loaring N S, Miller L, Smith R J 2004 *Mon. Not. Roy. Astron. Soc.* **348**, 745

[46] Anderson L, et al. 2013 *Mon. Not. Roy. Astron. Soc.* **427**(4), 3435

[47] Paris I, et al. 2012 *Astron. Astrophys.* **548**, A66

[48] Schneider D P, et al. 2010 *Astron. J.* **139**, 2360

[49] Metropolis N, Rosenbluth A W, Rosenbluth M N, Teller A H, Teller E, 1953 *J. Chem. Phys.* **21**, 1087

[50] Hastings W K 1970 *Biometrika* **57**, 97

9th International Workshop DICE2018 : Spacetime - Matter - Quantum Mechanics IOP Publishing
IOP Conf. Series: Journal of Physics: Conf. Series **1275** (2019) 012041 doi:10.1088/1742-6596/1275/1/012041

Influence functionals, decoherence and conformally coupled scalars

C Burrage[1], C Käding[1], P Millington[1] and J Minář[2,3,1,4]

[1] School of Physics and Astronomy, University of Nottingham, Nottingham NG7 2RD, UK
[2] Institute for Theoretical Physics, University of Amsterdam, Science Park 904, 1098 XH Amsterdam
[3] Department of Physics, Lancaster University, Lancaster, LA1 4YB, UK
[4] Centre for the Mathematics and Theoretical Physics of Quantum Non-Equilibrium Systems, School of Physics and Astronomy, University of Nottingham, Nottingham, NG7 2RD, UK

E-mail: clare.burrage@nottingham.ac.uk, christian.kading@nottingham.ac.uk, p.millington@nottingham.ac.uk, jminar@uva.nl

Abstract. Some of the simplest modifications to general relativity involve the coupling of additional scalar fields to the scalar curvature. By making a Weyl rescaling of the metric, these theories can be mapped to Einstein gravity with the additional scalar fields instead being coupled universally to matter. The resulting couplings to matter give rise to scalar fifth forces, which can evade the stringent constraints from local tests of gravity by means of so-called screening mechanisms. In this talk, we derive evolution equations for the matrix elements of the reduced density operator of a toy matter sector by means of the Feynman-Vernon influence functional. In particular, we employ a novel approach akin to the LSZ reduction more familiar to scattering-matrix theory. The resulting equations allow the analysis, for instance, of decoherence induced in atom-interferometry experiments by these classes of modified theories of gravity.

1. Introduction

There are many pointers to a need to extend the Standard Model of particle physics (SM) and/or general relativity, not least the dark matter and dark energy problems, and the simplest extensions often involve additional scalar fields. If these new fields are neutral under the SM gauge groups, they may couple to the SM via the Higgs portal or through a non-minimal coupling to the Ricci scalar (of Brans-Dicke type [1]), mediating interactions with hidden sectors. In fact, for the SM, Higgs-portal and non-minimal gravitational couplings are related to one another (up to and including dimension-four operators) by a change of conformal frame [2].

If the scalar fields are heavy on astrophysical scales and stable on cosmological time-scales, they may compose the relic density of dark matter, produced thermally in the early universe. However, with the absence of a direct or indirect detection of a particulate dark matter candidate, interest has been growing in new light fields, which can yield additional long-range forces relevant on astrophysical scales [3, 4, 5] and be constrained by terrestrial atom-interferometry experiments [6, 7, 8, 9, 10, 11, 12]. Cosmologically light fields may also play a role in the late-time accelerated expansion, where, e.g., quintessence scenarios of dark energy can be realized through the dynamics of a pseudo-Goldstone boson of a spontaneously broken scale symmetry [13].

Content from this work may be used under the terms of the Creative Commons Attribution 3.0 licence. Any further distribution of this work must maintain attribution to the author(s) and the title of the work, journal citation and DOI.
Published under licence by IOP Publishing Ltd

IOP Conf. Series: Journal of Physics: Conf. Series **1275** (2019) 012041 doi:10.1088/1742-6596/1275/1/012041

Any additional scalar forces operative on Solar System scales are heavily constrained by local tests of gravity, as well as terrestrial metrology experiments. In order to evade these constraints, and without fine-tuning couplings to matter, we can appeal to so-called screening mechanisms, wherein the fifth force is suppressed dynamically in regions of high ambient density. Around a point source of unit mass, the scalar perturbations δX above some (approximately) constant background $\bar{X}$ satisfy an equation of motion schematically of the form [14]

$$-\,Z(\bar{X})\big(\delta\ddot{X}-c_s^2(\bar{X})\nabla^2\delta X\big)\;-\;M^2(\bar{X})\delta X\;=\;\frac{\mathrm{d}A(\bar{X})}{\mathrm{d}\bar{X}}\,\delta(\mathbf{x})\,, \tag{1}$$

giving rise to a Yukawa potential

$$U(r)\;\sim\;-\,\frac{1}{Z(\bar{X})c_s^2(\bar{X})}\bigg[\frac{\mathrm{d}A(\bar{X})}{\mathrm{d}\bar{X}}\bigg]^2\frac{1}{4\pi r}\exp\bigg[-\,\frac{M(\bar{X})\,r}{Z^{1/2}(\bar{X})c_s(\bar{X})}\bigg]\,. \tag{2}$$

The fifth force can therefore be suppressed by dynamically modifying the mass $M(\bar{X})$, as in the chameleon mechanism [15, 16]; the coupling function $A(\bar{X})$, as in the symmetron mechanism [17, 18] (for related models, see references [19, 20, 21, 22, 23, 24, 25]), or the kinetic structure $Z(\bar{X})c_s^2(\bar{X})$, as in the Vainshtein [26] mechanism (see also references [27, 28, 29, 30, 31]).

In this note, we consider the impact of an environment composed of a light, screened scalar field on a probe system, reporting the results of reference [32]. Proceeding from first principles by means of the Feynman-Vernon influence functional [33] (see also reference [34]), we derive the quantum master equation for the reduced density matrix of a toy 'atomic' system (having in mind an atom-interferometry experiment). Assuming that 'multi-atom' states are suppressed, we project into the single-particle subspace by means of an LSZ-like reduction technique [35], before commenting on how the renormalization procedure must be adapted to account for finite-time effects. While the impact of the screened scalar is beyond the reach of current experiments, as one might expect, screened scalar fields nevertheless provide a solid context within which to develop this general and robust approach to studying decoherence in quantum field theory.

2. Conformally coupled scalars

The action of the models that we have in mind can be written in the general form

$$S\;=\;\int_x\sqrt{-g}\,\bigg[\frac{1}{2}\,M_{\rm Pl}^2\mathcal{R}\;-\;\frac{1}{2}\,g^{\mu\nu}\,\partial_\mu X\,\partial_\nu X\;-\;V(X)\bigg]\;+\;\int_x\sqrt{-\tilde{g}}\;\tilde{\mathcal{L}}_m(\{\tilde{\phi}_i\},\tilde{g}_{\mu\nu})\,, \tag{3}$$

where $M_{\rm Pl}$ is the reduced Planck mass, $\mathcal{R}$ is the Ricci scalar with respect to the Einstein-frame metric $g_{\mu\nu}$ (see reference [36]) and $\tilde{\mathcal{L}}_m$ is the matter Lagrangian. The matter fields $\{\tilde{\phi}_i\}$ move on geodesics of the Jordan-frame metric $\tilde{g}_{\mu\nu}=A^2(X)g_{\mu\nu}$, coupling universally to the scalar field X. We employ the $(-,+,+,+)$ signature convention and use the shorthand notation $\int_x\equiv\int\mathrm{d}^4x$.

As a proxy for our 'atom', we take a massive scalar field theory with Lagrangian

$$\tilde{\mathcal{L}}_m\;=\;-\,\frac{1}{2}\,\tilde{g}^{\mu\nu}\,\partial_\mu\tilde{\phi}\,\partial_\nu\tilde{\phi}\;-\;\frac{1}{2}\,\tilde{m}^2\,\tilde{\phi}^2\,. \tag{4}$$

The matter action then has the following form in terms of the Einstein-frame metric:

$$S_m\;=\;\int_x\sqrt{-g}\,\bigg[-\,\frac{1}{2}\,A^2(X)\,g^{\mu\nu}\,\partial_\mu\tilde{\phi}\,\partial_\nu\tilde{\phi}\;-\;\frac{1}{2}\,A^4(X)\,\tilde{m}^2\tilde{\phi}^2\bigg]\,. \tag{5}$$

After defining $\phi\equiv A(X)\tilde{\phi}$, so as to obtain a near-canonical kinetic term, we have

$$S_m\;=\;\int_x\sqrt{-g}\,\bigg[-\,\frac{1}{2}\,g^{\mu\nu}\,\partial_\mu\phi\,\partial_\nu\phi\;-\;\frac{1}{2}\,\phi^2\,g^{\mu\nu}\,\partial_\mu\ln A(X)\,\partial_\nu\ln A(X)$$
$$+\;\phi\,g^{\mu\nu}\,\partial_\mu\phi\,\partial_\nu\ln A(X)\;-\;\frac{1}{2}\,A^2(X)\,\tilde{m}^2\phi^2\bigg]\,. \tag{6}$$

The equation of motion for the expectation value of the universally coupled scalar $\langle X\rangle$ is

$$\Box\,\langle X\rangle\;-\;\frac{\mathrm{d}V(X)}{\mathrm{d}X}\bigg|_{X\,=\,\langle X\rangle}\;=\;-\,\frac{1}{2}\,\frac{\mathrm{d}A^2(X)}{\mathrm{d}X}\bigg|_{X\,=\,\langle X\rangle}\;\tilde{T}\;, \tag{7}$$

where, in our case, $\tilde{T}$ is the trace of the energy-momentum tensor of the vacuum chamber of the atom-interferometry experiment, taking the form $\tilde{T}=-A^{-1}(\langle X\rangle)\rho^{\mathrm{ext}}$, where ρ^{ext} is the covariantly conserved energy density (see, e.g., reference [14]).

We now specialize to the quartic chameleon [37] with potential and coupling function

$$V(X)\;=\;-\,\frac{\lambda}{4!}\,X^4\qquad\text{and}\qquad A(X)\;=\;e^{X/\mathcal{M}}\,. \tag{8}$$

While the parameter space for this model is heavily constrained by atom-interferometry and torsion-balance experiments [38], it provides a convenient context for our discussions.

Working to leading order in the expansion of the coupling function, and in a constant density environment, the chameleon field acquires a background value of

$$\langle X\rangle\;=\;-\left(\frac{6\rho^{\mathrm{ext}}}{\lambda\mathcal{M}}\right)^{1/3}\,, \tag{9}$$

with the fluctuations $\chi\equiv X-\langle X\rangle$ about the minimum of the effective potential $V^{\mathrm{eff}}=V+\rho^{\mathrm{ext}}X/\mathcal{M}$ having a density-dependent mass

$$M^2\;=\;\left(\frac{\lambda}{2}\right)^{1/3}\left(\frac{3\rho^{\mathrm{ext}}}{\mathcal{M}}\right)^{2/3}\,, \tag{10}$$

which vanishes in vacuum. Hence, in regions of high density, the mass of the fluctuations becomes large, and the Yukawa potential induced by the chameleon field is suppressed.

Working with operators up to dimension four only, the action for the 'atom' field ϕ and the chameleon fluctuations χ can be separated into free and (self-)interacting parts as

$$S_\phi[\phi]\;=\;\int_x\left[-\,\frac{1}{2}\,g^{\mu\nu}\,\partial_\mu\phi\,\partial_\nu\phi\;-\;\frac{1}{2}\,m^2\phi^2\right]\,, \tag{11a}$$

$$S_\chi[\chi]\;=\;\int_x\left[-\,\frac{1}{2}\,g^{\mu\nu}\,\partial_\mu\chi\,\partial_\nu\chi\;-\;\frac{1}{2}\,M^2\chi^2\right]\,, \tag{11b}$$

$$S_{\chi,\mathrm{int}}[\chi]\;=\;\int_{x\,\in\,\Omega_t}\left[-\,\frac{\lambda}{4!}\,(\chi^4+4\,\langle X\rangle\,\chi^3)\right]\,, \tag{11c}$$

$$S_{\mathrm{int}}[\phi,\chi]\;=\;\int_{x\,\in\,\Omega_t}\left[-\,\frac{1}{2}\left(2\,\frac{m^2}{\mathcal{M}}\right)\chi\phi^2\;-\;\frac{1}{4}\left(4\,\frac{m^2}{\mathcal{M}^2}\right)\chi^2\phi^2\right]\,, \tag{11d}$$

where we have defined (assuming $\langle X\rangle$ is constant)

$$m^2\;\equiv\;\tilde{m}^2\left(1\;+\;2\frac{\langle X\rangle}{\mathcal{M}}\;+\;2\,\frac{\langle X\rangle^2}{\mathcal{M}^2}\right)\,. \tag{12}$$

Our aim in the next two sections is to trace out the chameleon fluctuations and derive the quantum master equation describing the evolution of the density matrix of the remaining open scalar system. With this in mind, and in order to account for the fact that the initial state of the 'atom' system will be prepared and its final state measured over a finite interval of time, we have restricted the support of the spacetime integrals in the interaction parts to the hypervolume $\Omega_t=[0,t]\times\mathbb{R}^3$ (see references [39, 40]), approximating the vacuum chamber as infinitely large. Note that this amounts to an instantaneous switching on and off of the interactions.

IOP Conf. Series: Journal of Physics: Conf. Series **1275** (2019) 012041 doi:10.1088/1742-6596/1275/1/012041

3. Open dynamics and the Feynman-Vernon influence functional

Our starting point is the reduced density functional — the matrix element of the density operator in the basis of field eigenstates — of the 'atom' system (presented here in a simplified notation, cf. the full time-dependence and functional delta functions that appear in reference [32])

$$\rho_\phi[\phi^\pm;t_f] = \int \mathrm{d}\chi_f^\pm \, \rho[\phi^\pm,\chi^\pm;t_f]\,, \tag{13}$$

where $\int \mathrm{d}\chi_f^\pm \equiv \int \mathrm{d}\chi_f^+ \int \mathrm{d}\chi_f^-$ is an integral over the doubled field configurations at time t_f. This doubling of degrees of freedom necessarily arises when constructing a path-integral representation of a trace, leading to the Schwinger-Keldysh closed-time-path (CTP) formalism [41, 42].

Assuming that the initial state of the system (at time t_i) is factorizable, i.e.

$$\rho[\phi^\pm,\chi^\pm;t_i] = \rho_\phi[\phi^\pm;t_i]\rho_\chi[\chi^\pm;t_i]\,, \tag{14}$$

the reduced density functional evolves as

$$\rho_\phi[\phi^\pm;t_f] = \int \mathrm{d}\phi_i^\pm \, \mathcal{I}[\phi^\pm;t_f,t_i]\rho_\phi[\phi^\pm;t_i]\,, \tag{15}$$

where

$$\mathcal{I}[\phi^\pm;t_f,t_i] = \int_{\phi_i^\pm}^{\phi_f^\pm} \mathcal{D}\phi^\pm \, e^{i\widehat{S}_{\mathrm{eff}}[\phi;t_f,t_i]} \tag{16}$$

is the influence-functional propagator ($\hbar = 1$). The effective action $\widehat{S}_{\mathrm{eff}}[\phi;t_f,t_i]$ can be written

$$\widehat{S}_{\mathrm{eff}}[\phi;t_f,t_i] = \widehat{S}_\phi[\phi] + \widehat{S}_{\mathrm{IF}}[\phi;t_f,t_i]\,. \tag{17}$$

We use a $\widehat{\ }$ to identify actions that depend on both of the doubled degrees of freedom. Doubled actions corresponding to unitary evolution, i.e. obtained from equation (11), have, e.g., a form

$$\widehat{S}_\phi[\phi] = S_\phi[\phi^+] - S_\phi[\phi^-]\,. \tag{18}$$

The influence action $\widehat{S}_{\mathrm{IF}}[\phi;t_f,t_i]$ on the other hand, which arises after tracing out the χ field, mixes the $+$ and $-$ type fields. It is defined via the Feynman-Vernon influence functional

$$\mathcal{F}[\phi^\pm] = e^{i\widehat{S}_{\mathrm{IF}}[\phi]} = \int \mathrm{d}\chi_f^\pm \int \mathrm{d}\chi_i^\pm \, \rho_\chi[\chi^\pm;t_i] \int_{\chi_i^\pm}^{\chi_f^\pm} \mathcal{D}\chi^\pm \, e^{i(\widehat{S}_\chi[\chi]+\widehat{S}_{\chi,\mathrm{int}}[\chi]+\widehat{S}_{\mathrm{int}}[\phi,\chi])}\,, \tag{19}$$

wherein we have omitted time arguments for brevity. Taking the partial time derivative of the reduced density functional in equation (15), we obtain the functional expression of the master equation (with $t \equiv t_f$)

$$\partial_t\rho_\phi[\phi^\pm;t] = -i\big(-\partial_t S_\phi[\phi^+] + \partial_t S_\phi[\phi^-]\big)\rho_\phi[\phi^\pm;t] + i\partial_t\widehat{S}_{\mathrm{IF}}[\phi;t]\rho_\phi[\phi^\pm;t]\,. \tag{20}$$

4. LSZ-like reduction

The path-integral approach detailed in the previous section has the advantage that we can make use of the powerful and familiar diagrammatic techniques to evaluate the influence functional. However, we would like to work in a basis other than that of the field eigenstates. As in the case of scattering-matrix theory, we can project into the Fock space, and we will limit ourselves here to the single-particle subspace, by means of an LSZ-like reduction technique [35].

In order to motivate the form of this reduction, it is convenient to work in the canonical operator approach known as thermo field dynamics [43, 44, 45]. Therein, we work with a doubled Hilbert space $\widehat{\mathcal{H}} = \mathcal{H}^+ \otimes \mathcal{H}^-$ (cf. the notation of reference [46]), which allows the quantum Liouville equation (in the Schrödinger picture) to be recast in a Schrödinger-like form

$$\partial_t|\rho(t)\rangle\!\rangle \;=\; -\,i\widehat{H}|\rho(t)\rangle\!\rangle\,. \tag{21}$$

Herein, $\widehat{H}$ is the Liouvillian operator, which, for an isolated system, takes the form

$$\widehat{H} \;=\; \hat{H}\otimes\hat{\mathbb{I}} \,-\, \hat{\mathbb{I}}\otimes\hat{H}\,, \tag{22}$$

where $\hat{H}$ is the Hamiltonian and $\hat{\mathbb{I}}$ is the identity operator. The density operator $\hat{\rho}(t)$ is replaced by a state in the doubled Hilbert space, taking the form

$$|\rho(t)\rangle\!\rangle \;=\; \hat{\rho}^+(t)|1\rangle\!\rangle \;\equiv\; \big(\hat{\rho}(t)\otimes\hat{\mathbb{I}}\big)|1\rangle\!\rangle\,. \tag{23}$$

The state $|1\rangle\!\rangle$ can be written schematically as

$$|1\rangle\!\rangle \;=\; \sum_n |n^+\rangle\otimes|n^-\rangle\,, \tag{24}$$

where n labels a complete basis. The expectation value of an operator $\hat{O}(t)$ is then

$$\langle\hat{O}(t)\rangle \;=\; \mathrm{tr}\,\hat{O}(t)\hat{\rho}(t) \;=\; \langle\!\langle 1|\big(\hat{O}(t)\otimes\hat{\mathbb{I}}\big)\big(\hat{\rho}(t)\otimes\hat{\mathbb{I}}\big)|1\rangle\!\rangle \;=\; \langle\!\langle 1|\hat{O}^+(t)|\rho(t)\rangle\!\rangle\,. \tag{25}$$

Working in the interaction picture, we are interested in the single-particle matrix element

$$\rho(\mathbf{p},\mathbf{p}';t) \;=\; \langle\mathbf{p};t|\hat{\rho}(t)|\mathbf{p}';t\rangle \;=\; \mathrm{tr}\,|\mathbf{p}';t\rangle\langle\mathbf{p};t|\,\hat{\rho}(t) \;=\; \langle\!\langle 1(t)|\big(|\mathbf{p}';t\rangle\langle\mathbf{p};t|\otimes\hat{\mathbb{I}}\big)\hat{\rho}^+(t)|1(t)\rangle\!\rangle \tag{26}$$

in the basis of momentum eigenstates $|\mathbf{p};t\rangle$. Note that the states and operators are evaluated at equal times, ensuring that this matrix element is picture independent (see references [39, 40]). Taking the time derivative, recalling that the state and basis vectors evolve respectively with the interaction and free parts of the effective Liouvillian operator $\widehat{H}_{\rm eff}$, we obtain

$$\partial_t\rho(\mathbf{p},\mathbf{p}';t) \;=\; -\,i\int \mathrm{d}\Pi^\phi_{\mathbf{k}}\,\mathrm{d}\Pi^\phi_{\mathbf{k}'}\,\rho(\mathbf{k},\mathbf{k}';t)\langle\!\langle 0|\mathrm{T}[\hat{a}^+_{\mathbf{p}}(t)\hat{a}^-_{\mathbf{p}'}(t)\widehat{H}_{\rm eff}(t)\hat{a}^{+\dagger}_{\mathbf{k}}(t)\hat{a}^{-\dagger}_{\mathbf{k}'}(t)|0\rangle\!\rangle\,, \tag{27}$$

where $E^\phi_{\mathbf{k}} = \sqrt{\mathbf{k}^2+m^2}$ and we use the shorthand notation

$$\int \mathrm{d}\Pi^\phi_{\mathbf{k}} \;\equiv\; \int_{\mathbf{k}}\frac{1}{2E^\phi_{\mathbf{k}}} \;=\; \int\frac{\mathrm{d}^3\mathbf{k}}{(2\pi)^3 2E^\phi_{\mathbf{k}}} \tag{28}$$

for the Lorentz-invariant phase-space integral. By using

$$\hat{a}^+_{\mathbf{p}}(t) \;=\; +\,i\int_{\mathbf{x}} e^{-i\mathbf{p}\cdot\mathbf{x}}\,\partial_{t,E^\phi_{\mathbf{p}}}\hat{\phi}^+(t,\mathbf{x})\,,\qquad \hat{a}^-_{\mathbf{p}}(t) \;=\; -\,i\int_{\mathbf{x}} e^{+i\mathbf{p}\cdot\mathbf{x}}\,\partial^*_{t,E^\phi_{\mathbf{p}}}\hat{\phi}^-(t,\mathbf{x})\,, \tag{29}$$

where $\int_{\mathbf{x}} \equiv \int \mathrm{d}^3\mathbf{x}$ and

$$\partial_{t,E^\phi_{\mathbf{p}}} \;\equiv\; \overrightarrow{\partial}_t \,-\, iE^\phi_{\mathbf{p}}\,, \tag{30}$$

equation (27) can be rewritten entirely in terms of field operators as

$$\begin{aligned}\partial_t\rho(\mathbf{p},\mathbf{p}';t) \;=\; -\,i\lim_{\substack{x^{0(\prime)}\to t^+\\ y^{0(\prime)}\to 0^-}}\int \mathrm{d}\Pi^\phi_{\mathbf{k}}\,\mathrm{d}\Pi^\phi_{\mathbf{k}'}\,e^{i(E^\phi_{\mathbf{k}}-E^\phi_{\mathbf{k}'})t}\rho(\mathbf{k},\mathbf{k}';t)\int_{\mathbf{x}\mathbf{x}'\mathbf{y}\mathbf{y}'} e^{-i(\mathbf{p}\cdot\mathbf{x}-\mathbf{p}'\cdot\mathbf{x}')+i(\mathbf{k}\cdot\mathbf{y}-\mathbf{k}'\cdot\mathbf{y}')}\\ \times\,\partial_{x^0,E^\phi_{\mathbf{p}}}\partial^*_{x^{0\prime},E^\phi_{\mathbf{p}'}}\partial^*_{y^0,E^\phi_{\mathbf{k}}}\partial_{y^{0\prime},E^\phi_{\mathbf{k}'}}\langle\!\langle 0|\mathrm{T}[\hat{\phi}^+(x)\hat{\phi}^-(x')\widehat{H}_{\rm eff}(t)\hat{\phi}^+(y)\hat{\phi}^-(y')]|0\rangle\!\rangle\,,\end{aligned} \tag{31}$$

where $x^{0(\prime)} \to t$ from above and $y^{0(\prime)} \to 0$ from below so that the operator time-ordering reproduces that in equation (27). This can be expressed in the path-integral language as

$$\partial_t \rho(\mathbf{p},\mathbf{p}';t) = i \lim_{\substack{x^{0(\prime)}\to t^+ \\ y^{0(\prime)}\to 0^-}} \int \mathrm{d}\Pi^\phi_{\mathbf{k}}\,\mathrm{d}\Pi^\phi_{\mathbf{k}'}\, e^{i(E^\phi_{\mathbf{k}}-E^\phi_{\mathbf{k}'})t} \rho(\mathbf{k},\mathbf{k}';t) \int_{\mathbf{x}\mathbf{x}'\mathbf{y}\mathbf{y}'} e^{-i(\mathbf{p}\cdot\mathbf{x}-\mathbf{p}'\cdot\mathbf{x}')+i(\mathbf{k}\cdot\mathbf{y}-\mathbf{k}'\cdot\mathbf{y}')}$$
$$\times\, \partial_{x^0,E^\phi_{\mathbf{p}}}\partial^*_{x^{0\prime},E^\phi_{\mathbf{p}'}}\partial^*_{y^0,E^\phi_{\mathbf{k}}}\partial_{y^{0\prime},E^\phi_{\mathbf{k}'}} \int \mathcal{D}\phi^\pm\, e^{i\widehat{S}_\phi[\phi]}\phi^+(x)\phi^-(x')\Big(\partial_t \widehat{S}_{\text{eff}}[\phi;t]\Big)\phi^+(y)\phi^-(y')\,, \quad (32)$$

where we have replaced $\widehat{H}_{\text{eff}} \to -\,\partial_t \widehat{S}_{\text{eff}}$. Up to factors of the wavefunction normalization, the right-hand side contains the LSZ reduction [35] of a CTP four-point correlation function.

5. Quantum master equation

Inserting the explicit form of the influence-functional action for the model of section 2 (see reference [32]) into equation (32), we obtain the master equation

$$\partial_t\rho(\mathbf{p},\mathbf{p}';t) = -i(E^\phi_{\mathbf{p}}-E^\phi_{\mathbf{p}'})\,\rho(\mathbf{p},\mathbf{p}';t) - \frac{im^2}{\mathcal{M}}\left(\frac{1}{E^\phi_{\mathbf{p}}}-\frac{1}{E^\phi_{\mathbf{p}'}}\right)\rho(\mathbf{p},\mathbf{p}';t)$$
$$\times\left\{\frac{\Delta^{\mathrm{F}}_{xx}}{\mathcal{M}} + \left[\frac{m^2}{\mathcal{M}}D^{\mathrm{F}}_{xx} + \frac{\lambda}{2}\langle X\rangle\,\Delta^{\mathrm{F}}_{xx}\right]\frac{\cos(Mt)-1}{M^2}\right\}$$
$$+\, i\,\frac{4m^4}{\mathcal{M}^2}\sum_{s=\pm}\int_{\mathbf{k}}\Bigg\{\Bigg[\rho(\mathbf{p},\mathbf{p}';t)\,\frac{1}{E^\phi_{\mathbf{p}}2E^\chi_{\mathbf{k}}2E^\phi_{\mathbf{p}-\mathbf{k}}}\,\frac{s}{\big(sE^\chi_{\mathbf{k}}+E^\phi_{\mathbf{p}-\mathbf{k}}\big)^2-\big(E^\phi_{\mathbf{p}}\big)^2}$$
$$\times\Big[\Big(sE^\chi_{\mathbf{k}}+E^\phi_{\mathbf{p}-\mathbf{k}}\Big)\Big(1-\exp\big[-i(sE^\chi_{\mathbf{k}}+E^\phi_{\mathbf{p}-\mathbf{k}})t\big]\cos\big(E^\phi_{\mathbf{p}}t\big)\Big)$$
$$-\,iE^\phi_{\mathbf{p}}\exp\big[-i(sE^\chi_{\mathbf{k}}+E^\phi_{\mathbf{p}-\mathbf{k}})t\big]\sin\big(E^\phi_{\mathbf{p}}t\big)\Big]\big[1+f(sE^\chi_{\mathbf{k}})\big]$$
$$+\,\rho(\mathbf{p}-\mathbf{k},\mathbf{p}'-\mathbf{k};t)\,\frac{1}{2E^\chi_{\mathbf{k}}2E^\phi_{\mathbf{p}-\mathbf{k}}2E^\phi_{\mathbf{p}'-\mathbf{k}}}\,\frac{s}{sE^\chi_{\mathbf{k}}+E^\phi_{\mathbf{p}-\mathbf{k}}-E^\phi_{\mathbf{p}}}$$
$$\times\Big(1-\exp\big[i(sE^\chi_{\mathbf{k}}+E^\phi_{\mathbf{p}-\mathbf{k}}-E^\phi_{\mathbf{p}})t\big]\Big)f(sE^\chi_{\mathbf{k}})\Bigg] - (\mathbf{p}\longleftrightarrow\mathbf{p}')^*\Bigg\}$$
$$-\,\frac{4m^4}{\mathcal{M}^2}\,\rho(\mathbf{p},\mathbf{p}';t)\sum_{s=\pm}\int_{\mathbf{x}}\int_{\mathbf{k}_1\mathbf{k}_2}\frac{\sin\big[(E^\phi_{\mathbf{k}_1}+E^\phi_{\mathbf{k}_1-\mathbf{k}_2}+sE^\chi_{\mathbf{k}_2})t\big]}{E^\phi_{\mathbf{k}_1}+E^\phi_{\mathbf{k}_1-\mathbf{k}_2}+sE^\chi_{\mathbf{k}_2}}\,\frac{s\big[1+f(sE^\chi_{\mathbf{k}_2})\big]}{2E^\phi_{\mathbf{k}_1}2E^\phi_{\mathbf{k}_1-\mathbf{k}_2}2E^\chi_{\mathbf{k}_2}}\,. \quad (33)$$

Herein, Δ^{F}_{xy} and D^{F}_{xy} are the Feynman propagators of the chameleon and 'atom' fields, respectively, and $E^\chi_{\mathbf{p}} = \sqrt{\mathbf{p}^2+M^2}$ and $E^\phi_{\mathbf{p}} = \sqrt{\mathbf{p}^2+m^2}$ are their on-shell energies. We have also accounted for thermal fluctuations in the chameleon field (which is assumed to be in thermal equilibrium with the walls of the vacuum chamber), giving rise to the factors of the Bose-Einstein distribution $f(E) = 1/(e^{\beta E}-1)$, where $\beta = 1/T$ is the inverse thermodynamic temperature (see, e.g., reference [47]). We recognize the first term on the right-hand side of equation (33) as the free-phase evolution, the remaining terms of the first and second lines as tadpole self-energy insertions, and the terms of the third to seventh lines as the contributions from the one-loop bubble diagram. The final line contains a disconnected vacuum diagram that we omit hereafter; for further details, see reference [32]. The self-energies of the second to fifth lines contain quadratic and logarithmic divergences. Noticing, however, that all but the first of the self-energy insertions vanish in the limit $t \to 0$, it follows that the contributions from any counterterms must vanish in the same limit. As a result, the renormalization procedure must be modified to reflect the fact that the external switching on and off of the interactions leads to violation of both time-translational invariance and Lorentz invariance, as described in detail in reference [32].

Subtracting the divergences at a reference three-momentum $\bar{\mathbf{p}}$ (see reference [32]), the master equation can be written in the schematic form

$$\partial_t \rho(\mathbf{p},\mathbf{p}';t) = -\left[iu(\mathbf{p},\mathbf{p}';t) + \Gamma(\mathbf{p},\mathbf{p}';t)\right]\rho(\mathbf{p},\mathbf{p}';t) + \int_{\mathbf{k}} \gamma(\mathbf{p},\mathbf{p}',\mathbf{k};t)\rho(\mathbf{p}-\mathbf{k},\mathbf{p}'-\mathbf{k};t)\ , \quad (34)$$

where the various terms are given as follows:

$$u(\mathbf{p},\mathbf{p}';t) \equiv E^{\phi}_{\mathbf{p}} - E^{\phi}_{\mathbf{p}'} + \frac{m^2}{\mathcal{M}}\left(\frac{1}{E^{\phi}_{\mathbf{p}}} - \frac{1}{E^{\phi}_{\mathbf{p}'}}\right)\Delta^{\mathrm{F}(T\neq 0)}_{xx}\left\{\frac{1}{\mathcal{M}} + \frac{\lambda}{2}\langle X\rangle\frac{\cos(Mt)-1}{M^2}\right\}$$
$$- \left\{\frac{1}{E^{\phi}_{\mathbf{p}}}\int_{p^0}\frac{\sin\left[(p^0-E^{\phi}_{\mathbf{p}})t\right]}{p^0-E^{\phi}_{\mathbf{p}}}\mathrm{Re}\Big[\Pi(-p^2) - \Pi^{(T=0)}(-p^2)\big|_{\mathbf{p}=\bar{\mathbf{p}}}\Big] - (\mathbf{p}\longleftrightarrow\mathbf{p}')\right\}, \quad (35a)$$

$$\Gamma(\mathbf{p},\mathbf{p}';t) \equiv \frac{1}{E^{\phi}_{\mathbf{p}}}\int_{p^0}\frac{\sin\left[(p^0-E^{\phi}_{\mathbf{p}})t\right]}{p^0-E^{\phi}_{\mathbf{p}}}\mathrm{Im}\,\Pi(-p^2) + (\mathbf{p}\longleftrightarrow\mathbf{p}')\ , \quad (35b)$$

$$\gamma(\mathbf{p},\mathbf{p}',\mathbf{k};t) \equiv i\frac{4m^4}{\mathcal{M}^2}\sum_{s=\pm}\left\{\frac{1}{2E^{\chi}_{\mathbf{k}}2E^{\phi}_{\mathbf{p}-\mathbf{k}}2E^{\phi}_{\mathbf{p}'-\mathbf{k}}}\frac{s}{sE^{\chi}_{\mathbf{k}}+E^{\phi}_{\mathbf{p}-\mathbf{k}}-E^{\phi}_{\mathbf{p}}}\right.$$
$$\left.\times\left(1-\exp\left[i(sE^{\chi}_{\mathbf{k}}+E^{\phi}_{\mathbf{p}-\mathbf{k}}-E^{\phi}_{\mathbf{p}})t\right]\right)f(sE^{\chi}_{\mathbf{k}}) - (\mathbf{p}\longleftrightarrow\mathbf{p}')^*\right\}, \quad (35c)$$

where the t-dependence should be compared with the diagrammatics of references [39, 40] and

$$i\Pi(-p^2) = \left(-\frac{2im^2}{\mathcal{M}}\right)^2\int_k\left[\frac{-i}{k^2+M^2-i\epsilon} + 2\pi f(|k^0|)\delta(k^2+M^2)\right]\frac{-i}{(k-p)^2+m^2-i\epsilon}\ . \quad (36)$$

The superscripts $T=0$ and $T\neq 0$ indicate respectively the zero-temperature and finite-temperature parts of the various propagators and self-energies. The coefficient $u\in\mathbb{R}$ describes coherent evolution, and $\Gamma\in\mathbb{R}$ and $\gamma\in\mathbb{C}$ describe decoherence and momentum diffusion.

The dominant effect comes from the phase shift that results from the change in the effective mass m of the 'atom' due to the ambient value of the chameleon field [cf. equation (12)]. (In order to be sensitive to the difference $m-\tilde{m}$, we might imagine a setup in which the ambient value of the chameleon field is different along the two paths of the interferometer.) For a spherical vacuum chamber of radius 1 m, the value of the chameleon field at the centre of the chamber is $\langle X\rangle = 1.287/\sqrt{\lambda L^2}$ [6, 16]. Assuming the thermal effects are subleading, and taking $\lambda = 1/10$, $\mathcal{M} = M_{\mathrm{Pl}}$ (the reduced Planck mass), $\tilde{m} = 87m_u$ (the mass of a ^{87}Rb atom in atomic mass units) and a characteristic velocity scale of the atom in the atom interferometer of $v = 10$ ms^{-1}, we find an upper bound on the magnitude of the effects

$$\Delta u \sim \langle X\rangle\,\tilde{m}v^2/(\mathcal{M}c) \sim 10^{-23}\ \mathrm{Hz}\ , \quad (37)$$

which is far below the current sensitivity $\sim 10^{-8}$ Hz (as inferred from reference [48]).

6. Concluding remarks

We have reported a novel way of deriving the quantum master equation for matrix elements of the reduced density operator from first principles in quantum field theory, as presented in reference [32], which yields cut-off independent and quantitative predictions that are crucial for ongoing precision searches for dark matter and dark energy. While the context presented here has been screened scalar fifth forces, our approach is general, and its extension to higher-spin fields may find use, for example, in studies of gravitational decoherence.

IOP Conf. Series: Journal of Physics: Conf. Series **1275** (2019) 012041 doi:10.1088/1742-6596/1275/1/012041

Acknowledgments

This work was supported by a Royal Society University Research Fellowship [CB], a Leverhulme Trust Research Leadership Award [CB and PM], the University of Nottingham Vice-Chancellor's Scholarship for Research Excellence [CK], the UK Science and Technology Facilities Council (STFC) grant no. ST/L000393/1 [PM], the UK Engineering and Physical Sciences Research Council (EPSRC) grant no. EP/P026133/1 [JM] and a European Research Council (ERC) grant, agreement no. 335266 (ESCQUMA) [JM].

References

[1] Brans C and Dicke R H 1961 *Phys. Rev.* **124** 925–35
[2] Burrage C, Copeland E J, Millington P and Spannowsky M 2018 *J. Cosmol. Astropart. Phys.* JCAP11(2018)036
[3] Burrage C, Copeland E J and Millington P 2017 *Phys. Rev. D* **95** 064050 [Erratum: *Phys. Rev. D* **95** 129902]
[4] O'Hare C A J and Burrage C 2018 *Phys. Rev. D* **98** 064019
[5] Burrage C, Copeland E J, Käding C and Millington P 2018 *Preprint* arXiv:1811.12301 [astro-ph.CO]
[6] Burrage C, Copeland E J and Hinds E A 2015 *J. Cosmol. Astropart. Phys.* JCAP03(2015)042
[7] Hamilton P, Jaffe M, Haslinger P, Simmons Q, Müller H and Khoury J 2015 *Science* **349** 849–51
[8] Schlögel S, Clesse S and Füzfa A 2016 *Phys. Rev. D* **93** 104036
[9] Elder B, Khoury J, Haslinger P, Jaffe M, Müller H and Hamilton P 2016 *Phys. Rev. D* **94** 044051
[10] Jaffe M, Haslinger P, Xu V, Hamilton P, Upadhye A, Elder B, Khoury J and Müller H 2017 *Nat. Phys.* **13** 938–42
[11] Brax P and Davis A-C 2016 *Phys. Rev. D* **94** 104069
[12] Brax P, Fichet S and Pignol G 2018 *Phys. Rev. D* **97** 115034
[13] García-Bellido J, Rubio J, Shaposhnikov M and Zenhäusern D 2011 *Phys. Rev. D* **84** 123504
[14] Joyce A, Jain B, Khoury J and Trodden M 2015 *Phys. Rep.* **568** 1–98
[15] Khoury J and Weltman A 2004 *Phys. Rev. Lett.* **93** 171104
[16] Khoury J and Weltman A 2004 *Phys. Rev. D* **69** 044026
[17] Hinterbichler K and Khoury J 2010 *Phys. Rev. Lett.* **104** 231301
[18] Hinterbichler K, Khoury J, Levy A and Matas A 2011 *Phys. Rev. D* **84** 103521
[19] Dehnen H, Frommert H and Ghaboussi F 1992 *Int. J. Theor. Phys.* **31** 109–14
[20] Gessner E 1992 *Astrophys. Space Sci.* **196** 29–43
[21] Damour T and Polyakov A M 1994 *Nucl. Phys. B* **423** 532–58
[22] Pietroni M 2005 *Phys. Rev. D* **72** 043535
[23] Olive K A and Pospelov M 2008 *Phys. Rev. D* **77** 043524
[24] Brax P, van de Bruck C, Davis A-C and Shaw D 2010 *Phys. Rev. D* **82** 063519
[25] Burrage C, Copeland E J and Millington P 2016 *Phys. Rev. Lett.* **117** 211102
[26] Vainshtein A I 1972 *Phys. Lett. B* **39** 393–4
[27] Nicolis A and Rattazzi R 2004 *J. High Energy Phys.* JHEP06(2004)059
[28] Nicolis A, Rattazzi R and Trincherini E 2009 *Phys. Rev. D* **79** 064036
[29] Babichev E, Deffayet C and Ziour R 2009 *Int. J. Mod. Phys. D* **18** 2147–54
[30] Brax P, Burrage C and Davis A-C 2013 *J. Cosmol. Astropart. Phys.* JCAP01(2013)020
[31] Burrage C and Khoury J 2014 *Phys. Rev. D* **90** 024001
[32] Burrage C, Käding C, Millington P and Minář J 2018 *Preprint* arXiv:1812.08760 [hep-th]
[33] Feynman R P and Vernon Jr F L 1963 *Ann. Phys.* **24** 118–73 [2000 *Ann. Phys.* **281** 547–607]
[34] Calzetta E A and Hu B-L 2008 *Nonequilibrium Quantum Field Theory* (Cambridge: Cambridge Univ. Press)
[35] Lehmann H, Symanzik K and Zimmermann W 1955 *Nuovo Cimento* **1** 205–25
[36] Fujii Y and Maeda K-I 2003 *The Scalar-Tensor Theory of Gravitation* (Cambridge: Cambridge Univ. Press)
[37] Gubser S S and Khoury J 2004 *Phys. Rev. D* **70** 104001
[38] Burrage C and Sakstein J 2018 *Living Rev. Relativ.* **21** 1
[39] Millington P and Pilaftsis A 2013 *Phys. Rev. D* **88** 085009
[40] Millington P and Pilaftsis A 2013 *Phys. Lett. B* **724** 56–62
[41] Schwinger J S 1961 *J. Math. Phys.* **2** 407–32
[42] Keldysh L V 1964 *Zh. Eksp. Teor. Fiz.* **47** 1515–27 [1965 *Sov. Phys. JETP* **20** 1018–26]
[43] Takahasi Y and Umezawa H 1975 *Collective Phenomena* **2** 55–80 [1996 *Int. J. Mod. Phys. B* **10** 1755–805]
[44] Arimitsu T and Umezawa H 1985 *Prog. Theor. Phys.* **74** 429–32
[45] Arimitsu T and Umezawa H 1987 *Prog. Theor. Phys.* **77** 32–52
[46] Khanna F C, Malbouisson A P C, Malbouisson J M C and Santana A E 2009 *Thermal Quantum Field Theory: Algebraic Aspects and Applications* (Singapore: World Scientific)
[47] Le Bellac M 1996 *Thermal Field Theory* (Cambridge: Cambridge Univ. Press)
[48] Estey B, Yu C, Müller H, Kuan P-C and Lan S-Y 2015 *Phys. Rev. Lett.* **115** 083002

IOP Conf. Series: Journal of Physics: Conf. Series **1275** (2019) 012042 doi:10.1088/1742-6596/1275/1/012042

Linearized metric solutions in ghost-free nonlocal gravity

Luca Buoninfante

Dipartimento di Fisica "E.R. Caianiello", Università di Salerno, I-84084 Fisciano (SA), Italy
INFN - Sezione di Napoli, Gruppo collegato di Salerno, I-84084 Fisciano (SA), Italy
Van Swinderen Institute, University of Groningen, 9747 AG, Groningen, The Netherlands

E-mail: lbuoninfante@sa.infn.it

Abstract. In this manuscript we review some aspects of linearized metric solutions in ghost-free nonlocal gravity, in which the action is made up of non-polynomial differential operators containing covariant derivatives of infinite order. By working with the simplest model of such a wide class of infinite derivative theories of gravity, we will first compute the spacetime metric generated by a static point-like source and show that all curvature invariants are nonsingular at the origin. Secondly, a similar computation is performed for an electrically charged source and also in this case the regularizing feature of nonlocality plays a crucial role. As a third case, we consider the spacetime metric generated by a Dirac delta distribution on a ring and show that, at least in the linear regime, Kerr-like singularities can be avoided in ghost-free non-local gravity.

1. Introduction

Einstein's general relativity (GR) has been the most successful theory of gravity so far, indeed its predictions have been tested to very high precision in the infrared (IR) regime, i.e. at large distances and late times [1]. For instance, we can think of the recent observation of gravitational wave emission [2], which can represents one of the greatest triumph of theoretical physics. However, despite its great achievements, there are still unsolved problems which suggest that Einstein's GR can be only seen as an effective field theory of gravitational interaction, which works very well at low energy but breaks down in the ultraviolet (UV) regime. In fact, at the classical level the Einstein-Hilbert Lagrangian, $\sqrt{-g}\mathcal{R}$, suffers from the presence of blackhole and cosmological singularities (short-distance regime), while at the quantum level it turns out to be non-renormalizable from a perturbative point of view (high-energy regime).

It is worthwhile emphasizing that our knowledge of the gravitational interaction at short distances is really limited. It suffices to think that, from a pure experimental point of view, Newton's $1/r$ law has been tested only up to a distance of 5.6×10^{-5} meters, which in terms of energies means 0.01eV [3]. Therefore, there exist a very wide desert of roughly thirty orders of magnitude, from 0.01eV to the Planck scale $M_p \sim 10^{19}$GeV [1], in which we do not know anything about gravity. This is the place where UV extensions of GR can play a crucial role.

[1] In this paper we work with Natural Units in which $c = 1 = \hbar$ and the Coulomb constant is $k_e = 1$. Moreover, we adopt the mostly positive convention for the metric signature, $\eta = \text{diag}(-1, +1, +1, +1)$.

Content from this work may be used under the terms of the Creative Commons Attribution 3.0 licence. Any further distribution of this work must maintain attribution to the author(s) and the title of the work, journal citation and DOI.
Published under licence by IOP Publishing Ltd

One way to extend GR is to add terms quadratic in the curvatures in the gravitational action, like for example $\mathcal{R}^2$ and $\mathcal{R}_{\mu\nu}\mathcal{R}^{\mu\nu}$. This kind of action was shown to be power counting renormalizable in Ref. [4, 5], but still non-physical because of the presence of a massive spin-2 ghost degree of freedom which causes Hamiltonian instabilities, classically, and breaking of the unitarity condition of the S-matrix, quantum mechanically[2]. See Refs. [10, 11, 12, 13, 14, 15, 16, 17, 18, 19] for other relevant works on applications involving gravitational actions quadratic in the curvature invariants.

The emergence of ghost modes is related to the presence of higher order time derivatives in the field equations [22]. In the last four decades, it was realized that unhealthy degrees of freedom can be still avoided in higher derivative theories if the order of the derivatives is not finite but *infinite*. Indeed, by introducing certain non-polynomial differential operators in the action, like for example $e^{\Box}$, one can prevent the appearance of extra poles in the physical spectrum, which was already noticed in the last century by [23, 24, 25, 26]. The presence of non-polynomial derivatives makes the action *nonlocal*, and this kind of nonlocal models were already studied in the early fifties in relation with the improved UV behavior of loop integrals, see Refs. [27]. This possibility turned out to be very promising and has motivated a deeper investigation of this unexplored sector of *nonlocal (or infinite derivative) field theories.*

First relevant applications of nonlocal field theory in a gravitational context were made in Refs. [28, 29, 30, 31] were the authors explicitly show the possibility to construct a quadratic curvature theory of gravity which is classically stable and unitary at the quantum level. It was also noticed that the presence of nonlocality can regularize infinities and many efforts have been made in order to resolve blackhole [29, 30, 32, 33, 34, 35, 36, 37, 38, 39, 40, 41, 42, 43, 44, 45, 46, 47] and cosmological [28, 48, 49, 50] singularities. Further applications appear in the context of inflation [51] and thermal field theory [52, 53, 54].

This class of nonlocal field theories shows an improved UV behavior [55, 58, 57, 58] but a rigorous proof of renormalizability is only available for some peculiar non-polynomial operators [57]. See also Refs. [60, 61, 62, 63, 64] for recent progresses on infinite derivative field theories, and in particular Refs. [65, 66, 67, 68] where the authors prove perturbative unitarity.

It is worthwhile emphasizing that similar infinite derivative operators also appear in string field theory and p-adic string [69, 70, 71, 72, 73].

Our aim is to review some classical aspects of ghost-free infinite derivative gravity (IDG); in particular, we will find linearized metric solutions for several sources and discuss the physical implications of nonlocality. The manuscript is organized as follows. In Section 2, we introduce the gravitational action and, we will find the propagator and the linearized field equations around Minkowski background. In Section 3, we will find the linearized metric solution for a static neutral point-like source. In Section 4, we will do the same for an electrically charged point-like source. Moreover, in Section 5 we will find the spacetime metric for a Delta Dirac distribution on a ring. Finally, in Section 6 we will draw our conclusions and discuss the outlook.

2. Nonlocal gravitational action

The class of ghost-free nonlocal gravitational actions is very wide but we will only consider one particular model for our analysis, which will capture all the main classical features due to nonlocality. Let us consider the following infinite derivative action [29, 30, 32]

$$S=\frac{1}{2\kappa^2}\int d^4x\sqrt{-g}\left\{\mathcal{R}+G_{\mu\nu}\frac{e^{-\Box/M_s^2}-1}{\Box}\mathcal{R}^{\mu\nu}\right\}, \tag{1}$$

[2] See Refs. [6, 7, 8, 9] for recent interesting works, in which the authors introduce a new quantization prescription through which the ghost is converted into a fake degree of freedom (*fakeon*), in such a way that the optical theorem is still satisfied and so perturbative unitarity is maintained.

where $G_{\mu\nu} = \mathcal{R}_{\mu\nu} - g_{\mu\nu}\mathcal{R}/2$ is the Einstein tensor, $\kappa^2 \equiv 8\pi G$ with $G = 1/M_p^2$ being the Newton constant, while M_s is a new fundamental energy scale at which nonlocal effects should manifest and is mathematically needed to make the exponent of the exponential dimensionless.

Note that the non-polynomial form-factor in Eq. (1) is

$$\mathcal{F}(\Box) \equiv \frac{e^{-\Box/M_s^2} - 1}{\Box} = -\sum_{n=1}^{\infty} \frac{1}{n!} \frac{(-\Box)^{n-1}}{M_s^{2n}}, \tag{2}$$

which is analytic and modify the short-distance (UV) behavior of Einstein's GR as only positive power of $\Box$ appear. It is worthwhile mentioning that in the literature there are also examples of non-analytic form-factors which are responsible for IR modifications; see for instance Refs. [74, 75, 76, 77, 78, 79].

2.1. Graviton propagator and linearized field equations

Let us now consider linear perturbations around the Minkowski background,

$$g_{\mu\nu} = \eta_{\mu\nu} + \kappa h_{\mu\nu}\,, \tag{3}$$

where $h_{\mu\nu}$ is the graviton perturbation, so that by expanding the action up to order $\mathcal{O}(h_{\mu\nu}^2)$ [30] we obtain:

$$S^{(2)} = \frac{1}{4}\int d^4x\, h_{\mu\nu}\, e^{-\Box/M_s^2}\mathcal{O}^{\mu\nu\rho\sigma}\, h_{\rho\sigma}, \tag{4}$$

where

$$\mathcal{O}^{\mu\nu\rho\sigma} := \frac{1}{4}\left(\eta^{\mu\rho}\eta^{\nu\sigma} + \eta^{\mu\sigma}\eta^{\nu\rho}\right)\Box - \frac{1}{2}\eta^{\mu\nu}\eta^{\rho\sigma}\Box + \frac{1}{2}\left(\eta^{\mu\nu}\partial^\rho\partial^\sigma + \eta^{\rho\sigma}\partial^\mu\partial^\nu - \eta^{\mu\rho}\partial^\nu\partial^\sigma - \eta^{\mu\sigma}\partial^\nu\partial^\rho\right) \tag{5}$$

is a four-rank operator $\mathcal{O}^{\mu\nu\rho\sigma}$ which is totally symmetric in all its indices.

By inverting the kinetic operator $e^{-\Box/M_s^2}\mathcal{O}^{\mu\nu\rho\sigma}$ one obtains the graviton propagator around Minkowski, and its gauge independent and saturated part is given by [23, 26, 29, 30]

$$\Pi_{\mu\nu\rho\sigma}(k) = e^{-k^2/M_s^2}\left(\frac{\mathcal{P}^2_{\mu\nu\rho\sigma}}{k^2} - \frac{\mathcal{P}^0_{s,\,\mu\nu\rho\sigma}}{2k^2}\right) \equiv e^{-k^2/M_s^2}\Pi_{GR,\,\mu\nu\rho\sigma}(k), \tag{6}$$

where $\Pi_{GR} = \mathcal{P}^2/k^2 - \mathcal{P}^0_s/2k^2$ is the graviton propagator in Einstein's GR, while $\mathcal{P}^2$ and $\mathcal{P}^0_s$ are the spin-2 and spin-0 projection operators; see Refs. [80, 81] for more details.

Very interestingly, the nonlocal modification in the propagator turns out to be a simple extra factor e^{-k^2/M_s^2}, which does not introduce any new pole in the theory other than the standard massless spin-2, $k^2 = 0$.

The linearized field equations corresponding the the action in Eq. (4) are given by [30]

$$e^{-\Box/M_s^2}\mathcal{O}_{\mu\nu}{}^{\rho\sigma}\, h_{\rho\sigma} = -16\pi G T_{\mu\nu}, \tag{7}$$

where

$$T_{\mu\nu} = -\frac{2}{\sqrt{-g}}\frac{\delta(\sqrt{-g}\mathcal{L}_m)}{\delta g^{\mu\nu}} \tag{8}$$

is the stress-energy tensor of the matter sector.

In the following sections we will solve the field equations in Eq. (7) for three different choices of $T_{\mu\nu}$.

IOP Conf. Series: Journal of Physics: Conf. Series **1275** (2019) 012042 doi:10.1088/1742-6596/1275/1/012042

3. Spacetime metric for a static point-like source

In this Section we wish to find a spherically symmetric static metric solution in presence of a Delta Dirac distribution placed at $r = 0$. By working in the conformal Newtonian gauge, we can write the perturbed metric in Eq. (3) in isotropic coordinates as follows:

$$ds^2 = -(1 + 2\Phi(r))dt^2 + (1 - 2\Psi(r))(dr^2 + r^2 d\Omega^2), \tag{9}$$

where $r = \sqrt{x^2 + y^2 + z^2}$ is the isotropic radial coordinate, so that the field equations in Eq. (7) become

$$\begin{aligned} e^{-\nabla^2/M_s^2}\nabla^2\Phi = &\ 4\pi G(T + 2T_{00}), \\ e^{-\nabla^2/M_s^2}\nabla^2\Psi = &\ 4\pi G T_{00}. \end{aligned} \tag{10}$$

where $T \equiv \eta^{\mu\nu}T_{\mu\nu}$, and we have used $\kappa h_{00} = -2\Phi$, $\kappa h_{ij} = -2\Psi\delta_{ij}$, $\kappa h = 2(\Phi - 3\Psi)$ and $\Box \simeq \nabla^2$.

In the case of a neutral static point-like source, the stress energy tensor is given by

$$T_{\mu\nu} = m\delta_\mu^0\delta_\nu^0\delta^{(3)}(\vec{r}), \tag{11}$$

so that the modified Poisson equations in Eq. (10) reduce to

$$e^{-\nabla^2/M_s^2}\nabla^2\Phi = e^{-\nabla^2/M_s^2}\nabla^2\Psi = 4\pi m\delta^{(3)}(\vec{r}), \tag{12}$$

implying that the two metric potentials are equal, $\Phi = \Psi$. Note that we have to deal with a differential equation of infinite order due to the presence of the exponential operator $e^{-\nabla^2/M_s^2}$; however it can be easily solved by going to Fourier space and then anti-transforming back to coordinate space. Indeed, by doing so we obtain the following solution:

$$\Phi(r) = -4\pi Gm\int d^3k\frac{e^{-k^2/M_s^2}}{k^2}e^{i\vec{k}\cdot\vec{r}} = -\frac{2Gm}{\pi}\frac{1}{r}\int_0^\infty dk\frac{\sin(kr)}{k}e^{-k^2/M_s^2} = -\frac{Gm}{r}\text{Erf}\left(\frac{M_s r}{2}\right), \tag{13}$$

where

$$\text{Erf}(x) := \frac{2}{\sqrt{\pi}}\int_0^x e^{-t^2}dt, \tag{14}$$

is the so called error-function.

Very interestingly, note that the gravitational potential in Eq. (13) is non-singular at $r = 0$, indeed it tends to the finite constant value $\Phi(0) = 2GmM_s/\sqrt{\pi}$; while for large distances we recover the $1/r$ behavior of the Newtonian potential, as expected. Note that the linearized regime is valid as long as the inequality $2GmM_s/\sqrt{\pi} < 1$ holds true.

From a physical point of view, nonlocality is able to regularize the singularity, indeed the point-like source at $r = 0$ is smeared out on a region of size $1/M_s$ due to the presence of infinite order derivative. Not only the linearized metric potentials but also all the linearized curvature invariants turn out to be non-singular. Furthermore, at $r = 0$ the metric becomes conformally-flat, since all the components of the Weyl tensor vanish at the origin [38].

4. Spacetime metric for an electrically charged point-like source

Let us now consider the case of a static electric charge as a gravitational source and find the corresponding metric potentials for the spherically symmetric metric in Eq. (9). The stress-energy tensor in this case is given by electro-magnetic one which reads

$$T_{\mu\nu} = \frac{1}{4\pi}\left(\eta_{\rho\nu}F_{\mu\sigma}F^{\rho\sigma} - \frac{1}{4}\eta_{\mu\nu}F_{\rho\sigma}F^{\rho\sigma}\right), \tag{15}$$

where $F_{\mu\nu} = \partial_\mu A_\nu - \partial_\nu A_\mu$ is the electro-magnetic field strength defined in terms of the potential-vector A_μ. In the simplest case of an electric charge, only the components related to the electric field are non-vanishing:

$$F_{10} = -F_{01} = E_r \quad \text{and} \quad E_r = \frac{Q}{r^2}, \tag{16}$$

with E_r being the radial component of the electric field. Given the stress-energy tensor in Eq. (15) with the components in Eq. (16), the infinite derivative differential equations in Eq. (10) become [41]:

$$\begin{aligned} e^{-\nabla^2/M_s^2}\nabla^2\Phi &= \frac{GQ^2}{r^4}, \\ e^{-\nabla^2/M_s^2}\nabla^2\Psi &= \frac{GQ^2}{2r^4}. \end{aligned} \tag{17}$$

Also in this case we can solve the modified Poisson equations by using the Fourier transform method. By making the field redefinitions $\tilde{\Phi} := e^{-\nabla^2/M_s^2}\Phi$, $\tilde{\Psi} := e^{-\nabla^2/M_s^2}\Psi$, we obtain the following two solutions from Eq. (17):

$$\begin{aligned} \tilde{\Phi}(r) &= -\frac{C_1}{r} + \frac{GQ^2}{2r^2} + C_2 \\ \tilde{\Psi}(r) &= -\frac{C_1}{r} + \frac{GQ^2}{4r^2} + C_2, \end{aligned} \tag{18}$$

where we fix $C_1 = Gm$ by requiring that we want to recover the static neutral case in Eq. (13) when $Q = 0$, and $C_2 = 0$ since we want asymptotic flatness (for $r \longrightarrow \infty$). We can now go back to the fields Φ and Ψ, which are given by

$$\begin{aligned} \Phi(r) &= -Gme^{\nabla^2/M_s^2}\left(\frac{1}{r}\right) + \frac{GQ^2}{2}e^{\nabla^2/M_s^2}\left(\frac{1}{r^2}\right), \\ \Psi(r) &= -Gme^{\nabla^2/M_s^2}\left(\frac{1}{r}\right) + \frac{GQ^2}{4}e^{\nabla^2/M_s^2}\left(\frac{1}{r^2}\right). \end{aligned} \tag{19}$$

By using the fact that $4\pi/k^2$ is the Fourier transform of $1/r$, we can write

$$e^{\nabla^2/M_s^2}\left(\frac{1}{r}\right) = \int \frac{d^3k}{(2\pi)^3}\frac{4\pi}{k^2}e^{-k^2/M_s^2}e^{i\vec{k}\cdot\vec{r}} = \frac{2}{\pi}\int_0^\infty dk\,\frac{\sin(kr)}{kr}e^{-k^2/M_s^2} = \frac{1}{r}\mathrm{Erf}\left(\frac{M_s r}{2}\right), \tag{20}$$

which gives the neutral part of the potentials; then, by using the Fourier transform $\frac{2\pi^2}{k}\mathrm{sign}(k)$ of $1/r^2$, we can write

$$e^{\nabla^2/M_s^2}\left(\frac{1}{r^2}\right) = \int \frac{d^3k}{(2\pi)^3}\frac{2\pi^2}{k}\mathrm{sign}(k)e^{-k^2/M_s^2}e^{i\vec{k}\cdot\vec{r}} = \int_0^\infty dk\,\frac{\sin(kr)}{r}e^{-k^2/M_s^2} = \frac{M_s}{r}\mathrm{F}\left(\frac{M_s r}{2}\right), \tag{21}$$

where

$$\mathrm{F}(x) := e^{-x^2}\int_0^x e^{t^2}dt \tag{22}$$

is the so called Dawson function. Therefore, the two metric potentials in Eq. (19) read [41]

$$\begin{aligned} \Phi(r) &= -\frac{Gm}{r}\mathrm{Erf}\left(\frac{M_s r}{2}\right) + \frac{GQ^2M_s}{2r}\mathrm{F}\left(\frac{M_s r}{2}\right), \\ \Psi(r) &= -\frac{Gm}{r}\mathrm{Erf}\left(\frac{M_s r}{2}\right) + \frac{GQ^2M_s}{4r}\mathrm{F}\left(\frac{M_s r}{2}\right). \end{aligned} \tag{23}$$

As expected, for large distances $M_s r \gg 1$, we recover the metric potentials of the linearized Reissner-Nordström metric of GR [41], while in the limit $r \longrightarrow 0$ the two metric potentials tend to finite values, $\Phi(0) = -GmM_s/\sqrt{\pi} + GQ^2M_s^2/4$ and $\Psi(0) = -GmM_s/\sqrt{\pi} + GQ^2M_s^2/8$, as can be easily checked. The same regularized behavior can be shown for all linearized curvature invariants; in particular, the Weyl tensor vanishes implying that the metric is conformally-flat at the origin [41], as it also happens for the case of a neutral source studied in Section 3. Therefore, also for a point-like electric charge nonlocality is able to regularize the singularity at $r = 0$.

The linearized regime holds all the way from $r = \infty$ up to $r = 0$, as long as the inequalities $2|\Phi(0)|, 2|\Psi(0)| < 1$ are satisfied, which means $mM_s < M_p^2$ and $|Q|M_s < M_p$, where we have neglected constant factors of order one [41].

5. Spacetime metric for a Dirac delta distribution on a ring

In this Section, we wish to determine the spacetime metric generated by a Dirac Delta distribution on a ring and show how nonlocality can regularize Kerr-like ring singularities. It is well known that the Kerr metric suffers from the presence of a ring singularity which in Boyer-Lindquist coordinates is described by the equation $r^2 + a^2\cos^2\vartheta = 0$, or in Cartesian coordinates by $z = 0$, $x^2 + y^2 = a^2$, where a is the radius of the ring [82].

To mimic such a ring distribution, we consider a ring of radius a rotating with constant angular velocity ω, described by following stress-energy tensor:

$$T_{00} = m\delta(z)\frac{\delta(x^2 + y^2 - a^2)}{\pi}, \qquad T_{0i} = T_{00}v_i, \tag{24}$$

where v_i is the tangential velocity and its magnitude is related to the angular velocity through the relation $v = \omega\, a$; moreover, by assuming that the rotation happens around the z-axis, we can write $v_x = -y\,\omega$, $v_y = x\,\omega$ and $v_z = 0$. The stress-energy tensor in Eq. (24) will source the following linearized non-diagonal metric:

$$ds^2 = -(1 + 2\Phi)dt^2 + 2\vec{h}\cdot d\vec{x}dt + (1 - 2\Psi)d\vec{x}^2, \tag{25}$$

where $h_i \equiv h_{0i}$ and the coordinate r is the isotropic radius which should not be confused with the Boyer-Lindquist radial coordinate used above. The metric components in Eq. (25) can be found by solving a set of decoupled infinite order differential equations [42]:

$$\begin{aligned} e^{-\nabla^2/M_s^2}\nabla^2\Phi(\vec{r}) = e^{-\nabla^2/M_s^2}\nabla^2\Psi(\vec{r}) &= 4Gm\,\delta(z)\delta(x^2 + y^2 - a^2), \\ e^{-\nabla^2/M_s^2}\nabla^2 h_{0x}(\vec{r}) &= -16Gm\,\omega\, y\delta(z)\delta(x^2 + y^2 - a^2), \\ e^{-\nabla^2/M_s^2}\nabla^2 h_{0y}(\vec{r}) &= 16Gm\,\omega\, x\delta(z)\delta(x^2 + y^2 - a^2). \end{aligned} \tag{26}$$

Also in this case we can solve the modified Poisson equations by using the Fourier transform method. Note that, by going to cylindrical coordinates, we can Fourier transform the stress-energy tensor components as follows [42]:

$$\begin{aligned} \mathcal{F}[\delta(z)\delta(x^2 + y^2 - a^2)] &= \int_{-\infty}^{\infty} dz\delta(z)e^{ik_z z}\int_0^{\infty} d\rho\rho\delta(\rho^2 - a^2)\int_0^{2\pi} d\varphi e^{ik_x\rho\cos\varphi}e^{ik_y\rho\sin\varphi} \\ &= \pi\int_0^{\infty} d(\rho^2)\delta(\rho^2 - a^2)I_0\left(i\rho\sqrt{k_x^2 + k_y^2}\right) \\ &= \pi I_0\left(ia\sqrt{k_x^2 + k_y^2}\right), \end{aligned} \tag{27}$$

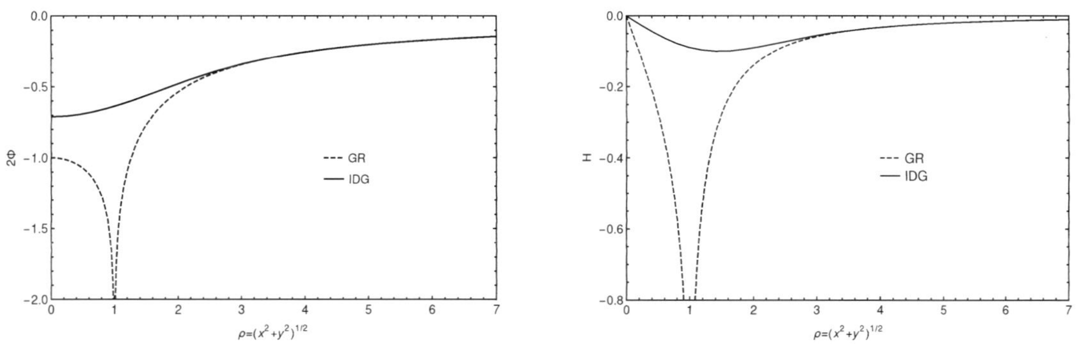

(a) Metric component $-h_{00}$ (b) Metric component h_{0i}

Figure 1: In this plot we have shown the behavior of the components $-h_{00}$ and $h_{0i} \sim H$ as functions of the cylindrical radius ρ, in both cases of IDG (orange line) and IDG (blue line). For convenience, we have set $M_s = 1.5$, $G = 1$, $m = 0.5$ and $a = 1$. It is clear that, while the metric potentials in GR have a singularity for $\rho = a$, they turn out to be regular in IDG.

$$\begin{aligned}\mathcal{F}[\delta(z)\delta(x^2+y^2-a^2)] &= \int_{-\infty}^{\infty} dz\delta(z)e^{ik_z z}\int_0^{\infty} d\rho\rho^2\delta(\rho^2-a^2)\int_0^{2\pi} d\varphi e^{ik_x\rho\cos\varphi}e^{ik_y\rho\sin\varphi}\cos\varphi \\ &= \pi\frac{k_x}{\sqrt{k_x^2+k_y^2}}\int_0^{\infty} d(\rho^2)\rho\delta(\rho^2-a^2)I_1\left(i\rho\sqrt{k_x^2+k_y^2}\right) \\ &= \pi a\frac{k_x}{\sqrt{k_x^2+k_y^2}}I_1\left(ia\sqrt{k_x^2+k_y^2}\right),\end{aligned} \tag{28}$$

and

$$\mathcal{F}[y\delta(z)\delta(x^2+y^2-a^2)] = \pi a\frac{k_y}{\sqrt{k_x^2+k_y^2}}I_1\left(ia\sqrt{k_x^2+k_y^2}\right), \tag{29}$$

where $\mathcal{F}[\cdots]$ stands for the Fourier transform operation, I_0 and I_1 are two Modified Bessel functions. For simplicity, let us only consider the plane $z = 0$ and work with the cylindrical radial coordinate $\rho = \sqrt{x^2+y^2}$, as we know that in GR the ring singularity lies in the x-y plane. Therefore, by anti-transforming to coordinate space, the metric potentials will be given by [42]

$$\Phi(\rho) = \Psi(\rho) = -Gm\int_0^{\infty} d\zeta I_0\left(ia\zeta\right)I_0\left(i\zeta\rho\right)\mathrm{Erfc}\left(\frac{\zeta}{M_s}\right), \tag{30}$$

$$h_{0x}(x,y) = 4Gm\,\omega\,a\frac{y}{\rho}H(\rho), \qquad h_{0y}(x,y) = -4Gm\,\omega\,a\frac{x}{\rho}H(\rho), \tag{31}$$

with

$$H(\rho) := \int_0^{\infty} d\zeta I_1(ia\zeta)I_1(i\zeta\rho)\mathrm{Erfc}\left(\frac{\zeta}{M_s}\right), \tag{32}$$

which in the limit $M_s \longrightarrow \infty$ all reduce to GR case.

The above three integrals cannot be solved analytically but we can do it numerically as shown in Fig. 1. We can explicitly see that in GR the metric potentials suffers from the presence of a ring singularity at $\rho = a$, while in infinite derivative gravity such a singularity is regularized. Also in this case, one can show that all the curvature invariants are non-singular in the entire spacetime and that the Weyl tensor vanishes at $r = 0$.

Note that the linearized regime holds from $r = \infty$ all the way to $r = 0$ as long as the inequalities $2|\Phi| < 1$ and $|h_{0i}| < 1$ are satisfied for any r, which, by neglecting constant factors of order one, also means $GmM_s < 1$ and $GmM_s^2\omega a^2 < 1$, respectively [42].

6. Conclusions and outlook

In this manuscript, we have studied linearized metric solutions for one model of infinite derivative theories of gravity, in which the choice of nonlocal form-factor is the one in Eq. (2). We have considered three different configurations, neutral static point-like source, static electrically charged point-like source and Dirac delta distribution on a ring rotating with constant angular velocity. Our aim was to show how the presence of nonlocality through infinite order derivatives can regularize both point-like and ring singularities, at least at the linearized level.

The next important step is to explore the non-linear regime and understand whether singularities can be also avoided when dealing with the full set of non-linear field equations. Some progress has been made in Ref. [40], where physical arguments have been presented in support of singularity avoidance. However, a rigorous treatment is still lacking and no full non-linear spherically symmetric solution is known in the literature. Such a program is challenging but at the same time very promising, and it will subject of future works.

Acknowledgments

The author would like to thank A. S. Cornell, G. Harmsen, A. S. Koshelev, G. Lambiase, S. Maheswari, J. Marto and A. Mazumdar, for the fruitful collaboration and all the intensive discussions.

References

[1] Will C M 2014 Living Rev. Rel. 17 4
[2] Abbott B P et al. [LIGO Scientific and Virgo Collaborations] 2016 Phys. Rev. Lett. **116** no.6 061102
[3] Kapner D J, Cook T S, Adelberger E G, Gundlach J H, Heckel B R, Hoyle C D and Swanson H E 2007 Phys. Rev. Lett. **98** 021101
[4] Stelle K S 1977 Phys. Rev. D **16** 953
[5] Stelle K S 1978 Gen. Rel. Grav. **9** 353-371
[6] Anselmi D and Piva M 2017 JHEP **1706** 066; Anselmi D and Piva M 2017 Phys. Rev. D **96** no. 4 045009
[7] Anselmi D 2018 JHEP **1802** 141
[8] Anselmi D 2017 JHEP **1706** 086
[9] Anselmi D and Piva M 2018 JHEP **1805** 027; Anselmi D and Piva M 2018 JHEP **1811** 021; Anselmi D *Preprint* arXiv:1809.05037 [hep-th]
[10] Starobinski A A 1980 Phys. Lett. B **91** 99-102; 1983 Sov. Astron. Lett. **9** 302
[11] Capozziello S, Lambiase G, Sakellariadou M and Stabile An. 2015 Phys. Rev. D **91** 044012
[12] Lambiase G, Sakellariadou M, Stabile A and Stabile An. 2015 JCAP **1507** 003
[13] Lambiase G, Sakellariadou M and Stabile A 2013 JCAP **1312** 020
[14] Capozziello S and Lambiase G 2003 Int. J. Mod. Phys. D **12** 843-852
[15] Calchi Novati S, Capozziello S and Lambiase G 2000 Grav. Cosmol. **6** 173-180
[16] Capozziello S, Lambiase G and Schmidt H J 2000 Annalen Phys. **9** 39-48
[17] Capozziello S, Lambiase G, Papini G and Scarpetta G 1999 Phys. Lett. A **254** 11-17
[18] Lambiase G, Stabile A and Stabile An. 2017 Phys. Rev. D **95** 084019. Buoninfante L, Lambiase G, Petruzziello L and Stabile A. 2019 Eur. Phys. J. C **79** no. 1 41
[19] Blasone M, Capozziello S, Lambiase G and Petruzziello L *Preprint* arXiv:1812.08029 [gr-qc]
[20] Giacchini B L and de Paula Netto T *Preprint* arXiv:1806.05664 [gr-qc]
[21] Accioly A, Giacchini B L and Shapiro I L 2017 Phys. Rev. D **96** no. 10, 104004. Giacchini B L 2017 Phys. Lett. B **766** 306
[22] Oströgradski M 1850 *Mémoires sur les Équations Différentielles, relatives au Problème des Isopérimètres* Mem. Ac. St. Petersbourg 6 385-517

[23] Krasnikov N V 1987 Theor Math. Phys. **73** 1184; 1987 Teor. Mat. Fiz. **73** 235
[24] Kuzmin Y V 1989 Yad. Fiz. **50** 1630-1635
[25] Moffat J W 1990 Phys. Rev. D **41** 1177; Evens D, Moffat J W, Kleppe G and Woodard R P 1991 Phys. Rev. D **43** 499
[26] Tomboulis E T 1997 *Preprint* hep-th/9702146
[27] Feynman R P 1948 Phys. Rev. **74** 939. Pais A and Uhlenbeck G E 1950 Phys. Rev. **79** 145; Efimov V 1967 Comm. Math. Phys. **5** 42; Efimov V 1968 ibid 7 138; Alebastrov V A, Efimov V 1973 Comm. Math. Phys. **31** 1 1-24; Alebastrov V A, V. Efimov 1974 Comm. Math. Phys. **38** 1 11-28; Kirzhnits D A 1967 Sov. Phys. Usp. **9** 692
[28] Biswas T, Mazumdar A and Siegel W 2006 JCAP **0603** 009
[29] Modesto L 2012 Phys. Rev. D **86** 044005
[30] Biswas t, Gerwick E, Koivisto T and Mazumdar A 2012 Phys. Rev. Lett. **108** 031101
[31] Biswas T, Koshelev A S and Mazumdar A 2016 Fundam. Theor. Phys. **183** 97 Biswas T, Koshelev A S and Mazumdar A 2017 Phys. Rev. D **95** no. 4 043533
[32] Biswas T, Conroy A, Koshelev A S and Mazumdar A 2014 Class. Quant. Grav. **31** 015022, Erratum: Class. Quant. Grav. **31** 159501
[33] Edholm J, Koshelev A S and Mazumdar A 2016 Phys. Rev. D **94** no. 10 104033; Frolov V P and Zelnikov A 2016 Phys. Rev. D **93** no. 6 064048
[34] Frolov V P, Zelnikov A and de Paula Netto T 2015 JHEP **1506** 107
[35] Frolov V P 2015 Phys. Rev. Lett. **115**, no. 5 051102 (2015)
[36] Frolov V p and Zelnikov A 2016 Phys. Rev. D **93** no. 6 064048
[37] Koshelev A S and Mazumdar A 2017 Phys. Rev. D **96** no. 8 084069
[38] Buoninfante L, Koshelev A S, Lambiase G and Mazumdar A 2018 JCAP **1809** no.09 034
[39] Koshelev A S, Marto J and Mazumdar A 2018 Phys. Rev. D **98** no.6 064023
[40] Buoninfante L, Koshelev A S, Lambiase G, Marto J and Mazumdar A 2018 JCAP **1806** no.06 014
[41] Buoninfante L, Harmsen G, Maheshwari S and Mazumdar A 2018 Phys. Rev. D **98** no.8 084009
[42] Buoninfante L, Cornell A S, Harmsen G, Koshelev A S, Lambiase G, Marto J and Mazumdar A 2018 Phys. Rev. D **98** no. 8 084041
[43] Edholm J 2018 Phys. Rev. D **97** no.6 064011
[44] Boos J, Frolov V P and Zelnikov A 2018 Phys. Rev. D **97** no. 8 084021
[45] Kilicarslan E 2018 Phys. Rev. D **98** no. 6 064048
[46] de la Cruz-Dombriz A, Torralba F J M and Mazumdar A *Preprint* arXiv:1812.04037 [gr-qc]
[47] Mazumdar A and Stettinger G *Preprint* arXiv:1811.00885 [hep-th]
[48] Biswas T, Koivisto T and Mazumdar A 2010 JCAP **1011** 008
[49] Biswas T, Koshelev A S, Mazumdar A and Vernov S Y 2012 JCAP **1208** 024
[50] Koshelev A S and Vernov S Y 2012 Phys. Part. Nucl. **43** 666
[51] Joukovskaya L 2007 Phys. Rev. D **76** 105007; Calcagni G, Montobbio M and Nardelli G 2007 Phys. Rev. D **76** 126001; Biswas T, Brandenberger R, Mazumdar A and Siegel W 2007 JCAP **0712** 011; Biswas T and Mazumdar A 2014 Class. Quant. Grav. **31** 025019; Chialva D and Mazumdar A 2015 Mod. Phys. Lett. A **30** no. 03n04 1540008; Koshelev A S, Modesto L, Rachwal L and Starobinsky A A 2016 JHEP **1611** 067; Koshelev A S, Kumar S K and Starobinsky A A 2018 JHEP **1803** 071; Kumar S K and Modesto L *Preprint* arXiv:1810.02345 [hep-th]
[52] Biswas T, Cembranos J A R and Kapusta J I 2010 Phys. Rev. Lett. **104** 021601
[53] Biswas T, Cembranos J A R and Kapusta J I 2010 JHEP **1010** 048
[54] Biswas T, Cembranos J A R and Kapusta J I 2010 Phys. Rev. D **82** 085028
[55] Biswas T and Okada N 2015 Nucl. Phys. B **898** 113
[56] Tomboulis E T 2015 Phys. Rev. D **92** no. 12 125037
[57] Modesto L and Rachwal L 2014 Nucl. Phys. B **889** 228; Modesto L and Rachwał L 2015 Nucl. Phys. B **900** 147
[58] Talaganis S, Biswas T and Mazumdar A 2015 Class. Quant. Grav. **32** no. 21 215017
[59] Ghoshal A, Mazumdar A, Okada N and Villalba D 2018 Phys. Rev. D **97** no. 7 076011
[60] Gama F S, Nascimento J R, Petrov A Y and Porfirio P J, *Preprint* arXiv:1804.04456 [hep-th]
[61] Hashi M N, Isono H, Noumi T, Shiu G and Soler P 2018 JHEP **1808** 064
[62] Buoninfante L, Lambiase G and Mazumdar A *Preprint* arXiv:1805.03559 [hep-th]
[63] Buoninfante L, Ghoshal A, Lambiase G and Mazumdar A *Preprint* arXiv:1812.01441 [hep-th]
[64] Buoninfante L, Lambiase G and Yamaguchi M *Preprint* arXiv:1812.10105 [hep-th]
[65] Pius R and Sen A 2016 JHEP **1610** 024, Erratum: 2018 JHEP **1809** 122; Sen A 2016 JHEP **1611** 050
[66] Carone C D 2017 Phys. Rev. D **95** 045009
[67] Chin P and Tomboulis E T 2018 JHEP **1806** 014

IOP Conf. Series: Journal of Physics: Conf. Series **1275** (2019) 012042 doi:10.1088/1742-6596/1275/1/012042

[68] Briscese F and Modesto L *Preprint* arXiv:1803.08827 [gr-qc]
[69] Witten E 1986 Nucl. Phys. B **268** 253
[70] Eliezer D A, Woodard R P 1989 Nucl. Phys. B 325 389
[71] Tseytlin A A 1995 Phys. Lett. B **363** 223
[72] Siegel W *Preprint* hep-th/0309093
[73] Freund P G O and Olson M 1987 Phys. Lett. B **199** 186; Brekke L, Freund P G O, Olson M and Witten E 1988 Nucl. Phys. B **302** 365; Freund P G O and Witten E 1987 Phys. Lett. B **199** 191; Frampton P H and Okada Y 1988 Phys.Rev. D **37** 3077-3079; Barnaby N, Biswas T and Cline J M 2007 JHEP 0704 056
[74] Barvinsky A O and Vilkovisky G A 1985 Phys. Rept. **119** 1-74
[75] Deser S and Woodard R P 2007 Phys. Rev. Lett. **99** 111301
[76] Conroy A, Koivisto T, Mazumdar A and Teimouri A 2015 Class. Quant. Grav. **32** no. 1 015024
[77] Belgacem E, Dirian Y, Foffa S and Maggiore M 2018 JCAP **1803** no. 03 002
[78] Woodard R P 2018 Universe **4** no. 8 88
[79] Belenchia A, Benincasa D M T, Liberati S 2015 JHEP 1503 036. Belenchia A, Benincasa D M T, Martin-Martinez E and Saravani M 2016 Phys. Rev. D **94** 061902(R)
[80] Biswas T, Koivisto T and Mazumdar A *Preprint* arXiv:1302.0532 [gr-qc]
[81] Buoninfante L *Preprint* arXiv:1610.08744 [gr-qc]
[82] Hawking S W and Ellis R G 2011 *The Large Scale Structure of Space-Time* Cambridge: Cambridge University Press

IOP Conf. Series: Journal of Physics: Conf. Series **1275** (2019) 012043 doi:10.1088/1742-6596/1275/1/012043

How to obtain initial degrees of freedom for the early universe of about 1000, instead of 100, while preserving a uniformly increasing arrow of time due to entropy increase

Andrew Walcott Beckwith

Physics Department, Chongqing University,
Chongqing 401331, People's Republic of China
english.cqu.edu.cn

E-mail: `Rwill9955b@gmail.com; abeckwith@uh.edu`

Abstract. Using the Klauder enhanced quantization to specify the cosmological constant as a baseline for the mass of a graviton, we eventually develop with an initial degrees of freedom argument of about 1000, instead of 100 or so, implying no quantum fluctuations such as reversals in the arrow of time due to the size of initial black holes, even if the cosmological constant as outlined may be due to quantum processes, as outlined initially. Classicality in black holes, which may shed gravitons, avoids quantum-induced reversals in the arrow of time, and keeps time flow increasing rather than initially decreasing.

1. The general relativity first integral

We use Padmanabhan's first integral [1, 2] ((1)–(3)), with a Ricci scalar in (3) [3] and usually the curvature, $\aleph$, set as extremely small, with general relativity [4, 5]

$$S_1 = \frac{1}{2\kappa}\int \sqrt{-g}\,\mathrm{d}^4x(\Re - 2\Lambda) \tag{1}$$

$$-g = -\det g_{\mathrm{uv}} \tag{2}$$

$$\Re = 6\left[\frac{\ddot{a}}{a} + \left(\frac{\dot{a}}{a}\right)^2 + \frac{\aleph}{a^2}\right]. \tag{3}$$

Also, the variation of $\delta g_{tt} \approx a_{\min}^2\phi$ [6, 7] will have an inflaton, ϕ, [2, 7, 3, 2, 8] which is combined into other procedures for a solution to the cosmological constant problem. Here, $a_{\min}$ is a minimum value of the scale factor. [8, 3]

1.1. Next for the idea from Klauder

We go to a restricted quantum action principle [5, p. 78] with a one–one equivalence, [1, 4]

$$S = \int_0^T \mathrm{d}t[p(t)\dot{q}(t) - H\ (p(t), q(t))] \approx S = \frac{1}{2\kappa}\int \sqrt{-g}\,\mathrm{d}^4x(\Re - 2\Lambda). \tag{4}$$

Content from this work may be used under the terms of the Creative Commons Attribution 3.0 licence. Any further distribution of this work must maintain attribution to the author(s) and the title of the work, journal citation and DOI.
Published under licence by IOP Publishing Ltd

IOP Conf. Series: Journal of Physics: Conf. Series **1275** (2019) 012043 doi:10.1088/1742-6596/1275/1/012043

We assume Λ is a constant. Hence, we assume the following approximation, [4]

$$\frac{p_0^2}{2} = \frac{p_0^2(N)}{2} + N; \quad \text{for } 0 < N \leq \infty \text{ and } q = q_0 \pm p_0 t \tag{5}$$

$$V_N = \begin{cases} 0; & \text{for } 0 < x < 1 \\ N; & \text{otherwise} \end{cases} \tag{6}$$

$$H_N(p(t), q(t)) = \frac{p_0^2}{2} + \frac{(\hbar\pi)^2}{2} + N; \quad \text{for } 0 < N \leq \infty. \tag{7}$$

Our innovation is to then equate $q = q_0 \pm p_0 t \sim \phi$ and to assume small time step values. Then,

$$\Lambda \approx \frac{-\left[\frac{V_0}{3\gamma-1} + 2N + \frac{\gamma(3\gamma-1)}{8\pi G\tilde{t}^2}\right]}{\frac{1}{\kappa}\int\sqrt{-g}\,\mathrm{d}^3x} + \left\{6\left[\frac{\ddot{a}}{a} + \left(\frac{\dot{a}}{a}\right)^2\right]\right\}\Bigg|_{t=\tilde{t}} \tag{8}$$

These terms within the bubble of space–time [1] use the same inflaton potential. The scale factor is presumed to obey the value of that in Camarade et al. [3]

2. Why this is linked to gravity/massive gravitons

Klauder's program [5] is to embed via (10)–(13) as a quantum mechanical well for a pre-Planckian system for inflaton physics as given by (5)–(7). In a treatment of the action integral, [5, p. 87] Klauder refers to the weak-correspondence principle, where an enhanced classical Hamiltonian is given one–one correspondence with quantum effects in a nonvanishing fashion. If so [4, 9]

$$m_g = \frac{\hbar\sqrt{\Lambda}}{c} \tag{9}$$

Therefore, we link this value of the cosmological constant and the production of gravitons by mini black holes. Klauder's idea of enhanced quantization for Λ and the possible decay of mini black holes gives [10]

$$T_{\mathrm{H}} = m_D \cdot \left[\frac{m_D \cdot (n+2)}{m_{\mathrm{bh}} \cdot 8 \cdot \Gamma \cdot \left(\frac{n+3}{2}\right)}\right]^{\frac{1}{n+1}} \cdot \left(\frac{n+1}{4\sqrt{\pi}}\right) \tag{10}$$

$$m_D \doteq 1 \text{ TeV} \equiv 10^{12} \text{ eV} = 1.783 \times 10^{-30} \text{ g} \tag{11}$$

$$m_D \doteq 1.22 \times 10^{21} \text{ TeV} = 2.175 \times 10^{-9} \text{ g} \tag{12}$$

$$T_{\mathrm{H}} \propto 1 \text{ TeV} \cdot \left[\frac{1 \text{ TeV} \cdot (n+2)}{1.22 \times 10^{21} \cdot 8 \cdot \Gamma \cdot \left(\frac{n+3}{2}\right)}\right]^{\frac{1}{n+1}} \cdot \left(\frac{n+1}{4\sqrt{\pi}}\right) \text{ gravitons,} \tag{13}$$

after a comparison with the conditions for which the Hawkings temperature is created. [11] We then can, if we have this Hawkings temperature, as given in (14)–(16), consider the question whether the black holes have classical or quantum behavior as well as Γ being a gamma function. That is, the idea is based upon the formation of a finite number of black holes that decay. [12]

In Corda's recent work, [13] we have a so-called horizon volume, where $\tilde{n}_{qm}$ is the so-called quantum number put in where the four-dimensional Planck mass is normalized to one. So then, if there are 10^2 black holes of mass 10^2 times the Planck mass (setting the four-dimensional Planck mass to 1), there is a change in volume of early space–time for which we have, if the Planck length is normalized to one and if n of (10) is equal to zero initially for a four-dimensional space–time volume [12, 13]

$$16\pi\frac{\widetilde{n}_{qm}}{4\times10^4}\left|\sqrt{1-\left(\frac{\widetilde{n}_{qm}-1}{\widetilde{n}_{qm}}\right)}\right|\approx10^6 \tag{14}$$

$$m_{\mathrm{bh}}\approx10^2\cdot m_{\mathrm{4D-Planck}} \tag{15}$$

$$\Delta V_{\mathrm{total}}\simeq10^2\cdot\Delta V_{\widetilde{n}_{qm}-1\to\widetilde{n}_{qm}}\approx10^2\cdot16\pi\frac{\widetilde{n}_{qm}}{4\times10^4}\left|\sqrt{1-\left(\frac{\widetilde{n}_{qm}-1}{\widetilde{n}_{qm}}\right)}\right|\approx10^8 \tag{16}$$

This puts a serious restriction on the number of allowed quantization levels $\widetilde{n}_{qm}$, but it also means that, within this horizon space, we may be seeing mini black holes created that could release gravitons. We then discuss what may be pertinent to characterizing if the black holes are behaving classically or quantum mechanically. Note, if $n=0$ (4 dimensions) in a litteral interpretation of (10) and (13) with a 10^2 Planck-mass black hole (in four dimensions), we would have

$$T_{\mathrm{H}}\propto1\ \mathrm{TeV}\cdot\left(\frac{1}{1.22\times10^{21}\cdot2\sqrt{\pi}}\right)\frac{1}{4\sqrt{\pi}}. \tag{17}$$

Notably, this has a few reality problems to attend to. Using more standard four-dimensional theory for the Hawkings temperature, [11] we have Hawking radiation from a blackbody (Planck) spectrum with a normal, four-dimensional temperature, T_{H}, given by

$$T_{\mathrm{H}}=\begin{cases}\frac{\hbar c^3}{8\pi G m_{\mathrm{bh}}k_B}=6.17\times10^{-8}\ \mathrm{K} & \text{if } m_{\mathrm{bh}}=1\ \text{solar mass}\\ \frac{\hbar c^3}{8\pi G m_{\mathrm{bh}}k_B}=6.17\times10^{28}\ \mathrm{K} & \text{if } m_{\mathrm{bh}}=10^2 m_{\mathrm{4D-Planck}}\\ \frac{6.17\times10^{28}}{11604}\mathrm{eV} & \text{if } m_{\mathrm{bh}}=10^2 m_{\mathrm{4D-Planck}}\\ 5.317\times10^{12}\ \mathrm{TeV} & \text{if } m_{\mathrm{bh}}=10^2 m_{\mathrm{4D-Planck}}\end{cases} \tag{18}$$

The sun is 9.8×10^{37} Planck masses, so the Hawkings temperature of a 10^2 four-dimensional Planck-mass black hole, as postulated here, would be then about 10^{36} times higher, 1 eV is 11604 K, hence if we have only $n=0$ we really would prefer not to use (17). Hence, using (18), we can then move on to entropy: [14]

$$\begin{aligned}S_{\mathrm{bh}}=\frac{1+n}{2+n}\frac{m_{\mathrm{bh}}}{T_{\mathrm{bh}}}&=\frac{1+n}{2+n}\frac{10^2\cdot m_{\mathrm{4D-Planck}}}{5.317\times10^2\ \mathrm{TeV}}\\ &\approx\frac{1+n}{2+n}\frac{10^2\cdot1.22\times10^{16}\ \mathrm{TeV}}{5.317\times10^{12}\ \mathrm{TeV}}\approx\frac{1+n}{4+2n}\cdot10^4\xrightarrow[n\to0]{}10^3\end{aligned} \tag{19}$$

That is, if we had 10^2 black holes, of mass about 10^2 times a four-dimensional black hole, we probably would be looking at an initial entropy of about 10^5. Then, we would see the entropy density at[15]

$$s=\frac{2\pi^2}{45}g_*\left(\frac{T_{\mathrm{universe}}}{T_{\mathrm{Planck}}}\right)^2 \tag{20}$$

This suggests that relic, initially formed, black holes would be formed, say in a one-meter-radius ball about 10^{-27} s after the onset of inflation and that we would see the rapid decay of these micro black holes in less than 10^{-27} s: $T_{\mathrm{Planck}}=1.416833(85)\times10^{32}\ \mathrm{K}\approx10^8$ TeV, so if the entropy density is $s\equiv10^2S_{\mathrm{bh}}$,

$$10^5=\frac{2\pi^2}{45}g_*\left(\frac{T_{\mathrm{universe}}}{T_{\mathrm{Planck}}}\right)^2\Leftrightarrow g_*\approx10^5\left(\frac{2\pi^2}{45}\right)^{-1}\left(\frac{T_{\mathrm{universe}}}{T_{\mathrm{Planck}}}\right)^{-2} \tag{21}$$

IOP Conf. Series: Journal of Physics: Conf. Series **1275** (2019) 012043 doi:10.1088/1742-6596/1275/1/012043

3. Conclusions

We speculate that this initial degree of freedom increase from 100 to at least 1000 would be part of the new physics which would, when explored, explain the different temperatures given in (18) and (17), a topic we will explore in future publications. Additionally, in (17) and (14)–(16) the $n > 0$ Planck mass is stated to be about 1 TeV versus its four-dimensional value of about 10^{12} TeV. The resolution of these glaring inconsistencies is likely, with the DOF increase implied in (21). Furthermore, if 10^2 Planck-mass is, indeed, the mass of a graviton, we have the necessity of semiclassical treatment of the black holes, [14] whereas if the black holes are of the order of a Planck mass, we have quantum effects. Now for a quote from 't Hooft.

> One may observe however that another source of time reversal asymmetry can be contemplated. As explained in previous sections, this author does not believe that "quantum mechanics" will be the last and permanent framework for the ultimate laws of nature. If we drop it, to be replaced by some classical ideas, the need for time reversal symmetry also subsides. We could opt for an underlying theory where information, in the classical sense, can disappear. Considering cellular automata, systems where information does get lost are much more general than the ones where information is conserved, so that switching the direction of time brings about much more dramatic changes. [16]

Our 10^2 Planck-mass black holes, creating the decay leading to graviton production, which we use as a creation of time's arrow, should be bridged to the following. We wish to go to five or more dimensions and reconcile the contradictions inherent in (17) and (18) as well as to possibly explain the oddness of the 1 TeV higher dimensional Planck mass, if $n > 0$, versus the four-dimensional value of Planck mass, which is many times larger. Hence having said that, we allude to the possibility of a template of double inflation that may be useful, what may be a long-term research resolution that may lead to explaining the difference between (17) and (18). The following is speculative, and if confirmed through additional research, would be a major step toward a cosmological linkage between initial inflation, and reacceleration of the universe one billion years ago. Look at A. Yurov's double-inflation hypothesis. [17] There exist one emergent complex scalar field Φ, and its evolution is linked to both initial inflation and reacceleration. He states that this scalar field would account for both first and second inflation potential in both cases of chaotic inflation. [17] That is, if $H^2 = \frac{\kappa^2}{3}\left(\rho + \frac{\rho^2}{2\lambda}\right) + \frac{m^2}{a^4}$ [18, 19] and a potential $V = m^2\Phi \cdot \Phi$, which was used in part to obtain the reacceleration of the universe, [20] the behavior of massive gravitons one billion years ago to mimic DE in terms of the reacceleration parameter. In any case, the following must be verified to make the linkage. note, a is a scale factor, whereas $\mathrm{d}S_{5\mathrm{D}} = \frac{l^2}{z^2}(\eta_{\mu\nu}\,\mathrm{d}x^\mu\,\mathrm{d}x^\nu + \mathrm{d}z^3)$, for certain Brane-theory ideas, [17] with the result that in line with the data, [21] we must have

$$\frac{3H^2}{4\pi G} \gg V(t)\Big|_{\text{time} \sim 10^{-44}\ \text{s}} . \tag{22}$$

That is, the potential energy, V, of initial inflation is overshadowed by the contributions of the Friedman equation, H, at the onset of inflation. We should note that the potential energy as stated would be assuming that (22) has consistency with $H \sim 1.66\lfloor\sqrt{\tilde{g}_*}\rfloor \frac{T^2}{m^2_{4\mathrm{D-Planck}}}$ for very large temperatures. [22] We need to have a thorough investigation into the emergent scalar field, such that one emergent complex scalar field Φ is identified and the degree-of-freedom issue is made consistent with (17) and (18). That is, perhaps resolving their contradictory result, while having m of (21) and $V = m^2\Phi \cdot \Phi$ of the value of 100 initial black holes, which would be part of our linkage resolving the contradiction given in (17) and (18), while also having $\frac{|m| \leq l^2}{4}$. The evaluative process means that the density function, ρ, would be more than m which we identify as, here, being of the value of 100 initial black holes. This complexity, and some of the dynamics may already have been partly demonstrated in a laboratory, [23] which is something to keep in mind when we look for analogues as to this thought experiment. Keep in mind the following executive summary of what we have, prior to these would-be extensions of this research program. Even if the cosmological constant may be due to quantum processes, classicality in black holes, which may shed gravitons, avoids quantum-induced reversals in the arrow of time. We do not have time reversal in the initial onset of time evolution. We have yet to extend our analysis, with regard to gravitons, to a proof of concept linked to the future of our universe. That is the bridge we must set in place. Also see Appendix Appendix A which outlines the counting algorithm and gives us the entropy counting for gravitons we use here.

Acknowledgments
This work is supported in part by National Natural Science Foundation of China grant No. 11375279.

References

[1] Padmanabhan T 2005 *100 Years of Relativity, Space–Time, Structure: Einstein and Beyond* ed Ashtekar A (Singapore: World Press Scientific) pp 175–204 https://doi.org/10.1142/9789812700988_0007
[2] Padmanabhan T 2005 *An Invitation to Astrophysics* (*World Scientific Series in Astronomy and Astrophysics* vol 8) (Singapore: World Press Scientific) https://doi.org/10.1142/9789812700988_0007
[3] Camara C S, de Garcia Maia M R, de Carvalho J C and de Lima J A 2004 *Phys. Rev. D* **69** Article 063501
[4] Beckwith A 2018 Using "enhanced quantization" to bound the cosmological constant, (for a bound on graviton mass), by comparing two action integrals (one being from general relativity) at the start of inflation http://vixra.org/pdf/1802.0305v1.pdf
[5] Klauder J 2015 *Enhanced Quantization, Particles, Fields and Gravity* (Singapore: World Press Scientific)
[6] Beckwith A 2016 *J. High Energy Phys. Gravitation Cosmol.* **2** 106–124 https://doi.org/10.4236/jhepgc.2016.21012
[7] Giovannini M 2008 *A Primer on the Physics of the Cosmic Microwave Background* (Hackensack, NJ: World Press Scientific) https://doi.org/10.1142/6730
[8] Beckwith A 2017 Creating a (quantum?) constraint, in pre Planckian space–time early universe via the Einstein cosmological constant in a one to one and onto comparison between two action integrals text of Talk for FFP 15, Spain November 30, 11 am-11:30 am, http://vixra.org/abs/1711.0355
[9] Novello M 2005 The mass of the graviton and the cosmological constant puzzle https://arxiv.org/abs/astro-ph/0504505
[10] Hajdukovic D S 2011 Concerning production and decay of mini black holes https://arxiv.org/abs/gr-qc/0702142
[11] Hawking S W 1974 *Nature* **248** 30–31 https://doi.org/10.1038/248030a0
[12] Beckwith A 2018 *J. High Energy Phys. Gravitation Cosmol.* **4** 549–566 https://doi.org/10.4236/jhepgc.2018.43033
[13] Corda C 2018 *Mod. Phys. Lett. A* **33** Article 1850069 https://doi.org/10.1142/S0217732318500694
[14] Calmet X 2013 Quantum black holes Tech. rep. University of Sussex https://www.mpifr-bonn.mpg.de/1311186/Calmet.pdf
[15] Kolb E W and Turner M S 1991 *The Early Universe* (Redwood City, CA: Addison and Wesley) http://power.itp.ac.cn/~guozk/books/The_Early_Universe(Kolb_Turner_1988).pdf
[16] 'T Hooft G 2018 Time, the arrow of time, and quantum mechanics https://arxiv.org/pdf/1804.01383.pdf
[17] Yurov A V 2002 Complex field as inflaton and quintessence https://arxiv.org/abs/hep-th/0208129
[18] Maartens R 2004 *Living Rev. Relativ.* **7** Article 7 https://doi.org/10.12942/lrr-2004-7
[19] Maartens R 2005 *The Physics of the Early Universe* (*Lecture Notes in Physics* vol 653) ed Papantonopoulos E (Amsterdam: Springer Verlag) pp 213–247
[20] Beckwith A 2010 Could gravitons from prior universe survive quantum bounce to reappear in present universe http://vixra.org/abs/1008.0050
[21] Kirshner R 1999 *Proc. Natl. Acad. Sci. USA* **96** 4224–4227 https://doi.org/10.1073/pnas.96.8.4224
[22] Balazs C, Carena M, Menon A, Morrissey D E and Wagner C E M 2005 Overview of electroweak baryogenesis http://www.slac.stanford.edu/econf/C0508141/proc/papers/ALCPG0333.PDF

[23] Ruutu V, Eltsov V, Gill A J, Kibble T W B, Krusius M, Makhlin Y G, Plaçais B, Volvik G E and Xu W D 1996 *Nature* **382** 334–336 `https://doi.org/10.1038/382334a0`
[24] Ng Y J 2008 *Entropy* **10** 441–461 `https://doi.org/10.3390/e10040441`

Appendix A. Infinite quantum statistics as given by Jack Ng

Jack Ng changes conventional statistics: He outlines how to get $S \approx N$, which, with additional arguments, we refine to be $S \approx \langle n \rangle$ (where $\langle n \rangle$ is graviton density). Begin with a partition function [24]

$$Z_N \sim \frac{1}{N!}\left(\frac{V}{\lambda^3}\right)^N . \tag{A.1}$$

This, according to Ng, leads to entropy of the limiting value of, if $S = (\log[Z_N]$

$$S \approx N\left(\log\left[\frac{V}{N\lambda^3}\right] + \frac{5}{2}\right) \xrightarrow[\text{Ng infinite quantum statistics}]{} N\left(\log\left[\frac{V}{\lambda^3}\right] + \frac{5}{2}\right) \approx N. \tag{A.2}$$

IOP Conf. Series: Journal of Physics: Conf. Series **1275** (2019) 012044 doi:10.1088/1742-6596/1275/1/012044

Reconciling reality, space and time: A graviton driven quantum mechanism of cosmic expansion and CMB radiation

G. Doyen[1], **D. Drakova**[2]

[1] Ludwig-Maximilians Universität, München, Germany
[2] Faculty of Chemistry, University of Sofia, Bulgaria

E-mail: gerold@gerold-doyen.de

Abstract. The theory of emerging quantum mechanics (EQM) is a quantum field theory in flat 11 dimensional spacetime, quantizing gravity in the weak interaction limit. In EQM the quantum fields materialize (i.e., they become real) if they entangle with the gravonons, i.e. localized gravitons, thereby forming beables. If not entangled with gravonons, the quantum field is in a limbo state as e.g. exemplified by the state inbetween source and screen in the double slit experiment. Quantum diffusion proceeds via repeated limbo - beable transitions. This leads to the impression that particles having been measured at a certain separation in space suddenly disappear and reappear at a different separation. For any cosmological experiment this is consistent with the interpretation that space has expanded. The rate of cosmic expansion is then equal to the rate of beable - limbo transitions. This rate is calculated from first principles and equals the experimentally determined Hubble parameter. Explicit calculations on the generation of the cosmic microwave radiation (CMB) require to consider the beabling process of the electromagnetic quantum field. The beabling condition is fulfilled for light-atom-lattices (LAL). Temperature emerges in EQM by escape of the particle out of the warp resonance (beable). Without fitting any free model parameter the CMB radiation temperature is 2.2 K, which is to be compared to the experimental value of 2.7 K.

1. Introduction

The generation of structure from quantum fluctuations is postulated in inflationary cosmology where, however, the measurement problem (the so called collapse or reduction of the wave function which creates reality from potentiality) remains unsolved. Further problems refer to the unspecified physical nature of dark energy and dark matter, the coincidence problem, etc. Recently a solution of this problem has been suggested based on the ADD hypothesis [1] (i.e., existence of large extra dimensions) which leads to the formation of localized gravitons, so called gravonons [2, 3]. This theory of emerging quantum mechanics (EQM) is a quantum field theory in flat 11 dimensional spacetime. The gravonons create effects as postulated within the dark energy - dark matter scenario of cosmology [3]. Whether dark energy or dark matter effects appear, depends on the gravonon density in the considered spacial region.

In EQM the quantum fields materialize (i.e., they become real) if they entangle with the gravonons, thereby forming so called beables. If not entangled with gravonons, the quantum field is in a limbo state as e.g. exemplified by the state inbetween source and screen in the

Content from this work may be used under the terms of the Creative Commons Attribution 3.0 licence. Any further distribution of this work must maintain attribution to the author(s) and the title of the work, journal citation and DOI.
Published under licence by IOP Publishing Ltd

IOP Conf. Series: Journal of Physics: Conf. Series **1275** (2019) 012044 doi:10.1088/1742-6596/1275/1/012044

double slit experiment. Quantum diffusion proceeds via repeated limbo - beable transitions. In the limbo state the quantum fields are not real (i.e. not measureable). This leads then to the impression that particles, having been measured at a certain separation in space, suddenly disappear and reappear at a different separation, there being no possibility to trace the particles inbetween. For any cosmological experiment this is consistent with the interpretation that space has expanded. The rate of cosmic expansion is then equal to the rate of beable - limbo transitions. This rate can be calculated from first principles and equals the experimentally determined Hubble parameter.

The formal definitions of the introduced terms is as follows:

(i) A limbo state is a tensor product of a solution of Schrödinger's equation in four dimensional spacetime without gravitation with a wave functional of gravitons.

(ii) A gravonon is a linear combination of graviton modes which is localized over a region of several molecules or a complex of atoms in a piece of matter.

(iii) A beable is a state describing an entanglement between four dimensional brane states (not including gravitation) and gravitational states in 10 dimensional space, i.e. it is a (non-trivial) linear combination of limbo states:

$$\mid beable\rangle = \sum_{limbo} C_{limbo} \mid limbo\rangle. \tag{1}$$

(iv) A limbo - beable transition is a time-dependent development of a wave packet of field configurations from a dominant limbo character to a dominant beable character.

As all measurement devices are made up of beables defined by eq. (1) (meter sticks and clocks) EQM has to distinguish between background parameter spacetime and physical space and time as they appear in experiment. Physical space and time are fixed by the density of beables (gravonons) and by the rate of beable-limbo transitions. Of course, this means that if in different regions of the universe we have different gravonon density and gravonon frequency, physical space and time will differ. With this definition of physical spacetime applied to intergalactic matter the Einstein-de Sitter result can be derived. Within EQM the gravitational constant is taken as a universal invariant. The velocity of light (being constant within any local Lorentz frame) and Planck's constant (relating the energy of quanta to their wave properties) depend on the specification of what is a clock, which changes with the expansion of the universe: time is what is read on the clock, length is what is read on the meter stick. In this connection it is worth mentioning that recently the meter and the second have been defined by the velocity of light as measured in our reference system here on earth and the kilogram is defined via Planck's constant [4]. The velocity of light and Planck's constant are no longer physical quantities the values of which can be improved by improved experiments, the numerical values assigned to these quantities are fixed by a definition of the human race.

The cosmic frequency redshift of light has to be explained within physical spacetime. Energy is conserved in EQM. Explicit calculations require to consider the beabling process of the electromagnetic quantum field. For this purpose light-atom-lattices are constructed in the spirit of experimentally investigated atomic aggregates in optical lattices, where recently light mediated collective atomic oscillations and lattice instabilities have been reported [5, 6, 7]. This allows to analyze Cosmic Microwave Background (CMB) radiation. Anisotropies in the CMB temperature distribution can only arise, if quantum fluctuations materialize (i.e. become real) in form of beables which then constitute "classical" primordial density variations (see [8] for an attempt to invoke decoherence theory for this purpose).

IOP Conf. Series: Journal of Physics: Conf. Series **1275** (2019) 012044 doi:10.1088/1742-6596/1275/1/012044

2. Outline of the physics

There can be no reconciliation of realism, space and time as long as quantum mechanics is interpreted in terms of probability amplitudes. The behaviour of hydrogen atoms in ultrahigh vacuum chambers is extremly well described by EQM [9]. This has also to be true for hydrogen atoms in the ultrahigh vacuum of intergalactic space. To assume that hydrogen atoms behave like small classical billard balls is in no way justified.

The connections and differences between the common Copenhagen interpretation based on Born's rule and probabilities and the realistic beable formation are best illustrated with the example of the double slit experiment [10]. The limbo wave function has amplitude everywhere on the screen, but in the Copenhagen interpretation collapse is postulated to occur only at a single point. In EQM the wave function develops amplitude only at a single point on the screen. It is absolutely central to the theory developed here to have in mind why the matter field localizes at a particular spot on the screen [11]. It is determined by energy conservation and by extremely weak coupling between the matter field and gravitational degrees of freedom near the point of impact. Beables are configurations where a quantum field is localized because it is entangled to gravonons.

The *beabling condition* is that

- the initial energy of the wave packet when it was sent off from the oven matches exactly (with accuracy of typically 10^{-12} eV) the energy of the "warp resonance", where the particle resides when it entangles with the gravonons;
- in the "warp resonance" the matter field entangles with gravitional degrees of freedom of high density of states.

If experimentally we could prepare the emitted wave packet with an infinitely precise and predetermined value of the energy then there might be only a single (or none) site of the screen where beabling is possible. Experimentally we then can identify the localized matter field only at this particular site on the screen or not at all. If there is only one particular site on the screen where entanglement to gravitational degrees of freedom is possible, then we have to try very many precise energies for our wave packet to fulfil the beabling condition.

The experimental findings for hydrogen quantum diffusion on metal surfaces can be rationalized in an analogous way by calculating the position of the warp resonance and the corresponding gravonon mode [9]. The consequence is that hydrogen atoms on metal surfaces can only be observed at very specific sites where the beabling condition is fulfilled. Inbetween these sites the H-atoms are not observable. The rate with which the H-atoms hop between those specific sites depend on the vibrational properties of the gravonons.

From a quantum mechanical point of view it is therefore clear that diffusion of H-atoms in the intergalactic space can only occur if such very specific beabling conditions are fulfilled. As the density is very sparse here the beabling conditions are much harder to fulfil than in the case of the double slit or for H-diffusion on metal surfaces. Quantum diffusion in the intergalactic space will therefore be extremely slow. It is the aim of this contribution to calculate the rate of H-diffusion in the intergalatic space and to put this in relation to the experimentally determined rate of expansion of the universe.

3. Model universe in EQM: universe expansion is quantum diffusion

Various estimates yield for the intergalactic density of hydrogen atoms 5 to 10 H-atoms per m^3. Our model universe consists of eight hydrogen atoms per m^3 interacting via Newton's

IOP Conf. Series: Journal of Physics: Conf. Series **1275** (2019) 012044 doi:10.1088/1742-6596/1275/1/012044

gravitational law and forming a loose soft matter lattice. Softness and enormous lattice constant of matter in the universe imply that phonon Debye energy is enormously small and that therefore the phonon density of states near energy zero is enormously large. This means that configurations describing phonon deexcitation plus hydrogen excitation into plane wave limbo states are practically on-shell for an enormous number of such configurations. All these on-shell configurations mix and entangle with gravonon configurations.

3.1. Estimate of change of potential for vibration of atoms around equilibrium position by Taylor expansion

A hydrogen atom in the loose crystal feels gravitational attraction with all its neighbours and its equilibrium is unstable at a maximum of the energy. A local weak potential minimum arises because when hydrogen shifts away from the equilibrium position a weak repulsive wall appears. Imagine a hydrogen atom in the center of a bipyramid with 6 other hydrogen atoms at the apices surrounding it at distance a. If the central hydrogen atom shifts by a small distance Δ towards one of them the change of the potential equals:

$$\begin{aligned} \delta V &= GM^2\left(-\frac{1}{(a-\Delta)}-\frac{1}{(a+\Delta)}-\frac{4}{a(1+\frac{1}{2}\frac{\Delta^2}{a^2})}+\frac{6}{a}\right) \\ &= \frac{2GM^2}{a^3}\Delta^2 \end{aligned} \tag{2}$$

The expression for δV allows to derive the dependence of ω_{grav} on the size of the crystal a. $\omega_{grav} = \sqrt{\frac{K}{M}}$ where K is the force constant and M the mass. The force constant is the force per unit displacement, hence $K = 2\delta V$ if inserting $\Delta = 1$ in $\delta V = 1/2K\Delta^2$. It follows that

$$\omega_{grav} = \sqrt{\frac{4GM}{a^3}}. \tag{3}$$

For eight hydrogen atoms per m^3 the lattice constant equals $a = 0.81 \times 10^{10}$ bohr. The Hubble parameter in EQM is the gravonon mode frequency eq. (3) which turns out to be of the order of 10^{-10} per year in our theory. The inverse of this is the diffusion jump time which equals the Hubble time.

The calculated Hubble parameter is decisively determined by the averaged density of hydrogen atoms in the intergalactic space.

3.1.1. Expansion dynamics from diffusion model.

Given that real space and time, as measured in experiment, emerge from gravonon formation, a diffusion equation can be set up as

$$\dot{a}(t) = C_{jump}a(t)\omega_{grav}. \tag{4}$$

This diffusion equation is explained by the fact that the lattice constant a of the intergalactic cluster can only change by discrete multiples of the present lattice constant a. This follows from the quantum diffusion model of expansion. The atom always hops to a predetermined possible lattice site. The factor $C_{jump}a(t)$ specifies an average jump length in terms of the lattice constant $a(t)$. Inserting the dependence of ω_{grav} on the lattice constant a from eq. (3) into eq. (4), yields with $C_\omega = \sqrt{4GM}$ and initial condition $a(0) = 0$

$$\dot{a}(t) = a(t)C_{jump}C_\omega a(t)^{-3/2} \tag{5}$$

$$a(t) = (3/2C_{jump}C_\omega t)^{2/3} \tag{6}$$

IOP Conf. Series: Journal of Physics: Conf. Series **1275** (2019) 012044 doi:10.1088/1742-6596/1275/1/012044

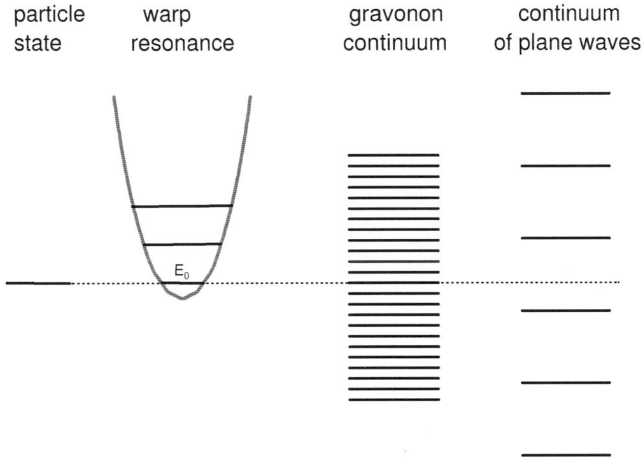

Figure 1. Level scheme illustrating the interaction between the warp resonance and the gravonons.

This is the Einstein-de Sitter result for a flat universe ($\Lambda = 0, k = 0$). Hence EQM predicts a flat universe in terms of an emerging classical spacetime. This is not a consequence of the flat background spacetime inherent in the Lagrangian, but is a consequence of spacetime expansion being interpreted as a diffusion process.

An atom in the intergalactic soft lattice is in limbo for a time ω_{grav}^{-1}. We can only observe it at time steps of ω_{grav}^{-1}, i.e. the clock ticks with an interval of ω_{grav}^{-1}. The possible clocks in the intergalactic space (which have to be constructed from the intergalactic matter) are a factor $\omega_{grav}^{-1}/\omega_{grav-today}^{-1}$ slower than the clocks we have presently here on earth. We would then conclude that for times, measured by the classical clocks constructed from matter here on earth, the velocity of atom diffusion is given by the above diffusion equation.

This way of deriving the Einstein - de Sitter equation is similar in spirit to chemical reaction kinetics. Here the elementary rate constants and pre-exponential factors are calculated from quantum mechanics, but the macroscopic (quasiclassical) reaction velocities as observed in experiment are obtained by feeding the quantum mechanical quantities into kinetic differential equations.

3.2. The beabling condition

This subsection summarizes the EQM-theory in ref. [2] with respect to cosmic diffusion. The interaction potential between atoms is Newton's gravitational potential:

$$V_{grav} = -GM_1M_2/r. \tag{7}$$

The time dependence of the beabling process is calculated by solving Schrödinger's equation.

The gravitational interaction which acts between two atoms separated by distances larger than millimeters is the ordinary Newton gravitation in 4 dimensional spacetime, which turns out to be much too small to induce a dimensionality-like coupling to vibrations with a frequency corresponding to the measured Hubble constant.

IOP Conf. Series: Journal of Physics: Conf. Series **1275** (2019) 012044 doi:10.1088/1742-6596/1275/1/012044

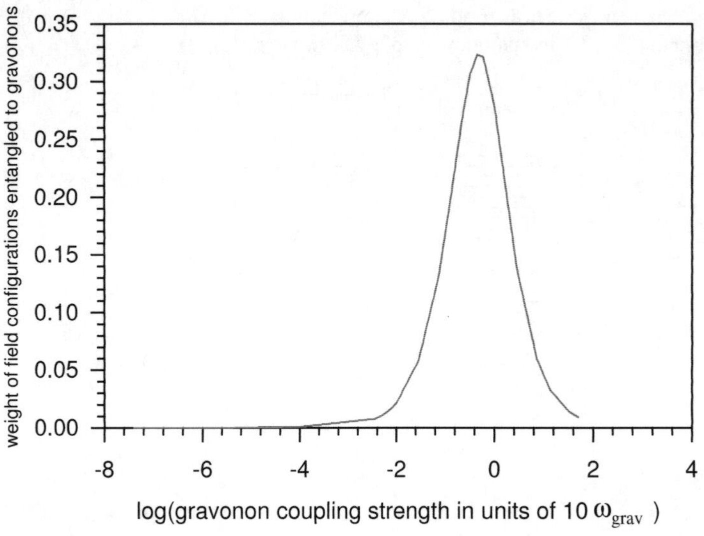

Figure 2. Summed weight of all field configurations entangled to gravonons versus the logarithm of the gravonon coupling strength in units of $10\omega_{grav}$.

Then how does the beabling process work? The condition for beabling is a local and very weak interaction between the warp resonance and the gravonons, see fig. 1. If the interaction is zero we have no beabling process. Turning on a very weak interaction with the gravonons means involving just those environmental configurations which are resonant or near-resonant with the initial configuration. (Perturbation theory shows that their involvement is inversely proportional to the energy difference with the initial configuration.)

If the interaction with the gravonons is localized and strong then environmental excitations of different nature over a broad energy interval will be involved, states describing particles in limbo reflection back into the vacuum included. So the particle wave will be preferentially in a limbo state rather than in beable.

Even if the warp resonances are exactly on the energy shell, the coupling to the gravonons must not become arbitrarily small for beabling to occur. The reason is that the gravonons represent a local spacetime structure exhibiting few modes of extremely small but non-zero frequency. If the matrix elements for gravonon coupling become smaller than the softest gravonon mode, gravonons will no longer entangle with the particle motion. We have studied this by calculating the degree of entanglement as a function of gravonon coupling strength. In fig. 2 the summed weight of all field configurations entangled to gravonons is plotted versus the logarithm of the gravonon coupling strength. Entanglement to gravonons is effective over two orders of magnitude and becomes unimportant for weaker and stronger interaction. A variation of this interaction strength by a factor 10^2 or even less from site to site can decide, whether the particle sticks here or somewhere else. The results of fig. 2 are, of course, only valid for the model structure and the parameters applied here. They depend on the nature of the particle (shape, mass, electronic structure) and on the nature of the gravonon continuum.

This means that in EQM a matter assembly with a density of the intergalactic matter as existing today (a few atoms per cubic meter) in an arrangement where single H-atoms have a distance from each other of half a meter or so will not beable due to the short range of ADD

IOP Conf. Series: Journal of Physics: Conf. Series **1275** (2019) 012044 doi:10.1088/1742-6596/1275/1/012044

high dimensional gravitation.

This shortcoming and problem can be solved, if we observe that the Newton gravonon frequency does only dependent on the matter density in the intergalactic space and is not bound to the condition that single hydrogen atoms, spaced at a definite distance a, form a kind of soft lattice:

$$\omega_{Newton\ mode} = \sqrt{4GM/a^3} = \sqrt{4G\rho_{int.galac.}} \tag{8}$$

It is possible to have e.g. 10^6 H-atoms grouped together at a separation $a_{group} << a$ and then to have these groups vibrate against each other with the same frequency $\omega_{Newton\ mode}$ given by eq. (8) if the distance $a_{group-group}$ between these groups is such that $M_{group}/a^3_{group-group} = \rho_{int.galac.}$.

The beabling condition then requires

$$V_{grav} \approx 10 * \omega_{Newton\ mode}. \tag{9}$$

3.2.1. Rearrangement of intergalactic matter: Atom clouds forming a soft Newton lattice. If we group N_{group} hydrogen atoms together and separate these groups at distances of

$$a_{group-group} = N^{1/3}_{group} * a \tag{10}$$

so that the average intergalactic matter density stays the same, the gravitational interaction energy increases to

$$\begin{aligned} V_{grav} &= -GM^2_{group}/a_{group-group} \\ &= -GM^2 N^{5/3}_{group}/a \end{aligned} \tag{11}$$

where we used eq. (10) and $M_{group} = MN_{group}$.

For example, enlarging the lattice constant by a factor 100 and increasing the number of atoms per group to $N_{group} = 10^6$ leaves the average matter density the same. The gravitational interaction between clouds increases according to eq. (11) by a factor $N^{5/3}_{group} = 10^{10}$, the bridge law might be fulfilled and the intergalactic lattice might beable.

3.2.2. Construct an example of a beabled form of intergalactic matter. The van der Waals force can pull the H-atoms together so that they form H_2-molecules which then combine to a H_2 van der Waals solid. This beabling process is known from a long chemical experience to occur. When van der Waals solids of H_2 form and in a period of millions or billions of years arrange themselves at a lateral separation of 10 meters or so, the beabling condition is fulfilled and this form of matter is manifestly accessible to experimental observation. These matter aggregates expand then by quantum diffusion which is the phenomenon we refer to as spacetime expansion.

4. Light - Atom - Lattices (LAL)

There is, however, the possibility of another beabling process leading to a different form of beabled intergalactic matter. We will call this form a Light - Atom - Lattice (LAL).

Light in an optomechanical lattice generates vibrations and coupling between this optomechanical motion and the Newton mode vibrations in the lattice leads to modification of the latter [3].

IOP Conf. Series: Journal of Physics: Conf. Series **1275** (2019) 012044 doi:10.1088/1742-6596/1275/1/012044

4.1. Mechanism of beabling a light - atom - lattice
In the case of the LAL we need a new (non-adiabatic) kind of coupling in order that the light, which perturbs the atoms, makes the atoms beable. According to the principles of EQM, beabling can only occur if matter fields couple to gravonon modes of low frequency. The only low frequency gravonons that are around in our cosmological situation are the Newton mode vibrations. The optomechanical vibrations have a much too large vibrational energy ($\omega_{LAL} = 10^{-5}$ Hartree) for providing a beabling process.

The mechanism can be imagined as follows. A photon arrives and polarizes a H-atom. For an arbitrary photon wave length and an arbitrary distance between the intergalactic hydogen atoms the effects are vanisingly small and the atoms remain in limbo. But now imagine that the wave length of the photon matches the distance between two intergalactic hydrogen atoms. Then according to the optomechanical experiments and their interpretation by Domokos et al. [5, 6, 7] we have a resonance like effect which makes the hydrogen atoms move due to the force exerted by the light. The frequency of the atomic vibrations is, according to the Domokos theory and the experiments, the same as the photon frequency. This is still a limbo description without any gravitational forces involved.

In order to come to beabling of the light - atom complex we have to include the Newton mode vibration which, according to the above calculations, coincides with the experimental Hubble parameter $\omega_{Newton\ mode} = 5.76 \times 10^{-35}$ a.u. time^{-1} and hence is 29 orders of magnitude smaller than the photon frequency and the frequency of the light induced atomic vibrations ω_{LAL}.

The frequency of the light and the resonant atomic vibrations have to be 29 orders of magnitude smaller to match the frequency of the Newton mode. This would mean a wavelength of order 10^{29} cm $\approx 10^{27}$ m, which is 27 orders of magnitude larger than the average distance between two intergalactic H-atoms today.

Let us think more carefully and consider the two kinds of forces that act between two H-atoms at intergalactic matter densities today. These are the graviton mediated gravitational force and the photon mediated van der Waals force. The van der Waals potential varies with the inverse sixth power of the distance and hence at some large distance the gravitational potential becomes strongest. At these large distances the photon mediated forces, whether resonant or not, can no more determine the motions of the H-atoms, these motions are dictated here by Newton's law.

At small distances, where the van der Waals potential dominates, the gravitational force can no longer establish a Newton mode kind of vibration of the H-atoms, because the van der Waals force will pull away the hydrogen atoms from their symmetric positions, required to form the Newton mode.

So only at the distance where the van der Waals potential and gravitational potential have equal values there is a possibility that light induced interactions influence the atomic motion, without at the same time destroying the Newton mode vibrations, which are necessary for the beabling process. The coefficients for a hydrogen van der Waals crystal have been measured and can be found in the literature [12]. The two potentials are plotted in fig. 3.

4.1.1. Character of the non-adiabatic coupling between LAL and the global Newton crystal. In order that beabling of the light - atom lattice via the four dimensional Newton mode works, the following bridge law has to be satisfied:

$$V_{atom-Newton\ vibration} \approx 10\hbar\omega_{Newton\ mode} \qquad (12)$$

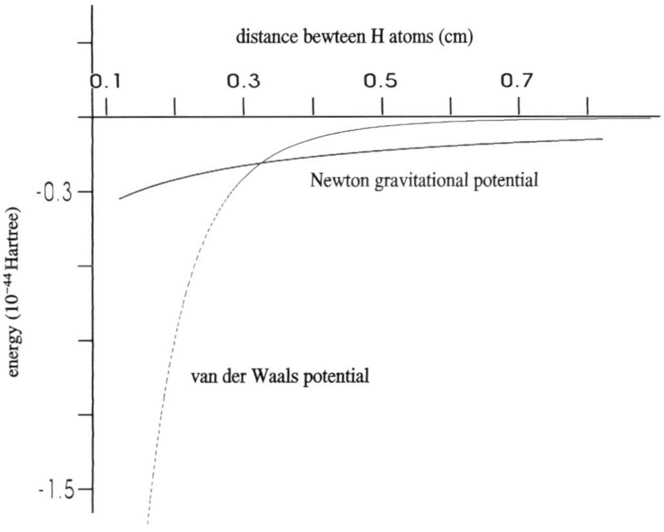

Figure 3. Van der Waals potential (dotted violet curve) and Newton's gravitational potential (full green curve) as function of distance.

It is not at all obvious what the origin of the interaction $V_{atom-Newton\ vibration}$ should be. We suggest to investigate the model in ref. [13] with the following assignments:

(i) The many atoms in the cluster of model [13] are the atoms performing the Newton vibration. ω_S in [13] is equal to $\omega_{Newton\ mode}$ and is of order 10^{-34} Hartree.

(ii) The adatom in [13] is one arbitrarily chosen atom in the light - atom lattice performing a "phonon" vibration of frequency $\omega_{LAL} = 10^{-5}$ Hartree.

(iii) The surface atom "x" in [13] is an atom out of the many atoms in the cluster, performing the Newton vibration. It is adjacent to the adatom and vibrates with $\omega_S = \omega_{Newton\ mode} = 10^{-34}$ Hartree.

We apply eq. (13) on page 110 in [13]. The first term in eq. (13) vanishes, because $q_A = 0$. We assume this, because the vibration of the "adatom" is so much faster (10^{29} times faster) than the Newton vibration so that its position can be replaced by its expectation value. The atoms of the Newton mode effectively experience only the average of q_A, which is zero.

The second term in eq. (13) gives a renormalization of the frequency $\omega_S = \omega_{Newton\ mode}$ of the surface atom "x" (see e.g. [14])

$$\begin{aligned} \omega_S &\rightarrow \sqrt{\omega_S^2 + \eta\gamma\omega_S^2} \\ &\approx \omega_S(1 + \frac{1}{2}\eta\gamma) \end{aligned} \tag{13}$$

which has to be small so that it does not spoil the results with respect to CMB, velocity of light, etc., i.e. $\eta\gamma \approx 1$. This frequency renormalization has to be small, it must not be larger than ω_S. Only then do we get the correct Hubble parameter, which equals the frequency of the Newton mode. This implies $\eta\gamma \approx 1$.

[h]

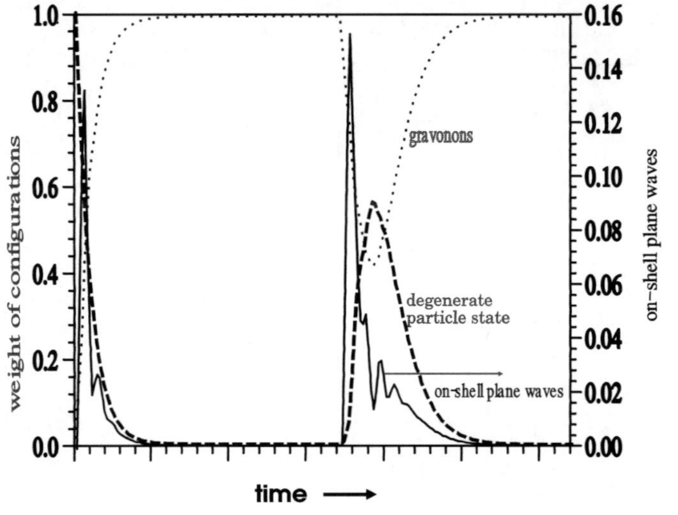

Figure 4. Partial sums of weights (i.e. squared coefficient of the field configurations) of configurations as a function of time as a solution of Schrödinger's equation. The dotted curve comprises all configurations entangled to gravonons (Newton mode). The dashed curve is the sum over all configurations with the LALs in their harmonic oscillator ground state. The full curve represents all configurations where H-atoms are in a plane wave state.

η has to be of order unity (i.e. ≈ 0.5 or so) in order to retain the phonon character of the light - atom lattice arising from the photon mediated van der Waals interactions and the graviton mediated gravitational interactions in the previous section 4.1. γ has to be larger than unity so that the soft mode resulting from the model in ref. [13] can be of order 10^{-34} Hartree.

The third and fourth terms of eq. (13) in [13] are then proportional to $(1-\eta)\gamma$. The third term is $(1-\eta)\gamma\omega_S^2 q_x q_n$. The frequency change due to this term can by means of perturbation theory be estimated as $\Delta\omega = \frac{V^2}{\omega_S}$, where V is the perturbing potential which has to be calculated from the $\gamma - \eta$ model. The frequency change due to the second term of eq. (13) in [13] has already been estimated as $\Delta\omega = \frac{1}{2}\eta\gamma\omega_S$. An analysis analogous to that of the second term of eq. (13), which has the form $\eta\gamma\omega_S^2(distance)^2$, can be performed for the third term, which has the form $(1-\eta)\gamma\omega_S^2(distance)^2$. Comparing these two forms, we realize that the only difference is $\eta \to (1-\eta)$. Therefore if the first form yields the frequency change $\Delta\omega = \frac{1}{2}\eta\gamma\omega_S$ then the second form yields $\Delta\omega = \frac{1}{2}(1-\eta)\gamma\omega_S$. The factor $1/2$ is without significance for an order of magnitude estimate. Therefore: $\Delta\omega \approx (1-\eta)\gamma\omega_S$, $V^2 = (1-\eta)\gamma\omega_S^2$, and $V = \sqrt{(1-\eta)\gamma}\omega_S$. This is the potential with which the Newton mode couples to the optomechanical phononic vibrations and which has to satisfy the bridge law $V = \sqrt{1-\eta)\gamma}\omega_S \approx 10^{-34}$. Because ω_S is of order 10^{-34} we have $(1-\eta)\gamma \approx 1$.

According to the spring model the force constant K of the Newton mode vibrations equals $M\omega_S^2 \approx 10^{-67}$ and the frequency of the 'adatom' mode equals $\omega_A = \omega_{LAL} \approx 10^{-5}$. Hence γ can be calculated as

$$\gamma = \frac{m\omega_A^2}{M\omega_S^2} \tag{14}$$

$$= \frac{m10^{-10}}{M10^{-70}}$$
$$= \frac{m}{M}10^{60}$$

Because γ has to be of order $\gamma \approx 1$ we have $\frac{m}{M} = 10^{-60}$.

This signifies that the interaction mediating the entanglement of the hydrogen atoms to the Newton vibrations has a kind of non-adiabatic character. If the mass of the "adatom" would be zero there would be no entanglement to the Newton vibrations, no dimensionality effect [15] and no beabling. The mass of the "adatom" is a factor 10^{60} smaller than the mass of the atoms of the Newton mode which is just the right magnitude for the dimensionality effect [15] to get a grip.

η und γ can now be calculated as follows:

$$\eta\gamma = (1-\eta)\gamma \rightarrow \eta = 0.5 \tag{15}$$
$$(1-\eta)\gamma = 1 \rightarrow \gamma = 2 \tag{16}$$

Concluding we see that the lattice constant of the LAL (0.33 cm), the number of atoms in the LAL (10^6) and the distance separating the LAL crystallites (10 m) is determined by the beabling condition for the global Newton crystal.

4.2. Light emission from the Light - Atom - Lattice: emerging temperature

If a cosmic ray particle hits the LAL and forces a H-atom out of its harmonic oscillator ground state $\mid \psi_0\rangle$, it will make a transition into a plane wave. Because of the weakness of the potentials involved plane waves are to a very good approximation excited eigenstates of the unperturbed LAL. A transient perturbing potential V_{pert} due to a cosmic ray particle will hence induce transitions to plane waves with rate
$\frac{2\pi}{\hbar} \mid \langle\psi_0 \mid V_{pert} \mid e^{-ikx}\rangle \mid^2 \rho_{pl.waves} \approx \frac{2\pi}{\hbar} \mid \langle\psi_0 \mid e^{-ikx}\rangle \mid^2 \overline{V}^2_{pert}\, \rho_{pl.waves}$.
The transition amplitude will be proportional to the overlap of a gaussian (the harmonic oscillator ground state) and the plane wave $\langle\psi_0(x) \mid e^{ikx}\rangle \propto \mathrm{e}^{-\frac{k^2}{4\alpha^2}} = \mathrm{e}^{-\frac{E_k}{2E_0}} \rightarrow \mathrm{e}^{-\frac{E_k}{2k_BT}}$, where α is the exponent of the gaussian and $E_0 = \frac{\alpha^2}{m} \rightarrow \alpha^2 = mE_0$ ($\hbar = 1$). The energy E_0 can be substituted by the thermal energy k_BT, introducing the temperature T.

When squaring the overlap we see that the coefficients squared of the hydrogen plane waves in the limbo functional have a distribution $e^{-E_k*const}$ and the constant can then be set to define a temperature. This is the emergence of a temperature. It turns out to be just the temperature that is measured for the CMB radiation.

The cosmic particle has transferred the H-atom into a limbo state. When it rebeables from a particular plane wave k into its harmonic oscillator ground state, the energy E_k is regained. Because the hydrogen vibration couples to the photon field it means that the energy difference excites photons, i.e., the wave packet state decribing the re-beabled hydrogen atom contains configurations with the excited photon with coefficients proportional to the Boltzmann factor. These photons can then be beabled in a measurement here on earth.

When then calculating the Planck distribution we have to take into account that the Boltzmann factor enters for the coefficients, with which the configurations having the excited photon with a special k-vector, are contained in the wave packet. When from this the expectation value of the photon number operator is calculated, we obtain the Bose-Einstein distribution. This has

IOP Conf. Series: Journal of Physics: Conf. Series **1275** (2019) 012044 doi:10.1088/1742-6596/1275/1/012044

then to be multiplied by the photon mode density to get the full Planck distribution.

Fig. 4 shows the solution of Schrödinger's equation of a model hamiltonian describing the dynamics of the intergalatic hydrogen matter. The weight of field configurations with plane wave component is given by the full curve. *During a beable - limbo transition the plane wave components of the total wave functional acquire weight according to the Boltzmann factor* $e^{-\frac{E_k}{k_B T_{CMB}}}$. $\omega_{LAL} = 2k_B T_{CMB}$. The wave vector k determines the energy of the respective plane wave $\frac{k^2}{2m} = E_k$ and the exponent $\alpha^2 = \frac{m\omega_{LAL}}{2}$ of the gaussian characterizes the oscillating H-atom. It scales with the kinetic energy of the H-atom $E_0 = k_B T$ in the vibrational ground state.

During a beable - limbo transion the LAL develops into a black body radiator, which emits the CMB radiation. In EQM temperature emerges as quantum mechanical phenomenon.

5. Conclusions

In this contribution a purely quantum mechanical model of the intergalactic hydrogen matter has been suggested where only three experimental quantities enter: the mass density of intergalactic hydrogen atoms, the experimentally determined parameter describing van der Waals hydrogen crystals and Newton's gravitational constant. The determistic EQM theory has been applied to this model which leads to localization (beabling) of the hydrogen atoms, if the beabling condition is fulfilled. If the beabling condition is not fulfilled the H atoms remain non-observable. Without fitting any free model parameters the following quantities have been calculated:

quantity	EQM	experiment
Hubble time	$1.3 * 10^{10}$ years	$1.36 * 10^{10}$ years
		(standard candles)[16]
CMB radiation temp.	2.2 Kelvin	2.725 Kelvin

The important physical insights gained are:

- Quantum diffusion of hydrogen atoms simulates expansion of three dimensional space.
- Numbers emerge by satisfying the beabling conditions.
- Einstein-de Sitter expansion with zero curvature results.
- Temperature emerges from the model, a horizon problem does not exist.

References

[1] Arkani-Hamed N, Dimopoulos S and Dvali G 1998 *Physics Letters* **429** 263
[2] Doyen G and Drakova D 2015 *Found Physics* **45** 959; 2014 *Preprint* arXiv: quant-ph/1408.2716
[3] Drakova D and Doyen G this volume
[4] Rothleitner Ch, Becker P, Bettin H and Knopf D 2019 *Bunsenmagazin* **1** 13
[5] Vochezer A, Kampschulte T, Hammerer K and Treutlein Ph 2018 *Phys. Rev. Lett.* **120** 073602
[6] Asboth J K, Ritch H and Domokos P 2007 *Phys. Rev. Lett.* **98** 203008
[7] Asboth J K, Ritch H and Domokos P 2008 *Phys. Rev.* A **77** 063424
[8] Kiefer C 2007 *Quantum Gravity* (Oxford Univ. Press)
[9] Drakova D and Doyen G 2013 *J. Physics: Conf. Series* **442** 012049
[10] Ananthaswamy A 2018 *Through Two Doors at Once* (Penguin Random House LLC)
[11] Doyen G and Drakova D 2015 *J. Physics: Conf. Series* **442** 012063
[12] Atkins P W 1986 *Physical Chemistry,* third edition (Oxford University Press) p 814, table 1.3
[13] Doyen G 1982 in *Vibrations at Surfaces* eds R Caudano, J-M Gilles and A A Lucas, (Plenum Press) p 105
[14] Cochran W 1973 *The dynamics of atoms in crystals* (Arnold)
[15] Doyen G 1980 *Phys. Rev.* B **22** 497; Doyen G and Drakova D 2011 *J. Physics: Conf. Series* **306** 012033
[16] Riess A G et al. 2016 *Astrophys. J.* **826** 56

The distance formula in algebraic spacetime theories

D Canarutto and E Minguzzi

Dipartimento di Matematica e Informatica "U. Dini",
Università degli Studi di Firenze, Via S. Marta 3, I-50139 Firenze, Italy

E-mail: daniel.canarutto@unifi.it, ettore.minguzzi@unifi.it

Abstract. The Lorentzian distance formula, conjectured several years ago by Parfionov and Zapatrin, has been recently proved by the second author. In this work we focus on the derivation of an equivalent expression in terms of the geometry of 2-spinors by using a partly original approach due to the first author. Our calculations clearly show the independence of the algebraic distance formula of the observer.

1. Introduction

The *distance* between any two points of a Riemannian manifold M can be expressed through *Connes' formula* [1, 2]: for any $x, y \in M$

$$\mathrm{d}(x,y) = \sup_{\|\nabla f\| \leq 1} \big\{ f(y) - f(x) \big\} , \tag{R1}$$

where the supremum is over the continuously differentiable 1-Lipschitz functions. Moreover, as shown by Connes it admits the equivalent expression

$$\mathrm{d}(x,y) = \sup_{f \in \mathcal{A}} \big\{ |f(y) - f(x)| : \|[D, f]\| \leq 1 \big\} , \tag{R2}$$

where the space $\mathcal{A} \equiv \mathrm{C}^1(M, \mathbb{R})$ is regarded as an algebra of operators (with pointwise multiplication) and D is the Dirac operator, acting on sections of a spinor bundle over M. The latter expression is related to the so-called 'algebraic' approach, in which geometric information about an underlying manifold M can be derived from a *spectral triple*

$$\big(\mathcal{A}, \mathcal{H}, D\big) ,$$

where $\mathcal{H}$ is a Hilbert space, $\mathcal{A}$ is an algebra of operators on it and D is a special operator.

Analogous results in Lorentzian geometry were established only recently. Let (M, g) be a time-oriented Lorentzian manifold, namely a *spacetime*. Our choice for the Lorentzian metric signature is $(+, -, -, -)$. The Lorentzian distance $\mathrm{d}_{\mathrm{Lor}}(x, y)$ is defined on a spacetime as the supremum of the lengths of all piecewise C^1, future-oriented causal curves γ joining x and y

$$\mathrm{d}_{\mathrm{Lor}}(x,y) = \sup_{\gamma} \int_{\gamma} \sqrt{g(\dot\gamma, \dot\gamma)}\, dt.$$

One sets $\mathrm{d}_{\mathrm{Lor}}(x, y) = 0$ if x and y are not causally related [3].

Content from this work may be used under the terms of the Creative Commons Attribution 3.0 licence. Any further distribution of this work must maintain attribution to the author(s) and the title of the work, journal citation and DOI.
Published under licence by IOP Publishing Ltd

IOP Conf. Series: Journal of Physics: Conf. Series **1275** (2019) 012045 doi:10.1088/1742-6596/1275/1/012045

The Lorentzian analog of (R1) reads

$$\mathrm{d}_{\mathrm{Lor}}(x,y)=\inf_{f\in\mathcal{S}}\Bigl\{[f(x)-f(y)]^{+}\Bigr\}\,,\qquad [r]^{+}\equiv\max\{0,r\}\,,\ r\in\mathbb{R}\,,\tag{L1}$$

where $\mathcal{S}$ is the space of all *steep* functions, i.e. C^1 functions $M\to\mathbb{R}$ such that $\mathrm{d}f$ fulfills $\langle\mathrm{d}f,\mathrm{Y}\rangle\geq\|\mathrm{Y}\|\equiv\sqrt{g(\mathrm{Y},\mathrm{Y})}$ for every future-oriented causal vector Y. In fact, the family $\mathcal{S}$ can be further reduced to the strictly steep functions, i.e. those for which the mentioned inequality is strict, $\langle\mathrm{d}f,\mathrm{Y}\rangle>\|\mathrm{Y}\|$, in which case f has timelike gradient. We shall not use the stronger version.

This formula was conjectured by Parfionov & Zapatrin [2], used by Franco & Eckstein [4] as an hypothesis for their spectral triple formulation (see below and also Franco [5]), and eventually proved by the second author [6, 7] under the hypothesis of global hyperbolicity (and also by weaker assumptions). Although there were previous works devoted to the proof of (L1), they failed to show that the steep functions in the formula could be taken to be defined all over of M, cf. [8] and C^1, cf. [9].

The spectral triple reformulation of this formula, due to Franco and Eckstein [4] applies to spacetimes of arbitrary dimension but is subjected to the choice of an observer. In this work our aim is to obtain a spectral triple formulation in the 4-dimensional case, by using the formalism of 2-spinors. Our approach has the advantage that the independence of the observer and its equivalence with Eq. (L1) will be manifest.

2. Two-spinor algebra and spacetime

We give a sketchy account of a partly original approach to spinors and spacetime geometry, studied by the first author [10–12], that has some differences with the well-known Penrose-Rindler formalism [13, 14]. We start by noting that a finite-dimensional complex vector space U yields its complex dual and anti-dual spaces, $U^\star$ and $\overline{U}^\star$, respectively constituted of all linear and anti-linear maps $U\to\mathbb{C}$. The space $\overline{U}\equiv(\overline{U}^\star)^\star$ is then called the *conjugate space* of U. Complex conjugation determines anti-isomorphisms $U^\star\leftrightarrow\overline{U}^\star$ and $U\leftrightarrow\overline{U}$. In the standard way of writing coordinate expressions, conjugate spaces typically yield 'dotted indices'.

The space $U\otimes\overline{U}$ is endowed with the anti-linear involution characterized by the rule $u\otimes\bar{v}\mapsto v\otimes\bar{u}$; accordingly we get its splitting

$$U\otimes\overline{U}=\mathrm{H}(U\otimes\overline{U})\oplus\mathrm{iH}(U\otimes\overline{U})$$

into the real subspaces (of the same dimensions) of all *Hermitian* and anti-Hermitian tensors. A Hermitian 2-form on U is an element in $\mathrm{H}(\overline{U}^\star\otimes U^\star)\cong(\mathrm{H}(U\otimes\overline{U}))^*$.

Now we consider the case when $\dim U=2$ and the 1-dimensional space $\wedge^2 U$ (not U) is endowed with a positive Hermitian metric. This yields, up to a phase factor, a unique normalized 'complex symplectic' tensor $\epsilon\in\wedge^2U^\star$. Accordingly one gets 'index moving' isomorphisms $\epsilon^\flat:U\to U^\star:u\mapsto u^\flat\equiv\epsilon(u,\_)$ and $\epsilon^\#:U^\star\to U:\lambda\mapsto\lambda^\#\equiv\epsilon^{-1}(\lambda,\_)$ that are unique up to a phase factor.

The object $\mathrm{g}\equiv\epsilon\otimes\bar{\epsilon}\in\wedge^2U^\star\otimes\wedge^2\overline{U}^\star$ is then well-defined independently of the choice of a normalized ϵ, and the rule $\mathrm{g}(u\otimes\bar{v},u'\otimes\bar{v}')\equiv\epsilon(u,u')\bar{\epsilon}(\bar{v},\bar{v}')$ makes it a bilinear form on $U\otimes\overline{U}$. Its restriction to the 4-dimensional real vector space

$$H\equiv\mathrm{H}(U\otimes\overline{U})$$

turns out to have signature $(+---)$. Thus (H,g) is a Minkowski space. Isotropic elements in H are of the form $\pm u\otimes\bar{u}$ with $u\in U$, so that there is a natural way of fixing a time orientation in H.

Now the 4-dimensional complex space

$$W \equiv U \oplus \overline{U}^\star$$

can be identified with the space of Dirac spinors, that may be represented as $\psi \equiv (u, \bar{\lambda})$. In fact the linear map $\gamma : U \otimes \overline{U} \to \mathrm{End}(W)$ characterized by

$$\gamma[r \otimes \bar{s}](u, \bar{\lambda}) \equiv \sqrt{2}\left(\langle \bar{\lambda}, \bar{s}\rangle\, r\,,\ \epsilon(r, u)\, \bar{\epsilon}^\flat(\bar{s})\right)$$

restricts to a Clifford map $H \to \mathrm{End}(W)$. Its image generates the *Dirac algebra*, that has the distinguished element γ_η , corresponding to the volume form η of H, acting as $\gamma_\eta(u, \bar{\lambda}) = \mathrm{i}\,(u, -\bar{\lambda})$ and determining the projections $\frac{1}{2}\,(1 \mp \mathrm{i}\gamma_\eta)$ onto the *chiral subspaces* $U \oplus \{0\}$ and $\{0\} \oplus \overline{U}^\star$.

The *Dirac adjunction* is the anti-linear involution

$$W \to W^\star \equiv U^\star \oplus \overline{U} : (u, \bar{\lambda}) \mapsto (\lambda, \bar{u}) ,$$

and is associated with the Hermitian product

$$\mathrm{k} : \overline{W} \times W \to \mathbb{C} : \big((\bar{u}, \lambda), (v, \bar{\mu})\big) \mapsto \langle \lambda, v\rangle + \langle \bar{\mu}, \bar{u}\rangle$$

that turns out to have signature $(+ + - -)$.

It should be noticed that we assume no *positive* Hermitian metric on U . Indeed such object is a timelike, future-oriented element $\mathrm{h} \in H^*$, so that its assignment is essentially equivalent to fixing an observer. It can always be expressed as

$$\mathrm{h} = \lambda \otimes \bar{\lambda} + \mu \otimes \bar{\mu}\ , \qquad \lambda, \mu \in U^\star\ .$$

Similarly a timelike, future-oriented vector can be written as

$$Y = \tfrac{1}{\sqrt{2}}\left(u \otimes \bar{u} + \lambda^\# \otimes \bar{\lambda}^\#\right)\ , \qquad u \in U\,,\ \lambda \in U^\star\ .$$

This particular combination fulfills $g(Y, Y) = |\langle \lambda, u\rangle|^2$, and we get

$$\gamma[Y](u, \bar{\lambda}) = \Big(\langle \bar{\lambda}, \bar{u}\rangle\, u\ ,\ \langle \lambda, u\rangle\, \bar{\lambda}\Big)\ .$$

Thus the Dirac spinor $\psi \equiv (u, \bar{\lambda})$ fulfills $\gamma[Y]\psi = \pm m\,\psi$ if and only if $\langle \lambda, u\rangle = \pm m$.

The standard matricial formalism of Dirac algebra can be recovered by choosing a basis $(\mathrm{z}_1\,, \mathrm{z}_2)$ of U such that $\mathrm{z}_1 \wedge \mathrm{z}_2$ is normalized. This yields the g-orthonormal basis

$$(\tau_\lambda) \equiv \left(\tfrac{1}{\sqrt{2}}\sigma_\lambda^{AA'}\, \mathrm{z}_A \otimes \bar{\mathrm{z}}_{A'}\right) \subset H$$

(where (σ_λ) is the λ-th Pauli matrix), the *Weyl* and *Dirac bases* (upper indices label elements of dual bases)

$$(\zeta_\alpha) \equiv \left(\mathrm{z}_1\,, \mathrm{z}_2\,, -\bar{\mathrm{z}}^1\,, -\bar{\mathrm{z}}^2\right)\ , \qquad \alpha = 1, 2, 3, 4\ ,$$

$$(\zeta'_\alpha) \equiv \left(\tfrac{1}{\sqrt{2}}(\zeta_1 - \zeta_3)\,,\ \tfrac{1}{\sqrt{2}}(\zeta_2 - \zeta_4)\,,\ \tfrac{1}{\sqrt{2}}(\zeta_2 - \zeta_4)\,,\ \tfrac{1}{\sqrt{2}}(\zeta_2 + \zeta_4)\right)$$

The usual Weyl and Dirac representations of the Dirac algebra are then recovered as the matrices of the endomorphisms $\gamma_\lambda \equiv \gamma[\tau_\lambda]$.

* * *

Let now $U \to M$ be a 2-spinor bundle over the 4-dimensional basis manifold M ; then, according to the above constructions, we get bundles $H \to M$ and $W \to M$ with induced fiber algebraic structures. A *tetrad-affine approach* to spacetime geometry can be formulated by describing the gravitational field as a couple (ω, θ) , where ω is a linear connection of $U \to M$ (*spinor connection*) and $\theta : \mathrm{T}M \to H$ is a *soldering form* (or *tetrad*), i.e. a linear invertible fibered morphism; this yields a Lorentz metric $\theta^* g$ on M (still denoted as g if no confusion arises). The condition $\nabla\theta = 0$ then determines a metric (possibly non-symmetric) connection of $\mathrm{T}M \to M$, and one eventually gets all the geometric structures needed for the theory of Einstein-Cartan-Maxwell-Dirac fields [10].

IOP Conf. Series: Journal of Physics: Conf. Series **1275** (2019) 012045 doi:10.1088/1742-6596/1275/1/012045

3. Two-spinors and Lorentzian distance

Let us come to the spectral triple reformulation of the Lorentzian distance formula. The spectral triple's elements in this case are [4]

- The algebra $\mathcal{A} \equiv \mathrm{C}^1(M, \mathbb{R})$ with pointwise multiplication.
- The Hilbert space $\mathcal{H} \equiv \mathrm{L}^2(M, W)$ of square integrable sections of the spinor bundle $W \to M$, associated with the choice of a positive Hermitian metric h on W.
- The Dirac operator $\nabla\!\!\!/ \equiv -\mathrm{i}\,\theta^a_\lambda\,\gamma^\lambda \nabla_a \equiv -\mathrm{i}\,\gamma^a \nabla_a$ associated with the spin connection.

Moreover one uses

- The endomorphism $\mathrm{i}\gamma^0$, referred in [4] as the *fundamental symmetry.*
- The chirality operator $\chi \equiv \pm\mathrm{i}\gamma_\eta = \pm\mathrm{i}\gamma_0\gamma_1\gamma_2\gamma_3$.

Taking different notations and conventions into account—in particular opposite metric signatures—we can rewrite the Lorentzian distance in the spectral triple formulation as presented by Franco and Eckstein as

$$\mathrm{d}_{\mathrm{Lor}}(x, y) = \inf_{f \in \mathcal{A}} \Big\{ [f(x) - f(y)]^+ : \mathrm{h}(\bar{\psi}, \gamma^0([\nabla\!\!\!/, f] \pm \gamma_\eta)\psi) \geq 0, \ \forall \psi \in \mathcal{H} \Big\} . \tag{L2}$$

We wish to show, through the elaboration of the right-hand side, that this equation is equivalent to the Lorentzian distance formula. Since the latter has been proved, this equation will also be proved.

We observe that the above use of γ^0 implies the choice of an observer (identified with the element τ_0 of the used Pauli frame), and that the positive Hermitian metric h depends on the choice of an observer, too. It is then natural to assume that these are the same observer; in that case we have $\mathrm{h}(\bar{\psi}, \gamma^0\phi) = \mathrm{k}(\bar{\psi}, \phi)$, where k is the (observer-independent) Dirac adjuction metric. Thus we obtain the more satisfactory version

$$\mathrm{d}_{\mathrm{Lor}}(x, y) = \inf_{f \in \mathcal{A}} \Big\{ [f(x) - f(y)]^+ : \mathrm{k}(\bar{\psi}, ([\nabla\!\!\!/, f] \pm \gamma_\eta)\psi) \geq 0, \ \forall \psi \in \mathcal{H} \Big\} . \tag{L2'}$$

In this expression the observer is still present though, since it is needed in the choice of h, which is used in the definition of $\mathcal{H}$. By the way, we note that the Hilbert spaces associated with different observers may not coincide, since a given section can be square-integrable with respect to an observer and not with respect to the other (as it can be easily shown by a suitable counterexample).

We have

$$[\nabla\!\!\!/, f]\psi = \nabla\!\!\!/(f\psi) - f\nabla\!\!\!/\psi = \gamma^a\nabla_a(f\psi) - f\nabla\!\!\!/\psi = (\gamma^a\partial_a f)\,\psi = \gamma^\#[\mathrm{d}f]\psi \equiv \mathrm{d}\!\!\!/ f\,\psi ,$$

whence for $\psi \equiv (u, \bar{\lambda}) : M \to U \oplus \overline{U}^\star \equiv W$ we get

$$\mathrm{k}(\bar{\psi}, ([\nabla\!\!\!/, f] \pm \gamma_\eta)\psi) = \mathrm{k}(\bar{\psi}, (\mathrm{d}\!\!\!/ f \pm \gamma_\eta)\psi) = \mp 2\,|\langle\lambda, u\rangle|\,\sin\alpha + \sqrt{2}\,\langle \overleftarrow{\theta}^*(\mathrm{d}f), u \otimes \bar{u} + (\lambda \otimes \bar{\lambda})^\#\rangle) ,$$

where $\alpha \equiv \arg\langle\lambda, u\rangle$ that is

$$\sin\alpha = \frac{\langle\lambda, u\rangle - \langle\bar{\lambda}, \bar{u}\rangle}{2\,\mathrm{i}\,|\langle\lambda, u\rangle|} .$$

Now setting

$$Y \equiv \tfrac{1}{\sqrt{2}}\,(u \otimes \bar{u} + \lambda^\flat \otimes \bar{\lambda}^\flat) , \qquad |Y|^2 \equiv g(Y, Y) = |\langle\lambda, u\rangle|^2 ,$$

we get

$$\mathrm{k}(\bar{\psi}, (\mathrm{d}\!\!\!/ f \pm \gamma_\eta)\psi) = 2\,\langle \mathrm{d}f, Y\rangle \mp 2\,|Y|\,\sin\alpha .$$

IOP Conf. Series: Journal of Physics: Conf. Series **1275** (2019) 012045 doi:10.1088/1742-6596/1275/1/012045

At each spacetime point we obtain an arbitrary timelike future-pointing vector $Y \in H$ by the above expression (this is essentially the contravariant 4-momentum associated with an appropriate Dirac spinor $\psi \equiv (u, \bar{\lambda})$). Thus (L2) can be equivalently written as

$$d_{\text{Lor}}(x, y) = \inf_{f \in \mathcal{A}} \left\{ [f(x) - f(y)]^+ : \langle df, Y \rangle \mp 2\,|Y|\, \sin\alpha \geq 0 \;\; \forall Y \in I^+,\; \alpha \in \mathbb{R} \right\},$$

where $I^+ \subset H$ denotes the interior of the future cone. Moreover we observe that, since α is real, for any $Y \in I^+$ both conditions

$$\langle df, Y \rangle \geq |Y| \sin\alpha \quad \forall \alpha \in \mathbb{R},$$

$$\langle df, Y \rangle \geq -|Y| \sin\alpha \quad \forall \alpha \in \mathbb{R},$$

are equivalent to

$$\langle df, Y \rangle \geq |Y| .$$

Hence eventually we get

$$d_{\text{Lor}}(x, y) = \inf_{f \in \mathcal{A}} \left\{ [f(x) - f(y)]^+ : \langle df, Y \rangle \geq |Y|, \;\; \forall Y \in I^+ \right\},$$

which is essentially the expression (L1) of the Lorentzian distance proved by the first author.

Finally we remark that, in the above expression, the observer and the spinor ψ disappeared, as well as any spinor-related objects; in particular, the condition that ψ belong to a Hilbert space has no role at all.

References

[1] Connes A 1995 *J. Math. Phys.* 6194–6231

[2] Parfionov G N and Zapatrin R R 2000 *J. Math. Phys.* **41** 7122–7128

[3] Beem J K, Ehrlich P E and Easley K L 1996 *Global Lorentzian Geometry* (New York: Marcel Dekker Inc.)

[4] Franco N and Eckstein M 2013 *Class. Quantum Grav.* **30** 135007

[5] Franco N 2018 *J. Phys.: Conf. Ser.* **968** 012005

[6] Minguzzi E 2019 *Rev. Math. Phys.* **31** 1930001 arXiv:1709.06494

[7] Minguzzi E 2018 *J. of Phys.: Conf. Ser.* **968** 012009

[8] Moretti V 2003 *Rev. Math. Phys.* **15** 1171–1217

[9] Franco N 2010 *SIGMA Symmetry Integrability Geom. Methods Appl.* **6** Paper 064

[10] Canarutto D 1998 *J. Math. Phys.* **39** 4814–4823

[11] Canarutto D 2000 *Acta Appl. Math.* **62** 187–224

[12] Canarutto D 2007 *Int. J. Geom. Met. Mod. Phys.* **4** 1005–1040

[13] Penrose R and Rindler W 1984 *Spinors and Space-Time. I: Two-spinor calculus and relativistic fields* (Cambridge: Cambridge University Press)

[14] Penrose R and Rindler W 1986 *Spinors and Space-Time. II: Spinor and twistor methods in space-time geometry* (Cambridge: Cambridge University Press)

IOP Conf. Series: Journal of Physics: Conf. Series **1275** (2019) 012046 doi:10.1088/1742-6596/1275/1/012046

Non-Cauchy surface foliations and their expected connection to quantum gravity theories

Merav Hadad

Department of Natural Sciences, The Open University of Israel P.O.B. 808, Raanana 43107, Israel

E-mail: meravha@openu.ac.il

Abstract. We argue that quantum gravity theories should involve constructing a quantum theory on non-Cauchy hypersurfaces and suggest that the hypersurface direction should be the same as the direction of the effective non-gravitational force field at a point. We start we a short review of works which support this idea and then we foliate spacetime along a effective non-gravitational force field direction. Next we discuss the implication of this foliation on the expected properties of quantum gravity. We also discuss the vagueness caused by constructing any quantum theory when using non-Cauchy foliation.

1. Introduction

Obtaining a gravitational theory from microscopic objects or quantum fields is extremely important and challenging. It is important because it is expected to help us understand problems of gravity at very high energy, such as the behavior of black holes and the origin of the universe. It is challenging because this kind of theory is as yet unknown. Attempts to describe gravity using microscopic objects, such as strings, loops or triangles, have not yet led us to the desired Einstein equations. Whereas attempts to treat the gravitational metric as simply another quantum field are problematic. They result in a theory that involves spin-2 massless fields and this kind of theory is not renormalizable in more than $2 + 1$ dimensions. Moreover, combining quantum mechanics with general relativity is also problematic since time plays a different role within these two frameworks: whereas general relativity treats time as a dynamical variable, quantum theories use the Hamiltonian formalism which causes time to act as an independent parameter through which states evolve. Moreover, attempting to use the Hamiltonian formalism in order to quantize the gravitational theory, leads to a non renormalizable theory and to the problem of time in the ADM formalism.

Instead of giving up the powerful Hamiltonian formalism when general relativity theories are concerned, we suggest to use this formalism differently. We suggest to consider the symmetry breaking caused by an effective non-gravitational force field, and to use its direction as an independent parameter through which states evolve. This mean that instead of singling out the direction of a time vector field in the Hamiltonian formalism, we single out the direction of the effective non-gravitational force.

As we will show in the next section this idea is supported by several works which all single out the direction of a non-gravitational force in order to obtain different aspects of quantum gravity. It is also in agreement with holography and Verlinde's suggestion. Moreover, not only it uses the natural symmetry breaking caused by an effective non gravitational force field, it also eliminates the problem

Content from this work may be used under the terms of the Creative Commons Attribution 3.0 licence. Any further distribution of this work must maintain attribution to the author(s) and the title of the work, journal citation and DOI.
Published under licence by IOP Publishing Ltd

IOP Conf. Series: Journal of Physics: Conf. Series **1275** (2019) 012046 doi:10.1088/1742-6596/1275/1/012046

of time in the ADM formalism. Finally, this foliation enables us to obtain the conditions which are needed in order to derive an effective (2+1)D gravitational theory instead of (3+1)D theory, with can be renormelized in the quantum limit. However, since the direction of any force field is space-like, and not time-like this suggestion leads to vagueness regarding the basic concepts of relativistic quantum field theories. We suggest ways to overcome some of these vagueness in the last section.

This paper is organized as follows: we start with a review of works which support the idea that quantum gravity theories should involve constructing a quantum theory on non-Cauchy hypersurfaces and that the hypersurface direction at any point should be the same as the direction of the effective non-gravitational force field at this point. Then we foliate spacetime along this direction, and discus the implication of this foliation on the expected properties of quantum gravity. Finally, we discuss the vagueness caused by constructing any quantum theory when using non-Cauchy foliation, and discuss the implications on quantum gravity.

2. Review on the relevance of non-Cauchy foliation to quantum gravity

The main idea of this paper is that in order to find the proper way to quantize the gravitational theory, we need to consider the symmetry breaking caused by an effective non-gravitational force field, and to use the direction of that field as the direction through which states evolve. This suggestion is supported by several examples which relate non-Cauchy surfaces to different aspects of quantum gravity.

Before listing these examples, we first mention three known properties relating quantum gravity to light-like hypersurfaces. The first is the holographic principle [1] , first proposed by 't Hooft [2], which states that in quantum gravity, the description of a volume of space can be encoded on a lower-dimensional boundary to the region. Next is the AdS/CFT [3] correspondence, which uses a non-perturbative formulation of string theory to obtain a realization of the holographic principle. And finally, the third is Verlinde's suggestion [4] that gravity is an entropic force which arises from the statistical behavior of microscopic degrees of freedom encoded on a holographic screen. We discuss the way these are related to our suggestion and reinforce it, in the next section.

However, in these descriptions the holographic screen is a light-like surface and therefore a Cauchy surface. Conversely, treating any force direction as the direction which states "evolve", involves non-Cauchy hypersurfaces. Fortunately, non-Cauchy hypersurfaces in general, and that resulting from singling out the direction of a non-gravitational force in particular, have also been found useful when dealing with aspects of quantum gravity. We will discuss in detail some such foliation.

2.1. The thermodynamics properties of gravity can also be derived by non-Cauchy foliation

The first example we discus in details involves the surface density of space time degrees of freedom (DoF). These are expected to be observed by an accelerating observer in curved spacetime, i.e. whenever an external non-gravitational force field is introduced [5]. This DoF surface density was first derived by Padmanabhan for a static spacetime using thermodynamic considerations. We found that this density can also be constructed from specific canonical conjugate pairs as long as they are derived in a unique way. Here we briefly summaries the assumptions and conclusions, for the detailed calculation see [6].

To see this one starts by defining the direction of the space-like vector field in a stationary D-dimensional spacetime. We take accelerating detectors that have a D velocity unit vector field u^a and acceleration $a^a = u^b\nabla_b u^a \equiv an^a$ (where n^a is a unit vector and $u^a n_a = 0$). Next one foliate spacetime with respect to the unit vector field n_a by defining a $(D-1)$- hyper-surface which is normal to n_a. The lapse function M and shift vector W_a satisfy $r_a = Mn_a + W_a$ where $r^a\nabla_a r = 1$ and r is constant on Σ_{D-1}. The Σ_{D-1} hyper-surfaces metric h_{ab} is given by $g_{ab} = h_{ab} + n_a n_b$. The extrinsic curvature of the hyper-surfaces is given by $K_{ab} = -\frac{1}{2}\mathcal{L}_n h_{ab}$ where $\mathcal{L}_n$ is the Lie derivative along n^a. The Σ_{D-2} hyper-surfaces metric σ_{ab} is given by $h_{ab} = \sigma_{ab} - u_a u_b$.

As was first noted by Brown [7] for generalized theories of gravity, the canonical conjugate variable of the extrinsic curvature K_{bc} is $4\sqrt{-h}n_a n_d U_0^{abcd}$. Where U_0^{abcd} is an auxiliary variable, which equals $\frac{\partial \mathcal{L}}{\partial R_{abcd}}$ when the equations of motion hold. We found in [6] that the relevant phase space for detectors with D-velocity u^a at point P can be identified by projection of the extrinsic curvature tensor and its canonical conjugate variable on the vector field u^a:

$$\left\{K^{nm}u_m u_n, 4\sqrt{h}U_0^{abcd}n_a u_b u_c n_d\right\} \tag{1}$$

The gravitational density degrees of freedom detected by an accelerating detector with D-velocity u^a at point P is constructed from multiplying these special stationary canonically conjugate variables. Thus, using $K^{ab}u_b u_a = n^a a_a = a$, the gravitational $D-2$ surface density of the spacetime DoF observed by an accelerating observer Δn per unit time Δt is

$$\frac{\Delta n}{\Delta t} = 4a\sqrt{h}U_0^{abcd}n_a u_b u_c n_d. \tag{2}$$

In order to derive Δn one integrate this term during some natural period of time. In this case we can use the assumption that an accelerated observer detects her environment as an equilibrium system at temperature T even in curved spacetime[9]. This kind of canonical ensemble gives thermal Greens functions which are periodic in the Euclidean time with period $\beta = 1/T$ [8]. Assuming this is also the case for fields which represent the spacetime degrees of freedom we integrate eq. (2) over the Euclidian period β and find that the surface density of the spacetime degrees of freedom:

$$\Delta n = 4\beta\sqrt{h}U_0^{abcd}a_a u_b u_c n_d. \tag{3}$$

Next, using the assumption that the temperature $T = 1/\beta$ seen by an accelerated observer equals $Na/2\pi$ even in curved spacetime [9], and $\sqrt{h} = N\sqrt{\sigma}$,we deduce that the degrees of freedom surface density in a one period Euclidean time detected by an accelerating observer is:

$$\Delta n = 8\pi U_0^{abcd}\epsilon_{ab}\boldsymbol{\epsilon}_{\mathbf{cd}}. \tag{4}$$

(Where $\epsilon_{ab} = \frac{1}{2}(n_a u_b - n_b u_a)$ and $\boldsymbol{\epsilon}_{\mathbf{ab}} = \sqrt{\sigma}\epsilon_{ab}$). Up to factor of 4, this is precisely the expression for entropy density of spacetime found by Padmanabhan using the equipartition law for a static metric.

To conclude: we see that entropy density of spacetime can also be constructed from specific canonical conjugate pairs as long as they are derived by foliating spacetime with respect to the direction of the non-gravitational vector force field. That this aspect reinforces the importance of singling out a very unique spatial direction: the direction of a non-gravitational force.

2.2. The effects of string vibrations on the metric can also be derived by non-Cauchy foliation

The second example we discuss in details involves string theory excitations. String theory succeeds in describing a kind of quantum gravity, and with its aid one can identify the entropy of a BPS black hole as string theory excitations [11, 12]. Some of these vibrations create metrics with quantized conical singularities in the $r-y$ surface [13]. Thus we expect that a BPS black hole metric should be a kind of superposition of metrics, some of which have a conical singularity in the $r-y$ surface. This creates an uncertainty at the opening angle at the $r-y$ surface. It turns out that these singularities can also be explained using the uncertainty principle, as long as the variables in the uncertainty principle are obtained in a unique way. Here we briefly summaries the assumptions and conclusions for the detailed calculation see [14].

In order to see this we need to begin with the same foliation as before. Only now the Σ_{D-1} hypersurfaces metric h_{ab} is given by $g_{ab} = h_{ab} + (-1)^{s_u} u_a u_b$, where $s_u = 0$ if u_a is space-like and $s_u = 1$ if u_a is time-like. We also define a $D-2$ hypersurfaces defined by $t = const$ and $r = const$ are thus normal to the given vector n_a and to u_a. The D-2 hypersurfaces metric σ_{ab} is given by $h_{ab} = \sigma_{ab} + (-1)^{s_n} n_a n_b$, where $s_n = 0$ if n_a is space-like and $s_n = 1$ if n_a is time-like. Moreover, as was found out in [14] , for the $D1D5$ black hole, instead of one unit normal vector n_a, one should deal with a several vectors. It is a special case where we can find n $(n < D-1)$ unit vectors: $n_{(i)a}$ $(i = 1...n)$ which are normal to each other and to the same unit vector u_a which is defined by $n^b_{(i)} \nabla_b n_{(i)a} = a_{(i)} u_a$ and $a_{(i)} \neq 0$. In this case the $D-n-1$ metric $\tilde{\sigma}_{ab}$ is defined by $h_{ab} = \tilde{\sigma}_{ab} + \sum_{i=1}^{n} (-1)^{si} n_{(i)a} n_{(i)b}$ and one obtains (see details in [14]):

$$\left[K^{ab} n_{(i)b} n_{(i)a}(x), \sqrt{\sigma_{(i)}} U_0^{abcd} u_a n_{(i)b} n_{(i)c} u_d(\tilde{x})\right] = \hbar \delta^{D-1}(x - \tilde{x}) \tag{5}$$

where $\sqrt{\sigma_{(i)}} = \prod_{j \neq i} N_j \sqrt{-\tilde{\sigma}(-1)^{s_u + \sum_l sl}}$.

For a $D1D5$ black hole we examine the static metric

$$ds^2 = f(r)(-dt^2 + dy^2) + f(r)^{-1}(dr^2 + r^2 d\Omega^2) + g(r) dz_i^2, \tag{6}$$

where y and z_i are compact. We will use the 'naive' geometry [12] where: $f(r) = (1 + \frac{Q_1}{r^2})^{-1/2}(1 + \frac{Q_5}{r^2})^{-1/2}$ and $g(r) = (1 + \frac{Q_1}{r^2})^{1/2}(1 + \frac{Q_5}{r^2})^{-1/2}$.

Choosing the two vectors:

$$\begin{aligned} n^a_{(1)} &= (f(r)^{-1/2}, 0, 0, ..., 0) \\ n^a_{(2)} &= (0, f(r)^{-1/2}, 0, 0, ..., 0) \end{aligned} \tag{7}$$

and define a $D-3$ matric $\tilde{\sigma}_{ab}$ as $h_{ab} = \tilde{\sigma}_{ab} + (-1)^{s1} n_{(1)a} n_{(1)b} + (-1)^{s(2)} n_{(2)a} n_{(2)b}$, then

$$\left[K_{tt}(x), N\sqrt{\tilde{\sigma}} U_0^{abcd} u_a n_{(1)b} n_{(1)c} u_d(\tilde{x})\right] = \hbar \delta^{D-1}(x - \tilde{x}). \tag{8}$$

$$\left[K_{yy}(x), N\sqrt{\tilde{\sigma}} U_0^{abcd} u_a n_{(2)b} n_{(2)c} u_d(\tilde{x})\right] = \hbar \delta^{D-1}(x - \tilde{x}). \tag{9}$$

where $K_{tt} \equiv K^{ab} n_{(1)b} n_{(1)a}$, $K_{yy} \equiv K^{ab} n_{(2)b} n_{(2)a}$ and $N = N_t = N_y$. Integrating over a closed $D-1$ hypersurface we find

$$[\Theta_{r-t_E}, S_{Wr-t}] = \hbar \tag{10}$$

$$[\Theta_{r-y}, S_{Wr-y}] = \hbar \tag{11}$$

where
$\Theta_{r-t_E} \equiv \oint ds_1 K_{tt} = \oint N dt_E K_{tt}$ is the opening angle at the $r - t_E$ surface,
$S_{Wr-t} \equiv \oint N dy d^{D-3} x \sqrt{\tilde{\sigma}} u_a n_{(1)b} n_{(1)c} u_d U_0^{abcd}$
$\Theta_{r-y} \equiv \oint ds_2 K_{yy} = \oint N dy K_{yy}$ is the opening angle at the $r - y$ surface,
$S_{Wr-y} \equiv \oint N dt_E d^{D-3} x \sqrt{\tilde{\sigma}} u_a n_{(2)b} n_{(2)c} u_d U_0^{abcd}$.

IOP Conf. Series: Journal of Physics: Conf. Series **1275** (2019) 012046 doi:10.1088/1742-6596/1275/1/012046

Finally we get the following uncertainty relation on the horizon of a $D1D5$ black hole:

$$\Delta\Theta_{r-t_E}\Delta S_W \geq \hbar \tag{12}$$

$$\Delta\Theta_{r-y}\Delta S_{Wr-y} \geq \hbar \tag{13}$$

The uncertainty at the opening angle in the $r - y$ surface for $D1D5$ black hole is in agreement with the fuzzball proposal, where it was found that some specific string vibrations form metrics with conical singularities at the $r - y$ surface [12, 13, 15, 16].

To conclude: we see that these singularities can be explained not only by string vibrations but also from quantum properties: the uncertainty principle. Note that the variables in the commutation relations must be canonical conjugate pairs which are obtained by singling out the radial direction. The radial direction can be regarded as the direction of a non gravitational force that causes observers to "stand steel" in these coordinate frame. Thus, these singularities, which according to string theory are expected in quantum gravity theories, are derived by the uncertainty principle only when singling out the non-gravitational force direction.

2.3. Several expected quantum black hole properties can also be derived by non-Cauchy foliation

Finally lets mention shortly several example regarding the benefit of the non-Cauchy surface foliation to the research of quantum black hole.

The first example comes long ago from the membrane paradigm [17]. The membrane models a black hole as a thin, classically radiating membrane vanishingly close to the black hole's event horizon. It turns out that this non-Cauchy surface is useful for visualizing and calculating the effects predicted by quantum mechanics for the exterior physics of black holes.

The second example involves the Wheeler-De Witt metric probability wave equation. Recently, in [18], foliation in the radial direction was used to obtain Wheeler-De Witt metric probability wave equation on the apparent horizon hypersurface of the Schwarzschild-de Sitter black hole. By solving this equation, the authors found that a quantized Schwarzschild-de Sitter black hole has a nonzero value for the mass in its ground state. This property of quantum black holes leads to stable black hole remnants, which is also an expected property of quantum gravity.

The last example we mention shortly for the benefit of non-Cauchy foliation to quantum gravity involves "holographic quantization". The holographic quantization uses spatial foliation in order to quantize the gravitational fields for different backgrounds in Einstein theory. This is carried out by singling out one of the spatial directions in a flat background [19], and also singling out the radial direction for a Schwarzschild metric [20]. Moreover, other works [21] even suggest that the holographic quantization causes the (3+1)D Einstein gravity to become effectively reduced to (2+1)D after solving the Lagrangian analogues of the Hamiltonian and momentum constraints.

All these examples support the idea that quantum gravity theories should involve constructing a quantum theory on non-Cauchy hypersurfaces and that the hypersurface direction should be the same as the direction of the effective non-gravitational force field at a point. In the next section we foliate spacetime along this direction, and discuss the implication of this foliation on Einstein equations the expected properties of quantum gravity.

3. Expected advantages of non-Cauchy foliation to quantum gravity.

In order to discuss the implication of the foliation of spacetime along the direction of an effective non-gravitational force field to the expected properties of quantum gravity we need first to describe properly some terms. In this section we foliate spacetime along this direction and then discuss its

IOP Conf. Series: Journal of Physics: Conf. Series **1275** (2019) 012046 doi:10.1088/1742-6596/1275/1/012046

expected benefits regarding the problems of quantum gravity.

3.1. Foliate spacetime along the direction of an effective non-gravitational force field

In order to single out the direction of the gravitational force, we first define its direction. We define it as follow: for a given metric in a D-dimensional spacetime, we take accelerating detectors that have a D velocity unit vector field u^a and acceleration $a^a = u^b\nabla_b u^a \equiv an^a$ (where n^a is a unit vector and $u^a n_a = 0$). Note also that this direction is a space-like vector field. Next, we foliate spacetime with respect to the unit vector field n_a by defining a $(D-1)$ hyper-surface which is normal to n_a and constant r on Σ_{D-1}. The lapse function M and shift vector W_a satisfy $r_a = Mn_a + W_a$ where $r^a\nabla_a r = 1$ and r is constant on Σ_r. The Σ_r hyper-surfaces metric h_{ab} is given by $g_{ab} = h_{ab} + n_a n_b$. The extrinsic curvature of the hyper-surfaces is given by $K_{ab} = -\frac{1}{2}\mathcal{L}_n h_{ab}$ where $\mathcal{L}_n$ is the Lie derivative along n^a. The intrinsic curvature $R^{(3)}_{ab}$ is then given by the 2+1 Christoffel symbols: $\Gamma^k_{ab} = \frac{1}{2}h^{kl}\left(\frac{\partial h_{lb}}{\partial x^a} + \frac{\partial h_{al}}{\partial h^b} - \frac{\partial h_{ab}}{\partial x^l}\right)$ so that $R^{(3)}_{ab} = \frac{\partial \Gamma^k_{ab}}{\partial x^k} - \frac{\partial \Gamma^k_{ak}}{\partial x^b} + \Gamma^k_{ab}\Gamma^l_{kl} - \Gamma^l_{al}\Gamma^k_{lb}$. Thus, instead of 3+1 Einstein equation,

$$R^{(4)}_{ab} = 8\pi\left(T_{ab} - \frac{1}{2}Tg_{ab}\right),$$

one finds [26] that the 3+1 Einstein equation, can be written in terms of the Σ_r hyper-surfaces metric h_{ab}, it's extrinsic curvature K_{ab} and the intrinsic curvature $R^{(3)}_{ab}$:

$$R^{(3)}_{ab} + KK_{ab} - 2K_{ai}K^i_b - M^{-1}\left(\mathcal{L}_r K_{ab} + D_a D_b M\right) = 8\pi\left(S_{ab} - \frac{1}{2}\left(S - P\right)h_{ab}\right),$$

$$R^{(3)} + K^2 - K_{ab}K^{ab} = 16\pi P,$$

$$D_b K^b_a - D_a K = 8\pi F_a.$$

where D_a is the 2+1 covariant derivatives, $S_{ab} = h_{ac}h_{ad}T^{cd}$, $P = n_c n_d T^{cd}$ and $F_a = h_{ac}n_d T^{cb}$.

Note that whenever the extrinsic curvature does vanish, even only on the hypersurface r_0, this foliation leads to an interesting observation. Though a (2+1)D gravitational theory is believed to be a toy model for quantum gravity, this foliation suggests that a (3+1)D gravitational theory can be regard as an "evolution" of a (2+1)D gravitational theory along a non-gravitational force direction. Thus, given the fact that a renormalized (2+1)D quantum gravity theory can be obtained, this leads to a (3+1)D quantum gravity originated from a (2+1)D renormalized theory. Whether or not this renormalized construction leads to an effectively renormalized gravitational theory on the (3+1)D is remain to be seen.

3.2. Implication of this foliation on the expected properties of quantum gravity

Now, lets discuss the implication of this foliation on the expected properties of quantum gravity. We expect this foliation to contribute in three main topics. The first relates the equivalence principle and the insights of this equivalence in the quantum limit. The second is regarding the connection between quantum gravity and holography. Finally the third validates Verlinde's suggestion which relates the origin of a third spatial dimension to the existence and direction of acceleration in a gravitational force. We expend on these issues.

1) The equivalence principle in the quantum limit. We argue that the foliation along the non gravitational force, is related to the equivalence principle and can contribute impotent insights of this equivalence in the quantum limit. To see this, first note that any effective non-gravitational force causes the observers to accelerate in a spatial direction and thus breaks the symmetry between the three spatial directions. As shown by the Unruh effect, this symmetry breaking is extremely important when dealing with quantum theories in flat space. Given the equivalence principle, we expect the same symmetry breaking to occur for accelerating observers in curved spacetime. Though it is extremely complicated (and maybe even impossible) to derive Unruh radiation in curved space, this means that accelerating detectors are expected to detect a kind of Unruh radiation [25] of a gravitational field as well. Moreover, as was shown in [6], singling out the effective non-gravitational field enables identification of the relevant canonical term and calculation of the expected surface density DoF obtained by acceleration observers. Thus singling out the effective non-gravitational force is expected to be a powerful tool in identifying and even eliminating the extra quantum gravitational fields obtained by accelerating observers in curved space.

2) The procedure reinforces the connection between quantum gravity and holography. We argue that this procedure reinforces the connection between quantum gravity and holography. This can be seen in Hamiltonian formalism, when singling out the effective non-gravitational force direction instead of a direction of time. In general the Hamiltonian formalism singles out the time direction and thus the time act as an independent parameter through which states evolve. We suggest to consider the symmetry breaking caused by an effective non-gravitational force field, and to use the direction of that field as the direction through which states evolve. This suggestion supports holography because using the Hamiltonian formalism in this case, leads to equations which determine the evolution of the gravitational fields along the effective non-gravitational force direction, i.e. along the space-like vector field r_a. Moreover, they are determined not only by the Hamiltonian equation, but also by the "initial" condition on the unique non-Cauchy hyper-surface $r = r_0$. Thus when singling out the effective non-gravitational force direction instead of a direction of time, we expect all the information needed to describe the gravitational field in the $r \neq r_0$ is encoded on a non-Cauchy hyper-surface, as suggested by holography.

3) This procedure validates Verlinde's suggestion. This procedure validates Verlinde's suggestion which relates the origin of a third spatial dimension to the existence and direction of acceleration in a gravitational force. To clarify this, note that when we write the 3+1 Einstein equation, in terms of the Σ_r hyper-surfaces metric h_{ab}, its extrinsic curvature $K_{ab} = -\frac{1}{2}\mathcal{L}_n h_{ab}$ and the intrinsic curvature $R^{(3)}_{ab}$,we find that if on $r = r_0$: $K_{ab} = 0$, and also using gauge fixing in order to eliminate the term $M^{-1}\left(\mathcal{L}_r K_{ab} + D_a D_b M\right)$, then all the observers who are restricted to be on $r = r_0$ are not accelerating. This is because $a = n^a a_a = K^{ab} u_b u_a = 0$. Thus in this case gravity can be describe as a $(2+1)D$ theory. On the other hand, if on some $r = r_0$: $K_{ab} \neq 0$, then all the observers on the surface do accelerate and the $(3+1)D$ Einstein equation cannot be reduced to any kind of$(2+1)D$ Einstein equation. Thus this foliation reinforces Verlinde's suggestion relating the direction of the acceleration direction to an extra 3th spatial dimension.

3.3. The implication of this foliation on the obstacles of quantum gravity

Next we discuss the implication of this foliation on the possibility of solving the obstacles of quantum gravity which are the problem of time in the ADM formalism and the renormalization problem.

Eliminate the problem of time in the ADM formalism. Our suggestion is strongly related to a problem with time in the ADM formalism [3]. This problem results from the conceptual difficulty that arises when attempting to combine quantum mechanics with general relativity. It arises from the contrasting role of time within these two frameworks. In quantum theories time acts as an independent parameter through which states evolve, with the Hamiltonian operator acting as the generator of infinitesimal translations of quantum states through time. In contrast, general relativity treats time as a dynamical variable which interacts directly with matter. The problem of time occurs because in general relativity the Hamiltonian is a constraint that must vanish in vacuum. However, in any ordinary canonical theory, the Hamiltonian generates time translations. Therefore, we arrive at the conclusion that "nothing moves" ("there is no time") in general relativity. This so-called 'frozen formalism' caused much confusion when it was first discovered, since it seems to imply that nothing happens in a quantum theory of gravity.

The problem of time vanishes if, instead of the ordinary formulation which splits between one dimension of time and three dimensions of space, one splits between one dimension of space and the 2+1 dimensions of spacetime. This is because in this case, thou the new Hamiltonian vanish, the theory does still depend on the time coordinate even in the vacuum.

Solving the non-renormalization problem Moreover, if we consider a non-Cauchy foliation in the vacuum, then the vanishing of the new Hamiltonian turns into an advantage. This accrue because it leads to the possibility of describing the vacuum case as a renormalized (2+1)D quantum gravity theory. To see this, note that if instead of time direction, we foliate along one of the spatial directions, we arrive at the conclusion that, instead of "nothing changes along the time direction", it seems that "nothing changes along this spatial direction". In other words "there is no third spatial dimension" and the (3+1)D gravitational theory becomes effectively a (2+1)D theory. This is excellent from the point of view of quantum gravity since it is possible to quantize a (2+1)D gravitational theory.

4. Remarks on the expected causality problem due to the non-Cauchy foliation

This paper deals with the possibility of employing the Hamiltonian formalism in a unique way. Instead of singling out the direction of a time vector field, we single out the direction of the effective non-gravitational force. We expect this to be a powerful tool in order to obtain a quantum gravity theory that does not diverge in some cases, and to understand the origin of its divergence in others. However, since this involves singling out a spatial direction instead of a time direction, this also leads to vagueness regarding the basic concepts of relativistic quantum field theories.

Moreover, this is expected to suffer from problems such as causality, probability, conservation and unitarity and thus all these properties should be reexamined in this case. In order to learn how to overcome these difficulties we need to set aside the gravitational theory and focus on investigating known quantum field theories in flat spacetime. We need to investigate the possibility of obtaining relativistic quantum theories by singling out a spatial direction. This is expected to be useful since in this case we know what are the quantum properties that should be obtained and the fact that these theories do not have any of the expected problems.

In order to do that we repeat the known process of quantization, but instead of singling out time and thus Cauchy surface which can be defined as an equal time surface $x_0 = 0$, we use a non-Cauchy surface. A non-Cauchy surface can be defined by constant spatial coordinate surface $x_1 = 0$. In this case we expect that the quantization will be as follows. For a given Lagrangian density of a relativistic field theory $\mathscr{L}(\phi(x), \partial_\mu\phi(x))$, we define a new canonically conjugate momentum to the field variable $\phi(x)$: $\Pi_1(x) = \frac{\partial\mathscr{L}}{\partial\phi'(x)}$ where $\phi' = \partial_1\phi$. The new Hamiltonian density will be $\mathscr{H}_1 = \Pi_1(x)\phi' - \mathscr{L}$. Next we need to verify that the dynamical equation derived by Euler-Lagrange equation, can be written in the new Hamiltonian form as $\phi(x)' = \{\phi(x), H_1\}$ and $\Pi_1(x)' = \{\Pi_1(x), H_1\}$ (where $H_1 = \int d^3\tilde{x}\mathscr{H} \equiv \int dx^0dx^2dx^3\mathscr{H}$). For this purpose we need to assume equal x_1 bracket relations between the field variable $\phi(x)$ and the new conjugate momentum $\Pi_1(x)$. However, if we use equal x_1 canonical Poisson bracket relations: $\{\phi(x),\phi(y)\}_{x^1=y^1} = 0 = \{\Pi(x),\Pi(y)\}_{x^1=y^1}$ and $\{\phi(x),\Pi(y)\}_{x^1=y^1} = \delta^3(\widetilde{x}-\widetilde{y}) = \delta(x^0-y^0)\delta(x^2-y^2)\delta(x^3-y^3)$ we get a non causal relation. Thus equal spatial coordinate bracket cannot be defined by the ordinary canonical Poisson brackets. We must identify correctly the equal x_1 classical bracket relations between the field variable $\phi(x)$ and the new conjugate momentum $\Pi_1(x)$, in order to find out whether the dynamical equation derived by Euler-Lagrange equation can be derived by using the Hamiltonian-like form. However, it seems that in order to identify correctly the equal x_1 classical bracket relations between the field variable $\phi(x)$ and the new conjugate momentum $\Pi_1(x)$ we need to have an extension of the canonical Poisson brackets in such a way that it will be causally defined even when foliating spacetime to non-Cauchy surfaces. Unfortunately, we do not have this kind of extension, and it may be that this kind of extension can not even be derived. For a scalar field, we suggested [27]. a way to identify the equal x_1 canonical relation without using extended Poisson bracket and found that these relations are

$$\{\phi(x),\phi(y)\}_{x^1=y^1} = -i\int \frac{d^3\widetilde{k}}{(2\pi)^3}\epsilon(\pm K^0, P_x^2)\frac{1}{P_x}e^{i\widetilde{k}\cdot(\widetilde{x}-\widetilde{y})} \tag{14}$$

$$\{\Pi_1(x),\Pi_1(y)\}_{x^1=y^1} = -i\int \frac{d^3\widetilde{k}}{(2\pi)^3}\epsilon(\pm K^0, P_x^2)P_x e^{i\widetilde{k}\cdot(\widetilde{x}-\widetilde{y})} \tag{15}$$

$$\{\phi(x),\Pi_1(y)\}_{x^1=y^1} = 0. \tag{16}$$

These relations are causal on the hypersurface $x^1 = y^1$ and it was found that the equation of motion derived by them and the new Hamiltonian leads to the expected equations of motion. This means that deriving a relativistic quantum gravity theory by singling out a spatial direction instead of a time direction is possible, at least for scalars in Minkowski space. However, in order to prove this generally one needs to extend this work to other cases such as different kind of fields, and hopefully in curved spacetime.

5. Summary

In this paper we suggest to use the Hamilton formalism in a unique way. We suggest to consider the symmetry breaking caused by an effective non-gravitational force field, and to use its direction as an independent parameter through which states evolve.

After a short introduction, we examined in the second section several works which support this idea. All these works, which comes from different areas in physics, single out the direction of a non-gravitational force in order to obtain different aspects of quantum gravity. We extended our discussion in two different areas: thermodynamics and string theory. In the third section we found that in some cases the (3+1)D gravitational theory can be regard as an "evolution" of a (2+1)D gravitational theory along a non-gravitational force direction. Then, given the fact that a renormalized (2+1)D quantum gravity theory can be obtained, this leads naturally to a (3+1)D quantum gravity which is originated from a renormalized theory. In the third section we also explained how this idea is in agreement with general expected concepts relating quantum gravity such as holography and Verlinde's suggestion. Moreover, we discussed the option that this unique foliation is useful in order to eliminate the problem of time in the ADM formalism. Finally, in the forth section we discussed the causality problem expected from such non-Cauchy foliation. In this context we mention our latest work which solved the expected causality problem expected from such non-Cauchy foliation for scalars in Minkowski space.

To conclude, it seems that singling out the direction of an effective non-gravitational force field, instead of the direction of time, may leads to a (3+1)D quantum gravitational theory that can be regard as an "evolution" of a (2+1)D renormalized quantum gravitational theory along a non-gravitational

IOP Conf. Series: Journal of Physics: Conf. Series **1275** (2019) 012046 doi:10.1088/1742-6596/1275/1/012046

force direction. Whether or not this renormalized construction leads to an effectively renormalized gravitational theory on the (3+1)D is remain to be seen. However, even if this way of construction does not lead to a renormalized quantum gravitational field theory in the (3+1)D, this construction can be considered as a proof that our inability to renormalized the (3+1)D theory is related directly to the existence of non-gravitational forces in our everyday life.

References

[1] Bousso R 2002 Rev. Mod. Phys. **74** 825
[2] 't Hooft G 1993 Conf. Proc. C930308 284-296
[3] Maldacena J 1998 Adv. Theor. Math. Phys. 2 231-252
[4] Verlinde E P 2011 JHEP 1104 029
[5] Padmanabhan T 2010 Phys. Rev. D **81** 124040
[6] Hadad M 2018 Phys. Lett. B **782** 566–569
[7] Brown J D 1995 Phys. Rev. D **52** 7011
[8] Birrell N D and Davies P C W 1982 (Cambridge: Cambridge University Press) chapter 2 pp 23-28
[9] Padmanabhan T 2010 Phys. Rev. D **81** 124040
[10] Giusto S, Lunin O, Mathur S D and Turton D 2013 JHEP **1302** 050
[11] Strominger A and Vafa C 1996 Phys. Lett. B **379** 99
[12] Mathur S D 2005 Fortsch. Phys. **53** 793
[13] Giusto S, Lunin O, Mathur S D and Turton D 2013 JHEP **1302** 050
[14] Hadad M and Rosenblum L 2015 Phys. Rev. D **92** 12 124025
[15] Lunin O, Maldacena J M and Maoz L 2002 *Preprint* hep-th/0212210
[16] Balasubramanian V, de Boer J, Keski-Vakkuri E and Ross S F 2001 Phys. Rev. D **64** 064011
[17] Price R H. and Thorne K 1988 Scientific American. **258** (4) 69 - 77
[18] Ghaffarnejad H 2018*Preprint* arXiv:1805.01283
[19] Park I Y 2014 *Preprint* arXiv:1404.5066
[20] Park I Y 2016 Math. Phys. **57** 022305
[21] Park I Y 2015 Eur. Phys. J. C **75** 9 459
[22] Carlip S, Korean J 1995 Phys. Soc. **28** S447
[23] Arnowitt R, Deser S, Misner C 1959 Phys. Rev. **116** (5) 1322 - 1330
[24] Isham C J 1992 *Preprint* gr-qc/9210011
[25] Unruh W G 1976 Phys. Rev. D **14** 870
[26] Gourgoulhon E 2007 *Preprint* gr-qc/0703035
[27] Hadad M and Rosenblum L 2019 in preparation

IOP Conf. Series: Journal of Physics: Conf. Series **1275** (2019) 012047 doi:10.1088/1742-6596/1275/1/012047

Generally covariant quantum theory

Johan Noldus
Goorweg 24/73, 3191 Hever, Belgium
and
Noldus Research "Horusiem", Tropicana 5, la Herradura, Andalucia, Spain
E-mail: johan.noldus@gmail.com

Abstract. This paper represents a concise introduction to the quantum theory of point particles in a time orientable curved spacetime part of which was presented in the DICE conference in Castiglioncello, Italy. It is manifest covariant with a unique section of vacuum states as part of a Hilbert bundle.

1. Introduction

There is only one quantum theory on Minkowski and that is the one presented by Weinberg proceeding upon work started by Wigner and Von Neumann. It is axiomatic, starts from a clear definition of a particle and constrains the dynamics as such that the notion of a field becomes useful. Weinberg wrote his summary after all the Evil happened and the beast was baptised "quantum field theory" instead of relativistic particle dynamics. Often, it is useful to attribute the correct name to something as it must reflect its deepest inner workings such as man, woman and hermaphrodite although it is kind of embarrassing to see the last one as a convex combination of the previous extremal cases (it is much more sexy than both of them).

The idea of this paper is to give two distinct proper introductions to RQT (relativistic quantum theory), a Weinbergian one - which we will end with and was not presented on the conference - and a divine one, starting from the most simple of considerations, having nothing to do a priori with probability theory and Hilbert bundles (instead of spaces). Both approaches provide one with a different view on classical and quantum mechanics; they are geometrical and entirely devoid of a coordinatized language as well as symplectic approaches due to globally hyperbolic foliations.

I thank the organizers of the conference for the opportunity to present my viewpoint on the matter -as well as other things- and will return in due time with further elaborations on this work.

2. Foundational arguments

In this section, I see nature as a communist reflects upon society; the foundational quantity of everything is contained in an action signifying "work" or "rabota". That is, consider $\phi(\gamma(s), p(\gamma(s))) \in \mathbf{B}$ where $\gamma : [a, b] \to \mathcal{M}$ is a curve joining an event x to an event y in spacetime $\mathcal{M}$ in affine parametrization with respect to a Lorentzian or Riemannian metric where, moreover,

Content from this work may be used under the terms of the Creative Commons Attribution 3.0 licence. Any further distribution of this work must maintain attribution to the author(s) and the title of the work, journal citation and DOI.
Published under licence by IOP Publishing Ltd

IOP Conf. Series: Journal of Physics: Conf. Series **1275** (2019) 012047 doi:10.1088/1742-6596/1275/1/012047

p is a field on that line associated with the physical quantity of "momentum". We do not really know yet what momentum is but it represents a kind of weight or importance given to that motion, p must not be proportional to $\dot{\gamma}$ as weight might sometimes be opposant to the current motion. Given that rednecks love calculus, **B** is a division algebra over the real numbers with standard operations $+, .$, that is $\mathbb{R}, \mathbb{C}$ or $\mathbb{Q}$ disregarding the non-associative octonions.

A frictionless theory is a dreamworld as no waste is produced. Mathematically, this translates as follows: there exists an involution $\dagger$ and operation $\star$ such that $\phi(\gamma(b-s), p(\gamma(b-s))) \star \phi(\gamma(s), p(\gamma(s)) = 1_\star$ and $\phi(\gamma(b-s), p(\gamma(b-s))) = \phi(\gamma(s), p(\gamma(s)))^\dagger$. It is worthwhile to comment upon these; the first one means that that reversing the process is arithmetically equivalent to taking the inverse whereas the second one says that the inverse has a metrical significance. This last stance is useful as inverting two processes must preserve the distance between them. No discussion about this is allowed for.

As a consequence, the constant curve $\gamma_e(s) = x = y$ satisfies

$$\phi(\gamma_e(s), p(\gamma_e(s)))^2 = 1_\star$$

which for $\star = +$ and $\mathbf{B} = \mathbb{R}$ gives $\phi(\gamma(b-s), p(\gamma(b-s)) = -\phi(\gamma(s), p(\gamma(s))$ and $\phi(\gamma_e(s), p(\gamma_e(s))) = 0$. These simple observations give rise to the notion of work *and* classical physics. For, $\mathbf{B} = \mathbb{C}$, we have that

$$\phi(\gamma(b-s), p(\gamma(b-s)) = \overline{\phi(\gamma(s), p(\gamma(s))}$$

and $|\phi(\gamma(s), p(\gamma(s))|^2 = 1$ what leads to the $U(1)$ Fourier waves in quantum theory.

2.1. Classical theory

The idea is to write down a first order differential equation for the quantity of labor. Reparametrization invariance forces $\frac{d}{ds}\phi(\gamma(s), p(\gamma(s))$ to be proportional to $\frac{d}{ds}\gamma(s)$. Furthermore, the reversion property implies

$$\frac{d}{ds}\phi(\gamma(s), p(\gamma(s)) = \frac{d}{ds}\gamma(s) \cdot \mathbf{F}(\gamma(s), p(\gamma(s)))$$

which is the old Newtonian expression with **F** having the meaning of force. Note that this is an expression local in spacetime as the change of work does not depend upon its previous history. Theories with radiative effects do *not* satisfy this expression but nevertheless a more complicated one. To complete the dynamics, Newton supposed that $p(\gamma(s))$ must maximally stimulate the direction in which the particle is moving and that, therefore

$$p(\gamma(s)) = m\dot{\gamma}(s)$$

where $m > 0$ and expresses the weight attached to persistence of motion, called *physical mass*. Another observation was of an Einsteinian nature, namely that the change of work should reflect the change in an inherent physical property of the particle and not depend upon external forces at all. This would mean that, in a way, a particle is free infinitesimally; the lowest order, in the derivatives of the world line, such invariant is given by the momentum squared

$$h(p(\gamma(s)), p(\gamma(s)))$$

which suggests something like

$$\frac{d}{ds}\left(\frac{m}{2}\left(\frac{d}{ds}\gamma(s)\right)^2\right) = \frac{d}{ds}\phi(\gamma(s), \frac{d}{ds}\gamma(s))$$

and bestows $\phi(\gamma(s), \frac{d}{ds}\gamma(s))$ with the dimension of mass which it should be given that the notion of force must be associated to something intrinsic which is change of momentum

$$\mathbf{F}(\gamma(s), p(\gamma(s))) := \frac{d}{ds}p(\gamma(s)).$$

This is the simplest idea possible, given that the kinetic term is the lowest order invariant and m can be thought of as some material based constant. This leads to

$$\frac{m}{2}\left(\frac{d}{ds}\gamma(b)\right)^2 - \frac{m}{2}\left(\frac{d}{ds}\gamma(a)\right)^2 = \phi(\gamma(b), p(\gamma(b))) - \phi(\gamma(a), p(\gamma(a)))$$

and in a way generalizes a concerved quantity given that ϕ depends upon the entire path and not just the endpoints in general.

One could make higher derivative theories also in this way and allow for Newtonian laws with third order derivatives. These naturally appear in the context of back reactions in electromagnetism for example and allow for "unphysical" solutions with causality going backwards in time. For example, an electron would accelerate prior to turning on a light bulb. Note also that the interpretation of γ as the physical path of the particle naturally emerges given that Newtons law fixes it entirely given two "initial data".

2.2. Quantum theory

Now, we derive quantum theory, as well as the probability interpretation, in the same vein. One notices that the *obvious*, but not only, candidate for an equation of motion is given by

$$\hbar\frac{d}{ds}\phi(\gamma(s), p(\gamma(s))) = ig(p(\gamma(s)), \dot{\gamma}(s))\phi(\gamma(s), p(\gamma(s)))$$

where p is the energy momentum vector and $\dot{\gamma}(s)$ dimensionless. Notice that $\hbar$ is needed for dimensional reasons to get a nontrivial theory given that ϕ must be dimensionless as any physical quantity is a real and not unitary number. On flat space times $\phi(\gamma(s), p(\gamma(s)))$ is topological as it just depends upon the homotopy class or winding number. That is,

$$\phi(\gamma(b), p(\gamma(b))) = e^{ip.(y-x)}$$

which is the standard Fourier wave in y with base point, or origin x. Given that $e^{ip.(y-x)}$ provides for a trivial unitary mapping between $e^{ip.(z-y)}$ and $e^{ip.(z-x)}$, the waves are identical up to a momentum dependent constant multiplicative $U(1)$ factor. In traditional RQT, this is precisely the impact of the translation symmetry in Minkowski QFT. To arrive at the QFT propagator for a quantum field on Minkowski, we notice that the total "propagator"

$$D(x, y) = \alpha \int_{\mathbb{R}^4} d^4p\theta(p^0)\delta(g(p, p) - m^2)\phi(\gamma(b), p)$$

is the expression we are looking after. Indeed, in $\phi(x, w, p)$, as well w as p are uncertain, which is kind of logical given that the momentum is *not* necessarily the maximal forward one but is dragged over the curve as to indicate the initial direction of preference. In order to recuperate the classical bi-functional way of thinking, we have to integrate over p on mass-shell. It utters nothing but the Heisenberg uncertainty principle that if the positions x, y are known sharply, then the momentum is totally uncertain apart from the fact that it needs to be forwards pointing in time and have mass energy mc^2. Actually, what this integration says is that all preferences are

IOP Conf. Series: Journal of Physics: Conf. Series **1275** (2019) 012047 doi:10.1088/1742-6596/1275/1/012047

taken into account democratically unless some higher intelligence, due to a spiritual interaction, desires differently. This is actually a classical Bayesian way of thinking except that the weights or probabilities are here given by complex numbers. In this vein, quantum theory is one of reality w and desire p which do not need to coincide as happens in the classical case; the theory may get more psychological than this by having more complex "momenta" and Einsteinian constraints as happens in (an appropriate version of) string theory. x is here interpreted as the point of birth of a particle and y as the point of dissapearance, death or annihilation. Actually, this is all there is to free QFT on Minkowski; we dispose of no Hamiltonian operator but the Wightman function, from which the Feynman propagator can be uniquely defined, has the standard singularity structure which makes implementation of interactions difficult.

To arrive at the full expression in curved spacetime, we notice that

$$D(x, y) \in \mathbb{R}$$

if x is space like to y written by $x \sim y$ and given that $D(x, y) = \overline{D(y, x)}$ in general we arrive at the conclusion that for space like separated events, the creation and annihilation processes at x and y can be swapped without altering the "propagator". This is the expression of Bose-Einstein statistics, a desirable property in the general theory. In a general curved spacetime, ϕ depends upon curve and not just the homotopy class due to the existence of local gravitational degrees of freedom. In light of Bose statistics, only geodesics count given that the scalar product is preserved. In the light of constructing regularized propagators with a smooth structure, consider the "Schrödinger" equation,

$$\frac{d}{ds}\phi(x, y = \exp_x(w); w, k, s)$$
$$= (ig(w(s), k(s)) - \kappa\sqrt{h(w(s), w(s))} - \frac{2}{L^2\sigma^3(x, \exp_x(ws))}\sigma_{\alpha'}(x, \exp_x(ws))w^{\alpha'}(s))\phi(x, y$$
$$= \exp_x(w); w, k, s)$$

where $s \in [0, 1]$ and $w, k \in T_x\mathcal{M}$, h is a Riemannian metric delivering an energy μ and L is a huge mass smoothening out the light cone. Finally,

$$\frac{D}{ds}k(s) = \frac{D}{ds}w(s) = 0.$$

Notice that violation of unitarity occurs by means of κ, L: they represent irreducible *imaginary* friction terms, meaning that every process has a "cost" which cannot be undone -no perfect "Carnot cycle". It endows spacetime with a kind of effective granularity in the metrics determined by h and σ^2. The solution to the equation reads

$$\phi(x, y = \exp_x(w); w, k) := \phi(x, y = \exp_x(w); w, k, 1) = e^{ik.w}e^{-\kappa\int_0^1\sqrt{h(w(s),w(s))}ds - \frac{1}{L^2\sigma^2(x,\exp_x(w))}}$$

which produces a generalized wave given by

$$\phi(x, y; k) := \sum_{w \in T_x\mathcal{M} : \exp_x(w) = y} \phi(x, y = \exp_x(w); w, k).$$

This calls for the following definition of the propagator

$$D(x, y) = \alpha \sum_{w \in T_x\mathcal{M} : \exp_x(w) = y} \int_{\mathbb{R}^4} d^4p\theta(p^0)$$

$$e^{-\mu h(p(0),p(0))-\mu h(p(1),p(1))-\mu(1-\theta(g(w,w)))(h(R_{w(0)}p(0),R_{w(0)}p(0))-h(R_{-w(1)}p(1),R_{-w(1)}p(1)))}$$
$$\delta(g(p,p)-m^2)\phi(x,y;w,p)$$

where the μ terms express that the creation and annihilation processes come at a cost.

Under reasonable conditions, this regularized propagator is smooth everywhere and has exponential falloff behavior towards infinity. The Feynman propagator gets the following universal prescription:

$$\Delta_{F,\mu,\kappa,L}(x,y) = \sum_{w:\exp_x(w)=y \text{ and w is in the future lightcone of x}} D_{\mu,\kappa,L}(x,w)$$
$$+ \sum_{w':\exp_y(w')=x \text{ and w' is in the future lightcone of y}} D_{\mu,\kappa,L}(y,w')$$
$$+ \sum_{w:\exp_x(w)=y \text{ and w is space like at x}} D_{\mu,\kappa,L}(x,w)$$

meaning that all "information" has to travel towards the future which constitutes clearly the right function to study interactions with.

This theory has been worked out in a book [1] published on Amazon. It turns out that all Feynman diagrams are finite in all known interaction theories for particles of spin less than $\frac{5}{2}$. Moreover, they are suitably bounded and show "exponential falloff behavior" even on Riemannian spaces with negative sectional curvature when friction κ is large enough. All theories need a modification of the standard Dyson expression for diagrams with a large number of internal vertices in order to procure analytical results and make the whole power series well defined. The interested reader is referred to that book for an entire elaboration of this theory.

3. Haute Weinbergian cuisine

Whereas this previous section procured extremely deep connections between different branches of physics from an elementary point of view, this section is somewhat more traditional but no less profound. It is just so that in the end, the same formalism is recovered in all known cases but a different looking avenue is opened up. The latter might be completely isomorphic to the previous one however.

3.1. Classical physics revisited

Consider a particle moving in a bundle $\mathcal{E}$ over a Lorentzian spacetime $(\mathcal{M},g)$ where the fibers are equipped with a metric field and the associated connection preserves the total metric (which is usually a product metric). Regard the word line as an immersion $\gamma:\mathbb{R}\to\mathcal{E}$ and the momentum as its the push forward of ∂_t with equals

$$\frac{D}{dt} := \nabla_{\frac{d}{dt}\gamma(t)}$$

where ∇ is the bundle connection. Given that we shall only work with functions $f:\mathcal{E}\to\mathbb{R}$, the latter expression can be taken for $(\partial_t)_\star$ as an ordinary vectorfield instead of a general derivative operator. To every curve γ and function f we can attach a function $\gamma_f:\mathbb{R}\to\mathbb{R}:t\to f(\gamma(t))$. We can now define a $C^\infty(\mathbb{R})$ algebra of operators $\mathbb{L}$ on the function space $f:\mathcal{E}\to\mathbb{R}$ mapping them to functions from $\mathbb{R}$ to $\mathbb{R}$. Concretely

$$[(\gamma_f)(g)](t) := f(\gamma(t))g(\gamma(t))$$

and

$$[p_\gamma f](t) := \frac{d}{dt} f(\gamma(t)).$$

We have moreover,

$$\gamma_f(g+h) = \gamma_f(g) + \gamma_f(h)$$

and

$$[(\partial_t)(\gamma_f g)](t) := [(\partial_t)_\star f](t) g(\gamma(t)) + f(\gamma(t))[(\partial_t)_\star(g)](t).$$

This suggests to extend the definition of the momentum in this way to functions $\mathbb{R} \to \mathbb{R}$. The same comment holds for γ_f. In this vein,

$$[\gamma_g \gamma_f h](t) = g(\gamma(t)) f(\gamma(t)) h(\gamma(t))$$

and

$$[p_\gamma \gamma_f h](t) := \partial_t(f(\gamma(t))h(\gamma(t)))$$

as well as

$$[\gamma_f p_\gamma h](t) := f(\gamma(t))\partial_t h(\gamma(t)).$$

Finally,

$$[p_\gamma p_\gamma h](t) = (\partial_t)^2 h(\gamma(t))$$

which induces a real algebra generated by

$$\gamma_g, p_\gamma$$

where γ varies over all immersions. This algebra is represented by means of linear operators on the function algebra

$$\mathcal{B} := C^\infty(\mathbb{R}) \otimes C^\infty(\mathcal{E})$$

which may be given the structure of an Hilbert algebra in the usual L^2 sense by introducing an einbein on the "time line" $\mathbb{R}$. Concretely

$$[\gamma_f, \gamma_h]\,(g) = 0 = [p_\gamma, p_\gamma]\,(g), [p_\gamma, \gamma_f]\,(g) = p_\gamma(f)\gamma_\star(g) = \gamma_{p_\gamma(f)}(g)$$

where $\gamma_\star$ is the pull back defined by the immersion γ. Here, the commutation relations employ the full $\mathcal{B}$ action but are understood to apply on $f, g, h \in C^\infty(\mathcal{E})$ and result in an element of $C^\infty(\mathbb{R})$.

Covariant dynamics requires dynamics without potential energy terms; therefore, any force has to be implemented in the momentum what explains the bundle $\mathcal{E}$. Moreover, according to Einstein himself, every force, including the gravitational one, can be gauged away in some point so that locally and physically every particle is a free one meaning that the correct equation is the geodesic bundle equation. Therefore, the classical Hamiltonian is a constraint and moreover, commuting it with a vector leaves a covector if it were an invariant energy so that

$$[\mathcal{H}(\gamma_f, p_\gamma), p_\gamma]$$

cannot represent $\frac{D}{dt}p_\gamma$ unless we would make an extra metric contraction. Actually, the whole Hamiltonian edifice is kind of meaningless as we shall see now. Indeed, taking $\mathcal{H}(\gamma_f, p_\gamma)$ to be p_γ with equations of motion given by

$$[\frac{D}{dt}\triangle\gamma_f](g) := [p_\gamma, \gamma_f](g) = \gamma_{p_\gamma(f)}(g)$$

IOP Conf. Series: Journal of Physics: Conf. Series **1275** (2019) 012047 doi:10.1088/1742-6596/1275/1/012047

and

$$[\frac{D}{dt}(\partial_t)_\star](g) = [\frac{D}{dt}\triangle p_\gamma](dg) := [p_\gamma, p_\gamma](dg) = [p_\gamma, p_\gamma](g) = 0$$

where

$$[\frac{D}{dt}\triangle\zeta](g) = [\frac{D}{dt}, \zeta](g).$$

There is nothing more to say really apart from the constraint $g(p_\gamma, p_\gamma) = \frac{m^2c^4}{\hbar^2}$ which is the mass energy relation. This is all what is allowed in classical physics of point particles really and we now proceed to quantum theory. Notice that the dynamical content is completely implied by the commutator algebra which constitutes a total unison between dynamics and kinematics. Physically, this is entirely trivial and completely justified given that the momentum just corresponds to the energy in a rest frame. Note also the presence of $\hbar$ in the latter formula which is there for dimensional reasons; alas, it does not do anything else apart from setting a time scale given that the covariant derivative does not depend upon it.

3.2. Relativistic quantum theory

As we have shown in the previous section, the Poisson Bracket really is a commutator and the Hamiltonian formulation is rather void given that the total free momentum, constrained by the quantum mechanical mass formula is the only real quantity of interest. Unlike in classical physics, quantum mechanics cannot use an external time in a sense given that a particle is not specified anymore by a world line but by a wave. In a way, it is the complex dual of the classical situation where "world lines" correspond to functions $\psi : \mathcal{E} \to \mathbb{C}$ which are C^∞. The operators γ_f and p_γ are replaced then by x_f and $i\nabla_V$ where V is a real vectorfield over $\mathcal{E}$ and f is a real valued function over $\mathcal{E}$. Here, $[x_f](g)(x) = f(x)g(x)$ and

$$P(V)(g) := i\nabla_V(g) = iV(g).$$

They obey the algebra

$$[x_f, x_h] = 0, [i\nabla_V, i\nabla_W] = -R(V,W)(\cdot) - \nabla_{[V,W]}$$

and finally

$$[i\nabla_V, x_f] = x_{iV(f)}.$$

The momentum commutation relations have been put in this exotic form because the covariant derivative can work on vectorfields and higher objects too. The i is just there to ensure that the momentum operator is real given that the commutator of two real operators is imaginary. The situation here is very different as one cannot just pick a Hamiltonian linear in the momenta given that one would as thus preselect a non dynamical arrow of time. Hence our only choice is given by

$$H = \sum_{i,j=1}^{n} \eta^{ij}\nabla_{E_i}\nabla_{E_j}$$

where the E_i correspond to local vielbeins and η^{ij} is the inverse of the standard flat metric. In order for this to work ∇ must be extended to the spin connection to digest local boost transformations. Furthermore, one has

$$H = m^2$$

as constraint. It is clear one has no Heisenberg type dynamics here as the vectorfields really are spacetime vectorfields; hence, the entire theory is encapsulated by the constraint and the geometry of the bundle $\mathcal{E}$. It has been shown by Ashtekar and Magnon that this theory only

IOP Conf. Series: Journal of Physics: Conf. Series **1275** (2019) 012047 doi:10.1088/1742-6596/1275/1/012047

works out fine in stationary space times with Minkowski as the prime example due to the existence of scalar products on leafs of a foliation for which the latter is preserved in "time".

In 2011, I wrote a book about an operational approach to quantum theory with local vacua delineating a Fock-Hilbert bundle $\otimes_{x\in\mathcal{M}}\mathcal{H}_x$ over the space-time manifold $\mathcal{M}$. However, the approach was troublesome and muddled with two "fundamental errors" of mine, not due to a lack of mathematical precision, but being the consequence of a poor understanding of what curved spacetime really signifies. This error found a natural solution in [1] written on generally covariant quantum theory from the point of view of the Feynman series.

Concretely, we assumed $\mathcal{H}_x$ to be constructed by means of a cyclic quasi-free vacuum state $|0\rangle_x$ and multi particle states showing Bose or Fermi statistics constructed in the Fock way. The dynamical object was a unitary bi-field $U(x,y)$ mapping $\mathcal{H}_y \to \mathcal{H}_x$ and obeying a Schroedinger like differential equation

$$\frac{d}{dt}U(t,s) = iHU(t,s)$$

but then with the times t,s replaced by x,y. The two errors in the book originated from the mathematical implementation of this idea I conceived; first of all $U(t,s) = U(t)U^{\dagger}(s)$ and moreover the only covariant first order differential operator *homogeneous* in the spacetime coordinates is given by the covariant Dirac operator D. The first condition is equivalent to a "cohomology" condition

$$U(x,y)U(y,z)U(z,x) = 1$$

which turns out to hold in Minkowski or any maximally symmetric spacetime only and reflects the absence of local gravitational degrees of freedom. Consequently, the only solution I was able to find of my field equations was free quantum field theory on Minkowski in a way I shall explain later. The Dirac operator gives all sorts of trouble meaning we have to replace the complex numbers by an appropriate Clifford algebra of signature $(1,3)$ or $(3,1)$. This gives rise to negative probabilities and huge problems with the spectral theorem even for finite dimensional Clifford bi-modules. The approach was clearly dead as it stood which I realized later on.

3.3. Taking bi-fields seriously

As pointed out in [1], the idea of a Hilbert bundle *is* adequate, but the correct differential equation for $U(y,x)$ needs to run over geodesics connecting x with y in a fully reparametrization invariant way. The obvious candidate is given by

$$\frac{d}{ds}U(\gamma(s),x) = i\dot{\gamma}(s)^a P_a U(\gamma(s),x)$$

where $\gamma(s)$ is the unique geodesic connecting x with y and P_a equals the free momentum generator, given by the expression

$$P_a = \sum_{\text{particles j, internal degrees}\,\sigma_j} \int \frac{d^3k}{\sqrt{k_0}} k_a a^{\dagger}_{k;j,\sigma_j} a_{k,j,\sigma_j}$$

at the point z with respect to the dragged vierbein in x along the geodesic. The coincidence limit is fixed by $U(x,x) = 1$; this suggests one to enlarge the notation to $U(y,x;e_a(y),e_b(x))$ as well as a unitary action $T(\Lambda, e_b(z))$ of the ortochronous Lorentz group $\Lambda \in O^{+}(1,3)$ on $U(z,x;{e'}_b(z),e_a(x))$ by means of conjugation $TUT^{\dagger}$. All this has been explained in [2]; in order for $T(\Lambda(s), e_a(\gamma(x))$ to shift through $\frac{d}{ds}$ we need a Lorentz covariant derivative and, henceforth, an anti hermitian connection $L_\mu(z, e_b(z))$ such that

$$\left(\frac{d}{ds} + \dot{\gamma}^\mu(s)L_\mu(\gamma(s), \Lambda_a^b(s)e_b(\gamma(s)))\right) T(\Lambda(s), e_b(\gamma(s)))U(\gamma(s), x; e_b(\gamma(s)), e_a(x))T^\dagger(\Lambda(s), e_b(\gamma(s)))$$

$$-T(\Lambda(s), e_b(\gamma(s)))U(\gamma(s), x; e_b(\gamma(s)), e_a(x))T^\dagger(\Lambda(s), e_b(\gamma(s)))\dot{\gamma}^\mu(s)L_\mu(\gamma(s), \Lambda_a^b(s)e_b(\gamma(s))) =$$

$$T(\Lambda(s), e_b(\gamma(s)))\left[\left(\frac{d}{ds} + \dot{\gamma}^\mu(s)L_\mu(\gamma(s), e_b(\gamma(s)))\right) U(\gamma(s), x; e_b(\gamma(s)), e_a(x))\right] T^\dagger(\Lambda, e_b(\gamma(s)))$$

$$-T(\Lambda, e_b(\gamma(s)))U(\gamma(s), x; e_b(\gamma(s)), e_a(x))\dot{\gamma}^\mu(s)L_\mu(\gamma(s), e_b(\gamma(s)))T^\dagger(\Lambda(s), e_b(\gamma(s))).$$

In case we dispose of multiple geodesics connecting x with y, we just multiply the corresponding unitary operators in the *same* vierbein at y, the order of which does not matter given that all P_a commute and because the action of the Lorentz group acts by boosting the momenta. Therefore, we can just sum up the momenta which can accommodate for topology change of Minkowski into a flat space like cylinder giving rise to the correct field picture.

There is however a small caveat here in case multiple geodesics connect x and y in the sense that the gauge field might acquire a nontrivial significance due to multi valuedness of $e_a(y)$, where the latter is the dragged vielbein from x to y. Hence, it is better to replace the argument y by a tangent vector V in $T\mathcal{M}_x$ and take the x perspective where $\exp_x(v) = y$. In that case, we set $L_\mu(sv, \Lambda_a^b(s)e_b(\exp_x(sv)))$ to zero in case

$$\frac{D}{ds}e_b(\exp_x(sv)) = 0$$

for $s = 0 \ldots 1$. In other words, the vielbein in the warped point in that direction must be the dragged one; this makes both formalisms entirely equivalent what the free theory is concerned. Notice that by construction, $U(y, x) = U^\dagger(x, y)$ due to the minus sign caused by flipping $\dot{\gamma}^\mu(s)$. Given that the connection $L_\mu(v, e_b(\exp_x(v)))$ is a new object defined on

$$T\mathcal{M}_x(v) \times V\mathcal{M}_g(\exp_x(v))$$

where $V\mathcal{M}_g(\exp_x(v))$ is the nonlinear space of g vierbeins over $\mathcal{M}$, which is equivalent to the group manifold $O^+(1,3)$ regarded as a homogeneous space with a hyperbolic Cartan metric of signature $(3,3)$; it might be opportune to make it more dynamical and invent a new type of non abelian Yang-Mills theory over $T\mathcal{M}_x$. This author tried this also in 2011 but failed to recognize the bundle perspective as well was stuck with Clifford modules for replacements of Hilbert spaces. The easiest thing is to see L_μ as $L_{\partial_{vj}\exp_x(v)}(v, e_b(\exp_x(v)))$ where e_b varies independently and refers to $y = \exp_x(v)$ and subsequently write out a Yang-Mills equation of the kind

$$\left(D_v L_{\partial_{vj}\exp_x(v)}(v, e_b(\exp_x(v)))\right)_{kl} = \left(\partial_{v[k} - L_{[k}\right) L_{l]} = (dL)_{kl} - (L \wedge L)_{kl} = 0$$

where d is the Hodge operator on flat tangent space. Life could be more exciting as to pick out the zero solution in parallel transport gauge and we leave this new piece of physics for further examination of the bored ones.

So far, we have determined only our quantum connection; now, we develop bi-fields which are nothing but the warps of coincidence fields meaning

$$\Phi(y, x) : \left(\prod_{v \in T\mathcal{M}_x : \exp_x(v) = y} U(y = \exp_x(v), x; (\exp_x(v))_\star e_a(x), e_a(x))\right) \Phi(x, x)$$

IOP Conf. Series: Journal of Physics: Conf. Series **1275** (2019) 012047 doi:10.1088/1742-6596/1275/1/012047

$$\left(\prod_{v \in T\mathcal{M}_x : \exp_x(v) = y} U(y = \exp_x(v), x; (\exp_x(v))_\star e_a(x), e_a(x)) \right)^\dagger .$$

Here,

$$\Phi(x,x) = \sum_{\text{particles j with internal quantum numbers}\,\sigma_j} \int_{\mathbb{R}^3} \frac{d^3k}{k_0} (v_{\sigma_j} \sqrt{k_0} a^\dagger_{k,j,\sigma_j} + \overline{v}_{\sigma_j} \sqrt{k_0} a_{k,j,\sigma_j})$$

where v_{σ_j} is an internal field vector associated to the internal particle degrees of freedom. They are needed to obtain different physical behavior, $\frac{d^3k}{k_0}$ is the on shell relativistic measure in Fourier space on Minkowski and finally, $\sqrt{k_0} a_{k,j,\sigma_j}$ is relativistic normalization of the creation annihilation algebra. I leave it as an elementary exercise to find out principles determining v_{σ_j}. So $\Phi(x,x)$ is the proper democratic relativistic expression taking into all matter degrees of freedom in the universe.

3.4. Interaction theory

So far, we have delineated the free theory from an operational bi-field formalism which reduces in Minkowski to a single field formalism due to the remarkable "cohomology" property

$$U(x,y)U(y,z)U(z,x) = 1$$

where we have dropped the vielbeins and assumed dragging all along which is logical given that dragging is trivial and hence consistent along closed paths due to the vanishing of the Riemann tensor. The trick now is to work directly into an interaction picture and forget about a closed bi-field equation. That is, we write down spacetime interaction densities of the kind

$$i\lambda \int_{\mathcal{M}} \sqrt{g(y)} \Phi(y,x)\Phi(y,z)\Phi(y,p)\Phi(y,q).$$

This is an obvious exercise leading to a completely equivalent formalism as in the book.

References

[1] Noldus J 2018 Generally covariant quantum theory (Amazon)
[2] Noldus J 2011 Foundations of a theory of quantum gravity (Shiny World Corporation); *do. Preprint*, arXiv:1101.5113

Some aspects of the canonical analysis of Reuter-Weyer RG improved Einstein-Hilbert action

Gabriele Gionti, S.J.

Specola Vaticana, V-00120 Vatican City, Vatican City State
and Vatican Observatory Research Group Steward Observatory, The University Of Arizona, 933 North Cherry Avenue, Tucson, Arizona 85721, USA.
INFN, Laboratori Nazionali di Frascati, Via E. Fermi 40, 00044 Frascati, Italy.

E-mail: ggionti@specola.va

Abstract. A canonical analysis of RG improved action of the Einstein-Hilbert functional is performed. The gravitational and cosmological constants as function of the space-time coordinates are treated as external non-geometrical fields. Dirac's constraint analysis is performed, in the general case, up to secondary constraints. The constraints are second class and, in general, the problem appears to be technically complicated. This fact suggests studying the Dirac's constraint analysis of the related Brans-Dicke theory, which shows the Poisson Brackets between Hamiltonian-Hamiltonian constraints do not close.
A simplified FLRW minisuperspace model based on the RG improved Einstein Hilbert action contains Bouncing and Emergent Universes for values of $K = -1, 0, 1$

1. Introduction

Einstein General Relativity appears to be a successful phenomenological theory at laboratory, solar system, galactic and in general at distances bigger than the Planck length $l >> l_{Pl} \equiv \frac{1}{\sqrt{G}} \approx 10^{-33} cm.$. All the classical tests, precession of Mercury, bending of the light rays close to massive bodies [1], and the recent detection of the Gravitational Waves [2], attest Einstein General Relativity is a sound classical theory. But as it is even known at the popular level [3], Einstein General Relativity has an initial singularity. This fact means that General Relativity is no longer predictive around the singularity. People have speculated that this breakdown of the physical laws signals the emergence of a new physics. Although the previous statement is still matter of debate, certainly it is quite well known that matter at atomic and subatomic level behaves according to the laws of Quantum Mechanics. The attempt to formulate a sound Quantum Theory of Gravity (Quantum Gravity) is as yet unrealized. There are many different attempts: String Theory, Loop Quantum Gravity, Non-Commutative Geometry, Dynamical Triangulation and Causal Dynamical Triangulations, Asymptotic Safety etc. None of these approaches have come yet to a final theory of Quantum Gravity which satisfies the expectations of the entire scientific community.

Quantized General Relativity, along the guidelines of Quantum Field theory, is perturbatively non-renormalizable. The Newton constant G has the dimension of the inverse of a square length. Therefore, we have to add a number of counter terms which increase as the loop orders do. The renormalizzation process introduces infinitely many parameters so that the resulting theory

Content from this work may be used under the terms of the Creative Commons Attribution 3.0 licence. Any further distribution of this work must maintain attribution to the author(s) and the title of the work, journal citation and DOI.
Published under licence by IOP Publishing Ltd

IOP Conf. Series: Journal of Physics: Conf. Series **1275** (2019) 012048 doi:10.1088/1742-6596/1275/1/012048

does not have any predictive power [4]. In general a theory is considered "fundamental" if it is perturbatively renormalizable. This means that its infinities can be absorbed by redefining a finite number of parameters. It follows that Quantum General Relativity is not a fundamental theory in this sense. But this is not the end of the story, because there exist fundamental theories which are non-perturbatively renormalizable. This non-perturbative renormalizability, introduced by K. Wilson [5], is related to the existence of a Non-Gaussian Fixed Point (NGFP) in the space of the parameters which guarantee the finiteness of the theory in the ultraviolet limit[6].

Stephen Weinberg [7] proposed the Asymptotic Safety conjecture. He suggested that Einstein General Relativity might be defined non-perturbatively at the non-Gaussian fixed point . He himself proved that NGFP exists in 2+ϵ dimensions [7]. In d=4 there has been no progress because of the lack of a calculation scheme. Recently, [8], progress has been made using the "effective average action". Implementing the "Einstein Hilbert Truncation"[9], it has been shown that there is a NGFP. There is strong evidence that the fixed point exists in the exact theory as well.

2. Renormalization Group approach

A Wilson-type, coarse-grained, free energy functional $\Gamma_k\left[g_{\mu\nu}\right]$ is defined in the following way: $\Gamma_k\left[g_{\mu\nu}\right]$ contains all the quantum fluctuations with momenta $p > k$ and not yet of those with $p < k$ [10]. The modes $p < k$ are suppressed in the path-integral by a mass-square type term $R_k(p^2)$. The behavior of the free-energy functional interpolates between $\Gamma_{k\mapsto\infty} = S$, S being the classical (bare) action, and $\Gamma_{k\mapsto 0} = \Gamma$, Γ being the standard effective action. Γ_k satisfies the RG-equation, called also the Wetterich equation [11],

$$k\partial_k\Gamma_k = \frac{1}{2}Tr\left[(\delta^2\Gamma_k + R_k)^{-1}k\partial_k R_k\right] \tag{1}$$

In general, since this RG-equation is very complicated, one adopts a powerlul non perturbative approximation scheme: truncates the space of the action functional and projects the RG flow onto a finite dimensional space. That is to say, one considers that the free energy functional Γ_k, formally, can be expanded in the following way

$$\Gamma_k[\cdot] = \sum_{i=0}^{N} g_i(k)k^{d_i}I_i[\cdot] \quad , \tag{2}$$

where $I_i[\cdot]$ are given local or non local functionals" of the fields and a-dimensional coefficients $g_i(k)$. In the case of gravity, the following truncation antsatz is usually made:

$$I_0[g] = \int d^4x\sqrt{g} \quad , I_1[g] = \int d^4x\sqrt{g}R \quad , I_2[g] = \int d^4x\sqrt{g}R^2 \quad ,\text{etc.} \tag{3}$$

The simplest truncation is the Einstein-Hilbert truncation which looks like

$$\Gamma_k = -\frac{1}{16\pi G_k}\int d^4x\left(R - 2\bar{\lambda}_k\right) + \text{g.f.} + \text{g.t.} \quad , \tag{4}$$

here g.f. are classical gauge fixing terms, while g.h. are ghost terms. There are two running parameters G_k, the Newton constant, which can be written in a dimensionless way as $g(k) = k^2G_k$. In the same manner, the cosmological constant $\bar{\lambda}_k$ becomes $\lambda(k) = \bar{\lambda}_k/k^2$.

Inserting this ansatz into the flow (Wetterich) equation, one obtains "a projection" onto e finite dimensional space [6]

$$Tr[...] = (...)\int\sqrt{g} + (...)\int\sqrt{g}R + ... \quad , \tag{5}$$

IOP Conf. Series: Journal of Physics: Conf. Series **1275** (2019) 012048 doi:10.1088/1742-6596/1275/1/012048

and then the following finite-dimensional RG equations

$$k\partial_k g(k) = \beta_g(g,\lambda)$$
$$k\partial_k \lambda(k) = \beta_\lambda(g,\lambda) \ . \qquad (6)$$

The solutions of these equations provide the scaling relation for the a-dimensional gravitational constant $g(k)$ and the a-dimensional cosmological constant $\lambda(k)$.

A point $(g_\star, \lambda_\star)$ is a NGFP if it is a non trivial zero of the beta-functions, that is $\beta_g(g_\star,\lambda_\star) = 0$ $\beta_\lambda(g_\star,\lambda_\star) = 0$ and $(g_\star,\lambda_\star) \neq 0$.

3. Reuter-Weyer action proposal and its Hamiltonian

Reuter and Weyer [12] proposed a modified Einstein-Hilbert action with a non geometrical field $G(x)$ and $\Lambda(x)$.

$$S_{mEH}[g, G(x), \Lambda(x)] \equiv \frac{1}{16\pi}\int d^4x\sqrt{-g}\left(\frac{R}{G(x)} - 2\frac{\Lambda(x)}{G(x)}\right) \quad . \qquad (7)$$

They made the hypothesis that the functional form of the gravitational constant, as function of the Space-Time coordinates, and of the cosmological constant are determined completely by the Renormalization Group and are independent of the metric tensor g. In other words, the variation of the metric tensor g in the action functional S_{mEH} does not affect $G(x)$ and $\Lambda(x)$.

Looking carefully to the action (7) and positing $\phi(x) = \frac{1}{G(x)}$, it looks like a Brans-Dicke[13] theory without the kinetic term for the scalar field $\phi(x)$. Reference [12] discusses the fact that the equation of motion of this theory should impose integrability condition on $G(x)$ and $\Lambda(x)$. Since the functional form of $G(x)$ and $\Lambda(x)$ is fixed once we make the cut-off identification $k(x)$, the equation of motion finally will put constraints on the cut-off identification.

We want to study the Hamiltonian Theory (see also [14] for a parallel study) derived by this action.

The first step of this process is to consider a split of the Space-Time (M, g) in which the manifold M becomes ,topologically, $M = R\times\Sigma$: R is a one dimensional space, the time direction, Σ is a three dimensional space-like surface embedded in M,and g becomes the so called ADM [15] metric

$$g = -(N^2 - N_iN^i)dt\otimes dt + N_i(dx^i\otimes dt + dt\otimes dx^i) + h_{ij}dx^i\otimes dx^j \quad , \qquad (8)$$

$N = N(t,x)$ is the so called lapse function and $N^i = N^i(t,x)$ are the shift functions.

The York-boundary term [16] is introduced in the action (7) to make it a differential functional under the variation δg af the metric g, then, implementing the ADM metric, the action functional becomes

$$S_{ADM}[h_{ij}, N, N^i] = \frac{1}{16\pi}\int_{R\times\Sigma} dtd^3x\sqrt{h}N\frac{1}{G(t,x)}\left({}^4R - 2\Lambda(t,x)\right) + \frac{1}{8\pi}\int_{\partial M}\frac{K\sqrt{h}}{G(t,x)}d^3x \ . \qquad (9)$$

Now we introduce some identities as in [17]

$$\frac{1}{G}\left(K\sqrt{h}\right)_{,0} = \frac{G_{,0}}{G^2}K\sqrt{h} + \left(\frac{K\sqrt{h}}{G}\right)_{,0} \qquad (10)$$

$$\frac{1}{G}\frac{\partial f^i}{\partial x^i} = \frac{G_{,i}}{G^2}f^i + \frac{\partial}{\partial x^i}\left(\frac{f^i}{G}\right) \ . \qquad (11)$$

Here K is the trace, performed via the three dimensional metric h_{ij} on the three surface Σ, of the extrinsic curvature tensor

$$K_{ij} = \frac{1}{2}(-\frac{\partial h_{ij}}{\partial t} + \bar{\nabla}_i N_j + \bar{\nabla}_j N_i) \tag{12}$$

$\bar{\nabla}_i N_j$ is the covariant derivative on the space-like surface defined through the three-dimensional metric h_{ij} and f^i is a vector function defined in the following way [18] [17]

$$f^i \equiv \sqrt{h}\left(K N^i - h^{ij} N_{,j}\right) \quad . \tag{13}$$

The ADM Lagrangian density $\mathcal{L}_{ADM}$ is then (for all details see [19])

$$\mathcal{L}_{ADM} \equiv \frac{1}{16\pi}\left[\frac{N\sqrt{h}}{G}(K_{ij}K^{ij} - K^2 + {}^{(3)}R - 2\Lambda) - 2\frac{G_{,0}}{G^2}K\sqrt{h} + 2\frac{G_{,i}f^i}{G^2}\right] \quad . \tag{14}$$

from this Lagrangian density, we can compute the spatial momentum π^{ij} and get

$$\pi^{ij} = \frac{\partial \mathcal{L}_{ADM}}{\partial \dot{h}_{ij}} = -\frac{\sqrt{h}}{16\pi G}\left(K^{ij} - h^{ij}K\right) + \frac{\sqrt{h}\, h^{ij}}{16\pi N G^2}\left(G_{,0} - G_{,k}N^k\right) \quad . \tag{15}$$

In a straight-forward, [19], it is possible to see that the following re-definition of the spatial momenta

$$\tilde{\pi}^{ij} = \pi^{ij} - \frac{\sqrt{h}\, h^{ij}}{16\pi N G^2}\left(G_{,0} - G_{,k}N^k\right) = -\frac{\sqrt{h}}{16\pi G}\left(K^{ij} - h^{ij}K\right) \tag{16}$$

allows the definition of the following transformation of coordinates

$$\left(N, N^i, h_{ij}, \pi, \pi_i, \pi^{ij}\right) \mapsto \left(N, N^i, h_{ij}, \pi, \pi^i, \tilde{\pi}^{ij}\right) \quad , \tag{17}$$

which can be shown to be canonical [19]. In these coordinates, the Hamiltonian density is

$$\begin{aligned}\mathcal{H}_{ADM} &= N\left((16\pi G)G_{abcd}\tilde{\pi}^{ab}\tilde{\pi}^{cd} - \frac{\sqrt{h}({}^3R - 2\Lambda)}{16\pi G}\right) + 2\tilde{\pi}^{ab}\bar{\nabla}_a N_b \\ &+ \frac{\sqrt{h}(G_{,0} - G_{,k}N^k)\bar{\nabla}_a N^a}{8\pi G^2 N} + \frac{G_{,i}\sqrt{h}h^{ij}}{8\pi G^2}N_{,j} \quad ,\end{aligned} \tag{18}$$

G_{abcd} is the DeWitt supermetric

It is straightforward to show that the primary constraint are $\pi \approx 0$ and $\pi_i \approx 0$. The Hamiltonian constraint $\mathcal{H}$ and the momentum constraint $\mathcal{H}_)$ are, respectively,

$$\mathcal{H} = (16\pi G)G_{abcd}\tilde{\pi}^{ab}\tilde{\pi}^{cd} - \frac{\sqrt{h}({}^3R - 2\Lambda)}{16\pi G} - \frac{\sqrt{h}(G_{,0} - G_{,k}N^k)\bar{\nabla}_a N^a}{8\pi G^2 N^2} - \nabla_j\left(\frac{G_{,i}\sqrt{h}h^{ij}}{8\pi G^2}\right) \tag{19}$$

$$\mathcal{H}_i = -2\bar{\nabla}^a\tilde{\pi}_{ai} + \frac{\sqrt{h}(-G_{,i})\bar{\nabla}_a N^a}{8\pi G^2 N} - \sqrt{h}\bar{\nabla}_i\left(\frac{G_{,0} - G_{,k}N^k}{8\pi G^2 N}\right) \quad . \tag{20}$$

The constraints analysis does not close at the secondary level. It easy to notice that the constraints are second class. At this step the invariance under diffeomorphism appears broken since we have introduced $G(x)$ and $\Lambda(x)$ which allows us to distinguish space-time points (we thank M. Reuter for this remark). In fact, for example, if we compute the variation of the three metric h_{ij} and its momentum $\tilde{\pi}_{ij}$ generated by the momentum constraints $\mathcal{H}_i$

$$\{h_{ij}, \int d^3x \tilde{N}^i \mathcal{H}_i\} = \mathcal{L}_{\tilde{\mathbf{N}}} h_{ij} \quad , \tag{21}$$

$$\left\{\tilde{\pi}^{ij}, \int d^3x \tilde{N}^i \mathcal{H}_i\right\} = \mathcal{L}_{\tilde{\mathbf{N}}} \tilde{\pi}^{ij} + \bar{\nabla}_a \left[\frac{\tilde{N}^s}{2}\left(\frac{G_{,s}}{8\pi G^2 N}\right) N^a h^{ij} \sqrt{h}\right] \quad , \tag{22}$$

and it is clear that the momenta constraints are not the generators of the space symmetries. A way to get out from all these technical difficulties is to consider instead of a general ADM metric (8), an ADM metric in Gaussian normal coordinates [20][21]

$$g = -N^2(t) dt \otimes dt + h_{ij} dx^i \otimes dx^j \quad . \tag{23}$$

Basically we are putting the shift functions N^i to zero, that is we are doing a gauge fixing on the spatial diffeomorphisms. The Hamiltonian density $\mathcal{H}_{\mathcal{ADM}}$ then reduces to

$$\mathcal{H}_{ADM} = N\left((16\pi G) G_{abcd} \tilde{\pi}^{ab} \tilde{\pi}^{cd} - \frac{\sqrt{h}(^{(3)}R - 2\Lambda)}{16\pi G}\right) \quad , \tag{24}$$

and the Hamiltonian constraint $\mathcal{H}$ is just the quantity multiplied by the shift function N,

$$\mathcal{H} = \left((16\pi G) G_{abcd} \tilde{\pi}^{ab} \tilde{\pi}^{cd} - \frac{\sqrt{h}(^{(3)}R - 2\Lambda)}{16\pi G}\right) \quad . \tag{25}$$

Momenta constraints are absent and the constraint algebra easily closes.

4. Hamiltonian analysis of Brans-Dicke theory

The difficulties faced in the previous Hamiltonian Analysis of the Reuter-Weyer modified Einstein-Hilbert action suggest exploring the analogous Brans-Dicke theory in the "non singular case" when Brans-Dicke is equivalent to tree level effective gravity from String Theory coupled to a dilaton field [22]. In this case, the field $\phi(x)$ is dynamical with respect to the Reuter-Weyer action functional, and it has a kinetic term and a potential $U(\phi)$. We also included a York-boundary term.

$$S = \frac{1}{4q^2}\left[\int_M d^4x \sqrt{-g}\left(\phi^2\, {}^{(4)}R + 4g^{\mu\nu}\partial_\mu\phi\partial_\nu\phi - U(\phi)\right) + 2\int_{\partial M} d^3x \sqrt{h}\phi^2 K\right] \quad . \tag{26}$$

Its ADM decomposition, following the lines of the previous Reuter-Weyer ADM decomposition, is

$$S_{ADM} = \int_{t\times\Sigma} dt d^3x N\sqrt{h}\Big(\phi^{2(3)}R + \phi^2 K_{ij}K^{ij} - \phi^2 K^2 - \frac{4}{N^2}(\partial_0\phi)^2 \tag{27}$$

$$\frac{8}{N^2} N^i \partial_0 \partial_i \phi + 4\partial_i\phi\partial^i\phi - 4\frac{N^i N^j}{N^2}\partial_i\partial_j\phi + \frac{4}{N}\phi\phi_{,0} K - \frac{4}{\sqrt{h}N}\phi\phi_{,i} f^i - U(\phi)\Big) \quad .$$

We can now compute the momenta associated to N, N^i, h^{ij} and ϕ, and so we get

$$\pi_N \equiv \frac{\partial\mathcal{L}_{ADM}}{\partial\dot{N}} \approx 0 \tag{28}$$

$$\pi_{N^i} = \frac{\partial \mathcal{L}_{ADM}}{\partial \dot{N}^i} \approx 0 \tag{29}$$

$$\pi^{ij} = \frac{\partial \mathcal{L}_{ADM}}{\partial \dot{h}_{ij}} = -\frac{\sqrt{h}}{4q^2}\phi^2 K^{ij} + \frac{1}{4}\phi\pi_\phi h^{ij} \tag{30}$$

$$\pi_\phi = \frac{\partial \mathcal{L}_{ADM}}{\partial \dot{\phi}} = \frac{\sqrt{h}}{2q^2 N}\left(\partial_0\phi - N^i\partial_i\phi - \frac{N}{2}\phi K\right) \tag{31}$$

the first four momenta are primary constraints in close analogy to Einstein General Relativity. The total Hamiltonian H_T is

$$H_T = \int d^3x \left(\lambda_{\pi_N}\pi_N + \lambda^i_{\pi_{N^i}}\pi_{N^i} + N\mathcal{H} + N^i\mathcal{H}_i\right) \tag{32}$$

where the Hamiltonian Constraint $\mathcal{H}$ and the momentum constraints $\mathcal{H}^i$ are, respectively,

$$\mathcal{H} = \frac{4q^2}{\sqrt{h}\phi^2}\pi^{ij}\pi_{ij} - \frac{2q^2}{\sqrt{h}\phi}\pi\pi_\phi - \frac{\sqrt{h}}{4q^2}\phi^2\,{}^{(3)}R + \frac{q^2}{2\sqrt{h}}\pi_\phi^2 - \frac{\sqrt{h}}{4q^2}\partial^i\phi\partial_i\phi + \frac{\sqrt{h}}{q^2}\nabla^i(\phi\phi_{,i}) + \frac{\sqrt{h}}{4q^2}U(\phi), \tag{33}$$

$$\mathcal{H}_i = -2\nabla_j\pi_i^j + \pi_\phi\partial_i\phi \quad . \tag{34}$$

Here with π we have indicated the trace of π^{ij}. Following [23] one can show that the momentum constraints are the generators of the space diffeomorphisms on the three-dimensional surfaces

$$\{h_{ij}, \int d^3y N^l\mathcal{H}_l\} = (\mathcal{L}_{\mathbf{N}}h)_{ij} \quad , \tag{35}$$

$$\{\pi^{ij}, \int d^3y N^l\mathcal{H}_l\} = (\mathcal{L}_{\mathbf{N}}\pi)^{ij} \quad . \tag{36}$$

These observations [23] allow, very easily, the computation of the following commutators

$$\{\mathcal{H}_i(x), \mathcal{H}_j(x')\} = \mathcal{H}_i(x')\partial_j\delta(x,x') - \mathcal{H}_j(x)\partial_i\delta(x,x') \tag{37}$$

$$\{\mathcal{H}(x), \mathcal{H}_j(x')\} = -\mathcal{H}(x')\partial'_j\delta(x',x) \quad . \tag{38}$$

The difficult part of this calculation is, as usual in Hamiltonian theories of General Relativity, the evaluation of the Poisson brackets between the Hamiltonian-Hamiltonian constraints. We have found the following result

$$\{\int d^3x\mathcal{H}(x), \int d^3x'\mathcal{H}(x')\} = \chi^i(x)\partial_i\delta(x,x') - \chi^i(x')\partial'_i\delta(x,x') \;, \tag{39}$$

where $\chi^i(x)$ is defined as

$$\chi^i(x) \equiv \left(\frac{8\nabla_k log\phi(x)}{\sqrt{h(x)}}\right)\left(\pi^{ik}(x) + \left(\pi(x) - \frac{5}{4}\pi_\phi(x)\phi(x)\right)h^{ik}(x)\right) \quad . \tag{40}$$

Einstein General Relativity formulated into the Hamiltonian formalism through the ADM picture (3 + 1 decomposition) has secondary constraints, the momenta and the Hamiltonian constraints, which are first class (see ref. [23] [24]). The very fact that this non-singular Brans-Dicke theory does not close at the secondary level of the constraint analysis shows it is completely different with respect to Einstein General Relativity. This feature was already highlighted by a seminal paper, in canonical analysis of Einstein General Relativity, by Hojman, Kuchar and Teitelboim [25] (we are grateful to A.Kamenshchik for pointing out this reference to us). They started from a generic scenario of a Hamiltonian geometrodynamic. Here the hypothesis is to consider a three-dimesional space-like surface Σ embedded in the four dimensional Lorentzian manifolds M. Σ evolves varing the shifts functions N^i and the Lapse function N. Let's call these variations, respectively, δN^i and δN. The three metric h_{ij} and its conjugated momenta π^{ij} are the sole canonical variables. The momenta constraints $\mathcal{H}^i$ and the Hamiltonian constraint $\mathcal{H}$ are functions only of the canonical variables h_{ij} and π^{ij} and they are the generators , respectively, of the variations induced by the shifts δN^i and the lapse δN. The four constraints obey the following commutations relations

$$\begin{aligned}\{\mathcal{H}_i(x),\mathcal{H}_j(x')\} &= \mathcal{H}_i(x')\partial_j\delta(x,x') - \mathcal{H}_j(x)\partial_i\delta(x,x')\\ \{\mathcal{H}(x),\mathcal{H}_j(x')\} &= -\mathcal{H}(x')\partial_j'\delta(x',x)\\ \{\mathcal{H}(x),\mathcal{H}(x')\} &= \mathcal{H}^i(x)\partial_i\delta(x,x') - \mathcal{H}^i(x')\partial_i'\delta(x,x')\end{aligned} \quad (41)$$

then Einstein geometrodynamics is the only theory which satisfies all the above conditions. This results remains the same in case we consider Einstein General Relativity coupled to a field ϕ^A in a "non derivative gravitational way" [26]. This signifies that if $H^{(T)}$ is the total Hamiltonian and $H^{(M)}$ the matter part containing ϕ^A, then a "non derivative gravitational coupling" means the total Hamiltonian $H^{(T)}$ can be decomposed as a part that depends only on the geometrical variable h_{ij} and π^{ij}, and another one, $H^{(M)}$, that depends by ϕ_A, π^A and g_{ij} as below [26].

$$H^{(T)} = H(g_{ij},\pi^{ij}) + H^{(M)}(g_{ij},\phi_A,\pi^A) \quad (42)$$

In [25] page 131 the authors themselves recognise that the previous condition does not hold for Brans-Dicke theory.

5. Cosmological application to the sub-Planck Era

We are now ready to apply the previous considerations to a specific Friedman Lemaitre Robertson Walker (FLRW) minisuperspace model of RG improved Reuter-Weyer action with matter following [27] (the reader is advised to look to this reference for all the technical details). Just to fix the ideas, we start from the Einstein-Hilbert action (7) now in presence of matter $\mathcal{L}_m$ with the York boundary term on the boundary ∂M of a four dimensional Lorentian Manifold (M,g)

$$S = \int_M d^4x\sqrt{-g}\left\{\frac{R-2\Lambda(k)}{16\pi G(k)} + \mathcal{L}_m\right\} + \frac{1}{8\pi}\int_{\partial M}\frac{K\sqrt{h}}{G(k)}d^3x \quad , \quad (43)$$

and consider a FLRW metric with a lapse function $N(t)$

$$ds^2 = -N^2(t)dt^2 + \frac{a(t)^2}{1-Kr^2}dr^2 + a(t)^2(r^2d\theta^2 + r^2\sin\theta d\phi^2) \quad . \quad (44)$$

Let's suppose matter made of a barotropic perfect fluid, with density ρ, pressure p and equation of state $p = w\rho$, w being constant. Imposing the conservation of the matter stress energy-momentum tensor $T^{\mu\nu}{}_{;\nu} = 0$ we get $\rho = ma^{-3-3w}$, m being an integration constant. The matter Lagrangian density is then [28] $\mathcal{L}_m = -mNa^{-3w}$. Manrique et al. [29] have proposed a cut off identification which is proportional to the eigenvalue of the Laplacian on the there dimensional ADM spatial-like surfaces $k \sim \frac{1}{a}$. Implementing these considerations into the Lagrangian (43), the corresponding point Lagrangian $\mathcal{L}_g$ is

$$\mathcal{L}_g = -\frac{3\,a\dot{a}^2}{8\pi N(t)G(a)} + \frac{3\,aNK}{8\pi G(a)} - \frac{a^3 N\Lambda(a)}{8\pi G(a)} - \frac{2Nm}{a^{3w}} + \frac{3\,,a^2\dot{a}^2 G'(a)}{8\pi N G(a)^2} \quad . \tag{45}$$

We highlight FLRW metric (44) is a particular case of ADM-metric tensor in Gaussian normal coordinates (23). Therefore the Hamiltonian structure should be the same described by (24). This is not completely true, since $G(a)$ and $\Lambda(a)$ are now functionally dependent on the dynamical variable $a(t)$ and not determined completely by the Renormalization Group. Therefore, strictly speaking, it is not a particular case of the Reuter-Weyer improved Einstein-Hilbert action. An "ad hoc"Dirac's constraint analysis is needed. Performing it [27] [19], a primary constraint $\pi \approx 0$, related to the lapse N, and a secondary constrain $\mathcal{H}$, that turns to be the Hamiltonian constraint, they have been found. Both are first class (see [27]) for all the technical details). The Hamiltonian constraint $\mathcal{H} = 0$ gives the RG-improved Quantum Freedman equation

$$\frac{K}{a^2 H^2} - \frac{8\pi G(a)\,\rho + \Lambda(a)}{3H^2} + \eta(a) + 1 = 0 \; . \tag{46}$$

The main difference of this equation with respect to the Freedman equation is the appearance of a factor $\eta(a) = -\frac{a\,G'(a)}{G(a)}$, the anomalous dimension. K takes values $-1, 0, 1$ according to hyperbolic, flat, closed universes. $H = \left(\frac{\dot{a}}{a}\right)$ is the Hubble term. The equation of evolution for $a(t)$ becomes

$$\dot{a}^2 = -\tilde{V}_K(a) \equiv -\frac{K + V(a)}{\eta(a) + 1} \quad \text{where } V(a) = \frac{a^2}{3}(8\pi G(a)\,\rho + \Lambda(a)) \tag{47}$$

Notice the allowed region for dynamical evolution is $\tilde{V}_K(a) \leq 0$. Close to the non-Gaussian fixed point, using the cut off $k \sim \frac{1}{a}$, the following approximate solutions for RG-equation are deduced [30]

$$\begin{aligned} G(a) &\simeq G_0 \left(1 + G_0\, g_*^{-1} a^{-2}\right)^{-1} \\ \Lambda(a) &\simeq \Lambda_0 + \lambda_* a^{-2} \ , \end{aligned} \tag{48}$$

where (G_*, Λ_*) is the value of the Gravitational and Cosmological constant at the non-Gaussian fixed point. (G_0, Λ_0) are the infrared values of the Gravitation and Cosmological constants and coincide with the observed values.

In the radiation dominated era $w = \frac{1}{3}$, the condition for Bouncing Universes $\tilde{V}_K(a) = 0$ and then $\dot{a}(t_b) \equiv \dot{a}_b = 0$ (see [27]), ([19]) for all the details) has two real solution with non negative real part

$$a_b^2 = -\frac{G_0\Lambda_0 + g_*(\lambda_* - 3K)}{2g_*\Lambda_0} \pm \sqrt{\left(\frac{G_0\Lambda_0 - g_*(\lambda_* - 3K)}{2g_*\Lambda_0}\right)^2 - \frac{8\pi m\, G_0}{\Lambda_0}} \quad . \tag{49}$$

The special condition for having "Emergent Universes "[31], [27] is $\dot{a}_b = \ddot{a}_b = 0$. Of course a_b^2 has to be positive, that is

$$\lambda_* - 3K < -\frac{G_0\Lambda_0}{g_*} \quad . \tag{50}$$

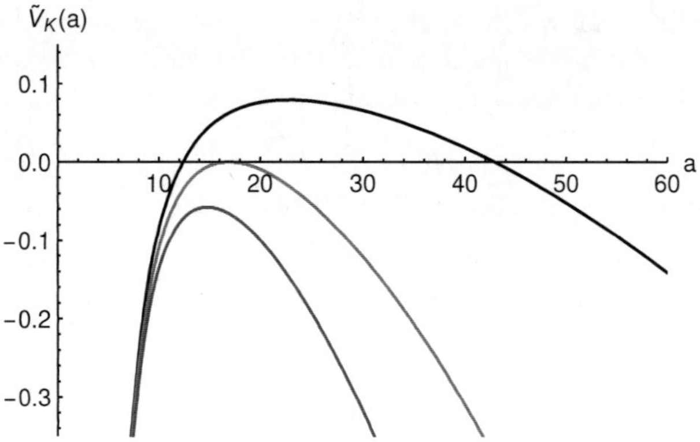

Figure 1. The effective potential $\tilde{V}_K(a)$ for a bouncing universe (black), emergent universe (red), singular universe (blue), for $K = 0$, $w = 1/3$, $g_* = 0.1$, $\lambda_* = -0.5$ and $m = 3$. Black, red and blue correspond to $\Lambda_0 = 2 \times 10^{-4}$, $\Lambda_0 = 8.3 \times 10^{-4}$ and $\Lambda_0 = 1.5 \times 10^{-3}$ respectively.

In Classical General Relativity $\lambda_* = 0$ implies only closed universes K=1 are possible. The sub-Planckian regime in Asymptotic Safety allows, in some coupling of gravity with matter [32], λ_* to have a negative value so that K=0 and K=-1 are possible as well, which is a feature only of quantum era.

We can now study the behaviour of the early emergent universe close to a_b linearizing the quantum equation (47) around a_b. The approximate equation is then:

$$\dot{a}^2 = \frac{4 g_* a_b^2 \Lambda_0}{3\left(g_* a_b^2 - G_0\right)} (a - a_b)^2 \ , \tag{51}$$

then the general solution is

$$a(t) = a_b + \epsilon \exp\left\{ \sqrt{\frac{4 g_* a_b^2 \Lambda_0}{3\left(g_* a_b^2 - G_0\right)}}\, t \right\} \ , \tag{52}$$

ϵ being an integration constant. It is evident that (52) exibits an emergent universe scenario: there is a minimal radius a_b and an everlasting (eternal) exponential evolution of the scale factor $a(t)$. There is no need of a model with an "*ad hoc*"inflation. The density parameter can be written

$$\Omega - 1 = \frac{3\left(g_* a_b^2 - G_0\right) K}{4 g_* a_b^4 \Lambda_0} e^{-2N_e} \ . \tag{53}$$

The number N_e of e-folds is

$$N_e \simeq \log\left(\frac{\epsilon}{a_b} \exp\left\{ \sqrt{\frac{4 g_* a_b^2 \Lambda_0}{3\left(g_* a_b^2 - G_0\right)}}\, t_e \right\} \right) \ , \tag{54}$$

where t_e is the cosmic time at the inflation exit.

6. Conclusions

Hamiltonian analysis of the Reuter-Weyer improved Einstein-Hilbert action has been performed. G and Λ have been treated as external non-geometrical fields. The Dirac's constraint analysis of Reuter-Weyer improved Einstein Hilbert action has been carried out up to the secondary constraint level of the Dirac's constraint analysis. It looks quite complicated and the constraints appear to be second class. This theory behaves much better if we use "Gaussian Normal Coordinates". There exists a Hamiltonian constraint and it is first class. The difficulties emerged in the constraint analysis of the Reuter-Weyer improved Einstein-Hilbert action suggest the Dirac's constraint analysis of Brans-Dicke theory. This shows the Poisson brackets of the Hamiltonian-Hamiltonian constraint generate tertiary constraint.

The simple FLRW minisuperspace model based on the Reuter-Weyer improved Einstein-Hilbert action functional exibits bouncing and emergent universes for $K = 0$ and $K = -1$, which are impossible to get in classical Einstein General Relativity.

Future directions to explore are a better study of Dirac's constraint analysis of the improved Reuter-Weyer Einstein-Hilbert action functional, in particular the role of the $G(x)$ and $\Lambda(x)$ as external fields. A better understanding of the Dirac's constraint analysis of Brans-Dicke theory and the "questio disputata" of the equivalence between the Jordan frame and the Einstein frame. Finally, proceeding along these lines, it could be also quite interesting to study ADM analysis of a Black Hole minisupersapce model based on the improved Reuter-Weyer Einstein Hilbert action functional, which could indicate interesting features as the case of Bouncing and Emergent Universes just shown.

Acknowledgments
I am very grateful to A. Bonanno for suggesting this topic of research, for discussions and encouragment during my visits to OACT in Catania. I would like also to thank G. Esposito, A.Kamenshchik and M. Reuter for enlightening conversations on this topic.

References

[1] Weinberg S 1972 *Gravitation and Cosmology* (New York: John Wiley and Sons)
[2] Abbott B P *et al.* (LIGO Scientific, Virgo) 2016 *Phys. Rev. Lett.* **116** 221101 [Erratum: Phys. Rev. Lett.121,no.12,129902(2018)] (*Preprint* gr-qc/1602.03841)
[3] Hawking S W and Ellis G F R 2011 *The Large Scale Structure of Space-Time* Cambridge Monographs on Mathematical Physics (Cambridge University Press)
[4] Lauscher O and Reuter M 2002 *Classical and Quantum Gravity* **19** 483–492 (*Preprint* hep-th/0110021)
[5] Wilson K G 1973 *Phys. Rev.* **D7** 2911–2926
[6] Niedermaier M and Reuter M 2006 *Living Reviews in Relativity* **9** 5
[7] Weinberg S 1979 *General Relativity: An Einstein centenary survey* ed Hawking S W and Israel W pp 790–831
[8] Reuter M 1998 *Phys. Rev.* **D57** 971–985 (*Preprint* hep-th/9605030)
[9] Reuter M and Saueressig F 2002 *Phys. Rev.* **D65** 065016 (*Preprint* hep-th/0110054)
[10] Reuter M and Saueressig F 2010 *Geometric and topological methods for quantum field theory* pp 288–329 (*Preprint* hep-th/0708.1317)
[11] Berges J, Tetradis N and Wetterich C 2002 *Phys. Rept.* **363** 223–386 (*Preprint* hep-ph/0005122)
[12] Reuter M and Weyer H 2004 *Phys. Rev.* **D69** 104022 (*Preprint* hep-th/0311196)
[13] Brans C and Dicke R H 1961 *Physical Review* **124** 925–935
[14] Olmo G J and Sanchis-Alepuz H 2011 *Phys. Rev.* **D83** 104036 (*Preprint* gr-qc/1101.3403)
[15] Arnowitt R, Deser S and Misner C W 1960 *Phys. Rev.* **117**(6) 1595–1602 URL https://link.aps.org/doi/10.1103/PhysRev.117.1595
[16] York J W 1986 *Foundations of Physics* **16** 249–257

IOP Conf. Series: Journal of Physics: Conf. Series **1275** (2019) 012048 doi:10.1088/1742-6596/1275/1/012048

[17] Bonanno A, Esposito G and Rubano C 2004 *Classical and Quantum Gravity* **21** 5005–5016 (*Preprint* gr-qc/0403115)

[18] DeWitt B S 1967 *Phys. Rev.* **160** 1113–1148 [3,93(1987)]

[19] Gionti S J G 2018 *PoS* **CORFU2017** 192 (*Preprint* gr-qc/1805.02318)

[20] Wiltshire D L 1995 *Cosmology: The Physics of the Universe. Proceedings, 8th Physics Summer School, Canberra, Australia, Jan 16-Feb 3, 1995* pp 473–531 (*Preprint* gr-qc/0101003)

[21] Christodoulou D and Klainerman S 1993 *The global nonlinear stability of the Minkowski space* Princeton mathematical series ; 41 (Princeton University Press)

[22] Garay L J and Garcia-Bellido J 1993 *Nucl. Phys.* **B400** 416–434 (*Preprint* gr-qc/9209015)

[23] Menotti P 2017 *ArXiv e-prints* (*Preprint* gr-qc/1703.05155)

[24] Esposito G 1992 *Lect. Notes Phys. Monogr.* **12** 1–326

[25] Hojman S A, Kuchař K and Teitelboim C 1976 *Annals of Physics* **96** 88–135

[26] Kuchař K 1974 *Journal of Mathematical Physics* **15** 708–715

[27] Bonanno A, Gionti S J G and Platania A 2018 *Classical and Quantum Gravity* **35** 065004 (*Preprint* gr-qc/1710.06317)

[28] Dimakis N, Christodoulakis T and Terzis P A 2014 *Journal of Geometry and Physics* **77** 97–112 (*Preprint* gr-qc/1311.4358)

[29] Manrique E, Rechenberger S and Saueressig F 2011 *Physical Review Letters* **106** 251302 (*Preprint* hep-th/1102.5012)

[30] Bonanno A and Reuter M 2000 *Phys. Rev.* **D62** 043008 (*Preprint* hep-th/0002196)

[31] Ellis G F R and Maartens R 2004 *Class. Quant. Grav.* **21** 223–232 (*Preprint* gr-qc/0211082)

[32] Biemans J, Platania A and Saueressig F 2017 *JHEP* **05** 093 (*Preprint* hep-th/1702.06539)

IOP Conf. Series: Journal of Physics: Conf. Series **1275** (2019) 012049 doi:10.1088/1742-6596/1275/1/012049

Born rule and Schrödinger equation from first principles

Robert Oeckl

Centro de Ciencias Matemáticas,
Universidad Nacional Autónoma de México,
C.P. 58190, Morelia, Michoacán, Mexico

E-mail: robert@matmor.unam.mx

Abstract. A concise review of the derivation of the Born rule and Schrödinger equation from first principles is provided. The starting point is a formalization of fundamental notions of measurement and composition, leading to a general framework for physical theories known as the positive formalism. Consecutively adding notions of spacetime, locality, absolute time and causality recovers the well established convex operational framework for classical and quantum theory. Requiring the partially ordered vector space of states to be an anti-lattice one obtains quantum theory in its standard formulation. This includes Born rule and Schrödinger equation.

1. Introduction

While many of the mathematical ingredients to describe classical physics are accepted as natural, this is much less so in the case of quantum physics. In particular, a complex Hilbert space, the notion of operators as observables, the Schrödinger equation and the Born rule might initially appear as lacking motivation from first principles. Indeed, many of these structures were originally introduced in a somewhat ad hoc fashion. However, as our understanding of the foundations of physics and quantum theory in particular has advanced, much of the mystery surrounding these structures has disappeared. I think it is fair to say that any mystery that remains today may be reduced to the appearance of the complex Hilbert space alone, with all other structures following naturally. While outside the scope of this note, I should add that for this last mysterious ingredient there has been a concerted effort, particularly in the last 20 years or so to explain it. These attempts, known as *reconstructions of quantum theory* succeed in predicting the appearance of a complex Hilbert space by introducing some additional assumption that might appear more or less plausible.

The purpose of the present note is to review very briefly how the mathematical structures and interpretational rules of quantum theory arise naturally. I hope that especially non-experts in the foundations of quantum theory might find this account useful amid sometimes confusing and contradictory claims in the literature. The starting point is a minimalist account of notions of measurement and composition, leading to a framework for encoding theories of physics, known as the *positive formalism* [1], in its *abstract* version. Adding a notion of spacetime and implementing the principle of locality leads to the *local* version of the positive formalism (which is the starting point in [1]). Rigidifying a *metric background time* and implementing the principle of *causality* leads to the *convex operational framework*. The latter has been used at least since 1970 to

Content from this work may be used under the terms of the Creative Commons Attribution 3.0 licence. Any further distribution of this work must maintain attribution to the author(s) and the title of the work, journal citation and DOI.
Published under licence by IOP Publishing Ltd

IOP Conf. Series: Journal of Physics: Conf. Series **1275** (2019) 012049 doi:10.1088/1742-6596/1275/1/012049

describe classical and quantum dynamical systems in a unified fashion [2]. The remaining steps as presented (starting with Section 5) to recover either a rudimentary formulation of classical statistical physics or the standard formulation of quantum theory are well known. However, I have not found a presentation elsewhere in the literature that is equally condensed. Many technical details are simplified or omitted, as are most references. The interested reader is directed instead to consider [1] and the references therein.

2. The abstract positive formalism

Let us start by subsuming notions of experiment, measurement, observation, preparation and intervention into a single notion, called *probe*. (In the recent quantum foundations literature the word *process* is often used with a similar meaning.) A probe is a fine-grained notion in the sense that it allows, put does not require, to distinguish different measurement outcomes. For example, suppose a measurement has two distinct possible outcomes A and B. Then, there is a probe encoding each of these, call them $P(\mathsf{A})$ and $P(\mathsf{B})$. But there is also a probe $P(*)$ encoding the experiment without outcome, i.e., where no knowledge of a specific outcome is taken into account. In general, an experiment may involve a number of output channels, each of which may or may not be observed in a given run. There is then a probe encoding each of these different possibilities in addition to each of the possible outcomes of the observed output channels. This leads to a *hierarchy of generality* in the space of probes that gives rise to a binary *relation* of relative generality. It is not difficult to see that the properties of this binary relation (reflexive, anti-symmetric and transitive) are precisely those of a *partial order*. Hence we shall say that a probe P that is more special than a probe Q, i.e., implies Q, is less or equal to Q, $P \le Q$. In the simple example with two outcomes we have $P(\mathsf{A}) \le P(*)$ and $P(\mathsf{B}) \le P(*)$.

It is not sufficient to consider single experiments or their outcomes in isolation. Meaningful predictions arise in terms of correlations between different observations and interventions. To capture this, let us introduce a notion of *composition* of probes, denoted $P \diamond Q$, for probes P and Q. For example, if the component probes encode certain outcomes of different experiments, then the composite probe encodes the joint occurrence of all of these outcomes. Not all probes can be meaningfully composed. For example, a probe involving a certain apparatus at a certain place and time cannot be composed with a probe involving a different (or even a copy of the same) apparatus at the same place and time.

In general it is not practical to consider explicitly for a given experiment all observations or interventions taking place that are in some way correlated to this experiment. Instead, let us introduce a notion of *boundary condition* that parametrizes all potential correlations between the experiment and the rest of the world. So, associated to any given probe is a set of boundary conditions. This set may be different for different probes, but it is the same for probes that encode different aspects (such as different outcomes) of the same run of the same experiment. Let us say that probes that share the same set of boundary conditions are of the same *type*. In particular, there are partial order relations only between probes of the same type. On the other hand, probes of the same type cannot be composed with each other.

The most elementary prediction one might wish to make is the *probability* for a certain outcome of an observation or measurement. Consider the previous example of a single experiment with two possible outcomes A and B. Say, we are interested in the probability Π for outcome A. Clearly we need the probe $P(\mathsf{A})$ to encode this outcome. However, and this is less obvious, we also need the probe $P(*)$ as a reference for comparison. Moreover, we need a boundary condition b that parametrizes our knowledge about the influence of all the external degrees of freedom on our experiment. We may then loosely paraphrase the question we ask as follows: How probable is the observation of $P(\mathsf{A})$ subject to boundary condition b, given that we have set up experiment $P(*)$ subject to boundary condition b? We translate this into a

IOP Conf. Series: Journal of Physics: Conf. Series **1275** (2019) 012049 doi:10.1088/1742-6596/1275/1/012049

mathematical formula as follows,

$$\Pi = \frac{[\![P(\mathsf{A}), b]\!]}{[\![P(*), b]\!]}. \tag{1}$$

Here we have introduced a pairing $[\![\cdot,\cdot]\!]$ between probes of a given type and corresponding boundary conditions. The pairing takes non-negative real values. These values are not in general probabilities, but can be thought of as "relative probabilities" or *compatibilities.* So, $[\![P(*), b]\!]$ quantifies the compatibility between the experimental setup and the boundary condition. $[\![P(\mathsf{A}), b]\!]$ quantifies this same compatibility, but where in addition the occurrence of outcome A is imposed. Crucially, the inequality $P(\mathsf{A}) \leq P(*)$ encoding relative generality translates precisely into a corresponding inequality of compatibilities:

$$[\![P(\mathsf{A}), b]\!] \leq [\![P(*), b]\!]. \tag{2}$$

This guarantees that the quotient (1) is less or equal to one, as a probability must be. Note that this inequality will hold for any boundary condition b.

Conversely, we may use this to define the partial order relation in the space of probes of a given type. This amounts to viewing probes, via the pairing, as real valued functions on a set (the set of boundary conditions). The real vector space of real valued functions on a set naturally carries a partial order compatible with the vector space structure. It is thus called a *partially ordered vector space.* There holds $f \leq g$ for functions f and g if $f(x) \leq g(x)$ for all elements x of the set. We denote the partially ordered vector space spanned by all probes of a given type by $\mathcal{P}$. We denote the subset of *positive elements* of $\mathcal{P}$ by $\mathcal{P}^+$. We call an element *positive* if it satisfies $0 \leq P$. This subset is in fact a *convex cone.* In particular, all probes are positive since compatibilities are non-negative. Returning to the example, we also have a completeness condition that assures that probabilities add up to one. Here, for any boundary condition b,

$$[\![P(\mathsf{A}), b]\!] + [\![P(\mathsf{B}), b]\!] = [\![P(*), b]\!]. \tag{3}$$

This translates to, $P(\mathsf{A}) + P(\mathsf{B}) = P(*)$.

What is more, we shall assume that all elements of $\mathcal{P}^+$ are probes. This may be justified as follows. Firstly, we claim that performing a probabilistic mixture of experiments is also a valid experiment. Thus, probabilistic mixtures of probes are probes. Secondly, we notice a rescaling freedom for probes in expression (1). (Only relative scalings are important.) Thus a positive multiple of a probe is again a probe. Consequently, linear combinations of probes with positive coefficients are probes. In this way we obtain the whole cone $\mathcal{P}^+$. In contrast, we call general elements of $\mathcal{P}$ *generalized probes.*[1] Similarly, it is legitimate to form probabilistic mixtures of boundary conditions and expression (1) reveals a scaling freedom also for these. In this way, the boundary conditions form a positive cone $\mathcal{B}^+$ in a partially ordered vector space $\mathcal{B}$ of *generalized boundary conditions.* The pairing $[\![\cdot,\cdot]\!]$ thus becomes a map $\mathcal{P}^+ \times \mathcal{B}^+ \to \mathbb{R}_0^+$ and we extend it bilinearly to a map $\mathcal{P} \times \mathcal{B} \to \mathbb{R}$.

We take a closer look at the interplay of composition and boundary conditions. Suppose we consider the composition of a probe P with a probe Q. We recall that the boundary conditions for the probe P encode knowledge about correlations between P and the rest of the world. Now Q is part of this rest of the world. Thus, we should expect part of these boundary conditions to encode correlations with Q and another part correlations with the rest of the world excluding Q. Corresponding considerations hold for Q. To formalize this we introduce a notion of decomposition of the spaces of boundary conditions. On the level of generalized boundary conditions which are vector spaces, this decomposition should take the form of a tensor product. This is again because we permit probabilistic combinations as well as rescalings.

[1] What is called a probe here is called a *primitive probe* in [1]. A generalized probe here is simply a probe in [1].

IOP Conf. Series: Journal of Physics: Conf. Series **1275** (2019) 012049 doi:10.1088/1742-6596/1275/1/012049

In the example with probes P and Q, call the corresponding spaces of generalized boundary conditions $\mathcal{B}_P$ and $\mathcal{B}_Q$. We denote the decompositions by $\mathcal{B}_P = \mathcal{B}_{PQ} \otimes \mathcal{B}'_P$ and $\mathcal{B}_Q = \mathcal{B}_{PQ} \otimes \mathcal{B}'_Q$. Crucially, the space $\mathcal{B}_{PQ}$ of boundary conditions encoding correlations between P and Q appears in both decompositions. What is more, P and Q are composed "along" $\mathcal{B}_{PQ}$. The joint probe $P \diamond Q$, arising from this composition has a space of boundary conditions that decomposes as $\mathcal{B} = \mathcal{B}'_P \otimes \mathcal{B}'_Q$.

We allow a special *transparent* probe that does not correspond to any actual intervention, measurement or observation, but just "lets signals pass through" from any space $\mathcal{B}$ of boundary conditions to a copy of this same space. Using this we can derive an inner product $(\!|\cdot,\cdot|\!)$: $\mathcal{B} \times \mathcal{B} \to \mathbb{R}$ that is positive-definite and non-negative on pairs of positive elements (proper boundary conditions). This in turn leads to the *composition rule* of probes as follows. We return to the example of probes P and Q with corresponding spaces of boundary conditions $\mathcal{B}_P = \mathcal{B}_{PQ} \otimes \mathcal{B}'_P$ and $\mathcal{B}_Q = \mathcal{B}_{PQ} \otimes \mathcal{B}'_Q$. Let $b \in \mathcal{B}'_P$ and $c \in \mathcal{B}'_Q$. Let $\{\xi_k\}_{k\in I}$ be an orthonormal basis of $\mathcal{B}_{PQ}$. Then,

$$[\![P \diamond Q, b \otimes c]\!] = \sum_{k\in I} [\![P, b \otimes \xi_k]\!][\![Q, c \otimes \xi_k]\!]. \tag{4}$$

We arrive at an *axiomatic* framework for physical theories, called the *abstract positive formalism*. Its main mathematical ingredients, called probes and boundary conditions, are *partially ordered vector spaces*. Crucially this framework includes rules for the *prediction of physically measurable quantities*, the simplest of those taking the form of formula (1). A *physical theory* now consists of a model satisfying the axioms, i.e., an explicit specification of spaces of probes, boundary conditions, types etc. along with a correspondence to physical phenomena for some of these objects.

3. Spacetime and locality

The framework as considered so far is meant to be general and minimalistic, involving as few assumptions about our concrete experience of the physical world as possible. However, adding suitable structure that codifies (perhaps a simplified version of) this experience can greatly improve the practical usefulness and predictive power of the framework when applicable. We proceed here to integrate a notion of *spacetime* into the framework, which allows to implement the powerful principle of *locality*.

We suppose that measurements can be localized in *spacetime regions*. In fact, we identify the *types* of probes with spacetime regions. In this way we assign to each spacetime region M its corresponding space $\mathcal{P}_M$ of (generalized) probes. We permit the composition of probes only if the underlying spacetime regions have no intersection, except in their boundaries. That is, the composition of probes corresponds to a composition of the underlying spacetime regions in terms of a *gluing* operation. Similarly, we associate the space of (generalized) boundary conditions for probes in M to the boundary hypersurface ∂M of the region M, with notation $B_{\partial M}$. More generally, we allow the association of a space $\mathcal{B}_\Sigma$ of boundary conditions to any hypersurface Σ. Moreover, we bring into correspondence decompositions of spaces of boundary conditions with decompositions of the underlying hypersurfaces. Thus, a decomposition of a hypersurface Σ is its presentation as a union $\Sigma = \Sigma_1 \cup \cdots \cup \Sigma_n$ of hypersurfaces such that the components may intersect only in their boundaries. To this corresponds then the tensor product decomposition of associated spaces of boundary conditions, $\mathcal{B}_\Sigma = \mathcal{B}_{\Sigma_1} \otimes \cdots \otimes \mathcal{B}_{\Sigma_n}$. The composition of probes in adjacent spacetime regions involves the decomposition of the boundary hypersurface of each involved region into pieces. Each piece arises as the intersections with the boundaries of one of the other regions. Returning to the example of composing probes P and Q, these are now associated to underlying spacetime regions M and N. The boundaries ∂M and ∂N intersect in some hypersurface Σ_{PQ}. This leads to decompositions $\partial M = \Sigma_{PQ} \cup \Sigma_P$ and $\partial N = \Sigma_{PQ} \cup \Sigma_Q$, where Σ_P and Σ_Q are the non-intersecting pieces of the boundaries. We recover the previously

IOP Conf. Series: Journal of Physics: Conf. Series **1275** (2019) 012049 doi:10.1088/1742-6596/1275/1/012049

considered decomposition the spaces of boundary conditions as follows, $\mathcal{B}_P = \mathcal{B}_{\partial M}$, $\mathcal{B}_Q = \mathcal{B}_{\partial N}$, $\mathcal{B}_{PQ} = \mathcal{B}_{\Sigma_{PQ}}$, $\mathcal{B}'_P = \mathcal{B}_{\Sigma_P}$, $\mathcal{B}'_Q = \mathcal{B}_{\Sigma_Q}$.

Crucially, two probes only share a space of boundary conditions if the underlying regions are adjacent, that is if parts of their boundaries intersect. This implements the powerful principle of *locality* which dramatically reduces the possible direct connections between probes in a composition. The underlying physical idea is that interaction is exclusively mediated by signals traversing spacetime and thereby necessarily crossing interfacing hypersurfaces. For each spacetime region there is a special kind of probe that corresponds to not making any intervention, measurement or observation in this region. We call this the *nul probe*. Clearly, the composition of nul probes is a nul probe. We arrive at the *local positive formalism*, also called *spacetime positive formalism* (and previously simply referred to as the *positive formalism*). It is at this level that we may make contact with classical field theory or quantum field theory [1]. However, as we are interested in recovering the standard formulation of quantum theory we proceed to specialize to a non-relativistic picture of spacetime.

4. Time, evolution and causality

We drastically simplify the notion of spacetime to a linear notion of time, measured in terms of real numbers. Spacetime regions are reduced to time-intervals $[t_1, t_2]$ and finite unions of these. Hypersurfaces are reduced to instants of time (and finite unions thereof). In particular, we have a space of boundary conditions $\mathcal{B}_t$ for each time $t \in \mathbb{R}$. Any other space of boundary conditions arises as a tensor product of these. We also call $\mathcal{B}_t$ the *state space* at time t and its elements *(generalized) states*. Intuitively speaking, a state parametrizes knowledge about the past relevant for making predictions about the future. But it also parametrizes information about the future relevant for making predictions about the past. What is more, by construction, any possibly relevant information about the past (or future) may be encoded in this way.

Probes are associated to time intervals and admit a pairing with boundary conditions that arise from tensor products of initial and final states. However, it turns out to be convenient to view probes instead as maps from the initial to the final state space. This is equivalent via the inner product. Consider a time interval $[t_1, t_2]$ and a corresponding probe $P \in \mathcal{P}^+_{[t_1,t_2]}$. Set $\tilde{P} : \mathcal{B}_{t_1} \to \mathcal{B}_{t_2}$ via,

$$[\![P, b_1 \otimes b_2]\!]_{[t_1,t_2]} = (\!(b_2, \tilde{P} b_1)\!)_{t_2} \quad \forall b_1 \in \mathcal{B}_{t_1}, \forall b_2 \in \mathcal{B}_{t_2}. \tag{5}$$

$\tilde{P}$ is uniquely determined by P and vice versa. Note crucially that the positivity of P implies that $\tilde{P}$ maps positive elements to positive elements, i.e., restricts to a map $\mathcal{B}^+_{t_1} \to \mathcal{B}^+_{t_2}$. (The converse is not necessarily true, see [1] for more details.) From this point onwards we exclusively use the presentation of probes as maps between state spaces and refer to them as *operations*. (We also drop the tilde in the notation, i.e., write P to mean $\tilde{P}$.) This has the advantage that the somewhat cumbersome composition rule for probes (4) translates into a simple composition of maps. Recall the special nul probe (or operation) for any time interval $[t_1, t_2]$ which represents the absence of any intervention, measurement or observation during that time. We denote this by $T_{[t_1,t_2]} : \mathcal{B}_{t_1} \to \mathcal{B}_{t_2}$ and call it the *time evolution* map. Time evolution maps compose as $T_{[t_1,t_3]} = T_{[t_2,t_3]} \circ T_{[t_1,t_2]}$ for $t_1 \leq t_2 \leq t_3$ because nul probes compose to nul probes.

We introduce a further simplification at his point, namely we suppose that all the state spaces $\mathcal{B}_t$ are isomorphic and can be canonically identified with a single state space $\mathcal{B}$ via a time translation symmetry. Operations are then positive operators on $\mathcal{B}$, i.e., linear maps $\mathcal{B} \to \mathcal{B}$ that preserve positivity. Further, we refer from here onwards to a generic space of operations $\mathcal{P}^+ \subseteq \mathcal{P}$ without necessarily specifying an initial and final time for each operation. Let us consider the form the probability formula (1) takes with the newly introduced notation. Thus, we consider a measurement with two possible outcomes A and B. The operations $P(\mathsf{A})$ and $P(*)$ represent the

IOP Conf. Series: Journal of Physics: Conf. Series **1275** (2019) 012049 doi:10.1088/1742-6596/1275/1/012049

measurement with specific outcome A and unspecified outcome respectively. We need to specify an initial state $b_1 \in \mathcal{B}^+$ and a final state $b_2 \in \mathcal{B}^+$ to obtain a probability Π for the occurrence of outcome A,

$$\Pi = \frac{(\!|b_2, P(\mathsf{A})\, b_1|\!)}{(\!|b_2, P(*)\, b_1|\!)}. \tag{6}$$

Here, b_1 encodes knowledge about the past of the measurement and b_2 knowledge about its future. The choice of b_1 is usually referred to as a *preparation.* A choice of the state b_2 on the other hand is usually called *post-selection.*

But what if we do not have any knowledge about the future (or past)? To codify this situation we distinguish a special state $\mathbf{e} \in \mathcal{B}^+$. Recall how the partial order structure of $\mathcal{B}^+$ corresponds to a hierarchy of generality. What we want is a "maximally general" state, a *state of maximal uncertainty.* In terms of the partial order this should mean something like $b \le \mathbf{e}$ for any other state b. Of course this cannot be literally true, because we have rescaling freedom. But taking this into account we can demand the inequality after suitable rescaling. That is, we demand that for any state $b \in \mathcal{B}^+$ there exist $\lambda > 0$ such that $b \le \lambda\mathbf{e}$. Such an element of a partially ordered vector space is called an *order unit.* Our distinguished state $\mathbf{e}$ is thus a particular order unit (in general there are many). We are now in a position to talk about a measurement without post-selection. This is achieved by setting the final state to be the state of maximal uncertainty, $b_2 = \mathbf{e}$ in formula (6). This is also referred to as *discarding* the system that has been measured. Note that we could equally contemplate using the state of maximal uncertainty as an initial state or even introducing this concept already at the level of the local or abstract positive formalism, but this is outside of the scope of this short piece.

The concept of the state of maximal uncertainty gives us the opportunity to hard-code another well established empirical principle into the formalism, namely *causality.* By this we mean that a future choice of measurement (but not outcome!) does not influence outcomes in the past. To this end we distinguish *non-selective* from *selective* operations. Thus, consider again a measurement with two possible outcomes A and B and corresponding operations $P(\mathsf{A})$, $P(\mathsf{B})$, $P(*)$. Then, the operation $P(*)$ is non-selective as it only represents performing the measurement without knowledge of the outcome, while the operations $P(\mathsf{A})$ and $P(\mathsf{B})$ are selective as they select outcome A or B.

Consider now the consecutive execution of two measurements of this type, first P, then Q. Given a fixed initial state $b \in \mathcal{B}^+$, the probability Π to measure outcome A via P and then discarding should be the same as when performing Q (without getting to know its outcome) after P and only then discarding. That is,

$$\Pi = \frac{(\!|\mathbf{e}, Q(*)P(\mathsf{A})\, b|\!)}{(\!|\mathbf{e}, Q(*)P(*)\, b|\!)} = \frac{(\!|\mathbf{e}, P(\mathsf{A})\, b|\!)}{(\!|\mathbf{e}, P(*)\, b|\!)}. \tag{7}$$

For this to hold with arbitrary b and P requires the *normalization condition* for non-selective operations,

$$(\!|\mathbf{e}, Q(*)c|\!) = (\!|\mathbf{e}, c|\!), \quad \forall c \in \mathcal{B}^+. \tag{8}$$

For selective probes this implies a corresponding inequality, e.g., $Q(\mathsf{A}) \le Q(*)$ implies, $(\!|\mathbf{e}, Q(\mathsf{A})c|\!) \le (\!|\mathbf{e}, Q(*)c|\!) = (\!|\mathbf{e}, c|\!)$. From this point onwards we implement the causality principle by requiring non-selective operations to satisfy the normalization condition (8). Note that this condition is *time asymmetric*, in contrast to all other elements of the formalism considered so far.

Non-degeneracy and positivity of the inner product imply that the inner product between the state of maximal uncertainty $\mathbf{e}$ and an arbitrary state $b \in \mathcal{B}^+$ that is non-zero, $b \neq 0$, is strictly positive $(\!|\mathbf{e}, b|\!) > 0$. We can take advantage of this to introduce a *normalization condition* for states. We say that a state $b \in \mathcal{B}^+$ is *normalized* if $(\!|\mathbf{e}, b|\!) = 1$. Thus, any non-zero state can be

IOP Conf. Series: Journal of Physics: Conf. Series **1275** (2019) 012049 doi:10.1088/1742-6596/1275/1/012049

normalized by multiplying it with a positive scalar. We denote the subset of normalized states by $\mathcal{B}^{\mathrm{n}} \subseteq \mathcal{B}^{+} \subseteq \mathcal{B}$. Using normalized states when the normalization condition for non-selective operations is enforced leads to simplifications of some formulas. Most notably, the probability for the outcome of a sequence of measurements takes the form of a linear expression. However, this only works if the system is discarded at the end and no conditioning on outcomes is considered. For a single measurement P the probability Π of outcome A with initial state $b \in \mathcal{B}^{\mathrm{n}}$ takes the simple form,

$$\Pi = (\!|\mathbf{e}, P(\mathsf{A})\, b|\!). \tag{9}$$

This arises from formula (6) with $b_2 = \mathbf{e}$, $b_1 = b \in \mathcal{B}^{\mathrm{n}}$ because the denominator takes the value 1, as $P(*)$ satisfies (8).

If we perform consecutive measurements we can ask how we should "update" a state to codify the outcome of a previous measurement. Starting with a state $b \in \mathcal{B}^{+}$ and performing a measurement P with outcome A, the probability for an outcome B in a subsequent measurement Q is,

$$\Pi = \frac{(\!|\mathbf{e}, Q(\mathsf{B})P(\mathsf{A})b|\!)}{(\!|\mathbf{e}, Q(*)P(\mathsf{A})b|\!)} = \frac{(\!|\mathbf{e}, Q(\mathsf{B})c|\!)}{(\!|\mathbf{e}, Q(*)c|\!)}. \tag{10}$$

Here we have set $c = P(\mathsf{A})b$. This can be seen as the updated state. However, if we want to take advantage of simplified formulas such as (9) we have to work with normalized state. This yields the *state update rule*,

$$b \mapsto \frac{P(\mathsf{A})\, b}{(\!|\mathbf{e}, P(\mathsf{A})\, b|\!)}. \tag{11}$$

5. The convex operational framework

The framework we have arrived at is the well known *convex operational framework*, designed to describe classical and quantum theory in a unified way [2]. The main ingredients of this framework are thus a partially ordered vector space $\mathcal{B}$ called (generalized) *state space* and a partially ordered vector space $\mathcal{P}$ of (generalized) *operations*. The elements of $\mathcal{P}$ are linear operators on $\mathcal{B}$ with the subset $\mathcal{P}^{+}$ of (proper) operations mapping states to states, i.e., $\mathcal{B}^{+}$ to itself. Moreover, there is a special *state of maximal uncertainty* $\mathbf{e} \in \mathcal{B}^{+}$ which is an order unit. We have a subset $\mathcal{B}^{\mathrm{n}} \subseteq \mathcal{B}^{+}$ of *normalized states* b that satisfy $(\!|\mathbf{e}, b|\!) = 1$. Operations are non-selective or selective with the former satisfying the normalization condition (8). Probabilities for outcomes of single measurements with out post-selection take the form (9) and come with the state update rule (11).

The evolution of states in the absence of any measurement or intervention is encoded in the time evolution maps $T_{[t_1,t_2]} \in \mathcal{P}^{+}$, associated to time intervals $[t_1, t_2]$. These compose in the obvious way, as already remarked. We shall assume now in addition that the time evolution maps $T_{[t_1,t_2]}$ preserve the structure of the state space completely. This is reasonable if we are considering a complete description of physics in the sense that we are not neglecting any interaction with an "exterior" system. Thus, $T_{[t_1,t_2]} : \mathcal{B} \to \mathcal{B}$ is an isomorphism of vector spaces and reduces to a bijection of positive cones $\mathcal{B}^{+} \to \mathcal{B}^{+}$. Moreover, it preserves the inner product $(\!|\cdot,\cdot|\!)$ and the state of maximal uncertainty, $T_{[t_1,t_2]}\mathbf{e} = \mathbf{e}$ which implies that it also preserves the subset of normalized states $\mathcal{B}^{\mathrm{n}}$. We make now another simplifying assumption about the physics, namely that time evolution is itself time translation invariant. That is, $T_{[t_1,t_1+\Delta]} = T_{[t_2,t_2+\Delta]}$ for any t_1, t_2, Δ. Consequently, we simply write T_Δ instead of $T_{[t_1,t_1+\Delta]}$. In this way we obtain a one-parameter group of isomorphisms of $\mathcal{B}$, i.e., satisfying $T_{\Delta_1}T_{\Delta_2} = T_{\Delta_1+\Delta_2}$ and $T_0 = \mathrm{id}$.

6. Classical theory

After this lightning review of the convex operational framework as derived from first principles via the positive formalism we come to the question of how we recover the details of the standard

IOP Conf. Series: Journal of Physics: Conf. Series **1275** (2019) 012049 doi:10.1088/1742-6596/1275/1/012049

formulation of quantum theory from it. The answer is that, of course, we cannot do so yet as we have not introduced any principle that would exclude, say, classical physics. Indeed, not all partially ordered vector spaces are alike, even with the additional structures they carry in our case. (This is in notable contrast to, say, infinite-dimensional separable complex Hilbert spaces.) It is thus not surprising that radically different models of physics can be captured with this framework.

We start with a particular type of partially ordered vector space as state space, a *lattice.* Roughly speaking this means that for any two positive elements $b, c \in \mathcal{B}^+$ there is a positive element that represent the minimum $\min(b, c)$ and one that represents the maximum $\max(b, c)$ of the two elements. This turns out to imply that $\mathcal{B}$ is the vector space $\mathcal{C}(L)$ of functions over a set L. This set L is nothing but the *phase space* of a *classical theory.* $\mathcal{B}^+$ is the subset of positive functions, i.e., the functions that have everywhere non-negative values. The physical interpretation of the states is that of *statistical distributions* on phase space. Without explaining details let us quickly go over the other structures. There is also a *measure* μ on L and the inner product on $\mathcal{B}$ is the corresponding L^2-inner product,

$$(\!|b, c|\!) = \int b(x)c(x)\mathrm{d}\mu(x). \tag{12}$$

The state $\mathbf{e}$ of maximal uncertainty is the constant function of unit value. (We assume that all functions are bounded.) The normalization condition $(\!|\mathbf{e}, b|\!)$ for a state $b \in \mathcal{B}^+$ means that it has unit integral. The operations are positive maps $\mathcal{B} \to \mathcal{B}$ which fall into classes with different physical interpretations. An interesting class is given by the particular *observables* that are characteristic functions χ_S for subsets $S \subseteq L$ of the phase space. This yields a selective operation as the map $b \mapsto \chi_S \cdot b$. The corresponding non-selective operation is the identity operator. This map encodes a measurement to determine whether the system is in the subset S. The probability for an affirmative answer is given by formula (9) and the corresponding update of the statistical distribution is the state update rule (11). Another interesting class of operations contains those that preserve the structure of $\mathcal{B}$ completely. It turns out that these are necessarily induced by measure preserving bijections $v : L \to L$ and take the form, $b \mapsto b \circ v^{-1}$. The time evolution maps are of this type. In particular, they arise thus from a one-parameter group of measure preserving bijections v_Δ in phase space. This flow in phase space can be described infinitesimally by a vector field. Usually there is a symplectic structure on phase space so that this is the Hamiltonian vector field of the Hamiltonian function.

7. Quantum theory

Another type of partially ordered vector space arises as the space of *self-adjoint operators* on a complex Hilbert space $\mathcal{H}$ with inner product denoted by $\langle\cdot,\cdot\rangle$. One way to characterize this type of space is as an *anti-lattice.* That is, for any two positive elements a minimum or maximum exists only in the trivial case when the elements are linearly dependent. Thus, the generalized state space $\mathcal{B}$ consists of self-adjoint operators on $\mathcal{H}$. $\mathcal{B}^+$ is the subset of *positive operators*, i.e., operators b such that $\langle b\psi, \psi\rangle \geq 0$ for all $\psi \in \mathcal{H}$. The inner product on $\mathcal{B}$ is then necessarily the Hilbert-Schmidt inner product,

$$(\!|b, c|\!) = \mathrm{tr}(bc), \tag{13}$$

and the state of maximal uncertainty is $\mathbf{e} = \mathrm{id}$, the identity operator (both up to suitable equivalence). The state normalization condition acquires the simple form $(\!|\mathbf{e}, b|\!) = \mathrm{tr}(b) = 1$ for $b \in \mathcal{B}^+$. In particular, the subset $\mathcal{B}^{\mathbf{n}} \subseteq \mathcal{B}^+$ is thus precisely the set of positive operators on $\mathcal{H}$ of unit trace. The set $\mathcal{P}^+$ of operations is the set of *completely positive maps* $\mathcal{B} \to \mathcal{B}$. (In the language of Section 4: A generalized probe is positive if and only if the associated probe map is completely positive.) In the present context they are called *quantum operations.*

IOP Conf. Series: Journal of Physics: Conf. Series **1275** (2019) 012049 doi:10.1088/1742-6596/1275/1/012049

Such an operation Q can be written in terms of Kraus operators $\{K_i\}_{i\in I}$ as, $Qb = \sum_i K_i b K_i^\dagger$. The normalization condition for non-selective operations (8), is easily seen to be equivalent to $\mathrm{tr}(\sum_i K_i^\dagger K_i) = 1$. The special case of a measurement given by an *observable* A, i.e., a self-adjoint operator on $\mathcal{H}$ is as follows. Let $\sum_i a_i P_i$ be the spectral decomposition of A in terms of orthogonal projectors P_i. (For simplicity we suppose a finite sum.) a_i are the distinct real numbers labeling the distinct measurement outcomes. The non-selective operation $Q(*)$ associated with this measurement is the map $b \mapsto \sum_i P_i b P_i$ while the selective operation $Q(k)$ selecting outcome k is, $b \mapsto P_k b P_k$. With this the probability (9) for outcome k in state $b \in \mathcal{B}^{\mathrm{n}}$ becomes,

$$\Pi = (\!|\mathbf{e}, Q(k)b|\!) = \mathrm{tr}(P_k b). \tag{14}$$

If b is a pure state, i.e., can be written as $b = |\psi\rangle\langle\psi|$ for $\psi \in \mathcal{H}$ in the usual bra-ket notation, then this is also the same as,

$$\Pi = \langle\psi, P_k\psi\rangle. \tag{15}$$

If P_k is a one-dimensional projector, i.e., of the form $P_k = |\eta_k\rangle\langle\eta_k|$, then this in turn can be written as,

$$\Pi = |\langle\eta_k, \psi\rangle|^2. \tag{16}$$

This is precisely the Born rule. The state transformation rule (11) in case of outcome k is,

$$b \mapsto \frac{P_k b P_k}{\mathrm{tr}(P_k b)}, \qquad \text{or} \qquad \psi \mapsto \frac{P_k\psi}{\sqrt{\langle\psi, P_k\psi\rangle}}, \tag{17}$$

in terms of pure states if applicable. The latter expression is justified by reproducing the former and involves a convenient, but irrelevant choice of phase. In particular, we recover the Lüders rule.

We proceed to consider operations that preserve the structure of the state space, including its inner product and state of maximal uncertainty. These have necessarily the form $b \mapsto UbU^\dagger$, where U is a unitary operator on $\mathcal{H}$. In particular, for the one-parameter group of time-evolution maps T_Δ this means that (fixing phases as necessary) there is a corresponding one-parameter group of unitary maps U_Δ such that $T_\Delta b = U_\Delta b U_\Delta^\dagger$. This satisfies $U_{\Delta_1} U_{\Delta_2} = U_{\Delta_1+\Delta_2}$ and $U_0 = \mathrm{id}$. By Stone's theorem there exists then a self-adjoint operator H on $\mathcal{H}$ such that $U_\Delta = \exp(-\mathrm{i}\Delta H)$, called the *Hamiltonian operator*. Adopting the notation $b(t) = T_t b = U_t b U_t^{-1}$ we can then write the time evolution of the state $b(t)$ as a differential equation,

$$\frac{\partial}{\partial t} b(t) = -\mathrm{i}[H, b(t)], \tag{18}$$

where $[\cdot,\cdot]$ denotes the commutator. This is precisely the *quantum Liouville equation*, i.e., the special case of the *Lindblad equation* for closed systems. For a pure state $b(t) = |\psi(t)\rangle\langle\psi(t)|$ we can set $|\psi(t)\rangle = U_t|\psi(0)\rangle$. In differential form,

$$\frac{\partial}{\partial t}|\psi(t)\rangle = -\mathrm{i}H|\psi(t)\rangle. \tag{19}$$

This is the Schrödinger equation.

Acknowledgments
This work was partially supported by CONACYT project grant 259258 and UNAM-PAPIIT project grant IA-106418.

References
[1] Oeckl R A local and operational framework for the foundations of physics to appear in Adv. Theor. Math. Phys. (*Preprint* 1610.09052)
[2] Davies E B and Lewis J T 1970 *Commun. Math. Phys.* **17** 239–260

IOP Conf. Series: Journal of Physics: Conf. Series **1275** (2019) 012050 doi:10.1088/1742-6596/1275/1/012050

Mathematics and physics of consistency of the Born rule with normal probability distribution

Alexey A. Kryukov

Department of Mathematics & Natural Sciences, University of Wisconsin-Milwaukee, USA

E-mail: kryukov@uwm.edu

Abstract. Newtonian and Scrödinger dynamics can be formulated in a physically meaningful way within the same Hilbert space framework. The resulting unexpected relation between classical and quantum motions goes beyond the results provided by the Ehrenfest theorem. The normal probability distribution and the Born rule turn out to be related. A dynamical mechanism responsible for the latter formula is proposed and applied to measurements of macroscopic and microscopic systems.

1. Introduction

In recent papers [1] and [2] an important new connection between classical and quantum dynamics was derived. The starting point was a realization of classical and quantum mechanics on an equal footing within the same Hilbert space framework, and identification of observables with vector fields on the sphere of normalized states. This resulted in a physically meaningful interpretation of the components of the velocity of state. Newtonian dynamics was shown to be the Schrödinger dynamics of a system whose state is constrained to the classical phase space submanifold in the Hilbert space of states. This also resulted in a formula relating the normal probability distribution and the Born rule, and interpretation of quantum collapse in terms of diffusion of the state on the projective space of states. Simply put, the classical space and classical phase space of a system of particles can be identified with a submanifold of the space of states of the corresponding quantum system. When the system is constrained to the submanifold, it behaves classically. Otherwise, it behaves quantum-mechanically. The velocity of the state at any point of the classical space submanifold can be decomposed into the classical (velocity, acceleration) and non-classical (phase velocity, spreading) components. The curvature of the sphere of states is determined from the canonical commutation relations.

In this paper, we continue to explore the implications of the proposed geometric framework. It has been known since Einstein that the thermal motion of molecules in a liquid results in the Brownian motion of pollen grain in the liquid, with probability distribution of the position of the grain satisfying the diffusion equation. It is shown here that the state of a microscopic particle exposed to the same random potential as the one experienced by the pollen grain is equally likely to get displaced in any direction tangent to the projective space of states. The relationship between the normal probability distribution and the Born rule that was established earlier signifies, then, that the probability density for the state to reach a particular point in the space of states is given by the Born rule. A diffusion equation for the motion of the state in these conditions is then obtained.

Content from this work may be used under the terms of the Creative Commons Attribution 3.0 licence. Any further distribution of this work must maintain attribution to the author(s) and the title of the work, journal citation and DOI.
Published under licence by IOP Publishing Ltd

IOP Conf. Series: Journal of Physics: Conf. Series **1275** (2019) 012050 doi:10.1088/1742-6596/1275/1/012050

In this model, the role of the measuring device in measurements on macroscopic and microscopic particles is similar. In both cases, the device is designed to record values of the measured quantity and is responsible for a noise and the resulting distribution of these values. The effect of the noise can be modeled by a diffusion equation. In the case of a measurement on a macroscopic particle, the diffusion equation is the familiar equation on the classical space $\mathbb{R}^3$. In the case of a measurement on a microscopic particle, the equation describes the random motion of the state of the particle and represents an extension of the diffusion equation from the classical space onto the space of states. Under the diffusion, the state of a microscopic particle can reach the classical space submanifold in the space of states and trigger a detector. The probability of reaching a particular point of the submanifold is given by the Born rule.

To make the paper self-contained, we begin with a review of the results reported in [1] and [2].

2. The classical mechanics in the language of quantum states

Experience shows that macroscopic bodies have a well-defined position in space at any time. In quantum mechanics the state of a spinless particle with a known position $\mathbf{a} \in \mathbb{R}^3$ is described by the Dirac delta function $\delta^3_{\mathbf{a}}(\mathbf{x}) = \delta^3(\mathbf{x}-\mathbf{a})$. The map $\omega : \mathbf{a} \longrightarrow \delta^3_{\mathbf{a}}$ provides a one-to-one correspondence between points $\mathbf{a} \in \mathbb{R}^3$ and state "functions" $\delta^3_{\mathbf{a}}$. This allows us to describe points in $\mathbb{R}^3$ in functional terms and identify the set $\mathbb{R}^3$ with the set M_3 of all delta functions in the space of state functions of the particle.

The common Hilbert space $L_2(\mathbb{R}^3)$ of state functions of a particle does not contain delta functions. By writing the inner product of functions $\varphi, \psi \in L_2(\mathbb{R}^3)$ as

$$(\varphi, \psi)_{L_2} = \int \delta^3(\mathbf{x}-\mathbf{y})\varphi(\mathbf{x})\overline{\psi}(\mathbf{y})d^3\mathbf{x}d^3\mathbf{y} \tag{1}$$

and approximating the kernel $\delta^3(\mathbf{x}-\mathbf{y})$ with a Gaussian function, one obtains a new inner product in $L_2(\mathbb{R}^3)$

$$(\varphi, \psi)_{\mathbf{H}} = \int e^{-\frac{(\mathbf{x}-\mathbf{y})^2}{8\sigma^2}}\varphi(\mathbf{x})\overline{\psi}(\mathbf{y})d^3\mathbf{x}d^3\mathbf{y}. \tag{2}$$

The Hilbert space $\mathbf{H}$ obtained by completing $L_2(\mathbb{R}^3)$ with respect to this inner product contains delta functions and their derivatives. In particular,

$$\int e^{-\frac{(\mathbf{x}-\mathbf{y})^2}{8\sigma^2}}\delta^3(\mathbf{x}-\mathbf{a})\delta^3(\mathbf{y}-\mathbf{a})d^3\mathbf{x}d^3\mathbf{y} = 1. \tag{3}$$

It follows that the set M_3 of all delta functions $\delta^3_{\mathbf{a}}(\mathbf{x})$ with $\mathbf{a} \in \mathbb{R}^3$ form a submanifold of the unit sphere in the Hilbert space $\mathbf{H}$, diffeomorphic to $\mathbb{R}^3$.

The kernel $\delta^3(\mathbf{x}-\mathbf{y})$ of the metric on $L_2(\mathbb{R}^3)$ is analogous to the Kronecker delta δ_{ik}, representing Euclidean metric in orthogonal coordinates. The "skewed" kernel $e^{-\frac{(\mathbf{x}-\mathbf{y})^2}{8\sigma^2}}$ of the metric on $\mathbf{H}$ is then analogous to the Euclidean metric represented in linear coordinates with skewed axes by a constant non-diagonal matrix g_{ik}.

The map $\rho_\sigma : \mathbf{H} \longrightarrow L_2(\mathbb{R}^3)$ that relates L_2 and $\mathbf{H}$-representations is given by the Gaussian kernel

$$\rho_\sigma(\mathbf{x}, \mathbf{y}) = \left(\frac{1}{2\pi\sigma^2}\right)^{3/4} e^{-\frac{(\mathbf{x}-\mathbf{y})^2}{4\sigma^2}}. \tag{4}$$

In fact, multiplying the operators (integrating the product of the corresponding kernels) one can see that

$$k(\mathbf{x}, \mathbf{y}) = (\rho^*_\sigma\rho_\sigma)(\mathbf{x}, \mathbf{y}) = e^{-\frac{(\mathbf{x}-\mathbf{y})^2}{8\sigma^2}}, \tag{5}$$

IOP Conf. Series: Journal of Physics: Conf. Series **1275** (2019) 012050 doi:10.1088/1742-6596/1275/1/012050

which is consistent with (2). The map ρ_σ transforms delta functions $\delta^3_{\mathbf{a}}$ to Gaussian functions $\widetilde{\delta}^3_{\mathbf{a}} = \rho_\sigma(\delta^3_{\mathbf{a}})$, centered at $\mathbf{a}$. The image M_3^σ of M_3 under ρ_σ is an embedded submanifold of the unit sphere in $L_2(\mathbb{R}^3)$ made of the functions $\widetilde{\delta}^3_{\mathbf{a}}$. The map $\omega_\sigma = \rho_\sigma \circ \omega : \mathbb{R}^3 \longrightarrow M_3^\sigma$ is a diffeomorphism. Here ω is the same as before. In what follows, the obtained realizations will be used interchangeably.

Let $\mathbf{r} = \mathbf{a}(t)$ be a path with values in $\mathbb{R}^3$ and let $\varphi = \delta^3_{\mathbf{a}(t)}$ be the corresponding path in M_3. It is easy to see that the norm $\left\|\frac{d\varphi}{dt}\right\|^2_H$ of the velocity in the space $\mathbf{H}$ is given by

$$\left\|\frac{d\varphi}{dt}\right\|^2_H = \left.\frac{\partial^2 k(\mathbf{x},\mathbf{y})}{\partial x^i \partial y^k}\right|_{\mathbf{x}=\mathbf{y}=\mathbf{a}} \frac{da^i}{dt}\frac{da^k}{dt}. \tag{6}$$

Here $k(\mathbf{x},\mathbf{y}) = e^{-\frac{(\mathbf{x}-\mathbf{y})^2}{8\sigma^2}}$ as in (5), so that

$$\left.\frac{\partial^2 k(\mathbf{x},\mathbf{y})}{\partial x^i \partial y^k}\right|_{\mathbf{x}=\mathbf{y}=\mathbf{a}} = \frac{1}{4\sigma^2}\delta_{ik}, \tag{7}$$

where δ_{ik} is the Kronecker delta symbol. Assuming now that the distance in $\mathbb{R}^3$ is measured in the units of 2σ, we obtain

$$\left\|\frac{d\varphi}{dt}\right\|_H = \left\|\frac{d\mathbf{a}}{dt}\right\|_{\mathbb{R}^3}. \tag{8}$$

It follows that the map $\omega : \mathbb{R}^3 \longrightarrow \mathbf{H}$ is an isometric embedding. Furthermore, the set M_3 is complete in $\mathbf{H}$ so that there is no vector in $\mathbf{H}$ orthogonal to all of M_3. By defining the operations of addition $\oplus$ and multiplication by a scalar $\lambda\odot$ via $\omega(\mathbf{a})\oplus\omega(\mathbf{b}) = \omega(\mathbf{a}+\mathbf{b})$ and $\lambda\odot\omega(\mathbf{a}) = \omega(\lambda\mathbf{a})$ with ω as before, we obtain M_3 as a vector space isomorphic to the Euclidean space $\mathbb{R}^3$.

The projection of velocity and acceleration of the state $\delta^3_{\mathbf{a}(t)}$ onto the Euclidean space M_3 yields correct Newtonian velocity and acceleration of the classical particle:

$$\left(\frac{d}{dt}\delta^3_{\mathbf{a}}(\mathbf{x}), -\frac{\partial}{\partial x^i}\delta^3_{\mathbf{a}}(\mathbf{x})\right)_{\mathbf{H}} = \frac{da^i}{dt} \tag{9}$$

and

$$\left(\frac{d^2}{dt^2}\delta^3_{\mathbf{a}}(\mathbf{x}), -\frac{\partial}{\partial x^i}\delta^3_{\mathbf{a}}(\mathbf{x})\right)_{\mathbf{H}} = \frac{d^2a^i}{dt^2}. \tag{10}$$

The Newtonian dynamics of the classical particle can be derived from the principle of least action for the action functional S on paths in $\mathbf{H}$, defined by

$$\int k(\mathbf{x},\mathbf{y})\left[\frac{m}{2}\frac{d\varphi_t(\mathbf{x})}{dt}\frac{d\overline{\varphi}_t(\mathbf{y})}{dt} - V(\mathbf{x})\varphi_t(\mathbf{x})\overline{\varphi}_t(\mathbf{y})\right]d^3\mathbf{x}d^3\mathbf{y}dt, \tag{11}$$

where m is the mass of the particle, V is the potential and $k(\mathbf{x},\mathbf{y}) = e^{-\frac{1}{2}(\mathbf{x}-\mathbf{y})^2}$ (same as in (5) with $2\sigma = 1$; see (8)). Namely, under the constraint $\varphi_t(\mathbf{x}) = \delta^3(\mathbf{x}-\mathbf{a}(t))$ the action (11) becomes

$$S = \int\left[\frac{m}{2}\left(\frac{d\mathbf{a}}{dt}\right)^2 - V(\mathbf{a})\right]dt, \tag{12}$$

which is the classical action functional for the particle.

This shows that a classical particle can be considered a constrained dynamical system with the state φ of the particle and the velocity $\frac{d\varphi}{dt}$ of the state as dynamical variables. A similar

IOP Conf. Series: Journal of Physics: Conf. Series **1275** (2019) 012050 doi:10.1088/1742-6596/1275/1/012050

realization exists for mechanical systems consisting of any number of classical particles. For example, the map $\omega \otimes \omega : \mathbb{R}^3 \times \mathbb{R}^3 \longrightarrow \mathbf{H} \otimes \mathbf{H}$, $\omega \otimes \omega(\mathbf{a}, \mathbf{b}) = \delta^3_{\mathbf{a}} \otimes \delta^3_{\mathbf{b}}$ identifies the configuration space $\mathbb{R}^3 \times \mathbb{R}^3$ of a two particle system with the embedded submanifold $M_6 = \omega \otimes \omega(\mathbb{R}^3 \times \mathbb{R}^3)$ of the Hilbert space $\mathbf{H} \otimes \mathbf{H}$. Consider a path $(\mathbf{a}(t), \mathbf{b}(t))$ in $\mathbb{R}^3 \times \mathbb{R}^3$ and the corresponding path $\delta^3_{\mathbf{a}(t)} \otimes \delta^3_{\mathbf{b}(t)}$ with values in M_6. For any t, the vectors $\frac{d}{dt}\delta^3_{\mathbf{a}(t)} \otimes \delta^3_{\mathbf{b}(t)}$ and $\delta^3_{\mathbf{a}(t)} \otimes \frac{d}{dt}\delta^3_{\mathbf{b}(t)}$ are tangent to M_6 at the point $\delta^3_{\mathbf{a}(t)} \otimes \delta^3_{\mathbf{b}(t)}$ and orthogonal in $\mathbf{H} \otimes \mathbf{H}$. The space M_6 with the induced metric is isometric to the direct product $\mathbb{R}^3 \times \mathbb{R}^3$ with the natural Euclidean metric. Projection of velocity and acceleration of the state $\varphi(t) = \delta^3_{\mathbf{a}(t)} \otimes \delta^3_{\mathbf{b}(t)}$ onto the basis vectors $\left(-\frac{\partial}{\partial x^i}\delta^3_{\mathbf{a}(t)}\right) \otimes \delta^3_{\mathbf{b}(t)}$ and $\delta^3_{\mathbf{a}(t)} \otimes \left(-\frac{\partial}{\partial x^k}\delta^3_{\mathbf{b}(t)}\right)$ yields the velocity and acceleration of the particles by means of the formulas similar to (9) and (10).

3. Observables as vector fields

Quantum observables can be identified with vector fields on the space of states. Given a self-adjoint operator $\widehat{A}$ on a Hilbert space L_2 of square-integrable functions (it could in particular be the tensor product space of a many body problem) one can introduce the associated linear vector field A_φ on L_2 by

$$A_\varphi = -i\widehat{A}\varphi. \tag{13}$$

If D is the domain of the operator $\widehat{A}$, then A_φ maps D into the vector space L_2. Because $\widehat{A}$ is self-adjoint, the field A_φ, being restricted to the sphere S^{L_2} of unit normalized states, is tangent to the sphere. The commutator of observables and the commutator (Lie bracket) of the corresponding vector fields are related in a simple way:

$$[A_\varphi, B_\varphi] = [\widehat{A}, \widehat{B}]\varphi. \tag{14}$$

Furthermore, a Hilbert metric on the space of states yields a Riemannian metric on the sphere. For this consider the realization $L_{2\mathbb{R}}$ of the Hilbert space L_2, i.e., the real vector space of pairs $X = (\mathrm{Re}\psi, \mathrm{Im}\psi)$ with ψ in L_2. If ξ, η are vector fields on S^{L_2}, define a Riemannian metric $G_\varphi : T_{\mathbb{R}\varphi}S^{L_2} \times T_{\mathbb{R}\varphi}S^{L_2} \longrightarrow \mathbb{R}$ on the sphere by

$$G_\varphi(X, Y) = \mathrm{Re}(\xi, \eta). \tag{15}$$

Here $X = (\mathrm{Re}\xi, \mathrm{Im}\xi)$, $Y = (\mathrm{Re}\eta, \mathrm{Im}\eta)$ and (ξ, η) denotes the L_2-inner product of ξ, η.

The Riemannian metric on S^{L_2} yields a Riemannian (Fubini-Study) metric on the projective space CP^{L_2}, which is the base of the fibration $\pi : S^{L_2} \longrightarrow CP^{L_2}$. For this an arbitrary tangent vector $X \in T_{R\varphi}S^{L_2}$ is decomposed into two components: tangent and orthogonal to the fibre $\{\varphi\}$ through φ (i.e., to the plane C^1 containing the circle $S^1 = \{\varphi\}$). The differential $d\pi$ maps the tangential component to zero-vector. The orthogonal component of X can be then identified with $d\pi(X)$. If two vectors X, Y are orthogonal to the fibre $\{\varphi\}$, the inner product of $d\pi(X)$ and $d\pi(Y)$ in the Fubini-Study metric is equal to the inner product of X and Y in the metric G_φ:

$$(d\pi(X), d\pi(Y))_{FS} = G_\varphi(X, Y). \tag{16}$$

The resulting metrics can be used to find physically meaningful components of vector fields A_φ associated with observables. Since A_φ is tangent to S^{L_2}, it can be decomposed into components tangent and orthogonal to the fibre $\{\varphi\}$ of the fibre bundle $\pi : S^{L_2} \longrightarrow CP^{L_2}$. These components have a simple physical meaning, justifying the use of the projective space CP^{L_2}. From

$$\overline{A} \equiv (\varphi, \widehat{A}\varphi) = (-i\varphi, -i\widehat{A}\varphi), \tag{17}$$

one can see that the expected value of an observable $\widehat{A}$ in state φ is the projection of the vector $-i\widehat{A}\varphi \in T_\varphi S^{L_2}$ onto the fibre $\{\varphi\}$. Because

$$(\varphi, \widehat{A}^2\varphi) = (\widehat{A}\varphi, \widehat{A}\varphi) = (-i\widehat{A}\varphi, -i\widehat{A}\varphi), \tag{18}$$

the term $(\varphi, \widehat{A}^2\varphi)$ is the norm of the vector $-i\widehat{A}\varphi$ squared. The vector $-i\widehat{A}_\perp\varphi = -i\widehat{A}\varphi - (-i\overline{A}\varphi)$ associated with the operator $\widehat{A} - \overline{A}I$ is orthogonal to the fibre $\{\varphi\}$. Accordingly, the variance

$$\Delta A^2 = (\varphi, (\widehat{A} - \overline{A}I)^2\varphi) = (\varphi, \widehat{A}_\perp^2\varphi) = (-i\widehat{A}_\perp\varphi, -i\widehat{A}_\perp\varphi) \tag{19}$$

is the norm squared of the component $-i\widehat{A}_\perp\varphi$. Recall that the image of this vector under $d\pi$ can be identified with the vector itself. It follows that the norm of $-i\widehat{A}_\perp\varphi$ in the Fubini-Study metric coincides with its norm in the Riemannian metric on S^{L_2} and in the original L_2-metric.

The Schrödinger equation

$$\frac{d\varphi}{dt} = -i\widehat{h}\varphi \tag{20}$$

is an equation for the integral curves of the vector field $-i\widehat{h}\varphi$ on the sphere S^{L_2}. Let's decompose $-i\widehat{h}\varphi$ onto the components parallel and orthogonal to the fibre. The parallel component of $\frac{d\varphi}{dt}$ is numerically

$$Re(-i\varphi, -i\widehat{h}\varphi) = \overline{E}, \tag{21}$$

i.e., the expected value of the energy. The decomposition of the velocity vector $\frac{d\varphi}{dt}$ into the parallel and orthogonal components is then given by

$$\frac{d\varphi}{dt} = -i\overline{E}\varphi + -i(\widehat{h} - \overline{E})\varphi = -i\overline{E}\varphi - i\widehat{h}_\perp\varphi. \tag{22}$$

The orthogonal component of the velocity $\frac{d\varphi}{dt}$ is equal to $-i\widehat{h}_\perp\varphi$. From this and equation (19) we conclude that: *The velocity of evolution of state in the projective space is equal to the uncertainty of energy.* Equation (22) also demonstrates that the physical state is driven by the operator $\widehat{h}_\perp$, associated with the uncertainty in energy rather than the energy itself.

The uncertainty relation

$$\Delta A \Delta B \geq \frac{1}{2}\left|\left(\varphi, [\widehat{A}, \widehat{B}]\varphi\right)\right| \tag{23}$$

follows geometrically from the comparison of areas of rectangle $A_{|XY|}$ and parallelogram A_{XY} formed by vectors $X = -i\widehat{A}_\perp\varphi$ and $Y = -i\widehat{B}_\perp\varphi$:

$$A_{|XY|} \geq A_{XY}. \tag{24}$$

There is also an uncertainty identity, [3]:

$$\Delta A^2 \Delta B^2 = A_{XY}^2 + G_\varphi^2(X, Y). \tag{25}$$

4. Components of the velocity of state under the Schrödinger evolution

From (22) we know that for any state $\varphi \in S^{L_2}$ the velocity of state $\frac{d\varphi}{dt}$ in the Schrödinger equation can be decomposed onto the components parallel and orthogonal to the fibre $\{\varphi\}$ of the bundle $\pi : S^{L_2} \longrightarrow CP^{L_2}$:

$$\frac{d\varphi}{dt} = -i\overline{E}\varphi - i\widehat{h}_\perp\varphi. \tag{26}$$

The norm of the parallel component $-i\overline{E}\varphi$ is the expected value of energy $\overline{E}$. It represents the phase velocity of state. The norm of the orthogonal component $-i\widehat{h}_\perp\varphi$ is equal to the uncertainty of energy ΔE on the state φ. It represents the velocity of motion of the fibre $\{\varphi\}$. In particular, from (26) it follows that under the Schrödinger evolution the speed of evolution of state in the projective space is equal to the uncertainty in energy.

The orthogonal component $-i\widehat{h}_{\perp}\varphi$ of the velocity can be further decomposed into physically meaningful components. Suppose that at $t=0$, a microscopic particle is prepared in the state $\varphi_0 \equiv \varphi_{\mathbf{a},\mathbf{p}}$ given by

$$\varphi_{\mathbf{a},\mathbf{p}}(\mathbf{x}) = \left(\frac{1}{2\pi\sigma^2}\right)^{3/4} e^{-\frac{(\mathbf{x}-\mathbf{a})^2}{4\sigma^2}} e^{i\frac{\mathbf{p}(\mathbf{x}-\mathbf{a})}{\hbar}}, \tag{27}$$

where σ is the same as in (4) and $\mathbf{p} = m\mathbf{v}_0$ with $\mathbf{v}_0$ being the initial group-velocity of the packet. The set $M^{\sigma}_{3,3}$ of all initial states $\varphi_{\mathbf{a},\mathbf{p}}$ given by (27) form a 6-dimensional embedded submanifold in $L_2(\mathbb{R}^3)$. Consider the set of all fibres of the bundle $\pi : S^{L_2} \longrightarrow CP^{L_2}$ through the points of $M^{\sigma}_{3,3}$. The resulting bundle $\pi : M^{\sigma}_{3,3} \times S^1 \longrightarrow M^{\sigma}_{3,3}$ identifies $M^{\sigma}_{3,3}$ with a submanifold of CP^{L_2}, denoted by the same symbol. The map $\Omega : \mathbb{R}^3 \times \mathbb{R}^3 \longrightarrow M^{\sigma}_{3,3}$,

$$\Omega(\mathbf{a},\mathbf{p}) = \left(\frac{1}{2\pi\sigma^2}\right)^{3/4} e^{-\frac{(\mathbf{x}-\mathbf{a})^2}{4\sigma^2}} e^{i\frac{\mathbf{p}(\mathbf{x}-\mathbf{a})}{\hbar}} \tag{28}$$

is a diffeomorphism from the classical phase space of the particle onto $M^{\sigma}_{3,3}$. For $\Omega = re^{i\theta}$, where r is the modulus and θ is the argument of Ω, the vectors $\frac{\partial r}{\partial a^{\alpha}}e^{i\theta}$ and $i\frac{\partial \theta}{\partial p^{\beta}}re^{i\theta}$ are orthogonal to the fibre $\{\varphi_0\}$ at the point φ_0 in $L_2(\mathbb{R}^3)$. These vectors can be then identified with vectors tangent to $M^{\sigma}_{3,3}$, considered as a submanifold of CP^{L_2}. They form a basis in the tangent space $T_{\{\varphi_0\}}M^{\sigma}_{3,3}$. Furthermore, the induced Riemannian metric is the usual Fubini-Study metric on CP^{L_2}, restricted to $M^{\sigma}_{3,3}$.

For any path $\{\varphi\} = \{\varphi_\tau\}$ with values in $M^{\sigma}_{3,3} \subset CP^{L_2}$ the norm of velocity vector $\frac{d\{\varphi\}}{d\tau}$ in the Fubini-Study metric is given by

$$\left\|\frac{d\{\varphi\}}{d\tau}\right\|^2_{FS} = \frac{1}{4\sigma^2}\left\|\frac{d\mathbf{a}}{d\tau}\right\|^2_{\mathbb{R}^3} + \frac{\sigma^2}{\hbar^2}\left\|\frac{d\mathbf{p}}{d\tau}\right\|^2_{\mathbb{R}^3}. \tag{29}$$

It follows that under a proper choice of units, the map Ω is an isometry that identifies the Euclidean phase space $\mathbb{R}^3 \times \mathbb{R}^3$ of the particle with the submanifold $M^{\sigma}_{3,3} \subset CP^{L_2}$ furnished with the induced metric. The map Ω is an extension to the phase space of the isometric embedding $\omega_\sigma = \rho_\sigma \circ \omega$ of the space $\mathbb{R}^3$, introduced in the section entitled "The classical mechanics in the language of quantum states".

The obtained embedding of the classical phase space into the space of quantum states is physically meaningful. To see this let us first project the orthogonal component $-\frac{i}{\hbar}\widehat{h}_{\perp}\varphi$ of the velocity $\frac{d\varphi}{dt}$ onto vectors tangent to the curves of constant values of $\mathbf{p}$ and $\mathbf{a}$ (classical space and momentum space components) in the projective manifold $M^{\sigma}_{3,3}$. Calculation of the projection of the velocity $\frac{d\varphi}{dt}$ onto the unit vector $-\widehat{\frac{\partial r}{\partial a^{\alpha}}e^{i\theta}}$ (i.e., the classical space component of $\frac{d\varphi}{dt}$) for an arbitrary Hamiltonian of the form $\widehat{h} = -\frac{\hbar^2}{2m}\Delta + V(\mathbf{x})$ yields

$$\mathrm{Re}\left(\frac{d\varphi}{dt}, -\widehat{\frac{\partial r}{\partial a^{\alpha}}e^{i\theta}}\right)\Bigg|_{t=0} = \left(\frac{dr}{dt}, -\widehat{\frac{\partial r}{\partial a^{\alpha}}}\right)\Bigg|_{t=0} = \frac{v_0^{\alpha}}{2\sigma}. \tag{30}$$

Calculation of the projection of velocity $\frac{d\varphi}{dt}$ onto the unit vector $i\widehat{\frac{\partial\theta}{\partial p^{\alpha}}\varphi}$ (momentum space component) gives

$$\mathrm{Re}\left(\frac{d\varphi}{dt}, i\widehat{\frac{\partial\theta}{\partial p^{\alpha}}\varphi}\right)\Bigg|_{t=0} = \frac{mw^{\alpha}\sigma}{\hbar}, \tag{31}$$

IOP Conf. Series: Journal of Physics: Conf. Series **1275** (2019) 012050 doi:10.1088/1742-6596/1275/1/012050

where

$$mw^{\alpha} = -\left.\frac{\partial V(\mathbf{x})}{\partial x^{\alpha}}\right|_{\mathbf{x}=\mathbf{x}_0} \tag{32}$$

and σ is assumed to be small enough for the linear approximation of $V(\mathbf{x})$ to be valid within intervals of length σ.

The velocity $\frac{d\varphi}{dt}$ also contains component due to the change in σ (spreading), which is orthogonal to the fibre $\{\varphi\}$ and the phase space $M^{\sigma}_{3,3}$, and is equal to

$$\mathrm{Re}\left(\frac{d\varphi}{dt}, i\widehat{\frac{d\varphi}{d\sigma}}\right)\Big|_{t=0} = \frac{\sqrt{2}\hbar}{8\sigma^2 m}. \tag{33}$$

Calculation of the norm of $\frac{d\varphi}{dt} = \frac{i}{\hbar}\widehat{h}\varphi$ at $t=0$ gives

$$\left\|\frac{d\varphi}{dt}\right\|^2 = \frac{\overline{E}^2}{\hbar^2} + \frac{\mathbf{v}_0^2}{4\sigma^2} + \frac{m^2\mathbf{w}^2\sigma^2}{\hbar^2} + \frac{\hbar^2}{32\sigma^4 m^2}, \tag{34}$$

which is the sum of squares of the found components. This completes a decomposition of the velocity of state at any point $\varphi_{\mathbf{a},\mathbf{p}} \in M^{\sigma}_{3,3}$.

For a closed system the norm of $\frac{d\varphi}{dt} = \frac{i}{\hbar}\widehat{h}\varphi$ is preserved in time. For a system in a stationary state, this amounts to conservation of energy. In fact, in this case $\varphi_t(\mathbf{x}) = \psi(\mathbf{x})e^{-\frac{iEt}{\hbar}}$, which is a motion along the phase circle, and

$$\left\|\frac{d\varphi}{dt}\right\|^2 = \frac{E^2}{\hbar^2}. \tag{35}$$

For a closed system in any initial state the norm of the phase component (expected energy) and orthogonal component (energy uncertainty) of the velocity $\frac{d\varphi}{dt}$ are both preserved.

From (30) and (31) and a simple consistency check, showing that the rate of change of the projection in (30) is given by acceleration $\mathbf{w}$, one can see that the phase space components of the velocity of state $\frac{d\varphi}{dt}$ assume correct classical values at any point $\varphi_{\mathbf{a},\mathbf{p}} \in M^{\sigma}_{3,3}$. This remains true for the time dependent potentials as well. The immediate consequence of this and the linear nature of the Schrödinger equation is that under the Schrödinger evolution with the Hamiltonian $\widehat{h} = -\frac{\hbar^2}{2m}\Delta + V(\mathbf{x},t)$, the state constrained to $M^{\sigma}_{3,3} \subset CP^{L_2}$ moves like a point in the phase space representing a particle in Newtonian dynamics. More generally, Newtonian dynamics of n particles is the Schrödinger dynamics of n-particle quantum system whose state is constrained to the phase-space submanifold $M^{\sigma}_{3n,3n}$ of the projective space for $L_2(\mathbb{R}^3)\otimes \ldots \otimes L_2(\mathbb{R}^3)$, formed by tensor product states $\varphi_1 \otimes \ldots \otimes \varphi_n$ with φ_k of the form (27). On the contrary, there exists a unique extension of the Newtonian dynamics on the classical phase space submanifold to a unitary dynamics in the full Hilbert space [2].

Note again that the velocity and acceleration terms in (34) are orthogonal to the fibre $\{\varphi_{\mathbf{a},\mathbf{p}}\}$ of the fibration $\pi : S^{L_2} \longrightarrow CP^{L_2}$, showing that these Newtonian variables have to do with the motion in the projective space CP^{L_2}. The velocity of spreading is orthogonal to the fibre and to the phase space submanifold $M^{\sigma}_{3,3}$.

5. The Born rule and the normal probability distribution

The isometric embedding of the classical space M^{σ}_3 into the space of states $L_2(\mathbb{R}^3)$ results in a relationship between distances in $\mathbb{R}^3$ and in the projective space CP^{L_2}. The distance between two points $\mathbf{a}$ and $\mathbf{b}$ in $\mathbb{R}^3$ is $\|\mathbf{a}-\mathbf{b}\|_{\mathbb{R}^3}$. Under the embedding of the classical space into the space

IOP Conf. Series: Journal of Physics: Conf. Series **1275** (2019) 012050 doi:10.1088/1742-6596/1275/1/012050

of states, the variable $\mathbf{a}$ is represented by the state $\tilde{\delta}^3_{\mathbf{a}}$. The set of states $\tilde{\delta}^3_{\mathbf{a}}$ form a submanifold M^σ_3 in the Hilbert spaces of states $L_2(\mathbb{R}^3)$, which is "twisted" in $L_2(\mathbb{R}^3)$. It belongs to the sphere S^{L_2} and spans all dimensions of $L_2(\mathbb{R}^3)$. The distance between the states $\tilde{\delta}^3_{\mathbf{a}}$, $\tilde{\delta}^3_{\mathbf{b}}$ on the sphere S^{L_2} or in the projective space CP^{L_2} is not equal to $\|\mathbf{a}-\mathbf{b}\|_{\mathbb{R}^3}$. In fact, the former distance measures length of a geodesic between the states while the latter is obtained using the same metric on the space of states, but applied along a geodesic in the twisted manifold M^σ_3. The precise relation between the two distances is given by

$$e^{-\frac{(\mathbf{a}-\mathbf{b})^2}{4\sigma^2}} = \cos^2\theta(\tilde{\delta}^3_{\mathbf{a}}, \tilde{\delta}^3_{\mathbf{b}}), \tag{36}$$

where θ is the Fubini-Study distance between states in CP^{L_2}. The distance θ in the projective space of states CP^{L_2} appears here for a good reason: the fibres of the fibration $\pi : S^{L_2} \longrightarrow CP^{L_2}$ through the points of the classical space M^σ_3 are orthogonal to this space. This is why the distance in M^σ_3 can be expressed in terms of the distance in CP^{L_2}. Despite the non-trivial geometry contained in (36), the formula itself is easy to derive. The left hand side is the result of integration in $|(\tilde{\delta}^3_{\mathbf{a}}, \tilde{\delta}^3_{\mathbf{b}})|^2$. On the other hand, the same expression is equal to the right side of (36) by definition of the Fubini-Study metric.

The relation (36) has an immediate implication onto the form of probability distributions of random variables over M^σ_3 and CP^{L_2}. In particular, consider a random variable over CP^{L_2}. Suppose that the restricted random variable defined over $M^\sigma_3 = \mathbb{R}^3$ is distributed normally on $\mathbb{R}^3$. Then the direction-independent probability of transition between arbitrary quantum states must satisfy the Born rule. So, *The normal distribution law on* M^σ_3 *implies the Born rule on* CP^{L_2}. *Conversely, the Born rule implies the normal distribution law for states in* M^σ_3.

The fact that the Born rule implies the normal distribution on M^σ_3 is straightforward. According to the Born rule, the probability density $f_{\mathbf{a},\sigma}(\mathbf{b})$ to find the particle in a state $\tilde{\delta}^3_{\mathbf{a}}$ at a point $\mathbf{b}$ is equal to

$$f_{\mathbf{a},\sigma}(\mathbf{b}) = |\tilde{\delta}^3_{\mathbf{a}}(\mathbf{b})|^2 = |(\tilde{\delta}^3_{\mathbf{a}}, \delta^3_{\mathbf{b}})|^2 = \left(\frac{1}{2\pi\sigma^2}\right)^{3/2} e^{-\frac{(\mathbf{a}-\mathbf{b})^2}{2\sigma^2}}, \tag{37}$$

which is the normal distribution function. It follows that on the elements of M^σ_3, the Born rule *is* the rule of normal distribution.

The Born rule on M^σ_3 can be also written in term of the probability $P(\tilde{\delta}^3_{\mathbf{a}}, \tilde{\delta}^3_{\mathbf{b}})$ of transition between the states $\tilde{\delta}^3_{\mathbf{a}}, \tilde{\delta}^3_{\mathbf{b}}$ in M^σ_3:

$$P(\tilde{\delta}^3_{\mathbf{a}}, \tilde{\delta}^3_{\mathbf{b}}) = |(\tilde{\delta}^3_{\mathbf{a}}, \tilde{\delta}^3_{\mathbf{b}})|^2. \tag{38}$$

Assuming $\tilde{\delta}^3_{\mathbf{b}}$ is sufficiently sharp, the formulas (37) and (38) mean the same thing. In fact,

$$|(\tilde{\delta}^3_{\mathbf{a}}, \tilde{\delta}^3_{\mathbf{b}})|^2 = f_{\mathbf{a},\sqrt{2}\sigma}(\mathbf{b})(\Delta x)^3, \tag{39}$$

where $f_{\mathbf{a},\sqrt{2}\sigma}$ is the normal distribution function with variance $\sqrt{2}\sigma$ and $\Delta x = \sqrt{4\pi\sigma^2}$. This relates the probability in (38) to the normal probability density in (37) and identifies $P(\tilde{\delta}^3_{\mathbf{a}}, \tilde{\delta}^3_{\mathbf{b}})$ with the probability of finding the macroscopic particle near the point $\mathbf{b}$.

Conversely, suppose we have a rule for probability of transition between states in CP^{L_2} which gives the normal distribution law for the states in M^σ_3 and depends only on the distance between states. Let's show that this must be the Born rule. In fact, the Fubini-Study distance between the states $\tilde{\delta}^3_{\mathbf{a}}$, $\tilde{\delta}^3_{\mathbf{b}}$ takes on all values from 0 to $\pi/2$, which is the largest possible distance between points in CP^{L_2}. By assumption, the probability $P(\varphi, \psi)$ of transition between any states φ and

IOP Conf. Series: Journal of Physics: Conf. Series **1275** (2019) 012050 doi:10.1088/1742-6596/1275/1/012050

ψ depends only on the Fubini-Study distance $\theta(\pi(\varphi), \pi(\psi))$ between the states. Given arbitrary states $\varphi, \psi \in S^{L_2}$, let then $\tilde{\delta^3_{\mathbf{a}}}$, $\tilde{\delta^3_{\mathbf{b}}}$ be two states in M^3_σ, such that

$$\theta(\pi(\varphi), \pi(\psi)) = \theta(\tilde{\delta^3_{\mathbf{a}}}, \tilde{\delta^3_{\mathbf{b}}}). \quad (40)$$

It then follows that

$$P(\varphi, \psi) = P(\tilde{\delta^3_{\mathbf{a}}}, \tilde{\delta^3_{\mathbf{b}}}) = \cos^2 \theta(\tilde{\delta^3_{\mathbf{a}}}, \tilde{\delta^3_{\mathbf{b}}}) = \cos^2 \theta(\pi(\varphi), \pi(\psi)), \quad (41)$$

which yields the Born rule for arbitrary states. This proves the claim and completes a review of [1] and [2].

6. Measurements on macroscopic and microscopic particles

We are now in a position to compare the process of measurement in the classical and quantum physics. First of all, the classical space and phase space are now submanifolds in the Hilbert space of states. This allows us to use the same language when analyzing both types of measurement. Second, the Newtonian dynamics is now a restriction of the Schrödinger dynamics to the classical phase space submanifold. Conversely, the Schrödinger dynamics is a unique unitary extension of the Newtonian dynamics from the classical phase space to the Hilbert space. This allows us to begin with a model of measurement satisfying Newton laws and extend it to a model consistent with the rules of quantum mechanics. Finally, the normal probability law is the restriction of the Born rule to the classical space submanifold. Conversely, the Born rule is the unique isotropic extension of the normal probability law from the classical space to the space of states. In particular, a classical model of measurement with a normal distribution of the measured quantity should lead us to a model consistent with the Schrödinger dynamics and the Born rule.

Measurements performed on a macroscopic particle satisfy generically the normal distribution law for the measured observable. This is consistent with the central limit theorem and indicates that a specific way in which the observable was measured is not important. To be specific, consider measurements of position of a particle. A common way of finding the position of a macroscopic particle is to expose it to light of sufficiently short wavelength and to observe the scattered photons. Due to the unknown path of the incident photons, multiple scattering events on the particle, random change in position of the particle, etc., the process of observation can be described by the diffusion equation with the observed position of the particle experiencing Brownian motion from an initial point during the time of observation. This results in the normal distribution of observed position of the particle.

The ability to describe measurement of position of a macroscopic particle as a diffusion seems to be a general feature of measurements in the macro-world, independent of a particular measurement set up. The averaging process making the central limit theorem applicable and leading to the normal distribution of the position random variable can be seen as the result of random hits experienced by the particle from the surrounding particles participating in the measurement. These random hits are equally likely to come from any direction, independent of the initial position of the particle, leading to Brownian motion and the validity of the diffusion equation for the probability density of the position random variable for the particle.

It is claimed now that at any time t the initial state ψ of a microscopic particle undergoing a position measurement is equally likely to shift in any direction in the tangent space $T_{\{\psi\}}CP^{L_2}$. Suppose, for example, that the particle is exposed to a stream of photons of sufficiently high frequency and number density. The scattered photons are then observed to determine the position of the particle. The field of photons in the experiment will be treated classically, as a fluctuating potential in a region surrounding the source. The assumptions about the potential that will be made will determine to what extent the derived results can be applied to different experiments of measuring position of a microscopic particle.

Recall that the space $M^{\sigma}_{3,3}$ is complete in $L_2(\mathbb{R}^3)$. Consider the subset of $M^{\sigma}_{3,3}$ formed by the states

$$\varphi_{\mathbf{mn}}(\mathbf{x}) = \left(\frac{1}{2\pi\sigma^2}\right)^{3/4} e^{-\frac{(\mathbf{x}-\alpha\mathbf{n})^2}{4\sigma^2}} e^{i\frac{\beta\mathbf{mx}}{\hbar}}, \tag{42}$$

where $\alpha = \sqrt{2\pi}\sigma$, $\beta = \frac{h}{\sqrt{2\pi}\sigma}$ and $\mathbf{m}, \mathbf{n}$ take values on the lattice $\mathbb{Z}^3 \times \mathbb{Z}^3$ of points with integer coordinates in $\mathbb{R}^3 \times \mathbb{R}^3$. The set of functions (42) is known to be also complete in $L_2(\mathbb{R}^3)$. Any state in $L_2(\mathbb{R}^3)$ can be then represented by a linear combination of states $\varphi_{\mathbf{mn}}$. (For $\alpha\beta < h$ the system of functions $\varphi_{\mathbf{mn}}$ is called the Gabor or Weil-Heisenberg frame.) In particular, the initial state ψ of the particle can be represented by a sum

$$\psi = \sum_{\mathbf{m},\mathbf{n}} C_{\mathbf{mn}}\varphi_{\mathbf{mn}}. \tag{43}$$

The set M^{σ}_3 is also complete in $L_2(\mathbb{R}^3)$. Here too there exist countable subsets of M^{σ}_3 that are complete in $L_2(\mathbb{R}^3)$. Moreover, an arbitrary initial state ψ in $L_2(\mathbb{R}^3)$ can be approximated as well as necessary by a finite discrete sum

$$\psi \approx \sum_{\mathbf{n}} C_{\mathbf{n}}\widetilde{\delta}^3_{\mathbf{a}-\gamma\mathbf{n}}, \tag{44}$$

where $\mathbf{a}$ is arbitrary, $\mathbf{n} \in \mathbb{Z}^3$, and the value of $\gamma > 0$ together with the number of terms in the sum depend on ψ and the needed approximation. Taking γ sufficiently small, let's partition the space $\mathbb{R}^3$ into the cubes of edge γ centered at the lattice points $\gamma\mathbf{n}$ and consider the indicator function $1_{\mathbf{n}}$ for each cube. The potential $\widehat{V}$ can be written as a sum $\sum_{\mathbf{n}} 1_{\mathbf{n}}\widehat{V}_{\mathbf{n}}$. The components $\widehat{V}_{\mathbf{n}}$ for different $\mathbf{n}$ can be assumed to be independent, identically distributed random variables. In the case of position measurement by scattering photons off the particle, the components $\widehat{V}_{\mathbf{n}}$ can be associated with a single photon at time t.

For simplicity, let's neglect the kinetic energy term in the Hamiltonian $\widehat{h}$. Let us denote the solution of the Shrödinger equation with the initial state ψ by $\chi(t)$ and let us write $\chi(t) = e^{-\frac{i\overline{V}t}{\hbar}}\psi(t)$, where $\overline{V} = (\widehat{V}\psi, \psi)$ and $\psi(0) = \psi$. We then have at $t = 0$:

$$\frac{d\psi}{dt} = -\frac{i}{\hbar}\widehat{V}_{\perp}\psi, \tag{45}$$

where $\widehat{V}_{\perp} = \widehat{V} - \overline{V}$, as before. This equation gives the velocity of the state $\chi(t)$ in the projective space CP^{L_2} at $t = 0$. To prove that under the action of $\widehat{V}_{\perp}$ all directions of velocity of state in $T_{\psi}CP^{L_2}$ are equally likely, consider the components of the velocity in the basis $-i\widetilde{\delta}^3_{\mathbf{m}} \equiv -i\widetilde{\delta}^3_{\mathbf{a}-\gamma\mathbf{m}}$

$$\left(\frac{d\psi}{dt}, -i\widetilde{\delta}^3_{\mathbf{m}}\right) = \frac{1}{\hbar}(\widehat{V}_{\perp}\psi, \widetilde{\delta}^3_{\mathbf{m}}). \tag{46}$$

For any given potential and a given ψ, the form in (46) is a function of the distance between the points ψ and $\widetilde{\delta}^3_{\mathbf{m}}$ in the Fubini-Study metric and, possibly, of the vector $\eta \in T_{\psi}CP^{L_2}$ of a fixed norm, tangent to the geodesic from $\{\psi\}$ to $\widetilde{\delta}^3_{\mathbf{m}}$ in CP^{L_2}. We have $\widehat{V} = \sum_{\mathbf{n}} 1_{\mathbf{n}}\widehat{V}_{\mathbf{n}}$, where each $\widehat{V}_{\mathbf{n}}$ is a random variable. Accordingly, (46) defines a random variable for each $\mathbf{m}$. Dividing these variables by $|(\psi, \widetilde{\delta}^3_{\mathbf{m}})|$ we obtain a new set of random variables

$$\frac{\frac{1}{\hbar}(\widehat{V}_{\perp}\psi, \widetilde{\delta}^3_{\mathbf{m}})}{|(\psi, \widetilde{\delta}^3_{\mathbf{m}})|}. \tag{47}$$

Because $|(\psi, \widetilde{\delta}^3_{\mathbf{m}})|$ depends only on the distance between $\{\psi\}$ and $\widetilde{\delta}^3_{\mathbf{m}}$, the probability distributions of the random variables given by (46) and (47) are either both dependent or both independent of η. Provided the potential does not change much within each cube, up to a constant phase factor, the expression (47) is equal to

$$\frac{1}{\hbar}(V_{\mathbf{m}} - \overline{V}). \tag{48}$$

From the decomposition (44), the near-orthogonality of the functions $\widetilde{\delta}^3_{\mathbf{m}}$ and the definition of $\overline{V}$, we have

$$\overline{V} = \sum_{\mathbf{n}} V_{\mathbf{n}} |C_{\mathbf{n}}|^2. \tag{49}$$

Because $\sum |C_{\mathbf{n}}|^2 = 1$, the mean value of the random variable in (48) is zero:

$$E(V_{\mathbf{m}} - \overline{V}) = E(V_{\mathbf{m}}) - E(V_{\mathbf{m}}) \sum_{\mathbf{n}} |C_{\mathbf{n}}|^2 = 0. \tag{50}$$

As discussed, the random variables $V_{\mathbf{m}}$ at different cubes, i.e., for different values of $\mathbf{m}$ can be considered independent and identically distributed. It follows that the probability distributions of the random variables $V_{\mathbf{m}} - \overline{V}$ have a zero mean and are identical for all values of $\mathbf{m}$. With the help of the central limit theorem one can also claim that these distributions are normal. So the random vector with components (47) has an isotropic multivariate Gaussian distribution (the covariance matrix is proportional to the identity).

The standard deviation ΔV for the distribution of $V_{\mathbf{n}}$ in time satisfies the uncertainty relation $\Delta V \tau > h$, where τ is the time interval of observation. Therefore, the phase $-\frac{1}{\hbar}\overline{V}\tau$ in the phase factor of $\chi(t)$ should be considered uniformly distributed on $[0, 2\pi)$ so that any value of the phase factor is equally likely. It follows that the components of $\frac{d\chi}{dt}$ may have arbitrary constant phase factors at any given time and so the complex random vector made of these components has an isotropic normal circularly symmetric distribution. Accordingly, the vector $\frac{d\chi}{dt}\tau$ is equally likely to point in any direction in $T_{\psi}CP^{L_2}$ at $t = 0$. To make this result valid at an arbitrary moment of time, it remains to assume that the distribution of potentials is stationary and that changes in potentials over time are independent random variables.

Under a measurement of position of a macroscopic particle the observed particle is exposed to a random potential that is responsible for the normal distribution of the position random variable. We now see that the state of a microscopic particle undergoing a similar measurement and exposed to the same potential will experience a random motion on the sphere of states under which any direction η of displacement of the state is equally likely. From the section entitled "The Born Rule and the Normal Probability Distribution" we know that the normal distribution on M_3^{σ} and equal probability of any direction of displacement of the state result in the Born rule. That is, under the random potential produced by the measuring device the state ψ of the measured microscopic particle performs a random walk on the sphere of states and the probability for the state of reaching a neighborhood of any point φ on the sphere is given by the Born rule: $P(\varphi, \psi) = |(\varphi, \psi)|^2$.

Given the lack of Lebesgue measure on an infinite-dimensional Hilbert space, one may wonder how the state would have any chance of reaching a neighborhood of a given point in S^{L_2}. However, a realistic measuring device occupies a finite volume in the classical space. So the potential created by it can only affect a region $Q \subset \mathbb{R}^3$ of a finite volume V. The initial state ψ of the particle can be split onto the state $\psi_Q = \psi|_Q$ with support in Q (restriction of ψ to Q) and the leftover state $\chi = \psi - \psi_Q$. The state χ is not going to change under the potential and does not participate in the measurement (the probability for it of reaching a detector in Q is zero). Furthermore, possible group-velocity $\mathbf{v}_g$ of the measured particle in the given potential is also

IOP Conf. Series: Journal of Physics: Conf. Series **1275** (2019) 012050 doi:10.1088/1742-6596/1275/1/012050

bounded. The corresponding momentum $m\mathbf{v}_g$ of the particle belongs then to a bounded region $P \subset \mathbb{R}^3$. Therefore the state ψ_Q of the particle is limited to a superposition of states in the region $P \times Q$ in the phase space $\mathbb{R}^3 \times \mathbb{R}^3 = M^{\sigma}_{3,3}$. But there are only finitely many elements of the Weil-Heisenberg frame in such a region. Therefore, under the measurement the state ψ_Q evolves in a finite-dimensional subspace $V_{P\times Q}$ of the Hilbert space $L_2(\mathbb{R}^3)$. In particular, the Lebesgue volume of a neighborhood $Q_{\mathbf{a}}$ of any point $\widetilde{\delta}^3_{\mathbf{a}}$ (the state of a particle with a known position) in $V_{P\times Q}$ is well defined. Accordingly, the state has a non-vanishing probability of reaching $Q_{\mathbf{a}}$ and the relative probabilities of reaching neighborhoods of different points are given by the Born rule.

7. The motion of a state under a measurement

Let's now look into details of the motion of a state under a measurement. Note that in the non-relativistic quantum mechanics the particle, and, therefore, its state in a single particle Hilbert space, cannot disappear or get created. The unitary property of evolution means that the state can only evolve along the unit sphere in the space of states $L_2(\mathbb{R}^3)$. To express this conservation of states in the case of observation of position of the particle, consider the density of states functional $\rho_t[\varphi;\psi_Q]$. This functional measures the number of states that belong to a neighborhood of a point φ on the sphere of states $S^{V_{P\times Q}} \subset V_{P\times Q}$ at time t for an ensemble of particles, prepared in the initial state ψ_Q each. Under the isometric embedding $\omega : \mathbb{R}^3 \longrightarrow M^{\sigma}_3 \subset L_2(\mathbb{R}^3)$ the states in M^{σ}_3 are identified with (positions of) particles. So the density of states functional $\rho_t[\varphi;\psi_Q]$ must be an extension of the density of macroscopic particles $\rho_t(\mathbf{a};\mathbf{b})$ on $\mathbb{R}^3$. Here $\mathbf{b}$ is the initial position of the particle. In other words, $\rho_t(\mathbf{a};\mathbf{b}) = \rho_t[\widetilde{\delta}^3_{\mathbf{a}};\widetilde{\delta}^3_{\mathbf{b}}]$.

In the case of macroscopic particles the conservation of the number of particles is expressed in differential form by the continuity equation. For instance, if $\rho_t(\mathbf{a};\mathbf{b})$ is the density at a point $\mathbf{a} \in \mathbb{R}^3$ of an ensemble of Brownian particles with initial position $\mathbf{b}$ and $\mathbf{j}_t(\mathbf{a};\mathbf{b})$ is the current density at $\mathbf{a}$ of the particles, then

$$\frac{\partial \rho_t(\mathbf{a};\mathbf{b})}{\partial t} + \nabla \mathbf{j}_t(\mathbf{a};\mathbf{b}) = 0. \tag{51}$$

The conservation of states of an ensemble of microscopic particles is expressed by the continuity equation that follows from the Schrödinger dynamics. This is the same equation (51) with

$$\rho_t = |\psi|^2, \quad \text{and} \quad \mathbf{j}_t = \frac{i\hbar}{2m}(\psi\nabla\overline{\psi} - \overline{\psi}\nabla\psi). \tag{52}$$

For the states $\psi \in M^{\sigma}_{3,3}$ we obtain

$$\mathbf{j}_t = \frac{\mathbf{p}}{m}|\psi|^2 = \mathbf{v}\rho_t. \tag{53}$$

Because the restriction of Schrödinger evolution to $M^{\sigma}_{3,3}$ is the corresponding Newtonian evolution, the function ρ_t in (53) must be the density of particles. This density was denoted earlier by $\rho_t(\mathbf{a};\mathbf{b})$. It gives the number of Brownian particles that start at $\mathbf{b}$ and by the time t reach a neighborhood of a point $\mathbf{a}$. The relation $\rho_t(\mathbf{a};\mathbf{b}) = \rho_t[\widetilde{\delta}^3_{\mathbf{a}};\widetilde{\delta}^3_{\mathbf{b}}]$ tells us that ρ_t in (52) must be then the density of states $\rho_t[\widetilde{\delta}^3_{\mathbf{a}};\psi]$. It gives the number of particles in initial state ψ found under the measurement at time t in a state $\widetilde{\delta}^3_{\mathbf{a}}$ (i.e., on a neighborhood of the point $\mathbf{a}$ in $\mathbb{R}^3$).

In the integral form the conservation of states in $L_2(\mathbb{R}^3)$ can be written in the following form:

$$\rho_{t+\tau}[\varphi;\psi] = \int \rho_t[\varphi + \eta;\psi]\gamma[\eta]D\eta, \tag{54}$$

IOP Conf. Series: Journal of Physics: Conf. Series **1275** (2019) 012050 doi:10.1088/1742-6596/1275/1/012050

where $\gamma[\eta]$ is the probability functional of the variation η in the state φ and integration goes over all variations η such that $\varphi+\eta \in S^{L_2}$. When the state of the particle is constrained to $M_3^\sigma = \mathbb{R}^3$ this equation must imply the usual diffusion on $\mathbb{R}^3$. The restriction of (54) to M_3^σ means that $\varphi = \tilde{\delta}^3_{\mathbf{a}}$ and $\eta = \tilde{\delta}^3_{\mathbf{a}+\epsilon} - \tilde{\delta}^3_{\mathbf{a}}$, where ϵ is a displacement vector in $\mathbb{R}^3$. As we already know, the function $\rho_t[\tilde{\delta}^3_{\mathbf{a}}; \tilde{\delta}^3_{\mathbf{b}}] = \rho_t(\mathbf{a};\mathbf{b})$ is the usual density of particles in space. Let us substitute this into (54), replace $\gamma[\eta]$ with the corresponding probability density function $\gamma(\epsilon) \equiv \gamma[\tilde{\delta}^3_{\mathbf{a}+\epsilon} - \tilde{\delta}^3_{\mathbf{a}}]$ and integrate over the space $\mathbb{R}^3$ of all possible vectors ϵ. As in the Einstein derivation of the Brownian motion, assume that $\gamma(\epsilon)$ is the same for all $\mathbf{a}$ and independent of the direction of ϵ (space symmetry). Therefore, the terms $\int \epsilon^k \gamma(\epsilon) d\epsilon$ and $\int \epsilon^k \epsilon^l \gamma(\epsilon) d\epsilon$ with $k \neq l$ vanish. It follows by the Einstein derivation that

$$\frac{\partial \rho_t(\mathbf{a};\mathbf{b})}{\partial t} = k\Delta \rho_t(\mathbf{a};\mathbf{b}), \tag{55}$$

where $k = \frac{1}{2\tau}\int \epsilon^2 \gamma(\epsilon) d\epsilon$ is a constant.

The diffusion equation (55) describes the dynamics of an ensemble of particles in the classical space M_3^σ. If initially all particles in the ensemble are at the origin, then the density of the particles at a point $x \in \mathbb{R}$ (one dimensional case) at time t is given by

$$\rho_t(x;0) = \frac{1}{\sqrt{4\pi k t}} e^{-\frac{x^2}{4kt}}. \tag{56}$$

The path of a Brownian particle under random hits from the surrounding particles is a particular path of the state ψ in the space of states under the corresponding random potential. Since the distribution of displacement of ψ is isotropic, the distribution (56) can be expressed in terms of the Fubini-Study distance between states. From (36) we have

$$e^{-\frac{x^2}{4\sigma^2}} = \cos^2\theta, \tag{57}$$

where θ is the Fubini-Study distance between the state $\tilde{\delta}_x(u) = \left(\frac{1}{2\pi\sigma^2}\right)^{1/4} e^{-\frac{(u-x)^2}{4\sigma^2}}$ and the like state centered at $x=0$. Therefore,

$$x^2 = -4\sigma^2 \ln(\cos^2\theta). \tag{58}$$

Equating the probability density for the Brownian particle initially at the origin to be found at time t at the point x with the probability density for transition between the corresponding states (see (39)), we have

$$\frac{1}{\sqrt{4\pi k t}} e^{-\frac{x^2}{4kt}} = \frac{1}{\sqrt{4\pi\sigma^2}} \rho_t(\theta), \tag{59}$$

where by (36) the function $\rho_t(\theta)$ is equal to $\cos^2\theta$ for $t = \frac{\sigma^2}{k}$. Denoting this value by τ and substituting (58) into (59), we have

$$\rho_t(\theta) = \sqrt{\frac{\tau}{t}} \left(\cos^2\theta\right)^{\frac{\tau}{t}}. \tag{60}$$

As t increases, $\left(\cos^2\theta\right)^{\tau/t}$ approaches 1 for all $\theta \in [0,\pi/2)$, while $\rho_t(\pi/2) = 0$. Note that the integral of ρ_t with respect to $dx = \frac{dx}{d\theta}d\theta$ is, of course, 1.

From (58) we have

$$\left(\frac{d\theta}{dx}\right)^2 = -\frac{1}{2\sigma^2}\ln(\cos\theta)\cdot\cot^2\theta \tag{61}$$

and

$$\frac{d^2\theta}{dx^2} = \frac{\cot\theta}{4\sigma^2}\left[1 + 2\ln(\cos\theta)\cdot\csc^2\theta\right]. \tag{62}$$

This yields the second derivative $\frac{d^2}{dx^2}$ in the the form of the following operator Δ_θ:

$$\Delta_\theta = \frac{\cot\theta}{4\sigma^2}\left[1 - 2\ln(\cos\theta)\cdot\frac{d}{d\theta}\left(\cot\theta\frac{d}{d\theta}\right)\right]. \tag{63}$$

The diffusion equation takes the form

$$\frac{\partial\rho_t(\theta)}{\partial t} = k\Delta_\theta\rho_t(\theta), \tag{64}$$

where $\rho_t(\theta) = \rho_t[\varphi;\psi]$ with θ being the Fubini-Study distance between the initial state ψ and the variable point φ. The corresponding fundamental solution is the function in (60).

The parameter $\tau = \frac{\sigma^2}{k}$ should be interpreted as the time interval of observation. It is the time from the beginning of the diffusion process of the initial state ψ to the moment when the state has reached the end point φ (the moment of observation). Of course, this time may vary in different trials. However, the same is true of the time interval of observation of a Brownian particle in a particular experiment. By the central limit theorem the normal distribution on the left side of (59) is correct for the averaged time interval of observation τ. Because the left hand side of (59) is valid, the density (60) is accurate as well, giving the Born rule at $t = \tau$. One can say that a specific measuring device has an associated time parameter τ and diffusion coefficient k, and, therefore, the variance of the normal distribution of the position $\sigma^2 = k\tau$. The value of σ is then used in the isometric embedding $\omega : \mathbf{a} \longrightarrow \widetilde{\delta}^3_{\mathbf{a}}$, giving the relationship (57) and leading to the validity of the Born rule at the time of observation.

Note that if the diffusion of a state takes place but an observation is not made, the state continues its random walk on the sphere of states. Because of this additional (Schrödinger) evolution, the Born rule for any future observation has to be applied to the evolved state, rather than the original state ψ.

8. Collapse of a quantum state

It was shown that under the action of a random potential typically experienced by a Brownian particle, the state of a microscopic particle is equally likely to fluctuate in any direction tangent to the projective space of states. This, together with the normal distribution of position of the Brownian particle, signifies the validity of the Born rule for the motion of state. The motion satisfies the diffusion equation (64), whose fundamental solution is (60). It was shown that the state under the random potential evolves in the finite-dimensional subspace of the space of states. As a result, there is a non-vanishing probability for the initial state ψ of reaching a neighborhood of any non-orthogonal state φ.

The presented process of transition between states is very different from what is usually understood by the collapse. The fact that a noise may lead to random fluctuation of state is rather simple and goes against of what one normally tries to achieve when explaining collapse under a measurement. The collapse models utilize various ad hoc additions to Schrödinger equation with the goal of explaining why the state under the resulting stochastic process "concentrates" to an eigenstate of the measured observable (usually, position or energy) [6]-[17]. Instead, it is argued here that under a generic measurement an ensemble of states in the same initial position ψ "diffuses" isotropically into the space of states. Whenever a particular state in the ensemble reaches a neighborhood of an eigenstate of the measured observable, we say that the "collapse" has occurred. In this case the measuring device can record the value of the measured physical quantity.

IOP Conf. Series: Journal of Physics: Conf. Series **1275** (2019) 012050 doi:10.1088/1742-6596/1275/1/012050

The role of the measuring device in this scenario is reduced to initiating the diffusion(creating a "noise") and to registering a particular location of the diffused state. For instance, the "noise" in the position measuring device under consideration is due to the stream of photons. The device then registers the state reaching a point in M_3^σ. In a similar way, a momentum measuring device registers the states that reach under the diffusion the eigen-manifold of the momentum operator (which is the image of M_3^σ under the Fourier transformation). Note also the similarity in the role of measuring devices in quantum and classical mechanics: in both cases the devices are designed to measure a particular physical quantity and inadvertently create a "noise", which results in a distribution of values of the measured quantity.

It follows, in particular, that the measuring device in quantum mechanics is not responsible for creating a basis into which the state is to be expanded. If several measuring devices are present, they are not "fighting" for the basis. When the eigen-manifolds of the corresponding observables don't overlap, only one of them can "click" for the measured particle as the state can reach only one of the eigen-manifolds at a time.

What does it all say about measurement of position of macroscopic and microscopic particles? During the period of observation of position of a macroscopic particle, the position is a random variable with normal distribution. Normally, observation happens during a short enough interval of time and the spread of the probability density is sufficiently small. A particular value of position variable during the observation is simply a realization of one of the possible outcomes. The change in observed position of the particle can be equivalently thought of as either a stochastic process $\mathbf{b}_t$ with values in $\mathbb{R}^3$ or a process $\tilde{\delta}^3_{\mathbf{b},t}$ with values in M_3^σ. The advantage of the latter representation is that the position random variable $\tilde{\delta}^3_{\mathbf{b}}$ gives both, the position of the particle in $M_3^\sigma = \mathbb{R}^3$ and the probability density to find it in a different location $\mathbf{a}$ (in the state $\tilde{\delta}^3_{\mathbf{a}}$), due to uncontrollable interactions with the surroundings under the observation.

Measuring position of a microscopic particle has, in essence, a very similar nature. Under observation the state ψ is a random variable with values in the space of states CP^{L_2}. To measure position is to observe the state on the submanifold M_3^σ or $M_{3,3}^\sigma$ in CP^{L_2}. In this case, the random variable ψ assumes one of the values $\tilde{\delta}^3_{\mathbf{a}}$, with the uniquely defined probability density compatible with the normal density in the space $\mathbb{R}^3$. This probability density is given by the Born rule. Here too the random variable ψ gives both, the position of the state of the particle in CP^{L_2} and the probability density to find the particle in a different state $\tilde{\delta}^3_{\mathbf{a}}$.

So the difference between the measurements is two-fold. First, under a measurement the state ψ of a microscopic particle is a random variable over the entire space of states CP^{L_2} and not just over the submanifold M_3^σ. Second, unless ψ is already constrained to M_3^σ (the case which would mimic the measurement of position of a macroscopic particle), to measure position is to observe the state that "diffused" enough to reach the submanifold M_3^σ. To put it differently, the measuring device is not where the initial state was. Assuming the state has reached M_3^σ, the probability density of reaching a particular point in M_3^σ is given, as we saw, by the Born rule.

We don't use the term collapse of position random variable when measuring position of a macroscopic particle. Likewise, there seems to be no physics in the term collapse of the state of a microscopic particle. Instead, due to the diffusion of state, there is a probability density to find the particle in various locations on CP^{L_2}. In particular, the state may reach the space manifold $M_3^\sigma = \mathbb{R}^3$. If that happens and we have detectors spread over the space, then one of them clicks. If the detector at a point $\mathbf{a} \in \mathbb{R}^3$ clicks, that means the state is at the point $\tilde{\delta}^3_{\mathbf{a}} \in CP^{L_2}$ (that is, the state *is* $\tilde{\delta}^3_{\mathbf{a}}$). The number of clicks at different points $\mathbf{a}$ when experiment is repeated is given by the Born rule. The state is not a "cloud" in $\mathbb{R}^3$ that shrinks to a point under observation. Rather, the state is a point in CP^{L_2} which may or may not be on $\mathbb{R}^3 = M_3^\sigma$. When the detector clicks we know that the state is on M_3^σ.

Note once again that there is no need in any new mechanism of "collapse" in the model. An observation is not about a "concentration" of state and the stochastic process initiated by

IOP Conf. Series: Journal of Physics: Conf. Series **1275** (2019) 012050 doi:10.1088/1742-6596/1275/1/012050

the observation is in agreement with the conventional Schrödinger equation with a randomly fluctuating potential ("noise"). The origin of the potential depends on the type of measuring device or properties of the environment, capable of "measuring" the system. Fluctuation of the potential can be traced back to thermal motion of molecules, atomic vibrations in solids, vibrational and rotational molecular motion, and the surrounding fields.

9. Summary

The dynamics of a classical n-particle mechanical system can be identified with the Schrödinger dynamics constrained to the classical phase space submanifold $M^{\sigma}_{3n,3n}$ in the space of states. Conversely, there is a unique unitary time evolution on the space of states of a quantum system that yields Newtonian dynamics when constrained to the classical phase space. This results in a tight, previously unnoticed relationship between classical and quantum physics. In particular, under a measurement of the position of a macroscopic particle, the position random variable generically obeys the normal distribution law. The normal distribution law for the position variable implies the Born rule for transitions between arbitrary quantum states. Therefore, any classical (i.e., based on Newtonian dynamics) model of measurement of a macroscopic particle that predicts the normal distribution of the position random variable extends in a unique way to the corresponding quantum (i.e., satisfying Schrödinger dynamics) model that enforces the Born rule for probability of transition between states. The central limit theorem makes it easy for the outcomes of a measurement of a classical system to satisfy the normal distribution law. It follows that the Born rule in measurements of a quantum system is as generic as the normal distribution law in classical measurements.

In this paper, the proved relationship between classical and quantum concepts is taken to mean that physical laws, which govern the behavior of macroscopic and microscopic bodies, are fundamentally the same. For instance, there exists a unique extension of the classical Brownian motion from the classical space submanifold M^{σ}_{3} to the space of states CP^{L_2} of the particle. Because the Brownian motion can model the process of measurement in classical physics, its unique extension is taken to be adequate for the description of measurements on microscopic systems.

With this understood, the process of measurement on a quantum system can be described in terms of a diffusion of state of the measured system in which the state has equal probability of being displaced in any direction in the space of states CP^{L_2}. The role of the measuring device is reduced to creating a "noise" that triggers the diffusion in CP^{L_2} and in recording the diffused state when it reaches a particular region in CP^{L_2}. The conclusion is that the so-called collapse of the wave function in the framework is not about an instantaneous "concentration" of state near an eigenstate of the measured observable. Instead, it is about diffusion on the space of states under interaction with the measuring device and the environment. The "collapse" to an eigenstate of an observable happens when the state under the diffusion reaches the eigenmanifold of that observable. In the case of position measurements, the state must reach the classical space or phase space submanifolds in CP^{L_2}. Due to the enormous amount of collisions between a macroscopic body and the particles in the surroundings, position of the body is constantly measured. As a result, the diffusion process for macroscopic bodies can trivialize, which may explain why they remain in the classical space and, therefore, have a definite position.

We see that macroscopic and microscopic particles may not be so different after all. The only important distinction is that microscopic systems within the proposed framework live in the space of states while their macroscopic counterparts live in the classical space submanifold. Since our own life happens primarily in the macro-world, it is difficult for us to understand the infinite-dimensional quantum world around us. As soon as the classical-space-centered point of view is extended to its Hilbert-space-centered counterpart, the new, clearer view of the classical-quantum relationship emerges.

IOP Conf. Series: Journal of Physics: Conf. Series **1275** (2019) 012050 doi:10.1088/1742-6596/1275/1/012050

Acknowledgments
I would like to express my deep gratitude to Thomas Elze for organizing my talk at DICE2018 conference. I am also grateful to John Klauder and other conference participants for their sincere interest and great questions.

References
[1] Kryukov A 2017 *J. Math. Phys.* **52** 082103
[2] Kryukov A 2018 *J. Math. Phys.* **59** 052103
[3] Kryukov A 2007 *Phys. Lett.* A **370** 419
[4] Sanz A and Miret-Artés S 2012 *Dynamics of Open Classical Systems* in: Lecture Notes in Physics vol 850 (Berlin: Springer) p 47
[5] Percival I 1998 *Quantum State Diffusion* (Cambridge: Cambridge University Press)
[6] Pearle P 1976 *Phys. Rev.* D **13** 857
[7] Pearle P 1999 *Collapse Models* in: Lecture Notes in Physics vol 526 (Berlin: Springer) p 195
[8] Ghirardi G, Rimini A and Weber T 1986 *Phys. Rev.* D **34** 470
[9] Ghirardi G, Pearle P and Rimini A 1990 *Phys. Rev.* A **42** 78
[10] Diósi L 1989 *Phys. Rev.* D **40** 1165
[11] Adler S 2002 *J. Phys.* A **35** 841
[12] Adler S 2004 *Quantum theory as an emergent phenomenon* (Cambridge: Cambridge University Press)
[13] Adler S, Brody D, Brun T and Hughston L 2001 *J. Phys.* A **34** 8795
[14] Adler S and Brun T 2001 *J. Phys.* A **34** 4797
[15] Adler S and Horwitz L 2000 *J. Math. Phys.* **41** 2485
[16] Hughston L 1996 *Proc. Roy. Soc. London* A 953
[17] Bassi A, Lochan K, Satin S, Singh T and Ulbricht H 2013 *Rev. Mod. Phys.* **85** 471
[18] Joos E, Giulini D, Kiefer C, Kupsch J and Stamatescu I 2003 *Decoherence and the appearance of a classical world in quantum theory* (Berlin: Springer)

How an induced Kerr–Newman black hole releases gravitational waves without a mini black hole explosion

Andrew Walcott Beckwith
Physics Department, Chongqing University,
Chongqing 401331, People's Republic of China
english.cqu.edu.cn

E-mail: Rwill9955b@gmail.com; abeckwith@uh.edu

Abstract. Our task is to create circumstances in which the no-hair theorem of black holes no longer applies. First, we must determine if the mini black hole bomb would spontaneously occur. In all, the main end result is to try to avoid the so called black hole bomb effect, where a mini black hole explodes in a laboratory setting within say 10^{-16} s or so. That is, the idea would be to have a reasonably stable configuration with input laser energy, but a small mass, and to do it over hopefully 10^{-15}, or more, times longer than the 10^{-16} s preevaporation life of the mini black hole. That is, a duration of say up to 10^{-1} s which would provide a baseline as to astrophysical modeling of a Kerr–Newman black hole.

Keywords: Kerr–Newman black hole, high-frequency gravitational waves (HGW), causal discontinuity, PACS: 98.80.-k.

1. Introduction

To begin, a Kerr–Newman black-hole event horizon, with a charge Q and a constant angular momentum J as an induced state of affairs, is fed by laser-induced energy to generate gravitational waves and gravitons. Therefore, we use the formation of an event horizon [1] at the outer boundary of a matter-energy *bubble* of space–time in a laboratory setting and also using criteria for spatial resolution of a graviton [2] within a confined metric geometry. In addition, we generalize the entropy, [2] depending upon graviton production, due to infinite quantum statistics, [2] where we assume that graviton count, equivalent to a particle count, N, is equivalent to an entropy count, for reasons we go into in our manuscript. Furthermore, we use background as to the Kerr and Kerr–Newman metrics, [3, 4] which is important for our write up, and, in addition, we use Appell's 1887 nonstandard treatment of electrodynamics, [5] which is part and parcel of what is implied elsewhere. [6, 7] Note that the treatment of the ergosphere, and the question of a nonzero angular momentum associated with a black hole, [8] means that we have far more detail as to black hole physics than usual. [4] We do *not* call the angular momentum a constant in time. That is, we have torque in our model. Park's description [9] of how a rod rotating at a given high frequency, ω, gives a distinct GW and would be a graviton creator if the ends of the spinning rod were tapped by a laser. In a different way, we can use the idea of an artificial Kerr black hole generating GW as also another cosmological window into relic conditions of cosmology. Finally, our laboratory test, if initiated properly, may falsify or

Content from this work may be used under the terms of the Creative Commons Attribution 3.0 licence. Any further distribution of this work must maintain attribution to the author(s) and the title of the work, journal citation and DOI.
Published under licence by IOP Publishing Ltd

IOP Conf. Series: Journal of Physics: Conf. Series **1275** (2019) 012051 doi:10.1088/1742-6596/1275/1/012051

give credence to the 7.7×10^{-23} eV/c^2 upper bound to a massive graviton [10, p. 320], which may clarify if there is, say a difference between relic gravitons and later versions of what gravitons are, well after the onset of inflation.

2. A brief recap of Kerr–Newman black hole physics

A complete derivation of Lens and Thirring's study of how a spinning sphere of uniform density creates a gravitational field, [11, p. 257] leads to a metric of

$$\mathrm{d}S^2 = \left(1 - \frac{2m}{r}\right) c^2\, \mathrm{d}t^2 - \left(1 - \frac{2m}{r}\right) \mathrm{d}\sigma^2 + \frac{4\kappa J}{c^2 r}(\sin^2\theta) c\, \mathrm{d}\varphi\, \mathrm{d}t, \tag{1}$$

where J is the angular momentum of the sources, $\mathrm{d}\sigma^2$ is the three-dimensional Lens–Thirring flat-space line element, iff $J = -c^3 m \frac{a}{\kappa}$, and $r = \rho$. This gives us a rotating Kerr black hole.

In our consideration, we set J to a constant for ease of calculation and—with the caveat that, if a is a measure of the angular momentum per mass and if m in (1) is mass, we can say that m is the "geometric mass"—this simplification leads to a Coriolis-like force: [8, 11]

$$\rho\ddot{\varphi} + \frac{-2ma}{\rho^3}\dot{\rho} = 0. \tag{2}$$

If we replace ρ with the angular velocity ω, the above is the Coriolis force. [11, p. 130] If we add a charge, Q, into this business, we get, in part, [12]

$$\mathrm{d}S^2 = \frac{\widetilde{\rho}^2}{c^2}\left(\frac{-\,\mathrm{d}r^2}{\Delta} + \mathrm{d}\theta^2\right) = \frac{\Delta(c\,\mathrm{d}t - \widetilde{a}(\sin^2\theta)\,\mathrm{d}\varphi^2)}{\widetilde{\rho}^2} - \frac{(\sin^2\theta)(c\,\mathrm{d}t - \widetilde{a}(\sin^2\theta)\,\mathrm{d}\varphi^2)}{\widetilde{\rho}^2}$$

where $\widetilde{a} = \frac{J}{mc}$, $\Delta = r^2 - r_S r + \widetilde{a}^2 + r_Q^2$, $\widetilde{\rho}^2 = r^2 + \widetilde{a}^2(\cos^2\theta)$, $r_S = \frac{2Gm}{c^2}$, and $r_Q^2 = \frac{Q^2 G}{4\pi\varepsilon_0 c^4}$. We will, for the sake of simplicity, approximate J as a constant when we do our calculations.

3. What a charge, Q, in a rotating-black-hole solution gives

With an extremal condition on the mass of a Kerr–Newman black hole bounded below by angular momentum J, and charge, Q, [8] and if we have a mass m redefined as the Christodoulou–Ruffini mass, [6, p. 12] which we could set as M, [13] then we have $M = m$, $S = S_{\mathrm{ext}}$, $J = J_{\mathrm{ext}}$,

$$S = \pi\sqrt{Q_{\mathrm{ext}}^4 + 4J_{\mathrm{ext}}^2} = \pi\sqrt{Q^4 + 4J^2} \text{ and } M^2 = m^2 = \frac{1}{2}(Q^2 + \sqrt{Q^4 + 4J^2}). \tag{3}$$

The last two parts of (3) can be interpreted using the ideas of infinite quantum statistics, as a way of making a linkage between entropy and counting numbers of emitted particles, using $S = S_{\mathrm{ext}} \approx n$(particle count). [3] We can then, using (3), make the following statement as to number of particles stimulated by a laser hitting an artificial black hole, which we will in this first reading equate with Gravitons (massive) and the matter–energy input into the artificial black hole is

$$E_{\mathrm{ext}} = \frac{k_B}{2} T_{\mathrm{applied}} \approx Mc^2 \Leftrightarrow \left(\frac{k_B}{2c^2} T_{\mathrm{applied}}\right)^2 = \frac{1}{2}\left(q^2 + \frac{c_1 n}{\pi}\right)$$

$$\Rightarrow Q^2 = 2\left(\frac{k_B}{2c^2} T_{\mathrm{applied}}\right)^2 + \frac{c_1 n}{\pi}. \tag{4}$$

The particle count—in this case, stimulated graviton emission from the black hole—and the temperature, T_{applied}, from a laser smashing into a target will influence an effective charge, Q.

IOP Conf. Series: Journal of Physics: Conf. Series **1275** (2019) 012051 doi:10.1088/1742-6596/1275/1/012051

4. Specifying conditions for the production of gravitons, from the artificial Kerr–Newman black hole

We can consider working with the induced Kerr-Newman black hole assuming that there is a stimulated emission of particles from the artificial black hole and assuming that there is a method of input from lasers, or possibly thermonuclear fusion:

$$Q = \sqrt{2\left(\frac{k_B}{2c^2} T_{\text{applied}}\right)^2 + \frac{c_1 n}{\pi}} \text{ and } E_{\text{ext}} = \frac{k_B}{2} T_{\text{applied}}. \tag{5}$$

We will be examining what would be possible input energy into this *induced Kerr–Newman* black hole.

We go back to optimizing [8]

$$Q = \sqrt{2\left(\frac{k_B}{2c^2} T_{\text{applied}}\right)^2 + \frac{c_1 n}{\pi}} \tag{6}$$

$$E_{\text{ext}} = \frac{k_B}{2} T_{\text{applied}} \approx Mc^2 \tag{7}$$

$$2M^2 \geq Q^2 + \sqrt{Q^4 + 4J^2} \tag{8}$$

and

$$2M^2 \geq 2\left(\frac{k_B}{2c^2} T_{\text{applied}}\right)^2 + \frac{c_1 n}{\pi} + \sqrt{\left[2\left(\frac{k_B}{2c^2} T_{\text{applied}}\right)^2 + \frac{c_1 n}{\pi}\right]^2 + 4J^2} \tag{9}$$

and

$$B = \frac{+2Q}{\left(\frac{rJ}{mc}\right)} \tag{10}$$

Note that the expression for B, magnetic field, is commensurate with a specific value of r, such that we have E effectively disappear.

5. What conditions permit $T_E(\omega) \equiv \frac{1}{4\pi(2\widetilde{M}-\omega)} \cong T_{\text{applied}}$?

We submit that this is not a trivial question, and answering it would lead to perhaps successful implementation of our idea for forming a Kerr–Newman artificial black hole. Answering it will require well-posed modeling and experimental constraint conditions, which we will try to bring up in this section, First, to do this identification of $T_E(\omega) \equiv \frac{1}{4\pi(2\widetilde{M}-\omega)} \cong T_{\text{applied}}$, we have to have the fix put in. This is basic. That is, a mathematical investigation may, indeed, yield conditions in which one can establish $T_E(\omega) \cong T_{\text{applied}}$. Furthermore, we must also investigate the strength of gravitational waves from the formula for GW *strain.* [6, p. 505] Namely, in the case of laser light implosion, we have $h \sim$ the strain strength of GW, which may be measured from an induced black hole, [6] which were initially for a laser interferometer system, in LIGO, with the following comparisons. [8]

$$T_E(\omega) \equiv \frac{1}{4\pi(2\widetilde{M} - \omega)} \cong T_{\text{applied}}, \tag{11}$$

where $\omega = \omega_{\text{GW}}$, $\widetilde{M}$ is the mass of the induced black hole, ω_{laser} is the frequency of the laser, and W_{laser} is the laser light power. Then

$$h \sim GW - strength = \left(\frac{h\omega_{\text{GW}}}{4\pi\omega_{\text{laser}} W_{\text{laser}}}\right)^{\frac{1}{2}}. \tag{12}$$

6. Conclusion: what (11) portends for emitted GW (graviton?) radiation from the artificial black hole

Equation (11) is for laser-induced implosions on a black hole, Kerr–Newman style, which would, in the case of the national ignition facility, have an enormous power behind it. This assumes a signal-to-noise ratio of about one. Note that the original form of (11) [6] was for laser interferometry in a LIGO-style system. To get what we are seeking, we are likely assuming that the laser light would be very high frequency, and that both ω_{laser} and ω_{GW} would be very high, likely in the 10^8 to 10^{12} Hz range. Furthermore, we would be looking at an induced ring singularity of at least an angstrom in *width,* likely much larger. Our working assumption would be then that the emitted GW from the induced black hole would scale roughly as [8]

$$\omega_{\text{GW}} \approx 2\widetilde{M} = \frac{1}{4\pi T_{\text{applied}}}, \tag{13}$$

if $\omega = \omega_{\text{GW}}$ and $\widetilde{M}$ is the mass of the induced black hole.

This should be seen against the usual dimensional analysis, assuming that $k_B = h = c \to 1$ in dimensional analysis which would be seen as akin to the more usual [8]

$$E_{\text{ext}} = \frac{k_B}{2} T_{\text{applied}} \approx Mc^2 \xrightarrow[k_B=h=c\to 1]{} \omega_{\text{GW}}. \tag{14}$$

That is, as the applied temperature grows, the system approaches that required for GW production. In addition, if we are referring to a ring singularity [8] in an induced Kerr–Newman black hole, we would have [8]

$$\lambda_{\text{GW}}\nu_{\text{GW}} \equiv 2\pi\lambda_{\text{GW}}\omega_{\text{GW}} \approx c \equiv 1 \Rightarrow \lambda_{\text{GW}} \approx \frac{1}{2\pi\omega_{\text{GW}}}, \tag{15}$$

if $\lambda_{\text{GW}} \geq$ the radius of the ring singularity $\approx \frac{1}{2\pi\omega_{\text{GW}}}$.

If the radius of the black hole singularity ring is not on an angstrom scale, it is easy to postulate that one is having at least a 10^{10} Hz frequency in emitted radiation, and the strength of the GW, can be easily made, with adjustment in input parameters, so $h \approx 10^{23}$ is probable. That is, this should be seen in light of having a suitable temperature, T_{applied}, applied to the artificial Kerr–Newman black hole provided that we are looking at, say [8]

$$h \sim GW - strength = \left(\frac{h\omega_{\text{GW}}}{4\pi\omega_{\text{laser}}W_{\text{laser}}}\right)^{\frac{1}{2}} \propto 10^{-23}. \tag{16}$$

Leading to, in terms of laboratory conditions,

$$\begin{aligned} T_{\text{applied}} \xrightarrow[\text{derived}]{} (T_{\text{applied}})_{\text{derived}} \Leftrightarrow (\Delta E)_{\text{derived}} &= \frac{k_B}{2}(T_{\text{applied}})_{\text{derived}} \\ &\approx \frac{h}{(\Delta t_{\text{applied}})_{\text{derived}}} \approx \hbar(\omega_{\text{applied}})_{\text{derived}}. \end{aligned} \tag{17}$$

If this is done and we then wind up with $(\omega_{\text{applied}})_{\text{derived}}$ on the order of 10^{10} Hz, we are then in terms of experimental input well on our way toward setting parameterization of a quantum theory of gravity, which so far has eluded experimentalists. [8, 14, 15, 16, 17, 18, 19, 20, 21]

Acknowledgments
This work is supported in part by National Natural Science Foundation of China grant No. 11375279.

References

[1] Ohanian H C and Ruffini R 2013 *Gravitation and Spacetime* 3rd ed (New York: Cambridge University Press)
[2] Ng Y J 2008 *Phys. Lett. B* **657** 10–14 `https://doi.org/10.1016/j.physletb.2007.09.052`
[3] Newman E T, Couch E, Chinnapared K, Exton A, Prakash A and Torrence R 1965 *J. Math. Phys.* **6** 918–919 `https://doi.org/10.1063/1.1704351`
[4] Kerr R P 1963 *Physs Rev. Lett.* **11** 237–238
[5] Whittaker E and Watson G 1927 *A Course of Modern Analysis* (New York: Cambridge University Press)
[6] Thorne K S and Blandford R D 2017 *Modern Classical Physics* (Princeton, NJ: Princeton University Press)
[7] Kieffer C 2001 *Cosmology and Particle Physics, CAPP 2000, Verbier, Switzerland* ed Duerrer R, Garcia-Bellido J and Shaposhnikov M (Melville, NY: American Institute of Physics) pp 499–504
[8] Beckwith A 2018 *J. High Energy Phys. Gravitation Cosmol.* **4** 743–778 `https://doi.org/10.4236/jhepgc.2018.44042`
[9] Park D 1955 *Phys. Rev.* **99** 1324–1325 `https://doi.org/10.1103/PhysRev.99.1324`
[10] Maggiore M 2018 *Gravitational Waves* (*Astrophysics and Cosmology* vol 2) (Oxford, England: Oxford University Press)
[11] Adler R, Bazin M and Schiffer M 1975 *Introduction to General Relativity* 2nd ed (San Francisco: McGraw–Hill)
[12] Hajicek P 2008 *An Introduction to the Relativistic Theory of Gravitation* (Berlin: Springer-Verlag)
[13] Christodoulou D and Ruffini R 1971 *Phys. Rev. D* **4** 3552–3555
[14] Schoenlein R W, Chattopadhyay S, Chong H H W, Glover T E, Heimann P A, Leemans W P, Shank C V and Zholents A 2000 *Appl. Phys. B* **71** 1–10 `https://doi.org/10.1007/PL00021152`
[15] Beckwith A 2017 *J. High Energy Phys. Gravitation Cosmol.* **3** 558–563 `https://doi.org/10.4236/jhepgc.2017.34042`
[16] Corda C 2011 *J. High Energy Phys.* **2011** Article 101 `https://doi.org/10.1007/JHEP08(2011)101`
[17] Corda C, Hendi S H, Katebi R and Schmidt N O 2014 *Adv. High Energy Phys.* **2014** Article 527874
[18] Corda C 2012 *J. Mod. Phys. D* **21** Article 1242023 `https://doi.org/10.1142/S0218271812420230`
[19] Corda C, Hendi S H, Katebi R and Schmidt N O 2013 *J. High Energy Phys.* **2013** Article 008
[20] LIGO Scientific Collaboration and Virgo Collaboration 2017 *Phys. Rev. Lett.* **118** Article 221101 `https://doi.org/10.1103/PhysRevLett.118.221101`
[21] Sun D Q, Wang Z L, He M, Hu X R and Deng J B 2017 *Adv. High Energy Phys.* **2017** Article 4817948 `https://doi.org/10.1155/2017/4817948`

IOP Conf. Series: Journal of Physics: Conf. Series **1275** (2019) 012052 doi:10.1088/1742-6596/1275/1/012052

Quantum field theory of axion-photon mixing and vacuum polarization

Antonio Capolupo
Dipartimento di Fisica E.R.Caianiello and INFN Gruppo Collegato di Salerno, Universitá di Salerno, Fisciano (SA) - 84084, Italy

Abstract. We report on recent results obtained by analyzing axion–photon mixing in the framework of quantum field theory. We obtain corrections to the oscillation formulae and we reveal a new effect of the vacuum polarization due to the non-zero value of the vacuum energy for the component of the photon polarization mixed with the axion. The study of axion–photon mixing in curved space is also presented. Numerical analysis show that some quantum field theory effect of axion–photon mixing, in principle, could be detected experimentally.

1. Introduction

Axions are light neutral particles with mass $10^{-6} - 10^{-2}$ eV [1, 2] that, together with ultra-light axions (ULAs) with masses of $\sim 10^{-22}$ eV [3]–[8] are considered as possible candidates for dark matter [9]–[28]. An interesting property of axions and axion-like particles (ALPs) is axion–photon mixing and oscillation in the presence of strong magnetic fields [29]. This phenomenon (which is negligible for ULAs) could be used to detect axions with masses of 10^{-3} and 10^{-2} eV. However, at the moment there is no experimental demonstration of the existence of axions [30]–[40]. Therefore the study of possible experimental setups that could reveal axions and ULAs, and the complete theoretical understandings of the axions and of axion–photon mixing are needed.

In this paper we report the results presented in Ref.[41], obtained by analyzing axion–photon mixing in the framework of quantum field theory (QFT). We present new oscillation formulae for the axion–photon system and demonstrate the presence of QFT vacuum polarization induced by the condensate structure of the vacuum for mixed particles. We show that the QFT corrections to the amplitudes of the oscillation formulae, which are negligible for neutrinos and kaons, are detectable in principle in the case of axion–photon mixing. We also study axion–photon mixing in curved space and we provide numerical estimates of the energy density of the vacuum for axions and photons.

It should be noted that axion–photon mixing, at the difference of the mixing of neutrinos and bosons such as kaons, is characterized by the mixing of particles with different nature. Moreover, only one component of the photon polarization state couples and oscillates with the axion, generating a vacuum polarization that is absent in the other mixed systems. Therefore the results achieved by the QFT analysis of the mixing of neutrinos and mesons [42]–[55] cannot be applied immediately to axion–photon mixing and specific adjustments are required.

The paper is organized as follows. In Section 2, we present the QFT treatment of axion–photon mixing and we show new oscillation formulae. In Section 3, we analyze axion–photon

Content from this work may be used under the terms of the Creative Commons Attribution 3.0 licence. Any further distribution of this work must maintain attribution to the author(s) and the title of the work, journal citation and DOI.
Published under licence by IOP Publishing Ltd

mixing in curved space–time and we show a new polarization effect of the vacuum and in Section 4 we give our conclusions.

2. Axion–photon mixing

By neglecting the Heisenberg–Euler term due to loop correction in QED [56]-[58], $\frac{\alpha^2}{90m_e^4}\left[(F_{\mu\nu}F^{\mu\nu})^2 + \frac{7}{4}(F_{\mu\nu}\tilde{F}^{\mu\nu})^2\right]$, the Lagrangian density describing the ALP–photon system is

$$L = -\frac{1}{4}F_{\mu\nu}F^{\mu\nu} + \frac{1}{2}(\partial_\mu a\partial^\mu a - m_a^2 a^2) + \frac{g_{a\gamma\gamma}}{4}\, a\, F_{\mu\nu}\tilde{F}^{\mu\nu}\,. \tag{1}$$

Here the first two terms are the Lagrangian densities of the free photon and axion, respectively, and the last term is the interaction of two photons with the axion pseudoscalar field a in the presence of a magnetic field, which is responsible for axion–photon mixing [59]. In Eq.(1), $\tilde{F}_{\mu\nu} = \frac{1}{2}\epsilon_{\mu\nu\rho\sigma}F^{\rho\sigma}$ is the dual electromagnetic tensor and $g_{a\gamma\gamma} \equiv g \equiv g_\gamma\alpha/\pi f_a$ is the axion–photon coupling, with $g_\gamma \sim 1$, $\alpha = 1/137$ and f_a decay constant for ALPs.

We consider a laser beam propagating through a magnetic field $\mathbf{B}$ in the direction perpendicular to $\mathbf{B}$. In this case, the photon polarization state $\gamma_\perp$ perpendicular to $\mathbf{B}$ and to the direction of propagation of the beam decouples. Then, the mixing characterizes only the axion and the photon polarization state $\gamma_\parallel$ parallel to the magnetic field and the propagation equations can be written as: $(\omega - i\partial_z + M)\begin{pmatrix}\gamma_\parallel \\ a\end{pmatrix} = 0$, where, M is the mixing matrix $M = -\frac{1}{2\omega}\begin{pmatrix}\omega_P^2 & -g\omega B_T \\ -g\omega B_T & m_a^2\end{pmatrix}$. Here, the magnetic field $\mathbf{B}$ coincides with the purely transverse field B_T, and $\omega_P = (4\pi\alpha N_e/m_e)^{1/2}$ is the plasma frequency that is equal to zero in the case of propagation in the vacuum. Moreover, N_e is the electron density and m_a is the axion mass. The matrix M can be diagonalized by means of a rotation

$$\begin{pmatrix}\gamma_\parallel'(z) \\ a'(z)\end{pmatrix} = \begin{pmatrix}\cos\theta & \sin\theta \\ -\sin\theta & \cos\theta\end{pmatrix}\begin{pmatrix}\gamma_\parallel(z) \\ a(z)\end{pmatrix}, \tag{2}$$

where $\gamma_\parallel'$ and a' are the "free" fields with definite masses, $\gamma_\parallel$ and a are the fields of the mixed particles and $\theta = \frac{1}{2}\arctan\left(\frac{2g\omega B_T}{m_a^2 - \omega_P^2}\right)$ is the mixing angle. Since any polarization component of a neutral vectorial field can be represented by a neutral scalar field, then the decoupling of the photon polarization state $\gamma_\perp$ allows us to treat $\gamma_\parallel$ as a neutral scalar field. Thus axion–photon mixing can be described by the formalism presented in Ref.[44] used for the mixing of neutral particles in the QFT framework.

We invert Eqs.(2), we quantize $\gamma_\parallel'(x)$ and $a'(x)$ in the usual way and we express the mixed fields by means of the mixing generator $G_\vartheta(t)$ as $\gamma_\parallel(z) = G_\vartheta^{-1}(t)\ \gamma_\parallel'(z)\ G_\vartheta(t)$ and $a(z) = G_\vartheta^{-1}(t)\ a'(z)\ G_\vartheta(t)$, where $\vartheta = -\theta$. In a similar way, we can define the annihilators for $\gamma_\parallel$ and a as $\alpha_{\mathbf{k},\gamma_\parallel}(\vartheta, t) \equiv G_\vartheta^{-1}(t)\ \alpha_{\mathbf{k},\gamma_\parallel'}(t)\ G_\vartheta(t)$ and $\alpha_{\mathbf{k},a}(\vartheta, t) \equiv G_\vartheta^{-1}(t)\ \alpha_{\mathbf{k},a'}(t)\ G_\vartheta(t)$, such that $\alpha_{\mathbf{k},\gamma_\parallel}(\vartheta, t)|0(\vartheta, t)\rangle_{\gamma_\parallel,a} = \alpha_{\mathbf{k},a}(\vartheta, t)|0(\vartheta, t)\rangle_{\gamma_\parallel,a} = 0$. Here $|0(\vartheta, t)\rangle_{\gamma_\parallel,a} \equiv G_\vartheta^{-1}(t)\ |0\rangle_{\gamma_\parallel',a'}$ is the vacuum for mixed fields which in the infinite volume limit is orthogonal to the vacuum for free fields $|0\rangle_{\gamma_\parallel',a'}$. Moreover, $\alpha_{\mathbf{k},\gamma_\parallel'}(t) = \alpha_{\mathbf{k},\gamma_\parallel'}\, e^{-i\omega_{k,\gamma_\parallel'}t}$, $\alpha_{\mathbf{k},a'}(t) = \alpha_{\mathbf{k},a'}\, e^{i\omega_{k,a'}t}$, are the annihilators for the free fields $\gamma_\parallel'$ and a', respectively, where $\omega_{k,\gamma_\parallel'} = \omega_k + \Delta_+$, and $\omega_{k,a'} = \omega_k + \Delta_-$, with ω_k photon energy and $\Delta_\pm = -\frac{\omega_P^2 + m_a^2}{4\omega_k} \pm \frac{1}{4\omega_k}\sqrt{(\omega_P^2 - m_a^2)^2 + (2g\omega_k B_T)^2}$. Explicitly, we have:

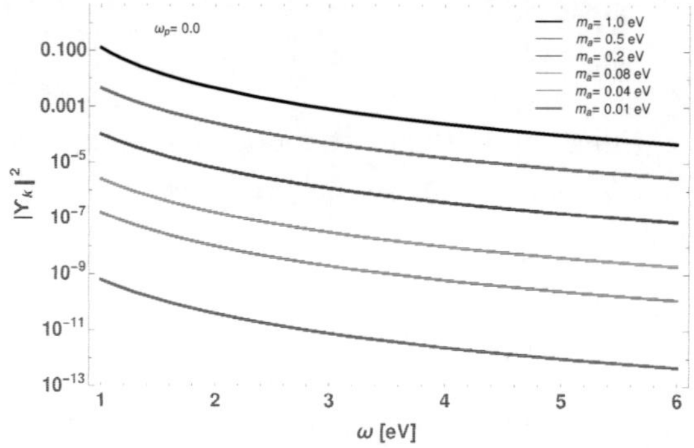

Figure 1. *(Color online) Plots of* $|\Upsilon_{\mathbf{k}}|^2$ *as a function of the photon energy for* $\omega_p = 0$.

$\alpha_{\mathbf{k},\gamma_{||}}(t) = \cos\vartheta \alpha_{\mathbf{k},\gamma'_{||}}(t) + \sin\vartheta\left(|\Sigma_{\mathbf{k}}|\alpha_{\mathbf{k},a'}(t) + |\Upsilon_{\mathbf{k}}|\alpha^{\dagger}_{\mathbf{k},a'}(t)\right)$, and similar for $\alpha_{\mathbf{k},a}(t)$, where the Bogoliubov coefficients are $|\Upsilon_{\mathbf{k}}| \equiv \frac{1}{2}\left(\sqrt{\frac{\omega_{k,\gamma'_{||}}}{\omega_{k,a'}}} - \sqrt{\frac{\omega_{k,a'}}{\omega_{k,\gamma'_{||}}}}\right)$, and $|\Sigma_{\mathbf{k}}| = \sqrt{1 + |\Upsilon_{\mathbf{k}}|^2}$.

Notice that, the vacuum for axions and photons is given by the following product:

$$|0(\vartheta,t)\rangle_{\gamma,a} = |0(\vartheta,t)\rangle_{\gamma_{||},a} \otimes |0\rangle_{\gamma'_{\perp}}, \tag{3}$$

where $|0(\vartheta,t)\rangle_{\gamma_{||},a}$ is the vacuum for $\gamma_{||}$ and a, and $|0\rangle_{\gamma'_{\perp}}$ is the vacuum for the unmixed component $\gamma'_{\perp}$. The state $|0(\vartheta,t)\rangle_{\gamma_{||},a}$ has a structure of condensed particles, with the condensation density given by: ${}_{\gamma_{||},a}\langle 0(\vartheta,t)|\alpha^{\dagger}_{\mathbf{k},\gamma'_{||}}\alpha_{\mathbf{k},\gamma'_{||}}|0(\vartheta,t)\rangle_{\gamma_{||},a} = \sin^2\vartheta|\Upsilon_{\mathbf{k}}|^2$. By contrast, $|0\rangle_{\gamma'_{\perp}}$ has a trivial structure and it does not exhibit a condensed structure. The presence of a condensate only for the component $\gamma_{||}$ produces the polarization of the vacuum for axions and photons.

The QFT oscillation formulae for the axion–photon system are given by the expectation value of the momentum operator for mixed fields $\mathcal{P}_\sigma(t) = \int d^3x\, \pi_\sigma(x) \nabla \phi_\sigma(x)$ at $t \neq 0$ on the initial state of the photon $|\alpha_{\mathbf{k},\gamma_{||}}\rangle = \alpha^{\dagger}_{\mathbf{k},\gamma_{||}}(0)|0(\vartheta,0)\rangle_{\gamma_{||},a}$, normalized to its initial value: $\mathcal{P}^{\mathbf{k}}_{\gamma\to\sigma}(t) \equiv \frac{\langle\alpha_{\mathbf{k},\gamma_{||}}|\mathcal{P}_\sigma(t)|\alpha_{\mathbf{k},\gamma_{||}}\rangle}{\langle\alpha_{\mathbf{k},\gamma_{||}}|\mathcal{P}_\sigma(0)|\alpha_{\mathbf{k},\gamma_{||}}\rangle}$, with $\sigma = \gamma_{||}, a$. Explicitly, we have:

$$\mathcal{P}^{\mathbf{k}}_{\gamma\to\gamma}(t) = 1 - \sin^2(2\theta)\Big[|\Sigma_{\mathbf{k}}|^2 \sin^2\left(\Omega^-_k t\right) - |\Upsilon_{\mathbf{k}}|^2 \sin^2\left(\Omega^+_k t\right)\Big], \tag{4}$$

$$\mathcal{P}^{\mathbf{k}}_{\gamma\to a}(t) = \sin^2(2\theta)\Big[|\Sigma_{\mathbf{k}}|^2 \sin^2\left(\Omega^-_k t\right) - |\Upsilon_{\mathbf{k}}|^2 \sin^2\left(\Omega^+_k t\right)\Big], \tag{5}$$

where $\Omega^{\pm}_k = \frac{\omega_{k,a'} \pm \omega_{k,\gamma'_{||}}}{2}$. In Fig.1 we plot the coefficient $|\Upsilon_{\mathbf{k}}|^2$ which multiplies the high-frequency oscillation term in Eqs. (4) and (5). We assume a photon energy $\omega \simeq 1.5eV$, a coupling constant $g_{a\gamma} \sim 1.4 \times 10^{-11} GeV^{-1}$ [60] and an axion mass $m_a \in [1 - 10^{-2}]eV$. The plots show that values of $|\Upsilon_{\mathbf{k}}|^2$ which are negligible in the case of very low axion masses, in principle, are detectable in the case of "heavy" axions, (axions with masses up to $0.2eV$ are expected in some models [61]). The values of $|\Upsilon_{\mathbf{k}}|^2$ are much higher than the corresponding field-theoretical corrections for the other mixed systems such as kaons and neutrinos [44]. Therefore, in principle, the axion–photon system could allow to test the QFT effects. Finally, we note that the presence of a gas in the conversion region increases the oscillation rate because the gas induces the effective mass of photon $m_\gamma = \frac{\hbar}{c^2}\omega_P$. However, the plasma presence reduces the value of $|\Upsilon_{\mathbf{k}}|^2$, which in any case remain detectable [41].

3. Axion–photon mixing in curved space–time and vacuum polarization

We analyze axion–photon mixing in curved space-time, by considering a homogeneous magnetic field which coincides with its transverse component, $B = B_T$ and we study the energy density and the pressure of the vacuum for mixed fields.

The unmixed scalar fields $\psi_i^C = \gamma'^C$, a'^C in curved space-time are

$$\psi_i^C(\mathbf{x},t) = J^{-1}(\mathbf{x},t)\,\psi_i(\mathbf{x},t)\,J(\mathbf{x},t) = \frac{1}{\sqrt{V}}\sum_{\mathbf{k}}\left[u_{\mathbf{k},i}(\mathbf{x},t)\alpha_{\mathbf{k},i} + u^*_{\mathbf{k},i}(\mathbf{x},t)\alpha^\dagger_{\mathbf{k},i}\right], \tag{6}$$

where J is the generator of the curvature (which depends on the metric considered), $\psi_i = \gamma'$, a' are the free fields operator in flat space–time, and $u_{\mathbf{k},i}(\mathbf{x},t) = \zeta_{\mathbf{k},i}(t)e^{i\mathbf{k}\cdot\mathbf{x}}$ are the mode functions that can be expressed analytically only in particular cases. The vacuum state for ψ_i^C is $|0(\mathbf{x},t)\rangle^C_{\gamma'_\parallel,a'} = J^{-1}(\mathbf{x},t)|0\rangle_{\gamma'_\parallel,a'}$, and the curved mixed vacuum is $|0(\vartheta,x)\rangle^C_{\gamma_\parallel,a} = (G^C_\vartheta(x))^{-1}\ J^{-1}(x)|0\rangle_{\gamma'_\parallel,a'}$, where $G^C_\vartheta(x)$ is the mixing generator for fields in curved space-time [41]. We take into account that for any curved space, the covariant derivative for scalar fields is equal to the ordinary derivative, and we compute the expectation value of the energy momentum tensor density $T_{\mu\nu}(x)$ for $\gamma'_\parallel, a'$ on the mixed vacuum $|0(\vartheta,x)\rangle^C_{\gamma_\parallel,a}$, i.e., $\Xi^\parallel_{\mu\nu} \equiv {}^C_{\gamma_\parallel,a}\langle 0(\vartheta,x)| : T_{\mu\nu}(x) : |0(\vartheta,x)\rangle^C_{\gamma_\parallel,a}$. Here : ... : indicates the normal ordering with respect to $|0(\mathbf{x},t)\rangle^C_{\gamma'_\parallel,a'}$. As shown in Ref.[41], the off-diagonal components of $\Xi^\parallel_{\mu\nu}(x)$ are zero for a homogeneous and isotropic universe, as well as for diagonal metrics, thus the energy density and pressure are $\rho_\parallel = g^{00}\ \Xi^\parallel_{00}$, and $p_\parallel = -g^{jj}\ \Xi^\parallel_{jj}$ (no summation on the index j is intended), respectively. Then, we obtain the state equation of the cosmological constant: $p_\parallel = -\rho_\parallel$, where

$$\rho_\parallel = \frac{\Delta m^2 \sin^2\theta}{2V}\sum_{\mathbf{k}}\left(|\zeta_{\mathbf{k},1}(t)|^2 - |\zeta_{\mathbf{k},2}(t)|^2\right), \tag{7}$$

and $\Delta m^2 = |m^2_{a'} - m^2_{\gamma'}|$.

Astrophysical systems that produce strong magnetic fields are pulsar and neutron stars. In order to study of Eq.(7) for these object we should consider the Schwarzschild metric. In this case, the explicit form of $\rho_\parallel$ is difficult. An estimate of $\rho_\parallel$, can be obtained by considering the Minkowski metric for simplicity. This means that we neglect the gravitational effects induced by the Schwarzschild metric. This assumption is reasonable for many systems generating magnetic fields, because their radius is very small. In the Minkowski metric, the explicit form of the energy density is:

$$\begin{aligned}\rho_\parallel &= \frac{\Delta m^2\sin^2\theta}{8\pi^2}\Big\{\left(m^2_{a'} - \sqrt{m^4_{a'} + 4g^2B_T^2K^2}\right) - \left(m^2_{a'} + g^2B_T^2\right)\Big[\tanh^{-1}\left(\frac{m^2_{a'}}{m^2_{a'} + g^2B_T^2}\right) \\ &- \tanh^{-1}\left(\frac{\sqrt{m^4_{a'} + 4g^2B_T^2K^2}}{m^2_{a'} + g^2B_T^2}\right)\Big]\Big\},\end{aligned} \tag{8}$$

where K is the cut-off on the momenta. Since, for $x \in R$, the function $\tanh^{-1}(x)$ is defined in the domain: $-1 < x < 1$, then Eq.(8) imposes a condition on the cut-off K given by: $-1 < \frac{\sqrt{m^4_{a'}+4g^2B_T^2K^2}}{m^2_{a'}+g^2B_T^2} < 1$, which implies, $K < \frac{\sqrt{2m^2_{a'}+g^2B_T^2}}{2}$.

We now give a numerical estimation of Eq.(8) for astrophysical objects and for terrestrial experiments.

- *Astrophysical systems:* we consider the following values of the parameters: $m_{a'} \sim 2 \times 10^{-2} eV$, $g \sim 10^{10} GeV^{-1}$, $B_T \sim (10^{15} - 10^{16})G$ (which can be obtained for different astrophysical objects), $\omega \sim 100eV$, $K \sim 10^{-2}eV$ (which satisfies the upper bound on K). We have: $\rho_{\parallel} \sim 10^{-47} GeV^4$, which is of the same order as the estimated value of the dark energy. However, $\rho_{\parallel}$ does not contribute to the dark energy, because it appears only in the regions where the magnetic field is localized. Moreover, in these regions other fields would have a larger influence on the energy momentum tensor.

- *Terrestrial experiments:* We consider $B = 45T$, $\omega \sim 10eV$ and $g \sim 10^{10} GeV^{-1}$. For these values of the parameter, the term $\tanh^{-1}\left(\frac{m_{a'}^2}{m_{a'}^2 + g^2 B_T^2}\right)$ in Eq.(8), imposes that $\rho_{\parallel}$ has a finite value only for values of axion mass less than $3 \times 10^{-7}eV$. For example, for $m_{a'} \sim 2 \times 10^{-7}eV$, we have $\rho \sim 10^{-66} GeV^4$.

It should be noted that the vacuum energy density and pressure are non zero only for the component parallel to the magnetic field of the photon polarization. On the contrary, the contributions $\rho_{\perp}$ and $p_{\perp}$ of the component perpendicular to the magnetic field of the photon polarization $\gamma_{\perp}$ are equal to zero: $\Xi^{\perp}_{\mu\nu}(x) \equiv_{\gamma_\perp} \langle 0(\mathbf{x},t)| : T_{\mu\nu}(x) : |0(\mathbf{x},t)\rangle_{\gamma_\perp} = 0$. This fact leads to a QFT polarization of the vacuum. The QFT vacuum polarization here presented could be detected in next experiments on axion–photon mixing.

4. Conclusions

We analyzed axion–photon mixing in the QFT framework. We derived new oscillation formulae for axion–photon transitions and revealed a vacuum polarization due to the condensate structure of the vacuum for mixed fields. These effects are not expected in previous studies of axion–photon mixing in the framework of quantum mechanics. Our numerical analysis shows that, in principle, the QFT effects can be detected in laser beam experiments.

Acknowledgments

I acknowledge partial financial support from MIUR and INFN and the COST Action CA1511 Cosmology and Astrophysics Network for Theoretical Advances and Training Actions (CANTATA) supported by COST (European Cooperation in Science and Technology).

References

[1] Peccei R D and Quinn H 1997 *Phys. Rev. Lett.* **38**, 1440
[2] Peccei R D and Quinn H 1977 *Phys. Rev. D* **16**, 1791
[3] Witten E 1984 *Phys. Lett. B* **149**, 351
[4] Hu W, Barkana R and Gruzinov A 2000 *Phys. Rev. Lett.* **85**, 1158
[5] Svrcek P and Witten E 2006 *J. High Energy Phys.* **06**, 051
[6] Arvanitaki A, Dimopoulos S, Dubovsky S, Kaloper N and March-Russell J 2010 *Phys. Rev. D* **81**, 123530
[7] Marsh D J E 2016 *Phys. Rep.* **643**, 1
[8] Hui L, Ostriker J P, Tremaine S and Witten E 2017 *Phys. Rev. D* **95**, no. 4, 043541
[9] Khmelnitsky A and Rubakov V 2014 *JCAP*, **2**, 019
[10] De Martino I, Broadhurst T, Tye S H H, Chiueh T, Schive H Y and Lazkoz R 2017 *Phys. Rev. Lett.* , **119**, 221103
[11] De Martino I, Broadhurst T, Tye S H H, Chiueh T, Schive H Y and Lazkoz R 2018 *Galaxies*, **6**, 10
[12] De Martino I, Broadhurst T, Tye S H H, Chiueh T, Schive H Y and Lazkoz R 2018 arXiv:1807.08153
[13] Barkana R. 2018 *Nature* **555**, no. 7694, 71
[14] Yoshiura S, Takahashi K and Takahashi T 2018 *Phys.Rev. D* **98**, 063529
[15] Moroi T, Nakayama K and Tang Y 2018 *Phys. Lett. B* **783**, 301
[16] Preskill J, Wise M B and Wilczek F 1983 *Phys. Lett. B* **120**, 127
[17] Abbott L F and Sikivie P 1983 *Phys. Lett. B* **120**, 133

[18] Dine M and Fischler W 1983 *Phys. Lett. B* **120**, 137
[19] Woo T-P and Chiueh T 2009 *Ap. J.* **697**, 850–861
[20] Schive H Y, Chiueh T and Broadhurst T 2014 *Nature Physics*, **10**, 7, 496–499
[21] Schive H Y et al. 2014 *Phys. Rev. Lett.* **113**, 261302
[22] Mocz P et al. 2017 *MNRAS*, **471**, 4559–4570
[23] Veltmaat J, Niemeyer J C and Schwabe B 2018 arXiv:1804.09647
[24] De Rocco W and Hook A 2018 *Phys. Rev. D* **98**, 035021
[25] Capolupo A, Lambiase G and Vitiello G 2015 *Adv. High Energy Phys.* **2015**, 826051
[26] Capolupo A, Lambiase G and Vitiello G 2015 *J. Phys. Conf. Ser.* **626**, 012059
[27] Tam H and Yang Q 2012 *Phys. Lett. B* **716**, 435
[28] Emami R, Broadhurst T, Smoot G, Chiueh T and Hoang Nhan L 2018 arXiv:1806.04518
[29] Raffelt G, 2008 *Lectures Notes Physics* **741**, 51.
[30] Zavattini E et al. (PVLAS Collaboration) 2008 *Phys. Rev. D* **77**, 032006
[31] Aune S et al. (CAST Collaboration) 2011 *Phys. Rev. Lett.* **107**, 261302
[32] van Bibber K, McIntyre P M, Morris D E and Raffelt G G 1989 *Phys. Rev. D* **39**, 2089
[33] Arik M et al. 2011 *Phys. Rev. Lett.* **107**, 261302
[34] Asztalos S et al. (ADMX Collaboration) 2004 *Phys. Rev. D* **69**, 011101
[35] Asztalos S et al. (ADMX Collaboration) 2010 *Phys. Rev. Lett.* **104**, 041301
[36] Cameron R et al. 1993 *Phys. Rev. D* **47**, 3707
[37] Fouche M. et al. (BMV Collab.) 2008 *Phys. Rev. D* **78**, 032013
[38] Pugnat P et al. (OSQAR Collab.) 2008 *Phys. Rev. D* **78**, 092003
[39] Pugnat P et al. (OSQAR Collab.) 2014 *Eur. Phys. J. C* **74**, 3027
[40] Ehret K. et al. (ALPS Collab.) 2010 *Phys. Lett. B* **689**, 149
[41] Capolupo A, De Martino I, Lambiase G and Stabile A 2019 *Axion–photon mixing in quantum field theory and vacuum energy, Phys. Lett. B*, in press, https://doi.org/10.1016/j.physletb.2019.01.056.
[42] Blasone M, Capolupo A and Vitiello G 2002 *Phys. Rev. D* **66**, 025033 and references therein.
[43] Blasone M, Capolupo A, Romei O , and Vitiello G 2001 *Phys. Rev. D* **63**, 125015
[44] Capolupo A, Ji C R, Mishchenko Y and Vitiello G 2004 *Phys. Lett. B* **594**, 135
[45] Blasone M, Capolupo A, Terranova F and Vitiello G 2005 *Phys. Rev. D* **72**, 013003
[46] Blasone M, Capolupo A, Ji C-R and Vitiello G 2010 *Int. J. Mod. Phys. A* **25**, 4179
[47] Capolupo A. 2016 *Adv. High Energy Phys.* **2016**, 8089142, 10
[48] Capolupo A. 2018 *Adv. High Energy Phys.* **2018**, 9840351
[49] Capolupo A, Capozziello S and Vitiello G 2009 *Phys. Lett. A* **373**, 601
[50] Capolupo A, Capozziello S and Vitiello G 2007 *Phys. Lett. A* **363**, 53
[51] Capolupo A, Capozziello S and Vitiello G 2008 *Int. J. Mod. Phys. A* **23**, 4979
[52] Blasone M, Capolupo A, Capozziello S and Vitiello G. 2008 *Nucl. Instrum. Meth. A* **588**, 272
[53] Blasone M, Capolupo A and Vitiello G. 2010 *Prog. Part. Nucl. Phys.* **64**, 451
[54] Blasone M, Capolupo A , Capozziello S, Carloni S and Vitiello G 2004 *Phys. Lett. A* **323**, 182
[55] Capolupo A, Di Mauro M and Iorio A 2011 *Phys. Lett. A* **375**, 3415
[56] Heisenberg W and Euler H 1936 *Z. Phys.* **98**, 714
[57] Schwinger J S 1951 *Phys. Rev.* **82** 664
[58] Dobrich B and Gies H 2009 *Europhys. Lett.* **87**, 21002
[59] Olive K A et al. (Particle Data Group) 2014 *Chin. Phys. C* **38**, 090001
[60] Giannotti M et al. 2017 *JCAP* **1710**, 010
[61] Di Luzio L et al. 2018 *Phys. Rev. Lett.* **120**, 261803

IOP Conf. Series: Journal of Physics: Conf. Series **1275** (2019) 012053 doi:10.1088/1742-6596/1275/1/012053

Neutrino nature, total and geometric phase

Antonio Capolupo
Dipartimento di Fisica E.R.Caianiello and INFN Gruppo Collegato di Salerno, Universitá di Salerno, Fisciano (SA) - 84084, Italy

S.M. Giampaolo
Division of Theoretical Physics, Ruder Bošković Institute, Bijenčka cesta 54, 10000 Zagreb, Croatia

Abstract. We study the total and the geometric phase associated with neutrino mixing and we show that the phases produced by the neutrino oscillations have different values depending on the representation of the mixing matrix and on the neutrino nature. Therefore the phases represent a possible probe to distinguish between Dirac and Majorana neutrinos.

1. Introduction

The phenomenon of neutrino mixing and oscillation, that has been proved experimentally [1]-[6], implies that the neutrino has a mass. Then the neutrino, being a neutral particle, can be a Majorana particle (a fermion that is its own antiparticle), or a Dirac particle (a fermion different from its antiparticle). At the moment the neutrino nature is not established.

A Majorana field is characterized by the presence of the Majorana phases ϕ_i in the mixing matrix which violate the CP symmetry. These phases cannot be eliminated since the Lagrangian of Majorana neutrinos is not invariant under $U(1)$ global transformation. By contrast, for Dirac neutrino, the Lagrangian is invariant under $U(1)$ global transformation and the ϕ_i phases can be removed. The mixing matrices for Majorana U_M and for Dirac neutrinos U_D can be related for example by the equation, $U_M = U_D \cdot diag(1, e^{i\phi_1}, e^{i\phi_2}, ..., e^{i\phi_{n-1}})$, where $i = 1, ..., n-1$. Other representations of U_M can be obtained by the rephasing the lepton charge fields in the charged current weak-interaction Lagrangian [7]. For example, in two flavor neutrino mixing case, one can consider the following mixing matrices for Majorana neutrinos

$$U_1 = \begin{pmatrix} \cos\theta & \sin\theta\, e^{i\phi} \\ -\sin\theta & \cos\theta\, e^{i\phi} \end{pmatrix}, \qquad or \qquad U_2 = \begin{pmatrix} \cos\theta & \sin\theta\, e^{-i\phi} \\ -\sin\theta\, e^{i\phi} & \cos\theta \end{pmatrix}, \tag{1}$$

where θ is the mixing angle, and ϕ is the Majorana phase. It should be noted that, neglecting the dissipation [8], the Majorana phases do not affect the neutrino oscillation formulae, being such formulae equivalent for Majorana and for Dirac neutrinos [9]. Therefore, the oscillation formulae are not useful in the study of the neutrino nature.

Recently, the study of the geometric phase has attracted also a great attention. The geometric phase appears in the evolution of any quantum state describing a system characterized by a Hamiltonian defined on a parameter space [10]–[25]. This phase arises in many physical systems [26]–[41] and it has been observed experimentally.

Content from this work may be used under the terms of the Creative Commons Attribution 3.0 licence. Any further distribution of this work must maintain attribution to the author(s) and the title of the work, journal citation and DOI.
Published under licence by IOP Publishing Ltd

IOP Conf. Series: Journal of Physics: Conf. Series **1275** (2019) 012053 doi:10.1088/1742-6596/1275/1/012053

In this paper, we report the results of the study on the total and geometric phases of neutrino presented in Ref.[42] and we show that, unlike the oscillation formulae, the total phase (and the dynamical one), generated by the transition between different flavors, depends on the choice of the matrix U. Indeed, different choices of U lead to different values of the total phases. In particular, considering the two flavor neutrino mixing case, we show that the use of the matrix U_2 in Eq.(1) (and of that corresponding to oscillations in a medium), generates values of the phases which are different for Majorana and for Dirac neutrinos. By contrast, if we consider the U_1 matrix, all the phases are independent from ϕ and Majorana neutrinos cannot be distinguished from Dirac neutrinos.

The paper is organized as follows. In Section 2 we analyze the total and the geometric phase for neutrinos by using different mixing matrces. In Section 3 we report a numerical analysis on the neutrino phases and in Section 4 we give our conclusions.

2. Total and geometric phases for neutrinos

We analyze the neutrinos propagation in vacuum and through a medium. The matter effects, are taken into account by replacing in the flavor states in vacuum, Δm^2 with $\Delta m_m^2 = \Delta m^2 R_\pm$, and $\sin 2\theta$ with $\sin 2\theta_m = \sin 2\theta / R_\pm$. The coefficients $R_\pm$ are, $R_\pm = \sqrt{\left(\cos 2\theta \pm \frac{2\sqrt{2} G_F n_e E}{\Delta m^2}\right)^2 + \sin^2 2\theta}$, with $+$ for oscillation of antineutrinos and $-$ for oscillations of neutrinos [43, 44]. In the following, we consider the flavor states $|\nu_e(z)\rangle$ and $|\nu_\mu(z)\rangle$ at the z distance given by the mixing matrix U_2, with θ replaced by θ_m,

$$\begin{aligned} |\nu_e(z)\rangle &= \cos\theta_m e^{i\frac{\Delta m_m^2}{4E}z}|\nu_1\rangle + e^{-i\phi}\sin\theta_m e^{-i\frac{\Delta m_m^2}{4E}z}|\nu_2\rangle, \\ |\nu_\mu(z)\rangle &= -e^{i\phi}\sin\theta_m e^{i\frac{\Delta m_m^2}{4E}z}|\nu_1\rangle + \cos\theta_m e^{-i\frac{\Delta m_m^2}{4E}z}|\nu_2\rangle. \end{aligned} \tag{2}$$

and we derive the total and the non–cyclic geometric phase [18]. For a quantum system whose state vector is $|\psi(s)\rangle$, the geometric phase is defined as the difference between the total phase $\Phi_\psi^{tot} = \arg\langle\psi(s_1)|\psi(s_2)\rangle$ and the dynamic phase $\Phi_\psi^{dyn} = \Im\int_{s_1}^{s_2}\langle\psi(s)|\dot\psi(s)\rangle ds$, i.e. $\Phi^g = \Phi_\psi^{tot} - \Phi_\psi^{dyn}$. Here, s is a real parameter such that $s \in [s_1, s_2]$, and the dot denotes the derivative with respect to s. For electron neutrino, the geometric phase is

$$\begin{aligned} \Phi_{\nu_e}^g(z) &= \arg\left[\langle\nu_e(0)|\nu_e(z)\rangle\right] - \Im\int_0^z \langle\nu_e(z')|\dot\nu_e(z')\rangle dz' \\ &= \arg\left[\cos\left(\frac{\Delta m_m^2 z}{4E}\right) + i\cos 2\theta_m \sin\left(\frac{\Delta m_m^2 z}{4E}\right)\right] - \frac{\Delta m_m^2 z}{4E}\cos 2\theta_m\,. \end{aligned} \tag{3}$$

For muon neutrino we have $\Phi_{\nu_\mu}^g(z) = -\Phi_{\nu_e}^g(z)$. Eq.(3) holds both for Majorana and for Dirac neutrinos, indeed it does not depend on the CP violating phase ϕ and thus it is independent on the choice of the mixing matrix. However, we can also consider the following phases due to the neutrino transitions between different flavors,

$$\Phi_{\nu_e\to\nu_\mu}(z) = \arg\left[\langle\nu_e(0)|\nu_\mu(z)\rangle\right] - \Im\int_0^z \langle\nu_e(z')|\dot\nu_\mu(z')\rangle dz'\,, \tag{4}$$

$$\Phi_{\nu_\mu\to\nu_e}(z) = \arg\left[\langle\nu_\mu(0)|\nu_e(z)\rangle\right] - \Im\int_0^z \langle\nu_\mu(z')|\dot\nu_e(z')\rangle dz'\,. \tag{5}$$

Eqs.(4) and (5) represent the differences between the total and the dynamic phases generated by the transitions $\nu_e \to \nu_\mu$ and $\nu_\mu \to \nu_e$, respectively. By using the Majorana neutrino states in

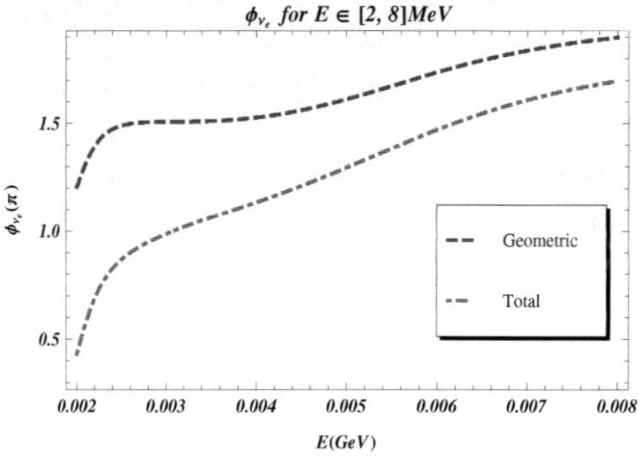

Figure 1. *(Color online) Plots of the total (the red dot dashed line) and the geometric phases (the blue dashed line) of ν_e, as a function of the neutrino energy E, for a distance length $z = 100km$.*

Eqs.(2), we have

$$\Phi_{\nu_e\to\nu_\mu}(z) = \frac{3\pi}{2} + \phi + \left(\frac{\Delta m_m^2}{4E}\sin 2\theta_m\, \cos\phi\right) z\,, \tag{6}$$

$$\Phi_{\nu_\mu\to\nu_e}(z) = \frac{3\pi}{2} - \phi + \left(\frac{\Delta m_m^2}{4E}\sin 2\theta_m\, \cos\phi\right) z\,. \tag{7}$$

Then, $\Phi_{\nu_e\to\nu_\mu} \neq \Phi_{\nu_\mu\to\nu_e}$. Although both the total and the dynamic phases depend on ϕ, the asymmetry between the transitions $\nu_e \to \nu_\mu$ and $\nu_\mu \to \nu_e$ is due to the total phases. Indeed, we have $\Phi^{tot}_{\nu_e\to\nu_\mu} = \frac{3\pi}{2} + \phi$ and $\Phi^{tot}_{\nu_\mu\to\nu_e} = \frac{3\pi}{2} - \phi$ (whereas $\Phi^{dyn}_{\nu_e\to\nu_\mu} = \Phi^{dyn}_{\nu_\mu\to\nu_e} = \left(\frac{\Delta m_m^2}{4E}\sin 2\theta_m\, \cos\phi\right) z$). By contrast, for Dirac neutrinos we have

$$\Phi_{\nu_e\to\nu_\mu}(z) = \Phi_{\nu_\mu\to\nu_e}(z) = \frac{3\pi}{2} + \left(\frac{\Delta m_m^2}{4E}\sin 2\theta_m\right) z\,, \tag{8}$$

and the total phases reduce to $\Phi^{tot}_{\nu_e\to\nu_\mu}(z) = \Phi^{tot}_{\nu_\mu\to\nu_e}(z) = \frac{3\pi}{2}$. The phases defined in Eqs.(4) and (5) and the total phases depend on the choice of the mixing matrix. Indeed, if we consider the mixing matrix obtained by U_1 by replacing θ with θ_m, the result of Eq.(8) is obtained also for Majorana neutrinos. Similar results are found for oscillation in vacuum. Therefore the phases $\Phi_{\nu_e\to\nu_\mu}$ and $\Phi_{\nu_\mu\to\nu_e}$ and the total phases $\Phi^{tot}_{\nu_e\to\nu_\mu}$ and $\Phi^{tot}_{\nu_\mu\to\nu_e}$ discriminate between the two matrices U_1 and U_2.

3. Numerical analysis.

In order to connect of our results with experiments, we plot in Figs.1 and 2 the total, the geometric phases and the phases defined in Eqs.(4) and (5) by using the characteristic values of experiments such as RENO [2] and T2K [4]. In Fig.1 we report the total and geometric phases associated with the evolution of ν_e. We consider the neutrino propagation through the matter and the values of the parameters of RENO experiment [2]: neutrino energy $E \in [2-8]MeV$, electron earth density $n_e = 10^{24}cm^{-3}$, $\Delta m^2 = 7.6\times10^{-3}eV^2$ and distance $z = 100km$. In Fig.2 we report the phases $\Phi_{\nu_e\to\nu_\mu}$ and $\Phi_{\nu_\mu\to\nu_e}$, by assuming $E \sim 1GeV$ and $z = 300km$, which are values compatible with the parameters of $T2K$ experiment [4]. Moreover we consider $\phi = 0.3$, and the values of n_e and Δm^2 considered above.

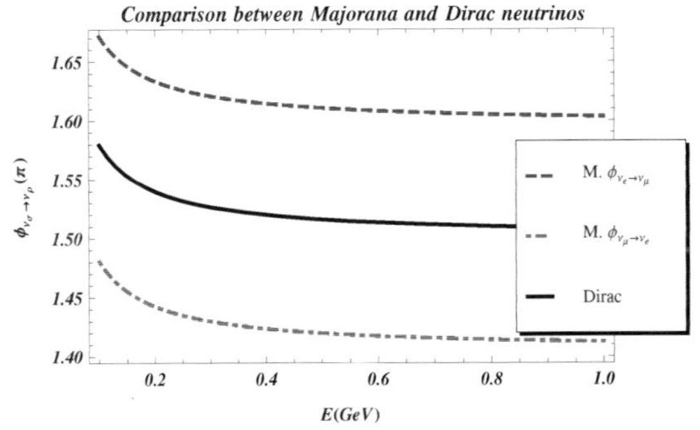

Figure 2. *(Color online) Plot of the phases* $\Phi_{\nu_e \to \nu_\mu}$ *(the blue dashed line) and* $\Phi_{\nu_\mu \to \nu_e}$ *(the red dot dashed line) for Majorana neutrinos as a function of the neutrino energy* E*, for a distance length* $z = 300km$*. The phases* $\Phi_{\nu_e \to \nu_\mu} = \Phi_{\nu_\mu \to \nu_e}$ *for Dirac neutrinos is represented by the black solid line.*

4. Conclusions.

We analyzed the total and the geometric phases generated in the evolution of the neutrino. We have shown that for Majorana neutrinos the phases due to a transition between different neutrino flavors take different values depending on the representation of the mixing matrix and on the nature of neutrinos. By considering the mixing matrix U_2, we obtained for Majorana neutrinos, $\Phi_{\nu_e \to \nu_\mu} \neq \Phi_{\nu_\mu \to \nu_e}$ (and $\Phi^{tot}_{\nu_e \to \nu_\mu} \neq \Phi^{tot}_{\nu_\mu \to \nu_e}$), that reveals an asymmetry in the transitions $\nu_e \to \nu_\mu$ and $\nu_\mu \to \nu_e$. This asymmetry disappears for Dirac neutrinos. On the contrary, by using U_1, we have $\Phi_{\nu_e \to \nu_\mu} = \Phi_{\nu_\mu \to \nu_e}$, (and $\Phi^{tot}_{\nu_e \to \nu_\mu} = \Phi^{tot}_{\nu_\mu \to \nu_e}$) both for Dirac and Majorana neutrinos and nothing can be said on the neutrino natures. We presented a numerical analysis by using the characteristic parameters of RENO and T2K experiments and we have obtained values for the neutrino phases which, in principle, are detectable. Our results pave the way for a completely new method to study the nature of neutrinos. In our discussion, the quantum field theory effects on particle mixing [45]-[60], can be safely neglected [47].

Acknowledgements

A.C. acknowledges partial financial support from MIUR and INFN and the COST Action CA1511 Cosmology and Astrophysics Network for Theoretical Advances and Training Actions (CANTATA) supported by COST (European Cooperation in Science and Technology). S.M.G. acknowledge support by the H2020 CSA Twinning project No. 692194, "RBI-T-WINNING".

References

[1] An F P et al. [Daya-Bay Collaboration] 2012 *Phys. Rev. Lett.* **108**, 171803
[2] Ahn J K et al. [RENO Collaboration] 2012 Experiment, *Phys. Rev. Lett.***108**, 191802
[3] Abe Y et al. [Double Chooz Collaboration] 2012 *Phys. Rev. Lett.* **108**, 131801
[4] Abe K et al. [T2K Collaboration] 2011 *Phys. Rev. Lett.* **107**, 041801
[5] Adamson P et al. [MINOS Collaboration] 2011 *Phys. Rev. Lett.* **107**, 181802
[6] Nakamura K and Petcov S T 2012 *Phys. Rev. D* **86**, 010001
[7] Giunti C 2010 *Phys. Lett. B* **686**, 41
[8] Capolupo A, Giampaolo S M and Lambiase G 2018 arXiv:1807.07823 [hep-ph]
[9] Bilenky S M and Pontecorvo B 1978 Phys. Rep. **41**, 225
[10] Berry M V 1984 *Proc. Roy. Soc. Lond. A* **392**, 45
[11] Aharonov Y and Anandan J 1987 *Phys. Rev. Lett.* **58**, 1593
[12] Samuel J and Bhandari R 1988 *Phys. Rev. Lett.* **60**, 2339
[13] Pancharatnam S 1956 Proc. Indian Acad. Sci. A **44**, 1225

[14] Shapere A and Wilczek F 1989 Geometric Phases in Physics, *World Scientific*, Singapore.
[15] Garrison J C and Wright E M 1988 *Phys. Lett. A* **128**, 177
[16] Pati A K 1995 *J. Phys. A* **28**, 2087
[17] Pati A K 1995 *Phys. Rev. A* **52**, 2576
[18] Mukunda N and Simon R 1993 *Ann. Phys.*(N.Y) **228**, 205
[19] Mostafazadeh A 1999 *J. Phys. A* **32**, 8157
[20] Anandan J 1988 *Phys. Lett. A* **133**, 171
[21] Tomita A and Chiao R Y 1986 *Phys. Rev. Lett.* **57**, 937
[22] Jones J A, Vedral V, Ekert A and Castagnoli G 2000 *Nature* **403**, 869
[23] Leek P J et. al 2007 *Science* **318**, 1889
[24] Neeley M et al. 2009 *Science* **325**, 722
[25] Pechal M et al. 2012 *Phys. Rev. Lett.* **108**, 170401
[26] Zhang Y, Tan Y W, Stormer H L and Kim P 2005 *Nature* **438**, 201-204
[27] Falci G et al. 2000 *Nature* **407**, 355-358
[28] Mottonen M, Vartiainen J J and Pekola J P 2008 *Phys. Rev. Lett.* **100**, 177201
[29] Murakawa H et al. 2013 *Science*, **342**, Issue 6165, 1490-1493
[30] Xiao D et al. 2010 *Rev. Mod. Phys.* **82**, 1959
[31] Capolupo A and Vitiello G 2013 *Adv. High Energy Phys.* **2013**, 850395
[32] Capolupo A and Vitiello G 2013 *Phys. Rev. D* **88**, 024027
[33] Capolupo A and Vitiello G 2015 *Adv. High Energy Phys.* **2015**, 878043
[34] Bruno A, Capolupo A, Kak S, Raimondo G and Vitiello G 2011 *Mod. Phys. Lett. B* **25**, 1661
[35] Hu J and Yu J 2012 *Phys. Rev. A* **85**, 032105
[36] Blasone M, Capolupo A, Celeghini E and Vitiello G 2009 *Phys. Lett. B* **674**, 73
[37] Joshi S and Jain S R 2016 *Phys. Lett. B* **754**, 135
[38] Johns L and Fuller G M 2017 *Phys. Rev. D* **95**, 043003
[39] Capolupo A, Lambiase G and Vitiello G 2015 *Adv. High Energy Phys.* **2015**, 826051
[40] Bertlmann R A, Durstberger K, Hasegawa Y and Hiesmayr B C 2004 *Phys.Rev.A* **69**, 032112
[41] Capolupo A 2011 *Phys. Rev. D* **84**, 116002
[42] Capolupo A, Giampaolo S M, Hiesmayr B C and Vitiello G 2018 *Phys. Lett. B* **780**, 216
[43] Mikheev S P and Smirnov A Yu 1985 *Sov. J. Nuc. Phys.* **42** (6): 913917
[44] Wolfenstein L 1978 *Phys. Rev. D* **17** (9): 2369
[45] Blasone M, Capolupo A and Vitiello G 2002 *Phys. Rev. D* **66**, 025033 and references therein
[46] Blasone M, Capolupo A, Romei O and Vitiello G 2001 *Phys. Rev. D* **63**, 125015
[47] Capolupo A, Ji C-R, Mishchenko Y and Vitiello G 2004 *Phys. Lett. B* **594**, 135
[48] Blasone M, Capolupo A, Terranova F and Vitiello G 2005 *Phys. Rev. D* **72**, 013003
[49] Blasone M, Capolupo A, Ji C-R and Vitiello G 2010 *Int. J. Mod. Phys. A* **25**, 4179
[50] Capolupo A 2018 *Adv. High Energy Phys.* **2018**, 9840351
[51] Capolupo A 2016 *Adv. High Energy Phys.* **2016**, 8089142
[52] Capolupo A, Capozziello S and Vitiello G 2009 *Phys. Lett. A* **373**, 601
[53] Capolupo A, Capozziello S and Vitiello G 2007 *Phys. Lett. A* **363**, 53
[54] Capolupo A, Capozziello S and Vitiello G 2008 *Int. J. Mod. Phys. A* **23**, 4979
[55] Blasone M, Capolupo A, Capozziello S and Vitiello G 2008 *Nucl. Instrum. Meth. A* **588**, 272
[56] Blasone M, Capolupo A and Vitiello G 2010 *Prog. Part. Nucl. Phys.* **64**, 451
[57] Blasone M, Capolupo A, Capozziello S, Carloni S and Vitiello G 2004 *Phys. Lett. A* **323**, 182
[58] Capolupo A, De Martino I, Lambiase G and Stabile A 2019 *Axion–photon mixing in quantum field theory and vacuum energy, Phys. Lett. B*, in press, https://doi.org/10.1016/j.physletb.2019.01.056.
[59] Capolupo A and Di Mauro M 2013 *Acta Phys. Polon. B* **44**, 81
[60] Capolupo A, Di Mauro M and Iorio A 2011 *Phys. Lett. A* **375**, 3415

IOP Conf. Series: Journal of Physics: Conf. Series **1275** (2019) 012054 doi:10.1088/1742-6596/1275/1/012054

Physics in one dimension with perpendicular non-locality

E Cerwen
Krokbacksv. 11, 68063 Likenäs, Sweden
E-mail: erik_cervensci@hotmail.com

Abstract. A single momentum-carrying dimension connected by Lorentz transformations to a perpendicular non-local dimension having time but lacking spatial measures is applied to electromagnetic radiation, thermal radiation, the Schrödinger equation, Kepler's 3:rd law, the rotation curve of spiral galaxies, and the universe as a whole represented by the linear part of its apparent expansion. This is made possible by identifying and making local terms equal to non-local ones as prescribed by the geometry. For example, 1-D momentum, the oscillating orbited radius of a system and baryonic mass are local whereas field components, oscillation periods and the electron cloud of an atom are non-local. Accordingly, in thermal radiation the physical process of a single quantum transfer replaces field radiation intensity as a function value. The Schrödinger equation can be rearranged and split into factors of circulating current carried by electrons surrounding magnetic charge. The latter result derives from a suitable factorization of the Bohr atom. Based on the assumption that the geometrical framework is valid generally for the rest frame the so-called 'dark matter' of galaxies can be identified by analogy with black body radiation and with the electron cloud as being non-local as might be expected of a 'massive field'. The basic theory yields a radius that is the inverse of a line increment per unit length and per unit time as in Hubble's constant. A tentative numerical value of this line increment appears as a residue when the Bohr atom is factorized for the purpose of providing a circular current to the Schrödinger equation.

1. Introduction

Ever since the experimental discovery of Compton scattering it is agreed in the physics community that electromagnetic radiation has a forward-directed momentum component, notably above the transition from Thompson to Compton scattering at about $4 \times 10^{17}\ Hz \approx 10^{26}\ m^{-1}$ where the radiation starts to behave as particles. In the photon picture of electromagnetic radiation the forward Abraham and Minkowsky momenta now form the basis of quantitative applications [1] [2].The existence of a forward momentum component can also be anticipated from the cross-product-Poynting vector based on Maxwell's description of the propagation of light [3] [4]. Since the latter is consistent with special relativity by the Faraday tensor most contemporary descriptions of electromagnetic radiation are based on Maxwell's equations. Nevertheless, many ambiguities are still unresolved and actively discussed in the literature. Maxwell's equations have been criticized for not providing a physical interpretation of the displacement current [5], for ambiguity regarding Coulomb and Lorentz gauge and for not dealing with signal source and sink events. Furthermore, they lack a conceptualization of the non-locality of the wave-front inferred from the self-interference of single quantum signals passing a double-slit. Non-locality in signal transmission has also been discussed in the context of polarization correlations and

Content from this work may be used under the terms of the Creative Commons Attribution 3.0 licence. Any further distribution of this work must maintain attribution to the author(s) and the title of the work, journal citation and DOI.
Published under licence by IOP Publishing Ltd

IOP Conf. Series: Journal of Physics: Conf. Series **1275** (2019) 012054 doi:10.1088/1742-6596/1275/1/012054

dismissed in the wave picture of the signal [6] whereas it persists in the photon picture. It is not known how the signal turns into local momentum upon absorption and then disappears from the entire wave-front, which in principle may span the entire galaxy or more. Any kind of non-locality which leads to instantaneous communication is considered unphysical in the geometrical framework of special relativity, which leads to still unresolved problems of fundamental importance in physics besides the need for a comprehensive quantitative picture of signal transmission.

These fundamental problems can only be solved from a broad perspective that includes the ontology of space-time, in other words, evaluating whether space-time exists on its own or if it appears as a consequence of its physical contents. It is not necessarily the case that the Cartesian coordinate system, invented in the 17:th century and later by special relativity adapted to the constant velocity of light, is the best geometrical framework for most processes in nature. In fact, the propagation of electromagnetic radiation carrying forward momentum and perpendicular electric and magnetic fields is not rotationally symmetric, adding to its wave-front non-locality which is unaccounted for in physics' descriptions even to this day. Even though electromagnetic radiation spans most of the universe's volume and would be a good candidate to determine the geometry of space-time the foundations of contemporary physics were laid in the 17:th and 18:th centuries based on our subjective impressions of the classical material 3-dimensional world. At the elementary level, however, the material world is made visible by the atom through its quantum interactions with light. The quantum transfers are discontinuous, which implies that on each quantum transfer the primordial atom redefines its momentum axis irrespective of whether the momentum is radial (head-on) or axial (curl or spin). In a typical quantum transfer the atom only accepts the proper head-on energy transfer in a probabilistic fashion without negotiating the energy of the signal based on incoming angle as is the case in classical momentum transfer. This means that the atom sees its surrounding on a one-dimensional axis different from our 3-dimensional world and all other dimensions will necessarily be non-local from *its* perspective. Hence, non-locality by reference to the momentum axis characterizes both the electromagnetic radiation and the atom and seems to be a fundamental property of space dictated by its physical contents. This idea will now be examined quantitatively by setting up a geometry that exhibits a momentum axis as well as non-locality, and applying this geometry at the quantum level in electromagnetic signal transfer, thermal radiation and the Schrödinger equation. It will also be shown that the geometry applies in the classical world based on examples from Kepler's 3:rd law, galaxy star rotation and even the entire universe.

2. A geometrical construct exhibiting a momentum axis and perpendicular non-locality

A geometry with the required properties is obtained by performing a Lorentz transformation of the inverse of the number-flux vector at discrete time coordinates -1 and 0 defining an interval of observation:

$$(q_0, t_0) = \left(\frac{\sqrt{1-\frac{v^2}{c^2}}}{v}\frac{m^2}{s}, 0\right); \qquad (\overline{q}_0, \overline{t}_0) = \left(\frac{1}{v}\frac{m^2}{s}, -s\right) \tag{1}$$

$$(q_r, t_r) = \left(\frac{\sqrt{1-\frac{v^2}{c^2}}}{v}\frac{m^2}{s}, s\sqrt{1-\frac{v^2}{c^2}}\right); \qquad (\overline{q}_r, \overline{t}_r) = \left(\frac{1}{v}\frac{m^2}{s} - vs, 0\right) \tag{2}$$

$$\overline{\Delta q} = -vs\ , \qquad \overline{\Delta t} = \overline{t}_r - \overline{t}_0 = s \quad \Rightarrow \frac{\overline{\Delta q}}{\overline{\Delta t}} = -v \tag{3}$$

$$\Delta q = 0 \ , \qquad \Delta t = t_r - t_0 = s\sqrt{1 - \frac{v^2}{c^2}} \ . \tag{4}$$

Here, m is the unit of length and s the geometrized unit of time [1]. This system of equations defines two observers located at the origin (un-barred) and at radius distance from the origin (barred observer). The latter observer is capable of observations along the momentum axis, $\overline{\Delta q}$, and of measuring the unit of time while the observer at the origin only is aware of time and recognizes an angular velocity v. The two observers are space-like separated.

The directions of the axes is defined by analogy with the unit circle, $(\cos x)^2 + (\sin y)^2 = 1$, as

$$q_r{}^2 + \frac{1}{c^2}\frac{m^4}{s^2} = \frac{1}{v^2}\frac{m^4}{s^2} = \overline{q}_r{}^2 \tag{5}$$

or

$$(\frac{\Delta t}{s})^2 + (\frac{\overline{\Delta q}}{m})^2 = 1 \tag{6}$$

so that line increment and time interval are perpendicular. The time interval measured by the momentum observer is also perpendicular to the momentum frame where it defines the tangential velocity as shown in eq. 3c.

The sign of the line increment (cf. eq. 3) shows that for $v > 0$ the radius of the observed object decreases in the time interval. An observer at the origin computes a contracted radius $\overline{q}_0$ similarly to the Lorentz-Fitzgerald case, $q_0 = \overline{q}_0\sqrt{1 - v^2/c^2}$. Hence, the geometry can be understood as a circle space-like separated from a peripheral observer who detects it in the form of a line increment in the direction of observation after the passage of one unit of time. In physics, line increments in the direction of observation are known from the Bohr atom and the cosmological expansion which hints at concrete applications.

For observations towards the origin along the radius in the local frame of observation, the magnitude of the line increment is summed from $\overline{\Delta q}$ per unit length, m, to the unit length proper, m, per radius, $\overline{q}_0$

$$\frac{-\overline{\Delta q}}{m} = \frac{m}{\overline{q}_0} \quad \Leftrightarrow \quad \overline{q}_0 \ \overline{\Delta q} = -m^2 \tag{7}$$

as can be derived from eqs. 1b and 3c. Here, the velocity of light, m/s, limits the radial extension of the geometry to $|\overline{q}_0|$ ($v \leq c$ as required by $\sqrt{1 - v^2/c^2}$). In eq. 7 the radius and the line increment define the unit of length so that each unit length comes with a line increment as well as a radius whereas eq. 3a shows that the line increment also can be interpreted as a rate. Because of eq. 3 and 4, observations can only be made from the laboratory frame at the periphery towards the origin of space and time coordinates. The observer at the origin is non-local in the sense of performing all observations solely on the time axis (eq. 4b)

Summarizing the above,
1. Each length in the local frame comes with a line increment, a momentum.
2. This line increment is the inverse of a radial length in the momentum frame.

[1] using non-standard notation for the purpose of distinguishing the two units

IOP Conf. Series: Journal of Physics: Conf. Series **1275** (2019) 012054 doi:10.1088/1742-6596/1275/1/012054

3. Time is perpendicular to the momentum axis. The time axis harbors the non-local observer. Because of the absence of spatial measures in the non-local frame the geometry is not rotationally symmetric like the Euclidean-Cartesian one. Besides the absence of spatial measure and the two frames being perpendicular the non-locality is also evident from the space-like separation of the local and the non-local observer.

4. The two observers are located either at the origin or at the end of a radial length.

5. The line increment per unit time equals a perpendicular velocity (eq 3).

6. The time dilatation of special relativity is preserved (eqs. 2a and 4b).

7. There is a length contraction of the same magnitude as a Lorentz-Fitzgerald contraction applied to the radius, which is confined to the non-local observer's frame (eq. 2). This contraction is not visible in the local frame of observation where it is known from special relativity theory that Lorentz contractions can indeed not be observed [7] [8].

8. There is a relativistic horizon obtained by adding line increments linearly to each unit length until the sum of these increments adds up to the velocity of light. However, there are no relativistic effects in the local frame (eq. 3) since the atom either responds to or does not respond to the signal depending on its internal structure in its rest frame without taking into account relativistic alternatives. The relativistic distortions take place prior to the quantum transfer and make the signal fit or not fit to transfer its momentum coming from a moving or stationary source.

3. From quantum to classical in a one-dimensional material world with perpendicular non-locality

In as much as known physical processes can be interpreted within the geometrical framework defined above it is a valid framework for these processes and the framework consequently has 'physicality' with regard to these processes. The property 5 above is most useful for evaluating this since it implies that some factors assigned to the local frame are equal to other factors in the non-local frame. A quantitative example to start with can then easily be obtained by rearranging the thermal Planck distribution [9], placing local factors to the left and non-local ones to the right, as

$$h\nu \; exp\Big(\frac{-h\nu}{kT}\Big) = \frac{c^3}{8\pi}\; U(\nu)d\nu^{-1}\; \tau^2\; \Big(1 - exp\Big(\frac{-h\nu}{kT}\Big)\Big) \quad . \tag{8}$$

The left side contains the momentum transfer to and from the matter, $h\nu$, and the exponential term which is interpreted classically as a probability that the excited state of the matter decays. The exponential term was extensively characterized in the early 20:th century, e.g. [9] [10] [12] [13]. The right side contains the complementary probability that the electromagnetic field transfers energy to the matter. Furthermore, the right side contains the field intensity per frequency component, which is known to result from a squared probability amplitude, and the oscillation period τ of the radiation squared. In accordance with the present geometrical framework and with the well-established directions of the electric and magnetic fields of electromagnetic radiation these components are perpendicular to the momentum axis. Hence, the geometry embeds in a natural way thermal radiation. It is noteworthy that in the present case the field intensity appears as a function variable rather than as a function value like in the well-established approaches to Planck's equation. Letting the amplitude work jointly with the squared oscillation period to establish the radiation field like in eq. 8 is consistent with many known exceptions from the Bohr-Einstein energy level rules [14] [15]. In the present case the actual physical event, the momentum transfer, appears as a function value, supporting the conjecture that the geometrical framework is a 'natural' or 'physical' one. The form and interpretation of eq 8 given here applies at equilibrium when any absorption of radiation is immediately emitted or the radiation intensity is low in comparison with spontaneous emission. A conventional analysis of stimulated emission similarly to eq. 8 can be found in [16].

This form of the Planck distribution with the quantum-momentum left and non-local factors right agrees geometrically with the trivial example of electromagnetic radiation *per se* mentioned in the Introduction, which can be illustrated by the Poynting vector[2]

$$\mathbf{S} = \mathbf{B} \times \mathbf{E} \tag{9}$$

The Schrödinger equation for a free particle, at left in

$$\frac{p^2}{2M_e}\Psi = -i\hbar\frac{\partial}{\partial t}\Psi \quad \Rightarrow \quad \frac{\hbar}{2}\Big(\frac{\partial}{\partial x}\Big)^2\Psi = -iM_e\frac{\partial}{\partial t}\Psi\ , \tag{10}$$

yields the form at right where local terms (length measures) are on the left side and non-local ones (electron cloud and time) on the right side, which agrees with the notion of Property 5 in Section 2 that physical processes in general are composed of equally large local and non-local terms. Furthermore, the imaginary term, $-i$, makes the non-local terms perpendicular to the momentum frame as prescribed in Section 2.

The Bohr atom in the ground state with its oscillating radius defined by the circulating non-local electron provides further justification for the proposed geometry. The Bohr atom can be factorized such as to give a concrete physical-geometrical meaning to the Schrödinger equation. First, write the Bohr atom, $a_0 = 4\pi\epsilon_0\hbar/(e^2m_e)$, as

$$\Big[a_0\frac{\alpha m_e}{\hbar}\Big]\Big(\frac{e^2}{4\pi\epsilon_0\alpha}\Big) = \hbar \Rightarrow 2.4113\times10^{-53}\ \frac{e^2c^2}{4\alpha^2} = \hbar \Rightarrow \frac{2\overline{\Delta q}}{\pi\ Ampere}\frac{ec}{2\alpha} = \sqrt{\hbar} \tag{11}$$

where geometrized units have been used except in the case of the factor c of magnetic charge because of charge invariance. The square-bracketed term becomes unity and can be omitted whereas the number in the middle is interpreted as $4\overline{\Delta q}^2/\pi^2$ where $\overline{\Delta q} = 7.714\times10^{-27}\ m$ [3]. Use is thus made of the substitution

$$\sqrt{\hbar} = \frac{2\overline{\Delta q}}{\pi\ Ampere}\ \frac{ec}{2\alpha} \tag{12}$$

which is inserted into the linear Schrödinger equation for a free particle in the form at the right side of eq. 10 to give

$$(\overline{\Delta q})^2\ \Big(\frac{ec}{2\alpha}\Big)^2\ \Big(\frac{\partial}{\partial x}\Big)^2\ \Psi = i\ 2M_e\ \frac{\partial}{\partial t}\ \Psi\Big(\frac{\pi}{2}Ampere\Big)^2 s^{-2}\ . \tag{13}$$

Here the non-local right side has a circular ($\pi/2$) electric current carried by a pair of electrons surrounding on the left local side a squared magnetic charge[4] (possibly two charges of opposite sign) that are associated with a squared line increment interpretable as angular momentum.

[2] The ambiguity of the Poynting vector and Maxwell's equations, much discussed in the literature, cf. e.g. [3] [4], is left aside here.

[3] The numerical value of $\overline{\Delta q} = 7.714\times10^{-27}s^{-1}$ (equal to 71.36 $km/second/Mparsec$), agrees within errors with the Hubble constant in astrophysics obtained by two independent measuring methods [17] [18] [19]. The inverse of the line increment prescribed by Property 2 in Section 2 is $1.296\times10^{26}\ m$ and a light signal would take 13.7 billion years to travel between the endpoints of this radial length. Since the line increment refers to the unit length by eq. 7 and is taken per unit time by eq. 3c it has the same dimension as Hubble's constant, which is m/ms. In the present paper though the chosen numerical value has a quantum-physical justification on its own as shown subsequently.

[4] The magnetic charge monopole was originally identified as a one-dimensional entity [11]

IOP Conf. Series: Journal of Physics: Conf. Series **1275** (2019) 012054 doi:10.1088/1742-6596/1275/1/012054

While the Schrödinger equation previously has been criticized for being purely mathematical the substitution using eq. 12 shows that it may actually *define* the structure of the atom in terms of a non-local charged electron pair surrounding the nucleus. The physical interpretation of eq. 13 with several but not all terms squared invites further theory.

The examples given show that several quantum processes of fundamental importance can be advantageously interpreted in the proposed geometrical framework composed of a single spatial dimension and a non-local frame. Subsequently it will be shown that many classical systems also are compatible with the proposed geometry such that quantum and classical worlds exist on an equal footing within the same framework. This is also a result that has been obtained with conventional geometry and quantum mechanics [20] [21]. For example, Kepler's 3:rd law supplemented with the gravitational constant may be rearranged and written (T = orbital period, r = orbital radius = an oscillating length with period T, local factors left, non-local ones (M, T and v^{-1}) right

$$r = G\ M_{sun}\ \frac{T^2}{4\pi^2 r^2} \tag{14}$$

whereas Galilee's acceleration may be written (L = length displacement at the observed velocity, a = acceleration, t = time passed, s = unit of time)

$$L = a\ t\ s\ . \tag{15}$$

The latter may be adapted to the apparent cosmological expansion, which *a priori* is linear in the present geometry, so that by analogy with eq. 15 (m = meter, r = radius) the universe accelerates onto the local unit length,

$$m = \frac{\overline{\Delta q}}{s^2}\left(\frac{r_{universe}}{m}s\right)\ s\ , \tag{16}$$

where, tentatively, $\overline{\Delta q}/ms = H_0$. As is well known the linearity prevails over quite long distances before the measures are obscured by relativistic effects and by the long time intervals ΔT between astrophysical observation and actual event leading to cosmological model-dependence. If one ignores that only radial endpoints are allowed in this geometry it basically holds that

$$\sum \overline{\Delta q} = H_0\ \Delta T \tag{17}$$

since each unit length comes with a line increment and the universe is assumed to have the same space-time structure everywhere along its radius unaffected by relativistic effects. This agrees with the trivial case in standard cosmology where, however, the present acceleration of the distant origin onto the local unit length is seen in the reverse as a literal expansion from a local origin. In eqs. 14 - 16 the acceleration acts like an operator from the non-local to the local frame whereby the former has a squared term like in the quantum world (eqs. 8, 10, 13).

Finally, the rotation of stars in spiral galaxies adapted from the baryonic Tully-Fischer relation [22] [23] and rearranged by analogy with thermal radiation as in eq. 8 yields

$${v_{bar}}^2 = {v_{obs}}^2\left(1 - exp\left(-\sqrt{\frac{v_{bar}^2}{Rg^\dagger}}\right)\right) \tag{18}$$

indicating by analogy with the previous examples in this paper that the robust baryons contained in the term v_{bar} (baryonic velocity) are local and stable (no exponential term) whereas

IOP Conf. Series: Journal of Physics: Conf. Series **1275** (2019) 012054 doi:10.1088/1742-6596/1275/1/012054

so called 'dark matter' contained in the term v_{obs} (observed velocity) is non-local.

These examples from the classical world show that it too can be seen from the perspective of a one-dimensional observer immersed in a non-local dimension, not just in the manner that the tiny atom sees a signal but on a macroscopic scale as well.

4. Discussion

In this paper, the physicality of a geometrical construct composed of one spatial dimension embedded in a perpendicular non-local field, like in electromagnetic signal propagation, has been evaluated. It has been shown that such a geometry naturally accommodates thermal radiation and provides physical meaning to the Schrödinger equation including its complex terms based on a suitable factorization of the Bohr atom. Furthermore, the solar system given by Kepler's 3:rd law, the apparent cosmological expansion and the velocity distribution of stars in spiral galaxies conform to the geometry. Besides providing a tentative general framework for the rest frame evidenced by these applications the coexistence of our 3-dimensional world with a hidden, quantitatively well defined non-local frame of observation implies a kind of manifest 'duality' which has not previously been documented[5]. The non-local frame may in a very concrete fashion harbor all kinds of physics descriptions that have previously only been thought of as merely mathematics, like e.g. particle-field duality, superpositions, path integrals, number theory, permutations and the like.

References

[1] Partanen M, Häyrynen T, Oksanen J and Tulkki J 2017 Photon momentum and optical forces in cavities *Preprint* arXiv:1707.04505v1 [quant-ph]
[2] Partanen M, Häyrynen T, Oksanen J and Tulkki J 2017 Photon mass drag and the momentum of light in a medium *Phys. Rev. A* **95** 063850
[3] Chen X-S, Chen X-B, Sun W-M and Wang F 2007 Testing the correctness of the Poynting vector E x B as the momentum density of gauge fields *Preprint* arXiv:0710.1427v1 [hep-ph]
[4] Puthoff H E 2009 Electromagnetic potentials basis for energy density and power flux *Preprint* arXiv:0904.1617 [physics.gen-ph]
[5] Selvan K T 2009 A revisiting of scientific and philosophical perspectives on Maxwell's displacement current *IEEE Antennas and Propagation Magazine* **51(3)** 36-46
[6] Jung K 2017 Violation of Bell's inequality: Must the Einstein locality really be abandoned? *J. Phys.: Conf. Ser.* **880** 012065
[7] Terrell J 1959 Invisibility of the Lorentz contraction *Phys. Rev.* **116(4)** 1041-1045
[8] Penrose R 1958 The apparent shape of a relativistically moving sphere *Proc. Cambridge Phil. Soc.* **55(1)** 137-139
[9] Planck. M 1901, 1921 *Vorlesungen über die Theorie der Wärmestrahlung.* (Johann Ambrosius Barth, Leipzig)
[10] Bose S N 1924 Planck's Gesetz und Lichtquantenhypothese *Zeitschr. f. Phys.* **26** 178-181
[11] Dirac P A P 1948 The theory of magnetic poles *Phys. Rev.* **74(7)** 817-830
[12] Einstein A 1916 Absorption nach der Quantentheorie *Verh. Deutsch. Phys. Ges.* **18** 318-323
[13] Pauli W 1923 Über das thermische Gleichgewicht zwischen Strahlung und freie Elektronen *Zeitschr. f. Physik* **18** 272-286
[14] Moore C I, Knauer J P and Meyerhofer D D 1995 Observation of the transition from Thomson to Compton scattering in multiphoton interactions with low-energy electrons *Phys. Rev. Lett.* **74(13)** 2439-2442
[15] Cole D C and Zou Y 2004 Analysis of orbital decay time for the classical hydrogen atom interacting with circularly polarized electromagnetic radiation *Phys. Rev. E* **69** 016601(1) - 016601(12)
[16] Dirac P 1924 The conditions for statistical equilibrium between atoms, electrons and radiation *Proc. Royal Soc. London, Ser. A* **106** 581-596

[5] The word 'duality' in the sense intended here has previously been used in philosophy (Descartes) and physics (de Broglie). In quantum physics various 'pictures' usually denote how the same phenomenon can be described with different mathematical approaches.

[17] Riess A G *et. al.* 2018 Milky way Cepheid standards for measuring cosmic distances and application to Gaia DR2: Implications for the Hubble constant *Preprint* arXiv:1804:10655v2 [astro-ph.CO]
[18] Riess A G *et. al.* 2016 A 2.4 % determination of the local value of the Hubble constant *Preprint* arXiv:1604.01424v3 [astro-ph.CO]
[19] Birrer S *et. al.* 2019 HoLiCOW - IX. Cosmographic analysis of the doubly imaged quasar SDSS 1206+4332 and a new measurement of the Hubble constant *MNRAS* https://doi.org/10.1093/mnras/stz200
[20] Batygin K 2018 Schrödinger evolution of self-gravitating discs *MNRAS* **475** 5070-5084
[21] Klauder J R 2018 Solving the insoluble: A new rule for quantization *Preprint* arXiv:1710.05085v3 [quant-ph]
[22] McGaugh S S 2014 The third law of galactic rotation *Preprint* arXiv:1412.3767v1 [astro-ph.GA]
[23] Lelli F, McGaugh S S, Schombert J M and Pawlowski M S 2017 One law to rule them all: The radial acceleration relation of galaxies *Preprint* arXiv:1610.08981v2 [astro-ph.GA]

IOP Conf. Series: Journal of Physics: Conf. Series **1275** (2019) 012055 doi:10.1088/1742-6596/1275/1/012055

Localized gravitons as essence of dark energy: Implications for accelerated cosmic expansion

D. Drakova[1], **G. Doyen**[2]

[1] Faculty of Chemistry, University of Sofia, Bulgaria
[2] Ludwig-Maximilians Universität, München, Germany

E-mail: Drakova@chem.uni-sofia.bg

Abstract. A theory of Emerging Quantum Mechanics (EQM) has been developed [1] contributing to the understanding of the measurement problem by assigning a quantum dynamical process to collapse within Schrödinger's theory.

EQM treats cosmological problems in a quantum field theory in high dimensional spacetime (11D) including matter and the gravitational field. Gravity is quantized in the weak interaction limit. The EQM Lagrangian contains terms up to fourth order in the gravitational field. Second and fourth order terms are associated with repulsive dark energy and third order terms with attractive dark matter effects. Gravonon-gravonon repulsion, i.e. dark energy in EQM, modifies the structure of the gravitational field and leads to accelerated expansion of the universe.

Quantum diffusion of the matter particles in three dimensional space is suggested as the mechanism of the expansion of the universe. The rate of expansion is given by the frequency of the soft gravitational modes. A lower limit of the deceleration parameter for the expansion of the universe, derived in EQM, is in the range between -0.6 and -0.8, an observed value of -0.55 has been reported.

1. Introduction: Some unexplained experimental observations

The measurement problem is the most prominent unresolved problem of standard quantum mechanics. Observations unexpained by standard quantum mechanics range from atomic to cosmological scales, some of them are summarized in table 1. The collapse process is not the result of any well defined quantum dynamics in four dimensional spacetime. Furthermore in cosmology the physical nature of dark energy and dark matter, needed to reproduce numerous cosmological observations, is not clarified. The coincidence problem presents a further difficulty for quantum cosmology.

In a recent publication a quantum field theory was presented [1], called Emerging Quantum Mechanics (EQM), which uses high dimensional spacetime and quantizes gravity in the weak interaction limit, leading to the solution of the measurement problem [2]. The idea in EQM is that the entanglement with gravonons, i.e. localized gravitons, leads to the localization of quantum particles as beables. This happens if and only if the coupling is weak and nearly degenerate with a continuum of states of high density. This is a kind of dimensionality effect warranted only by the continuum of gravonons which, in addition to our three dimensional space, live in

Content from this work may be used under the terms of the Creative Commons Attribution 3.0 licence. Any further distribution of this work must maintain attribution to the author(s) and the title of the work, journal citation and DOI.
Published under licence by IOP Publishing Ltd

IOP Conf. Series: Journal of Physics: Conf. Series **1275** (2019) 012055 doi:10.1088/1742-6596/1275/1/012055

Table 1. Summary of some experimental observations and their understanding within standard quantum mechanics

Experiment	Experimental result	Prediction by standard quantum mechanics	Judgement
expansion of universe	small value of dark energy (cosmological constant)	extremely large value of cosmological constant	quantum theory fails by 123 orders of magnitude
diffraction of mesoscopic molecules	wave $\rightarrow$ particle transition	collapse, measurement problem	heuristic postulate instead of a theory
sticking of gas atoms at $T \rightarrow 0$K	everything is stuck at $T \rightarrow 0$K	nothing is stuck at $T \rightarrow 0$K	quantum theory fails by an infinite order of magnitude
Xe atom localization	Xe atom on a metal surface does not move for days and weeks	Xe atom wave packet decays within 10^{-10} s	theory fails by twenty orders of magnitude
hydrogen diffusion on metal surfaces	diffusion rate depends strongly on experimental setup	no explanation	theory fails by nine orders of magnitude

hidden high spacial dimensions. A beabling condition must be fulfilled which is a requirement for the magnitude of the coupling strength between the quantum particle and the gravonons [5]. Because of the weakness of the coupling, particle localization and beabling via gravonons is effective on the energy shell defined by the incident wave packet. Therefore the site of particle localization, e.g. on the detector surface in the double slit experiment, is determined by the presence of gravonons, degenerate with the initial state of the incident particle wave. Schrödinger's equation leads deterministically to the chosen site of particle localization if the initial state is known.

We use the EQM approach to solve the cosmological problems of the nature and the effects of dark energy and dark matter, which emerge in the theory (section 2). Quantum diffusion of hydrogen atoms in a crystal in intergalactic space is the mechanism of the expansion of the universe in EQM [5]. The accelerated expansion of the universe is due to gravonon-gravonon repulsive interactions resulting in a range of values for the deceleration parameter of the order of the experimentally determined value [3] (section 4.1).

2. Features of EQM. Emerging dark energy and dark matter.

Inspired by string theory in EQM we assume 11 dimensional spacetime, with the additional spacial dimensions being hidden and compactified. A three dimensional brane exists and is embedded in 10 dimensional space. The size of the compactified hidden dimensions is stabilized by the presence of branes. It is of the order of $10^3 - 10^4$ bohr and is determned by comparison with the experimental rate of adsorbate quantum diffusion on solid surfaces [1]. Gravitons, the quanta of the gravitational field, are the only particles living in all spacial dimensions. All other particles (fields) live only on the three dimensional brane.

The mathematical implications of the features of EQM refer to:

- Localization of particles out of delocalized wave functionals.
- Telegraph-like transitions between localized particles states.
- Transitions between localized and delocalized particle states (tiny bursts).

IOP Conf. Series: Journal of Physics: Conf. Series **1275** (2019) 012055 doi:10.1088/1742-6596/1275/1/012055

- "Strong" gravitational force on the molecular scale.
- Existence of localized graviton states (gravonons).
- Reproduction of the experimental facts in table 1.
- Single world wave functional, which is a superposition of delocalized and localized configurations, and develops with time via Schrödinger's equation.

As it is derived in ref. [1], the gravitational field is quantized in the limit of weak gravity. In the non-relativistic limit new particles emerge named gravonons, which represent local versions of the gravitons.

2.1. *Objects in EQM*
New objects are defined:

- Gravonons are linear combinations of gravitons of various wave lengths. They are localized over a region encompassing a molecule or a micro-, meso- or macroscopic piece of matter.
- Beables: matter fields entangled to gravonons constitute beables. A matter particle loosing entanglement to gravonons can escape into plane waves and then be recaptured on a localized place.
- Limbo state is any solution of Schrödinger's equation in 4 dimensional spacetime in tensor product with gravitational configurations in 10 spacial dimensions, out of which 7 are hidden. In the limbo state matter particles are not accessible in experiment.
- Limbo-beable transition is a change of the wave functional with time, from a dominating tensor product of 4 dimensional components times gravitational components, to a wave functional, for which linear combination of such tensor products has essential physical consequences.

The world wave functional $\mid \Psi(t)\rangle$ is a solution of Schrödinger's equation:

$$\mid \Psi(t)\rangle = \sum_{limbo} C_{limbo}(t) \mid limbo\rangle. \tag{1}$$

It is a superposition of many-particle states where $\mid limbo\rangle$ is a four dimensional solution of Schrödinger's equation, not involving gravitational interaction, in tensor product with a state describing gravitons and/or gravonons.

Equation (1) is a sum of tensor products, which is not a tensor product itself. If this superposition is dominated by a sum of components, which describe a matter particle in tensor product with gravonons, then this is a beable where the matter particle is entangled with the gravonons. If the world wave functional is dominated by a single tensor product of a single particle state and gravonons, it corresponds to an independent particle and gravonon continuum, hence there is no entanglement between them. This state we call the limbo state.

2.2. *Interaction Lagrangian. Dark energy and dark matter emerge in EQM.*
Einstein's field equation describes classical physics where particles move along trajectories (geodesics):

$$E_{\mu\nu} = 8\pi G(T_{\mu\nu} + \Lambda g_{\mu\nu}) \tag{2}$$

$E_{\mu\nu}$: Einstein's tensor; G: Newton's constant; $T_{\mu\nu}$: energy-momentum tensor (T_{00} contains the density of classical particles); Λ: the cosmological constant, which together with the metric $g_{\mu\nu}$

IOP Conf. Series: Journal of Physics: Conf. Series **1275** (2019) 012055 doi:10.1088/1742-6596/1275/1/012055

represents the so called dark energy term.

In the weak gravitational limit the Euler-Lagrange equation in EQM [1] is the analogue of Einstein's field equation and reads for the 00-component:

$$\partial^2 h_{00} = -8\pi \mathrm{G}^{(N)}(T_{00} + \frac{1}{2}T_{00}h_{00}) \tag{3}$$

$$T_{00} = m\psi^+\psi + \frac{1}{2}m\psi^+\psi h_{00} \tag{4}$$

$$\begin{aligned}\partial^2 h_{00} &= -8\pi \mathrm{G}^{(N)}(m\psi^+\psi + m\psi^+\psi h_{00}) \\ &= -8\pi \mathrm{G}^{(N)}(\rho_{matter} + \rho_{matter}h_{00}).\end{aligned} \tag{5}$$

This is the equation of motion of the gravitational field h_{00}. (G^N: gravitational constant in N-dimensional space.) Because of the existence of additional hidden dimensions the term proportional to h_{00} becomes only significant for short separations between matter particles.

The (negative) interaction Lagrangian in EQM is derived from the Einstein-Hilbert action:

$$\begin{aligned}&\frac{1}{2}h_{00}T^{00}\left(1 - \frac{1}{2}h^{00}\right) \approx \\ &\frac{1}{2}m^2\psi^+\psi\left(h_{00} + \zeta^+_{00}\zeta_{00} + \frac{1}{4}\zeta^+_{00}\zeta_{00}h_{00} + \frac{3}{8}(\zeta^+_{00}\zeta_{00})^2\right).\end{aligned} \tag{6}$$

The graviton field h_{00} is defined in the non-relativistic limit with the gravonon operators ζ^+, ζ as: $h_{00}(x) = \mathrm{e}^{-ikct}\zeta_{00}(x)$, $h^{00}(x) = -\mathrm{e}^{ikct}\zeta^+_{00}(x)$. The first term in eq. (6) is Newton's gravity for mass m with density distribution $\psi^+\psi$, corresponding to the first term on the r.h.s of eq. (2). Terms of second, third and fourth order in the gravitational field emerge. The second and the fourth order terms in the gravitational field are repulsive terms. They are associated with dark energy. The attractive term in eq. (6) is due to third order terms in the gravonon field and it is associated with dark matter effects.

2.3. Cosmological constant problem solved in EQM

The analogy with Einstein's field equation allows to associate the dark energy term in eq. (2) with:

$$\Lambda g_{00} \rightarrow \frac{1}{2}\rho_{matter}\left(\zeta^+_{00}\zeta_{00} + \frac{3}{8}(\zeta^+_{00}\zeta_{00})^2\right) \tag{7}$$

hence Λ is an operator:

$$\Lambda \rightarrow \frac{1}{2}\rho_{matter}\left(\zeta^+_{00} + \frac{3}{8}(\zeta^+_{00})^2\zeta_{00}\right) \tag{8}$$

involving the density of matter ρ_{matter}, which is not a constant in space and in time. The "cosmological coincidence" $\Lambda \propto \rho_{matter}$ comes out of EQM. Our quantum gravity explains the value of Λ which is proportional to the density of observable matter for all cosmological times.

The cosmological term eq. (8) is a product of three operators and has enormous consequence for the solution of Schrödinger's equation, for instance it leads to the accelerated expansion of the universe. The cosmological constant paradox does not exist within EQM, no fit to anything is needed. This is in contrast with standard QFT, where the cosmological constant has a value 10^{123} times the experimental value and has the meaning of the vacuum energy. In EQM, also a QFT theory, which includes gravity as a quantum field, a cosmological constant with the correct value emerges.

IOP Conf. Series: Journal of Physics: Conf. Series **1275** (2019) 012055 doi:10.1088/1742-6596/1275/1/012055

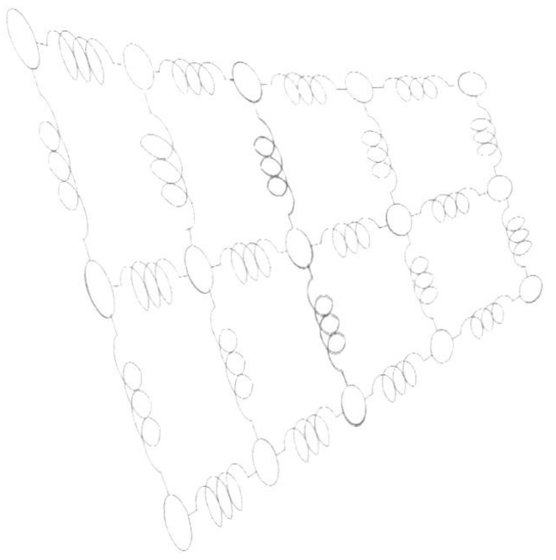

Figure 1. EQM model of the universe: a fraction of a light-atom-lattice, consisting of hydrogen atoms and photons, in intergalactic space. It might as well represent a Newton crystal [5] where the circles represent LALs.

3. The model universe in EQM

The model universe in EQM is a collection of hydrogen-atom lattices in free space and is described in detail in ref. [5]. A fraction of a hydrogen atom lattice is shown in fig. 1. In article [5] on cosmic expansion it is shown that the interactions are Newton-like, leading to soft gravonon modes with frequency $\omega_{Newton\ mode}$. Beabling via the Newton gravonons is only possible for a hydrogen-atom crystal with approximtely 10^6 atoms on the lattice sites and lattice constant of the order of 10 m in interaction with the photon field. The gravonon mode of the crystal is only negligibly influenced by the atom vibrations induced by the optical field but the strength of gravitational interaction in this light-atom-lattice (LAL) is enhanced so that the beabling condition is fulfilled.

4. Accelerated expansion of the universe in EQM

In order to study accelerated expansion one has to study interpenetrating LALs, where the separation between the hydrogen atoms might become smaller than the compactification radius. Then the higher order terms in eq. (5) become effective, i.e. dark energy emerges and modifies the unperturbed expansion of the universe as described in ref. [5]. This is symbolically illustrated in the upper panel of fig. 2 where the two mattresses represent now two LALs, i.e. the circles represent now single hydrogen atoms, in contrast to fig. 1.

In EQM the accelerated expansion of the universe is due to gravonon-gravonon repulsion between intergalactic hydrogen-atom crystals (dark energy repulsion). According to the $\gamma - \eta$ theory [6] this leads to the generation of harder gravonon modes of the Newton crystal $\omega_{Newton\ mode}$ and hence to acceleration of the expansion of the universe.

4.1. Deceleration parameter in EQM. $\gamma - \eta$ theory.

Accelerated expansion of the universe in EQM is due to dark energy repulsion, i.e. gravonon-gravonon repulsion between hydrogen-atom lattices (fig. 2, upper panel). These are the terms of second and fourth order in the gravitational field in the (negative) interaction Lagrangian of EQM (eq. 6).

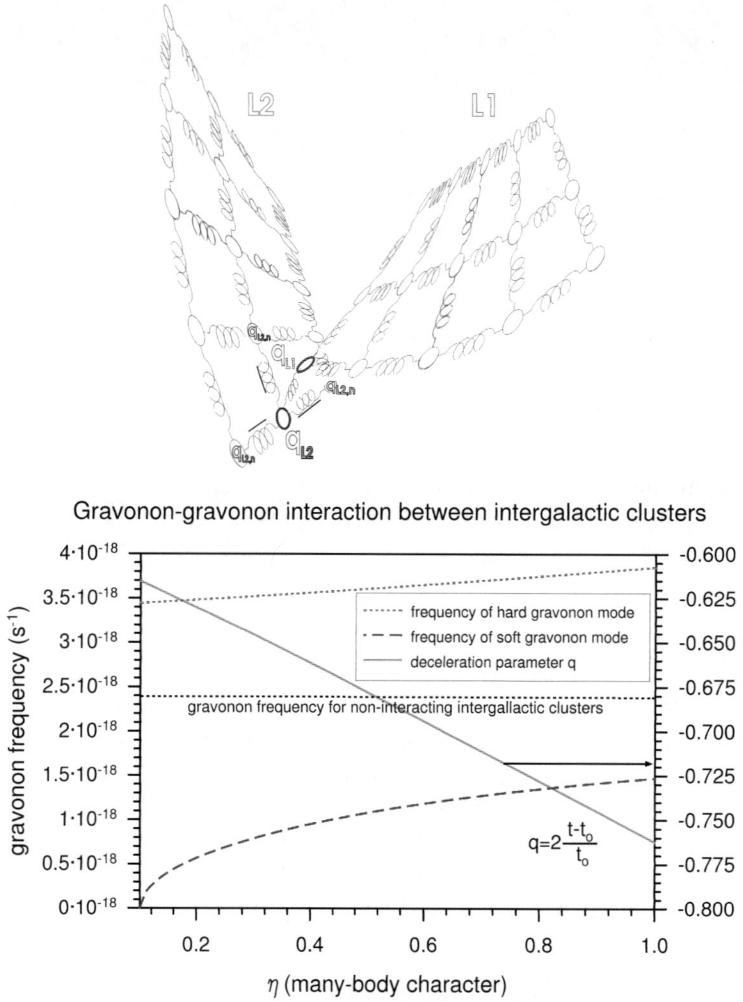

Figure 2. Upper panel: fractions of a two hydrogen atom crystals $L1$ and $L2$. One selected atom q_{L1} in $L1$ interacts directly with one atom q_{L2} in $L2$, which is involved in the Newton gravonon mode and couples to its neighbours $\{q_{L2_i}\}$ via gravonon-gravonon interactions. Lower panel: variation of the frequency of a soft (dashes, blue curve) and hard Newton gravonon modes (dots, red curve) in $L2$, resulting as a solution of eq. (9). This is the effect of the direct gravonon interaction between the two hydrogen atoms in the two intergalactic hydrogen-atom lattices, shown in the upper panel, and the gravonon-gravonon interactions induced by atom q_{L1} among the neighbours $\{q_{L2_n}\}$ of atom q_{L2} in the second lattice. The deceleration parameter $q(\eta)$ resulting from the faster expansion of $L2$ due to the gravonon-gravonon repulsion induced among its atoms, is plotted with the full green curve which may serve as a lower bound for the experimental value of q.

The Hamiltonian is that of the $\gamma - \eta$ theory [6], reformulated for a cosmological model of two intergalactic clusters ($\omega_{Newton} = \omega_{Newton\ mode}$ in ref. [5]):

$$\begin{aligned} H &= H_o + V \\ &= H_o - \gamma\sqrt{\frac{m_{L2}}{m_{L1}}}\omega^2_{Newton}q_{L1}q_{L2} + \frac{\eta}{2}\gamma\omega^2_{Newton}q^2_{L2} + \frac{1-\eta}{K}\gamma\omega^2_{Newton}q_{L2}\sum_n q_{L2,n} \\ &- \frac{1-\eta}{2K}\gamma\omega^2_{Newton}\sum_n q^2_{L2,n} \\ &= H_o - \omega^2_{Newton}q_{L1}q_{L2} + \frac{\eta}{2}\omega^2_{Newton}q^2_{L2} + \frac{1-\eta}{K}\omega^2_{Newton}q_{L2}\sum_n q_{L2,n} \\ &- \frac{1-\eta}{2K}\omega^2_{Newton}\sum_n q^2_{L2,n}. \end{aligned} \tag{9}$$

$\gamma = 1$, because it is the ratio of the force constants of the harmonic oscillators q_{L2} and q_{L1}, which are equal ($\omega_{L1} = \omega_{L2}$ and $m_{L1} = m_{L2}$). η is a measure of the many-body character of the interactions in lattice $L2$. $\eta = 0$ means extreme many-body character, $\eta = 1$ refers to a spring model. The many-body interactions in $L2$ are perturbed by the presence of the intergalactic lattice $L1$ and new vibrational modes are generated: a softer and a harder one compared to the non-perturbed Newton vibrational mode of a single hydrogen-atom lattice.

The result of the solution is the generation of a softer and a harder vibration modes compared to the Newton mode for a single lattice (fig. 2, lower panel). The softer vibration (blue curve, dashes) is not important for expansion because it is just an in-phase vibration of the atoms in lattice $L2$. The harder vibration (red curve, dots) is the counter-phase vibration of the hydrogen atoms in $L2$ and leads to a faster beable-limbo transition, hence to accelerated expansion of the lattice. Varying the weight of the many-body character of the interactions within lattice $L2$ changes the frequency of the hard vibration and as a consequence the Hubble time changes. A pure spring model of the gravonon-gravonon interactions in $L2$ yields harder Newton-like vibration than a vibration resulting from many-body interactions because in the first case the whole interaction force is concentrated between each pair of interacting atoms in $L2$, whereas in the latter case it is distributed among the surrounding atoms.

If just a single atom of one lattice interacts with a single atom of the second lattice, then nothing will change. Only if the two lattices penetrate into each other there will be an effect. But this is the extreme situation of strongest gravonon-gravonon repulsion and it yields a lower limit for the value of the deceleration parameter:

$$q = 2\frac{t - t_o}{t_o}, \tag{10}$$

t_o being the Hubble time. The deceleration parameter $q(\eta)$ varies with η in the range from -0.6 to -0.8 (full green curve in fig. 2). It can be compared to the value of -0.55 based on Planck spacecraft data [3]. The comparison of our result with the Planck data assumes that (nearly) every hydrogen atom in crystal $L1$ interacts with another hydrogen atom in crystal $L2$ in the way indicated in the upper panel of fig. 2. In fact it requires that every LAL in a Newton-mode-lattice penetrates a LAL in a different Newton-mode-lattice. Because this will hardly be realized in the intergalactic matter it is clear that the theoretical values of q are lower limits for the experimental Planck data.

Choosing one value for the frequency of the vibrational mode $\omega_{Newton} = 8.64 \times 10^{-35}$ a.u.$^{-1}$ of the interacting lattices yields, solving Scrödingers equation, a faster transition beable-limbo,

IOP Conf. Series: Journal of Physics: Conf. Series **1275** (2019) 012055 doi:10.1088/1742-6596/1275/1/012055

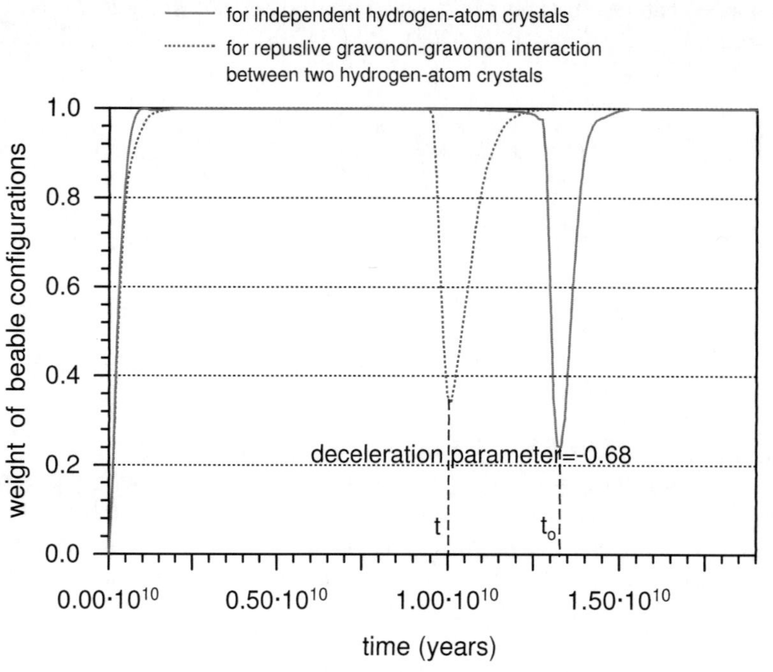

Figure 3. Development with time of the weight of the beable of a single hydrogen atom in one crystal entangled with non-perturbed gravonons of frequency of the order of 5.68×10^{-35} a.u.$^{-1}$, i.e. $t_o = 4.25 \times 10^{17}$ seconds, typical for a single crystal (red full curve). The weight of the beable due to entanglement of a hydrogen atom with the modified harder gravonons of frequency $\omega_{Newton} = 8.64 \times 10^{-35}$ a.u.$^{-1}$, i.e. $t = 2.80 \times 10^{17}$ s, one typical value for a modified Newton mode of the hydrogen atom crystals due to interaction with a second hydrogen atom crystal, is plotted with the blue dotted curve.

hence faster expansion of the interacting hydrogen-atom lattices (fig. 3).

The $\gamma - \eta$ model applies only if the distance between atoms on the different clusters are smaller than the compactification radius.

5. Conclusion

Terms associated with dark matter attraction and dark energy repulsion emerge in the Lagrangian of the quantum field theory EQM, starting from Einstein-Hilbert action in high dimensional spacetime. New particles, called gravonons, emerge in the non-relativistic limit. Gravonon-gravonon interactions are the physical background of dark matter and dark energy. Non-adiabatic effects due to interaction between gravonons and the Newton vibrational mode in the hydrogen atom crystal and gravonon-gravonon repulsive interaction are the physical reason for the accelerated expansion of the universe.

In EQM there is no cosmological coincidence problem, the cosmological "constant" being an operator, defined by the gravonon creation and destruction operators and the matter density. They differ in space and time.

Some typical problems of standard quantum mechanics, collected in table 1, have been resolved within the present theory and the reader can find the solutions in refs. [1, 2, 7]. Within EQM experimental observations in a range of time scales of 10^{-1} seconds to 10^{10} years and in a range of distances from atomic to cosmological scales are reproduced and explained.

References

[1] Doyen G and Drakova D 2015 *Found Physics* **45**, 959; 2014 arXiv: quant-ph/1408.2716
[2] Doyen G and Drakova D 2015 *J. Phys.: Conf. Series* **626** 012063
[3] for the parameters Ω_m and Ω_Λ estimated from the Planck spacecraft data: Planck collaboration 2015 *Astronomy and Astrophys.* **594** A13; 2016 arXiv: astro-ph/1502.01589; 2018 arXiv: astro-ph/1807.06209
[4] Riess A G et al. 2016 *Astrophys. J.* **826** 56
[5] Doyen G and Drakova D this book
[6] Doyen G 1982 in *Vibrations at Surfaces* eds R Caudano, J-M Gilles and A A Lucas, (Plenum Press) p 105
[7] Doyen G and Drakova D 2011 *J. Phys.: Conf. Series* **306** 012033; Drakova D and Doyen G 2012 arXiv: quant-ph/1204.5606; Drakova D and Doyen G 2013 *J. Phys.: Conf. Series* **442** 012049; Doyen G and Drakova D 2014 arXiv: quant-ph/1408.2716; Drakova D and Doyen G 2015 arxiv: quant-ph/1220200

Quantum thermodynamic instabilities in compact stars

A Lavagno[1,4], **D Pigato**[2] **and G Gervino**[3,4]

[1]Department of Applied Science and Technology, Politecnico di Torino, I-10129 Torino, Italy
[2]U.B.A., I-13897 Occhieppo Inferiore, Italy
[3]Dipartimento di Fisica, Università di Torino, I-10126 Torino, Italy
[4]INFN, Sezione di Torino, I-10125 Torino, Italy

Abstract. We study the existence of thermodynamic instabilities in the nuclear equation of state relative to the high density regime reached in the central core of compact stars. In the framework of a relativistic mean-field theory, we analyze the asymmetric nuclear properties in beta-equilibrium, including hyperons and Delta-isobar degrees of freedom. We investigate a finite density phase transition characterized by pure hadronic matter with the presence of mechanical instability (relative to the fluctuation of baryon number) and of chemical-diffusive instability (relative to the fluctuation of electric charge concentration). We find that, in the presence of thermodynamic instabilities, two hadronic phases with different values of electric charge content may coexist, with several phenomenological consequences in the physics of compact stars.

The new accumulating data from X-ray satellites provide important information on the structure and formation of compact stellar objects giving a unique opportunity to explore the behavior of high density nuclear matter [1]. Comparing the experimental data with different theoretical models, it is possible to study fundamental aspects of nuclear and hadron physics at different regimes [2, 3, 4, 5, 6, 7, 8]. This demands a detailed knowledge of the bulk thermodynamics and the equation of state of strongly interacting matter.

The behavior of matter at large densities is still poorly known but, on general grounds, new degrees of freedom are expected to appear and they should generate a softening of the equation of state. In this direction, many efforts have been focused on searching for possible phase transitions in heavy-ion collisions. At low temperatures and subnuclear densities, a liquid-gas type of phase transition was first predicted theoretically and later observed experimentally in a nuclear multifragmentation phenomenon at intermediate-energy nuclear reactions [9, 10].

A relevant aspect of a system with two conserved charges (baryon and electric charge numbers) is that the phase transition is of second order from the viewpoint of Ehrenfests definition. At variance with the so-called Maxwell construction for one conserved charge, the pressure is not constant in the mixed phase and, therefore, the incompressibility does not vanish [11]. Moreover, for a binary system with two phases, the binodal coexistence surface is two dimensional and the instabilities in the mixed phase arise from quantum fluctuations in the electric charge concentration (chemical instability) and in the baryon density (mechanical instability) [12, 13, 14].

In the present paper we study β-stable nuclear matter at zero temperature by requiring the Gibbs conditions on the global conservation of baryon number and zero net electric charge in the

Content from this work may be used under the terms of the Creative Commons Attribution 3.0 licence. Any further distribution of this work must maintain attribution to the author(s) and the title of the work, journal citation and DOI.
Published under licence by IOP Publishing Ltd

framework of a quantum relativistic mean-field model in which the interaction between baryons is mediated by the exchange of a scalar meson σ, an isoscalar vector meson ω, and a isovector vector ρ. We consider the recently introduced parametrization of the nuclear equation of state called SFHo which takes into account of the experimental observation related to the symmetry energy and of constraints from terrestrial and astrophysical data relative to the density derivative of the symmetry energy [15].

A crucial contribution to the formation of thermodynamic instabilities is that a state of high density resonance $\Delta(1232)$-isobar matter may be formed in the core of the compact star. Transport model calculations and experimental results indicate that an excited state of baryonic matter is dominated by the Δ-resonance at the energy from AGS to RHIC [12, 16, 17, 18, 19]. It has been pointed out that the existence of Δ-isobars can be very relevant also in the core of neutron stars [20, 21, 22]. Moreover, in symmetric nuclear matter and in the framework of a non-linear Walecka model, it has been predicted that a phase transition from nucleonic matter to Δ-excited nuclear matter can take place and the occurrence of this transition sensibly depends on the Δ-meson coupling constants [23, 24].

Qualitatively, it has been possible to establish that the Δs inside a nucleus feel an attractive potential. There are several purely theoretical studies on the properties of the isobars in the nuclear medium: for instance, in Ref. [25], from QCD sum rules, it has been found that the coupling ratio $\omega - \Delta$ with respect to ω-nucleons is significantly smaller than one. From phenomenological analysis of data from electron-nucleus, photoabsorption and pion nucleus scattering can be extracted different constraints related to the values of the Δ-meson coupling constants $x_{\omega\Delta} = g_{\omega\Delta}/g_{\omega N}$ and $x_{\sigma\Delta} = g_{\sigma\Delta}/g_{\sigma N}$, used in the nuclear equation of state [21].

In Fig. 1, we report the relation between the coupling ratios obtained by considering the experimental constraints (see Ref.[21] for details). Furthermore, for coupling ratios in the lower region of the red line the equation of state contains thermodynamic instabilities. Such a region of values is compatible with the experimental constraints and can be very interesting to investigate how this matter of fact can influence the bulk properties of compact stars.

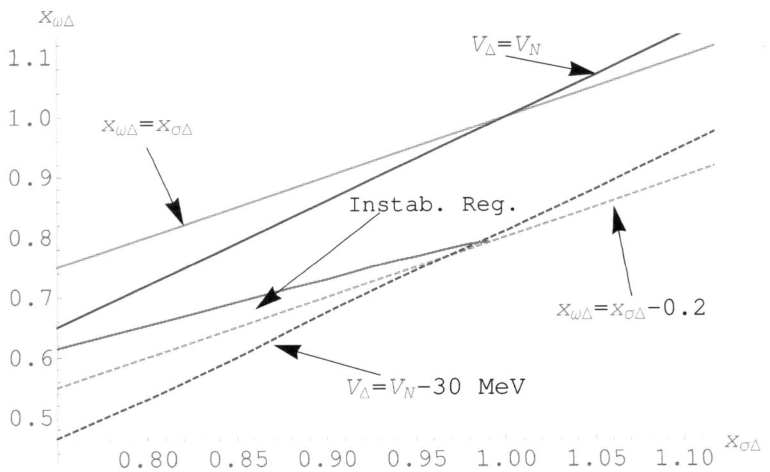

Figure 1. Relation between the coupling ratios $x_{\omega\Delta}$ and $x_{\sigma\Delta}$ in terms of experimental constraints related to pion and electron scattering and from photoabsorption on nuclei. Below the red line, thermodynamic instabilities are present in the equation of state.

Let us remember that we are dealing with the study of a multi-component system with two conserved charges: baryon number and electric charge. The baryon chemical potential μ_i of the

IOP Conf. Series: Journal of Physics: Conf. Series **1275** (2019) 012056 doi:10.1088/1742-6596/1275/1/012056

ith baryon particle are obtained by the β-equilibrium conditions: $\mu_i = \mu_B + c_i\,\mu_C$,where μ_B and μ_C are the chemical potentials associated with conservation of the baryon number and electric charge, respectively, and c_i is the electric charge of the ith baryon.

Assuming the presence of two phases (denoted as I and II, respectively), the system is stable against the separation in two phases if the free energy of a single phase is lower than the free energy in all two phases configuration. The phase coexistence is given by the Gibbs conditions

$$\mu_B^I = \mu_B^{II}\,, \qquad \mu_C^I = \mu_C^{II}\,, \tag{1}$$

$$P^I(T,\mu_B,\mu_C) = P^{II}(T,\mu_B,\mu_C)\,. \tag{2}$$

Therefore, at a given baryon density ρ_B and at a given net electric charge density $\rho_C = r_c\,\rho_B$ (with $r_c = Z/A$), the chemical potentials μ_B are μ_C are univocally determined. An important feature of this conditions is that, unlike the case of a single conserved charge, the pressure in the mixed phase is not constant and, although the total ρ_B and ρ_C are fixed, baryon and charge densities can be different in the two phases. For such a system in thermal equilibrium, the possible phase transition can be characterized by mechanical and chemical instabilities [11, 12, 26]. As usual the condition of the mechanical stability implies

$$\rho_B \left(\frac{\partial P}{\partial \rho_B}\right)_{T,\rho_C} > 0\,. \tag{3}$$

Furthermore, the chemical stability condition is satisfied if [12]

$$\left(\frac{\partial \mu_C}{\partial r_c}\right)_{T,P} > 0 \;\; \text{or} \;\; \begin{cases} \left(\dfrac{\partial \mu_B}{\partial r_c}\right)_{T,P} < 0\,, & \text{if } r_c > 0\,, \\ \left(\dfrac{\partial \mu_B}{\partial r_c}\right)_{T,P} > 0\,, & \text{if } r_c < 0\,. \end{cases} \tag{4}$$

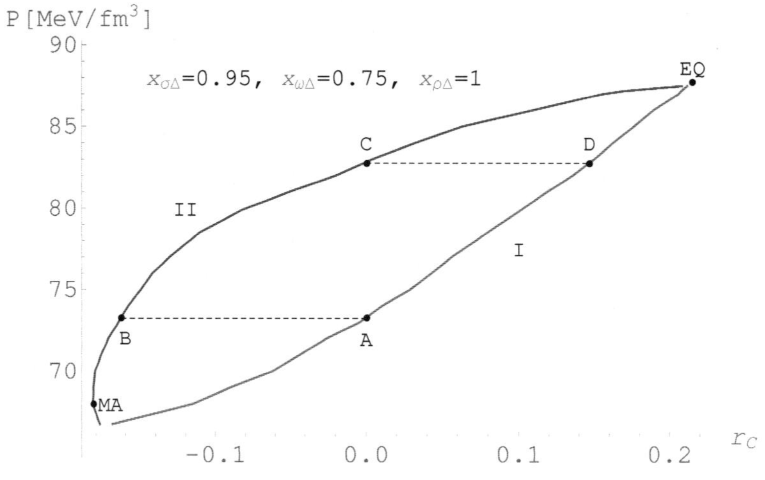

Figure 2. Binodal section with in evidence the point of maximum asymmetry (MA) and the point of equal equilibrium (EQ). In the mixed phase (from the points A to C) the system has different electric charge concentration r_c in the two phases.

In Fig. 2, we show the binodal section for a value of the Δ-meson coupling ratios where both mechanical and chemical-diffusive instabilities are present in the nuclear equation of state.

The right branch (phase I at lower density) corresponds to the initial phase, where the dominant component of the system is given by nucleons. The left branch (phase II at higher density) is related to the final phase, where the system is composed primarily by Δ-isobar degrees of freedom. In presence of Δ-isobars the phase coexistence region extends up to regions of negative electric charge fraction r_c due to the formation of Δ^- particles.

During the isothermal compression from the initial point A, the system becomes unstable and starts to be energetically favorable the separation into two phases, therefore an infinitesimal Δ-dominant phase appears in B, at the same temperature and pressure but at higher baryon density and different electric charge content. In the phase transition, each phase evolves towards a configuration with increasing r_c until the value of pressure in the point C in the phase II where the system becomes stable.

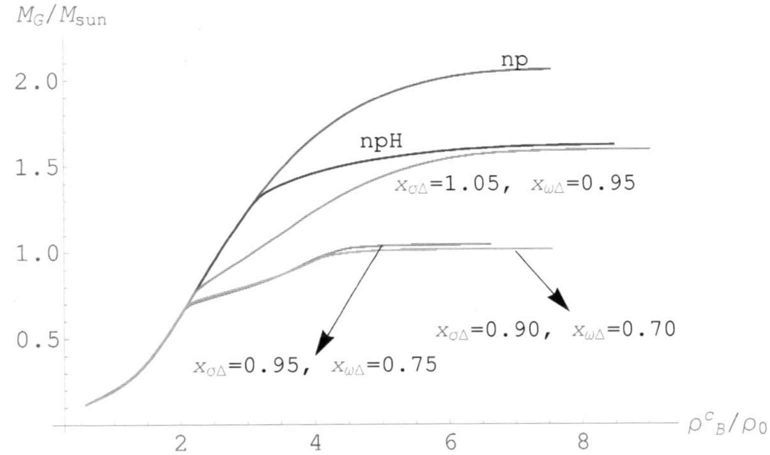

Figure 3. Gravitational mass of the stars (in units of the solar mass) as a function of the central baryon density (in units of the nuclear saturation density ρ_0) for different parameter sets. The symbol np stands for stars without hyperons and Δ degrees of freedom, npH with hyperons but without Δ particles. The two lower curves correspond to values of the meson-Δ coupling constants where thermodynamic instabilities are present.

We are going now to investigate the relevance of the above considered thermodynamic instabilities in the core of the compact stars. In Fig. 3, we report the gravitational mass as a function of central baryon density ρ_B^c of the star in absence (np) and in presence (npH) of hyperons, without Δ-isobar degrees of freedom. Otherwise, the three lower curves contain Δ particles. For the coupling ratios: $x_{\sigma\Delta} = 1.05$, $x_{\omega\Delta} = 0.95$, the equation of state is always stable and the core of the star is composed by nucleons, hyperons and Δ-isobars. Instead for the values: $x_{\sigma\Delta} = 0.95$, $x_{\omega\Delta} = 0.75$ and $x_{\sigma\Delta} = 0.90$, $x_{\omega\Delta} = 0.70$, mechanical and chemical-diffusion instability are present and the Gibbs construction must be applied to equation of state.

The presence of Δ-isobar degrees of freedom smooths the equation of state. This effect is much more evident when thermodynamic instabilities are present with a strong reduction of the maximum gravitational mass. On the other hand, very compact object with smaller radii can be formed. These features can be very relevant in the phenomenological interpretation of compact star objects.

References

[1] Antoniadis J *et al.* 2013 *Science* **340** 1233232
[2] Drago A, Lavagno A, Metzger B D and Pagliara G 2016 *Phys. Rev.* D **93** 103001

[3] Drago A, Lavagno A, Pagliara G and Pigato D 2016 *Eur. Phys. J* A **52** 40
[4] Torres J R, Gulminelli F, Menezes D P 2017 *Phys. Rev.* C **95** 025201
[5] Lavagno A, Pigato D and Quarati P 2010 *J. Phys. G: Nucl. Part. Phys.* **37** 115102
[6] Christian J -E, Zacchi A and Schaffner-Bielich J 2019 *Phys. Rev.* D **99** 023009
[7] Lavagno A and Quarati P 2001 *Phys. Lett.* B **498** 47
[8] Drago A and Lavagno A 2001 *Phys. Lett.* B **511** 229
[9] Goodman A L, Kapusta J I and Mekjian A Z 1984 *Phys. Rev.* C **30** 851
[10] Pochodzalla J *et al.* 1995 *Phys. Rev. Lett.* **75** 1040
[11] Müller H and Serot B D 1995 *Phys. Rev.* C **52** 2072
[12] Lavagno A and Pigato D 2012 *Phys. Rev.* C **86** 024917
[13] Lavagno A and Narayana Swamy P 2010 *Found. Phys.* **40** 814
[14] Lavagno A, Scarfone A M and Narayana Swamy P 2006 *Eur. Phys. J.* C **47** 253
[15] Lattimer J M and Lim Y 2013 *Astrophys. J.* **771** 51
[16] Hofmann M, Mattiello R, Sorge H, Stocker H and Greiner W 1995 *Phys. Rev.* C **51** 2095
[17] Li Z, Mao G, Zhuo Y and Greiner W 1997 *Phys. Rev.* C **56** 1570
[18] Lavagno A 2010 *Phys. Rev.* C **81** 044909
[19] Lavagno A 2013 *Eur. Phys. J.* A **49** 102
[20] Schurhoff T, Schramm S and Dexheimer V 2010 *Astrophys. J.* **724** L74
[21] Drago A, Lavagno A, Pagliara G and Pigato D 2014 *Phys. Rev.* C **90** 065809
[22] Drago A, Lavagno A and Pagliara G 2014 *Phys. Rev.* D **89** 043014
[23] Waldhauser B W *et al.* 1987 *Phys. Rev.* C **36** 1019
[24] Li Z *et al.* 1997 *Phys. Rev.* C **56** 1570
[25] Jin X 1995 *Phys. Rev.* C **51** 2260
[26] Lavagno A, Narayana Swamy P 2010 *Physica* A **389** 993

IOP Conf. Series: Journal of Physics: Conf. Series **1275** (2019) 012057 doi:10.1088/1742-6596/1275/1/012057

Simulation of Quantum Universe

Sang Pyo Kim

Department of Physics, Kunsan National University, Kunsan 54150, Korea

E-mail: `sangkim@kunsan.ac.kr`

Abstract. Quantum simulation provides quantum systems under study with analogous controllable quantum systems and has wide applications from condensed-matter physics to high energy physics and to cosmology. The quantum system of a homogeneous and isotropic field in the Friedmann-Robertson-Walker universe can be simulated by a charge in an electrically modulated ion trap. The quantum states of these time-dependent oscillators are constructed by quantum invariants. Further, we propose simulation of quantum Friedmann-Robertson-Walker universe with a minimal massive scalar field by a charged scalar field in a homogeneous, time-dependent, magnetic field in quantum electrodynamics and investigate the Cauchy problem of how the wave functions evolve.

1. Introduction

Understanding the evolution of quantum systems has been one of unsolved and challenging problems since the advent of quantum theory. In particular, the information about the evolution of a quantum system increases exponentially or geometrically with the degrees of freedom and requires huge data. Quantum simulation is an alternative to large or uncontrollable quantum systems and has wide applications from condensed-matter physics to high energy physics and to cosmology [1]. The advancement of technology in separating and confining a single charged or neutral particle within ion traps makes it possible precisely measure physical quantities with the unprecedented precision [2, 3].

In gravity there have been many proposals to simulate or imitate nonperturbative nature of quantum fields in curved spacetimes. An acoustic analogue of a black hole is the most well known model [4]. A sonic black hole can be formed in Bose-Einstein condensates [5] and the Unruh effect can be measured by an accelerating electron in intense lasers [6]. Ion traps can be used to simulate and measure the Gibbons-Hawking radiation of a de Sitter space [7].

Simulation of cosmology is another interesting arena because the homogeneous Bianchi models undergo Belinskii-Khalatnikov-Lifshitz's chaotic motions and infinite oscillations near the singularity (for a review and references, see refs. [8, 9] and also for numerical approaches, see ref. [10]). Though the Friedmann-Robertson-Walker (FRW) universe exhibits a simpler behavior near the singularity than the homogeneous Bianchi models, the quantum FRW cosmology with a minimal massive scalar field shows complicated and interesting motions that undergo infinite oscillations [11, 12] in contrast to the classical counterpart [13].

The main purpose of this paper is to propose simulations of quantum universe of the FRW geometry with a minimal massive scalar field both in the second quantized theory and the third quantized theory in the superspace of the geometry and scalar field. In the second quantized theory, the wave function of the universe obeys the Wheeler-DeWitt (WDW) equation,

Content from this work may be used under the terms of the Creative Commons Attribution 3.0 licence. Any further distribution of this work must maintain attribution to the author(s) and the title of the work, journal citation and DOI.
Published under licence by IOP Publishing Ltd

IOP Conf. Series: Journal of Physics: Conf. Series **1275** (2019) 012057 doi:10.1088/1742-6596/1275/1/012057

a relativistic wave equation, in the superspace of the three-geometry and the scalar field. In the third quantized theory, the WDW equation follows from an action of quantum fields in the superspace and the wave function has the status of operators that create or annihilate the universe of a given wave function just as a second quantized field creates or annihilates a particle and antiparticles. In the second quantized or third quantized theory, the universal interaction of gravity manifests itself through parametric interactions of the underlying geometry of general relativity. The quantum universe of the FRW geometry with massless fields is equivalent to a charged scalar field in time-dependent electric fields and that with a massive complex field to the charged field in time-dependent magnetic fields.

The organization of this paper is as follows. In section 2, we explore the quantum states of a charge modulated by time-dependent electric fields in an ion trap and compare those with the quantum system of a homogeneous and isotropic field (inflaton) in the FRW universe. By using linear quantum invariants acting as the annihilation and creation operators, the Fock space of all excited harmonic states can be constructed and the most general coherent-squeezed states and their dispersion relations can also be found. In section 3, we explain the third quantized theory of the FRW universe with a minimal massive complex scalar field and compare the theory with quantum electrodynamics (QED) of a charged scalar field in a homogeneous, time-dependent, magnetic field. The simulation of the quantized universe is proposed using a charged field in an electromagnetic field in QED. In section 4, we study the evolution of wave functions of the WDW equation and compare these with Landau levels in the time-dependent magnetic field.

2. Ion traps and simulation of quantum fields in FRW Universe

A Paul trap or an ion trap is a device of generating two- or three-dimensional quadrupole electromagnetic fields that confine a charged or neutral particle [2, 3]. The charge with mass m in properly arrayed and oriented fields has decoupled quadratic Hamiltonian

$$H(t) = \sum_{i=1}^{3} H_{(i)}(t) := \sum_{i=1}^{3} \frac{1}{2m} p_{(i)}^2 + \frac{m}{2} \omega_{(i)}^2(t) x_{(i)}^2, \tag{1}$$

where the frequencies squared are an interaction of the charge q ($q > 0$) with an electric potential

$$\omega_{(i)}^2(t) = \frac{qU_0}{m}\alpha_i + \frac{qf(t)}{m}\tilde{\alpha}_i. \tag{2}$$

Here, $x_{(i)}$ denotes x, y and z, and $p_{(i)} = m\dot{x}_{(i)}$ and $f(t)$ is a modulating, time-dependent, electric field. The equations of motion are the harmonic motions with time-dependent frequencies squared

$$\ddot{x}_{(i)} + \omega_{(i)}^2(t) x_{(i)} = 0. \tag{3}$$

The charge neutrality of the Paul trap demands the Gauss law, $\nabla^2 \Phi(\vec{x}, t) = 0$, which leads to the condition for any time

$$\sum_{i=1}^{3} \alpha_i = 0, \qquad \sum_{i=1}^{3} \tilde{\alpha}_i = 0. \tag{4}$$

The condition (4) is the only constraint for, otherwise, decoupled oscillators (1). Depending on the parameters α_i, $\tilde{\alpha}_i$ and the modulation field $f(t)$, at least one of frequencies squared can take a negative value.

The FRW universe with the metric

$$ds^2 = -dt^2 + a^2(t) d\mathbf{x}_3^2, \tag{5}$$

by redefining a new time variable $d\eta = dt/a^3(t)$, can have another metric

$$ds^2 = -a^6(\eta)d\eta^2 + a^2(\eta)d\mathbf{x}_3^2. \tag{6}$$

For instance, a de Sitter space $a = e^{Ht}$, $\eta = -e^{-3Ht}/3H$ has a range $(-\infty, 0)$. Then the Hamiltonian for a complex scalar field with mass μ in the new metric is

$$H_{\mathrm{M}} = \int \frac{d^3\mathbf{k}}{(2\pi)^3}\Big[\pi^*_{\mathbf{k}}\pi_{\mathbf{k}} + \omega^2_{\mathbf{k}}(\eta)\phi^*_{\mathbf{k}}\phi_{\mathbf{k}}\Big], \tag{7}$$

where

$$\omega^2_{\mathbf{k}}(\eta) = a^6(\eta)\Big(\mu^2 + \frac{\mathbf{k}^2}{a^2(\eta)}\Big). \tag{8}$$

The complex scalar field is equivalent to two real scalar fields. The equation of motion for each momentum is given by

$$\ddot{\phi}_{\mathbf{k}}(\eta) + \omega^2_{\mathbf{k}}(\eta)\phi_{\mathbf{k}}(\eta) = 0, \tag{9}$$

where dots denote derivatives with respect to η, and the same equation holds for $\phi^*_{\mathbf{k}}$. A homogeneous and isotropic field (inflaton) suppresses all momenta to $\mathbf{k} = 0$ and the Hamiltonian reduces to a quantum system of complex variable $\phi(\eta)$.

The exact quantum states for the time-dependent oscillators (1) and (7) can be found in terms of a quadratic quantum invariant by Lewis and Riesenfeld [14]. Following refs. [15, 16], we use two linear quantum invariants that act as the annihilation and creation operators (units of $\hbar = c = l_{\mathrm{P}}^2 = 16\pi/m_{\mathrm{P}}^2 = 1$ will be used)

$$\begin{aligned} \hat{a}_{(i)}(t) &= i[u^*_{(i)}(t)\hat{p}_{(i)} - m\dot{u}^*_{(i)}(t)\hat{x}_{(i)}], \\ \hat{a}^\dagger_{(i)}(t) &= -i[u_{(i)}(t)\hat{p}_{(i)} - m\dot{u}_{(i)}(t)\hat{x}_{(i)}], \end{aligned} \tag{10}$$

where $u_{(i)}(t)$ is a complex solution to eqs. (3), (9) and (16) below that satisfies the Wronskian condition

$$\mathrm{Wr}[u_{(i)}(t), u^*_{(i)}(t)] = \frac{i}{m}. \tag{11}$$

Note that the complex solution is not unique since any linear superposition also satisfies eq. (11)

$$u_{(i\nu)}(t) = \mu u_{(i)}(t) + \nu u^*_{(i)}(t), \qquad |\mu|^2 - |\nu|^2 = 1. \tag{12}$$

In quantum field theory, the choice of $u_{(i\nu)}(t)$ corresponds to the selection of the vacuum state: the vacuum constructed from $u_{(i\nu)}(t)$ is a squeezed vacuum of $u_{(i)}(t)$ and vice versa, as shown below. The most general coherent(displaced)-squeezed states take the form [16]

$$\begin{aligned} \Psi_{(i\nu)n}(x_{(i)}; x_{(i)c}, p_{(i)c}; u_{(i\nu)}(t)) &= \Big(\frac{1}{\sqrt{2\pi}2^n n!|u_{(i\nu)}|}\Big)^{\frac{1}{2}}\Big(\frac{u_{(i\nu)}}{|u_{(i\nu)}|}\Big)^{n+\frac{1}{2}} e^{ip_{(i)c}x_{(i)}} \\ &\times H_n\Big(\frac{x_{(i)} - x_{(i)c}}{\sqrt{2}|u_{(i\nu)}|}\Big)\exp\Big[i\frac{m}{2}\frac{\dot{u}^*_{(i\nu)}}{u^*_{(i\nu)}}(x_{(i)} - x_{(i)c})^2\Big], \end{aligned} \tag{13}$$

where H_n is the Hermite polynomial and the centroid of the wave packets are

$$\begin{aligned} x_{(i)c}(t) &= \langle \Psi_{(i\nu)n}(x_{(i)}; u_{(i\nu)}(t)) | \hat{x}_{(i)} | \Psi_{(i\nu)n}(x_{(i)}; u_{(i\nu)}(t)) \rangle, \\ p_{(i)c}(t) &= \langle \Psi_{(i\nu)n}(x_{(i)}; u_{(i\nu)}(t)) | \hat{p}_{(i)} | \Psi_{(i\nu)n}(x_{(i)}; u_{(i\nu)}(t)) \rangle. \end{aligned} \tag{14}$$

Their dispersion relations are

$$\begin{aligned} \Delta x_{(i)}(t) &= \langle \Psi_{(i\nu)n}(x_{(i)}; u_{(i\nu)}(t)) | (\hat{x}_{(i)} - x_{(i)c}(t))^2 | \Psi_{(i\nu)n}(x_{(i)}; u_{(i\nu)}(t)) \rangle \\ &= (2n+1) u^*_{(i\nu)}(t) u_{(i\nu)}(t) \\ \Delta p_{(i)}(t) &= \langle \Psi_{(i\nu)n}(x_{(i)}; u_{(i\nu)}(t)) | (\hat{p}_{(i)} - p_{(i)c}(t))^2 | \Psi_{(i\nu)n}(x_{(i)}; u_{(i\nu)}(t)) \rangle \\ &= (2n+1) m \dot{u}^*_{(i\nu)}(t) \dot{u}_{(i\nu)}(t) \rangle. \end{aligned} \tag{15}$$

The scalar field model (7) has $m = 1$ from eqs. (10) to (15).

The Paul trap has a sinusoidal field $f(t) = \tilde{U}_0 \cos(\omega_{\rm rf} t)$ with a radio frequency to confine the charge. In real experiments, a pure oscillating electric field with $\alpha_i = 0$ for $i = 1, 2, 3$ and $\tilde{\alpha}_3 = -(\tilde{\alpha}_1 + \tilde{\alpha}_2)$ dynamically confines the charge while $\alpha_3 = -(\alpha_1 + \alpha_2) > 0$ and $\tilde{\alpha}_1 = -\tilde{\alpha}_2$ provides also a dynamical confinement [3]. An experimental design of three-dimensional trap, $U_0 = 0 \sim 50\,\mathrm{V}$, $\tilde{U}_0 = 100 \sim 500\,\mathrm{V}$ and $\omega_{\rm rf}/2\pi = 100\,\mathrm{kHz} \sim 100\,\mathrm{MHz}$ [3], implies that at least one of the equations of motion has a negative frequency squared

$$x''_{(i)} + [a_i - 2q_i \cos(2z)] x_{(i)} = 0, \tag{16}$$

where the prime denotes derivative with respect to z and

$$a_i = \frac{4qU_0}{m\omega_{\rm rf}^2} \alpha_i, \qquad q = -\frac{2q\tilde{U}_0}{m\omega_{\rm rf}^2} \tilde{\alpha}_i, \qquad z = \frac{\omega_{\rm rf}}{2} t. \tag{17}$$

The Mathieu equation (16) has either stability regions or instability regions depending on parameters a_i and q_i, whose stability regions lead to confinement of the charge [17]. Two independent real solutions with $\mathrm{Wr}[w_{(i)0}, w_{(i)1}] = 1$ combine to yield a complex solution to eq. (16) satisfying the Wronskian condition (11)

$$u_{(i)}(t) = \frac{1}{\sqrt{2}} (w_{(i)0}(t) - i w_{(i)1}(t)). \tag{18}$$

An ion trap experiment was proposed to observe by the Unruh-DeWitt's monopole detector the analogue of quantum effects in cosmological spacetimes, such as the Gibbons-Hawking effect and inflationary structure [7]. Phonon excitations of an ion in a trap with a trap frequency exponentially modulated with κ exhibit a thermal spectrum with the Unruh temperature $T_{\rm U} = \kappa / k_{\rm B}$ [18].

3. Third Quantized Universe vs Second Quantized scalar QED

We study the quantum cosmology of the FRW geometry with a minimal complex scalar field ϕ with mass μ, which has the extended supermetric

$$ds^2 = -da^2 + a^2 d\phi^* d\phi = e^{2\alpha}(-d\alpha^2 + d\phi^* d\phi). \tag{19}$$

Then, the Hamiltonian constraint in the Arnowitt-Deser-Misner (ADM) formalism and the WDW equation take the form (for a review and references, see ref. [19])

$$\mathrm{H}(\alpha, \phi, \phi^*) = -\underbrace{(\pi_\alpha^2 + V_{\rm G}(\alpha))}_{\text{gravity part } \mathrm{H_G}} + \underbrace{(\pi^*_\phi \pi_\phi + \mu^2 e^{6\alpha} \phi^* \phi)}_{\text{scalar field part } \mathrm{H_M}} \tag{20}$$

and

$$\Big[\Box + e^{6\alpha}\mu^2\phi^*\phi - V_{\rm G}(\alpha)\Big]\Psi(\alpha,\phi,\phi^*) = 0, \tag{21}$$

where the d'Alembertian in the superspace and the gravitational potential are, respectively,

$$\Box = \frac{\partial^2}{\partial\alpha^2} - \frac{\partial^2}{\partial\phi\partial\phi^*}, \qquad V_{\rm G}(\alpha) = ke^{4\alpha} - 2\Lambda e^{6\alpha}. \tag{22}$$

Here, $k = 1, 0, -1$ for a closed, flat and open universe and Λ denotes a cosmological constant. In terms of the real and imaginary components ($\phi = \phi_1 + i\phi_2$), the complex scalar field equals to two real scalar fields $\phi^*\phi = \phi_1^2 + \phi_2^2$ and the d'Alembertian becomes a three-dimensional wave operator

$$\Box = \frac{\partial^2}{\partial\alpha^2} - \frac{\partial^2}{\partial\phi_1^2} - \frac{\partial^2}{\partial\phi_2^2}. \tag{23}$$

The WDW equation is a relativistic wave equation for scalar fields with α-dependent mass $e^{3\alpha}\mu$ and with an α-dependent effective mass $V_{\rm G}(\alpha)$ for the field Ψ.

There have been a few approaches to understand and study the WDW equation (21). One approach is to look for the wave function with specific boundary conditions necessary for describing the present universe. Hartle-Hawing's no-boundary wave function is such a wave function that matches the present Lorentzian geometry to a Euclidean compact manifold [20, 21] and thus satisfies a specific boundary condition. The other approach is to interpret eq. (21) as a relativistic wave equation for scalar fields with α-dependent mass and an α-dependent effective mass $V_{\rm G}$ in the superspace (α, ϕ_1, ϕ_2) with supermetric (19). The Cauchy problem has been developed in refs. [12, 22] that evolves the initial data at a given hypersurface $\Sigma = \alpha_0$ to any hypersurface $\Sigma = \alpha$ of the superspace in the same way as the ADM formalism does in classical general relativity.

Along the line of the second approach of relativistic wave equation, we may third-quantize the WDW equation in the superspace. Then the WDW equation follows from the third quantized action

$$S = \int d\alpha d\phi_1 d\phi_2\Big[-\Psi\frac{\partial^2}{\partial\alpha^2}\Psi + \Psi\Big(\frac{\partial^2}{\partial\phi_1} + \frac{\partial^2}{\partial\phi_2}\Big)\Psi - \Psi\Big(e^{6\alpha}\mu^2(\phi_1^2+\phi_2^2) - V_{\rm G}(\alpha)\Big)\Psi\Big]. \tag{24}$$

The super-Hamiltonian from the action (24) is

$$H = \int d\phi_1 d\phi_2\Big[\Pi^2 + \Psi\Big(-\frac{\partial^2}{\partial\phi_1} - \frac{\partial^2}{\partial\phi_2} + e^{6\alpha}\mu^2(\phi_1^2+\phi_2^2)\Big)\Psi - \Psi V_{\rm G}(\alpha)\Psi\Big], \tag{25}$$

where $\Pi = \partial\Psi/\partial\alpha$ is the conjugate to the wave function Ψ. Massless fields lead to a super-Hamiltonian of α-dependent oscillators [23, 24, 25, 26, 27, 28] and the tachyonic state was studied for the closed universe [11]. We note that for the massless fields the wave function separates by $e^{i\pi_i\phi_i}$ $(i = 1, 2)$ and the super-Hamiltonian (25) is an α-dependent oscillator, which can be simulated by ion traps in section 2.

We now compare the WDW equation (21) for the massive complex field and the Hamiltonian (25) with those of a charged scalar field in a magnetic field in scalar QED. We note that the third quantization of a massive field is analogous to the second quantized Klein-Gordon field with charge q in a homogeneous, time-dependent, magnetic field

$$\mathbf{A}(t) = \frac{1}{2}\mathbf{B}(t)\times\mathbf{x}. \tag{26}$$

In fact, the transverse motion of a charged scalar field with mass m in a temporal, homogeneous, magnetic field along the z-direction is given by [29]

$$\Big[\frac{\partial^2}{\partial t^2} + \Big(\mathbf{p}_\perp^2 + \Big(\frac{qB(t)}{2}\Big)^2 \mathbf{x}_\perp^2 - qB(t)L_z\Big) + (m^2 + k_z^2)\Big]\Phi_\perp(t, \mathbf{x}_\perp) = 0. \tag{27}$$

The Hamiltonian of the transverse part of the field has the form [30]

$$H_\perp(t) = \int d^2\mathbf{x}_\perp^2 \Big[\pi_\perp^* \pi_\perp + \Phi_\perp^* (\mathbf{p}_\perp^2 + \omega_L^2(t)\mathbf{x}_\perp^2 - 2\omega_L(t)L_z)\Phi_\perp\Big]. \tag{28}$$

Thus the analog of quantum universe (25) and the second quantized scalar QED (28) is manifest:

$$\nabla_\perp^2 \Leftrightarrow \frac{\partial^2}{\partial\phi_1} + \frac{\partial^2}{\partial\phi_2}, \qquad \omega_L(t) \Leftrightarrow e^{3\alpha}\mu, \qquad m^2 + k_z^2 \Leftrightarrow -V_{\rm G}(\alpha). \tag{29}$$

The difference due to L_z becomes a technical point when the quantum state of the charged scalar and the wave function of the universe are expanded by the Landau levels and harmonic wave functions of the scalar fields, respectively, as will be shown in section 4. Table 1 compares characteristics of quantum FRW universe with a minimal scalar and a scalar field in QED: massless scalar fields of quantum FRW universe correspond to a charged scalar field in a time-dependent electric field and a massive complex scalar to the charged scalar field in a time-dependent magnetic field.

Table 1. Quantum Simulation of Universe

Quantum Universe	Scalar QED
Universes	Charged scalars
Wheeler-DeWitt equation	Klein-Gordon equation
Superspace of spacetime	Electromagnetic fields
Massive scalar in the early universe	Scalar in a time-dependent magnetic field
Coupling of harmonic wave functions	Coupling of Landau levels
Gravitational potential	Scalar in a time-dependent electric field
Wave functions of the universe	Quantum states of charged field

4. Landau levels and wave functions of WDW equation

The Landau levels in a constant magnetic field are well known in nonrelativistic and relativistic theory. However, the time-dependency of the magnetic field makes Landau levels changing in time and instantaneous Landau levels are a good adiabatic approximation only for a slowly varying field. Another issue is the couplings of instantaneous Landau levels due to L_z, which can be diagonalized in some new Landau levels. The essential complication originates from the time-dependent Landau levels, which make the task of finding the exact Landau levels nontrivial in relativistic theory. The Cauchy problem has been proposed for the charged scalar field in a time-dependent magnetic field in the new Landau levels [29].

The WDW equation and the super-Hamiltonian contain the Hamiltonian of α-dependent oscillators

$$\hat{H}_{(i)}(\phi_i; \alpha) = -\frac{\partial^2}{\partial\phi_i^2} + e^{6\alpha}\mu^2\phi_i^2. \tag{30}$$

Each oscillator has the α-dependent harmonic wave functions

$$\hat{H}_{(i)}(\phi_i;\alpha)|\Phi_{(i)n}(\phi_i;\alpha)\rangle = e^{3\alpha}\mu(2n+1)|\Phi_{(i)n}(\phi_i;\alpha)\rangle. \tag{31}$$

However, the α-dependency makes the eigenfunctions continuously transit to neighbor eigenfunctions preserving the parity [12, 22]

$$\frac{\partial}{\partial\alpha}|\Phi_{(i)n}(\phi_i;\alpha)\rangle = \mathbf{\Omega}_{(i)}|\Phi_{(i)n}(\phi_i;\alpha)\rangle, \tag{32}$$

where the transition matrix operator takes the form

$$\mathbf{\Omega}_{(i)} := \frac{3}{4}(\hat{A}^2_{(i)} - \hat{A}^{\dagger 2}_{(i)}). \tag{33}$$

Here, $\hat{A}_{(i)}$ ($\hat{A}^{\dagger}_{(i)}$) is an annihilation (creation) operator for the harmonic oscillator $H_{(i)}$ that lowers (raises) the quantum numbers by one unit.

The stratagem for the Cauchy initial value problem that evolves the wave function $\Psi(\alpha_0,\phi_1,\phi_2)$ to $\Psi(\alpha,\phi_1,\phi_2)$ is to expand the wave function by the eigenfunctions on each hypersurface $\Sigma = \alpha$

$$\Psi(\alpha,\phi_1,\phi_2) = \sum_{l,n=0}^{\infty} \psi_{l,n}(\alpha)\Phi_{(1)n}(\alpha)\rangle\Phi_{(2)n}(\alpha)\rangle, \tag{34}$$

and to solve the ordinary second-order differential equation for the multi-indexed column vector $\psi_{l,n}(\alpha)$ with multi-indexed off-diagonal transition matrices $\mathbf{\Omega}_{(i)}$ and the diagonal energy matrix $\mathbf{E} = 2e^{3\alpha}(\hat{A}^{\dagger}_{(1)}\hat{A}_{(1)} + \hat{A}^{\dagger}_{(2)}\hat{A}_{(2)} + 1)$. The detailed procedure can be found in refs. [12, 22].

5. Conclusion

The quantum system of a charge in an electrically modulated ion trap is analogous to the quantum system of a homogeneous and isotropic scalar field (inflaton) in the FRW universe. Thus an ion trap that is properly modulated by a time-dependent electric field can simulate the quantum evolution of the inflaton in the universe. The WDW equation, the second quantized theory, for the FRW universe with a minimal complex scalar field and the super-Hamiltonian, the third quantized theory, in the superspace of the geometry and scalar field, are analogous to a charged scalar field in QED. Massless scalar fields correspond to the charged scalar field in a time-dependent electric field while a massive complex scalar field corresponds to the charged scalar field in a time-dependent magnetic field.

These analogies between a quantum field in the FRW universe and a charge in an electrically modulated ion trap, and also between the wave functions of the WDW equation and the quantum states of a charged scalar field either in time-dependent electric fields or magnetic fields open a new window for quantum simulation of the universe. Another challenging simulation of quantum universe is the homogeneous Bianchi models that exhibit chaotic motions and infinite oscillations [8, 9] and it would be interesting to see how a minimal scalar field changes the characteristic of the system.

Acknowledgments
The author would like to thank Christian Schubert for the warm hospitality at Instituto de Física y Matemáticas (ifm), Universidad Michoacana de San Nicolás de Hidalgo (UMSNH), where this paper was initiated, and Rong-Gen Cai for the warm hospitality at Institute of Theoretical Physics (ITP), Chinese Academy of Sciences (CAS), where this paper was

IOP Conf. Series: Journal of Physics: Conf. Series **1275** (2019) 012057 doi:10.1088/1742-6596/1275/1/012057

completed. This work was supported by the Basic Science Research Program through the National Research Foundation of Korea (NRF) funded by the Ministry of Education (NRF-2015R1D1A1A01060626) and also in part by Institute of Basic Science (IBS) under IBS-R012-D1.

References

[1] Georgescu I M, Ashhab S and Nori F 2014 Quantum Simulation *Rev. Mod. Phys.* **86** 153
[2] Paul W 1990 Electromagnetic traps for charged and neutral particles *Rev. Mod. Phys.* **62** 531
[3] Leibfried D, Blatt R, Monroe C and Wineland D 2003 Quantum dynamics of single trapped ions *Rev. Mod. Phys.* **75** 281
[4] Unruh W G 1981 Experimental black hole evaporation *Phys. Rev. Lett.* **46** 1351
[5] Garay L J, Anglin J R, Cirac J I and Zoller P 2001 Sonic black holes in dilute Bose-Einstein condensates *Phys. Rev.* A **63** 023611
[6] Schutzhold R, Schaller G and Habs D 2006 Signatures of the Unruh effect from electrons accelerated by ultra-strong laser fields *Phys. Rev. Lett.* **97** 121302 [Erratum 2006 *Phys. Rev. Lett.* **97** 139902]
[7] Menicucci N C, Olson S J and Milburn G J 2010 Simulating quantum effects of cosmological expansion using a static ion trap *New J. Phys.* **12** 095019
[8] Belinskii V A, Lifshitz E M and Khalatnikov I M 1970 Oscillatory approach to a singular point in the relativistic cosmology *Adv. Phys.* **19** 525
[9] Belinskii V A, Lifshitz E M and Khalatnikov I M 1982 A general solution of the Einstein equations with a time singularity *Adv. Phys.* **31** 639
[10] Berger B K 1998 Numerical approaches to space-time singularities *Living Rev. Rel.* **1** 7
[11] Kim S P 2014 Third Quantization and Quantum Universes *Nucl. Phys. Proc. Suppl.* **246-247** 68
[12] Kim S P 2013 Massive Scalar Field Quantum Cosmology *The Universe* **1** no. 1 11
[13] Belinskii V A, Lifshitz E M and Khalatnikov I M 1973 Effect of scalar and vector fields on the nature of the cosmological singularity *Sov. Phys. JETP* **36** 591
[14] Lewis H R and Riesenfeld W B 1969 An Exact quantum theory of the time dependent harmonic oscillator and of a charged particle time dependent electromagnetic field *J. Math. Phys.* **10** 1458
[15] Kim S P and Page D N 2001 Classical and quantum action phase variables for time dependent oscillators *Phys. Rev.* A **64** 012104
[16] Kim S P 2004 Time-dependent Displaced and Squeezed Number States *J. Korean Phys. Soc.* **44** 446
[17] Olver F W J, Lozier D W, Boisvert R F and Clark C W 2010 *The NIST Handbook of Mathematical Functions* (New York: Cambridge University Press)
[18] Alsing P M, Dowling J P and Milburn G J 2005 Ion Trap Simulations of Quantum Fields in an Expanding Universe *Phys. Rev. Lett.* **94** 220401
[19] Kiefer C 2004 *Quantum gravity, Int. Ser. Monogr. Phys.* **124** 1 (Oxford: Oxford University Press)
[20] Hartle J B and S. W. Hawking S W 1983 Wave Function of the Universe *Phys. Rev.* D **28** 2960
[21] Hawking S W 1984 The Quantum State of the Universe *Nucl. Phys.* B **239** 257
[22] Kim S P 1992 Quantum mechanics of conformally and minimally coupled Friedmann-Robertson-Walker cosmology *Phys. Rev.* D **46** 3403
[23] Banks T 1985 T C P, Quantum Gravity, the Cosmological Constant and All That... *Nucl. Phys.* B **249** 332 (1985)
[24] McGuigan M 1988 Third Quantization and the Wheeler-dewitt Equation *Phys. Rev.* D **38** 3031
[25] McGuigan M 1989 Universe Creation From the Third Quantized Vacuum *Phys. Rev.* D **39** 2229
[26] Hosoya A and Morikawa M 1989 Quantum Field Theory of Universe *Phys. Rev.* D **39** 1123
[27] Abe S 1993 Fluctuations around the Wheeler-DeWitt trajectories in third quantized cosmology *Phys. Rev.* D **47** 718
[28] Buonanno A, Gasperini M, Maggiore M and Ungarelli C 1997 Expanding and contracting universes in third quantized string cosmology *Class. Quant. Grav.* **14** L97
[29] Kim S P 2014 Landau Levels of Scalar QED in Time-Dependent Magnetic Fields *Annals Phys.* **344** 1
[30] Kim S P 2014 Second Quantized Scalar QED in Homogeneous Time-Dependent Electromagnetic Fields *Annals Phys.* **351** 54

IOP Conf. Series: Journal of Physics: Conf. Series **1275** (2019) 012058 doi:10.1088/1742-6596/1275/1/012058

No-go theorems for ekpyrosis from ten-dimensional supergravity

Kunihito Uzawa

Department of Physics, School of Science and Technology, Kwansei Gakuin University, Sanda, Hyogo 669-1337, Japan

E-mail: kunihito.uzawa@gmail.com

Abstract. We study whether the new ekpyrotic scenario can be embedded into ten-dimensional supergravity. We use that the scalar potential obtained from flux compactifications of type II supergravity with sources has a universal scaling with respect to the dilaton and the volume mode. Similar to the investigation of inflationary models, we obtain very strong constraints ruling out ekpyrosis from analysing the fast-roll conditions. We conclude that flux compactifications tend to provide potentials that are neither too flat and positive (inflation) nor too steep and negative (ekpyrosis).

1. Introduction

The strong no-go theorems which exclude tree-level de Sitter compactifications under a few simple assumptions with or without negative tension objects such as orientifold planes have been much explored because of the possible cosmological and phenomenological interests. However, the no-go theorems for ekpyrotic scenario which is alternative to inflation model in string theory is much less extensive. Our motivation for the present work is to improve this situation.

The ekpyrosis inspired by string theory and brane world model suggests alternative solutions to the early universe puzzles such as inflation and dark energy, and assumes that two four-dimensional boundary branes which are located at the endpoints of orbifold in the higher-dimensional bulk spacetime [1, 2, 3, 4, 5, 6]. When we assume that there is an attractive force between two branes, these branes approach to each other, which gives big bang. Since the big bang is described as a collision of branes there is not the beginning of time in the ekpyrotic scenario. Although two branes move through each other at once after collision, we can get a model so that the branes become closer again.

The motion of branes is described by the potential of scalar field in a four-dimensional effective theory. Non-perturbative effects result in a potential which attracts one brane towards the other brane [1]. In order to resolve a horizon and flatness problem, the potential during a period of slow contraction before the big bang is negative and steeply falling. There was plenty of time before the big bang for the universe to be in causal contact over large regions, and in this way the horizon and flatness problem is automatically solved [4]. Then, the usual statement of the ekpyrosis is that the universe slowly contracts before the big bang with the equation of state $w = p/\rho \gg 1$. The scalar potential is expected to turn up towards zero at large negative value of scalar field in the ekpyrotic or cyclic models.

Content from this work may be used under the terms of the Creative Commons Attribution 3.0 licence. Any further distribution of this work must maintain attribution to the author(s) and the title of the work, journal citation and DOI.
Published under licence by IOP Publishing Ltd

IOP Conf. Series: Journal of Physics: Conf. Series **1275** (2019) 012058 doi:10.1088/1742-6596/1275/1/012058

As two branes approach each other, the branes are rippled because of generating quantum fluctuations. Since the collision of branes cannot be happened at exactly the same time, the branes collide slightly earlier or later in some places. Our universe thus has a little bit more time to cool or hot. When we consider the curvature perturbation in the ekpyrotic scenario, it occurs a strong blueshift, which is in sharp contradiction with the small redshift of the scalar fluctuation in the CMB [7, 8, 9, 10, 11, 12, 13, 14, 15]. The occurrence of blueshift scalar perturbation is nonetheless a minor flaw of ekpyrotic theory which can be easily corrected. Hence, new ekpyrotic theory considers multiple scalar field and successfully generated a scalar spectrum which is scale invariant and slightly redshifted [16, 17, 18, 19, 20, 21, 22, 23, 24, 25, 26, 27, 28, 29, 30, 31]. Producing primordial gravitational waves sourced by the gauge field in the ekpyrotic scenario was also studied in [32].

It is the purpose of this work to give a No-Go theorem of the ekpyrotic scenario in a ten-dimensional supergravity model which is low energy limit of a string theory. We study the dynamics of two scalar fields in the four-dimensional effective theory after a compactification in string theory. There are a dilaton and the volume modulus of the internal manifold in the effective theory, which is a four-dimensional theory of gravity minimally coupled to two-scalar fields. We will derive the two moduli fields with negative exponential potentials. Since this potential is steep, the scalar field should be satisfied by "fast-roll" condition instead of slow-roll parameter in inflation model [33, 34]. Although the terms coupling the scalar fields to the scalar curvature of the internal space and orientifold plane contribute the negative value to the potential in the string theory, we find that the potential does not satisfy the fast-roll condition in general. Therefore, it is not possible for us to realize the ekpyrotic phase in a string theory.

Section 2 describes the potential of scalar field for ekpyrotic scenario and the way it derives the four-dimensional effective theory. We discuss the approach to the effective action in more detail. The No-Go theorem of the ekpyrosis thus given by the string theory is discussed. We also investigate the detailed properties of these models, their embedding in a string theory and their viability. For simplicity, we do not consider D-branes and the associated moduli except for the volume modulus (breathing mode) of internal space although the analysis would not be different. Finally, section 3 provides a brief summary and an outlook to future developments.

2. No-Go theorem of the ekpyrotic scenario in the type II theory

In this section, we consider compactifications of the type II theory to four-dimensional spacetime on compact manifold Y. The ten-dimensional low-energy effective action for the type II theory takes the form [35, 36]

$$\begin{aligned} S &= \frac{1}{2\bar{\kappa}^2}\int d^{10}x\sqrt{-g}\left[e^{-2\phi}\left(R+4g^{MN}\partial_M\phi\partial_N\phi-\frac{1}{2}|H|^2\right)-\frac{1}{2}\sum_p|F_p|^2\right] \\ &\quad -\sum_p\left(T_{\mathrm{D}p}+T_{\mathrm{O}p}\right)\int d^{p+1}x\sqrt{-g_{p+1}}\,e^{-\phi}\,, \end{aligned} \tag{1}$$

where $\bar{\kappa}^2$ is the ten-dimensional gravitational constant, R denotes the ten-dimensional Ricci scalar, ϕ is the scalar field, H is the NS-NS 3-form field strength, F_p are the R-R p-form field strengths ($p = 0, 2, 4, 6, 8$ for type IIA, and $p = 1, 3, 5, 7, 9$ for type IIB) that are sourced by D-branes, and $T_{\mathrm{D}p}$ ($T_{\mathrm{O}p}$) is the Dp-brane (Op-plane) charge and tension. Although there are Chern-Simons terms in the ten-dimensional action, these are essentially independent of the dilaton and the scale of the background metric. Hence, we will not consider them.

To compactify the theory to four dimensions, we consider the a metric ansatz of the form [35, 36]

$$\begin{aligned} ds^2 &= g_{MN}dx^M dx^N = q_{\mu\nu}dx^\mu dx^\nu + g_{ij}dy^i dy^j \\ &= q_{\mu\nu}dx^\mu dx^\nu + \rho\, u_{ij}(\mathrm{Y})dy^i dy^j\,, \end{aligned} \tag{2}$$

where ρ is breathing mode (volume modulus of the compact space), x^μ denote the coordinates of four-dimensional spacetime, y^i are local coordinates on the internal space Y, and g_{MN}, $q_{\mu\nu}$, $u_{ij}(\mathrm{Y})$ are the metrices of ten-dimensional spacetime, four-dimensional spacetime, six-dimensional internal space, respectively. We assume that $q_{\mu\nu}$, $u_{ij}(\mathrm{Y})$ depend only on the coordinates x^μ, y^i, respectively. Since we factored out the overall volume modulus ρ of the internal space in the ten-dimensional metric (2), the modulus ρ is related to the total physical volume of the internal space v_6 and the volume of Y space $v(\mathrm{Y})$ as

$$\rho = \left[\frac{v_6}{v(\mathrm{Y})}\right]^{1/3}, \qquad v_6 = \int d^6y\sqrt{g_6}\,, \qquad v(\mathrm{Y}) = \int d^6y\sqrt{u}\,. \tag{3}$$

Here, g_6, u denote the determinant of the metric g_{ij}, $u_{ij}(\mathrm{Y})$, respectively. The volume modulus ρ is chosen such that the metric $u_{ij}(\mathrm{Y})$ of the internal space is normalized $v(\mathrm{Y}) = 1$.

After we integrate over the internal space Y, the four-dimensional effective action S_{E} in the Einstein frame is given by

$$\begin{aligned} S_{\mathrm{E}} &= \frac{1}{2\kappa^2}\int d^4x\sqrt{-\bar{q}}\left[\bar{R} - 2\bar{q}^{\mu\nu}\partial_\mu \ln\tau\,\partial_\nu\ln\tau - \frac{3}{2}\bar{q}^{\mu\nu}\partial_\mu\ln\rho\,\partial_\nu\ln\rho\right.\\ &\quad + \left(\frac{\bar{\kappa}}{\tau\kappa}\right)^2\rho^{-1}R(\mathrm{Y}) - \frac{1}{2}\left(\frac{\bar{\kappa}}{\tau\kappa}\right)^2\rho^{-3}|H|^2 - \frac{1}{2}\sum_p\left(\frac{\bar{\kappa}}{\tau^2\kappa}\right)^2\rho^{3-p}|F_p|^2\\ &\quad \left. -2\sum_p\left(T_{\mathrm{D}p} + T_{\mathrm{O}p}\right)\left(\frac{\bar{\kappa}^2}{\kappa}\right)^2\tau^{-3}\rho^{(p-6)/2}\int d^{p-3}x\sqrt{g_{p-3}}\,\right], \end{aligned} \tag{4}$$

where $R(\mathrm{Y})$ denotes the Ricci scalar constructed from the metric $u_{ij}(\mathrm{Y})$ and κ^2 is the four-dimensional gravitational constant. Orientifold planes occupy $(p-3)$-dimensional internal space due to extending our four-dimensional universe. Then, the contribution of Op-plane ($p \geq 3$) to moduli potential will survive.

In the four-dimensional action, we have defined the dilaton modulus [35, 36]

$$\tau = \mathrm{e}^{-\phi}\rho^{3/2}\,, \tag{5}$$

and performed a conformal transformation on the four-dimensional metric

$$q_{\mu\nu} = \left(\frac{\bar{\kappa}}{\tau\kappa}\right)^2\bar{q}_{\mu\nu}\,. \tag{6}$$

Here, $\bar{q}_{\mu\nu}$ is the four-dimensional metric in the Einstein frame. $\bar{R}$ and $\bar{q}$ in the four-dimensional action (4) are the Ricci scalar and the determinant constructed from the metric $\bar{q}_{\mu\nu}$, respectively.

Because of the conformal transformation (6), the kinetic term of the fields ρ and τ is diagonal in the four-dimensional effective action. Since these fields do not have canonical kinetic energies, they are redefined as [35]

$$\bar{\rho} = \sqrt{\frac{3}{2}}\kappa^{-1}\ln\rho\,, \qquad \bar{\tau} = \sqrt{2}\kappa^{-1}\ln\tau\,. \tag{7}$$

The moduli potential arises from the compactification of the terms in ten-dimensional action (1) associated with the various field strengths, Dp-branes and Op-planes as well as the gravity and the dilaton. The 3-form H and p-form field strengths F_p can have a non-vanishing integral over any closed three-, p-dimensional internal manifold of the compact space Y, and have to obey generalized Dirac charge quantization conditions, respectively

$$\int_\Sigma H = h_\Sigma\,, \qquad \int_{\mathcal{C}_p} F_p = f^{(p)}_{\mathcal{C}_p}\,, \tag{8}$$

IOP Conf. Series: Journal of Physics: Conf. Series **1275** (2019) 012058 doi:10.1088/1742-6596/1275/1/012058

where h_Σ and $f^{(p)}_{\mathcal{C}_p}$ are integers associated with number of quanta of H and F_p through each three-, p-dimensional homology cycles Σ, $\mathcal{C}_p$ in the internal manifold, respectively. We derive the potential energy in the four-dimensional Einstein frame arising from a three-form field strength H, and a p-form field strength F_p coming from the terms in the ten-dimensional action (1) which is proportional to $|H|^2$, $|F_p|^2$, respectively. We contract with three and p factors of the internal space metric g^{ij} so that

$$V_H \propto \exp\left[-\kappa\left(\sqrt{2}\bar{\tau}+\sqrt{6}\bar{\rho}\right)\right], \quad \text{for } H, \tag{9}$$

$$V_p \propto \exp\left[-\kappa\left\{2\sqrt{2}\bar{\tau}+\frac{\sqrt{6}}{3}(p-3)\bar{\rho}\right\}\right], \quad \text{for } F_p. \tag{10}$$

Then we obtain the four-dimensional effective action in the Einstein frame:

$$S_{\rm E} = \int d^4x\sqrt{-\bar{q}}\left[\frac{1}{2\kappa^2}\bar{R}-\frac{1}{2}\bar{q}^{\mu\nu}\partial_\mu\bar{\tau}\,\partial_\nu\bar{\tau}-\frac{1}{2}\bar{q}^{\mu\nu}\partial_\mu\bar{\rho}\,\partial_\nu\bar{\rho}-V(\bar{\tau},\bar{\rho})\right], \tag{11}$$

where the moduli potential of four-dimensional effective theory is given by

$$V(\bar{\tau},\bar{\rho}) = V_{\rm Y}+V_{\rm H}+\sum_p V_p+\sum_p V_{{\rm D}p}+\sum_p V_{{\rm O}p}. \tag{12}$$

Here, each components of the moduli potential can be expressed as [35]

$$V_{\rm Y}(\bar{\tau},\bar{\rho}) = -A_{\rm Y}\exp\left[-\kappa\left(\sqrt{2}\bar{\tau}+\frac{\sqrt{6}}{3}\bar{\rho}\right)\right]R({\rm Y}), \tag{13}$$

$$V_{\rm H}(\bar{\tau},\bar{\rho}) = A_{\rm H}\exp\left[-\kappa\left(\sqrt{2}\bar{\tau}+\sqrt{6}\bar{\rho}\right)\right], \tag{14}$$

$$V_p(\bar{\tau},\bar{\rho}) = A_p\exp\left[-\kappa\left\{2\sqrt{2}\bar{\tau}+\frac{\sqrt{6}}{3}(p-3)\bar{\rho}\right\}\right], \tag{15}$$

$$V_{{\rm D}p}(\bar{\tau},\bar{\rho}) = A_{{\rm D}p}\exp\left[-\kappa\left\{\frac{3\sqrt{2}}{2}\bar{\tau}+\frac{\sqrt{6}}{6}(6-p)\bar{\rho}\right\}\right]\int d^{p-3}x\sqrt{g_{p-3}}, \tag{16}$$

$$V_{{\rm O}p}(\bar{\tau},\bar{\rho}) = -A_{{\rm O}p}\exp\left[-\kappa\left\{\frac{3\sqrt{2}}{2}\bar{\tau}+\frac{\sqrt{6}}{6}(6-p)\bar{\rho}\right\}\right]\int d^{p-3}x\sqrt{g_{p-3}}, \tag{17}$$

where $A_{\rm Y}$, $A_{\rm H}$, A_p, $A_{{\rm D}p}$, and $A_{{\rm O}p}$ are coefficients to scale with fluxes and numbers of Op-planes and Dp-branes. These coefficients in general depend on the choice of flux integers h_Σ, $f^{(p)}_{\mathcal{C}_p}$, and also the function of the moduli of the internal space Y.

When the potential form for the ekpyrotic scenario gives the negative and steep, the fast-roll parameters for the ekpyrosis have to obey [33, 34]

$$\varepsilon_{\rm f} \equiv \kappa^2\frac{V^2}{\left(\partial_{\bar{\tau}}V\right)^2+\left(\partial_{\bar{\rho}}V\right)^2} \ll 1, \qquad |\eta_{\rm f}| \equiv \left|1-\frac{V\left(\partial^2_{\bar{\tau}}V+\partial^2_{\bar{\rho}}V\right)}{\left(\partial_{\bar{\tau}}V\right)^2+\left(\partial_{\bar{\rho}}V\right)^2}\right| \ll 1. \tag{18}$$

This is analogy with the standard slow-roll parameters in inflation. The potential form satisfying the condition (18) gives the ekpyrotic period of slow contraction before the big bang.

In the following, we illustrate how the above ingredients may be useful from the point of view of excluding the ekpyrosis. We focus the discussion on the contribution of negative energy, which appear particularly promising. Our goal is only to show that simple, available ingredients in the type II theory have energy densities which scale with the volume and dilaton moduli in a way which suffices to obey our no-go theorem, which was based purely on scaling of energy densities. This should act as a guide to model building, but should be taken in the heuristic spirit it is offered.

It is especially interesting to understand the dynamics of moduli at negative potential energy. If there are non-trivial fluxes in the background (12), one notes that these make uplifting the moduli potential to positive energy. In this work, we consider the moduli potential (11) without flux:

$$\begin{aligned} V(\bar{\tau}\,,\bar{\rho}) &= V_{\rm Y} + \sum_p V_{{\rm O}p} \\ &= -A_{\rm Y}\exp\left[-\kappa\left(\sqrt{2}\bar{\tau} + \frac{\sqrt{6}}{3}\bar{\rho}\right)\right] R({\rm Y}) \\ &\quad -\sum_p A_{{\rm O}p}\exp\left[-\kappa\left\{\frac{3\sqrt{2}}{2}\bar{\tau} + \frac{\sqrt{6}}{6}(6-p)\bar{\rho}\right\}\right]\int d^{p-3}x\sqrt{g_{p-3}}\,. \end{aligned} \tag{19}$$

Before discussing the No-Go theorem, we comment about the other moduli of the theory associated with the compactification. The four-dimensional effective action (11) includes in general kinetic energy terms of so-called Kähler moduli, complex structure moduli, and axions. As we have mentioned above, the various coefficients $A_{\rm Y}$, $A_{\rm H}$, A_p, $A_{{\rm D}p}$, and $A_{{\rm O}p}$ in the potential (12) are complicated functions of these moduli. Although there are kinetic terms of these moduli in the four-dimensional effective action, their contributions will always be positive [35]. For simplicity, we do not consider the dynamics and fixing of these moduli in the following. However, if four-dimensional effective action is described by Kähler moduli, complex structure moduli, and axions as well as volume moduli (breathing mode), the moduli potential will be modified. We will discuss these in the end of this section. The stabilization mechanisms of all the geometric moduli and many axions in type II string theory have been discussed in [37, 38, 39, 40, 41] (see also recent analysis [42]).

2.1. Type IIA compactification

There are No-Go theorems which exclude slow roll inflation and de Sitter vacua in the simple IIA compactifications with orientifold planes [1] [35, 44]. For the IIA flux compactifications on Calabi-Yau manifolds with O6-planes in the four-dimensional potential (12), the slow-roll inflation is forbidden [35]. The construction classical de Sitter vacua in IIA compactifications on internal manifolds with negative curvature and orientifold planes were studied in [36, 45, 46].

In this section, we consider IIA compactifications on an internal space (2), namely positive curvature and Ricci flat spaces, involving orientifold planes, and discuss the No-Go theorem for ekpyrotic scenario. The analysis will focus on the behavior of the moduli potential in the volume modulus and dilaton. In order to present the no-go theorem using these fields, we have to still make sure that there are no steep directions of the scalar potential in the $(\bar{\tau}\,,\bar{\rho})$-plane.

In such cases one can then study directions involving $\bar{\rho}$, $\bar{\tau}$ and finds that the scalar potential satisfies

$$\varepsilon_{\rm f} = \frac{V^2}{2}\left[\left\{V_{\rm Y} + \frac{3}{2}\left(V_{\rm O4} + V_{\rm O6} + V_{\rm O8}\right)\right\}^2 + \frac{1}{3}\left(V_{\rm Y} + V_{\rm O4} - V_{\rm O8}\right)^2\right]^{-1} > \frac{6}{31}\,. \tag{20}$$

[1] There are no large volume flux compactifications in type IIA theory involving fluxes other than F_2 after compactifying on a circle in M-theory because we have no method for canceling the flux tadpole on an 8-fold with trivial Euler characteristic (For a treatment of this situation in type IIA theory, see [43].)

IOP Conf. Series: Journal of Physics: Conf. Series **1275** (2019) 012058 doi:10.1088/1742-6596/1275/1/012058

The fast-roll parameter $\varepsilon_{\rm f}$ has the bound $\varepsilon_{\rm f} > 6/31$. This result does not depend on the choice of coefficients $A_{\rm Y}$ and $A_{{\rm O}p}$. The value of parameter $\varepsilon_{\rm f}$ is not much less than one, which is the contradiction with the fast-roll condition for ekpyrosis (18). Hence, ekpyrosis is not allowed. This can also be seen from results in [47, 48].

2.2. Type IIB compactification

There are many no-go theorems and exclude most concrete examples of de Sitter solution [38, 49, 50, 51, 52]. It is possible to evade simple no-go theorems of inflationary scenario for SU(2)×SU(2) with an SU(2)-structure and O5- and O7-planes. Although there are de Sitter critical points, these have at least one tachyonic direction with a large η (slow-roll) parameter [52].

On the other hand, for type IIB compactifications in the ekpyrotic model, we have also seen that it is possible to obtain simple no-go theorems in the $(\bar{\tau}\,,\,\bar{\rho})$-plane if one includes orientifold planes and the curvature of the internal space.

From the Eq. (18), we find a constraint of the fast-roll parameter $\varepsilon_{\rm f}$ [48]

$$\begin{aligned}\varepsilon_{\rm f} &= V^2\left[2\left\{V_{\rm Y}+\frac{3}{2}\left(V_{\rm O3}+V_{\rm O5}+V_{\rm O7}+V_{\rm O9}\right)\right\}^2\right.\\&\left.+\frac{1}{6}\left(2V_{\rm Y}+3V_{\rm O3}+V_{\rm O5}-V_{\rm O7}-3V_{\rm O9}\right)^2\right]^{-1} > \frac{1}{6}\,.\end{aligned} \tag{21}$$

Unfortunately, the form moduli potential is not steep again as $\varepsilon_{\rm f}$ and $\eta_{\rm f}$ parameters turn out to be large value. Just as in the IIA analogue, one obtains the bound $\varepsilon_{\rm f} > 1/6$. If we choose different values for $A_{\rm Y}$ and $A_{{\rm O}p}$ in the moduli potential (12), we can find again the same bound.

If we consider the dynamics of remaining moduli such as Kähler moduli, complex structure moduli and axions, the potential is in general a function of hundreds of fields, for large number of infinite families of possible flux combinations on each of the many available internal manifolds. When we treat dynamics of these moduli or the coupling between moduli and fluxes or D-brane, O-plane, they contribute the potential. There is a possibility that the no-go theorems using these fields are circumvented.

We have used the criterion for ekpyrosis that the fast-roll parameter defined in equation (18) should be very much smaller than unity. If the definition of ekpyrosis is as a phase during which the pressure is larger than the energy density, it is not necessary for it to be much larger, and there even exist models where the perturbations can be generated with an equation of state where the pressure is only a little bit larger than the energy density [28, 53]. Now let us redefine moduli

$$\varphi_1 = \kappa\tau\,, \qquad \varphi_2 = \kappa\rho\,, \tag{22}$$

and set $\kappa = 1$ to have canonical kinetic energy, the four-dimensional action for moduli can be expressed as

$$S_\varphi = \sum_{i=1}^{2}\int d^4x\sqrt{-\bar{q}}\left(-\frac{1}{2}\,\bar{q}^{\mu\nu}\,\partial_\mu\varphi_i\,\partial_\mu\varphi_i + \sum_p A_{{\rm O}p}\,{\rm e}^{c_i\varphi_i}\right), \tag{23}$$

where c_i $(i = 1, 2)$ are constants. In order to ensure that $w > 1$, we need $c_i > \sqrt{6}$ [4]. In the absence of a concrete model predicting it in this work, it is therefore impossible to realize the ekpyrosis.

3. Discussions

In this work, we have studied the No-Go theorem of the ekpyrosis for string theory in a spacetime of ten dimensions. We gave a potential of the scalar fields in four-dimensional effective theory, in terms of the compactification with smooth manifold.

The effective potential of two scalar fields can be constructed by postulating suitable emergent gravity, orientifold planes, and vanishing fluxes on the ten-dimensional background. The construction of four-dimensional effective action was developed in [35], mainly but not entirely in the context of a new ekpyrotic scenario.

We have used results of section 2 to analyze the dynamics of moduli that can be constructed using a simple compactification. The scalar potential depends only on two moduli: $\bar{\rho}$ and $\bar{\tau}$. In such a simple setting, one can show that $\varepsilon_{\rm f} > 6/31$ for IIA theory and $\varepsilon_{\rm f} > 1/6$ for IIB theory whenever $V(\bar{\tau}, \bar{\rho}) < 0$. It has been known for some time [4] that the effective potential of scalar fields requires the fast-roll parameter to be small during the ekpyrotic phase. However, with the help of the tools developed in section 2, this is prohibited in a string theory with a compactification. Hence, the explicit nature of the dynamics has made it impossible to realize the ekpyrotic phase in the present work. This is consistent with results in [47].

As we have commented in Sec. 2 , we have not considered the dynamics for moduli other than the volume modulus of the six-dimensional internal space in this work. If these moduli have dynamics, kinetic terms of moduli appear in the four-dimensional effective theory. Moreover, since moduli couple to orientifolds, it is possible that there are other steep directions of the scalar potential in directions outside the $(\bar{\rho}, \bar{\tau})$-plane.

Let us comment about the No-Go theorem of ekpyrotic scenario in heterotic M-theory. The original ekpyrotic model was inspired by heterotic M-theory which is obtained by going from eleven to five dimensions first, and only at even lower energies to 4 dimensions. It arose precisely from the realization that in heterotic M-theory a realistic phenomenology is obtained by introducing non-perturbative effects in the scalar field potential. Otherwise, in the Heterotic theory with a simple Calabi-Yau compactification we have considered in this work, the four-dimensional effective potential form of scalar fields φ_i is given by $V(\varphi_i) \propto e^{c_i\varphi_i}$, where $c_i < \sqrt{6}$. Then, it is impossible for us to construct the ekpyrotic scenario in the heterotic M-theory without non-perturbative correction. Unfortunately, no one has obtained an exact calculation of the potential involving all non-perturbative effects at present.

The compactification we have considered in this work also gives the No-Go theorem of the inflationary scenario as well as the ekpyrosis because the moduli potential cannot satisfy the slow-roll condition [35]. There are constrains or No-Go theorems to construct the inflationary or de Sitter model in the string theory with a few simple assumptions [35, 46, 54, 55, 56, 57] (see also Minkowski vacua for compactifications in type II theory [58]). There are also the strongest constraints on inflationary or de Sitter models in the heterotic string where both higher derivative interactions can be precisely analyzed from a space-time perspective in [59] and from a world-sheet perspective in [60].

We have not considered the warped compactification because it does not make the modification of fast roll condition even if it couples to matter field in the ten-dimensional theory. The effect of warped factor gives only the change of coefficients for each term in the four-dimensional effective action after integration of internal space.

In order to embed ekpyrotic or cyclic models in a ten-dimensional supergravity we have investigated in this work, we may consider some ingredients, for instance, the dynamics of remaining moduli, higher curvature correction other than orientifold and flux. We have not attempted an explicit construction here, since that will take us beyond the scope of this work. A lot of study remains to be done in string theory before a cosmologically realistic case is treated.

Acknowledgments
We would like to thank Thomas Van Riet for numerous valuable discussions and careful reading of the manuscript, and thank Savdeep Sethi, Tetsuya Shiromizu, Masato Minamitsuji, Shuntaro Mizuno, Toshifumi Noumi, David Benoit Andriot, Yusuke Yamada, Taishi Ikeda, Asuka Ito for discussions and valuable comments. We would like to also thank the Yukawa Institute for

IOP Conf. Series: Journal of Physics: Conf. Series **1275** (2019) 012058 doi:10.1088/1742-6596/1275/1/012058

Theoretical Physics at Kyoto University. Discussions during the YITP symposium YKIS2018a "General Relativity – The Next Generation –" and the workshop YITP-T-17-02 "Gravity and Cosmology 2018" were useful to complete this work. We are grateful to Shinji Mukohyama and Atsushi Watanabe for their warm hospitalities in the Yukawa Institute for Theoretical Physics. We also thank the Yukawa Institute for Theoretical Physics at Kyoto University for hospitality during the YITP workshop on "Gravity and Cosmology for Young Researchers" (Workshop No. YITP-X-17-08) which was supported by Grant-in-Aid for Scientific Research on Innovative Areas No. 17H06359. This work is supported by Grants-in-Aid from the Scientific Research Fund of the Japan Society for the Promotion of Science, under Contracts No. 16K05364.

References

[1] Khoury J, Ovrut B A, Steinhardt P J and Turok N 2001 *Phys. Rev.* D **64** 123522
[2] Steinhardt P J and Turok N 2002 *Science* **296** 1436
[3] Erickson J K, Gratton S, Steinhardt P J and Turok N 2007 *Phys. Rev.* D **75** 123507
[4] Lehners J L 2008 *Phys. Rept.* **465** 223
[5] Lehners J L, Steinhardt P J and Turok N 2009 *Int. J. Mod. Phys.* D **18** 2231
[6] Khoury J and Steinhardt P J 2010 *Phys. Rev. Lett.* **104** 091301
[7] Lyth D H 2002 *Phys. Lett.* B **524** 1
[8] Brandenberger R and Finelli F 2001 *J. High Energy Phys.* JHEP11(2001)056
[9] Hwang J C 2002 *Phys. Rev.* D **65** 063514
[10] Khoury J, Ovrut B A, Steinhardt P J and Turok N 2002 *Phys. Rev.* D **66** 046005
[11] Lyth D H 2002 *Phys. Lett.* B **526** 173
[12] Tsujikawa S 2002 *Phys. Lett.* **526** 179
[13] Notari A and Riotto A 2002 *Nucl. Phys.* B **644** 371
[14] Tsujikawa S, Brandenberger R and Finelli F 2002 *Phys. Rev.* D **66** 083513
[15] Lehners J L, McFadden P, Turok N and Steinhardt P J 2007 *Phys. Rev.* D **76** 103501
[16] Buchbinder E I, Khoury J and Ovrut B A 2007 "New Ekpyrotic cosmology," *Phys. Rev.* D **76** 123503
[17] Koyama K, Mizuno S and Wands D 2007 *Class. Quant. Grav.* **24** 3919
[18] Koyama K, Mizuno S, Vernizzi F and Wands D 2007 *J. Cosmol. Astropart. Phys.* JCAP11(2007)024
[19] Buchbinder E I, Khoury J and Ovrut B A 2008 *Phys. Rev. Lett.* **100** 171302
[20] Lehners J L and Steinhardt P J 2008 *Phys. Rev.* D **77** 063533 Erratum: [2009 *Phys. Rev.* D **79** 129903]
[21] Lehners J L and Steinhardt P J 2008 *Phys. Rev.* D **78** 023506 Erratum: [2009 *Phys. Rev.* D **79** 129902]
[22] Mizuno S, Koyama K, Vernizzi F and Wands D 2008 *AIP Conf. Proc.* **1040** 121
[23] Linde A, Mukhanov V and Vikman A 2010 *J. Cosmol. Astropart. Phys.* JCAP02(2010)006
[24] Lehners J L 2010 *Adv. Astron.* **2010** 903907
[25] Li M 2013 *Phys. Lett.* B **724** 192
[26] Battarra L and Lehners J L 2014 *Phys. Rev.* D **89** no.6, 063516
[27] Fertig A, Lehners J L and Mallwitz E 2014 *Phys. Rev.* D **89** no.10, 103537
[28] Ijjas A, Lehners J L and Steinhardt P J 2014 *Phys. Rev.* D **89** no.12, 123520
[29] Levy A M, Ijjas A and Steinhardt P J 2015 *Phys. Rev.* D **92** no.6, 063524
[30] Fertig A and Lehners J L 2016 *J. Cosmol. Astropart. Phys.* JCAP01(2016) no.01, 026
[31] Lehners J L 2018 *J. Cosmol. Astropart. Phys.* JCAP11(2018) no.11, 001
[32] Ito A and Soda J 2017 *Phys. Lett.* B **771** 415
[33] Gratton S, Khoury J, Steinhardt P J and Turok N 2004 *Phys. Rev.* D **69** 103505
[34] Khoury J, Steinhardt P J and Turok N 2004 *Phys. Rev. Lett.* **92** 031302
[35] Hertzberg M P, Kachru S, Taylor W and Tegmark M 2007 *J. High Energy Phys.* JHEP12(2007)095
[36] Haque S S, Shiu G, Underwood B and Van Riet T 2009 *Phys. Rev.* D **79** 086005
[37] Dasgupta K, Rajesh G and Sethi S 1999 *J. High Energy Phys.* JHEP08(1999) 023
[38] Giddings S B, Kachru S and Polchinski J 2002 *Phys. Rev.* D **66** 106006
[39] Kachru S, Kallosh R, Linde A D and Trivedi S P 2003 *Phys. Rev.* D **68** 046005
[40] Villadoro G and Zwirner F 2005 *J. High Energy Phys.* JHEP06(2005)047
[41] DeWolfe O, Giryavets A, Kachru S and Taylor W 2005 *J. High Energy Phys.* JHEP07(2005)066
[42] Sethi S 2018 *J. High Energy Phys.* JHEP10(2018)022
[43] McOrist J and Sethi S 2012 *J. High Energy Phys.* JHEP12(2012)122
[44] de Carlos B, Guarino A and Moreno J M 2010 *J. High Energy Phys.* JHEP01(2010)012
[45] Silverstein E 2008 *Phys. Rev.* D **77** 106006
[46] Danielsson U H, Haque S S, Shiu G and Van Riet T 2009 *J. High Energy Phys.* JHEP09(2009)114

[47] Meeus E 2016 *A nogo-theorem for Ekpyrosis from 10D supergravity* (master's thesis, KU Leuven)
[48] Uzawa K 2018 *J. High Energy Phys.* JHEP06(2018)041
[49] Gibbons G W 1985 *Aspects Of Supergravity Theories* (*Preprint* Print-85-0061 (CAMBRIDGE))
[50] de Wit B, Smit D J and Hari Dass N D 1987 *Nucl. Phys.* B **283** 165
[51] Maldacena J M and Nunez C 2001 *Int. J. Mod. Phys.* A **16** 822
[52] Caviezel C, Wrase T and Zagermann M 2010 *J. High Energy Phys.* JHEP04(2010)011
[53] Koehn M, Lehners J L and Ovrut B A 2014 *Phys. Rev.* D **90** no.2, 025005
[54] Wrase T and Zagermann M 2010 *Fortsch. Phys.* **58** 906
[55] Shiu G and Sumitomo Y 2011 *J. High Energy Phys.* JHEP09(2011)052
[56] Van Riet T 2012 *Class. Quant. Grav.* **29** 055001
[57] Andriot D 2018 *J. High Energy Phys.* JHEP03(2018)054
[58] Andriot D, Blåbäck J and Van Riet T 2017 *Phys. Rev. Lett.* **118** no.1, 011603
[59] Green S R, Martinec E J, Quigley C and Sethi S 2012 *Class. Quant. Grav.* **29** 075006
[60] Kutasov D, Maxfield T, Melnikov I and Sethi S 2015 *Phys. Rev. Lett.* **115** no.7, 071305

IOP Conf. Series: Journal of Physics: Conf. Series **1275** (2019) 012059 doi:10.1088/1742-6596/1275/1/012059

Dark matter in modified cosmologies

G. Lambiase

Dipartimento di Fisica "E.R. Caianiello" Università di Salerno, I-84084 Fisciano (Sa), Italy, and
INFN - Gruppo Collegato di Salerno, Italy.

E-mail: lambiase@sa.infn.it

Abstract. The IceCube experiment discovered a high-energy astrophysical neutrino flux with energies of the order of PeV. To explain this phenomenon one may consider the minimal 4-dimensional operator $\sim y_{\alpha\chi}\overline{L_{L_\alpha}}H\chi$. An interesting possibility is to also explain the relic abundance of DM in the Universe. Assuming that the cosmological background evolves according to the standard cosmological model, this model fails since the rate of DM decay $\Gamma_\chi \sim |y_{\alpha\chi}|^2$ needed to get the correct DM relic abundance differs by many orders of magnitude with respect that one needed to explain the IceCube data. We show that such a discrepancy can be solved if the Universe evolution is governed by a modified cosmology.

1. Introduction

In the recent years, the IceCube Collaboration [1, 2] has reported several neutrino-induced cascade events with energies ~ 1 PeV [2]. Candidates for the generation of such neutrino high energy events could be various astrophysical sources [3, 4, 5, 6], although till now there are no clear correlations with the known astrophysical sources, such as supernova remnants or Active Galactic Nuclei [7]. Other possibilities rely on the idea that neutrinos could arise from the decay of PeV mass Dark Matter (DM) (see [8] and references therein).

An interesting possibility raised in Ref. [8] is whether it is possible to explain both the PeV DM relic density and the decay rate required for IceCube with only one operator. The minimal DM-neutrino 4-dimensional interaction is given by $y\,\bar{L}\cdot H\chi$ [8]. Such a minimal dimension four operator however fails to account for both the PeV dark matter relic abundance and the decay rate required to explain IceCube. The DM models analyzed in [8] are based on the standard cosmological. On the other hand, observations from Type Ia Supernovae [9], CMB radiation [10], and the large scale structure [11, 12], have given evidence that the present cosmic expansion of the Universe is accelerating, and that therefore a form of unknown energy/matter, generally called Dark Energy (DE), is needed [13]. The latter could be also ascribed to an extension of Einstein's theory, and studies in this direction are till now under investigation [14, 15, 16, 17, 18, 19, 20]. From a cosmological point of view, one of the consequences of dealing with the cosmology based on modified gravity is that the thermal history of particles gets modified and the expansion rates H of the Universe can be written in terms of the expansion rate H_{GR} of General Relativity (GR) [21, 22]

$$H_{MC}(T) = A(T)H_{GR}(T)\,, \tag{1}$$

where $A(T)$ is the so called amplification factor. To preserve the successful predictions of BBN, one refers to the pre-BBN epoch since it is not directly constrained by cosmological observations.

Content from this work may be used under the terms of the Creative Commons Attribution 3.0 licence. Any further distribution of this work must maintain attribution to the author(s) and the title of the work, journal citation and DOI.
Published under licence by IOP Publishing Ltd

This means $A(T) \neq 1$ at early time, and $A(T) \to 1$ before BBN begins. Typically one has

$$A(T) = \eta \left(\frac{T}{T_*}\right)^{\nu}, \tag{2}$$

where T_* is a reference temperature, and $\{\eta, \nu\}$ free parameters that depend on the cosmological model under consideration [21, 22, 23]. The parameter ν labels cosmological models: $\nu = 2$ in Randall-Sundrum type II brane cosmology [24], $\nu = 1$ in kination models [25], $\nu = 0$ in cosmologies with an overall boost of the Hubble expansion rate [21], $\nu = -0.8$ in scalar-tensor cosmology [26, 21], $\nu = 2/n - 2$ in $f(R)$ cosmology, with $f(R) = R + \alpha R^n$ [27].

In this paper we consider generalizations of the standard cosmology with the aim to get a consistent minimal model of PeV DM and IceCube neutrinos. We calculate the freeze-in abundance of DM in a modified cosmology and show that the couplings required for obtaining the required relic abundance depend upon the modified gravity parameters [28]. In the following, we shall assume that the Universe is described by a flat Friedman-Robertson-Walker metric

$$ds^2 = dt^2 - a^2(t)(dx^2 + dy^2 + dz^2), \tag{3}$$

where $a(t)$ is the scale factor. We refer to a Universe radiation dominated, so that the energy density is given by $\rho = \frac{\pi^4 g_* T^4}{30}$, $g_* = 106$, while the pressure is $p = \rho/3$ (the adiabatic index is $w = 1/3$). The equation of continuity $\dot{\rho} + 3H(\rho + p = 0$ is assumed to hold (the dot will stand for the derivative with respect to the cosmic time t).

2. PeV neutrinos and Ice Cube data

The simplest 4-dimensional operator that allows to explain the IceCube high energy signal, is

$$\mathcal{L}_{d=4} = y_{\alpha\chi} \overline{L_{L_\alpha}} H \chi, \qquad \alpha = e, \mu, \tau, \tag{4}$$

where χ is the DM particle that transforms as $\chi \sim (1, 1, 0)$ of SM, $H \sim (1, 2, +1/2)$ is the Higgs doublet, $L_{L_\alpha} \sim (1, 2, -1/2)$ is the left-handed lepton doublet corresponding to the generation $\alpha (= e, \mu \tau)$, and finally $y_{\alpha\chi}$ are the Yukawa couplings. We consider the freeze-in production [29, 8], i.e. the DM particles are never in thermal equilibrium since they interact very weakly, but are gradually produced from the hot thermal bath. As a consequence, a sizable DM abundance is allowed until the temperature falls down to $T \sim m_\chi$ (temperatures below m_χ are such that DM particles phase-space is kinematically difficult to access).

Denoting with $Y_\chi = n_\chi / s$ the DM abundance, where n_χ is the number density of the DM particles and $s = \frac{2\pi^2}{45} g_*(T) T^3$ the entropy density, the evolution of the DM particle is governed by the Boltzmann equation

$$\frac{dY_\chi}{dT} = -\frac{1}{HTs}\left[\frac{g_\chi}{(2\pi)^3} \int C \frac{d^3 p_\chi}{E_\chi}\right], \tag{5}$$

where H is the expansion rate of the Universe and C the general collision term. The DM relic abundance is given by [8]

$$\Omega_{DM} h^2 = \frac{2 m_\chi^2 s_0 h^2}{\rho_{cr}} \int_0^\infty \frac{dx}{x^2} \left(-\left.\frac{dY_\chi}{dT}\right|_{T=\frac{m_\chi}{x}}\right), \tag{6}$$

where $x = m_\chi / T$, $s_0 \simeq 2891.2/\text{cm}^3$ is the present value of the entropy density, and $\rho_{cr} = 1.054 \times 10^{-5} h^2 \text{GeV}/\text{cm}^3$ the critical density. Equation (6) must reproduce the observed DM abundance [30]

$$\left.\Omega_{DM} h^2\right|_{obs} = 0.1188 \pm 0.0010, \tag{7}$$

and at the same time, explain the IceCube data. Concerning the operator (4), the dominant contributions to DM production are *a*) the *inverse decay* processes $\nu_\alpha + H^0 \to \chi$ and $l_\alpha + H^+ \to \chi$, that occurs when $m_\chi > m_H + m_{\nu,l}$ proportional to factor $|y_{\alpha\chi}|^2$, and *b*) the *Yukawa production* processes, such as $t + \bar{t} \to \bar{\nu}_\alpha + \chi$ is proportional to $|y_{\alpha\chi} y_t|^2$, where t represents the quark top. After solving the Boltzamnn equation one gets

$$\Omega_{DM} h^2|_{inv.dec.} = 0.1188 \frac{\sum_\alpha |y_{\alpha\chi}|^2}{7.5 \times 10^{-25}} . \tag{8}$$

From (8) immediately follows that to have the correct DM relic abundance (7) one has to require

$$\sum_{\alpha=e,\mu,\tau} |y_{\alpha\chi}|^2 = 7.5 \times 10^{-25} . \tag{9}$$

However, Eq. (9) is in conflict with the value of $\sum_{\alpha=e,\mu,\tau} |y_{\alpha\chi}|^2$ needed to explain the IceCube data. To see that, first note that the DM lifetime $\tau_\chi = \Gamma_\chi^{-1}$ has to be larger that the age of the Universe, $\tau_\chi > t_U \simeq 4.35 \times 10^{17}$sec. Moreover, IceCube spectrum sets a constraints on lower bounds of DM lifetime $\tau_\chi^b \simeq 10^{28}$sec, i.e. $\tau_\chi \gtrsim \tau_\chi^b$, which is (approximatively) model-independent (see [8]). Using (9) one obtains

$$\Gamma_\chi = \sum_{\alpha=e,\mu,\tau} \frac{|y_{\alpha\chi}|^2}{8\pi} m_\chi \simeq 4.5 \times 10^4 \frac{m_\chi}{1\text{PeV}} \text{sec}^{-1} \quad \to \quad \tau_\chi \simeq 2.2 \times 10^{-5} \frac{1\text{PeV}}{m_\chi} \text{sec} \ll t_U .$$

The observations of IceCube require the DM decay lifetime $\tau_\chi \sim 10^{28}$ sec which implies

$$\sum_\alpha |y_{\alpha\chi}|^2 \simeq 10^{-58} , \tag{10}$$

which is many order of magnitudes smaller than the value needed to explain the DM relic abundance, see (9). As a consequence, the IceCube high energy events and the DM relic abundance are not compatible with the DM production if the latter is ascribed to the 4-dimensional operator $\overline{L_{L_\alpha}} H \chi$.

3. PeV neutrinos in modified cosmologies

As seen in the previous Sections, the 4-dimensional operator fails to explaining both the IceCube data and DM relic abundance. We face this problem in terms of modified cosmology. Using (2), (5) and (6), it follows (for $\nu > -3$)

$$\begin{aligned} \Omega_{DM} h^2 &= \frac{45 h^2}{1.66 \pi^2 g_*^{3/2}} \frac{s_0 M_{Pl}}{\rho_{cr}} \frac{\Gamma_\chi}{m_\chi} \frac{2^{2+\nu}}{\eta} \left(\frac{T_*}{m_\chi}\right)^\nu \Gamma\left(\frac{5+\nu}{2}\right) \Gamma\left(\frac{3+\nu}{2}\right) \\ &\simeq 0.1188 \frac{\sum_\alpha |y_{\alpha\chi}|^2}{7.5 \times 10^{-24}} \Pi , \quad \text{where} \quad \Pi \equiv \frac{2^{3+\nu}}{3\pi\eta} \left(\frac{T_*}{m_\chi}\right)^\nu \Gamma\left(\frac{5+\nu}{2}\right) \Gamma\left(\frac{3+\nu}{2}\right) . \end{aligned} \tag{11}$$

Π accounts for all corrections induced by modified cosmology. To explain the DM relic abundance and the IceCube data, we require

$$\frac{\sum_\alpha |y_{\alpha\chi}|^2}{7.5 \times 10^{-24}} \Pi \lesssim \mathcal{O}(1) . \tag{12}$$

The parameters $\{\eta, \nu, T_*\}$ in (11) are arbitrary and can be chosen such that the condition (12) is fulfilled. Therefore we may have $-3 < \nu < 0$ for $T_* < M_\chi$ or $\nu > 0$ for $T_* > m_\chi$.

4. Examples of modified cosmologies

In this Section we discuss the previous issues for two specific example of modified cosmologies.

4.1. Scalar Tensor Theories (STTs)

The total action of a STT of gravity is given by $S = S_{STT} + S_m$ [21], where

$$S_{STT} = \frac{1}{16\pi}\int d^4x\sqrt{-\tilde{g}}\left[\Phi^2\tilde{R}(\tilde{g}) + 4\omega(\Phi)\tilde{g}^{\mu\nu}\partial_\mu\Phi\partial_\nu\Phi - 4\tilde{V}(\Phi)\right], \tag{13}$$

and $S_m = S_m[\Psi, \tilde{g}_{\mu\nu}]$ is the matter action (the matter fields Ψ_m couple to the metric tensor $\tilde{g}_{\mu\nu}$). The action (13) encodes the Brans-Dicke theory of gravity for $\omega(\Phi) = \omega = constant$. In the form (13), the STT action is refereed as Jordan frame. By means of the conformal transformation $\tilde{g}_{\mu\nu} = A_C(\phi)g_{\mu\nu}$ (A_C is the conformal factor that depends on $\phi(x)$) and setting $\Phi^2 = 8\pi M_*/A_C^2$, $V(\phi) = A_C^4(\phi)\tilde{V}(\phi)/4\pi$, and $\alpha(\phi) = \frac{d\log A_C(\phi)}{d\phi}$ $(= (\omega(\Phi)+3)^{-1})$, the action (13) can be casted in the so-called Einstein frame

$$S_{STT} = \frac{M_*^2}{2}\int d^4x\sqrt{-g}\left[R(g) + g^{\mu\nu}\partial_\mu\phi\partial_\nu\phi - \frac{2}{M_*^2}V(\phi)\right], \tag{14}$$

while the action of matter fields assumes the form $S_m = S_M[\Psi_m, A_C^2(\phi)g_{\mu\nu}]$. M_* accounts for the fact that the gravitational *constant* may vary with the scalar field, $M_* = M_{Pl}(\phi) = G^{-1}(\phi)$. In the FRW flat Universe (3), the cosmological field equations read

$$H^2 \equiv \frac{\dot{a}^2}{a^2} = \frac{1}{3M_*^2}\left[\rho + \frac{M_*^2}{2}\dot{\phi}^2 + V(\phi)\right], \tag{15}$$

$$\frac{\ddot{a}}{a} = -\frac{1}{6M_*^2}\left[\rho + 3p + 2M_*^2\dot{\phi}^2 - 2V(\phi)\right], \tag{16}$$

$$\ddot{\phi} + 3H\dot{\phi} + \frac{1}{M_*^2}\left[\frac{\alpha(\phi)}{\sqrt{2}}(\rho - 3p) + V_\phi\right] = 0, \tag{17}$$

where $V_\phi = \partial V/\partial\phi$. The Bianchi identity (conservation of the energy momentum tensor) $d(\rho a^3) + pd(a^3) = (\rho - 3p)d\log A_C(\phi)$ implies $Ta = constant$ for $p = \rho/3$. From Eqs. (15)-(17), one gets [21]

$$H^2 \equiv H_{MC}^2 = \frac{A_C^2(\phi)[1+\alpha(\phi)\phi']^2}{1-\phi'^2/6}H_{GR}^2. \tag{18}$$

The prime indicates the derivative with respect to $N \equiv \ln a$ ($\phi' \equiv \frac{d\phi}{dN} = \frac{d\phi}{d\ln a} = -T\phi_T$, where $\phi_T \equiv d\phi/dT$), while for the scalar field equation one gets (setting $\lambda = V(\phi)/\rho$)

$$\frac{2(1+\lambda)}{3(1-\phi'^2/6)}\phi'' + [(1-w)+2\lambda]\phi' + \sqrt{2}\alpha(\phi)(1-3w) + 2\lambda\frac{V_\phi}{V} = 0. \tag{19}$$

The form of the factor $A(T)$ in (2) for a STT follows from (18)

$$A(T) \equiv \frac{A_C(\phi)[1+\alpha(\phi)\phi']}{(1-\phi'^2/6)^{1/2}}. \tag{20}$$

Assuming that $A(T)$ is of the form (2), Eq. (20) can be rewritten in the form

$$\left[A_C^2\alpha^2T^2 + \frac{\eta^2T_*^2}{6}\left(\frac{T}{T_*}\right)^{2(\nu+1)}\right]\phi_T^2 - 2A_C^2\alpha\phi_T + \left[A_C^2 - \eta^2\left(\frac{T}{T_*}\right)^{2\nu}\right] = 0. \tag{21}$$

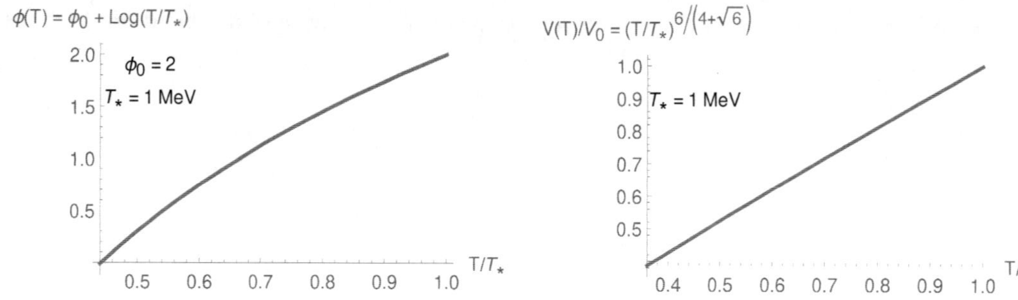

Figure 1. ϕ and $V(\phi)$ vs T.

To solve Eq. (21) one has to specify the form of $A_C(\phi)$. Since the problem is quite involved, here we discuss the particular case $\alpha, A_C \ll 1$. Equation (21) reduces to $\left(\frac{d\phi}{dz}\right)^2 = \frac{6}{z^2}$, where $z \equiv \frac{T}{T_*}$. The solution is $\phi(z) = \phi^{(0)} + \sqrt{6}\log z \simeq \phi^{(1)} - \sqrt{6}N$, with $\phi^{(0)}, \phi^{(1)}$ constants. Noting that $\phi'' = 0$, Eq. (19) assumes the form $-\frac{\sqrt{6}}{3} + \frac{V}{\rho}(1 + \log V_\phi) = 0$. Writing the energy density in terms of the field ϕ, $\rho = K_* e^{4\phi/\sqrt{6}}$ with $K_* \equiv \frac{\pi^2 g_* T_*^4}{30}$, Eq. (19) allows to derive the potential V (the integration constant is set equal to zero) $V(\phi) = \gamma^{-1} K_* e^{\gamma\phi}$, where $\gamma \equiv \frac{\sqrt{6}}{\sqrt{6}+4}$. The potential V is suppressed for temperatures $T < T_*$ (see Fig. 1). Results are independent on ν and η. To obtain the correct DM relic abundance and explain the IceCube results, hence to fulfill the condition (12), the parameters $\{\eta, \nu, T_*\}$ must fine tuned. Setting $\eta \sim \mathcal{O}(1)$ and $T_* = 10^q$GeV, and for $T_* \gg m_\chi$, Eq. (12) implies $\nu = \frac{34}{q-6}$. For example, if the transition temperature occurs at $\sim 10^{12}$GeV, i.e. $q \sim 12$, then it follows $\nu \sim 5 - 6$.

4.2. $f(\mathcal{T})$ cosmology

Another interesting model is provided by the theory of gravity based on the Weitzenböck connection (instead of the usual Levi-Civita connection). Here the gravitational field is described by the torsion (instead of the curvature tensor). This model is referred as the Teleparallel Equivalent of General Relativity (TEGR), that is equivalent to GR at the level of field equations [31, 31, 32, 33, 34, 35]. In TEGR, one adopts the curvatureless Weitzenböck connection

$$\mathcal{T}^\lambda_{\mu\nu} = \hat{\Gamma}^\lambda_{\nu\mu} - \hat{\Gamma}^\lambda_{\mu\nu} = e^\lambda_i(\partial_\mu e^i_\nu - \partial_\nu e^i_\mu), \tag{22}$$

where $e^i_\mu(x)$ are the vierbein fields defined as $g_{\mu\nu}(x) = \eta_{ij} e^i_\mu(x) e^j_\nu(x)$. The action is given by $S^0_I = \frac{1}{16\pi G}\int d^4x e\mathcal{T}$, where $\mathcal{T} = S_\rho{}^{\mu\nu}\mathcal{T}^\rho{}_{\mu\nu}$ is the torsion scalar, $e = det(e^i_\mu) = \sqrt{-g}$, and

$$S_\rho{}^{\mu\nu} = \frac{1}{2}\left[\frac{1}{4}(\mathcal{T}^{\mu\nu}{}_\rho - \mathcal{T}^{\nu\mu}{}_\rho - \mathcal{T}_\rho{}^{\mu\nu}) + \delta^\mu_\rho \mathcal{T}^{\theta\nu}{}_\theta - \delta^\nu_\rho \mathcal{T}^{\theta\mu}{}_\theta\right]. \tag{23}$$

We shall consider the simplest generalization of the action S^0_I, i.e.

$$S_I = \frac{1}{16\pi G}\int d^4x e\,[\mathcal{T} + f(\mathcal{T})], \tag{24}$$

IOP Conf. Series: Journal of Physics: Conf. Series **1275** (2019) 012059 doi:10.1088/1742-6596/1275/1/012059

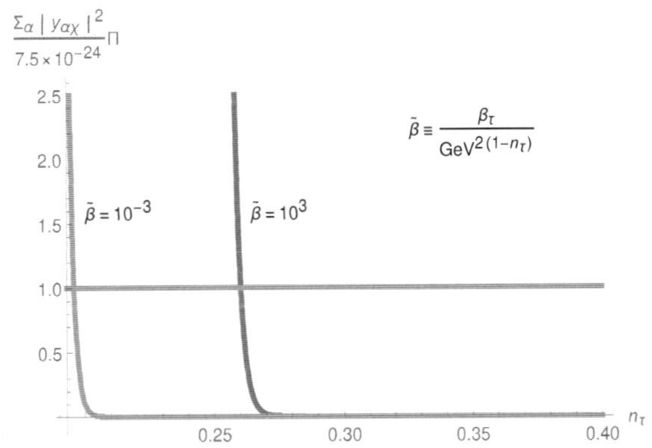

Figure 2. For fixed values of $\eta = 1$ and $\beta_{\mathcal{T}}$, the values of Π needed to explain DM relic abundance and IceCube data follow for $n_{\mathcal{T}} \lesssim 0.25$.

where in the present work we take $f(\mathcal{T}) = \beta_{\mathcal{T}}|\mathcal{T}|^{n_{\mathcal{T}}}$ (in general $f(\mathcal{T})$ is a generic function of the torsion) [36, 37, 38]. Equations (22) and (23) imply $\mathcal{T} = -6H^2$. The cosmological field equations are [34]

$$12H^2[1+f_{\mathcal{T}}] + [\mathcal{T} + f] = 16\pi G\rho, \tag{25}$$

$$48H^2 f_{\mathcal{T}\mathcal{T}}\dot{H} - (1+f_{\mathcal{T}})[12H^2 + 4\dot{H}] - (\mathcal{T} - f) = 16\pi G p,$$

where $f_{\mathcal{T}} = df/d\mathcal{T}$. By rewriting (25) in the form $H^2 + H_{\mathcal{T}}^2 = \frac{8\pi}{3M_{Pl}^2}\rho$, where $H_{\mathcal{T}}^2 \equiv \frac{f}{6} - \frac{\mathcal{T} f_{\mathcal{T}}}{3} = 6^{n-1}\beta_{\mathcal{T}}(2n_{\mathcal{T}}+1)H^{2n}$, and assuming $H_{\mathcal{T}} \gg H$, one gets $H_{\mathcal{T}} \equiv H_{MC} = A(T)H_{GR}$, where $A(T)$ is of the form (2) with

$$T_* \equiv \left(\frac{24\pi^3 g_*}{45}\right)^{\frac{1}{4}} (2n_{\mathcal{T}}+1)^{\frac{1}{4(1-n_{\mathcal{T}})}} \left(\frac{\beta_{\mathcal{T}}}{\mathrm{GeV}^{2(1-n_{\mathcal{T}})}}\right)^{\frac{1}{4(1-n_{\mathcal{T}})}} \left(\frac{M_{Pl}}{\mathrm{GeV}}\right)^{\frac{1}{2}} \mathrm{GeV}, \tag{26}$$

$\eta = 1$ and $\nu = \frac{2}{n_{\mathcal{T}}} - 2$. For $a(t) = a_0 t^{\delta}$, i.e. $H = \frac{\delta}{t}$, it follows $T(t)a(t) = constant$. The transition temperature T_* given in (26) has to be used into Eq. (12). In Fig. 2 is plotted (12) for the $f(\mathcal{T})$ model. The value of $\beta_{\mathcal{T}}$ is obtained by fixing the transition temperature at $T_* \sim 10^{11} - 10^9$GeV, that is $T_* \gg m_\chi$ (the parameter δ enters into the expression of p; we shall not present explicitly it being not relevant for our analysis).

5. Conclusions

In this paper we have studied the possibility to reconcile the current bound on DM relic abundance with IceCube data in terms of the 4-dimensional operator. Motivated by cosmological observations by Type Ia Supernovae, CMB radiation, and the large scale structure, according to which the present Universe is in an accelerating phase, new theories beyond GR have been proposed. We have shown that modified gravity models can explain the IceCube outputs and at the same time the DM relic abundance observed in a minimal particle physics model. This because the cosmological field equations based on modified gravity models change the thermal history of particles, so that the expansion rate of the Universe can be written in the form (1). As cosmological models, we have considered the STTs and models related to torsion $f(\mathcal{T})$. We have solved the Boltzmann equation to get the abundance of DM particles. The latter turns out to be modified by a quantity that does only depend on parameters of the modified cosmological models, and allows to explain, consistently, both the IceCube data and the correct DM abundance $\Omega_D M h^2 \sim 0.11$.

References

[1] Aartsen M G *et al* 2015 *Phys. Rev. Lett* **115** 081102
[2] Aartsen M G *et al* 2015 *Phys. Rev.* D **91** 072004
[3] Cholis I and Hooper D 2013 *JCAP* **06** 030
[4] Anchordoqui L A *et al* 2014 *JHEAp* **1-2** 1
[5] Murase K 2015 *AIP Conf. Proc.* **1666** 040006
[6] Sahu S and Miranda L S 2015 *Eur. Phys. J.* C **75** 273
[7] Aartsen M G *et al* 2017 *Astrophys. J.* **835** 151
[8] Chianese M and Merle A 2017 *JCAP* **04** 017
[9] Riess A G *et al* 1998 *Astron. J.* **116** 1009
Perlmutter S *et al* 1999 *Astrophys. J.* **517** 565
[10] Spergel D N *et al* 2007 *ApJS* **170** 377
[11] Tegmark M *et al* 2004 *Phys. Rev.* D **69** 103501
[12] Eisenstein D J it et al 2005 *Astrophys. J.* **633** 560
[13] Capozziello S and De Laurentis M 2011 *Phys. Rept.* **509** 167
Chakravarty G, Mohanty S and Lambiase G 2017 *Int. J. Mod. Phys.* D **26** 1730023
Amendola L and Tsujikawa S 2010 *Dark Energy: Theory and Observations* (Cambridge University Press)
Nojiri S and Odintsov S D 2011 *Phys. Rep.* **505** 59
Durrer R and Maartens R 2008 *Gen. Relat. Grav.* **40** 301
Sami M 2007 *Lect. Notes Phys.* **720** 219
Copeland E J, Sami M and Tsujikawa S 2006 *Int. J. Mod. Phys.* D **15** 1753
Capolupo A 2017 *Galaxies* **5** 98
Capolupo A 2016 *Adv. High Energy Phys.* **2016** 8089142
[14] De Felice A and Tsujikawa S 2010 *Living Rev. Rel.* **13** 3
[15] Starobinsky A A 1980 *Phys. Lett.* B **91** 99
[16] Ellis J, Nanopoulos D V and Olive K A 2013 *Phys. Rev. Lett.* **111** 111301
[17] Hu W and Sawicki I 2007 *Phys. Rev.* D **76** 064004
[18] Starobinsky A A 2007 *JETP Lett.* **86** 157
[19] Clifton T, Ferreira P G, Padilla A and C. Skordis 2012 *Phys. Rept.* **513** 1
[20] Brans C and Dicke R H 1961 *Phys. Rev.* D **124** 925
[21] Catena R, Fornengo N, Pato M, Pieri L and Masiero A 2010 *Phys. Rev.* D **81** 123522
[22] Schelke M, Catena R, Fornengo N, Masiero A and Pietroni M 2006 *Phys. Rev.* D **74** 083505
Kamionkowski M and Turner M S 1990 *Phys. Rev.* D **42** 3310
Salati P 2003 *Phys. Lett.* B **571** 121
Lambiase G 2012 *JCAP* **10** 028
Iorio A and Lambiase G 2015 *Eur. Phys. J.* C **75** 115
[23] Gelmini G and Gondolo P 2010 DM production mechanisms (*Preprint* arXiv:1009.3690 [astro-ph.CO])
D'Amico G, Kamionkowski M and Sigurdson K 2009 Dark Matter Astrophysics (*Preprint* arXiv:0907.1912 [astro-ph.CO])
[24] Randal L and Sundrum R 1991 *Phys. Rev. Lett.* **83** 4690
[25] Profumo F and Ullio P 2003 *JCAP* **0311** 006
Rosati F 2003 *Phys. Lett.* B **570** 5
Pallis C 2005 ĴCAP **0510** 015
[26] Catena R, Fornengo N, Masiero A, Pieroni M and Rosati F 2004 *Phys. Rev.* D **70** 063519
[27] Capozziello S, Galluzzi V, Lambiase G and Pizza L 2015 *Phys. Rev.* D **92** 084006
Capozziello S, De Laurentis M and Lambiase G 2012 *Phys. Lett.* B **715** 1
[28] Lambiase G, Mohanty S and Stabile A 2018 *Eur. Phys. J.* C **78** 350
[29] Hall L J, Jedamzik K, March-Russell J and West S M 2010 *JHEP* **03** 080
[30] Ade P A R *et al* 2016 *Astron. Astrophys.* **594** 13
[31] Ferraro R and Fiorini F 2007 *Phys. Rev.* D **75** 084031
[32] Linder E V 2010 *Phys. Rev.* D **81** 127301
Bahamonde S, Böhmer C G and Wright M 2015 *Phys. Rev.* D **92** 104042
Capozziello S, Lambiase G and Saridakis E N 2017 *Eur. Phys. J.* C **77** 576
Fazlpour B 2016 *Gen. Rel. Grav.* **48** 159
[33] Aldrovandi R and Pereira J G 2013 *Teleparallel Gravity: An Introduction* (Springer, Dordrecht)
[34] Cai Y F, Capozziello S, De Laurentis M and Saridakis E N 2016 *Rept. Prog. Phys.* **79** 106901
[35] Cai Y, Li C, Saridakis E N and Xue L Q 2018 *Phys. Rev.* D **97** 103513
[36] Nesseris S, Basilakos S, Saridakis E N and Perivolaropoulos L 2013 *Phys. Rev.* D **88** 103010
[37] Ferraro R and Fiorini F 2008 *Phys. Rev.* D **78** 124019

[38] Nunes R C, Pan C and Saridakis E N 2016 *JCAP* **1608** 011

IOP Conf. Series: Journal of Physics: Conf. Series **1275** (2019) 012060 doi:10.1088/1742-6596/1275/1/012060

Matter-as-constrained-gravity quanta on multiple Rindler observer grounds

P A Mandrin

IMSD, Lavaterstrasse 103, 8002 Zürich, Switzerland

E-mail: pierre.mandrin@imsd.ch

Abstract. By considering matter as a constraint on the availability of gravitational degrees of freedom and accounting for the statistical interpretation of Rindler horizons, the freedom to construct quantum gravity theories reproducing General Relativity and Quantum Field Theory (QFT) as special cases is considerably reduced. On one hand, the mathematical structure of quantum gravity is restricted by the properties of Quantum Mechanics. On the other hand, one can predict a value for the fundamental quantum constant of gravity which is related to the Planck area via the Planck constant. These findings are compatible with spin-less particles of matter. In the context of canonical ensemble statistics, the von Neumann entropy concept is found to extend from matter to gravity. An important motivation and pillar for this development is the concept of multiple observer statistics and the total entropy perceived by Rindler observers with a particular spacing, which has a one-to-one correspondence to the Gibbons-Hawking-York boundary term.

1. Introduction

So far, experimental evidence is lacking of which of the approaches to quantum gravity (if any) should be applicable. Because of its smallness, the value of the fundamental constant of quantum gravity is also unknown. However, there might be some profound relation between gravity and matter. This possibility has received little attention in the past, but could help reduce the realm of admissible theories. We shall consider here matter as a constraint which restricts the gravitational degrees of freedom. The goal of this article is to focus on a gravity-matter relation instead of quantising some set of canonical variables.

As a skeleton of our formalism, we use the ideas of black hole thermodynamics [1, 2, 3] which have been extended to Rindler space-times, where the Unruh radiation [4] corresponds to the Hawking radiation of black holes. In the limit of stationarity, Einstein's field equations appear as a manifestation of the First Law applied to the null-horizon [5, 6], so that any null-surface acquires a local, observer-dependent temperature $T = \kappa/(2\pi)$ with surface gravity κ, and an entropy $S_{BHJP} = A/(4L_p^2)$, where A is the horizon surface area and L_p is the Planck length.

This concept has recently been extended to describe multiple Rindler observers as required for observing a horizon temperature, this yields a Boltzmann-like 3d-space entropy – we call it m-entropy S – which is equivalent (up to signatures) to the Gibbons-Hawking-York boundary term [7, 8] if the density of observers is proportional to the external curvature [9]. In contrast to S_{BHJP}, the existence of S does not require the underlying microscopic model to be in local equilibrium, since S is not paired with a temperature. This allows for hypothetical departures

Content from this work may be used under the terms of the Creative Commons Attribution 3.0 licence. Any further distribution of this work must maintain attribution to the author(s) and the title of the work, journal citation and DOI.
Published under licence by IOP Publishing Ltd

from General Relativity and opens a door to quantum behaviour on a space of reduced dimension $(2+1)$, in correspondence to the holographic principle [10].

The present article investigates departures from the vacuum geometry. These departures are interpreted in terms of matter fields, the latter play the role of constraints. We derive some formal properties of matter (in the context of GR, we only consider spin-0 fields) and investigate the resulting fundamental quanta of gravity.

For the convenience of the reader, we also summarise in this article:

- the concept of m-entropy which has been formerly derived using [6],
- how the Gibbons-Hawking boundary term can be subject to the variation principle without involving the bulk action (cf Appendix A) and
- the multiple observer description of gravity (cf Appendix B); its statistical interpretation is summarized in Appendix C.

For the quantum properties of the m-entropy, we refer to [9], where we have sketched the derivation of quantum uncertainties in a system with a small number of quantum bits of gravity, as well as the derivation of some typical transition probabilities between quantum states.

2. Synopsis of the multiple observer formulation of GR

The starting point is the Gibbons-Hawking boundary term for a compact 4-volume V; we omit any terms free of derivatives of the metric for simplicity, and no matter term is added:

$$S_{\partial 1} = \frac{c^4}{8\pi G} \sum_A \varepsilon_A \int_{B_A} d^3x \, \sqrt{|\gamma|} \, K. \tag{1}$$

In (1), the integration is performed over the piece-wise smooth boundary $\partial V = \sum_A B_A$ of V, where every component B_A is time- or space-like. Furthermore, $K = K_{ab}\gamma^{ab}$, where γ_{ab} is the induced metric on ∂V with determinant γ, K_{ab} is the exterior curvature on ∂V with unit normal vector n^a, namely $K_{ab} = -\frac{1}{2}\mathcal{L}_\perp \gamma_{ab} = -g^{cd}\gamma_{da}\nabla_c n_b$, $\mathcal{L}_\perp = \mathcal{L}_n$, and $\varepsilon_A = n^a n_a$ on B_A. Equivalently, following the notation of [6], if $g_{\perp k \neq \perp} = 0$, (1) can be written as

$$S_{\partial 1} = \frac{c^4}{16\pi G} \int_{\partial V} N^c_{ab} \, f^{ab} \, d^3\Sigma_c, \qquad f^{ab} = \sqrt{-g}\, g^{ab}, \qquad N^c_{ab}\, d^3\Sigma_c = \varepsilon_A \, N^{ab} \, d^3x, \tag{2}$$

where N^c_{ab} is defined as the momentum conjugate to f^{ab}, $d^3\Sigma_c$ is the the 3-surface element covector and $N_{ab} = K_{ab} - K\gamma_{ab}$. $S_{\partial 1}$ can be subject to the variation principle (without involving the bulk, cf Appendix A). We divide the boundary ∂V into rectangular cuboids $\Delta\partial V$ with edges short enough for geometric variations to be negligible across $\Delta\partial V$ and large enough for quantum gravity effects to be negligible. For the portion of boundary term (1) contained within $\Delta\partial V$, it can be shown (cf Appendix B) [9, 11]:

$$\begin{aligned}
S_{\partial 1\ \Delta\partial V} &= \frac{c^4\varepsilon_A}{32\pi G} \int_{\Delta\partial V} d^3x \, \sqrt{|\gamma|}\, \gamma^{ab}\, \mathcal{L}_\perp \gamma_{ab} \\
&= \frac{c^4\varepsilon_A}{32\pi G} \sum_{k=1}^{N} \int_{A(x_k^{||})} d^2\tilde{x} \left[\sqrt{|\gamma_{(u)}||\frac{e_{||}^2}{e_u^2}|}\ \Delta u \ \gamma^{ab}_{(u)}\ \mathcal{L}_v \gamma^{(u)}_{ab} + \sqrt{|\gamma_{(v)}||\frac{e_{||}^2}{e_v^2}|}\ \Delta v\ \gamma^{ab}_{(v)}\ \mathcal{L}_u \gamma^{(v)}_{ab} \right] \\
&= \frac{c^4\varepsilon_A}{32\pi G} \sum_{k=1}^{N} \sum_{w=u,v} \int_{\Delta w} \int_{A(x_k^{||})} d^3\tilde{x}\ \sqrt{|\gamma_{(w)}||\frac{e_{||}^2}{e_w^2}|}\ \gamma^{ab}_{(w)}\ \mathcal{L}_v \gamma^{(w)}_{ab} \qquad (3) \\
&= \sum_{k=1}^{N} \left[\int_{\Sigma_{(u)}(x_k^{||})} \tilde{T}_u \ s\ du\ d^2x + \int_{\Sigma_{(v)}(x_k^{||})} \tilde{T}_v\ s\ dv\ d^2x \right], \qquad (4)
\end{aligned}$$

where e^J_a is the induced triad on ∂V, $\gamma^{(w)}_{ab}$ and $\gamma^{ab}_{(w)}$ are the metric and inverse metric induced on the hypersurface $w =$ constant, respectively, $w = u, v$ are null coordinates, $\tilde{x}^\mu = (u, v, x^2, x^3)$, $\tilde{T}_w = |\frac{e^2_{||}}{e^2_w}|\, T_w$ and T_w is the horizon temperature observed on the hypersurface $w =$ constant. If we choose precisely $\Delta w = \tilde{T}_w^{-1}$, the temperatures disappear together with the null distances, i.e. $S_{\partial 1\ \Delta\partial V}$ can be interpreted as a Boltzmann-like entropy if the spacings between the observers are wright at the limit of the classical spatial resolution, i.e. the density of observers is proportional to the resolvable normal change of γ_{ab}. Since $S_{\partial 1\ \Delta\partial V}$ is not time-dependent, it does not depend on any thermalization condition, but exists beyond stationarity. Only in the case of stationarity does $S_{\partial 1\ \Delta\partial V}$ reduce to a sum of products of horizon entropies and temperatures. From the point of view of the Rindler observers, the total entropy within $\Delta\partial V$ is the m-entropy, i.e. the sum of the entropies of all the null-strips contained in (4) – this is the lack of information related to a thermodynamic measurement process which takes some time and/or distance.

As a consequence, the boundary space naturally splits into multiple layers of 2d-surfaces (like a foliation across a given direction), and every layer consists of a collection of ordered surface microstates $|q\rangle$ (cf Appendix C). The layer structure is similar to a computer hard disk, where every layer stores a certain number of bits. For gravity, however, there is no a-priori restriction about which fraction of the states is on which layer. A stationary point of $S_{\partial 1\ \Delta\partial V}$ is a macroscopic geometry, for which the number of microstate configurations is (locally) maximum.

Multiple observer statistical gravity is not the first development for which some kind of analogy between the gravitational action and the notion of "entropy" is suggested [12, 13, 14]. However, it is the first derivation of such a relation on the basis of horizon thermodynamics. For quantum mechanics, such an analogy has been conjectured even earlier [15].

3. Quasi-stationarity condition as a matter generating mechanism

Although the stationary point of $S_{\partial 1}$ is the geometry of vacuum (γ^{vac}_{ab}), there is enough room for non-vacuum metrics since the fraction of vacuum $(N_K, \{n_{Ki}|i \leq N_K\})$-states is small against the non-vacuum states. Our goal is:

(i) constrain γ_{ab} to non-vacuum configurations with fixed "distance" from γ^{vac}_{ab} and
(ii) solve for a stationary point $\delta S_{\partial 1}|_{constr} = 0$ under the constraint (i) (quasi-stationarity).

Procedure: Be $\{\Gamma_{d^I_a(\tau)}\}$ with $\Gamma_{d^I_a(\tau)} = (e^{vac\ I}_a + d^I_a(\tau))\eta_{IJ}(e^{vac\ J}_b + d^J_b(\tau))$ a family of distinct curves in the neighbourhood of $\gamma^{vac}_{ab} = e^{vac\ I}_a \eta_{IJ} e^{vac\ J}_b$ in the space of γ_{ab} with $0 \leq \tau \leq \epsilon$, $\Gamma_{d^I_a(0)} = \gamma^{vac}_{ab}$ and $\Gamma_{d^I_a(\epsilon)} \neq \gamma^{vac}_{ab}$. We impose the stationarity condition $\delta(S|_{\gamma_{ab}=\Gamma_{d^I_a(\epsilon)}}) = 0$, where $d^I_a(\epsilon)$ is varied, up to gauge transformations of γ_{ab} leaving S invariant. To impose condition (i), we introduce a compensation term $S_m(\Gamma_{d^I_a(\epsilon)}, \varphi)$ such that $\delta(S + S_m) = 0$, and the independent object φ ensures a fixed "distance" between γ_{ab} and γ^{vac}_{ab}. We call this the "compensation method". We look for a general ansatz for S_m. A special case with simplified treatment is

$$|\Gamma_{d^I_a(\epsilon)} - \gamma^{vac}_{ab}|/|\gamma^{vac}_{ab}| \ll 1, \tag{5}$$

which we call the tenuous limit, where $|\gamma_{ab}|$ denotes the largest absolute component of γ_{ab}.

3.1. Tenuous limit

We expand the induced metric up to the first order in the real "distance“ parameter τ:

$$\begin{aligned} \gamma_{ab}(\tau) &\approx \eta_{ab} + \zeta_{ab}\tau, \\ \gamma^{ab}(\tau) &\approx \eta^{ab} - \bar{\zeta}^{ab}\tau, \end{aligned} \tag{6}$$

IOP Conf. Series: Journal of Physics: Conf. Series **1275** (2019) 012060 doi:10.1088/1742-6596/1275/1/012060

where we define $\bar{\zeta}^{ad} = \eta^{ab}\zeta_{bc}\eta^{cd}$ so that $\gamma^{ab}\gamma_{bc} = \delta^a_c$ holds to first order in τ. Next, we write the integrand σ_g of the boundary term (1) to lowest order in τ:

$$\sigma_g = \frac{c^4\varepsilon_A}{16\pi G}\sqrt{|\gamma|}\gamma^{ab}\mathcal{L}_\perp\gamma_{ab} \approx -\frac{c^4\varepsilon_A}{16\pi G}\sqrt{|\gamma|}\bar{\zeta}^{ab}\mathcal{L}_\perp\zeta_{ab}\tau^2. \tag{7}$$

Notice that the first order term in τ vanishes because $\gamma_{ab}(0)$ is a stationary point for S. Because σ_g is of second order in τ, so must be the compensation term integrand σ_m. We reduce (7) to

$$\sigma_g \approx -\frac{c^4\varepsilon_A}{32\pi G}\sqrt{|\gamma|}\mathcal{L}_\perp(\bar{\zeta}^{ab}\zeta_{ab})\tau^2. \tag{8}$$

In the expression for (8), we replace $-\bar{\zeta}^{ab}\zeta_{ab}\tau^2$ by the square of a constraining scalar function $\varphi \sim \tau$ to construct σ_m:

$$\sigma_m \approx \vartheta\sqrt{-g}\nabla_\perp\varphi^2 = 2\vartheta\sqrt{-g}\varphi\nabla_\perp\varphi, \tag{9}$$

where ϑ is an arbitrary constant (still to be determined). A form like (9) generates gravity-independent dynamical degrees of freedom: a field. Choosing at least $\sigma_m \sim \nabla_\perp\varphi^2$ rather than $\sim \nabla_\perp\varphi$ is necessary in order that a non-trivial bulk Lagrangian for φ with boundary term S_m and non-trivial equations of motion exists (we ignore theories without a Lagrangian). More generally, φ can be allowed to be complex, $\sigma_m \sim \nabla_\perp\varphi^*\varphi$. Notice that the objects $\bar{\zeta}^{ab}$ and ζ_{ab} at the stationary points of $S + S_m$ ($\tau = \epsilon$ fixed) do not depend on the choice of ϵ.

Because our starting point was the boundary term for GR, we expect that φ is a spin-less matter field, namely the Klein Gordon (KG) field or some transformation of it. To examine this, we compute the boundary KG term S_{bKG} from the bulk KG Lagrangian $\mathcal{L}_{KG}$ or action S_{KG} (restricting ourselves to a real φ), and use Gauss' Theoreme:

$$\begin{aligned}
\mathcal{L}_{KG} &= \frac{1}{2}(\partial^\mu\varphi\partial_\mu\varphi - \frac{m^2c^2}{\hbar}\varphi^2), \\
\delta S_{KG}/\delta\varphi &= \int_V d^4x\sqrt{-g}(-\nabla^\mu\nabla_\mu - \frac{m^2c^2}{\hbar})\varphi + \delta S_{bKG}/\delta\varphi, \qquad (10)\\
S_{bKG} &= \int_V d^4x\sqrt{-g}\nabla^\mu(\varphi\nabla_\mu\varphi) = \int_{\partial V} d^3x\sqrt{-g}\varphi\nabla_\perp\varphi. \qquad (11)
\end{aligned}$$

We see that S_m and S_{bKG} are of the same mathematical form, so that they compensate each other (and the Klein Gordon Equation holds) if φ in (9) equals φ in (11) and $\vartheta = -\frac{1}{2}$.

3.2. Higher density

If (5) is not satisfied (higher density regime), it follows from the non-linearity of S_g that a metric description similar to (6) would cause ζ_{ab} at the stationary points (while τ is fixed) to depend on τ in a more complex way:

$$\begin{aligned}
\gamma_{ab}(\tau) &\approx \eta_{ab} + \zeta_{ab}(\tau), \\
\gamma^{ab}(\tau) &\approx \eta^{ab} - \bar{\zeta}^{ab}(\tau),
\end{aligned} \tag{12}$$

where $\bar{\zeta}^{ab}(\eta_{bc} + \zeta_{bc}) = \eta^{ab}\zeta_{bc}$, namely, the tensorial structure of ζ_{ab} itself would depend on τ as well. Nevertheless, we can still proceed in analogy to the tenuous limit, as for (8,9).

IOP Conf. Series: Journal of Physics: Conf. Series **1275** (2019) 012060 doi:10.1088/1742-6596/1275/1/012060

4. Quanta of matter, quanta of gravity, Planck area and canonical ensemble

Consider a cuboid $\Delta\partial V$ (on ∂V) which is foliated across x^K (K is a Minkowski index) into N_K pieces of 2d-surfaces of aria A_K representing n_K bits each, where N_K is defined according to Appendix C. If n_q defines the number of possible states of every bit and if every microstate of $\Delta\partial V$ is given by a pair (N_K, n_K), the m-entropy of $\Delta\partial V$ is [9] (see also Appendix C)

$$s_K = N_K n_K \ln n_q, \tag{13}$$
$$s_K = \sum_i n_{Ki} \ln n_q. \tag{14}$$

(K is not used as a tensorial index in (13)). (13) generalizes to (14) if we allow the number n_{Ki} of bits on every ith surface to vary individually. Equations (13,14) describe the ordered quantum bit structure for GR. We analyse now how (14) can be related to quantum matter in the tenuous limit. Consider a curve $\Gamma_{d^I_a(\tau)}$ in γ_{ab}-space, made of points satisfying the stationarity condition $\delta(S|_{\gamma_{ab}=\Gamma_{d^I_a(\tau)}}) = 0$ (for τ fixed). We change $s_K/(\ln n_q)$ along $\Gamma_{d^I_a(\tau)}$ by a number of bits $\Delta(\sum_i n_{Ki})$ so that the change of γ_{ab} is no less than the smallest classically observable difference. To maintain stationarity, we modify $\varphi = \varphi(\tau)$ (by the same number of quanta as bits) via

$$\Delta(N_K n_K) \ln n_q = \frac{d\sigma_g}{d\tau}\Delta\tau = -\frac{d\sigma_m}{d\tau}\Delta\tau \approx -\frac{\partial\sigma_m}{\partial\varphi}\Delta\varphi. \tag{15}$$

In (15), the stationarity condition has been required to hold in physical equivalence to the quantum mechanical commutation relations which reflect the equations of motion. Subjecting φ to second quantisation, $\varphi \to \hat{\varphi}$, converts (15) into an operator equation (i.e. γ_{ab} is not sharp)

$$\frac{d\sigma_g}{d\tau}\hat{\tau} = -\frac{\partial\sigma_m}{\partial\varphi}\hat{\varphi}, \qquad \hat{\tau} = \sum_i \tau_{i-} a_i^\dagger + \tau_{i+} a_i, \tag{16}$$

with $a_i|n_{Ki}\rangle = n_{Ki}^{1/2}|n_{Ki}-1\rangle$, $a_i^\dagger|n_{Ki}\rangle = (n_{Ki}+1)^{1/2}|n_{Ki}+1\rangle$. (15) establishes how every quantum change of gravity yields a quantum change in matter. Conversely, $\hbar$ fixes the fundamental quantum constant of gravity, using (9, 11, 15) and $\int_{\partial V} d^3x\sqrt{-g}\varphi\nabla_\perp\varphi = \hbar c$:

$$\frac{1}{32\pi}\left|\int_{\partial V} d^3x\sqrt{|\gamma|}\mathcal{L}_\perp(\bar{\zeta}^{ab}\zeta_{ab})\Delta(\tau^2)\right| = \frac{{}^-hG}{c^3} = L_p^2, \tag{17}$$

which relates the quantum of "surface curving" to the Planck area. Notice that L_p^2 is not just an ad-hoc scale of breakdown of GR, but fixes the elementary quantum constant.

Equation (17) simplifies further if we consider the example of a scalar particle in the form of a monochromatic planar wave function with wave covector $k_\mu = (k_0, k_\perp, 0, 0)$. The non-zero elements of $(\zeta_{ab}\tau)$ also yield a monochromatic plane wave function. If we choose V to be a hypercuboid with edges parallel to the coordinate axes and the two space-like hypersurfaces $\Sigma_\pm$ separated by π/k_0 (half a period) and the time-like hypersurfaces $\Pi_\pm$ perpendicular to $(0, k_\perp, 0, 0)$ very near to each other (distance $\ll \pi/k_\perp$), we obtain

$$|\bar{\zeta}^{ab}\zeta_{ab}|_{max}\Delta(\tau^2) = 16\pi\frac{L_p^2}{|\Sigma_+|k_0}, \tag{18}$$

where $|\Sigma_+|$ is the 3-volume of Σ_+, since only $\Sigma_\pm$ contribute significantly.

Finally, a cuboid $\Delta\partial V$ can be considered to be immersed in a "bath" of matter field (φ). It can therefore be described quantum mechanically as a canonical ensemble. The ensemble average entropy s_N of all possible multiple observer states of $\Delta\partial V$, with bit-number eigenstates $|n_{lKi}\rangle$, density matrix $\hat{\rho}$ and matter compensation term $s_m(\Delta\partial V, \varphi)$, is the von Neumann entropy:

IOP Conf. Series: Journal of Physics: Conf. Series **1275** (2019) 012060 doi:10.1088/1742-6596/1275/1/012060

$$s_N = \mathrm{tr}\,(\hat{\rho}\ln\hat{\rho}), \tag{19}$$

$$\rho_{lm} = |n_{lKi}\rangle \frac{\exp\left[s_K(n_{lKi}) + s_m\right]}{\sum_k \exp\left[s_K(n_{kKi}) + s_m\right]} \left[\prod_j \delta_{n_{lKj}n_{mKj}}\right] \langle n_{mKi}| \quad (i \geq 1). \tag{20}$$

5. Conclusions

By constraining the Gibbons-Hawking-York boundary term to lowest physically nontrivial order (quadratic) in the constraint "parameter" φ while taking the first derivative over to the constraint term, we have obtained the emergence of a spin-less matter field (which is compatible with the torsionlesness of GR), at least in the tenuous limit. This fact suggests a profound relation between gravity and matter. Constraining the boundary term (and not the bulk action) is largely motivated by the multiple observer statistical interpretation of GR. We infer that quantum matter might be intimately related to the quantum structure of gravity as well, so that the full quantum mechanical treatment can be transferred from quantum matter to the gravitational field. Under this condition, the fundamental area of one bit of "surface curving" has been found to be 32π times the Planck area. If we take this computation literally, that would mean that the quantisation of gravity has indirectly already been quantified since $\hbar$ has been determined. Finally, von Neumann's quantum canonical ensemble statistics straight-forwardly extends to the "matter bath" description of quantum gravity. Since our approach does not rely on the quantisation prescription of an a priori model of canonically conjugated variables, but instead on the concept of "gravity constrained by matter" with a generalised statistical mechanical background, the applicability of our investigations might be quite general.

Appendix A. The boundary term and the variation principle

For a compact 4d-region V with boundary ∂V, consider the Hilbert-Einstein action S_H and the Gibbons-Hawking-York boundary term $S_{\partial 1}$, with the symbols defined as in Section 2,

$$S_H = \frac{c^4}{16\pi G}\int_V d^4x\,\sqrt{-g}\,R, \tag{A.1}$$

$$S_{\partial 1} = \frac{c^4}{8\pi G}\sum_A \varepsilon_A \int_{B_A} d^3x\,\sqrt{|\gamma|}\,K, \tag{A.2}$$

ignoring terms which do not depend on derivatives of the metric. In order for the variation principle to correctly yield Einstein's Equations [7, 8], we either need to fix $N_{ab} = K_{ab} - K\gamma_{ab}$ while varying S_H or to fix γ_{ab} while varying $S_H + S_{\partial 1}$. We also can vary the difference $S_{\partial 1}$ separately, with respect to both γ_{ab} and N_{ab} on ∂V (but neither $g_{\alpha\beta}$ nor the bulk is involved),

$$\delta S_{\partial 1} = 0, \tag{A.3}$$

i.e. the Gibbons-Hawking-York term is subject to the variational principle.

Appendix B. Multiple horizon thermodynamics

Following [10], we consider a 2+1-dimensional microscopic concept, namely on the non-null boundary ∂V of a compact space-time region V. We restrict ourselves to the gauge $g_{\perp k \neq \perp} = 0$ ($\perp$ is the component normal to ∂V). The boundary term $S_{\partial 1}$ (A.2) can also be written as [6]

$$S_{\partial 1} = \frac{c^4}{16\pi G}\int_{\partial V} N^c_{ab}\,f^{ab}\,d^3\Sigma_c, \qquad f^{ab} = \sqrt{-g}\,g^{ab}, \qquad N^c_{ab}\,d^3\Sigma_c = \varepsilon_A\,N^{ab}\,d^3x, \tag{B.1}$$

IOP Conf. Series: Journal of Physics: Conf. Series **1275** (2019) 012060 doi:10.1088/1742-6596/1275/1/012060

where N^c_{ab} is the momentum conjugate to f^{ab} and $d^3\Sigma_c$ is the 3-surface element covector. We divide ∂V into rectangular cuboids $\Delta\partial V \subset B_A \subset \partial V$ with edge length L_e for each edge e so that $L_p \ll L_e$ and L_e is smaller than the typical variation scale of γ_{ab} and K_{ab}. $S_{\partial 1}$ then reads:

$$S_{\partial 1} = \sum_{\Delta\partial V \subset \partial V} S_{\partial 1\ \Delta\partial V} = \frac{c^4}{32\pi G} \sum_{\Delta\partial V \subset \partial V} \varepsilon_A \int_{\Delta\partial V} d^3x\ \sqrt{|\gamma|}\ \gamma^{ab}\ \mathcal{L}_\perp \gamma_{ab}. \tag{B.2}$$

We diagonalise γ_{ab} via an orthogonal transformation which is constant across $\Delta\partial V$, and we rewrite γ_{ab} with induced triads $e^I_{(a)}$. We then replace $(x^a) = (x^\perp, x^{||}, x^3, x^4)$, where $x^{||}$ is tangent to ∂V, by null coordinates $(\tilde{x}^a) = (\tilde{x}^1 = u, \tilde{x}^2 = v, x^3, x^4)$. We split the integration range in (B.2) into an interval $\mathcal{I} = [x^{||}_-, x^{||}_+]$ and a piece of 2-surface $A(x^{||})$. The integral over $\mathcal{I}$ can be written as a sum over $N = (x^{||}_+ - x^{||}_-)/\Delta x^{||}$ strips, with strip width $\Delta x^{||}$ and locations $x^{||}_k = x^{||}_- + k\Delta x^{||}$, where $\Delta x^{||}$ spatially resolves $\mathcal{L}_\perp \gamma_{ab}$. It can be shown [9] that (B.2) is equal to

$$S_{\partial 1\ \Delta\partial V} = \frac{c^4 \varepsilon_A}{16\pi G} \sum_{k=1}^{N} \int_{A(x^{||}_k)} d^2x\ \sqrt{|\gamma|}\ \Delta x^{||}\ \sigma, \tag{B.3}$$

$$\Delta x^{||}\ \sigma = \Delta u\ e^{(a)_u}_I\ \mathcal{L}_v e^I_{(a)_u} + \Delta v\ e^{(a)_v}_I\ \mathcal{L}_u e^I_{(a)_v}, \tag{B.4}$$

where $A(x^{||}_k)$ is the intersection of $\Delta\partial V$ with the (x^3, x^4)-surface at $x^{||} = x^{||}_k$, $e^I_{(a)_w}$ are the triads induced on the hypersurface w = constant with $w = u, v$. Diagonalising in turn $e^I_{(a)_w}$ and changing back to metric notation yields

$$\begin{aligned} S_{\partial 1\ \Delta\partial V} &= \frac{c^4 \varepsilon_A}{32\pi G} \sum_{k=1}^{N} \int_{A(x^{||}_k)} d^2\tilde{x} \left[\sqrt{|\gamma_{(u)}||\frac{e^2_{||}}{e^2_u}|}\ \Delta u\ \gamma^{ab}_{(u)}\ \mathcal{L}_v \gamma^{(u)}_{ab} + \sqrt{|\gamma_{(v)}||\frac{e^2_{||}}{e^2_v}|}\ \Delta v\ \gamma^{ab}_{(v)}\ \mathcal{L}_u \gamma^{(v)}_{ab}\right] \\ &= \frac{c^4 \varepsilon_A}{32\pi G} \sum_{k=1}^{N} \left[\int_{\Delta u} \int_{A(x^{||}_k)} d^3\tilde{x}\ \sqrt{|\gamma_{(u)}||\frac{e^2_{||}}{e^2_u}|}\ \gamma^{ab}_{(u)}\ \mathcal{L}_v \gamma^{(u)}_{ab} \right. \\ &\left. + \int_{\Delta v} \int_{A(x^{||}_k)} d^3\tilde{x}\ \sqrt{|\gamma_{(v)}||\frac{e^2_{||}}{e^2_v}|}\ \gamma^{ab}_{(v)}\ \mathcal{L}_u \gamma^{(v)}_{ab}\right], \end{aligned} \tag{B.5}$$

where $\gamma^{(w)}_{ab}$ and $\gamma^{ab}_{(w)}$ are the metric and inverse metric induced on the hypersurface w = constant, respectively. Equation (B.5) expresses $S_{\partial 1\ \Delta\partial V}$ with respect to multiple null surfaces (layers) which can be interpreted using horizon thermodynamics [5, 6] on null hypersurfaces Σ, via [11]

$$\frac{c^4}{16\pi G} \int_\Sigma N^c_{ab}\ f^{ab}\ d^3\Sigma_c = \int_\Sigma T\ s\ d\lambda\ d^2x, \tag{B.6}$$

where s is the surface density of the entropy S_{BHJP} and T the temperature perceived by the Rindler observer on the horizon with parameter $\lambda = \bar{u}$ or $\bar{v}$. Therefore, (B.5) can be written as

$$\begin{aligned} S_{\partial 1\ \Delta\partial V} &= \frac{1}{2} \sum_{k=1}^{N} \left[\int_{\Sigma_{(u)}(x^{||}_k)} \tilde{T}_u\ s\ du\ d^2x + \int_{\Sigma_{(v)}(x^{||}_k)} \tilde{T}_v\ s\ dv\ d^2x\right] \\ &= \frac{1}{2} \sum_{k=1}^{N} \int_{A(x^{||}_k)} [T_u\ s\ \Delta u + T_v\ s\ \Delta v]\ d^2x, \end{aligned} \tag{B.7}$$

where $\tilde{T}_w = |\frac{e_{||}^2}{e_w^2}|\ T_w$ and $w = u, v$. The multiple observer interpretation of (B.7) in terms of hidden information has been discussed in [9], and this justifies that $S_{\partial 1\ \Delta\partial V}$ be denoted as multiple observer entropy or m-entropy, in contrast to the single observer entropy S_{BHJP}.

Appendix C. Statistical gravity

Because of the thermodynamic property of the boundary space, the quantum bits of the multiple layers can be interpreted as the building blocks of a statistical theory as follows. The cuboid $\Delta\partial V$ is foliated across the direction defined by an arbitrary Minkowski index vector x^K into N_K pieces of 2d-surfaces of aria A_K restricted on $\Delta\partial V$. The number N_K is given by the best resolution achievable classically. A_K represents n_K quantum bits with $n_q \geq 2$ different possible states $|q\rangle$ each, so that $\Delta\partial V$ counts

$$\Omega_K(\Delta\partial V) = ({n_q}^{n_K})^{N_K} = \exp{(s_K)} \tag{C.1}$$

microscopic states and the m-entropy per volume $\Delta\partial V$ (s_K) and m-entropy (S) are

$$s_K = N_K n_K \ln n_q, \qquad S = \sum_{\Delta\partial V_m \subset \partial V} s_{Km} \Delta\partial V_m, \tag{C.2}$$

while γ_{ab} is obtained via $A_K \approx 4L_p^2 n_K$ and $\gamma^{ab}\mathcal{L}_\perp \gamma_{ab} \approx s_K$. Further relations and a more detailed description can be found in [9]. As a generalisation, the microstates are described by states labeled $s_{(j)}^{(i)}$ for every ith out of n_{Kj} quantum bits on every jth out of N_K layers:

$$|(s_{(1)}^{(1)}, s_{(1)}^{(2)}, \ldots, s_{(1)}^{(n_{K1})}); (s_{(2)}^{(1)}, \ldots, s_{(2)}^{(n_{K2})}); \ldots; (s_{(N_K)}^{(1)}, \ldots, s_{(N_K)}^{(n_{KN_K})})\rangle. \tag{C.3}$$

A classical geometry corresponds to the macroscopic state with the highest probability. More precisely, if we divide the space of γ_{ab} into small cells of equal size defined by a range $\Delta\gamma_{ab}$, the classical solution of GR lies in the cell which contains the largest number of microstates, or equivalently, at the stationary point of the boundary term.

References

[1] Bekenstein J D 1973 *Phys. Rev.* D **7** 2333
[2] Hawking S W 1975 *Commun. math. Phys.* **43** 199
[3] Wald R M 2001 *Liv. Rev. Rel.* **4** 2001-6wald
[4] Unruh W G 1976 *Phys. Rev.* D **14** 870
[5] Jacobson T 1995 *Phys. Rev. Lett.* **75** 1260
[6] Padmanabhan T 2014 *Gen. Rel. Grav.* **46** 1673
[7] York J W 1986 *Found. Phys.* **16** 249
[8] Gibbons Q W and Hawking S W 1977 *Phys. Rev.* D **15** 2752
[9] Mandrin P A 2017 *Proceedings of the 8th international workshop DICE2016: spacetime - matter - quantum mechanics, J. Phys.: Conf. Series* **880** 012028
[10] G. 't Hooft 1993 Dimensional reduction in quantum gravity *Preprint* gr-qc/9310026
[11] B. R. Majhi B R and Padmanabhan T 2013 *Eur. Phys. J.* C **73** 2651
[12] Mandrin P A 2014 *Int. J. Theor. Phys.* **53** 4250
[13] Mandrin P A 2017 *Proceedings of the MG14 meeting on general relativity* (Rome 2015) 3902.
[14] Munkhammar J 2016 in Beyond peaceful coexistence, Imperial College Press
[15] Lisi A G 2006 Quantum mechanics from a universal action reservoir *Preprint* physics.pop-ph/0605068

IOP Conf. Series: Journal of Physics: Conf. Series **1275** (2019) 012061 doi:10.1088/1742-6596/1275/1/012061

Generalized uncertainty principle in graphene

Alfredo Iorio and Pablo Pais

Faculty of Mathematics and Physics, Charles University
V Holešovičkách 2, 18000 Prague 8, Czech Republic

E-mail: iorio@ipnp.troja.mff.cuni.cz, pais@ipnp.troja.mff.cuni.cz

Abstract. We show that, by going beyond the low-energy approximation for which the dispersion relations of graphene are linear, the corresponding emergent field theory is a specific generalization a Dirac field theory. The generalized Dirac Hamiltonians one obtains are those compatible with specific generalizations of the uncertainty principle. We also briefly comment on the compatibility of the latter with noncommuting positions, $[x_i, x_j] \neq 0$, and on their possible physical realization.

1. Introduction

At low energy, the quasiparticles of graphene responsible for its transport properties, have a well know description in terms of an emergent Dirac field theory, both in flat and curved spacetimes. See [1] for the original theory, [2] for the first experiments, and the reviews [3, 4]. Recently [5], the possibility that a *deformed* Dirac structure could still exist was explored, in a range of energies beyond such linear regime.

Indeed, as we shall recall later, it was found in [5] that a Dirac structure of a generalized kind is present, beyond the linear approximation, and it is precisely of the specific kind obtained earlier in quantum gravity scenarios with a *minimal length*, see [6] for a review. This is as it must be: the higher the energy (or, equivalently, the shorter the wavelength) of the π-electron quasiparticles, the more the space they see while they propagate looses its smoothness, and the effects of the discrete honeycomb lattice (see Fig. 1) show-up[1]. This is also consistent with doubly special relativity [6] (DSR), with its maximal Planck energy scale, where the smooth manifold structure of spacetime breaks down, along with the local Lorentz invariance.

These scenarios, in fundamental research, are also the stage for certain generalized uncertainty principles (GUPs) stemming from the presence in the theory of a fundamental length scale [8, 9, 10, 11]. Those differ considerably from other GUPs in the literature, see, e.g. [12]. Noticeably, the latter GUPs would not apply here, while the former do.

We shall also show that noncommutativity, either of coordinates or of momenta, is supported by such GUPs. In particular, we shall comment on the fact that a specific kind of $[x_i, x_j]$ (or of $[p_i, p_j]$) can be accommodated in this context, and shall point to the role of an external magnetic field for its realization.

It is important to appreciate that these features of graphene can be useful for high energy theoretical research, because there are many open questions there that could be addressed in

[1] In fact, even in the continuum/large-wavelength regime, memories of the lattice are not completely gone. On this see, e.g., [7].

Content from this work may be used under the terms of the Creative Commons Attribution 3.0 licence. Any further distribution of this work must maintain attribution to the author(s) and the title of the work, journal citation and DOI.
Published under licence by IOP Publishing Ltd

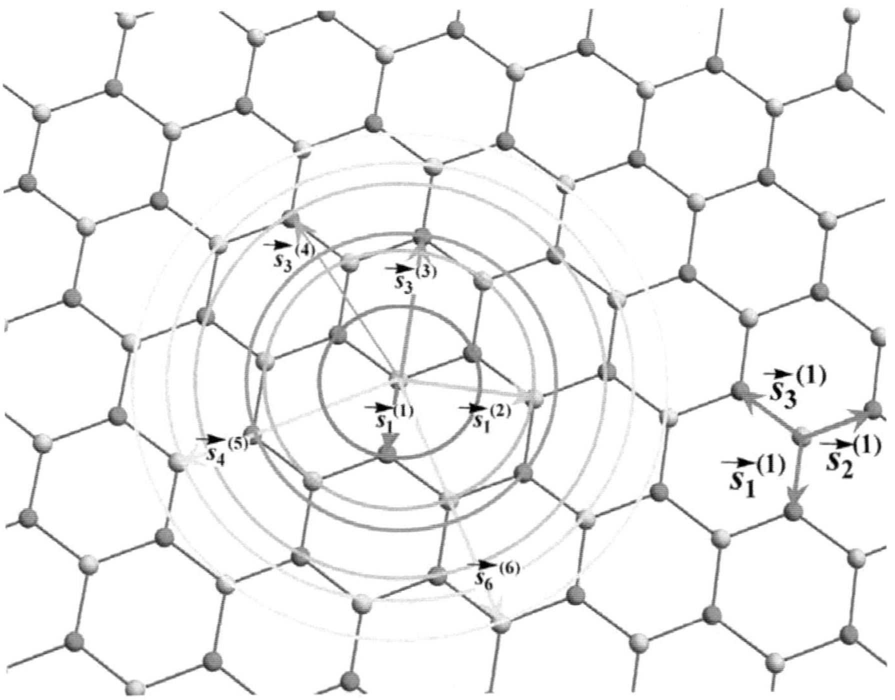

Figure 1. The honeycomb lattice of carbons making graphene. The two Bravais sublattices, L_A and L_B, are identified by different colors (green and yellow). The three vectors connecting nearest neighbors are explicitly shown, $\vec{s}_1^{(1)} = \ell(0,-1)$, $\vec{s}_2^{(1)} = \frac{\ell}{2}(\sqrt{3},1)$, $\vec{s}_3^{(1)} = \frac{\ell}{2}(-\sqrt{3},1)$ with $\ell \approx 1.42$Å the carbon-to-carbon distance, whereas only one representative per m of the $\vec{s}_i^{(m)}$s, till $m = 6$, is shown. See [5].

this way. For instance, previous work [13, 14, 15, 16, 17], has extensively explored the possibility to realize on graphene black hole physics within the standard description of a quantum field on (nearly everywhere) continuous spacetimes. The present approach could open the way to explore what happens to such physics when the granular effects of space(time) are taken into account, see, e.g., [18]. One key example that comes to the mind is the experimental realization of quantum or noncommutative (or both) corrections to the Bekenstein formula for the black-hole entropy.

In the following, we shall recall how the Dirac structure survives beyond the linear regime, in the hypothesis that $[x_i, x_j] = 0 = [p_i, p_j]$. We then briefly comment on the compatibility of such structures with noncommutativity, and close with some conclusions.

2. Dirac structure beyond the linear regime

The honeycomb lattice of graphene, see Fig. 1, has *two* atoms per unit cell. The quantum state associated to such configuration is $|\Phi_{\vec{k}}\rangle = a_{\vec{k}}|\phi^A_{\vec{k}}\rangle + b_{\vec{k}}|\phi^B_{\vec{k}}\rangle$. Here $\vec{k}$ labels the crystal quasi-momentum, $a_{\vec{k}}$ and $b_{\vec{k}}$ are complex functions, and $\phi^I_{\vec{k}}(\vec{r}) = \langle\vec{r}|\phi^I_{\vec{k}}\rangle = e^{i\vec{k}\cdot\vec{r}}\varphi^I(\vec{r})$, with $I = A, B$, are Block functions, with periodic orbital wave-functions $\varphi^I(\vec{r})$ localized at the sites of the Bravais sublattices, either L_A or L_B.

From the Schrödinger equation $\hat{H}_{\vec{k}}|\Phi_{\vec{k}}\rangle = E(\vec{k})|\Phi_{\vec{k}}\rangle$, we obtain

$$\mathcal{H}_{\vec{k}} \equiv \langle\Phi_{\vec{k}}|\hat{H}_{\vec{k}}|\Phi_{\vec{k}}\rangle = \int d^2\vec{r}d^2\vec{r}'\langle\Phi_{\vec{k}}|\vec{r}\rangle\langle\vec{r}|\hat{H}_{\vec{k}}|\vec{r}'\rangle\langle\vec{r}'|\Phi_{\vec{k}}\rangle\,. \quad (1)$$

We rewrite the overall interaction as a sum of interactions between subsequent orders of neighbors, and go up to the $m^{\rm th}$-near neighbors. It can be shown [5] that the field Hamiltonian (1) becomes

$$\mathcal{H}_{\vec{k}} = \sum_{m\in \mathrm{diag}} (\epsilon^{(0)}\varsigma_m + \eta_m)\mathcal{F}_m(\vec{k})(a^*_{\vec{k}}a_{\vec{k}} + b^*_{\vec{k}}b_{\vec{k}}) + \left(\sum_{m\in \mathrm{off}} (\epsilon^{(0)}\varsigma_m + \eta_m)\mathcal{F}^*_m(\vec{k})a^*_{\vec{k}}b_{\vec{k}} + h.c.\right),$$

where ς_m and η_m are the overlapping and the hopping parameters, respectively, and $\mathcal{F}_m(\vec{k}) \equiv \sum_{i=1}^{n_m} \exp\{i\vec{k}\cdot\vec{s}_i^{(m)}\}$. We have

$$\mathcal{F}_1 = \sum_{i=1}^{3} e^{i\vec{k}\cdot\vec{s}_i^{(1)}} = e^{-i\ell k_y}[1 + 2e^{i\frac{3}{2}\ell k_y}\cos(\frac{\sqrt{3}}{2}\ell k_x)], \tag{2}$$

where ℓ is the lattice spacing, and the important relation $\mathcal{F}_2 = |\mathcal{F}_1|^2 - 3$, holds [5].

By setting the zero of $E(\vec{k})$ to $\epsilon^{(0)}$, the Schrödinger equation becomes $\psi^\dagger_{\vec{k}}\widehat{\mathcal{H}}_{\vec{k}}\psi_{\vec{k}} = E(\vec{k})\psi^\dagger_{\vec{k}}\widehat{\mathcal{S}}_{\vec{k}}\psi_{\vec{k}}$ where $\psi^\dagger_{\vec{k}} \equiv (a^*_{\vec{k}}, b^*_{\vec{k}})$, and the Hamiltonian matrix is $\widehat{\mathcal{H}}_{\vec{k}} = h^{\rm diag}_{\vec{k}}\mathbf{1}_2 + {\rm Re}h^{\rm off}_{\vec{k}}\sigma_x + {\rm Im}h^{\rm off}_{\vec{k}}\sigma_y$, while the overlapping matrix is $\widehat{\mathcal{S}}_{\vec{k}} = \mathcal{S}^{\rm diag}_{\vec{k}}\mathbf{1}_2 + {\rm Re}\mathcal{S}^{\rm off}_{\vec{k}}\sigma_x + {\rm Im}\mathcal{S}^{\rm off}_{\vec{k}}\sigma_y$, with σ_x and σ_y the standard Pauli matrices. Evidently, it is the concurrence of two atoms per unit cell, and of the Hermiticity of the Hamiltonian, that are behind the emergence of the Pauli matrices, although this is not enough to have a Dirac structure, as we need to focus on the two solutions, $E_\pm$, to the secular equation $\det\left(\widehat{\mathcal{H}}_{\vec{k}} - E\widehat{\mathcal{S}}_{\vec{k}}\right) = 0$.

This gives an involved expression in terms of the $\mathcal{F}_m$s, and in general there is no Dirac structure behind such dispersion relations, but the occurrence that $\mathcal{F}_2 = |\mathcal{F}_1|^2 - 3$, allows us to write $E_\pm = \eta_1\left(\pm|\mathcal{F}_1| - \tilde{A}|\mathcal{F}_1|^2\right)$, where $\tilde{A} = \epsilon'_1/\eta_1$ is a dimensionless parameter[2]. By defining $\vec{P} \equiv (\hbar/\ell)\,(Re\mathcal{F}_1, Im\mathcal{F}_1)$, $A \equiv (\ell/\hbar)\tilde{A}$, and[3] $V_F \equiv \eta_1\ell/\hbar$ we arrive at

$$E_\pm \equiv \ V_F\left(\pm|\vec{P}| - A\,|\vec{P}|^2\right). \tag{3}$$

This, including the important minus sign in front of A, is precisely what obtained in [9, 10, 11] when generalizing the Dirac dispersion relations, to accommodate the effects of a GUP with a minimal fundamental length $A \propto \ell$, the honeycomb lattice spacing here.

By introducing a new generalized momentum $\vec{Q} \equiv \vec{P}(1 - A|\vec{P}|)$, of which the former, $\vec{P}$, is the low-energy approximation, we have the Dirac Hamiltonian

$$H(Q) = V_F\sum_{\vec{k}} \psi^\dagger_{\vec{k}}\, \not{Q}\psi_{\vec{k}}, \tag{4}$$

where we used the Pauli matrices properties to write $\not{P}\,\not{P} = |\vec{P}|^2\mathbf{1}_2$. When

$$[X^P_i, X^P_j] = 0 = [P_i, P_j], \quad \text{and} \quad [X^P_i, P_j] = i\hbar\delta_{ij}, \tag{5}$$

then

$$[X^P_i, Q_j] = i\hbar\left[\delta_{ij} - A\left(Q\delta_{ij} + \frac{Q_iQ_j}{Q}\right)\right] \tag{6}$$

[2] Here we are entitled to consider one Dirac point only [16], we shift the zero of the energy again, $\epsilon_1 \to \epsilon'_1 \equiv \epsilon_1 - \epsilon_0$, and we drop terms of the kind $O(\varsigma_1^2)$ and $O(\varsigma_1\eta_2)$ or higher.

[3] Notice that the Fermi velocity is usually defined as $v_F = 3/2\eta_1\ell/\hbar$, that is $v_F = 3/2V_F$. See [3].

IOP Conf. Series: Journal of Physics: Conf. Series **1275** (2019) 012061 doi:10.1088/1742-6596/1275/1/012061

that is precisely the GUP of [9, 10, 11]. As long those phase space variables are concerned, this is it. Nonetheless, it is crucial to notice that here we have *three* regimes, with a different kind of momentum at any given regime: two generalized momenta, $\vec{Q}$ and $\vec{P}$ (with $\vec{P}$ low-energy with respect to $\vec{Q}$), and the standard momentum $\vec{p}$ (that is low-energy with respect to $\vec{P}$, hence with respect to $\vec{Q}$).

At the first level of the ladder, (X_i^Q, Q_j), full Lorentz symmetry is preserved, see (4). At the second level, (X_i^P, Q_j), the Hamiltonian preserves rotation symmetry, $SO(2)$ in this case, but full Lorentz is gone (see $Q_i Q_j / Q$ in (6)). At the last level in the ladder, (x_i, Q_j), not even rotation symmetry of the Hamiltonian survives (see the so-called "trigonal warping" [3], that here we naturally reproduce).

Therefore, in the hypothesis that $[x_i, x_j] = 0 = [p_i, p_j]$, and since the low energy momenta are the p_is (i.e., $[x_i, p_j] = i\delta_{ij}$), by expanding $\vec{P}(\vec{p})$ to $O(\ell)$ we have

$$\begin{aligned}
[x, Q_x] &= i\left(1 - \frac{\ell}{2} Q_x - \frac{\ell {\epsilon'}_1}{\eta_1}\left(|\vec{Q}| + \frac{Q_x^2}{|\vec{Q}|}\right)\right), \\
[x, Q_y] &= i\left(\frac{\ell}{2} Q_y - \frac{\ell {\epsilon'}_1}{\eta_1 |\vec{Q}|} Q_x Q_y\right), \\
[y, Q_x] &= i\left(\frac{\ell}{2} Q_y - \frac{\ell {\epsilon'}_1}{\eta_1 |\vec{Q}|} Q_x Q_y\right), \\
[y, Q_y] &= i\left(1 + \frac{\ell}{2} Q_x - \frac{\ell {\epsilon'}_1}{\eta_1}\left(|\vec{Q}| + \frac{Q_y^2}{|\vec{Q}|}\right)\right).
\end{aligned} \tag{7}$$

In [5] the result up to order $O(\ell^2)$ is given. This is a rich algebraic structure, worth investigating, especially in relation to the possibility for noncommutativity.

3. GUP algebra and noncommutativity in graphene

Noncommutativity of position and momentum is at the core of quantum mechanics, hence, as such, a well accepted fact of Nature. On the other hand, noncommutativity of positions among themselves still seems quite a radical and fully theoretical hypothesis.

In fact, such a possibility has been studied from many perspectives, starting from the early days of quantum mechanics with the Lorentz-symmetry-preserving model of Snyder [19]. It then went through different stages of fortune, from the ground-breaking mathematical framework of Connes's Noncommutative Geometry [20], to Doplicher et al's hypothesis of a necessary outcome of the reconciliation between general relativity and quantum mechanics [21], to the noncommutative gauge theory as a possible low-energy limit of string theory, obtained via the Seiberg-Witten map[4] [23]. The latter results, in the past decade, ignited research on the emergence of such phenomenon in condensed matter systems [24], especially impacts on Lorentz symmetry, e.g., in electrodynamics [25, 26].

Let us then comment here on whether this radical idea can find some room in these settings, and what could this possibly mean from the physical point of view. For the sake of simplicity and generality of the discussion, here we do not follow too closely the notation of the previous Section, and we call the coordinates x_i, and the momenta p_i, without explicit reference to what is high-energy and what is low-energy. It should not be difficult to reconstruct the right correspondence, though.

[4] String theory is actually not necessary to obtain the Seiberg-Witten map, as shown by Wess and coworkers [22].

IOP Conf. Series: Journal of Physics: Conf. Series **1275** (2019) 012061 doi:10.1088/1742-6596/1275/1/012061

In the previous Section, we saw that, by considering terms beyond the linear approximation, a modified Heisenberg algebra emerges. In there, we assumed that the coordinates (and momenta) commute among themselves. Indeed, the algebra (6) is computed by employing Jacobi identities where the commutators $[x_i, x_j]$ and $[p_i, p_j]$ are demanded to be zero [11]. If we allow for noncommutativity, we have many more possibilities. Here we merely explore which noncommutative algebra is compatible with the GUP emerged in graphene, see (6), up to quadratic contributions.

Suppose we have a general Weyl-Heisenberg algebra of the kind studied in [9, 10, 11]

$$[x_i, p_j] = i\hbar \left(\delta_{ij} \left(1 + \alpha_1 p + \beta_1 p^2\right) + p_i p_j \left(\frac{\alpha_2}{p} + \beta_2 \right) \right) , \tag{8}$$

where $p \equiv \sqrt{p^2} = \sqrt{p_i p^i}$, and αs and βs must be dimensionfull constants, $[\alpha_a] \sim 1/p$, $[\beta_a] \sim 1/p^2$, for $a = 1, 2$. Imposing the Jacobi identity,

$$[[x_i, x_j], p_k] + [[x_j, p_k], x_i] + [[p_k, x_i], x_j] = 0 , \tag{9}$$

leads to the relation [11]

$$-[[x_i, x_j], p_k] = \left((\alpha_1 - \alpha_2)p^{-1} + (\alpha_1^2 + 2\beta_1 - \beta_2)\right) \Delta_{ijk} , \tag{10}$$

with $\Delta_{ijk} \equiv \hbar^2 (p_i \delta_{jk} - p_j \delta_{ik})$.

Equation (10) can be solved by demanding both its sides to be zero, which gives [9, 10, 11] $\alpha_1 = \alpha_2 \equiv -\alpha$ and $\beta_2 = 3\alpha^2$, with $\beta_1 = \alpha^2$, on the one hand, and commuting coordinates, on the other hand. The final expression for the GUP is then

$$[x_i, p_j] = i\hbar \left(\delta_{ij} - \alpha \left(p\delta_{ij} + \frac{p_i p_j}{p} \right) + \alpha^2 (p^2 \delta_{ij} + 3 p_i p_j) \right) , \tag{11}$$

that, at order α, is what we obtain in graphene too, see (6).

Noticeably, this GUP does not rule out the possibility of a noncommutative algebra, when

$$[x_i, x_j] = i\hbar\theta_{ij} , \tag{12}$$

and θ_{ij} is a *constant*, antisymmetric, $c-$number valued quantity. Indeed, in this case, the left side of (10) is still zero, so then is the right side, and the previous argument leading to (11) just goes through.

On the other hand, a more general noncommutativity needs more drastic changes. For instance, one well known model is the *Lie-algebra model* [22],

$$[x_i, x_j] = i\hbar C_{ij}^k x_k , \tag{13}$$

where C_{ij}^k are $c-$number valued structure constants. With a noncommutativity of the kind (13), the left side of (10) is no longer zero, and one cannot find a set of constant αs and βs satisfying the equation.

Another popular realization of noncommutativity is the q-deformed algebra [22], based on deformed operators x_q and commutators

$$x_q^i x_q^j = \frac{1}{q} R_{kl}^{ij} x_q^k x_q^l , \tag{14}$$

with $q(\hbar)$, and R_{kl}^{ij} a $c-$number valued quantity. This approach lead, for instance, to the interesting *twisted coproduct* model of Balachandran et al [27]. For this kind of noncommutativity

IOP Conf. Series: Journal of Physics: Conf. Series **1275** (2019) 012061 doi:10.1088/1742-6596/1275/1/012061

based on q-deformation, all the commutators need be, at least in principle, modified. Hence, to study whether the latter approach can find room in the graphene scenario, an analysis deeper than the one carried on here is necessary.

Similarly, one can work-out from the Jacobi for $[[p_i, p_j], x_k]$, that noncommutativity of the momenta is also allowed along a similar pattern as for coordinates.

Therefore, we conclude that there is indeed room for noncommutativity in this graphene scenario, surely in the settings provided by (12). The questions are then: What physical situation are we describing this way? What θ_{ij} corresponds to in graphene?

We shall not answer fully these questions here, but shall only point to a scenario that could be easily realized with graphene, and that is known to have noncommutativity of the kind we are looking for. The scenario is that of planar electrons in the lowest Landau level. The energy levels of such electrons, moving within an external magnetic field of strength B, perpendicular to the plane, are the well-known Landau levels. In graphene, Landau levels have been realized in experiments and studied, see, e.g., [28].

When the magnetic field is strong the dynamics is effectively projected to the lowest such levels[5], giving rise to [24]

$$[x_i, x_j] = -i\hbar \frac{c}{eB} \epsilon_{ij} \,, \tag{15}$$

that in our context would mean $\theta_{ij} = \frac{c}{eB}\epsilon_{ij}$.

4. Conclusions

We have shown that, by going beyond the linear dispersion relations of graphene, a generalized Dirac field theory survives, and it is compatible with GUPs that require a minimal fundamental length scale. This is also consistent with Amelino-Camelia's DSR, where the smooth manifold structure of spacetime breaks down, along with the local Lorentz invariance.

While it may be surprising that some sort of field theory description still holds in graphene at higher energies (a fact due to the fortuitus occurrence $\mathcal{F}_2 = |\mathcal{F}_1|^2 - 3$), the correspondence of the latter with fields in a granular spacetime scenarios is something one should expect. In fact, the shorter the wavelength of the π-electron quasiparticles, the more the granular structure of the honeycomb lattice affects their dynamics.

It is important for high energy theory to have found here these features, because there are many open theoretical questions that could be addressed this way. To cite a key one: the experimental realization of quantum corrections to the Bekenstein formula for the black-hole entropy. Many more are, of course, possible to envisage.

We also briefly commented here on the compatibility of noncommutativity with the obtained GUP. We concluded that the simplest compatibility with the GUP modified by the graphene granular structure is given by $[x_i, x_j] = \theta_{ij}$, with θ_{ij} a constant, antisymmetric, $c-$number valued quantity. The Lie-algebra noncommutativity is not compatible, while compatibility with q-deformation needs further analysis. We then pointed to a possible physical set-up, where the presence of a strong perpendicular external magnetic field B could create the conditions for such noncommutativity, with $\theta_{ij} = (c/eB)\epsilon_{ij}$, by projecting the dynamics of the π-electron quasiparticles on the lowest Landau level.

Interesting would be to compare the Lorentz violation emerging in graphene in the commutative case of Section 2 (see the violation in (6), and the even stronger violation in (7)), with the Lorentz violation due to noncommutativity in the approach of [29, 30, 31].

[5] This is because the Landau levels have a separation proportional to $\sqrt{B}$ or to B, depending on whether we are dealing with relativistic or non-relativistic fermions.

Acknowledgments
The authors are indebted to Mir Faizal, Ahmed Ali and Ibrahim Elmashad for their collaboration on these exciting matters, and thank Gaetano Lambiase and Fabio Scardigli for lively and informative discussions. P. P. acknowledges the warm hospitality of the Centro de Estudios Científicos (CECs) in Valdivia, Chile, at the final stage of this work.

References
[1] Semenoff G W 1984 *Phys. Rev. Lett.* **53** 2449
[2] Novoselov K S, Geim A K, Morozov S V, Jiang D, Katsnelson M I, Grigorieva I V, Dubonos S V and Firsov A A 2005 *Nature* **438** 197 (*Preprint* `cond-mat/0509330`)
[3] Castro Neto A H, Guinea F, Peres N M R, Novoselov K S and Geim A K 2009 *Rev. Mod. Phys.* **81** 109–162 (*Preprint* `cond-mat/0709.1163`)
[4] Iorio A 2015 *Int. J. Mod. Phys.* D **24** 1530013 (*Preprint* `hep-th/1412.4554`)
[5] Iorio A, Pais P, Elmashad I A, Ali A F, Faizal M and Abou-Salem L I 2018 *Int. J. Mod. Phys.* D **27** 1850080 (*Preprint* `physics.gen-ph/1706.01332`)
[6] Amelino-Camelia G 2013 *Living Rev. Relat.* **16** 5
[7] Iorio A and Pais P 2015 *Phys. Rev.* D **92** 125005 (*Preprint* `hep-th/1508.00926`)
[8] Das S and Vagenas E C 2008 *Phys. Rev. Lett.* **101** 221301
[9] Ali A F, Das S and Vagenas E C 2009 *Phys. Lett.* B **678** 497
[10] Das S, Vagenas E C and Ali A F 2010 *Phys. Lett.* B **690** 407
[11] Ali A F, Das S and Vagenas E C 2011 *Phys. Rev.* D **84** 044013
[12] Maggiore M 1993 *Phys. Lett.* B **304** 65
[13] Iorio A 2011 *Ann. Phys. (N.Y.)* **326** 1334 (*Preprint* `hep-th/1007.5012`)
[14] Iorio A and Lambiase G 2012 *Phys. Lett.* B **716** 334 (*Preprint* `cond-mat.mtrl-sci/1108.2340`)
[15] Iorio A and Lambiase G 2014 *Phys. Rev.* D **90** 025006 (*Preprint* `hep-th/1308.0265`)
[16] Iorio A and Pais P 2018 *Ann. Phys. (N.Y.)* **398** 265
[17] Taioli S, Gabbrielli R, Simonucci S, Pugno N M and Iorio A 2016 *J. Phys. Condens. Matter* **28** 13LT01
[18] Scardigli F, Lambiase G and Vagenas E C 2017 *Phys. Lett.* B **767** 242
[19] Snyder H S 1947 *Phys. Rev.* **71** 38
[20] Connes A 1994 *Noncommutative geometry* (Academic Press)
[21] Doplicher S, Fredenhagen K and Roberts J E 1995 *Commun. Math. Phys.* **172** 187–220
[22] Madore J, Schraml S, Schupp P and Wess J 2000 *Eur. Phys. J.* **C16** 161 (*Preprint* `hep-th/0001203`)
[23] Seiberg N and Witten E 1999 *JHEP* **09** 032 (*Preprint* `hep-th/9908142`)
[24] Jackiw R 2002 *Nucl. Phys. B - Proceedings Supplements* B **108** 30
[25] Guralnik Z, Jackiw R, Pi S and Polychronakos A 2001 *Phys. Lett.* B **517** 450
[26] Castorina P, Iorio A and Zappalà D 2004 *Phys. Rev.* D **69**(6) 065008
[27] Balachandran A, Pinzul A and Qureshi B 2006 *Phys. Lett.* B **634** 434
[28] Jiang Z, Henriksen E A, Tung L C, Wang Y J, Schwartz M E, Han M Y, Kim P and Stormer H L 2007 *Phys. Rev. Lett.* **98**(19) 197403
[29] Iorio A 2008 *Phys. Rev.* D **77** 048701
[30] Iorio A and Sykora T 2002 *Int. J. Mod. Phys.* A **17** 2369 (*Preprint* `hep-th/0111049`)
[31] Álvarez-Gaumé L and Vázquez-Mozo M A 2003 *Nucl. Phys.* B **668** 293

Some algebraically solvable two-dimensional dynamical systems with polynomial interactions

Francesco Calogero[a,b,1] **and Farrin Payandeh**[a,c,2]

[a] Physics Department, University of Rome "La Sapienza", Rome, Italy
[b] INFN, Sezione di Roma 1
[c] Department of Physics, Payame Noor University (PNU), PO BOX 19395-3697 Tehran, Iran

E-mail: [1] francesco.calogero@roma1.infn.it, francesco.calogero@uniroma1.it

E-mail: [2] f_payandeh@pnu.ac.ir

Abstract. We tersely review a recently introduced technique to identify systems of two nonlinearly-coupled Ordinary Differential Equations (ODEs) *solvable by algebraic operations*; and we report some specific examples of this kind, namely systems of 2 first-order ODEs with polynomial right-hand sides,

$$\dot{x}_n = P^{(n)}(x_1, x_2) \ , \quad n = 1, 2 \ ,$$

satisfied by the 2 (possibly *complex*) dependent variables $x_n \equiv x_n(t)$. Here $P^{(n)}(x_1, x_2)$ indicates some specific *polynomial*. These examples are analogous, but different, from those previously reported.

1. Introduction

A technique to identify dynamical systems characterized by systems of *nonlinearly-coupled* Ordinary Differential Equations (ODEs) *solvable by algebraic operations* is based on the relations among the N *coefficients* $y_m(t)$ and the N *zeros* $x_n(t)$ of a time-dependent (monic) polynomial of degree N in the (complex) variable z:

$$p_N(z;t) = z^N + \sum_{m=1}^{N} \left[y_m(t) \ z^{N-m} \right] = \prod_{n=1}^{N} \left[z - x_n(t) \right] \ . \tag{1}$$

The basic idea is to consider "simple" time evolutions of the N *coefficients* $y_m(t)$—possibly *explicitly solvable* evolutions, maybe featuring remarkable properties such as *isochrony* (implying that *all* the N coefficients $y_m(t)$ are periodic in t with the same fixed period independent of the initial data); and to then consider the corresponding time evolutions of the N zeros $x_n(t)$, which are generally characterized by "more nonlinear" equations of motions, the solutions of which are then obtainable via the *algebraic* operation of computing the N zeros $x_n(t)$ of the polynomial $p_N(z;t)$ defined in terms of its N *coefficients* $y_m(t)$. This generally entails that the time evolution of the *zeros* $x_n(t)$ *inherits* properties (such as *isochrony*) of the evolution of the coefficients $y_m(t)$. This idea has a long history [1], and it has produced many developments, see for instance [2] [3] [4] [5] [6] and references therein.

Content from this work may be used under the terms of the Creative Commons Attribution 3.0 licence. Any further distribution of this work must maintain attribution to the author(s) and the title of the work, journal citation and DOI.
Published under licence by IOP Publishing Ltd

IOP Conf. Series: Journal of Physics: Conf. Series **1275** (2019) 012062 doi:10.1088/1742-6596/1275/1/012062

An additional recent development has explored the modifications of the approach outlined above which emerge if, rather than considering *generic* time-dependent polynomials, one focusses on *special* time-dependent polynomials featuring (for all time) *multiple zeros*: in particular, as a first step in this direction, results have been reported for the case of a polynomial featuring *one double zero* [7], then some significant progress has been made on the case of a polynomial featuring *one zero of arbitrary multiplicity* [8], and finally the most general case has been treated of a polynomial $p_M(z;t)$ of degree M featuring an *arbitrary* number N of *zeros* each of them of *arbitrary multiplicity* μ_n [9]:

$$p_M(z;t) = z^N + \sum_{m=1}^{M} \left[y_m(t)\ z^{M-m} \right] = \prod_{n=1}^{N} \left\{ \left[z - x_n(t) \right]^{\mu_n} \right\} , \tag{2}$$

of course with the N positive integers μ_n related to the positive integer parameters M and N by the relation

$$M = \sum_{n=1}^{N} (\mu_n) . \tag{3}$$

A second twist of this development restricted attention to the $N = 2$ case—i. e., polynomials (see (2)) featuring only 2 *zeros* [10] [9]. This stringent limitation has opened the way to the identification of *algebraically solvable two-dimensional dynamical systems with polynomial interactions*, namely systems characterized by the following systems of 2 nonlinearly coupled ODEs:

$$\dot{x}_n = P^{(n)}(x_1, x_2) , \quad n = 1, 2 . \tag{4}$$

Notation 1.1. Here and hereafter superimposed dots denote differentiations with respect to the dependent variable t ("time"), $x_n \equiv x_n(t)$ are the 2 (generally *complex*) dependent variables, and $P^{(n)}(x_1, x_2)$ are generally 2 polynomials (or "almost polynomials": see below) in these 2 dependent variables x_1, x_2. Parameters such a, b, c, α, β, γ (possibly equipped with subscripts) are time-independent; they can be arbitrarily assigned (up to explicitly indicated relations among them). Note that often, above and hereafter, the t-dependence of variables is *not* explicitly indicated (when this is unlikely to generate any misunderstanding). □

A few such systems have been identified and tersely discussed in [10] by taking as point of departure the case of a polynomial such as (2) with $M = 3$, and several such cases have been treated in [9] by focussing on the case with $M = 4$. In the present paper—after a terse reminder of the methodology to obtain these results—we focus again on the $M = 3$ case, reporting several solvable models of type (4) not previously identified. Analogous treatments of cases with $M > 4$ shall be performed by ourselves or by others in the future.

In the following **Section 2** we tersely review our notation and some results which are basic for the following treatment. In **Section 3** and its subsections several specific examples are reported (with some related results confined to **Appendix A**). The last **Section 4** outlines possible future developments.

2. Preliminaries

A basic tool of our treatment are the following definitions of the 3 variables $y_m \equiv y_m(t)$, $m = 1, 2, 3$ in terms of the 2 variables $x_n \equiv x_n(t)$, $n = 1, 2$:

$$y_1 = -(2x_1 + x_2) , \quad y_2 = x_1(x_1 + 2x_2) , \quad y_3 = -(x_1)^2 x_2 ; \tag{5}$$

they of course imply (see (2)) that x_1 and x_2 are the 2 *zeros*—with respective multiplicities 2 and 1—of the (monic) third-degree polynomial $p_3(z;t)$ with coefficients y_1, y_2, y_3:

$$p_3(z;t) = z^3 + \sum_{m=1}^{3} \left[y_m(t)\ z^{3-m} \right] = \left[z - x_1(t) \right]^2 \ \left[z - x_2(t) \right] . \tag{6}$$

IOP Conf. Series: Journal of Physics: Conf. Series **1275** (2019) 012062 doi:10.1088/1742-6596/1275/1/012062

Note that this entails that x_1 and x_2 can be obtained—in terms of y_1 and y_2, or y_1 and y_3, or y_2 and y_3—by solving the following algebraic equations:

$$3\left(x_1\right)^2 + 2y_1x_1 + y_2 = 0 \; , \quad x_2 = -\left(2x_1 + y_1\right) \; , \tag{7}$$

or

$$2\left(x_1\right)^3 + y_1\left(x_1\right)^2 + y_3 = 0 \; , \quad x_2 = -\left(2x_1 + y_1\right) \; , \tag{8}$$

or

$$\left(x_1\right)^3 - y_2x_1 - 2y_3 = 0 \; , \quad x_2 = \frac{-\left(x_1\right)^2 + y_2}{2x_1} \; . \tag{9}$$

It can moreover be easily shown—or see [7] or [9] or [10]—that these formulas imply the following differential relations:

$$\dot{x}_1 = -\frac{2x_1\dot{y}_1 + \dot{y}_2}{2\left(x_1 - x_2\right)} \; , \quad \dot{x}_2 = \frac{\left(x_1 + x_2\right)\dot{y}_1 + \dot{y}_2}{x_1 - x_2} \; , \tag{10}$$

$$\dot{x}_1 = \frac{-\left(x_1\right)^2\dot{y}_1 + \dot{y}_3}{2x_1\left(x_1 - x_2\right)} \; , \quad \dot{x}_2 = \frac{x_1x_2\dot{y}_1 - \dot{y}_3}{x_1\left(x_1 - x_2\right)} \; , \tag{11}$$

$$\dot{x}_1 = \frac{x_1\dot{y}_2 + 2\dot{y}_3}{2x_1\left(x_1 - x_2\right)} \; , \quad \dot{x}_2 = -\frac{x_1x_2\dot{y}_2 + \left(x_1 + x_2\right)\dot{y}_3}{\left(x_1\right)^2\left(x_1 - x_2\right)} \; . \tag{12}$$

It is plain from these formulas that if the two variables $y_1 \equiv y_1(t)$ and $y_2 \equiv y_2(t)$, or $y_1 \equiv y_1(t)$ and $y_3 \equiv y_3(t)$, or $y_2 \equiv y_2(t)$ and $y_3 \equiv y_3(t)$ satisfy an *algebraically solvable* system of 2 first-order ODEs, say

$$\dot{y}_{m_1} = f_{m_1}\left(y_{m_1}, y_{m_2}\right) \; , \quad \dot{y}_{m_2} = f_{m_2}\left(y_{m_1}, y_{m_2}\right) \tag{13}$$

with $m_1 = 1$, $m_2 = 2$ or $m_1 = 1$, $m_2 = 3$ or $m_1 = 2$, $m_2 = 3$, then the corresponding system of 2—generally nonlinearly-coupled, first-order—ODEs satisfied by $x_1 \equiv x_1(t)$ and $x_2 \equiv x_2(t)$ is as well *algebraically solvable.* Moreover, if the functions $f_{m_1}\left(y_{m_1}, y_{m_2}\right)$ and $f_{m_2}\left(y_{m_1}, y_{m_2}\right)$ are *appropriately chosen*, then the right-hand sides of the ODEs (10), (11), (12) become *polynomial* (or "*almost polynomial*": see below). In the following **Section 3** we report several *algebraically solvable* systems satisfied by the 2 dependent variables $x_1 \equiv x_1(t)$ and $x_2 \equiv x_2(t)$ obtained in this manner, i. e. by first replacing in the relevant equations (10), (11) and (12), the expressions of the time-derivatives of the relevant 2 variables y_m via (13) and subsequently replacing the expressions of these 2 variables y_m in terms of the 2 variables x_n via the relevant equations (5); of course with the functions $f_{m_1}\left(y_{m_1}, y_{m_2}\right)$ and $f_{m_2}\left(y_{m_1}, y_{m_2}\right)$ *appropriately chosen polynomials*(see below).

3. Results

Our first step is to identify 3 *solvable* systems of 2 evolution equations satisfied by the dependent variables $y_{m_1} \equiv y_{m_1}(t)$ and $y_{m_2} \equiv y_{m_2}(t)$. Their solvability is discussed in **Appendix A** of Ref. [10] and tersely reviewed below in **Appendix A**.

The first of these 3 systems—hereafter identified as **A1** (see **Appendix A**)—is characterized by the following 2 *uncoupled* ODEs:

$$\dot{y}_{m_1} = \sum_{\ell=0}^{L}\left[\alpha_\ell\left(y_{m_1}\right)^{\ell m_2+1}\right] \; , \quad \dot{y}_{m_2} = \sum_{\ell=0}^{L}\left[\beta_\ell\left(y_{m_2}\right)^{\ell m_1+1}\right] \; . \tag{14}$$

The second of these 3 systems—hereafter identified as **A2** (see **Appendix A**)—is characterized by the following 2 *coupled* ODEs:

$$\dot{y}_{m_1} = \sum_{\ell=0}^{L} \left[\alpha_\ell \left(y_{m_1}\right)^\ell\right] , \quad \dot{y}_{m_2} = y_{m_2} \sum_{\ell=1}^{L} \left[\beta_\ell \left(y_{m_1}\right)^{\ell-1}\right] + \sum_{\ell=0}^{L} \left[\gamma_\ell \left(y_{m_1}\right)^{\ell-1+(m_2/m_1)}\right] . \tag{15}$$

The third of these 3 systems—hereafter identified as **A3** (see **Appendix A**)—is characterized by the following 2 coupled ODEs:

$$\begin{aligned} \dot{y}_{m_1} &= \alpha_0 + \alpha_1 y_{m_2} , \quad \dot{y}_{m_2} = \beta_0 \left(y_{m_1}\right)^{-1+m} + \beta_1 \left(y_{m_1}\right)^{-1+2m} , \\ m &= m_2/m_1 , \quad m_1 = 1 , \quad m_2 = 2, 3 . \end{aligned} \tag{16}$$

Our next step is to list in the following 11 subsections 11 *algebraically solvable* systems of 2 nonlinearly-coupled ODEs (4) with *polynomial* (or "*almost polynomial*": see below) right-hand sides; in each case we identify the corresponding *appropriately chosen algebraically solvable* system of ODEs—see above and **Appendix A**—satisfied by the corresponding functions $y_{m_1}(t)$ and $y_{m_2}(t)$. But note that the algebraically solvable systems (4) thus identified are only 9, because 2 pairs of the systems identified below—although arrived at differently—are in fact *identical* (a phenomenon already noted in [9]).

*3.1. Models of type **A1***

In the 3 cases listed in this **Subsection** 3.1 the variables $y_{m_1}(t)$ and $y_{m_2}(t)$ are supposed to satisfy the system **A1** of 2 uncoupled ODEs (see (14) and **Appendix A**), with the indicated assignments of the various parameters.

Model A1.1: $m_1 = 1,\ m_2 = 2;\ L = 1;\ \alpha_0 = a,\ \alpha_1 = b;\ \beta_0 = 2a,\ \beta_1 = 6b;$

$$\begin{aligned} \dot{x}_1 &= a x_1 + b x_1 \left[5 \left(x_1\right)^2 + 5 x_1 x_2 - \left(x_2\right)^2\right] , \\ \dot{x}_2 &= a x_2 - b \left[2 \left(x_1\right)^3 - 2 \left(x_1\right)^2 x_2 - 8 x_1 \left(x_2\right)^2 - \left(x_2\right)^3\right] . \end{aligned} \tag{17}$$

Model A1.2: $m_1 = 1,\ m_2 = 3;\ L = 1;\ \alpha_0 = a,\ \alpha_1 = -2b;\ \beta_0 = 3a,\ \beta_1 = -162b;$

$$\begin{aligned} \dot{x}_1 &= x_1 \left\{a + b \left[16 \left(x_1\right)^3 + 48 \left(x_1\right)^2 x_2 - 9 x_1 \left(x_2\right)^2 - \left(x_2\right)^3\right]\right\} , \\ \dot{x}_2 &= x_2 \left\{a - 2b \left[16 \left(x_1\right)^3 - 33 \left(x_1\right)^2 x_2 - 9 x_1 \left(x_2\right)^2 - \left(x_2\right)^3\right]\right\} . \end{aligned} \tag{18}$$

Model A1.3: $m_1 = 2,\ m_2 = 3;\ L = 1;\ \alpha_0 = 2a,\ \alpha_1 = 2b;\ \beta_0 = 3a,\ \beta_1 = 81b;$

$$\begin{aligned} \dot{x}_1 &= x_1 \left\{a + b \left(x_1\right)^3 \left[\left(x_1\right)^3 + 9 \left(x_1\right)^2 x_2 + 33 x_1 \left(x_2\right)^2 - 16 \left(x_2\right)^3\right]\right\} , \\ \dot{x}_2 &= x_2 \left\{a - b \left(x_1\right)^3 \left[2 \left(x_1\right)^3 + 18 \left(x_1\right)^2 x_2 - 15 x_1 \left(x_2\right)^2 - 32 \left(x_2\right)^3\right]\right\} . \end{aligned} \tag{19}$$

*3.2. Models of type **A2***

In the 6 cases listed in this **Subsection** 3.2 the variables $y_{m_1}(t)$ and $y_{m_2}(t)$ are supposed to satisfy the system **A2** of 2 coupled ODEs (see (15) and **Appendix A**), with the indicated assignments of the various parameters.

Model A2.1: $m_1 = 1,\ m_2 = 2;\ L = 3;\ \alpha_0 = 3b_0,\ \alpha_1 = a_0 - 3b_1,\ \alpha_2 = -a_1 + 3b_2,\ \alpha_3 = a_2 - 3b_3;$ $\beta_\ell = (-1)^{\ell-1} 2a_{\ell-1},\ \ell = 1, 2, 3;\ \gamma_\ell = (-1)^\ell 2b_\ell,\ \ell = 0, 1, 2, 3;$

$$\begin{aligned} \dot{x}_n &= x_n \left(a_0 + a_1 X + a_2 X^2\right) - \left[b_0 + b_1 X + b_2 X^2 + b_3 X^3\right] , \\ n &= 1, 2;\ X = 2x_1 + x_2 . \end{aligned} \tag{20}$$

Model A2.2: $m_1 = 1,\ m_2 = 2;\ L = 3;\ \alpha_0 = 0,\ \alpha_\ell = c_\ell;\ \beta_\ell = 2c_\ell;\ \ell = 1, 2, 3$;

$$\dot{x}_n = x_n \left(c_1 + c_2 X + c_3 X^2\right) , \quad n = 1, 2, \quad X = x_1 (x_1 + 2x_2) . \tag{21}$$

Model A2.3: $m_1 = 1,\ m_2 = 3;\ L = 3;\ \alpha_0 = 9b_0,\ \alpha_\ell = (-1)^{\ell-1} (a_0 - 9b_\ell),\ \beta_\ell = (-1)^{\ell-1} 3a_{\ell-1},\ \ell = 1, 2, 3;\ \gamma_\ell = (-1)^\ell b_\ell,\ \ell = 0, 1, 2, 3.$

$$\begin{aligned} \dot{x}_1 &= x_1 \left(a_0 + a_1 X + a_2 X^2\right) - \left(\frac{5x_1 + x_2}{2x_1}\right) \left(b_0 + b_1 X + b_2 X^2 + +b_3 X^3\right) , \\ \dot{x}_2 &= x_2 \left(a_0 + a_1 X + a_2 X^2\right) - \left(\frac{4x_1 - x_2}{x_1}\right) \left(b_0 + b_1 X + b_2 X^2 + +b_3 X^3\right) , \\ X &= 2x_1 + x_2 . \end{aligned} \tag{22}$$

Note that the right hand sides of these ODEs (22) are *polynomial* only if *all* the coefficients b_ℓ vanish.

Model A2.4: $m_1 = 3,\ m_2 = 1;\ L = 3;\ \alpha_0 = 0,\ \alpha_\ell = (-1)^{\ell-1} 3c_\ell,\ \beta_\ell = (-1)^{\ell-1} c_{\ell-1},\ \ell = 1, 2, 3;\ \gamma_\ell = 0,\ \ell = 0, 1, 2, 3.$

$$\dot{x}_n = x_n \left(c_0 + c_1 X + c_2 X^2\right) , \quad n = 1, 2, \quad X = (x_1)^2 x_2 . \tag{23}$$

Model A2.5: $m_1 = 2,\ m_2 = 3;\ L = 3;\ \alpha_0 = 0,\ \alpha_\ell = 2c_{\ell-1},\ \beta_\ell = 3c_{\ell-1},\ \ell = 1, 2, 3;\ \gamma_\ell = 0,\ \ell = 0, 1, 2, 3.$

$$\dot{x}_n = x_n \left(c_0 + c_1 X + c_2 X^2\right) , \quad n = 1, 2, \quad X = x_1 (x_1 + 2x_2) . \tag{24}$$

Model A2.6: $m_1 = 3,\ m_2 = 2;\ L = 3;\ \alpha_0 = 0,\ \alpha_\ell = (-1)^{\ell-1} 3c_{\ell-1},\ \beta_\ell = (-1)^{\ell-1} 2c_{\ell-1},\ \ell = 1, 2, 3;\ \gamma_\ell = 0,\ \ell = 0, 1, 2, 3.$

$$\dot{x}_n = x_n \left(c_0 + c_1 X + c_2 X^2\right) , \quad n = 1, 2, \quad X = (x_1)^2 x_2 . \tag{25}$$

Remark 3.2-1. Note that the systems of ODEs (21) and (24) are identical, and likewise the systems of ODEs (23) and (25) are identical. □

*3.3. Models of type **A3***

In the 2 cases listed in this **Subsection** the variables $y_{m_1}(t)$ and $y_{m_2}(t)$ are supposed to satisfy the system **A3** of 2 coupled ODEs (see (16) and **Appendix A**), with the indicated assignments of the various parameters.

Model A3.1: $m_1 = 1,\ m_2 = 2;\ \alpha_0 = -3a,\ \alpha_1 = -9b;\ \beta_0 = -2a,\ \beta_1 = -2b;$

$$\begin{aligned} \dot{x}_1 &= a + b \left[(x_1)^2 + 7x_1 x_2 + (x_2)^2\right] , \\ \dot{x}_2 &= a + b \left[7 (x_1)^2 + 4x_1 x_2 - 2 (x_2)^2\right] . \end{aligned} \tag{26}$$

Model A3.2: $m_1 = 1,\ m_2 = 3;\ \alpha_0 = -18a,\ \alpha_1 = -486b;\ \beta_0 = -2a,\ \beta_1 = -2b;$

$$\begin{aligned} \dot{x}_1 &= (x_1)^{-1} \{a (5x_1 + x_2) \\ &\quad +b \left[32 (x_1)^4 - 131 (x_1)^3 x_2 - 51 (x_1 x_2)^2 - 11x_1 (x_2)^3 - (x_2)^4\right]\} , \\ \dot{x}_2 &= (x_1)^{-1} \{2a (4x_1 - x_2) \\ &\quad -2b \left[32 (x_1)^4 + 112 (x_1)^3 x_2 - 51 (x_1 x_2)^2 - 11x_1 (x_2)^3 - (x_2)^4\right]\} . \end{aligned} \tag{27}$$

4. Outlook

Space limitations do not allow to report here some natural developments of the results reported above, such as the identification of the *more general solvable* systems which obtain from those reported in the preceding **Section 3** via *linear* transformations of the dependent variables $x_n(t)$, say

$$x_1(t) = A_1 + A_{11}z_1(t) + A_{12}z_2(t) \ , \quad x_2(t) = A_2 + A_{21}z_1(t) + A_{22}z_2(t) \ , \tag{28}$$

with the 6 parameters A_1, A_2, A_{11}, A_{12} , A_{21}, A_{22} *arbitrarily assigned*: the interested reader is referred to the analogous developments discussed in **Section 5** of Ref. [9].

And of course the investigation in various fields of applied mathematics of the *algebraically solvable* systems of 2 nonlinearly-coupled ODEs reported above is an interesting open task.

Acknowledgments

FP likes to thank the Physics Department of the University of Rome "La Sapienza" for the hospitality from April 2018 to January 2019 (during her sabbatical, when the results reported in this paper were obtained), and the organizers of DICE2018 (where these results were reported by her), especially Prof. Hans Thomas Elze.

Appendix A.

The 2 ODEs of the system **A1** (see (14)) are uncoupled, and each of them is clearly solvable by quadratures. Note in particular that the initial-value problem of the ODE

$$\dot{y} = ay + by^{m+1} \tag{A.1}$$

is given by the *explicit* formula

$$y(t) = y(0)\exp(at)\{1 + (b/a)[y(0)]^m[1 - \exp(mat)]\}^{-1} \ . \tag{A.2}$$

To solve the system **A2** one treats firstly the first of the 2 ODEs (15)—essentially as just above. Then one notes—see **Appendix A** of Ref. [10]—that the solution of the second ODE of the system (15) reads

$$y_{m_2}(t) = F(t)\left[y_{m_2}(0) + \int_0^t dt' \ [F(t')]^{-1}\sum_{\ell=0}^{L}\left\{\gamma_\ell\left[y_{m_1}(t')\right]^{\ell-1+(m_2/m_1)}\right\}\right] \tag{A.3}$$

with

$$F(t) = \exp\left\{\int_0^t \left[dt'\sum_{\ell=1}^{L}\left\{\beta_\ell\left[y_{m_1}(t')\right]^{\ell-1}\right\}\right]\right\} \ . \tag{A.4}$$

Explicit solutions can be easily obtained in the following cases:

$$m_1 = 1 \ , \quad m_2 = 2, 3 \ , \quad m = m_2/m_1 \ , \tag{A.5}$$

$$\dot{y}_{m_1} = \alpha_0 + \alpha_1 y_{m_1} + \alpha_2 (y_{m_1})^2 \ , \tag{A.6}$$

$$\dot{y}_{m_2} = y_{m_2}(\beta_0 + \beta_1 y_{m_1}) + \gamma_0 (y_{m_1})^{-1+m} + \gamma_1 (y_{m_1})^m + \gamma_2 (y_{m_1})^{1+m} \ ; \tag{A.7}$$

$$\begin{aligned} y_{m_1}(t) &= \frac{y_{m_1}(0)\left[1 + (\Delta/\alpha_1)\tanh(\Delta t)\right] - 2(\alpha_0/\alpha_1)\tanh(\Delta t)}{1 - \left\{\left[2\alpha_2 y_{m_1}(0) + \Delta\right]/\alpha_1\right\}\tanh(\Delta t)} \ , \\ \Delta^2 &= (\alpha_1)^2 - 4\alpha_0\alpha_2 \ , \end{aligned} \tag{A.8}$$

IOP Conf. Series: Journal of Physics: Conf. Series **1275** (2019) 012062 doi:10.1088/1742-6596/1275/1/012062

$$y_{m_2}(t) = f(t) \left[y_{m_2}(0) + \int_0^t dt' \ \left[f(t') \right]^{-1} \sum_{\ell=0}^{L} \left\{ \gamma_\ell \left[y_{m_1}(t') \right]^{\ell-1+m} \right\} \right] , \tag{A.9}$$

$$f(t) = \exp \left\{ \int_0^t \left[dt' \sum_{\ell=1}^{L} \left\{ \beta_\ell \left[y_{\tilde{m}_1}(t') \right]^{\ell-1} \right\} \right] \right\} . \tag{A.10}$$

Finally, to identify the solution of the system **A3** (see (16)) we note that time-differentiation of the first of its 2 ODEs entails that $y_{m_1}(t)$ satisfies the decoupled second-order ODE

$$\ddot{y}_{m_1} = \alpha_1 \left[\beta_0 (y_{m_1})^{-1+m} + \beta_1 (y_{m_1})^{-1+2m} \right] , \quad m = m_2/m_1 ; \tag{A.11}$$

hence (see (16)) $y_1(t)$ is an elliptic (for $m = 2$) or hyperelliptic (for $m = 3$) function; and likewise for $y_2(t)$ (see the first of the 2 ODEs (16)).

References

[1] Calogero F 1978, Motion of Poles and Zeros of Special Solutions of Nonlinear and Linear Partial Differential Equations, and Related "Solvable" Many Body Problems, *Nuovo Cimento* **43B** 177-241

[2] Calogero F 2001, *Classical many-body problems amenable to exact treatments*, Lecture Notes in Physics Monograph **m66**, Springer, Heidelberg, (749 pages)

[3] Calogero F 2008, *Isochronous systems*, Oxford University Press, Oxford, U. K. ; paperback (2012).

[4] Calogero F 2018, *Zeros of Polynomials and Solvable Nonlinear Evolution Equations*, Cambridge University Press, Cambridge, U. K., 2018 (168 pages)

[5] Calogero F 2018, Systems of nonlinearly-coupled differential equations solvable by algebraic operations, in *Nonlinear Systems and Their Remarkable Mathematical Structures* (edited by N. Euler), CRC Press (Taylor & Francis), pages 1-14

[6] Gómez-Ullate D and Sommacal M 2005, Periods of the Goldfish Many-Body Problem, *J. Nonlinear Math. Phys.* **12**, Suppl. **1**, 351-362

[7] Bihun O and Calogero F 2018, Time-dependent polynomials with *one double* root, and related new solvable systems of nonlinear evolution equations, *Qual. Theory Dyn. Syst.* doi.org/10.1007/s12346-018-0282-3; http://arxiv.org/abs/1806.07502

[8] Bihun O 2018, Time-dependent polynomials with one multiple root and new solvable dynamical systems, arXiv:1808.00512v1 [math-ph] 1 Aug 2018

[9] Calogero F and Payandeh F 2018, Polynomials with multiple zeros and solvable dynamical systems including models in the plane with polynomial interactions, *J. Math. Phys.* (submitted 16.12.2018)

[10] Calogero F and Payandeh F 2018, Solvable dynamical systems in the plane with polynomial interactions, to be published as a chapter in a collective book to celebrate the 65th birthdate of Emma Previato (in press)

IOP Conf. Series: Journal of Physics: Conf. Series **1275** (2019) 012063 doi:10.1088/1742-6596/1275/1/012063

Remarks on the Unruh effect with mixed neutrinos

M Blasone[1,2], G Lambiase[1,2], G G Luciano[2] and L Petruzziello[1,2]

[1] Dipartimento di Fisica, Università di Salerno, Via Giovanni Paolo II, 132 I-84084 Fisciano (SA), Italy.

[2] INFN Sezione di Napoli, Gruppo collegato di Salerno, Italy

E-mail: blasone@sa.infn.it, lambiase@sa.infn.it, gluciano@sa.infn.it, lpetruzziello@na.infn.it

Abstract. We discuss some recent results on the inverse β-decay of an accelerated proton in the presence of neutrino mixing. By comparing different approaches, we conclude that a consistent treatment preserving general covariance should be based on flavor neutrino states. We also comment on the issue of thermality of the Unruh effect for mixed neutrinos.

1. Introduction

The analysis of decay processes in accelerated frames [1] provides a useful framework to test the consistency of Quantum Field Theory (QFT) in non-trivial backgrounds. In particular, the inverse β-decay of accelerated protons has been addressed, showing the necessity of Unruh effect to preserve the general covariance of QFT [2, 3].

Recently, it has been argued that the introduction of neutrino mixing in the above process could lead to theoretical ambiguities [4]. In Refs. [5, 6], it has been shown that the equality of decay rates calculated in the inertial and comoving frames holds also in the presence of mixing, at least in a specific approximation. Subsequently, in the paper by Cozzella *et al.* [7], it has been claimed that no problem arises at all when neutrino mixing is involved, and that the two decay rates exactly match.

We remark that the fundamental difference between the approach of Refs. [5, 6] and the one of Ref. [7] relies in the choice of asymptotic states for mixed neutrinos. Indeed, whilst in Refs. [5, 6] such a rôle is played by flavor eigenstates, in Ref. [7] asymptotic neutrinos are required to be mass eigenstates. Therefore, it appears to be an irremediable dichotomy in the above two treatments, which somehow needs to be cured.

In the present paper, we show that the statements of Ref. [7] are unjustified, being based on incorrect assumptions. Before studying this, in Sec. 2 we review some general considerations on the quantization of neutrino fields with flavor mixing, which are useful for the subsequent analysis. In Sec. 3 we discuss the results of Ref. [7] in light of our approach [5]. Sec. 4 is devoted to conclusions.

2. Neutrino mixing in QFT

Neutrino mixing [8] is embedded in the Standard Model in analogy to quark mixing, at least for the case of Dirac neutrinos. Considering for simplicity the case of two flavors, we have that the Lagrangian associated to neutrino fields with definite flavors ν_e, ν_μ exhibits off-diagonal terms

Content from this work may be used under the terms of the Creative Commons Attribution 3.0 licence. Any further distribution of this work must maintain attribution to the author(s) and the title of the work, journal citation and DOI.
Published under licence by IOP Publishing Ltd

in the mass matrix, which can be diagonalized by means of the mixing transformations:

$$\begin{aligned} \nu_e(x) &= \cos\theta\,\nu_1(x) + \sin\theta\,\nu_2(x), \\ \nu_\mu(x) &= -\sin\theta\,\nu_1(x) + \cos\theta\,\nu_2(x), \end{aligned} \tag{1}$$

where ν_1, ν_2 are fields with definite masses.

Quantization of flavor fields is achieved [9] by identifying the algebraic generator of the rotation Eq. (1)

$$\begin{aligned} \hat{\nu}_e(x) &= \hat{G}^{-1}\,\hat{\nu}_1(x)\,\hat{G}, \\ \hat{\nu}_\mu(x) &= \hat{G}^{-1}\,\hat{\nu}_2(x)\,\hat{G}, \end{aligned} \tag{2}$$

with

$$\hat{G} = \exp\left\{\theta \int d^3x \left[\hat{\nu}_1^\dagger(x)\hat{\nu}_2(x) - \hat{\nu}_2^\dagger(x)\hat{\nu}_1(x)\right]\right\}, \tag{3}$$

which is easy to check using the canonical anticommutation relations. The expansions for the flavor fields are obtained by inserting into Eq. (2) the corresponding expansions for the free fields. A straightforward calculation leads to

$$\hat{\nu}_\alpha(x) = \sum_r \int d^3k\, e^{ik\cdot x}\left[\hat{a}^r_{k,\alpha}\, u^r_{k,\sigma}(t) + \hat{b}^{r\dagger}_{-k,\alpha}\, v^r_{-k,\sigma}(t)\right], \tag{4}$$

where $(\alpha,\sigma) = (e,1),(\mu,2)$ and r is the polarization. In this way, flavor ladder operators are consistently defined [9].

A subtle point in the above quantization procedure arises when considering the action of the mixing generator $\hat{G}$ on the free field vacuum $|0\rangle_m$ (mass vacuum). This indeed generates a non-trivial state $|0\rangle_f$ (flavor vacuum), defined as:

$$|0\rangle_f \equiv \hat{G}^{-1}\,|0\rangle_m\,. \tag{5}$$

It is possible to show that the Fock spaces for neutrinos with definite mass and definite flavor ($\mathcal{H}_m$ and $\mathcal{H}_f$, respectively) are unitarily *inequivalent* to each other in the infinite volume limit [9]. Indeed, one has

$$\lim_{V\to\infty}\ {}_m\langle 0|0\rangle_f = 0\,. \tag{6}$$

The flavor vacuum has the following properties:

- it is an eigenstate of the flavor charge operators $\hat{Q}_\alpha$, obtained by means of Noether's theorem [10];
- it is a generalized Gilmore-Perelomov coherent state [11];
- it is a condensate of neutrino-antineutrino pairs both with same and different masses [9].

Flavor neutrino states are then consistently defined as eigenstates of flavor charges

$$\hat{Q}_\alpha|\nu_\alpha\rangle = |\nu_\alpha\rangle, \quad \alpha = e,\mu. \tag{7}$$

In the relativistic limit, these states reduce to the usual Pontecorvo flavor states [8],

$$|\nu_\alpha\rangle_P = \sum_i U_{\alpha,i}|\nu_i\rangle, \tag{8}$$

where $U_{\alpha,i}$ is the generic element of the mixing matrix

$$U = \begin{pmatrix} \cos\theta & \sin\theta \\ -\sin\theta & \cos\theta \end{pmatrix}. \tag{9}$$

3. Remarks on the Unruh effect with mixed neutrinos

In what follows, we discuss the results of Ref. [7] in view of the framework developed in Sec. 2.

In the Introduction of Ref. [7], the authors consider the recent investigation of Ref. [4] about the inequality of the inertial and the comoving decay rates for the process $p \to n\, e^+\, \nu_e$ with mixed neutrinos, asserting that in principle no discrepancy should exist between the calculations done in the two reference frames. On the basis of this, they then affirm that "*either the Unruh effect is wrong (contradicting several previous results [7], including what we consider to be a virtual observation of it [8]) or some mistake was made in the previously mentioned analysis*".

We basically agree with the authors of Ref. [7] on the fact that general covariance should be taken as a guiding principle in the analyzed problem and thus the decay rates in the two frames should coincide also when mixed neutrinos are involved. On the other hand, we do not agree that a mismatch of the above decay rates in the presence of neutrino mixing undermines the validity of Unruh effect as proved in Refs. [2], for the simple reason that in these works mixing is not taken into account at all.

Rather, we think that it is possible that corrections to the Unruh effect (in the form of non-thermal contributions) may arise in connection with neutrino mixing. This has indeed been shown explicitly in Refs. [12, 13], where the quantization of both bosonic and fermionic mixed fields was performed in accelerated frame.

In this connection, we stress that our analysis of the proton decay rates with mixed neutrinos carried out in Ref. [5], demonstrates that general covariance holds in the approximation in which Pontecorvo states may be used to represent flavor neutrinos. Beyond such an approximation, general covariance must hold, but Unruh effect may be somehow modified.

* * *

We now turn the attention to another important debated point. Although in the Appendix B of Ref. [7] the authors show that Pontecorvo states are well-defined in the limit of small mass difference with respect to the intrinsic uncertainties in the neutrino momenta, they never use such states in their treatment, even in the aforementioned approximation. This allows us to conclude that their result should be equal to our outcome in the regime in which Pontecorvo states are correct, since we only deal with such states[1]. However, this is not the case: indeed, the use of Pontecorvo states produces "off-diagonal" terms (namely, contributions proportional to $\cos^2\theta \sin^2\theta$) in the expressions of the inertial ($\Gamma^{p\to n}_{in}$) and comoving ($\Gamma^{p\to n}_{acc}$) decay rates [5, 6],

$$\Gamma^{p\to n}_{in} = \cos^4\theta\, \Gamma^{p\to n}_{1} + \sin^4\theta\, \Gamma^{p\to n}_{2} + \cos^2\theta \sin^2\theta\, \Gamma^{p\to n}_{12}, \tag{10}$$

$$\Gamma^{p\to n}_{acc} = \cos^4\theta\, \widetilde{\Gamma}^{p\to n}_{1} + \sin^4\theta\, \widetilde{\Gamma}^{p\to n}_{2} + \cos^2\theta \sin^2\theta\, \widetilde{\Gamma}^{p\to n}_{12}. \tag{11}$$

Notice that these off-diagonal contributions are completely absent in Eqs. (49) and (53) of Ref. [7], thus showing the internal inconsistency of that formulation.

The origin of this discrepancy can be readily understood: in Ref. [7], the authors consider the decay process

$$p \to n\, \bar{l}_\alpha\, \nu_i, \tag{12}$$

with $l_\alpha = \{e, \mu, \tau\}$ and $\nu_i = \{\nu_1, \nu_2, \nu_3\}$. It is clear that, with such a choice, the aforementioned off-diagonal terms can never appear in any calculation of the decay rates. Thus, the outcome of Ref. [7] - which is claimed to be exact - cannot reproduce the one of Refs. [5, 6] within the relativistic approximation, where Pontecorvo states are universally recognized to be well-defined.

[1] Note that these states can be efficiently generalized also to the case of curved backgrounds, and in particular of accelerated frames [14, 15].

We further point out that in Ref. [7] mixing is correctly taken into account at the level of fields in the weak interaction action

$$\hat{S}_I = \int d^4x \sqrt{-g} \left(\sum_\alpha \hat{\bar{\nu}}_\alpha \gamma^\mu \hat{P}_L \hat{l}_\alpha \hat{j}_\mu + \text{h.c.} \right), \tag{13}$$

but not at the level of the states[2], not even in the particular situations in which Pontecorvo states are known to be valid.

Furthermore, it must be emphasized that the combined use in Ref. [7] of the action Eq. (13) with the neutrino states for the process Eq. (12) does not satisfy normalization conditions. Fields and states adopted in the evaluation of transition amplitudes should indeed be "compatible" in the sense of normalization. Namely, if the action is written in terms of a generic field $\hat{\phi}$ as[3]

$$\hat{S} = \int d^4x \, \mathcal{L}\left(\hat{\phi}, \partial\hat{\phi}\right), \tag{14}$$

then one expects to work with states $|\phi\rangle$ that satisfy the relation [16, 17]

$$\langle 0| \hat{\phi}(0) |\phi(p)\rangle = 1 . \tag{15}$$

Expanding the field $\hat{\phi}$ as

$$\hat{\phi}(x) = \int \frac{d^3k}{(2\pi)^3 2\,\omega_k} \left[\hat{a}_k \, e^{-ik\cdot x} + \hat{b}_k^\dagger \, e^{ik\cdot x}\right], \tag{16}$$

then the commutation relations for the ladder operators read [16, 18]

$$\left[\hat{a}_k, \hat{a}_p^\dagger\right] = \left[\hat{b}_k, \hat{b}_p^\dagger\right] = (2\pi)^3 2\,\omega_k \, \delta(\mathbf{k}-\mathbf{p}), \tag{17}$$

which validate the use of Eq. (15).

If now we consider the case involving mixing, we have

$$\begin{aligned} |\phi_A\rangle &= \cos\theta \, |\phi_1\rangle + \sin\theta \, |\phi_2\rangle, \\ |\phi_B\rangle &= -\sin\theta \, |\phi_1\rangle + \cos\theta \, |\phi_2\rangle, \end{aligned} \tag{18}$$

with a similar relation holding for fields, in analogy to Eq. (1). Then, in the expansion Eq. (16), for each field $\hat{\phi}_i$, $i = \{1, 2\}$, we have $\omega_k \to \omega_{k,i}$, $\hat{a}_k \to \hat{a}_{k,i}$ and $\hat{b}_k^\dagger \to \hat{b}_{k,i}^\dagger$.

Assuming for the mixed fields ϕ_a, $a = \{A, B\}$, an action analogous to that in Eq. (13) and considering a decay process involving $|\phi_i\rangle$ similar to the one in Eq. (12), we get

$$\sum_a \langle 0| \hat{\phi}_a |\phi_i\rangle = \sum_{a,i} U_{a,i} \neq 1 . \tag{19}$$

thus showing that the states $|\phi_i\rangle$ are at odds with the fields $\hat{\phi}_a$ in the sense of the normalization Eq. (15).

[2] This is the reason why the total decay rate Eq. (48) in Ref. [7] contains second powers of the mixing matrix elements, while in Eq. (38) of Ref. [5] they appear with the fourth power.

[3] To make the analysis as transparent as possible, in what follows we refer to scalar fields. Analogous considerations hold for fermionic fields.

IOP Conf. Series: Journal of Physics: Conf. Series **1275** (2019) 012063 doi:10.1088/1742-6596/1275/1/012063

A proper choice for the process Eq. (12) should instead be

$$p \to n\,\bar{l}_\alpha\,\nu_\alpha\,, \tag{20}$$

with

$$\langle 0|\,\hat{\phi}_\alpha\,|\phi_\alpha\rangle = \cos^2\theta + \sin^2\theta = 1\,, \tag{21}$$

which is the result one would expect in analogy with Eq. (15).

We stress that, with this normalization, the transition probabilities between the states $|\phi_A\rangle$ and $|\phi_B\rangle$ are slightly different from the usual ones. In fact, it is possible to derive that

$$|\langle\phi_A(t)|\phi_A\rangle|^2 = 4(2\pi)^6\omega_A^2\left[1 - \frac{\omega_1\omega_2}{\omega_A^2}\sin^2 2\theta\sin^2\left(\frac{\Delta\omega\, t}{2}\right)\right], \tag{22}$$

with $\omega_A = \omega_1\cos^2\theta + \omega_2\sin^2\theta$ and $\Delta\omega = \omega_2 - \omega_1$. By defining

$$|\tilde{\phi}(p)\rangle = \frac{|\phi(p)\rangle}{(2\pi)^3 2\,\omega_p}\,, \tag{23}$$

the expression Eq. (22) becomes

$$\left|\langle\tilde{\phi}_A(t)|\tilde{\phi}_A\rangle\right|^2 = 1 - \frac{\omega_1\omega_2}{\omega_A^2}\sin^2 2\theta\sin^2\left(\frac{\Delta\omega\, t}{2}\right), \tag{24}$$

which, in the relativistic limit, reduces to the usual Pontecorvo oscillation formula.

* * *

Finally, we comment on the effort of the authors of Ref. [7] to prove that mixed neutrinos do not spoil the thermality of the Unruh effect. Similarly to what done in Ref. [4], they assume a thermal bath of neutrinos with definite masses. The probability of detecting a neutrino with flavor α per unit proper time is then given by Eq. (34) of Ref. [7]:

$$\left.\frac{dP_\alpha}{d\tau}\right|_{\Delta m_{ij}^2\sim 0} = \sum_i |U_{\alpha,i}|^2 \left.\frac{dP_{exc,i}}{d\tau}\right|_{\Delta m_{ij}^2\sim 0} \approx \left.\frac{dP_{exc,i}}{d\tau}\right|_{m_i\approx const}, \tag{25}$$

with $\Delta m_{ij}^2 = |m_i^2 - m_j^2|$.

The last step of Eq. (25) practically erases mixing, which is not equivalent to consider it in the relativistic regime. Hence, their claim that the thermality of the Unruh effect is preserved also for flavor neutrino is unfounded. A rigorous analysis of field mixing in accelerated frame has been given in Refs. [12, 13], where nonthermal contributions have been found, which correctly vanish when the relativistic limit is taken into account. Of course, such contributions also vanish in the trivial equal-masses limit considered by Cozzella *et al.*

We want to remark once again that the assumption of Ref. [7] about thermal states for mixed neutrinos is the same adopted in Ref. [4] (where it is motivated by the requirement of KMS condition). Such an observation has been shown to violate general covariance in Refs. [5, 6].

4. Conclusions

In this paper we have discussed some claims of Ref. [7], where it is stated that the standard Unruh effect is perfectly valid for mixed neutrinos and that the decay rates in the inertial and comoving frames do indeed match also in the presence of mixing.

We share with the authors of Ref. [7] the belief that general covariance should act as a guiding principle in the comparison of the two decay rates. However, we have shown that the analysis performed by Cozzella *et al.* fails to pinpoint the peculiar aspects of mixing involved in the problem under consideration. In particular, this last statement is supported by the fact that calculations of Ref. [7] do not recover the outcome of Ref. [5] (where Pontecorvo states are employed to investigate the inverse β-decay) in the limit there considered. Indeed, there is no approximation that can provide the decay rates of Ref. [7] with the off-diagonal contributions arising from the study conducted in Ref. [5].

As a final remark, we would like to stress that the problem considered in Refs. [5, 7] has to do with the internal consistency of quantum field theory, and thus it is necessary to proceed as far as possible with an exact approach without assumptions based on phenomenological considerations. In our analysis [5, 6], by resorting to the concept of flavor neutrino states defined as exact eigenstates of flavor charges obtained via Noether's theorem, we consider an approximation in which usual Pontecorvo states for flavor neutrinos can be safely used. In Ref. [5], we have shown that, in such an approximation, the decay rates in the two frames do indeed agree. It remains an open question if corrections to the Unruh effect are necessary in the more general case of exact flavor states. In any case, corrections will be small and this is to be regarded as a virtue of the formalism which allows to highlight novel, unpredicted, effects (as, for example, in the context of the Casimir effect [19], where the non-trivial nature of flavor vacuum is responsible for an unusual behavior of the attractive strength between the binding plates [20]). Further investigation is required along this line [21].

Appendix: Meaning of the approximation used in Ref. [5]

In Ref. [5], we have adopted an approximation in the evaluation of the decay rates, in particular the interference terms Γ_{12} and $\widetilde{\Gamma}_{12}$ in Eqs. (10) and (11), by taking small $\epsilon \equiv \frac{\delta m}{m_{\nu_1}}$ and then letting $m_{\nu_1} \to 0$. Thus, one may think that it is necessary to carry out a "controlled expansion" of such terms, considering m_{ν_1} and ϵ on the same footing.

In this regard, we would like here to clarify that, although the expansions in Eqs. (41) and (56) of Ref. [5] are formally performed around $\epsilon = 0$, the contributions we actually compare in the two frames are the first-order terms of the expansions in δm (see Eqs. (42) and (57) of Ref. [5], where $\Gamma^{(1)}$ and $\tilde{\Gamma}^{(1)}$ are divided by m_{ν_1}), as it can be seen by simple calculations. These terms are clearly well-behaved for $m_{\nu_1} = 0$.

Similar considerations hold if one considers an expansion in terms of a dimensionless parameter such as $\frac{\delta m}{a}$ rather than δm.

We also emphasize that the particular approximation considered in Ref. [5], namely the expansion in δm and the subsequent requirement of vanishing m_{ν_1}, is adopted not only for bypassing technicalities, but also for physical reasons. Indeed, as shown in Ref. [6], in such a regime Pontecorvo states (which we use in Ref. [5]) can be identified with the exact QFT flavor states [9, 22], up to corrections of the order $\mathcal{O}(\delta m^2)$. This can be seen by observing that, in the QFT treatment of mixing, flavor neutrino states (coherent states) have schematically the form [9, 22]

$$|\nu_e\rangle = \left[\cos\theta\, a_1^\dagger + \sin\theta\, |U|\, a_2^\dagger + \sin\theta\, |V|\, a_1^\dagger\, a_2^\dagger\, b_1^\dagger\right] |0\rangle_{1,2}\,, \tag{26}$$

where $|0\rangle_{1,2}$ is the vacuum for free fields, $a_i^\dagger$ $(b_i^\dagger)$, $i = 1, 2$, is the creation operator for a neutrino (antineutrino) with mass m_i, and U, V are the Bogoliubov coefficients arising from the non-orthogonality of Dirac spinors with different masses, i.e.

$$|U| = \left(\frac{\omega_{\nu_1} + m_1}{2\omega_{\nu_1}}\right)^{\frac{1}{2}} \left(\frac{\omega_{\nu_2} + m_2}{2\omega_{\nu_2}}\right)^{\frac{1}{2}} \left(1 + \frac{k^2}{(\omega_{\nu_1} + m_1)(\omega_{\nu_2} + m_2)}\right), \tag{27}$$

$$|U|^2 + |V|^2 = 1. \tag{28}$$

IOP Conf. Series: Journal of Physics: Conf. Series **1275** (2019) 012063 doi:10.1088/1742-6596/1275/1/012063

Expanding in δm and then setting $m_{\nu_1} = 0$ yields

$$|U| \approx 1 - \frac{\delta m^2}{8k_\nu^2} + O(\delta m^4), \tag{29}$$

From Eq. (26), it then follows that the exact QFT flavor eigenstates are equal to the Pontecorvo states up to the leading order, thus justifying the employment of the latter in Ref. [5].

References

[1] Muller R 1997 *Phys. Rev. D* **56** 953
[2] Matsas G E A and Vanzella D A T 1999 *Phys. Rev. D* **59** 094004; Matsas G E A and Vanzella D A T 2000 *Phys. Rev. D* **63** 014010; Matsas G E A and Vanzella D A T 2001 *Phys. Rev. Lett.* **87** 151301
[3] Suzuki H and Yamada K 2003 *Phys. Rev. D* **67** 065002
[4] Ahluwalia D V, Labun L and Torrieri G 2016 *Eur. Phys. J. A* **52** 189
[5] Blasone M, Lambiase G, Luciano G G and Petruzziello L 2018 *Phys. Rev. D* **97** 105008
[6] Blasone M, Lambiase G, Luciano G G and Petruzziello L 2018 PoS CORFU **2017** 198
[7] Cozzella G, Fulling S A, Landulfo A G S, Matsas G E A and Vanzella D A T 2018 *Phys. Rev. D* **97** 105022
[8] Bilenky S M and Pontecorvo B 1978 *Phys. Rept.* **41** 225
[9] Blasone M and Vitiello G 1995 *Annals Phys.* **244** 283
[10] Blasone M, Jizba P and Vitiello G 2001 *Phys. Lett. B* **517** 471
[11] Gilmore R 1972 *Ann. Phys. NY* **74** 391; Perelomov A M 1972 *Commun. Math. Phys.* **26** 222
[12] Blasone M, Lambiase G and Luciano G G 2017 *Phys. Rev. D* **96** 025023
[13] Blasone M, Lambiase G and Luciano G G 2018 *J. Phys. Conf. Ser.* **956** 012021
[14] Cardall C Y and Fuller G M 1997 *Phys. Rev. D* **55** 7960
[15] Blasone M, Lambiase G, Luciano G G and Petruzziello L 2018 *EPL* **124** 51001
[16] Schwartz M D 2014 *Quantum Field Theory and the Standard Model* Cambridge UK: Cambridge University Press
[17] Cheng T P and Li L F 1984 *Gauge Theory Of Elementary Particle Physics* Oxford UK: Clarendon
[18] Greiner W and Reinhardt J 1996 *Field quantization* Berlin Germany: Springer
[19] Casimir H B G 1948 Proc. K. Ned. Akad. Wet. **51** 793; Casimir H B G and Polder D 1948 Phys. Rev. **73** 360; Milton K A 2001 *The Casimir effect: physical manifestations of zero-point energy* River Edge: World Scientific; Nesterenko V V, Lambiase G and Scarpetta G 2001 Phys. Rev. D **64** 025013 (2001); Nesterenko V V, Lambiase G and Scarpetta G 2002 Ann. Phys. **298** 403; Blasone M, Lambiase G, Petruzziello L and Stabile A 2018 Eur. Phys. J. C **78** 976 (2018); Buoninfante L, Lambiase G, Petruzziello L and Stabile A 2019 Eur. Phys. J. C **79** 41.
[20] Blasone M, Luciano G G L, Petruzziello L and Smaldone L 2018 *Phys. Lett. B* **786** 278
[21] Blasone M, Luciano G G L, Petruzziello L and Smaldone L *In preparation*
[22] Blasone M, Capolupo A, Romei O and Vitiello G 2001 *Phys. Rev. D* **63** 125015; Blasone M and Palmer J 2004 *Phys. Rev. D* **69** 057301; Blasone M, Capolupo A, Terranova F and Vitiello G 2005 *Phys. Rev. D* **72** 013003; Blasone M, Gargiulo M V and Vitiello G 2016 *Phys. Lett. B* **761** 104; Blasone M, Henning P A and Vitiello G 1999 *Phys. Lett. B* **451** 140; Blasone M, Capolupo A and Vitiello G 2002 *Phys. Rev. D* **66** 025033; Blasone M, Pires Pacheco P and Wan Chan Tseung H 2003 *Phys. Rev. D* **67** 073011; Blasone M, Jizba P and Luciano G G 2018 *Annals Phys.* **397** 213

IOP Conf. Series: Journal of Physics: Conf. Series **1275** (2019) 012064 doi:10.1088/1742-6596/1275/1/012064

On the time continuous evolution of the universe if time is discrete and irreversible in nature

Roland Riek

Laboratory of Physical Chemistry, ETH Zurich, Switzerland

E-mail: `roland.riek@phys.chem.ethz.ch`

Abstract. The time evolution of the universe is usually mathematically described under a continuous time and thus time reversible. Here, the consequences of studying the evolution of a homogenous isotropic universe by time continuous reversible physics are studied if time is actually discrete and irreversible in nature. The discrete dynamical time concept of Lee and its continuous time limit to the continuous time case is applied to the Newtonian limit of the general relativity theory. By doing so, the cosmic constant as well as the inflation of the universe arise and are predicted quantitatively well by assuming the smallest time step to be the Planck time and by using the current size of the universe.

1. Introduction

Because experimentally measured time is composed of an array of events, which can be exemplified in physics only by an energy-consuming clock measurement, and thus can not be measured continuously because of the uncertainty principle between energy and time ($\Delta E \Delta t > h/2$ with h being the Planck constant) time may be regarded discrete in nature (see for example [1-10]). It is however not a very popular concept, because the continuity of time and space enables powerful mathematical tools. Nonetheless, several approaches on discrete time physics have been elaborated on ([6]- [12]). Here, the evolution of a homogenous isotropic universe is explored under a discrete dynamical time and its consequences are discussed if time continuous reversible physics is applied to an universe that expands in time steps. The discrete dynamical time concept of Lee [6] is thereby used and its limit to the continuous time case is applied to the Newtonian limit of the general relativity theory [13]. This approach is therefore not a quantum mechanical-based or a quantum-loop gravity-based theory [12], while it circumvents principle issues on discrete time physics such as the Lorentz transformation, but yields an alternate explanation on the apparent inflation of the universe and the cosmic constant. After a short summary on several definitions and laws in cosmology (2.1) and the introduction of a discrete dynamical time (2.2), the request on time reversibility to Newton's equation is discussed by introducing a scaling of time (2.2). In 2.3 and 2.4, this concept is applied to the gravitational force yielding the cosmic constant as ad hoc introduced by Einstein [14]. In 2.5 the evolution of the universe is described yielding an evolutionary phase of the universe as postulated by Guth [15] and in 2.6 the apparent acceleration of the universe is discussed.

Content from this work may be used under the terms of the Creative Commons Attribution 3.0 licence. Any further distribution of this work must maintain attribution to the author(s) and the title of the work, journal citation and DOI.
Published under licence by IOP Publishing Ltd

2. Theory

2.1. The standard universe evolution under a continuous time

In the classical Newtonian description of gravity, the gravitational force $\mathbf{F}_G(\mathbf{r})$ on a symbolic point particle of mass m at radius r from the center of the mass M of a point-like particle or a spherically symmetric object (with size smaller than r) is given by

$$\mathbf{F}_G(\mathbf{r}) = m\,\ddot{\mathbf{r}} = -\frac{4G}{3\pi}\rho\,\mathbf{r}\,m = -\frac{G\,M\,m}{\mathbf{r}^2}\,\frac{\mathbf{r}}{|\mathbf{r}|} \tag{1}$$

with G the gravitational constant and $\mathbf{r}$ the vector with length r from the origin of reference at the center of mass M, which is considered here the mass of the universe, to the coordinates of the point particle (please note, vectors are written in bold) yielding for $\mathbf{r} = \begin{pmatrix} r^x \\ r^y \\ r^z \end{pmatrix} = \begin{pmatrix} r^x \\ r^x \\ r^x \end{pmatrix}$ with $r_x = \frac{r}{\sqrt{3}}$. With the theorem of Gauss and the gravitational field potential $\Phi(\mathbf{r}) = -\frac{G\,M}{|\mathbf{r}|}$ the following Poisson equation can be obtained:

$$\nabla^2\Phi(\mathbf{r}) = \Phi\,(\mathbf{r}) = 4\,\pi\,G\,\rho(\mathbf{r}) \tag{2}$$

with $\rho(\mathbf{r})$ being the mass density of the system with mass M. Thus, a Newtonian gravitational field can be derived from a scalar potential Φ that follows the Poisson equation.

In contrast, the Poisson equation derived from the Einstein field equations of the theory of general relativity at the Newtonian limit [16] is

$$\Phi(\mathbf{r}) = 4\,\pi\,G\,\rho(\mathbf{r}) - \Lambda \tag{3}$$

with Λ defined as the cosmic constant, which is constant by definition. Nowakowski et al. [13] showed using the Schwarzschild solution that the potential for a point-like and/or spherical symmetric object with mass M at the Newtonian limit is

$$\Phi(\mathbf{r}) = -\frac{G\,M}{|\mathbf{r}|} - \frac{1}{6}\Lambda\,\mathbf{r}^2 \tag{4}$$

This yields the following gravitational force acting on the point particle of mass m at radius r:

$$\mathbf{F}_G(\mathbf{r}) = m\,\ddot{\mathbf{r}} = -\frac{G\,M\,m}{\mathbf{r}^2}\,\frac{\mathbf{r}}{|\mathbf{r}|} + \frac{m}{3}\Lambda\,\mathbf{r} \tag{5}$$

Another equation of interest here is the space time metric in case of a flat space time geometry (restricted to a flat space time geometry for simplicity reasons) given by

$$dx^2 = (cdt)^2 - R(t)^2 dr^2 \tag{6}$$

with dx the space time interval, which is the distance between two events, $c$the speed of light, and $R(t)$ the scaling factor. It describes the radial evolution of an isotropic homogenous universe under a continuous time $t$under the theory of general relativity [14, 16]. We further mention the Hubble law [21] with the time-dependent Hubble constant given by

$$H(t) = \frac{\dot{R}(t)}{R(t)} \tag{7}$$

2.2. Under a dynamic discrete time

If time is a discrete dynamical variable $\hat{\mathbf{t}}_n$ the continuous space function r(t) of a homogenous isotropic universe is replaced by a sequence of discrete values $\mathbf{r}_n = \mathbf{r}\,(\hat{\mathbf{t}}_n)$ with [6]:

$$(\mathbf{r}_0,\, \hat{\mathbf{t}}_0),\, (\mathbf{r}_1,\, \hat{\mathbf{t}}_1), \ldots\ldots, (\mathbf{r}_n,\, \hat{\mathbf{t}}_n), \ldots\ldots, (\mathbf{r}_{N+1},\, \hat{\mathbf{t}}_{N+1}) \tag{8}$$

with $(\mathbf{r}_0,\, \hat{\mathbf{t}}_0)$ the initial and $(\mathbf{r}_{N+1},\, \hat{\mathbf{t}}_{N+1})$ the final position. In this description $\mathbf{r}_n$ is still continuous, while $\hat{\mathbf{t}}_n$ is discrete and as requested a dynamic variable described by a diagonal tensor. It is noted here, that the dynamic part of the time and its tensor character just introduced is only needed to show the effect of the use of a continuous time in describing the time evolution of the universe under the assumption that time is discrete in nature with a constant step size $\Delta t = const$ (for exampleΔ t could be the Planck time $5.4\,10^{-44}$ s). With other words, it is assumed here that the sequence of events happening are in general time irreversible and and can be described by

$$(\mathbf{r}_0,\, t_0),\, (\mathbf{r}_1,\, t_1), \ldots\ldots, (\mathbf{r}_n,\, t_n), \ldots\ldots, (\mathbf{r}_{N+1},\, t_{N+1})$$

with t_n being a scalar and the step size between time points is constant withΔ t. Thus, the exercise that follows with a dynamic tensor description of time and the scaling variable is regarded only a mathematical trick and not real in nature.

The dynamic part of the time can be described by a time scaling tensor variable denoted $\mathbf{s}_n\mathbf{1}$ defined by

$$\mathbf{s}_n\mathbf{1}\,\Delta t = \hat{\mathbf{t}}_n - \hat{\mathbf{t}}_{n-1} \tag{9}$$

withΔ $t = const.$

For simplicity reasons the system of interest is the gravitational force of an object of mass M that acts on a test particle at a distance of radius r, which enables by proper selection of the coordinate system the reduction of the dynamic part of the time to only a single variable s_n within the tensor notation as elaborated on more detailed in the Appendix (please note, the Appendix also includes the more general case).

In discrete mechanics there are many possible definitions of the velocity $\dot{\mathbf{r}}_n$. The following definition at time point n follows ref [11]:

$$\dot{\mathbf{r}}_n = \frac{\mathbf{r}_n - \mathbf{r}_{n-1}}{\Delta t} \tag{10}$$

It seems to be a plausible one because it is in line with the causality argument and our daily experience that the presence is determined by the past and presence (i.e. to describe a present state of a system information can experimentally-derived only from the past and presence). With eq. (10) the velocity is defined backward in time and thus time asymmetric, permitting a forward progressing description of the system from past and present information. Please note, that this definition contrasts other theories as for example quantum loop gravity theory [12]. In presence of a discrete dynamic time variable described by the additional time scaling variable s_n a Lagrangian must be derived that is able to describe adequately the system. Such a Lagrangian has been introduced by Nosé and Hoover [17]-[20] and its discrete analog has been introduced in ref. [11]:

$$L_n^N = L_n^N(\mathbf{r}_n,\, \mathbf{r}_{n-1};\, \dot{s}_n,\, s_n) = s_n\left(\frac{1}{2}\, m\, \dot{\mathbf{r}}_n^2 - V(\mathbf{r}_n) + \frac{1}{2} Q \frac{\dot{s}_n^2}{s_n^2} - N_{df}\, k_B T\, ln\, s_n\right) \tag{11}$$

where $V(\mathbf{r}_n)$ is the acting potential (in the case of a Newtonian gravitational potential $V(\mathbf{r}_n) = \Phi m$), k_B is the Boltzmann constant, Q is a constant with units Js^2 (energy*seconds*seconds), which has been described as a "mass"-like term for the motion of s_n with Q ¿ 0, N_{df} is the degree of freedom of the system, $\dot{s}_n = \frac{s_{n+1} - s_n}{\Delta t}$ is the velocity of the scaling factor with the unit s^{-1}, and T is the temperature of the system defined through the kinetic energy of the system. A corresponding Hamiltonian without the last two terms of eq. 11, which are not relevant here, has been described by Elze et al. [10].

If the discrete analog of the Lagrangian equation for each coordinate component i is defined as in ref. [11]

$$\frac{1}{\Delta t}(\frac{\partial L^N_{n+1}(\mathbf{r}_{\mathrm{n+1}},\dot{\mathbf{r}}_{\mathrm{n+1}})}{\partial \dot{r}^{\mathbf{i}}_{\mathrm{n+1}}} - \frac{\partial L^N_n(\mathbf{r}_{\mathrm{n}},\dot{\mathbf{r}}_{\mathrm{n}})}{\partial \dot{r}^{\mathbf{i}}_{\mathrm{n}}}) = \frac{\partial L^N_n}{\partial r^i_n} \tag{12}$$

the discrete Newton' law can be obtained using the discrete Lagrangian from eq. 10

$$\frac{1}{\Delta t}(\dot{\mathbf{r}}_{n+1}\frac{s_{n+1}}{s_n} - \dot{\mathbf{r}}_{n+1} + \dot{\mathbf{r}}_{n+1} - \dot{\mathbf{r}}_n) = \frac{1}{m}\mathbf{F}(\mathbf{r}_n) \tag{13}$$

$$\ddot{\mathbf{r}}_n = \frac{1}{\Delta t}(\dot{\mathbf{r}}_{n+1} - \dot{\mathbf{r}}_n) = \frac{1}{m}\mathbf{F}(\mathbf{r}_n) - \dot{\mathbf{r}}_{n+1}(\frac{\mathbf{s}_{n+1}}{\mathbf{s}_n} - 1)\frac{1}{\Delta t} \tag{14}$$

$$\ddot{\mathbf{r}}_n = \frac{1}{m}\mathbf{F}(\mathbf{r}_n) - \gamma_n \dot{\mathbf{r}}_{n+1} \tag{15}$$

with $\gamma_n = \frac{\dot{s}_{n+1}}{s_n}$ and $\dot{s_n} = \frac{s_{n+1}-s_n}{\Delta t}$.

If there is no scaling of time (i.e. $s_i = 1$ for all $i = 1...N+1$) the Newton's law under a discrete time with constant time steps is of the form $m\,\ddot{\mathbf{r}}_n = \mathbf{F}(\mathbf{r}_n)$

which resembles its continuous analog. In presence of a scaling of time (unequal to 1) however, the Newton's law has an additional term, which can be regarded an acceleration or a friction term in dependence on the sign of γ_n. It is this friction term, which enables discrete time physics to be time reversible with the following request (see ref. [11]) that holds for the symmetric case selected.

$$\frac{s_n}{s_{n+1}} = \frac{F^i(r_{n+1})}{F^i(r_n)} \quad i = x, y, z \tag{16}$$

with $\mathbf{F}(\mathbf{r}_n) = \begin{bmatrix} F^x(\mathbf{r}_n) \\ F^y(\mathbf{r}_n) \\ F^z(\mathbf{r}_n) \end{bmatrix} = \begin{bmatrix} F^x(\mathbf{r}_n) \\ F^x(\mathbf{r}_n) \\ F^x(\mathbf{r}_n) \end{bmatrix}$.

This equation (eq. 16) and its more general analog in the Supplementary Material (eq. 54) are called the reversibility axiom. If the reversibility axiom is fulfilled, the introduced discrete time physics is time reversible [11].

By incorporating the reversibility axiom into the discrete Newtonian equation (eq. 15) the following expression is obtained:

$$\ddot{\mathbf{r}}_n = \frac{1}{\Delta t}(\dot{\mathbf{r}}_{n+1} - \dot{\mathbf{r}}_n) = \frac{1}{m}\mathbf{F}(\mathbf{r}_n) + \dot{\mathbf{r}}_{n+1}\frac{\dot{F^x}(r_{n+1})}{F^x(r_{n+1})} \tag{17}$$

with $\dot{F^x}(r_{n+1}) = \frac{F^x(r_{n+1})-F^x(r_n)}{\Delta t}$. Eq. 17 reflects thereby the time reversible evolution of a system under a discrete time. It is highlighted that due to the dynamic nature of the discrete time with the variable s_n time reversibility is guaranteed.

As in the case of Newton's law (eq. 17) the introduction of a dynamic discrete time has an impact also on other time-dependent formulas such as the well known differential equation $\frac{\dot{x}}{x} = b$ with $b = const$ having as solution the exponential function $x(t) = x_0\, e^{b\,t}$ with x_0 the value at time 0. The corresponding differential equation under a discrete dynamic time (dependent on eq. 16) is given by

$$\frac{\dot{x}_n}{x_n} = b\, s_n \tag{18}$$

This can be derived from $x_n(t) = x_0 \prod_{i=1}^{n}(1 + bs_i\Delta t) \approx x_0\, e^{b\sum_{i=1}^{n} s_i\Delta t}$ with the discrete definition of $\dot{x}_n$ given above for $\dot{\mathbf{r}}_n$ (eq. 10). The latter equation is true at the limit $lim\,\Delta t \to 0$.

2.3. Under a dynamic continuous time
In order to elaborate on the consequences of using a time continuous description of the evolution of the universe if time is actually discrete and non dynamic in nature, the above description with a dynamic discrete time is set to the limit with $lim\,\Delta t \to 0$.

For this, the derived discrete Newtonian equation (eq. 17) is transformed into its corresponding continuous analog:

$$\ddot{\mathbf{r}} = \frac{1}{m}\mathbf{F}(\mathbf{r}) - \gamma\dot{\mathbf{r}} \tag{19}$$

with $g = \frac{\dot{s}}{s} = -\frac{\dot{F^x}}{F^x}$ with $s(t)$.
Correspondingly, in this time continuous frame the differential equation of the exponential function is then given by $\frac{\dot{x}}{x} = b\,s(t)$ and thus the Hubble law of eq. 5 is given by

$$\frac{\dot{R}}{R} = H(t)\,s(t) \tag{20}$$

Please note, within the system of interest, which is homogenous isotropic expanding universe with the argumentation put forward in eq. 22 below, the $s$in eq. 19 and eq. 20 are the same. The introduction of the dynamic continuous time resolves also the problem of the immediate request for a discrete space that comes along with a discrete time through the Lorentz transformation as has been done for example by the quantum loop gravity theory [12]. However, by the present translation from the discrete time to a dynamic continuous time with $lim\,\Delta t \to 0$ the space can be described as usual because no need for a dynamic continuous space is required as space has not the problem of irreversibility.

2.4. From the Poisson equation to the Evolution of the Universe
Applying the continuous formulation of the modified Newton's equation of eq. 17 to the gravitational force with $\mathbf{F}(\mathbf{r}) = \mathbf{F}_G(\mathbf{r})$ for the system of interest here (using $F_G^i(r) = -2\frac{GMm}{r^{i3}}\dot{r^i}$) the following equation is obtained:

$$\ddot{\mathbf{r}} = \frac{1}{m}\mathbf{F}(\mathbf{r}) - 2\frac{\dot{r_x}}{r_x}\dot{\mathbf{r}} = -\frac{GM}{r^2}\frac{\mathbf{r}}{|\mathbf{r}|} + 2(\frac{\dot{r_x}}{r_x})^2\,\mathbf{r} \tag{21}$$

In an isotropic homogenous and spherically expanding universe at the Newtonian limit (eq. 5: $\ddot{\mathbf{r}} = -\frac{GM}{\mathbf{r}^2}\frac{\mathbf{r}}{|\mathbf{r}|} + \frac{1}{3}\Lambda\,\mathbf{r}$) the scaling factor R is proportional to $r$and $\dot{R}$ proportional to $\dot{r}$ and thus we obtain for the second term of eq. 19 (using also eq. 20)

$$2\,(\frac{\dot{r_x}}{r_x})^2 = 2\,(\frac{\dot{R}}{R})^2 = 2\,H(t)^2\,s(t)^2 \tag{22}$$

which in combination with the Newtonian limit of the poisson equation derived for the general relativity theory (eq. 3) yields

$$\frac{1}{3}\Lambda = 2\,H(t)^2 s(t)^2 \tag{23}$$

This results under a slow changing force limit (i.e. $s \approx 1$) to

$$\Lambda = 6\,H(t)^2 \tag{24}$$

or under a constant changing force with $s(t) = \sqrt{1/2}$

$$H = \sqrt{\frac{\Lambda}{3}} \tag{25}$$

usually obtained under standard descriptions [16] (obviously $s(t) = \sqrt{1/2}$ was chosen by purpose albeit meaningless and not required for the further derivations; alternatively, the geometry of the expanding universe can be altered such that $2\,(\frac{\dot{r_x}}{r_x})^2 = (\frac{\dot{R}}{R})^2$).

Concluding the first part of the presented theory, when time continuous physics at the Newtonian limit is applied to a flat isotropic homogenous universe that evolves in discrete time steps under slow changing force, it appears that the universe is expanding with a positive cosmic constant of $\Lambda = 6\,\mathrm{H}^2$. With other words, if time is discrete, the cosmic constant is a consequence of applying time reversible continuous physics to the expansion of the universe. It is further interesting to note, that under the given condition with the cosmic constant being constant by definition, the Hubble constant is independent of time if the gravitational force does not and never did change fast.

2.5. *The expansion of the universe with its apparent inflation*

The condition that the gravitational force is not changing fast is of course not correct in the early time of the expansion of the universe, which is studied next. For this purpose we use the following relationship $\gamma = \frac{\dot{s}}{s} = -\frac{\dot{F}}{F} = \frac{\dot{r}}{r} = \frac{\dot{R}(t)}{R(t)}$ yielding

$$\frac{\dot{s}(t)}{s(t)} = \frac{\dot{R}(t)}{R(t)} = H(t)\,s(t) = \sqrt{\frac{\Lambda}{6}} \tag{26}$$

and correspondingly

$$s(t) = s_0\,e^{\sqrt{\frac{\Lambda}{6}}\,t} \tag{27}$$

with s_0 the scaling at time 0 (please note, that this s_0 is from the continuous description of time, while in the discrete description s_n i used with n an integer starting from 1)

Since

$$H(t) = \frac{\sqrt{\frac{\Lambda}{6}}}{s(t)} = \sqrt{\frac{\Lambda}{6}}\frac{1}{s_0}e^{-\sqrt{\frac{\Lambda}{6}}\,t} \tag{28}$$

If one now takes the Hubble constant of eq. 36 and applies it for the determination of the scaling factor by using the standard Hubble law of eq. 5 we obtain the following scaling factor of the metric of the universe

$$R(t) = R_0 e^{\int_0^t \frac{\sqrt{\frac{\Lambda}{6}}}{s(t)}dt} = R_0 e^{\frac{\sqrt{\frac{\Lambda}{6}}}{s_0}\int_0^t e^{-\sqrt{\frac{\Lambda}{6}}\,t}dt} = R_0 e^{\frac{1}{s_0}[1-e^{-\sqrt{\frac{\Lambda}{6}}\,t}]} \tag{29}$$

Hence, when time continuous physics is applied to a flat isotropic homogenous universe that evolves in discrete time steps there is an apparent expansion of the universe following eq. 29. This artifact is constant for large time $t lim_{t\to\infty} R(t) = R_0\,e^{\frac{1}{s_0}}$, while for small times using a Taylor expansion 1. order of the exponential function the universe appears to expand exponentially with $R(t) \approx R_0\,e^{\frac{\sqrt{\frac{\Lambda}{6}}}{s_0}\,t}$ as postulated by Guth [15].

There are three so far three undetermined constants left in eq. 29 (i.e. R_0, s_0, and Λ) that need to be discussed for a quantitative time continuous description of the evolution of

IOP Conf. Series: Journal of Physics: Conf. Series **1275** (2019) 012064 doi:10.1088/1742-6596/1275/1/012064

the universe taking into account that time is actually discrete and irreversible. First, R_0 is considered. Because the scaling factor $R(t)$ is in frame of the reference at time point 0, $R_0 = 1$. Next, the unknown constant s_0 is determined by assuming that the first time step of the universe is the Planck time$\Delta\, t_p = \sqrt{\frac{hG}{c^5}}$ =5.3 $10^{-44}s$ (with h Planck's Wirkungsquantum and $c$the velocity of light in vacuum) and that the universe was expanding with maximal velocity, i.e. light velocity c. Furthermore, it is assumed that the universe at time point $t = 0$ had the minimal size of a system possible, which is regarded the Planck length l_p. Since by definition $l_p = c\,\Delta t_p$ at time point t_1 the size of the universe was thus $l_p + c\,\Delta t_p = 2\,l_p = R(t_1)\,l_p$ yielding $R(t_1) = 2$. We further describe $R(t_1)$ by a Taylor expansion first order and using $\dot{R}(t) = R(t)\,e^{\sqrt{\frac{\Lambda}{6}}t}\sqrt{\frac{\Lambda}{6}}\frac{1}{s_0}$ starting with time $t = 0$ yielding

$$R(t_1) \approx R_0 + \dot{R}(0)\Delta t_p = 1 + \sqrt{\frac{\Lambda}{6}}\frac{1}{s_0}\Delta t_p \tag{30}$$

which results in

$$s_0 = \sqrt{\frac{\Lambda}{6}}\,\Delta t_p \tag{31}$$

since $R(t_1) = 2$. Following these assumptions, the quantitative expression of the expansion of the universe is given by

$$R(t) = e^{\frac{1}{\sqrt{\frac{\Lambda}{6}}\Delta t_p}[1-e^{-\sqrt{\frac{\Lambda}{6}}t}]} \tag{32}$$

The scaling factor for large time t is thus given by $lim_{t\to\infty} R(t) = R_{inf} = e^{\frac{\sqrt{\frac{6}{\Lambda}}}{\Delta t_p}}$ and constant, while for small times using a Taylor expansion 1. order of the exponential function the universe appears to expand exponentially with $R(t) \approx e^{\frac{t}{\Delta t_p}}$ as requested by Guth [15]. The last unknown is the cosmic constant Λ, which can be gathered from measurements and the Hubble law with an Λ in the order of $10^{-35}s^{-2}$ assumed in the following to be $3{*}10^{-35}s^{-2}$[16]. This allows to calculate the evolution of the universe using eq. 32 as demonstrated in Figure 1. Figure 1 shows that when time continuous physics is applied to a flat isotropic homogenous universe that evolves in discrete time steps, it appears that the universe is expanding through an inflationary phase during the first 3 $10^{-35}s$ after which it approximates its maximal expansion at an R(t) of ca $10^{60} = R_{inf}$. Please note, that the presented approach does neither take special relativity nor quantum effects into account, which in particular at early time points such as during the grand unification theory (GUT) time, may have to be considered. Nevertheless, assuming a universe size of $1\,l_p$ at time point 0 the expansion of the universe follows closely accepted inflationary universe models [15],[16]. Furthermore, the model predicts well the current size of the universe having $R(t_{present}) \approx R_{inf} = 4\,10^{60}$ yielding a size of the universe of $R_{inf}\,l_p = R_{inf}\,c\,\Delta t_p \approx 1\,10^{27}m$. Of course, the return process starting on the measured current size of the universe in combination with eq. 32 is also possible and results in the cosmic constant with expected size.

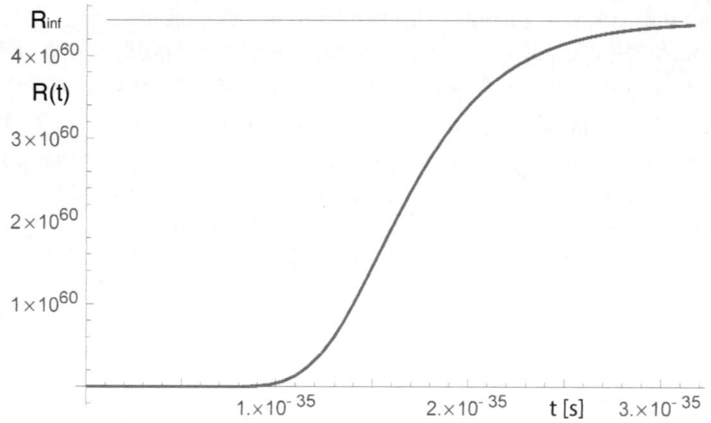

Figure 1: The scaling factor $R(t)$ versus time t using eq. 32. In addition, the maximal expansion factor $lim_{t\to\infty}R(t) = e^{\frac{\sqrt{\frac{6}{\Lambda}}}{\Delta t_p}}$ is indicated.

2.6. The apparent acceleration of the universe at late time points

Using the standard model of general relativity in combination with experimental measures it is often suggested that the universe is currently in an accelerating phase [16]. The presented description of the universe assuming a continuous time while time is postulated discrete in nature yields however a maximal scale factor of $lim_{t\to\infty}R(t) = e^{\frac{\sqrt{\frac{6}{\Lambda}}}{\Delta t_p}}$ (see also Figure 1). While it depends on the cosmic constant Λ, within the description presented the size of the cosmic constant does not determine whether the universe is accelerating in size, expanding forever, or may collapse. Within this model, there is thus no support and no requirement for the proclaimed current accelerated universe.

3. Conclusion

The present work studied the artifacts that appear when time continuous reversible physics is applied to a universe that evolves irreversibly under a discrete time. The artifacts are (i) a cosmic constant and (ii) an inflationary evolution of the universe at early time - two prominent postulates in cosmology of which physical origins and requests are puzzling. Several recent attempts to resolve these issues have been suggested as for example [22-24]. The presented work suggests a simple solution by demanding a discrete time. The consequences are manyfold such as that the present universe is not expanding under acceleration as believed, nor that an inflationary expansion of the universe is necessary to describe the evolution of the universe. We invite the reader to apply the concept of a discrete time to further areas in physics and cosmology.

Acknowledgments

Acknowledgments We would like to thank the ETH for unrestrained financial support and Dr. Yaron Harad, Dr. Alexander Sobol, Dr. Witek Kwiatkowski, and Dr. Jürg Fröhlich for helpful discussions.

Appendix

The following time reversible discrete time physics has been published in ref [11] and is repeated here for the example of interest.

If time is a discrete dynamical variable $\hat{\mathbf{t}}_n$ the continuous space function $\mathrm{r}(t)$ of a homogenous isotropic universe is replaced by a sequence of discrete values $\mathbf{r}_n = \mathbf{r}(\hat{\mathbf{t}}_n)$ with [6]:

$$(\mathbf{r}_0,\, \hat{\mathbf{t}}_0),\, (\mathbf{r}_1,\, \hat{\mathbf{t}}_1), \ldots\ldots\ldots,\, (\mathbf{r}_n,\, \hat{\mathbf{t}}_n), \ldots\ldots,\, (\mathbf{r}_{N+1},\, \hat{\mathbf{t}}_{N+1}) \tag{A.1}$$

with $(\mathbf{r}_0,\, \hat{\mathbf{t}}_0)$ the initial and $(\mathbf{r}_{N+1},\, \hat{\mathbf{t}}_{N+1})$ the final position. In this description $\mathbf{r}_n$ is still continuous, while $\hat{\mathbf{t}}_n$ is discrete and as requested a dynamic variable described by a diagonal tensor second order. It is noted here, that the dynamic part of the time and its tensor character just introduced is only needed to show the effect of the use of a continuous time in describing the time evolution of the universe under the assumption that time is discrete in nature with a constant step size$\Delta\ t\ = const$ (for example$\Delta\ t$ could be the Planck time $5.4\, 10^{-44}$ s). With other words, it is assumed here that the sequence of events happening are in general time irreversible and and can be described by

$$(\mathbf{r}_0,\, t_0),\, (\mathbf{r}_1,\, t_1), \ldots\ldots\ldots,\, (\mathbf{r}_n,\, t_n), \ldots\ldots,\, (\mathbf{r}_{N+1},\, t_{N+1})$$

with t_n being a scalar and the step size between time points is constant with$\Delta\ t$. Thus, the exercise that follows with a dynamic tensor description of time and the scaling variable is regarded only a mathematical trick and not real in nature.

The dynamic part of the time can be described by a time scaling tensor variable denoted $\mathbf{s}_n\mathbf{1}$ defined by

$$\mathbf{s}_n\mathbf{1}\,\Delta t\ = \hat{\mathbf{t}}_n - \hat{\mathbf{t}}_{n-1} \tag{A.2}$$

with$\Delta\ t\ = const$. In matrix notation this reads:

$$\begin{bmatrix} s_n^x \\ s_n^y \\ s_n^z \end{bmatrix} \begin{bmatrix} 1 & 0 & 0 \\ 0 & 1 & 0 \\ 0 & 0 & 1 \end{bmatrix} \Delta t\ = \begin{bmatrix} t_n^x - t_{n-1}^x & 0 & 0 \\ 0 & t_n^y - t_{n-1}^y & 0 \\ 0 & 0 & t_n^z - t_{n-1}^z \end{bmatrix} \tag{A.3}$$

For simplicity reasons the system of interest is the gravitational force of an object of mass M that acts on a test particle at a distance of radius r, which enables by proper selection of the coordinate system to reduce the dynamic part of the time to only a single variable s_n within the tensor notation.

$$\begin{bmatrix} s_n^x \\ s_n^y \\ s_n^z \end{bmatrix} \begin{bmatrix} 1 & 0 & 0 \\ 0 & 1 & 0 \\ 0 & 0 & 1 \end{bmatrix} \Delta t\ = s_n 1 \Delta t \tag{A.4}$$

The more general case is treated below.

In discrete mechanics there are many possible definitions of the velocity $\dot{\mathbf{r}}_n$. The following definition at time point n seems to be a plausible definition:

$$\dot{\mathbf{r}}_n = \frac{\mathbf{r}_n - \mathbf{r}_{n-1}}{\Delta t} \tag{A.5}$$

because it is in line with the causality argument and our daily experience that the presence is determined by the past and presence (i.e. to describe a present state of a system information can experimentally-derived only from the past and presence). With eq. (37) the velocity is defined backward in time and thus time asymmetric, permitting a forward progressing description of the system from past and present information. Please note, that this definition contrasts other theories as for example quantum loop gravity theory [12]. In presence of a discrete dynamic time variable described by the additional time scaling variable s_n a Lagrangian must be derived that is able to describe adequately the system. Such a Lagrangian under an artificial continuous scaled time s has been introduced by Nosé and Hoover ($L^N = s(\frac{1}{2}\, m\, \dot{r}^2 - V(r) + \frac{1}{2} Q \frac{\dot{s}^2}{s^2} - N_{df}\, k_B T\, ln\, s)$)[17]-[20]. The discrete analog of the Nosé-Hoover Lagrangian (in real space) is given by the following expression [11]:

IOP Conf. Series: Journal of Physics: Conf. Series **1275** (2019) 012064 doi:10.1088/1742-6596/1275/1/012064

$$L_n^N = L_n^N(\mathbf{r}_\mathrm{n},\, \mathbf{r}_{n-1};\, \dot{s}_n,\, s_n) = s_n(\frac{1}{2}\, m\, \dot{\mathbf{r}}_n^2 - V(\mathbf{r}_n) + \frac{1}{2}Q\frac{\dot{s}_n^2}{s_n^2} - N_{df}\, k_B T\, ln\, s_n) \tag{A.6}$$

where $V(\mathbf{r}_n)$ is the acting potential (in the case of a Newtonian gravitational potential $V(\mathbf{r}_n) \;=\; Fm\;\,)$, k_B is the Boltzmann constant, Q is a constant with units Js^2 (energy*seconds*seconds), which has been described as a "mass"-like term for the motion of s_n with Q ¿ 0, N_{df} is the degree of freedom of the system, $\dot{s}_n = \frac{s_{n+1}-s_n}{\Delta t}$ is the velocity of the scaling factor with the unit s^{-1}, and T is the temperature of the system defined through the kinetic energy of the system. A corresponding Hamiltonian without the last two terms of eq. 38 has been described by Elze et al. [10].

If the discrete analog of the Lagrangian equation for each coordinate component i is defined as in ref. [11]

$$\frac{1}{\Delta t}(\frac{\partial L_{n+1}^N(\mathbf{r}_{\mathrm{n}+1}, \dot{\mathbf{r}}_{\mathrm{n}+1})}{\partial \dot{r}_{\mathrm{n}+1}^{\mathbf{i}}} - \frac{\partial L_n^N(\mathbf{r}_\mathrm{n}, \dot{\mathbf{r}}_\mathrm{n})}{\partial \dot{r}_\mathrm{n}^{\mathbf{i}}}) = \frac{\partial L_n^N}{\partial r_n^i} \tag{A.7}$$

the discrete Newton' law can be obtained using the discrete Lagrangian from eq. 10

$$\frac{1}{\Delta t}(\,\dot{\mathbf{r}}_{n+1}\frac{s_{n+1}}{s_n} - \dot{\mathbf{r}}_{n+1} + \dot{\mathbf{r}}_{n+1} - \dot{\mathbf{r}}_n) = \frac{1}{m}\mathbf{F}(\mathbf{r}_n) \tag{A.8}$$

$$\ddot{\mathbf{r}}_n = \frac{1}{\Delta t}(\dot{\mathbf{r}}_{n+1} - \dot{\mathbf{r}}_n) = \frac{1}{m}\mathbf{F}(\mathbf{r}_n) - \dot{\mathbf{r}}_{n+1}(\frac{\mathbf{s}_{n+1}}{\mathbf{s}_n} - 1)\frac{1}{\Delta t} \tag{A.9}$$

$$\ddot{\mathbf{r}}_n = \frac{1}{m}\mathbf{F}(\mathbf{r}_n) - g_n\dot{\mathbf{r}}_{n+1} \tag{A.10}$$

with $g_n = \frac{\dot{s}_{n+1}}{s_n}$ and $\dot{s_n} = \frac{s_{n+1}-s_n}{\Delta t}$.
An alternative derivation results in

$$\ddot{\mathbf{r}}_n = \frac{s_n}{m\, s_{n+1}}\mathbf{F}(\mathbf{r}_n) - \dot{\mathbf{r}}_n(1 - \frac{s_n}{s_{n+1}})\frac{1}{\Delta t} \tag{A.11}$$

If there is no scaling of time (i.e. $s_i = 1$ for all $i = 1...N+1$) the Newton's law under a discrete time with constant time steps is of the form $m\,\ddot{\mathbf{r}}_n = \mathbf{F}(\mathbf{r}_n)$

which resembles its continuous analog. In presence of a scaling of time unequal to 1 however, the Newton's law has an additional term, which can be regarded an acceleration or a friction term in dependence on the sign of g_n. It is this friction term, which enables discrete time physics to be time reversible as demonstrated in the following.

Time reversibility can be described by a two step process having one step forward followed by a step backward. Let us consider the evolution of the discrete Newton's law with two steps forwards. Following eq. 14 by solving $\ddot{\mathbf{r}}_n = \frac{1}{\Delta t}(\dot{\mathbf{r}}_{n+1} - \dot{\mathbf{r}}_n)$ for

(i)

$$\dot{\mathbf{r}}_{n+1} = \dot{\mathbf{r}}_n + \frac{s_n}{m\, s_{n+1}}\Delta t\, \mathbf{F}(\mathbf{r}_n) - (1 - \frac{s_n}{s_{n+1}})\dot{\mathbf{r}}_n \tag{A.12}$$

For the second time step we take the second expression of the discrete Newton's law (eq. 13)

(ii)

$$\dot{\mathbf{r}}_{n+2} = \dot{\mathbf{r}}_{n+1} + \frac{1}{m}\Delta t\, \mathbf{F}(\mathbf{r}_{n+1}) - \dot{\mathbf{r}}_{n+2}(\frac{s_{n+2}}{s_{n+1}} - 1) \tag{A.13}$$

IOP Conf. Series: Journal of Physics: Conf. Series **1275** (2019) 012064 doi:10.1088/1742-6596/1275/1/012064

If the second step is now backward in time

$$\dot{\mathbf{r}}_{n+2} = \dot{\mathbf{r}}_{n+1} - \frac{1}{m}\Delta t\,\mathbf{F}(\mathbf{r}_{n+1}) - \dot{\mathbf{r}}_{n+2}(\frac{s_{n+2}}{s_{n+1}} - 1) \tag{A.14}$$

and if time reversibility is requested (i.e. $\dot{\mathbf{r}}_n = \dot{\mathbf{r}}_{n+2}$ and $s_{n+2} = s_n$)

$$\dot{\mathbf{r}}_n = \dot{\mathbf{r}}_{n+2} = \dot{\mathbf{r}}_{n+1} - \frac{1}{m}\Delta t\,\mathbf{F}(\mathbf{r}_{n+1}) - \dot{\mathbf{r}}_n(\frac{s_n}{s_{n+1}} - 1) \tag{A.15}$$

$$\dot{\mathbf{r}}_n = \dot{\mathbf{r}}_n + \frac{s_n}{m\,s_{n+1}}\Delta t\,\mathbf{F}(\mathbf{r}_n) - (1 - \frac{s_n}{s_{n+1}})\dot{\mathbf{r}}_n - \frac{1}{m}\Delta t\,\mathbf{F}(\mathbf{r}_{n+1}) - \dot{\mathbf{r}}_n(\frac{s_n}{s_{n+1}} - 1) \tag{A.16}$$

$$\frac{s_n}{s_{n+1}}\mathbf{F}(\mathbf{r}_n) = \mathbf{F}(\mathbf{r}_{n+1}) \tag{A.17}$$

and thus

$$\frac{s_n}{s_{n+1}} = \frac{F^i(r_{n+1})}{F^i(r_n)}\; i = x, y, z \tag{A.18}$$

with $\mathbf{F}(\mathbf{r}_n) = \begin{bmatrix} F^x(\mathbf{r}_n) \\ F^y(\mathbf{r}_n) \\ F^z(\mathbf{r}_n) \end{bmatrix} = \begin{bmatrix} F^x(\mathbf{r}_n) \\ F^x(\mathbf{r}_n) \\ F^x(\mathbf{r}_n) \end{bmatrix}$.

This equation holds for the symmetric case selected.

This equation (eq. 50) and its more general analog below (eq. 55) are called the reversibility axiom. If the reversibility axiom is fulfilled, the introduced discrete time physics is time reversible.

By incorporating the reversibility axiom on the symmetric case selected (eq. 50) into the discrete Newtonian equation (eq. 42) the following expression is obtained:

$$\ddot{\mathbf{r}}_n = \frac{1}{\Delta t}(\dot{\mathbf{r}}_{n+1} - \dot{\mathbf{r}}_n) = \frac{1}{m}\mathbf{F}(\mathbf{r}_n) + \dot{\mathbf{r}}_{n+1}\frac{\dot{F^x}(r_{n+1})}{F^x(r_{n+1})} \tag{A.19}$$

with $\dot{F^x}(r_{n+1}) = \frac{F^x(r_{n+1}) - F^x(r_n)}{\Delta t}$. Eq. 51 reflects thereby the time reversible evolution of a system under a discrete time. It is thereby highlighted that due to the dynamic nature of the discrete time with the variable s_n time reversibility was obtained.

The approach taken above has been focusing on a spherical potential that acts along all the coordinates the same for simplicity. Here, the more general case is highlighted of an acting force on a single particle starting with time to be a discrete dynamical tensor second order $\hat{\mathbf{t}}_n$ describing a series of events as follows

$$(\mathbf{r}_0,\, \hat{\mathbf{t}}_0),\, (\mathbf{r}_1,\, \hat{\mathbf{t}}_1), \ldots\ldots, (\mathbf{r}_n,\, \hat{\mathbf{t}}_n), \ldots\ldots, (\mathbf{r}_{N+1},\, \hat{\mathbf{t}}_{N+1}) \tag{A.20}$$

with $(\mathbf{r}_0,\, \hat{\mathbf{t}}_\mathbf{0})$ the initial and $(\mathbf{r}_{N+1},\, \hat{\mathbf{t}}_{N+1})$ the final position. the dynamical part of time can then be described by the scaling factor

$$\hat{\mathbf{s}}_n\Delta t = \mathbf{s}_n\mathbf{1}\,\Delta t = \begin{bmatrix} s_n^x & 0 & 0 \\ 0 & s_n^y & 0 \\ 0 & 0 & s_n^z \end{bmatrix}\Delta t = \begin{bmatrix} t_n^x & 0 & 0 \\ 0 & t_n^y & 0 \\ 0 & 0 & t_n^z \end{bmatrix} - \begin{bmatrix} t_{n-1}^x & 0 & 0 \\ 0 & t_{n-1}^y & 0 \\ 0 & 0 & t_{n-1}^z \end{bmatrix} = \hat{\mathbf{t}}_n - \hat{\mathbf{t}}_{n-1} \tag{A.21}$$

This yields the following reversibility axiom

$$\frac{s_n^x}{s_{n+1}^x} = \frac{F^x(r_{n+1})}{F^x(r_n)}, \frac{s_n^y}{s_{n+1}^y} = \frac{F^y(r_{n+1})}{F^y(r_n)}, \frac{s_n^z}{s_{n+1}^z} = \frac{F^z(r_{n+1})}{F^z(r_n)} \tag{A.22}$$

By incorporating the reversibility axiom into the discrete Newtonian equation (eq. 42) the following expression is obtained:

$$\ddot{\mathbf{r}}_n = \frac{1}{\Delta t}(\dot{\mathbf{r}}_{n+1}-\dot{\mathbf{r}}_n) = \frac{1}{m}\mathbf{F}(r_n) - \dot{\mathbf{r}}_{n+1}\begin{bmatrix} (\frac{F^x(r_n)}{F^x(r_{n+1})}-1)\frac{1}{\Delta t} & 0 & 0 \\ 0 & (\frac{F^y(r_n)}{F^y(r_{n+1})}-1)\frac{1}{\Delta t} & 0 \\ 0 & 0 & (\frac{F^z(r_n)}{F^z(r_{n+1})}-1)\frac{1}{\Delta t} \end{bmatrix} \tag{A.23}$$

Next, the derived discrete Newtonian equation is transformed into its corresponding continuous analog by $lim\,\Delta t \rightarrow 0$:

$$\ddot{\mathbf{r}} = \frac{1}{m}\mathbf{F}(\mathbf{r}) - \hat{\gamma}\,\dot{\mathbf{r}} \tag{A.24}$$

with

$$\hat{\gamma} = \begin{bmatrix} -\frac{\dot{F^x}}{F^x} & 0 & 0 \\ 0 & -\frac{\dot{F^y}}{F^y} & 0 \\ 0 & 0 & -\frac{\dot{F^z}}{F^z} \end{bmatrix}$$

References

[1] Farias RAH and Recami E 2007 arXiv:quant-ph/9706059
[2] Thomson JJ 1925 *Proc. Roy. Soc. of Edinburgh* **46** 90
[3] Yang CN 1947 *Phys. Rev.* **72** 874
[4] Levi R 1927 *Journal de Physique et le Radium* **8** 182
[5] Caldirola P 1953 *Supplemento al Nuovo Cimento* **10** 1747
[6] Lee TD 1983 *Physics Lett.* **122B** 217
[7] 't Hooft G 2010 *Int. J. Mod. Phys.* **A 25** 4385
[8] 't Hooft G 2014 arXiv:1405.1548
[9] Elze HT 2013 *Eur. Phys. J. Web of Conferences* **58** 01013
[10] Elze HT 2014 *Phys. Rev. A* **89** 012111
[11] Riek R 2014 *Entropy* **16** 3149
[12] Rovelli C 2011 arXiv:1102.3660
[13] Nowakowski M 2001 *Int. J. Mod. Phys. D* **10** 649
[14] Einstein A 1917 *Sitzungsberichte der Königlich Preussischen Akademie der Wissenschaften Berlin* **1** 142
[15] Guth A 1981 *Phys. Rev. D* **23** 347
[16] Straumann N 2013 *General Relativity*, Springer, Dordrecht
[17] Nosé S 1984 *J. Chem. Phys.* **81** 511
[18] Nosé S 1986 *Mol. Phys.* **57** 187
[19] Hoover WG 1985 *Phys. Rev. A* **31** 1695
[20] Hünenberger PH 2005 *Adv. Polym. Sci.* **173** 105
[21] Hubble E 1929 *Proc. Natl. Acad. Sci.* **15** 168
[22] Verlinde E 2016 arXiv:1611.02269v2
[23] Hu W and Sawicki I 2007 *Phys. Rev. D* **76**. 064004
[24] Davoudiasl H, Hooper D, and McDermott SD 2016 *Phys Rev Lett.* **116** 031303

AUTHOR INDEX

Allen, Roland E 222, 227
Amirkhani, A. 274
Andersen, T C 356
Bartalucci, S. 274
Bazzi, M. 274
Beckwith, Andrew Walcott 405, 485
Bernardini, Alex E. 308
Bertolami, Orfeu 308
Bertolucci, S. 274
Bittencourt, Victor A. S. V. 256
Blasone, Massimo 256
Blasone, M 230, 241, 570
Bragadireanu, M. 274
Bryan, K L H 350
Buoninfante, Luca 395
Burinskii, Alexander 294
Burrage, C 387
Butterfield, J 20
Calogero, Francesco 563
Canarutto, D 423
Capolupo, Antonio 490, 496
Cargnelli, M. 274
Celeghini, E 266
Cerwen, E 501
Chang, Lay Nam 342
Clozza, A. 274
Cocciaro, B 331
Curceanu, C. 274
De Paolis, L. 274
Degiovanni, I. 215
Del Grande, R. 274
Diósi, Lajos 77
Doyen, G. 411, 509
Drakova, D. 411, 509
Dunningham, JA 49
Egger, J.-P. 274
Eichhorn, Astrid 108
Ellis, G F R 1
Faetti, S 331
Finster, Felix 94
Fiorini, C. 274
Flachi, Antonino 289
Fronzoni, L 331
Gabriele, S. J. Gionti 448
García-Calderón, Gastón 280
Genovese, M. 215
Gervino, G 518
Geszti, Tamás 163
Giampaolo, S. M. 496
Guaraldo, C. 274
Hadad, Merav 428
Hall, M J W 364
Halliwell, J J 82
Hayes, A 49
Iliescu, M. 274
Iorio, Alfredo 146, 556
Jizba, P 49
Kelly, C 180
Khrennikov, Andrei 200
Kiefer, Claus 123
Kim, Sang Pyo 523
Klauder, John R 34
Klinkhamer, F R 130
Kryukov, Alexey A. 468
Kwidzinski, Nick 123
Käding, C 387
Lambiase, G. 540
Lambiase, G 241, 570
Laubenstein, M. 274
Lavagno, A 518
Lidström, Suzy 222
Losero, E. 215
Luciano, G G 241, 570
Mandrin, P A 548
Marsh, B 20
Marton, J. 274
Medved, A J M 350
Meyer, Gabriel 187
Millington, P 387
Milotti, E. 274
Milucci, M. 274
Minguzzi, E 423
Minic, Djordje 342
Minář, J 387
Nisticò, Giuseppe 323
Noldus, Johan 438
Oeckl, Robert 459
Pais, Pablo 556
Payandeh, Farrin 563
Petruzziello, L 241, 570
Pichler, A. 274
Pietreanu, D. 274
Pigato, D 518
Piscicchia, K. 274
Porotti, Riccardo 206
Pradyumna, S. T. 215

Prati, Enrico 206
Pye, J. 248
Reginatto, M 364
Restelli, Marcello 206
Riek, Roland 577
Romo, Roberto 280
Ruo-Berrchera, I. 215
Scardigli, Fabio 37
Scardigli, F. 241
Schuch, Dieter 315
Scordo, A. 274
Shi, H. 274
Sirghi, D. 274
Sirghi, F. 274
Smaldone, L 230
Sperandio, L. 274
Stachowski, A. 374
Szydłowski, M. 374
Takeuchi, Tatsu 342
Tamascelli, Dario 206
Terán, Miguel Ángel 280
Tilloy, Antoine 60
Traina, P. 215
Trugenberger, C A 180
Urbanowski, K. 374
Uzawa, Kunihito 531
Vazquez-Doce, O. 274
Vinante, Andrea 171
Vitagliano, Vincenzo 289
Vitiello, Giuseppe 187
Vitiello, G 230
Zmeskal, J. 274
Zucco, M. 215

Institute of Physics
Dirac House, Temple Back
Bristol BS1 6BE UK

ISSN: 1742-6588
ISBN 978-1-5108-9950-6
9 781510 899506

CURRAN ASSOCIATES INC.
proceedings.com